For my mother, Kit
with love

Contents

Preface xi

1 Algebra and Problem Solving

1.1 Some Basics of Algebra 2
1.2 Operations and Properties of Real Numbers 11
1.3 Solving Equations 22
Connecting the Concepts 25
1.4 Introduction to Problem Solving 29
1.5 Formulas, Models, and Geometry 39
1.6 Properties of Exponents 46
1.7 Scientific Notation 56
SUMMARY AND REVIEW 64
TEST 68

2 Graphs, Functions, and Linear Equations

2.1 Graphs 70
2.2 Functions 78
2.3 Linear Functions: Slope, Graphs, and Models 93
2.4 Another Look at Linear Graphs 106
2.5 Other Equations of Lines 117
Connecting the Concepts 122
2.6 The Algebra of Functions 126
SUMMARY AND REVIEW 135
TEST 138

3 Systems of Linear Equations and Problem Solving

3.1 Systems of Equations in Two Variables 142
3.2 Solving by Substitution or Elimination 151
Connecting the Concepts 156
3.3 Solving Applications: Systems of Two Equations 159
3.4 Systems of Equations in Three Variables 172
Connecting the Concepts 172
3.5 Solving Applications: Systems of Three Equations 181
3.6 Elimination Using Matrices 186
3.7 Determinants and Cramer's Rule 191
3.8 Business and Economic Applications 197
SUMMARY AND REVIEW 203
TEST 206
Cumulative Review: Chapters 1–3 207

4 Inequalities and Problem Solving

4.1 Inequalities and Applications 212
4.2 Intersections, Unions, and Compound Inequalities 224
4.3 Absolute-Value Equations and Inequalities 233
4.4 Inequalities in Two Variables 242
Connecting the Concepts 249
4.5 Applications Using Linear Programming 252
SUMMARY AND REVIEW 260
TEST 263

5 Polynomials and Polynomial Functions

5.1 Introduction to Polynomials and Polynomial Functions 266
Connecting the Concepts 266
5.2 Multiplication of Polynomials 277
5.3 Common Factors and Factoring by Grouping 287
Connecting the Concepts 287
5.4 Factoring Trinomials 293
5.5 Factoring Perfect-Square Trinomials and Differences of Squares 303
5.6 Factoring Sums or Differences of Cubes 309
5.7 Factoring: A General Strategy 313
5.8 Applications of Polynomial Equations 317
SUMMARY AND REVIEW 329
TEST 332

6 Rational Expressions, Equations, and Functions

6.1 Rational Expressions and Functions: Multiplying and Dividing 334
6.2 Rational Expressions and Functions: Adding and Subtracting 343
6.3 Complex Rational Expressions 352
6.4 Rational Equations 361
Connecting the Concepts 361
6.5 Solving Applications Using Rational Equations 368
6.6 Division of Polynomials 378
6.7 Synthetic Division 383
6.8 Formulas, Applications, and Variation 388
SUMMARY AND REVIEW 401
TEST 404
Cumulative Review: Chapters 1–6 405

7 Exponents and Radicals

7.1 Radical Expressions and Functions 408
7.2 Rational Numbers as Exponents 416
7.3 Multiplying Radical Expressions 423
7.4 Dividing Radical Expressions 429
7.5 Expressions Containing Several Radical Terms 435
7.6 Solving Radical Equations 443
Connecting the Concepts 443
7.7 Geometric Applications 450
7.8 The Complex Numbers 458
SUMMARY AND REVIEW 466
TEST 469

8 Quadratic Functions and Equations

8.1 Quadratic Equations 472
8.2 The Quadratic Formula 483
8.3 Applications Involving Quadratic Equations 490
8.4 Studying Solutions of Quadratic Equations 498
8.5 Equations Reducible to Quadratic 502
8.6 Quadratic Functions and Their Graphs 508
Connecting the Concepts 514
8.7 More About Graphing Quadratic Functions 517
8.8 Problem Solving and Quadratic Functions 523
8.9 Polynomial and Rational Inequalities 532
SUMMARY AND REVIEW 540
TEST 544

9 Exponential and Logarithmic Functions

9.1 Composite and Inverse Functions 548
9.2 Exponential Functions 560
Connecting the Concepts 560
9.3 Logarithmic Functions 569
9.4 Properties of Logarithmic Functions 576
9.5 Common and Natural Logarithms 583
9.6 Solving Exponential and Logarithmic Equations 590
9.7 Applications of Exponential and Logarithmic Functions 596
Summary and Review 610
Test 613
Cumulative Review: Chapters 1–9 615

10 Conic Sections

10.1 Conic Sections: Parabolas and Circles 620
10.2 Conic Sections: Ellipses 631
10.3 Conic Sections: Hyperbolas 639
Connecting the Concepts 643
10.4 Nonlinear Systems of Equations 649
Summary and Review 658
Test 660

11 Sequences, Series, and the Binomial Theorem

11.1 Sequences and Series 662
11.2 Arithmetic Sequences and Series 668
11.3 Geometric Sequences and Series 677
11.4 The Binomial Theorem 688
Connecting the Concepts 688
Summary and Review 696
Test 698
Cumulative Review: Chapters 1–11 699

Appendix: The Graphing Calculator 703

Answers A-1

Index I-1

Index of Applications

Videotape and CD Index

Preface

We are pleased to present the sixth edition of *Intermediate Algebra: Concepts and Applications*. Each time we work on a new edition, it's a balancing act. On the one hand, we want to preserve the features, applications, and explanations that faculty have come to rely on and expect. On the other hand, we want to blend our own ideas for improvement with the many insights that we receive from faculty and students throughout North America. The result is a living document in which new features and applications are developed while successful features and popular applications from previous editions are updated and refined. Our goal, as always, is to present content that is easy to understand and has the depth required for success in this and future courses.

Appropriate for a one-term course in intermediate algebra, this text is intended for those students who have a firm background in elementary algebra. It is one of three texts in an algebra series that also includes *Elementary Algebra: Concepts and Applications*, Sixth Edition, by Bittinger/Ellenbogen and *Elementary and Intermediate Algebra: Concepts and Applications, A Combined Approach*, Third Edition, by Bittinger/Ellenbogen/Johnson.

Approach

Our goal, quite simply, is to help today's students both learn and retain mathematical concepts. To achieve this goal, we feel that we must prepare developmental-mathematics students for the transition from "skills-oriented" elementary and intermediate algebra courses to more "concept-oriented" college-level mathematics courses. This requires that we teach these same students critical thinking skills: to reason mathematically, to communicate mathematically, and to identify and solve mathematical problems. Following are some aspects of our approach that are used in this revision to help meet the challenges we all face teaching developmental mathematics.

Problem Solving

One distinguishing feature of our approach is our treatment of and emphasis on problem solving. We use problem solving and applications to motivate the material wherever possible, and we include real-life applications and problem-solving techniques throughout the text. Problem solving not only encourages students to think about how mathematics can be used, it helps to prepare them for more advanced material in future courses.

- In Chapter 1, we introduce the five-step process for solving problems: (1) Familiarize, (2) Translate, (3) Carry out, (4) Check, and (5) State the answer. These steps are then used consistently throughout the text whenever we encounter a problem-solving situation. Repeated use of this problem-solving strategy gives students a sense that they have a starting point for any type of problem they encounter, and frees them to focus on the mathematics necessary to successfully translate the problem situation. We often use estimation and carefully checked guesses to help with the Familiarize and Check steps (see pp. 33, 166, 373, and 524).

Applications

Interesting applications of mathematics help motivate both students and instructors. Solving applied problems gives students the opportunity to see their conceptual understanding put to use in a real way. In the Sixth Edition of *Intermediate Algebra: Concepts and Applications*, not only have we increased the total number of applications and real-data problems overall, nearly 20 percent of our applications are new, and we have increased the number of source lines to better highlight the real-world data. As in the past, art is integrated into the applications and exercises to aid the student in visualizing the mathematics. (See pp. 119, 128, 351, and 393.)

Pedagogy

New! **Connecting the Concepts.** To help students understand the "big picture," Connecting the Concepts subsections within each chapter (and highlighted in the table of contents) relate the concept at hand to previously learned and upcoming concepts. Because students may occasionally "lose sight of the forest because of the trees," we feel confident that this feature will help them keep better track of their bearings as they encounter new material. (See pp. 25, 172, 249, and 266.)

New! **Study Tips.** Most plentiful in the first three chapters when students are still establishing their study habits, Study Tips are found in the margins and interspersed throughout the first seven chapters. Our Study Tips range from how to approach assignments, to reminders of the various study aids that are available, to strategies for preparing for a final exam. (See pp. 93, 157, and 212.)

New! ***Aha!* Exercises.** Designated by Aha!, these exercises can be solved quickly if the student has the proper insight. The Aha! designation is used the first time a new insight can be used on a particular type of exercise and indicates to the student that there is a simpler way to complete the exercise that requires less lengthy computation. It's then up to the student to find the simpler approach and, in subsequent exercises, to determine if and when that particular insight can be used again. Occasionally the *Aha!* exercise is easily answered by looking at the preceding odd-numbered exercise. Our hope is that the *Aha!* exercises will discourage rote learning and reward students who "look before they leap" into a problem. (See pp. 122, 240, 285, 301, and 341.)

Technology Connections. Throughout each chapter, optional Technology Connection boxes help students use graphing calculator technology to better visualize a concept that they have just learned. To connect this feature to the exercise sets, certain exercises are marked with a graphing calculator icon and reinforce the use of this optional technology. (See pp. 94, 96, 120, 239, and 269.)

Skill Maintenance Exercises. Retaining mathematical skills is critical to a student's success in future courses. To this end, nearly every exercise set includes six to eight Skill Maintenance exercises that review skills and concepts from preceding chapters of the text. In this edition, not only have the Skill Maintenance exercises been increased by 50 percent, but they are now designed to provide extra practice with the specific skills needed for the very next section of the text. We also now list answers to both odd- and even-numbered Skill Maintenance exercises, along with their section references, in the answers at the back of the book. (See pp. 91, 158, 250, and 316.)

Synthesis Exercises. Following the Skill Maintenance section, each exercise set ends with a group of Synthesis exercises designated by their own heading. These exercises offer opportunities for students to synthesize skills and concepts from earlier sections with the present material, and often provide students with deeper insights into the current topic. Synthesis exercises are generally more challenging than those in the main body of the exercise set. (See pp. 150, 241, 435, and 442.)

Writing Exercises. In this edition, nearly every set of exercises includes at least four writing exercises. Two of these are more basic and appear just before the Skill Maintenance exercises. The other writing exercises are more challenging and appear as Synthesis exercises. All writing exercises are marked with an icon and require answers that are one or more complete sentences. This type of problem has been found to aid in student comprehension, critical thinking, and conceptualization. Because some instructors may collect answers to writing exercises, and because more than one answer may be correct, answers to writing exercises are not listed at the back of the text. (See pp. 231, 302, 342, and 351.)

Collaborative Corners. In today's professional world, teamwork is essential. We continue to provide optional Collaborative Corner features throughout the text that require students to work in groups to explore and solve mathematical problems. There are one to three Collaborative Corners per chapter, each one appearing after the appropriate exercise set. (See pp. 39, 223, 277, and 423.)

Cumulative Review. After Chapters 3, 6, 9, and 11, we have included a Cumulative Review, which reviews skills and concepts from preceding chapters of the text. (See pp. 207, 405, 615, and 699.)

What's New in the Sixth Edition?

We have rewritten many key topics in response to user and reviewer feedback and have made significant improvements in design, art, pedagogy, and an expanded supplements package. Detailed information about the content changes is available in the form of a conversion guide. Please ask your local Addison-Wesley sales consultant for more information. Following is a list of the major changes in this edition.

New Design

You will see that the page dimension for this edition is larger, which allows for an open look and a typeface that is easier to read. In addition, we continue to pay close attention to the pedagogical use of color to make sure that it is used to present concepts in the clearest possible manner.

Content Changes

A variety of content changes have been made throughout the text. Some of the more significant changes are listed below.

- Chapter 2 now includes a brief introduction to interpolation and extrapolation. The concept of slope is now closely linked with the idea of rate of change, beginning in Section 2.3.
- The topic of variation has been removed from Section 8.6 and moved into Section 6.8. As a result, Chapter 8 has been shortened to 9 sections.
- Chapter 7 has been rewritten so that Section 7.3 is now strictly on multiplying radical expressions. Section 7.4 is now strictly on division of radical expressions. Section 7.5 is now devoted to expressions with two or more radical terms.
- Chapter 9 now begins with Composite and Inverse Functions (formerly Section 9.2) and then moves to Exponential Functions (formerly Section 9.1). This makes for more flow of topics and facilitates coverage of Composite and Inverse Functions as a stand-alone topic if desired.

Supplements for the Instructor

New! Annotated Instructor's Edition

(ISBN 0-201-65873-9)

The *Annotated Instructor's Edition* includes all the answers to the exercise sets, usually right on the page where the exercises appear, and Teaching Tips in the margins that give insights and classroom discussion suggestions that will be especially useful for new instructors. These handy answers and ready Teaching Tips will help both new and experienced instructors save classroom preparation time.

New! MyMathLab

MyMathLab.com is a complete, on-line course for Addison-Wesley mathematics textbooks that integrates interactive, multimedia instruction correlated to the textbook content. MyMathLab can be easily customized to suit the needs of students and instructors and provides a comprehensive and efficient on-line course-management system that allows for diagnosis, assessment, and tracking of students' progress.

MyMathLab features the following:

- Fully interactive multimedia textbooks are built in CourseCompass, a version of Blackboard™ designed specifically for Addison-Wesley.
- Chapter and section folders from the textbook contain a wide range of instructional content: videos, software tools, audio clips, animations, and electronic supplements.
- Hyperlinks take you directly to on-line testing, diagnosis, tutorials, and gradebooks in MathXL—Addison-Wesley's tutorial and testing system for mathematics and statistics.
- Instructors can create, copy, edit, assign, and track all tests for their course as well as track student tutorial and testing performance.
- With push-button ease, instructors can remove, hide, or annotate Addison-Wesley preloaded content, add their own course documents, or change the order in which material is presented.
- Using the communication tools found in MyMathLab, instructors can hold on-line office hours, host a discussion board, create communication groups within their class, send e-mails, and maintain a course calendar.
- Print supplements are available on-line, side by side with their textbooks.

For more information, visit our Web site at www.mymathlab.com or contact your Addison-Wesley sales representative for a live demonstration.

Printed Test Bank/ Instructor's Resource Guide (ISBN 0-201-73490-7)

The Instructor's Resource Guide portion of this supplement contains the following:

- Extra practice problems and answers
- Black-line masters of grids and number lines for transparency masters or test preparation

- A videotape index and section cross-references to our tutorial software packages
- A syllabus conversion guide from the Fifth Edition to the Sixth Edition

The Printed Test Bank portion of this supplement contains the following:

- Six new alternate free-response test forms for each chapter, organized with the same topic order as the chapter tests in the main text. Each form includes synthesis questions, as appropriate, at the end of each test.
- Two new multiple-choice versions of each chapter test
- Eight new alternate test forms for the final examination: Alternate Test Forms A, B, and C of the final examinations are organized by chapter and D, E, and F are organized by problem type.
- Answers to all tests

Instructor's Solutions Manual

(ISBN 0-201-73489-3)

The *Instructor's Solutions Manual* contains fully worked-out solutions to the odd-numbered exercises and brief solutions to the even-numbered exercises in the exercise sets.

Answer Book (ISBN 0-201-73727-2)

The *Answer Book* includes answers to all even-numbered and odd-numbered exercises.

TestGen-EQ/QuizMaster-EQ

(ISBN 0-201-73732-9)

Available on a dual-platform Windows/Macintosh CD-ROM, this fully networkable software enables instructors to build, edit, print, and administer tests using a computerized test bank of questions organized according to the contents of each chapter. Tests can be printed or saved for on-line testing via a network on the Web, and the software can generate a variety of grading reports for tests and quizzes.

InterAct Math Plus

(ISBN 0-201-72140-6)

Available to Windows users of *Intermediate Algebra: Concepts and Applications*, Sixth Edition, this networkable software provides course management and on-line administration for Addison-Wesley's InterAct Math Tutorial Software (see "Supplements for the Student"). InterAct Math Plus enables instructors to create and administer on-line tests, summarize students' results, and monitor students' progress in the tutorial software, providing an invaluable teaching and tracking resource.

InterAct MathXL: www.mathxl.com

(12-month registration ISBN 0-201-71111-7, stand-alone)
The MathXL Web site provides diagnostic testing and tutorial help, all on-line using InterAct Math® tutorial software and TestGen-EQ testing software. Students can take chapter tests correlated to the text, receive individualized study plans based on those test results, work practice problems and receive tutorial instruction for areas in which they need improvement, and take further tests to gauge their progress. Instructors can customize tests and track all student test results, study plans, and practice work.

Supplements for the Student

New! Web Site: www.MyMathLab.com

Ideal for lecture-based, lab-based, and on-line courses, this state-of-the-art Web site provides students with a centralized point of access to the wide variety of on-line resources available with this text. The pages of the actual book are loaded into MyMathLab.com, and as students work through a section of the on-line text, they can link directly from the pages to supplementary resources (such as tutorial software, interactive animations, and audio and video clips) that provide instruction, exploration, and practice beyond what is offered in the printed book. MyMathLab.com generates personalized study plans for students and allows instructors to track all student work on tutorials, quizzes, and tests. Complete course-management capabilities, including a host of communication tools for course participants, are provided to create a user-friendly and interactive on-line learning environment.

Student's Solutions Manual

(ISBN 0-201-65874-7)
The *Student's Solutions Manual* by Judith A. Penna contains completely worked-out solutions with step-by-step annotations for all the odd-numbered exercises in the text, with the exception of the Writing exercises. This manual also lists, without complete solutions, the answers for even-numbered text exercises.

InterAct Math® Tutorial CD-ROM

(ISBN 0-201-74624-7)
This interactive tutorial software provides Windows users with algorithmically generated practice exercises that correlate at the objective level to the odd-numbered exercises in the text. Each practice exercise is accompanied by both an example and a guided solution designed to involve students in the solution process. Selected problems also include a video clip that helps students visualize concepts. The software recognizes common student errors and provides appropriate feedback. Instructors can use InterAct Math Plus course management software to create, administer, and track on-line tests and monitor student performance during practice sessions.

InterAct MathXL www.mathxl.com

(12-month registration ISBN 0-201-71630-5, stand-alone)
The MathXL Web site provides diagnostic testing and tutorial help, all on-line, using InterAct Math® tutorial software and TestGen-EQ testing software. Students can take chapter tests correlated to the text, receive individualized study plans based on those test results, work practice problems and receive tutorial instruction for areas in which they need improvement, and take further tests to gauge their progress.

Videotapes (ISBN 0-201-74209-8)

Developed and produced especially for this text, the videotapes feature an engaging team of instructors, including the authors. These instructors present material and concepts by using examples and exercises from every section of the text in a format that stresses student interaction.

Digital Video Tutor

(ISBN 0-201-74641-7, stand-alone)
The videotapes for this text are now available on CD-ROM, making it easy and convenient for students to watch video segments from a computer at home or on campus. The complete digitized video set, now affordable and portable for students, is ideal for distance learning or supplemental instruction.

AW Math Tutor Center

(ISBN 0-201-72170-8, stand-alone)
The AW Math Tutor Center is staffed by qualified mathematics instructors who provide students with tutoring on examples and odd-numbered exercises from the textbook. Tutoring is available via toll-free telephone, fax, or e-mail.

Acknowledgments

No book can be produced without a team of professionals who take pride in their work and are willing to put in long hours. Barbara Johnson, in particular, deserves special thanks for her work as development editor. Barbara's tireless devotion to all aspects of this project and her many fine suggestions have contributed immeasurably to the quality of this text. Laurie A. Hurley also deserves special thanks for her careful accuracy checks, well-thought-out suggestions, and uncanny eye for detail. Judy Penna's outstanding work in organizing and preparing the printed supplements and the indexes amounts to an inspection of the text that goes far beyond the call of duty and for which we will always be extremely grateful. Thanks to Tom Schicker for authoring the *Printed Test Bank*. Dawn Mulheron not only served as an accuracy checker, but was terrifically helpful in posting and double-checking the "catches" found by the checkers. Daphne Bell of Motlow State Community College, Cassidy Ferraro, and Donald Carlson provided enormous help, often in the face of great time pressure, as accuracy checkers. We are also indebted to Chris Burditt and Jann

MacInnes for their many fine ideas that appear in our Collaborative Corners and Janet Wyatt for her recommendations for Teaching Tips featured in the *Annotated Instructor's Edition*.

Martha Morong, of Quadrata, Inc., provided editorial and production services of the highest quality imaginable—she is simply a joy to work with. Geri Davis, of the Davis Group, Inc., performed superb work as designer, art editor, and photo researcher, and always with a disposition that can brighten an otherwise gray day. Network Graphics generated the graphs, charts, and many of the illustrations. Not only are the people at Network reliable, but they clearly take pride in their work. The many hand-drawn illustrations appear thanks to Jim Bryant, a gifted artist with true mathematical sensibilities. Tom and Pam Hansen, of Copy Ship Fax Plus, consistently went the extra yard in providing the best in copying services.

Our team at Addison-Wesley deserves special thanks. Assistant Editor Greg Erb coordinated all the reviews, tracked down countless pieces of information, and managed many of the day-to-day details—always in a pleasant and reliable manner. Executive Project Manager Kari Heen expertly provided a steadying influence along with gentle prodding at just the right moments. Senior Acquisitions Editor Jenny Crum provided many fine suggestions along with unflagging support. Senior Production Supervisor Kathy Manley exhibited patience when others would have shown frustration. Designer Dennis Schaefer's willingness to listen and then creatively respond resulted in a book that is beautiful to look at. Marketing Manager Dona Kenly skillfully kept us in touch with the needs of faculty; Executive Technology Producer Lorie Reilly provided us with the technological guidance so necessary for our many supplements; and Media Producer Tricia Mescall remains the steady hand responsible for our fine video series. Our publisher, Jason Jordan, deserves credit for assembling this fine team and remaining accessible to us on both a professional and personal level. To all of these people we owe a real debt of gratitude.

A special thanks to the students at the Community College of Vermont, Professor Tony Julianelle of the University of Vermont, and Sybil MacBeth of Tidewater Community College for their thoughtful comments and suggestions. We also thank the following professors for their thoughtful reviews and insights.

Prerevision Diary Reviewers (Fifth Edition)

Ray Brinker, *Western Illinois University*
Michael Divinia, *San Jose City College*
Richard Kern, *Luzerne County Community College*
Kamilia Nemri, *Spokane Community College*
Brenda Santistevan, *Salt Lake Community College*
Carol Satkowiak, *Florida Community College at Jacksonville*

Manuscript Reviewers (Sixth Edition)

Dianne Adams, *Hazard Community College*
Sonya Armstrong, *West Virginia State College*
James Ball, *Indiana State University*

Mark Bates, *Oxnard College*
Bob Bohac, *North Idaho College*
Paulette Callahan, *College of San Mateo, San Francisco State University*
Al Coons, *Pima Community College*
Stephen DeLong, *Tidewater Community College—Virginia Beach Campus*
Barbara Elzey, *Lexington Community College*
Laura Ferguson, *Weatherford College*
Ed Gallo, *Ivy Tech State College*
Chris Gardiner, *Eastern Michigan University*
Cheryl Gregory, *College of San Mateo*
Margret Hathaway, *Kansas City Community College*
Nancy Lehmann, *Austin Community College*
Linda Lohman, *Jefferson Community College*
Rachel Malucci, *San Francisco State University*
Perla Myers, *University of San Diego*
Kamilia Nemri, *Spokane Community College*
Irene Palacios, *Grossmont College*
Jane Pinnow, *University of Wisconsin—Parkside*
Gina Reed, *Gainesville College*
Karen Robinson, *Aims Community College*
Don Rose, *College of the Sequoias*
Greg Rosik, *Century College*
Slav Sharapov, *Quinebaug Valley Community Technical College*
Annette Smith, *South Plains College*
William Steed, *Los Angeles City College*
Jim Stewart, *Jefferson Community College*
Denise Widup, *University of Wisconsin—Parkside*

Finally, a special thank you to all those who so generously agreed to discuss their professional use of mathematics in our chapter openers. These dedicated people, none of whom we knew prior to writing this text, all share a desire to make math more meaningful to students. We cannot imagine a finer set of role models.

M.L.B.
D.J.E.

Feature Walkthrough

2 Graphs, Functions, and Linear Equations

2.1 Graphs
2.2 Functions
2.3 Linear Functions: Slope, Graphs, and Models
2.4 Another Look at Linear Graphs
2.5 Other Equations of Lines
Connecting the Concepts
2.6 The Algebra of Functions
Summary and Review
Test

AN APPLICATION

More and more Americans are making their travel arrangements using the Internet. The numer of on-line travel buyers has grown from 48 million in 1999 to approximately 60 million in 2001 (*Source*: Travel Industry Association of America). Find the rate at which this number is growing.

This problem appears as Example 10 in Section 2.3.

I use math in many different ways. Initially, I must estimate the hours and cost of the project for my customer. During development, I use math to create pages, frames, tables, automation objects, and graphics.

TAMMY GROSS
Webmaster
Indianapolis, IN

CHAPTER OPENERS
Each chapter opens with a list of the sections covered and a real-life application that includes a testimonial from a person in that field to show how integral mathematics is in problem solving. Real data are often used in these applications, as well as in many other exercises, and in "on the job" examples (like those students might find in the workplace) to increase student interest.

ANNOTATED INSTRUCTOR'S EDITION

A new *Annotated Instructor's Edition* provides both new and experienced instructors with helpful, timesaving classroom-preparation tools. In addition to including the answers to the exercise sets right on the page where the exercises appear, this special edition incorporates Teaching Tips that give new content insights and classroom discussion suggestions. These tips will save preparation time for each class and provide guidance for new instructors.

NEW!

Teaching Tip

If students struggle with solving formulas, you may wish to substitute numbers first. For example, in $A = lw$, if the area is 200 ft^2 and the width is 10 ft, what is the length?

Solving Formulas

Suppose we remember the area and the w... want to find the length. To do so, we coul... (Area = Length · Width) for l, using the sam... ing equations.

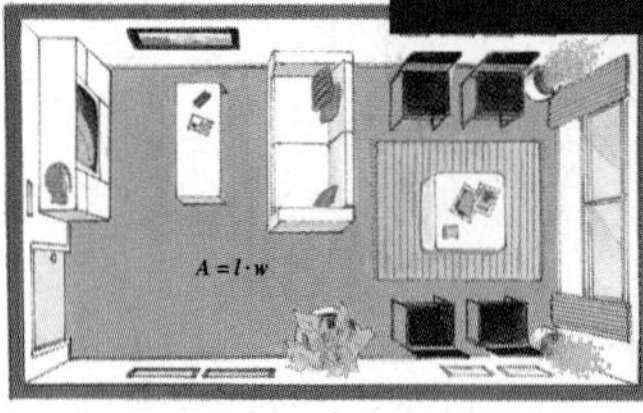

STUDY TIPS

Study Tips can be found throughout the first seven chapters of the text to help students develop sound study habits and to remind them to use the extra tutorial resources that are available.

NEW!

Study Tip

The AW Math Tutor Center provides free tutoring to students using this text. Assisted by qualified mathematics instructors via telephone, fax, or e-mail, you can receive live tutoring on examples and exercises. For more information on the AW Math Tutor Center, see the "To the Student" portion of the preface.

Negative Integers as Exponen...

Later in this text we will explain what num... Until then, integer exponents will suffice.

To develop a definition for negative ... two ways. First we proceed as in arithmetic:

$$\frac{5^3}{5^7} = \frac{5\cdot5\cdot5}{5\cdot5\cdot5\cdot5\cdot5\cdot5\cdot5} = \frac{5\cdot5\cdot5\cdot1}{5\cdot5\cdot5\cdot5\cdot5\cdot5\cdot5}$$
$$= \frac{5\cdot5\cdot5}{5\cdot5\cdot5}\cdot\frac{1}{5\cdot5\cdot5\cdot5}$$
$$= \frac{1}{5^4}.$$

Were we to apply the quotient rule, we would have

$$\frac{5^3}{5^7} = 5^{3-7} = 5^{-4}.$$

These two expressions for $5^3/5^7$ suggest that

$$5^{-4} = \frac{1}{5^4}.$$

This leads to the definition of negative exponents.

technology connection

To visualize Example 6, note that the graph of $y_1 = |x|$ (which is entered $y_1 = \text{abs}(x)$, using the NUM option of the MATH menu) appears without interruption for any piece of the x-axis that we examine.

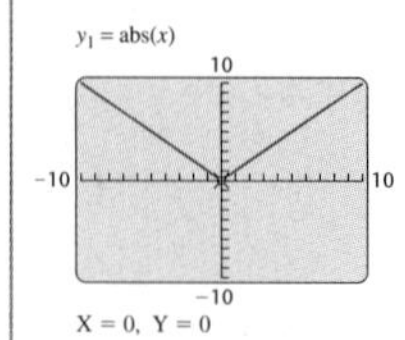

In contrast, the graph of $y_2 = \dfrac{7}{2x-6}$ has a break at $x = 3$.

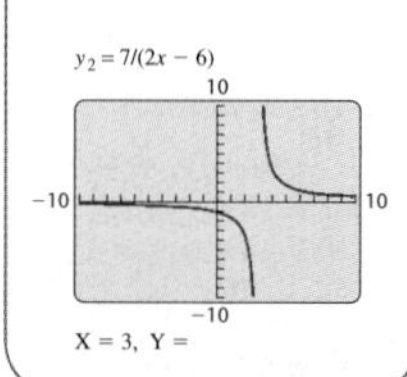

a) $f(x) = |x|$ **b)** $f(x) = \dfrac{7}{2x-6}$

Solution

a) We ask ourselves, "Is there any number x for which we cannot compute $|x|$?" Since we can find the absolute value of *any* number, the answer is no. Thus the domain of f is $\mathbb{R}$, the set of all real numbers.

b) Is there any number x for which $\dfrac{7}{2x-6}$ cannot be computed? Since $\dfrac{7}{2x-6}$ cannot be computed when $2x - 6$ is 0, the answer is yes. To determine what x-value causes the denominator to be 0, we solve an equation:

$2x - 6 = 0$ Setting the denominator equal to 0
$2x = 6$ Adding 6 to both sides
$x = 3.$ Dividing both sides by 2

Thus, 3 is *not* in the domain of f, whereas all other real... domain of f is $\{x\,|\,x \text{ is a real number } \textit{and } x \neq 3\}$.

Applications: Interpolation and Extrapolat...

Function notation is often used in formulas. For example, ... the area A of a circle is a function of its radius r, instead of

$$A = \pi r^2,$$

we can write

$$A(r) = \pi r^2.$$

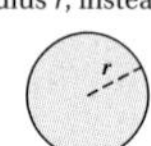

When a function is given as a graph in a problem-solving situation, we are often asked to determine certain quantities on the basis of the graph. Later in this text, we will develop models that can be used for calculations. For now we simply use the graph to estimate the coordinates of an unknown point by using other points with known coordinates. When the unknown point is *between* the known points, this process is called **interpolation**. If the unknown point extends *beyond* the known points, the process is called **extrapolation**.

TECHNOLOGY CONNECTION FEATURE

Optional Technology Connections appear throughout each chapter to help students visualize, through the use of technology, a concept that they have just learned. This feature is reinforced in many exercise sets through exercises marked with a graphing calculator icon.

COLLABORATIVE CORNER

Reduce, Reuse, and Recycle

Focus: Inequalities and problem solving
Time: 15–20 minutes
Group size: 2

In the United States, the amount of solid waste (rubbish) being recycled is slowly catching up to the amount being generated. In 1991, each person generated, on average, 4.3 lb of solid waste every day, of which 0.8 lb was recycled. In 2000, each person generated, on average, 4.4 lb of solid waste, of which 1.3 lb was recycled. (*Sources*: U.S. Census 2000 and EPA Municipal Solid Waste Factbook)

ACTIVITY

Assume that the amount of solid waste being generated and the amount recycled are both increasing linearly. One group member should find a linear function w for which $w(t)$ represents the number of pounds of waste generated per person per day t years after 1991. The other group member should find a linear function r for which $r(t)$ represents the number of pounds recycled per person per day t years after 1991. Finally, working together, the group should determine those years for which the amount recycled will meet or exceed the amount generated.

COLLABORATIVE CORNER FEATURE
A popular feature from the previous edition, optional Collaborative Corners are inserted throughout the text. Collaborative Corners give students the opportunity to work as a group to solve problems or to perform specially designed activities. There are two or three Collaborative Corners per chapter, each one appearing after the appropriate exercise set.

Example 3 ***Tattoo removal.*** In 1996, an estimated 275,000 Americans visited a doctor for tattoo removal. That figure was expected to grow to 410,000 in 2000 (*Source*: Mike Meyers, staff writer, Star-Tribune Newspaper of the Twin Cities Minneapolis–St. Paul, copyright 2000). Assuming constant growth since 1995, how many people will visit a doctor for tattoo removal in 2005?

Solution

1. **Familiarize.** Constant growth indicates a constant rate of change, so a linear relationship can be assumed. If we let n represent the number of people, in thousands, who visit a doctor for tattoo removal and t the number of years since 1995, we can form the pairs $(1, 275)$ and $(5, 410)$. After choosing suitable scales on the two axes, we draw the graph. Note that the jagged "break" on the vertical axis is used to avoid including a large portion of unused grid.

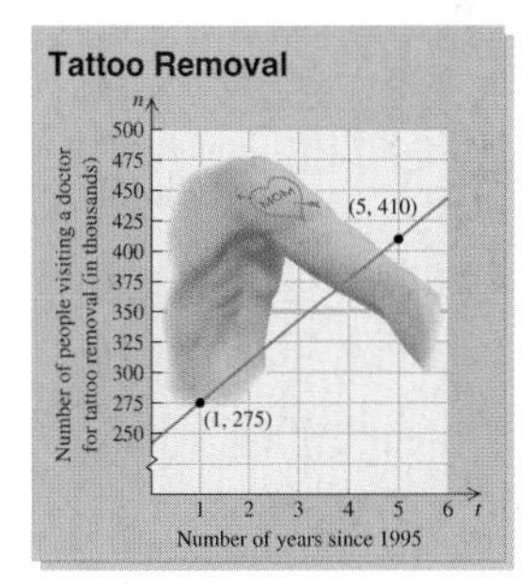

REAL-DATA APPLICATIONS
Applications have always been a strength of this text, and now the authors bring you even more of a good thing. This edition includes 20% new application and real-data problems, along with an increase in the total number of applications overall.

2. **Translate.** To find an equation relating n and t, we first find the slope of the line. This corresponds to the *growth rate*:

$$m = \frac{410 \text{ thousand people} - 275 \text{ thousand people}}{5 \text{ years} - 1 \text{ year}}$$
$$= \frac{135 \text{ thousand people}}{4 \text{ years}}$$
$$= 33.75 \text{ thousand people per year.}$$

Next, we use the point–slope equation and solve for n:

$n - 275 = 33.75(t - 1)$ Writing point–slope form
$n - 275 = 33.75t - 33.75$ Using the distributive law
$n = 33.75t + 241.25.$ Adding 275 to both sides

3. **Carry out.** Using function notation, we have

$$n(t) = 33.75t + 241.25.$$

CONNECTING THE CONCEPTS

Let's briefly summarize our work up to this point in the text: After a review of the basics of algebra and problem solving in Chapter 1, we turned our attention in Chapter 2 to equations in two variables for which graphs are used to represent the solution sets. Graphs also enabled us to visualize the solutions of systems of equations in Chapter 3. In Chapter 4, we continued finding solutions, but this time our work included absolute-value functions and inequalities.

Here in Chapter 5, we will take a break from solving equations and inequalities and concentrate on finding equivalent expressions. Our work with equivalent expressions will ultimately allow us to solve a new type of equation at the end of this chapter.

Chapters 6 and 7 will follow a similar pattern: After learning new ways of writing equivalent expressions, we will learn to solve new types of equations toward the end of each chapter.

CONNECTING THE CONCEPTS
This feature highlights the importance of connecting concepts and invites students to pause and check that they understand the "big picture." This helps assure that students understand how concepts work together in several sections at once. For example, students are alerted to shifts made from solving equations to writing equivalent expressions. The pacing of this feature helps students increase their comprehension and maximize their retention of key concepts.

NEW!

AHA! EXERCISES
In many exercise sets, students will see a new icon, *Aha!*. This icon indicates to students that there is a simpler way to complete the exercise without going through a lengthy computation. It's then up to the student to discover that simpler approach. The *Aha!* icon appears the first time a new insight can be used on a particular type of exercise. After that, it's up to the student to determine if and when that particular insight can be reused.

SYNTHESIS EXERCISES
Synthesis exercises in this new edition guarantee an extensive and wide-ranging variety of problems in every exercise set. The Synthesis exercises allow students to combine concepts from more than one section and provide challenge for even the strongest students. Mixed in with these problems are occasional *Aha!* exercises (described above).

EXERCISES

5.2 MULTIPLICATION OF POLYNOMIALS **285**

47. $(2x^3 - 3y^2)^2$
48. $(3s^2 + 4t^3)^2$
49. $(a^2b^2 + 1)^2$
50. $(x^2y - xy^2)^2$
51. Let $P(x) = 4x - 1$. Find $P(x) \cdot P(x)$.
52. Let $Q(x) = 3x^2 + 1$. Find $Q(x) \cdot Q(x)$.
53. Let $F(x) = 2x - \frac{1}{3}$. Find $[F(x)]^2$.
54. Let $G(x) = 5x - \frac{1}{2}$. Find $[G(x)]^2$.

Multiply.
55. $(c + 2)(c - 2)$
56. $(x - 3)(x + 3)$
57. $(4x + 1)(4x - 1)$
58. $(3 - 2x)(3 + 2x)$
59. $(3m - 2n)(3m + 2n)$
60. $(3x + 5y)(3x - 5y)$
61. $(x^3 + yz)(x^3 - yz)$
62. $(4a^3 + 5ab)(4a^3 - 5ab)$
63. $(-mn + m^2)(mn + m^2)$
64. $(-3b + a^2)(3b + a^2)$
65. $(x + 1)(x - 1)(x^2 + 1)$
66. $(y - 2)(y + 2)(y^2 + 4)$
67. $(a - b)(a + b)(a^2 - b^2)$
68. $(2x - y)(2x + y)(4x^2 - y^2)$
Aha! **69.** $(a + b + 1)(a + b - 1)$
70. $(m + n + 2)(m + n - 2)$
71. $(2x + 3y + 4)(2x + 3y - 4)$
72. $(3a - 2b + c)(3a - 2b - c)$
73. *Compounding interest.* Suppose that P dollars is invested in a savings account at interest rate i, compounded annually, for 2 yr. The amount A in the account after 2 yr is given by
$$A = P(1 + i)^2.$$
Find an equivalent expression for A.
74. *Compounding interest.* Suppose that P dollars is invested in a savings account at interest rate i, compounded semiannually, for 1 yr. The amount A in the account after 1 yr is given by
$$A = P\left(1 + \frac{i}{2}\right)^2.$$
Find an equivalent expression for A.
75. Given $f(x) = x^2 + 5$, find and simplify.
a) $f(t - 1)$
b) $f(a + h) - f(a)$
c) $f(a) - f(a - h)$
76. Given $f(x) = x^2 + 7$, find and simplify.
a) $f(p + 1)$
b) $f(a + h) - f(a)$
c) $f(a) - f(a - h)$
77. Find two binomials whose product is $x^2 - 25$ and explain how you decided on those two binomials.
78. Find two binomials whose product is $x^2 - 6x + 9$ and explain how you decided on those two binomials.

SKILL MAINTENANCE

Solve.
79. $ab + ac = d$, for a
80. $xy + yz = w$, for y
81. $mn + m = p$, for m
82. $rs + s = t$, for s
83. *Value of coins.* There are 50 dimes in a roll of dimes, 40 nickels in a roll of nickels, and 40 quarters in a roll of quarters. Kacie has 13 rolls of coins, which have a total value of \$89. There are three more rolls of dimes than nickels. How many of each type of roll does she have?
84. *Wages.* Takako worked a total of 17 days last month at her father's restaurant. She earned \$50 a day during the week and \$60 a day during the weekend. Last month Takako earned \$940. How many weekdays did she work?

SYNTHESIS

85. We have seen that $(a - b)(a + b) = a^2 - b^2$. Explain how this result can be used to develop a fast way of multiplying $95 \cdot 105$.
86. A student incorrectly claims that since $2x^2 \cdot 2x^2 = 4x^4$, it follows that $5x^5 \cdot 5x^5 = 25x^{25}$. What mistake is the student making?

Multiply. Assume that variables in exponents represent natural numbers.
87. $[(-x^ay^b)^4]^a$
88. $(z^{n^2})^{n^3}(z^{4n^3})^{n^2}$
89. $(a^xb^{2y})(\frac{1}{2}a^{3x}b)^2$
90. $(a^xb^y)^{w+z}$
91. $y^3z^n(y^{3n}z^3 - 4yz^{2n})$

WRITING EXERCISES
Writing exercises, indicated by [icon], provide opportunities for students to answer problems with one or more sentences. Often, these questions have more than one correct response and ask students to explain *why* a certain concept works as it does. In this new edition, two Writing exercises now precede the Skill Maintenance exercises, indicating that they are somewhat less challenging than those that follow the Skill Maintenance exercises. This allows for Writing exercises to be assigned to a wider cross section of the student body.

SKILL MAINTENANCE EXERCISES
As in the past, Skill Maintenance exercises appear in all exercise sets as a means of keeping past concepts fresh and previously covered skills sharp. Two changes to the Skill Maintenance exercises now improve this already popular feature: The number has been increased nearly 50% and they are now designed to provide extra practice with the specific skills needed for the very next section of the text.

Intermediate Algebra

CONCEPTS AND APPLICATIONS

1 Algebra and Problem Solving

1.1 Some Basics of Algebra
1.2 Operations and Properties of Real Numbers
1.3 Solving Equations
Connecting the Concepts
1.4 Introduction to Problem Solving
1.5 Formulas, Models, and Geometry
1.6 Properties of Exponents
1.7 Scientific Notation
Summary and Review
Test

AN APPLICATION

The Delta Queen is a paddleboat that tours the Mississippi River near New Orleans, Louisiana. It is not uncommon for the Delta Queen to run 7 mph in still water and for the river to flow at a rate of 3 mph (*Source*: Delta Queen information). At these rates, how long will it take the boat to cruise 2 mi upstream?

This problem appears as Exercise 3 in Section 1.4.

A river pilot uses math in a variety of ways, from piloting a radar course when meeting other vessels to calculating the estimated time of arrival at a river location.

RANDY SVEUM
Licensed Pilot
Winona, MN

The principal theme of this text is problem solving in algebra. An overall strategy for solving problems is presented in Section 1.4. Additional and increasing emphasis on problem solving appears throughout the book.

This chapter begins with a short review of algebraic symbolism and properties of numbers. As you will see, the manipulations of algebra, such as simplifying expressions and solving equations, are based on the properties of numbers.

1.1 Some Basics of Algebra

Algebraic Expressions and Their Use • Translating to Algebraic Expressions • Evaluating Algebraic Expressions • Sets of Numbers

The primary difference between algebra and arithmetic is the use of *variables*. In this section, we will see how variables can be used to represent various situations. We will also examine the different types of numbers that will be represented by variables throughout this text.

Algebraic Expressions and Their Use

We are all familiar with expressions like

$$95 + 21, \qquad 57 \times 34, \qquad 9 - 4, \quad \text{and} \quad \frac{35}{71}.$$

In algebra, we use these as well as expressions like

$$x + 21, \qquad l \cdot w, \qquad 9 - s, \quad \text{and} \quad \frac{d}{t}.$$

When a letter is used to stand for various numbers, it is called a **variable**. If a letter represents one particular number, it is called a **constant**. Let $d =$ the number of hours it takes the moon to orbit the earth. Then d is a constant. If $a =$ the age of a baby chick, in minutes, then a is a variable since a changes as time passes.

An **algebraic expression** consists of variables, numbers, and operation signs. All of the expressions above are examples of algebraic expressions. When an equals sign is placed between two expressions, an **equation** is formed.

Algebraic expressions and equations arise frequently in problem-solving situations. Suppose, for example, that we want to determine by how much the gas mileage of the most fuel-efficient cars increased between 1999 and 2000.

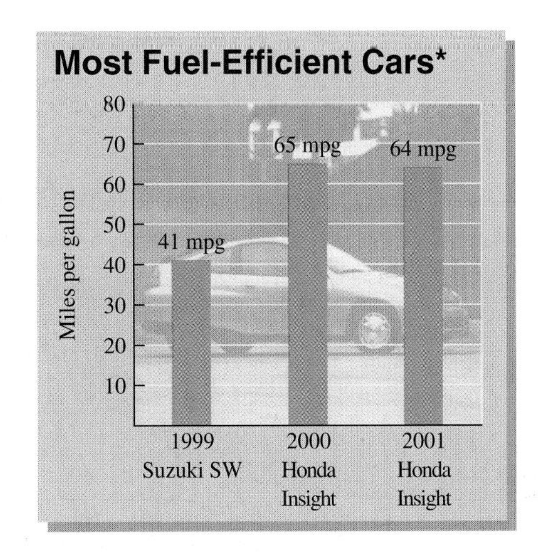

By using x to represent the increase in mileage, we can form an equation:

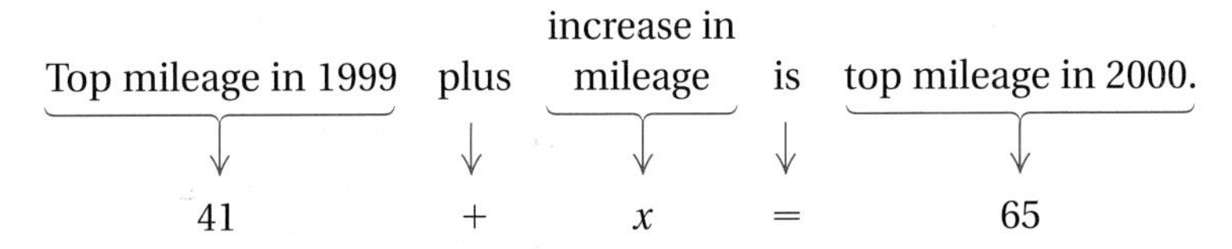

To find a **solution**, we can subtract 41 on both sides of the equation:

$$x = 65 - 41$$
$$x = 24.$$

We see that the mileage increased by 24 miles per gallon (mpg) from 1999 to 2000.

Translating to Algebraic Expressions

To translate problems to equations, we need to know which words correspond to which symbols:

Key Words

Addition	Subtraction	Multiplication	Division
add sum of plus increased by more than	subtract difference of minus decreased by less than	multiply product of times twice of	divide divided by quotient of ratio per

**Source*: U.S. Department of Energy; figures represent a combination of highway and city driving.

When the value of a number is unknown, we represent it with a variable.

Phrase	Algebraic Expression
Five *more than* some number	$n + 5$
Half *of* a number	$\frac{1}{2}t$ or $\frac{t}{2}$
Five *more than* three *times* some number	$3p + 5$
The *difference of* two numbers	$x - y$
Six *less than* the *product of* two numbers	$rs - 6$
Seventy-six percent *of* some number	$0.76z$ or $\frac{76}{100}z$

Note that an expression like rs represents a product and can also be written as $r \cdot s$, $r \times s$, or $(r)(s)$. The multipliers r and s are also called **factors**.

Example 1

Translate to an algebraic expression:

Five less than forty-three percent of the quotient of two numbers.

Solution We let r and s represent the two numbers.

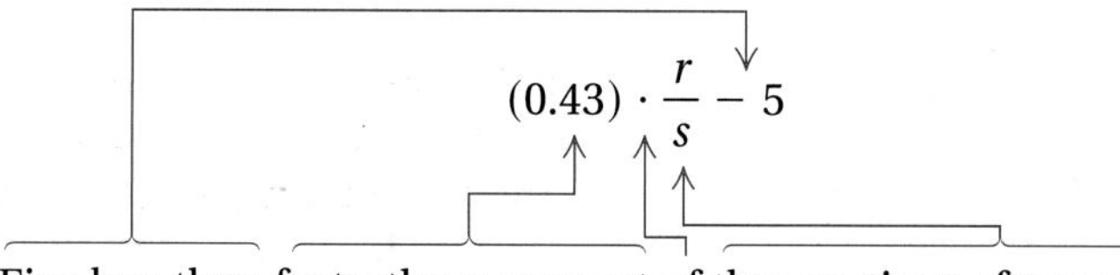

Five less than forty-three percent of the quotient of two numbers

Evaluating Algebraic Expressions

When we replace a variable with a number, we say that we are **substituting** for the variable. The calculation that follows the substitution is called **evaluating the expression**.

Example 2

Evaluate the expression $3xy + z$ for $x = 2$, $y = 5$, and $z = 7$.

Solution We substitute and carry out the multiplication and addition:

$$3xy + z = 3 \cdot 2 \cdot 5 + 7$$ We use color to highlight the substitution.

$$= 30 + 7$$

$$= 37.$$

Geometric formulas are often evaluated. In the next example, we use the formula for the area A of a triangle with a base of length b and a height of length h. This is an important formula that is worth remembering:

$$A = \tfrac{1}{2} \cdot b \cdot h.$$

Example 3 The base of a triangular sail is 4 m and the height is 6.3 m. Find the area of the sail.

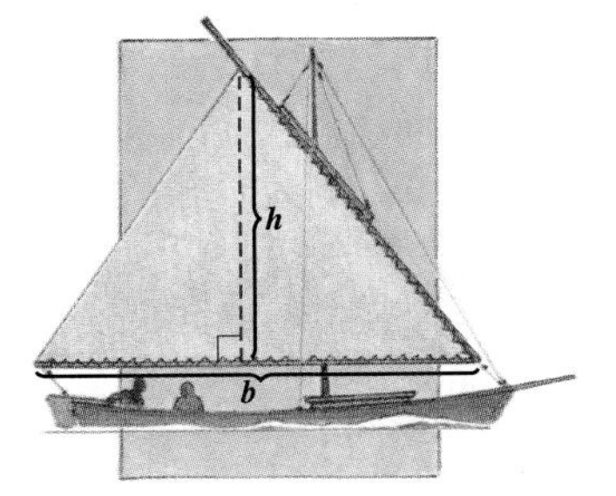

Solution We substitute 4 for b and 6.3 for h and multiply:

$$\begin{aligned}\tfrac{1}{2} \cdot b \cdot h &= \tfrac{1}{2} \cdot 4 \cdot 6.3 \\ &= 12.6 \text{ square meters (sq m).}\end{aligned}$$

Before evaluating other algebraic expressions, we need to develop *exponential notation.* Many different kinds of numbers can be used as *exponents.* Here we establish the meaning of a^n when n is a counting number, $1, 2, 3, \ldots$.

Exponential Notation

The expression a^n, in which n is a counting number, means

$$\underbrace{a \cdot a \cdot a \cdot \cdots \cdot a \cdot a}_{n \text{ factors.}}$$

In a^n, a is called the *base* and n is the *exponent,* or *power.* When no exponent appears, it is assumed to be 1. Thus, $a^1 = a$.

The expression a^n is read "a raised to the nth power" or simply "a to the nth." We read s^2 as "s-squared" and x^3 as "x-cubed." This terminology comes from the fact that the area of a square of side s is $s \cdot s = s^2$ and the volume of a cube of side x is $x \cdot x \cdot x = x^3$.

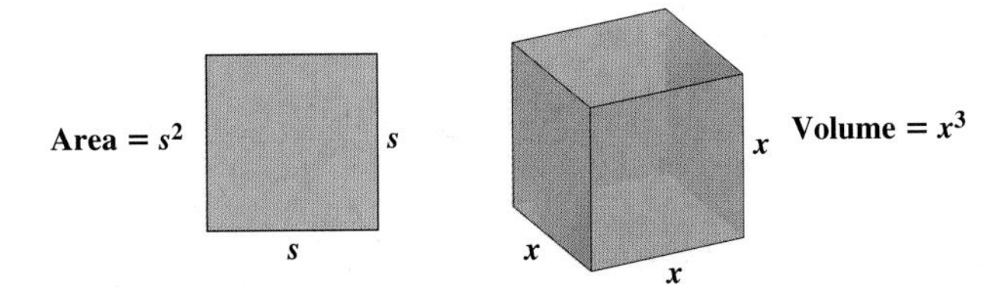

Exponential notation tells us that 5^2 means $5 \cdot 5$, or 25, but what does $1 + 2 \cdot 5^2$ mean? If we add 1 and 2 and multiply by 25, we get 75. If we multiply 2 times 5^2, or 25, and add 1, we get 51. A third possibility is to square $2 \cdot 5$ to get 100 and then add 1. The following convention indicates that only the second of these approaches is correct: We square 5, then multiply, and then add.

Rules for Order of Operations

1. Simplify within any grouping symbols.
2. Simplify all exponential expressions.
3. Perform all multiplication and division working from left to right.
4. Perform all addition and subtraction working from left to right.

Example 4

Evaluate $5 + 2(a - 1)^2$ for $a = 4$.

Solution

$5 + 2(a - 1)^2 = 5 + 2(4 - 1)^2$	Substituting
$= 5 + 2(3)^2$	Working within parentheses first
$= 5 + 2(9)$	Simplifying 3^2
$= 5 + 18$	Multiplying
$= 23$	Adding

Step (3) in the rules for order of operations tells us to divide before we multiply when division appears first, reading left to right. Similarly, if subtraction appears before addition, we subtract before we add.

Example 5

Evaluate $9 - x^3 + 6 \div 2y^2$ for $x = 2$ and $y = 5$.

Solution

$9 - x^3 + 6 \div 2y^2 = 9 - 2^3 + 6 \div 2(5)^2$	Substituting
$= 9 - 8 + 6 \div 2 \cdot 25$	Simplifying 2^3 and 5^2
$= 9 - 8 + 3 \cdot 25$	Dividing
$= 9 - 8 + 75$	Multiplying
$= 1 + 75$	Subtracting
$= 76$	Adding

Sets of Numbers

When evaluating algebraic expressions, and in problem solving in general, we often must examine the *type* of numbers used. For example, if a formula is used to determine an optimal class size, any fractional results must be rounded up or down, since it is impossible to have a fractional part of a student. Three frequently used sets of numbers are listed below.

Natural Numbers, Whole Numbers, and Integers

Natural Numbers (or Counting Numbers)

Those numbers used for counting: $\{1, 2, 3, \dots\}$

Whole Numbers

The set of natural numbers with 0 included: $\{0, 1, 2, 3, \dots\}$

Integers

The set of all whole numbers and their opposites:

$$\{\dots, -4, -3, -2, -1, 0, 1, 2, 3, 4, \dots\}$$

The dots are called ellipses and indicate that the pattern continues without end.

Study Tip

The examples in each section are designed to prepare you for success with the exercise set. Study the step-by-step solutions of the examples, noting that substitutions are highlighted in red. The time you spend studying the examples will save you valuable time when you do your assignment.

The integers correspond to the points on a number line as follows:

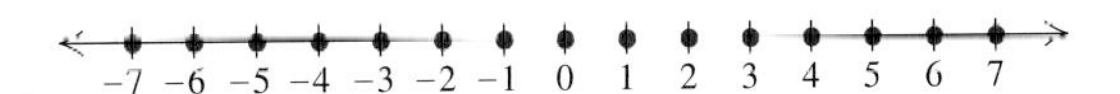

To fill in the numbers between these points, we must describe two more sets of numbers. This requires us to first discuss set notation.

The set containing the numbers −2, 1, and 3 can be written $\{-2, 1, 3\}$. This way of writing a set is known as **roster notation**. Roster notation was used for the sets listed above. A second type of set notation, **set-builder notation**, specifies conditions under which a number is in the set. The following example of set-builder notation is read as shown:

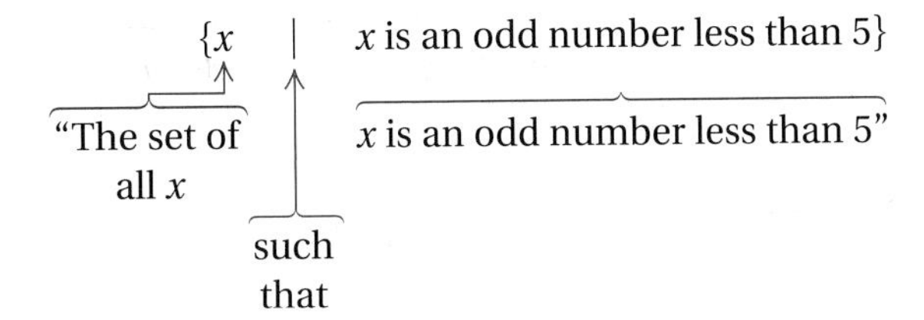

Set-builder notation is generally used when it is difficult to list a set using roster notation.

Example 6 Using both roster notation and set-builder notation, represent the set consisting of the first 15 even natural numbers.

Solution

Using roster notation: $\{2, 4, 6, 8, 10, 12, 14, 16, 18, 20, 22, 24, 26, 28, 30\}$

Using set-builder notation: $\{n \mid n$ is an even number between 1 and 31$\}$

The symbol $\in$ is used to indicate that an element belongs to a set. Thus if $A = \{2, 4, 6, 8\}$, we can write $4 \in A$ to indicate that 4 *is an element of* A. We can also write $5 \notin A$ to indicate that 5 *is not an element of* A.

Example 7 Classify the statement $8 \in \{x \mid x$ is an integer$\}$ as true or false.

Solution Since 8 *is* an integer, the statement is true. In other words, since 8 is an integer, it belongs to the set of all integers.

With set-builder notation, we can describe the set of all *rational numbers.*

Rational Numbers

Numbers that can be expressed as an integer divided by a nonzero integer are called *rational numbers*:

$$\left\{ \frac{p}{q} \,\middle|\, p \text{ is an integer, } q \text{ is an integer, and } q \neq 0 \right\}.$$

Rational numbers can be written using fraction or decimal notation. *Fractional notation* uses symbolism like the following:

$$\frac{5}{8},\quad \frac{12}{-7},\quad \frac{-17}{15},\quad -\frac{9}{7},\quad \frac{39}{1},\quad \frac{0}{6}.$$

In *decimal notation,* rational numbers either *terminate* (end) or *repeat.*

Example 8 When written in decimal form, does each of the following numbers terminate or repeat? **(a)** $\frac{5}{8}$; **(b)** $\frac{6}{11}$.

Solution

a) Since $\frac{5}{8}$ means $5 \div 8$, we perform long division to find that $\frac{5}{8} = 0.625$, a decimal that ends. Thus, $\frac{5}{8}$ can be written as a terminating decimal.

b) Using long division, we find that $6 \div 11 = 0.5454\ldots$, so we can write $\frac{6}{11}$ as a repeating decimal. Repeating decimal notation can be abbreviated by writing a bar over the repeating part—in this case, $0.\overline{54}$.

Many numbers, like π, $\sqrt{2}$, and $-\sqrt{15}$, can be only approximated by rational numbers. For example, $\sqrt{2}$ is the number for which $\sqrt{2} \cdot \sqrt{2} = 2$. A calculator's representation of $\sqrt{2}$ as 1.414213562 is an approximation since $(1.414213562)^2$ is not exactly 2.

To see that $\sqrt{2}$ is a "real" point on the number line, it can be shown that when a right triangle has two legs of length 1, the remaining side has length $\sqrt{2}$. Thus we can "measure" $\sqrt{2}$ and locate it precisely on a number line.

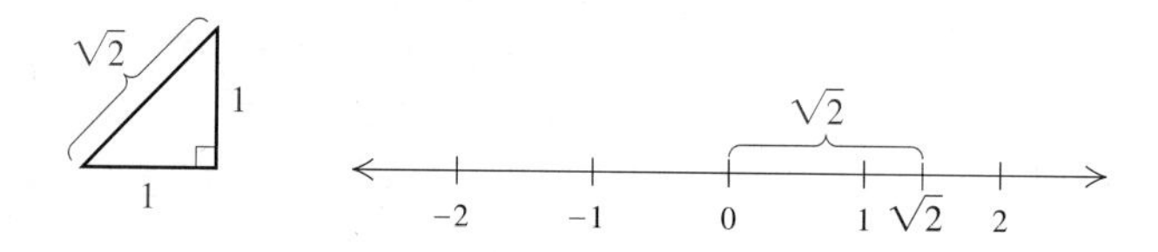

Numbers like π, $\sqrt{2}$, and $-\sqrt{15}$ are said to be **irrational**. Decimal notation for irrational numbers neither terminates nor repeats.

The set of all rational numbers, combined with the set of all irrational numbers, gives us the set of all **real numbers**.

Real Numbers

Numbers that are either rational or irrational are called *real numbers*:

$$\{x \mid x \text{ is rational or } x \text{ is irrational}\}.$$

Every point on the number line represents some real number and every real number is represented by some point on the number line.

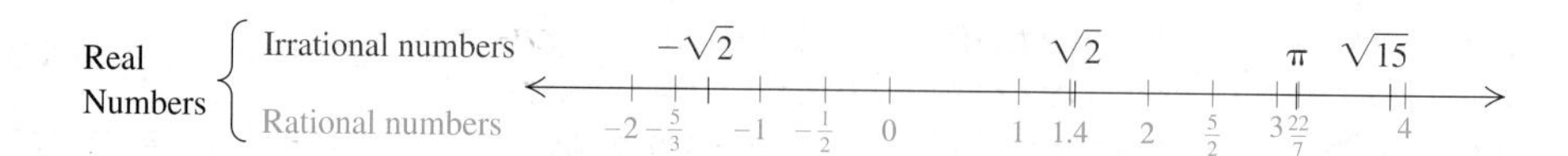

The following figure shows the relationships among various kinds of numbers, along with examples of how real numbers can be sorted.

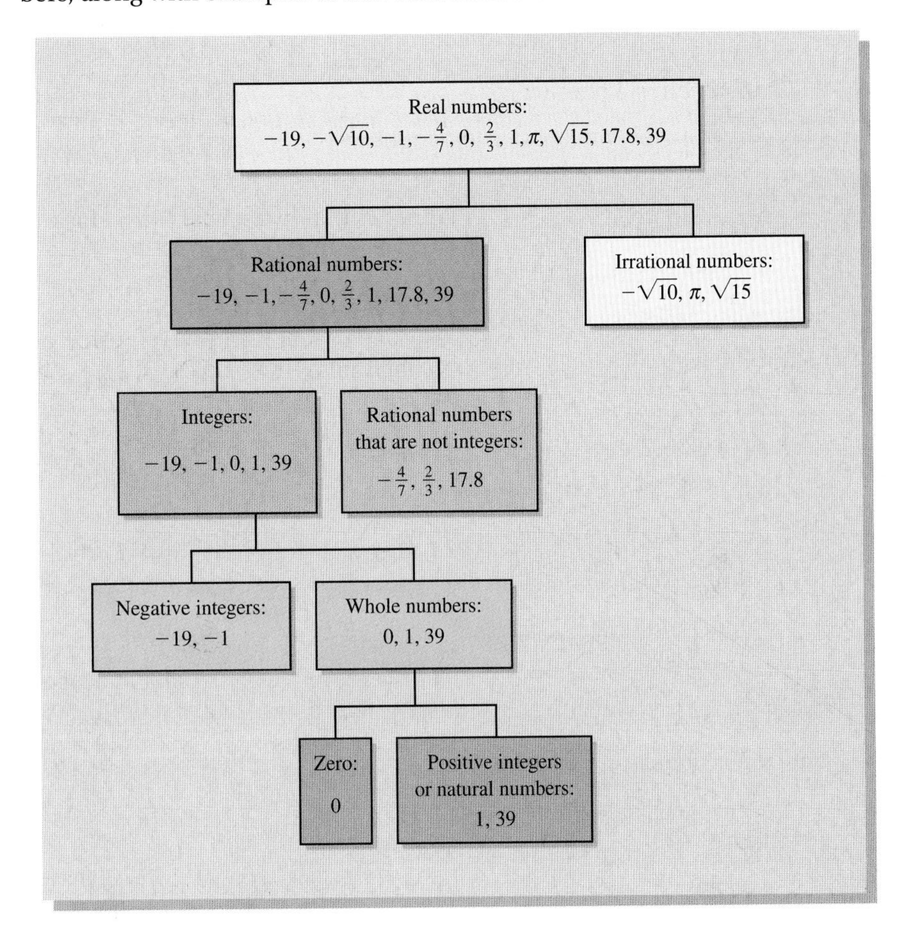

When all members of one set are found in a second set, the first set is a **subset** of the second set. Thus if $A = \{2, 4, 6\}$ and $B = \{1, 2, 4, 5, 6\}$, we write $A \subseteq B$ to indicate that *A is a subset of B*. Similarly, if $\mathbb{N}$ represents the set of all natural numbers and $\mathbb{Z}$ the set of all integers, we can write $\mathbb{N} \subseteq \mathbb{Z}$. Additional statements can be made using other sets in the diagram above.

Exercise Set 1.1

FOR EXTRA HELP

Digital Video Tutor CD 1
Videotape 1

InterAct Math

Math Tutor Center

MathXL

MyMathLab.com

To the student and the instructor: Throughout this text, selected exercises are marked with the icon Aha! *. These "Aha!" exercises can be answered quite easily if the student pauses to inspect the exercise rather than proceed mechanically. This is done to discourage rote memorization. Some "Aha!" exercises are left unmarked to encourage students to* always *pause before working a problem.*

Use mathematical symbols to translate each phrase.

1. Three less than some number

2. Four more than some number

3. Twelve times a number

4. Twice a number

5. Sixty-five percent of some number

6. Thirty-nine percent of some number

7. Ten more than twice a number

8. Six less than half of a number

9. Eight more than ten percent of some number

10. Five less than six percent of some number

11. One less than the difference of two numbers

12. Two more than the product of two numbers

13. Ninety miles per every four gallons of gas

14. One hundred words per every sixty seconds

Evaluate each expression using the values provided.

15. $7x + y$, for $x = 3$ and $y = 4$

16. $6a - b$, for $a = 5$ and $b = 3$

17. $2c \div 3b$, for $b = 4$ and $c = 6$

18. $3z \div 2y$, for $y = 1$ and $z = 6$

19. $25 + r^2 - s$, for $r = 3$ and $s = 7$

20. $n^3 + 2 - p$, for $n = 2$ and $p = 5$

Aha! **21.** $3n^2p - 3pn^2$, for $n = 5$ and $p = 9$

22. $2a^3b - 2b^2$, for $a = 3$ and $b = 7$

23. $5x \div (2 + x - y)$, for $x = 6$ and $y = 2$

24. $3(m + 2n) \div m$, for $m = 7$ and $n = 0$

25. $[10 - (a - b)]^2$, for $a = 7$ and $b = 2$

26. $[17 - (x + y)]^2$, for $x = 4$ and $y = 1$

27. $m + [n(3 + n)]^2$, for $m = 9$ and $n = 2$

28. $a^2 - [3(a - b)]^2$, for $a = 7$ and $b = 5$

In Exercises 29–32, find the area of a triangular window with the given base and height.

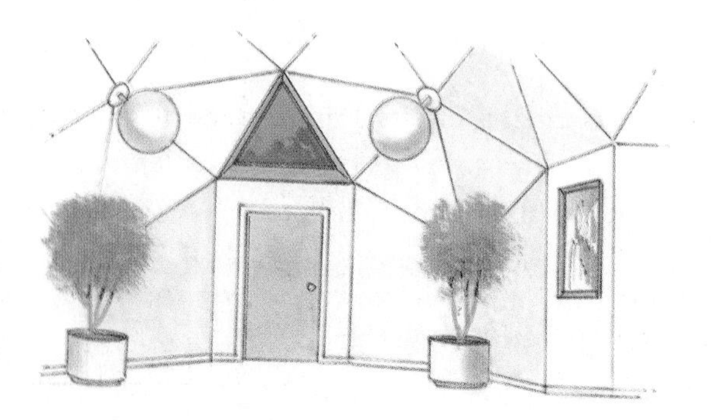

29. Base = 5 ft, height = 7 ft

30. Base = 2.9 m, height = 2.1 m

31. Base = 7 m, height = 3.2 m

32. Base = 3.6 ft, height = 4 ft

Use roster notation to write each set.

33. The set of all vowels in the alphabet

34. The set of all days of the week

35. The set of all odd natural numbers

36. The set of all even natural numbers

37. The set of all natural numbers that are multiples of 5

38. The set of all natural numbers that are multiples of 10

Use set-builder notation to write each set.

39. The set of all odd numbers between 10 and 20

40. The set of all multiples of 4 between 22 and 35

41. $\{0, 1, 2, 3, 4\}$

42. $\{-3, -2, -1, 0, 1, 2\}$

43. The set of all multiples of 5 between 7 and 79

44. The set of all even numbers between 9 and 99

Classify each statement as true or false. The following sets are used:

$\mathbb{N}$ = the set of natural numbers;
$\mathbb{W}$ = the set of whole numbers;
$\mathbb{Z}$ = the set of integers;
$\mathbb{Q}$ = the set of rational numbers;
$\mathbb{H}$ = the set of irrational numbers;
$\mathbb{R}$ = the set of real numbers.

45. $9 \in \mathbb{N}$

46. $5.1 \in \mathbb{N}$

47. $\mathbb{N} \subseteq \mathbb{W}$

48. $\mathbb{W} \subseteq \mathbb{Z}$

49. $\sqrt{8} \in \mathbb{Q}$

50. $\frac{2}{3} \in \mathbb{H}$

51. $\mathbb{H} \subseteq \mathbb{R}$

52. $\sqrt{10} \in \mathbb{R}$

53. $4.3 \notin \mathbb{Z}$

54. $\mathbb{Z} \not\subseteq \mathbb{N}$

55. $\mathbb{Q} \subseteq \mathbb{R}$

56. $\mathbb{Q} \subseteq \mathbb{Z}$

To the student and the instructor: The icon ▪ is used to denote writing exercises. These exercises are meant to be answered with one or more English sentences. Because many writing exercises have a variety of correct answers, these solutions are not listed in the answers at the back of the book.

57. What is the difference between rational numbers and integers?

58. Werner insists that $15 - 4 + 1 \div 2 \cdot 3$ is 2. What error is he making?

SYNTHESIS

To the student and the instructor: Synthesis exercises are designed to challenge students to extend the concepts or skills studied in each section. Many synthesis exercises require the assimilation of skills and concepts from several sections.

59. Is the following true or false, and why?

$$\{2, 4, 6\} \subseteq \{2, 4, 6\}$$

60. On a quiz, Francesca answers $6 \in \mathbb{Z}$ while Jacob writes $\{6\} \in \mathbb{Z}$. Jacob's answer does not receive full credit while Francesca's does. Why?

Translate to an algebraic expression.

61. The quotient of the sum of two numbers and their difference

62. Three times the sum of the cubes of two numbers

63. Half of the difference of the squares of two numbers

64. The product of the difference of two numbers and their sum

Use roster notation to write each set.

65. The set of all whole numbers that are not natural numbers

66. The set of all integers that are not whole numbers

67. $\{x \mid x = 5n, n \text{ is a natural number}\}$

68. $\{x \mid x = 3n, n \text{ is a natural number}\}$

69. $\{x \mid x = 2n + 1, n \text{ is a whole number}\}$

70. $\{x \mid x = 2n, n \text{ is an integer}\}$

71. Draw a right triangle that could be used to measure $\sqrt{13}$ units.

1.2 Operations and Properties of Real Numbers

Absolute Value • Inequalities • Addition, Subtraction, and Opposites • Multiplication, Division, and Reciprocals • The Commutative, Associative, and Distributive Laws

In this section, we review addition, subtraction, multiplication, and division of real numbers. We also study important rules for manipulating algebraic expressions. First, however, it is important to discuss absolute value and inequalities.

Absolute Value

It is convenient to have a notation that represents a number's distance from zero on the number line.

Absolute Value

The notation $|a|$, read "the absolute value of a," represents the number of units that a is from zero.

Example 1

Find the absolute value: **(a)** $|-3|$; **(b)** $|2.5|$; **(c)** $|0|$.

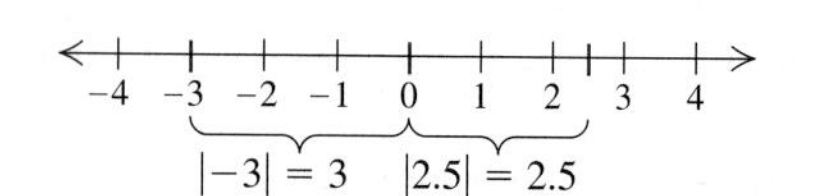

Solution

a) $|-3| = 3$ — -3 is 3 units from 0.
b) $|2.5| = 2.5$ — 2.5 is 2.5 units from 0.
c) $|0| = 0$ — 0 is 0 units from itself.

Note that whereas the absolute value of a nonnegative number is the number itself, the absolute value of a negative number is its opposite.

Inequalities

For any two numbers on the number line, the one to the left is said to be less than, or smaller than, the one to the right. The symbol $<$ means "is less than" and the symbol $>$ means "is greater than." The symbol $\leq$ means "is less than or equal to" and the symbol $\geq$ means "is greater than or equal to." These symbols are used to form **inequalities.**

In the figure below, we have $-6 < -1$ (since -6 is to the left of -1) and $|-6| > |-1|$ (since 6 is to the right of 1).

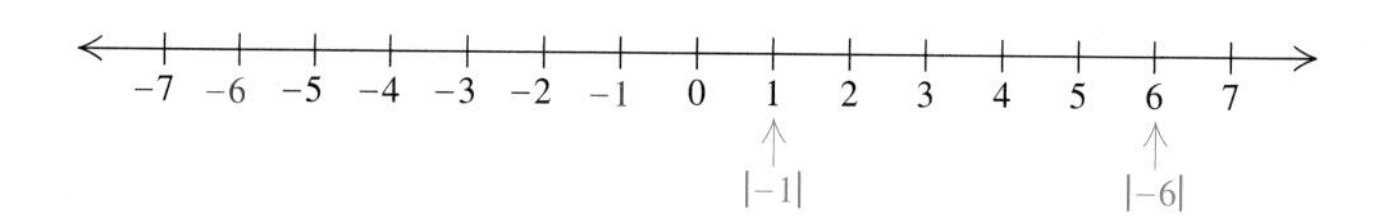

Example 2

Write out the meaning of each inequality and determine whether it is a true statement.

a) $-7 < -2$ **b)** $4 > -1$ **c)** $-3 \geq -2$
d) $5 \leq 6$ **e)** $6 \leq 6$

Solution

Inequality	*Meaning*
a) $-7 < -2$	"-7 is less than -2" is true because -7 is to the left of -2.
b) $4 > -1$	"4 is greater than -1" is true because 4 is to the right of -1.
c) $-3 \geq -2$	"-3 is greater than or equal to -2" is false because -3 is to the left of -2.
d) $5 \leq 6$	"5 is less than or equal to 6" is true because $5 < 6$ is true.
e) $6 \leq 6$	"6 is less than or equal to 6" is true because $6 = 6$ is true.

Addition, Subtraction, and Opposites

We are now ready to review the addition of real numbers.

Addition of Two Real Numbers

1. *Positive numbers*: Add the numbers. The result is positive.
2. *Negative numbers*: Add absolute values. Make the answer negative.
3. *A negative and a positive number*: If the numbers have the same absolute value, the answer is 0. Otherwise, subtract the smaller absolute value from the larger one:
   a) If the positive number is further from 0, make the answer positive.
   b) If the negative number is further from 0, make the answer negative.
4. *One number is zero*: The sum is the other number.

Example 3 Add: **(a)** $-9 + (-5)$; **(b)** $-3.2 + 9.7$; **(c)** $-\frac{3}{4} + \frac{1}{3}$.

Solution

a) $-9 + (-5)$ We add the absolute values, getting 14. The answer is *negative*, -14.

b) $-3.2 + 9.7$ The absolute values are 3.2 and 9.7. Subtract 3.2 from 9.7 to get 6.5. The positive number is further from 0, so the answer is *positive*, 6.5.

c) $-\frac{3}{4} + \frac{1}{3} = -\frac{9}{12} + \frac{4}{12}$ The absolute values are $\frac{9}{12}$ and $\frac{4}{12}$. Subtract to get $\frac{5}{12}$. The negative number is further from 0, so the answer is *negative*, $-\frac{5}{12}$.

When numbers like 7 and -7 are added, the result is 0. Such numbers are called **opposites**, or **additive inverses**, of one another.

The Law of Opposites

For any two numbers a and $-a$,

$$a + (-a) = 0.$$

(When opposites are added, their sum is 0.)

Example 4 Find the opposite: **(a)** -17.5; **(b)** $\frac{4}{5}$; **(c)** 0.

Solution

a) The opposite of -17.5 is 17.5 because $-17.5 + 17.5 = 0$.
b) The opposite of $\frac{4}{5}$ is $-\frac{4}{5}$ because $\frac{4}{5} + \left(-\frac{4}{5}\right) = 0$.
c) The opposite of 0 is 0 because $0 + 0 = 0$.

To name the opposite, we use the symbol "$-$" and read the symbolism $-a$ as "the opposite of a."

> ***Caution!*** $-a$ does not necessarily denote a negative number. In particular, when a is *negative*, $-a$ is *positive*.

Example 5

Find $-x$ for the following: **(a)** $x = -2$; **(b)** $x = \frac{3}{4}$.

Solution

a) If $x = -2$, then $-x = -(-2) = 2$. The opposite of -2 is 2.
b) If $x = \frac{3}{4}$, then $-x = -\frac{3}{4}$. The opposite of $\frac{3}{4}$ is $-\frac{3}{4}$.

Technology Connections highlight situations in which calculators (primarily graphing calculators) or computers can be used to enrich the learning experience. Most Technology Connections present information in a generic form—consult an outside reference for specific keystrokes. (Henceforth in the text we will refer to all graphing utilities as graphers.)

Graphers have different keys for subtracting and writing negatives. The key labeled $\boxed{(-)}$ is used to create a negative sign, whereas $\boxed{-}$ is used for subtraction.

1. Use a grapher to check Example 6.

Using the notation of opposites, we can formally define absolute value.

Absolute Value

$$|x| = \begin{cases} x & \text{if } x \geq 0, \\ -x & \text{if } x < 0 \end{cases}$$

(When x is nonnegative, the absolute value of x is x. When x is negative, the absolute value of x is the opposite of x. Thus, $|x|$ is never negative.)

A negative number is said to have a negative "sign" and a positive number a positive "sign." To subtract, we can add an opposite. Thus we sometimes say that we "change the sign of the number being subtracted and then add."

Example 6

Subtract: **(a)** $5 - 9$; **(b)** $-1.2 - (-3.7)$; **(c)** $-\frac{4}{5} - \frac{2}{3}$.

Solution

a) $5 - 9 = 5 + (-9)$ Change the sign and add.
$= -4$

b) $-1.2 - (-3.7) = -1.2 + 3.7$ Instead of *subtracting* -3.7, we *add* 3.7.
$= 2.5$

c) $-\frac{4}{5} - \frac{2}{3} = -\frac{4}{5} + \left(-\frac{2}{3}\right)$
$= -\frac{12}{15} + \left(-\frac{10}{15}\right)$ Finding a common denominator
$= -\frac{22}{15}$

Multiplication, Division, and Reciprocals

Multiplication of real numbers can be regarded as repeated addition or as repeated subtraction that begins at 0. For example,

$$3 \cdot (-2) = 0 + (-2) + (-2) + (-2) = -6$$

and

$$(-2)(-5) = 0 - (-5) - (-5) = 0 + 5 + 5 = 10.$$

Multiplication of Two Real Numbers

1. To multiply two numbers with *unlike signs,* multiply their absolute values. The answer is *negative.*
2. To multiply two numbers having the *same sign,* multiply their absolute values. The answer is *positive.*

Thus we have $(-4)9 = -36$ and $\left(-\frac{2}{3}\right)\left(-\frac{3}{7}\right) = \frac{2}{7}$.

To divide, recall that the quotient $a \div b$ (also denoted a/b) is that number c for which $c \cdot b = a$. For example, $10 \div (-2) = -5$ since $(-5)(-2) = 10$; $(-12) \div 3 = -4$ since $(-4)3 = -12$; and $-18 \div (-6) = 3$ since $3(-6) = -18$. Thus the rules for division are just like those for multiplication.

Division of Two Real Numbers

1. To divide two numbers with *unlike signs,* divide their absolute values. The answer is *negative.*
2. To divide two numbers having the *same sign,* divide their absolute values. The answer is *positive.*

Thus we have

$$\frac{-45}{-15} = 3 \quad \text{and} \quad \frac{20}{-4} = -5.$$

Note that since

$$\frac{-8}{2} = \frac{8}{-2} = -\frac{8}{2} = -4,$$

we have the following generalization.

The Sign of a Fraction

For any number a and any nonzero number b,

$$\frac{-a}{b} = \frac{a}{-b} = -\frac{a}{b}.$$

Recall that

$$\frac{a}{b} = \frac{a}{1} \cdot \frac{1}{b} = a \cdot \frac{1}{b}.$$

That is, if we prefer, we can multiply by $1/b$ rather than divide by b. Provided that b is not 0, the numbers b and $1/b$ are called **reciprocals**, or **multiplicative inverses**, of each other.

The Law of Reciprocals

For any two numbers a and $1/a$ $(a \neq 0)$,

$$a \cdot \frac{1}{a} = 1.$$

(When reciprocals are multiplied, their product is 1.)

Example 7

Find the reciprocal: **(a)** $\frac{7}{8}$; **(b)** $-\frac{3}{4}$; **(c)** -8.

Solution

a) The reciprocal of $\frac{7}{8}$ is $\frac{8}{7}$ because $\frac{7}{8} \cdot \frac{8}{7} = 1$.

b) The reciprocal of $-\frac{3}{4}$ is $-\frac{4}{3}$.

c) The reciprocal of -8 is $\frac{1}{-8}$, or $-\frac{1}{8}$.

To divide, we can multiply by a reciprocal. We sometimes say that we "invert and multiply."

Example 8

Divide: **(a)** $-\frac{1}{4} \div \frac{3}{5}$; **(b)** $-\frac{6}{7} \div (-10)$.

Solution

a) $-\frac{1}{4} \div \frac{3}{5} = -\frac{1}{4} \cdot \frac{5}{3}$ "Inverting" $\frac{3}{5}$ and changing division to multiplication

$= -\frac{5}{12}$

b) $-\frac{6}{7} \div (-10) = -\frac{6}{7} \cdot \left(-\frac{1}{10}\right) = \frac{6}{70}$, or $\frac{3}{35}$

Thus far, we have never divided by 0 or, equivalently, had a denominator of 0. There is a reason for this. Suppose 5 were divided by 0. The answer would have to be a number that, when multiplied by 0, gave 5. But any number times 0 is 0. Thus we cannot divide 5 or any other nonzero number by 0.

What if we divide 0 by 0? In this case, our solution would need to be some number that, when multiplied by 0, gave 0. But then *any* number would work as a solution to $0 \div 0$. This could lead to contradictions so we agree to exclude division of 0 by 0 also.

Division by Zero

We never divide by 0. If asked to divide a nonzero number by 0, we say that the answer is *undefined*. If asked to divide 0 by 0, we say that the answer is *indeterminate*.

The rules for order of operations discussed in Section 1.1 apply to *all* real numbers, regardless of their signs.

Example 9 Simplify: $7 - 5^2 + 6 \div 2(-5)^2$.

Solution

$$7 - 5^2 + 6 \div 2(-5)^2 = 7 - 25 + 6 \div 2 \cdot 25 \quad \text{Simplifying } 5^2 \text{ and } (-5)^2$$
$$= 7 - 25 + 3 \cdot 25 \quad \text{Dividing}$$
$$= 7 - 25 + 75 \quad \text{Multiplying}$$
$$= -18 + 75 \quad \text{Subtracting}$$
$$= 57 \quad \text{Adding}$$

Besides parentheses, brackets, and braces, groupings may be indicated by a fraction bar, absolute-value symbol, or radical sign ($\sqrt{\ }$).

Example 10 Calculate: $\dfrac{12|7 - 9| + 4 \cdot 5}{(-3)^4 + 2^3}$.

Solution We simplify the numerator and the denominator and divide the results:

$$\frac{12|7 - 9| + 4 \cdot 5}{(-3)^4 + 2^3} = \frac{12|-2| + 20}{81 + 8}$$
$$= \frac{12(2) + 20}{89}$$
$$= \frac{44}{89}. \quad \text{Multiplying and adding}$$

Study Tip

Take the time to include all the steps when working your homework problems. Doing so will help you organize your thinking and avoid computational errors. It will also give you complete, step-by-step solutions of the exercises that will make sense when studying for quizzes and tests.

The Commutative, Associative, and Distributive Laws

When a pair of real numbers are added or multiplied, the order in which the numbers are written does not affect the result.

The Commutative Laws

For any real numbers a and b,

$a + b = b + a$; (for Addition)

$a \cdot b = b \cdot a$. (for Multiplication)

The commutative laws provide one way of writing *equivalent expressions.*

Equivalent Expressions

Two expressions that have the same value for all possible replacements are called *equivalent expressions.*

Much of this text is devoted to finding equivalent expressions.

Example 11 Use a commutative law to write an expression equivalent to $7x + 9$.

Solution Using the commutative law of addition, we have

$$7x + 9 = 9 + 7x.$$

We can also use the commutative law of multiplication to write

$$7x + 9 = x7 + 9.$$

The expressions $7x + 9$, $9 + 7x$, and $x7 + 9$ are all equivalent. They name the same number for any replacement of x.

The *associative laws* also enable us to form equivalent expressions.

The Associative Laws

For any real numbers a, b, and c,

$a + (b + c) = (a + b) + c$; (for Addition)

$a \cdot (b \cdot c) = (a \cdot b) \cdot c$. (for Multiplication)

Example 12 Write an expression equivalent to $(3x + 7y) + 9z$, using the associative law of addition.

Solution We have

$$(3x + 7y) + 9z = 3x + (7y + 9z).$$

The expressions $(3x + 7y) + 9z$ and $3x + (7y + 9z)$ are equivalent. They name the same number for any replacements of x, y, and z.

Example 13 Use the commutative and associative laws to write an expression equivalent to

$$\frac{5}{x} \cdot (yz).$$

Solution Answers may vary. We use the associative law first, then the commutative law.

$$\frac{5}{x} \cdot (yz) = \left(\frac{5}{x} \cdot y\right) \cdot z \quad \text{Using the associative law of multiplication}$$

$$= \left(y \cdot \frac{5}{x}\right) \cdot z \quad \text{Using the commutative law inside the parentheses}$$

The *distributive law* that follows provides still another way of forming equivalent expressions. In essence, the distributive law allows us to rewrite the *product* of a and $b + c$ as the *sum* of ab and ac.

The Distributive Law

For any real numbers a, b, and c,

$$a(b + c) = ab + ac.$$

Example 14 Obtain an expression equivalent to $5x(y + 4)$ by multiplying.

Solution We use the distributive law to get

$$5x(y + 4) = 5x \cdot y + 5x \cdot 4 \quad \text{Using the distributive law}$$

$$= 5xy + 5 \cdot 4 \cdot x \quad \text{Using the commutative law of multiplication}$$

$$= 5xy + 20x. \quad \text{Simplifying}$$

The expressions $5x(y + 4)$ and $5xy + 20x$ are equivalent. They name the same number for any replacements of x and y.

When we reverse what we did in Example 14, we say that we are **factoring** an expression. This allows us to rewrite a sum or a difference as a product.

Example 15 Obtain an expression equivalent to $3x - 6$ by factoring.

Solution We use the distributive law to get

$$3x - 6 = 3(x - 2).$$

In Example 15, since the product of 3 and $x - 2$ is $3x - 6$, we say that 3 and $x - 2$ are **factors** of $3x - 6$. Thus the word "factor" can act as a noun or as a verb.

Exercise Set 1.2

FOR EXTRA HELP

Digital Video Tutor CD 1
Videotape 1

InterAct Math

Math Tutor Center

MathXL

MyMathLab.com

Find each absolute value.

1. $|-9|$
2. $|-7|$
3. $|6|$
4. $|47|$
5. $|-6.2|$
6. $|-7.9|$
7. $|0|$
8. $|3\frac{3}{4}|$
9. $|1\frac{7}{8}|$
10. $|7.24|$
11. $|-4.21|$
12. $|-5.309|$

Write the meaning of each inequality, and determine whether it is a true statement.

13. $-6 \le -2$
14. $-1 \le -5$
15. $-9 > 1$
16. $7 \ge -2$
17. $3 \ge -5$
18. $9 \le 9$
19. $-8 < -3$
20. $7 \ge -8$
21. $-4 \ge -4$
22. $2 < 2$
23. $-5 < -5$
24. $-2 > -12$

Add.

25. $4 + 7$
26. $8 + 3$
27. $-4 + (-7)$
28. $-8 + (-3)$
29. $-3.9 + 2.7$
30. $-1.9 + 7.3$
31. $\frac{2}{7} + \left(-\frac{3}{5}\right)$
32. $\frac{3}{8} + \left(-\frac{2}{5}\right)$
33. $-4.9 + (-3.6)$
34. $-2.1 + (-7.5)$
35. $-\frac{1}{9} + \frac{2}{3}$
36. $-\frac{1}{2} + \frac{4}{5}$
37. $0 + (-4.5)$
38. $-3.19 + 0$
39. $-7.24 + 7.24$
40. $-9.46 + 9.46$
41. $15.9 + (-22.3)$
42. $21.7 + (-28.3)$

Find the opposite, or additive inverse.

43. 3.14
44. 5.43
45. $-4\frac{1}{3}$
46. $2\frac{3}{5}$
47. 0
48. $-2\frac{3}{4}$

Find $-x$ for each of the following.

49. $x = 7$
50. $x = 3$
51. $x = -2.7$
52. $x = -1.9$
53. $x = 1.79$
54. $x = 3.14$
55. $x = 0$
56. $x = -1$

Subtract.

57. $9 - 2$
58. $10 - 3$
59. $2 - 9$
60. $3 - 10$
61. $-6 - (-10)$
62. $-3 - (-9)$
63. $-5 - 14$
64. $-7 - 8$
65. $2.7 - 5.8$
66. $3.7 - 4.2$
67. $-\frac{3}{5} - \frac{1}{2}$
68. $-\frac{2}{3} - \frac{1}{5}$
69. $-3.9 - (-6.8)$
70. $-5.4 - (-4.3)$
71. $0 - (-7.9)$
72. $0 - 5.3$

Multiply.

73. $(-5)6$
74. $(-4)7$
75. $(-3)(-8)$
76. $(-7)(-8)$
77. $(4.2)(-5)$
78. $(3.5)(-8)$
79. $\frac{3}{7}(-1)$
80. $-1 \cdot \frac{2}{5}$
81. $(-17.45) \cdot 0$
82. 15.2×0
83. $(-3.2) \times (-1.7)$
84. $(1.9) \cdot (4.3)$

Divide.

85. $\frac{-10}{-2}$
86. $\frac{-15}{-3}$
87. $\frac{-100}{20}$
88. $\frac{-50}{5}$
89. $\frac{73}{-1}$
90. $\frac{-62}{1}$
91. $\frac{0}{-7}$
92. $\frac{0}{-11}$

Find the reciprocal, or multiplicative inverse.

93. 4
94. 3
95. -9
96. -5
97. $\frac{2}{3}$
98. $\frac{4}{7}$
99. $-\frac{3}{11}$
100. $-\frac{7}{3}$

Divide.

101. $\frac{2}{3} \div \frac{4}{5}$
102. $\frac{2}{7} \div \frac{6}{5}$
103. $-\frac{3}{5} \div \frac{1}{2}$
104. $\left(-\frac{4}{7}\right) \div \frac{1}{3}$
105. $\left(-\frac{2}{9}\right) \div (-8)$
106. $\left(-\frac{2}{11}\right) \div (-6)$
107. *Aha!* $-\frac{12}{7} \div \left(-\frac{12}{7}\right)$
108. $\left(-\frac{2}{7}\right) \div (-1)$

Calculate using the rules for order of operations.

109. $7 - (8 - 3 \cdot 2^3)$
110. $19 - (4 + 2 \cdot 3^2)$

111. $\dfrac{5 \cdot 2 - 4^2}{27 - 2^4}$

112. $\dfrac{7 \cdot 3 - 5^2}{9 + 4 \cdot 2}$

113. $\dfrac{3^4 - (5 - 3)^4}{8 - 2^3}$

114. $\dfrac{4^3 - (7 - 4)^2}{3^2 - 7}$

115. $\dfrac{(2 - 3)^3 - 5|2 - 4|}{7 - 2 \cdot 5^2}$

116. $\dfrac{8 \div 4 \cdot 6|4^2 - 5^2|}{9 - 4 + 11 - 4^2}$

117. $|2^2 - 7|^3 + 1$

118. $|-2 - 3| \cdot 4^2 - 1$

119. $28 - (-5)^2 + 15 \div (-3) \cdot 2$

120. $43 - (-9 + 2)^2 + 18 \div 6 \cdot (-2)$

121. $12 - \sqrt{11 - (3 + 4) \div [-5 - (-6)]^2}$

122. $15 - 1 + \sqrt{5^2 - (3 + 1)^2(-1)}$

Write an equivalent expression using a commutative law. Answers may vary.

123. $4a + 7b$

124. $6 + xy$

125. $(7x)y$

126. $-9(ab)$

Write an equivalent expression using an associative law.

127. $(3x)y$

128. $-7(ab)$

129. $x + (2y + 5)$

130. $(3y + 4) + 10$

Write an equivalent expression using the distributive law.

131. $3(a + 7)$

132. $8(x + 1)$

133. $4(x - y)$

134. $9(a - b)$

135. $-5(2a + 3b)$

136. $-2(3c + 5d)$

137. $9a(b - c + d)$

138. $5x(y - z + w)$

Find an equivalent expression by factoring.

139. $5x + 25$

140. $7a + 7b$

141. $3p - 9$

142. $15x - 3$

143. $7x - 21y + 14z$

144. $6y - 9x - 3w$

Aha! **145.** $255 - 34b$

146. $132a + 33$

147. Describe in your own words a method for determining the sign of the sum of a positive number and a negative number.

148. What is the difference between the associative law of multiplication and the distributive law?

SKILL MAINTENANCE

To the student and the instructor: *Exercises included for Skill Maintenance review skills previously studied in the text. Usually these exercises provide preparation for the next section of the text. The answers to all Skill Maintenance exercises appear at the back of the book, along with a section number indicating where that type of problem first appeared.*

Evaluate.

149. $2(x + 5)$ and $2x + 10$, for $x = 3$

150. $2a - 3$ and $a - 3 + a$, for $a = 7$

SYNTHESIS

151. Explain in your own words why $7/0$ is undefined.

152. Write a sentence in which the word "factor" appears once as a verb and once as a noun.

Insert one pair of parentheses to convert each false statement into a true statement.

153. $8 - 5^3 + 9 = 36$

154. $2 \cdot 7 + 3^2 \cdot 5 = 104$

155. $5 \cdot 2^3 \div 3 - 4^4 = 40$

156. $2 - 7 \cdot 2^2 + 9 = -11$

157. Find the greatest value of a for which $|a| \geq 6.2$ and $a < 0$.

158. Use the commutative, associative, and distributive laws to show that $5(a + bc)$ is equivalent to $c(b5) + a5$. Use only one law in each step of your work.

159. Are subtraction and division commutative? Why or why not?

160. Are subtraction and multiplication associative? Why or why not?

1.3 Solving Equations

Equivalent Equations • The Addition and Multiplication Principles • Combining Like Terms • Types of Equations

Solving equations is an essential part of problem solving in algebra. In this section, we review and practice solving simple equations.

Equivalent Equations

In Section 1.1, we saw that the solution of $41 + x = 65$ is 24. That is, when x is replaced with 24, the equation $41 + x = 65$ is a true statement. Although this solution may seem obvious, it is important to know how to find such a solution using the principles of algebra. These principles can produce *equivalent equations* from which solutions are easily found.

Equivalent Equations

Two equations are *equivalent* if they have the same solution(s).

Example 1

Determine whether $4x = 12$ and $10x = 30$ are equivalent equations.

Solution The equation $4x = 12$ is true only when x is 3. Similarly, $10x = 30$ is true only when x is 3. Since both equations have the same solution, they are equivalent.

Example 2

Determine whether $x + 4 = 7$ and $x = 3$ are equivalent equations.

Solution Each equation has only one solution, the number 3. Thus the equations are equivalent.

Example 3

Determine whether $3x = 4x$ and $3/x = 4/x$ are equivalent equations.

Solution Note that 0 is a solution of $3x = 4x$. Since neither $3/x$ nor $4/x$ is defined for $x = 0$, the equations $3x = 4x$ and $3/x = 4/x$ are *not* equivalent.

The Addition and Multiplication Principles

Suppose that a and b represent the same number and that some number c is added to a. If c is also added to b, we will get two equal sums, since a and b are the same number. The same is true if we multiply both a and b by c. In this manner, we can produce equivalent equations.

The Addition and Multiplication Principles for Equations

For any real numbers a, b, and c:

a) $a = b$ is equivalent to $a + c = b + c$;
b) $a = b$ is equivalent to $a \cdot c = b \cdot c$, provided $c \neq 0$.

Example 4

Solve: $y - 4.7 = 13.9$.

Solution

$$y - 4.7 = 13.9$$

$$y - 4.7 + 4.7 = 13.9 + 4.7 \quad \text{Using the addition principle; adding 4.7}$$

$$y + 0 = 13.9 + 4.7 \quad \text{The law of opposites}$$

$$y = 18.6$$

Check:

$$\begin{array}{c|l} y - 4.7 & 13.9 \\ \hline 18.6 - 4.7 \; ? & 13.9 \quad \text{Substituting 18.6 for } y \\ 13.9 & 13.9 \quad \text{TRUE} \end{array}$$

The solution is 18.6.

In Example 4, why did we add 4.7 to both sides? Because 4.7 is the opposite of -4.7 and we wanted y alone on one side of the equation. Adding 4.7 gave us $y + 0$, or just y, on the left side. This led to the equivalent equation, $y = 18.6$, from which the solution, 18.6, is immediately apparent.

Example 5

Solve: $\frac{2}{5}x = \frac{9}{10}$.

Solution We have

$$\frac{2}{5}x = \frac{9}{10}$$

$$\frac{5}{2} \cdot \frac{2}{5}x = \frac{5}{2} \cdot \frac{9}{10} \quad \text{Using the multiplication principle, we multiply by } \tfrac{5}{2}, \text{ the reciprocal of } \tfrac{2}{5}.$$

$$1x = \frac{45}{20} \quad \text{The law of reciprocals}$$

$$x = \frac{9}{4}. \quad \text{Simplifying}$$

The check is left to the student. The solution is $\frac{9}{4}$.

In Example 5, why did we multiply by $\frac{5}{2}$? Because $\frac{5}{2}$ is the reciprocal of $\frac{2}{5}$ and we wanted x alone on one side of the equation. When we multiplied by $\frac{5}{2}$, we got $1x$, or just x, on the left side. This led to the equivalent equation $x = \frac{9}{4}$, from which the solution, $\frac{9}{4}$, is clear.

There is no need for a subtraction or division principle because subtraction can be regarded as adding opposites and division can be regarded as multiplying by reciprocals.

Combining Like Terms

In an expression like $8a^5 + 17 + 4/b + (-6a^3b)$, the parts that are separated by addition signs are called *terms*. A **term** is a number, a variable, a product of numbers and/or variables, or a quotient of numbers and/or variables. Thus, $8a^5$, 17, $4/b$, and $-6a^3b$ are terms in $8a^5 + 17 + 4/b + (-6a^3b)$. When terms have variable factors that are exactly the same, we refer to those terms as **like**, or **similar, terms**. Thus, $3x^2y$ and $-7x^2y$ are similar terms, but $3x^2y$ and $4xy^2$ are not. We can often simplify expressions by **combining**, or **collecting, like terms**.

Example 6 Combine like terms: $3a + 5a^2 + 4a + a^2$.

Solution

$3a + 5a^2 + 4a + a^2 = 3a + 4a + 5a^2 + a^2$ Using the commutative law

$= (3 + 4)a + (5 + 1)a^2$ Using the distributive law. Note that $a^2 = 1a^2$.

$= 7a + 6a^2$

Sometimes we must use the distributive law to remove grouping symbols before combining like terms. Remember to remove the innermost grouping symbols first.

Example 7 Simplify: $3x + 2[4 + 5(x + 2y)]$.

Solution

$3x + 2[4 + 5(x + 2y)] = 3x + 2[4 + 5x + 10y]$ Using the distributive law

$= 3x + 8 + 10x + 20y$ Using the distributive law

$= 13x + 8 + 20y$ Combining like terms

The product of a number and -1 is its opposite, or additive inverse. For example,

$$-1 \cdot 8 = -8 \qquad \text{(the opposite of 8)}.$$

Thus we have $-8 = -1 \cdot 8$, and in general, $-x = -1 \cdot x$. We can use this fact along with the distributive law when parentheses are preceded by a negative sign or subtraction.

Example 8

Simplify $-(a - b)$, using multiplication by -1.

Solution We have

$-(a - b) = -1 \cdot (a - b)$	Replacing $-$ with multiplication by -1
$= -1 \cdot a - (-1) \cdot b$	Using the distributive law
$= -a - (-b)$	Replacing $-1 \cdot a$ with $-a$ and $(-1) \cdot b$ with $-b$
$= -a + b$, or $b - a$.	Try to go directly to this step.

The expressions $-(a - b)$ and $b - a$ are equivalent. They name the same number for all replacements of a and b.

Example 8 illustrates a useful shortcut worth remembering:

The opposite of $a - b$ is $-a + b$, or $b - a$.

Example 9

Simplify: $9x - 5y - (5x + y - 7)$.

Solution

$9x - 5y - (5x + y - 7) = 9x - 5y - 5x - y + 7$	Using the distributive law
$= 4x - 6y + 7$	Combining like terms

CONNECTING THE CONCEPTS

It is important to distinguish between forming *equivalent expressions* and writing *equivalent equations.*

In Examples 4 and 5, we used the addition and multiplication principles to write a series of equivalent equations that led to an equation for which the solution is clear. Because the equations were equivalent, the solution of the last equation was also a solution of the original equation.

In Examples 6–9, we used the commutative, associative, and distributive laws to write equivalent expressions that take on the same value when the variables are replaced with numbers. This is how we "simplify" an expression.

Equivalent Equations	*Equivalent Expressions*
$y - 4.7 = 13.9$	$3x + 2[4 + 5(x + 2y)]$
$y - 4.7 + 4.7 = 13.9 + 4.7$	$= 3x + 2[4 + 5x + 10y]$
$y + 0 = 13.9 + 4.7$	$= 3x + 8 + 10x + 20y$
$y = 18.6$	$= 13x + 8 + 20y$

Often, as in Example 10, we merge these ideas by forming an equivalent equation by replacing part of an equation with an equivalent expression.

Example 10

Solve: $5x - 2(x - 5) = 7x - 2$.

Solution

$5x - 2(x - 5) = 7x - 2$	
$5x - 2x + 10 = 7x - 2$	Using the distributive law
$3x + 10 = 7x - 2$	Combining like terms
$3x + 10 - 3x = 7x - 2 - 3x$	Using the addition principle; adding $-3x$, the opposite of $3x$, to both sides
$10 = 4x - 2$	Combining like terms
$10 + 2 = 4x - 2 + 2$	Using the addition principle
$12 = 4x$	Simplifying
$\frac{1}{4} \cdot 12 = \frac{1}{4} \cdot 4x$	Using the multiplication principle; multiplying both sides by $\frac{1}{4}$, the reciprocal of 4
$3 = x$	The law of reciprocals; simplifying

Check:

$5x - 2(x - 5)$	$= 7x - 2$	
$5 \cdot 3 - 2(3 - 5)$?	$7 \cdot 3 - 2$	
$15 - 2(-2)$	$21 - 2$	
$15 + 4$	19	
19	19	TRUE

The solution is 3.

Types of Equations

In Examples 4, 5, and 10, we solved *linear equations*. A **linear equation** in one variable—say, x—is an equation equivalent to one of the form $ax = b$ with a and b constants and $a \neq 0$.

Don't rush to solve equations in your head. Work neatly, keeping in mind that the number of steps in a solution is less important than producing a simpler, yet equivalent, equation in each step.

Every equation falls into one of three categories. An **identity** is an equation that is true for all replacements that can be used on both sides of the equation (for example, $x + 3 = 2 + x + 1$). A **contradiction** is an equation, like $n + 5 = n + 7$, that is *never* true. A **conditional equation**, like $2x + 5 = 17$, is sometimes true and sometimes false, depending on what the replacement of x is. Most of the equations examined in this text are conditional.

Example 11

Solve each of the following equations and classify the equation as an identity, a contradiction, or a conditional equation.

a) $2x + 7 = 7(x + 1) - 5x$
b) $3x - 5 = 3(x - 2) + 4$
c) $3 - 8x = 5 - 7x$

Solution

a) $2x + 7 = 7(x + 1) - 5x$

$2x + 7 = 7x + 7 - 5x$ Using the distributive law

$2x + 7 = 2x + 7$ Combining like terms

The equation $2x + 7 = 2x + 7$ is true regardless of what x is replaced with, so all real numbers are solutions. Note that $2x + 7 = 2x + 7$ is equivalent to $2x = 2x$, $7 = 7$, or $0 = 0$. All real numbers are solutions and the equation is an identity.

b) $3x - 5 = 3(x - 2) + 4$

$3x - 5 = 3x - 6 + 4$ Using the distributive law

$3x - 5 = 3x - 2$ Combining like terms

$-3x + 3x - 5 = -3x + 3x - 2$ Using the addition principle

$-5 = -2$

Since our original equation is equivalent to $-5 = -2$, which is false for any choice of x, there is no solution to this problem. There is no choice of x that will solve the original equation. The equation is a contradiction.

c) $3 - 8x = 5 - 7x$

$3 - 8x + 7x = 5 - 7x + 7x$ Using the addition principle

$3 - x = 5$ Simplifying

$-3 + 3 - x = -3 + 5$ Using the addition principle

$-x = 2$ Simplifying

$x = \frac{2}{-1}$, or -2 Dividing both sides by -1 or multiplying both sides by $\frac{1}{-1}$

There is one solution, -2. For other choices of x, the equation is false. This equation is conditional since it can be true or false, depending on the replacement for x.

We will sometimes refer to the set of solutions, or **solution set**, of a particular equation. Thus the solution set for Example 11(c) is $\{-2\}$. The solution set for Example 11(a) is simply $\mathbb{R}$, the set of all real numbers, and the solution set for Example 11(b) is the **empty set**, denoted $\varnothing$ or $\{\ \}$. As its name suggests, the empty set is the set containing no elements.

Exercise Set 1.3

FOR EXTRA HELP

Digital Video Tutor CD 1
Videotape 1

InterAct Math

Math Tutor Center

MathXL

MyMathLab.com

Determine whether the two equations in each pair are equivalent.

1. $3x = 15$ and $5x = 25$

2. $6x = 24$ and $15x = 60$

3. $t + 5 = 11$ and $3t = 18$

4. $t - 3 = 7$ and $3t = 24$

5. $12 - x = 3$ and $2x = 20$

6. $3x - 4 = 8$ and $3x = 12$

7. $5x = 2x$ and $\dfrac{4}{x} = 3$

8. $6 = 2x$ and $5 = \dfrac{2}{3 - x}$

Solve. Be sure to check.

9. $x - 2.9 = 13.4$

10. $y + 4.3 = 11.2$

11. $8t = 72$

12. $9t = 63$

13. $4x - 12 = 60$

14. $4x - 6 = 70$

15. $3n + 5 = 29$

16. $7t + 11 = 74$

17. $2y - 11 = 37$

18. $3x - 13 = 29$

Simplify by combining like terms. Use the distributive law as needed.

19. $3x + 7x$

20. $9x + 3x$

21. $7rt - 9rt$

22. $3ab + 7ab$

23. $9t^2 + t^2$

24. $7a^2 + a^2$

25. $12a - a$

26. $15x - x$

27. $n - 8n$

28. $x - 6x$

29. $5x - 3x + 8x$

30. $3x - 11x + 2x$

31. $4x - 2x^2 + 3x$

32. $9a - 5a^2 + 4a$

33. $6a + 7a^2 - a + 4a^2$

34. $9x + 2x^3 + 5x - 6x^2$

35. $4x - 7 + 18x + 25$

36. $13p + 5 - 4p + 7$

37. $-7t^2 + 3t + 5t^3 - t^3 + 2t^2 - t$

38. $-9n + 8n^2 + n^3 - 2n^2 - 3n + 4n^3$

39. $7a - (2a + 5)$

40. $x - (5x + 9)$

41. $m - (3m - 1)$

42. $5a - (4a - 3)$

43. $3d - 7 - (5 - 2d)$

44. $8x - 9 - (7 - 5x)$

45. $-2(x + 3) - 5(x - 4)$

46. $-9(y + 7) - 6(y - 3)$

47. $4x - 7(2x - 3)$

48. $9y - 4(5y - 6)$

49. $9a - [7 - 5(7a - 3)]$

50. $12b - [9 - 7(5b - 6)]$

51. $5\{-2a + 3[4 - 2(3a + 5)]\}$

52. $7\{-7x + 8[5 - 3(4x + 6)]\}$

53. $2y + \{7[3(2y - 5) - (8y + 7)] + 9\}$

54. $7b - \{6[4(3b - 7) - (9b + 10)] + 11\}$

Solve. Be sure to check.

55. $6x + 2x = 56$

56. $3x + 7x = 120$

57. $9y - 7y = 42$

58. $8t - 3t = 65$

59. $5t - 13t = -32$

60. $-9y - 5y = 28$

61. $2(x + 6) = 8x$

62. $3(y + 5) = 8y$

63. $70 = 10(3t - 2)$

64. $27 = 9(5y - 2)$

65. $180(n - 2) = 900$

66. $210(x - 3) = 840$

67. $5y - (2y - 10) = 25$

68. $8x - (3x - 5) = 40$

69. $7y - 1 = 23 - 5y$

70. $14t + 20 = 8t - 22$

71. $\frac{1}{5} + \frac{3}{10}x = \frac{4}{5}$

72. $-\frac{5}{2}x + \frac{1}{2} = -18$

73. $0.9y - 0.7 = 4.2$

74. $0.8t - 0.3t = 6.5$

75. $7r - 2 + 5r = 6r + 6 - 4r$

76. $9m - 15 - 2m = 6m - 1 - m$

77. $\frac{1}{4}(16y + 8) - 17 = -\frac{1}{2}(8y - 16)$

78. $\frac{1}{3}(6t + 48) - 20 = -\frac{1}{4}(12t - 72)$

79. $5 + 2(x - 3) = 2[5 - 4(x + 2)]$

80. $3[2 - 4(x - 1)] = 3 - 4(x + 2)$

Find each solution set. Then classify each equation as a conditional equation, an identity, or a contradiction.

81. $5x + 7 - 3x = 2x$

82. $3t + 5 + t = 5 + 4t$

83. $1 + 9x = 3(4x + 1) - 2$

84. $4 + 7x = 7(x + 1)$

Aha! **85.** $-9t + 2 = -9t - 7(6 \div 2(49) + 8)$

86. $-9t + 2 = 2 - 9t - 5(8 \div 4(1 + 3^4))$

87. $2\{9 - 3[-2x - 4]\} = 12x + 42$

88. $3\{7 - 2[7x - 4]\} = -40x + 45$

89. As the first step in solving

$$2x + 5 = -3,$$

Pat multiplies both sides by $\frac{1}{2}$. Is this incorrect? Why or why not?

90. Explain how an identity can be easily altered so that it becomes a contradiction.

SKILL MAINTENANCE

Translate to an algebraic expression.

91. Nine more than twice a number

92. Forty-two percent of half of a number

SYNTHESIS

93. Explain the difference between equivalent expressions and equivalent equations.

94. Explain how the distributive and commutative laws can be used to rewrite $3x + 6y + 4x + 2y$ as $7x + 8y$.

Solve and check. The symbol ▦ indicates an exercise designed to be solved with a calculator.

95. $4.23x - 17.898 = -1.65x - 42.454$

96. $-0.00458y + 1.7787 = 13.002y - 1.005$

97. $4x - \{3x - [2x - (5x - (7x - 1))]\} = 4x + 7$

98. $3x - \{5x - [7x - (4x - (3x + 1))]\} = 3x + 5$

99. $17 - 3\{5 + 2[x - 2]\} + 4\{x - 3(x + 7)\}$
$= 9\{x + 3[2 + 3(4 - x)]\}$

100. $23 - 2\{4 + 3[x - 1]\} + 5\{x - 2(x + 3)\}$
$= 7\{x - 2[5 - (2x + 3)]\}$

101. Create an equation for which it is preferable to use the multiplication principle *before* using the addition principle. Explain why it is best to solve the equation in this manner.

1.4 Introduction to Problem Solving

The Five-Step Strategy • Problem Solving

We now begin to study and practice the "art" of problem solving. Although we are interested mainly in using algebra to solve problems, much of what we say here applies to solving all kinds of problems.

What do we mean by a *problem*? Perhaps you've already used algebra to solve some "real-world" problems. What procedure did you use? Is there an approach that can be used to solve problems of a more general nature? These are some questions that we will answer in this section.

In this text, we do not restrict the use of the word "problem" to computational situations involving arithmetic or algebra, such as $589 + 437 = a$ or

$3x + 5x = 9$. We mean instead some question to which we wish to find an answer. Perhaps this can best be illustrated with some sample problems:

1. Can I afford to rent a bigger apartment?
2. If I exercise twice a week and eat 3000 calories a day, will I lose weight?
3. Do I have enough time to take 4 courses while working 20 hours a week?
4. My fishing boat travels 12 km/h in still water. How long will it take me to cruise 25 km upstream if the river's current is 3 km/h?

Although these problems are all different, there is a general strategy that can be applied to all of them.

The Five-Step Strategy

Since you have already studied some algebra, you have had some experience with problem solving. The following steps constitute a strategy that you may already have used and a good strategy for problem solving in general.

Five Steps for Problem Solving with Algebra

1. *Familiarize* yourself with the problem.
2. *Translate* to mathematical language.
3. *Carry out* some mathematical manipulation.
4. *Check* your possible answer in the original problem.
5. *State* the answer clearly.

Of the five steps, probably the most important is the first: becoming familiar with the problem situation. Here are some ways in which this can be done.

The First Step in Problem Solving with Algebra

To familiarize yourself with the problem:

1. If the problem is written, read it carefully.
2. Reread the problem, perhaps aloud. Verbalize the problem to yourself.
3. List the information given and restate the question being asked. Select a variable or variables to represent any unknown(s) and clearly state what each variable represents. Be descriptive! For example, let $t =$ the flight time, in hours; let $p =$ Paul's weight, in kilograms; and so on.
4. Find additional information. Look up formulas or definitions with which you are not familiar. Geometric formulas appear on the inside back cover of this text; important words appear in the index. Consult an expert in the field or a reference librarian.
5. Create a table, using variables, in which both known and unknown information is listed. Look for possible patterns.
6. Make and label a drawing.
7. Estimate or guess an answer and check to see whether it is correct.

Example 1

How might you familiarize yourself with the situation of Problem 1: "Can I afford to rent a bigger apartment?"

Solution Clearly more information is needed to solve this problem. You might:

a) Estimate the rent of some apartments in which you are interested.
b) Examine what your savings are and how your income is budgeted.
c) Determine how much rent you can afford.

When enough information is known, it might be wise to make a chart or table to help you reach an answer.

Example 2

How might you familiarize yourself with Problem 4: "How long will it take the boat to cruise 25 km upstream?"

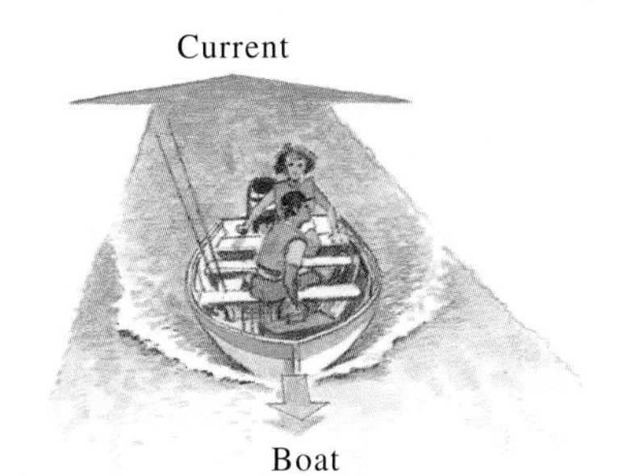

Solution First read the question *very* carefully. This may even involve speaking aloud. You may need to reread the problem several times to fully understand what information is given and what information is required. A sketch or table is often helpful.

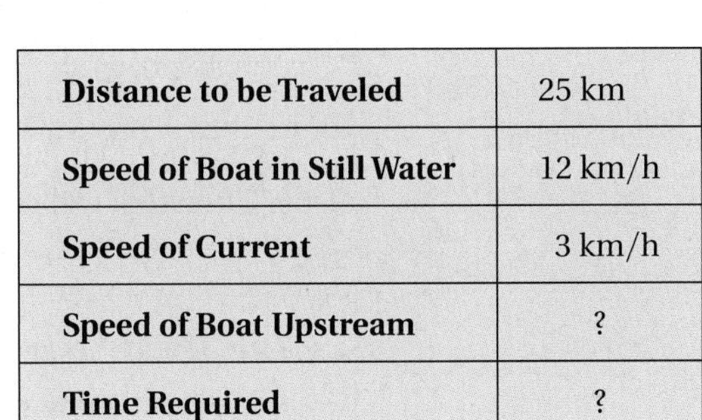

Distance to be Traveled	25 km
Speed of Boat in Still Water	12 km/h
Speed of Current	3 km/h
Speed of Boat Upstream	?
Time Required	?

To continue the familiarization process, we should determine, possibly with the aid of outside references, what relationships exist among the various quantities in the problem. With some effort it can be learned that the current's speed should be subtracted from the boat's speed in still water to determine the boat's speed going upstream. We also need to either find or recall an extremely important formula:

Distance = Speed × Time.

We rewrite part of the table, letting $t =$ the number of hours required for the boat to cruise 25 km upstream.

Distance to be Traveled	25 km
Speed of Boat Upstream	$12 - 3 = 9$ km/h
Time Required	t

At this point we might try a guess. Suppose the boat traveled upstream for 2 hr. The boat would have then traveled

$$9\,\frac{\text{km}}{\text{hr}} \times 2\text{ hr} = 18\text{ km}. \qquad \text{Note that } \frac{\text{km}}{\text{hr}} \cdot \text{hr} = \text{km}.$$

$$\uparrow \qquad \uparrow \qquad \uparrow$$

$$\text{Speed} \times \text{Time} = \text{Distance}$$

Since $18 \neq 25$, our guess is wrong. Still, examining how we checked our guess sheds light on how to translate the problem to an equation. Note that a better guess, when multiplied by 9, would yield a number closer to 25.

The second step in problem solving is to translate the situation to mathematical language. In algebra, this often means forming an equation.

The Second Step in Problem Solving with Algebra

Translate the problem to mathematical language. In some cases, translation can be done by writing an algebraic expression, but most problems in this text are best solved by translating to an equation.

In the third step of our process, we work with the results of the first two steps. Often this will require us to use the algebra that we have studied.

The Third Step in Problem Solving with Algebra

Carry out some mathematical manipulation. If you have translated to an equation, this means to solve the equation.

To complete the problem-solving process, we should always **check** our solution and then **state** the solution in a clear and precise manner. To check, we make sure that our answer is reasonable and that all the conditions of the original problem have been satisfied. If our answer checks, we write a complete English sentence stating the solution. The five steps are listed again below. Try to apply them regularly in your work.

Five Steps for Problem Solving with Algebra

1. *Familiarize* yourself with the problem.
2. *Translate* to mathematical language.
3. *Carry out* some mathematical manipulation.
4. *Check* your possible answer in the original problem.
5. *State* the answer clearly.

Problem Solving

At this point, our study of algebra has just begun. Thus we have few algebraic tools with which to work problems. As the number of tools in our algebraic "toolbox" increases, so will the difficulty of the problems we can solve. For now our problems may seem simple; however, to gain practice with the problem-solving process, you should try to use all five steps. Later some steps may be shortened or combined.

Example 3

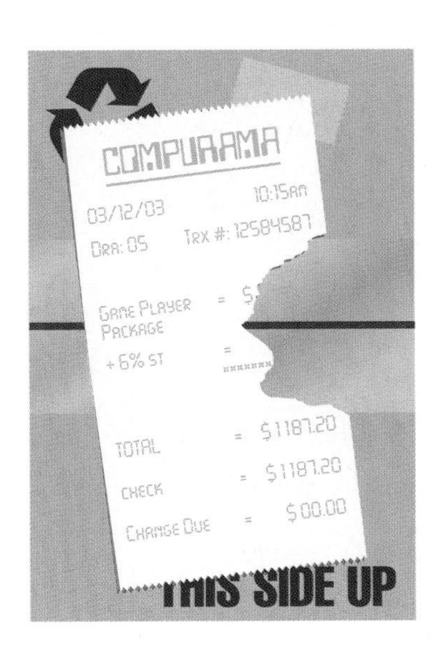

Purchasing. Elka pays \$1187.20 for a computer. If the price paid includes a 6% sales tax, what is the price of the computer itself?

Solution

1. **Familiarize.** First, we familiarize ourselves with the problem. Note that tax is calculated from, and then added to, the computer's price. Let's guess that the computer's price is \$1000. To check the guess, we calculate the amount of tax, $(0.06)(\$1000) = \$60$, and add it to \$1000:

$$\begin{aligned} \$1000 + (0.06)(\$1000) &= \$1000 + \$60 \\ &= \$1060. \qquad \$1060 \neq \$1187.20 \end{aligned}$$

Our guess was too low, but the manner in which we checked the guess will guide us in the next step. We let

$C =$ the computer's price, in dollars.

2. **Translate.** Our guess leads us to the following translation:

Rewording:	The computer's price	plus	6% sales tax	is	the price with sales tax.
	↓	↓	↓	↓	↓
Translating:	C	$+$	$(0.06)C$	$=$	\$1187.20

3. **Carry out.** Next, we carry out some mathematical manipulation:

$$\begin{aligned} C + (0.06)C &= 1187.20 \\ 1.06C &= 1187.20 && \text{Combining like terms} \\ \frac{1}{1.06} \cdot 1.06C &= \frac{1}{1.06} \cdot 1187.20 && \text{Using the multiplication principle} \\ C &= 1120. \end{aligned}$$

4. **Check.** To check the answer in the original problem, note that the tax on a computer costing \$1120 would be $(0.06)(\$1120) = \$67.20$. When this is added to \$1120, we have

$\$1120 + \$67.20$, or \$1187.20.

Thus, \$1120 checks in the original problem.

5. **State.** We clearly state the answer: The computer itself costs \$1120.

Example 4

Home maintenance. In an effort to make their home more energy-efficient, Alma and Drew purchased 200 in. of 3M Press-In-Place window glazing. This will be just enough to outline their two square skylights. If the length of the sides of the larger skylight is $1\frac{1}{2}$ times the length of the sides of the smaller one, how should the glazing be cut?

Solution

1. **Familiarize.** Note that the *perimeter* of (distance around) each square is four times the length of a side. Furthermore, if s is used to represent the length of a side of the smaller square, then $\left(1\frac{1}{2}\right)s$ will represent the length of a side of the larger square. We make a drawing and note that the two perimeters must add up to 200 in.

   Perimeter of a square $= 4 \cdot$ *length of a side*

2. **Translate.** Rewording the problem can help us translate:

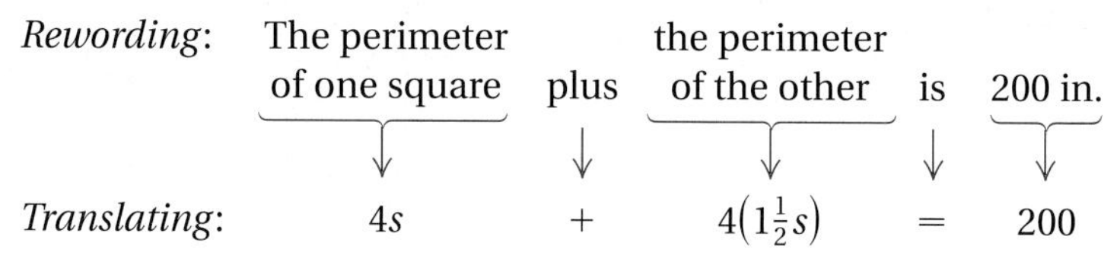
Rewording:
The perimeter of one square
plus
the perimeter of the other
is
200 in.
Translating:
4s
+
4(1½s)
=
200

3. **Carry out.** We solve the equation:

$$4s + 4\left(1\tfrac{1}{2}s\right) = 200$$
$$4s + 6s = 200 \qquad \text{Simplifying}$$
$$10s = 200 \qquad \text{Combining like terms}$$
$$s = \frac{1}{10} \cdot 200 \qquad \text{Multiplying both sides by } \tfrac{1}{10}$$
$$s = 20. \qquad \text{Simplifying}$$

4. **Check.** If 20 is the length of the smaller side, then $\left(1\frac{1}{2}\right)(20) = 30$ is the length of the larger side. The two perimeters would then be

   $4 \cdot 20 \text{ in.} = 80 \text{ in.}$ and $4 \cdot 30 \text{ in.} = 120 \text{ in.}$

   Since $80 \text{ in.} + 120 \text{ in.} = 200 \text{ in.}$, our answer checks.

5. **State.** The glazing should be cut into two pieces, one 80 in. long and the other 120 in. long.

We cannot stress too much the importance of labeling the variables in your problem. In Example 4, solving for s was not enough: We needed to find $4s$ and $4\left(1\frac{1}{2}s\right)$ to determine the numbers we were after.

Example 5

Three numbers are such that the second is 6 less than three times the first and the third is 2 more than two-thirds the first. The sum of the three numbers is 150. Find the largest of the three numbers.

Solution We proceed according to the five-step process.

1. **Familiarize.** We need to find the largest of three numbers. We list the information given in a table in which x represents the first number.

First Number	x
Second Number	6 less than 3 times the first
Third Number	2 more than $\frac{2}{3}$ the first

$$\text{First} + \text{Second} + \text{Third} = 150$$

Try to check a guess at this point. We will proceed to the next step.

2. **Translate.** Because we wish to write an equation in just one variable, we need to express the second and third numbers using x ("in terms of x"). To do so, we expand the table:

First Number	x	x
Second Number	6 less than 3 times the first	$3x - 6$
Third Number	2 more than $\frac{2}{3}$ the first	$\frac{2}{3}x + 2$

We know that the sum is 150. Substituting, we obtain an equation:

$$\underbrace{\text{First}} + \underbrace{\text{second}} + \underbrace{\text{third}} = 150.$$

$$x + (3x - 6) + \left(\tfrac{2}{3}x + 2\right) = 150$$

3. **Carry out.** We solve the equation:

$$x + 3x - 6 + \tfrac{2}{3}x + 2 = 150 \quad \text{Leaving off unnecessary parentheses}$$
$$\left(4 + \tfrac{2}{3}\right)x - 4 = 150 \quad \text{Combining like terms}$$
$$\tfrac{14}{3}x - 4 = 150$$
$$\tfrac{14}{3}x = 154 \quad \text{Adding 4 to both sides}$$
$$x = \tfrac{3}{14} \cdot 154 \quad \text{Multiplying both sides by } \tfrac{3}{14}$$
$$x = 33. \quad \text{Remember, } x \text{ represents the first number.}$$

Going back to the table, we can find the other two numbers:

Second: $3x - 6 = 3 \cdot 33 - 6 = 93$;

Third: $\frac{2}{3}x + 2 = \frac{2}{3} \cdot 33 + 2 = 24$.

4. **Check.** We return to the original problem. There are three numbers: 33, 93, and 24. Is the second number 6 less than three times the first?

$$3 \times 33 - 6 = 99 - 6 = 93$$

The answer is *yes*.

Is the third number 2 more than two-thirds the first?

$$\tfrac{2}{3} \times 33 + 2 = 22 + 2 = 24$$

The answer is *yes*.

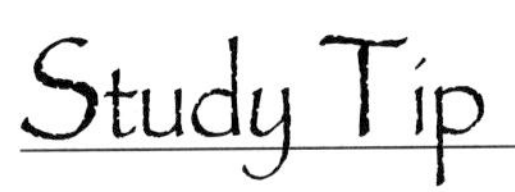

Consider forming a study group with some of your fellow students. Exchange telephone numbers, schedules, and any e-mail addresses so that you can coordinate study time for homework and tests.

Is the sum of the three numbers 150?

$33 + 93 + 24 = 150$

The answer is *yes*. The numbers do check.

5. **State.** The problem asks us to find the largest number, so the answer is: "The largest of the three numbers is 93."

> ***Caution!*** In Example 5, although the equation $x = 33$ enabled us to find the largest number, 93, the number 33 was *not* the solution to the problem. By carefully labeling our variable in the first step of problem solving, we may avoid the temptation of thinking that our variable always represents the solution of the problem.

Exercise Set 1.4

FOR EXTRA HELP

Digital Video Tutor CD 1
Videotape 2

InterAct Math

Math Tutor Center

MathXL

MyMathLab.com

For each problem, familiarize yourself with the situation. Then translate to mathematical language. You need not actually solve the problem; just carry out the first two steps of the five-step strategy. You will be asked to complete some of the solutions as Exercises 31–38.

1. The sum of two numbers is 65. One of the numbers is 7 more than the other. What are the numbers?

2. The sum of two numbers is 83. One of the numbers is 11 more than the other. What are the numbers?

3. *Boating.* The Delta Queen is a paddleboat that tours the Mississippi River near New Orleans, Louisiana. It is not uncommon for the Delta Queen to run 7 mph in still water and for the Mississippi to flow at a rate of 3 mph (*Source*: Delta Queen information). At these rates, how long will it take the boat to cruise 2 mi upstream?

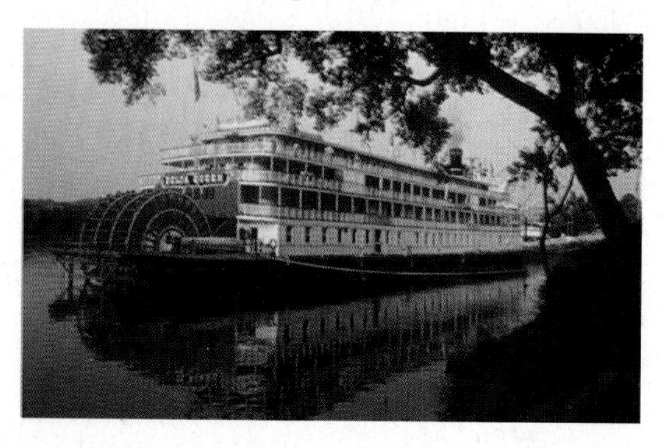

4. *Swimming.* Fran swims at a rate of 5 km/h in still water. The Lazy River flows at a rate of 2.3 km/h. How long will it take Fran to swim 1.8 km upstream?

5. *Moving sidewalks.* The moving sidewalk in O'Hare Airport is 300 ft long and moves at a rate of 5 ft/sec. If Alida walks at a rate of 4 ft/sec, how long will it take her to walk the length of the moving sidewalk?

6. *Aviation.* A Cessna airplane traveling 390 km/h in still air encounters a 65-km/h headwind. How long will it take the plane to travel 725 km into the wind?

7. *Angles in a triangle.* The degree measures of the angles in a triangle are three consecutive integers.

Find the measures of the angles.

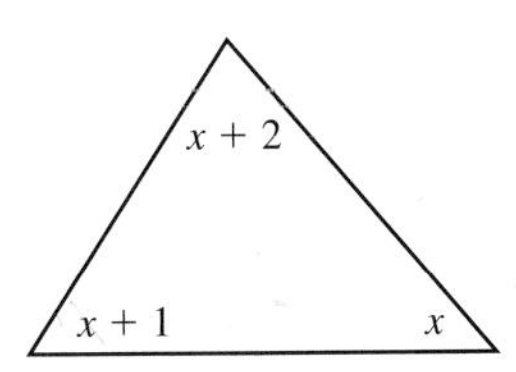

8. *Pricing.* The Sound Connection prices TDK 100-min blank audiotapes by raising the wholesale price 50% and adding 25 cents. What must a tape's wholesale price be if the tape is to sell for \$1.99?

9. *Pricing.* Becker Lumber gives contractors a 10% discount on all orders. After the discount, a contractor's order cost \$279. What was the original cost of the order?

10. *Pricing.* Miller Oil offers a 5% discount to customers who pay promptly for an oil delivery. The Blancos promptly paid \$142.50 for their December oil bill. What would the cost have been had they not promptly paid?

11. *Cruising altitude.* A Boeing 747 has been instructed to climb from its present altitude of 8000 ft to a cruising altitude of 29,000 ft. If the plane ascends at a rate of 3500 ft/min, how long will it take to reach the cruising altitude?

12. A piece of wire 10 m long is to be cut into two pieces, one of them $\frac{2}{3}$ as long as the other. How should the wire be cut?

13. *Angles in a triangle.* One angle of a triangle is three times as great as a second angle. The third angle measures 12° less than twice the second angle. Find the measures of the angles.

14. *Angles in a triangle.* One angle of a triangle is four times as great as a second angle. The third angle measures 5° more than twice the second angle. Find the measures of the angles.

15. Find two consecutive even integers such that two times the first plus three times the second is 76.

16. Find three consecutive odd integers such that the sum of the first, twice the second, and three times the third is 70.

17. A steel rod 90 cm long is to be cut into two pieces, each to be bent to make an equilateral triangle. The length of a side of one triangle is to be twice the length of a side of the other. How should the rod be cut?

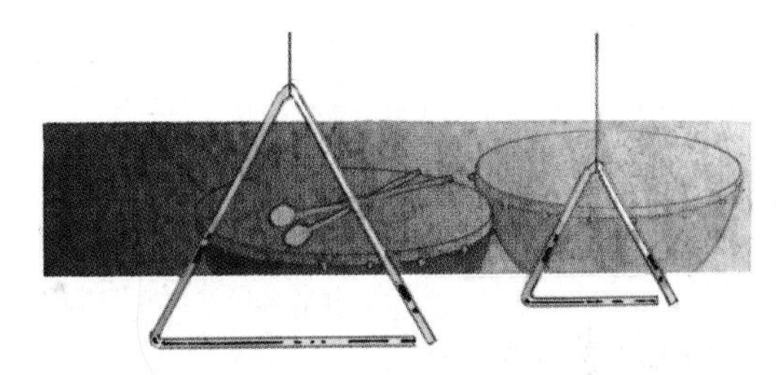

18. A piece of wire 100 cm long is to be cut into two pieces, and those pieces are each to be bent to make a square. The area of one square is to be 144 cm^2 greater than that of the other. How should the wire be cut? (*Remember*: Do not solve.)

19. *Test scores.* Deirdre's scores on five tests are 93, 89, 72, 80, and 96. What must the score be on her next test so that the average will be 88?

20. *Pricing.* Whitney's Appliances is having a sale on 13 TV sets. They are displayed in order of increasing price from left to right. The price of each set differs by \$20 from either set next to it. For the price of the set at the extreme right, a customer can buy both the second and seventh sets. What is the price of the least expensive set?

Solve each problem. Use all five problem-solving steps.

21. The number 38.2 is less than some number by 12.1. What is the number?

22. The number 173.5 is greater than a certain number by 16.8. What is the number?

23. The number 128 is 0.4 of what number?

24. The number 456 is $\frac{1}{3}$ of what number?

25. One number exceeds another by 12. The sum of the numbers is 114. What is the larger number?

26. One number is less than another by 65. The sum of the numbers is 92. What is the smaller number?

27. A rectangle's length is three times its width and its perimeter is 120 cm. Find the dimensions of the rectangle.

28. A rectangle's length is twice its width and its perimeter is 21 cm. Find the dimensions of the rectangle.

29. A rectangle's width is one-fourth its length and its perimeter is 130 m. Find the dimensions of the rectangle.

30. A rectangle's width is one-third its length and its perimeter is 32 m. Find the dimensions of the rectangle.

31. Solve the problem of Exercise 4.

32. Solve the problem of Exercise 3.

33. Solve the problem of Exercise 14.

34. Solve the problem of Exercise 13.

35. Solve the problem of Exercise 10.

36. Solve the problem of Exercise 9.

37. Solve the problem of Exercise 8.

38. Solve the problem of Exercise 15.

39. Write a problem for a classmate to solve. Devise the problem so that the solution is "The first angle is 40°, the second angle is 50°, and the third angle is 90°."

40. Write a problem for a classmate to solve. Devise the problem so that the solution is "The material should be cut into two pieces, one 30 cm long and the other 45 cm long."

SKILL MAINTENANCE

Solve.

41. $7 = \frac{2}{3}(x + 6)$

42. $9 = \frac{x}{4}$

43. $8 = \frac{5 + t}{3}$

44. $6t - 8 = 0$

SYNTHESIS

45. How can a guess or estimate help prepare you for the *Translate* step when solving problems?

46. Why is it important to check the solution from step 3 (*Carry out*) in the original wording of the problem being solved?

47. *Test scores.* Tico's scores on four tests are 83, 91, 78, and 81. How many points above his current average must Tico score on the next test in order to raise his average 2 points?

48. *Geometry.* The height and sides of a triangle are four consecutive integers. The height is the first integer, and the base is the third integer. The perimeter of the triangle is 42 in. Find the area of the triangle.

49. *Home prices.* Panduski's real estate prices increased 6% from 1998 to 1999 and 2% from 1999 to 2000. From 2000 to 2001, prices dropped 1%. If a house sold for $117,743 in 2001, what was its worth in 1998? (Round to the nearest dollar.)

50. *Adjusted wages.* Blanche's salary is reduced $n\%$ during a period of financial difficulty. By what number should her salary be multiplied in order to bring it back to where it was before the reduction?

COLLABORATIVE CORNER

Who Pays What?

Focus: Problem solving

Time: 15 minutes

Group size: 5

Suppose that two of the five members in each group are celebrating birthdays and the entire group goes out to lunch. Suppose further that each member whose birthday it is gets treated to his or her lunch by the other *four* members. Finally, suppose that all meals cost the same amount and that the total bill is $40.00.*

*This activity was inspired by "The Birthday-Lunch Problem," *Mathematics Teaching in the Middle School,* vol. 2, no. 1, September–October 1996, pp. 40–42.

ACTIVITY

1. Determine, as a group, how much each group member should pay for the lunch described above. Then explain how this determination was made.
2. Compare the results and methods used for part (1) with those of the other groups in the class.

1.5 Formulas, Models, and Geometry

Solving Formulas • Mathematical Models

A **formula** is an equation that uses letters to represent a relationship between two or more quantities. For example, in Section 1.4, we made use of the formula $P = 4s$, where P represents the perimeter of a square and s the length of a side. Other important geometric formulas are $A = \pi r^2$ (for the area A of a circle of radius r), $C = \pi d$ (for the circumference C of a circle of diameter d), and $A = b \cdot h$ (for the area A of a parallelogram of height h and base length b).* A more complete list of geometric formulas appears at the very end of this text.

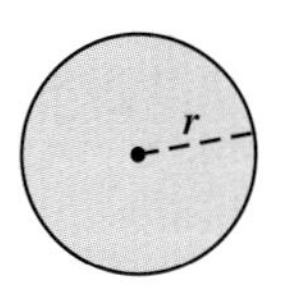

$A = \pi r^2$

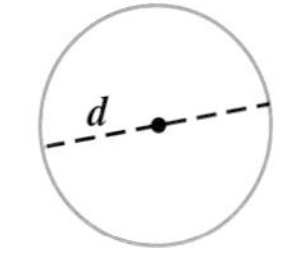

$C = \pi d$

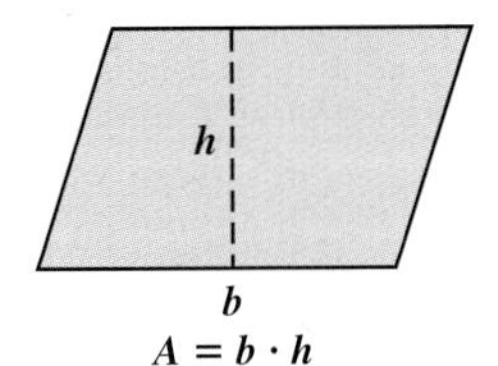

$A = b \cdot h$

*The Greek letter π, read "pi," is *approximately* 3.14159265358979323846264. Often 3.14 or 22/7 is used to approximate π when a calculator with a π key is unavailable.

Solving Formulas

Suppose we remember the area and the width of a rectangular room and want to find the length. To do so, we could "solve" the formula $A = l \cdot w$ (Area = Length · Width) for l, using the same principles that we use for solving equations.

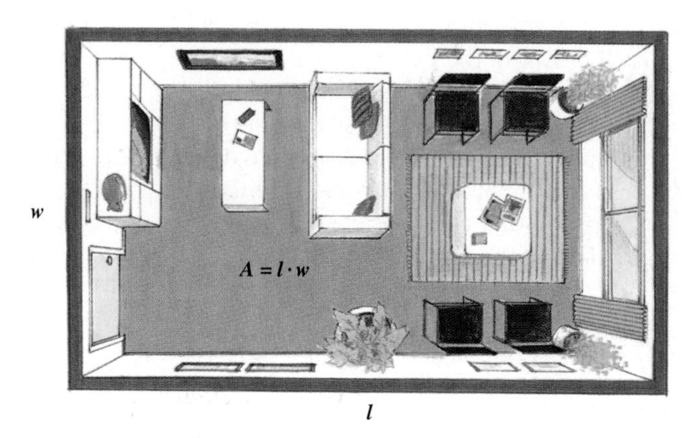

Example 1 ***Area of a rectangle.*** Solve the formula $A = l \cdot w$ for l.

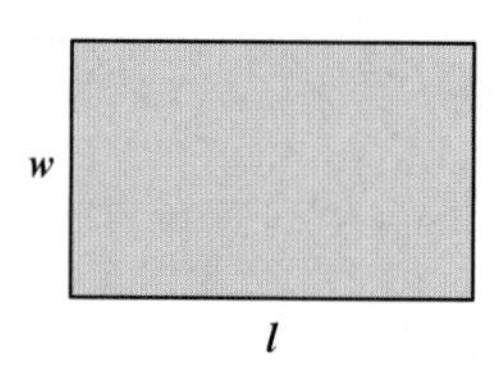

Solution

$$A = l \cdot w \quad \text{We want this letter alone.}$$

$$\frac{A}{w} = \frac{l \cdot w}{w} \quad \text{Dividing both sides by } w\text{, or multiplying both sides by } 1/w$$

$$\left.\begin{aligned} \frac{A}{w} &= l \cdot \frac{w}{w} \\ \frac{A}{w} &= l \end{aligned}\right\} \quad \text{Simplifying by removing a factor equal to 1: } \frac{w}{w} = 1$$

Thus to find the length of a rectangular room, we can divide the area of the room by its width. Were we to do this calculation for a variety of rectangular rooms, the formula $l = A/w$ would be more convenient than repeatedly substituting into $A = l \cdot w$.

Example 2 ***Simple interest.*** The formula $I = Prt$ is used to determine the simple interest I earned when a principal of P dollars is invested for t years at an interest rate r. Solve this formula for t.

Solution

$$I = Prt \qquad \text{We want this letter alone.}$$

$$\frac{I}{Pr} = \frac{Prt}{Pr} \qquad \text{Dividing both sides by } Pr\text{, or multiplying both sides by } \frac{1}{Pr}$$

$$\frac{I}{Pr} = t \qquad \text{Simplifying by removing a factor equal to 1: } \frac{Pr}{Pr} = 1$$

Example 3

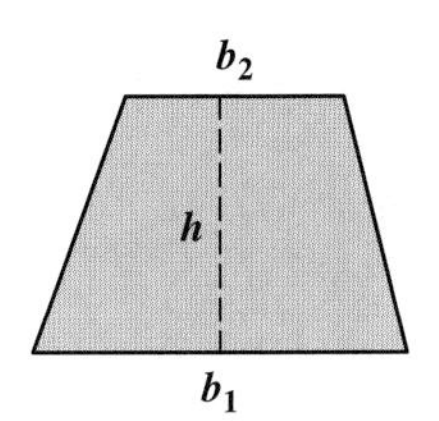

Area of a trapezoid. A trapezoid is a geometric shape with four sides, exactly two of which, the bases, are parallel to each other. The formula for calculating the area A of a trapezoid with bases b_1 and b_2 (read "b sub one" and "b sub two") and height h is given by

$$A = \frac{h}{2}(b_1 + b_2),$$

A derivation of this formula is outlined in Exercise 66 of this section.

where the *subscripts* 1 and 2 distinguish one base from the other. Solve for b_1.

Solution There are several ways to "remove" the parentheses. We could distribute $h/2$, but an easier approach is to multiply both sides by the reciprocal of $h/2$.

$$A = \frac{h}{2}(b_1 + b_2)$$

$$\frac{2}{h} \cdot A = \frac{2}{h} \cdot \frac{h}{2}(b_1 + b_2) \qquad \text{Multiplying both sides by } \frac{2}{h} \left(\text{or dividing by } \frac{h}{2}\right)$$

$$\frac{2A}{h} = b_1 + b_2 \qquad \text{Simplifying. The right side is "cleared" of fractions.}$$

$$\frac{2A}{h} - b_2 = b_1 \qquad \text{Adding } -b_2 \text{ on both sides}$$

The similarities between solving formulas and solving equations can be seen below. In (a), we solve as we did before; in (b), we do not carry out all calculations; and in (c), we cannot carry out all calculations because the numbers are unknown. The same steps are used each time.

a)

$$9 = \frac{3}{2}(x + 5)$$

$$\frac{2}{3} \cdot 9 = \frac{2}{3} \cdot \frac{3}{2}(x + 5)$$

$$6 = x + 5$$

$$1 = x$$

b)

$$9 = \frac{3}{2}(x + 5)$$

$$\frac{2}{3} \cdot 9 = \frac{2}{3} \cdot \frac{3}{2}(x + 5)$$

$$\frac{2 \cdot 9}{3} = x + 5$$

$$\frac{2 \cdot 9}{3} - 5 = x$$

c)

$$A = \frac{h}{2}(b_1 + b_2)$$

$$\frac{2}{h} \cdot A = \frac{2}{h} \cdot \frac{h}{2}(b_1 + b_2)$$

$$\frac{2A}{h} = b_1 + b_2$$

$$\frac{2A}{h} - b_2 = b_1$$

Example 4 ***Simple interest.*** The formula $A = P + Prt$ gives the amount A that a principal of P dollars will be worth in t years when invested at simple interest rate r. Solve the formula for P.

Solution

$A = P + Prt$	We want this letter alone.
$A = P(1 + rt)$	Factoring (using the distributive law) to combine like terms
$\dfrac{A}{1 + rt} = \dfrac{P(1 + rt)}{1 + rt}$	Dividing both sides by $1 + rt$, or multiplying both sides by $\dfrac{1}{1 + rt}$
$\dfrac{A}{1 + rt} = P$	Simplifying

This last equation can be used to determine how much should be invested at interest rate r in order to have A dollars t years later.

Note in Example 4 that the factoring enabled us to write P once rather than twice. This is comparable to combining like terms when solving an equation like $16 = x + 7x$.

You may find the following summary useful.

To Solve a Formula for a Specified Letter

1. Get all terms with the letter being solved for on one side of the equation and all other terms on the other side, using the addition principle. To do this may require removing parentheses.
   - To remove parentheses, either divide both sides by the multiplier in front of the parentheses or use the distributive law.
2. When all terms with the specified letter are on the same side, factor (if necessary) to combine like terms.
3. Solve for the letter in question by dividing both sides by the multiplier of that letter.

Mathematical Models

The above formulas from geometry and economics are examples of *mathematical models.* A **mathematical model** can be a formula, or set of formulas, developed to represent a real-world situation. In problem solving, a mathematical model is formed in the *Translate* step.

Example 5

Density. A collector suspects that a silver coin is not solid silver. The density of silver is 10.5 grams per cubic centimeter (g/cm^3) and the coin is 0.2 cm thick with a radius of 2 cm. If the coin is really silver, how much should it weigh?

Solution

1. **Familiarize.** From an outside reference, we find that density depends on mass and volume and that, in this setting, mass means weight. A formula for the volume of a right circular cylinder appears at the very end of this text.

2. **Translate.** We need to use two formulas:

$$D = \frac{m}{V} \quad \text{and} \quad V = \pi r^2 h,$$

where D represents the density, m the mass, V the volume, r the length of the radius, and h the height of a right circular cylinder. Since we need a model relating mass to the measurements of the coin, we solve for m and then substitute for V:

$$D = \frac{m}{V}$$

$$V \cdot D = V \cdot \frac{m}{V} \qquad \text{Multiplying by } V$$

$$V \cdot D = m \qquad \text{Simplifying}$$

$$\pi r^2 h \cdot D = m. \qquad \text{Substituting}$$

3. **Carry out.** The model $m = \pi r^2 hD$ can be used to find the mass of any right circular cylinder for which the dimensions and the density are known:

$$m = \pi r^2 hD$$

$$= \pi(2)^2(0.2)(10.5) \qquad \text{Substituting}$$

$$\approx 26.3894. \qquad \text{Using a calculator with a } \pi \text{ key}$$

4. **Check.** To check, we could repeat the calculations. We might also check the model by examining the units:

$$\pi r^2 h \cdot D = \text{cm}^2 \cdot \text{cm} \cdot \frac{\text{g}}{\text{cm}^3} = \text{cm}^3 \cdot \frac{\text{g}}{\text{cm}^3} = \text{g}. \qquad \pi \text{ has no unit.}$$

Since g (grams) is the unit in which the mass, m, is given, we have at least a partial check.

5. **State.** The coin, if it is indeed silver, should weigh about 26 grams.

Exercise Set 1.5

FOR EXTRA HELP

Digital Video Tutor CD 1
Videotape 2

InterAct Math

Math Tutor Center

MathXL

MyMathLab.com

Solve.

1. $d = rt$, for r (a distance formula)

2. $d = rt$, for t

3. $F = ma$, for a (a physics formula)

4. $A = lw$, for w (an area formula)

5. $W = EI$, for I (an electricity formula)

6. $W = EI$, for E

7. $V = lwh$, for h (a volume formula)

8. $I = Prt$, for r (a formula for interest)

9. $L = \dfrac{k}{d^2}$, for k
(a formula for intensity of sound or light)

10. $F = \dfrac{mv^2}{r}$, for m (a physics formula)

11. $G = w + 150n$, for n
(a formula for the gross weight of a bus)

12. $P = b + 0.5t$, for t (a formula for parking prices)

13. $2w + 2h + l = p$, for l
(a formula used when shipping boxes)

14. $2w + 2h + l = p$, for w

15. $5x + 2y = 8$, for y (a formula for a line)

16. $2x + 3y = 12$, for y

17. $Ax + By = C$, for y (a formula for graphing lines)

18. $P = 2l + 2w$, for l (a perimeter formula)

19. $C = \frac{5}{9}(F - 32)$, for F (a temperature formula)

20. $T = \frac{3}{10}(I - 12{,}000)$, for I (a tax formula)

21. $V = \frac{4}{3}\pi r^3$, for r^3 (a volume formula)

22. $V = \frac{4}{3}\pi r^3$, for π

23. $A = \dfrac{h}{2}(b_1 + b_2)$, for b_2 (an area formula)

24. $A = \dfrac{h}{2}(b_1 + b_2)$, for h (an area formula)

25. $A = \dfrac{q_1 + q_2 + q_3}{n}$, for n (a formula for averaging)
(*Hint*: Multiply by n to "clear" fractions.)

26. $g = \dfrac{km_1m_2}{d^2}$, for d^2 (Newton's law of gravitation)

27. $v = \dfrac{d_2 - d_1}{t}$, for t (a physics formula)

28. $v = \dfrac{s_2 - s_1}{m}$, for m

29. $v = \dfrac{d_2 - d_1}{t}$, for d_1

30. $v = \dfrac{s_2 - s_1}{m}$, for s_1

31. $r = m + mnp$, for m

32. $p = x - xyz$, for x

33. $y = ab - ac^2$, for a

34. $d = mn - mp^3$, for m

35. *Investing.* Janos has $2600 to invest for 6 months. If he needs the money to earn $156 in that time, at what rate of simple interest must Janos invest?

36. *Banking.* Yvonne plans to buy a one-year certificate of deposit (CD) that earns 7% simple interest. If she needs the CD to earn $110, how much should Yvonne invest?

37. *Geometry.* The area of a parallelogram is 78 cm^2. The base of the figure is 13 cm. What is the height?

38. *Geometry.* The area of a parallelogram is 72 cm^2. The height of the figure is 6 cm. How long is the base?

39. *Weight of a coin.* The density of gold is 19.3 g/cm^3. If the coin in Example 5 were made of gold instead of silver, how much more would it weigh?

40. *Weight of salt.* The density of salt is 2.16 g/cm^3 (grams per cubic centimeter). An empty cardboard salt canister weighs 28 g, is 13.6 cm tall, and has a 4-cm radius. How much will a filled canister weigh? Use the model developed in Example 5.

Projected birth weight. *Ultrasonic images of 29-week-old fetuses can be used to predict weight. One model, developed by Thurnau,* is* $P = 9.337da - 299$; *a second model, developed by Weiner,*† *is* $P = 94.593c + 34.227a - 2134.616$. *For both formulas, P represents the estimated fetal weight in grams, d the diameter of the fetal head in centimeters, c the circumference of the fetal head in centimeters, and a the circumference of the fetal abdomen in centimeters.*

41. Use Thurnau's model to estimate the diameter of a fetus' head at 29 weeks when the estimated weight is 1614 g and the circumference of the fetal abdomen is 24.1 cm.

42. Use Weiner's model to estimate the circumference of a fetus' head at 29 weeks when the estimated weight is 1277 g and the circumference of the fetal abdomen is 23.4 cm.

43. *Gardening.* A garden is being constructed in the shape of a trapezoid. The dimensions are as shown in the figure. The unknown dimension is to be such that the area of the garden is 90 ft^2. Find that unknown dimension.

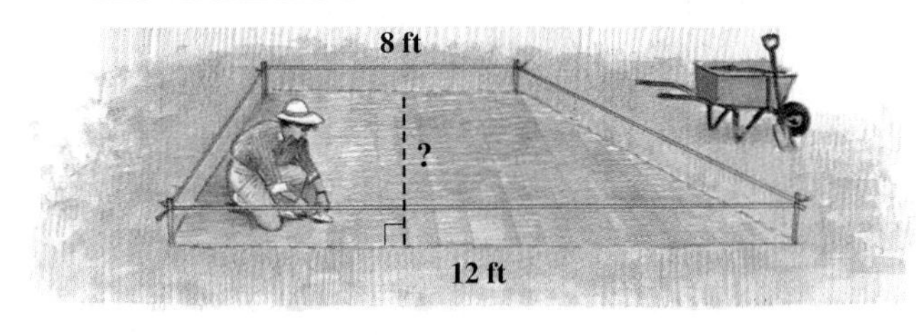

*Thurnau, G. R., R. K. Tamura, R. E. Sabbagha, et al. *Am. J. Obstet Gynecol* 1983; **145**:557.

†Weiner, C. P., R. E. Sabbagha, N. Vaisrub, et al. *Obstet Gynecol* 1985; **65**:812.

44. *Fencing.* A rectangular garden is being constructed, and 76 ft of fencing is available. The width of the garden is to be 13 ft. What should the length be, in order to use just 76 ft of fence?

Aha! **45.** *Investing.* Bok Lum Chan is going to invest \$1000 at simple interest at 9%. How long will it take for the investment to be worth \$1090?

46. Rik is going to invest \$950 at simple interest at 7%. How long will it take for his investment to be worth \$1349?

Waiting time. *In an effort to minimize waiting time for patients at a doctor's office without increasing a physician's idle time, Michael Goiten of Massachusetts General Hospital has developed a model. Goiten suggests that the interval time I, in minutes, between scheduled appointments be related to the total number of minutes T that a physician spends with patients in a day and the number of scheduled appointments N according to the formula* $I = 1.08(T/N)$.‡

47. Dr. Cruz determines that she has a total of 8 hr a day to see patients. If she insists on an interval time of 15 min, according to Goiten's model, how many appointments should she make in one day?

48. A doctor insists on an interval time of 20 min and must be able to schedule 25 appointments a day. According to Goiten's model, how many hours a day should the doctor be prepared to spend with patients?

49. Is every rectangle a trapezoid? Why or why not?

50. Predictions made using the models of Exercises 41 and 42 are often off by as much as 10%. Does this mean the models should be discarded? Why or why not?

SKILL MAINTENANCE

Use the associative and commutative laws to write two equivalent expressions for each of the following. Answers may vary.

51. $(7a)(3a)$

52. $(4y)(xy)$

SYNTHESIS

53. Which would you expect to have the greater density, and why: cork or steel?

‡*New England Journal of Medicine,* 30 August 1990, pp. 604–608.

54. Both of the models used in Exercises 41 and 42 have P alone on one side of the equation. Why?

55. The density of platinum is 21.5 g/cm^3. If the ring shown in the figure below is crafted out of platinum, how much will it weigh?

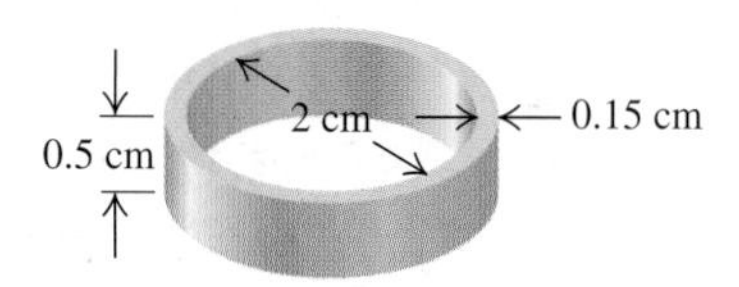

56. The density of a penny is 8.93 g/cm^3. The mass of a roll of pennies is 177.6 g. If the diameter of a penny is 1.85 cm, how tall is a roll of pennies?

57. The density of copper is 8.93 g/cm^3. How long must a copper wire be if it is 1 cm thick and has a mass of 4280 g?

Solve.

58. $s = v_i t + \frac{1}{2}at^2$, for a

59. $A = 4lw + w^2$, for l

60. $\dfrac{P_1V_1}{T_1} = \dfrac{P_2V_2}{T_2}$, for T_2

61. $\dfrac{P_1V_1}{T_1} = \dfrac{P_2V_2}{T_2}$, for T_1

62. $\dfrac{b}{a-b} = c$, for b

63. $m = \dfrac{(d/e)}{(e/f)}$, for d

64. $\dfrac{a}{a+b} = c$, for a

Aha! **65.** $s + \dfrac{s+t}{s-t} = \dfrac{1}{t} + \dfrac{s+t}{s-t}$, for t

66. To derive the formula for the area of a trapezoid, consider the area of two trapezoids, one of which is upside down.

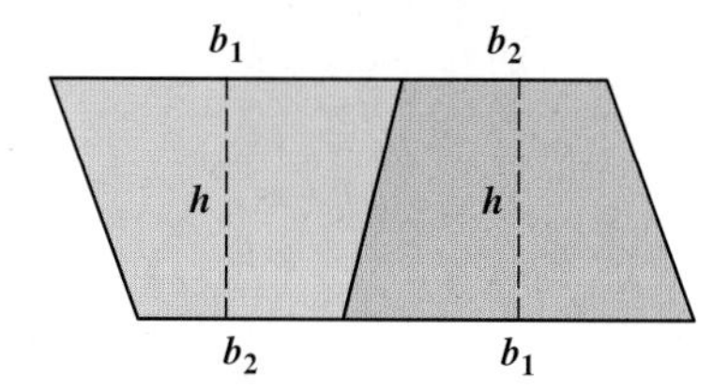

Explain why the total area of the two trapezoids is given by $h(b_1 + b_2)$. Then explain why the area of a trapezoid is given by $\dfrac{h}{2}(b_1 + b_2)$.

Properties of Exponents

1.6

The Product and Quotient Rules • The Zero Exponent • Negative Integers as Exponents • Raising Powers to Powers • Raising a Product or a Quotient to a Power

In Section 1.1, we discussed how whole-number exponents are used. We now develop rules for manipulating exponents and determine what zero and negative integers will mean as exponents.

The Product and Quotient Rules

Note that the expression $x^3 \cdot x^4$ can be rewritten as follows:

$$x^3 \cdot x^4 = \underbrace{x \cdot x \cdot x}_{\text{3 factors}} \cdot \underbrace{x \cdot x \cdot x \cdot x}_{\text{4 factors}}$$
$$= \underbrace{x \cdot x \cdot x \cdot x \cdot x \cdot x \cdot x}_{\text{7 factors}}$$
$$= x^7.$$

This result is generalized in the *product rule.*

Multiplying with Like Bases: The Product Rule

For any number a and any positive integers m and n,

$$a^m \cdot a^n = a^{m+n}.$$

(When multiplying powers, if the bases are the same, keep the base and add the exponents.)

Example 1 Multiply and simplify: **(a)** $m^5 \cdot m^7$; **(b)** $(5a^2b^3)(3a^4b^5)$.

Solution

a) $m^5 \cdot m^7 = m^{5+7} = m^{12}$ — Multiplying powers by adding exponents

b) $(5a^2b^3)(3a^4b^5) = 5 \cdot 3 \cdot a^2 \cdot a^4 \cdot b^3 \cdot b^5$ — Using the associative and commutative laws

$= 15a^{2+4}b^{3+5}$ — Multiplying; using the product rule

$= 15a^6b^8$

Caution! $5^8 \cdot 5^6 = 5^{14}$; $\quad 5^8 \cdot 5^6 \neq 25^{14}$.

Next, we simplify a quotient:

$\dfrac{x^8}{x^3} = \dfrac{x \cdot x \cdot x \cdot x \cdot x \cdot x \cdot x \cdot x}{x \cdot x \cdot x}$ — Using the definition of exponential notation

$= \dfrac{x \cdot x \cdot x}{x \cdot x \cdot x} \cdot x \cdot x \cdot x \cdot x \cdot x$

$= x \cdot x \cdot x \cdot x \cdot x$ — Removing a factor equal to 1: $\dfrac{x \cdot x \cdot x}{x \cdot x \cdot x} = 1$

$= x^5.$

The generalization of this result is the *quotient rule.*

Dividing with Like Bases: The Quotient Rule

For any nonzero number a and any positive integers m and n, $m > n$,

$$\frac{a^m}{a^n} = a^{m-n}.$$

(When dividing powers, if the bases are the same, keep the base and subtract the exponent of the denominator from the exponent of the numerator.)

Example 2 Divide and simplify: **(a)** $\frac{r^9}{r^3}$; **(b)** $\frac{10x^{11}y^5}{2x^4y^3}$.

Solution

a) $\frac{r^9}{r^3} = r^{9-3} = r^6$ Using the quotient rule

b) $\frac{10x^{11}y^5}{2x^4y^3} = 5 \cdot x^{11-4} \cdot y^{5-3}$ Dividing; using the quotient rule

$= 5x^7y^2$

Caution! $\frac{7^8}{7^2} = 7^6$; $\frac{7^8}{7^2} \neq 7^4$.

The Zero Exponent

Suppose now that the bases in the numerator and the denominator are both raised to the same power. On the one hand, any (nonzero) expression divided by itself is equal to 1. For example,

$$\frac{t^5}{t^5} = 1 \quad \text{and} \quad \frac{6^4}{6^4} = 1.$$

On the other hand, if we continue subtracting exponents when dividing powers with the same base, we have

$$\frac{t^5}{t^5} = t^{5-5} = t^0 \quad \text{and} \quad \frac{6^4}{6^4} = 6^{4-4} = 6^0.$$

This suggests that t^5/t^5 equals both 1 *and* t^0. It also suggests that $6^4/6^4$ equals both 1 *and* 6^0. This leads to the following definition.

The Zero Exponent

For any nonzero real number a,

$$a^0 = 1.$$

(Any nonzero number raised to the zero power is 1. 0^0 is undefined.)

Example 3 Evaluate each of the following for $x = 2.9$: **(a)** x^0; **(b)** $-x^0$; **(c)** $(-x)^0$.

Solution

a) $x^0 = 2.9^0 = 1$ Using the definition of 0 as an exponent

b) $-x^0 = -2.9^0 = -1$ The exponent 0 pertains only to the 2.9.

c) $(-x)^0 = (-2.9)^0 = 1$ The base here is -2.9.

Parts (b) and (c) of Example 3 illustrate an important result:

Since $-a^n$ means $-1 \cdot a^n$, in general, $-a^n \neq (-a)^n$.

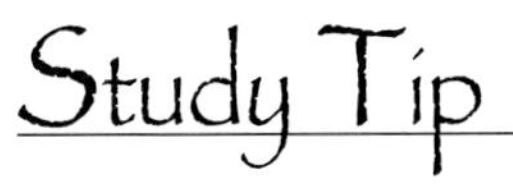

The AW Math Tutor Center provides free tutoring to students using this text. Assisted by qualified mathematics instructors via telephone, fax, or e-mail, you can receive live tutoring on examples and exercises. For more information on the AW Math Tutor Center, see the "To the Student" portion of the preface.

Negative Integers as Exponents

Later in this text we will explain what numbers like $\frac{2}{9}$ or $\sqrt{2}$ mean as exponents. Until then, integer exponents will suffice.

To develop a definition for negative integer exponents, we simplify $5^3/5^7$ two ways. First we proceed as in arithmetic:

$$\frac{5^3}{5^7} = \frac{5 \cdot 5 \cdot 5}{5 \cdot 5 \cdot 5 \cdot 5 \cdot 5 \cdot 5 \cdot 5} = \frac{5 \cdot 5 \cdot 5 \cdot 1}{5 \cdot 5 \cdot 5 \cdot 5 \cdot 5 \cdot 5 \cdot 5}$$
$$= \frac{5 \cdot 5 \cdot 5}{5 \cdot 5 \cdot 5} \cdot \frac{1}{5 \cdot 5 \cdot 5 \cdot 5}$$
$$= \frac{1}{5^4}.$$

Were we to apply the quotient rule, we would have

$$\frac{5^3}{5^7} = 5^{3-7} = 5^{-4}.$$

These two expressions for $5^3/5^7$ suggest that

$$5^{-4} = \frac{1}{5^4}.$$

This leads to the definition of negative exponents.

Negative Exponents

For any real number a that is nonzero and any integer n,

$$a^{-n} = \frac{1}{a^n}.$$

(The numbers a^{-n} and a^n are reciprocals of each other.)

The definitions above preserve the following pattern:

$$4^3 = 4 \cdot 4 \cdot 4,$$
$$4^2 = 4 \cdot 4, \quad \text{Dividing both sides by 4}$$
$$4^1 = 4, \quad \text{Dividing both sides by 4}$$
$$4^0 = 1, \quad \text{Dividing both sides by 4}$$
$$4^{-1} = \frac{1}{4}, \quad \text{Dividing both sides by 4}$$
$$4^{-2} = \frac{1}{4 \cdot 4} = \frac{1}{4^2}. \quad \text{Dividing both sides by 4}$$

> ***Caution!*** A negative exponent does not, in itself, indicate that an expression is negative. As shown at the bottom of p. 48,
>
> $$4^{-2} \neq 4(-2).$$

Example 4 Express using positive exponents and, if possible, simplify.

a) 3^{-2} **b)** $5x^{-4}y^3$ **c)** $\dfrac{1}{7^{-2}}$

Solution

a) $3^{-2} = \dfrac{1}{3^2} = \dfrac{1}{9}$

b) $5x^{-4}y^3 = 5\left(\dfrac{1}{x^4}\right)y^3 = \dfrac{5y^3}{x^4}$

c) Since $\dfrac{1}{a^n} = a^{-n}$, we have

$$\frac{1}{7^{-2}} = 7^{-(-2)} = 7^2, \text{ or } 49.$$ Remember: n can be a negative integer.

The result from part (c) above can be generalized.

> ***Factors and Negative Exponents***
>
> For any nonzero real numbers a and b and any integers m and n,
>
> $$\frac{a^{-n}}{b^{-m}} = \frac{b^m}{a^n}.$$
>
> (A factor can be moved to the other side of the fraction bar if the sign of the exponent is changed.)

Example 5 Write an equivalent expression without negative exponents:

$$\frac{vx^{-2}y^{-5}}{z^{-4}w^{-3}}.$$

Solution We can move each factor to the other side of the fraction bar if we change the sign of each exponent:

$$\frac{vx^{-2}y^{-5}}{z^{-4}w^{-3}} = \frac{vz^4w^3}{x^2y^5}.$$

The product and quotient rules apply for all integer exponents.

Example 6 Simplify: **(a)** $7^{-3} \cdot 7^8$; **(b)** $\dfrac{b^{-5}}{b^{-4}}$.

Solution

a) $7^{-3} \cdot 7^8 = 7^{-3+8}$ ← Using the product rule

$= 7^5$

b) $\dfrac{b^{-5}}{b^{-4}} = b^{-5-(-4)} = b^{-1}$ ← Using the quotient rule

$= \dfrac{1}{b}$ Writing the answer without a negative exponent

Example 6(b) can also be simplified as follows:

$$\frac{b^{-5}}{b^{-4}} = \frac{b^4}{b^5} = b^{4-5} = b^{-1} = \frac{1}{b}.$$

Raising Powers to Powers

Next, consider an expression like $(3^4)^2$:

$$(3^4)^2 = (3^4)(3^4) \quad \text{We are raising } 3^4 \text{ to the second power.}$$

$$= (3 \cdot 3 \cdot 3 \cdot 3)(3 \cdot 3 \cdot 3 \cdot 3)$$

$$= 3 \cdot 3 \cdot 3 \cdot 3 \cdot 3 \cdot 3 \cdot 3 \cdot 3 \quad \text{Using the associative law}$$

$$= 3^8.$$

Note that in this case, we could have multiplied the exponents:

$$(3^4)^2 = 3^{4 \cdot 2} = 3^8.$$

Likewise, $(y^8)^3 = (y^8)(y^8)(y^8) = y^{24}$. Once again, we get the same result if we multiply the exponents:

$$(y^8)^3 = y^{8 \cdot 3} = y^{24}.$$

Raising a Power to a Power: The Power Rule

For any real number a and any integers m and n,

$$(a^m)^n = a^{mn}.$$

(To raise a power to a power, multiply the exponents.)

technology connection

To simplify expressions like 4^7, most calculators are equipped with an exponentiation key. Such a key is often labeled [x^y] or [^]. To enter 4^7 on most scientific calculators, we press

[4] [x^y] [7] [=].

On other calculators, we press

[4] [^] [7] [ENTER]

1. List keystrokes that could be used to simplify 2^{-5} on a scientific or graphing calculator.
2. How could 2^{-5} be simplified on a calculator lacking an exponentiation key?

Example 7

Simplify: **(a)** $(3^5)^4$; **(b)** $(y^{-5})^7$; **(c)** $(a^{-3})^{-7}$.

Solution

a) $(3^5)^4 = 3^{5 \cdot 4} = 3^{20}$

b) $(y^{-5})^7 = y^{-5 \cdot 7} = y^{-35} = \dfrac{1}{y^{35}}$

c) $(a^{-3})^{-7} = a^{(-3)(-7)} = a^{21}$

Raising a Product or a Quotient to a Power

When an expression inside parentheses is raised to a power, the inside expression is the base. Let's compare $2a^3$ and $(2a)^3$.

$$2a^3 = 2 \cdot a \cdot a \cdot a; \qquad (2a)^3 = (2a)(2a)(2a)$$
$$= 2 \cdot 2 \cdot 2 \cdot a \cdot a \cdot a$$
$$= 2^3a^3 = 8a^3$$

We see that $2a^3$ and $(2a)^3$ are *not* equivalent. Note also that to simplify $(2a)^3$ we can raise each factor to the power 3. This leads to the following rule.

Raising a Product to a Power

For any integer n, and any real numbers a and b for which $(ab)^n$ exists,

$$(ab)^n = a^nb^n.$$

(To raise a product to a power, raise each factor to that power.)

Example 8

Simplify: **(a)** $(-2x)^3$; **(b)** $(-3x^5y^{-1})^{-4}$.

Solution

a) $(-2x)^3 = (-2)^3 \cdot x^3$ Raising each factor to the third power

$= -8x^3$

b) $(-3x^5y^{-1})^{-4} = (-3)^{-4}(x^5)^{-4}(y^{-1})^{-4}$ Raising each factor to the negative fourth power

$= \dfrac{1}{(-3)^4} \cdot x^{-20}y^4$ Multiplying powers; writing $(-3)^{-4}$ as $\dfrac{1}{(-3)^4}$

$= \dfrac{y^4}{81x^{20}}$ Note that $x^{-20} = \dfrac{1}{x^{20}}$.

There is a similar rule for raising a quotient to a power.

Raising a Quotient to a Power

For any integer n, and any real numbers a and b for which a/b, a^n, and b^n exist,

$$\left(\frac{a}{b}\right)^n = \frac{a^n}{b^n}.$$

(To raise a quotient to a power, raise both the numerator and the denominator to that power.)

Example 9 Simplify: **(a)** $\left(\frac{x^2}{2}\right)^4$; **(b)** $\left(\frac{y^2z^3}{5}\right)^{-3}$.

Solution

a) $\left(\frac{x^2}{2}\right)^4 = \frac{(x^2)^4}{2^4} = \frac{x^8}{16}$ $\leftarrow 2 \cdot 4 = 8$; $\leftarrow 2^4 = 16$

b) $\left(\frac{y^2z^3}{5}\right)^{-3} = \frac{(y^2z^3)^{-3}}{5^{-3}}$

$= \frac{5^3}{(y^2z^3)^3}$ Moving factors to the other side of the fraction bar and changing each −3 to 3

$= \frac{125}{y^6z^9}$

The rule for raising a quotient to a power allows us to derive a useful result for manipulating negative exponents:

$$\left(\frac{a}{b}\right)^{-n} = \frac{a^{-n}}{b^{-n}} = \frac{b^n}{a^n} = \left(\frac{b}{a}\right)^n.$$

Using this result, we can simplify Example 9(b) as follows:

$$\left(\frac{y^2z^3}{5}\right)^{-3} = \left(\frac{5}{y^2z^3}\right)^3$$ Taking the reciprocal of the base and changing the exponent's sign

$$= \frac{5^3}{(y^2z^3)^3} = \frac{125}{y^6z^9}.$$

Definitions and Properties of Exponents

The following summary assumes that no denominators are 0 and that 0^0 is not considered. For any integers m and n,

1 as an exponent:	$a^1 = a$
0 as an exponent:	$a^0 = 1$
Negative exponents:	$a^{-n} = \frac{1}{a^n}$
	$\frac{a^{-n}}{b^{-m}} = \frac{b^m}{a^n}$
	$\left(\frac{a}{b}\right)^{-n} = \left(\frac{b}{a}\right)^n$
The Product Rule:	$a^m \cdot a^n = a^{m+n}$
The Quotient Rule:	$\frac{a^m}{a^n} = a^{m-n}$
The Power Rule:	$(a^m)^n = a^{mn}$
Raising a product to a power:	$(ab)^n = a^n b^n$
Raising a quotient to a power:	$\left(\frac{a}{b}\right)^n = \frac{a^n}{b^n}$

Exercise Set 1.6

FOR EXTRA HELP

Digital Video Tutor CD 1
Videotape 2

InterAct Math

Math Tutor Center

MathXL

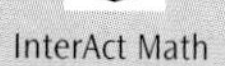

MyMathLab.com

Multiply and simplify. Leave the answer in exponential notation.

1. $2^7 \cdot 2^4$ **2.** $7^3 \cdot 7^4$

3. $5^6 \cdot 5^3$ **4.** $6^3 \cdot 6^5$

5. $t^0 \cdot t^8$ **6.** $x^0 \cdot x^5$

7. $6x^5 \cdot 3x^2$ **8.** $4a^3 \cdot 2a^7$

9. $(-3m^4)(-7m^9)$ **10.** $(-2a^5)(7a^4)$

11. $(x^3y^4)(x^7y^6z^0)$ **12.** $(m^6n^5)(m^4n^7p^0)$

Divide and simplify.

13. $\frac{a^9}{a^3}$ **14.** $\frac{x^{12}}{x^3}$

15. $\frac{12t^7}{4t^2}$ **16.** $\frac{20a^{20}}{5a^4}$

17. $\frac{m^7n^9}{m^2n^5}$ **18.** $\frac{m^{12}n^9}{m^4n^6}$

19. $\frac{45x^8y^5}{5x^2y}$ **20.** $\frac{35x^7y^8}{7xy^2}$

21. $\frac{28x^{10}y^9z^8}{-7x^2y^3z^2}$ **22.** $\frac{18x^8y^6z^7}{-3x^2y^3z}$

Evaluate each of the following for $x = -2$.

23. $-x^0$ **24.** $(-x)^0$

25. $(4x)^0$ **26.** $4x^0$

Simplify.

27. $(-3)^4$ **28.** $(-2)^6$ **29.** -3^4

30. -2^6 **31.** $(-4)^{-2}$ **32.** $(-5)^{-2}$

33. -4^{-2} **34.** -5^{-2} **35.** -2^{-4}

36. -5^{-3} **37.** -2^{-6} **38.** -1^{-8}

Write an equivalent expression without negative exponents and, if possible, simplify.

39. a^{-3} **40.** n^{-6} **41.** $\frac{1}{5^{-3}}$

42. $\frac{1}{2^{-6}}$ **43.** $4x^{-3}$ **44.** $7x^{-3}$

45. $2a^3b^{-6}$ **46.** $5a^{-7}b^4$ **47.** $\frac{z^{-4}}{3x^5}$

48. $\frac{y^{-5}}{x^{-3}}$ **49.** $\frac{x^{-2}y^7}{z^{-4}}$ **50.** $\frac{y^4z^{-3}}{x^{-2}}$

Write an equivalent expression with negative exponents.

51. $\frac{1}{3^4}$ **52.** $\frac{1}{9^2}$ **53.** $\frac{1}{(-16)^2}$

54. $\frac{1}{(-8)^6}$ **55.** x^5 **56.** n^3

57. $6x^2$ **58.** $-4y^5$ **59.** $\frac{1}{(5y)^3}$

60. $\frac{1}{(5x)^5}$ **61.** $\frac{1}{3y^4}$ **62.** $\frac{1}{4b^3}$

Simplify. Should negative exponents appear in the answer, write a second answer using only positive exponents.

63. $8^{-2} \cdot 8^{-4}$ **64.** $9^{-1} \cdot 9^{-6}$

65. $b^2 \cdot b^{-5}$ **66.** $a^4 \cdot a^{-3}$

67. $a^{-3} \cdot a^4 \cdot a^2$ **68.** $x^{-8} \cdot x^5 \cdot x^3$

69. $(9mn^3)(-2m^3n^2)$ **70.** $(6x^5y^{-2})(-3x^2y^3)$

71. $(-2x^{-3})(7x^{-8})$ **72.** $(6x^{-4}y^3)(-4x^{-8}y^{-2})$

73. $(5a^{-2}b^{-3})(2a^{-4}b)$ **74.** $(3a^{-5}b^{-7})(2ab^{-2})$

75. $\frac{10^{-3}}{10^6}$ **76.** $\frac{12^{-4}}{12^8}$

77. $\frac{2^{-7}}{2^{-5}}$ **78.** $\frac{9^{-4}}{9^{-6}}$

79. $\frac{y^4}{y^{-5}}$ **80.** $\frac{a^3}{a^{-2}}$

81. $\frac{24a^5b^3}{-8a^4b}$ **82.** $\frac{9a^2}{3ab^3}$

83. $\frac{14a^4b^{-3}}{-8a^8b^{-5}}$ **84.** $\frac{-24x^6y^7}{18x^{-3}y^9}$

85. $\frac{-5x^{-2}y^4z^7}{30x^{-5}y^6z^{-3}}$ **86.** $\frac{9a^6b^{-4}c^7}{27a^{-4}b^5c^9}$

87. $(x^4)^3$ **88.** $(a^3)^2$

89. $(9^3)^{-4}$ **90.** $(8^4)^{-3}$

91. $(t^{-8})^{-5}$ **92.** $(x^{-4})^{-3}$

93. $(6xy)^2$ **94.** $(5ab)^3$

95. $(a^3b)^4$ **96.** $(x^3y)^5$

97. $5(x^2y^2)^{-7}$ **98.** $7(a^3b^4)^{-5}$

99. $(a^{-5}b^2)^3(a^4b^{-1})^2$

100. $(x^2y^{-3})^3(x^{-4}y)^2$

Aha! **101.** $(5x^{-3}y^2)^{-4}(5x^{-3}y^2)^4$

102. $(2a^{-1}b^3)^{-2}(2a^{-1}b^3)^{-2}$

103. $\frac{(3x^3y^4)^3}{6xy^3}$ **104.** $\frac{(5a^3b)^2}{10a^2b}$

105. $\left(\frac{-4x^4y^{-2}}{5x^{-1}y^4}\right)^{-4}$ **106.** $\left(\frac{2x^3y^{-2}}{3y^{-3}}\right)^3$

107. $\left(\frac{3a^{-2}b^5}{9a^{-4}b^0}\right)^{-2}$ **108.** $\left(\frac{30x^5y^{-7}}{6x^{-2}y^{-6}}\right)^0$

Aha! **109.** $\left(\frac{4a^3b^{-9}}{2a^{-2}b^5}\right)^0$ **110.** $\left(\frac{5x^0y^{-7}}{2x^{-2}y^4}\right)^{-2}$

111. Explain why $(-1)^n = 1$ for any even number n.

112. Explain why $(-17)^{-8}$ is positive.

SKILL MAINTENANCE

Evaluate.

113. $4.9t^2 + 3t$, for $t = -3$

114. $16t^2 + 10t$, for $t = -2$

SYNTHESIS

115. Explain in your own words why a^0 is defined to be 1. Assume that $a \neq 0$.

116. Is the following true or false, and why?

$$5^{-6} > 4^{-9}$$

Simplify. Assume that all variables represent nonzero integers.

117. $\frac{12a^{x-2}}{3a^{2x+2}}$

118. $\frac{-12x^{a+1}}{4x^{2-a}}$

119. $[7y(7-8)^{-2} - 8y(8-7)^{-2}]^{(-2)^2}$

120. $\{[(8^{-a})^{-2}]^{b}\}^{-c} \cdot [(8^0)^a]^c$

121. $(3^{a+2})^a$

122. $(12^{3-a})^{2b}$

123. $\dfrac{4x^{2a+3}y^{2b-1}}{2x^{a+1}y^{b+1}}$

124. $\dfrac{25x^{a+b}y^{b-a}}{-5x^{a-b}y^{b+a}}$

125. $\dfrac{(2^{-2})^a \cdot (2^b)^{-a}}{(2^{-2})^{-b}(2^b)^{-2a}}$

126. $\dfrac{-28x^{b+5}y^{4+c}}{7x^{b-5}y^{c-4}}$

127. $\dfrac{3^{q+3} - 3^2(3^q)}{3(3^{q+4})}$

128. $\left[\left(\dfrac{a^{-2c}}{b^{7c}}\right)^{-3}\left(\dfrac{a^{4c}}{b^{-3c}}\right)^{2}\right]^{-a}$

1.7 Scientific Notation

Conversions • Significant Digits and Rounding • Scientific Notation in Problem Solving

There is a variety of symbolism, or *notation*, for numbers. You are already familiar with fraction notation, decimal notation, and percent notation. We now study **scientific notation**, so named because of its usefulness in work with the very large and very small numbers that occur in science.

The following are examples of scientific notation:

7.2×10^5 means 720,000;

3.4×10^{-6} means 0.0000034;

4.89×10^{-3} means 0.00489.

Scientific Notation

Scientific notation for a number is an expression of the form $N \times 10^m$, where N is in decimal notation, $1 \le N < 10$, and m is an integer.

Conversions

Note that $10^b/10^b = 10^b \cdot 10^{-b} = 1$. To convert to scientific notation, we can multiply by 1, writing 1 in the form $10^b/10^b$ or $10^b \cdot 10^{-b}$.

Example 1

Population projections. It has been estimated that in the year 2010, the world population will be 6,832,000,000 (*Source*: *The Statistical Abstract of the United States*, 2000). Write scientific notation for this number.

Solution To write 6,832,000,000 as 6.832×10^m for some integer m, we must move the decimal point in 6,832,000,000 to the left 9 places. This can be accomplished by dividing—and then multiplying—by 10^9:

$$6{,}832{,}000{,}000 = \frac{6{,}832{,}000{,}000}{10^9} \cdot 10^9 \qquad \text{Multiplying by 1: } \frac{10^9}{10^9} = 1$$

$$= 6.832 \times 10^9. \qquad \text{This is scientific notation.}$$

Example 2

Write scientific notation for the mass of a grain of sand:

0.0648 gram (g).

Solution To write 0.0648 as 6.48×10^m for some integer m, we must move the decimal 2 places to the right. To do this, we multiply—and then divide—by 10^2:

$$0.0648 = \frac{0.0648 \times 10^2}{10^2} \qquad \text{Multiplying by 1: } \frac{10^2}{10^2} = 1$$

$$= \frac{6.48}{10^2}$$

$$= 6.48 \times 10^{-2} \text{ g.} \qquad \text{Writing scientific notation}$$

Try to make conversions to scientific notation mentally if possible. In doing so, remember that negative powers of 10 are used for small numbers and positive powers of 10 are used for large numbers.

Example 3

Convert mentally to scientific notation: **(a)** 82,500,000; **(b)** 0.0000091.

Solution

a) $82{,}500{,}000 = 8.25 \times 10^7$ *Check*: Multiplying 8.25 by 10^7 moves the decimal point 7 places to the right.

b) $0.0000091 = 9.1 \times 10^{-6}$ *Check*: Multiplying 9.1 by 10^{-6} moves the decimal point 6 places to the left.

Example 4

Convert mentally to decimal notation: **(a)** 4.371×10^7; **(b)** 1.73×10^{-5}.

Solution

a) $4.371 \times 10^7 = 43{,}710{,}000$ Moving the decimal point 7 places to the right

b) $1.73 \times 10^{-5} = 0.0000173$ Moving the decimal point 5 places to the left

Significant Digits and Rounding

In the world of science, it is important to know just how accurate a measurement is. For example, the measurement 5.12 cm is more precise than the measurement 5.1 cm. We say that the number 5.12 has three **significant digits** whereas 5.1 has only two significant digits. If the measurement 5.1 cm were precise to the nearest hundredth of a centimeter, it would be written as 5.10 cm. When two or more measurements are added, subtracted, multiplied, or divided, the result is only as accurate as the *least* precise measurement used in the computation. Thus scientists have agreed on the following conventions.

1. The sum or difference of two numbers should be rounded off so that it has the same number of significant digits to the right of the decimal as the number in the measurement with the fewest significant digits to the right of the decimal.

   For example,

   $$135.\underbrace{4}_{\text{1 digit}}\text{ cm} + 50.\underbrace{28}_{\text{2 digits}}\text{ cm} = 185.68\text{ cm}$$

   should be rounded off to

   $$185.\overbrace{7}^{\text{1 digit}}\text{ cm.}$$

2. The product or quotient of two numbers should be rounded off so that it contains the same number of significant digits as the measurement with the fewest significant digits.

   For example,

   $$\underbrace{2.1}_{\text{2 digits}}\text{ cm} \times \underbrace{6.45}_{\text{3 digits}}\text{ cm} = 13.545\text{ cm}^2$$

   should be rounded off to

   $$\overbrace{14}^{\text{2 digits}}\text{ cm}^2.$$

For future work in this text with scientific notation, results will be rounded off according to the above conventions.

Example 5 Multiply and write scientific notation for the answer:

$$(7.2 \times 10^5)(4.3 \times 10^9).$$

Solution We have

$$(7.2 \times 10^5)(4.3 \times 10^9) = (7.2 \times 4.3)(10^5 \times 10^9)$$ Using the commutative and associative laws

$$= 30.96 \times 10^{14}$$ Adding exponents

$$= 31 \times 10^{14}.$$ Rounding to 2 significant digits

To find scientific notation for the result, we convert 31 to scientific notation and simplify:

$$\begin{aligned} 31 \times 10^{14} &= (3.1 \times 10^{1}) \times 10^{14} \\ &= 3.1 \times 10^{15}. \end{aligned}$$

Example 6 Divide and write scientific notation for the answer:

$$\frac{3.48 \times 10^{-7}}{4.64 \times 10^{6}}.$$

Solution

$$\frac{3.48 \times 10^{-7}}{4.64 \times 10^{6}} = \frac{3.48}{4.64} \times \frac{10^{-7}}{10^{6}}$$ Separating factors. Our answer must have 3 significant digits.

$$= 0.75 \times 10^{-13}$$ Subtracting exponents; simplifying

$$= (7.5 \times 10^{-1}) \times 10^{-13}$$ Converting 0.75 to scientific notation

$$= 7.50 \times 10^{-14}$$ Adding exponents. We write 7.50 to indicate 3 significant digits.

Scientific Notation in Problem Solving

Scientific notation can be useful in problem solving.

Example 7 ***Astronomy.*** The largest known star, Betelgeuse, is about 700 times the size of the sun and about 3.06×10^{15} mi from Earth (*Sources*: *Guinness Book of Records* 2000 and *The Cambridge Factfinder*, 4th ed.). How many light years is it from Earth to Betelgeuse?

Solution

1. **Familiarize.** From an astronomy text, we learn that light travels about 5.88×10^{12} mi in one year. Thus 1 light year $= 5.88 \times 10^{12}$ mi. We will let y represent the number of light years from Earth to Betelgeuse. Let's guess that the answer is 6 light years. Then the distance in miles would be

$$(5.88 \times 10^{12}) \cdot 6 = 35.28 \times 10^{12} = 3.528 \times 10^{13}.$$

Our guess is incorrect, but it tells us that the distance is more than 6 light years. We are also better able to translate to an equation.

Both graphing and scientific calculators allow expressions to be entered using scientific notation. To do so, a key normally labeled EE or EXP is used. Often this is a secondary function and a key labeled SHIFT or 2nd must be pressed first. To check Example 5, we press 7.2 EE 5 × 4.3 EE 9. When we then press ENTER or =, the result 3.096E15 or 3.096 15 appears. We must interpret this result as 3.096×10^{15}.

2. **Translate.** Note that the distance to Betelgeuse is y light years, or $(5.88 \times 10^{12})y$ mi. Also, note that the distance is 3.06×10^{15} mi. Since the quantities $(5.88 \times 10^{12})y$ and 3.06×10^{15} both represent the number of miles to Betelgeuse, we form the equation

$$(5.88 \times 10^{12})y = 3.06 \times 10^{15}.$$

3. **Carry out.** We solve the equation:

$$(5.88 \times 10^{12})y = 3.06 \times 10^{15}$$

$$\frac{1}{5.88 \times 10^{12}}(5.88 \times 10^{12})y = \frac{1}{5.88 \times 10^{12}}(3.06 \times 10^{15})$$ Multiplying both sides by $1/(5.88 \times 10^{12})$

$$y = \frac{3.06 \times 10^{15}}{5.88 \times 10^{12}}$$ Simplifying

$$= \frac{3.06}{5.88} \times \frac{10^{15}}{10^{12}}$$ Factoring

$$\approx 0.5204 \times 10^{3}$$

$$\approx 5.20 \times 10^{2} \text{ or } 520. \text{ yr.}$$ Our answer has 3 significant digits because of the 3.06 and 5.88.

4. **Check.** Since light travels 5.88×10^{12} mi in one year, in 520. yr it will travel $520 \times 5.88 \times 10^{12} \approx 3.06 \times 10^{15}$ mi, which is approximately the distance from Earth to Betelgeuse. The decimal point in 520. indicates that we did not round to the nearest ten. A number like 527 would not need a decimal point

5. **State.** The distance from Earth to Betelgeuse is about 520. light years.

Example 8

Telecommunications. A fiber-optic wire will be used for 375 km of transmission line. The wire has a diameter of 1.2 cm. What is the volume of wire needed for the line?

Solution

1. **Familiarize.** Making a drawing, we see that we have a cylinder (a very *long* one). Its length is 375 km and the base has a diameter of 1.2 cm.

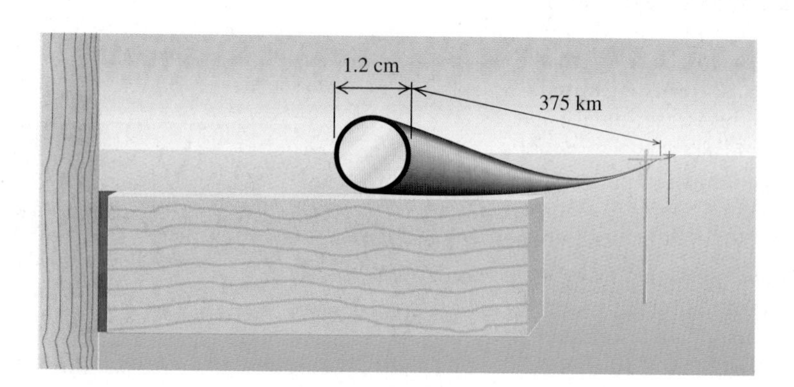

Recall that the formula for the volume of a cylinder is

$$V = \pi r^2 h,$$

where r is the radius and h is the height (in this case, the length of the wire).

2. **Translate.** We will use the volume formula, but it is important to make the units consistent. Let's express everything in meters:

Length: 375 km = 375,000 m, or 3.75×10^5 m;
Diameter: 1.2 cm = 0.012 m, or 1.2×10^{-2} m.

The radius, which we will need in the formula, is half the diameter:

Radius: 0.6×10^{-2} m, or 6×10^{-3} m.

We now substitute into the above formula:

$$V = \pi(6 \times 10^{-3})^2(3.75 \times 10^5).$$

3. **Carry out.** We do the calculations:

$$\begin{aligned} V &= \pi \times (6 \times 10^{-3})^2(3.75 \times 10^5) \\ &= \pi \times 6^2 \times 10^{-6} \times 3.75 \times 10^5 && \text{Using the properties of exponents} \\ &= (\pi \times 6^2 \times 3.75) \times (10^{-6} \times 10^5) \\ &= 423.9 \times 10^{-1} && \text{Using 3.14 for } \pi \text{ and rounding} \\ &\approx 42. && \text{Rounding 42.39 to 2 significant digits because of the 1.2} \end{aligned}$$

4. **Check.** About all we can do here is recheck the translation and calculations.

5. **State.** The volume of the wire is about 42 m^3 (cubic meters).

Exercise Set 1.7

FOR EXTRA HELP

Digital Video Tutor CD 1
Videotape 2

InterAct Math

Math Tutor Center

MathXL

MyMathLab.com

Convert to scientific notation.

1. 83,000,000,000
2. 2,600,000,000,000
3. 863,000,000,000,000,000
4. 572,000,000,000,000,000
5. 0.000000016
6. 0.000000263
7. 0.00000000007
8. 0.00000000009
9. 803,000,000,000
10. 3,090,000,000,000
11. 0.000000904
12. 0.00000000802
13. 431,700,000,000
14. 953,400,000,000

Convert to decimal notation.

15. 5×10^{-4}
16. 5×10^{-5}
17. 9.73×10^8
18. 9.24×10^7
19. 4.923×10^{-10}
20. 7.034×10^{-2}
21. 9.03×10^{10}
22. 1.01×10^{12}
23. 4.037×10^{-8}
24. 3.007×10^{-9}
25. 7.01×10^{12}
26. 9.001×10^{10}

Simplify and write scientific notation for the answer. Use the correct number of significant digits.

27. $(2.3 \times 10^6)(4.2 \times 10^{-11})$

28. $(6.5 \times 10^3)(5.2 \times 10^{-8})$

29. $(2.34 \times 10^{-8})(5.7 \times 10^{-4})$

30. $(4.26 \times 10^{-6})(8.2 \times 10^{-6})$

31. $(5.2 \times 10^6)(2.6 \times 10^4)$

32. $(6.11 \times 10^3)(1.01 \times 10^{13})$

33. $(7.01 \times 10^{-5})(6.5 \times 10^7)$

34. $(4.08 \times 10^{-10})(7.7 \times 10^5)$

Aha! **35.** $(2.0 \times 10^6)(3.02 \times 10^{-6})$

36. $(7.04 \times 10^{-9})(9.01 \times 10^{-7})$

37. $\dfrac{5.1 \times 10^6}{3.4 \times 10^3}$

38. $\dfrac{8.5 \times 10^8}{3.4 \times 10^5}$

39. $\dfrac{7.5 \times 10^{-9}}{2.5 \times 10^{-4}}$

40. $\dfrac{4.0 \times 10^{-6}}{8.0 \times 10^{-3}}$

41. $\dfrac{3.2 \times 10^{-7}}{8.0 \times 10^8}$

42. $\dfrac{12.6 \times 10^8}{4.2 \times 10^{-3}}$

43. $\dfrac{9.36 \times 10^{-11}}{3.12 \times 10^{11}}$

44. $\dfrac{2.42 \times 10^5}{1.21 \times 10^{-5}}$

45. $\dfrac{6.12 \times 10^{19}}{3.06 \times 10^{-7}}$

46. $\dfrac{4.7 \times 10^{-9}}{2.0 \times 10^{-9}}$

47. $\dfrac{780{,}000{,}000 \times 0.00071}{0.000005}$

48. $\dfrac{830{,}000{,}000 \times 0.12}{3{,}100{,}000}$

49. $5.9 \times 10^{23} + 2.4 \times 10^{23}$

50. $1.8 \times 10^{-34} + 5.4 \times 10^{-34}$

Solve.

51. *Astronomy.* Venus has a nearly circular orbit of the sun. If the average distance from the sun to Venus is 1.08×10^8 km, how far does Venus travel in one orbit?

52. *Office supplies.* A ream of copier paper weighs 2.25 kg. How much does a sheet of copier paper weigh?

53. *Printing and engraving.* A ton of five-dollar bills is worth \$4,540,000. How many pounds does a five-dollar bill weigh?

54. *Astronomy.* The average distance of the earth from the sun is about 9.3×10^7 mi. About how far does the earth travel in a yearly orbit about the sun? (Assume a circular orbit.)

55. *Astronomy.* The brightest star in the night sky, Sirius, is about 4.704×10^{13} mi from the earth. How many light years is it from the earth to Sirius?

56. *Astronomy.* The diameter of the Milky Way galaxy is approximately 5.88×10^{17} mi. How many light years is it from one end of the galaxy to the other?

Named in tribute to Anders Ångström, a Swedish physicist who measured light waves, 1 Å *(read "one Angstrom") equals* 10^{-10} *meters. One parsec is about* 3.26 *light years, and one light year equals* 9.46×10^{15} *meters.*

57. How many Angstroms are in one parsec?

58. How many kilometers are in one parsec?

For Exercises 59 and 60, approximate the average distance from the earth to the sun by 1.50×10^{11} *meters.*

59. Determine the volume of a cylindrical sunbeam that is 3 Å in diameter.

60. Determine the volume of a cylindrical sunbeam that is 5 Å in diameter.

61. *Biology.* An average of 4.55×10^{11} bacteria live in each pound of U.S. mud.* There are 60.0 drops in one teaspoon and 6.0 teaspoons in an ounce. How many bacteria live in a drop of U.S. mud?

**Harper's Magazine*, April 1996, p. 13.

62. *Astronomy.* If a star 5.9×10^{14} mi from the earth were to explode today, its light would not reach us for 100 years. How far does light travel in 13 weeks?

63. *Astronomy.* The diameter of Jupiter is about 1.43×10^5 km. A day on Jupiter lasts about 10 hr. At what speed is Jupiter's equator spinning?

64. *Finance.* A *mil* is one thousandth of a dollar. The taxation rate in a certain school district is 5.0 mils for every dollar of assessed valuation. The assessed valuation for the district is 13.4 million dollars. How much tax revenue will be raised?

65. Write a problem for a classmate to solve. Design the problem so the solution is "The area of the pinhead is 3.14×10^{-4} cm^2."

66. List two advantages of using scientific notation. Answers may vary.

SKILL MAINTENANCE

Evaluate.

67. $3x - 7y$, for $x = 5$ and $y = 1$

68. $2a - 5b$, for $a = 1$ and $b = -6$

SYNTHESIS

69. A criminal claims to be carrying \$5 million in twenty-dollar bills in a briefcase. Is this possible? Why or why not? (*Hint*: See Exercise 53.)

70. When a calculator indicates that $5^{17} = 7.629394531 \times 10^{11}$, an approximation is being made. How can you tell? (*Hint*: What should the ones digit be?)

71. The Sartorius Microbalance Model 4108 can weigh objects to an accuracy of 3.5×10^{-10} oz (*Source*: *Guinness World Records 2000*). A chemical compound weighing 1.2×10^{-9} oz is split in half and weighed on the microbalance. Give a weight range for the actual weight of each half.

72. Write the reciprocal of 2.57×10^{-17} in scientific notation.

73. Compare $8 \cdot 10^{-90}$ and $9 \cdot 10^{-91}$. Which is the larger value? How much larger is it? Write scientific notation for the difference.

74. Write the reciprocal of 8.00×10^{-23} in scientific notation.

75. Evaluate: $(4096)^{0.05}(4096)^{0.2}$.

76. What is the ones digit in 513^{128}?

77. A grain of sand is placed on the first square of a chessboard, two grains on the second square, four grains on the third, eight on the fourth, and so on. Without a calculator, use scientific notation to approximate the number of grains of sand required for the 64th square. (*Hint*: Use the fact that $2^{10} \approx 10^3$.)

COLLABORATIVE CORNER

Paired Problem Solving

Focus: Problem solving, scientific notation, and unit conversion

Time: 15–25 minutes

Group size: 3

ACTIVITY

Given that the earth's average distance from the sun is 1.5×10^{11} meters, determine the earth's orbital speed around the sun in miles per hour. Assume a circular orbit and use the following guidelines.

1. Two students should attempt to solve this problem while the third group member silently observes and writes notes describing the efforts of the other two.

2. After 10–15 minutes, all observers should share their observations with the class as a whole, answering these three questions:

a) What successful strategies were used?
b) What unsuccessful strategies were used?
c) What recommendations do the observers have to make for students working in pairs to solve a problem?

Summary and Review 1

Key Terms

Variable, p. 2
Constant, p. 2
Algebraic expression, p. 2
Equation, p. 2
Solution, p. 3
Factors, p. 4
Substituting, p. 4
Evaluating, p. 4
Exponent, or power, p. 5
Base, p. 5
Natural numbers, p. 6
Whole numbers, p. 6
Integers, p. 6
Roster notation, p. 7
Set-builder notation, p. 7
Element, p. 7
Rational numbers, p. 7
Fraction notation, p. 8
Decimal notation, p. 8
Irrational numbers, p. 8
Real numbers, p. 8
Subset, p. 9
Absolute value, p. 11
Inequality, p. 12
Opposite, p. 13
Additive inverse, p. 13
Reciprocal, p. 16
Multiplicative inverse, p. 16
Indeterminate, p. 17
Equivalent expressions, p. 18
Factoring, p. 19
Equivalent equations, p. 22
Term, p. 24
Like, or similar, terms, p. 24
Combine, or collect, like terms, p. 24
Linear equation, p. 26
Identity, p. 26
Contradiction, p. 26
Conditional equation, p. 26
Solution set, p. 27
Empty set, p. 27
Perimeter, p. 34
Formula, p. 39
Area, p. 40
Mathematical model, p. 42
Scientific notation, p. 56
Significant digits, p. 58

Important Properties and Formulas

Area of a rectangle:	$A = lw$
Area of a square:	$A = s^2$
Area of a parallelogram:	$A = bh$
Area of a trapezoid:	$A = \frac{h}{2}(b_1 + b_2)$
Area of a triangle:	$A = \frac{1}{2}bh$
Area of a circle:	$A = \pi r^2$
Circumference of a circle:	$C = \pi d$
Volume of a cube:	$V = s^3$
Volume of a right circular cylinder:	$V = \pi r^2 h$
Perimeter of a square:	$P = 4s$

Distance traveled:	$d = rt$
Simple interest:	$I = Prt$

Addition of Two Real Numbers

1. *Positive numbers*: Add the numbers. The result is positive.
2. *Negative numbers*: Add absolute values. Make the answer negative.
3. *A negative and a positive number*: If the numbers have the same absolute value, the answer is 0. Otherwise, subtract the smaller absolute value from the larger one:
   a) If the positive number is further from 0, make the answer positive.
   b) If the negative number is further from 0, make the answer negative.
4. *One number is zero*: The sum is the other number.

Multiplication of Two Real Numbers

1. To multiply two numbers with *unlike signs,* multiply their absolute values. The answer is *negative.*
2. To multiply two numbers with the *same sign,* multiply their absolute values. The answer is *positive.*

Division of Two Real Numbers

1. To divide two numbers with *unlike signs,* divide their absolute values. The answer is *negative.*
2. To divide two numbers with the *same sign,* divide their absolute values. The answer is *positive.*

The law of opposites: $a + (-a) = 0$

The law of reciprocals: $a \cdot \frac{1}{a} = 1, a \neq 0$

Absolute value: $|x| = \begin{cases} x, & \text{if } x \geq 0, \\ -x, & \text{if } x < 0 \end{cases}$

For any number a and any nonzero number b,

$$\frac{-a}{b} = \frac{a}{-b} = -\frac{a}{b}.$$

Rules for Order of Operations

1. Simplify within any grouping symbols.
2. Simplify all exponential expressions.
3. Perform all multiplication and division, working from left to right.
4. Perform all addition and subtraction, working from left to right.

Commutative laws:	$a + b = b + a$, $ab = ba$
Associative laws:	$a + (b + c) = (a + b) + c$, $a(bc) = (ab)c$
Distributive law:	$a(b + c) = ab + ac$

The Addition Principle for Equations

$a = b$ is equivalent to $a + c = b + c$.

The Multiplication Principle for Equations

For $c \neq 0$, $a = b$ is equivalent to $ac = bc$.

Five Steps for Problem Solving

1. *Familiarize* yourself with the problem.
2. *Translate* to mathematical language.
3. *Carry out* some mathematical manipulation.
4. *Check* your possible answer in the original problem.
5. *State* the answer clearly.

To solve a formula for a specified letter:

1. Get all terms with the letter being solved for on one side of the equation and all other terms on the other side, using the addition principle. To do this may require removing parentheses.
   - To remove parentheses, divide both sides by the multiplier in front of the parentheses or use the distributive law.
2. When all terms with the specified letter are on the same side, factor (if necessary) to combine like terms.
3. Solve for the specified letter by dividing both sides by the multiplier of that letter.

Definitions and Rules for Exponents

For any integers m and n (assuming 0 is not raised to a nonpositive power):

Zero as an exponent: $a^0 = 1$

Negative exponents:

$$a^{-n} = \frac{1}{a^n}; \frac{a^{-n}}{b^{-m}} = \frac{b^m}{a^n}; \left(\frac{a}{b}\right)^{-n} = \left(\frac{b}{a}\right)^n$$

Multiplying with like bases:

$a^m \cdot a^n = a^{m+n}$ (Product Rule)

Dividing with like bases:

$\frac{a^m}{a^n} = a^{m-n}; a \neq 0$ (Quotient Rule)

Raising a product to a power:

$(ab)^n = a^n b^n$

Raising a power to a power:

$(a^m)^n = a^{mn}$ (Power Rule)

Raising a quotient to a power:

$\left(\frac{a}{b}\right)^n = \frac{a^n}{b^n}; b \neq 0$

Scientific notation for a number is an expression of the form $N \times 10^m$, where $1 \leq N < 10$, N is in decimal notation, and m is an integer.

Review Exercises

The following review exercises are for practice. Answers are at the back of the book. If you need to, restudy the section indicated alongside the answer.

1. Translate to an algebraic expression: Three less than the quotient of two numbers.

2. Evaluate

$$7x^2 - 5y \div zx$$

for $x = -2$, $y = 3$, and $z = -5$.

3. Name the set consisting of the first six even natural numbers using both roster notation and set-builder notation.

4. Find the area of a triangular sign that has a base of 50 cm and a height of 70 cm.

Find the absolute value.

5. $|-7.3|$ **6.** $|4.09|$ **7.** $|0|$

Perform the indicated operation.

8. $-6.5 + (-3.7)$ **9.** $\left(-\frac{4}{5}\right) + \left(\frac{1}{7}\right)$

10. $\left(-\frac{1}{3}\right) + \frac{4}{5}$ **11.** $-7.9 - 3.6$

12. $-\frac{2}{3} - \left(-\frac{1}{2}\right)$ **13.** $12.5 - 17.9$

14. $(-4.2)(-3)$ **15.** $\left(-\frac{2}{3}\right)\left(\frac{5}{8}\right)$

16. $(1.2)(-4)$ **17.** $\frac{-18}{-3}$

18. $\frac{72.8}{-8}$ **19.** $-7 \div \frac{4}{3}$

20. Find $-a$ if $a = -4.01$.

Use a commutative law to write an equivalent expression.

21. $7 + a$ **22.** $7y$ **23.** $5x + y$

Use an associative law to write an equivalent expression.

24. $(4 + a) + b$ **25.** $(xy)7$

26. Obtain an expression that is equivalent to $7mn + 14m$ by factoring.

27. Combine like terms: $5x^3 - 8x^2 + x^3 + 2$.

28. Simplify: $7x - 4[2x + 3(5 - 4x)]$.

Solve. If the solution set is ∅ or ℝ, classify the equation as a contradiction or as an identity.

29. $x - 3.9 = 2.7$

30. $\frac{2}{3}a = 9$

31. $-9x + 4(2x - 3) = 5(2x - 3) + 7$

32. $3(x - 4) + 2 = x + 2(x - 5)$

33. $5t - (7 - t) = 4t + 2(9 + t)$

34. Translate to an equation but do not solve: 13 more than twice a number is 21.

35. A number is 17 less than another number. The sum of the numbers is 115. Find the smaller number.

36. One angle of a triangle measures three times the second angle. The third angle measures twice the second angle. Find the measures of the angles.

37. Solve for m: $P = m/S$.

38. Solve for x: $c = mx - rx$.

39. The volume of a film canister is 28.26 cm^3. If the radius of the canister is 1.5 cm, determine the height.

40. Multiply and simplify: $(5a^2b^7)(-2a^3b)$.

41. Divide and simplify: $\dfrac{12x^3y^8}{3x^2y^2}$.

42. Evaluate a^0, a^2, and $-a^2$ for $a = -5.3$.

Simplify. Do not use negative exponents in the answer.

43. $3^{-4} \cdot 3^7$

44. $(5a^2)^3$

45. $(-2a^{-3}b^2)^{-3}$

46. $\left(\dfrac{x^2y^3}{z^4}\right)^{-2}$

47. $\left(\dfrac{2a^{-2}b}{4a^3b^{-3}}\right)^4$

Simplify.

48. $\dfrac{4(9 - 2 \cdot 3) - 3^2}{4^2 - 3^2}$

49. $1 - (2 - 5)^2 + 5 \div 10 \cdot 4^2$

50. Convert 0.000000103 to scientific notation.

51. One *parsec* (a unit that is used in astronomy) is 30,860,000,000,000 km. Write scientific notation for this number.

Simplify and write scientific notation for each answer. Use the correct number of significant digits.

52. $(8.7 \times 10^{-9}) \times (4.3 \times 10^{15})$

53. $\dfrac{1.2 \times 10^{-12}}{6.1 \times 10^{-7}}$

54. A sheet of plastic has a thickness of 0.00015 mm. The sheet is 1.2 m by 79 m. Use scientific notation to find the volume of the sheet.

SYNTHESIS

55. Describe a method that could be used to write equations that have no solution.

56. Explain how the distributive law can be used when combining like terms.

57. If the smell of gasoline is detectable at 3 parts per billion, what percent of the air is occupied by the gasoline?

58. Evaluate $a + b(c - a^2)^0 + (abc)^{-1}$ for $a = 2$, $b = -3$, and $c = -4$.

59. What's a better deal: a 13-in. diameter pizza for \$8 or a 17-in. diameter pizza for \$11? Explain.

60. The surface area of a cube is 486 cm^2. Find the volume of the cube.

61. Solve for z: $m = \dfrac{x}{y - z}$.

62. Simplify: $\dfrac{(3^{-2})^a \cdot (3^b)^{-2a}}{(3^{-2})^b \cdot (9^{-b})^{-3a}}$.

63. Each of Ray's test scores counts three times as much as a quiz score. If after 4 quizzes Ray's average is 82.5, what score does he need on the first test in order to raise his average to 85?

64. Fill in the following blank so as to ensure that the equation is an identity.

$$5x - 7(x + 3) - 4 = 2(7 - x) + \_\_\_\_\_$$

65. Replace the blank with one term to ensure that the equation is a contradiction.

$$20 - 7[3(2x + 4) - 10] = 9 - 2(x - 5) + \_\_\_\_\_$$

66. Use the commutative law for addition once and the distributive law twice to show that

$$a2 + cb + cd + ad = a(d + 2) + c(b + d).$$

67. Find an irrational number between $\frac{1}{2}$ and $\frac{3}{4}$.

Chapter Test 1

1. Translate to an algebraic expression: Three more than the product of two numbers.
2. Evaluate $a^3 - 5b + b \div ac$ for $a = -2$, $b = 6$, and $c = 3$.
3. A triangular stamp's base measures 3 cm and its height 2.5 cm. Find its area.

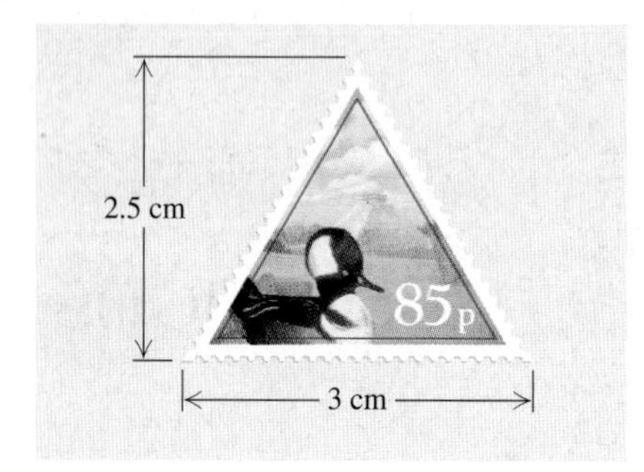

Perform the indicated operation.

4. $-15 + (-16)$
5. $-7.5 + 3.8$
6. $3.21 + (-8.32)$
7. $29.5 - 43.7$
8. $-16.8 - 26.4$
9. $-6.4(5.3)$
10. $-\frac{7}{3} - \left(-\frac{3}{4}\right)$
11. $-\frac{2}{7}\left(-\frac{5}{14}\right)$
12. $\frac{-42.6}{-7.1}$
13. $\frac{2}{5} \div \left(-\frac{3}{10}\right)$
14. Simplify: $7 + (1 - 3)^2 - 9 \div 2^2 \cdot 6$.
15. Use a commutative law to write an expression equivalent to $7x + y$.

Combine like terms.

16. $5y - 14y + 19y$
17. $6a^2b - 5ab^2 + ab^2 - 5a^2b + 2$
18. Simplify: $9x - 3(2x - 5) - 7$.

Solve. If the solution set is $\mathbb{R}$ or $\varnothing$, classify the equation as an identity or a contradiction.

19. $10x - 7 = 38x + 49$
20. $13t - (5 - 2t) = 5(3t - 1)$
21. Solve for P_2: $\frac{P_1V_1}{T_1} = \frac{P_2V_2}{T_2}$.
22. Greg's scores on five tests are 84, 80, 76, 96, and 80. What must Greg score on the sixth test so that his average will be 85?
23. Find three consecutive odd integers such that the sum of four times the first, three times the second, and two times the third is 167.

Simplify. Do not use negative exponents in the answer.

24. $-5(x - 4) - 3(x + 7)$
25. $6b - [7 - 2(9b - 1)]$
26. $(7x^{-4}y^{-7})(-6x^{-6}y)$
27. -3^{-2}
28. $(-6x^2y^{-4})^{-2}$
29. $\left(\frac{2x^3y^{-6}}{-4y^{-2}}\right)^2$
30. $(7x^3y)^0$

Simplify and write scientific notation for the answer. Use the correct number of significant digits.

31. $(9.05 \times 10^{-3})(2.22 \times 10^{-5})$
32. $\frac{5.6 \times 10^7}{2.8 \times 10^{-3}}$
33. $\frac{1.8 \times 10^{-4}}{4.8 \times 10^{-7}}$

Solve.

34. The average distance from the planet Venus to the sun is 6.7×10^7 mi. About how far does Venus travel in one orbit around the sun? (Assume a circular orbit.)

SYNTHESIS

Simplify.

35. $(4x^{3a}y^{b+1})^{2c}$
36. $\frac{-27a^{x+1}}{3a^{x-2}}$
37. $\frac{(-16x^{x-1}y^{y-2})(2x^{x+1}y^{y+1})}{(-7x^{x+2}y^{y+2})(8x^{x-2}y^{y-1})}$

2 Graphs, Functions, and Linear Equations

2.1 Graphs
2.2 Functions
2.3 Linear Functions: Slope, Graphs, and Models
2.4 Another Look at Linear Graphs
2.5 Other Equations of Lines
Connecting the Concepts
2.6 The Algebra of Functions
Summary and Review
Test

AN APPLICATION

More and more Americans are making their travel arrangements using the Internet. The number of on-line travel buyers has grown from 48 million in 1999 to approximately 60 million in 2001 (*Source*: Travel Industry Association of America). Find the rate at which this number is growing.

This problem appears as Example 10 in Section 2.3.

I use math in many different ways. Initially, I must estimate the hours and cost of the project for my customer. During development, I use math to create pages, frames, tables, automation objects, and graphics.

TAMMY J. GROSS
Webmaster
Indianapolis, IN

Graphs are important because they allow us to see relationships. For example, a graph of an equation in two variables helps us see how those two variables are related. Graphs are useful in a wide variety of settings. In this chapter, we emphasize graphs of equations.

A certain kind of relationship between two variables is known as a function. *Functions are very important in mathematics in general, and in problem solving in particular. In this chapter, you will learn what we mean by a function and then begin to use functions to solve problems.*

2.1 Graphs

Points and Ordered Pairs • Quadrants • Solutions of Equations • Nonlinear Equations

Study Tip

This text is accompanied by a set of videotapes in which the authors and other instructors present material and concepts for each section of the text. These videos are available on CD-ROM and in most campus' math labs. When missing a class is unavoidable, try to view the appropriate video lecture for that section. These videotapes can also be used as an aid whenever difficulty arises.

It has often been said that a picture is worth a thousand words. As we turn our attention to the study of graphs, we discover that in mathematics this is quite literally the case. Graphs are a compact means of displaying information and provide a visual approach to problem solving.

Points and Ordered Pairs

Whereas on a number line, each point corresponds to a number, on a plane, each point corresponds to a pair of numbers. The idea of using two perpendicular number lines, called **axes** (pronounced ak-sēz; singular, **axis**) to identify points in a plane is commonly attributed to the great French mathematician and philosopher René Descartes (1596–1650). In honor of Descartes, this representation is also called the **Cartesian coordinate system**. The variable x is normally represented on the horizontal axis and the variable y on the vertical axis, so we will also refer to the ***x*, *y*-coordinate system**.

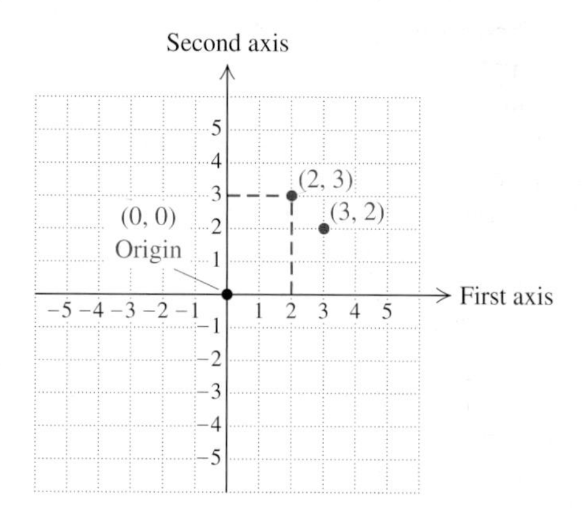

To identify a point that appears on the x, y-coordinate system, we write a pair of numbers in the form (x, y), where x is the number directly above or below the point on the horizontal axis and y is the number to the left or right of the point on the vertical axis (see the dashed lines in the figure on p. 70). Thus, as shown on p. 70, $(2,3)$ and $(3,2)$ are different points. Because the order in which the numbers are listed is important, these are called **ordered pairs**.

Example 1 Plot the points $(-4,3)$, $(-5,-3)$, $(0,4)$, and $(2.5,0)$.

Solution To plot $(-4,3)$, we note that the first number, -4, tells us the distance in the first, or horizontal, direction. We go 4 units *left* of the origin. The second number tells us the distance in the second, or vertical, direction. We go 3 units *up*. The point $(-4,3)$ is then marked, or "plotted."

The points $(-5,-3)$, $(0,4)$, and $(2.5,0)$ are also plotted below.

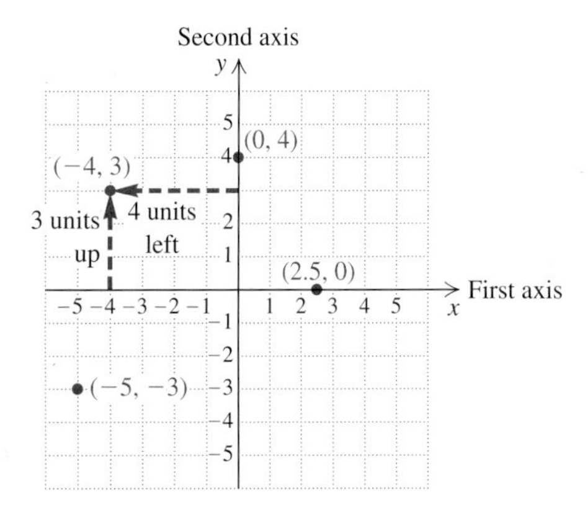

The numbers in an ordered pair are called **coordinates**. In $(-4,3)$, the *first coordinate* is -4 and the *second coordinate** is 3. The point with coordinates $(0,0)$ is called the **origin**.

Quadrants

The axes divide the plane into four regions called **quadrants**, as shown here.

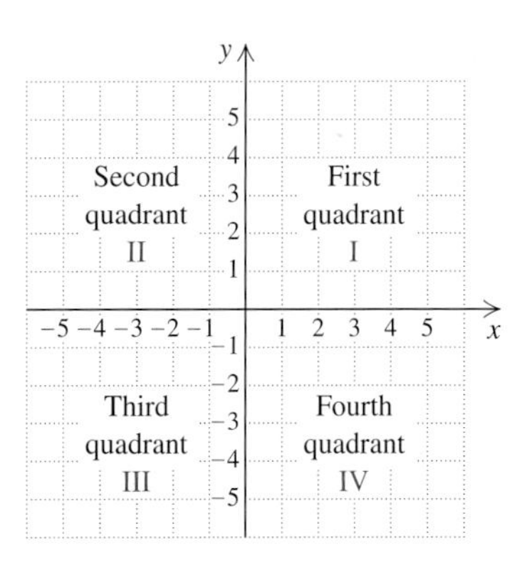

*The first coordinate is sometimes called the **abscissa** and the second coordinate the **ordinate**.

technology connection A

The window of a grapher is the rectangular portion of the screen in which a graph appears. Windows are described by four numbers of the form $[L, R, B, T]$, representing the left and right endpoints of the x-axis and the bottom and top endpoints of the y-axis. Below, we have graphed $y = -4x + 3$, using the "standard" $[-10, 10, -10, 10]$ window.

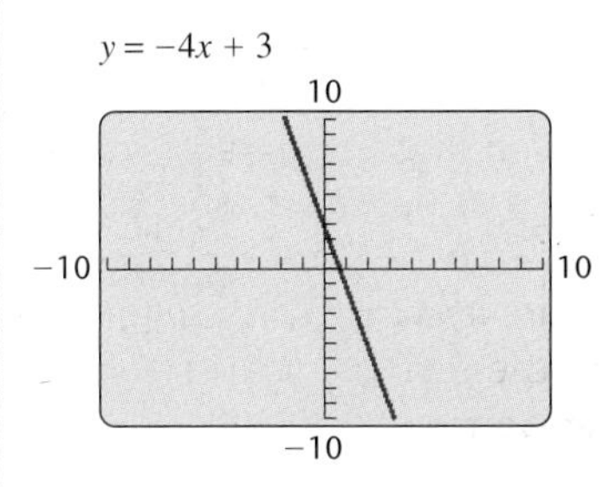

Once an equation has been graphed, we can investigate some of its points by using the TRACE feature. When the TRACE key is pressed, a cursor can be moved along the graph while its coordinates appear. To find the y-value that is paired with a particular x-value, we simply key in that x-value and press ENTER.

1. Graph $y = -4x + 3$ using a $[-10, 10, -10, 10]$ window. Then TRACE to find coordinates of several points, including the points with the x-values -1.5 and 1.

In region I (the *first* quadrant), both coordinates of a point are positive. In region II (the *second* quadrant), the first coordinate is negative and the second coordinate is positive. In the third quadrant, both coordinates are negative, and in the fourth quadrant, the first coordinate is positive while the second coordinate is negative.

Points with one or more 0's as coordinates, such as $(0, -6)$, $(4, 0)$, and $(0, 0)$, are on axes and *not* in quadrants.

Solutions of Equations

If an equation has two variables, its solutions are pairs of numbers. When such a solution is written as an ordered pair, the first number listed in the pair usually replaces the variable that occurs first alphabetically.

Example 2 Determine whether the pairs $(4, 2)$, $(-1, -4)$, and $(2, 5)$ are solutions of the equation $y = 3x - 1$.

Solution To determine whether each pair is a solution, we replace x with the first coordinate and y with the second coordinate. When the replacements make the equation true, we say that the ordered pair is a solution.

$y = 3x - 1$	$y = 3x - 1$	$y = 3x - 1$
$2 \;?\; 3(4) - 1$	$-4 \;?\; 3(-1) - 1$	$5 \;?\; 3(2) - 1$
$\mid\; 12 - 1$	$\mid\; -3 - 1$	$\mid\; 6 - 1$
$2 \mid 11$	$-4 \mid -4$	$5 \mid 5$
Since $2 = 11$ is *false*, the pair $(4, 2)$ is *not* a solution.	Since $-4 = -4$ is *true*, the pair $(-1, -4)$ *is* a solution.	Since $5 = 5$ is *true*, the pair $(2, 5)$ *is* a solution.

In fact, there is an infinite number of solutions of $y = 3x - 1$. We can use a graph as a convenient way of representing these solutions. Thus to **graph** an equation means to make a drawing that represents its solutions.

Example 3 Graph the equation $y = x$.

Solution We label the horizontal axis as the x-axis and the vertical axis as the y-axis.

Next, we find some ordered pairs that are solutions of the equation. In this case, it is easy. Here are a few pairs that satisfy the equation $y = x$:

$$(0, 0), \quad (1, 1), \quad (5, 5), \quad (-1, -1), \quad (-6, -6).$$

Plotting these points, we can see that if we were to plot a million solutions, the dots would merge into a solid line. Observing the pattern, we can draw the line with a ruler. The line is the graph of the equation $y = x$. We label the line $y = x$.

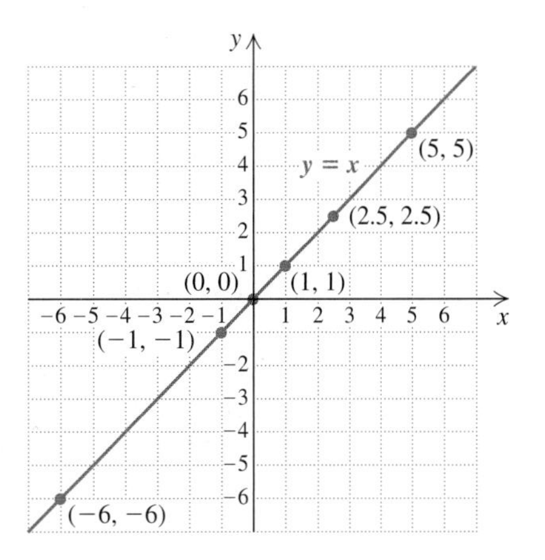

Note that the coordinates of *any* point on the line—for example, $(2.5, 2.5)$—satisfy the equation $y = x$. Note too that the line continues indefinitely in both directions—only part of it is shown.

Example 4 Graph the equation $y = 2x$.

Solution We find some ordered pairs that are solutions. This time we list the pairs in a table. To find an ordered pair, we can choose *any* number for x and then determine y. For example, if we choose 3 for x, then $y = 2 \cdot 3 = 6$ (substituting into the equation $y = 2x$). We choose some negative values for x, as well as some positive ones. If a number takes us off the graph paper, we generally do not use it. Next, we plot these points. If we plotted *many* such points, they would appear to make a solid line. We draw the line with a ruler and label it $y = 2x$.

x	y	(x, y)
0	0	$(0, 0)$
1	2	$(1, 2)$
3	6	$(3, 6)$
−2	−4	$(-2, -4)$
−3	−6	$(-3, -6)$

Choose any x.
Compute y.
Form the pair.
Plot the points and draw the line.

Example 5 Graph the equation $y = -\frac{1}{2}x$.

Solution By choosing even integers for x, we can avoid fractional values when calculating y. For example, if we choose 4 for x, we get $y = \left(-\frac{1}{2}\right)(4)$, or -2. When x is -6, we get $y = \left(-\frac{1}{2}\right)(-6)$, or 3. We find several ordered pairs, plot them, and draw the line.

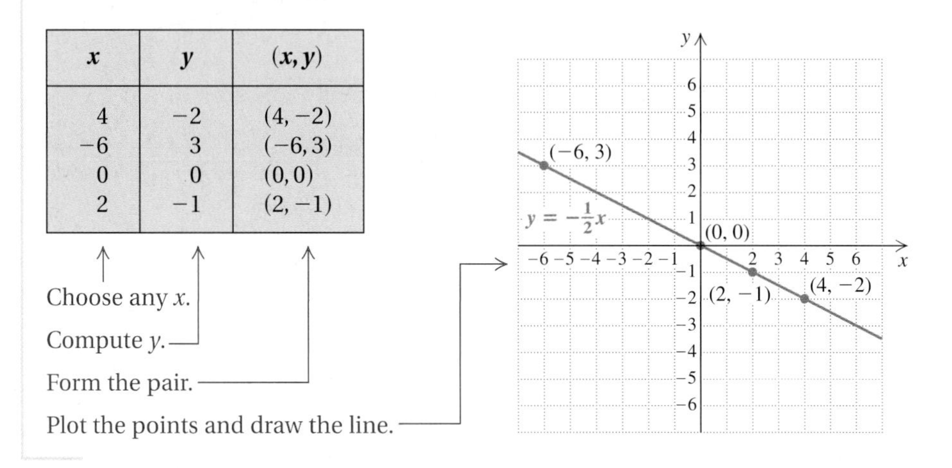

x	y	(x, y)
4	-2	$(4, -2)$
-6	3	$(-6, 3)$
0	0	$(0, 0)$
2	-1	$(2, -1)$

As you can see, the graphs in Examples 3–5 are straight lines. We will refer to any equation whose graph is a straight line as a **linear equation**. Linear equations will be discussed in more detail in Sections 2.3–2.5.

Nonlinear Equations

There are many equations for which the graph is not a straight line. Let's look at some of these **nonlinear equations**.

Example 6 Graph: $y = x^2 - 5$.

Solution We select numbers for x and find the corresponding values for y. For example, if we choose -2 for x, we get $y = (-2)^2 - 5 = 4 - 5 = -1$. The table lists several ordered pairs.

x	y
0	-5
-1	-4
1	-4
-2	-1
2	-1
-3	4
3	4

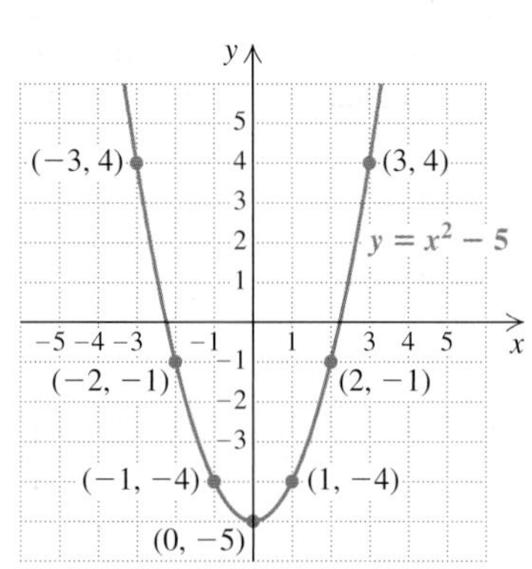

Next, we plot the points. The more points plotted, the clearer the shape of the graph becomes. Since the value of $x^2 - 5$ grows rapidly as x moves away from the origin, the graph rises steeply on either side of the y-axis.

Curves similar to the one in Example 6 are studied in detail in Chapter 8.

Example 7 Graph: $y = |x|$.

Solution We select numbers for x and find the corresponding values for y. For example, if we choose -1 for x, we get $y = |-1| = 1$. Several ordered pairs are listed in the table below.

x	y
-3	3
-2	2
-1	1
0	0
1	1
2	2
3	3

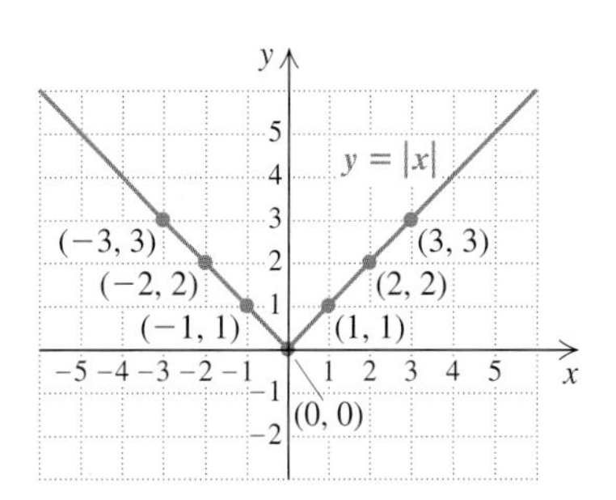

We plot these points, noting that the absolute value of a positive number is the same as the absolute value of its opposite. Thus the x-values 3 and -3 both are paired with the y-value 3. The graph is V-shaped, as shown above.

Often, graphs are drawn with the aid of a grapher. As shown above, determining just a few ordered pairs can be quite time-consuming. With a grapher, thousands of ordered pairs can be found at the touch of a button. Most graphers can display a table of ordered pairs for any equation that is entered. By adjusting the TABLE SETUP, we can control the smallest x-value listed (using Table Min) and the difference between successive x-values (using $\triangle$TBL). The up and down arrow keys allow us to scroll up and down the table. For the table shown, we used $y_1 = -4x + 3$.

TABLE MIN = 1.4 ΔTBL = .1

X	Y1	
1.4	−2.6	
1.5	−3	
1.6	−3.4	
1.7	−3.8	
1.8	−4.2	
1.9	−4.6	
2	−5	

X = 1.4

Graph each equation using a $[-10, 10, -10, 10]$ window. Then create a table of ordered pairs in which the x-values start at -1 and are 0.1 unit apart.

1. $y = 5x - 3$

2. $y = x^2 - 4x + 3$

3. $y = (x + 4)^2$

4. $y = \sqrt{x + 2}$

5. $y = |x + 2|$

Exercise Set 2.1

FOR EXTRA HELP

Digital Video Tutor CD 1
Videotape 3

InterAct Math

Math Tutor Center

MathXL

MyMathLab.com

To the student and the instructor: Throughout this text, selected exercises are marked with the icon Aha! *. These "Aha!" exercises can be answered quite easily if the student pauses to inspect the exercise rather than proceed mechanically. This is done to discourage rote memorization. Some "Aha!" exercises are left unmarked to encourage students to always pause before working a problem.*

Give the coordinates of each point.

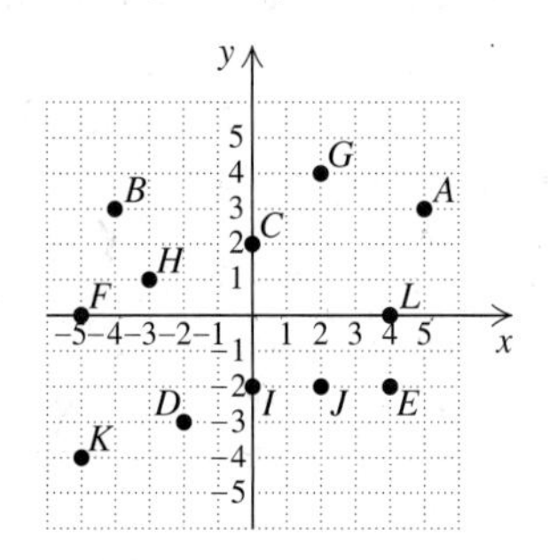

1. $A, B, C, D, E,$ and F

2. $G, H, I, J, K,$ and L

Plot the points. Label each point with the indicated letter.

3. $A(3,0)$, $B(4,2)$, $C(5,4)$, $D(6,6)$, $E(3,-4)$, $F(3,-3)$, $G(3,-2)$, $H(3,-1)$

4. $A(1,1)$, $B(2,3)$, $C(3,5)$, $D(4,7)$, $E(-2,1)$, $F(-2,2)$, $G(-2,3)$, $H(-2,4)$, $J(-2,5)$, $K(-2,6)$

5. Plot the points $M(2,3)$, $N(5,-3)$, and $P(-2,-3)$. Draw $\overline{MN}$, $\overline{NP}$, and $\overline{MP}$. ($\overline{MN}$ means the line segment from M to N.) What kind of geometric figure is formed? What is its area?

6. Plot the points $Q(-4,3)$, $R(5,3)$, $S(2,-1)$, and $T(-7,-1)$. Draw $\overline{QR}$, $\overline{RS}$, $\overline{ST}$, and $\overline{TQ}$. What kind of figure is formed? What is its area?

Name the quadrant in which each point is located.

7. $(-4,-9)$

8. $(2,17)$

9. $(-6,1)$

10. $(4,-8)$

11. $(3,\frac{1}{2})$

12. $(-1,-7)$

13. $(6.9,-2)$

14. $(-4,31)$

Determine if each ordered pair is a solution of the given equation.

15. $(1,-1)$; $y = 3x - 4$

16. $(2,5)$; $y = 4x - 3$

17. $(2,4)$; $5s - t = 8$

18. $(1,3)$; $4p + q = 5$

19. $(3,5)$; $4x - y = 7$

20. $(2,7)$; $5x - y = 3$

21. $(0,\frac{3}{5})$; $6a + 5b = 3$

22. $(0,\frac{3}{2})$; $3f + 4g = 6$

23. $(2,-1)$; $4r + 3s = 5$

24. $(2,-4)$; $5w + 2z = 2$

25. $(5,3)$; $x - 3y = -4$

26. $(1,2)$; $2x - 5y = -6$

27. $(-1,3)$; $y = 3x^2$

28. $(2,4)$; $2r^2 - s = 5$

29. $(2,3)$; $5s^2 - t = 7$

30. $(2,3)$; $y = x^3 - 5$

Graph.

31. $y = x + 1$

32. $y = x + 3$

33. $y = -x$

34. $y = 3x$

35. $y = 3x - 2$

36. $y = -4x + 1$

37. $y = -2x + 3$

38. $y = -3x + 1$

Aha! **39.** $y + 2x = 3$

40. $y + 3x = 1$

41. $y = -\frac{3}{2}x + 5$

42. $y = -\frac{2}{3}x - 2$

43. $y = \frac{3}{4}x + 1$

44. $y = \frac{3}{4}x + 2$

45. $y = x^2 + 1$

46. $y = x^2 + 2$

47. $y = x^2 - 3$

48. $y = x^2 - 1$

49. $y = 5 - x^2$

50. $y = 4 - x^2$

51. $y = |x| + 1$

52. $y = |x| + 2$

53. $y = |x| - 3$

54. $y = |x| - 1$

55. What can be said about the location of two points that have the same first coordinates and second coordinates that are opposites of each other?

56. Examine Example 7 and explain why it is unwise to draw a graph after plotting just two points.

SKILL MAINTENANCE

Evaluate.

57. $5s - 3t$, for $s = 2$ and $t = 4$

58. $2a + 7b$, for $a = 3$ and $b = 1$

59. $(3x - y)^2$, for $x = 4$ and $y = 2$

60. $(2m + n)^2$, for $m = 3$ and $n = 1$

Aha! **61.** $(5 - x)^4(x + 2)^3$, for $x = -2$

62. $2x^2 + 4x - 9$, for $x = -3$

SYNTHESIS

63. Without making a drawing, how can you tell that the graph of $y = x - 30$ passes through three quadrants?

64. At what point will the line passing through $(a, -1)$ and $(a, 5)$ intersect the line that passes through $(-3, b)$ and $(2, b)$? Why?

65. Graph $y = 6x$, $y = 3x$, $y = \frac{1}{2}x$, $y = -6x$, $y = -3x$, and $y = -\frac{1}{2}x$ using the same set of axes and compare the slants of the lines. Describe the pattern that relates the slant of the line to the multiplier of x.

66. Using the same set of axes, graph $y = 2x$, $y = 2x - 3$, and $y = 2x + 3$. Describe the pattern relating each line to the number that is added to $2x$.

67. Match each sentence with the most appropriate of the four graphs shown.

a) Roberta worked part time until September, full time until December, and overtime until Christmas.

b) Clyde worked full time until September, half time until December, and full time until Christmas.

c) Clarissa worked overtime until September, full time until December, and overtime until Christmas.

d) Doug worked part time until September, half time until December, and full time until Christmas.

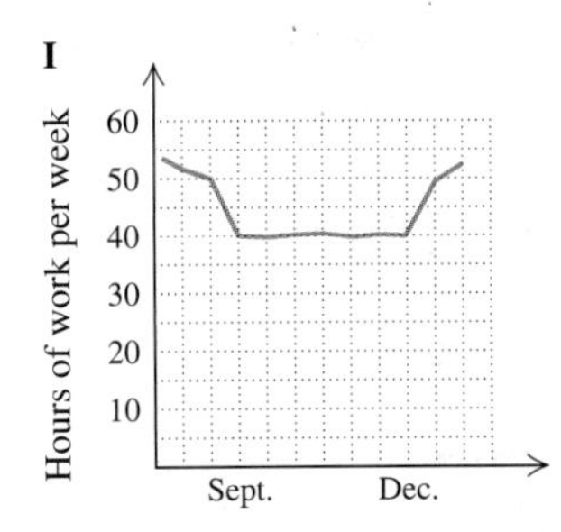

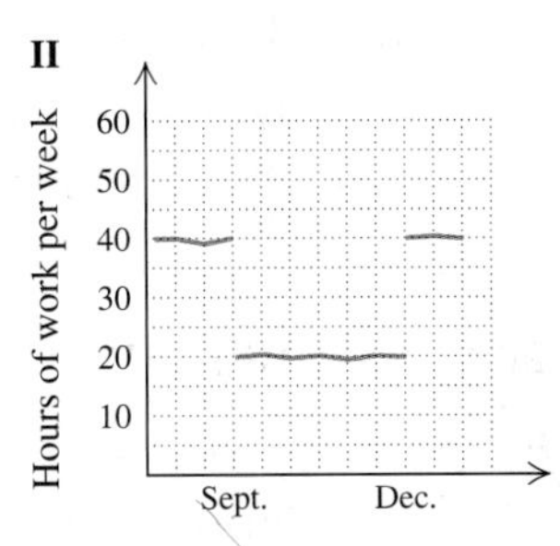

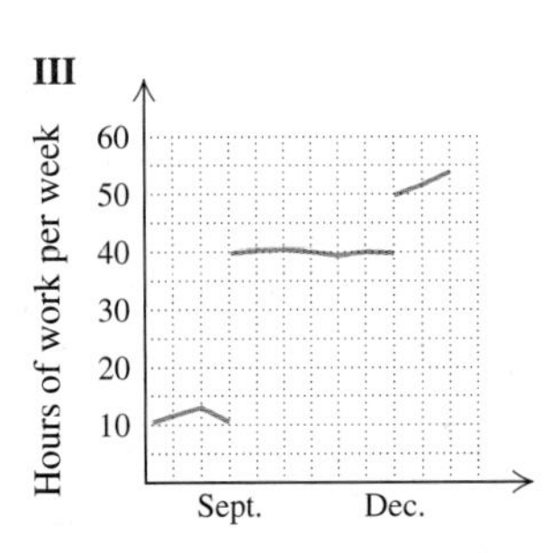

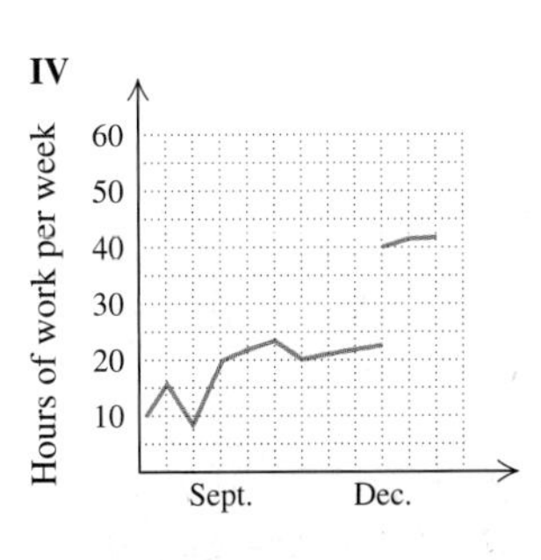

68. Match each sentence with the most appropriate of the four graphs shown below.

a) Carpooling to work, Terry spent 10 min on local streets, then 20 min cruising on the freeway, and then 5 min on local streets to his office.

b) For her commute to work, Sharon drove 10 min to the train station, rode the express for 20 min, and then walked for 5 min to her office.

c) For his commute to school, Roger walked 10 min to the bus stop, rode the express for 20 min, and then walked for 5 min to his class.

d) Coming home from school, Kristy waited 10 min for the school bus, rode the bus for 20 min, and then walked 5 min to her house.

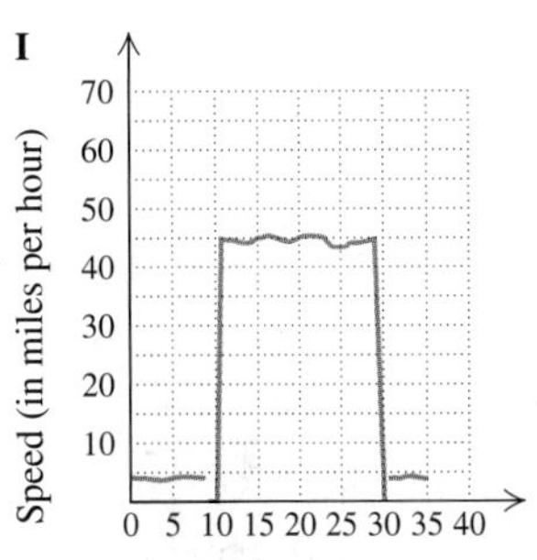

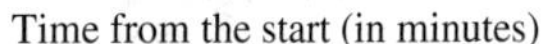

Time from the start (in minutes)

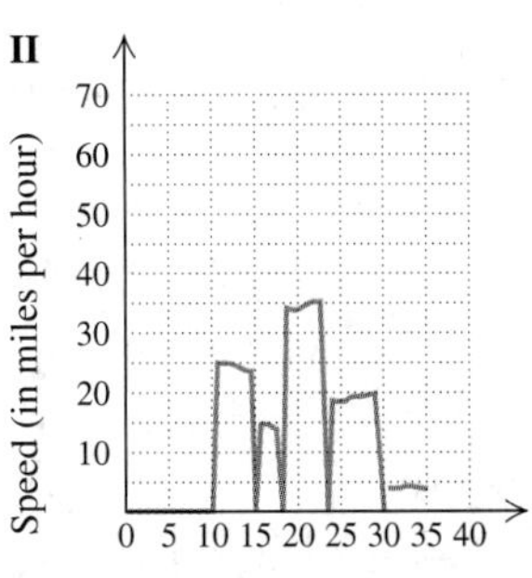

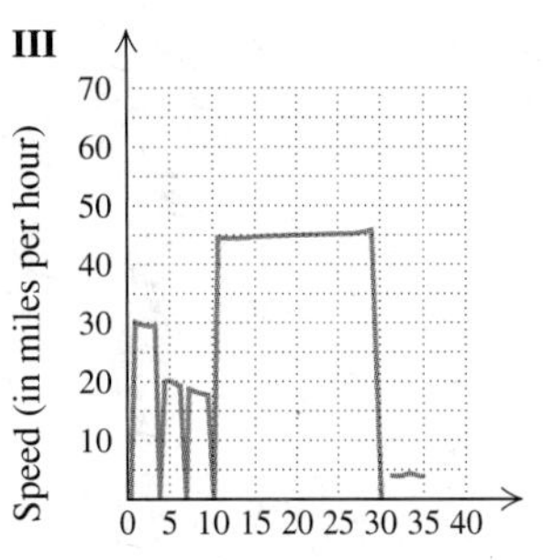

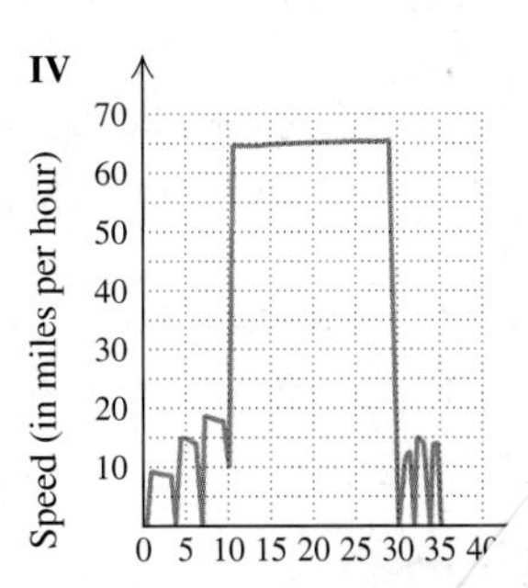

Time from the start (in minutes)

69. Which of the following equations have $\left(-\frac{1}{3}, \frac{1}{4}\right)$ as a solution?

a) $-\frac{3}{2}x - 3y = -\frac{1}{4}$
b) $8y - 15x = \frac{7}{2}$
c) $0.16y = -0.09x + 0.1$
d) $2(-y + 2) - \frac{1}{4}(3x - 1) = 4$

Graph each equation after plotting at least 10 points.

70. $y = 1/x$; use x-values from -4 to 4

71. $y = 1/(x - 2)$; use values of x from -2 to 6

72. $y = \sqrt{x}$; use values of x from 0 to 10

73. $y = 1/x^2$; use values of x from -4 to 4

74. If $(2, -3)$ and $(-5, 4)$ are the endpoints of a diagonal of a square, what are the coordinates of the other two vertices? What is the area of the square?

75. If $(-10, -2)$, $(-3, 4)$, and $(6, 4)$ are the coordinates of three vertices of a parallelogram, determine the coordinates of three different points that could serve as the fourth vertex.

Note: Throughout this text, the icon is used to indicate exercises designed for graphers (graphing calculators or computers).

In Exercises 76 and 77, use a grapher to draw the graph of each equation. For each equation, select a window that shows the curvature of the graph. Then, if possible, create a table of ordered pairs in which x-values extend, by tenths, from 0 to 0.6.

76. a) $y = -12.4x + 7.8$
b) $y = -3.5x^2 + 6x - 8$
c) $y = (x - 3.4)^3 + 5.6$

77. a) $y = 2.3x^4 + 3.4x^2 + 1.2x - 4$
b) $y = 12.3x - 3.5$
c) $y = 3(x + 2.3)^2 + 2.3$

2.2 Functions

Functions and Graphs • Function Notation and Equations • Applications: Interpolation and Extrapolation

We now develop the idea of a *function*—one of the most important concepts in mathematics. In much the same way that the ordered pairs of Section 2.1 formed correspondences between first coordinates and second coordinates, a function is a special kind of correspondence between two sets. For example,

To each person in a class	there corresponds	his or her biological mother.
To each item in a shop	there corresponds	its price.
To each real number	there corresponds	the cube of that number.

In each example, the first set is called the **domain**. The second set is called the **range**. For any member of the domain, there is *just one* member of the range to which it corresponds. This kind of correspondence is called a **function**.

Example 1

Determine whether each correspondence is a function.

a)

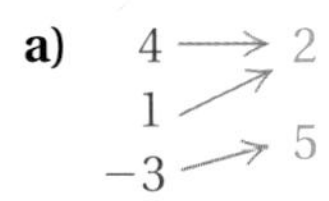

b)

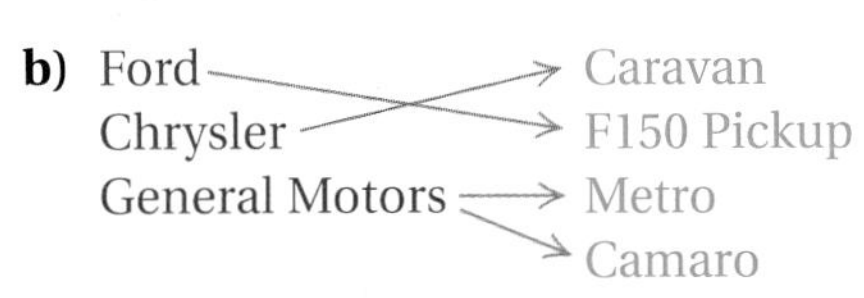

Solution

a) The correspondence *is* a function because each member of the domain corresponds to *just one* member of the range.

b) The correspondence *is not* a function because a member of the domain (General Motors) corresponds to more than one member of the range.

> **Function**
>
> A *function* is a correspondence between a first set, called the *domain,* and a second set, called the *range,* such that each member of the domain corresponds to *exactly one* member of the range.

Example 2

Determine whether each correspondence is a function.

Domain	*Correspondence*	*Range*
a) An elevator full of people	Each person's weight	A set of positive numbers
b) $\{-2, 0, 1, 2\}$	Each number's square	$\{0, 1, 4\}$
c) The books in a college bookstore	Each book's author	A set of people

Solution

a) The correspondence *is* a function, because each person has *only one* weight.

b) The correspondence *is* a function, because every number has *only one* square.

c) The correspondence *is not* a function, because some books have *more than one* author.

Study Tip

The *Student's Solutions Manual* is an excellent resource if you need additional help with an exercise in the exercise sets. It contains worked-out solutions to the odd-numbered exercises in each exercise set.

Functions and Graphs

The functions in Examples 1(a) and 2(b) can be expressed as sets of ordered pairs. Example 1(a) can be written $\{(-3,5),(1,2),(4,2)\}$ and Example 2(b) can be written $\{(-2,4),(0,0),(1,1),(2,4)\}$. We can graph these functions as follows.

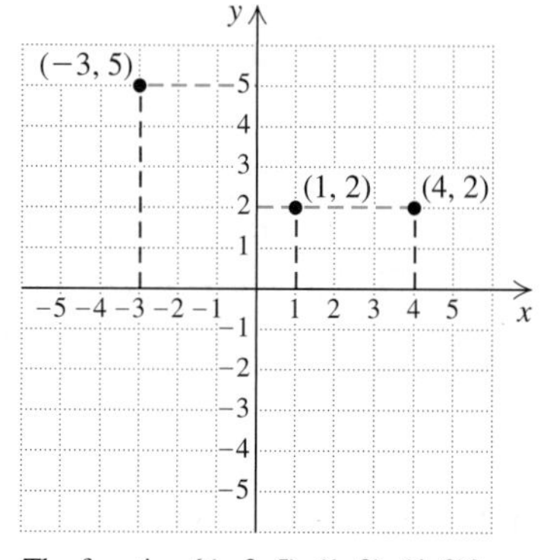

The function $\{(-3, 5), (1, 2), (4, 2)\}$
Domain is $\{-3, 1, 4\}$
Range is $\{5, 2\}$

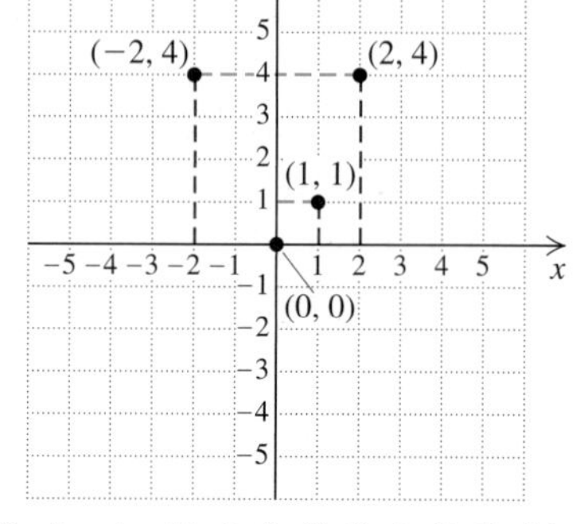

The function $\{(-2, 4), (0, 0), (1, 1), (2, 4)\}$
Domain is $\{-2, 0, 1, 2\}$
Range is $\{4, 0, 1\}$

When a function is given as a set of ordered pairs, the domain is the set of all first coordinates and the range is the set of all second coordinates. Functions are generally represented by lower- or upper-case letters.

Example 3 Find the domain and the range of the function f shown here.

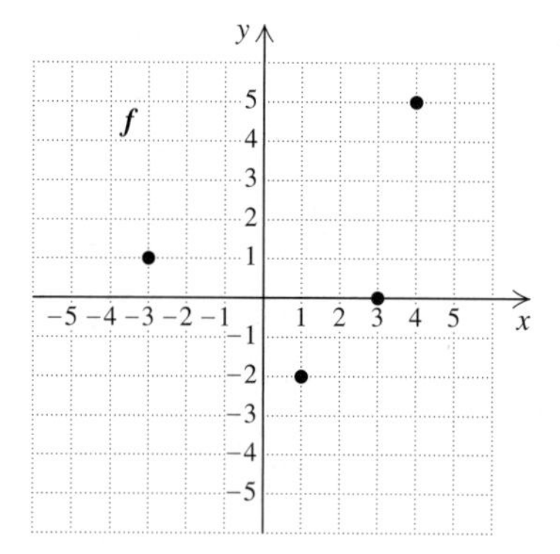

Solution Here f can be written $\{(-3,1),(1,-2),(3,0),(4,5)\}$. The domain is the set of all first coordinates, $\{-3,1,3,4\}$, and the range is the set of all second coordinates, $\{1,-2,0,5\}$. We can also find the domain and the range directly, without first writing f.

Example 4 For the function f shown here, determine each of the following.

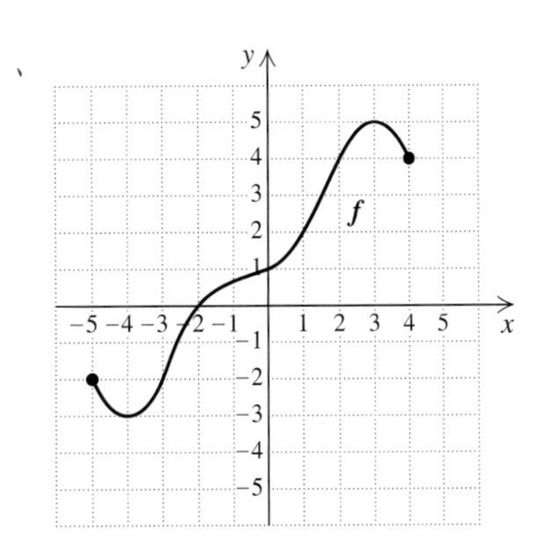

a) What member of the range is paired with 2
b) The domain of f
c) What member of the domain is paired with -3
d) The range of f

Solution

a) To determine what member of the range is paired with 2, we locate 2 on the horizontal axis (this is where the domain is located). Next, we find the point on the graph of f for which 2 is the first coordinate. From that point, we can look to the vertical axis to find the corresponding y-coordinate, 4. The "input" 2 has the "output" 4.

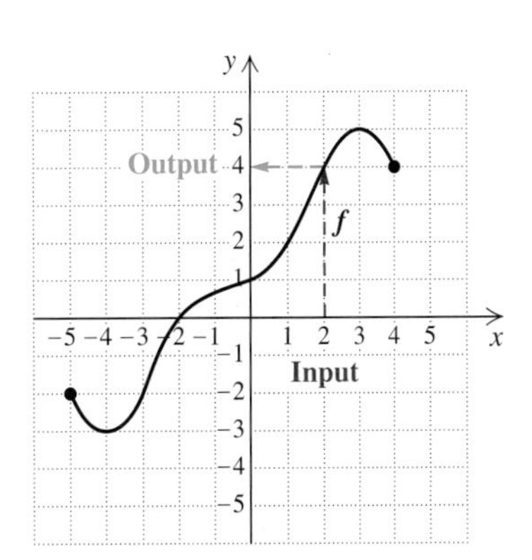

b) The domain of the function is the set of all x-values that are in the graph. Because there are no breaks in the graph of f, these extend continuously from -5 to 4 and can be viewed as the curve's shadow, or *projection*, on the x-axis. Thus the domain is $\{x \mid -5 \le x \le 4\}$.

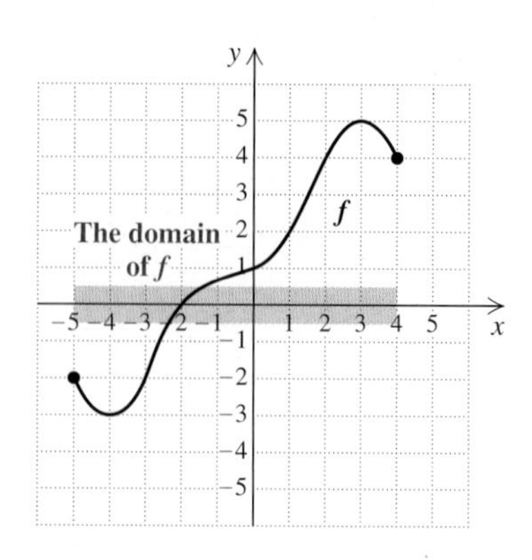

c) To determine what member of the domain is paired with -3, we locate -3 on the vertical axis. (This is where the range is located; see below.) From there we look left and right to the graph of f to find any points for which -3 is the second coordinate. One such point exists, $(-4, -3)$. We observe that -4 is the only element of the domain paired with -3.

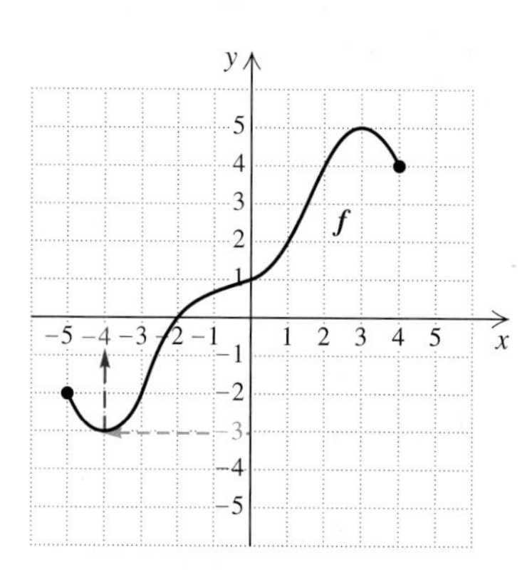

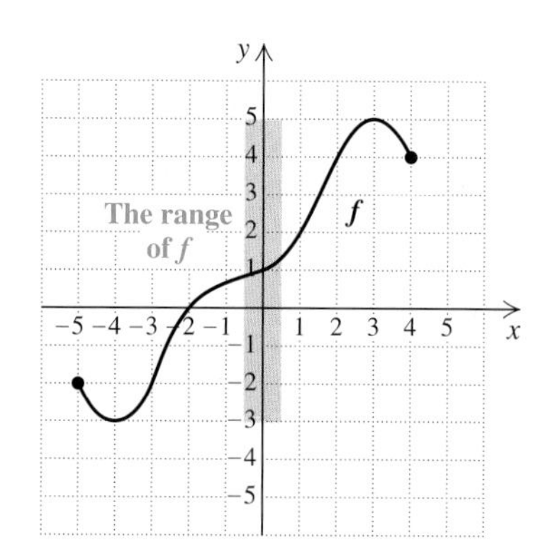

d) The range of the function is the set of all y-values that are in the graph. (See the graph on the right above.) These extend continuously from -3 to 5, and can be viewed as the curve's projection on the y-axis. Thus the range is $\{y \mid -3 \le y \le 5\}$.

Note that if a graph contains two or more points with the same first coordinate, that graph cannot represent a function (otherwise one member of the domain would correspond to more than one member of the range). This observation is the basis of the *vertical-line test.*

The Vertical-Line Test

If it is possible for a vertical line to cross a graph more than once, then the graph is not the graph of a function.

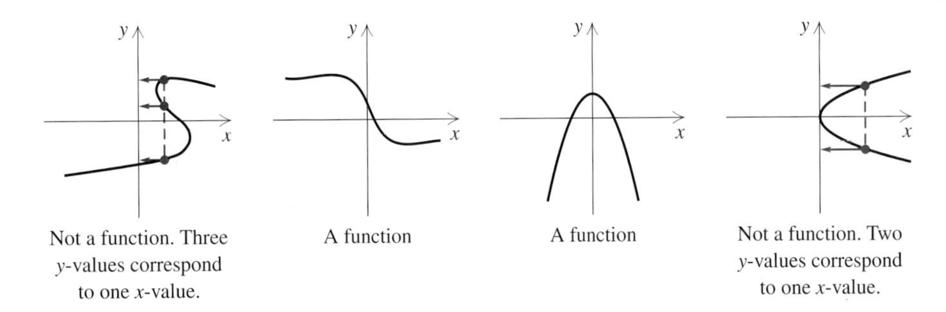

Not a function. Three y-values correspond to one x-value.

A function

A function

Not a function. Two y-values correspond to one x-value.

Graphs that do not represent functions still do represent *relations.*

> ***Relation***
>
> A *relation* is a correspondence between a first set, called the *domain,* and a second set, called the *range,* such that each member of the domain corresponds to *at least one* member of the range.

Thus, although the correspondences and graphs above are not all functions, they *are* all relations.

Function Notation and Equations

To understand function notation, it helps to imagine a "function machine." Think of putting a member of the domain (an *input*) into the machine. The machine is programmed to produce the appropriate member of the range (the *output*).

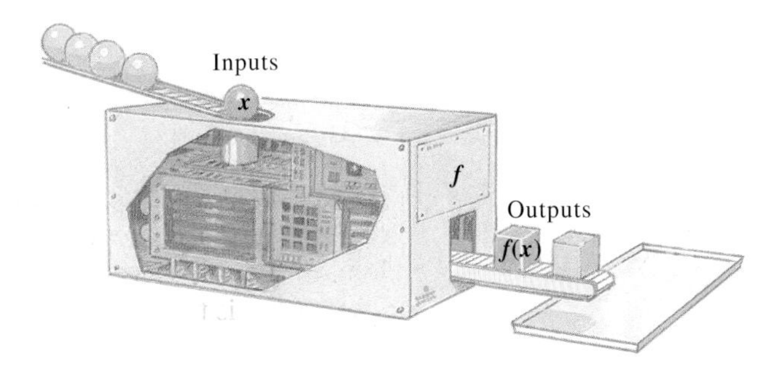

The function pictured has been named f. Here x represents an arbitrary input, and $f(x)$—read "f of x," "f at x," or "the value of f at x"—represents the corresponding output. You should check that in Example 3, $f(4)$ is 5, $f(-3)$ is 1 and so on. Similarly, in Example 4, $f(2) = 4$.

Note that $f(x)$ *does not mean* f times x.

Most functions are described by equations. For example, $f(x) = 2x + 3$ describes the function that takes an input x, multiplies it by 2, and then adds 3.

Input

$$f(x) = \underbrace{2x}_{\text{Double}} \underbrace{+\ 3}_{\text{Add 3}}$$

To calculate the output $f(4)$, we take the input 4, double it, and add 3 to get 11. That is, we substitute 4 into the formula for $f(x)$:

$$\begin{aligned} f(4) &= 2 \cdot 4 + 3 \\ &= 11. \end{aligned}$$

Sometimes, in place of $f(x) = 2x + 3$, we write $y = 2x + 3$, where it is understood that the value of y, the *dependent variable*, depends on our choice of x, the *independent variable*. To understand why $f(x)$ notation is so useful, consider two equivalent statements:

a) If $f(x) = 2x + 3$, then $f(4) = 11$.
b) If $y = 2x + 3$, then the value of y is 11 when x is 4.

The notation used in part (a) is far more concise.

Example 5

Find the indicated function value.

a) $f(5)$, for $f(x) = 3x + 2$
b) $g(-2)$, for $g(r) = 5r^2 + 3r$
c) $h(4)$, for $h(x) = 7$
d) $F(a + 1)$, for $F(x) = 3x + 2$

Solution Finding function values is much like evaluating an algebraic expression.

a) $f(5) = 3 \cdot 5 + 2 = 17$

b) $g(-2) = 5(-2)^2 + 3(-2)$
$\qquad = 5 \cdot 4 - 6 = 14$

c) For the function given by $h(x) = 7$, all inputs share the same output, 7. Therefore, $h(4) = 7$. The function h is an example of a *constant function*.

d) $F(a + 1) = 3(a + 1) + 2$
$\qquad = 3a + 3 + 2 = 3a + 5$

Note that whether we write $f(x) = 3x + 2$, or $f(t) = 3t + 2$, or $f(\square) = 3\square + 2$, we still have $f(5) = 17$. Thus the independent variable can be thought of as a *dummy variable*. The letter chosen for the dummy variable is not as important as the algebraic manipulations to which it is subjected.

When a function is described by an equation, the domain is often unspecified. In such cases, the domain is the set of all numbers for which function values can be calculated. If an x-value is not in the domain of a function, the graph of the function will not include any point above or below that x-value.

Example 6

To visualize Example 6, note that the graph of $y_1 = |x|$ (which is entered $y_1 = \text{abs}(x)$, using the NUM option of the MATH menu) appears without interruption for any piece of the x-axis that we examine.

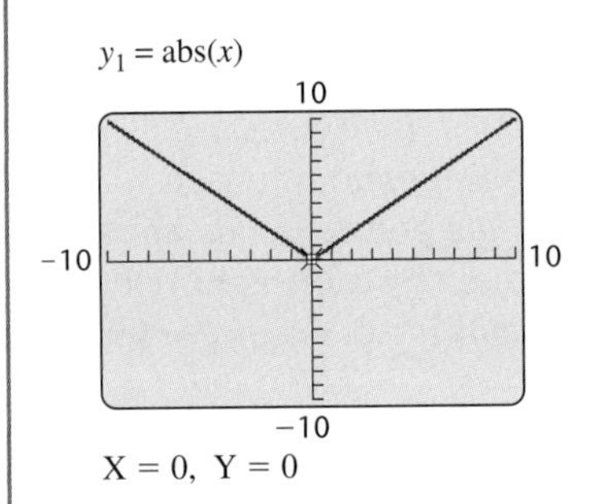

In contrast, the graph of

$y_2 = \dfrac{7}{2x - 6}$ has a break at $x = 3$.

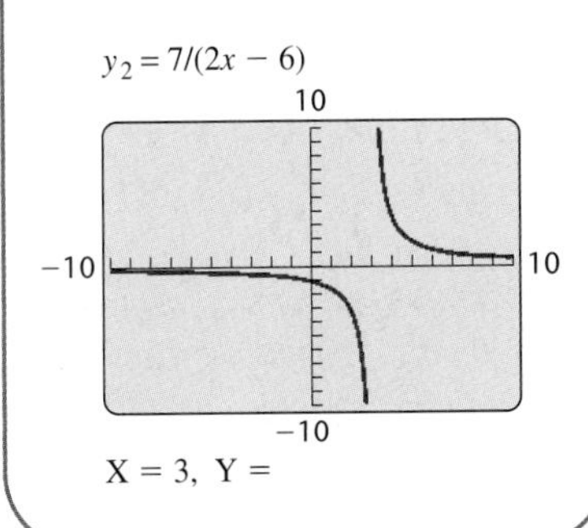

For each equation, determine the domain of f.

a) $f(x) = |x|$

b) $f(x) = \dfrac{7}{2x - 6}$

Solution

a) We ask ourselves, "Is there any number x for which we cannot compute $|x|$?" Since we can find the absolute value of *any* number, the answer is no. Thus the domain of f is $\mathbb{R}$, the set of all real numbers.

b) Is there any number x for which $\dfrac{7}{2x - 6}$ cannot be computed? Since $\dfrac{7}{2x - 6}$ cannot be computed when $2x - 6$ is 0, the answer is yes. To determine what x-value causes the denominator to be 0, we solve an equation:

$2x - 6 = 0$ Setting the denominator equal to 0

$2x = 6$ Adding 6 to both sides

$x = 3.$ Dividing both sides by 2

Thus, 3 is *not* in the domain of f, whereas all other real numbers are. The domain of f is $\{x \mid x \text{ is a real number } \textit{and } x \neq 3\}$.

Applications: Interpolation and Extrapolation

Function notation is often used in formulas. For example, to emphasize that the area A of a circle is a function of its radius r, instead of

$$A = \pi r^2,$$

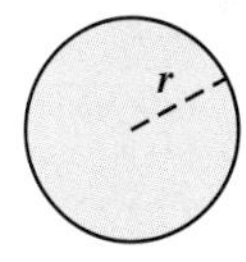

we can write

$$A(r) = \pi r^2.$$

When a function is given as a graph in a problem-solving situation, we are often asked to determine certain quantities on the basis of the graph. Later in this text, we will develop models that can be used for calculations. For now we simply use the graph to estimate the coordinates of an unknown point by using other points with known coordinates. When the unknown point is *between* the known points, this process is called **interpolation**. If the unknown point extends *beyond* the known points, the process is called **extrapolation**.

Example 7

Working mothers. More women than ever before are returning to the workforce within a year of giving birth. According to the U.S. Census Bureau, in 1980 38% of women who gave birth returned to work within a year. The figure grew to 43% in 1983, 50% in 1986, 54% in 1992, 55% in 1995, and 59% in 1998. Estimate the percentage of women who returned to work within a year of giving birth in 1989 (a year the Census Bureau did not include) and in 2001.

Solution

1. and **2.** **Familiarize** and **Translate.** The given information enables us to plot and connect six points. We let the horizontal axis represent the year and the vertical axis the percentage of women who gave birth and returned to work within one year. We label the function itself P.

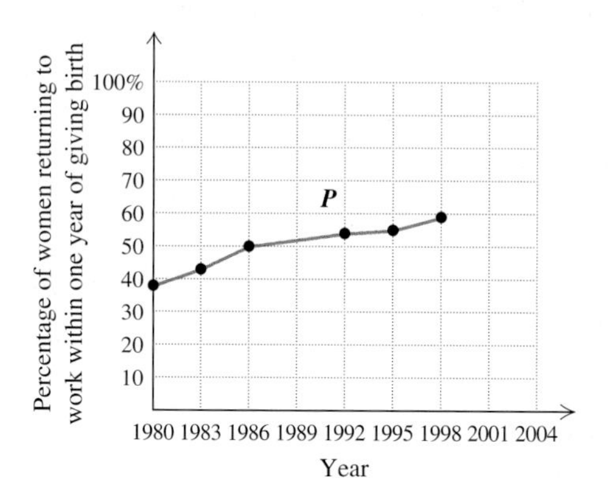

3. **Carry out.** To estimate the percentage of women who returned to work within a year of giving birth in 1989, we locate the point directly above the year 1989. We then estimate its second coordinate by moving horizontally from that point to the y-axis. Although our result is not exact, we see that $P(1989) \approx 52$.

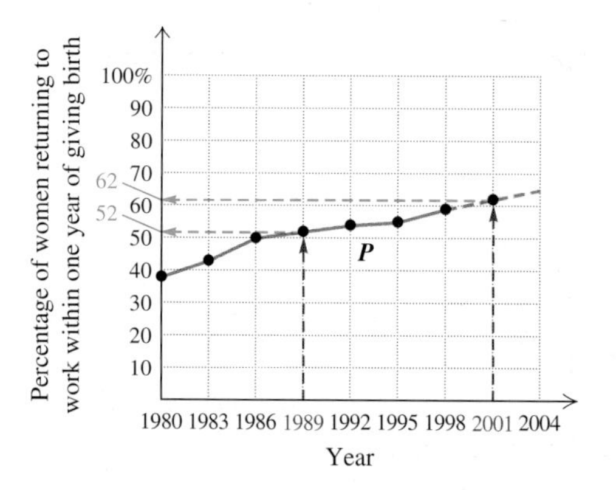

To estimate the percentage of women who returned to work within a year of giving birth in 2001, we extend the graph and extrapolate. It appears that $P(2001) \approx 62$.

4. **Check.** A precise check requires consulting an outside information source. Since 52% is between 50% and 54% and 62% is greater than 59%, our estimates seem plausible.

5. **State.** In 1989, about 52% of all women returned to work within a year of giving birth. By 2001, that percentage would be predicted to be about 62%.

Exercise Set 2.2

FOR EXTRA HELP

Digital Video Tutor CD 2
Videotape 3

InterAct Math

Math Tutor Center

MathXL

MyMathLab.com

Determine whether each correspondence is a function.

1.

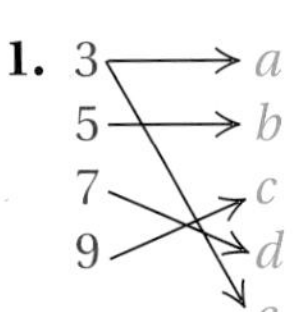

2.

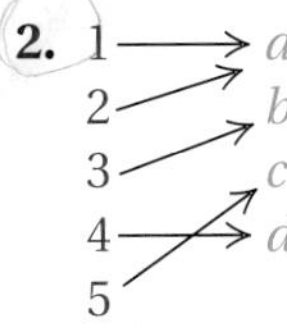

3. *Girl's Age (in months)* — *Average Daily Weight Gain (in grams)*

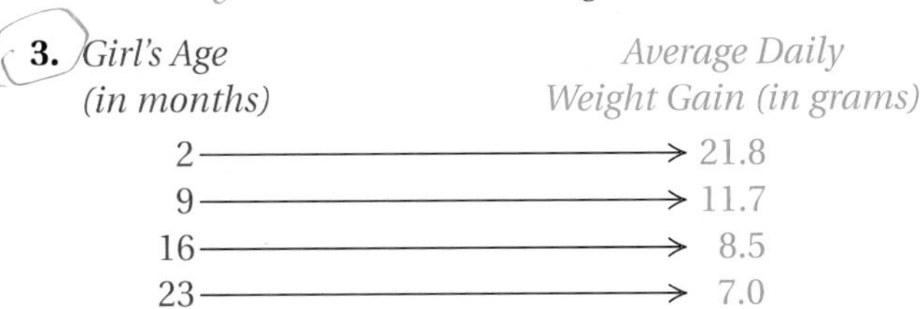

Source: *American Family Physician,* December 1993, p. 1435.

4. *Boy's Age (in months)* — *Average Daily Weight Gain (in grams)*

Source: *American Family Physician,* December 1993, p. 1435.

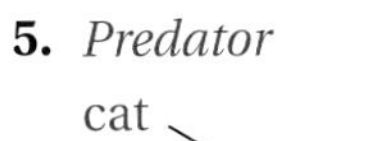

5. *Predator* — *Prey*

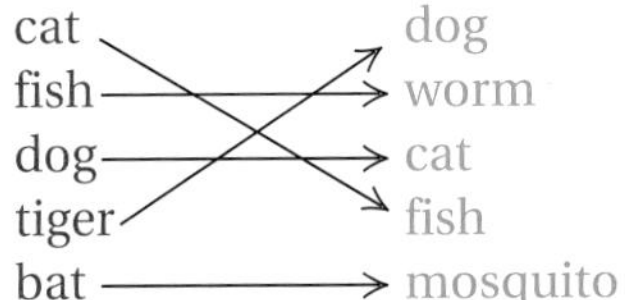

6. *Olympics Site* — *Year*

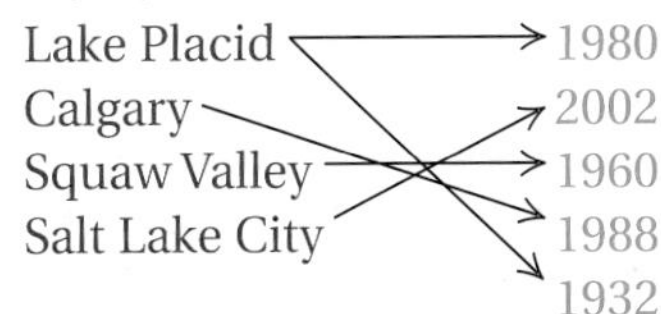

Determine whether each of the following is a function. Identify any relations that are not functions.

	Domain	*Correspondence*	*Range*
7.	A yard full of Christmas trees	Each tree's price	A set of prices
8.	The swordfish stored in a boat	Each fish's weight	A set of weights
9.	The members of a rock band	An instrument the person can play	A set of instruments
10.	The students in a math class	Each person's seat number	A set of numbers
11.	A set of numbers	Square each number and then add 4.	A set of numbers
12.	A set of shapes	The area of each shape	A set of numbers

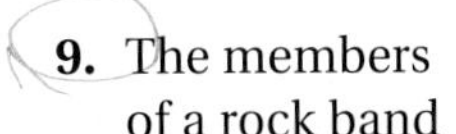

For each graph of a function, determine **(a)** $f(1)$; **(b)** *the domain;* **(c)** *any x-values for which* $f(x) = 2$; *and* **(d)** *the range.*

13.

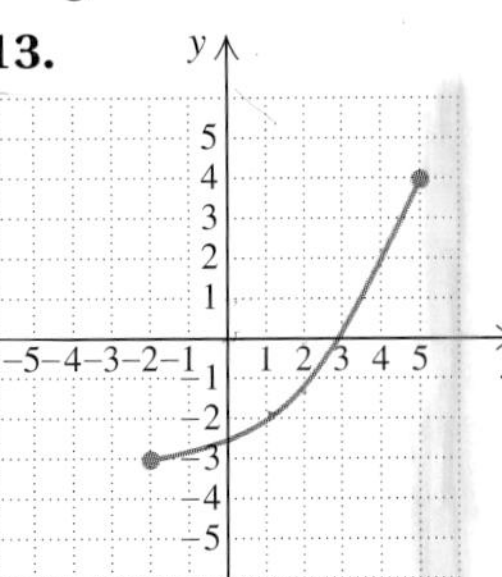

14.

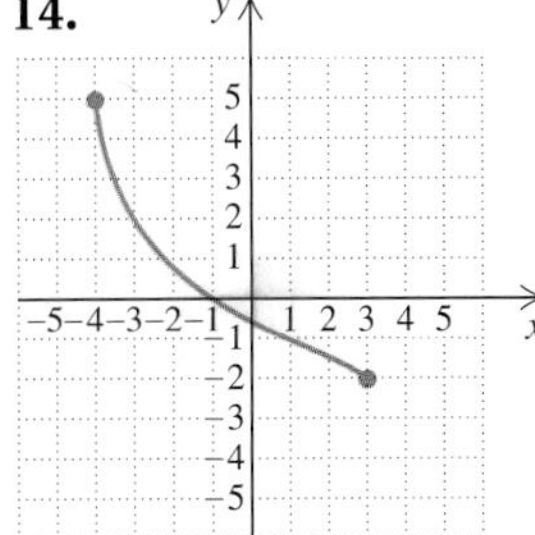

15.

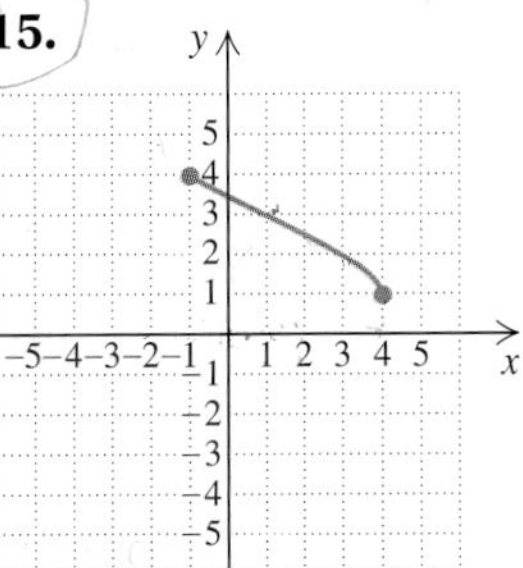

16.

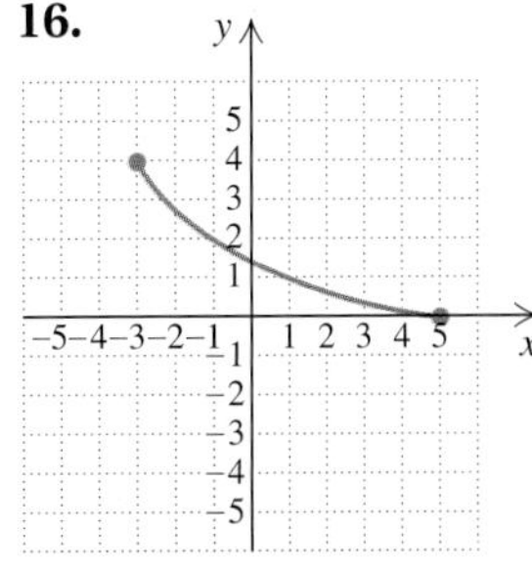

17.

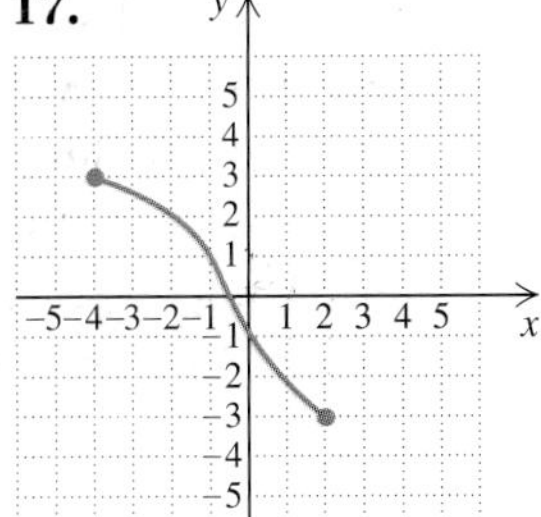

18.

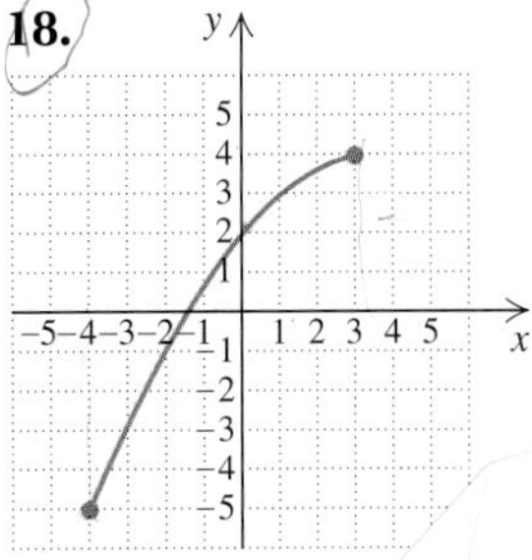

19.

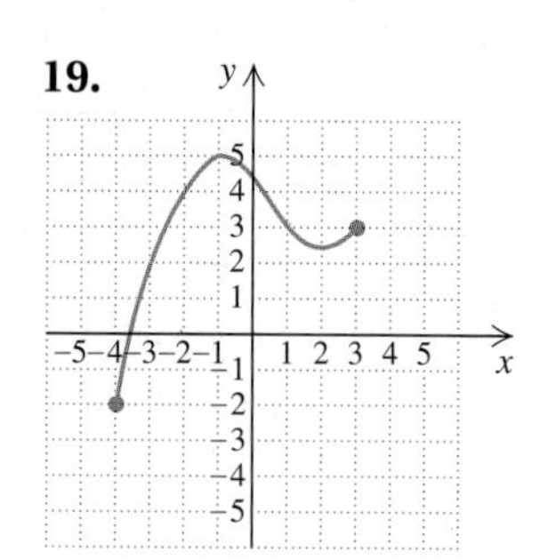

20.

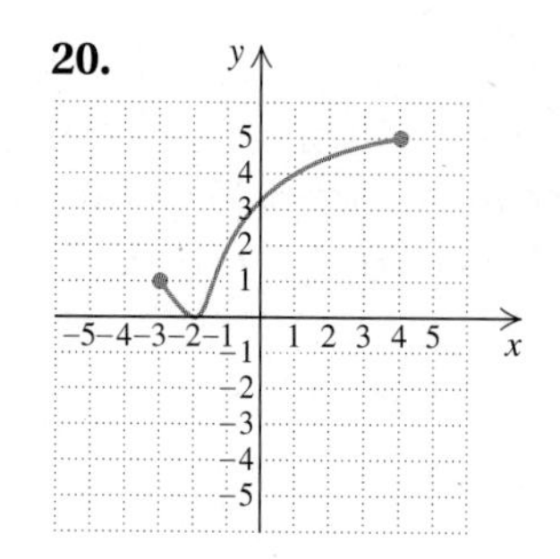

21.

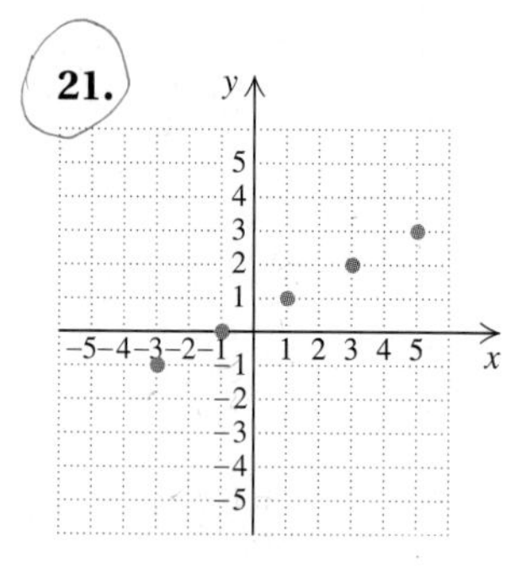

22.

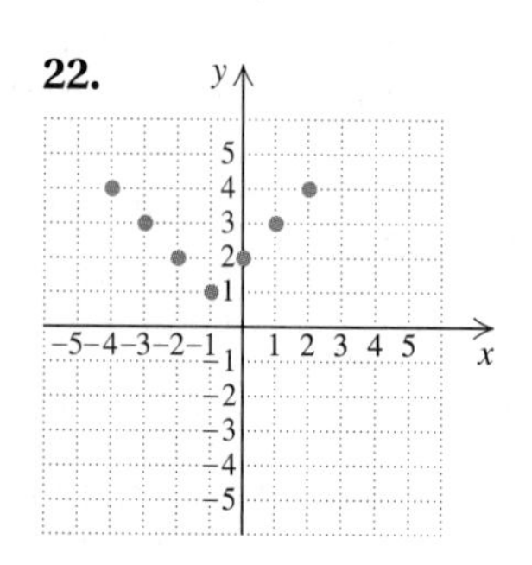

23.

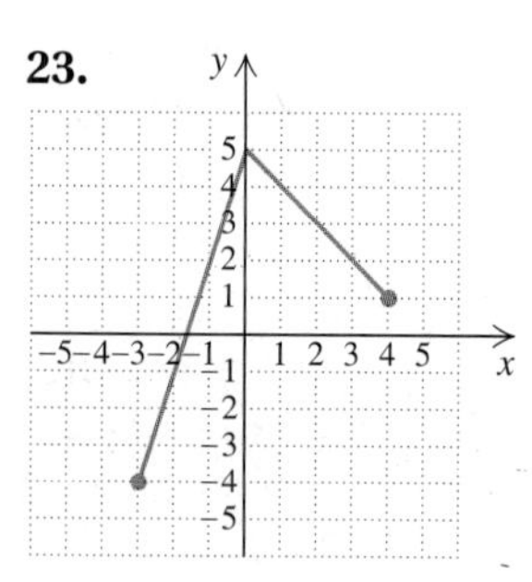

24.

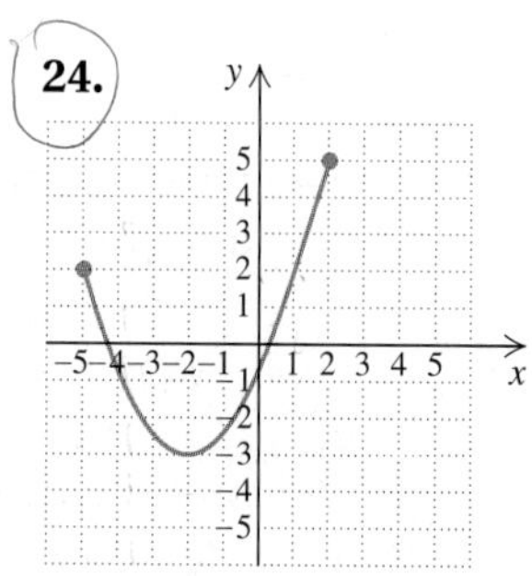

25.

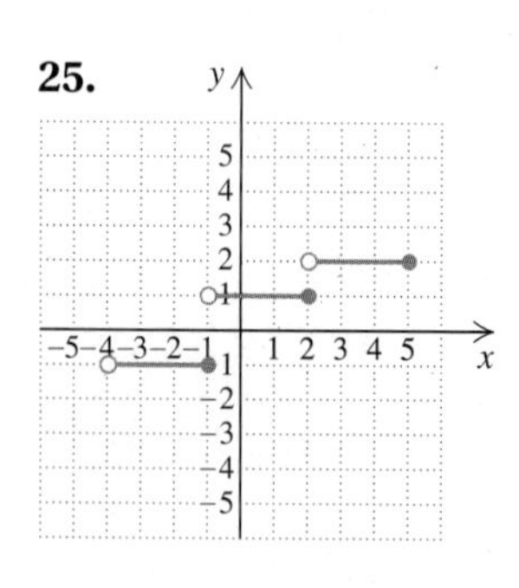

26.

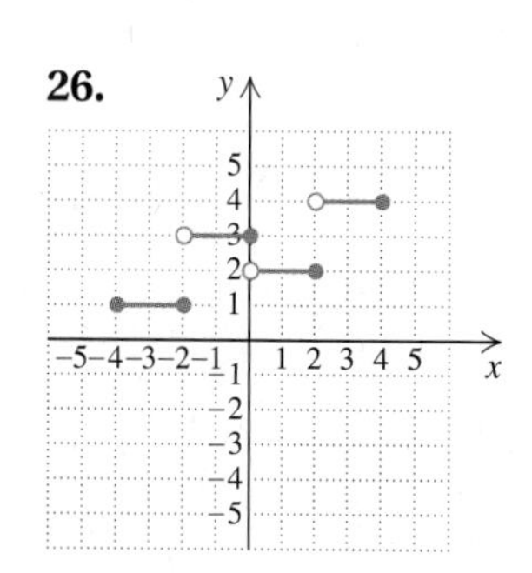

Determine whether each of the following is the graph of a function.

27.

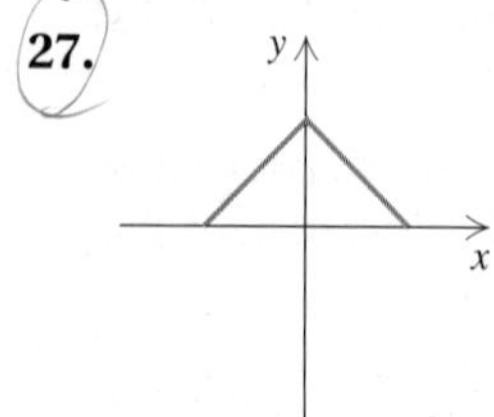

28.

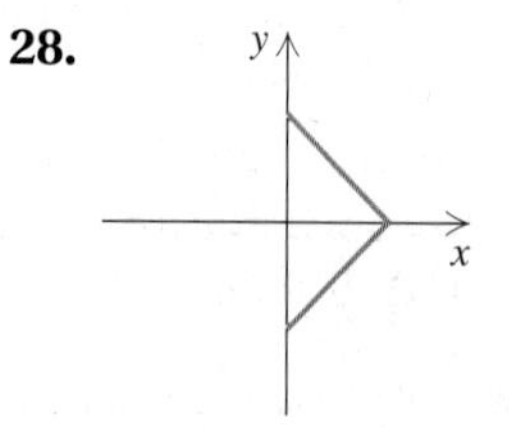

29.

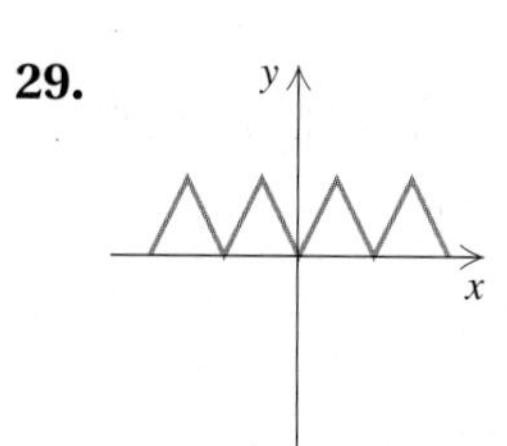

30.

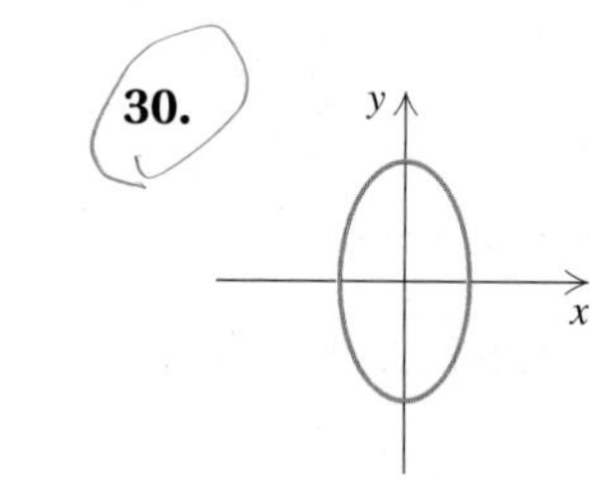

31.

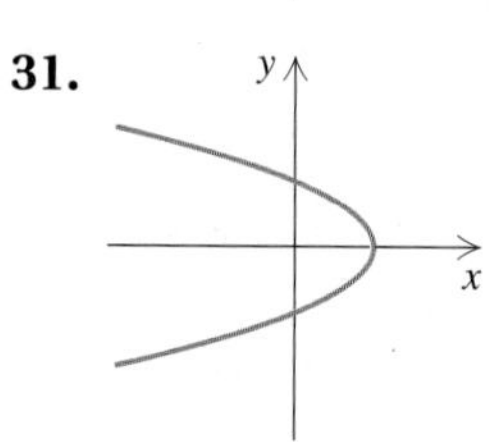

32.

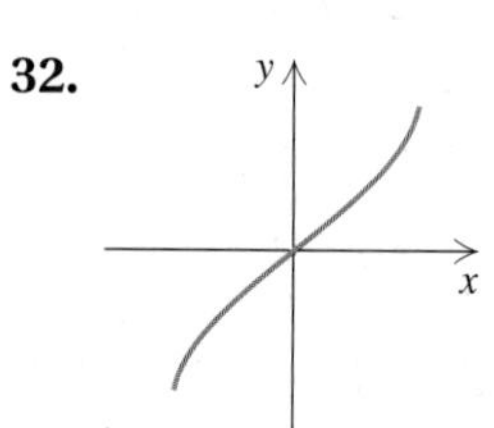

33.

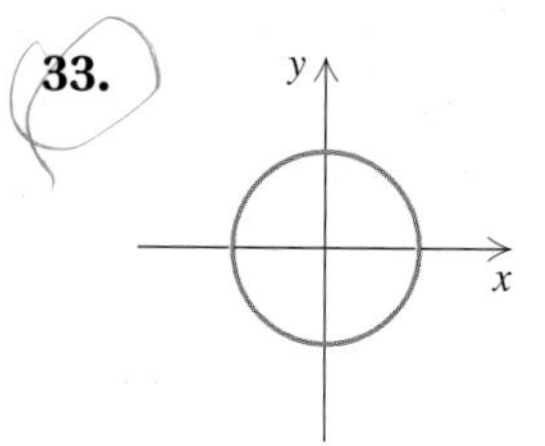

34.

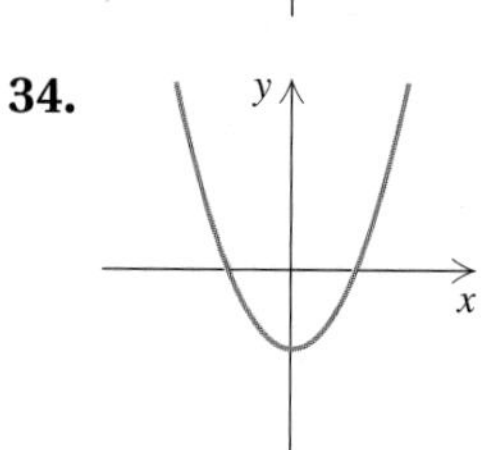

Find the function values.

35. $g(x) = x + 3$

a) $g(0)$ **b)** $g(-4)$ **c)** $g(-7)$
d) $g(8)$ **e)** $g(a + 2)$

36. $h(x) = x - 2$

a) $h(4)$ **b)** $h(8)$ **c)** $h(-3)$
d) $h(-4)$ **e)** $h(a - 1)$

37. $f(n) = 5n^2 + 4n$

a) $f(0)$ **b)** $f(-1)$ **c)** $f(3)$
d) $f(t)$ **e)** $f(2a)$

38. $g(n) = 3n^2 - 2n$

a) $g(0)$ **b)** $g(-1)$ **c)** $g(3)$
d) $g(t)$ **e)** $g(2a)$

39. $f(x) = \dfrac{x - 3}{2x - 5}$

a) $f(0)$ **b)** $f(4)$ **c)** $f(-1)$
d) $f(3)$ **e)** $f(x + 2)$

40. $s(x) = \dfrac{3x - 4}{2x + 5}$

a) $s(10)$ **b)** $s(2)$ **c)** $s(\frac{1}{2})$
d) $s(-1)$ **e)** $s(x + 3)$

41. Find the domain of f.

a) $f(x) = \dfrac{5}{x - 3}$ **b)** $f(x) = \dfrac{7}{6 - x}$

c) $f(x) = 2x + 1$

d) $f(x) = x^2 + 3$

e) $f(x) = \dfrac{3}{2x - 5}$

f) $f(x) = |3x - 4|$

42. Find the domain of g.

a) $g(x) = \dfrac{3}{x - 1}$

b) $g(x) = |5 - x|$

c) $g(x) = \dfrac{9}{x + 3}$

d) $g(x) = \dfrac{4}{3x + 4}$

e) $g(x) = x^3 - 1$

f) $g(x) = 7x - 8$

The function A described by $A(s) = s^2\dfrac{\sqrt{3}}{4}$ *gives the area of an equilateral triangle with side s.*

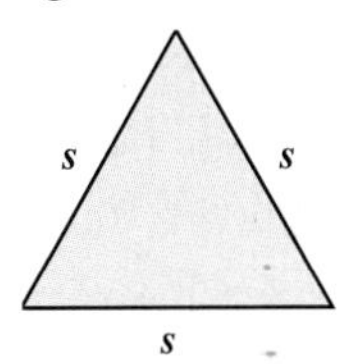

43. Find the area when a side measures 4 cm.

44. Find the area when a side measures 6 in.

The function V described by $V(r) = 4\pi r^2$ *gives the surface area of a sphere with radius r.*

45. Find the area when the radius is 3 in.

46. Find the area when the radius is 5 cm.

Chemistry. *The function F described by*

$$F(C) = \tfrac{9}{5}C + 32$$

gives the Fahrenheit temperature corresponding to the Celsius temperature C.

47. Find the Fahrenheit temperature equivalent to $-10°C$.

48. Find the Fahrenheit temperature equivalent to 5°C.

Archaeology. *The function H described by*

$$H(x) = 2.75x + 71.48$$

can be used to predict the height, in centimeters, of a woman whose humerus *(the bone from the elbow to the shoulder) is x cm long. Predict the height of a woman whose humerus is the length given.*

49. 32 cm

50. 35 cm

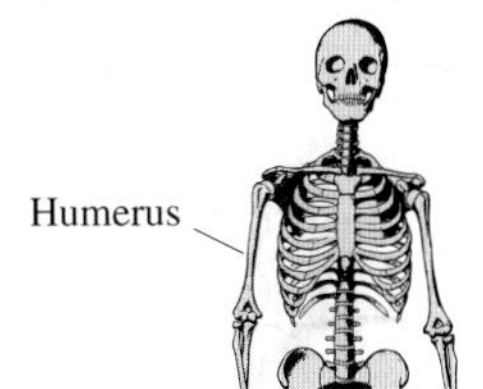

Heart attacks and cholesterol. *For Exercises 51 and 52, use the following graph, which shows the annual heart attack rate per 10,000 men as a function of blood cholesterol level.**

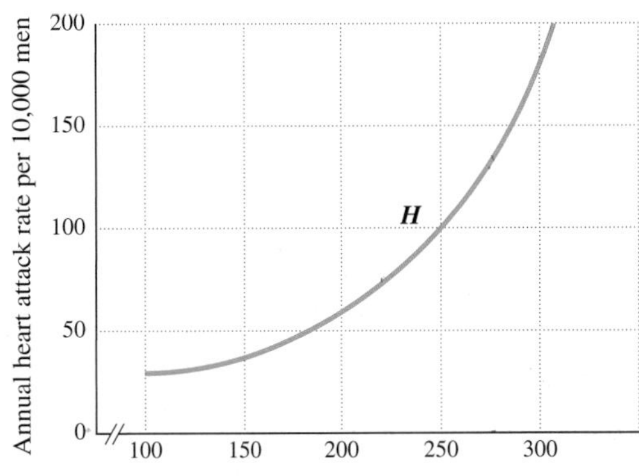

51. Approximate the annual heart attack rate for those men whose blood cholesterol level is 225 mg/dl. That is, find $H(225)$.

52. Approximate the annual heart attack rate for those men whose blood cholesterol level is 275 mg/dl. That is, find $H(275)$.

Voting attitudes. *For Exercises 53 and 54, use this graph, which shows the percentage of people responding yes to the question, "If your (political) party nominated a generally well-qualified person for president who happened to be a woman, would you vote for that person?" (Source: The New York Times, August 13, 2000)*

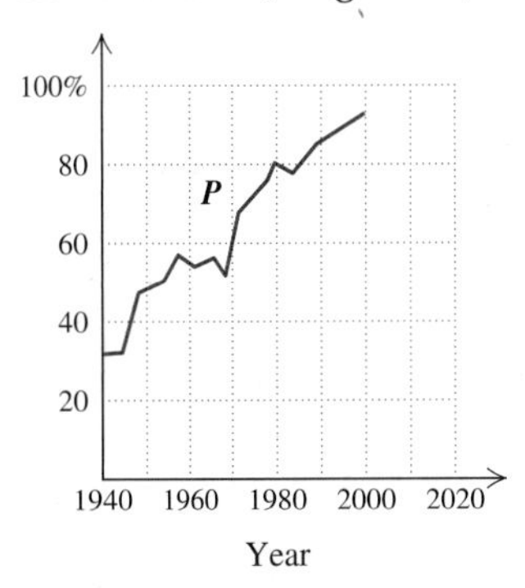

53. Approximate the percentage of Americans willing to vote for a woman for president in 1960. That is, find $P(1960)$.

54. Approximate the percentage of Americans willing to vote for a woman for president in 2000. That is, find $P(2000)$.

*Copyright 1989, CSPI. Adapted from *Nutrition Action Healthletter* (1875 Connecticut Avenue, N.W., Suite 300, Washington, DC 20009-5728. $24 for 10 issues).

Blood alcohol level. *The following table can be used to predict the number of drinks required for a person of a specified weight to be considered legally intoxicated (blood alcohol level of 0.08 or above). One 12-oz glass of beer, a 5-oz glass of wine, or a cocktail containing 1 oz of a distilled liquor all count as one drink. Assume that all drinks are consumed within one hour.*

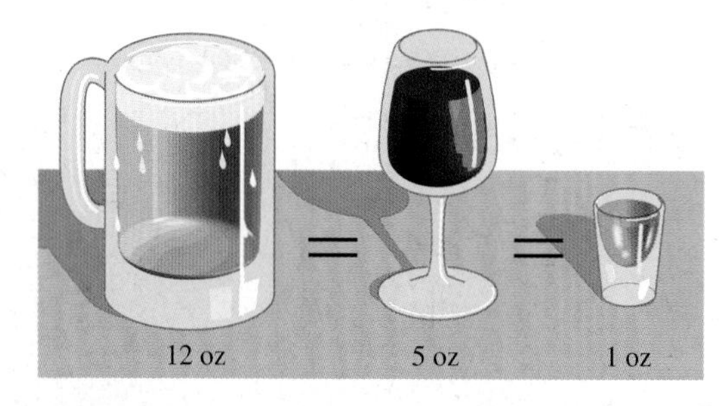

Input, Body Weight (in pounds)	Output, Number of Drinks
100	2.5
160	4
180	4.5
200	5

55. Use the data in the table above to draw a graph and to estimate the number of drinks that a 140-lb person would have to drink to be considered intoxicated.

56. Use the graph from Exercise 55 to estimate the number of drinks a 120-lb person would have to drink to be considered intoxicated.

Incidence of AIDS. *The following table indicates the number of cases of AIDS reported in each of several years (Source: U.S. Centers for Disease Control and Prevention).*

Input, Year	Output, Number of Cases Reported
1994	77,103
1996	66,497
1998	48,269

57. Use the data in the table above to draw a graph and to estimate the number of cases of AIDS reported in 2001.

58. Use the graph from Exercise 57 to estimate the number of cases of AIDS reported in 1997.

Population growth. *The town of Falconburg recorded the following dates and populations.*

Input, Year	Output, Population (in tens of thousands)
1995	5.8
1997	6
1999	7
2001	7.5

59. Use the data in the table above to draw a graph of the population as a function of time. Then estimate what the population was in 1998.

60. Use the graph in Exercise 59 to predict Falconburg's population in 2003.

61. *Retailing.* Shoreside Gifts is experiencing constant growth. They recorded a total of $250,000 in sales in 1996 and $285,000 in 2001. Use a graph that displays the store's total sales as a function of time to predict total sales for 2005.

62. Use the graph in Exercise 61 to estimate what the total sales were in 1999.

Researchers at Yale University have suggested that the following graphs may represent three different aspects of love.*

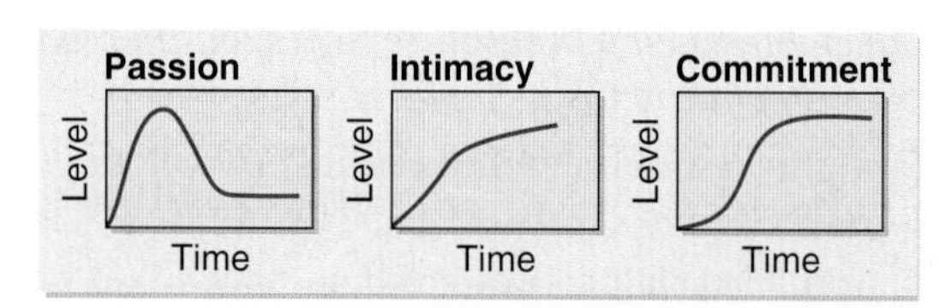

63. In what unit would you measure time if the horizontal length of each graph were ten units? Why?

64. Do you agree with the researchers that these graphs should be shaped as they are? Why or why not?

*From "A Triangular Theory of Love," by R. J. Sternberg, 1986, *Psychological Review,* **93**(2), 119–135. Copyright 1986 by the American Psychological Association, Inc. Reprinted by permission.

SKILL MAINTENANCE

Simplify.

65. $\dfrac{10 - 3^2}{9 - 2 \cdot 3}$

66. $\dfrac{2^4 - 10}{6 - 4 \cdot 3}$

67. The surface area of a rectangular solid of length l, width w, and height h is given by $S = 2lh + 2lw + 2wh$. Solve for l.

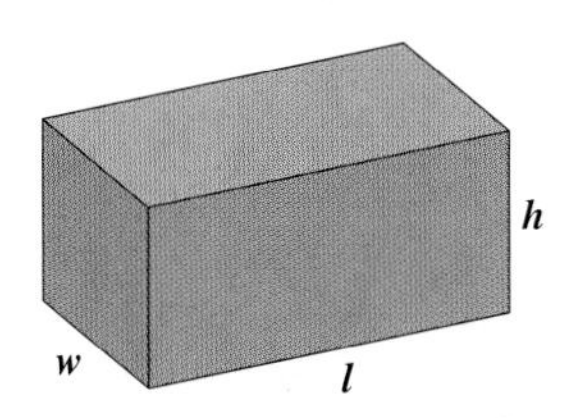

68. Solve the formula in Exercise 67 for w.

Solve for y.

69. $2x + 3y = 6$

70. $5x - 4y = 8$

SYNTHESIS

71. Which would you trust more and why: estimates made using interpolation or those made using extrapolation?

72. Explain in your own words why every function is a relation, but not every relation is a function.

For Exercises 73 and 74, let $f(x) = 3x^2 - 1$ *and* $g(x) = 2x + 5$.

73. Find $f(g(-4))$ and $g(f(-4))$.

74. Find $f(g(-1))$ and $g(f(-1))$.

Pregnancy. *For Exercises 75–78, use the following graph of a woman's "stress test." This graph shows the size of a pregnant woman's contractions as a function of time.*

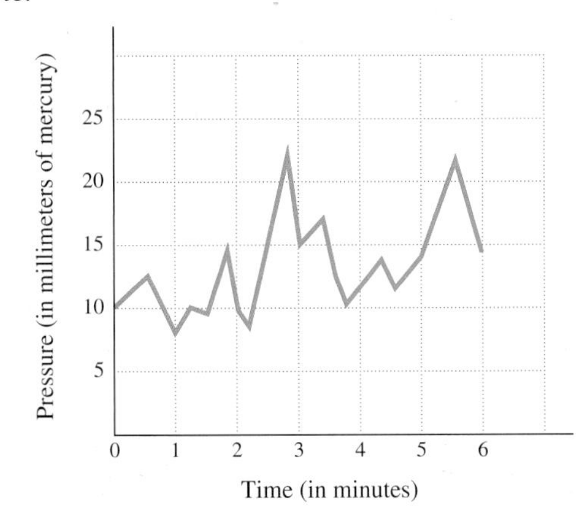

75. How large is the largest contraction that occurred during the test?

76. At what time during the test did the largest contraction occur?

77. On the basis of the information provided, how large a contraction would you expect 60 seconds after the end of the test? Why?

78. What is the frequency of the largest contraction?

79. The *greatest integer function* $f(x) = [\![x]\!]$ is defined as follows: $[\![x]\!]$ is the greatest integer that is less than or equal to x. For example, if $x = 3.74$, then $[\![x]\!] = 3$; and if $x = -0.98$, then $[\![x]\!] = -1$. Graph the greatest integer function for $-5 \le x \le 5$. (The notation $f(x) = \text{INT}[x]$ is used in many graphers and computer programs.)

80. Suppose that a function g is such that $g(-1) = -7$ and $g(3) = 8$. Find a formula for g if $g(x)$ is of the form $g(x) = mx + b$, where m and b are constants.

81. *Energy expenditure.* On the basis of the information given below, what burns more energy: walking $4\frac{1}{2}$ mph for two hours or bicycling 14 mph for one hour?

Approximate Energy Expenditure by a 150-Pound Person in Various Activities

Activity	Calories per Hour
Walking, $2\frac{1}{2}$ mph	210
Bicycling, $5\frac{1}{2}$ mph	210
Walking, $3\frac{3}{4}$ mph	300
Bicycling, 13 mph	660

Source: Based on material prepared by Robert E. Johnson, M.D., Ph.D., and colleagues, University of Illinois.

COLLABORATIVE

CORNER

Calculating License Fees

Focus: Functions

Time: 15–20 minutes

Group size: 3–4

The California Department of Motor Vehicles calculates automobile registration fees (VLF) according to the schedule shown below.

ACTIVITY

1. Determine the original sale price of the oldest vehicle owned by a member of your group. If necessary, use the price and age of a family member's vehicle. Be sure to note the year in which the car was purchased.
2. Use the schedule below to calculate the vehicle license fee (VLF) for the vehicle in part (1) above for each year from the year of purchase to the present. To speed your work, each group member can find the fee for a few different years.
3. Graph the results from part (2). On the x-axis, plot years beginning with the year of purchase, and on the y-axis, plot $V(x)$, the VLF as a function of year.
4. What is the lowest VLF that the owner of this car will ever have to pay, according to this schedule? Compare your group's answer with other groups' answers.
5. Does your group feel that California's method for calculating registration fees is fair? Why or why not? How could it be improved?
6. Try, as a group, to find an algebraic form for the function $y = V(x)$.
7. *Optional out-of-class extension:* Create a program for a grapher that accepts two inputs (initial value of the vehicle and year of purchase) and produces $V(x)$ as the output.

DMV VEHICLE LICENSE FEE INFORMATION

A Public Service Agency

The 2% **Vehicle License Fee (VLF)** is in lieu of a personal property tax on vehicles. Most VLF revenue is returned to City and County Local Governments (see reverse side). The license fee charged is based upon the sale price or vehicle value when initially registered in California. The vehicle value is adjusted for any subsequent sale or transfer, that occurred 8/19/91 or later, excluding sales or transfers between specified relatives.

The VLF is calculated by rounding the sale price to the nearest **odd** hundred dollar. That amount is reduced by a percentage utilizing an eleven year schedule (shown to the right), and 2% of that amount is the fee charged. See the accompanying example for a vehicle purchased last year for $9,199. This would be the second registration year following that purchase.

WHERE DO YOUR DMV FEES GO? SEE REVERSE SIDE.

PERCENTAGE SCHEDULE
Rev. & Tax. Code Sec. 10753.2
(Trailer coaches have a different schedule)

1st Year	100%	7th Year	40%
2nd Year	90%	8th Year	30%
3rd Year	80%	9th Year	25%
4th Year	70%	10th year	20%
5th Year	60%	11th Year onward	15%
6th Year	50%		

VLF CALCULATION EXAMPLE

Purchase Price:	$9,199
Rounded to:	$9,100
Times the Percentage:	90%
Equals Fee Basis of:	$8,190
Times 2% Equals:	$163.80
Rounded to:	$164

DMV77 8(REV.8/95) 95 30123

Linear Functions: Slope, Graphs, and Models

2.3

Slope–Intercept Form of an Equation • Applications

Different functions have different graphs. In this section, we examine functions and real-life models with graphs that are straight lines. Such functions and their graphs are called *linear* and can be written in the form $f(x) = mx + b$, where m and b are constants.

Slope–Intercept Form of an Equation

Examples 3–5 in Section 2.1 suggest that for any number m, the graph of $y = mx$ is a straight line passing through the origin. What will happen if we add a number b on the right side to get the equation $y = mx + b$?

Example 1 Graph $y = 2x$ and $y = 2x + 3$, using the same set of axes.

Solution We first make a table of solutions of both equations.

	y	y
x	$y = 2x$	$y = 2x + 3$
0	0	3
1	2	5
−1	−2	1
2	4	7
−2	−4	−1
3	6	9

We then plot these points. Drawing a blue line for $y = 2x + 3$ and a red line for $y = 2x$, we observe that the graph of $y = 2x + 3$ is simply the graph of $y = 2x$ shifted, or *translated*, 3 units up. The lines are parallel.

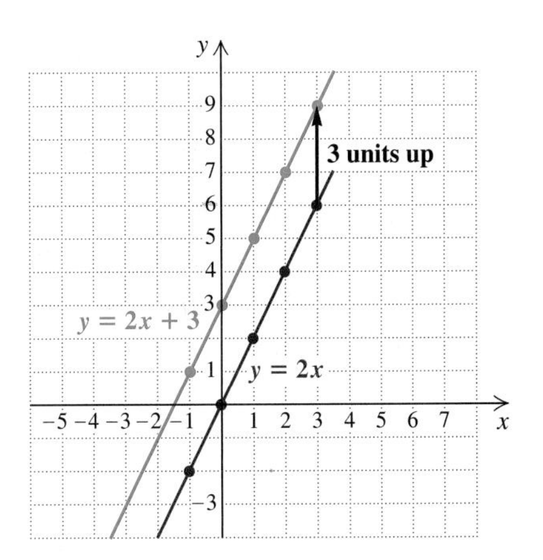

Study Tip

Make an effort to do your homework as soon as possible after each class. Make this part of your routine, choosing a time and a place where you can focus with a minimum of distractions.

Example 2 Graph $f(x) = \frac{1}{3}x$ and $g(x) = \frac{1}{3}x - 2$, using the same set of axes.

Solution We first make a table of solutions of both equations. By choosing multiples of 3, we can avoid fractions.

	$f(x)$	$g(x)$
x	$f(x) = \frac{1}{3}x$	$g(x) = \frac{1}{3}x - 2$
0	0	−2
3	1	−1
−3	−1	−3
6	2	0

We then plot these points. Drawing a blue line for $g(x) = \frac{1}{3}x - 2$ and a red line for $f(x) = \frac{1}{3}x$, we see that the graph of $g(x) = \frac{1}{3}x - 2$ is simply the graph of $f(x) = \frac{1}{3}x$ shifted, or translated, 2 units down.

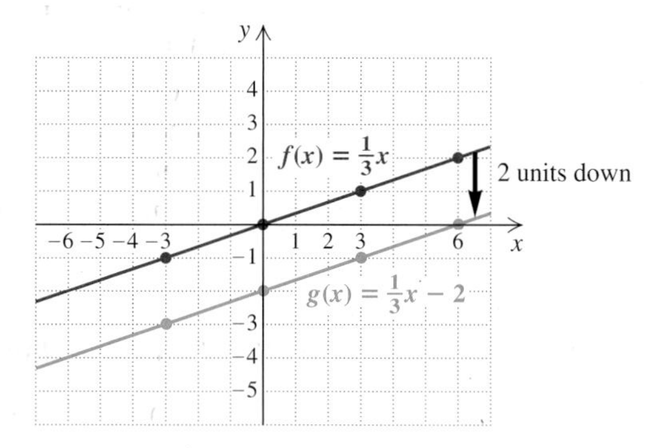

To explore the effect of b when graphing $y = mx + b$, begin with the graph of $y_1 = x$. On the same set of axes, graph the lines $y_2 = x + 3$ and $y_3 = x - 4$. How do these lines differ from $y_1 = x$? What do you think the line $y = x - 5$ will look like? Try drawing lines like $y = x + \frac{1}{4}$ and $y = x + (-3.2)$ and describe what happens to the graph of $y = x$ when a number b is added to x. By creating a table of values, explain how the values of y_2 and y_3 differ from y_1.

Note that the graph of $y = 2x + 3$ passed through the point $(0, 3)$ and the graph of $g(x) = \frac{1}{3}x - 2$ passed through the point $(0, -2)$. In general, the graph of $y = mx + b$ is a line parallel to $y = mx$, passing through the point $(0, b)$. **The point $(0, b)$ is called the y-intercept**. Often, to save time, we refer to the number b as the y-intercept.

Example 3 For each equation, find the y-intercept.

a) $y = -5x + 4$ **b)** $f(x) = 5.3x - 12$

Solution

a) The y-intercept is $(0, 4)$, or simply 4.
b) The y-intercept is $(0, -12)$, or simply -12.

In examining the graphs in Examples 1 and 2, note that the slant of the red lines seems to match the slant of the blue lines. Note too that the slant of the lines in Example 1 differs from the slant of the lines in Example 2. This leads us to suspect that it is the number m, in the equation $y = mx + b$, that is responsible for the slant of the line. The following definition enables us to visualize this slant, or *slope*, as a ratio of two lengths.

Slope

The *slope* of the line passing through (x_1, y_1) and (x_2, y_2) is given by

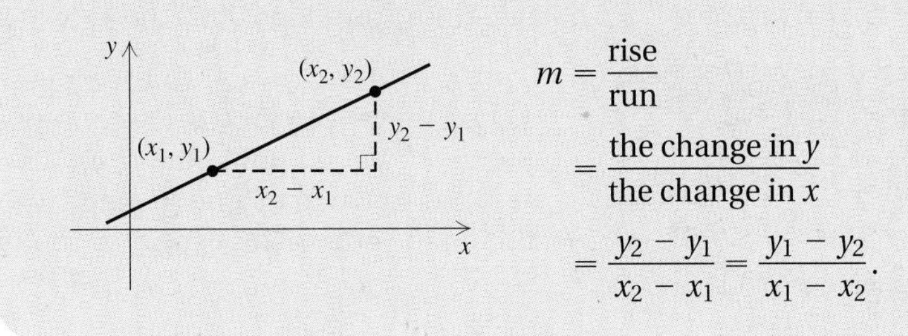

$$m = \frac{\text{rise}}{\text{run}}$$
$$= \frac{\text{the change in } y}{\text{the change in } x}$$
$$= \frac{y_2 - y_1}{x_2 - x_1} = \frac{y_1 - y_2}{x_1 - x_2}.$$

In the definition above, (x_1, y_1) and (x_2, y_2)—read "x sub-one, y sub-one and x sub-two, y sub-two"—represent two different points on a line. It does not matter which point is considered (x_1, y_1) and which is considered (x_2, y_2) so long as coordinates are subtracted in the same order in both the numerator and the denominator.

The letter m is traditionally used for slope. This usage has its roots in the French verb *monter*, to climb.

Example 4

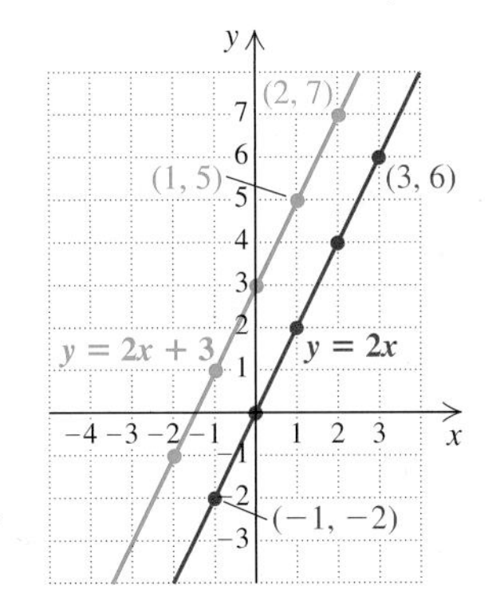

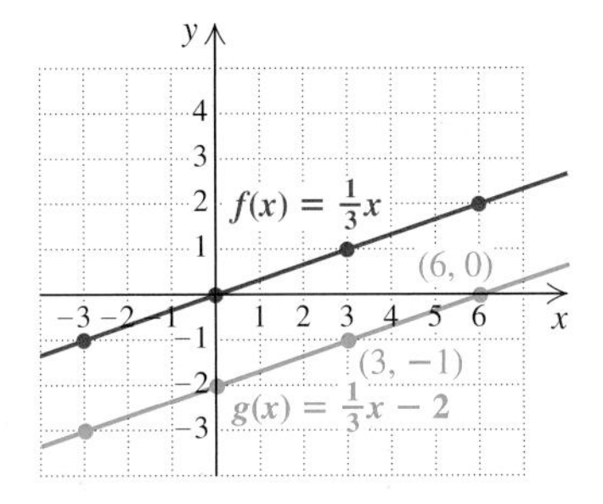

Find the slope of the lines drawn in Examples 1 and 2.

Solution To find the slope of a line, we can use the coordinates of any two points on that line. We use $(1, 5)$ and $(2, 7)$ to find the slope of the blue line in Example 1:

$$\text{Slope} = \frac{\text{rise}}{\text{run}} = \frac{\text{change in } y}{\text{change in } x}$$
$$= \frac{y_2 - y_1}{x_2 - x_1} = \frac{7 - 5}{2 - 1} = 2.$$

Any pair of points on the line will give the same slope.

To find the slope of the red line in Example 1, we use $(-1, -2)$ and $(3, 6)$:

$$\text{Slope} = \frac{\text{rise}}{\text{run}} = \frac{\text{change in } y}{\text{change in } x} = \frac{6 - (-2)}{3 - (-1)} = \frac{8}{4} = 2.$$

Any pair of points on the line will give the same slope.

We can use $(3, -1)$ and $(6, 0)$ to find the slope of the blue line in Example 2:

$$\text{Slope} = \frac{\text{rise}}{\text{run}} = \frac{\text{change in } y}{\text{change in } x} = \frac{0 - (-1)}{6 - 3} = \frac{1}{3}.$$

Any pair of points on the line will give the same slope.

You should confirm that the red line in Example 2 also has a slope of $\frac{1}{3}$.

In Example 4, we found that the lines given by $y = 2x + 3$, $y = 2x$, $g(x) = \frac{1}{3}x - 2$, and $f(x) = \frac{1}{3}x$ have slopes 2, 2, $\frac{1}{3}$, and $\frac{1}{3}$, respectively. This supports (but does not prove) the following:

The slope of any line written in the form $y = mx + b$ is m.

A proof of this result is outlined in Exercise 95 on p. 104.

Example 5

Determine the slope of the line given by $y = \frac{2}{3}x + 4$, and graph the line.

Solution Here $m = \frac{2}{3}$, so the slope is $\frac{2}{3}$. This means that from *any* point on the graph, we can locate a second point by simply going *up* 2 units (the *rise*) and *to the right* 3 units (the *run*). Where do we start? Because the y-intercept, $(0, 4)$, is known to be on the graph, we calculate that $(0 + 3, 4 + 2)$, or $(3, 6)$, is also on the graph. Knowing two points, we can draw the line.

technology connection B

To use a grapher to examine the effect of m when graphing $y = mx + b$, first graph $y_1 = x + 1$. Then on the same set of axes, graph $y_2 = 2x + 1$ and $y_3 = 3x + 1$. What do you think the graph of $y = 4x + 1$ looks like? Try graphing $y = 2.5x + 1$ and $y = \frac{3}{4}x + 1$ and describe how a positive multiplier of x affects the graph.

To see the effect of a negative value for m, graph the equations $y_1 = x + 1$ and $y_2 = -x + 1$ on the same set of axes. Then graph $y_3 = 2x + 1$ and $y_4 = -2x + 1$ on those same axes. How does a negative multiplier of x affect the graph?

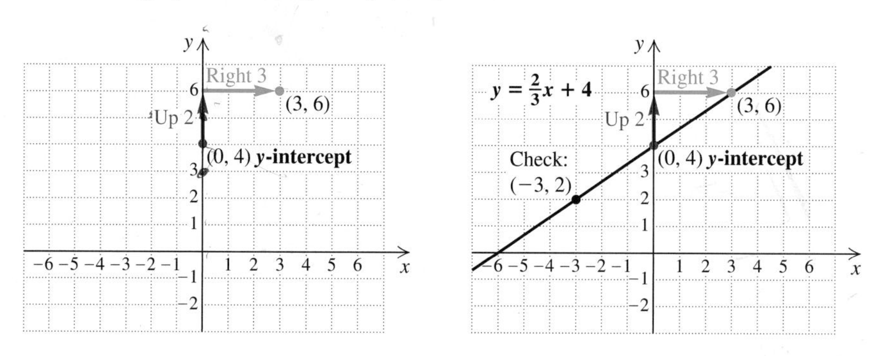

Important: To check the graph, we use some other value for x, say -3, and determine y (in this case, 2). We plot that point and see that it *is* on the line. Were it not, we would know that some error had been made.

The Slope-Intercept Equation

Any equation $y = mx + b$ has a graph that is a straight line. It goes through the *y-intercept* $(0, b)$ and has slope m. Any equation of the form $y = mx + b$ is said to be a *slope–intercept equation.*

Example 6

Determine the slope and the y-intercept of the line given by $y = -\frac{1}{3}x + 2$.

Solution The equation $y = -\frac{1}{3}x + 2$ is written in the form $y = mx + b$, with $m = -\frac{1}{3}$ and $b = 2$. Thus the slope is $-\frac{1}{3}$ and the y-intercept is $(0, 2)$.

Example 7

Find a linear function whose graph has slope $-\frac{2}{3}$ and y-intercept $(0, 4)$.

Solution We use the slope–intercept form, $f(x) = mx + b$:

$$f(x) = -\tfrac{2}{3}x + 4. \qquad \text{Substituting } -\tfrac{2}{3} \text{ for } m \text{ and 4 for } b$$

In Examples 1, 2, and 5, the lines have positive slopes and slant upward from left to right. When the slope of a line is negative, the line slants downward from left to right.

Example 8 Graph: $f(x) = -\frac{1}{2}x + 5$.

Solution The y-intercept is $(0, 5)$. The slope is $-\frac{1}{2}$, or $\frac{-1}{2}$. From the y-intercept, we go *down* 1 unit and *to the right* 2 units. That gives us the point $(2, 4)$. We can now draw the graph.

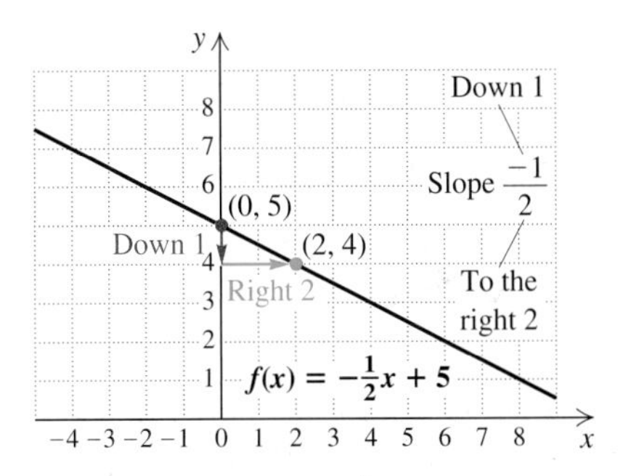

As a new type of check, we rename the slope and find another point:

$$-\frac{1}{2} = \frac{1}{-2}.$$

Thus we can go *up* 1 unit and then *to the left* 2 units. This gives the point $(-2, 6)$. Since $(-2, 6)$ is on the line, we have a check.

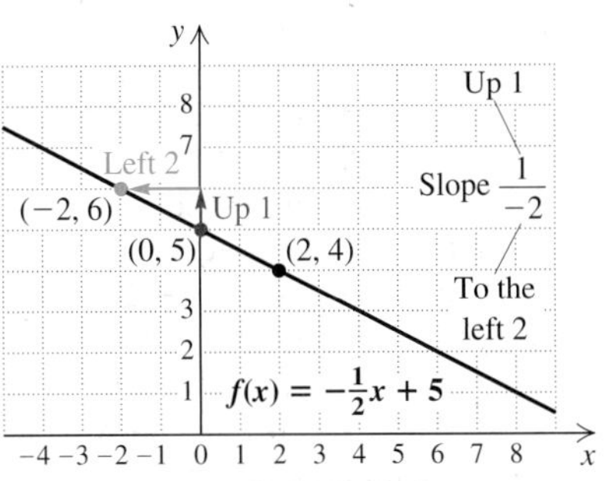

Often the easiest way to graph an equation is to rewrite it in slope–intercept form and then proceed as in Examples 5 and 8 above.

Example 9 Determine the slope and the y-intercept for the equation $5x - 4y = 8$. Then graph.

Solution We convert to a slope–intercept equation:

$$5x - 4y = 8$$
$$-4y = -5x + 8 \quad \text{Adding } -5x \text{ to both sides}$$
$$y = -\tfrac{1}{4}(-5x + 8) \quad \text{Multiplying both sides by } -\tfrac{1}{4}$$
$$y = \tfrac{5}{4}x - 2. \quad \text{Using the distributive law}$$

Because we have an equation of the form $y = mx + b$, we know that the slope is $\frac{5}{4}$ and the y-intercept is $(0, -2)$. We plot $(0, -2)$, and from there we go *up* 5 units and *to the right* 4 units, and plot a second point at $(4, 3)$. We then draw the graph.

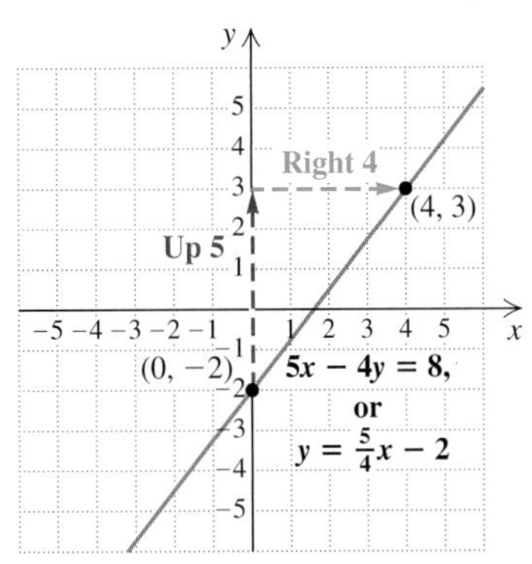

As a check, we calculate the coordinates of another point on the line. For $x = 2$, we have

$$\begin{aligned} 5 \cdot 2 - 4y &= 8 \\ 10 - 4y &= 8 \\ -4y &= -2 \\ y &= \frac{1}{2}. \end{aligned}$$

Thus, $\left(2, \frac{1}{2}\right)$ should appear on the graph. Since it *does* appear to be on the line (see the graph at the bottom of p. 97), we have a check.

Applications

Because slope is a ratio that indicates how a change in the vertical direction corresponds to a change in the horizontal direction, it has many real-world applications. Foremost is the use of slope to represent a *rate of change.*

Example 10 ***On-line travel plans.*** More and more Americans are making their travel arrangements using the Internet. Use the graph below to find the rate at which this number is growing (*Source*: Travel Industry Association of America). Note that the jagged "break" on the vertical axis is used to avoid including a large portion of unused grid.

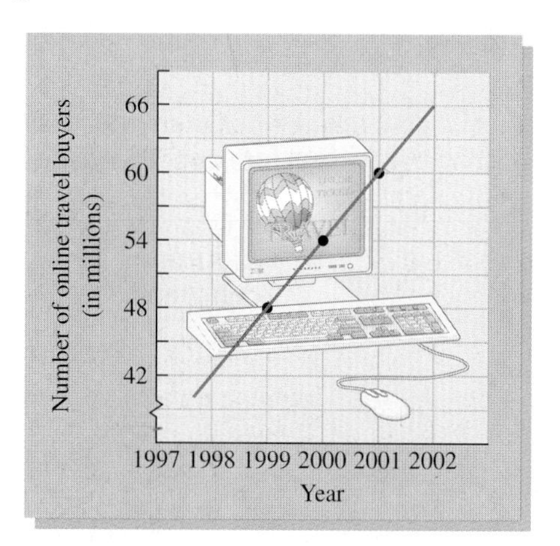

Solution Since the graph is linear, we can use any pair of points to determine the rate of change. We choose the points (1999, 48 million) and (2001, 60 million), which gives us

$$\begin{aligned} \text{Rate of change} &= \frac{60 \text{ million} - 48 \text{ million}}{2001 - 1999} \\ &= \frac{12 \text{ million}}{2 \text{ years}} = 6 \text{ million per year.} \end{aligned}$$

The number of Americans making on-line travel arrangements is growing at the rate of approximately 6 million users per year.

Example 11

Cellular phone use. In 1999, there were approximately 86 million cellular phone customers in the United States. By 2001, the figure had grown to 110 million (*Source*: based on data from the Cellular Telecommunications Industry Association). At what rate is the number of cellular phone customers changing?

Solution The rate at which the number of customers is changing is given by

$$\begin{aligned}\text{Rate of change} &= \frac{\text{change in number of customers}}{\text{change in time}}\\ &= \frac{110 \text{ million customers} - 86 \text{ million customers}}{2001 - 1999}\\ &= \frac{24 \text{ million customers}}{2 \text{ years}}\\ &= 12 \text{ million customers per year.}\end{aligned}$$

Between 1999 and 2001, the number of cellular phone customers grew at a rate of approximately 12 million customers per year.

Example 12

Running speed. Stephanie runs 10 km during each workout. For the first 7 km, her pace is twice as fast as it is for the last 3 km. Which of the following graphs best describes Stephanie's workout?

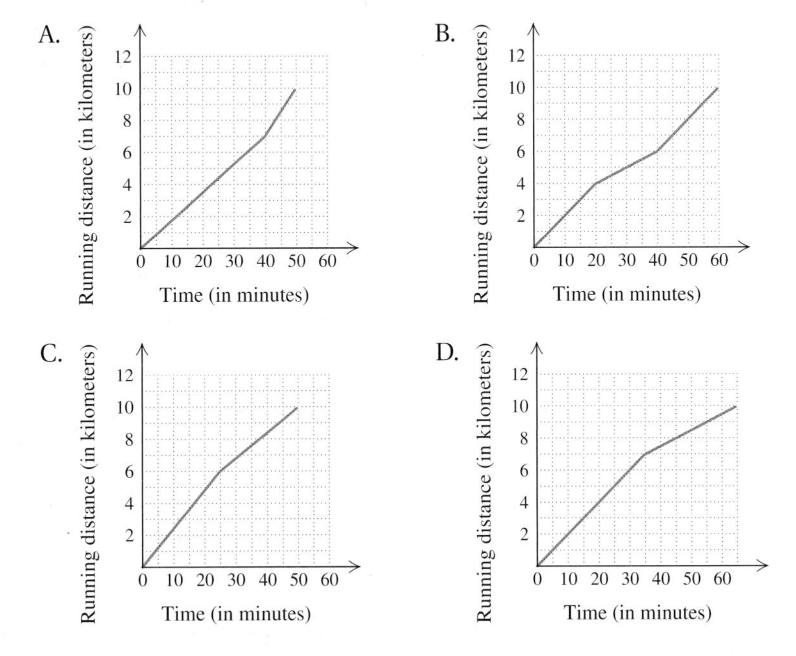

Solution The slopes in graph A increase as we move to the right. This would indicate that Stephanie ran faster for the *last* part of her workout. Thus graph A is not the correct one.

The slopes in graph B indicate that Stephanie slowed down in the middle of her run and then resumed her original speed. Thus graph B does not correctly model the situation either.

According to graph C, Stephanie slowed down not at the 7-km mark, but at the 6-km mark. Thus graph C is also incorrect.

Graph D indicates that Stephanie ran the first 7 km in 35 min, a rate of 0.2 km/min. It also indicates that she ran the final 3 km in 30 min, a rate of 0.1 km/min. This means that Stephanie's rate was twice as fast for the first 7 km, so graph D provides a correct description of her workout.

Linear functions arise continually in today's world. As with the rate problems appearing in Examples 10–12, it is critical to use proper units in all answers.

Example 13

Salvage value. Tyline Electric uses the function $S(t) = -700t + 3500$ to determine the *salvage value* $S(t)$, in dollars, of a color photocopier t years after its purchase.

a) What do the numbers -700 and 3500 signify?
b) How long will it take the copier to *depreciate* completely?
c) What is the domain of S?

Solution Drawing, or at least visualizing, a graph can be useful here.

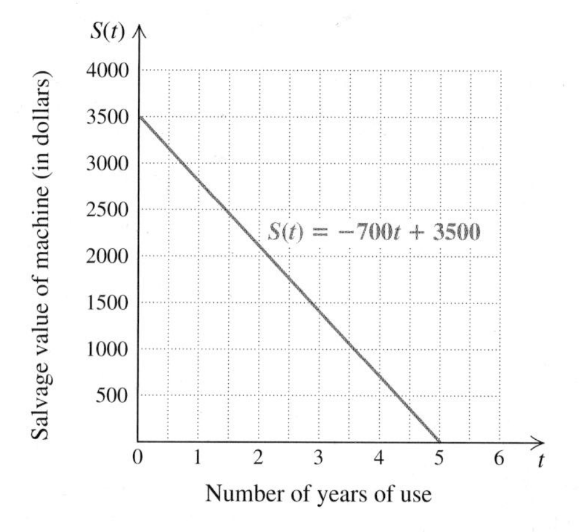

a) At time $t = 0$, we have $S(0) = -700 \cdot 0 + 3500 = 3500$. Thus the number 3500 signifies the original cost of the copier, in dollars.

This function is written in slope–intercept form. Since the output is measured in dollars and the input in years, the number -700 signifies that the value of the copier is declining at a rate of \$700 per year.

b) The copier will have depreciated completely when its value drops to 0. To learn when this occurs, we determine when $S(t) = 0$:

$S(t) = 0$	A graph is not always available.
$-700t + 3500 = 0$	Substituting $-700t + 3500$ for $S(t)$
$-700t = -3500$	Subtracting 3500 from both sides
$t = 5.$	Dividing both sides by -700

The copier will have depreciated completely in 5 yr.

c) Neither the number of years of service nor the salvage value can be negative. In part (b) we found that after 5 yr the salvage value will have dropped to 0. Thus the domain of S is $\{t \mid 0 \leq t \leq 5\}$. The graph above serves as a visual check of this result.

Exercise Set 2.3

FOR EXTRA HELP

Digital Video Tutor CD 2
Videotape 3

InterAct Math

Math Tutor Center

MathXL

MyMathLab.com

Graph.

1. $f(x) = 2x - 7$
2. $g(x) = 3x - 7$
3. $g(x) = -\frac{1}{3}x + 2$
4. $f(x) = -\frac{1}{2}x - 5$
5. $h(x) = \frac{2}{5}x - 4$
6. $h(x) = \frac{4}{5}x + 2$

Determine the y-intercept.

7. $y = 5x + 7$
8. $y = 4x - 9$
9. $f(x) = -2x - 6$
10. $g(x) = -5x + 7$
11. $y = -\frac{3}{8}x - 4.5$
12. $y = \frac{15}{7}x + 2.2$
13. $g(x) = 2.9x - 9$
14. $f(x) = -3.1x + 5$
15. $y = 37x + 204$
16. $y = -52x + 700$

For each pair of points, find the slope of the line containing them.

17. $(6, 9)$ and $(4, 5)$
18. $(8, 7)$ and $(2, -1)$
19. $(3, 8)$ and $(9, -4)$
20. $(17, -12)$ and $(-9, -15)$
21. $(-16.3, 12.4)$ and $(-5.2, 8.7)$
22. $(12.4, -5.8)$ and $(-14.5, -15.6)$
23. $(-9.7, 43.6)$ and $(4.5, 43.6)$
24. $(-2.8, -3.1)$ and $(-1.8, -2.6)$

Determine the slope and the y-intercept. Then draw a graph. Be sure to check as in Example 5 or Example 8.

25. $y = \frac{5}{2}x + 3$
26. $y = \frac{2}{5}x + 4$
27. $f(x) = -\frac{5}{2}x + 1$
28. $f(x) = -\frac{2}{5}x + 3$
29. $2x - y = 5$
30. $2x + y = 4$
31. $f(x) = \frac{1}{3}x + 2$
32. $f(x) = -3x + 6$
33. $7y + 2x = 7$
34. $4y + 20 = x$
35. Aha! $f(x) = -0.25x$
36. $f(x) = 1.5x - 3$
37. $4x - 5y = 10$
38. $5x + 4y = 4$
39. $f(x) = \frac{5}{4}x - 2$
40. $f(x) = \frac{4}{3}x + 2$
41. $12 - 4f(x) = 3x$
42. $15 + 5f(x) = -2x$
43. Aha! $g(x) = 2.5$
44. $g(x) = \frac{3}{4}x$

Find a linear function whose graph has the given slope and y-intercept.

45. Slope $\frac{2}{3}$, y-intercept $(0, -9)$
46. Slope $-\frac{3}{4}$, y-intercept $(0, 12)$
47. Slope -5, y-intercept $(0, 2)$
48. Slope 2, y-intercept $(0, -1)$
49. Slope $-\frac{7}{9}$, y-intercept $(0, 5)$
50. Slope $-\frac{4}{11}$, y-intercept $(0, 9)$
51. Slope 5, y-intercept $\left(0, \frac{1}{2}\right)$
52. Slope 6, y-intercept $\left(0, \frac{2}{3}\right)$

For each graph, find the rate of change. Remember to use appropriate units. See Example 10.

53.

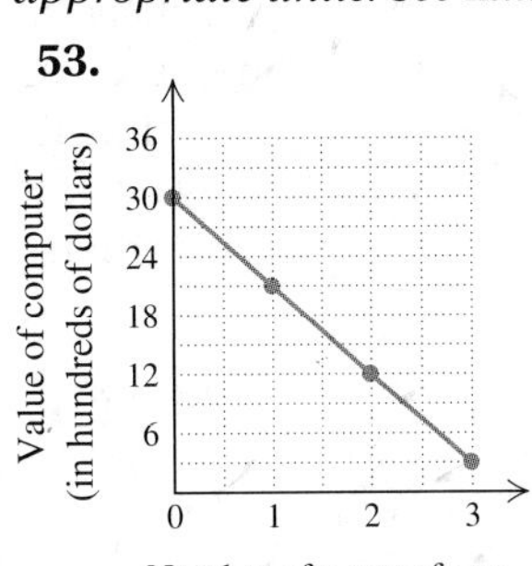

54.

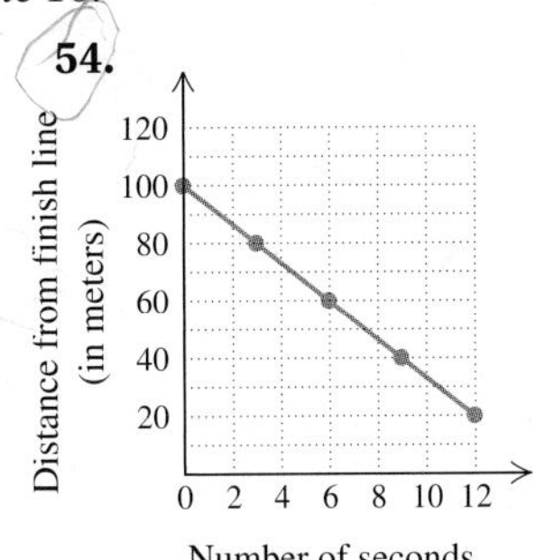

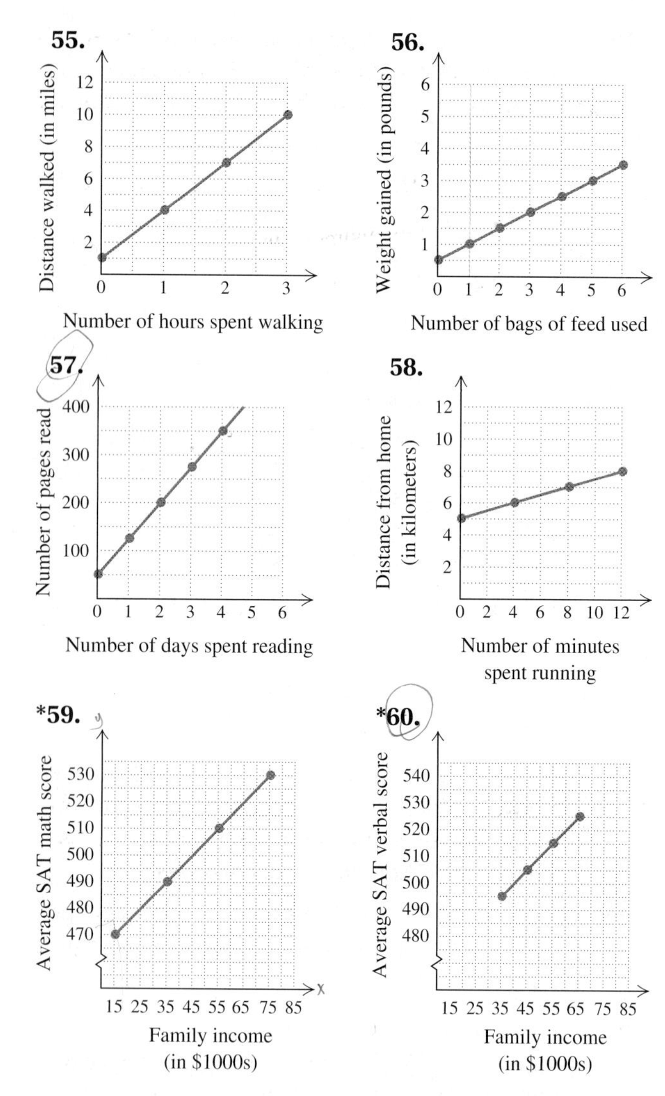

61. *Skiing rate.* A cross-country skier reaches the 3-km mark of a race in 15 min and the 12-km mark 45 min later. Find the speed of the skier.

62. *Running rate.* An Olympic marathoner passes the 5-mi point of a race after 30 min and reaches the 25-mi point 2 hr later. Find the speed of the marathoner.

*Based on data from the College Board Online.

63. *Rate of computer hits.* At the beginning of 1999, SciMor.com had already received 80,000 hits at their website. At the beginning of 2001, that number had climbed to 430,000. Calculate the rate at which the number of hits is increasing.

64. *Work rate.* As a painter begins work, one fourth of a house has already been painted. Eight hours later, the house is two-thirds done. Calculate the painter's work rate.

65. *Rate of descent.* A plane descends to sea level from 12,000 ft after being airborne for $1\frac{1}{2}$ hr. The entire flight time is 2 hr and 10 min. Determine the average rate of descent of the plane.

66. *Growth in overseas travel.* In 1988, the number of U.S. visitors overseas was about 11.6 million. In 1995, the number grew to 18.7 million (*Source: Statistical Abstract of the United States*). Determine the rate at which the number of U.S. visitors overseas was growing.

67. *Nursing.* Match each sentence with the most appropriate of the four graphs shown.

a) The rate at which fluids were given intravenously was doubled after 3 hr.

b) The rate at which fluids were given intravenously was gradually reduced to 0.

c) The rate at which fluids were given intravenously remained constant for 5 hr.

d) The rate at which fluids were given intravenously was gradually increased.

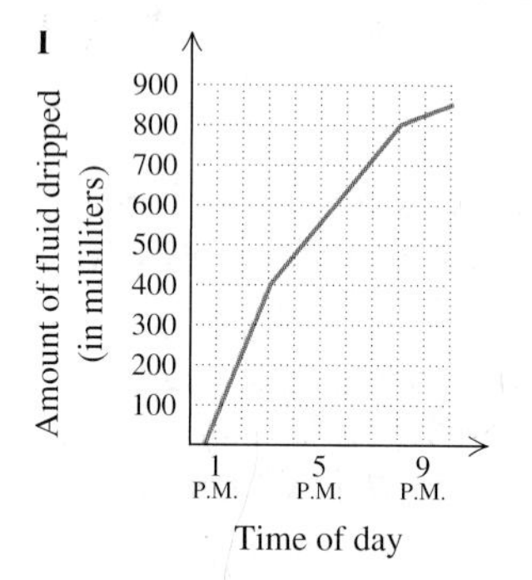

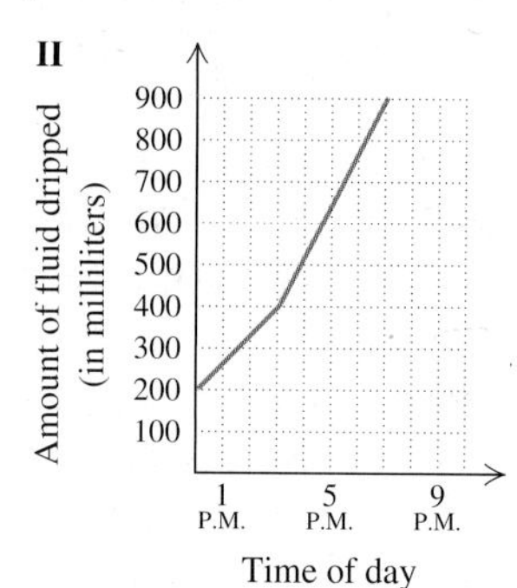

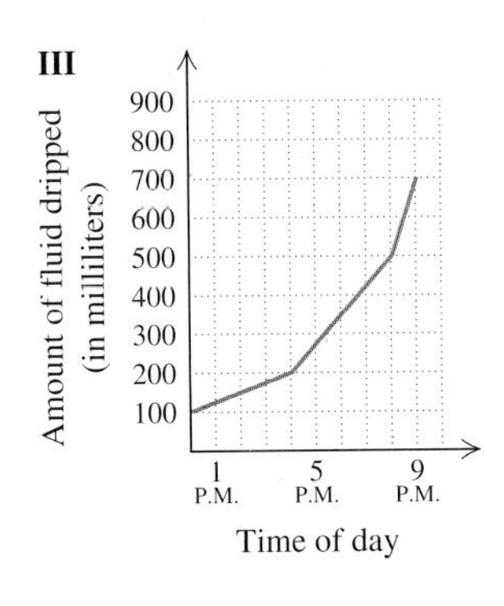

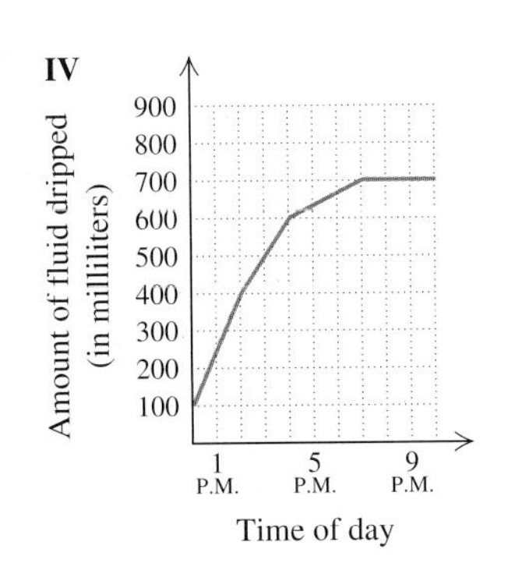

68. *Market research.* Match each sentence with the most appropriate graph below.

a) After January 1, daily sales continued to rise, but at a slower rate.

b) After January 1, sales decreased faster than they ever grew.

c) The rate of growth in daily sales doubled after January 1.

d) After January 1, daily sales decreased at half the rate that they grew in December.

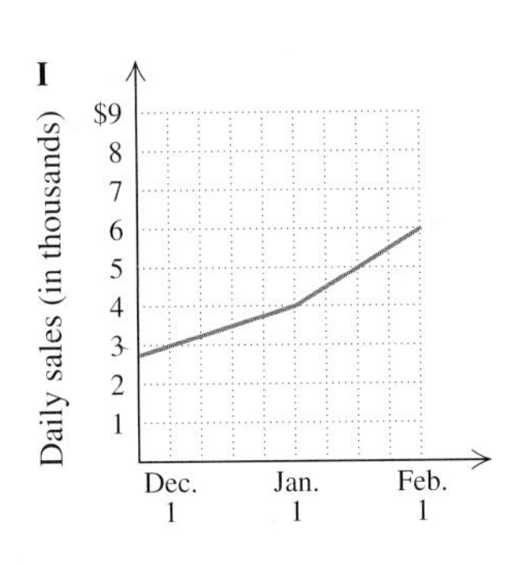

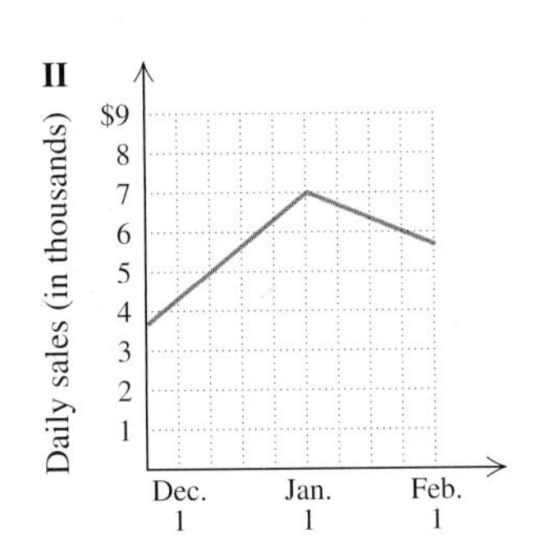

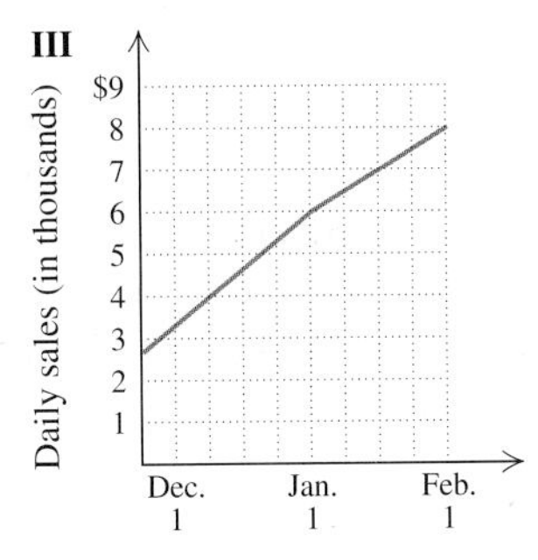

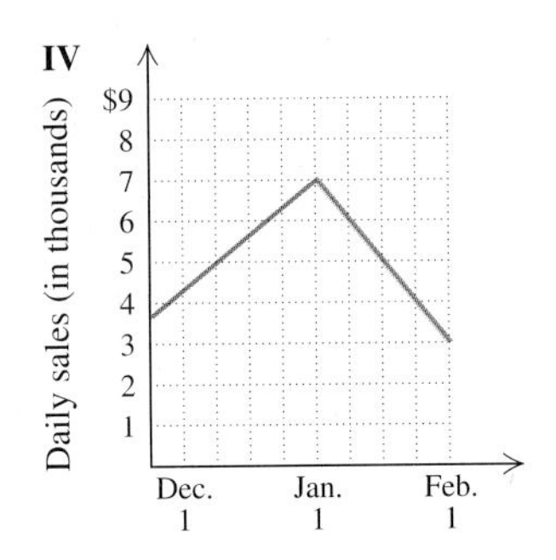

In Exercises 69–78, each model is of the form $f(x) = mx + b$. In each case, determine what m and b signify.

69. *Catering.* When catering a party for x people, Jennette's Catering uses the formula $C(x) = 25x + 75$, where $C(x)$ is the cost of the party, in dollars.

70. *Weekly pay.* Each salesperson at Knobby's Furniture is paid $P(x)$ dollars, where $P(x) = 0.05x + 200$ and x is the value of the salesperson's sales for the week.

71. *Hair growth.* After Tina gets a "buzz cut," the length $L(t)$ of her hair, in inches, is given by $L(t) = \frac{1}{2}t + 1$, where t is the number of months after she gets the haircut.

72. *Landscaping.* After being cut, the length $G(t)$ of the lawn, in inches, at Great Harrington Community College is given by $G(t) = \frac{1}{8}t + 2$, where t is the number of days since the lawn was cut.

73. *Life expectancy of American women.* The life expectancy of American women t years after 1950 is given by $A(t) = \frac{3}{20}t + 72$.

74. *Natural gas demand.* The demand, in quadrillions of joules, for natural gas is approximated by $D(t) = \frac{1}{5}t + 20$, where t is the number of years after 1960.

75. *Sales of cotton goods.* The function given by $f(t) = 2.6t + 17.8$ can be used to estimate the yearly sales of cotton goods, in billions of dollars, t years after 1975.

76. *Cost of a movie ticket.* The average price $P(t)$, in dollars, of a movie ticket is given by $P(t) = 0.1522t + 4.29$, where t is the number of years since 1990.

77. *Cost of a taxi ride.* The cost, in dollars, of a taxi ride in Pelham is given by $C(d) = 0.75d + 2$, where d is the number of miles traveled.

78. *Cost of renting a truck.* The cost, in dollars, of a one-day truck rental is given by $C(d) = 0.3d + 20$, where d is the number of miles driven.

79. *Salvage value.* Green Glass Recycling uses the function given by $F(t) = -5000t + 90{,}000$ to determine the salvage value $F(t)$, in dollars, of a waste removal truck t years after it has been put into use.

a) What do the numbers -5000 and 90,000 signify?
b) How long will it take the truck to depreciate completely?
c) What is the domain of F?

80. *Salvage value.* Consolidated Shirt Works uses the function given by $V(t) = -2000t + 15{,}000$ to determine the salvage value $V(t)$, in dollars, of a color separator t years after it has been put into use.

a) What do the numbers -2000 and 15,000 signify?
b) How long will it take the machine to depreciate completely?
c) What is the domain of V?

81. *Trade-in value.* The trade-in value of a Homelite snowblower can be determined using the function given by $v(n) = -150n + 900$. Here $v(n)$ is the trade-in value, in dollars, after n winters of use.

a) What do the numbers -150 and 900 signify?
b) When will the trade-in value of the snowblower be $300?
c) What is the domain of v?

82. *Trade-in value.* The trade-in value of a John Deere riding lawnmower can be determined using the function given by $T(x) = -300x + 2400$. Here $T(x)$ is the trade-in value, in dollars, after x summers of use.

a) What do the numbers -300 and 2400 signify?
b) When will the value of the mower be $1200?
c) What is the domain of T?

83. *Economics.* In 2000, the federal debt could be modeled using $D(t) = mt + 6000$, where $D(t)$ is in billions of dollars t years after 2000, and m is a constant. If you were president of the United States, would you want m to be positive or negative? Why?

84. *Economics.* Examine the function given in Exercise 83. What units of measure must be used for m? Why?

SKILL MAINTENANCE

Solve.

85. $2x - 5 = 7x + 3$

86. $4t + 9 = t - 6$

87. $\frac{1}{5}x + 7 = 2$

88. $-\frac{2}{3}t + 4 = t - 1$

89. $3 \cdot 0 - 2y = 9$

90. $4x - 7 \cdot 0 = 3$

SYNTHESIS

91. Hope Diswerkz claims that her firm's profits continue to go up, but the rate of increase is going down.

a) Sketch a graph that might represent her firm's profits as a function of time.
b) Explain why the graph can go up while the rate of increase goes down.

92. Belly Up, Inc., is losing $1.5 million per year while Spinning Wheels, Inc., is losing $170 an hour. Which firm would you rather own and why?

In Exercises 93 and 94, assume that r, p, and s are constants and that x and y are variables. Determine the slope and the y-intercept.

93. $rx + py = s$

94. $rx + py = s - ry$

95. Let (x_1, y_1) and (x_2, y_2) be two distinct points on the graph of $y = mx + b$. Use the fact that both pairs are solutions of the equation to prove that m is the slope of the line given by $y = mx + b$. (*Hint*: Use the slope formula.)

Given that $f(x) = mx + b$, classify each of the following as true or false.

96. $f(c + d) = f(c) + f(d)$

97. $f(cd) = f(c)f(d)$

98. $f(kx) = kf(x)$

99. $f(c - d) = f(c) - f(d)$

100. Find k such that the line containing $(-3, k)$ and $(4, 8)$ is parallel to the line containing $(5, 3)$ and $(1, -6)$.

101. Match each sentence with the most appropriate graph below.

a) Annie drove 2 mi to a lake, swam 1 mi, and then drove 3 mi to a store.

b) During a preseason workout, Rico biked 2 mi, ran for 1 mi, and then walked 3 mi.

c) James bicycled 2 mi to a park, hiked 1 mi over the notch, and then took a 3-mi bus ride back to the park.

d) After hiking 2 mi, Marcy ran for 1 mi before catching a bus for the 3-mi ride into town.

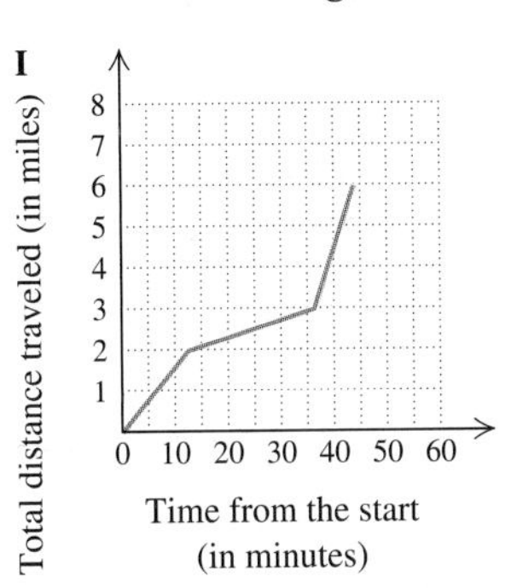

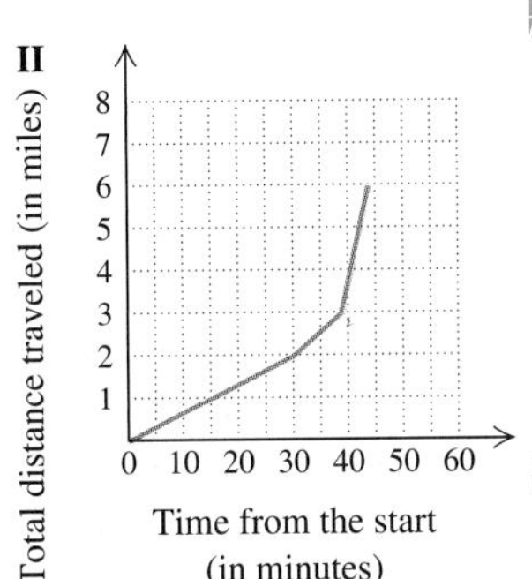

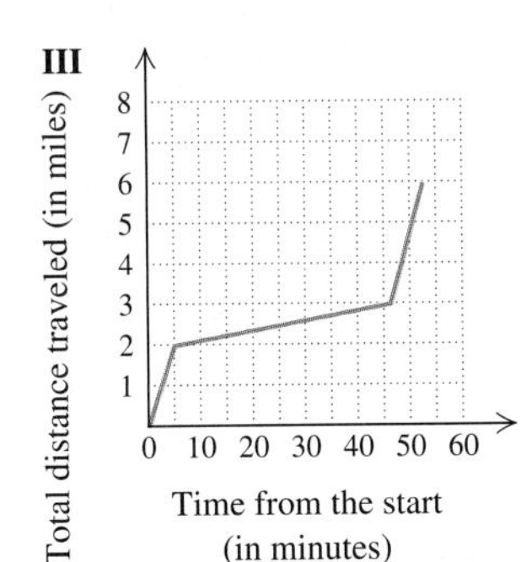

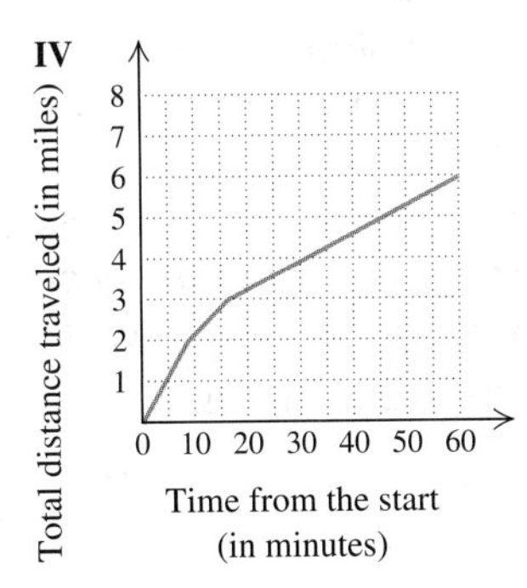

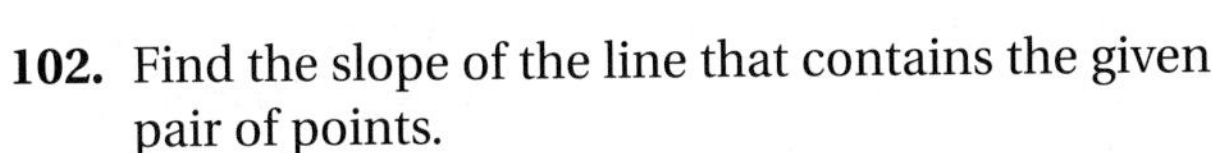

102. Find the slope of the line that contains the given pair of points.

a) $(5b, -6c), (b, -c)$

b) $(b, d), (b, d + e)$

c) $(c + f, a + d), (c - f, -a - d)$

103. *Cost of a speeding ticket.* The penalty schedule shown at the top of the next column is used to determine the cost of a speeding ticket in certain states. Use this schedule to graph the cost of a speeding ticket as a function of the number of miles per hour over the limit that a driver is going.

STATE POLICE
SPEEDING VIOLATION
FINES
1 – 10 mph over limit: $5.00/mph plus $17.50 surcharge
11 – 20 mph over limit: $6.00/mph plus $17.50 surcharge
21 – 30 mph over limit: $7.00/mph plus $17.50 surcharge
31+ mph over limit: $8.00/mph plus $17.50 surcharge
Officer will enter mph over limit in line 5a

104. Graph the equations

$$y_1 = 1.4x + 2, \quad y_2 = 0.6x + 2,$$
$$y_3 = 1.4x + 5, \quad \text{and} \quad y_4 = 0.6x + 5$$

using a grapher. If possible, use the SIMULTANEOUS mode so that you cannot tell which equation is being graphed first. Then decide which line corresponds to each equation.

105. A student makes a mistake when using a grapher to draw $4x + 5y = 12$ and the following screen appears.

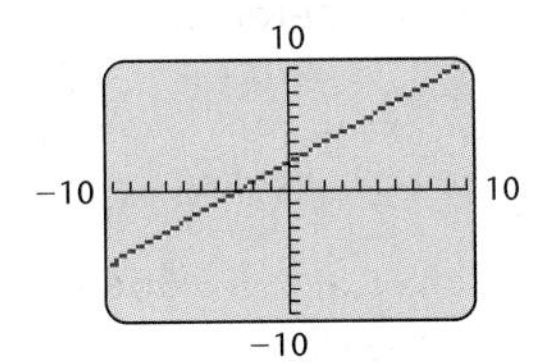

Use algebra to show that a mistake has been made. What do you think the mistake was?

106. A student makes a mistake when using a grapher to draw $5x - 2y = 3$ and the following screen appears.

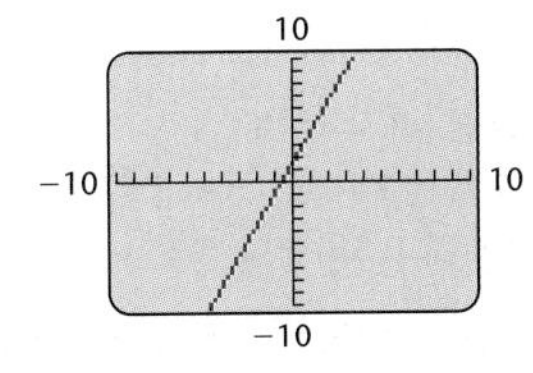

Use algebra to show that a mistake has been made. What do you think the mistake was?

Another Look at Linear Graphs

2.4

Zero Slope and Lines with Undefined Slope • Graphing Using Intercepts • Solving Equations Graphically • Recognizing Linear Equations

In Section 2.3, we graphed linear equations using slopes and y-intercepts. We now graph lines that have slope 0 or that have an undefined slope. We also graph lines using both x- and y-intercepts and learn how to use graphs to solve certain equations.

Zero Slope and Lines with Undefined Slope

If two different points have the same second coordinate, what is the slope of the line joining them? In this case, we have $y_2 = y_1$, so

$$m = \frac{y_2 - y_1}{x_2 - x_1} = \frac{0}{x_2 - x_1} = 0.$$

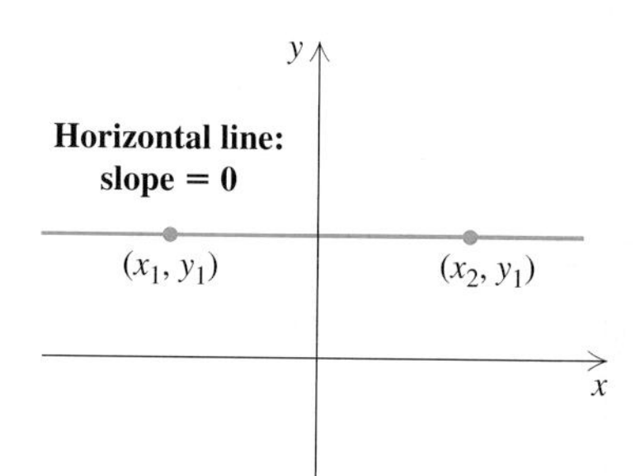

Slope of a Horizontal Line

Every horizontal line has a slope of 0.

Example 1

Graph: $f(x) = 3$.

Solution Recall from Example 5(c) in Section 2.2 that a function of this type is called a *constant function.* Writing slope–intercept form,

$$f(x) = 0 \cdot x + 3,$$

we see that the y-intercept is $(0, 3)$ and the slope is 0. Thus we can graph f by plotting the point $(0, 3)$ and, from there, determining a slope of 0. Because $0 = 0/2$ (any nonzero number could be used in place of 2), we can draw the graph by going up 0 units and to the right 2 units. As a check, we also find some ordered pairs. Note that for any choice of x-value, $f(x)$ must be 3.

x	$f(x)$
-1	3
0	3
2	3

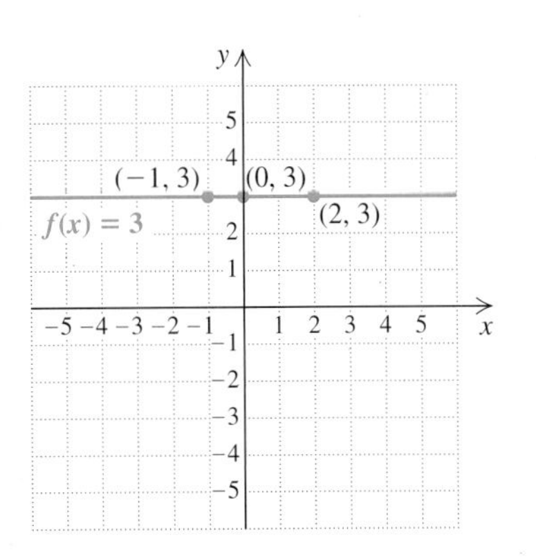

We see from Example 1 the following:

The graph of any constant function of the form $f(x) = b$ or $y = b$ is a horizontal line that crosses the y-axis at $(0, b)$.

Suppose that two different points are on a vertical line. They then have the same first coordinate. In this case, we have $x_2 = x_1$, so

$$m = \frac{y_2 - y_1}{x_2 - x_1} = \frac{y_2 - y_1}{0}.$$

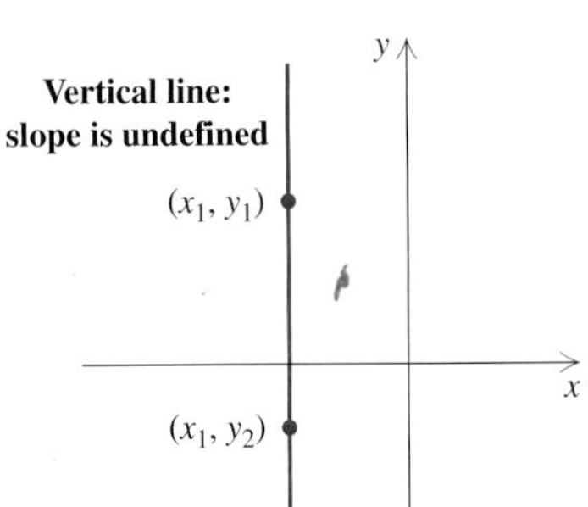

Since we cannot divide by 0, this is undefined. Note that when we say that $(y_2 - y_1)/0$ is undefined, it means that we have agreed to not attach any meaning to that expression.

Slope of a Vertical Line

The slope of a vertical line is undefined.

Example 2 Graph: $x = -2$.

Solution With y missing, no matter which value of y is chosen, x must be -2. Thus the pairs $(-2, 3)$, $(-2, 0)$, and $(-2, -4)$ all satisfy the equation. The graph is a line parallel to the y-axis. Note that since y is missing, this equation cannot be written in slope–intercept form.

x	y
-2	3
-2	0
-2	-4

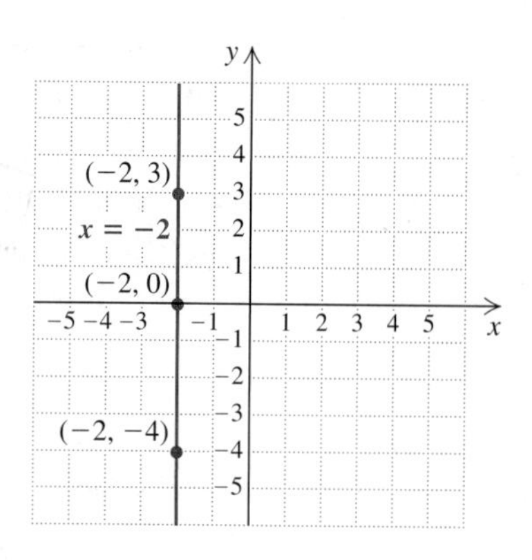

Example 2 shows the following:

The graph of any equation of the form $x = a$ is a vertical line that crosses the x-axis at $(a, 0)$.

Example 3 Find the slope of each given line. If the slope is undefined, state this.

a) $3y + 2 = 14$ **b)** $2x = 10$

Solution

a) We solve for y:

$$3y + 2 = 14$$
$$3y = 12 \quad \text{Subtracting 2 from both sides}$$
$$y = 4. \quad \text{Dividing both sides by 3}$$

The graph of $y = 4$ is a horizontal line. Since $3y + 2 = 14$ is equivalent to $y = 4$, the slope of the line $3y + 2 = 14$ is 0.

b) When y does not appear, we solve for x:

$$2x = 10$$
$$x = 5. \quad \text{Dividing both sides by 2}$$

The graph of $x = 5$ is a vertical line. Since $2x = 10$ is equivalent to $x = 5$, the slope of the line $2x = 10$ is undefined.

Graphing Using Intercepts

Any line that is not horizontal or vertical will cross both the x- and y-axes. We have already seen that the point at which a line crosses the y-axis is called the *y-intercept.* Similarly, the point at which a line crosses the x-axis is called the *x-intercept.* Any time the x- and y-intercepts are not both $(0, 0)$, they provide enough information to graph a linear equation. Recall that to find the y-intercept, we replace x with 0 and solve for y. To find the x-intercept, we replace y with 0 and solve for x.

To Determine Intercepts

The x-intercept is $(a, 0)$. To find a, let $y = 0$ and solve the original equation for x.

The y-intercept is $(0, b)$. To find b, let $x = 0$ and solve the original equation for y.

Example 4

Graph the equation $3x + 2y = 12$ by using intercepts.

Solution *To find the y-intercept, we let $x = 0$ and solve for y:*

$$3 \cdot 0 + 2y = 12$$
$$2y = 12$$
$$y = 6.$$

The y-intercept is $(0, 6)$.

To find the x-intercept, we let $y = 0$ and solve for x:

$$3x + 2 \cdot 0 = 12$$
$$3x = 12$$
$$x = 4.$$

The x-intercept is $(4, 0)$.

We plot the two intercepts and draw the line. A third point could be calculated and used as a check.

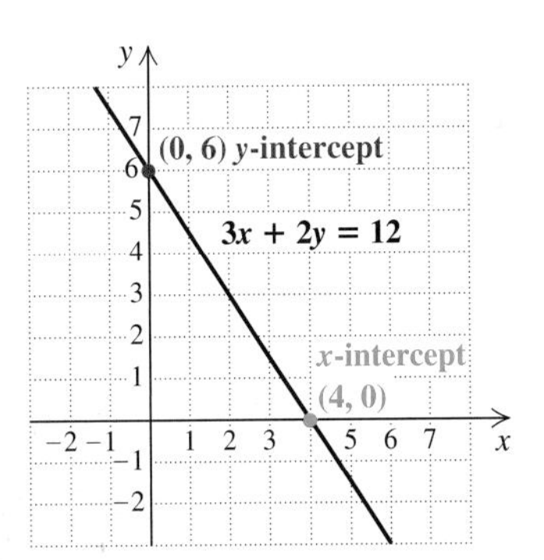

Example 5

Graph $f(x) = 2x + 5$ by using intercepts.

Solution Because the function is in slope–intercept form, we know that the y-intercept is $(0, 5)$. To find the x-intercept, we replace $f(x)$ with 0 and solve for x:

$$0 = 2x + 5$$
$$-5 = 2x$$
$$-\frac{5}{2} = x.$$

The x-intercept is $\left(-\frac{5}{2}, 0\right)$.

We plot the intercepts $(0, 5)$ and $\left(-\frac{5}{2}, 0\right)$ and draw the line. As a check, we can calculate the slope:

$$\begin{aligned} m &= \frac{5-0}{0-\left(-\frac{5}{2}\right)} \\ &= \frac{5}{\frac{5}{2}} \\ &= 5 \cdot \frac{2}{5} \\ &= 2. \end{aligned}$$

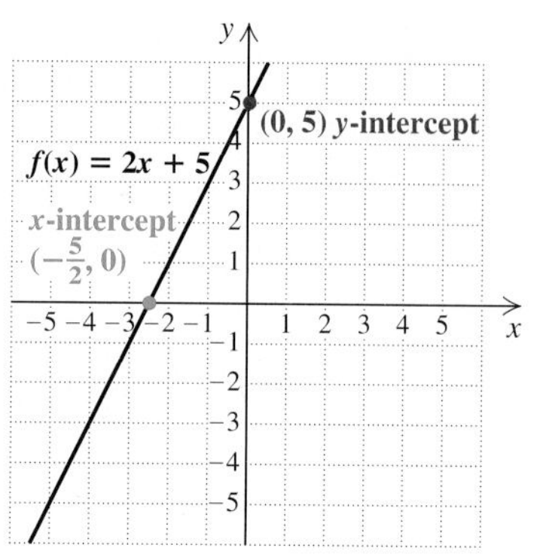

The slope is 2, as expected.

Solving Equations Graphically

Note in Example 5 that the x-intercept, $-\frac{5}{2}$, is the solution of $2x + 5 = 0$. Visually, $-\frac{5}{2}$ is the x-coordinate of the point at which the graphs of $f(x) = 2x + 5$ and $h(x) = 0$ intersect. Similarly, we can solve $2x + 5 = -3$ by finding the x-coordinate of the point at which the graphs of $f(x) = 2x + 5$ and $g(x) = -3$ intersect. Careful inspection suggests that -4 is that x-value. To check, note that $f(-4) = 2(-4) + 5 = -3$.

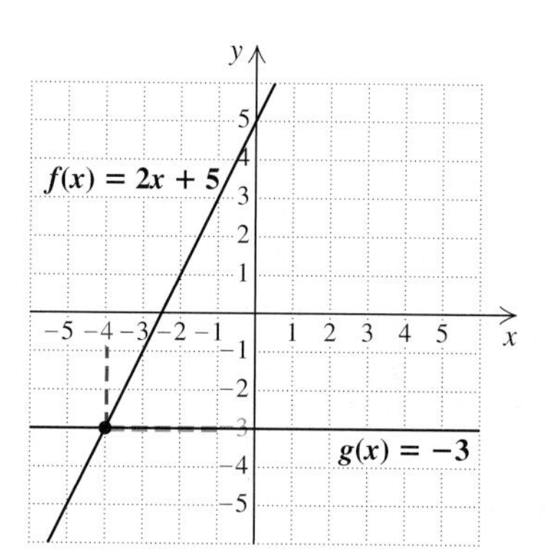

Example 6 Solve graphically: $\frac{1}{2}x + 3 = 2$.

Solution To find the x-value for which $\frac{1}{2}x + 3$ will equal 2, we graph $f(x) = \frac{1}{2}x + 3$ and $g(x) = 2$ on the same set of axes. Since the intersection appears to be $(-2, 2)$, the solution is apparently -2.

Check:

$$\begin{array}{r|l} \frac{1}{2}x + 3 = 2 & \\ \hline \frac{1}{2}(-2) + 3 \; ? & 2 \\ -1 + 3 & 2 \\ 2 & 2 \quad \text{TRUE} \end{array}$$

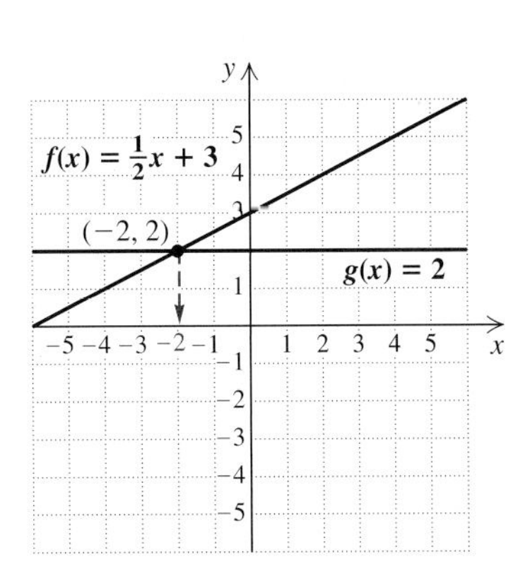

The solution is -2.

Example 7 ***Cost projections.*** Cleartone Communications charges \$50 for a cellular phone and \$40 per month for calls made under its Call Anywhere plan. Formulate and graph a mathematical model for the cost. Then use the model to estimate the time required for the total cost to reach \$250.

Solution

1. **Familiarize.** The problem describes a situation in which a monthly fee is charged after an initial purchase has been made. After 1 month of service, the total cost will be \$50 + \$40 = \$90. After 2 months, the total cost will be \$50 + \$40 · 2 = \$130. This can be generalized in a model if we let $C(t)$ represent the total cost, in dollars, for t months of service.

2. **Translate.** We rephrase and translate as follows:

Rephrasing:	The total cost	is	the cost of the phone	plus	\$40 per month.
Translating:	$C(t)$	$=$	50	$+$	$40 \cdot t$

where $t \geq 0$ (since there cannot be a negative number of months).

3. **Carry out.** Before graphing, we rewrite the model in slope–intercept form: $C(t) = 40t + 50$. We see that the vertical intercept is $(0, 50)$ and the slope—or rate—is \$40 per month.

 We plot $(0, 50)$ and, from there, count up \$40 and to the right 1 month. This takes us to $(1, 90)$. We then draw a line passing through both points.

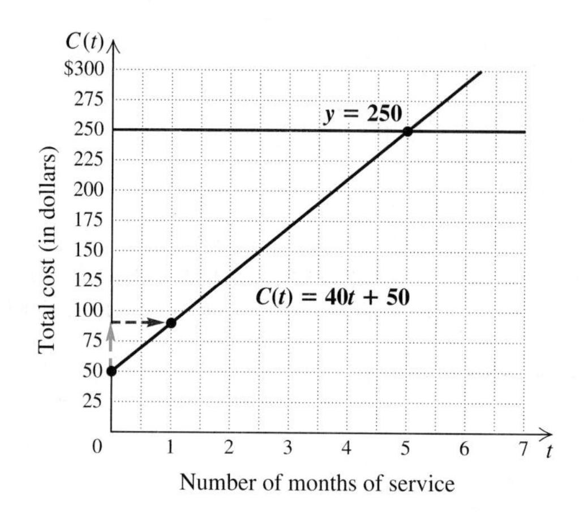

 To estimate the time required for the total cost to reach \$250, we are estimating the solution of

$$40t + 50 = 250. \qquad \text{Replacing } C(t) \text{ with } 250$$

 We do this by graphing $y = 250$ and looking for the point of intersection. This point appears to be $(5, 250)$. Thus we estimate that it takes 5 months for the total cost to reach \$250.

4. **Check.** We evaluate:

$$\begin{aligned} C(5) &= 40 \cdot 5 + 50 \\ &= 200 + 50 \\ &= 250. \end{aligned}$$

 Our estimate turns out to be precise.

5. **State.** It takes 5 months for the total cost to reach \$250.

There are limitations to solving equations graphically, as the next example illustrates.

Example 8 Solve graphically: $-\frac{3}{4}x + 6 = 2x - 1$.

Solution We graph $f(x) = -\frac{3}{4}x + 6$ and $g(x) = 2x - 1$ on the same set of axes. It *appears* that the lines intersect at $(2.5, 4)$. This would mean that $f(2.5)$ and $g(2.5)$ are identical. Let's check.

To solve Example 8 with a grapher, we can graph each side of the equation and then select the INTERSECT option of the CALC menu. Once this is done, we locate the cursor on each line and press ENTER. Finally, we enter a guess, and the calculator determines the coordinates of the intersection.

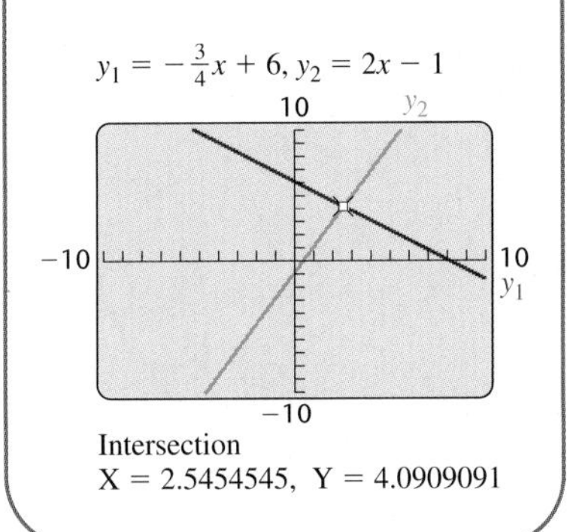

Check:

$-\frac{3}{4}x + 6 = 2x - 1$	
$-\frac{3}{4}(2.5) + 6$?	$2(2.5) - 1$
$-1.875 + 6$	$5 - 1$
4.125	4

FALSE

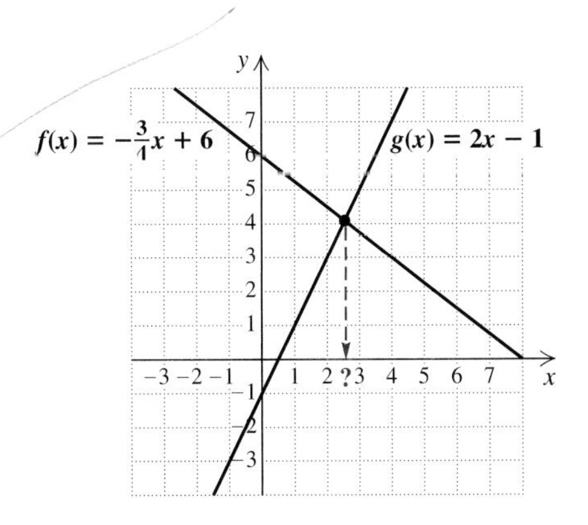

Our check shows that 2.5 is *not* the solution, although it may not be off by much. To find the exact solution, we need either a more precise way of determining the point of intersection (see the Technology Connection) or an algebraic approach:

$$-\tfrac{3}{4}x + 6 = 2x - 1$$

$$-\tfrac{3}{4}x + 7 = 2x \qquad \text{Adding 1 to both sides}$$

$$7 = \tfrac{11}{4}x \qquad \text{Adding } \tfrac{3}{4}x \text{ to both sides}$$

$$\tfrac{28}{11} = x. \qquad \text{Multiplying both sides by } \tfrac{4}{11}$$

The solution is $\frac{28}{11}$, or about 2.55. A check of this answer is left to the student.

> ***Caution!*** When using a graph to solve an equation, it is important to use graph paper and to work as neatly as possible.

Recognizing Linear Equations

Is every equation of the form $Ax + By = C$ linear? To find out, suppose that A and B are nonzero and solve for y:

$$Ax + By = C$$

$$By = -Ax + C \qquad \text{Adding } -Ax \text{ to both sides}$$

$$y = -\frac{A}{B}x + \frac{C}{B}. \qquad \text{Dividing both sides by } B$$

Since the last equation is a slope–intercept equation, we see that $Ax + By = C$ is a linear equation when $A \neq 0$ and $B \neq 0$.

But what if A or B (but not both) is 0? If A is 0, then $By = C$ and $y = C/B$. If B is 0, then $Ax = C$ and $x = C/A$. In the first case, the graph is a horizontal line; in the second case, the line is vertical. In either case, $Ax + By = C$ is a linear equation when A or B (but not both) is 0. We have now justified the following result.

The Standard Form of a Linear Equation

Any equation of the form $Ax + By = C$, where A, B, and C are real numbers and A and B are not both 0, is linear.

Any equation of the form $Ax + By = C$ is said to be a linear equation in *standard form.*

Example 9 Determine whether the equation $y = x^2 - 5$ is linear.

Solution We attempt to put the equation in standard form:

$$y = x^2 - 5$$

$$-x^2 + y = -5. \qquad \text{Adding } -x^2 \text{ to both sides}$$

This last equation is not linear because it has an x^2-term. We graphed this equation as Example 6 in Section 2.1.

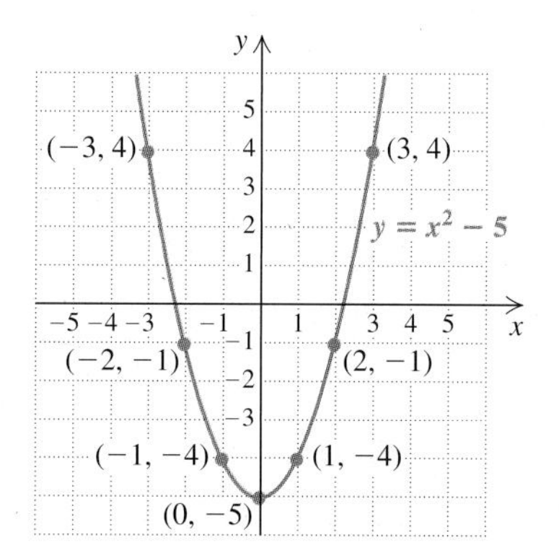

Only linear equations have graphs that are straight lines. Also, only linear graphs have a constant slope. Were you to try to calculate the slope between several pairs of points in Example 9, you would find that the slopes vary.

Exercise Set 2.4

FOR EXTRA HELP

Digital Video Tutor CD 2
Videotape 4

InterAct Math

Math Tutor Center

MathXL

MyMathLab.com

For each equation, find the slope. If the slope is undefined, state this.

1. $y - 7 = 5$
2. $x + 1 = 7$
3. $3x = 6$
4. $y - 3 = 5$
5. $4y = 20$
6. $19 = -6y$
7. $9 + x = 12$
8. $2x = 18$
9. $2x - 4 = 3$
10. $5y - 1 = 16$
11. $5y - 4 = 35$
12. $2x - 17 = 3$
13. $3y + x = 3y + 2$
14. $x - 4y = 12 - 4y$
15. $5x - 2 = 2x - 7$
16. $5y + 3 = y + 9$
17. Aha! $y = -\frac{2}{3}x + 5$
18. $y = -\frac{3}{2}x + 4$

Graph.

19. $y = 4$
20. $x = -1$
21. $x = 2$
22. $y = 5$
23. $4 \cdot f(x) = 20$
24. $6 \cdot g(x) = 12$
25. $3x = -15$
26. $2x = 10$
27. $4 \cdot g(x) + 3x = 12 + 3x$
28. $3 - f(x) = 2$

Find the intercepts. Then graph by using the intercepts and a third point as a check.

29. $x + y = 5$
30. $x + y = 4$
31. $y = 2x + 8$
32. $y = 3x + 9$
33. $3x + 5y = 15$
34. $5x - 4y = 20$
35. $2x - 3y = 18$
36. $3x + 2y = 12$
37. $7x = 3y - 21$
38. $5y = -15 + 3x$
39. $f(x) = 3x - 8$
40. $g(x) = 2x - 9$
41. $1.4y - 3.5x = -9.8$
42. $3.6x - 2.1y = 22.68$
43. $5x + 2g(x) = 7$
44. $3x - 4f(x) = 11$

Solve each equation graphically. Then check your answer by solving the same equation algebraically.

45. $x - 3 = 4$
46. $x + 4 = 6$
47. $3x - 4 = 1$
48. $2x + 1 = 7$
49. $\frac{1}{2}x + 3 = -1$
50. $\frac{1}{3}x - 2 = 1$
51. $x - 7 = 3x - 5$
52. $x + 3 = 5 - x$
53. $3 - x = \frac{1}{2}x - 3$
54. $5 - \frac{1}{2}x = x - 4$
55. $2x + 1 = -x + 7$
56. $-3x + 4 = 3x - 4$

Use a graph to estimate the solution in each of the following. Be sure to use graph paper.

57. *Telephone charges.* Skytone Calling charges \$50 for a telephone and \$25 per month under its economy plan. Estimate the time required for the total cost to reach \$150.

58. *Cellular phone charges.* The Cellular Connection charges \$60 for a cellular phone and \$40 per month under its economy plan. Estimate the time required for the total cost to reach \$260.

59. *Parking fees.* Karla's Parking charges \$3.00 to park plus 50¢ for each 15-min unit of time. Estimate how long someone can park for \$7.50.*

60. *Cost of a road call.* Dave's Foreign Auto Village charges \$35 for a road call plus \$10 for each 15-min unit of time. Estimate the time required for a road call that cost \$75.

61. *Cost of a FedEx delivery.* In 2001, for Standard delivery, within 150 mi, of packages weighing from 6 to 16 lb, FedEx charged \$18.75 for the first 6 lb plus \$0.75 for each additional pound. Estimate the weight of a package that cost \$24 to ship.

*More precise, nonlinear models of Exercises 59 and 60 appear in Exercises 94 and 95, respectively.

62. *Copying costs.* A local Mailboxes Etc.® store charges \$2.25 for binding plus 5¢ per page for each spiralbound copy of a town report. Estimate the length of a spiralbound report that cost \$3.50.

Determine whether each equation is linear. Find the slope of any nonvertical lines.

63. $5x - 3y = 15$

64. $3x + 5y + 15 = 0$

65. $16 + 4y = 10$

66. $3x - 12 = 0$

67. $3g(x) = 6x^2$

68. $2x + 4f(x) = 8$

69. $3y = 7(2x - 4)$

70. $2(5 - 3x) = 5y$

71. $g(x) - \frac{1}{x} = 0$

72. $f(x) + \frac{1}{x} = 0$

73. $\frac{f(x)}{5} = x^2$

74. $\frac{g(x)}{2} = 3 + x$

75. *Meteorology.* Wind chill is a measure of how cold the wind makes you feel. Below are some measurements of wind chill for a 15-mph breeze. How can you tell from the data that a linear function will give an approximate fit?

Temperature	15-mph Wind Chill
30°	9°
25°	2°
20°	−5°
15°	−11°
10°	−18°
5°	−25°
0°	−31°

Source: National Oceanic & Atmospheric Administration, as reported in the *Burlington Free Press,* 17 January 1992.

76. *Engineering.* Wind friction, or *air resistance,* increases with speed. At the top of the next column are some measurements made in a wind tunnel. Plot the data and explain why a linear function does or does not give an approximate fit.

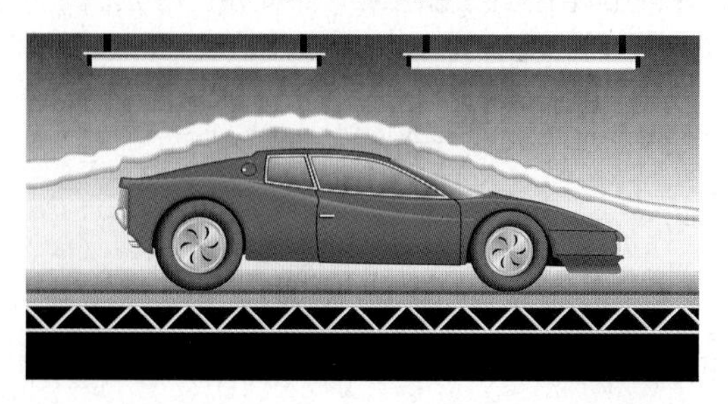

Velocity (in kilometers per hour)	Force of Resistance (in newtons)
10	3
21	4.2
34	6.2
40	7.1
45	15.1
52	29.0

SKILL MAINTENANCE

Simplify.

77. $-\frac{3}{7} \cdot \frac{7}{3}$

78. $\frac{5}{4}\left(-\frac{4}{5}\right)$

79. $-5[x - (-3)]$

80. $-2[x - (-4)]$

81. $\frac{2}{3}\left[x - \left(-\frac{1}{2}\right)\right] - 1$

82. $-\frac{3}{2}\left(x - \frac{2}{5}\right) - 3$

SYNTHESIS

83. Jim tries to avoid fractions as often as possible. Under what conditions will graphing using intercepts allow him to avoid fractions? Why?

84. Under what condition(s) will the x- and y-intercepts of a line coincide? What would the equation for such a line look like?

85. Give an equation, in standard form, for the line whose x-intercept is 5 and whose y-intercept is −4.

86. Find the x-intercept of $y = mx + b$, assuming that $m \neq 0$.

In Exercises 87–90, assume that r, p, and s are nonzero constants and that x and y are variables. Determine whether each equation is linear.

87. $rx + 3y = p - s$

88. $py = sx - ry + 2$

89. $r^2x = py + 5$

90. $\frac{x}{r} - py = 17$

91. Suppose that two linear equations have the same y-intercept but that equation A has an x-intercept that is half the x-intercept of equation B. How do the slopes compare?

Consider the linear equation

$$ax + 3y = 5x - by + 8.$$

92. Find a and b if the graph is a horizontal line passing through $(0, 2)$.

93. Find a and b if the graph is a vertical line passing through $(4, 0)$.

94. (Refer to Exercise 59.) It costs as much to park at Karla's for 16 min as it does for 29 min. Thus the linear graph drawn in the solution of Exercise 59 is not a precise representation of the situation. Draw a graph with a series of "steps" that more accurately reflects the situation.

95. (Refer to Exercise 60.) A 32-min road call with Dave's costs the same as a 44-min road call. Thus the linear graph drawn in the solution of Exercise 60 is not a precise representation of the situation. Draw a graph with a series of "steps" that more accurately reflects the situation.

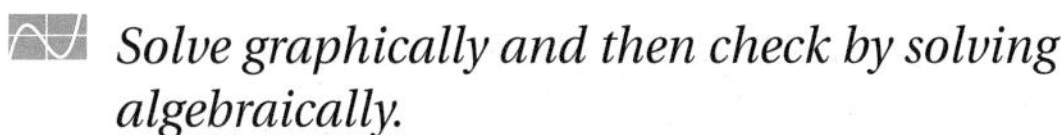

Solve graphically and then check by solving algebraically.

96. $5x + 3 = 7 - 2x$

97. $4x - 1 = 3 - 2x$

98. $3x - 2 = 5x - 9$

99. $8 - 7x = -2x - 5$

Solve using a grapher.

100. Weekly pay at Bikes for Hikes is \$219 plus a 3.5% sales commission. If a salesperson's pay was \$370.03, what did that salesperson's sales total?

101. It costs Bert's Shirts \$38 plus \$2.35 a shirt to print tee shirts for a day camp. Camp Weehawken paid Bert's \$623.15 for shirts. How many shirts were printed?

Other Equations of Lines

2.5

Point–Slope Equations • Parallel and Perpendicular Lines

Specifying a line's slope and one point through which the line passes enables us to draw the line. In this section, we study how this same information can be used to produce an *equation* of the line. The ability to do this is important in more advanced courses.

Point–Slope Equations

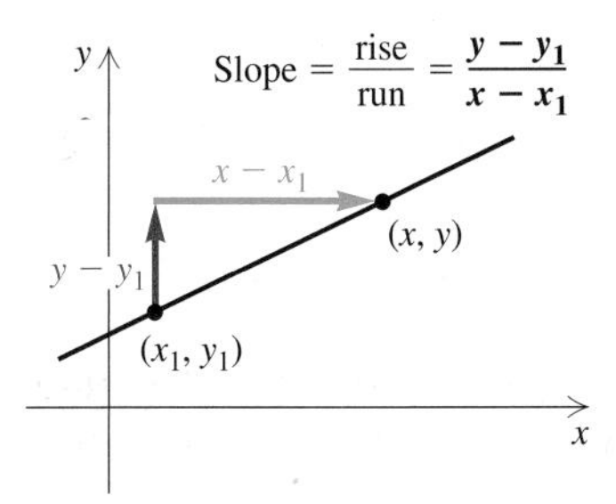

Suppose that a line of slope m passes through the point (x_1, y_1). For any other point (x, y) to lie on this line, we must have

$$\frac{y - y_1}{x - x_1} = m.$$

It is tempting to use this last equation as an equation of the line of slope m that passes through (x_1, y_1). The only problem with this form is that when x and y are replaced with x_1 and y_1, we have $\frac{0}{0} = m$, a false equation. To avoid this difficulty, we multiply both sides by $x - x_1$ and simplify:

$$(x - x_1)\frac{y - y_1}{x - x_1} = m(x - x_1) \qquad \text{Multiplying both sides by } x - x_1$$

$$y - y_1 = m(x - x_1). \qquad \text{Removing a factor equal to 1: } \frac{x - x_1}{x - x_1} = 1$$

This is the *point–slope* form of a linear equation.

Point–Slope Equation

The *point–slope equation* of a line with slope m, passing through (x_1, y_1), is

$$y - y_1 = m(x - x_1).$$

Example 1 Find and graph an equation of the line passing through $(3, 4)$ with slope $-\frac{1}{2}$.

Solution We substitute in the point–slope equation:

$$y - y_1 = m(x - x_1)$$

$$y - 4 = -\tfrac{1}{2}(x - 3). \qquad \text{Substituting}$$

To graph this point–slope equation, we count off a slope of $-\frac{1}{2}$, starting at $(3, 4)$. Then we draw the line.

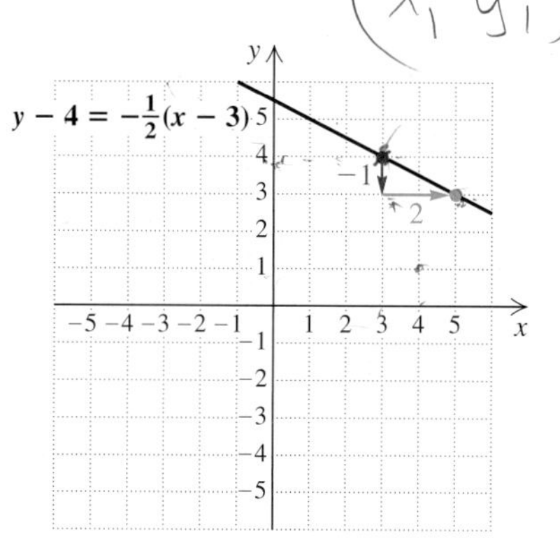

Study Tip

If you are finding it difficult to master a particular topic or concept, talk about it with a classmate. Verbalizing your questions about the material might help clarify it for you. If your classmate is also having difficulty, it is possible that a majority of your classmates are confused and you can ask your instructor to explain the concept again.

Example 2 Find a linear function that has a graph passing through the points $(-1, -5)$ and $(3, -2)$.

Solution We first determine the slope of the line and then use the point–slope equation. Note that

$$m = \frac{-5 - (-2)}{-1 - 3} = \frac{-3}{-4} = \frac{3}{4}.$$

Since the line passes through $(3, -2)$, we have

$$y - (-2) = \tfrac{3}{4}(x - 3) \qquad \text{Substituting into the point–slope equation}$$

$$y + 2 = \tfrac{3}{4}x - \tfrac{9}{4}. \qquad \text{Using the distributive law}$$

Before using function notation, we isolate y:

$$y = \tfrac{3}{4}x - \tfrac{9}{4} - 2 \qquad \text{Subtracting 2 from both sides}$$

$$y = \tfrac{3}{4}x - \tfrac{17}{4} \qquad -\tfrac{9}{4} - \tfrac{8}{4} = -\tfrac{17}{4}$$

$$f(x) = \tfrac{3}{4}x - \tfrac{17}{4}. \qquad \text{Using function notation}$$

You can check that substituting $(-1, -5)$ instead of $(3, -2)$ in the point–slope equation will yield the same expression for $f(x)$.

Example 3

Tattoo removal. In 1996, an estimated 275,000 Americans visited a doctor for tattoo removal. That figure was expected to grow to 410,000 in 2000 (*Source*: Mike Meyers, staff writer, Star-Tribune Newspaper of the Twin Cities Minneapolis–St. Paul, copyright 2000). Assuming constant growth since 1995, how many people will visit a doctor for tattoo removal in 2005?

Solution

1. **Familiarize.** Constant growth indicates a constant rate of change, so a linear relationship can be assumed. If we let n represent the number of people, in thousands, who visit a doctor for tattoo removal and t the number of years since 1995, we can form the pairs $(1, 275)$ and $(5, 410)$. After choosing suitable scales on the two axes, we draw the graph.

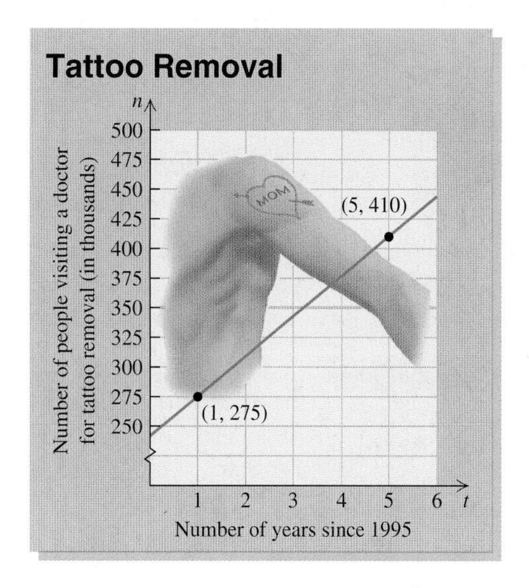

2. **Translate.** To find an equation relating n and t, we first find the slope of the line. This corresponds to the *growth rate*:

$$\begin{aligned} m &= \frac{410 \text{ thousand people} - 275 \text{ thousand people}}{5 \text{ years} - 1 \text{ year}} \\ &= \frac{135 \text{ thousand people}}{4 \text{ years}} \\ &= 33.75 \text{ thousand people per year.} \end{aligned}$$

Next, we use the point–slope equation and solve for n:

$$n - 275 = 33.75(t - 1) \quad \text{Writing point–slope form}$$

$$n - 275 = 33.75t - 33.75 \quad \text{Using the distributive law}$$

$$n = 33.75t + 241.25. \quad \text{Adding 275 to both sides}$$

3. **Carry out.** Using function notation, we have

$$n(t) = 33.75t + 241.25.$$

To predict the number of people who will visit a doctor for tattoo removal in 2005, we find

$$n(10) = 33.75 \cdot 10 + 241.25 \qquad \text{2005 is 10 years from 1995.}$$
$$= 578.75. \qquad \text{This represents 578,750 people.}$$

4. **Check.** To check, we can repeat our calculations. We could also extend the graph to see if (10, 578.75) appears to be on the line.
5. **State.** Assuming constant growth, there will be about 578,750 people visiting a doctor for tattoo removal in 2005.

Parallel and Perpendicular Lines

If two lines are vertical, they are parallel. How can we tell whether nonvertical lines are parallel? The answer is simple: We look at their slopes (see Examples 1 and 2 in Section 2.3).

> ***Slope and Parallel Lines***
>
> Two lines are parallel if they have the same slope.

Example 4

Determine whether the line passing through the points (1, 7) and (4, −2) is parallel to the line given by $f(x) = -3x + 4.2$.

Solution The slope of the line passing through (1, 7) and (4, −2) is given by

$$m = \frac{7 - (-2)}{1 - 4} = \frac{9}{-3} = -3.$$

Since the graph of $f(x) = -3x + 4.2$ also has a slope of −3, the lines are parallel.

To check that $y = \frac{7}{8}x - 3$ and $y = -\frac{8}{7}x + \frac{6}{7}$ are perpendicular, we can use a grapher. To do so, we use the ZSQUARE option of the ZOOM menu to create a "squared" window. This corrects distortion that would otherwise result from differing scales on the axes.

1. Use a grapher to check that $y = \frac{3}{4}x + 2$ and $y = -\frac{4}{3}x - 1$ are perpendicular.
2. Use a grapher to check that $y = -\frac{2}{5}x - 4$ and $y = \frac{5}{2}x + 3$ are perpendicular.
3. To see that this type of check is not foolproof, graph $y = \frac{31}{40}x + 2$ and $y = -\frac{40}{30}x - 1$. Are the lines perpendicular? Why or why not?

If one line is vertical and another is horizontal, they are perpendicular. There are other instances in which two lines are perpendicular.

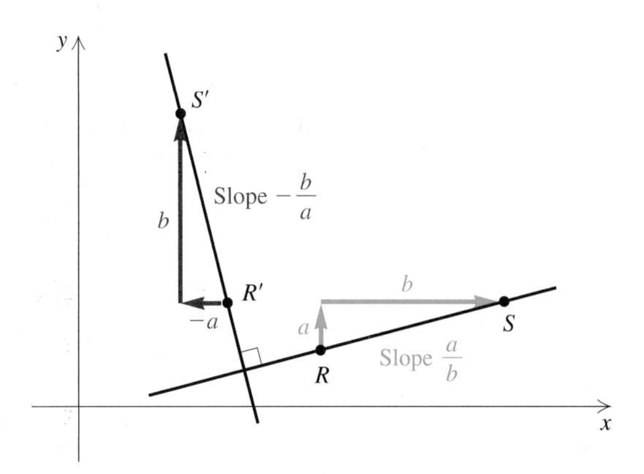

Consider a line $\overleftrightarrow{RS}$, as shown on p. 120, with slope a/b. Then think of rotating the figure 90° to get a line $\overleftrightarrow{R'S'}$ perpendicular to $\overleftrightarrow{RS}$. For the new line, the rise and the run are interchanged, but the run is now negative. Thus the slope of the new line is $-b/a$. Let's multiply the slopes:

$$\frac{a}{b}\left(-\frac{b}{a}\right) = -1.$$

This can help us determine which lines are perpendicular.

Slope and Perpendicular Lines

Two lines are perpendicular if the product of their slopes is -1. (If one line has slope m, the slope of a line perpendicular to it is $-1/m$. That is, we take the reciprocal and change the sign.) Lines are also perpendicular if one is vertical and the other is horizontal.

Example 5 Consider the line given by the equation $8y = 7x - 24$.

a) Find an equation for a parallel line passing through $(-1, 2)$.
b) Find an equation for a perpendicular line passing through $(-1, 2)$.

Solution To find the slope of the line given by $8y = 7x - 24$, we solve for y to find slope–intercept form:

$$8y = 7x - 24$$

$$y = \tfrac{7}{8}x - 3. \quad \text{Multiplying both sides by } \tfrac{1}{8}$$

The slope is $\frac{7}{8}$.

a) The slope of any parallel line will be $\frac{7}{8}$. The point–slope equation yields

$$y - 2 = \tfrac{7}{8}[x - (-1)] \quad \text{Substituting } \tfrac{7}{8} \text{ for the slope and } (-1, 2) \text{ for the point}$$

$$y - 2 = \tfrac{7}{8}[x + 1]$$

$$y = \tfrac{7}{8}x + \tfrac{7}{8} + 2 \quad \text{Using the distributive law and adding 2 to both sides}$$

$$y = \tfrac{7}{8}x + \tfrac{23}{8}.$$

b) The slope of a perpendicular line is given by the opposite of the reciprocal of $\frac{7}{8}$, or $-\frac{8}{7}$. The point–slope equation yields

$$y - 2 = -\tfrac{8}{7}[x - (-1)] \quad \text{Substituting } -\tfrac{8}{7} \text{ for the slope and } (-1, 2) \text{ for the point}$$

$$y - 2 = -\tfrac{8}{7}[x + 1]$$

$$y = -\tfrac{8}{7}x - \tfrac{8}{7} + 2 \quad \text{Using the distributive law and adding 2 to both sides}$$

$$y = -\tfrac{8}{7}x + \tfrac{6}{7}.$$

CONNECTING THE CONCEPTS

We have now studied the slope–intercept, point–slope, and standard forms of a linear equation. These are the most common ways in which linear equations are written. Depending on what information we are given and what information we are seeking, one form may be more useful than the others. A referenced summary is given below.

Slope–intercept form, $y = mx + b$ **or** $f(x) = mx + b$	▪ Useful when an equation is needed and the slope and y-intercept are given. See Example 7 on p. 96. ▪ Useful when a line's slope and y-intercept are needed. See Example 6 on p. 96. ▪ Useful when solving equations graphically. See Example 6 on p. 110. ▪ Commonly used for linear functions.
Standard form, $Ax + By = C$	▪ Allows for easy calculation of intercepts. See Example 4 on p. 109. ▪ Will prove useful in future work. See Sections 3.1–3.3.
Point–slope form, $y - y_1 = m(x - x_1)$	▪ Useful when an equation is needed and the slope and a point on the line are given. See Example 1 on p. 118. ▪ Useful when a linear function is needed and two points on its graph are given. See Example 2 on p. 118. ▪ Will prove useful in future work with curves and tangents in calculus.

Exercise Set 2.5

FOR EXTRA HELP

Digital Video Tutor CD 2 Videotape 4

InterAct Math

Math Tutor Center

MathXL

MyMathLab.com

Find an equation in point–slope form of the line having the specified slope and containing the point indicated. Then graph the line.

1. $m = -2, (2, 3)$

2. $m = 5, (7, 4)$

3. $m = 3, (4, 7)$

4. $m = 2, (7, 3)$

5. $m = \frac{1}{2}, (-2, -4)$

6. $m = 1, (-5, -7)$

7. $m = -1, (8, 0)$

8. $m = -3, (-2, 0)$

9. $m = \frac{2}{5}, (-3, 8)$

10. $m = \frac{3}{4}, (1, -5)$

For each point–slope equation listed, state the slope and a point on the graph.

11. $y - 4 = \frac{2}{7}(x - 1)$

12. $y - 3 = 9(x - 2)$

13. $y + 2 = -5(x - 7)$

14. $y - 1 = -\frac{2}{9}(x + 5)$

15. $y - 1 = -\frac{5}{3}(x + 2)$

16. $y + 7 = -4(x - 9)$

17. $y = \frac{4}{7}x$

18. $y = 3x$

Find an equation of the line having the specified slope and containing the indicated point. Write your final answer as a linear function in slope–intercept form.

19. $m = 4, (2, -3)$

20. $m = -4, (-1, 5)$

21. $m = -\frac{3}{5}, (4, -7)$

22. $m = -\frac{1}{5}, (-2, 1)$

23. $m = -0.6, (-3, -4)$

24. $m = 2.3, (4, -5)$

25. $m = \frac{2}{7}, (0, -5)$

26. $m = \frac{1}{4}, (0, 3)$

Find an equation of the line containing each pair of points. Write your final answer as a linear function in slope–intercept form.

27. $(1, 4)$ and $(5, 6)$

28. $(2, 6)$ and $(4, 1)$

29. $(2.5, -3)$ and $(6.5, 3)$

30. $(2, -1.3)$ and $(7, 1.7)$

Aha! **31.** $(1, 3)$ and $(0, -2)$

32. $(-3, 0)$ and $(0, -4)$

33. $(-2, -3)$ and $(-4, -6)$

34. $(-4, -7)$ and $(-2, -1)$

In Exercises 35–44, assume that a constant rate of change exists for each model formed.

35. *Records in the 400-meter run.* In 1930, the record for the 400-m run was 46.8 sec. In 1970, it was 43.8 sec. Let $R(t)$ represent the record in the 400-m run and t the number of years since 1930.

a) Find a linear function that fits the data.

b) Use the function of part (a) to predict the record in 2003; in 2006.

c) When will the record be 40 sec?

36. *Records in the 1500-meter run.* In 1930, the record for the 1500-m run was 3.85 min. In 1950, it was 3.70 min. Let $R(t)$ represent the record in the 1500-m run and t the number of years since 1930.

a) Find a linear function that fits the data.

b) Use the function of part (a) to predict the record in 2002; in 2006.

c) When will the record be 3.1 min?

37. *PAC contributions.* In 1992, Political Action Committees (PACs) contributed \$178.6 million to congressional candidates. In 2000, the figure rose to \$243.1 million (*Source*: Congressional Research Service and Federal Election Commission). Let $A(t)$ represent the amount of PAC contributions, in millions, and t the number of years since 1992.

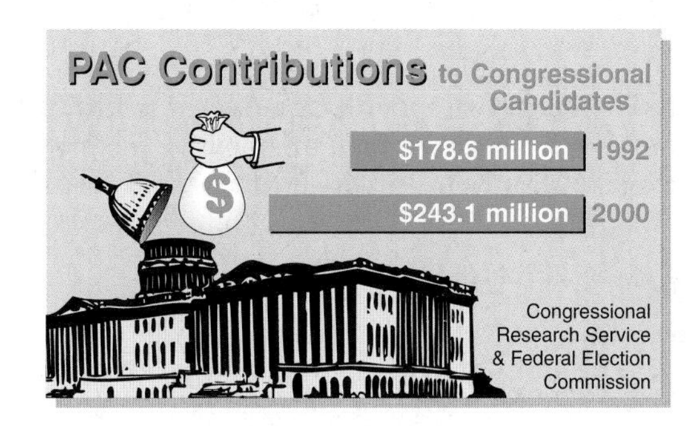

a) Find a linear function that fits the data.

b) Use the function of part (a) to predict the amount of PAC contributions in 2006.

38. *Consumer demand.* Suppose that 6.5 million lb of coffee are sold when the price is \$8 per pound, and 4.0 million lb are sold when it is \$9 per pound.

a) Find a linear function that expresses the amount of coffee sold as a function of the price per pound.

b) Use the function of part (a) to predict how much consumers would be willing to buy at a price of \$6 per pound.

39. *Recycling.* In 1993, Americans recycled 43.8 million tons of solid waste. In 1997, the figure grew to 60.8 million tons. (*Source*: *Statistical Abstract of the United States,* 1999) Let $N(t)$ represent the number of tons recycled, in millions, and t the number of years since 1993.

a) Find a linear function that fits the data.

b) Use the function of part (a) to predict the amount recycled in 2005.

40. *Seller's supply.* Suppose that suppliers are willing to sell 5.0 million lb of coffee at a price of \$8 per pound and 7.0 million lb at \$9 per pound.

a) Find a linear function that expresses the amount suppliers are willing to sell as a function of the price per pound.

b) Use the function of part (a) to predict how much suppliers would be willing to sell at a price of \$6 per pound.

41. *Life expectancy of females in the United States.* In 1990, the life expectancy of females was 78.8 yr. In 1997, it was 79.2 yr. (*Source*: *Statistical Abstract of the United States,* 1999) Let $E(t)$ represent life expectancy and t the number of years since 1990.

a) Find a linear function that fits the data.

b) Use the function of part (a) to predict the life expectancy of females in 2008.

42. *Life expectancy of males in the United States.* In 1990, the life expectancy of males was 71.8 yr. In 1997, it was 73.6 yr. Let $E(t)$ represent life expectancy and t the number of years since 1990.

a) Find a linear function that fits the data.

b) Use the function of part (a) to predict the life expectancy of males in 2007.

43. *National Park land.* In 1994, the National Park system consisted of about 74.9 million acres. By 1997, the figure had grown to 77.5 million acres. (*Source*: *Statistical Abstract of the United States*, 1999) Let $A(t)$ represent the amount of land in the National Park system, in millions of acres, t years after 1994.

a) Find a linear function that fits the data.

b) Use the function of part (a) to predict the amount of land in the National Park system in 2006.

44. *Pressure at sea depth.* The pressure 100 ft beneath the ocean's surface is approximately 4 atm (atmospheres), whereas at a depth of 200 ft, the pressure is about 7 atm.

a) Find a linear function that expresses pressure as a function of depth.

b) Use the function of part (a) to determine the pressure at a depth of 690 ft.

Without graphing, tell whether the graphs of each pair of equations are parallel.

45. $x + 8 = y$, $y - x = -5$

46. $2x - 3 = y$, $y - 2x = 9$

47. $y + 9 = 3x$, $3x - y = -2$

48. $y + 8 = -6x$, $-2x + y = 5$

49. $f(x) = 3x + 9$, $2y = 8x - 2$

50. $f(x) = -7x - 9$, $-3y = 21x + 7$

Write an equation of the line containing the specified point and parallel to the indicated line.

51. $(3, 7)$, $x + 2y = 6$

52. $(0, 3)$, $3x - y = 7$

53. $(2, -1)$, $5x - 7y = 8$

54. $(-4, -5)$, $2x + y = -3$

55. $(-6, 2)$, $3x - 9y = 2$

56. $(-7, 0)$, $5x + 2y = 6$

57. $(-3, -2)$, $3x + 2y = -7$

58. $(-4, 3)$, $6x - 5y = 4$

Aha! **59.** $(0, -7)$, $y = 2x + 3$

60. $(0, 4)$, $y = -x + 2$

Without graphing, tell whether the graphs of each pair of equations are perpendicular.

61. $f(x) = 4x - 3$, $4y = 7 - x$

62. $2x - 5y = -3$, $2x + 5y = 4$

63. $x + 2y = 7$, $2x + 4y = 4$

64. $y = -x + 7$, $f(x) = x + 3$

Write an equation of the line containing the specified point and perpendicular to the indicated line.

65. $(2, 5)$, $2x + y = -7$

66. $(4, 0)$, $x - 3y = 0$

67. $(3, -2)$, $3x + 4y = 5$

68. $(-3, -5)$, $5x - 2y = 4$

69. $(0, 9)$, $2x + 5y = 7$

70. $(-3, -4)$, $-3x + 6y = 2$

71. $(-4, -7)$, $3x - 5y = 6$

72. $(-4, 5)$, $7x - 2y = 1$

Aha! **73.** $(0, 6)$, $2x - 5 = y$

74. $(0, -4)$, $-x + 3 = y$

75. Rosewood Graphics recently promised its employees 6% raises each year for the next five years. Amy currently earns $30,000 a year. Can she use a linear function to predict her salary for the next five years? Why or why not?

76. If two lines are perpendicular, does it follow that the lines have slopes that are negative reciprocals of each other? Why or why not?

SKILL MAINTENANCE

Simplify.

77. $(3x^2 + 5x) + (2x - 4)$

78. $(5t^2 - 2t) - (4t + 3)$

Evaluate.

79. $\dfrac{2t-6}{4t+1}$, for $t = 3$

80. $(3x-1)(4t+20)$, for $t = -5$

81. $2x - 5y$, for $x = 3$ and $y = -1$

82. $3x - 7y$, for $x = -2$ and $y = 4$

SYNTHESIS

83. In 1986, Political Action Committees contributed \$132.7 million to congressional candidates. Does this information make your answer to Exercise 37(b) seem too low or too high? Why?

84. On the basis of your answers to Exercises 41 and 44, would you predict that at some point in the future the life expectancy of males will exceed that of females? Why or why not?

For Exercises 85–89, assume that a linear equation models each situation.

85. *Depreciation of a computer.* After 6 mos of use, the value of Pearl's computer had dropped to \$900. After 8 mos, the value had gone down to \$750. How much did the computer cost originally?

86. *Temperature conversion.* Water freezes at 32° Fahrenheit and at 0° Celsius. Water boils at 212°F and at 100°C. What Celsius temperature corresponds to a room temperature of 70° F?

87. *Cellular phone charges.* The total cost of Mel's cellular phone was \$230 after 5 mos of service and \$390 after 9 mos. What costs had Mel already incurred when his service just began?

88. *Operating expenses.* The total cost for operating Ming's Wings was \$7500 after 4 mos and \$9250 after 7 mos. Predict the total cost after 10 mos.

89. *Medical insurance.* In 1993, health insurance companies collected \$124.7 billion in premiums and paid out \$103.6 billion in benefits. In 1996, they collected \$137.1 billion in premiums and paid out \$113.8 billion in benefits. What percentage of premiums will be paid out in benefits in 2005?

90. Based on the information given in Exercises 38 and 40, at what price will the supply equal the demand?

91. Specify the domain of your answer to Exercise 38(a).

92. Specify the domain of your answer to Exercise 40(a).

93. For a linear function g, $g(3) = -5$ and $g(7) = -1$.

a) Find an equation for g.
b) Find $g(-2)$.
c) Find a such that $g(a) = 75$.

94. Find the value of k such that the graph of $5y - kx = 7$ and the line containing the points $(7, -3)$ and $(-2, 5)$ are parallel.

95. Find the value of k such that the graph of $7y - kx = 9$ and the line containing the points $(2, -1)$ and $(-4, 5)$ are perpendicular.

96. Use a grapher with a *squared* window to check your answers to Exercises 61–74.

97. When several data points are available and they appear to be nearly collinear, a procedure known as *linear regression* can be used to find an equation for the line that most closely fits the data.

a) Use a grapher with a LINEAR REGRESSION option and the table that follows to find a linear function that predicts a woman's life expectancy as a function of the year in which she was born. Compare this with the answer to Exercise 41.

b) Predict the life expectancy in 2008 and compare your answer with the answer to Exercise 41. Which answer seems more reliable? Why?

Life Expectancy of Women

Year, x	Life Expectancy, y (in years)
1920	54.6
1930	61.6
1940	65.2
1950	71.1
1960	73.1
1970	74.7
1980	77.5
1990	78.8
1999	79.2

Sources: *Statistical Abstract of the United States* and *The World Almanac* 1999.

The Algebra of Functions

2.6

The Sum, Difference, Product, or Quotient of Two Functions • Domains and Graphs

We now return to the idea of a function as a machine and examine four ways in which functions can be combined.

The Sum, Difference, Product, or Quotient of Two Functions

Suppose that a is in the domain of two functions, f and g. The input a is paired with $f(a)$ by f and with $g(a)$ by g. The outputs can then be added to get $f(a) + g(a)$.

Example 1 Let $f(x) = x + 4$ and $g(x) = x^2 + 1$. Find $f(2) + g(2)$.

Solution We visualize two function machines. Because 2 is in the domain of each function, we can compute $f(2)$ and $g(2)$.

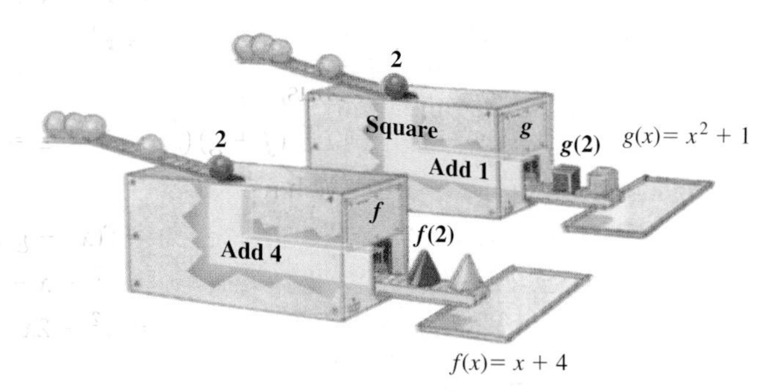

Since

$$f(2) = 2 + 4 = 6 \quad \text{and} \quad g(2) = 2^2 + 1 = 5,$$

we have

$$f(2) + g(2) = 6 + 5 = 11.$$

In Example 1, suppose that we were to write $f(x) + g(x)$ as $(x + 4) + (x^2 + 1)$, or $f(x) + g(x) = x^2 + x + 5$. This could then be regarded as a "new" function. The notation $(f + g)(x)$ is generally used to denote a function formed in this manner. Similar notations exist for subtraction, multiplication, and division of functions.

The Algebra of Functions

If f and g are functions and x is in the domain of both functions, then:

1. $(f+g)(x) = f(x) + g(x)$;
2. $(f-g)(x) = f(x) - g(x)$;
3. $(f \cdot g)(x) = f(x) \cdot g(x)$;
4. $(f/g)(x) = f(x)/g(x)$, provided $g(x) \neq 0$.

Example 2

For $f(x) = x^2 - x$ and $g(x) = x + 2$, find the following.

a) $(f+g)(3)$
b) $(f-g)(x)$ and $(f-g)(-1)$
c) $(f/g)(x)$ and $(f/g)(-4)$
d) $(f \cdot g)(3)$

Solution

a) Since $f(3) = 3^2 - 3 = 6$ and $g(3) = 3 + 2 = 5$, we have

$$\begin{aligned}(f+g)(3) &= f(3) + g(3)\\ &= 6 + 5 \qquad \text{Substituting}\\ &= 11.\end{aligned}$$

Alternatively, we could first find $(f+g)(x)$:

$$\begin{aligned}(f+g)(x) &= f(x) + g(x)\\ &= x^2 - x + x + 2\\ &= x^2 + 2. \qquad \text{Combining like terms}\end{aligned}$$

Thus,

$$(f+g)(3) = 3^2 + 2 = 11. \qquad \text{Our results match.}$$

b) We have

$$\begin{aligned}(f-g)(x) &= f(x) - g(x)\\ &= x^2 - x - (x+2) \qquad \text{Substituting}\\ &= x^2 - 2x - 2. \qquad \text{Removing parentheses and combining like terms}\end{aligned}$$

Thus,

$$\begin{aligned}(f-g)(-1) &= (-1)^2 - 2(-1) - 2 \qquad \text{Using } (f-g)(x) \text{ is faster than using } f(x) - g(x).\\ &= 1. \qquad \text{Simplifying}\end{aligned}$$

c) We have

$$\begin{aligned}(f/g)(x) &= f(x)/g(x)\\ &= \frac{x^2 - x}{x + 2}. \qquad \text{We assume that } x \neq -2.\end{aligned}$$

Thus,

$$\begin{aligned}(f/g)(-4) &= \frac{(-4)^2 - (-4)}{-4 + 2} \qquad \text{Substituting}\\ &= \frac{20}{-2} = -10.\end{aligned}$$

d) Using our work in part (a), we have

$$\begin{aligned}(f \cdot g)(3) &= f(3) \cdot g(3)\\ &= 6 \cdot 5\\ &= 30.\end{aligned}$$

It is also possible to compute $(f \cdot g)(3)$ by first multiplying $x^2 - x$ and $x + 2$ using methods we will discuss in Chapter 5.

Domains and Graphs

Although applications involving products and quotients of functions rarely appear in newspapers, situations involving sums or differences of functions often do appear in print. For example, the following graphs are similar to those published by the California Department of Education to promote breakfast programs in which students eat a balanced meal of fruit or juice, toast or cereal, and 2% or whole milk. The combination of carbohydrate, protein, and fat gives a sustained release of energy, delaying the onset of hunger for several hours.

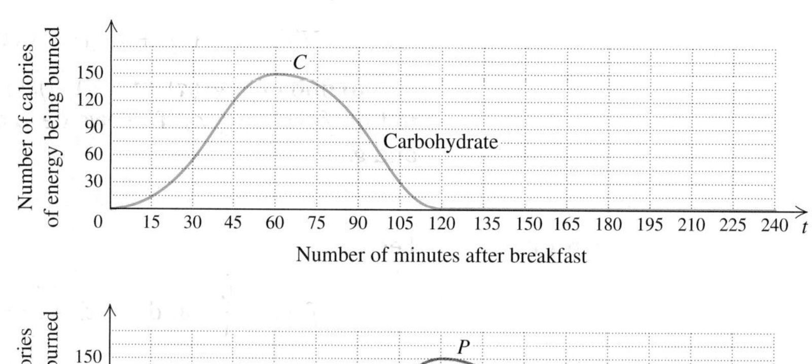

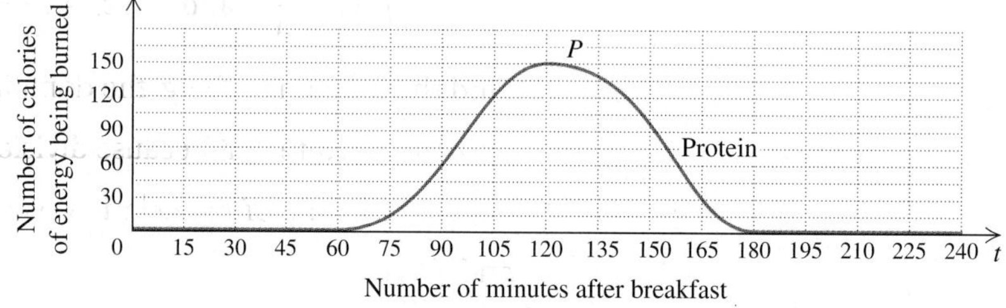

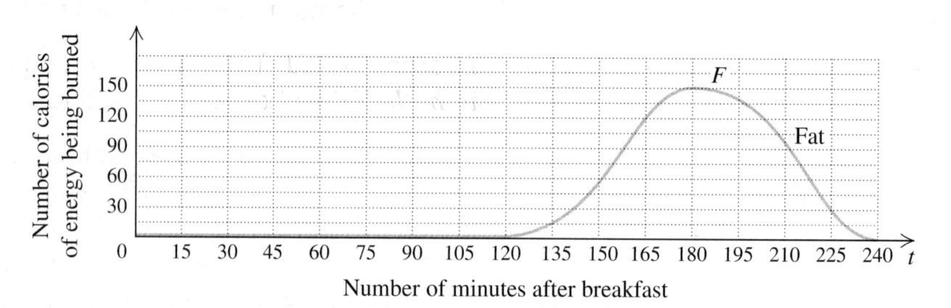

When the three graphs are superimposed, and the calorie expenditures added, it becomes clear that a balanced meal results in a steady, sustained supply of energy.

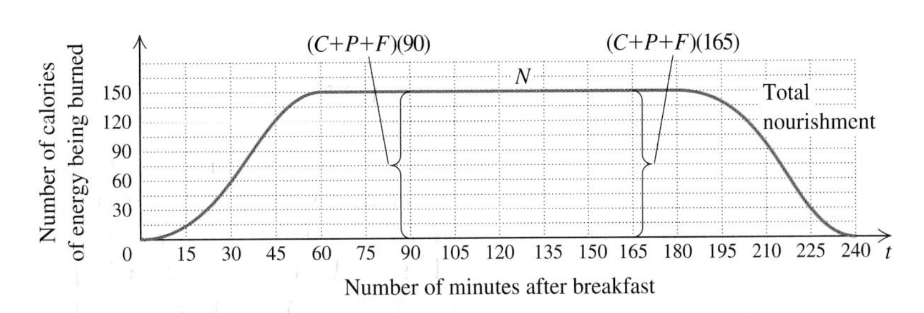

Note that for $t > 120$, we have $C(t) = 0$; for $t < 60$ or $t > 180$, we have $P(t) = 0$; and for $t < 120$, we have $F(t) = 0$. For any point on this last graph, we have

$$N(t) = (C + P + F)(t) = C(t) + P(t) + F(t).$$

To find $(f + g)(a)$, $(f - g)(a)$, $(f \cdot g)(a)$, or $(f/g)(a)$, we must first be able to find $f(a)$ and $g(a)$. Thus we need to ensure that a is in the domain of both f and g.

Example 3 Let

$$f(x) = \frac{5}{x} \quad \text{and} \quad g(x) = \frac{2x - 6}{x + 1}.$$

Find the domain of $f + g$, the domain of $f - g$, and the domain of $f \cdot g$.

Solution Note that because division by 0 is undefined, we have

$$\text{Domain of } f = \{x \mid x \text{ is a real number } \textit{and } x \neq 0\}$$

and

$$\text{Domain of } g = \{x \mid x \text{ is a real number } \textit{and } x \neq -1\}.$$

In order to find $f(a) + g(a)$, $f(a) - g(a)$, or $f(a) \cdot g(a)$, we must know that a is in *both* of the above domains. Thus,

$$\begin{aligned}\text{Domain of } f + g &= \text{Domain of } f - g = \text{Domain of } f \cdot g \\ &= \{x \mid x \text{ is a real number } \textit{and } x \neq 0 \textit{ and } x \neq -1\}.\end{aligned}$$

Suppose in Example 3 that we want to find $(f/g)(3)$. Finding $f(3)$ and $g(3)$ poses no problem:

$$f(3) = \frac{5}{3} \quad \text{and} \quad g(3) = \frac{2 \cdot 3 - 6}{3 + 1} = 0;$$

but then

$$\begin{aligned}(f/g)(3) &= f(3)/g(3)\\ &= \tfrac{5}{3}\ /\ 0\ . \qquad \text{Division by 0 is undefined.}\end{aligned}$$

Thus, although 3 is in the domain of both f and g, it is not in the domain of f/g.

Determining the Domain

The domain of $f + g$, $f - g$, or $f \cdot g$ is the set of all values common to the domains of f and g.

The domain of f/g is the set of all values common to the domains of f and g, excluding any values for which $g(x)$ is 0.

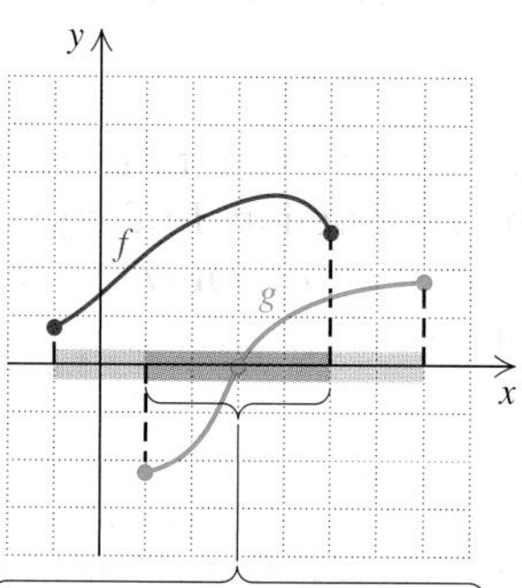

Domain of $f + g, f - g$, and $f \cdot g$

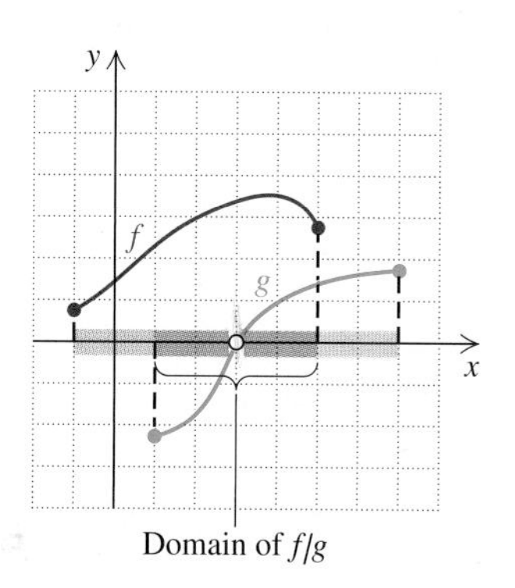

Domain of f/g

Example 4 Given $f(x) = 1/x$ and $g(x) = 2x - 7$, find the domains of $f + g$, $f - g$, $f \cdot g$, and f/g.

Solution The domain of f is $\{x | x \neq 0\}$ or $\{x | x \text{ is a real number } \textit{and } x \neq 0\}$. The domain of g is $\mathbb{R}$. Thus the domains of $f + g$, $f - g$, and $f \cdot g$ are the set of all elements common to both the domain of f and the domain of g. We have

$$\begin{aligned}\text{the domain of } f + g = \text{the domain of } f - g &= \text{the domain of } f \cdot g\\ &= \{x | x \text{ is a real number } \textit{and } x \neq 0\}.\end{aligned}$$

The domain of f/g is $\{x | x \text{ is a real number } \textit{and } x \neq 0\}$, with the additional restriction that $g(x) \neq 0$. To determine what x-values would make $g(x) = 0$, we solve:

$$\begin{aligned}2x - 7 &= 0 \qquad \text{Replacing } g(x) \text{ with } 2x - 7\\ 2x &= 7\\ x &= \tfrac{7}{2}.\end{aligned}$$

Since $g(x) = 0$ for $x = \frac{7}{2}$,

$$\text{the domain of } f/g = \left\{x | x \text{ is a real number } \textit{and } x \neq 0 \textit{ and } x \neq \tfrac{7}{2}\right\}.$$

A partial check of Example 4 can be performed by setting up a table so the TABLE MINIMUM is 0 and the increment of change (△Tbl) is 0.7. (Other choices, like 0.1, will also work.) Next, we let $y_1 = 1/x$ and $y_2 = 2x - 7$. Using the Y-VARS key to write $y_3 = y_1 + y_2$ and $y_4 = y_1/y_2$, we can create the table of values shown here. Note that when x is 3.5, a value for y_3 can be found, but y_4 is undefined. When setting up these functions, you may wish to enter y_1 and y_2 without "selecting" either. Otherwise, the table's columns must be scrolled to display y_3 and y_4.

X	Y3	Y4
0	ERROR	ERROR
.7	−4.171	−.2551
1.4	−3.486	−.1701
2.1	−2.324	−.1701
2.8	−1.043	−.2551
3.5	.28571	ERROR
4.2	1.6381	.17007
X = 0		

Use a similar approach to partially check Example 3.

Division by 0 is not the only condition that can force restrictions on the domain of a function. In Chapter 7, we will examine functions similar to that given by $f(x) = \sqrt{x}$, for which the concern is taking the square root of a negative number.

Exercise Set 2.6

FOR EXTRA HELP

Digital Video Tutor CD 2
Videotape 4

InterAct Math

Math Tutor Center

MathXL

MyMathLab.com

Let $f(x) = -3x + 1$ and $g(x) = x^2 + 2$. Find the following.

1. $f(2) + g(2)$

2. $f(-1) + g(-1)$

3. $f(5) - g(5)$

4. $f(4) - g(4)$

5. $f(-1) \cdot g(-1)$

6. $f(-2) \cdot g(-2)$

7. $f(-4)/g(-4)$

8. $f(3)/g(3)$

9. $g(1) - f(1)$

10. $g(2)/f(2)$

11. $(f + g)(x)$

12. $(g - f)(x)$

Let $F(x) = x^2 - 2$ and $G(x) = 5 - x$. Find the following.

13. $(F + G)(x)$

14. $(F + G)(a)$

15. $(F + G)(-4)$

16. $(F + G)(-5)$

17. $(F - G)(3)$

18. $(F - G)(2)$

19. $(F \cdot G)(-3)$

20. $(F \cdot G)(-4)$

21. $(F/G)(x)$

22. $(G - F)(x)$

23. $(F/G)(-2)$

24. $(F/G)(-1)$

The following graph shows the number of women, in millions, who had a child in the last year. Here $W(t)$ represents the number of women under 30 who gave birth in year t, $R(t)$ the number of women 30 and older who gave birth in year t, and $N(t)$ the total number of women who gave birth in year t.

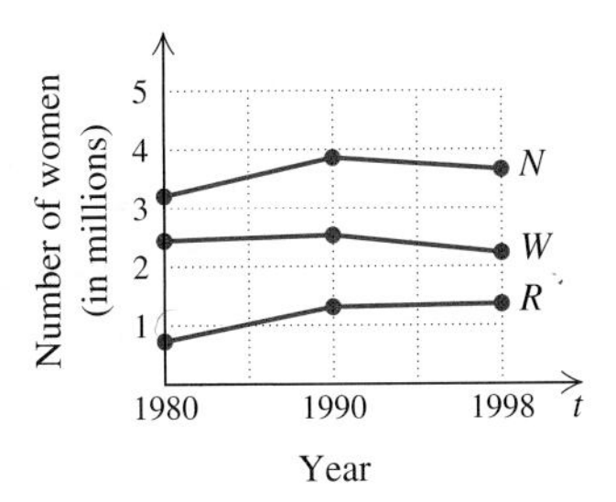

25. Use estimates of $R(1980)$ and $W(1980)$ to estimate $N(1980)$.

26. Use estimates of $R(1990)$ and $W(1990)$ to estimate $N(1990)$.

27. Which group of women was responsible for the drop in the number of births from 1990 to 1998?

28. Which group of women was responsible for the rise in the number of births from 1980 to 1990?

Often function addition is represented by stacking the individual functions directly on top of each other. The graph below indicates how the three major airports servicing New York City have been utilized. The braces indicate the values of the individual functions.

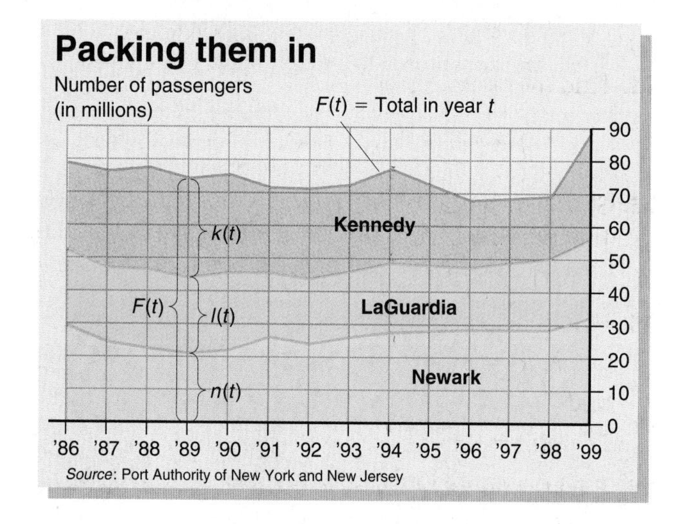

29. Estimate $(n + l)('98)$. What does it represent?

30. Estimate $(k + l)('98)$. What does it represent?

31. Estimate $(k - l)('94)$. What does it represent?

32. Estimate $(k - n)('94)$. What does it represent?

33. Estimate $(n + l + k)('99)$. What does it represent?

34. Estimate $(n + l + k)('98)$. What does it represent?

For each pair of functions f and g, determine the domain of the sum, difference, and product of the two functions.

35. $f(x) = x^2$, $g(x) = 7x - 4$

36. $f(x) = 5x - 1$, $g(x) = 2x^2$

37. $f(x) = \dfrac{1}{x - 3}$, $g(x) = 4x^3$

38. $f(x) = 3x^2$, $g(x) = \dfrac{1}{x - 9}$

39. $f(x) = \dfrac{2}{x}$, $g(x) = x^2 - 4$

40. $f(x) = x^3 + 1$, $g(x) = \dfrac{5}{x}$

41. $f(x) = x + \dfrac{2}{x - 1}$, $g(x) = 3x^3$

42. $f(x) = 9 - x^2$, $g(x) = \dfrac{3}{x - 6} + 2x$

43. $f(x) = \dfrac{3}{x - 2}$, $g(x) = \dfrac{5}{4 - x}$

44. $f(x) = \dfrac{5}{x - 3}$, $g(x) = \dfrac{1}{x - 2}$

For each pair of functions f and g, determine the domain of f/g.

45. $f(x) = x^4$, $g(x) = x - 3$

46. $f(x) = 2x^3$, $g(x) = 5 - x$

47. $f(x) = 3x - 2$, $g(x) = 2x - 8$

48. $f(x) = 5 + x$, $g(x) = 6 - 2x$

49. $f(x) = \dfrac{3}{x - 4}$, $g(x) = 5 - x$

50. $f(x) = \dfrac{1}{2 - x}$, $g(x) = 7 - x$

51. $f(x) = \dfrac{2x}{x + 1}$, $g(x) = 2x + 5$

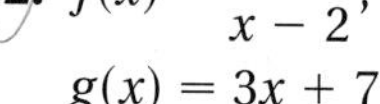

52. $f(x) = \dfrac{7x}{x - 2}$, $g(x) = 3x + 7$

For Exercises 53–60, consider the functions F and G as shown.

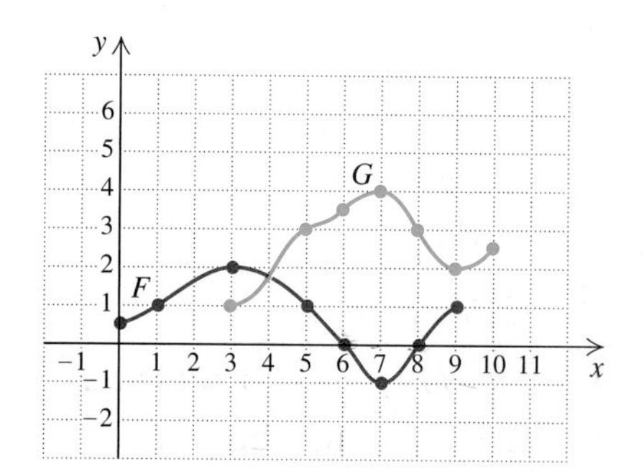

53. Determine $(F + G)(5)$ and $(F + G)(7)$.

54. Determine $(F \cdot G)(6)$ and $(F \cdot G)(9)$.

55. Determine $(G - F)(7)$ and $(G - F)(3)$.

56. Determine $(F/G)(3)$ and $(F/G)(7)$.

57. Find the domains of F, G, $F + G$, and F/G.

58. Find the domains of $F - G$, $F \cdot G$, and G/F.

59. Graph $F + G$.

60. Graph $G - F$.

*In the following graph, W(t) represents the number of gallons of whole milk, L(t) the number of gallons of low-fat milk, and S(t) the number of gallons of skim milk consumed by the average American in year t.**

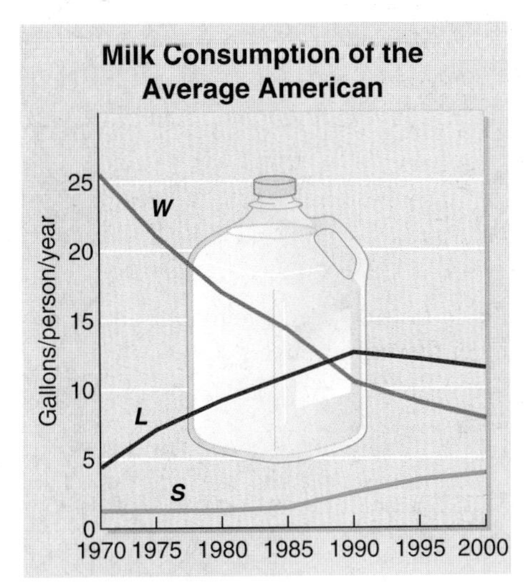

61. Explain in words what $(W - S)(t)$ represents and what it would mean to have $(W - S)(t) < 0$.

62. Consider $(W + L + S)(t)$ and explain why you feel that total milk consumption per person did or did not change over the years 1970–1998.

SKILL MAINTENANCE

Solve.

63. $4x - 7y = 8$, for x

64. $3x - 8y = 5$, for y

65. $5x + 2y = -3$, for y

66. $6x + 5y = -2$, for x

Translate each of the following. Do not solve.

67. Five more than twice a number is 49.

68. Three less than half of some number is 57.

69. The sum of two consecutive integers is 145.

70. The difference between a number and its opposite is 20.

*Sources: Copyright 1990, CSPI. Adapted from *Nutrition Action Healthletter* (1875 Connecticut Avenue, N.W., Suite 300, Washington, DC 20009-5728. $24.00 for 10 issues); USDA Agricultural Fact Book 2000, USDA Economic Research Service.

SYNTHESIS

71. If $f(x) = c$, where c is some positive constant, describe how the graphs of $y = g(x)$ and $y = (f + g)(x)$ will differ.

72. Examine the graphs following Example 2 and explain how they might be modified to represent the absorption of 200 mg of Advil® taken four times a day.

73. Find the domain of f/g, if

$$f(x) = \frac{3x}{2x + 5} \quad \text{and} \quad g(x) = \frac{x^4 - 1}{3x + 9}.$$

74. Find the domain of F/G, if

$$F(x) = \frac{1}{x - 4} \quad \text{and} \quad G(x) = \frac{x^2 - 4}{x - 3}.$$

75. Sketch the graph of two functions f and g such that the domain of f/g is

$$\{x \mid -2 \le x \le 3 \text{ and } x \ne 1\}.$$

76. Find the domain of m/n, if

$$m(x) = 3x \text{ for } -1 < x < 5$$

and

$$n(x) = 2x - 3.$$

77. Find the domains of $f + g$, $f - g$, $f \cdot g$, and f/g, if

$$f = \{(-2, 1), (-1, 2), (0, 3), (1, 4), (2, 5)\}$$

and

$$g = \{(-4, 4), (-3, 3), (-2, 4), (-1, 0), (0, 5), (1, 6)\}.$$

78. For f and g as defined in Exercise 77, find $(f + g)(-2)$, $(f \cdot g)(0)$, and $(f/g)(1)$.

79. Write equations for two functions f and g such that the domain of $f + g$ is

$$\{x \mid x \text{ is a real number and } x \ne -2 \text{ and } x \ne 5\}.$$

80. Let $y_1 = 2.5x + 1.5$, $y_2 = x - 3$, and $y_3 = y_1/y_2$. Depending on whether the CONNECTED or DOT mode is used, the graph of y_3 appears as follows.

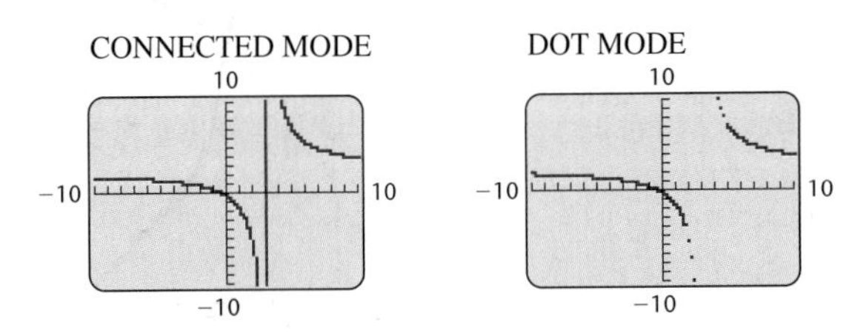

Use algebra to determine which graph more accurately represents y_3.

81. Using the window $[-5, 5, -1, 9]$, graph $y_1 = 5$, $y_2 = x + 2$, and $y_3 = \sqrt{x}$. Then predict what shape the graphs of $y_1 + y_2$, $y_1 + y_3$, and $y_2 + y_3$ will take. Use a grapher to check each prediction.

82. Use the TABLE feature on a grapher to check your answers to Exercises 37, 43, 45, and 51. (See the Technology Connection on p. 131.)

COLLABORATIVE CORNER

Time On Your Hands

Focus: The algebra of functions

Time: 10–15 minutes

Group size: 2–3

The graph and data at right chart the average retirement age $R(x)$ and life expectancy $E(x)$ of U.S. citizens in year x.

ACTIVITY

1. Working as a team, perform the appropriate calculations and then graph $E - R$.
2. What does $(E - R)(x)$ represent? In what fields of study or business might the function $E - R$ prove useful?
3. Should E and R really be calculated separately for men and women? Why or why not?
4. What advice would you give to someone considering early retirement?

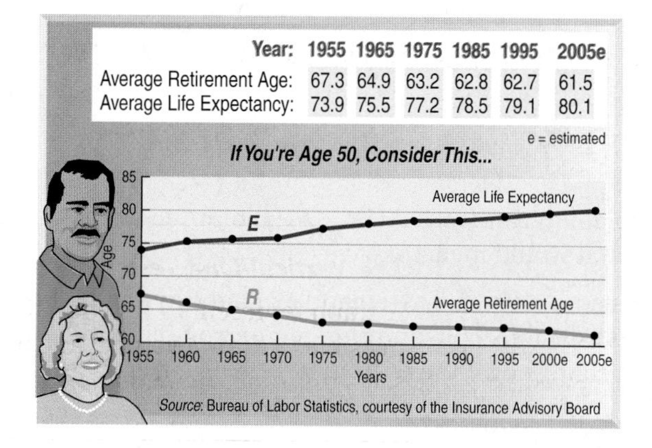

Year:	1955	1965	1975	1985	1995	2005e
Average Retirement Age:	67.3	64.9	63.2	62.8	62.7	61.5
Average Life Expectancy:	73.9	75.5	77.2	78.5	79.1	80.1

Summary and Review 2

Key Terms

Axes, axis, p. 70
x, y-coordinate system, p. 70
Cartesian coordinate system, p. 70
Ordered pair, p. 71
Origin, p. 71
Coordinates, p. 71
Quadrant, p. 71
Graph, p. 72
Linear equation, p. 74
Nonlinear equation, p. 74
Domain, p. 78
Range, p. 78
Function, p. 78
Projection, p. 81
Relation, p. 83
Input, p. 83
Output, p. 83
Dependent variable, p. 84
Independent variable, p. 84
Constant function, p. 84
Dummy variable, p. 84
Interpolation, p. 85
Extrapolation, p. 85
Linear function, p. 93
y-intercept, p. 94
Slope, p. 95
Rise, p. 95
Run, p. 95
Rate of change, p. 98
Salvage value, p. 100
Zero slope, p. 106
Undefined slope, p. 107
x-intercept, p. 108
Growth rate, p. 119

Important Properties and Formulas

The Vertical-Line Test

A graph represents a function if it is not possible to draw a vertical line that intersects the graph more than once.

$$\text{Slope} = m = \frac{\text{rise}}{\text{run}} = \frac{\text{change in } y}{\text{change in } x} = \frac{y_2 - y_1}{x_2 - x_1}$$

Every horizontal line has a slope of 0.
The slope of a vertical line is undefined.

The x-intercept is $(a, 0)$. To find a, let $y = 0$ and solve the original equation for x.

The y-intercept is $(0, b)$. To find b, let $x = 0$ and solve the original equation for y.

The slope–intercept equation of a line is

$$y = mx + b.$$

The point–slope equation of a line is

$$y - y_1 = m(x - x_1).$$

The standard form of a linear equation is

$$Ax + By = C.$$

Parallel lines: The slopes are equal.
Perpendicular lines: The product of the slopes is -1.

The Algebra of Functions

1. $(f + g)(x) = f(x) + g(x)$
2. $(f - g)(x) = f(x) - g(x)$
3. $(f \cdot g)(x) = f(x) \cdot g(x)$
4. $(f/g)(x) = f(x)/g(x)$, provided $g(x) \neq 0$

Review Exercises

Determine whether the ordered pair is a solution of the given equation.

1. $(3, 7)$, $4p - q = 5$

2. $(-2, 4)$, $x = 2y + 12$

3. $\left(0, -\frac{1}{2}\right)$, $3a - 4b = 2$

4. $(8, -2)$, $3c + 2d = 28$

Graph.

5. $y = -3x + 2$

6. $y = -x^2 + 1$

7. $8x + 32 = 0$

8. $y - 2 = 4$

9. For the following graph of f, determine **(a)** $f(2)$; **(b)** the domain of f; **(c)** any x-values for which $f(x) = 2$; and **(d)** the range of f.

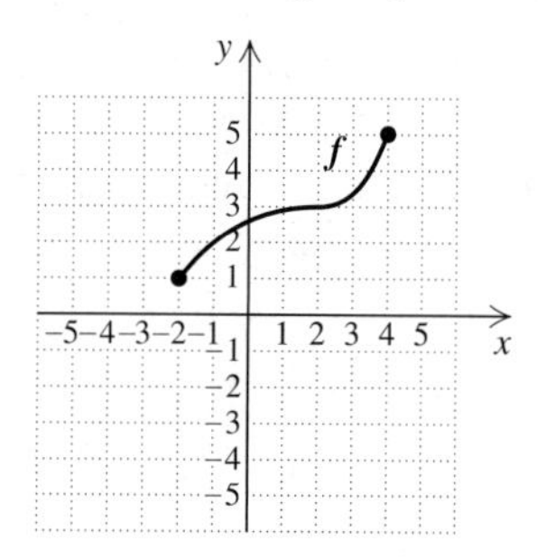

10. The function $A(t) = 0.233t + 5.87$ can be used to estimate the median age of cars in the United States t years after 1990 (*Source*: The Polk Co.). (In this context, a median age of 3 yr means that half the cars are more than 3 yr old and half are less.) Predict the median age of cars in 2010; that is, find $A(20)$.

Find the slope and the y-intercept.

11. $g(x) = -4x - 9$

12. $-6y + 2x = 7$

The following table shows the annual soft-drink production in the United States, in number of 12-oz cans per person (Sources: National Soft Drink Association; *Beverage World).*

Input, Year	Output, Number of 12-oz Cans per Person
1977	360
1987	480
1997	580

13. Use the data in the table at the bottom of the first column to draw a graph and to estimate the annual soft-drink production in 1990.

14. Use the graph from Exercise 13 to estimate the annual soft-drink production in 2005.

Find the slope of each line. If the slope is undefined, state this.

15. Containing the points $(4, 5)$ and $(-3, 1)$

16. Containing $(-16.4, 2.8)$ and $(-16.4, 3.5)$

17. Find the rate of change for the graph below. Use appropriate units.

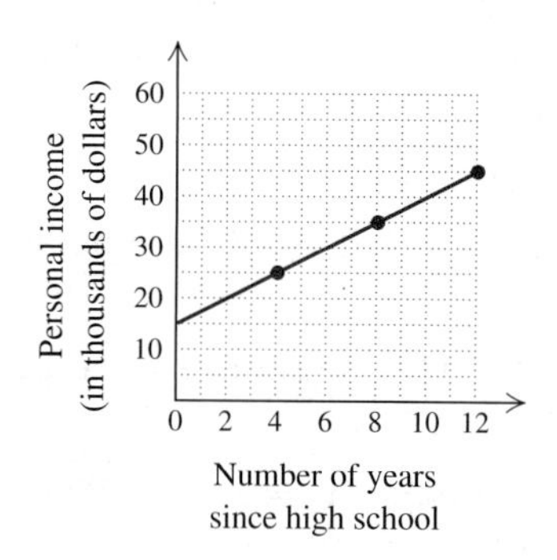

18. By March 1, 2000, U.S. builders had begun construction of 265,000 homes. By July 1, 2000, there were 795,000 begun (*Source*: U.S. Department of Commerce). Calculate the rate at which new homes were being started.

19. The average cost of tuition at a state university t years after 1997 can be estimated by $C(t) = 645t + 9800$. What do the numbers 645 and 9800 signify?

20. Find a linear function whose graph has slope $\frac{2}{7}$ and y-intercept $(0, -6)$.

21. Graph using intercepts: $-2x + 4y = 8$.

22. Solve $2 - x = 4 + x$ graphically. Then check your answer by solving the equation algebraically.

23. To join the Family Fitness Center, it costs \$75 plus \$15 a month. Use a graph to estimate the time required for the total cost to reach \$180.

Determine whether each of these is a linear equation.

24. $2x - 7 = 0$

25. $3x - 8f(x) = 7$

26. $2a + 7b^2 = 3$

27. $2p - \frac{7}{q} = 1$

28. Find an equation in point–slope form of the line with slope -2 and containing $(-3, 4)$.

29. Using function notation, write a slope–intercept equation for the line containing $(2, 5)$ and $(-4, -3)$.

Determine whether each pair of lines is parallel, perpendicular, or neither.

30. $y + 5 = -x,$
$x - y = 2$

31. $3x - 5 = 7y,$
$7y - 3x = 7$

32. In 1955, the U.S. minimum wage was \$0.75, and in 1997, it was \$5.15. Let W represent the minimum wage, in dollars, t years after 1955.

a) Find a linear function $W(t)$ that fits the data.

b) Use the function of part (a) to predict the minimum wage in 2005.

Find an equation of the line.

33. Containing the point $(2, -5)$ and parallel to the line $3x - 5y = 9$

34. Containing the point $(2, -5)$ and perpendicular to the line $3x - 5y = 9$

Let $g(x) = 3x - 6$ *and* $h(x) = x^2 + 1$. *Find the following.*

35. $g(0)$

36. $h(-5)$

37. $(g \cdot h)(4)$

38. $(g - h)(-2)$

39. $(g/h)(-1)$

40. $g(a + b)$

41. The domains of $g + h$ and $g \cdot h$

42. The domain of h/g

SYNTHESIS

43. Explain why every function is a relation, but not every relation is a function.

44. Explain why the slope of a vertical line is undefined whereas the slope of a horizontal line is 0.

45. Find the y-intercept of the function given by

$$f(x) + 3 = 0.17x^2 + (5 - 2x)^x - 7.$$

46. Determine the value of a such that the lines

$$3x - 4y = 12 \quad \text{and} \quad ax + 6y = -9$$

are parallel.

47. Homespun Jellies charges \$2.49 for each jar of preserves. Shipping charges are \$3.75 for handling, plus \$0.60 per jar. Find a linear function for determining the cost of shipping x jars of preserves.

48. Match each sentence with the most appropriate graph below.

a) Joni walks for 10 min to the train station, rides the train for 15 min, and then walks 5 min to the office.

b) During a workout, Phil bikes for 10 min, runs for 15 min, and then walks for 5 min.

c) Sam pilots his motorboat for 10 min to the middle of the lake, fishes for 15 min, and then motors for another 5 min to another spot.

d) Patti waits 10 min for her train, rides the train for 15 min, and then runs for 5 min to her job.

I

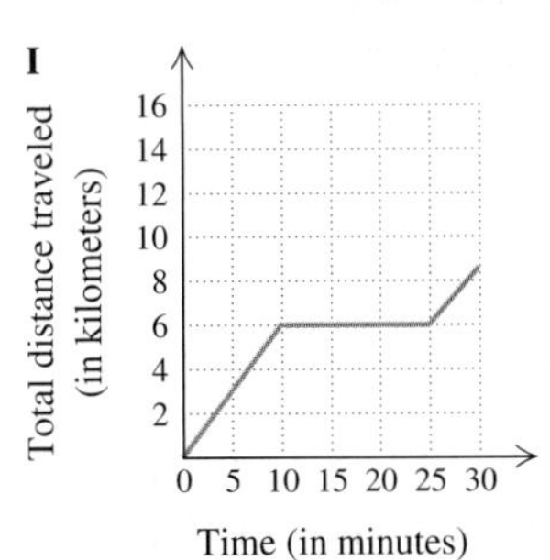

II

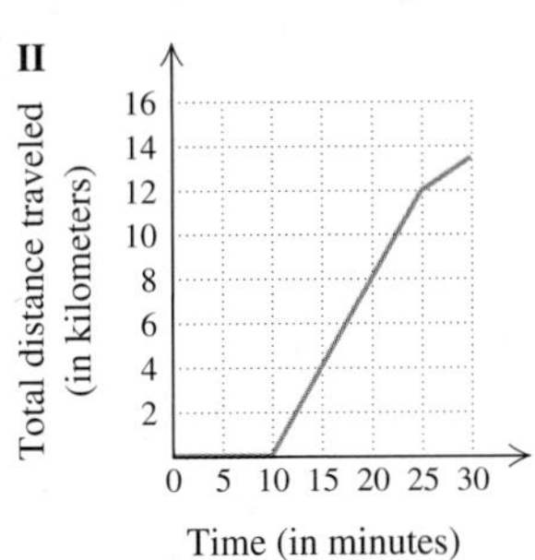

III

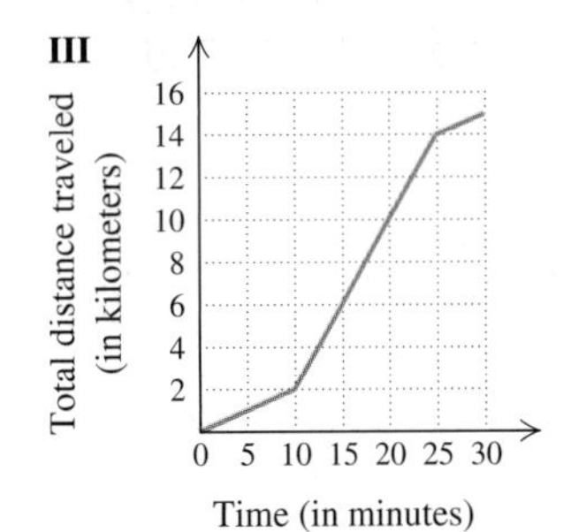

IV

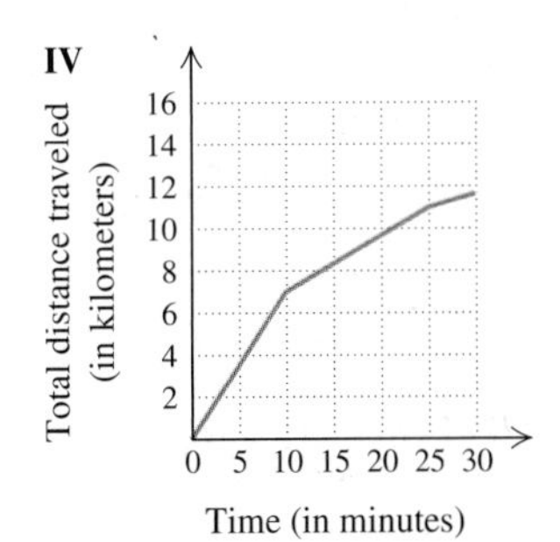

Chapter Test 2

Determine whether the ordered pair is a solution of the given equation.

1. $(0, -5)$, $x + 4y = -20$

2. $(1, -4)$, $-2p + 5q = 18$

Graph.

3. $y = -5x + 4$

4. $y = -2x^2 + 3$

5. $f(x) = 5$

6. $3 - x = 9$

7. For the following graph of f, determine **(a)** $f(-2)$; **(b)** the domain of f; **(c)** any x-value for which $f(x) = \frac{1}{2}$; and **(d)** the range of f.

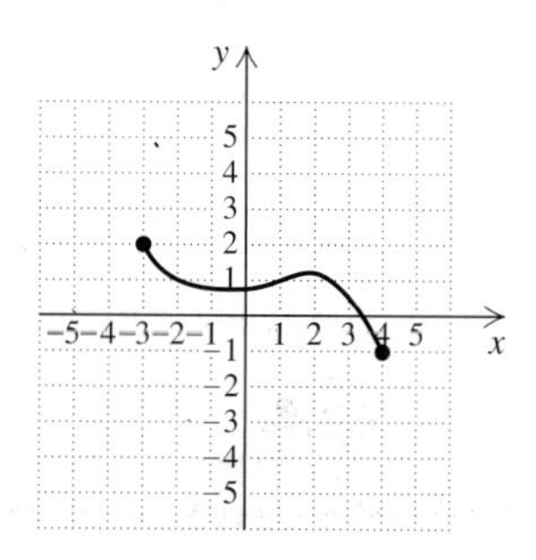

8. The function $S(t) = 1.2t + 21.4$ can be used to estimate the total U.S. sales of books, in billions of dollars, t years after 1992.

a) Predict the total U.S. sales of books in 2008.

b) What do the numbers 1.2 and 21.4 signify?

9. There were 43.3 million international visitors to the United States in 1995, and 48.5 million in 1999 (*Source*: Tourism Industries, International Trade Administration, Department of Commerce). Draw a graph and estimate the number of international visitors in 1997.

Find the slope and the y-intercept.

10. $f(x) = -\frac{3}{5}x + 12$

11. $-5y - 2x = 7$

Find the slope of the line containing the following points. If the slope is undefined, state this.

12. $(-2, -2)$ and $(6, 3)$

13. $(-3.1, 5.2)$ and $(-4.4, 5.2)$

14. Find the rate of change for the graph below. Use appropriate units.

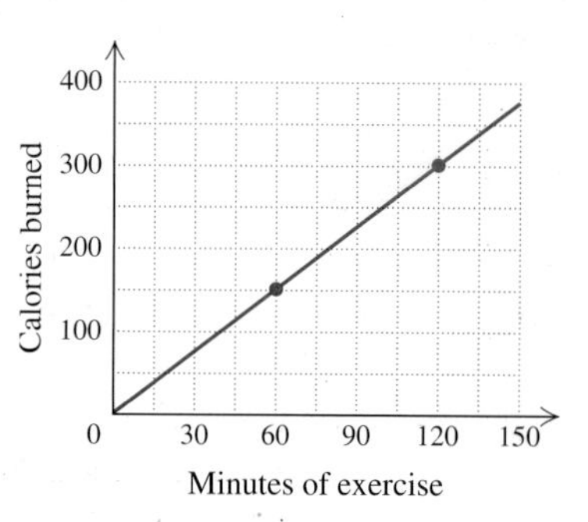

15. Find a linear function whose graph has slope -5 and y-intercept $(0, -1)$.

16. Graph using intercepts: $-2x + 5y = 12$.

17. Solve $x + 3 = 2x$ graphically. Then check your answer by solving the equation algebraically.

18. Which of these are linear equations?

a) $8x - 7 = 0$

b) $4b - 9a^2 = 2$

c) $2x - 5y = 3$

19. Find an equation in point–slope form of the line with slope 4 and containing $(-2, -4)$.

20. Using function notation, write a slope–intercept equation for the line containing $(3, -1)$ and $(4, -2)$.

Determine without graphing whether each pair of lines is parallel, perpendicular, or neither.

21. $4y + 2 = 3x$,
$-3x + 4y = -12$

22. $y = -2x + 5$,
$2y - x = 6$

Find an equation of the line.

23. Containing $(-3, 2)$ and parallel to the line $2x - 5y = 8$

24. Containing $(-3, 2)$ and perpendicular to the line $2x - 5y = 8$

25. Find the following, given that $g(x) = -3x - 4$ and $h(x) = x^2 + 1$.

a) $h(-2)$

b) $(g \cdot h)(3)$

c) The domain of h/g

26. If you rent a van for one day and drive it 250 mi, the cost is \$100. If you drive it 300 mi, the cost is \$115. Let $C(m)$ represent the cost, in dollars, of driving m miles.

a) Find a linear function that fits the data.

b) Use the function to find how much it will cost to rent the van for one day and drive it 500 mi.

SYNTHESIS

27. The function $f(t) = 5 + 15t$ can be used to determine a bicycle racer's location, in miles from the starting line, measured t hours after passing the 5-mi mark.

a) How far from the start will the racer be 1 hr and 40 min after passing the 5-mi mark?

b) Assuming a constant rate, how fast is the racer traveling?

28. The graph of the function $f(x) = mx + b$ contains the points $(r, 3)$ and $(7, s)$. Express s in terms of r if the graph is parallel to the line $3x - 2y = 7$.

29. Given that $f(x) = 5x^2 + 1$ and $g(x) = 4x - 3$, find an expression for $h(x)$ so that the domain of $f/g/h$ is

$$\{x \mid x \text{ is a real number } and\ x \neq \tfrac{3}{4}\ and\ x \neq \tfrac{2}{7}\}.$$

Answers may vary.

3 Systems of Linear Equations and Problem Solving

3.1 Systems of Equations in Two Variables
3.2 Solving by Substitution or Elimination
Connecting the Concepts
3.3 Solving Applications: Systems of Two Equations
3.4 Systems of Equations in Three Variables
Connecting the Concepts
3.5 Solving Applications: Systems of Three Equations
3.6 Elimination Using Matrices
3.7 Determinants and Cramer's Rule
3.8 Business and Economic Applications
Summary and Review
Test
Cumulative Review

AN APPLICATION

Industrial biochemists routinely use a machine to mix a buffer of 10% acetone by adding 100% acetone to water. One day, instead of adding 5 L of acetone to create a vat of buffer, a machine added 10 L. How much additional water was needed to bring the concentration down to 10%?

This problem appears as Exercise 63 in Section 3.3.

STEPHANI HARNICK
Environmental Engineer
McCordsville, IN

Chemical engineers often use math to calculate ratios in solutions and to evaluate formulas. As an environmental engineer, I need math to calculate air emissions to ensure acceptable air quality.

The most difficult part of problem solving is almost always translating the problem situation to mathematical language. Once a problem has been translated, the rest is usually straightforward. In this chapter, we study systems of equations *and how to solve them using graphing, substitution, elimination, and matrices. Systems of equations often provide the easiest way to model real-world situations in fields such as psychology, sociology, business, education, engineering, and science.*

3.1 Systems of Equations in Two Variables

Translating • Identifying Solutions • Solving Systems Graphically

Translating

Problems involving two unknown quantities are often solved most easily if we can first translate the situation to two equations in two unknowns.

Example 1

Real estate. Translate the following problem situation to mathematical language, using two equations.

In 1996, the Simon Property Group and the DeBartolo Realty Corporation merged to form the largest real estate company in the United States, owning 183 shopping centers in 32 states. Prior to merging, Simon owned twice as many properties as DeBartolo. How many properties did each company own before the merger?

DeBartolo shareholders would receive 0.68 share of Simon common stock for each share of DeBartolo common stock. Simon also would agree to repay $1.5 billion in DeBartolo debt. At Tuesday's closing price of $ 23.625 a share for common stock, the transaction is valued at roughly $3 billion.

Executives say the proposed company, Simon DeBartolo Group, would be the largest real estate company in the United States, worth $7.5 billion.

Not included in the deal: DeBartolo's ownership stake in the San Francisco 49ers, or the Indiana Pacers, owned separately by the Simon

Solution

1. **Familiarize.** We have already seen problems in which we need to look up certain formulas or the meaning of certain words. Here we need only observe that the words shopping centers and properties are being used interchangeably.

   Often problems contain information that has no bearing on the situation being discussed. In this case, the number 32 is irrelevant to the question being asked. Instead we focus on the number of properties owned and the phrase "twice as many." Rather than guess and check, let's proceed to the next step, using x for the number of properties originally owned by Simon and y for the number of properties originally owned by DeBartolo.

Study Tip

Don't be hesitant to ask questions in class at appropriate times. Most instructors welcome questions and encourage students to ask them. Other students in your class probably have the same questions you do.

2. **Translate.** There are two statements to translate. First we look at the total number of properties involved:

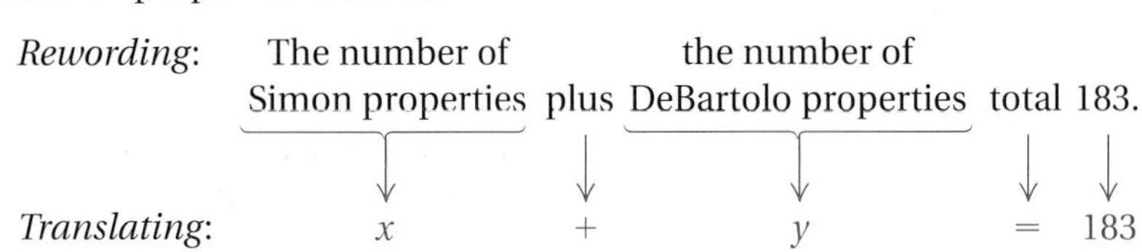

Rewording: The number of Simon properties plus the number of DeBartolo properties total 183.
Translating: x + y = 183

The second statement compares the number of properties that each company held before merging:

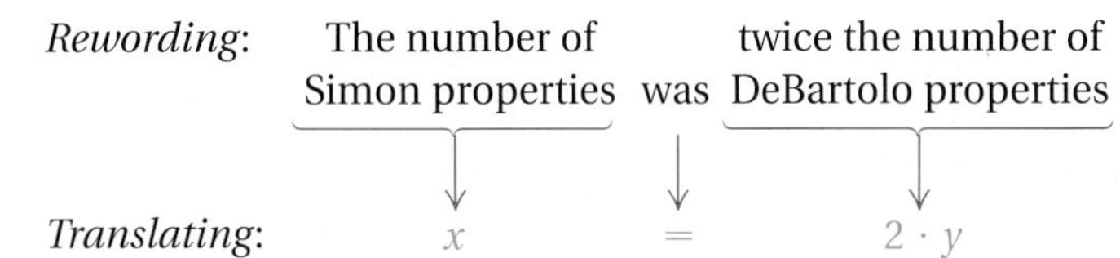

Rewording: The number of Simon properties was twice the number of DeBartolo properties
Translating: x = 2 · y

We have now translated the problem to a pair, or **system, of equations**:

$$x + y = 183,$$
$$x = 2y.$$

We will complete the solution of this problem in Section 3.3.

Problems like Example 1 *can* be solved using one variable; however, as problems become complicated, you will find that using more than one variable (and more than one equation) is often the preferable approach.

Example 2

Purchasing. Recently the Champlain Valley Community Music Center purchased 120 stamps for \$30.30. If the stamps were a combination of 20¢ postcard stamps and 34¢ first-class stamps, how many of each type were bought?

Solution

1. **Familiarize.** To familiarize ourselves with this problem, let's guess that the music center bought 60 stamps at 20¢ each and 60 stamps at 34¢ each. The total cost would then be

$$60 \cdot \$0.20 + 60 \cdot \$0.34 = \$12.00 + \$20.40, \text{ or } \$32.40.$$

Since $\$32.40 \neq \$30.30$, our guess is incorrect. Rather than guess again, let's see how algebra can be used to translate the problem.

2. **Translate.** We let p = the number of postcard stamps and f = the number of first-class stamps. The information can be organized in a table, which will help with the translating.

Type of Stamp	Postcard	First-class	Total	
Number Sold	p	f	120	→ $p + f = 120$
Price	\$0.20	\$0.34		
Amount	\$0.20p	\$0.34f	\$30.30	→ $0.20p + 0.34f = 30.30$

The first row of the table and the first sentence of the problem indicate that a total of 120 stamps were bought:

$$p + f = 120.$$

Since each postcard stamp cost \$0.20 and p stamps were bought, $0.20p$ represents the amount paid, in dollars, for the postcard stamps. Similarly, $0.34f$ represents the amount paid, in dollars, for the first-class stamps. This leads to a second equation:

$$0.20p + 0.34f = 30.30.$$

Multiplying both sides by 100, we can clear the decimals. This gives the following system of equations as the translation:

$$p + f = 120,$$
$$20p + 34f = 3030.$$

Identifying Solutions

A *solution* of a system of equations in two variables is an ordered pair of numbers that makes *both* equations true.

Example 3 Determine whether $(-4, 7)$ is a solution of the system

$$x + y = 3,$$
$$5x - y = -27.$$

Solution We use alphabetical order of the variables. Thus we replace x with -4 and y with 7:

$$\begin{array}{r|l} x + y & = 3 \\ \hline -4 + 7 & ?\ 3 \\ 3 & 3 \quad \text{TRUE} \end{array} \qquad \begin{array}{r|l} 5x - y & = -27 \\ \hline 5(-4) - 7 & ?\ -27 \\ -20 - 7 & \\ -27 & -27 \quad \text{TRUE} \end{array}$$

The pair $(-4, 7)$ makes both equations true, so it is a solution of the system. We can also describe the solution by writing $x = -4$ and $y = 7$. Set notation can also be used to list the solution set $\{(-4, 7)\}$.

Solving Systems Graphically

Recall that the graph of an equation is a drawing that represents its solution set. If we graph the equations in Example 3, we find that $(-4, 7)$ is the only point common to both lines. Thus one way to solve a system of two equations is to graph both equations and identify any points of intersection. The coordinates of each point of intersection represent a solution of that system.

$$x + y - 3,$$
$$5x - y = -27$$

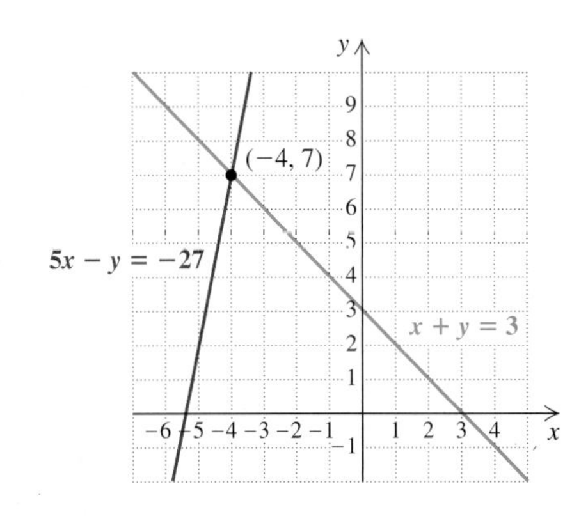

Most pairs of lines have exactly one point in common. We will soon see, however, that this is not always the case.

Example 4 Solve each system graphically.

a) $y - x = 1,$
$y + x = 3$

b) $y = -3x + 5,$
$y = -3x - 2$

c) $3y - 2x = 6,$
$-12y + 8x = -24$

Solution

a) We graph each equation using any method studied in Chapter 2. All ordered pairs from line L_1 are solutions of the first equation. All ordered pairs from line L_2 are solutions of the second equation. The point of intersection has coordinates that make *both* equations true. Apparently, (1, 2) is the solution. Graphing is not always accurate, so solving by graphing may yield approximate answers. Our check below shows that (1, 2) is indeed the solution.

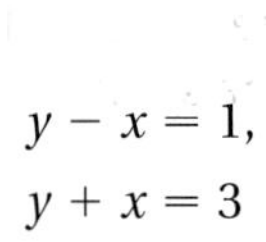

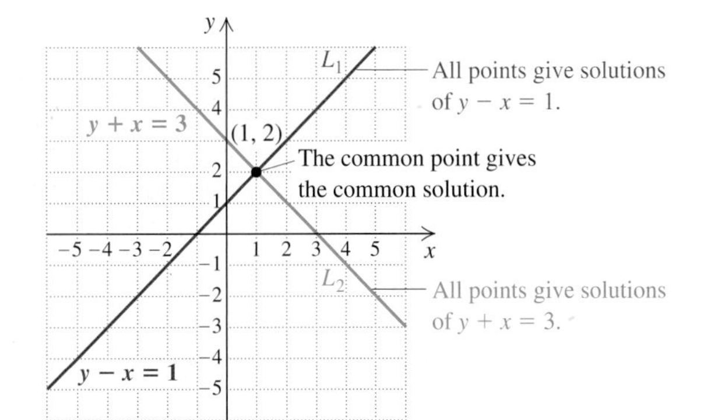

Check:

$y - x = 1$			$y + x = 3$		
$2 - 1$	?	1	$2 + 1$	?	3
1		1 TRUE	3		3 TRUE

b) We graph the equations. The lines have the same slope, -3, and different y-intercepts, so they are parallel. There is no point at which they cross, so the system has no solution.

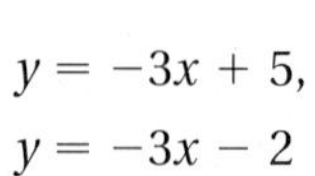

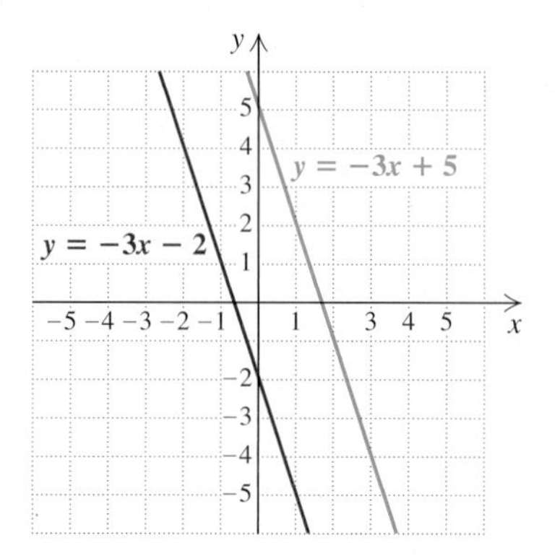

c) We graph the equations and find that the same line is drawn twice. Thus any solution of one equation is a solution of the other. Each equation has an infinite number of solutions, so the system itself has an infinite number of solutions. We check one solution, $(0, 2)$, which is the y-intercept of each equation.

$$3y - 2x = 6,$$
$$-12y + 8x = -24$$

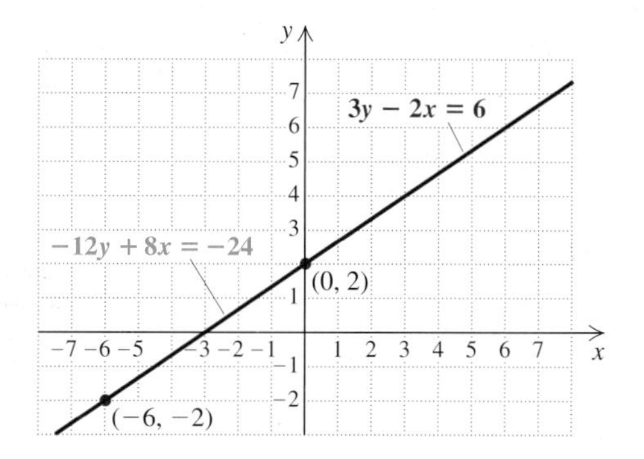

Check:

$$\begin{array}{c|c} 3y - 2x & 6 \\ \hline 3(2) - 2(0) & ?\ 6 \\ 6 - 0 & \\ 6 & 6 \end{array} \quad \text{TRUE}$$

$$\begin{array}{c|c} -12y + 8x & -24 \\ \hline -12(2) + 8(0) & ?\ -24 \\ -24 + 0 & \\ -24 & -24 \end{array} \quad \text{TRUE}$$

You can check that $(-6, -2)$ is another solution of both equations. In fact, any pair that is a solution of one equation is a solution of the other equation as well. Thus the solution set is

$$\{(x, y) \mid 3y - 2x = 6\}$$

or, in words, "the set of all pairs (x, y) for which $3y - 2x = 6$." Since the two equations are equivalent, we could have written instead $\{(x, y) \mid -12y + 8x = -24\}$.

On most graphers, an INTERSECT option allows you to find the coordinates of the intersection directly. This is especially useful when equations contain fractions or decimals or when the coordinates of the intersection are not integers. To illustrate, consider the following system:

$$3.45x + 4.21y = 8.39,$$
$$7.12x - 5.43y = 6.18.$$

Most graphers require equations to have y alone on one side. After solving for y in each equation, we can obtain the graph shown. Using INTERSECT, we see that, to the nearest hundredth, the coordinates of the intersection are (1.47, 0.79).

$y_1 = (8.39 - 3.45x)/4.21,$
$y_2 = (6.18 - 7.12x)/(-5.43)$

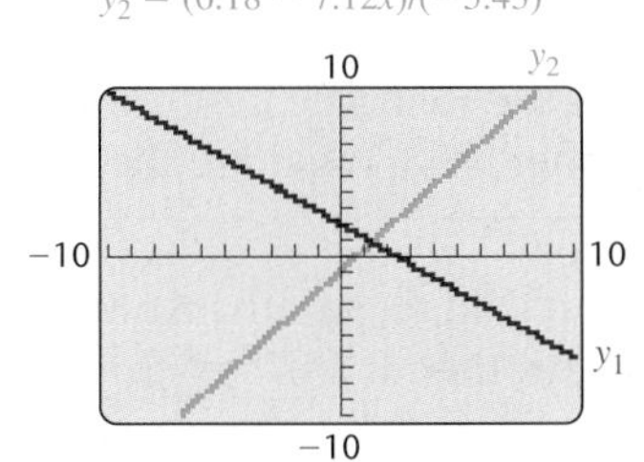

Use a grapher to solve each of the following systems. Make sure that all x- and y-coordinates are correct to the nearest hundredth.

1. $y = -5.43x + 10.89,$
   $y = 6.29x - 7.04$
2. $y = 123.52x + 89.32,$
   $y = -89.22x + 33.76$
3. $2.18x + 7.81y = 13.78,$
   $5.79x - 3.45y = 8.94$
4. $-9.25x - 12.94y = -3.88,$
   $21.83x + 16.33y = 13.69$

When we graph a system of two linear equations in two variables, one of the following three outcomes will occur.

1. The lines have one point in common, and that point is the only solution of the system (see Example 4a). Any system that has at least one solution is said to be **consistent.**
2. The lines are parallel, with no point in common, and the system has no solution (see Example 4b). This type of system is called **inconsistent**.
3. The lines coincide, sharing the same graph. Because every solution of one equation is a solution of the other, the system has an infinite number of solutions (see Example 4c). Since it has a solution, this type of system is consistent.

When one equation in a system can be obtained by multiplying both sides of another equation by a constant, the two equations are said to be **dependent**. Thus the equations in Example 4(c) are dependent, but those in Examples 4(a) and 4(b) are **independent**. For systems of three or more equations, the definitions of dependent and independent must be slightly modified.

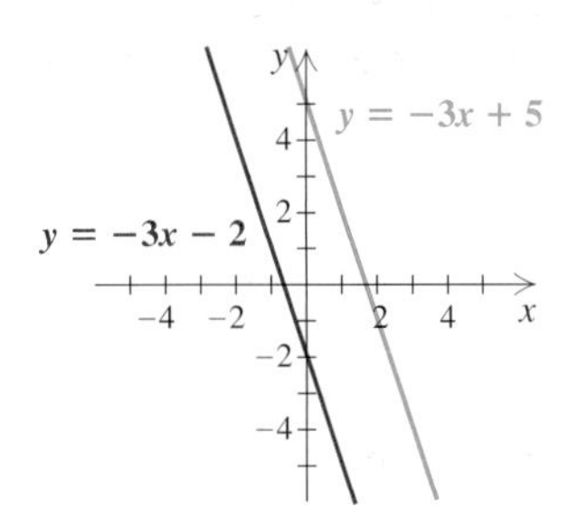

Graphs are parallel.
The system is *inconsistent* because there is no solution. Since the equations are not equivalent, they are *independent*.

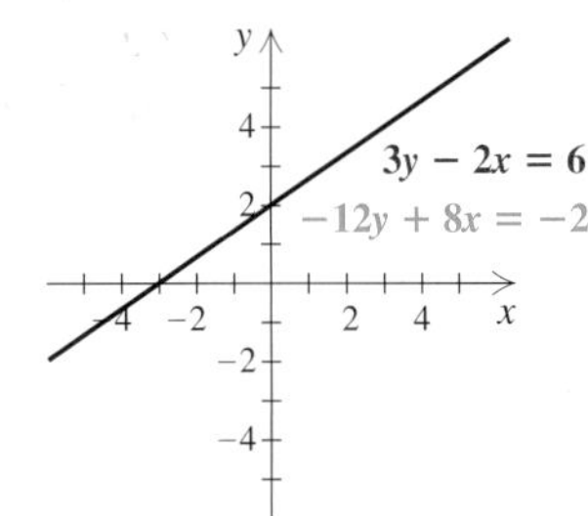

Equations have the same graph.
The system is *consistent* and has an infinite number of solutions. The equations are *dependent* since they are equivalent.

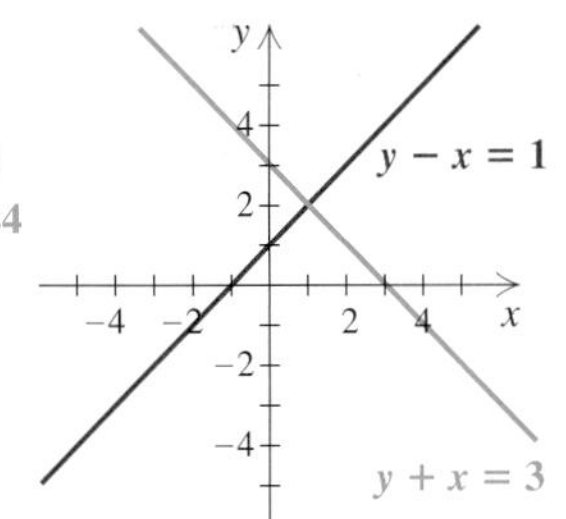

Graphs intersect at one point.
The system is *consistent* and has one solution. Since neither equation is a multiple of the other, they are *independent*.

Graphing is helpful when solving systems because it allows us to "see" the solution. It can also be used on systems of nonlinear equations, and in many applications, it provides a satisfactory answer. However, graphing often lacks precision, especially when fractional or decimal solutions are involved. In Section 3.2, we will develop two algebraic methods of solving systems. Both methods produce exact answers.

Exercise Set 3.1

FOR EXTRA HELP

Digital Video Tutor CD 2
Videotape 5

InterAct Math

Math Tutor Center

MathXL

MyMathLab.com

Determine whether the ordered pair is a solution of the given system of equations. Remember to use alphabetical order of variables.

1. $(1,2)$; $4x - y = 2$, $10x - 3y = 4$

2. $(-1,-2)$; $2x + y = -4$, $x - y = 1$

3. $(2,5)$; $y = 3x - 1$, $2x + y = 4$

4. $(-1,-2)$; $x + 3y = -7$, $3x - 2y = 12$

5. $(1,5)$; $x + y = 6$, $y = 2x + 3$

6. $(5,2)$; $a + b = 7$, $2a - 8 = b$

Aha! **7.** $(3,1)$; $3x + 4y = 13$, $6x + 8y = 26$

8. $(4,-2)$; $-3x - 2y = -8$, $8 = 3x + 2y$

Solve each system graphically. Be sure to check your solution. If a system has an infinite number of solutions, use set-builder notation to write the solution set. If a system has no solution, state this.

9. $x - y = 3$, $x + y = 5$

10. $x + y = 4$, $x - y = 2$

11. $3x + y = 5$, $x - 2y = 4$

12. $2x - y = 4$, $5x - y = 13$

13. $4y = x + 8$, $3x - 2y = 6$

14. $4x - y = 9$, $x - 3y = 16$

15. $x = y - 1$, $2x = 3y$

16. $a = 1 + b$, $b = 5 - 2a$

17. $x = -3$, $y = 2$

18. $x = 4$, $y = -5$

19. $t + 2s = -1$, $s = t + 10$

20. $b + 2a = 2$, $a = -3 - b$

21. $2b + a = 11$, $a - b = 5$

22. $y = -\frac{1}{3}x - 1$, $4x - 3y = 18$

23. $y = -\frac{1}{4}x + 1$, $2y = x - 4$

24. $6x - 2y = 2$, $9x - 3y = 1$

25. $y - x = 5$, $2x - 2y = 10$

26. $y = -x - 1$, $4x - 3y = 24$

27. $y = 3 - x$, $2x + 2y = 6$

28. $2x - 3y = 6$, $3y - 2x = -6$

29. For the systems in the odd-numbered exercises 9–27, which are consistent?

30. For the systems in the even-numbered exercises 10–28, which are consistent?

31. For the systems in the odd-numbered exercises 9–27, which contain dependent equations?

32. For the systems in the even-numbered exercises 10–28, which contain dependent equations?

Translate each problem situation to a system of equations. Do not attempt to solve, but save for later use.

33. The difference between two numbers is 11. Twice the smaller plus three times the larger is 123. What are the numbers?

34. The sum of two numbers is -42. The first number minus the second number is 52. What are the numbers?

35. *Retail sales.* Paint Town sold 45 paintbrushes, one kind at \$8.50 each and another at \$9.75 each. In all, \$398.75 was taken in for the brushes. How many of each kind were sold?

36. *Retail sales.* Mountainside Fleece sold 40 neckwarmers. Polarfleece neckwarmers sold for \$9.90 each and wool ones sold for \$12.75 each. In all, \$421.65 was taken in for the neckwarmers. How many of each type were sold?

37. *Geometry.* Two angles are supplementary.* One angle is 3° less than twice the other. Find the measures of the angles.

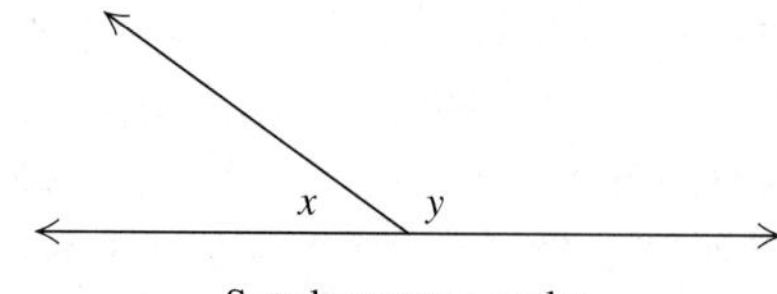

Supplementary angles

*The sum of the measures of two supplementary angles is 180°.

38. *Geometry.* Two angles are complementary.* The sum of the measures of the first angle and half the second angle is 64°. Find the measures of the angles.

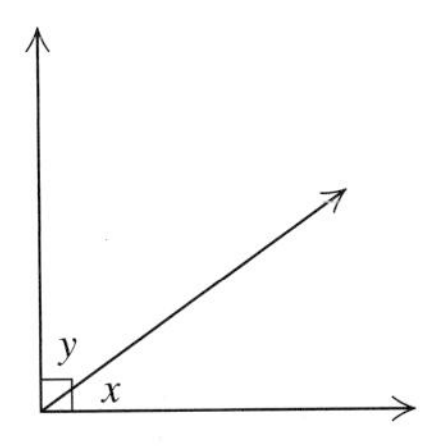

Complementary angles

39. *Basketball scoring.* Wilt Chamberlain once scored 100 points, setting a record for points scored in an NBA game. Chamberlain took only two-point shots and (one-point) foul shots and made a total of 64 shots. How many shots of each type did he make?

40. *Fundraising.* The St. Mark's Community Barbecue served 250 dinners. A child's plate cost $3.50 and an adult's plate cost $7.00. A total of $1347.50 was collected. How many of each type of plate was served?

41. *Sales of pharmaceuticals.* In 2001, the Diabetic Express charged $21.95 for a vial of Humulin insulin and $20.95 for a vial of Novolin insulin. If a total of $1077.50 was collected for 50 vials of insulin, how many vials of each type were sold?

42. *Court dimensions.* The perimeter of a standard basketball court is 288 ft. The length is 44 ft longer than the width. Find the dimensions.

$P = 288$ ft

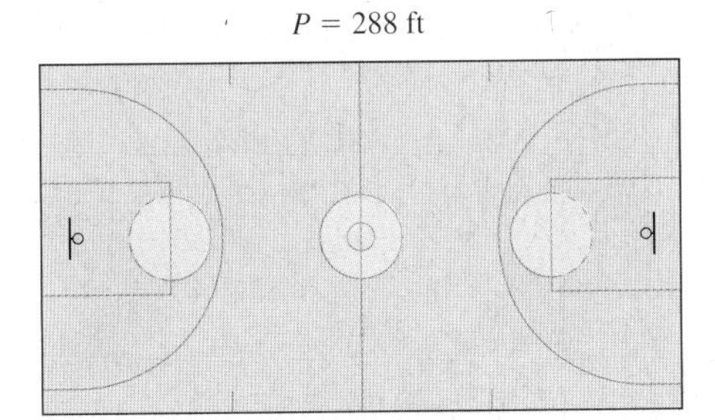

43. *Court dimensions.* The perimeter of a standard tennis court used for doubles is 228 ft. The width is 42 ft less than the length. Find the dimensions.

44. *Basketball scoring.* The Fenton College Cougars made 40 field goals in a recent basketball game, some 2-pointers and the rest 3-pointers. Altogether the 40 baskets counted for 89 points. How many of each type of field goal was made?

*The sum of the measures of two complementary angles is 90°.

45. *Hockey rankings.* Hockey teams receive 2 points for a win and 1 point for a tie. The Wildcats once won a championship with 60 points. They won 9 more games than they tied. How many wins and how many ties did the Wildcats have?

46. *Radio airplay.* Roscoe must play 12 commercials during his 1-hr radio show. Each commercial is either 30 sec or 60 sec long. If the total commercial time during that hour is 10 min, how many commercials of each type does Roscoe play?

47. *Nontoxic floor wax.* A nontoxic floor wax can be made from lemon juice and food-grade linseed oil. The amount of oil should be twice the amount of lemon juice. How much of each ingredient is needed to make 32 oz of floor wax? (The mix should be spread with a rag and buffed when dry.)

48. *Lumber production.* Denison Lumber can convert logs into either lumber or plywood. In a given day, the mill turns out 42 pallets of plywood and lumber. It makes a profit of $25 on a pallet of lumber and $40 on a pallet of plywood. How many pallets of each type must be produced and sold in order to make a profit of $1245?

49. *Video rentals.* J. P.'s Video rents general-interest films for $3.00 each and children's films for $1.50 each. In one day, a total of $213 was taken in from the rental of 77 videos. How many of each type of video was rented?

50. *Airplane seating.* An airplane has a total of 152 seats. The number of coach-class seats is 5 more than six times the number of first-class seats. How many of each type of seat are there on the plane?

51. Write a problem for a classmate to solve that requires writing a system of two equations. Devise the problem so that the solution is "The Lakers made 6 three-point baskets and 31 two-point baskets."

52. Write a problem for a classmate to solve that can be translated into a system of two equations. Devise the problem so that the solution is "Shelly gave 9 haircuts and 5 shampoos."

SKILL MAINTENANCE

Solve.

53. $2(4x - 3) - 7x = 9$

54. $6y - 3(5 - 2y) = 4$

55. $4x - 5x = 8x - 9 + 11x$

56. $8x - 2(5 - x) = 7x + 3$

Solve.

57. $3x + 4y = 7$, for y

58. $2x - 5y = 9$, for y

SYNTHESIS

Technology in U.S. schools. For Exercises 59–62, consider the following graph.

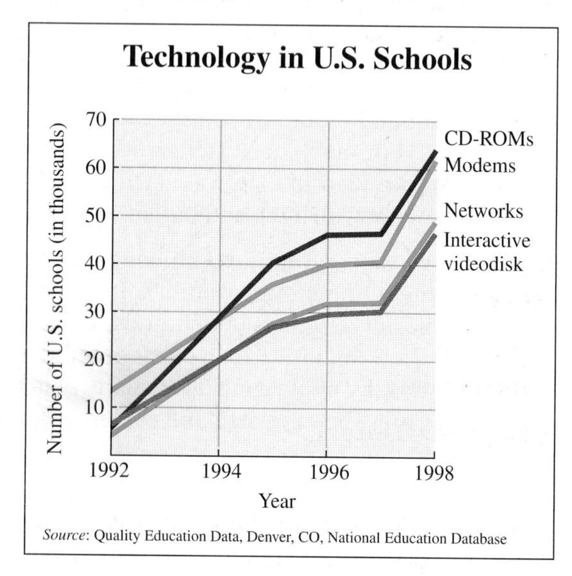

59. Is it accurate to state that there have always been more schools with CD-ROMs than with networks? Why or why not?

60. Is it accurate to state that there have always been more schools with networks than with interactive videodisks? Why or why not?

61. During which year did the number of schools with CD-ROMs first exceed the number of schools with modems?

62. During which year did the number of schools owning an interactive videodisk increase the most? At what rate did it increase?

63. For each of the following conditions, write a system of equations.

a) $(5, 1)$ is a solution.
b) There is no solution.
c) There is an infinite number of solutions.

64. A system of linear equations has $(1, -1)$ and $(-2, 3)$ as solutions. Determine:

a) a third point that is a solution, and
b) how many solutions there are.

65. The solution of the following system is $(4, -5)$. Find A and B.

$$Ax - 6y = 13,$$
$$x - By = -8.$$

Translate to a system of equations. Do not solve.

66. *Ages.* Burl is twice as old as his son. Ten years ago, Burl was three times as old as his son. How old are they now?

67. *Work experience.* Lou and Juanita are mathematics professors at a state university. Together, they have 46 years of service. Two years ago, Lou had taught 2.5 times as many years as Juanita. How long has each taught at the university?

68. *Design.* A piece of posterboard has a perimeter of 156 in. If you cut 6 in. off the width, the length becomes four times the width. What are the dimensions of the original piece of posterboard?

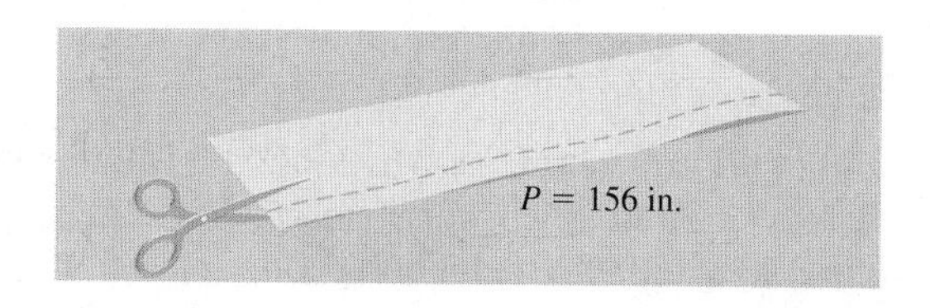

69. *Nontoxic scouring powder.* A nontoxic scouring powder is made up of 4 parts baking soda and 1 part vinegar. How much of each ingredient is needed for a 16-oz mixture?

Solve graphically.

70. $y = |x|$,
$x + 4y = 15$

71. $x - y = 0$,
$y = x^2$

In Exercises 72–75, use a grapher to solve each system of linear equations for x and y. Round all coordinates to the nearest hundredth.

72. $y = 8.23x + 2.11$,
$y = -9.11x - 4.66$

73. $y = -3.44x - 7.72$,
$y = 4.19x - 8.22$

74. $14.12x + 7.32y = 2.98$,
$21.88x - 6.45y = -7.22$

75. $5.22x - 8.21y = -10.21$,
$-12.67x + 10.34y = 12.84$

Solving by Substitution or Elimination

3.2

The Substitution Method • The Elimination Method • Comparing Methods

The Substitution Method

One algebraic (nongraphical) method for solving systems of equations, the *substitution method*, relies on having a variable isolated.

Example 1 Solve the system

$$x + y = 4, \quad (1)$$
$$x = y + 1. \quad (2)$$

For easy reference, we have numbered the equations.

Solution Equation (2) says that x and $y + 1$ name the same number. Thus we can substitute $y + 1$ for x in equation (1):

$$x + y = 4 \quad \text{Equation (1)}$$
$$(y + 1) + y = 4. \quad \text{Substituting } y + 1 \text{ for } x$$

We solve this last equation, using methods learned earlier:

$$(y + 1) + y = 4$$
$$2y + 1 = 4 \quad \text{Removing parentheses and combining like terms}$$
$$2y = 3 \quad \text{Subtracting 1 from both sides}$$
$$y = \tfrac{3}{2}. \quad \text{Dividing by 2}$$

We now return to the original pair of equations and substitute $\frac{3}{2}$ for y in either equation so that we can solve for x. For this problem, calculations are slightly easier if we use equation (2):

$$x = y + 1 \quad \text{Equation (2)}$$
$$= \tfrac{3}{2} + 1 \quad \text{Substituting } \tfrac{3}{2} \text{ for } y$$
$$= \tfrac{3}{2} + \tfrac{2}{2} = \tfrac{5}{2}.$$

We obtain the ordered pair $\left(\frac{5}{2}, \frac{3}{2}\right)$. A check ensures that it is a solution:

Check:

$x + y = 4$	
$\frac{5}{2} + \frac{3}{2}$?	4
$\frac{8}{2}$	
4	4 TRUE

$x = y + 1$	
$\frac{5}{2}$?	$\frac{3}{2} + 1$
	$\frac{3}{2} + \frac{2}{2}$
$\frac{5}{2}$	$\frac{5}{2}$ TRUE

Since $\left(\frac{5}{2}, \frac{3}{2}\right)$ checks, it is the solution.

x + y = 4
x = y + 1

A visualization of Example 1.

Note that the coordinates of the intersection are not obvious.

The exact solution to Example 1 is difficult to find graphically because it involves fractions. Despite this, the graph shown does serve as a check and provides a visualization of the problem.

If neither equation in a system has a variable alone on one side, we first isolate a variable in one equation and then substitute.

Example 2 Solve the system

$$2x + y = 6, \quad (1)$$
$$3x + 4y = 4. \quad (2)$$

Solution First, we select an equation and solve for one variable. To isolate y, we can subtract $2x$ from both sides of equation (1):

$$2x + y = 6 \quad (1)$$
$$y = 6 - 2x. \quad (3) \qquad \text{Subtracting } 2x \text{ from both sides}$$

Next, we proceed as in Example 1, by substituting:

$$3x + 4(6 - 2x) = 4 \qquad \text{Substituting } 6 - 2x \text{ for } y \text{ in equation (2). Use parentheses!}$$
$$3x + 24 - 8x = 4 \qquad \text{Distributing to remove parentheses}$$
$$3x - 8x = 4 - 24 \qquad \text{Subtracting 24 from both sides}$$
$$-5x = -20$$
$$x = 4. \qquad \text{Dividing both sides by } -5$$

Next, we substitute 4 for x in either equation (1), (2), or (3). It is easiest to use equation (3) because it has already been solved for y:

$$y = 6 - 2x$$
$$= 6 - 2(4)$$
$$= 6 - 8 = -2.$$

The pair $(4, -2)$ appears to be the solution. We check in equations (1) and (2).

Check:

$2x + y = 6$	
$2(4) + (-2)$	? 6
$8 - 2$	
6	6 TRUE

$3x + 4y = 4$	
$3(4) + 4(-2)$	? 4
$12 - 8$	
4	4 TRUE

Since $(4, -2)$ checks, it is the solution.

Study Tip

Immediately after each quiz or test, write out a step-by-step solution to any questions you missed. Visit your professor during office hours or consult with a tutor for help with problems that are still giving you trouble. Misconceptions tend to resurface if they are not corrected as soon as possible.

Some systems have no solution, as we saw graphically in Section 3.1. How do we recognize such systems if we are solving by an algebraic method?

Example 3 Solve the system

$$y = -3x + 5, \quad (1)$$
$$y = -3x - 2. \quad (2)$$

Solution We solved this system graphically in Example 4(b) of Section 3.1, and found that the lines are parallel and the system has no solution. Let's now try to solve the system by substitution. Proceeding as in Example 1, we substitute $-3x - 2$ for y in the first equation:

$$-3x - 2 = -3x + 5 \quad \text{Substituting } -3x - 2 \text{ for } y \text{ in equation (1)}$$
$$-2 = 5. \quad \text{Adding } 3x \text{ to both sides; } -2 = 5 \text{ is a contradiction.}$$

When we add $3x$ to get the x-terms on one side, the x-terms drop out and we end up with a contradiction—that is, an equation that is always false. When solving algebraically yields a contradiction, the system has no solution.

The Elimination Method

The *elimination method* for solving systems of equations makes use of the *addition principle*: If $a = b$, then $a + c = b + c$. Consider the following system:

$$2x - 3y = 0, \quad (1)$$
$$-4x + 3y = -1. \quad (2)$$

To see why the elimination method works well with this system, notice the $-3y$ in one equation and the $3y$ in the other. These terms are opposites. If we add all terms on the left side of the equations, $-3y$ and $3y$ add to 0, and in effect, the variable y is "eliminated."

To use the addition principle for equations, note that according to equation (2), $-4x + 3y$ and -1 are the same number. Thus we can work vertically and add $-4x + 3y$ to the left side of equation (1) and -1 to the right side:

$$\begin{array}{rll} 2x - 3y = & 0 & (1) \\ \underline{-4x + 3y =} & \underline{-1} & (2) \\ -2x + 0y = & -1. & \text{Adding} \end{array}$$

This eliminates the variable y, and leaves an equation with just one variable, x, for which we solve:

$$-2x = -1$$
$$x = \tfrac{1}{2}.$$

Next, we substitute $\frac{1}{2}$ for x in equation (1) and solve for y:

$$2 \cdot \tfrac{1}{2} - 3y = 0 \quad \text{Substituting. We also could have used equation (2).}$$
$$1 - 3y = 0$$
$$-3y = -1, \text{ so } y = \tfrac{1}{3}.$$

Check:

$$\begin{array}{c|l} 2x - 3y & 0 \\ \hline 2(\frac{1}{2}) - 3(\frac{1}{3}) \ ? & 0 \\ 1 - 1 & \\ 0 & 0 \quad \text{TRUE} \end{array} \qquad \begin{array}{c|l} -4x + 3y & -1 \\ \hline -4(\frac{1}{2}) + 3(\frac{1}{3}) \ ? & -1 \\ -2 + 1 & \\ -1 & -1 \quad \text{TRUE} \end{array}$$

Since $(\frac{1}{2}, \frac{1}{3})$ checks, it is the solution.

To eliminate a variable, we must sometimes multiply before adding.

Example 4 Solve the system

$$5x + 4y = 22, \quad (1)$$
$$-3x + 8y = 18. \quad (2)$$

Solution If we add the left sides of the two equations, we will not eliminate a variable. However, if the $4y$ in equation (1) were changed to $-8y$, we would. To accomplish this change, we multiply both sides of equation (1) by -2:

$$-10x - 8y = -44 \quad \text{Multiplying both sides of equation (1) by } -2$$
$$-3x + 8y = 18$$
$$-13x + 0 = -26 \quad \text{Adding}$$
$$x = 2. \quad \text{Solving for } x$$

Then

$$-3 \cdot 2 + 8y = 18 \quad \text{Substituting 2 for } x \text{ in equation (2)}$$
$$-6 + 8y = 18$$
$$\left.\begin{aligned} 8y &= 24 \\ y &= 3. \end{aligned}\right\} \quad \text{Solving for } y$$

We obtain $(2, 3)$, or $x = 2$, $y = 3$. We leave it to the student to confirm that this checks and is the solution.

Sometimes we must multiply twice in order to make two terms become opposites.

Example 5 Solve the system

$$2x + 3y = 17, \quad (1)$$
$$5x + 7y = 29. \quad (2)$$

Solution We multiply so that the x-terms are eliminated.

$$2x + 3y = 17, \longrightarrow \text{Multiplying both sides by 5} \longrightarrow 10x + 15y = 85$$
$$5x + 7y = 29 \longrightarrow \text{Multiplying both sides by } -2 \longrightarrow -10x - 14y = -58$$
$$0 + y = 27 \quad \text{Adding}$$
$$y = 27.$$

Next, we substitute to find x:

$$2x + 3 \cdot 27 = 17 \quad \text{Substituting 27 for } y \text{ in equation (1)}$$
$$2x + 81 = 17$$
$$\left.\begin{aligned} 2x &= -64 \\ x &= -32. \end{aligned}\right\} \quad \text{Solving for } x$$

Check:

$$\begin{array}{r|l} \multicolumn{2}{c}{2x + 3y = 17} \\ \hline 2(-32) + 3(27) & ?\ 17 \\ -64 + 81 & \\ 17 & 17 \text{ TRUE} \end{array} \qquad \begin{array}{r|l} \multicolumn{2}{c}{5x + 7y = 29} \\ \hline 5(-32) + 7(27) & ?\ 29 \\ -160 + 189 & \\ 29 & 29 \text{ TRUE} \end{array}$$

We obtain $(-32, 27)$, or $x = -32$, $y = 27$, as the solution.

Example 6 Solve the system

$$3y - 2x = 6, \qquad (1)$$
$$-12y + 8x = -24. \qquad (2)$$

Solution We graphed this system in Example 4(c) of Section 3.1, and found that the lines coincide and the system has an infinite number of solutions. Suppose we were to solve this system using the elimination method:

$$\begin{aligned} 12y - 8x &= 24 && \text{Multiplying both sides of equation (1) by 4} \\ -12y + 8x &= -24 && \\ \hline 0 &= 0. && \text{We obtain an identity; } 0 = 0 \text{ is always true.} \end{aligned}$$

Note that both variables have been eliminated and what remains is an identity—that is, an equation that is always true. Any pair that is a solution of equation (1) is also a solution of equation (2). The equations are dependent and the solution set is infinite:

$$\{(x, y) \mid 3y - 2x = 6\}.$$

Special Cases

When solving a system of two linear equations in two variables:

1. If an identity is obtained, such as $0 = 0$, then the system has an infinite number of solutions. The equations are dependent and, since a solution exists, the system is consistent.*
2. If a contradiction is obtained, such as $0 = 7$, then the system has no solution. The system is inconsistent.

Should decimals or fractions appear, it often helps to *clear* before solving.

Example 7 Solve the system

$$0.2x + 0.3y = 1.7,$$
$$\tfrac{1}{7}x + \tfrac{1}{5}y = \tfrac{29}{35}.$$

Solution We have

$$0.2x + 0.3y = 1.7, \longrightarrow \text{Multiplying both sides by 10} \longrightarrow 2x + 3y = 17$$
$$\tfrac{1}{7}x + \tfrac{1}{5}y = \tfrac{29}{35} \longrightarrow \text{Multiplying both sides by 35} \longrightarrow 5x + 7y = 29.$$

*Consistent systems and dependent equations are discussed in greater detail in Section 3.4.

We multiplied both sides of the first equation by 10 to clear the decimals. Multiplication by 35, the least common denominator, clears the fractions in the second equation. The problem is now identical to Example 5. The solution is $(-32, 27)$, or $x = -32, y = 27$.

Comparing Methods

The following table is a summary that compares the graphical, substitution, and elimination methods for solving systems of equations.

CONNECTING THE CONCEPTS

We now have three different methods for solving systems of equations. Each method has certain strengths and weaknesses, as outlined below.

Method	Strengths	Weaknesses
Graphical	Solutions are displayed graphically. Works with any system that can be graphed.	Inexact when solutions involve numbers that are not integers. Solution may not appear on the part of the graph drawn.
Substitution	Yields exact solutions. Easy to use when a variable is alone on one side.	Introduces extensive computations with fractions when solving more complicated systems. Solutions are not displayed graphically.
Elimination	Yields exact solutions. Easy to use when fractions or decimals appear in the system. The preferred method for systems of 3 or more equations in 3 or more variables (see Section 3.4).	Solutions are not displayed graphically.

(continued)

Before selecting a method to use, try to remember the strengths and weaknesses of each method. If possible, begin solving the system mentally before settling on the method that seems best suited for that particular system. Selecting the "best" method for a problem is a bit like selecting one of three different saws with which to cut a piece of wood. The "best" choice depends on what kind of wood is being cut and what type of cut is being made, as well as your skill level with each saw.

Note that each of the three methods was introduced using a rather simple example. As the examples became more complicated, additional steps were required in order to "turn" the new problem into a more familiar format. This is a common approach in mathematics: We perform one or more steps to make a "new" problem resemble a problem we already know how to solve.

Exercise Set 3.2

FOR EXTRA HELP

Digital Video Tutor CD 2
Videotape 5

InterAct Math

Math Tutor Center

MathXL

MyMathLab.com

For Exercises 1–48, if a system has an infinite number of solutions, use set-builder notation to write the solution set. If a system has no solution, state this.

Solve using the substitution method.

1. $y = 5 - 4x,$ $2x - 3y = 13$

2. $x = 8 - 4y,$ $3x + 5y = 3$

3. $2y + x = 9,$ $x = 3y - 3$

4. $9x - 2y = 3,$ $3x - 6 = y$

5. $3s - 4t = 14,$ $5s + t = 8$

6. $m - 2n = 16,$ $4m + n = 1$

7. $4x - 2y = 6,$ $2x - 3 = y$

8. $t = 4 - 2s,$ $t + 2s = 6$

9. $-5s + t = 11,$ $4s + 12t = 4$

10. $5x + 6y = 14,$ $-3y + x = 7$

11. $2x + 2y = 2,$ $3x - y = 1$

12. $4p - 2q = 16,$ $5p + 7q = 1$

13. $3a - b = 7,$ $2a + 2b = 5$

14. $5x + 3y = 4,$ $x - 4y = 3$

15. $2x - 3 = y,$ $y - 2x = 1$

16. $a - 2b = 3,$ $3a = 6b + 9$

Solve using the elimination method.

17. $x + 3y = 7,$ $-x + 4y = 7$

18. $x + y = 9,$ $2x - y = -3$

19. $2x + y = 6,$ $x - y = 3$

20. $x - 2y = 6,$ $-x + 3y = -4$

21. $9x + 3y = -3,$ $2x - 3y = -8$

22. $6x - 3y = 18,$ $6x + 3y = -12$

23. $5x + 3y = 19,$ $2x - 5y = 11$

24. $3x + 2y = 3,$ $9x - 8y = -2$

25. $5r - 3s = 24,$ $3r + 5s = 28$

26. $5x - 7y = -16,$ $2x + 8y = 26$

27. $6s + 9t = 12,$ $4s + 6t = 5$

28. $10a + 6b = 8,$ $5a + 3b = 2$

29. $\frac{1}{2}x - \frac{1}{6}y = 3,$ $\frac{2}{5}x + \frac{1}{2}y = 2$

30. $\frac{1}{3}x + \frac{1}{5}y = 7,$ $\frac{1}{6}x - \frac{2}{5}y = -4$

31. $\frac{x}{2} + \frac{y}{3} = \frac{7}{6},$ $\frac{2x}{3} + \frac{3y}{4} = \frac{5}{4}$

32. $\frac{2x}{3} + \frac{3y}{4} = \frac{11}{12},$ $\frac{x}{3} + \frac{7y}{18} = \frac{1}{2}$

Aha! **33.** $12x - 6y = -15,$ $-4x + 2y = 5$

34. $8s + 12t = 16,$ $6s + 9t = 12$

35. $0.2a + 0.3b = 1,$ $0.3a - 0.2b = 4$

36. $-0.4x + 0.7y = 1.3,$ $0.7x - 0.3y = 0.5$

Solve using any appropriate method.

37. $a - 2b = 16,$ $b + 3 = 3a$

38. $5x - 9y = 7,$ $7y - 3x = -5$

39. $10x + y = 306,$ $10y + x = 90$

40. $3(a - b) = 15,$ $4a = b + 1$

41. $3y = x - 2,$ $x = 2 + 3y$

42. $x + 2y = 8,$ $x = 4 - 2y$

43. $3s - 7t = 5,$
$7t - 3s = 8$

44. $2s - 13t = 120,$
$-14s + 91t = -840$

45. $0.05x + 0.25y = 22,$
$0.15x + 0.05y = 24$

46. $2.1x - 0.9y = 15,$
$-1.4x + 0.6y = 10$

47. $13a - 7b = 9,$
$2a - 8b = 6$

48. $3a - 12b = 9,$
$14a - 11b = 5$

49. Describe a procedure that can be used to write an inconsistent system of equations.

50. Describe a procedure that can be used to write a system that has an infinite number of solutions.

SKILL MAINTENANCE

51. The fare for a taxi ride from Johnson Street to Elm Street is \$5.20. If the rate of the taxi is \$1.00 for the first $\frac{1}{2}$ mi and 30¢ for each additional $\frac{1}{4}$ mi, how far is it from Johnson Street to Elm Street?

52. A student's average after 4 tests is 78.5. What score is needed on the fifth test in order to raise the average to 80?

53. *Home remodeling.* In a recent year, Americans spent \$35 billion to remodel bathrooms and kitchens. Twice as much was spent on kitchens as on bathrooms. (*Source*: *Indianapolis Star*) How much was spent on each?

54. A 480-m wire is cut into three pieces. The second piece is three times as long as the first. The third is four times as long as the second. How long is each piece?

55. *Car rentals.* Badger Rent-A-Car rents a compact car at a daily rate of \$34.95 plus 10¢ per mile. A businessperson is allotted \$80 for car rental. How many miles can she travel on the \$80 budget?

56. *Car rentals.* Badger rents midsized cars at a rate of \$43.95 plus 10¢ per mile. A tourist has a car-rental budget of \$90. How many miles can he travel on the \$90?

SYNTHESIS

57. Some systems are more easily solved by substitution and some are more easily solved by elimination. Write guidelines that could be used to help someone determine which method to use.

58. Explain how it is possible to solve Exercise 33 mentally.

59. If $(1, 2)$ and $(-3, 4)$ are two solutions of $f(x) = mx + b$, find m and b.

60. If $(0, -3)$ and $\left(-\frac{3}{2}, 6\right)$ are two solutions of $px - qy = -1$, find p and q.

61. Determine a and b for which $(-4, -3)$ is a solution of the system

$$ax + by = -26,$$
$$bx - ay = 7.$$

62. Solve for x and y in terms of a and b:

$$5x + 2y = a,$$
$$x - y = b.$$

Solve.

63. $\dfrac{x+y}{2} - \dfrac{x-y}{5} = 1,$
$\dfrac{x-y}{2} + \dfrac{x+y}{6} = -2$

64. $3.5x - 2.1y = 106.2,$
$4.1x + 16.7y = -106.28$

Each of the following is a system of nonlinear equations. However, each is reducible to linear, *since an appropriate substitution (say, u for $1/x$ and v for $1/y$) yields a linear system. Make such a substitution, solve for the new variables, and then solve for the original variables.*

65. $\dfrac{2}{x} + \dfrac{1}{y} = 0,$
$\dfrac{5}{x} + \dfrac{2}{y} = -5$

66. $\dfrac{1}{x} - \dfrac{3}{y} = 2,$
$\dfrac{6}{x} + \dfrac{5}{y} = -34$

67. A student solving the system

$$17x + 19y = 102,$$
$$136x + 152y = 826$$

graphs both equations on a grapher and gets the following screen.

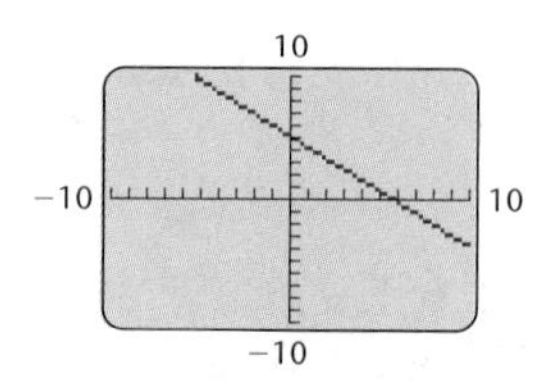

The student then (incorrectly) concludes that the equations are dependent and the solution set is infinite. How can algebra be used to convince the student that a mistake has been made?

COLLABORATIVE

CORNER

How Many Two's? How Many Three's?

Focus: Systems of linear equations

Time: 20 minutes

Group size: 3

The box score at right, from a basketball game in 2001 between the Philadelphia 76ers and the Indiana Pacers, contains information on how many field goals and free throws each player attempted and made. For example, the line "Rose 9–19 7–7 26" means that Indiana's Jalen Rose made 9 field goals out of 19 attempts and 7 free throws out of 7 attempts, for a total of 26 points. (Each free throw is worth 1 point and each field goal is worth either 2 or 3 points, depending on how far from the basket it was shot.)

ACTIVITY

1. Work as a group to develop a system of two equations in two unknowns that can be used to determine how many 2-pointers and how many 3-pointers were made by Philadelphia's Allen Iverson.
2. Each group member should solve the system from part (1) in a different way: one person algebraically, one person by making a table and methodically checking all combinations of 2- and 3-pointers, and one person by guesswork. Compare answers when this has been completed.
3. Determine, as a group, how many 2- and 3-pointers the Philadelphia 76ers made as a team.

■76ers 104, Pacers 93: At Philadelphia, Allen Iverson scored 37 points and had seven rebounds as Philadelphia beat Indiana, defeating the team that knocked the 76ers out of the playoffs the past two years. The victory was only the third in nine games for the 76ers, the team with the best record in the East. (*Source*: AP in the *Burlington Free Press*, 4/2/01)

Indiana (93)
Harrington 2-8 0-0 4, O'Neal 2-5 1-1 5, Perkins 0-4 0-0 0, Rose 9-19 7-7 26, R. Miller 3-14 4-4 11, Best 4-5 3-4 11, Croshere 3-9 15-16 21, Edney 2-6 4-4 8, Foster 1-2 1-4 3, Bender 2-2 0-0 4.
Totals 28-74 35-40 93.

Philadelphia (104)
Lynch 2-5 0-0 4, Hill 6-7 3-3 15, Mutombo 0-2 5-6 5, Snow 4-16 2-2 10, Iverson 12-27 10-15 37, Jones 4-11 3-4 11, Geiger 4-5 1-2 9, Ollie 2-4 4-6 8, Buford 1-3 0-0 2, Sanchez 0-0 0-0 0, MacCulloch 1-1 1-2 3.
Totals 36-81 29-40 104.

Indiana	**18**	**24**	**21**	**30**	**—**	**93**
Philadelphia	**25**	**29**	**23**	**27**	**—**	**104**

Solving Applications: Systems of Two Equations

3.3

Total-Value and Mixture Problems • Motion Problems

You are in a much better position to solve problems now that systems of equations can be used. Using systems often makes the translating step easier.

Example 1

Real estate. In 1996, the Simon Property Group and the DeBartolo Realty Corporation merged to form the largest real estate company in the United States, owning 183 shopping centers in 32 states. Prior to merging, Simon owned twice as many properties as DeBartolo. How many properties did each company own before the merger?

Solution The *Familiarize* and *Translate* steps have been done in Example 1 of Section 3.1. The resulting system of equations is

$$x + y = 183,$$
$$x = 2y,$$

where x is the number of properties originally owned by Simon and y is the number of properties originally owned by DeBartolo.

3. **Carry out.** We solve the system of equations. Since one equation already has a variable isolated, let's use the substitution method:

$$x + y = 183$$
$$2y + y = 183 \quad \text{Substituting } 2y \text{ for } x$$
$$3y = 183 \quad \text{Combining like terms}$$
$$y = 61.$$

We return to the second equation and substitute 61 for y and compute x:

$$x = 2y = 2 \cdot 61 = 122.$$

Apparently, Simon owned 122 properties and DeBartolo 61.

4. **Check.** The sum of 122 and 61 is 183, so the total number of properties is correct. Since 122 is twice 61, the numbers check.
5. **State.** Prior to merging, Simon owned 122 properties and DeBartolo owned 61.

Total-Value and Mixture Problems

Example 2

Purchasing. Recently the Champlain Valley Community Music Center purchased 120 stamps for \$30.30. If the stamps were a combination of 20¢ postcard stamps and 34¢ first-class stamps, how many of each type were bought?

Solution The *Familiarize* and *Translate* steps were completed in Example 2 of Section 3.1.

3. **Carry out.** We are to solve the system of equations

$$p + f = 120, \quad (1)$$
$$20p + 34f = 3030 \quad (2) \quad \text{Working in cents rather than dollars}$$

where p is the number of postcard stamps bought and f is the number of first-class stamps bought. Because both equations are in the form $Ax + By = C$, let's use the elimination method to solve the system. We can eliminate p by multiplying both sides of equation (1) by -20 and adding them to the corresponding sides of equation (2):

$$\begin{aligned} -20p - 20f &= -2400 && \text{Multiplying both sides of equation (1) by } -20 \\ 20p + 34f &= 3030 && \\ \hline 14f &= 630 && \text{Adding} \\ f &= 45. && \text{Solving for } f \end{aligned}$$

To find p, we substitute 45 for f in equation (1) and then solve for p:

$$\begin{aligned} p + f &= 120 && \text{Equation (1)} \\ p + 45 &= 120 && \text{Substituting 45 for } f \\ p &= 75. && \text{Solving for } p \end{aligned}$$

We obtain $(45, 75)$, or $f = 45$ and $p = 75$.

4. **Check.** We check in the original problem. Recall that f is the number of first-class stamps and p the number of postcard stamps.

   Number of stamps: $f + p = 45 + 75 = 120$

   Cost of first-class stamps: $\$0.34f = 0.34 \times 45 = \$15.30$

   Cost of postcard stamps: $\$0.20p = 0.20 \times 75 = \underline{\$15.00}$

   Total $= \$30.30$

   The numbers check.

5. **State.** The music center bought 45 first-class stamps and 75 postcard stamps.

Example 2 involved two types of items (first-class stamps and postcard stamps), the quantity of each type bought, and the total value of the items. We refer to this type of problem as a *total-value problem.*

Example 3

Blending teas. Tara's Tea Terrace sells loose Black tea for 95¢ an ounce and Lapsang Souchong for \$1.43 an ounce. Tara wants to make a 1-lb mixture of the two types, called Imperial Blend, that sells for \$1.10 an ounce. How much tea of each type should Tara use?

Solution

1. **Familiarize.** This problem is similar to Example 2. Rather than postcard stamps and first-class stamps, we have ounces of Black tea and ounces of Lapsang Souchong. Instead of a different price for each type of stamp, we have a different price per ounce for each type of tea. Finally, rather than knowing the total cost of the stamps, we know the weight and the price per

ounce of the Imperial Blend. It is important to note that we can find the total value of the blend by multiplying 16 ounces (1 lb) times \$1.10 per ounce. Although we could make and check a guess, we proceed to let $b =$ the number of ounces of Black tea and $l =$ the number of ounces of Lapsang Souchong.

2. **Translate.** Since a 16-oz batch is being made, we must have

$$b + l = 16.$$

To find a second equation, note that the total value of the 16-oz blend must match the combined value of the separate ingredients:

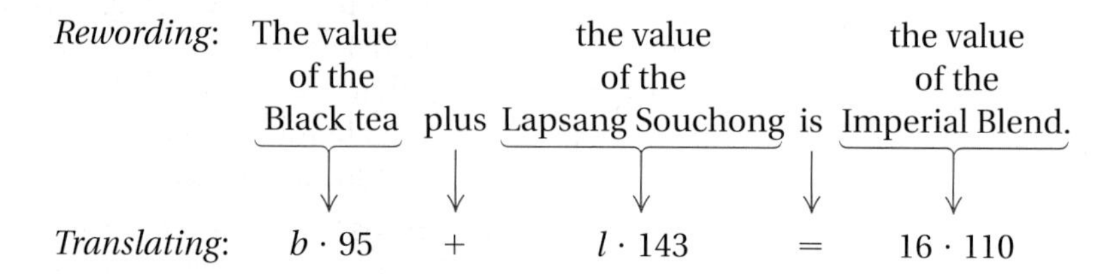

Rewording: The value of the Black tea plus the value of the Lapsang Souchong is the value of the Imperial Blend.
Translating: $b \cdot 95 + l \cdot 143 = 16 \cdot 110$

These equations can also be obtained from a table.

	Black Tea	Lapsang Souchong	Imperial Blend	
Number of Ounces	b	l	16	$\rightarrow b + l = 16$
Price per Ounce	95¢	143¢	110¢	
Value of Tea	$95b$	$143l$	$16 \cdot 110$, or 1760¢	$\rightarrow 95b + 143l = 1760$

We have translated to a system of equations:

$$b + l = 16, \quad (1)$$
$$95b + 143l = 1760. \quad (2)$$

3. **Carry out.** We can solve using substitution. When equation (1) is solved for b, we have $b = 16 - l$. Substituting $16 - l$ for b in equation (2), we find l:

$95(16 - l) + 143l = 1760$	Substituting
$1520 - 95l + 143l = 1760$	Using the distributive law
$48l = 240$	Combining like terms; subtracting 1520 from both sides
$l = 5.$	Dividing both sides by 48

We have $l = 5$ and, from equation (1) above, $b + l = 16$. Thus, $b = 11$.

4. **Check.** If 11 oz of Black tea and 5 oz of Lapsang Souchong are combined, a 16-oz, or 1-lb, blend will result. The value of 11 oz of Black tea is 11(\$0.95), or \$10.45. The value of 5 oz of Lapsang Souchong is 5(\$1.43), or \$7.15, so the combined value of the blend is \$10.45 + \$7.15 = \$17.60. A 16-oz batch, priced at \$1.10 an ounce, would also be worth \$17.60, so our answer checks.

5. **State.** The Imperial Blend should be made by combining 11 oz of Black tea with 5 oz of Lapsang Souchong.

Example 4

Student loans. Dawn's student loans totaled \$9600. Part was a Perkins loan made at 5% interest and the rest was a Federal Education Loan made at 8% interest. After one year, Dawn's loans accumulated \$633 in interest. What was the original amount of each loan?

Solution

1. **Familiarize.** We begin with a guess. If \$7000 was borrowed at 5% and \$2600 was borrowed at 8%, the two loans would total \$9600. The interest would then be 0.05(\$7000), or \$350, and 0.08(\$2600), or \$208, for a total of \$558 in interest. Our guess was wrong, but checking the guess familiarized us with the problem.

2. **Translate.** We let p = the amount of the Perkins loan and f = the amount of the Federal Education Loan. We then organize a table in which each column comes from the formula for simple interest:

   Principal · Rate · Time = Interest.

	Perkins Loan	Federal Loan	Total	
Principal	p	f	\$9600	→ $p + f = 9600$
Rate of Interest	5%	8%		
Time	1 yr	1 yr		
Interest	$0.05p$	$0.08f$	\$633	→ $0.05p + 0.08f = 633$

   The total amount borrowed is found in the first row of the table:

   $$p + f = 9600.$$

   A second equation, representing the accumulated interest, can be found in the last row:

   $$0.05p + 0.08f = 633, \quad \text{or} \quad 5p + 8f = 63{,}300. \qquad \text{Clearing decimals}$$

3. **Carry out.** The system can be solved by elimination:

$$\begin{aligned} p + f &= 9600, \longrightarrow \text{Multiplying both sides by } -5 \longrightarrow & -5p - 5f &= -48{,}000 \\ 5p + 8f &= 63{,}300. & 5p + 8f &= 63{,}300 \\ & & \hline 3f &= 15{,}300 \\ & & f &= 5100 \end{aligned}$$

$$\begin{aligned} p + f &= 9600 \longleftarrow f = 5100 \\ p + 5100 &= 9600 \\ p &= 4500. \end{aligned}$$

   We find that $p = 4500$ and $f = 5100$.

4. **Check.** The total amount borrowed is \$4500 + \$5100, or \$9600. The interest on \$4500 at 5% for 1 yr is 0.05(\$4500), or \$225. The interest on \$5100 at 8% for 1 yr is 0.08(\$5100), or \$408. The total amount of interest is \$225 + \$408, or \$633, so the numbers check.
5. **State.** The Perkins loan was for \$4500 and the Federal Education Loan was for \$5100.

Before proceeding to Example 5, briefly scan Examples 2–4 for similarities. Note that in each case, one of the equations in the system is a simple sum while the other equation represents a sum of products. Example 5 continues this pattern with what is commonly called a *mixture problem.*

Problem-Solving Tip

When solving a problem, see if it is patterned or modeled after a problem that you have already solved.

Example 5

Mixing fertilizers. Yardbird Gardening, Inc., carries two brands of fertilizer containing nitrogen and water. "Gently Green" is 5% nitrogen and "Sun Saver" is 15% nitrogen. Yardbird Gardening needs to combine the two types of solutions in order to make 100 L of a solution that is 12% nitrogen. How much of each brand should be used?

Solution

1. **Familiarize.** We make a drawing and then make a guess to gain familiarity with the problem.

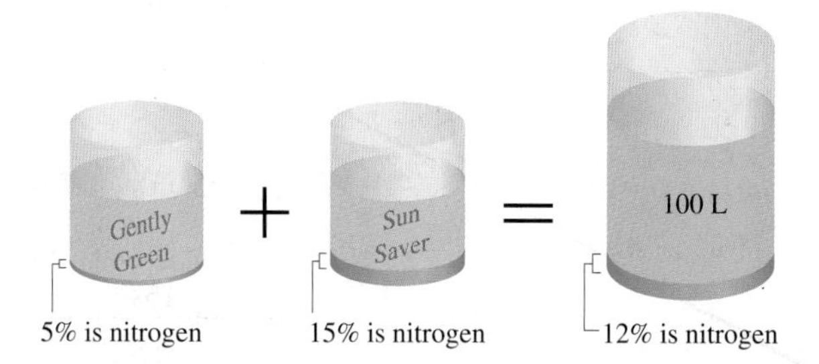

Suppose that 40 L of Gently Green and 60 L of Sun Saver are mixed. The resulting mixture will be the right size, 100 L, but will it be the right strength? To find out, note that 40 L of Gently Green would contribute $0.05(40) = 2$ L of nitrogen to the mixture while 60 L of Sun Saver would contribute $0.15(60) = 9$ L of nitrogen to the mixture. Altogether, 40 L of Gently Green and 60 L of Sun Saver would make 100 L of a mixture that has $2 + 9 = 11$ L of nitrogen. Since this would mean that the final mixture is only 11% nitrogen, our guess of 40 L and 60 L is incorrect. Still, the process of checking our guess has familiarized us with the problem.

2. **Translate.** Let g = the number of liters of Gently Green and s = the number of liters of Sun Saver. The information can be organized in a table.

	Gently Green	**Sun Saver**	**Mixture**	
Number of Liters	g	s	100	→ $g + s = 100$
Percent of Nitrogen	5%	15%	12%	
Amount of Nitrogen	$0.05g$	$0.15s$	0.12×100, or 12 liters	→ $0.05g + 0.15s = 12$

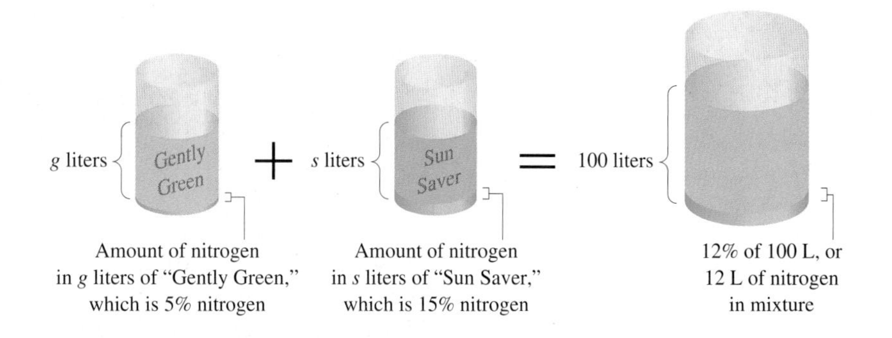

If we add g and s in the first row, we get one equation. It represents the total amount of mixture: $g + s = 100$.

If we add the amounts of nitrogen listed in the third row, we get a second equation. This equation represents the amount of nitrogen in the mixture: $0.05g + 0.15s = 12$.

After clearing decimals, we have translated the problem to the system

$$g + s = 100, \quad (1)$$
$$5g + 15s = 1200. \quad (2)$$

3. **Carry out.** We use the elimination method to solve the system:

$$-5g - 5s = -500 \quad \text{Multiplying both sides of equation (1) by } -5$$
$$5g + 15s = 1200$$
$$10s = 700 \quad \text{Adding}$$
$$s = 70; \quad \text{Solving for } s$$

$$g + 70 = 100 \quad \text{Substituting into equation (1)}$$
$$g = 30. \quad \text{Solving for } g$$

4. **Check.** Remember, g is the number of liters of Gently Green and s is the number of liters of Sun Saver.

Total amount of mixture: $g + s = 30 + 70 = 100$

Total amount of nitrogen: 5% of 30 + 15% of 70 = 1.5 + 10.5 = 12

Percentage of nitrogen in mixture: $\dfrac{\text{Total amount of nitrogen}}{\text{Total amount of mixture}} = \dfrac{12}{100} = 12\%$

The numbers check in the original problem.

5. **State.** Yardbird Gardening should mix 30 L of Gently Green with 70 L of Sun Saver.

Study Tip

Do not be surprised if your success rate drops some as you work through the exercises in this section. *This is normal.* Your success rate will increase as you gain experience with these types of problems and use some of the study tips already listed.

Motion Problems

When a problem deals with distance, speed (rate), and time, recall the following.

Distance, Rate, and Time Equations

If r represents rate, t represents time, and d represents distance, then:

$$d = rt, \qquad r = \frac{d}{t}, \quad \text{and} \quad t = \frac{d}{r}.$$

Be sure to remember at least one of these equations. The others can be obtained by using algebraic manipulations as needed.

Example 6

Train travel. A Vermont Railways freight train, loaded with logs, leaves Boston, heading to Washington D.C. at a speed of 60 km/h. Two hours later, an Amtrak® Metroliner leaves Boston, bound for Washington D.C., on a parallel track at 90 km/h. At what point will the Metroliner catch up to the freight train?

Solution

1. **Familiarize.** Let's make a guess—say, 180 km—and check to see if it is correct. The freight train, traveling 60 km/h, would reach a point 180 km from Boston in $\frac{180}{60} = 3$ hr. The Metroliner, traveling 90 km/h, would cover 180 km in $\frac{180}{90} = 2$ hr. Since 3 hr is *not* two hours more than 2 hr, our guess of 180 km is incorrect. Although our guess is wrong, we see that the time that the trains are running and the point at which they meet are both unknown. We let $t =$ the number of hours that the freight train is running before they meet and $d =$ the distance at which the trains meet. Since the freight train has a 2-hr head start, the Metroliner runs for $t - 2$ hours before catching up to the freight train, at which point both trains have traveled the same distance.

2. **Translate.** We can organize the information in a chart. Each row is determined by the formula *Distance* = *Rate* · *Time*.

	Distance	Rate	Time	
Freight Train	d	60	t	→ $d = 60t$
Metroliner	d	90	$t - 2$	→ $d = 90(t - 2)$

Using *Distance* = *Rate* · *Time* twice, we get two equations:

$$d = 60t, \qquad (1)$$
$$d = 90(t - 2). \qquad (2)$$

3. **Carry out.** We solve the system using substitution:

$$60t = 90(t - 2) \qquad \text{Substituting } 60t \text{ for } d \text{ in equation (2)}$$
$$60t = 90t - 180$$
$$-30t = -180$$
$$t = 6.$$

The time for the freight train is 6 hr, which means that the time for the Metroliner is 6 − 2, or 4 hr. Remember that it is distance, not time, that the problem asked for. Thus for $t = 6$, we have $d = 60 \cdot 6 = 360$ km.

4. **Check.** At 60 km/h, the freight train will travel $60 \cdot 6$, or 360 km, in 6 hr. At 90 km/h, the Metroliner will travel $90 \cdot (6 - 2) = 360$ km in 4 hr. The numbers check.

5. **State.** The freight train will catch up to the Metroliner at a point 360 km from Boston.

Example 7

Jet travel. An F16 jet flies 4 hr west with a 60-mph tailwind. Returning *against* the wind takes 5 hr. Find the speed of the jet with no wind.

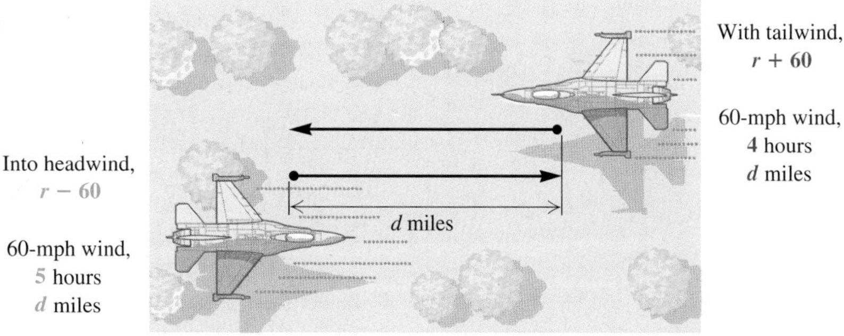

Solution

1. **Familiarize.** We imagine the situation and make a drawing. Note that the wind *speeds up* the jet on the outbound flight, but *slows down* the jet on the return flight. Since the distances traveled each way must be the same, we can check a guess of the jet's speed with no wind. Suppose the speed of the

jet with no wind is 400 mph. The jet would then fly $400 + 60 = 460$ mph with the wind and $400 - 60 = 340$ mph into the wind. In 4 hr, the jet would travel $460 \cdot 4 = 1840$ mi with the wind and $340 \cdot 5 = 1700$ mi against the wind. Since $1840 \neq 1700$, our guess of 400 mph is incorrect. Rather than guess again, let's have $r =$ the speed, in miles per hour, of the jet in still air. Then $r + 60 =$ the jet's speed with the wind and $r - 60 =$ the jet's speed against the wind. We also let $d =$ the distance traveled, in miles.

2. **Translate.** The information can be organized in a chart. The distances traveled are the same, so we use *Distance* = *Rate* (or *Speed*) · *Time*. Each row of the chart gives an equation.

	Distance	Rate	Time	
With Wind	d	$r + 60$	4	→ $d = (r + 60)4$
Against Wind	d	$r - 60$	5	→ $d = (r - 60)5$

The two equations constitute a system:

$$d = (r + 60)4, \quad (1)$$
$$d = (r - 60)5. \quad (2)$$

3. **Carry out.** We solve the system using substitution:

$$(r - 60)5 = (r + 60)4 \quad \text{Substituting } (r - 60)5 \text{ for } d \text{ in equation (1)}$$
$$5r - 300 = 4r + 240 \quad \text{Using the distributive law}$$
$$r = 540. \quad \text{Solving for } r$$

4. **Check.** When $r = 540$, the speed with the wind is $540 + 60 = 600$ mph, and the speed against the wind is $540 - 60 = 480$ mph. The distance with the wind, $600 \cdot 4 = 2400$ mi, matches the distance into the wind, $480 \cdot 5 = 2400$ mi, so we have a check.

5. **State.** The speed of the jet with no wind is 540 mph.

Tips for Solving Motion Problems

1. Draw a diagram using an arrow or arrows to represent distance and the direction of each object in motion.
2. Organize the information in a chart.
3. Look for times, distances, or rates that are the same. These often can lead to an equation.
4. Translating to a system of equations allows for the use of two variables.
5. Always make sure that you have answered the question asked.

Exercise Set 3.3

FOR EXTRA HELP

Digital Video Tutor CD 2
Videotape 5

InterAct Math

Math Tutor Center

MathXL

MyMathLab.com

1.–18. *For Exercises 1–18, solve Exercises 33–50 from pp. 148–149.*

19. *Sales.* Staples® recently sold a box of Flair® felt-tip pens for $12 and a four-pack of Sanford® Uni-ball® pens for $8. At the start of a recent fall semester, a combination of 40 boxes and four-packs of these pens was sold for a total of $372. How many of each type were purchased?

20. *Sales.* Staples recently sold a wirebound graph-paper notebook for $2.50 and a college-ruled notebook made of recycled paper for $2.30. At the start of a recent spring semester, a combination of 50 of these notebooks was sold for a total of $118.60. How many of each type were sold?

21. *Blending coffees.* The Coffee Counter charges $9.00 per pound for Kenyan French Roast coffee and $8.00 per pound for Sumatran coffee. How much of each type should be used to make a 20-lb blend that sells for $8.40 per pound?

22. *Mixed nuts.* The Nutty Professor sells cashews for $6.75 per pound and Brazil nuts for $5.00 per pound. How much of each type should be used to make a 50-lb mixture that sells for $5.70 per pound?

Aha! **23.** *Catering.* Casella's Catering is planning a wedding reception. The bride and groom would like to serve a nut mixture containing 25% peanuts. Casella has available mixtures that are either 40% or 10% peanuts. How much of each type should be mixed to get a 10-lb mixture that is 25% peanuts?

24. *Livestock feed.* Soybean meal is 16% protein and corn meal is 9% protein. How many pounds of each should be mixed to get a 350-lb mixture that is 12% protein?

25. *Ink remover.* Etch Clean Graphics uses one cleanser that is 25% acid and a second that is 50% acid. How many liters of each should be mixed to get 10 L of a solution that is 40% acid?

26. *Blending granola.* Deep Thought Granola is 25% nuts and dried fruit. Oat Dream Granola is 10% nuts and dried fruit. How much of Deep Thought and how much of Oat Dream should be mixed to form a 20-lb batch of granola that is 19% nuts and dried fruit?

27. *Student loans.* Lomasi's two student loans totaled $12,000. One of her loans was at 6% simple interest and the other at 9%. After one year, Lomasi owed $855 in interest. What was the amount of each loan?

28. *Investments.* An executive nearing retirement made two investments totaling $15,000. In one year, these investments yielded $1432 in simple interest. Part of the money was invested at 9% and the rest at 10%. How much was invested at each rate?

29. *Automotive maintenance.* "Arctic Antifreeze" is 18% alcohol and "Frost No-More" is 10% alcohol. How many liters of each should be mixed to get 20 L of a mixture that is 15% alcohol?

30. *Food science.* The following bar graph shows the milk fat percentages in three dairy products. How many pounds each of whole milk and cream should be mixed to form 200 lb of milk for cream cheese?

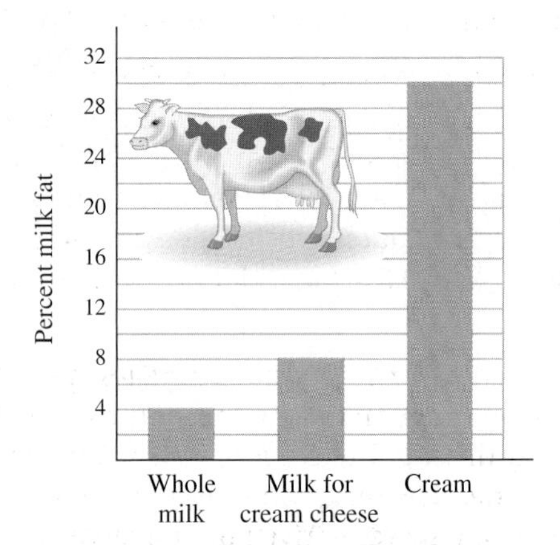

31. *Real estate.* The perimeter of an oceanfront lot is 190 m. The width is one fourth of the length. Find the dimensions.

32. *Architecture.* The rectangular ground floor of the John Hancock building has a perimeter of 860 ft. The length is 100 ft more than the width. Find the length and the width.

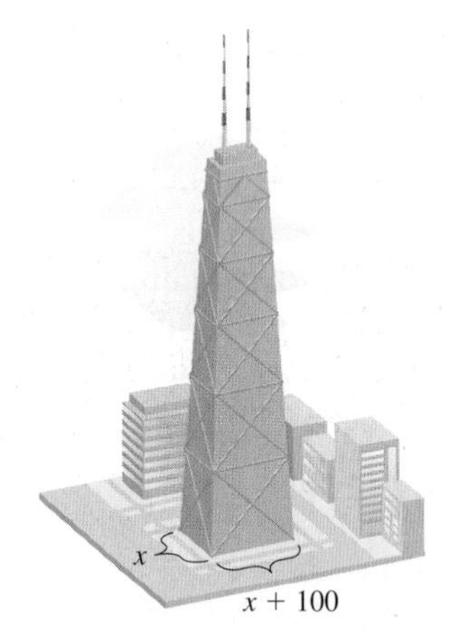

33. *Making change.* Cecilia makes a \$9.25 purchase at the bookstore with a \$20 bill. The store has no bills and gives her the change in quarters and fifty-cent pieces. There are 30 coins in all. How many of each kind are there?

34. *Teller work.* Ashford goes to a bank and gets change for a \$50 bill consisting of all \$5 bills and \$1 bills. There are 22 bills in all. How many of each kind are there?

35. *Train travel.* A train leaves Danville Junction and travels north at a speed of 75 km/h. Two hours later, an express train leaves on a parallel track and travels north at 125 km/h. How far from the station will they meet?

36. *Car travel.* Two cars leave Salt Lake City, traveling in opposite directions. One car travels at a speed of 80 km/h and the other at 96 km/h. In how many hours will they be 528 km apart?

37. *Canoeing.* Alvin paddled for 4 hr with a 6-km/h current to reach a campsite. The return trip against the same current took 10 hr. Find the speed of Alvin's canoe in still water.

38. *Boating.* Mia's motorboat took 3 hr to make a trip downstream with a 6-mph current. The return trip against the same current took 5 hr. Find the speed of the boat in still water.

39. *Point of no return.* A plane flying the 3458-mi trip from New York City to London has a 50-mph tailwind. The flight's *point of no return* is the point at which the flight time required to return to New York is the same as the time required to continue to London. If the speed of the plane in still air is 360 mph, how far is New York from the point of no return?

40. *Point of no return.* A plane is flying the 2553-mi trip from Los Angeles to Honolulu into a 60-mph headwind. If the speed of the plane in still air is 310 mph, how far from Los Angeles is the plane's point of no return? (See Exercise 39.)

41. Write at least three study tips of your own for someone beginning this exercise set.

42. In what ways are Examples 3 and 4 similar? In what sense are their systems of equations similar?

SKILL MAINTENANCE

Evaluate.

43. $2x - 3y + 12$, for $x = 5$ and $y = 2$

44. $7x - 4y + 9$, for $x = 2$ and $y = 3$

45. $5a - 7b + 3c$, for $a = -2$, $b = 3$, and $c = 1$

46. $3a - 8b - 2c$, for $a = -4$, $b = -1$, and $c = 3$

47. $4 - 2y + 3z$, for $y = \frac{1}{3}$ and $z = \frac{1}{4}$

48. $3 - 5y + 4z$, for $y = \frac{1}{2}$ and $z = \frac{1}{5}$

SYNTHESIS

49. Suppose that in Example 3 you are asked only for the amount of Black tea needed for the Imperial Blend. Would the method of solving the problem change? Why or why not?

50. Write a problem similar to Example 2 for a classmate to solve. Design the problem so that the solution is "The florist sold 14 hanging plants and 9 flats of petunias."

51.–54. *For Exercises 51–54, solve Exercises 66–69 from Exercise Set 3.1.*

55. *Retail.* Some of the world's best and most expensive coffee is Hawaii's Kona coffee. In order for coffee to be labeled "Kona Blend," it must contain at least 30% Kona beans. Bean Town Roasters has 40 lb of Mexican coffee. How much Kona coffee must they add if they wish to market it as Kona Blend?

56. *Automotive maintenance.* The radiator in Michelle's car contains 6.3 L of antifreeze and water. This mixture is 30% antifreeze. How much of this mixture should she drain and replace with pure antifreeze so that there will be a mixture of 50% antifreeze?

57. *Exercise.* Natalie jogs and walks to school each day. She averages 4 km/h walking and 8 km/h jogging. From home to school is 6 km and Natalie makes the trip in 1 hr. How far does she jog in a trip?

58. *Book sales.* A limited edition of a book published by a historical society was offered for sale to members. The cost was one book for $12 or two books for $20 (maximum of two per member). The society sold 880 books, for a total of $9840. How many members ordered two books?

59. The tens digit of a two-digit positive integer is 2 more than three times the units digit. If the digits are interchanged, the new number is 13 less than half the given number. Find the given integer. (*Hint*: Let $x =$ the tens-place digit and $y =$ the units-place digit; then $10x + y$ is the number.)

60. *Train travel.* A train leaves Union Station for Central Station, 216 km away, at 9 A.M. One hour later, a train leaves Central Station for Union Station. They meet at noon. If the second train had started at 9 A.M. and the first train at 10:30 A.M., they would still have met at noon. Find the speed of each train.

61. *Wood stains.* Williams' Custom Flooring has 0.5 gal of stain that is 20% brown and 80% neutral. A customer orders 1.5 gal of a stain that is 60% brown and 40% neutral. How much pure brown stain and how much neutral stain should be added to the original 0.5 gal in order to make up the order?*

62. *Fuel economy.* Grady's station wagon gets 18 miles per gallon (mpg) in city driving and 24 mpg in highway driving. The car is driven 465 mi on 23 gal of gasoline. How many miles were driven in the city and how many were driven on the highway?

63. *Biochemistry.* Industrial biochemists routinely use a machine to mix a buffer of 10% acetone by adding 100% acetone to water. One day, instead of adding 5 L of acetone to create a vat of buffer, a machine added 10 L. How much additional water was needed to bring the concentration down to 10%?

64. *Gender.* Phil and Phyllis are siblings. Phyllis has twice as many brothers as she has sisters. Phil has the same number of brothers as sisters. How many girls and how many boys are in the family?

65. See Exercise 61 above. Let $x =$ the amount of pure brown stain added to the original 0.5 gal. Find a function $P(x)$ that can be used to determine the percentage of brown stain in the 1.5-gal mixture. On a grapher, draw the graph of P and use INTERSECT to confirm the answer to Exercise 61.

*This problem was suggested by Professor Chris Burditt of Yountville, California.

Systems of Equations in Three Variables

3.4

Identifying Solutions • Solving Systems in Three Variables • Dependency, Inconsistency, and Geometric Considerations

CONNECTING THE CONCEPTS

As often happens in mathematics, once an idea is thoroughly understood, it can be extended to increasingly more complicated problems. This is precisely the situation for the material in Sections 3.4–3.7: We will extend the elimination method of Section 3.2 to systems of three equations in three unknowns. Although we will not do so in this text, the approach that we use can be further extended to systems with four equations in four unknowns, five equations in five unknowns, and so on.

Another common occurrence in mathematics is the streamlining of a sequence of steps that are used repeatedly. In Sections 3.6 and 3.7, we develop different notations that streamline the calculations of Sections 3.2 and 3.4.

Some problems translate directly to two equations. Others more naturally call for a translation to three or more equations. In this section, we learn how to solve systems of three linear equations. Later, we will use such systems in problem-solving situations.

Identifying Solutions

A **linear equation in three variables** is an equation equivalent to one in the form $Ax + By + Cz = D$, where A, B, C, and D are real numbers. We refer to the form $Ax + By + Cz = D$ as *standard form* for a linear equation in three variables.

A solution of a system of three equations in three variables is an ordered triple (x, y, z) that makes *all three* equations true.

Example 1

Determine whether $\left(\frac{3}{2}, -4, 3\right)$ is a solution of the system

$$\begin{aligned} 4x - 2y - 3z &= 5, \\ -8x - y + z &= -5, \\ 2x + y + 2z &= 5. \end{aligned}$$

Solution We substitute $\left(\frac{3}{2}, -4, 3\right)$ into the three equations, using alphabetical order:

$$\begin{array}{r|l} 4x - 2y - 3z & 5 \\ \hline 4 \cdot \frac{3}{2} - 2(-4) - 3 \cdot 3 & ?\ 5 \\ 6 + 8 - 9 & \\ 5 & 5 \quad \text{TRUE} \end{array} \qquad \begin{array}{r|l} -8x - y + z & -5 \\ \hline -8 \cdot \frac{3}{2} - (-4) + 3 & ?\ -5 \\ -12 + 4 + 3 & \\ -5 & -5 \quad \text{TRUE} \end{array}$$

$$\begin{array}{r|l} 2x + y + 2z & 5 \\ \hline 2 \cdot \frac{3}{2} + (-4) + 2 \cdot 3 & ?\ 5 \\ 3 - 4 + 6 & \\ 5 & 5 \quad \text{TRUE} \end{array}$$

The triple makes all three equations true, so it is a solution.

Solving Systems in Three Variables

Graphical methods for solving linear equations in three variables are problematic, because a three-dimensional coordinate system is required and the graph of a linear equation in three variables is a plane. The substitution method *can* be used but becomes very cumbersome unless one or more of the equations has only two variables. Fortunately, the elimination method allows us to manipulate a system of three equations in three variables so that a simpler system of two equations in two variables is formed. Once that simpler system has been solved, we can substitute into one of the three original equations and solve for the third variable.

Example 2 Solve the following system of equations:

$$\begin{aligned} x + y + z &= 4, && (1) \\ x - 2y - z &= 1, && (2) \\ 2x - y - 2z &= -1. && (3) \end{aligned}$$

Solution We select *any* two of the three equations and work to get one equation in two variables. Let's add equations (1) and (2):

$$\begin{aligned} x + y + z &= 4 && (1) \\ x - 2y - z &= 1 && (2) \\ \hline 2x - y \quad &= 5. && (4) \quad \text{Adding to eliminate } z \end{aligned}$$

Next, we select a different pair of equations and eliminate the *same variable* we did above. Let's use equations (1) and (3) to again eliminate z. Be careful here! A common error is to eliminate a different variable in this step.

$$\begin{aligned} x + y + z &= 4, \xrightarrow[\text{equation (1) by 2}]{\text{Multiplying both sides of}} & 2x + 2y + 2z &= 8 \\ 2x - y - 2z &= -1 & 2x - y - 2z &= -1 \\ & & \hline 4x + y \quad &= 7 \quad (5) \end{aligned}$$

Study Tip

When you feel confident in your command of a topic, don't hesitate to help classmates experiencing trouble. Your understanding and retention of a concept will deepen when you explain it to someone else and your classmate will appreciate your help.

Now we solve the resulting system of equations (4) and (5). That solution will give us two of the numbers in the solution of the original system.

$$\begin{aligned} 2x - y &= 5 \quad &(4) \\ 4x + y &= 7 \quad &(5) \\ \hline 6x \quad &= 12 \quad &\text{Adding} \\ x &= 2 \end{aligned}$$

Note that we now have two equations in two variables. Had we not eliminated the same variable in both of the above steps, this would not be the case.

We can use either equation (4) or (5) to find y. We choose equation (5):

$$\begin{aligned} 4x + y &= 7 \quad &(5) \\ 4 \cdot 2 + y &= 7 \quad &\text{Substituting 2 for } x \text{ in equation (5)} \\ 8 + y &= 7 \\ y &= -1. \end{aligned}$$

We now have $x = 2$ and $y = -1$. To find the value for z, we use any of the original three equations and substitute to find the third number, z. Let's use equation (1) and substitute our two numbers in it:

$$\begin{aligned} x + y + z &= 4 \quad &(1) \\ 2 + (-1) + z &= 4 \quad &\text{Substituting 2 for } x \text{ and } -1 \text{ for } y \\ 1 + z &= 4 \\ z &= 3. \end{aligned}$$

We have obtained the triple $(2, -1, 3)$. It should check in *all three* equations:

$$\begin{array}{c|l} x + y + z = 4 & \\ \hline 2 + (-1) + 3 \ ? & 4 \\ 4 & 4 \quad \text{TRUE} \end{array}$$

$$\begin{array}{c|l} x - 2y - z = 1 & \\ \hline 2 - 2(-1) - 3 \ ? & 1 \\ 1 & 1 \quad \text{TRUE} \end{array}$$

$$\begin{array}{c|l} 2x - y - 2z = -1 & \\ \hline 2 \cdot 2 - (-1) - 2 \cdot 3 \ ? & -1 \\ -1 & -1 \quad \text{TRUE} \end{array}$$

The solution is $(2, -1, 3)$.

Solving Systems of Three Linear Equations

To use the elimination method to solve systems of three linear equations:

1. Write all equations in the standard form $Ax + By + Cz = D$.
2. Clear any decimals or fractions.
3. Choose a variable to eliminate. Then select two of the three equations and work to get one equation in two variables.
4. Next, use a different pair of equations and eliminate the same variable that you did in step (3).
5. Solve the system of equations that resulted from steps (3) and (4).
6. Substitute the solution from step (5) into one of the original three equations and solve for the third variable. Then check.

Example 3 Solve the system

$$\begin{aligned} 4x - 2y - 3z &= 5, & (1)\\ -8x - y + z &= -5, & (2)\\ 2x + y + 2z &= 5. & (3) \end{aligned}$$

Solution

1., 2. The equations are already in standard form with no fractions or decimals.

3. Next, select a variable to eliminate. We decide on y because the y-terms are opposites of each other in equations (2) and (3). We add:

$$\begin{aligned} -8x - y + z &= -5 & (2)\\ 2x + y + 2z &= 5 & (3)\\ \hline -6x + 3z &= 0. & (4) \quad \text{Adding} \end{aligned}$$

4. We use another pair of equations to create a second equation in x and z. That is, we eliminate the same variable, y, as in step (3). We use equations (1) and (3):

$$\begin{aligned} 4x - 2y - 3z &= 5,\\ 2x + y + 2z &= 5 \end{aligned} \xrightarrow[\text{equation (3) by 2}]{\text{Multiplying both sides of}} \begin{aligned} 4x - 2y - 3z &= 5\\ 4x + 2y + 4z &= 10\\ \hline 8x + z &= 15. \quad (5) \end{aligned}$$

5. Now we solve the resulting system of equations (4) and (5). That allows us to find two parts of the ordered triple.

$$\begin{aligned} -6x + 3z &= 0,\\ 8x + z &= 15 \end{aligned} \xrightarrow[\text{equation (5) by } -3]{\text{Multiplying both sides of}} \begin{aligned} -6x + 3z &= 0\\ -24x - 3z &= -45\\ \hline -30x &= -45\\ x &= \tfrac{-45}{-30} = \tfrac{3}{2} \end{aligned}$$

We use equation (5) to find z:

$$\begin{aligned} 8x + z &= 15\\ 8 \cdot \tfrac{3}{2} + z &= 15 \quad \text{Substituting } \tfrac{3}{2} \text{ for } x\\ 12 + z &= 15\\ z &= 3. \end{aligned}$$

6. Finally, we use any of the original equations and substitute to find the third number, y. We choose equation (3):

$$\begin{aligned} 2x + y + 2z &= 5 \quad (3)\\ 2 \cdot \tfrac{3}{2} + y + 2 \cdot 3 &= 5 \quad \text{Substituting } \tfrac{3}{2} \text{ for } x \text{ and 3 for } z\\ 3 + y + 6 &= 5\\ y + 9 &= 5\\ y &= -4. \end{aligned}$$

The solution is $\left(\frac{3}{2}, -4, 3\right)$. The check was performed as Example 1.

Sometimes, certain variables are missing at the outset.

Example 4 Solve the system

$$\begin{aligned} x + y + z &= 180, && (1)\\ x \quad - z &= -70, && (2)\\ 2y - z &= 0. && (3) \end{aligned}$$

Solution

1., 2. The equations appear in standard form with no fractions or decimals.

3., 4. Note that there is no y in equation (2). Thus, at the outset, we already have y eliminated from one equation. We need another equation with y eliminated, so we use equations (1) and (3):

$$\begin{array}{l} x + y + z = 180, \xrightarrow[\text{equation (1) by } -2]{\text{Multiplying both sides of}} \\ 2y - z = 0 \end{array} \qquad \begin{array}{r} -2x - 2y - 2z = -360 \\ 2y - z = 0 \\ \hline -2x \quad - 3z = -360. \quad (4) \end{array}$$

5., 6. Now we solve the resulting system of equations (2) and (4):

$$\begin{array}{l} x - z = -70, \xrightarrow[\text{equation (2) by } 2]{\text{Multiplying both sides of}} \\ -2x - 3z = -360 \end{array} \qquad \begin{array}{r} 2x - 2z = -140 \\ -2x - 3z = -360 \\ \hline -5z = -500 \\ z = 100. \end{array}$$

Continuing as in Examples 2 and 3, we get the solution (30, 50, 100). The check is left to the student.

Dependency, Inconsistency, and Geometric Considerations

Each equation in Examples 2, 3, and 4 has a graph that is a plane in three dimensions. The solutions are points common to the planes of each system. Since three planes can have an infinite number of points in common or no points at all in common, we need to generalize the concept of *consistency.*

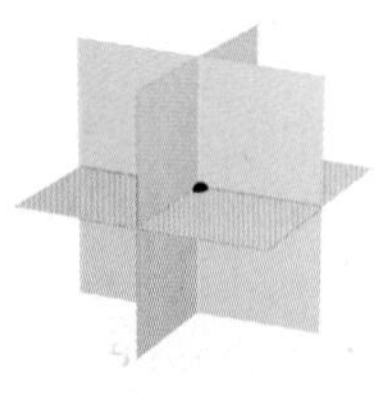

One solution: planes intersecting in exactly one point. System is consistent.

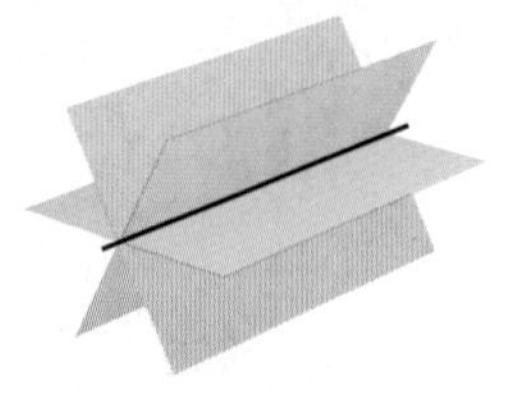

The planes intersect along a common line. An infinite number of points are common to the three planes. System is consistent.

Three parallel planes. There is no common point of intersection. System is inconsistent.

Planes intersect two at a time, but there is no point common to all three. System is inconsistent.

Consistency

A system of equations that has at least one solution is said to be **consistent.**

A system of equations that has no solution is said to be **inconsistent.**

Example 5 Solve:

$$\begin{aligned} y + 3z &= 4, \quad &(1)\\ -x - y + 2z &= 0, \quad &(2)\\ x + 2y + z &= 1. \quad &(3) \end{aligned}$$

Solution The variable x is missing in equation (1). By adding equations (2) and (3), we can find a second equation in which x is missing:

$$\begin{aligned} -x - y + 2z &= 0 \quad &(2)\\ x + 2y + z &= 1 \quad &(3)\\ \hline y + 3z &= 1. \quad &(4) \quad \text{Adding} \end{aligned}$$

Equations (1) and (4) form a system in y and z. We solve as before:

$$\begin{aligned} y + 3z &= 4, \xrightarrow[\text{equation (1) by } -1]{\text{Multiplying both sides of}} & -y - 3z &= -4\\ y + 3z &= 1 & y + 3z &= 1\\ \hline & \text{This is a contradiction.} \longrightarrow & 0 &= -3. \quad \text{Adding} \end{aligned}$$

Since we end up with a *false* equation, or contradiction, we know that the system has no solution. It is *inconsistent.*

The notion of *dependency* from Section 3.1 can also be extended.

Example 6 Solve:

$$\begin{aligned} 2x + y + z &= 3, \quad &(1)\\ x - 2y - z &= 1, \quad &(2)\\ 3x + 4y + 3z &= 5. \quad &(3) \end{aligned}$$

Solution Our plan is to first use equations (1) and (2) to eliminate z. Then we will select another pair of equations and again eliminate z:

$$\begin{array}{l} 2x + y + z = 3 \\ \underline{x - 2y - z = 1} \\ 3x - y \quad = 4. \qquad (4) \end{array}$$

Next, we use equations (2) and (3) to eliminate z again:

$$\begin{array}{l} x - 2y - z = 1, \\ 3x + 4y + 3z = 5 \end{array} \xrightarrow[\text{equation (2) by 3}]{\text{Multiplying both sides of}} \begin{array}{l} 3x - 6y - 3z = 3 \\ \underline{3x + 4y + 3z = 5} \\ 6x - 2y \quad = 8. \qquad (5) \end{array}$$

We now try to solve the resulting system of equations (4) and (5):

$$\begin{array}{l} 3x - y = 4, \\ 6x - 2y = 8 \end{array} \xrightarrow[\text{equation (4) by } -2]{\text{Multiplying both sides of}} \begin{array}{l} -6x + 2y = -8 \\ \underline{6x - 2y = 8} \\ 0 = 0. \qquad (6) \end{array}$$

Equation (6), which is an identity, indicates that equations (1), (2), and (3) are *dependent*. This means that the original system of three equations is equivalent to a system of two equations. One way to see this is to observe that two times equation (1), minus equation (2), is equation (3). Thus removing equation (3) from the system does not affect the solution of the system.* In writing an answer to this problem, we simply state that "the equations are dependent."

Recall that when dependent equations appeared in Section 3.1, the solution sets were always infinite in size and were written in set-builder notation. There, all systems of dependent equations were *consistent*. This is not always the case for systems of three or more equations. The following figures illustrate some possibilities geometrically.

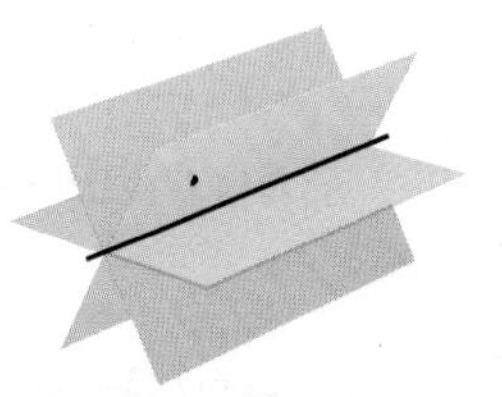

The planes intersect along a common line. The equations are dependent and the system is consistent. There is an infinite number of solutions.

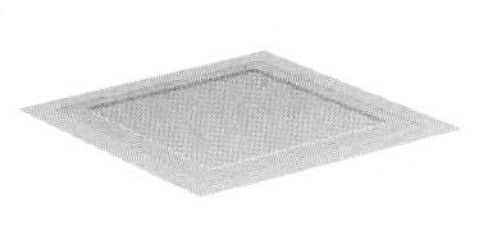

The planes coincide. The equations are dependent and the system is consistent. There is an infinite number of solutions.

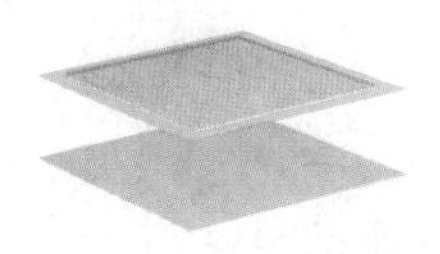

Two planes coincide. The third plane is parallel. The equations are dependent and the system is inconsistent. There is no solution.

*A set of equations is dependent if at least one equation can be expressed as a sum of multiples of other equations in that set.

Exercise Set 3.4

FOR EXTRA HELP

Digital Video Tutor CD 2
Videotape 5

InterAct Math

Math Tutor Center

MathXL

MyMathLab.com

1. Determine whether $(2, -1, -2)$ is a solution of the system

$$x + y - 2z = 5,$$
$$2x - y - z = 7,$$
$$-x - 2y + 3z = 6.$$

2. Determine whether $(1, -2, 3)$ is a solution of the system

$$x + y + z = 2,$$
$$x - 2y - z = 2,$$
$$3x + 2y + z = 2.$$

Solve each system. If a system's equations are dependent or if there is no solution, state this.

3. $x + y + z = 6$, $2x - y + 3z = 9$, $-x + 2y + 2z = 9$

4. $2x - y + z = 10$, $4x + 2y - 3z = 10$, $x - 3y + 2z = 8$

5. $2x - y - 3z = -1$, $2x - y + z = -9$, $x + 2y - 4z = 17$

6. $x - y + z = 6$, $2x + 3y + 2z = 2$, $3x + 5y + 4z = 4$

7. $2x - 3y + z = 5$, $x + 3y + 8z = 22$, $3x - y + 2z = 12$

8. $6x - 4y + 5z = 31$, $5x + 2y + 2z = 13$, $x + y + z = 2$

9. $3a - 2b + 7c = 13$, $a + 8b - 6c = -47$, $7a - 9b - 9c = -3$

10. $x + y + z = 0$, $2x + 3y + 2z = -3$, $-x + 2y - 3z = -1$

11. $2x + 3y + z = 17$, $x - 3y + 2z = -8$, $5x - 2y + 3z = 5$

12. $2x + y - 3z = -4$, $4x - 2y + z = 9$, $3x + 5y - 2z = 5$

13. $2x + y + z = -2$, $2x - y + 3z = 6$, $3x - 5y + 4z = 7$

14. $2x + y + 2z = 11$, $3x + 2y + 2z = 8$, $x + 4y + 3z = 0$

15. $x - y + z = 4$, $5x + 2y - 3z = 2$, $4x + 3y - 4z = -2$

16. $-2x + 8y + 2z = 4$, $x + 6y + 3z = 4$, $3x - 2y + z = 0$

17. $a + 2b + c = 1$, $7a + 3b - c = -2$, $a + 5b + 3c = 2$

18. $4x - y - z = 4$, $2x + y + z = -1$, $6x - 3y - 2z = 3$

19. $5x + 3y + \frac{1}{2}z = \frac{7}{2}$, $0.5x - 0.9y - 0.2z = 0.3$, $3x - 2.4y + 0.4z = -1$

20. $r + \frac{3}{2}s + 6t = 2$, $2r - 3s + 3t = 0.5$, $r + s + t = 1$

21. $3p + 2r = 11$, $q - 7r = 4$, $p - 6q = 1$

22. $4a + 9b = 8$, $8a + 6c = -1$, $6b + 6c = -1$

23. $x + y + z = 105$, $10y - z = 11$, $2x - 3y = 7$

24. $x + y + z = 57$, $-2x + y = 3$, $x - z = 6$

25. $2a - 3b = 2$, $7a + 4c = \frac{3}{4}$, $2c - 3b = 1$

26. $a - 3c = 6$, $b + 2c = 2$, $7a - 3b - 5c = 14$

Aha! **27.** $x + y + z = 182$, $y = 2 + 3x$, $z = 80 + x$

28. $l + m = 7$, $3m + 2n = 9$, $4l + n = 5$

29. $x + y = 0$, $x + z = 1$, $2x + y + z = 2$

30. $x + + z = 0$, $x + y + 2z = 3$, $y + z = 2$

31. $y + z = 1$, $x + y + z = 1$, $x + 2y + 2z = 2$

32. $x + y + z = 1$, $-x + 2y + z = 2$, $2x - y = -1$

33. Abbie recommends that a frustrated classmate double- and triple-check each step of work when attempting to solve a system of three equations. Is this good advice? Why or why not?

34. Describe a method for writing an inconsistent system of three equations in three variables.

SKILL MAINTENANCE

Translate each sentence to mathematics.

35. One number is twice another.

36. The sum of two numbers is three times the first number.

37. The sum of three consecutive numbers is 45.

38. One number plus twice another number is 17.

39. The sum of two numbers is five times a third number.

40. The product of two numbers is twice their sum.

SYNTHESIS

41. Is it possible for a system of three equations to have exactly two ordered triples in its solution set? Why or why not?

42. Describe a procedure that could be used to solve a system of four equations in four variables.

Solve.

43. $\frac{x+2}{3} - \frac{y+4}{2} + \frac{z+1}{6} = 0,$
$\frac{x-4}{3} + \frac{y+1}{4} - \frac{z-2}{2} = -1,$
$\frac{x+1}{2} + \frac{y}{2} + \frac{z-1}{4} = \frac{3}{4}$

44. $w + x + y + z = 2,$
$w + 2x + 2y + 4z = 1,$
$w - x + y + z = 6,$
$w - 3x - y + z = 2$

45. $w + x - y + z = 0,$
$w - 2x - 2y - z = -5,$
$w - 3x - y + z = 4,$
$2w - x - y + 3z = 7$

For Exercises 46 and 47, let u represent $1/x$*, v represent* $1/y$*, and w represent* $1/z$*. Solve for u, v, and w, and then solve for x, y, and z.*

46. $\frac{2}{x} - \frac{1}{y} - \frac{3}{z} = -1,$
$\frac{2}{x} - \frac{1}{y} + \frac{1}{z} = -9,$
$\frac{1}{x} + \frac{2}{y} - \frac{4}{z} = 17$

47. $\frac{2}{x} + \frac{2}{y} - \frac{3}{z} = 3,$
$\frac{1}{x} - \frac{2}{y} - \frac{3}{z} = 9,$
$\frac{7}{x} - \frac{2}{y} + \frac{9}{z} = -39$

Determine k so that each system is dependent.

48. $x - 3y + 2z = 1,$
$2x + y - z = 3,$
$9x - 6y + 3z = k$

49. $5x - 6y + kz = -5,$
$x + 3y - 2z = 2,$
$2x - y + 4z = -1$

In each case, three solutions of an equation in x, y, and z are given. Find the equation.

50. $Ax + By + Cz = 12;$
$(1, \frac{3}{4}, 3)$, $(\frac{4}{3}, 1, 2)$, and $(2, 1, 1)$

51. $z = b - mx - ny;$
$(1, 1, 2)$, $(3, 2, -6)$, and $(\frac{3}{2}, 1, 1)$

52. Write an inconsistent system of equations that contains dependent equations.

COLLABORATIVE CORNER

Finding the Preferred Approach

Focus: Systems of three linear equations
Time: 10–15 minutes
Group size: 3

Consider the six steps outlined on p. 174 along with the following system:

$$2x + 4y = 3 - 5z,$$
$$0.3x = 0.2y + 0.7z + 1.4,$$
$$0.04x + 0.03y = 0.07 + 0.04z.$$

ACTIVITY

1. Working independently, each group member should solve the system above. One person should begin by eliminating x, one should first eliminate y, and one should first eliminate z. Write neatly so that others can follow your steps.
2. Once all group members have solved the system, compare your answers. If the answers do not check, exchange notebooks and check each other's work. If a mistake is detected, allow the person who made the mistake to make the repair.
3. Decide as a group which of the three approaches above (if any) ranks as easiest and which (if any) ranks as most difficult. Then compare your rankings with the other groups in the class.

Solving Applications: Systems of Three Equations

3.5

Applications of Three Equations in Three Unknowns

Solving systems of three or more equations is important in many applications. Such systems arise in the natural and social sciences, business, and engineering. In mathematics, purely numerical applications also arise.

Example 1

The sum of three numbers is 4. The first number minus twice the second, minus the third is 1. Twice the first number minus the second, minus twice the third is -1. Find the numbers.

Solution

1. **Familiarize.** There are three statements involving the same three numbers. Let's label these numbers x, y, and z.
2. **Translate.** We can translate directly as follows.

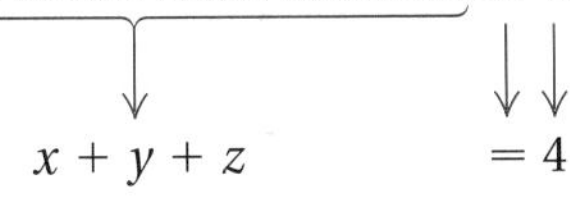
The sum of the three numbers is 4.
$x + y + z \quad = 4$

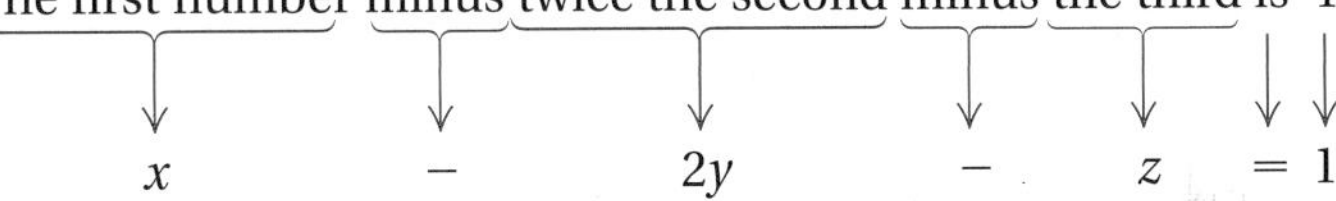
The first number minus twice the second minus the third is 1.
$x \quad - \quad 2y \quad - \quad z \quad = 1$

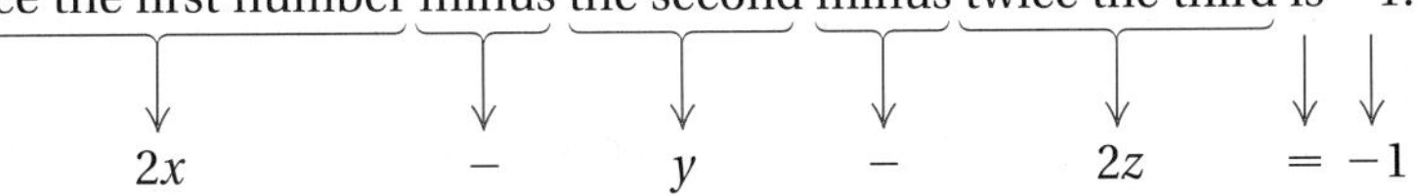
Twice the first number minus the second minus twice the third is -1.
$2x \quad - \quad y \quad - \quad 2z \quad = -1$

We now have a system of three equations:

$$\begin{aligned} x + y + z &= 4, \\ x - 2y - z &= 1, \\ 2x - y - 2z &= -1. \end{aligned}$$

3. **Carry out.** We need to solve the system of equations. Note that we found the solution, $(2, -1, 3)$, in Example 2 of Section 3.4.
4. **Check.** The first statement of the problem says that the sum of the three numbers is 4. That checks, because $2 + (-1) + 3 = 4$. The second statement says that the first number minus twice the second, minus the third is 1: $2 - 2(-1) - 3 = 1$. That checks. The check of the third statement is left to the student.
5. **State.** The three numbers are 2, -1, and 3.

Example 2

Architecture. In a triangular cross section of a roof, the largest angle is 70° greater than the smallest angle. The largest angle is twice as large as the remaining angle. Find the measure of each angle.

Solution

1. **Familiarize.** The first thing we do is make a drawing, or a sketch.

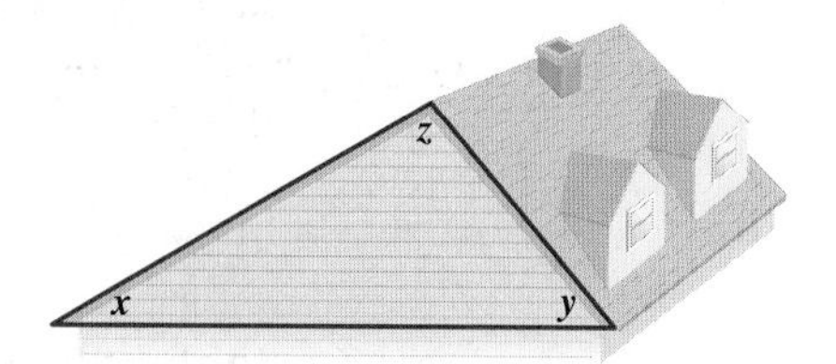
z
x
y

Since we don't know the size of any angle, we use x, y, and z to represent the three measures, from smallest to largest. Recall that the measures of the angles in any triangle add up to 180°.

2. **Translate.** This geometric fact about triangles gives us one equation:

$$x + y + z = 180.$$

Two of the statements can be translated almost directly.

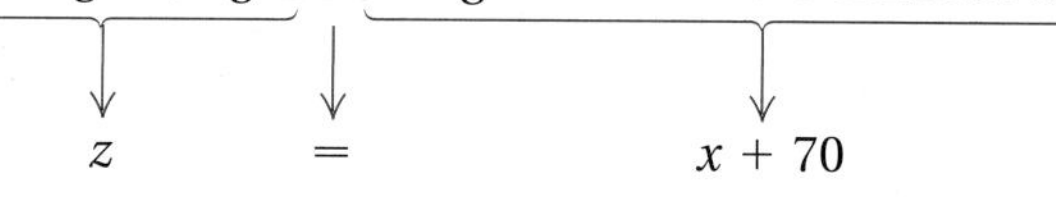
The largest angle is 70° greater than the smallest angle.
z
=
x + 70

The largest angle is twice as large as the remaining angle.
z
=
2y

We now have a system of three equations:

$$\begin{aligned} x + y + z &= 180, \\ x + 70 &= z, \\ 2y &= z; \end{aligned} \qquad \text{or} \qquad \begin{aligned} x + y + z &= 180, \\ x \quad\;\; - z &= -70, \\ 2y - z &= 0. \end{aligned} \qquad \begin{array}{l}\text{Rewriting in} \\ \text{standard form}\end{array}$$

3. **Carry out.** The system was solved in Example 4 of Section 3.4. The solution is $(30, 50, 100)$.

4. **Check.** The sum of the numbers is 180, so that checks. The measure of the largest angle, 100°, is 70° greater than the measure of the smallest angle, 30°, so that checks. The measure of the largest angle is also twice the measure of the remaining angle, 50°. Thus we have a check.

5. **State.** The angles in the triangle measure 30°, 50°, and 100°.

Example 3

Cholesterol levels. Americans have become very conscious of their cholesterol levels. Recent studies indicate that a child's intake of cholesterol should be no more than 300 mg per day. By eating 1 egg, 1 cupcake, and 1 slice of pizza, a child consumes 302 mg of cholesterol. If the child eats 2 cupcakes and 3 slices of pizza, he or she takes in 65 mg of cholesterol. By eating 2 eggs and 1 cupcake, a child consumes 567 mg of cholesterol. How much cholesterol is in each item?

Solution

1. **Familiarize.** After we have read the problem a few times, it becomes clear that an egg contains considerably more cholesterol than the other foods. Let's guess that one egg contains 200 mg of cholesterol and one cupcake contains 50 mg. Because of the third sentence in the problem, it would follow that a slice of pizza contains 52 mg of cholesterol since $200 + 50 + 52 = 302$.

   To see if our guess satisfies the other statements in the problem, we find the amount of cholesterol that 2 cupcakes and 3 slices of pizza would contain: $2 \cdot 50 + 3 \cdot 52 = 256$. Since this does not match the 65 mg listed in the fourth sentence of the problem, our guess was incorrect. Rather than guess again, we examine how we checked our guess and let e, c, and $s =$ the number of milligrams of cholesterol in an egg, a cupcake, and a slice of pizza, respectively.

2. **Translate.** By rewording some of the sentences in the problem, we can translate it into three equations.

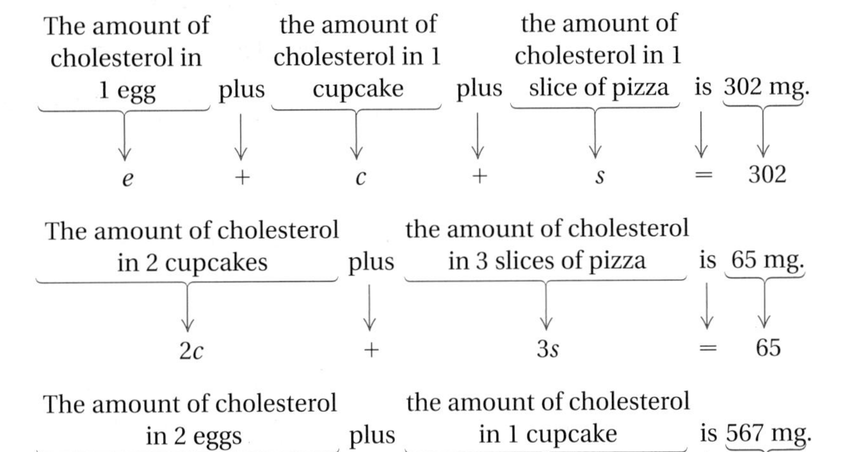

   We now have a system of three equations:

$$\begin{aligned} e + c + s &= 302, \\ 2c + 3s &= 65, \\ 2e + c \quad &= 567. \end{aligned}$$

3. **Carry out.** We solve and get $e = 274$, $c = 19$, $s = 9$, or $(274, 19, 9)$.

4. **Check.** The sum of 274, 19, and 9 is 302 so the total cholesterol in 1 egg, 1 cupcake, and 1 slice of pizza checks. Two cupcakes and three slices of pizza would contain $2 \cdot 19 + 3 \cdot 9 = 65$ mg, while two eggs and one cupcake would contain $2 \cdot 274 + 19 = 567$ mg of cholesterol. The answer checks.

5. **State.** An egg contains 274 mg of cholesterol, a cupcake contains 19 mg of cholesterol, and a slice of pizza contains 9 mg of cholesterol.

Exercise Set 3.5

FOR EXTRA HELP

Digital Video Tutor CD 3
Videotape 6

InterAct Math

Math Tutor Center

MathXL

MyMathLab.com

Solve.

1. The sum of three numbers is 57. The second is 3 more than the first. The third is 6 more than the first. Find the numbers.

2. The sum of three numbers is 5. The first number minus the second plus the third is 1. The first minus the third is 3 more than the second. Find the numbers.

3. The sum of three numbers is 26. Twice the first minus the second is 2 less than the third. The third is the second minus three times the first. Find the numbers.

4. The sum of three numbers is 105. The third is 11 less than ten times the second. Twice the first is 7 more than three times the second. Find the numbers.

5. *Geometry.* In triangle ABC, the measure of angle B is three times that of angle A. The measure of angle C is 20° more than that of angle A. Find the angle measures.

6. *Geometry.* In triangle ABC, the measure of angle B is twice the measure of angle A. The measure of angle C is 80° more than that of angle A. Find the angle measures.

7. *Automobile pricing.* A recent basic model of a particular automobile had a price of $12,685. The basic model with the added features of automatic transmission and power door locks was $14,070. The basic model with air conditioning (AC) and power door locks was $13,580. The basic model with AC and automatic transmission was $13,925. What was the individual cost of each of the three options?

8. *Lens production.* When Sight-Rite's three polishing machines, A, B, and C, are all working, 5700 lenses can be polished in one week. When only A and B are working, 3400 lenses can be polished in one week. When only B and C are working, 4200 lenses can be polished in one week. How many lenses can be polished in a week by each machine?

Aha! **9.** *Welding rates.* Elrod, Dot, and Wendy can weld 74 linear feet per hour when working together. Elrod and Dot together can weld 44 linear feet per hour, while Elrod and Wendy can weld 50 linear feet per hour. How many linear feet per hour can each weld alone?

10. *Telemarketing.* Sven, Tillie, and Isaiah can process 740 telephone orders per day. Sven and Tillie together can process 470 orders, while Tillie and Isaiah together can process 520 orders per day. How many orders can each person process alone?

11. *Restaurant management.* Kyle works at Dunkin® Donuts, where a 10-oz cup of coffee costs $1.05, a 14-oz cup costs $1.35, and a 20-oz cup costs $1.65. During one busy period, Kyle served 34 cups of coffee, emptying five 96-oz pots while collecting a total of $45. How many cups of each size did Kyle fill?

12. *Advertising.* In a recent year, companies spent a total of $84.8 billion on newspaper, television, and radio ads. The total amount spent on television and radio ads was only $2.6 billion more than the amount spent on newspaper ads alone. The amount spent on newspaper ads was $5.1 billion more than what was spent on television ads. How much was spent on each form of advertising?

13. *Investments.* A business class divided an imaginary investment of $80,000 among three mutual funds. The first fund grew by 10%, the second by 6%, and the third by 15%. Total earnings were $8850. The earnings from the first fund were $750 more than the earnings from the third. How much was invested in each fund?

14. *Restaurant management.* McDonald's® recently sold small soft drinks for 87¢, medium soft drinks for $1.08, and large soft drinks for $1.54. During a lunch-time rush, Chris sold 40 soft drinks for a total of $43.40. The number of small and large drinks, combined, was 10 fewer than the number of medium drinks. How many drinks of each size were sold?

15. *Nutrition.* A dietician in a hospital prepares meals under the guidance of a physician. Suppose that for a particular patient a physician prescribes a meal to have 800 calories, 55 g of protein, and 220 mg of vitamin C. The dietician prepares a meal of roast beef, baked potatoes, and broccoli according to the data in the following table.

	Calories	Protein (in grams)	Vitamin C (in milligrams)
Roast Beef, 3 oz	300	20	0
Baked Potato	100	5	20
Broccoli, 156 g	50	5	100

How many servings of each food are needed in order to satisfy the doctor's orders?

16. *Nutrition.* Repeat Exercise 15 but replace the broccoli with asparagus, for which a 180-g serving contains 50 calories, 5 g of protein, and 44 mg of vitamin C. Which meal would you prefer eating?

17. *Crying rate.* The sum of the average number of times a man, a woman, and a one-year-old child cry each month is 71.7. A one-year-old cries 46.4 more times than a man. The average number of times a one-year-old cries per month is 28.3 more than the average number of times combined that a man and a woman cry. What is the average number of times per month that each cries?

18. *Obstetrics.* In the United States, the highest incidence of fraternal twin births occurs among Asian-Americans, then African-Americans, and then Caucasians. Out of every 15,400 births, the total number of fraternal twin births for all three is 739, where there are 185 more for Asian-Americans than African-Americans and 231 more for Asian-Americans than Caucasians. How many births of fraternal twins are there for each group out of every 15,400 births?

19. *Basketball scoring.* The New York Knicks recently scored a total of 92 points on a combination of 2-point field goals, 3-point field goals, and 1-point foul shots. Altogether, the Knicks made 50 baskets and 19 more 2-pointers than foul shots. How many shots of each kind were made?

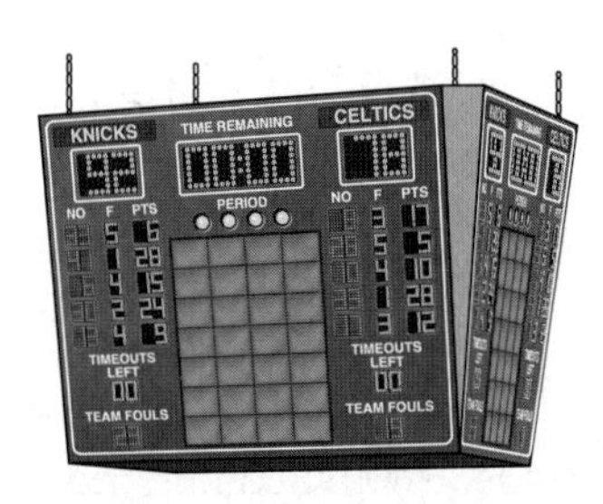

20. *History.* Find the year in which the first U.S. transcontinental railroad was completed. The following are some facts about the number. The sum of the digits in the year is 24. The ones digit is 1 more than the hundreds digit. Both the tens and the ones digits are multiples of 3.

21. Problems like Exercises 11 and 12 could be classified as total-value problems. How do these problems differ from the total-value problems of Section 3.3?

22. Write a problem for a classmate to solve. Design the problem so that it translates to a system of three equations in three variables.

SKILL MAINTENANCE

Simplify.

23. $5(-3) + 7$

24. $-4(-6) + 9$

25. $-6(8) + (-7)$

26. $7(-9) + (-8)$

27. $-7(2x - 3y + 5z)$

28. $-6(4a + 7b - 9c)$

29. $-4(2a + 5b) + 3a + 20b$

30. $3(2x - 7y) + 5x + 21y$

SYNTHESIS

31. Consider Exercise 19. Suppose there were no foul shots made. Would there still be a solution? Why or why not?

32. Consider Exercise 11. Suppose Kyle collected \$46. Could the problem still be solved? Why or why not?

33. Find a three-digit positive integer such that the sum of all three digits is 14, the tens digit is 2 more than the ones digit, and if the digits are reversed, the number is unchanged.

34. *Ages.* Tammy's age is the sum of the ages of Carmen and Dennis. Carmen's age is 2 more than the sum of the ages of Dennis and Mark. Dennis's age is four times Mark's age. The sum of all four ages is 42. How old is Tammy?

35. *Ticket revenue.* The Pops concert audience of 100 people consists of adults, students, and children. The ticket prices are \$10 for adults, \$3 for students, and 50¢ for children. The total amount of money taken in is \$100. How many adults, students, and children are in attendance? Does there seem to be some information missing? Do some more careful reasoning.

36. *Sharing raffle tickets.* Hal gives Tom as many raffle tickets as Tom first had and Gary as many as Gary first had. In like manner, Tom then gives Hal and Gary as many tickets as each then has. Similarly, Gary gives Hal and Tom as many tickets as each then has. If each finally has 40 tickets, with how many tickets does Tom begin?

37. Find the sum of the angle measures at the tips of the star in this figure.

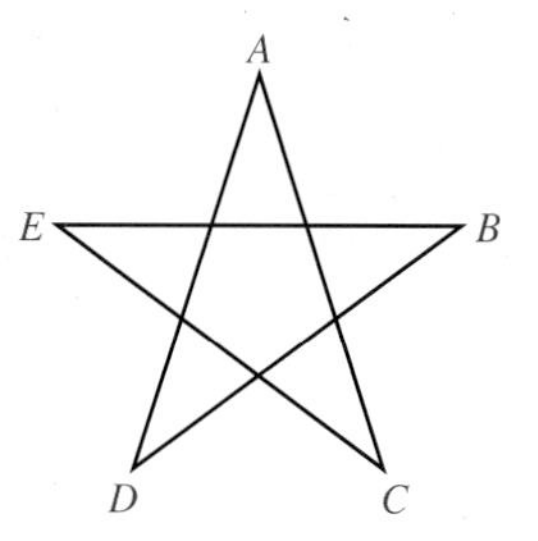

Elimination Using Matrices

3.6

Matrices and Systems • Row-Equivalent Operations

In solving systems of equations, we perform computations with the constants. The variables play no important role until the end. Thus we can simplify writing a system by omitting the variables. For example, the system

$$\begin{aligned} 3x + 4y &= 5, \\ x - 2y &= 1 \end{aligned} \quad \text{simplifies to} \quad \begin{array}{ccc} 3 & 4 & 5 \\ 1 & -2 & 1 \end{array}$$

if we do not write the variables, the operation of addition, and the equals signs.

Matrices and Systems

In the example above, we have written a rectangular array of numbers. Such an array is called a **matrix** (plural, **matrices**). We ordinarily write brackets around matrices. The following are matrices:

$$\begin{bmatrix} -3 & 1 \\ 0 & 5 \end{bmatrix}, \quad \begin{bmatrix} 2 & 0 & -1 & 3 \\ -5 & 2 & 7 & -1 \\ 4 & 5 & 3 & 0 \end{bmatrix}, \quad \begin{bmatrix} 2 & 3 \\ 7 & 15 \\ -2 & 23 \\ 4 & 1 \end{bmatrix}$$

The individual numbers are called *elements* or *entries.*

The **rows** of a matrix are horizontal, and the **columns** are vertical.

$$\begin{bmatrix} 5 & -2 & 2 \\ 1 & 0 & 1 \\ 0 & 1 & 2 \end{bmatrix} \begin{matrix} \longrightarrow \text{row 1} \\ \longrightarrow \text{row 2} \\ \longrightarrow \text{row 3} \end{matrix}$$

↑ column 1 ↑ column 2 ↑ column 3

Let's see how matrices can be used to solve a system.

Example 1 Solve the system

$$5x - 4y = -1,$$
$$-2x + 3y = 2.$$

As an aid for understanding, we list the corresponding system in the margin.

$5x - 4y = -1,$
$-2x + 3y = 2$

Solution We write a matrix using only coefficients and constants, listing x-coefficients in the first column and y-coefficients in the second. Note that in each matrix a dashed line separates the coefficients from the constants:

$$\left[\begin{array}{cc:c} 5 & -4 & -1 \\ -2 & 3 & 2 \end{array}\right].$$

Our goal is to transform

$$\left[\begin{array}{cc:c} 5 & -4 & -1 \\ -2 & 3 & 2 \end{array}\right] \quad \text{into the form} \quad \left[\begin{array}{cc:c} a & b & c \\ 0 & d & e \end{array}\right].$$

The variables x and y can then be reinserted to form equations from which we can complete the solution.

We do calculations that are similar to those that we would do if we wrote the entire equations. The first step is to multiply and/or interchange the rows so that each number in the first column below the first number is a multiple of that number. Here that means multiplying Row 2 by 5. This corresponds to multiplying both sides of the second equation by 5.

$5x - 4y = -1,$
$-10x + 15y = 10$

$$\left[\begin{array}{cc:c} 5 & -4 & -1 \\ -10 & 15 & 10 \end{array}\right] \quad \text{New Row 2} = 5(\text{Row 2 from above})$$

Next, we multiply the first row by 2, add this to Row 2, and write that result as the "new" Row 2. This corresponds to multiplying the first equation by 2 and adding the result to the second equation in order to eliminate a variable. Write out these computations as necessary—we perform them mentally.

$$5x - 4y = -1,$$
$$7y = 8$$

$$\left[\begin{array}{cc|c} 5 & -4 & -1 \\ 0 & 7 & 8 \end{array}\right]$$

$2(5 \quad -4 \mid -1) = (10 \quad -8 \mid -2)$ and
$(10 \quad -8 \mid -2) + (-10 \quad 15 \mid 10) = (0 \quad 7 \mid 8)$
New Row 2 = 2(Row 1) + (Row 2)

If we now reinsert the variables, we have

$$5x - 4y = -1, \quad (1)$$
$$7y = 8. \quad (2)$$

We can now proceed as before, solving equation (2) for y:

$$7y = 8 \quad (2)$$
$$y = \tfrac{8}{7}.$$

Next, we substitute $\frac{8}{7}$ for y in equation (1):

$5x - 4y = -1$ (1)
$5x - 4 \cdot \frac{8}{7} = -1$ Substituting $\frac{8}{7}$ for y in equation (1)
$x = \frac{5}{7}$. Solving for x

The solution is $\left(\frac{5}{7}, \frac{8}{7}\right)$. The check is left to the student.

Example 2 Solve the system

$$2x - y + 4z = -3,$$
$$x \quad\quad - 4z = 5,$$
$$6x - y + 2z = 10.$$

Solution We first write a matrix, using only the constants. Where there are missing terms, we must write 0's:

$2x - y + 4z = -3,$
$x - 4z = 5,$
$6x - y + 2z = 10$

$$\left[\begin{array}{ccc|c} 2 & -1 & 4 & -3 \\ 1 & 0 & -4 & 5 \\ 6 & -1 & 2 & 10 \end{array}\right]$$

Our goal is to transform the matrix to one of the form

$ax + by + cz = d,$
$ey + fz = g,$
$hz = i$

$$\left[\begin{array}{ccc|c} a & b & c & d \\ 0 & e & f & g \\ 0 & 0 & h & i \end{array}\right].$$

A matrix of this form can be rewritten as a system of equations that is equivalent to the original system, and from which a solution can be easily found.

The first step is to multiply and/or interchange the rows so that each number in the first column is a multiple of the first number in the first row. In this case, we do so by interchanging Rows 1 and 2:

$x - 4z = 5,$
$2x - y + 4z = -3,$
$6x - y + 2z = 10$

$$\left[\begin{array}{ccc|c} 1 & 0 & -4 & 5 \\ 2 & -1 & 4 & -3 \\ 6 & -1 & 2 & 10 \end{array}\right]$$

This corresponds to interchanging the first two equations.

Next, we multiply the first row by -2, add it to the second row, and replace Row 2 with the result:

$$\begin{aligned} x \quad\; - 4z &= 5, \\ -y + 12z &= -13, \\ 6x - y + 2z &= 10 \end{aligned}$$

$$\left[\begin{array}{rrr|r} 1 & 0 & -4 & 5 \\ 0 & -1 & 12 & -13 \\ 6 & -1 & 2 & 10 \end{array}\right].$$

$-2(1\ \ 0\ \ -4 \mid 5) = (-2\ \ 0\ \ 8 \mid -10)$ and $(-2\ \ 0\ \ 8 \mid -10) + (2\ \ -1\ \ 4 \mid -3) = (0\ \ -1\ \ 12 \mid -13)$

Now we multiply the first row by -6, add it to the third row, and replace Row 3 with the result:

$$\begin{aligned} x \quad\; - 4z &= 5, \\ -y + 12z &= -13, \\ -y + 26z &= -20 \end{aligned}$$

$$\left[\begin{array}{rrr|r} 1 & 0 & -4 & 5 \\ 0 & -1 & 12 & -13 \\ 0 & -1 & 26 & -20 \end{array}\right].$$

$-6(1\ \ 0\ \ -4 \mid 5) = (-6\ \ 0\ \ 24 \mid -30)$ and $(-6\ \ 0\ \ 24 \mid -30) + (6\ \ -1\ \ 2 \mid 10) = (0\ \ -1\ \ 26 \mid -20)$

Next, we multiply Row 2 by -1, add it to the third row, and replace Row 3 with the result:

$$\begin{aligned} x \quad\; - 4z &= 5, \\ -y + 12z &= -13, \\ 14z &= -7 \end{aligned}$$

$$\left[\begin{array}{rrr|r} 1 & 0 & -4 & 5 \\ 0 & -1 & 12 & -13 \\ 0 & 0 & 14 & -7 \end{array}\right].$$

$-1(0\ \ -1\ \ 12 \mid -13) = (0\ \ 1\ \ -12 \mid 13)$ and $(0\ \ 1\ \ -12 \mid 13) + (0\ \ -1\ \ 26 \mid -20) = (0\ \ 0\ \ 14 \mid -7)$

Reinserting the variables gives us

$$\begin{aligned} x \quad\; - 4z &= 5, \\ -y + 12z &= -13, \\ 14z &= -7. \end{aligned}$$

We now solve this last equation for z and get $z = -\frac{1}{2}$. Next, we substitute $-\frac{1}{2}$ for z in the preceding equation and solve for y: $-y + 12\left(-\frac{1}{2}\right) = -13$, so $y = 7$. Since there is no y-term in the first equation of this last system, we need only substitute $-\frac{1}{2}$ for z to solve for x: $x - 4\left(-\frac{1}{2}\right) = 5$, so $x = 3$. The solution is $\left(3, 7, -\frac{1}{2}\right)$. The check is left to the student.

Row-equivalent operations can be performed on a grapher. For example, to interchange the first and second rows of the matrix, as in step (1) of Example 2 above, we enter the matrix as matrix **A** and select "rowSwap" from the MATRIX MATH menu. Some graphers will not automatically store the matrix produced using a row-equivalent operation, so when several operations are to be performed in succession, it is helpful to store the result of each operation as it is produced. In the window at right, we see both the matrix produced by the rowSwap operation and the indication that this matrix is stored as matrix **B**.

```
rowSwap([A],1,2)→[B]
[[1  0 -4  5]
 [2 -1  4 -3]
 [6 -1  2 10]]
```

1. Use a grapher to proceed through all the steps in Example 2.

The operations used in the preceding example correspond to those used to produce equivalent systems of equations. We call the matrices **row-equivalent** and the operations that produce them **row-equivalent operations**.

Row-Equivalent Operations

> ***Row-Equivalent Operations***
>
> Each of the following row-equivalent operations produces a row-equivalent matrix:
>
> **a)** Interchanging any two rows.
> **b)** Multiplying all elements of a row by a nonzero constant.
> **c)** Replacing a row with the sum of that row and a multiple of another row.

The best overall method for solving systems of equations is by row-equivalent matrices; even computers are programmed to use them. Matrices are part of a branch of mathematics known as linear algebra. They are also studied in many courses in finite mathematics.

Exercise Set 3.6

FOR EXTRA HELP

Digital Video Tutor CD 3
Videotape 6

InterAct Math

Math Tutor Center

MathXL

MyMathLab.com

Solve using matrices.

1. $9x - 2y = 5,$
$3x - 3y = 11$

2. $4x + y = 7,$
$5x - 3y = 13$

3. $x + 4y = 8,$
$3x + 5y = 3$

4. $x + 4y = 5,$
$-3x + 2y = 13$

5. $6x - 2y = 4,$
$7x + y = 13$

6. $3x + 4y = 7,$
$-5x + 2y = 10$

7. $3x + 2y + 2z = 3,$
$x + 2y - z = 5,$
$2x - 4y + z = 0$

8. $4x - y - 3z = 19,$
$8x + y - z = 11,$
$2x + y + 2z = -7$

9. $p - 2q - 3r = 3,$
$2p - q - 2r = 4,$
$4p + 5q + 6r = 4$

10. $x + 2y - 3z = 9,$
$2x - y + 2z = -8,$
$3x - y - 4z = 3$

11. $3p + 2r = 11,$
$q - 7r = 4,$
$p - 6q = 1$

12. $4a + 9b = 8,$
$8a + 6c = -1,$
$6b + 6c = -1$

13. $2x + 2y - 2z - 2w = -10,$
$w + y + z + x = -5,$
$x - y + 4z + 3w = -2,$
$w - 2y + 2z + 3x = -6$

14. $-w - 3y + z + 2x = -8,$
$x + y - z - w = -4,$
$w + y + z + x = 22,$
$x - y - z - w = -14$

Solve using matrices.

15. *Coin value.* A collection of 34 coins consists of dimes and nickels. The total value is $1.90. How many dimes and how many nickels are there?

16. *Coin value.* A collection of 43 coins consists of dimes and quarters. The total value is \$7.60. How many dimes and how many quarters are there?

17. *Mixed granola.* Grace sells two kinds of granola. One is worth \$4.05 per pound and the other is worth \$2.70 per pound. She wants to blend the two granolas to get a 15-lb mixture worth \$3.15 per pound. How much of each kind of granola should be used?

18. *Trail mix.* Phil mixes nuts worth \$1.60 per pound with oats worth \$1.40 per pound to get 20 lb of trail mix worth \$1.54 per pound. How many pounds of nuts and how many pounds of oats should be used?

19. *Investments.* Elena receives \$212 per year in simple interest from three investments totaling \$2500. Part is invested at 7%, part at 8%, and part at 9%. There is \$1100 more invested at 9% than at 8%. Find the amount invested at each rate.

20. *Investments.* Miguel receives \$306 per year in simple interest from three investments totaling \$3200. Part is invested at 8%, part at 9%, and part at 10%. There is \$1900 more invested at 10% than at 9%. Find the amount invested at each rate.

21. Explain how you can recognize dependent equations when solving with matrices.

22. Explain how you can recognize an inconsistent system when solving with matrices.

SKILL MAINTENANCE

Simplify.

23. $5(-3) - (-7)4$

24. $8(-5) - (-2)9$

25. $-2(5 \cdot 3 - 4 \cdot 6) - 3(2 \cdot 7 - 15) + 4(3 \cdot 8 - 5 \cdot 4)$

26. $6(2 \cdot 7 - 3(-4)) - 4(3(-8) - 10) + 5(4 \cdot 3 - (-2)7)$

SYNTHESIS

27. If the matrices

$$\left[\begin{array}{cc|c} a_1 & b_1 & c_1 \\ d_1 & e_1 & f_1 \end{array}\right] \quad \text{and} \quad \left[\begin{array}{cc|c} a_2 & b_2 & c_2 \\ d_2 & e_2 & f_2 \end{array}\right]$$

share the same solution, does it follow that the corresponding entries are all equal to each other ($a_1 = a_2$, $b_1 = b_2$, etc.)? Why or why not?

28. Explain how the row-equivalent operations make use of the addition, multiplication, and distributive properties.

29. The sum of the digits in a four-digit number is 10. Twice the sum of the thousands digit and the tens digit is 1 less than the sum of the other two digits. The tens digit is twice the thousands digit. The ones digit equals the sum of the thousands digit and the hundreds digit. Find the four-digit number.

30. Solve for x and y:

$$ax + by = c,$$
$$dx + ey = f.$$

Determinants and Cramer's Rule

3.7

Determinants of 2 × 2 Matrices • Cramer's Rule: 2 × 2 Systems • Cramer's Rule: 3 × 3 Systems

Determinants of 2 × 2 Matrices

When a matrix has m rows and n columns, it is called an "m by n" matrix. Thus its *dimensions* are denoted by $m \times n$. If a matrix has the same number of rows and columns, it is called a **square matrix**. Associated with every square matrix is a number called its **determinant**, defined as follows for 2×2 matrices.

2 × 2 Determinants

The determinant of a two-by-two matrix $\begin{bmatrix} a & c \\ b & d \end{bmatrix}$ is denoted $\begin{vmatrix} a & c \\ b & d \end{vmatrix}$ and is defined as follows:

$$\begin{vmatrix} a & c \\ b & d \end{vmatrix} = ad - bc.$$

Example 1 Evaluate: $\begin{vmatrix} 2 & -5 \\ 6 & 7 \end{vmatrix}$.

Solution We multiply and subtract as follows:

$$\begin{vmatrix} 2 & -5 \\ 6 & 7 \end{vmatrix} = 2 \cdot 7 - 6 \cdot (-5) = 14 + 30 = 44.$$

Cramer's Rule: 2 × 2 Systems

One of the many uses for determinants is in solving systems of linear equations in which the number of variables is the same as the number of equations and the constants are not all 0. Let's consider a system of two equations:

$$a_1x + b_1y = c_1,$$
$$a_2x + b_2y = c_2.$$

If we use the elimination method, a series of steps can show that

$$x = \frac{c_1b_2 - c_2b_1}{a_1b_2 - a_2b_1} \quad \text{and} \quad y = \frac{a_1c_2 - a_2c_1}{a_1b_2 - a_2b_1}.$$

Determinants can be used in these expressions for x and y.

Cramer's Rule: 2 × 2 Systems

The solution of the system

$$a_1x + b_1y = c_1,$$
$$a_2x + b_2y = c_2,$$

if it is unique, is given by

$$x = \frac{\begin{vmatrix} c_1 & b_1 \\ c_2 & b_2 \end{vmatrix}}{\begin{vmatrix} a_1 & b_1 \\ a_2 & b_2 \end{vmatrix}}, \qquad y = \frac{\begin{vmatrix} a_1 & c_1 \\ a_2 & c_2 \end{vmatrix}}{\begin{vmatrix} a_1 & b_1 \\ a_2 & b_2 \end{vmatrix}}.$$

(continued)

The equations above make sense only if the determinant in the denominator is not 0. If the denominator *is* 0, then one of two things happens.

1. If the denominator is 0 and the other two determinants in the numerators are also 0, then the equations in the system are dependent.
2. If the denominator is 0 and at least one of the other determinants in the numerators is not 0, then the system is inconsistent.

To use Cramer's rule, we find the determinants and compute x and y as shown above. Note that the denominators are identical and the coefficients of x and y appear in the same position as in the original equations. In the numerator of x, the constants c_1 and c_2 replace a_1 and a_2. In the numerator of y, c_1 and c_2 replace b_1 and b_2.

Example 2 Solve using Cramer's rule:

$$2x + 5y = 7,$$
$$5x - 2y = -3.$$

Solution We have

$$x = \frac{\begin{vmatrix} 7 & 5 \\ -3 & -2 \end{vmatrix}}{\begin{vmatrix} 2 & 5 \\ 5 & -2 \end{vmatrix}} \quad \text{Using Cramer's rule}$$

$$= \frac{7(-2) - (-3)5}{2(-2) - 5 \cdot 5} = -\frac{1}{29}$$

and

$$y = \frac{\begin{vmatrix} 2 & 7 \\ 5 & -3 \end{vmatrix}}{\begin{vmatrix} 2 & 5 \\ 5 & -2 \end{vmatrix}} \quad \text{Using Cramer's rule}$$

$$= \frac{2(-3) - 5 \cdot 7}{-29} = \frac{41}{29}. \quad \text{The denominator is the same as in the expression for } x.$$

The solution is $\left(-\frac{1}{29}, \frac{41}{29}\right)$. The check is left to the student.

Cramer's Rule: 3 × 3 Systems

A similar method has been developed for solving systems of three linear equations in 3 unknowns. However, before stating the rule, we must extend our terminology.

3 × 3 Determinants

The determinant of a three-by-three matrix is defined as follows:

Subtract. Add.

$$\begin{vmatrix} a_1 & b_1 & c_1 \\ a_2 & b_2 & c_2 \\ a_3 & b_3 & c_3 \end{vmatrix} = a_1\begin{vmatrix} b_2 & c_2 \\ b_3 & c_3 \end{vmatrix} - a_2\begin{vmatrix} b_1 & c_1 \\ b_3 & c_3 \end{vmatrix} + a_3\begin{vmatrix} b_1 & c_1 \\ b_2 & c_2 \end{vmatrix}$$

Note that the a's come from the first column. Note too that the 2×2 determinants above can be obtained by crossing out the row and the column in which the a occurs.

For a_1: $\begin{vmatrix} \not{a_1} & \not{b_1} & \not{c_1} \\ \not{a_2} & b_2 & c_2 \\ \not{a_3} & b_3 & c_3 \end{vmatrix}$ For a_2: $\begin{vmatrix} \not{a_1} & b_1 & c_1 \\ \not{a_2} & \not{b_2} & \not{c_2} \\ \not{a_3} & b_3 & c_3 \end{vmatrix}$

For a_3: $\begin{vmatrix} \not{a_1} & b_1 & c_1 \\ \not{a_2} & b_2 & c_2 \\ \not{a_3} & \not{b_3} & \not{c_3} \end{vmatrix}$

Example 3 Evaluate:

$$\begin{vmatrix} -1 & 0 & 1 \\ -5 & 1 & -1 \\ 4 & 8 & 1 \end{vmatrix}.$$

Solution We have

Subtract. Add

$$\begin{vmatrix} -1 & 0 & 1 \\ -5 & 1 & -1 \\ 4 & 8 & 1 \end{vmatrix} = -1\begin{vmatrix} 1 & -1 \\ 8 & 1 \end{vmatrix} - (-5)\begin{vmatrix} 0 & 1 \\ 8 & 1 \end{vmatrix} + 4\begin{vmatrix} 0 & 1 \\ 1 & -1 \end{vmatrix}$$

$$= -1(1 + 8) + 5(0 - 8) + 4(0 - 1)$$ Evaluating the three determinants

$$= -9 - 40 - 4 = -53.$$

Cramer's Rule: 3 × 3 Systems

The solution of the system

$$a_1x + b_1y + c_1z = d_1,$$
$$a_2x + b_2y + c_2z = d_2,$$
$$a_3x + b_3y + c_3z = d_3$$

is found by considering the following determinants:

$$D = \begin{vmatrix} a_1 & b_1 & c_1 \\ a_2 & b_2 & c_2 \\ a_3 & b_3 & c_3 \end{vmatrix}, \quad D_x = \begin{vmatrix} d_1 & b_1 & c_1 \\ d_2 & b_2 & c_2 \\ d_3 & b_3 & c_3 \end{vmatrix},$$

$$D_y = \begin{vmatrix} a_1 & d_1 & c_1 \\ a_2 & d_2 & c_2 \\ a_3 & d_3 & c_3 \end{vmatrix}, \quad D_z = \begin{vmatrix} a_1 & b_1 & d_1 \\ a_2 & b_2 & d_2 \\ a_3 & b_3 & d_3 \end{vmatrix}.$$

D contains only coefficients.

In D_x, the d's replace the a's.

In D_y, the d's replace the b's.

In D_z, the d's replace the c's.

If a unique solution exists, it is given by

$$x = \frac{D_x}{D}, \quad y = \frac{D_y}{D}, \quad z = \frac{D_z}{D}.$$

Example 4

Solve using Cramer's rule:

$$x - 3y + 7z = 13,$$
$$x + y + z = 1,$$
$$x - 2y + 3z = 4.$$

Solution We compute D, D_x, D_y, and D_z:

$$D = \begin{vmatrix} 1 & -3 & 7 \\ 1 & 1 & 1 \\ 1 & -2 & 3 \end{vmatrix} = -10; \quad D_x = \begin{vmatrix} 13 & -3 & 7 \\ 1 & 1 & 1 \\ 4 & -2 & 3 \end{vmatrix} = 20;$$

$$D_y = \begin{vmatrix} 1 & 13 & 7 \\ 1 & 1 & 1 \\ 1 & 4 & 3 \end{vmatrix} = -6; \quad D_z = \begin{vmatrix} 1 & -3 & 13 \\ 1 & 1 & 1 \\ 1 & -2 & 4 \end{vmatrix} = -24.$$

Then

$$x = \frac{D_x}{D} = \frac{20}{-10} = -2;$$

$$y = \frac{D_y}{D} = \frac{-6}{-10} = \frac{3}{5};$$

$$z = \frac{D_z}{D} = \frac{-24}{-10} = \frac{12}{5}.$$

The solution is $\left(-2, \frac{3}{5}, \frac{12}{5}\right)$. The check is left to the student.

Determinants can be evaluated on most graphers using the MATRIX package. After entering a matrix on the grapher, we select the determinant operation from the MATRIX MATH menu and enter the name of the matrix. The grapher will return the value of the determinant of the matrix. For example, for

$$\mathbf{A} = \begin{bmatrix} 1 & 6 & -1 \\ -3 & -5 & 3 \\ 0 & 4 & 2 \end{bmatrix},$$

we have

```
det [A]
                26
```

Use a grapher to confirm the calculations in Example 4.

In Example 4, we need not have evaluated D_z. Once x and y were found, we could have substituted them into one of the equations to find z.

To use Cramer's rule, we divide by D, provided $D \neq 0$. If $D = 0$ and at least one of the other determinants is not 0, then the system is inconsistent. If *all* the determinants are 0, then the equations in the system are dependent.

Exercise Set 3.7

FOR EXTRA HELP

Digital Video Tutor CD 3
Videotape 6

InterAct Math

Math Tutor Center

MathXL

MyMathLab.com

Evaluate.

1. $\begin{vmatrix} 5 & 1 \\ 2 & 4 \end{vmatrix}$

2. $\begin{vmatrix} 3 & 2 \\ 2 & -3 \end{vmatrix}$

3. $\begin{vmatrix} 6 & -9 \\ 2 & 3 \end{vmatrix}$

4. $\begin{vmatrix} 3 & 2 \\ -7 & 5 \end{vmatrix}$

5. $\begin{vmatrix} 1 & 4 & 0 \\ 0 & -1 & 2 \\ 3 & -2 & 1 \end{vmatrix}$

6. $\begin{vmatrix} 3 & 0 & -2 \\ 5 & 1 & 2 \\ 2 & 0 & -1 \end{vmatrix}$

7. $\begin{vmatrix} -1 & -2 & -3 \\ 3 & 4 & 2 \\ 0 & 1 & 2 \end{vmatrix}$

8. $\begin{vmatrix} 1 & 2 & 2 \\ 2 & 1 & 0 \\ 3 & 3 & 1 \end{vmatrix}$

9. $\begin{vmatrix} -4 & -2 & 3 \\ -3 & 1 & 2 \\ 3 & 4 & -2 \end{vmatrix}$

10. $\begin{vmatrix} 2 & -1 & 1 \\ 1 & 2 & -1 \\ 3 & 4 & -3 \end{vmatrix}$

Solve using Cramer's rule.

11. $5x + 8y = 1,$
$3x + 7y = 5$

12. $3x - 4y = 6,$
$5x + 9y = 10$

13. $5x - 4y = -3,$
$7x + 2y = 6$

14. $-2x + 4y = 3,$
$3x - 7y = 1$

15. $3x - y + 2z = 1,$
$x - y + 2z = 3,$
$-2x + 3y + z = 1$

16. $3x + 2y - z = 4,$
$3x - 2y + z = 5,$
$4x - 5y - z = -1$

17. $2x - 3y + 5z = 27,$
$x + 2y - z = -4,$
$5x - y + 4z = 27$

18. $x - y + 2z = -3,$
$x + 2y + 3z = 4,$
$2x + y + z = -3$

19. $r - 2s + 3t = 6,$
$2r - s - t = -3,$
$r + s + t = 6$

20. $a - 3c = 6,$
$b + 2c = 2,$
$7a - 3b - 5c = 14$

21. What is it about Cramer's rule that makes it useful?

22. Which version of Cramer's rule do you find more useful: the version for 2×2 systems or the version for 3×3 systems? Why?

SKILL MAINTENANCE

Solve.

23. $0.5x - 2.34 + 2.4x = 7.8x - 9$

24. $5x + 7x = -144$

25. A piece of wire 32.8 ft long is to be cut into two pieces, and those pieces are each to be bent to make a square. The length of a side of one square is to be 2.2 ft greater than the length of a side of the other. How should the wire be cut?

26. *Inventory.* The Freeport College store paid \$1728 for an order of 45 calculators. The store paid \$9 for each scientific calculator. The others, all graphing calculators, cost the store \$58 each. How many of each type of calculator was ordered?

27. *Insulation.* The Mazzas' attic required three and a half times as much insulation as did the Kranepools'. Together, the two attics required 36 rolls of insulation. How much insulation did each attic require?

28. *Sales of food.* High Flyin' Wings charges \$12 for a bucket of chicken wings and \$7 for a chicken dinner. After filling 28 orders for buckets and dinners, High Flyin' Wings had collected \$281. How many buckets and how many dinners did they sell?

SYNTHESIS

29. Cramer's rule states that whenever the equations $a_1x + b_1y = c_1$ and $a_2x + b_2y = c_2$ are dependent, we have

$$\begin{vmatrix} a_1 & b_1 \\ a_2 & b_2 \end{vmatrix} = 0.$$

Explain why this occurs.

30. Under what conditions can a 3×3 system of linear equations be consistent but unable to be solved using Cramer's rule?

Solve.

31. $\begin{vmatrix} y & -2 \\ 4 & 3 \end{vmatrix} = 44$

32. $\begin{vmatrix} 2 & x & -1 \\ -1 & 3 & 2 \\ -2 & 1 & 1 \end{vmatrix} = -12$

33. $\begin{vmatrix} m+1 & -2 \\ m-2 & 1 \end{vmatrix} = 27$

34. Show that an equation of the line through (x_1, y_1) and (x_2, y_2) can be written

$$\begin{vmatrix} x & y & 1 \\ x_1 & y_1 & 1 \\ x_2 & y_2 & 1 \end{vmatrix} = 0.$$

3.8 Business and Economic Applications

Break-Even Analysis • Supply and Demand

Break-Even Analysis

When a company manufactures x units of a product, it spends money. This is **total cost** and can be thought of as a function C, where $C(x)$ is the total cost of producing x units. When the company sells x units of the product, it takes in money. This is **total revenue** and can be thought of as a function R, where $R(x)$ is the total revenue from the sale of x units. **Total profit** is the money taken in less the money spent, or total revenue minus total cost. Total profit from the production and sale of x units is a function P given by

$$\textbf{Profit} = \textbf{Revenue} - \textbf{Cost}, \quad \text{or} \quad P(x) = R(x) - C(x).$$

If $R(x)$ is greater than $C(x)$, the company makes money. If $C(x)$ is greater than $R(x)$, the company has a loss. When $R(x) = C(x)$, the company breaks even.

There are two kinds of costs. First, there are costs like rent, insurance, machinery, and so on. These costs, which must be paid whether a product is produced or not, are called *fixed costs.* When a product is being produced, there are costs for labor, materials, marketing, and so on. These are called *variable costs,* because they vary according to the amount being produced. The sum of the fixed cost and the variable cost gives the *total cost* of producing a product.

> ***Caution!*** Do not confuse "cost" with "price." When we discuss the *cost* of an item, we are referring to what it costs to produce the item. The *price* of an item is what a consumer pays to purchase the item and is used when calculating revenue.

Example 1 **_Manufacturing radios._** Ergs, Inc., is planning to make a new kind of radio. Fixed costs will be \$90,000, and it will cost \$15 to produce each radio (variable costs). Each radio sells for \$26.

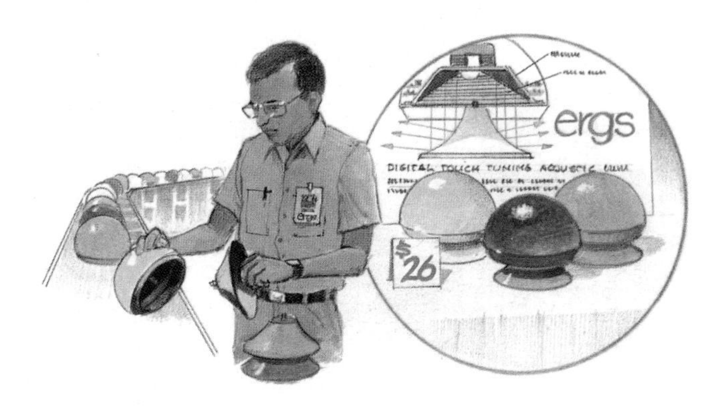

a) Find the total cost $C(x)$ of producing x radios.
b) Find the total revenue $R(x)$ from the sale of x radios.
c) Find the total profit $P(x)$ from the production and sale of x radios.
d) What profit or loss will the company realize from the production and sale of 3000 radios? of 14,000 radios?
e) Graph the total-cost, total-revenue, and total-profit functions using the same set of axes. Determine the break-even point.

Solution

a) Total cost is given by

$$C(x) = (\text{Fixed costs}) \text{ plus } (\text{Variable costs}),$$
$$\text{or } C(x) = 90{,}000 + 15x,$$

where x is the number of radios produced.

b) Total revenue is given by

$$R(x) = 26x.$$ \$26 times the number of radios sold. We assume that every radio produced is sold.

c) Total profit is given by

$$\begin{aligned} P(x) &= R(x) - C(x) \qquad \text{Profit is revenue minus cost.} \\ &= 26x - (90{,}000 + 15x) \\ &= 11x - 90{,}000. \end{aligned}$$

d) Profits will be

$$P(3000) = 11 \cdot 3000 - 90{,}000 = -\$57{,}000$$

when 3000 radios are produced and sold, and

$$P(14{,}000) = 11 \cdot 14{,}000 - 90{,}000 = \$64{,}000$$

when 14,000 radios are produced and sold. Thus the company loses money if only 3000 radios are sold, but makes money if 14,000 are sold.

e) The graphs of each of the three functions are shown below:

$R(x) = 26x,$ This represents the revenue function.

$C(x) = 90{,}000 + 15x,$ This represents the cost function.

$P(x) = 11x - 90{,}000.$ This represents the profit function.

$R(x)$, $C(x)$, and $P(x)$ are all in dollars.

The revenue function has a graph that goes through the origin and has a slope of 26. The cost function has an intercept on the \$-axis of 90,000 and has a slope of 15. The profit function has an intercept on the \$-axis of $-90{,}000$ and has a slope of 11. It is shown by the dashed line. The red dashed line shows a "negative" profit, which is a loss. (That is what is known as "being in the red.") The black dashed line shows a "positive" profit, or gain. (That is what is known as "being in the black.")

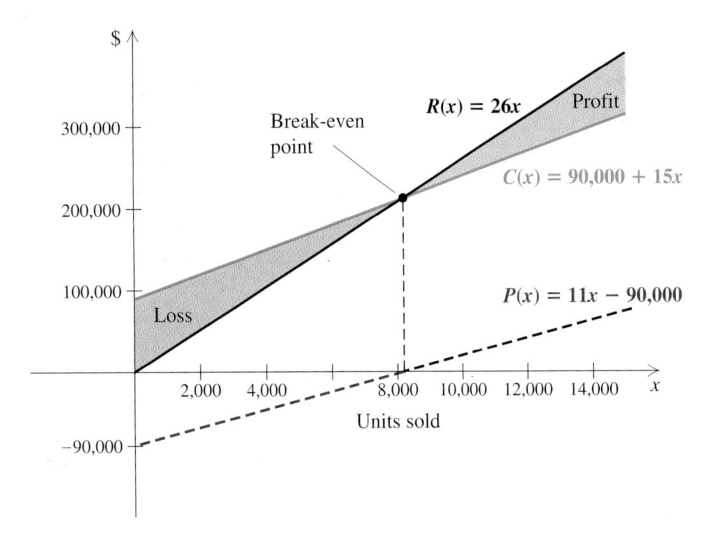

Profits occur where the revenue is greater than the cost. Losses occur where the revenue is less than the cost. The **break-even point** occurs where the graphs of R and C cross. Thus to find the break-even point, we solve a system:

$$R(x) = 26x,$$
$$C(x) = 90{,}000 + 15x.$$

Since both revenue and cost are in *dollars* and they are equal at the break-even point, the system can be rewritten as

$$d = 26x, \quad (1)$$
$$d = 90{,}000 + 15x \quad (2)$$

and solved using substitution:

$$26x = 90{,}000 + 15x \quad \text{Substituting } 26x \text{ for } d \text{ in equation (2)}$$
$$11x = 90{,}000$$
$$x \approx 8181.8.$$

The firm will break even if it produces and sells about 8182 radios (8181 will yield a tiny loss and 8182 a tiny gain), and takes in a total of $R(8182) = 26 \cdot 8182 = \$212{,}732$ in revenue. Note that the x-coordinate of the break-even point can also be found by solving $P(x) = 0$. The break-even point is (8182 radios, \$212,732).

Supply and Demand

As the price of coffee varies, the amount sold varies. The table and graph below show that consumers will demand less as the price goes up.

Demand Function, *D*

Price, *p*, per Kilogram	Quantity, *D*(*p*) (in millions of kilograms)
\$ 8.00	25
9.00	20
10.00	15
11.00	10
12.00	5

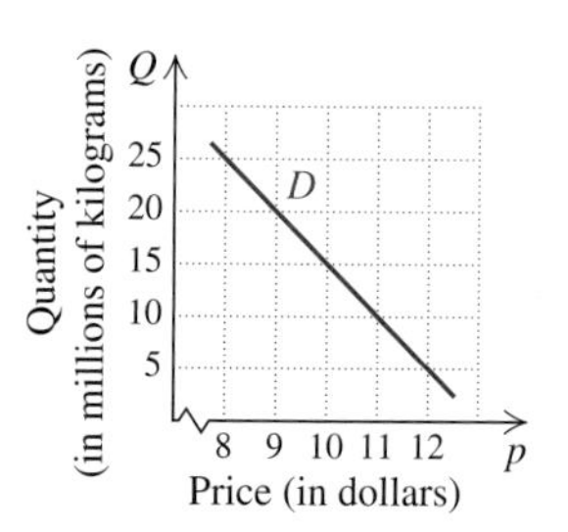

As the price of coffee varies, the amount available varies. The table and graph below show that sellers will supply less as the price goes down.

Supply Function, *S*

Price, *p*, per Kilogram	Quantity, *S*(*p*) (in millions of kilograms)
\$ 9.00	5
9.50	10
10.00	15
10.50	20
11.00	25

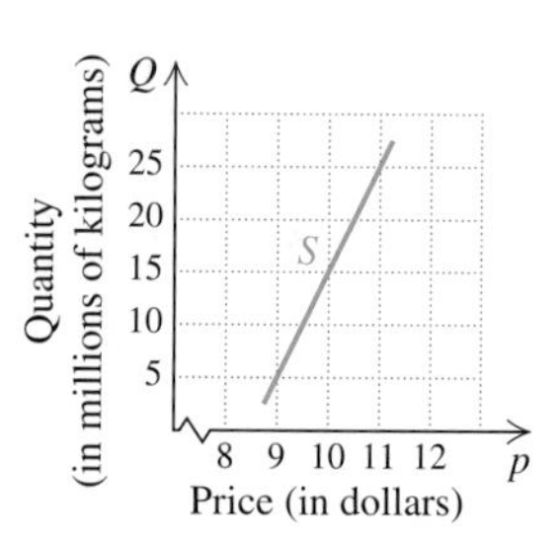

Let's look at the above graphs together. We see that as price increases, demand decreases. As price increases, supply increases. The point of intersection is called the **equilibrium point**. At that price, the amount that the seller will supply is the same amount that the consumer will buy. The situation is analogous to a buyer and a seller negotiating the price of an item. The equilibrium point is the price and quantity that they finally agree on.

Any ordered pair of coordinates from the graph is (price, quantity), because the horizontal axis is the price axis and the vertical axis is the quantity axis. If D is a demand function and S is a supply function, then the equilibrium point is where demand equals supply:

$$D(p) = S(p).$$

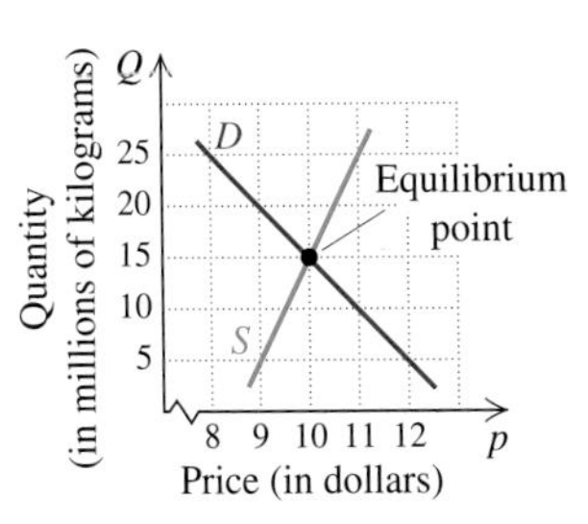

Example 2 Find the equilibrium point for the demand and supply functions given:

$$D(p) = 1000 - 60p, \quad (1)$$
$$S(p) = 200 + 4p. \quad (2)$$

Solution Since both demand and supply are *quantities* and they are equal at the equilibrium point, we rewrite the system as

$$q = 1000 - 60p, \quad (1)$$
$$q = 200 + 4p. \quad (2)$$

We substitute $200 + 4p$ for q in equation (1) and solve:

$200 + 4p = 1000 - 60p$ Substituting $200 + 4p$ for q in equation (1)

$200 + 64p = 1000$ Adding $60p$ to both sides

$64p = 800$ Adding -200 to both sides

$p = \frac{800}{64} = 12.5.$

Thus the equilibrium price is \$12.50 per unit.

To find the equilibrium quantity, we substitute \$12.50 into either $D(p)$ or $S(p)$. We use $S(p)$:

$$S(12.5) = 200 + 4(12.5) = 200 + 50 = 250.$$

Thus the equilibrium quantity is 250 units, and the equilibrium point is (\$12.50, 250).

Exercise Set 3.8

FOR EXTRA HELP

Digital Video Tutor CD 3
Videotape 6

InterAct Math

Math Tutor Center

MathXL

MyMathLab.com

For each of the following pairs of total-cost and total-revenue functions, find **(a)** *the total-profit function and* **(b)** *the break-even point.*

1. $C(x) = 45x + 300{,}000;$
$R(x) = 65x$

2. $C(x) = 25x + 270{,}000;$
$R(x) = 70x$

3. $C(x) = 10x + 120{,}000;$
$R(x) = 60x$

4. $C(x) = 30x + 49{,}500;$
$R(x) = 85x$

5. $C(x) = 40x + 22{,}500;$
$R(x) = 85x$

6. $C(x) = 20x + 10{,}000;$
$R(x) = 100x$

7. $C(x) = 22x + 16{,}000;$
$R(x) = 40x$

8. $C(x) = 15x + 75{,}000;$
$R(x) = 55x$

Aha! **9.** $C(x) = 75x + 100{,}000;$
$R(x) = 125x$

10. $C(x) = 20x + 120{,}000;$
$R(x) = 50x$

Find the equilibrium point for each of the following pairs of demand and supply functions.

11. $D(p) = 1000 - 10p,$
$S(p) = 230 + p$

12. $D(p) = 2000 - 60p,$
$S(p) = 460 + 94p$

13. $D(p) = 760 - 13p,$
$S(p) = 430 + 2p$

14. $D(p) = 800 - 43p,$
$S(p) = 210 + 16p$

15. $D(p) = 7500 - 25p,$
$S(p) = 6000 + 5p$

16. $D(p) = 8800 - 30p,$
$S(p) = 7000 + 15p$

17. $D(p) = 1600 - 53p,$
$S(p) = 320 + 75p$

18. $D(p) = 5500 - 40p,$
$S(p) = 1000 + 85p$

Solve.

19. *Computer manufacturing.* Biz.com Electronics is planning to introduce a new line of computers. The fixed costs for production are $125,300. The variable costs for producing each computer are $450. The revenue from each computer is $800. Find the following.

a) The total cost $C(x)$ of producing x computers
b) The total revenue $R(x)$ from the sale of x computers
c) The total profit $P(x)$ from the production and sale of x computers
d) The profit or loss from the production and sale of 100 computers; of 400 computers
e) The break-even point

20. *Manufacturing lamps.* City Lights, Inc., is planning to manufacture a new type of lamp. The fixed costs for production are $22,500. The variable costs for producing each lamp are estimated to be $40. The revenue from each lamp is to be $85. Find the following.

a) The total cost $C(x)$ of producing x lamps
b) The total revenue $R(x)$ from the sale of x lamps
c) The total profit $P(x)$ from the production and sale of x lamps
d) The profit or loss from the production and sale of 3000 lamps; of 400 lamps
e) The break-even point

21. *Manufacturing caps.* Martina's Custom Printing is planning on adding painter's caps to its product line. For the first year, the fixed costs for setting up production are $16,404. The variable costs for producing a dozen caps are $6.00. The revenue on each dozen caps will be $18.00. Find the following.

a) The total cost $C(x)$ of producing x dozen caps
b) The total revenue $R(x)$ from the sale of x dozen caps
c) The total profit $P(x)$ from the production and sale of x dozen caps
d) The profit or loss from the production and sale of 3000 dozen caps; of 1000 dozen caps
e) The break-even point

22. *Sport coat production.* Sarducci's is planning a new line of sport coats. For the first year, the fixed costs for setting up production are $10,000. The variable costs for producing each coat are $30. The revenue from each coat is to be $80. Find the following.

a) The total cost $C(x)$ of producing x coats
b) The total revenue $R(x)$ from the sale of x coats
c) The total profit $P(x)$ from the production and sale of x coats
d) The profit or loss from the production and sale of 2000 coats; of 50 coats
e) The break-even point

23. In Example 1, the slope of the line representing Revenue is the sum of the slopes of the other two lines. This is not a coincidence. Explain why.

24. Variable costs and fixed costs are often compared to the slope and the y-intercept, respectively, of an equation for a line. Explain why you feel this analogy is or is not valid.

SKILL MAINTENANCE

Solve.

25. $3x - 9 = 27$

26. $4x - 7 = 53$

27. $4x - 5 = 7x - 13$

28. $2x + 9 = 8x - 15$

29. $7 - 2(x - 8) = 14$

30. $6 - 4(3x - 2) = 10$

SYNTHESIS

31. Ian claims that since his fixed costs are $1000, he need sell only 20 birdbaths at $50 each in order to break even. Does this sound plausible? Why or why not?

32. In this section, we examined supply and demand functions for coffee. Does it seem realistic to you for the graph of D to have a constant slope? Why or why not?

33. *Yo-yo production.* Bing Boing Hobbies is willing to produce 100 yo-yo's at $2.00 each and 500 yo-yo's at $8.00 each. Research indicates that the public will buy 500 yo-yo's at $1.00 each and 100 yo-yo's at $9.00 each. Find the equilibrium point.

34. *Loudspeaker production.* Fidelity Speakers, Inc., has fixed costs of $15,400 and variable costs of $100 for each pair of speakers produced. If the speakers sell for $250 a pair, how many pairs of speakers must be produced (and sold) in order to have enough profit to cover the fixed costs of two additional facilities? Assume that all fixed costs are identical.

Use a grapher to solve.

35. *Dog food production.* Puppy Love, Inc., will soon begin producing a new line of puppy food. The marketing department predicts that the demand function will be $D(p) = -14.97p + 987.35$ and the supply function will be $S(p) = 98.55p - 5.13$.

a) To the nearest cent, what price per unit should be charged in order to have equilibrium between supply and demand?

b) The production of the puppy food involves $87,985 in fixed costs and $5.15 per unit in variable costs. If the price per unit is the value you found in part (a), how many units must be sold in order to break even?

36. *Computer production.* Number Cruncher Computers, Inc., is planning a new line of computers, each of which will sell for $970. The fixed costs in setting up production are $1,235,580 and the variable costs for each computer are $697.

a) What is the break-even point? (Round to the nearest whole number.) (

b) The marketing department at Number Cruncher is not sure that $970 is the best price. Their demand function for the new computers is given by $D(p) = -304.5p + 374{,}580$ and their supply function is given by $S(p) = 788.7p - 576{,}504$. To the nearest dollar, what price p would result in equilibrium between supply and demand?

Summary and Review 3

Key Terms

System of equations, p. 143
Solution of a system, p. 144
Consistent, p. 147
Inconsistent, p. 147
Dependent, p. 147
Independent, p. 147
Substitution method, p. 151
Elimination method, p. 153
Total-value problem, p. 160
Mixture problem, p. 164
Motion problem, p. 166
Matrix (plural, matrices), p. 187
Elements, p. 187
Entries, p. 187
Rows, p. 187
Columns, p. 187
Row-equivalent, p. 190
Dimensions, p. 191
Square matrix, p. 191
Determinant, p. 191
Cramer's rule, p. 192
Total cost, p. 197
Total revenue, p. 197
Total profit, p. 197
Fixed costs, p. 197
Variable costs, p. 197
Break-even point, p. 199
Demand function, p. 200
Supply function, p. 200
Equilibrium point, p. 200

Important Properties and Formulas

When solving a system of two linear equations in two variables:

1. If an identity is obtained, such as $0 = 0$, then the system has an infinite number of solutions. The equations are dependent and, since a solution exists, the system is consistent.
2. If a contradiction is obtained, such as $0 = 7$, then the system has no solution. The system is inconsistent.

To use the elimination method to solve systems of three linear equations:

1. Write all equations in the standard form $Ax + By + Cz = D$.
2. Clear any decimals or fractions.
3. Choose a variable to eliminate. Then select two of the three equations and work to get one equation in two variables.
4. Next, use a different pair of equations and eliminate the same variable that you did in step (3).
5. Solve the system of equations that resulted from steps (3) and (4).
6. Substitute the solution from step (5) into one of the original three equations and solve for the third variable. Then check.

Row-Equivalent Operations

Each of the following row-equivalent operations produces a row-equivalent matrix:

a) Interchanging any two rows.
b) Multiplying all elements of a row by a nonzero constant.
c) Replacing a row by the sum of that row and a multiple of another row.

Determinant of a 2 × 2 Matrix

$$\begin{vmatrix} a & c \\ b & d \end{vmatrix} = ad - bc$$

Determinant of a 3 × 3 Matrix

$$\begin{vmatrix} a_1 & b_1 & c_1 \\ a_2 & b_2 & c_2 \\ a_3 & b_3 & c_3 \end{vmatrix} = a_1\begin{vmatrix} b_2 & c_2 \\ b_3 & c_3 \end{vmatrix} - a_2\begin{vmatrix} b_1 & c_1 \\ b_3 & c_3 \end{vmatrix} + a_3\begin{vmatrix} b_1 & c_1 \\ b_2 & c_2 \end{vmatrix}$$

Cramer's Rule: 2 × 2 Systems

The solution of the system

$$a_1x + b_1y = c_1,$$
$$a_2x + b_2y = c_2$$

if it is unique, is given by

$$x = \frac{\begin{vmatrix} c_1 & b_1 \\ c_2 & b_2 \end{vmatrix}}{\begin{vmatrix} a_1 & b_1 \\ a_2 & b_2 \end{vmatrix}}, \qquad y = \frac{\begin{vmatrix} a_1 & c_1 \\ a_2 & c_2 \end{vmatrix}}{\begin{vmatrix} a_1 & b_1 \\ a_2 & b_2 \end{vmatrix}}.$$

Cramer's Rule: 3 × 3 Systems

The solution of the system

$$a_1x + b_1y + c_1z = d_1,$$
$$a_2x + b_2y + c_2z = d_2,$$
$$a_3x + b_3y + c_3z = d_3$$

is found by considering the following determinants:

$$D = \begin{vmatrix} a_1 & b_1 & c_1 \\ a_2 & b_2 & c_2 \\ a_3 & b_3 & c_3 \end{vmatrix}, \qquad D_x = \begin{vmatrix} d_1 & b_1 & c_1 \\ d_2 & b_2 & c_2 \\ d_3 & b_3 & c_3 \end{vmatrix},$$

$$D_y = \begin{vmatrix} a_1 & d_1 & c_1 \\ a_2 & d_2 & c_2 \\ a_3 & d_3 & c_3 \end{vmatrix}, \qquad D_z = \begin{vmatrix} a_1 & b_1 & d_1 \\ a_2 & b_2 & d_2 \\ a_3 & b_3 & d_3 \end{vmatrix}.$$

If a unique solution exists, it is given by

$$x = \frac{D_x}{D}, \qquad y = \frac{D_y}{D}, \qquad z = \frac{D_z}{D}.$$

Review Exercises

For Exercises 1–9, if a system has an infinite number of solutions, use set-builder notation to write the solution set. If a system has no solution, state this.

Solve graphically.

1. $3x + 2y = -4,$
$y = 3x + 7$

2. $2x + 3y = 12,$
$4x - y = 10$

Solve using the substitution method.

3. $9x - 6y = 2,$
$x = 4y + 5$

4. $y = x + 2,$
$y - x = 8$

5. $x - 3y = -2,$
$7y - 4x = 6$

Solve using the elimination method.

6. $8x - 2y = 10,$
$-4y - 3x = -17$

7. $4x - 7y = 18,$
$9x + 14y = 40$

8. $3x - 5y = -4,$
$5x - 3y = 4$

9. $1.5x - 3 = -2y,$
$3x + 4y = 6$

Solve.

10. Glynn bought two DVD's and one videocassette for $72. If he had purchased one DVD and two videocassettes, he would have spent $15 less. What is the price of a DVD? What is the price of a videocassette?

11. A freight train leaves Chicago at midnight traveling south at a speed of 44 mph. One hour later, a passenger train, going 55 mph, travels south from Chicago on a parallel track. How many hours will the passenger train travel before it overtakes the freight train?

12. Yolanda wants 14 L of fruit punch that is 10% juice. At the store, she finds punch that is 15% juice and punch that is 8% juice. How much of each should she purchase?

Solve. If a system's equations are dependent or if there is no solution, state this.

13. $x + 4y + 3z = 2,$
$2x + y + z = 10,$
$-x + y + 2z = 8$

14. $4x + 2y - 6z = 34,$
$2x + y + 3z = 3,$
$6x + 3y - 3z = 37$

15. $2x - 5y - 2z = -4,$
$7x + 2y - 5z = -6,$
$-2x + 3y + 2z = 4$

16. $-5x + 5y = -6,$
$2x - 2y = 4$

17. $3x + y = 2,$
$x + 3y + z = 0,$
$x + z = 2$

Solve.

18. In triangle *ABC*, the measure of angle *A* is four times the measure of angle *C*, and the measure of angle *B* is 45° more than the measure of angle *C*. What are the measures of the angles of the triangle?

19. Find the three-digit number in which the sum of the digits is 11, the tens digit is 3 less than the sum of the hundreds and ones digits, and the ones digit is 5 less than the hundreds digit.

20. Lynn has $159 in her purse, consisting of $20, $5, and $1 bills. The number of $20 bills is the same as the total number of $1 and $5 bills. If she has 14 bills in her purse, how many of each denomination does she have?

Solve using matrices. Show your work.

21. $3x + 4y = -13,$
$5x + 6y = 8$

22. $3x - y + z = -1,$
$2x + 3y + z = 4,$
$5x + 4y + 2z = 5$

Evaluate.

23. $\begin{vmatrix} -2 & 4 \\ -3 & 5 \end{vmatrix}$

24. $\begin{vmatrix} 2 & 3 & 0 \\ 1 & 4 & -2 \\ 2 & -1 & 5 \end{vmatrix}$

Solve using Cramer's rule. Show your work.

25. $2x + 3y = 6,$
$x - 4y = 14$

26. $2x + y + z = -2,$
$2x - y + 3z = 6,$
$3x - 5y + 4z = 7$

27. Find the equilibrium point for the demand and supply functions

$$S(p) = 60 + 7p$$

and

$$D(p) = 120 - 13p.$$

28. Robbyn is beginning to produce organic honey. For the first year, the fixed costs for setting up production are $18,000. The variable costs for producing each pint of honey are $1.50. The revenue from each pint of honey is $6. Find the following.

a) The total cost $C(x)$ of producing x pints of honey

b) The total revenue $R(x)$ from the sale of x pints of honey

c) The total profit $P(x)$ from the production and sale of x pints of honey

d) The profit or loss from the production and sale of 1500 pints of honey; of 5000 pints of honey

e) The break-even point

SYNTHESIS

29. How would you go about solving a problem that involves four variables?

30. Explain how a system of equations can be both dependent and inconsistent.

31. Robbyn is quitting a job that pays $27,000 a year to make honey (see Exercise 28). How many pints of honey must she produce and sell in order to make the same amount that she made in the job she left?

32. Solve graphically:

$$y = x + 2,$$
$$y = x^2 + 2.$$

33. The graph of $f(x) = ax^2 + bx + c$ contains the points $(-2, 3)$, $(1, 1)$, and $(0, 3)$. Find a, b, and c and give a formula for the function.

Chapter Test 3

1. Solve graphically:

$$2x + y = 8,$$
$$y - x = 2.$$

Solve, if possible, using the substitution method.

2. $x + 3y = -8,$
$4x - 3y = 23$

3. $2x + 4y = -6,$
$y = 3x - 9$

Solve, if possible, using the elimination method.

4. $4x - 6y = 3,$
$6x - 4y = -3$

5. $4y + 2x = 18,$
$3x + 6y = 26$

6. The perimeter of a rectangle is 96. The length of the rectangle is 6 less than twice the width. Find the dimensions of the rectangle.

7. Between her home mortgage (loan), car loan, and credit card bill (loan), Rema is $75,300 in debt. Rema's credit card bill accumulates 1.5% interest, her car loan 1% interest, and her mortgage 0.6% interest each month. After one month, her total accumulated interest is $460.50. The interest on Rema's credit card bill was $4.50 more than the interest on her car loan. Find the amount of each loan.

Solve. If a system's equations are dependent or if there is no solution, state this.

8. $-3x + y - 2z = 8,$
$-x + 2y - z = 5,$
$2x + y + z = -3$

9. $6x + 2y - 4z = 15,$
$-3x - 4y + 2z = -6,$
$4x - 6y + 3z = 8$

10. $2x + 2y = 0,$
$4x + 4z = 4,$
$2x + y + z = 2$

11. $3x + 3z = 0,$
$2x + 2y = 2,$
$3y + 3z = 3$

Solve using matrices.

12. $7x - 8y = 10,$
$9x + 5y = -2$

13. $x + 3y - 3z = 12,$
$3x - y + 4z = 0,$
$-x + 2y - z = 1$

Evaluate.

14. $\begin{vmatrix} 4 & -2 \\ 3 & 7 \end{vmatrix}$

15. $\begin{vmatrix} 3 & 4 & 2 \\ 2 & -5 & 4 \\ 4 & 5 & -3 \end{vmatrix}$

16. Solve using Cramer's rule:

$$8x - 3y = 5,$$
$$2x + 6y = 3.$$

17. An electrician, a carpenter, and a plumber are hired to work on a house. The electrician earns $21 per hour, the carpenter $19.50 per hour, and the plumber $24 per hour. The first day on the job, they worked a total of 21.5 hr and earned a total of $469.50. If the plumber worked 2 more hours than the carpenter did, how many hours did the electrician work?

18. Find the equilibrium point for the demand and supply functions

$$D(p) = 79 - 8p \quad \text{and} \quad S(p) = 37 + 6p.$$

19. Complete Communications, Inc., is producing a new family radio service model. For the first year, the fixed costs for setting up production are \$40,000. The variable costs for producing each radio are \$25. The revenue from each radio is \$70. Find the following.

a) The total cost $C(x)$ of producing x radios
b) The total revenue $R(x)$ from the sale of x radios
c) The total profit $P(x)$ from the production and sale of x radios
d) The profit or loss from the production and sale of 300 radios; of 900 radios
e) The break-even point

SYNTHESIS

20. The graph of the function $f(x) = mx + b$ contains the points $(-1, 3)$ and $(-2, -4)$. Find m and b.

21. At a county fair, an adult's ticket sold for \$5.50, a senior citizen's ticket for \$4.00, and a child's ticket for \$1.50. On opening day, the number of adults' and senior citizens' tickets sold was 30 more than the number of children's tickets sold. The number of adults' tickets sold was 6 more than four times the number of senior citizens' tickets sold. Total receipts from the ticket sales were \$11,219.50. How many of each type of ticket were sold?

Cumulative Review 1–3

Solve.

1. $-14.3 + 29.17 = x$

2. $x + 9.4 = -12.6$

3. $3.9(-11) = x$

4. $-2.4x = -48$

5. $4x + 7 = -14$

6. $-3 + 5x = 2x + 15$

7. $3n - (4n - 2) = 7$

8. $6y - 5(3y - 4) = 10$

9. $14 + 2c = -3(c + 4) - 6$

10. $5x - [4 - 2(6x - 1)] = 12$

Simplify. Do not leave negative exponents in your answers.

11. $x^4 \cdot x^{-6} \cdot x^{13}$

12. $(4x^{-3}y^2)(-10x^4y^{-7})$

13. $(6x^2y^3)^2(-2x^0y^4)^3$

14. $\dfrac{y^4}{y^{-6}}$

15. $\dfrac{-10a^7b^{-11}}{25a^{-4}b^{22}}$

16. $\left(\dfrac{3x^4y^{-2}}{4x^{-5}}\right)^4$

17. $(1.95 \times 10^{-3})(5.73 \times 10^8)$

18. $\dfrac{2.42 \times 10^5}{6.05 \times 10^{-2}}$

19. Solve $A = \frac{1}{2}h(b + t)$ for b.

20. Determine whether $(-3, 4)$ is a solution of $5a - 2b = -23$.

Graph.

21. $y = -2x + 3$

22. $y = x^2 - 1$

23. $4x + 16 = 0$

24. $-3x + 2y = 6$

25. Find the slope and the y-intercept of the line with equation $-4y + 9x = 12$.

26. Find the slope, if it exists, of the line containing the points $(2, 7)$ and $(-1, 3)$.

27. Find an equation of the line with slope -3 and containing the point $(2, -11)$.

28. Find an equation of the line containing the points $(-6, 3)$ and $(4, 2)$.

29. Determine whether the lines are parallel or perpendicular:

$$2x = 4y + 7,$$
$$x - 2y = 5.$$

30. Find an equation of the line containing the point $(2, 1)$ and perpendicular to the line $x - 2y = 5$.

31. For the graph of f shown, determine the domain, the range, $f(-3)$, and any value of x for which $f(x) = 5$.

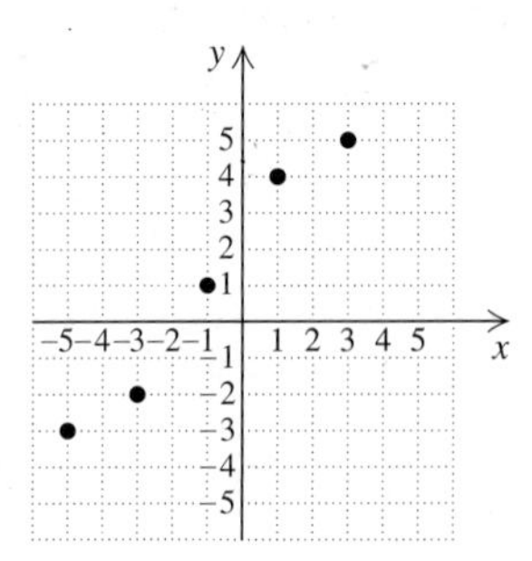

32. Determine the domain of the function given by

$$f(x) = \frac{7}{2x - 1}.$$

Given $g(x) = 4x - 3$ and $h(x) = -2x^2 + 1$, find the following function values.

33. $h(4)$

34. $-g(0)$

35. $(g \cdot h)(-1)$

36. $g(a) - h(2a)$

Solve.

37. $3x + y = 4,$
$6x - y = 5$

38. $4x + 4y = 4,$
$5x - 3y = -19$

39. $6x - 10y = -22,$
$-11x - 15y = 27$

40. $x + y + z = -5,$
$2x + 3y - 2z = 8,$
$x - y + 4z = -21$

41. $2x + 5y - 3z = -11,$
$-5x + 3y - 2z = -7,$
$3x - 2y + 5z = 12$

Evaluate.

42. $\begin{vmatrix} 2 & -3 \\ 4 & 1 \end{vmatrix}$

43. $\begin{vmatrix} 1 & 0 & 1 \\ -1 & 2 & 1 \\ 2 & 1 & 3 \end{vmatrix}$

44. The sum of two numbers is 26. Three times the smaller plus twice the larger is 60. Find the numbers.

45. In 1997, there were 652 endangered or threatened species of U.S. animals and plants that had recovery plans. By August 2000, there were 923 species with recovery plans. (*Source*: U.S. Fish and Wildlife Service, Department of the Interior) Find the rate at which recovery plans were being formed.

46. The number of U.S. pleasure trips, in millions, t years after 1994 can be estimated by $P(t) = 9t + 616$ (*Source*: Travel Industry Association of America). What do the numbers 9 and 616 signify?

47. In 1989, there were 6.6 million U.S. aircraft departures, and in 1999, there were 8.6 million departures (*Source*: Air Transport Association of America). Let $A(t)$ represent the number of departures, in millions, t years after 1989.

a) Find an equation for a linear function that fits the data.

b) Use the function of part (a) to predict the number of departures in 2010.

48. "Soakem" is 34% salt and the rest water. "Rinsem" is 61% salt and the rest water. How many ounces of each would be needed to obtain 120 oz of a mixture that is 50% salt?

49. Find three consecutive odd numbers such that the sum of four times the first number and five times the third number is 47.

50. Belinda's scores on four tests are 83, 92, 100, and 85. What must the score be on the fifth test so that the average will be 90?

51. The perimeter of a rectangle is 32 cm. If five times the width equals three times the length, what are the dimensions of the rectangle?

52. There are 4 more nickels than dimes in a bank. The total amount of money in the bank is \$2.45. How many of each type of coin are in the bank?

53. One month Lori and Jon spent \$680 for electricity, rent, and telephone. The electric bill was $\frac{1}{4}$ of the rent and the rent was \$400 more than the phone bill. How much was the electric bill?

54. A hockey team played 64 games one season. It won 15 more games than it tied and lost 10 more games than it won. How many games did it win? lose? tie?

55. Reggie, Jenna, and Achmed are counting calories. For lunch one day, Reggie ate two cookies and a banana, for a total of 260 calories. Jenna had a cup of yogurt and a banana, for a total of 245 calories. Achmed ate a cookie, a cup of yogurt, and two bananas, for a total of 415 calories. How many calories are in each item?

SYNTHESIS

56. Simplify: $(6x^{a+2}y^{b+2})(-2x^{a-2}y^{y+1})$.

57. An automotive dealer discovers that when \$1000 is spent on radio advertising, weekly sales increase by \$101,000. When \$1250 is spent on radio advertising, weekly sales increase by \$126,000. Assuming that sales increase according to a linear equation, by what amount would sales increase when \$1500 is spend on radio advertising?

58. Given that $f(x) = mx + b$ and that $f(5) = -3$ when $f(-4) = 2$, find m and b.

4 Inequalities and Problem Solving

4.1 Inequalities and Applications
4.2 Intersections, Unions, and Compound Inequalities
4.3 Absolute-Value Equations and Inequalities
4.4 Inequalities in Two Variables
Connecting the Concepts
4.5 Applications Using Linear Programming
Summary and Review
Test

AN APPLICATION

Slobberbone receives \$750 plus 15% of receipts over \$750 for playing a club date. If a club charges a \$6 cover charge, how many people must attend in order for the band to receive \$1200?

This problem appears as Exercise 63 in Section 4.1.

I use math in my job much more than I had thought I would. Bands are paid using various formulas, and I need to calculate capacity and admission prices to estimate income. I also use math to determine contracts, budgets, and royalty statements.

AMY POJMAN
Musician Management
New York, NY

Inequalities are mathematical sentences containing symbols such as < (is less than). Principles similar to those used for solving equations enable us to solve inequalities and the problems that translate to inequalities. In this chapter, we develop procedures for solving a variety of inequalities and systems of inequalities.

4.1 Inequalities and Applications

Solving Inequalities • Interval Notation • The Addition Principle for Inequalities • The Multiplication Principle for Inequalities • Using the Principles Together • Problem Solving

Study Tip

Try to at least glance at the next section of material that will be covered in class. This will make it easier to concentrate on your instructor's lecture instead of trying to write everything down.

Solving Inequalities

We can extend our equation-solving skills to the solving of inequalities. An **inequality** is any sentence containing $<, >, \leq, \geq$, or $\neq$ (see Section 1.2)—for example,

$$-2 < a, \qquad x > 4, \qquad x + 3 \leq 6, \qquad 6 - 7y \geq 10y - 4, \quad \text{and} \quad 5x \neq 10.$$

Any replacement for the variable that makes an inequality true is called a **solution**. The set of all solutions is called the **solution set**. When all solutions of an inequality are found, we say that we have **solved** the inequality.

Example 1 Determine whether the given number is a solution of the inequality.

a) $x + 3 < 6;\ \ 5$

b) $2x - 3 > -5;\ \ 1$

Solution

a) We substitute to get $5 + 3 < 6$, or $8 < 6$, a false sentence. Thus, 5 *is not* a solution.

b) We substitute to get $2 \cdot 1 - 3 > -5$, or $-1 > -5$, a true sentence. Thus, 1 *is* a solution.

The *graph* of an inequality is a drawing that represents its solutions. An inequality in one variable can be graphed on a number line. Inequalities in two variables can be graphed on a coordinate plane, and appear later in this chapter.

Example 2 Graph $x < 4$ on a number line.

Solution The solutions are all real numbers less than 4, so we shade all numbers less than 4. Since 4 is not a solution, we use an open dot at 4.

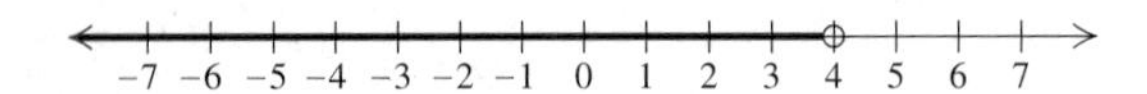

We can write the solution set using *set-builder notation* (see Section 1.1):

$$\{x \mid x < 4\}.$$

This is read

"The set of all x such that x is less than 4."

Interval Notation

Another way to write solutions of an inequality in one variable is to use **interval notation**. Interval notation uses parentheses, ( ), and brackets, [ ].

If a and b are real numbers such that $a < b$, we define the **open interval** (a, b) as the set of all numbers x for which $a < x < b$. Thus,

$$(a, b) = \{x \mid a < x < b\}.$$ Parentheses are used to exclude endpoints.

Its graph excludes the endpoints:

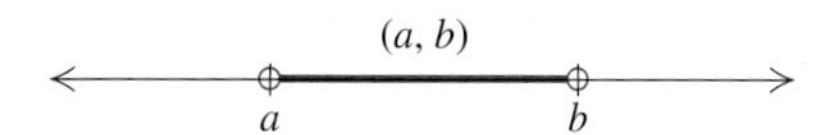

> ***Caution!*** Do not confuse the *interval* (a, b) with the *ordered pair* (a, b). The context in which the notation appears usually makes the meaning clear.

The **closed interval** $[a, b]$ is defined as the set of all numbers x for which $a \leq x \leq b$. Thus,

$$[a, b] = \{x \mid a \leq x \leq b\}.$$ Brackets are used to include endpoints.

Its graph includes the endpoints*:

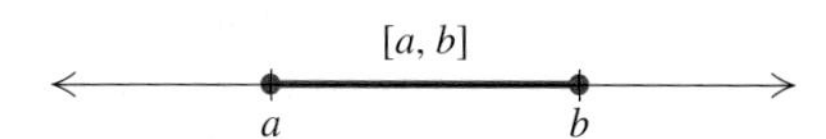

There are two kinds of **half-open intervals**, defined as follows:

1. $(a, b] = \{x \mid a < x \leq b\}$. This is open on the left. Its graph is as follows:

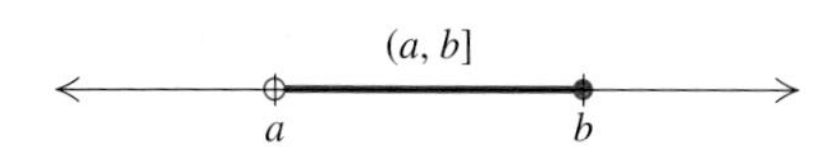

*Some books use the representations $\overset{(}{a}$———$\overset{)}{b}$ and $\overset{[}{a}$———$\overset{]}{b}$ instead of, respectively, $\underset{a}{\circ}$———$\underset{b}{\circ}$ and $\underset{a}{\bullet}$———$\underset{b}{\bullet}$.

2. $[a, b) = \{x \mid a \leq x < b\}$. This is open on the right. Its graph is as follows:

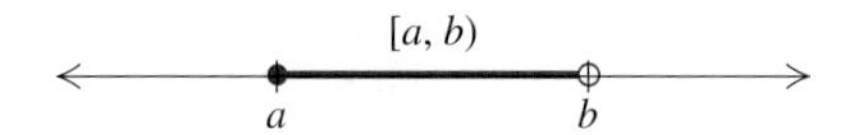

We use the symbols ∞ and $-\infty$ to represent positive and negative infinity, respectively. Thus the notation (a, ∞) represents the set of all real numbers greater than a, and $(-\infty, a)$ represents the set of all real numbers less than a.

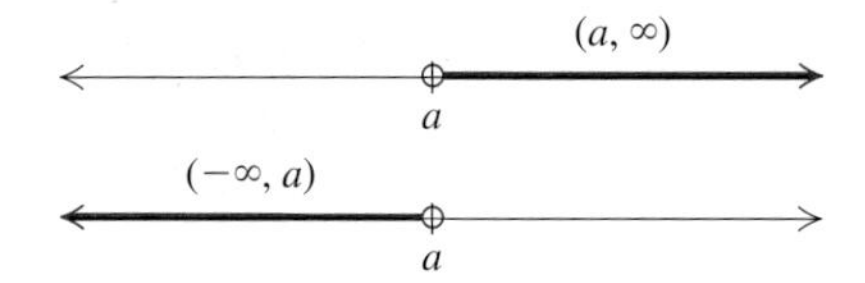

The notations $[a, \infty)$ and $(-\infty, a]$ are used when we want to include the endpoint a.

Example 3 Graph $y \geq -2$ on a number line and write the solution set using both set-builder and interval notations.

Solution Using set-builder notation, we write the solution set as $\{y \mid y \geq -2\}$.

Using interval notation, we write the solution set as $[-2, \infty)$.

To graph the solution, we shade all numbers to the right of -2 and use a solid dot to indicate that -2 is also a solution.

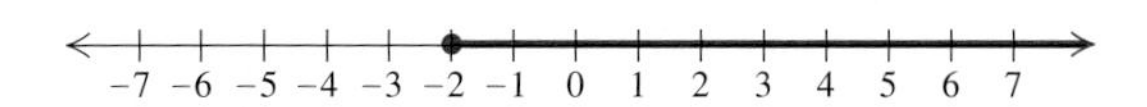

The Addition Principle for Inequalities

Two inequalities are *equivalent* if they have the same solution set. For example, the inequalities $x > 4$ and $4 < x$ are equivalent. Just as the addition principle for equations produces equivalent equations, the addition principle for inequalities produces equivalent inequalities.

> ***The Addition Principle for Inequalities***
>
> For any real numbers a, b, and c:
>
> $$a < b \text{ is equivalent to } a + c < b + c;$$
> $$a > b \text{ is equivalent to } a + c > b + c.$$
>
> Similar statements hold for $\leq$ and $\geq$.

As with equations, we try to get the variable alone on one side in order to determine solutions easily.

Example 4 Solve and graph: **(a)** $x + 5 > 1$; **(b)** $4x - 1 \geq 5x - 2$.

Solution

a)

$$x + 5 > 1$$

$$x + 5 + (-5) > 1 + (-5)$$ Using the addition principle to add -5 to both sides

$$x > -4$$

When an inequality—like this last one—has an infinite number of solutions, we cannot possibly check them all. Instead, we can perform a partial check by substituting one member of the solution set (here we use -1) into the original inequality:

$$\begin{array}{r|l} x + 5 > & 1 \\ \hline -1 + 5 \; ? & 1 \\ 4 & 1 \quad \text{TRUE} \end{array}$$

Since $4 > 1$ is true, we have our check. The solution set is $\{x \mid x > -4\}$, or $(-4, \infty)$. The graph is as follows:

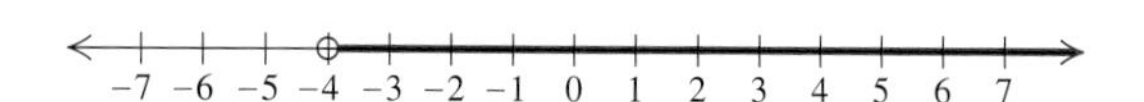

b)

$4x - 1 \geq 5x - 2$	
$4x - 1 + 2 \geq 5x - 2 + 2$	Adding 2 to both sides
$4x + 1 \geq 5x$	Simplifying
$4x + 1 - 4x \geq 5x - 4x$	Adding $-4x$ to both sides
$1 \geq x$	Simplifying

We know that $1 \geq x$ has the same meaning as $x \leq 1$. You can check that any number less than or equal to 1 is a solution. The solution set is $\{x \mid 1 \geq x\}$ or, more commonly, $\{x \mid x \leq 1\}$. Using interval notation, we write that the solution set is $(-\infty, 1]$. The graph is as follows:

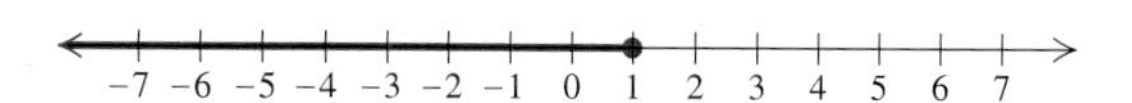

The Multiplication Principle for Inequalities

The multiplication principle for inequalities differs from the multiplication principle for equations.

Consider this true inequality:

$$4 < 9.$$

If we multiply both sides of $4 < 9$ by 2, we get another true inequality:

$$4 \cdot 2 < 9 \cdot 2, \quad \text{or} \quad 8 < 18.$$

If we multiply both sides of $4 < 9$ by -2, we get a false inequality:

$$\text{FALSE} \longrightarrow 4(-2) < 9(-2), \quad \text{or} \quad -8 < -18. \longleftarrow \text{FALSE}$$

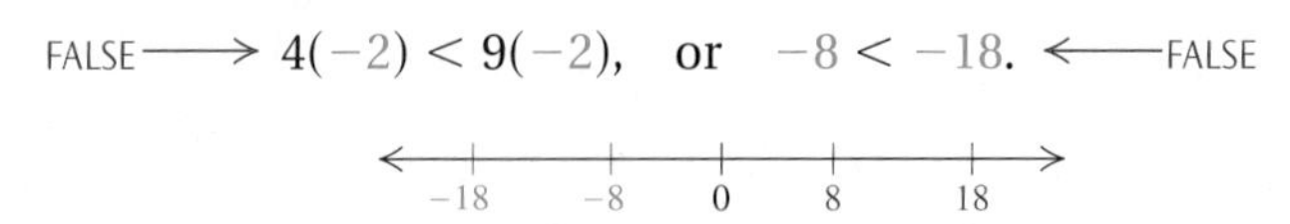

This is because multiplication (or division) by a negative number changes the sign of the number being multiplied (or divided). When the signs of both numbers in an inequality are changed, the position of the numbers with respect to each other is reversed.

$-8 > -18.$ ⟵ TRUE

↑ The $<$ symbol has been reversed!

The Multiplication Principle for Inequalities

For any real numbers a and b, and for any *positive* number c,

$a < b$ is equivalent to $ac < bc$;

$a > b$ is equivalent to $ac > bc$.

For any real numbers a and b, and for any *negative* number c,

$a < b$ is equivalent to $ac > bc$;

$a > b$ is equivalent to $ac < bc$.

Similar statements hold for $\leq$ and $\geq$.

Since division by c is the same as multiplication by $1/c$, there is no need for a separate division principle.

Caution! Remember that whenever we multiply or divide both sides of an inequality by a negative number, we must reverse the inequality symbol.

Example 5 Solve and graph: **(a)** $3y < \frac{3}{4}$; **(b)** $-5x \geq -80$.

Solution

a) $3y < \frac{3}{4}$

↕ The symbol stays the same.

$\frac{1}{3} \cdot 3y < \frac{1}{3} \cdot \frac{3}{4}$ Multiplying both sides by $\frac{1}{3}$ or dividing both sides by 3

$y < \frac{1}{4}$

Any number less than $\frac{1}{4}$ is a solution. The solution set is $\{y \mid y < \frac{1}{4}\}$, or $(-\infty, \frac{1}{4})$. The graph is as follows:

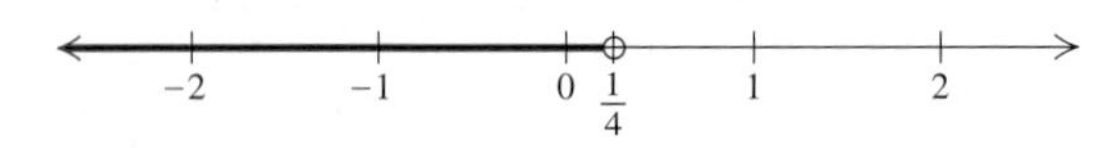

b) $-5x \geq -80$ — The symbol must be reversed.

$$\frac{-5x}{-5} \leq \frac{-80}{-5}$$ Dividing both sides by -5 or multiplying both sides by $-\frac{1}{5}$

$$x \leq 16$$

The solution set is $\{x \mid x \leq 16\}$, or $(-\infty, 16]$. The graph is as follows:

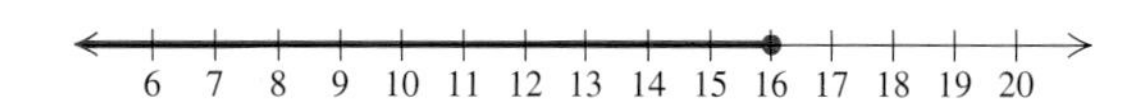

Using the Principles Together

Example 6

Solve: **(a)** $16 - 7y \geq 10y - 4$; **(b)** $-3(x + 8) - 5x > 4x - 9$.

Solution We use the addition and multiplication principles together in solving inequalities in much the same way as in solving equations.

a)

$$16 - 7y \geq 10y - 4$$

$$-16 + 16 - 7y \geq -16 + 10y - 4$$ Adding -16 to both sides

$$-7y \geq 10y - 20$$

$$-10y + (-7y) \geq -10y + 10y - 20$$ Adding $-10y$ to both sides

$$-17y \geq -20$$ The symbol must be reversed.

$$-\tfrac{1}{17} \cdot (-17y) \leq -\tfrac{1}{17} \cdot (-20)$$ Multiplying both sides by $-\frac{1}{17}$ or dividing both sides by -17

$$y \leq \tfrac{20}{17}$$

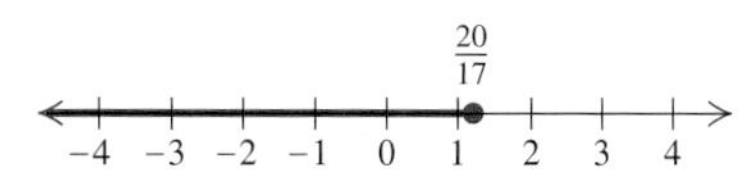

The solution set is $\left\{y \mid y \leq \frac{20}{17}\right\}$, or $\left(-\infty, \frac{20}{17}\right]$.

b)

$$-3(x + 8) - 5x > 4x - 9$$

$$-3x - 24 - 5x > 4x - 9$$ Using the distributive law

$$-24 - 8x > 4x - 9$$

$$-24 - 8x + 8x > 4x - 9 + 8x$$ Adding $8x$ to both sides

$$-24 > 12x - 9$$

$$-24 + 9 > 12x - 9 + 9$$ Adding 9 to both sides

$$-15 > 12x$$ The symbol stays the same.

$$-\tfrac{5}{4} > x$$ Dividing by 12 and simplifying

To check Example 6(b), graph $y_1 = -3(x + 8) - 5x$ and $y_2 = 4x - 9$ and identify those x-values for which $y_1 > y_2$.

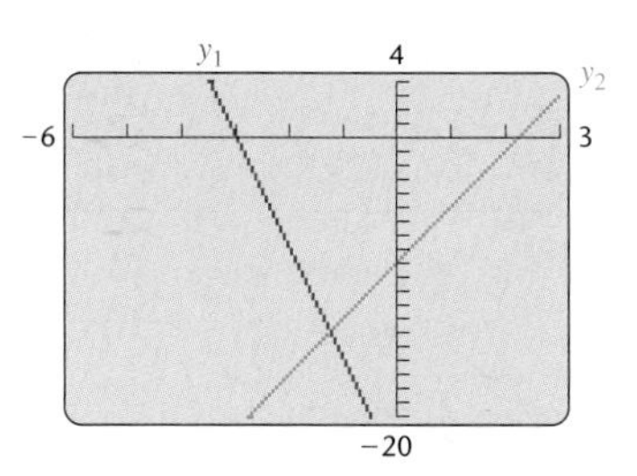

The INTERSECT option helps us find that $y_1 = y_2$ when $x = -1.25$. Note that $y_1 > y_2$ for x-values in the interval $(-\infty, -1.25)$.

On many graphers, the solution can be found by using the Y-VARS and TEST keys to enter $y_3 = y_1 > y_2$. The solution set is then displayed as an interval (shown by a horizontal line 1 unit above the x-axis).

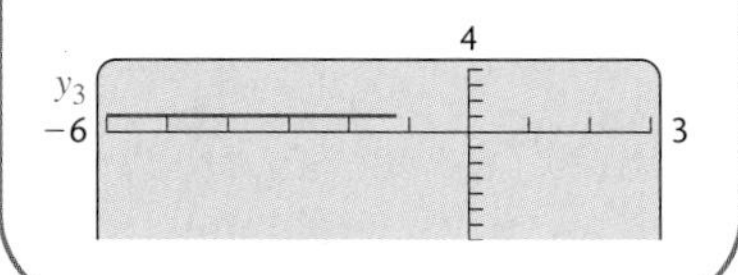

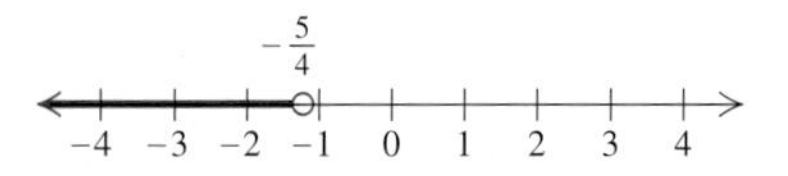

The solution set is $\{x \mid -\frac{5}{4} > x\}$, or $\{x \mid x < -\frac{5}{4}\}$, or $\left(-\infty, -\frac{5}{4}\right)$.

Problem Solving

Many problem-solving situations translate to inequalities. In addition to "is less than" and "is more than," other phrases are commonly used.

Important Words	Sample Sentence	Translation
is at least	Max is at least 5 years old.	$m \geq 5$
is at most	At most 6 people could fit in the elevator.	$n \leq 6$
cannot exceed	Total weight in the elevator cannot exceed 2000 pounds.	$w \leq 2000$
must exceed	The speed must exceed 15 mph.	$s > 15$
is between	Heather's income is between \$23,000 and \$35,000.	$23{,}000 < h < 35{,}000$
no more than	Bing weighs no more than 90 pounds.	$w \leq 90$
no less than	Saul would accept no less than 4 tickets for the show.	$t \geq 4$

Example 7

Records in the men's 200-m dash. Michael Johnson set a world record of 19.32 sec in the men's 200-m dash in the 1996 Olympics. If $R(t)$ is given in seconds, then the function given by

$$R(t) = -0.045t + 19.32$$

can be used to predict the world record in the men's 200-m dash t years after 1996. Determine (in terms of an inequality) those years for which the world record will be less than 19.0 sec.

Solution

1. **Familiarize.** We already have a formula. To become more familiar with it, we might make a substitution for t. Suppose we want to know the record after 20 years, in the year 2016. We substitute 20 for t:

$$R(20) = -0.045(20) + 19.32 = 18.42 \text{ sec.}$$

We see that by 2016, the record will be less than 19.0 sec. To predict the exact year in which the 19.0-sec mark will be broken, we could make other guesses that are less than 20. Instead, we proceed to the next step.

2. **Translate.** The record $R(t)$ is to be *less than* 19.0 sec. Thus we have

$$R(t) < 19.0.$$

We replace $R(t)$ with $-0.045t + 19.32$ to find the times t that solve the inequality:

$$-0.045t + 19.32 < 19.0. \quad \text{Substituting } -0.045t + 19.32 \text{ for } R(t)$$

3. **Carry out.** We solve the inequality:

$$-0.045t + 19.32 < 19.0$$
$$-0.045t < -0.32 \quad \text{Adding } -19.32 \text{ to both sides}$$
$$t > 7.1. \quad \text{Dividing both sides by } -0.045, \text{ reversing the symbol, and rounding}$$

4. **Check.** A partial check is to substitute a value for t greater than 7.1. We did that in the *Familiarize* step.

5. **State.** The record will be less than 19.0 sec for races occurring more than 7.1 years after 1996, or approximately all years after 2003.

Example 8

Earnings plans. On a new job, Rose can be paid in one of two ways:

Plan A: A salary of $600 per month, plus a commission of 4% of sales;

Plan B: A salary of $800 per month, plus a commission of 6% of sales in excess of $10,000.

For what amount of monthly sales is plan A better than plan B, if we assume that sales are always more than $10,000?

Solution

1. **Familiarize.** Listing the given information in a table will be helpful.

Plan A: Monthly Income	**Plan B: Monthly Income**
$600 salary 4% of sales *Total*: $600 + 4% of sales	$800 salary 6% of sales over $10,000 *Total*: $800 + 6% of sales over $10,000

Next, suppose that Rose sold a certain amount—say, $12,000—in one month. Which plan would be better? Under plan A, she would earn $600 plus 4% of $12,000, or

$$600 + 0.04(12{,}000) = \$1080.$$

Since with plan B commissions are paid only on sales in excess of $10,000, Rose would earn $800 plus 6% of ($12,000 − $10,000), or

$$800 + 0.06(2000) = \$920.$$

This shows that for monthly sales of $12,000, plan A is better. Similar calculations will show that for sales of $30,000 a month, plan B is better. To determine *all* values for which plan A earns more money, we must solve an inequality that is based on the calculations above.

2. **Translate.** We let S = the amount of monthly sales, in dollars. Examining the calculations in the *Familiarize* step, we see that monthly income from plan A is $600 + 0.04S$ and from plan B is $800 + 0.06(S - 10{,}000)$. We want to find all values of S for which

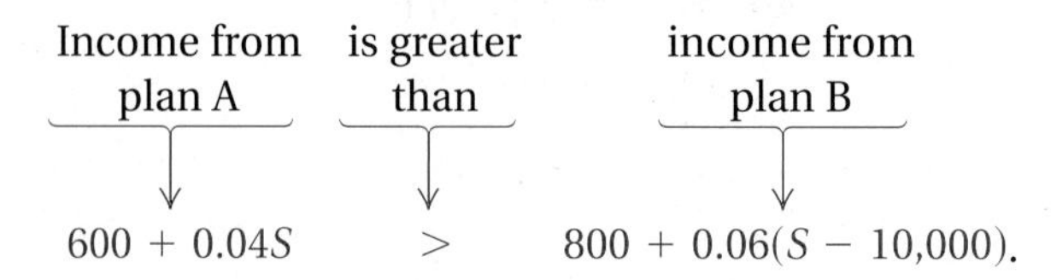

Income from plan A
is greater than
income from plan B
600 + 0.04S
>
800 + 0.06(S − 10,000).

3. **Carry out.** We solve the inequality:

$$600 + 0.04S > 800 + 0.06(S - 10{,}000)$$

$$600 + 0.04S > 800 + 0.06S - 600 \quad \text{Using the distributive law}$$

$$600 + 0.04S > 200 + 0.06S \quad \text{Combining like terms}$$

$$400 > 0.02S \quad \text{Subtracting 200 and } 0.04S \text{ from both sides}$$

$$20{,}000 > S, \text{ or } S < 20{,}000. \quad \text{Dividing both sides by 0.02}$$

4. **Check.** For $S = 20{,}000$, the income from plan A is

$600 + 4\% \cdot 20{,}000$, or \$1400.

The income from plan B is

$800 + 6\% \cdot (20{,}000 - 10{,}000)$, or \$1400.

This confirms that for sales totaling \$20,000, Rose's pay is the same under either plan.

In the *Familiarize* step, we saw that for sales of \$12,000, plan A pays more. Since $12{,}000 < 20{,}000$, this is a partial check. Since we cannot check all possible values of S, we will stop here.

5. **State.** For monthly sales of less than \$20,000, plan A is better.

Exercise Set 4.1

FOR EXTRA HELP

Digital Video Tutor CD 3 Videotape 7

InterAct Math

Math Tutor Center

MathXL

MyMathLab.com

Determine whether the given numbers are solutions of the inequality.

1. $x - 1 \ge 7$; $-4, 0, 8, 13$

2. $3x + 5 \le -10$; $-5, -10, 0, 27$

3. $t - 6 > 2t - 1$; $0, -8, -9, -3$

4. $5y - 9 < 3 - y$; $2, -3, 0, 3$

Graph each inequality, and write the solution set using both set-builder and interval notation.

5. $y < 6$

6. $x > 4$

7. $x \ge -4$

8. $t \le 6$

9. $t > -3$

10. $y < -3$

11. $x \le -7$

12. $x \ge -6$

Solve. Then graph.

13. $x + 9 > 4$

14. $x + 5 > 2$

15. $a + 7 \le -13$

16. $a + 9 \le -12$

17. $x - 5 \le 7$

18. $t + 14 \ge 9$

19. $y - 9 > -18$

20. $y - 8 > -14$

21. $y - 20 \le -6$

22. $x - 11 \le -2$

23. $9t < -81$

24. $8x \ge 24$

25. $0.5x < 25$

26. $0.3x < -18$

27. $-8y \leq 3.2$

28. $-9x \geq -8.1$

29. $-\frac{5}{6}y \leq -\frac{3}{4}$

30. $-\frac{3}{4}x \geq -\frac{5}{8}$

31. $5y + 13 > 28$

32. $2x + 7 < 19$

33. $-9x + 3x \geq -24$

34. $5y + 2y \leq -21$

35. Let $f(x) = 8x - 9$ and $g(x) = 3x - 11$. Find all values of x for which $f(x) < g(x)$.

36. Let $f(x) = 2x - 7$ and $g(x) = 5x - 9$. Find all values of x for which $f(x) < g(x)$.

37. Let $f(x) = 0.4x + 5$ and $g(x) = 1.2x - 4$. Find all values of x for which $g(x) \geq f(x)$.

38. Let $f(x) = \frac{3}{8} + 2x$ and $g(x) = 3x - \frac{1}{8}$. Find all values of x for which $g(x) \geq f(x)$.

Solve.

39. $4(3y - 2) \geq 9(2y + 5)$

40. $4m + 7 \geq 14(m - 3)$

41. $5(t - 3) + 4t < 2(7 + 2t)$

42. $2(4 + 2x) > 2x + 3(2 - 5x)$

43. $5[3m - (m + 4)] > -2(m - 4)$

44. $8x - 3(3x + 2) - 5 \geq 3(x + 4) - 2x$

45. $19 - (2x + 3) \leq 2(x + 3) + x$

46. $13 - (2c + 2) \geq 2(c + 2) + 3c$

47. $\frac{1}{4}(8y + 4) - 17 < -\frac{1}{2}(4y - 8)$

48. $\frac{1}{3}(6x + 24) - 20 > -\frac{1}{4}(12x - 72)$

49. $2[8 - 4(3 - x)] - 2 \geq 8[2(4x - 3) + 7] - 50$

50. $5[3(7 - t) - 4(8 + 2t)] - 20 \leq -6[2(6 + 3t) - 4]$

Solve.

51. *Truck rentals.* Campus Entertainment rents a truck for \$45 plus 20¢ per mile. A budget of \$75 has been set for the rental. For what mileages will they not exceed the budget?

52. *Truck rentals.* Metro Concerts can rent a truck for either \$55 with unlimited mileage or \$29 plus 40¢ per mile. For what mileages would the unlimited mileage plan save money?

53. *Insurance claims.* After a serious automobile accident, most insurance companies will replace the damaged car with a new one if repair costs exceed 80% of the NADA, or "blue-book," value of the car. Miguel's car recently sustained \$9200 worth of damage but was not replaced. What was the blue-book value of his car?

54. *Phone rates.* A long-distance telephone call using Down East Calling costs 10 cents for the first minute and 8 cents for each additional minute. The same call, placed on Long Call Systems, cost 15 cents for the first minute and 6 cents for each additional minute. For what length phone calls is Down East Calling less expensive?

Phone rates. *In Vermont, Verizon charges customers* \$13.55 *for monthly service plus* 2.2¢ *per minute for local phone calls between* 9 A.M. *and* 9 P.M. *weekdays. The charge for off-peak local calls is* 0.5¢ *per minute. Calls are free after the total monthly charges reach* \$39.40.

55. Assume that only peak local calls were made. For how long must a customer speak on the phone if the \$39.40 maximum charge is to apply?

56. Assume that only off-peak calls were made. For how long must a customer speak on the phone if the \$39.40 maximum charge is to apply?

57. *Checking-account rates.* The Hudson Bank offers two checking-account plans. Their Anywhere plan charges 20¢ per check whereas their Acu-checking plan costs \$2 per month plus 12¢ per check. For what numbers of checks per month will the Acu-checking plan cost less?

58. *Moving costs.* Musclebound Movers charges \$85 plus \$40 an hour to move households across town. Champion Moving charges \$60 an hour for cross-town moves. For what lengths of time is Champion more expensive?

59. *Wages.* Toni can be paid in one of two ways:

Plan A: A salary of \$400 per month, plus a commission of 8% of gross sales;

Plan B: A salary of \$610 per month, plus a commission of 5% of gross sales.

For what amount of gross sales should Toni select plan A?

60. *Wages.* Branford can be paid for his masonry work in one of two ways:

Plan A: \$300 plus \$9.00 per hour;

Plan B: Straight \$12.50 per hour.

Suppose that the job takes n hours. For what values of n is plan B better for Branford?

61. *Insurance benefits.* Bayside Insurance offers two plans. Under plan A, Giselle would pay the first \$50 of her medical bills and 20% of all bills after that. Under plan B, Giselle would pay the first \$250 of bills, but only 10% of the rest. For what amount of medical bills will plan B save Giselle money? (Assume that her bills will exceed \$250.)

62. *Wedding costs.* The Arnold Inn offers two plans for wedding parties. Under plan A, the inn charges \$30 for each person in attendance. Under plan B, the inn charges \$1300 plus \$20 for each person in excess of the first 25 who attend. For what size parties will plan B cost less? (Assume that more than 25 guests will attend.)

63. *Show business.* Slobberbone receives \$750 plus 15% of receipts over \$750 for playing a club date. If a club charges a \$6 cover charge, how many people must attend in order for the band to receive at least \$1200?

64. *Temperature conversion.* The function

$$C(F) = \tfrac{5}{9}(F - 32)$$

can be used to find the Celsius temperature $C(F)$ that corresponds to $F°$ Fahrenheit.

a) Gold is solid at Celsius temperatures less than 1063°C. Find the Fahrenheit temperatures for which gold is solid.

b) Silver is solid at Celsius temperatures less than 960.8°C. Find the Fahrenheit temperatures for which silver is solid.

65. *Manufacturing.* Ergs, Inc., is planning to make a new kind of radio. Fixed costs will be \$90,000, and variable costs will be \$15 for the production of each radio. The total-cost function for x radios is

$$C(x) = 90{,}000 + 15x.$$

The company makes \$26 in revenue for each radio sold. The total-revenue function for x radios is

$$R(x) = 26x.$$

(See Section 3.8.)

a) When $R(x) < C(x)$, the company loses money. Find the values of x for which the company loses money.

b) When $R(x) > C(x)$, the company makes a profit. Find the values of x for which the company makes a profit.

66. *Publishing.* The demand and supply functions for a locally produced poetry book are approximated by

$$D(p) = 2000 - 60p \quad \text{and}$$
$$S(p) = 460 + 94p,$$

where p is the price in dollars (see Section 3.8).

a) Find those values of p for which demand exceeds supply.

b) Find those values of p for which demand is less than supply.

67. Explain in your own words why the inequality symbol must be reversed when both sides of an inequality are multiplied by a negative number.

68. Why isn't roster notation used to write solutions of inequalities?

SKILL MAINTENANCE

Find the domain of f.

69. $f(x) = \dfrac{3}{x - 2}$

70. $f(x) = \dfrac{x - 5}{4x + 12}$

71. $f(x) = \dfrac{5x}{7 - 2x}$

72. $f(x) = \dfrac{x + 3}{9 - 4x}$

Simplify.

73. $9x - 2(x - 5)$

74. $8x + 7(2x - 1)$

SYNTHESIS

75. A Presto photocopier costs \$510 and an Exact Image photocopier costs \$590. Write a problem that involves the cost of the copiers, the cost per page of photocopies, and the number of copies for which the Presto machine is the more expensive machine to own.

76. Explain how the addition principle can be used to avoid ever needing to multiply or divide both sides of an inequality by a negative number.

Solve. Assume that a, b, c, d, and m are positive constants.

77. $3ax + 2x \geq 5ax - 4$; assume $a > 1$

78. $6by - 4y \leq 7by + 10$

79. $a(by - 2) \geq b(2y + 5)$; assume $a > 2$

80. $c(6x - 4) < d(3 + 2x)$; assume $3c > d$

81. $c(2 - 5x) + dx > m(4 + 2x)$; assume $5c + 2m < d$

82. $a(3 - 4x) + cx < d(5x + 2)$; assume $c > 4a + 5d$

Determine whether each statement is true or false. If false, give an example that shows this.

83. For any real numbers a, b, c, and d, if $a < b$ and $c < d$, then $a - c < b - d$.

84. For all real numbers x and y, if $x < y$, then $x^2 < y^2$.

85. Are the inequalities

$$x < 3 \quad \text{and} \quad x + \frac{1}{x} < 3 + \frac{1}{x}$$

equivalent? Why or why not?

86. Are the inequalities

$$x < 3 \quad \text{and} \quad 0 \cdot x < 0 \cdot 3$$

equivalent? Why or why not?

Solve. Then graph.

87. $x + 5 \leq 5 + x$

88. $x + 8 < 3 + x$

89. $x^2 > 0$

90. Assume that the graphs of $y_1 = -\frac{1}{2}x + 5$, $y_2 = x - 1$, and $y_3 = 2x - 3$ are as shown below. Solve each inequality, referring only to the figure.

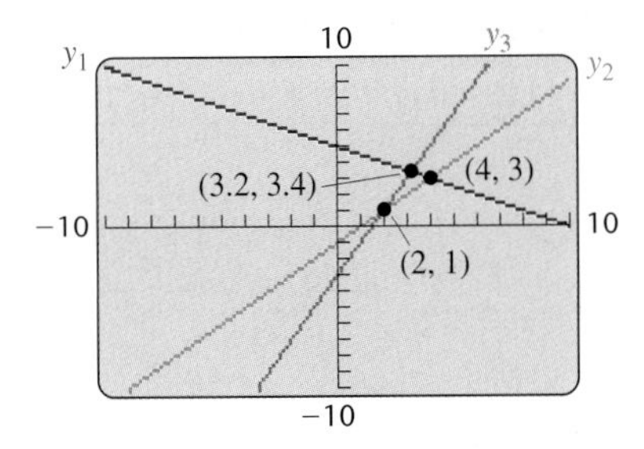

a) $-\frac{1}{2}x + 5 > x - 1$
b) $x - 1 \leq 2x - 3$
c) $2x - 3 \geq -\frac{1}{2}x + 5$

91. Using an approach similar to that in the Technology Connection on p. 217, use a grapher to check your answers to Exercises 13, 37, 53, and 61.

COLLABORATIVE CORNER

Reduce, Reuse, and Recycle

Focus: Inequalities and problem solving
Time: 15–20 minutes
Group size: 2

In the United States, the amount of solid waste (rubbish) being recycled is slowly catching up to the amount being generated. In 1991, each person generated, on average, 4.3 lb of solid waste every day, of which 0.8 lb was recycled. In 2000, each person generated, on average, 4.4 lb of solid waste, of which 1.3 lb was recycled. (*Sources*: U.S. Census 2000 and EPA Municipal Solid Waste Factbook)

ACTIVITY

Assume that the amount of solid waste being generated and the amount recycled are both increasing linearly. One group member should find a linear function w for which $w(t)$ represents the number of pounds of waste generated per person per day t years after 1991. The other group member should find a linear function r for which $r(t)$ represents the number of pounds recycled per person per day t years after 1991. Finally, working together, the group should determine those years for which the amount recycled will meet or exceed the amount generated.

Intersections, Unions, and Compound Inequalities

4.2

Intersections of Sets and Conjunctions of Sentences • Unions of Sets and Disjunctions of Sentences • Interval Notation and Domains

We now consider **compound inequalities**—that is, sentences like "$-2 < x$ *and* $x < 1$" or "$x < -3$ *or* $x > 3$" that are formed using the word *and* or the word *or*.

Intersections of Sets and Conjunctions of Sentences

The **intersection** of two sets A and B is the set of all elements that are common to both A and B. We denote the intersection of sets A and B as

$$A \cap B.$$

The intersection of two sets is often pictured as shown here.

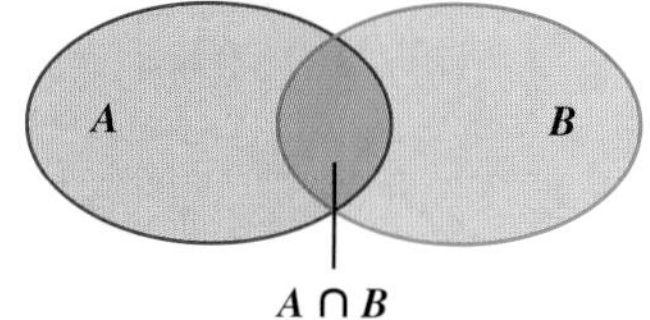

Example 1 Find the intersection: $\{1, 2, 3, 4, 5\} \cap \{-2, -1, 0, 1, 2, 3\}$.

Solution The numbers 1, 2, and 3 are common to both sets, so the intersection is $\{1, 2, 3\}$.

When two or more sentences are joined by the word *and* to make a compound sentence, the new sentence is called a **conjunction** of the sentences. The following is a conjunction of inequalities:

$$-2 < x \quad and \quad x < 1.$$

A number is a solution of a conjunction if it is a solution of both of the separate parts. For example, -1 is a solution because it is a solution of $-2 < x$ as well as $x < 1$.

Below we show the graph of $-2 < x$, followed by the graph of $x < 1$, and finally the graph of the conjunction $-2 < x$ *and* $x < 1$. *Note that the solution set of a conjunction is the intersection of the solution sets of the individual sentences.*

$\{x \mid -2 < x\}$ −7 −6 −5 −4 −3 −2 −1 0 1 2 3 4 5 6 7 $(-2, \infty)$

$\{x \mid x < 1\}$ −7 −6 −5 −4 −3 −2 −1 0 1 2 3 4 5 6 7 $(-\infty, 1)$

$\{x \mid -2 < x\} \cap \{x \mid x < 1\}$
$= \{x \mid -2 < x \text{ and } x < 1\}$ −7 −6 −5 −4 −3 −2 −1 0 1 2 3 4 5 6 7 $(-2, 1)$

Because there are numbers that are both greater than -2 and less than 1, the conjunction $-2 < x$ *and* $x < 1$ can be abbreviated by $-2 < x < 1$. Thus the interval $(-2, 1)$ can be represented as $\{x \mid -2 < x < 1\}$, the set of all numbers that are *simultaneously* greater than -2 *and* less than 1. Note that for $a < b$,

$$a < x \quad and \quad x < b \quad \textbf{can be abbreviated} \quad a < x < b;$$

and, equivalently,

$$b > x \quad and \quad x > a \quad \textbf{can be abbreviated} \quad b > x > a.$$

Example 2 Solve and graph: $-1 \le 2x + 5 < 13$.

Solution This inequality is an abbreviation for the conjunction

$$-1 \le 2x + 5 \quad and \quad 2x + 5 < 13.$$

The word *and* corresponds to set *intersection*. To solve the conjunction, we solve each of the two inequalities separately and then find the intersection of the solution sets:

$-1 \le 2x + 5$	*and*	$2x + 5 < 13$	
$-6 \le 2x$	*and*	$2x < 8$	Subtracting 5 from both sides of each inequality
$-3 \le x$	*and*	$x < 4.$	Dividing both sides of each inequality by 2

We now abbreviate the answer:

$$-3 \le x < 4.$$

The solution set is $\{x \mid -3 \le x < 4\}$, or, in interval notation, $[-3, 4)$. The graph is the intersection of the two separate solution sets.

$\{x \mid -3 \le x\}$ $[-3, \infty)$

$\{x \mid x < 4\}$ $(-\infty, 4)$

$\{x \mid -3 \le x\} \cap \{x \mid x < 4\} = \{x \mid -3 \le x < 4\}$ $[-3, 4)$

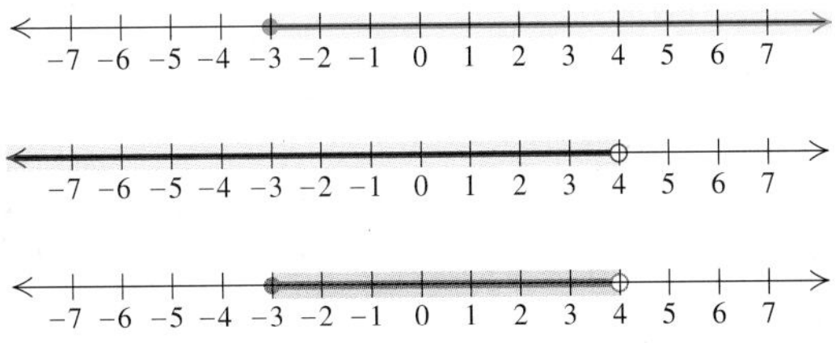

The steps in Example 2 are sometimes combined as follows:

$$-1 \le 2x + 5 < 13$$
$$-1 - 5 \le 2x + 5 - 5 < 13 - 5$$
$$-6 \le 2x < 8$$
$$-3 \le x < 4.$$

Such an approach saves some writing and will prove useful in Section 4.3.

Example 3 Solve and graph: $2x - 5 \geq -3$ *and* $5x + 2 \geq 17$.

Solution We first solve each inequality separately:

$$
\begin{aligned}
2x - 5 \geq -3 \quad &\textit{and} \quad 5x + 2 \geq 17\\
2x \geq 2 \quad &\textit{and} \quad 5x \geq 15\\
x \geq 1 \quad &\textit{and} \quad x \geq 3.
\end{aligned}
$$

Next, we find the intersection of the two separate solution sets.

$\{x \mid x \geq 1\}$ $\quad$ $[1, \infty)$

$\{x \mid x \geq 3\}$ $\quad$ $[3, \infty)$

$\{x \mid x \geq 1\} \cap \{x \mid x \geq 3\}$
$= \{x \mid x \geq 3\}$ $\quad$ $[3, \infty)$

The numbers common to both sets are greater than or equal to 3. Thus the solution set is $\{x \mid x \geq 3\}$, or, in interval notation, $[3, \infty)$. You should check that any number in $[3, \infty)$ satisfies the conjunction whereas numbers outside $[3, \infty)$ do not.

Mathematical Use of the Word "and"

The word "and" corresponds to "intersection" and to the symbol "$\cap$". Any solution of a conjunction must make each part of the conjunction true.

Sometimes there is no way to solve both parts of a conjunction at once.

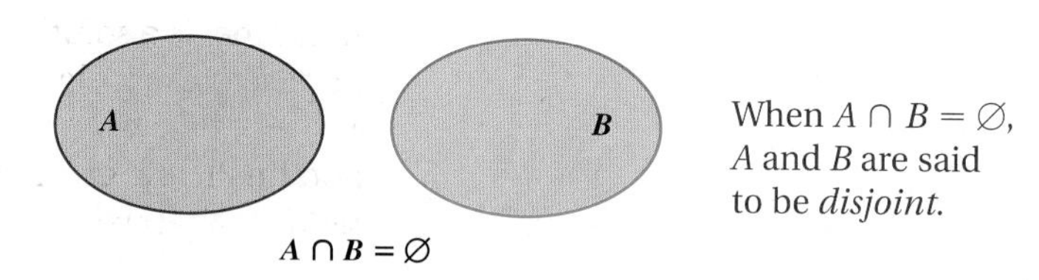

When $A \cap B = \varnothing$, A and B are said to be *disjoint*.

Example 4 Solve and graph: $2x - 3 > 1$ *and* $3x - 1 < 2$.

Solution We solve each inequality separately:

$$
\begin{aligned}
2x - 3 > 1 \quad &\textit{and} \quad 3x - 1 < 2\\
2x > 4 \quad &\textit{and} \quad 3x < 3\\
x > 2 \quad &\textit{and} \quad x < 1.
\end{aligned}
$$

The solution set is the intersection of the individual inequalities.

$\{x \mid x > 2\}$ — $(2, \infty)$

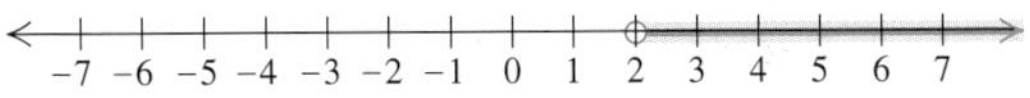

$\{x \mid x < 1\}$ — $(-\infty, 1)$

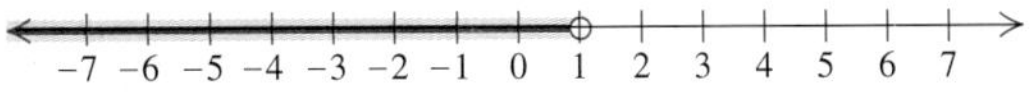

$\{x \mid x > 2\} \cap \{x \mid x < 1\}$
$= \{x \mid x > 2 \text{ and } x < 1\} = \varnothing$ — $\varnothing$

Since no number is both greater than 2 and less than 1, the solution set is the empty set, $\varnothing$.

Unions of Sets and Disjunctions of Sentences

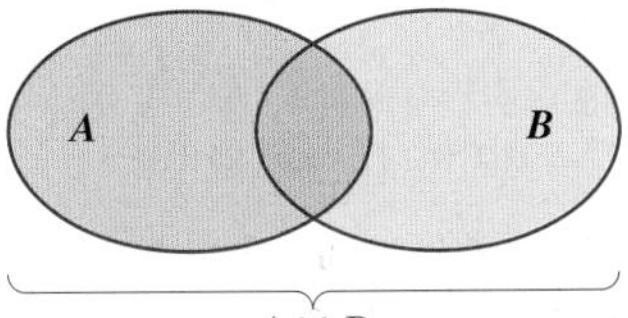

The **union** of two sets A and B is the collection of elements belonging to A and/or B. We denote the union of A and B by

$$A \cup B.$$

The union of two sets is often pictured as shown at left.

A B
$A \cup B$

Example 5 Find the union: $\{2, 3, 4\} \cup \{3, 5, 7\}$.

Solution The numbers in either or both sets are 2, 3, 4, 5, and 7, so the union is $\{2, 3, 4, 5, 7\}$.

When two or more sentences are joined by the word *or* to make a compound sentence, the new sentence is called a **disjunction** of the sentences. Here is an example:

$$x < -3 \quad or \quad x > 3.$$

A number is a solution of a disjunction if it is a solution of either of the separate parts. For example, -5 is a solution of this disjunction since -5 is a solution of $x < -3$. Below we show the graph of $x < -3$, followed by the graph of $x > 3$, and finally the graph of the disjunction $x < -3$ *or* $x > 3$. *Note that the solution set of a disjunction is the union of the solution sets of the individual sentences.*

$\{x \mid x < -3\}$ — $(-\infty, -3)$

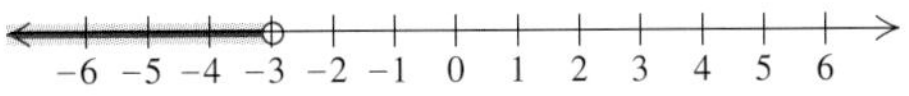

$\{x \mid x > 3\}$ — $(3, \infty)$

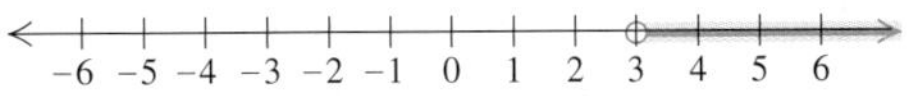

$\{x \mid x < -3\} \cup \{x \mid x > 3\}$
$= \{x \mid x < -3 \text{ or } x > 3\}$ — $(-\infty, -3) \cup (3, \infty)$

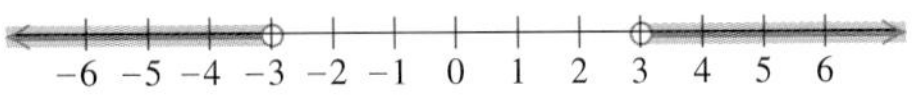

The solution set of $x < -3$ *or* $x > 3$ is $\{x \mid x < -3 \text{ or } x > 3\}$, or in interval notation, $(-\infty, -3) \cup (3, \infty)$. There is no simpler way to write the solution.

Mathematical Use of the Word "or"

The word "or" corresponds to "union" and to the symbol " $\cup$ ". For a number to be a solution of a disjunction, it must be in *at least one* of the solution sets of the individual sentences.

Example 6 Solve and graph: $7 + 2x < -1$ *or* $13 - 5x \le 3$.

Solution We solve each inequality separately, retaining the word *or*:

$$7 + 2x < -1 \quad or \quad 13 - 5x \le 3$$
$$2x < -8 \quad or \quad -5x \le -10$$

Reverse the symbol.

$$x < -4 \quad or \quad x \ge 2.$$

To find the solution set of the disjunction, we consider the individual graphs. We graph $x < -4$ and then $x \ge 2$. Then we take the union of the graphs.

$\{x \mid x < -4\}$

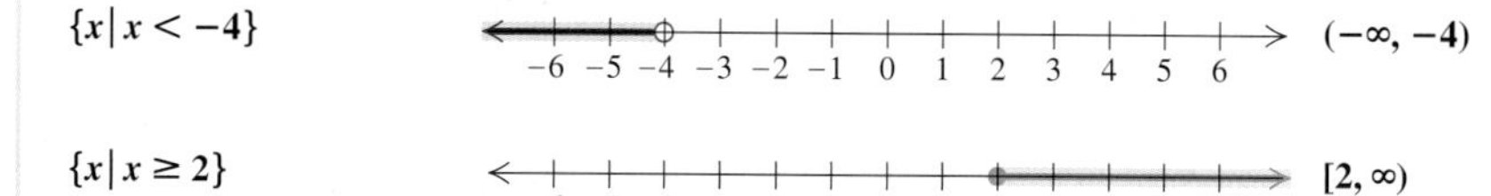

$(-\infty, -4)$

$\{x \mid x \ge 2\}$ $[2, \infty)$

$\{x \mid x < -4\} \cup \{x \mid x \ge 2\}$
$= \{x \mid x < -4 \text{ or } x \ge 2\}$ $(-\infty, -4) \cup [2, \infty)$

The solution set is $\{x \mid x < -4 \textit{ or } x \ge 2\}$, or $(-\infty, -4) \cup [2, \infty)$.

Caution! A compound inequality like

$$x < -4 \quad or \quad x \ge 2,$$

as in Example 6, *cannot* be expressed as $2 \le x < -4$ because to do so would be to say that x is *simultaneously* less than -4 and greater than or equal to 2. No number is both less than -4 *and* greater than 2, but many are less than -4 *or* greater than 2.

Example 7 Solve: $-2x - 5 < -2$ *or* $x - 3 < -10$.

Solution We solve the individual inequalities separately, retaining the word *or*:

$$-2x - 5 < -2 \quad or \quad x - 3 < -10$$
$$-2x < 3 \quad or \quad x < -7$$

Reverse the symbol.

$$x > -\tfrac{3}{2} \quad or \quad x < -7.$$

Keep the word "or."

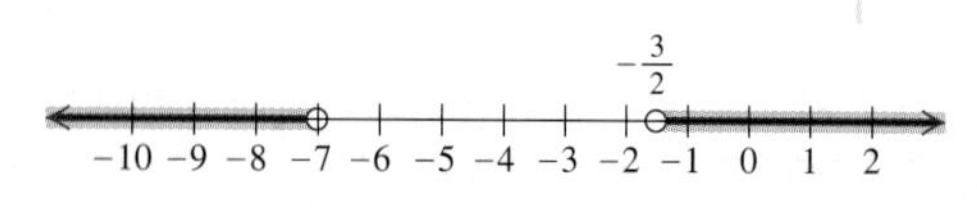

The solution set is $\left\{x \mid x < -7 \textit{ or } x > -\frac{3}{2}\right\}$, or $(-\infty, -7) \cup \left(-\frac{3}{2}, \infty\right)$.

Example 8 Solve: $3x - 11 < 4$ *or* $4x + 9 \geq 1$.

Solution We solve the individual inequalities separately, retaining the word *or*:

$$\begin{aligned} 3x - 11 < 4 \quad &or \quad 4x + 9 \geq 1 \\ 3x < 15 \quad &or \quad 4x \geq -8 \\ x < 5 \quad &or \quad x \geq -2. \end{aligned}$$

Keep the word "or."

To find the solution set, we first look at the individual graphs.

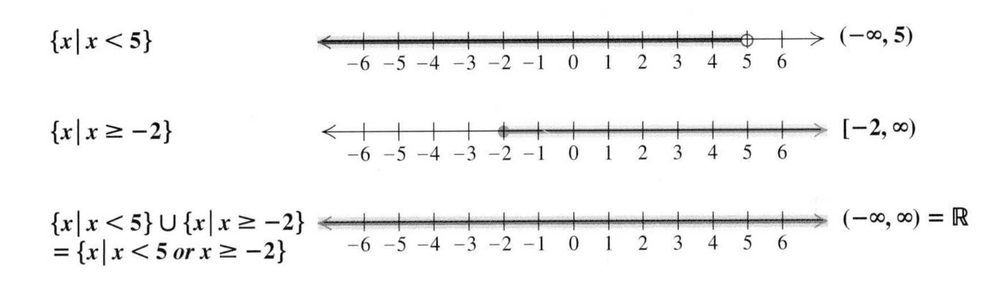

Since *all* numbers are less than 5 or greater than or equal to -2, the two sets fill the entire number line. Thus the solution set is $\mathbb{R}$, the set of all real numbers.

Interval Notation and Domains

In Section 2.2, we saw that if $g(x) = (5x - 2)/(3x - 7)$, then the domain of $g = \left\{x \mid x \text{ is a real number } and\ x \neq \frac{7}{3}\right\}$. We can now represent such a set using interval notation:

$$\left\{x \mid x \text{ is a real number } and\ x \neq \tfrac{7}{3}\right\} = \left(-\infty, \tfrac{7}{3}\right) \cup \left(\tfrac{7}{3}, \infty\right).$$

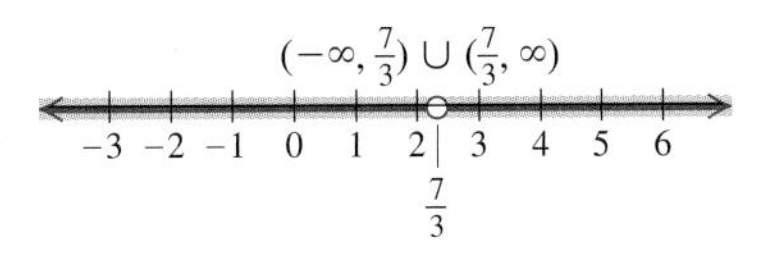

Example 9 Use interval notation to write the domain of f if $f(x) = \sqrt{x + 2}$.

Solution The expression $\sqrt{x + 2}$ is not a real number when $x + 2$ is negative. Thus the domain of f is the set of all x-values for which $x + 2 \geq 0$. Since $x + 2 \geq 0$ is equivalent to $x \geq -2$, we have

$$\text{Domain of } f = \{x \mid x \geq -2\} = [-2, \infty).$$

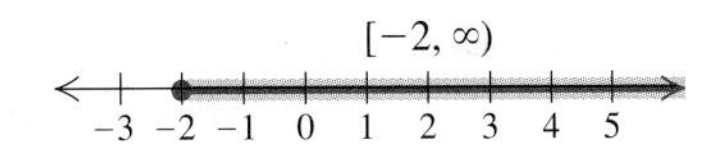

To visualize the domain of a sum, difference, or product of two functions, we have graphed below $y_1 = \sqrt{3 - x}$ and $y_2 = \sqrt{x + 1}$.

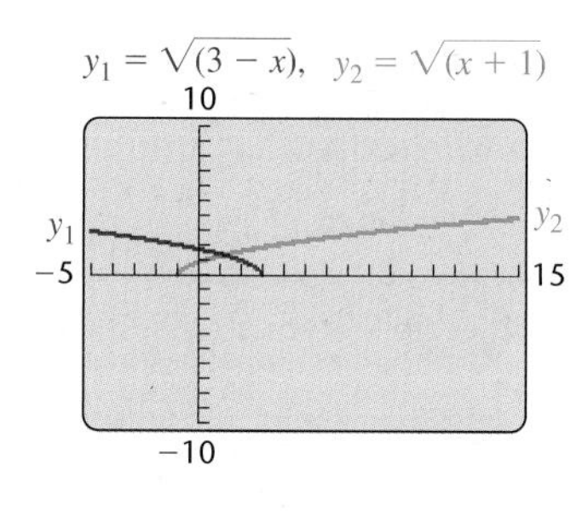

1. Determine algebraically the domains for y_1 and y_2 and TRACE each curve to verify your answer.
2. The Y-VARS option of the VARS key permits us to access the functions already entered. Use this feature to let $y_3 = y_1 + y_2$, $y_4 = y_1 - y_2$, $y_5 = y_2 - y_1$, and $y_6 = y_1 \cdot y_2$. Determine the domains of $y_1 + y_2$, $y_1 - y_2$, $y_2 - y_1$, and $y_1 \cdot y_2$ algebraically and then check by graphing.

Exercise Set 4.2

FOR EXTRA HELP

Digital Video Tutor CD 3 Videotape 7

InterAct Math

Math Tutor Center

MathXL

MyMathLab.com

Find each indicated intersection or union.

1. $\{7, 9, 11\} \cap \{9, 11, 13\}$
2. $\{2, 4, 8\} \cup \{8, 9, 10\}$
3. $\{1, 5, 10, 15\} \cup \{5, 15, 20\}$
4. $\{2, 5, 9, 13\} \cap \{5, 8, 10\}$
5. $\{a, b, c, d, e, f\} \cap \{b, d, f\}$
6. $\{a, b, c\} \cup \{a, c\}$
7. $\{r, s, t\} \cup \{r, u, t, s, v\}$
8. $\{m, n, o, p\} \cap \{m, o, p\}$
9. $\{3, 6, 9, 12\} \cap \{5, 10, 15\}$
10. $\{1, 5, 9\} \cup \{4, 6, 8\}$
11. $\{3, 5, 7\} \cup \varnothing$
12. $\{3, 5, 7\} \cap \varnothing$

Graph and write interval notation for each compound inequality.

13. $3 < x < 8$
14. $0 \leq y \leq 4$
15. $-6 \leq y \leq -2$
16. $-9 \leq x < -5$
17. $x < -2$ *or* $x > 3$
18. $x < -5$ *or* $x > 1$
19. $x \leq -1$ *or* $x > 5$
20. $x \leq -5$ *or* $x > 2$
21. $-4 \leq -x < 2$
22. $x > -7$ *and* $x < -2$
23. $x > -2$ *and* $x < 4$
24. $3 > -x \geq -1$
25. $5 > a$ *or* $a > 7$
26. $t \geq 2$ *or* $-3 > t$
27. $x \geq 5$ *or* $-x \geq 4$
28. $-x < 3$ *or* $x < -6$
29. $4 > y$ *and* $y \geq -6$
30. $6 > -x \geq 0$
31. $x < 7$ *and* $x \geq 3$
32. $x \geq -3$ *and* $x < 3$
33. Aha! $t < 2$ *or* $t < 5$
34. $t > 4$ *or* $t > -1$
35. $x > -1$ *or* $x \leq 3$
36. $4 > x$ *or* $x \geq -3$
37. $x \geq 5$ *and* $x > 7$
38. $x \leq -4$ *and* $x < 1$

Solve and graph each solution set.

39. $-1 < t + 2 < 7$
40. $-3 < t + 1 \leq 5$
41. $2 < x + 3$ *and* $x + 1 \leq 5$
42. $-1 < x + 2$ *and* $x - 4 < 3$
43. $-7 \leq 2a - 3$ *and* $3a + 1 < 7$

44. $-4 \le 3n + 2$ *and* $2n - 3 \le 5$

Aha! **45.** $x + 7 \le -2$ *or* $x + 7 \ge -3$

46. $x + 5 < -3$ *or* $x + 5 \ge 4$

47. $2 \le f(x) \le 8$, where $f(x) = 3x - 1$

48. $7 \ge g(x) \ge -2$, where $g(x) = 3x - 5$

49. $-21 \le f(x) < 0$, where $f(x) = -2x - 7$

50. $4 > g(t) \ge 2$, where $g(t) = -3t - 8$

51. $f(x) \le 2$ *or* $f(x) \ge 8$, where $f(x) = 3x - 1$

52. $g(x) \le -2$ *or* $g(x) \ge 10$, where $g(x) = 3x - 5$

53. $f(x) < -1$ *or* $f(x) > 1$, where $f(x) = 2x - 7$

54. $g(x) < -7$ *or* $g(x) > 7$, where $g(x) = 3x + 5$

55. $6 > 2a - 1$ *or* $-4 \le -3a + 2$

56. $3a - 7 > -10$ *or* $5a + 2 \le 22$

57. $a + 4 < -1$ *and* $3a - 5 < 7$

58. $1 - a < -2$ *and* $2a + 1 > 9$

59. $3x + 2 < 2$ *or* $4 - 2x < 14$

60. $2x - 1 > 5$ *or* $3 - 2x \ge 7$

61. $2t - 7 \le 5$ *or* $5 - 2t > 3$

62. $5 - 3a \le 8$ *or* $2a + 1 > 7$

For f(x) as given, use interval notation to write the domain of f.

63. $f(x) = \dfrac{9}{x + 7}$

64. $f(x) = \dfrac{2}{x + 3}$

65. $f(x) = \sqrt{x - 6}$

66. $f(x) = \sqrt{x - 2}$

67. $f(x) = \dfrac{x + 3}{2x - 5}$

68. $f(x) = \dfrac{x - 1}{3x + 4}$

69. $f(x) = \sqrt{2x + 8}$

70. $f(x) = \sqrt{8 - 4x}$

71. $f(x) = \sqrt{8 - 2x}$

72. $f(x) = \sqrt{10 - 2x}$

73. Why can the conjunction $2 < x$ *and* $x < 5$ be rewritten as $2 < x < 5$, but the disjunction $2 < x$ *or* $x < 5$ cannot be rewritten as $2 < x < 5$?

74. Can the solution set of a disjunction be empty? Why or why not?

SKILL MAINTENANCE

Graph.

75. $y = 5$

76. $y = -2$

77. $f(x) = |x|$

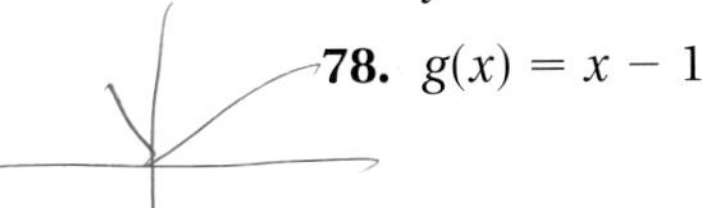

78. $g(x) = x - 1$

Solve each system graphically.

79. $y = x - 3,$
$y = 5$

80. $y = x + 2,$
$y = -3$

SYNTHESIS

81. What can you conclude about a, b, c, and d, if $[a, b] \cup [c, d] = [a, d]$? Why?

82. What can you conclude about a, b, c, and d, if $[a, b] \cap [c, d] = [a, b]$? Why?

83. Use the accompanying graph of $f(x) = 2x - 5$ to solve $-7 < 2x - 5 < 7$.

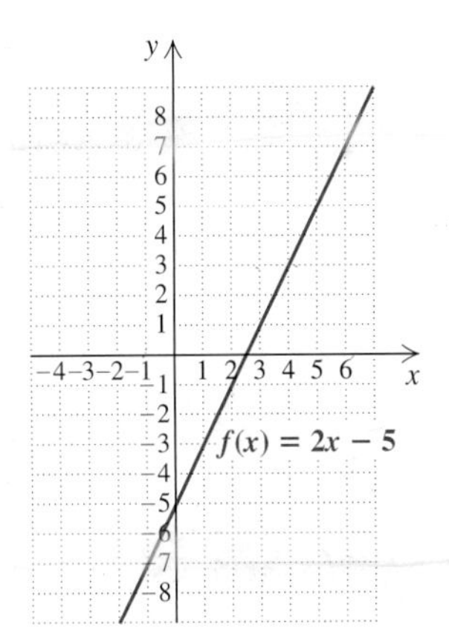

84. Use the accompanying graph of $g(x) = 4 - x$ to solve $4 - x < -2$ *or* $4 - x > 7$.

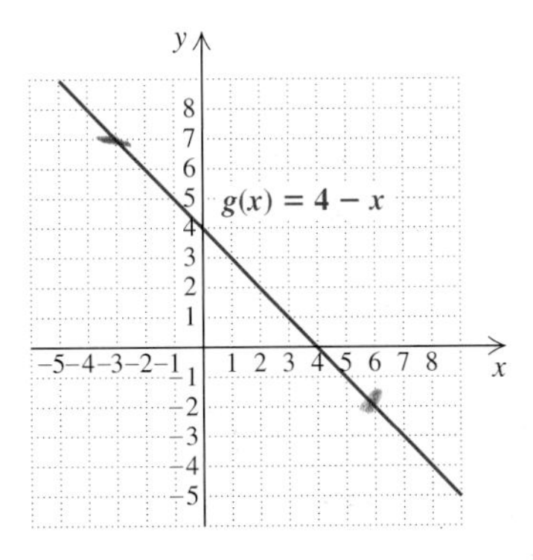

85. *Minimizing tolls.* A \$3.00 toll is charged to cross the bridge from Sanibel Island to mainland Florida. A six-month pass, costing \$15.00, reduces the toll to \$0.50. A one-year pass, costing \$150, allows for free crossings. How many crossings per year does it take, on average, for the two six-month passes to be the most economical choice? Assume a constant number of trips per month.

86. *Pressure at sea depth.* The function

$$P(d) = 1 + \frac{d}{33}$$

gives the pressure, in atmospheres (atm), at a depth of d feet in the sea. For what depths d is the pressure at least 1 atm and at most 7 atm?

87. *Converting dress sizes.* The function

$$f(x) = 2(x + 10)$$

can be used to convert dress sizes x in the United States to dress sizes $f(x)$ in Italy. For what dress sizes in the United States will dress sizes in Italy be between 32 and 46?

88. *Solid-waste generation.* The function

$$w(t) = 0.01t + 4.3$$

can be used to estimate the number of pounds of solid waste, $w(t)$, produced daily, on average, by each person in the United States, t years after 1991. For what years will waste production range from 4.5 to 4.75 lb per person per day?

89. *Temperatures of liquids.* The formula

$$C = \tfrac{5}{9}(F - 32)$$

can be used to convert Fahrenheit temperatures F to Celsius temperatures C.

a) Gold is liquid for Celsius temperatures C such that $1063° \le C < 2660°$. Find a comparable inequality for Fahrenheit temperatures.

b) Silver is liquid for Celsius temperatures C such that $960.8° \le C < 2180°$. Find a comparable inequality for Fahrenheit temperatures.

90. *Records in the women's 100-m dash.* Florence Griffith Joyner set a world record of 10.49 sec in the women's 100-m dash in 1988. The function

$$R(t) = -0.0433t + 10.49$$

can be used to predict the world record in the women's 100-m dash t years after 1988. Predict (in terms of an inequality) those years for which the world record was between 11.5 and 10.8 sec. (Measure from the middle of 1988.)

Solve and graph.

91. $4a - 2 \le a + 1 \le 3a + 4$

92. $4m - 8 > 6m + 5$ *or* $5m - 8 < -2$

93. $x - 10 < 5x + 6 \le x + 10$

94. $3x < 4 - 5x < 5 + 3x$

Determine whether each sentence is true or false for all real numbers a, b, and c.

95. If $-b < -a$, then $a < b$.

96. If $a \le c$ and $c \le b$, then $b > a$.

97. If $a < c$ and $b < c$, then $a < b$.

98. If $-a < c$ and $-c > b$, then $a > b$.

For f(x) as given, use interval notation to write the domain of f.

99. $f(x) = \dfrac{\sqrt{5 + 2x}}{x - 1}$

100. $f(x) = \dfrac{\sqrt{3 - 4x}}{x + 7}$

101. Let $y_1 = -1$, $y_2 = 2x + 5$, and $y_3 = 13$. Then use the graphs of y_1, y_2, and y_3 to check the solution to Example 2.

102. Let $y_1 = -2x - 5$, $y_2 = -2$, $y_3 = x - 3$, and $y_4 = -10$. Then use the graphs of y_1, y_2, y_3, and y_4 to check the solution to Example 7.

103. Use a grapher to check your answers to Exercises 33–36 and Exercises 53–56.

104. On many graphers, the TEST key provides access to inequality symbols, while the LOGIC option of that same key accesses the conjunction *and* and the disjunction *or*. Thus, if $y_1 = x > -2$ and $y_2 = x < 4$, Exercise 23 can be checked by forming the expression $y_3 = y_1$ *and* y_2. As in the Technology Connection on p. 217, the interval(s) in the solution set are shown as a horizontal line 1 unit above the x-axis. (Be careful to "deselect" y_1 and y_2 so that only y_3 is drawn.) Use this approach to check your answers to Exercises 29 and 60.

COLLABORATIVE

CORNER

Saving on Shipping Costs

Focus: Compound inequalities and solution sets

Time: 20–30 minutes

Group size: 2–3

At present (2001), the U.S. Postal Service charges 21 cents per ounce plus an additional 13-cent delivery fee (1 oz or less costs 34 cents; more than 1 oz, but not more than 2 oz, costs 55 cents; and so on). Rapid Delivery charges $1.05 per pound plus an additional $2.50 delivery fee (up to 16 oz costs $3.55; more than 16 oz, but less than 32 oz, costs $4.60; and so on). Let x be the weight, in ounces, of an item being mailed.*

*Based on an article by Michael Contino in *Mathematics Teacher*, May 1995.

ACTIVITY

One group member should determine the function p, where $p(x)$ represents the cost, in dollars, of mailing x ounces at a post office. Another group member should determine the function r, where $r(x)$ represents the cost, in dollars, of mailing x ounces with Rapid Delivery. The third group member should graph p and r on the same set of axes. Finally, working together, use the graph to determine those weights for which the Postal Service is less expensive. Express your answer using both set-builder and interval notation.

4.3 Absolute-Value Equations and Inequalities

Equations with Absolute Value • Inequalities with Absolute Value

Equations with Absolute Value

Recall from Section 1.2 the definition of absolute value.

Absolute Value

The absolute value of x, denoted $|x|$, is defined as

$$|x| = \begin{cases} x, & \text{if } x \geq 0, \\ -x, & \text{if } x < 0. \end{cases}$$

(When x is nonnegative, the absolute value of x is x. When x is negative, the absolute value of x is the opposite of x.)

To better understand this definition, suppose x is -5. Then $|x| = |-5| = 5$, and 5 is the opposite of -5. This shows that when x represents a negative number, we have $|x| = -x$.

Since distance is always nonnegative, we can think of a number's absolute value as its distance from zero on a number line.

Example 1 Find the solution set: **(a)** $|x| = 4$; **(b)** $|x| = 0$; **(c)** $|x| = -7$.

Solution

a) We interpret $|x| = 4$ to mean that the number x is 4 units from zero on a number line. There are two such numbers, 4 and -4. Thus the solution set is $\{-4, 4\}$.

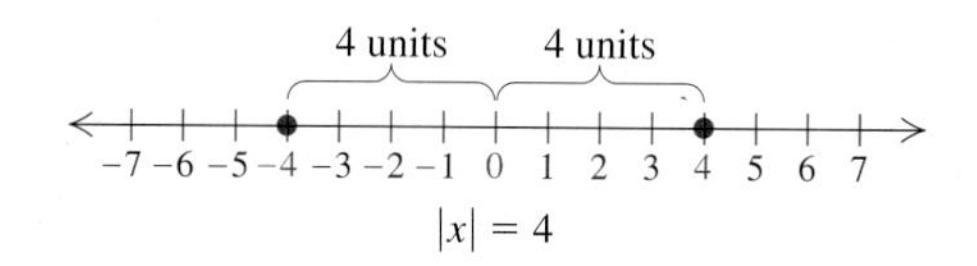

A second way to visualize this problem is to graph $f(x) = |x|$ (see Section 2.1). We also graph $g(x) = 4$. The x-values of the points of intersection are the solutions of $|x| = 4$.

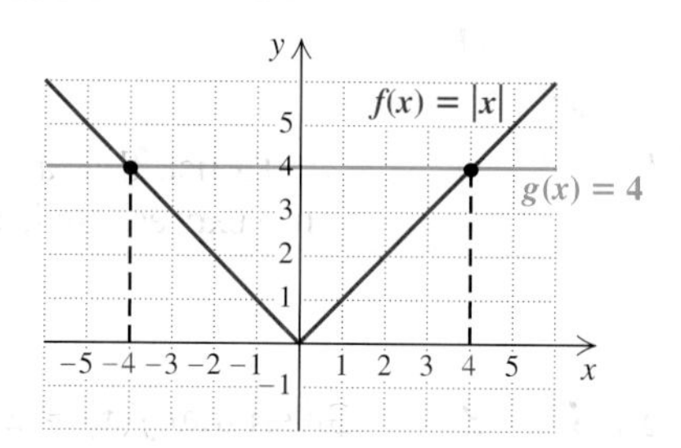

b) We interpret $|x| = 0$ to mean that x is 0 units from zero on a number line. The only number that satisfies this is 0 itself. Thus the solution set is $\{0\}$.

c) Since distance is always nonnegative, it doesn't make sense to talk about a number that is -7 units from zero. Remember: The absolute value of a number is never negative. Thus, $|x| = -7$ has no solution; the solution set is $\varnothing$.

Example 1 leads us to the following principle for solving equations.

Study Tip

The absolute-value principle will be used with a variety of replacements for X. Make sure that the principle, as stated here, makes sense before going further.

The Absolute-Value Principle for Equations

For any positive number p and any algebraic expression X:

a) The solutions of $|X| = p$ are those numbers that satisfy $X = -p$ or $X = p$.

b) The equation $|X| = 0$ is equivalent to the equation $X = 0$.

c) The equation $|X| = -p$ has no solution.

Example 2

A

To check Example 2(a), we let $y_1 = \textbf{abs}(2x + 5)$ and $y_2 = 13$. Using the window $[-12, 8, -2, 18]$, we use INTERSECT to find the points of intersection, $(-9, 13)$ and $(4, 13)$. The x-coordinates, -9 and 4, are the solutions.

$y_1 = \text{abs}(2x + 5), \; y_2 = 13$

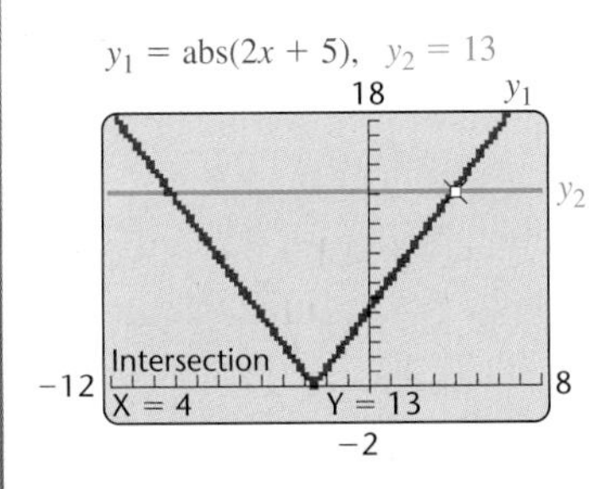

1. Use a grapher to show that Example 2(b) has no solution.

Find the solution set: **(a)** $|2x + 5| = 13$; **(b)** $|4 - 7x| = -8$.

Solution

a) We use the absolute-value principle, replacing X with $2x + 5$ and p with 13:

$$|X| = p$$
$$|2x + 5| = 13$$
$$2x + 5 = -13 \quad or \quad 2x + 5 = 13$$
$$2x = -18 \quad or \quad 2x = 8$$
$$x = -9 \quad or \quad x = 4.$$

Check: For -9:

$\|2x + 5\| = 13$	
$\|2(-9) + 5\|$	? 13
$\|-18 + 5\|$	
$\|-13\|$	
13	13 TRUE

For 4:

$\|2x + 5\| = 13$	
$\|2 \cdot 4 + 5\|$	? 13
$\|8 + 5\|$	
$\|13\|$	
13	13 TRUE

The number $2x + 5$ is 13 units from zero if x is replaced with -9 or 4. The solution set is $\{-9, 4\}$.

b) The absolute-value principle reminds us that absolute value is always nonnegative. The equation $|4 - 7x| = -8$ has no solution. The solution set is $\varnothing$.

To use the absolute-value principle, we must be sure that the absolute-value expression is alone on one side of the equation.

Example 3

Given that $f(x) = 2|x + 3| + 1$, find all x for which $f(x) = 15$.

Solution Since we are looking for $f(x) = 15$, we substitute:

$$f(x) = 15$$
$$2|x + 3| + 1 = 15 \qquad \text{Replacing } f(x) \text{ with } 2|x + 3| + 1$$
$$2|x + 3| = 14 \qquad \text{Subtracting 1 from both sides}$$
$$|x + 3| = 7 \qquad \text{Dividing both sides by 2}$$
$$x + 3 = -7 \quad or \quad x + 3 = 7 \qquad \text{Replacing } X \text{ with } x + 3 \text{ and } p \text{ with 7 in the absolute-value principle}$$
$$x = -10 \quad or \quad x = 4.$$

We leave it to the student to check that $f(-10) = f(4) = 15$. The solution set is $\{-10, 4\}$.

Example 4

Solve: $|x - 2| = 3$.

Solution Because this equation is of the form $|a - b| = c$, it can be solved in two different ways.

Method 1. We interpret $|x - 2| = 3$ as stating that the number $x - 2$ is 3 units from zero. Using the absolute-value principle, we replace X with $x - 2$ and p with 3:

$$|X| = p$$
$$|x - 2| = 3$$
$$x - 2 = -3 \quad or \quad x - 2 = 3 \qquad \text{Using the absolute-value principle}$$
$$x = -1 \quad or \quad x = 5.$$

Method 2. This approach is helpful in calculus. The expressions $|a - b|$ and $|b - a|$ can be used to represent the *distance between a* and *b* on the number line. For example, the distance between 7 and 8 is given by $|8 - 7|$ or $|7 - 8|$. From this viewpoint, the equation $|x - 2| = 3$ states that the distance between x and 2 is 3 units. We draw a number line and locate all numbers that are 3 units from 2.

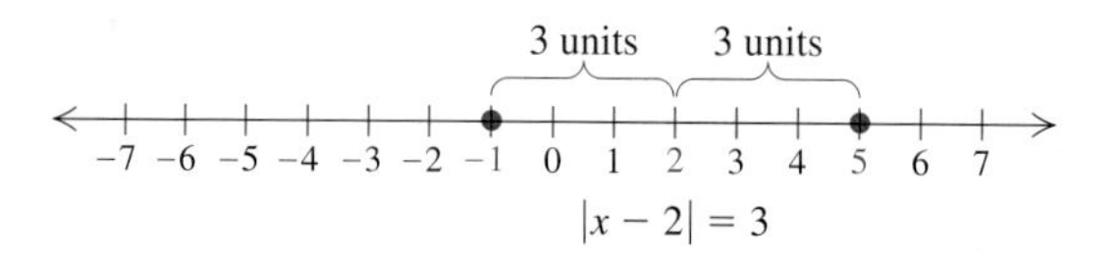

The solutions of $|x - 2| = 3$ are -1 and 5.

Check: The check consists of observing that both methods give the same solutions. The solution set is $\{-1, 5\}$.

Sometimes an equation has two absolute-value expressions. Consider $|a| = |b|$. This means that a and b are the same distance from zero.

If a and b are the same distance from zero, then either they are the same number or they are opposites.

Example 5 Solve: $|2x - 3| = |x + 5|$.

Solution Either $2x - 3 = x + 5$ (they are the same number) or $2x - 3 = -(x + 5)$ (they are opposites). We solve each equation separately:

This assumes these numbers are the same. | This assumes these numbers are opposites.

$$2x - 3 = x + 5 \quad or \quad 2x - 3 = -(x + 5)$$
$$x - 3 = 5 \quad or \quad 2x - 3 = -x - 5$$
$$x = 8 \quad or \quad 3x - 3 = -5$$
$$3x = -2$$
$$x = -\tfrac{2}{3}.$$

The check is left to the student. The solutions are 8 and $-\frac{2}{3}$ and the solution set is $\left\{-\frac{2}{3}, 8\right\}$.

Inequalities with Absolute Value

Our methods for solving equations with absolute value can be adapted for solving inequalities. Inequalities of this sort arise regularly in more advanced courses.

Example 6

Solve $|x| < 4$. Then graph.

Solution The solutions of $|x| < 4$ are all numbers whose *distance from zero is less than* 4. By substituting or by looking at the number line, we can see that numbers like $-3, -2, -1, -\frac{1}{2}, -\frac{1}{4}, 0, \frac{1}{4}, \frac{1}{2}, 1, 2$, and 3 are all solutions. In fact, the solutions are all the numbers between -4 and 4. The solution set is $\{x \mid -4 < x < 4\}$. In interval notation, the solution set is $(-4, 4)$. The graph is as follows:

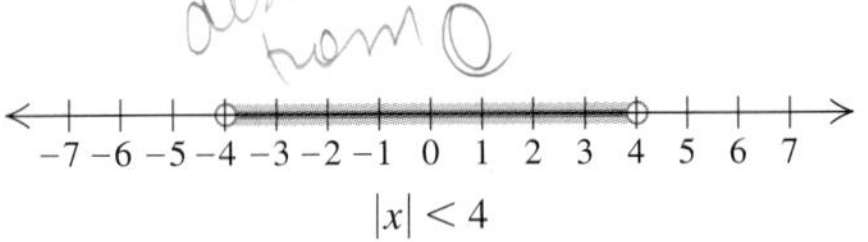

$|x| < 4$

We can also visualize Example 6 by graphing $f(x) = |x|$ and $g(x) = 4$, as in Example 1. The solution set consists of all x-values for which $(x, f(x))$ is below the horizontal line $g(x) = 4$. These x-values comprise the interval $(-4, 4)$.

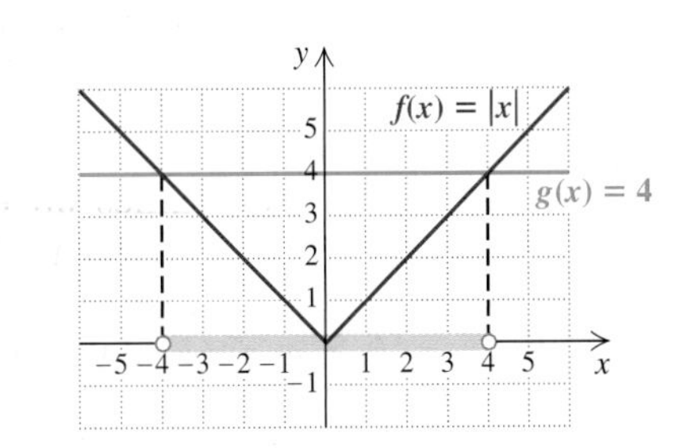

Example 7

Solve $|x| \geq 4$. Then graph.

Solution The solutions of $|x| \geq 4$ are all numbers that are at least 4 units from zero—in other words, those numbers x for which $x \leq -4$ or $4 \leq x$. The solution set is $\{x \mid x \leq -4 \text{ or } x \geq 4\}$. In interval notation, the solution set is $(-\infty, -4] \cup [4, \infty)$. We can check mentally with numbers like $-4.1, -5, 4.1$, and 5. The graph is as follows:

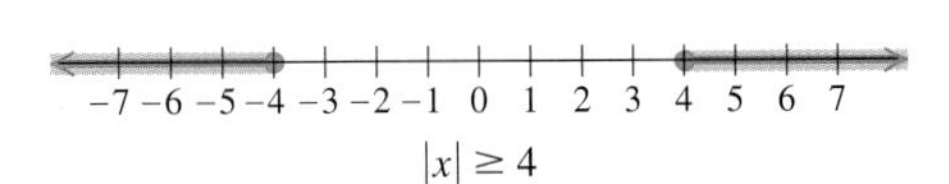

$|x| \geq 4$

As with Examples 1 and 6, Example 7 can be visualized by graphing $f(x) = |x|$ and $g(x) = 4$. The solution set of $|x| \geq 4$ consists of all x-values for which $(x, f(x))$ is on or above the horizontal line $g(x) = 4$. These x-values comprise $(-\infty, -4] \cup [4, \infty)$.

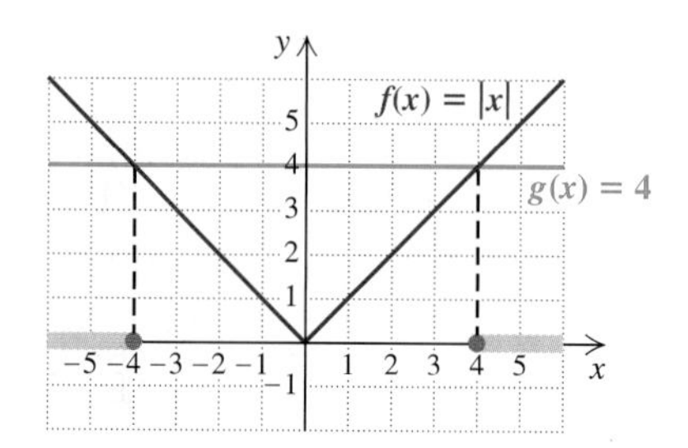

Examples 1, 6, and 7 illustrate three types of problems in which absolute-value symbols appear. The following is a general principle for solving such problems.

Principles for Solving Absolute-Value Problems

For any positive number p and any expression X:

a) The solutions of $|X| = p$ are those numbers that satisfy $X = -p$ *or* $X = p$.

b) The solutions of $|X| < p$ are those numbers that satisfy $-p < X < p$.

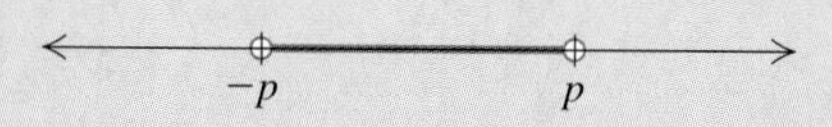

c) The solutions of $|X| > p$ are those numbers that satisfy $X < -p$ *or* $p < X$.

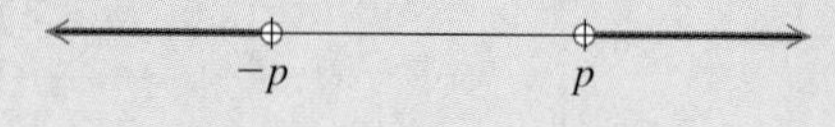

Of course, if p is negative, any value of X will satisfy the inequality $|X| > p$ because absolute value is never negative. By the same reasoning, $|X| < p$ has no solution when p is not positive. Thus, $|2x - 7| > -3$ is true for any real number x, and $|2x - 7| < -3$ has no solution.

Note that an inequality of the form $|X| < p$ corresponds to a *con*junction, whereas an inequality of the form $|X| > p$ corresponds to a *dis*junction.

Example 8 Solve $|3x - 2| < 4$. Then graph.

Solution We use part (b) of the principles listed above. In this case, X is $3x - 2$ and p is 4:

$\lvert X\rvert < p$	
$\lvert 3x - 2\rvert < 4$	Replacing X with $3x - 2$ and p with 4
$-4 < 3x - 2 < 4$	The number $3x - 2$ must be within 4 units of zero.
$-2 < 3x < 6$	Adding 2
$-\frac{2}{3} < x < 2.$	Multiplying by $\frac{1}{3}$

The solution set is $\{x \mid -\frac{2}{3} < x < 2\}$. In interval notation, the solution set is $(-\frac{2}{3}, 2)$. The graph is as follows:

−7 −6 −5 −4 −3 −2 −1 0 1 2 3 4 5 6 7

$|3x - 2| < 4$

Example 9 Given that $f(x) = |4x + 2|$, find all x for which $f(x) \geq 6$.

Solution We have

$$f(x) \geq 6,$$

or $\quad |4x + 2| \geq 6.\quad$ Substituting

To solve, we use part (c) of the principles listed above. In this case, X is $4x + 2$ and p is 6:

$\lvert X\rvert \geq p$	
$\lvert 4x + 2\rvert \geq 6$	Replacing X with $4x + 2$ and p with 6
$4x + 2 \leq -6$ *or* $6 \leq 4x + 2$	The number $4x + 2$ must be at least 6 units from zero.
$4x \leq -8$ *or* $4 \leq 4x$	Adding -2
$x \leq -2$ *or* $1 \leq x.$	Multiplying by $\frac{1}{4}$

The solution set is $\{x \mid x \leq -2 \text{ or } x \geq 1\}$. In interval notation, the solution is $(-\infty, -2] \cup [1, \infty)$. The graph is as follows:

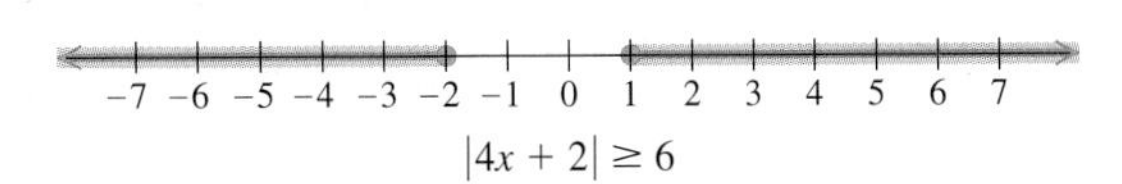

$|4x + 2| \geq 6$

technology connection B

To solve an inequality like $|4x + 2| \geq 6$ with a grapher, graph the equations $y_1 = \mathbf{abs}(4x + 2)$ and $y_2 = 6$. Then use INTERSECT to find where $y_1 = y_2$. The x-values on the graph of $y_1 = |4x + 2|$ that are *on or above* the line $y = 6$ solve the inequality.

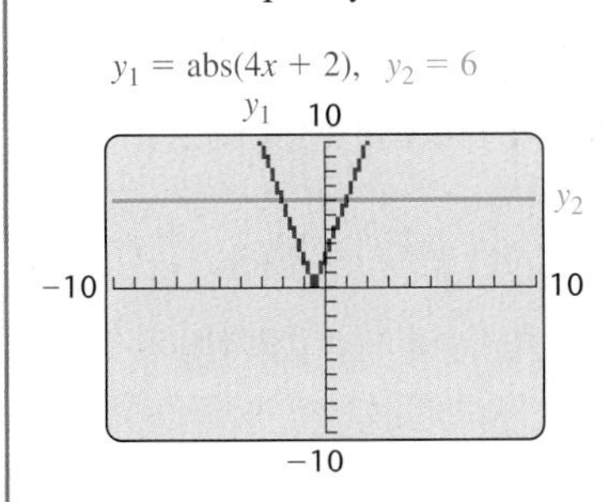

How can the same graph be used to solve the inequality $|4x + 2| < 6$ or the equation $|4x + 2| = 6$? Try using this procedure to solve Example 8 on a grapher.

Exercise Set 4.3

FOR EXTRA HELP

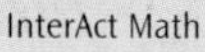
Digital Video Tutor CD 3
Videotape 8

InterAct Math

Math Tutor Center

MathXL

MyMathLab.com

Solve.

1. $|x| = 4$

2. $|x| = 9$

Aha! **3.** $|x| = -5$

4. $|x| = -3$

5. $|y| = 7.3$

6. $|p| = 0$

7. $|m| = 0$

8. $|t| = 5.5$

9. $|5x + 2| = 7$

10. $|2x - 3| = 4$

11. $|7x - 2| = -9$

12. $|3x - 10| = -8$

13. $|x - 3| = 8$

14. $|x - 2| = 6$

15. $|x - 6| = 1$

16. $|x - 5| = 3$

17. $|x - 4| = 5$

18. $|x - 7| = 9$

19. $|2y| - 5 = 13$

20. $|5x| - 3 = 37$

21. $7|z| + 2 = 16$

22. $5|q| - 2 = 9$

23. $\left|\frac{4 - 5x}{6}\right| = 3$

24. $\left|\frac{2x - 1}{3}\right| = 5$

25. $|t - 7| + 1 = 4$

26. $|m + 5| + 9 = 16$

27. $3|2x - 5| - 7 = -1$

28. $5 - 2|3x - 4| = -5$

29. Let $f(x) = |3x - 4|$. Find all x for which $f(x) = 8$.

30. Let $f(x) = |2x - 7|$. Find all x for which $f(x) = 10$.

31. Let $f(x) = |x| - 2$. Find all x for which $f(x) = 6.3$.

32. Let $f(x) = |x| + 7$. Find all x for which $f(x) = 18$.

33. Let $f(x) = \left|\frac{3x - 2}{5}\right|$. Find all x for which $f(x) = 2$.

34. Let $f(x) = \left|\frac{1 - 2x}{3}\right|$. Find all x for which $f(x) = 1$.

Solve.

35. $|x + 4| = |2x - 7|$

36. $|3x + 5| = |x - 6|$

37. $|x - 9| = |x + 6|$

38. $|x + 4| = |x - 3|$

39. $|5t + 7| = |4t + 3|$

40. $|3a - 1| = |2a + 4|$

Aha! **41.** $|n - 3| = |3 - n|$

42. $|y - 2| = |2 - y|$

43. $|7 - a| = |a + 5|$

44. $|6 - t| = |t + 7|$

45. $\left|\frac{1}{2}x - 5\right| = \left|\frac{1}{4}x + 3\right|$

46. $\left|2 - \frac{2}{3}x\right| = \left|4 + \frac{7}{8}x\right|$

Solve and graph.

47. $|a| \le 7$

48. $|x| < 2$

49. $|x| > 8$

50. $|a| \ge 3$

51. $|t| > 0$

52. $|t| \ge 1.7$

53. $|x - 3| < 5$

54. $|x - 1| < 3$

55. $|x + 2| \le 6$

56. $|x + 4| \le 1$

57. $|x - 3| + 2 > 7$

58. $|x - 4| + 5 > 2$

Aha! **59.** $|2y - 7| > -5$

60. $|3y - 4| > 8$

61. $|3a - 4| + 2 \ge 8$

62. $|2a - 5| + 1 \ge 9$

63. $|y - 3| < 12$

64. $|p - 2| < 3$

65. $9 - |x + 4| \le 5$

66. $12 - |x - 5| \le 9$

67. $|4 - 3y| > 8$

68. $|7 - 2y| < -6$

Aha! **69.** $|3 - 4x| < -5$

70. $7 + |4a - 5| \le 26$

71. $\left|\frac{2 - 5x}{4}\right| \ge \frac{2}{3}$

72. $\left|\frac{1 + 3x}{5}\right| > \frac{7}{8}$

73. $|m + 5| + 9 \le 16$

74. $|t - 7| + 3 \ge 4$

75. $25 - 2|a + 3| > 19$

76. $30 - 4|a + 2| > 12$

77. Let $f(x) = |2x - 3|$. Find all x for which $f(x) \le 4$.

78. Let $f(x) = |5x + 2|$. Find all x for which $f(x) \le 3$.

79. Let $f(x) = 2 + |3x - 4|$. Find all x for which $f(x) \ge 13$.

80. Let $f(x) = |2 - 9x|$. Find all x for which $f(x) \ge 25$.

81. Let $f(x) = 7 + |2x - 1|$. Find all x for which $f(x) < 16$.

82. Let $f(x) = 5 + |3x + 2|$. Find all x for which $f(x) < 19$.

83. Explain in your own words why -7 is not a solution of $|x| < 5$.

84. Explain in your own words why $[6, \infty)$ is only part of the solution of $|x| \ge 6$.

SKILL MAINTENANCE

Solve using substitution or elimination.

85. $2x - 3y = 7,$
$3x + 2y = -10$

86. $3x - 5y = 9,$
$4x - 3y = 1$

87. $x = -2 + 3y,$
$x - 2y = 2$

88. $y = 3 - 4x,$
$2x - y = -9$

Solve graphically.

89. $x + 2y = 9,$
$3x - y = -1$

90. $2x + y = 7,$
$-3x - 2y = 10$

SYNTHESIS

91. Is it possible for an equation in x of the form $|ax + b| = c$ to have exactly one solution? Why or why not?

92. Explain why the inequality $|x + 5| \geq 2$ can be interpreted as "the number x is at least 2 units from -5."

93. From the definition of absolute value, $|x| = x$ only when $x \geq 0$. Solve $|3t - 5| = 3t - 5$ using this same reasoning.

Solve.

94. $|x + 2| > x$

95. $2 \leq |x - 1| \leq 5$

96. $|5t - 3| = 2t + 4$

97. $t - 2 \leq |t - 3|$

Find an equivalent inequality with absolute value.

98. $-3 < x < 3$

99. $-5 \leq y \leq 5$

100. $x \leq -6$ *or* $6 \leq x$

101. $x < -4$ *or* $4 < x$

102. $x < -8$ *or* $2 < x$

103. $-5 < x < 1$

104. x is less than 2 units from 7.

105. x is less than 1 unit from 5.

Write an absolute-value inequality for which the interval shown is the solution.

106. 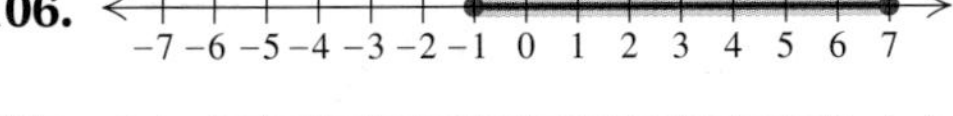

107. −5 −4 −3 −2 −1 0 1 2 3 4 5 6 7 8 9

108. −7 −6 −5 −4 −3 −2 −1 0 1 2 3 4 5 6 7

109. 0 1 2 3 4 5 6 7 8 9 10 11 12 13 14

110. *Motion of a spring.* A weighted spring is bouncing up and down so that its distance d above the ground satisfies the inequality $|d - 6 \text{ ft}| \leq \frac{1}{2} \text{ ft}$ (see the figure below). Find all possible distances d.

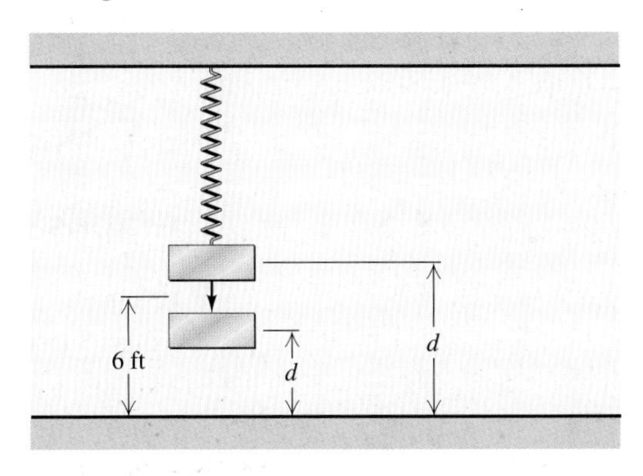

111. Use the accompanying graph of $f(x) = |2x - 6|$ to solve $|2x - 6| \leq 4$.

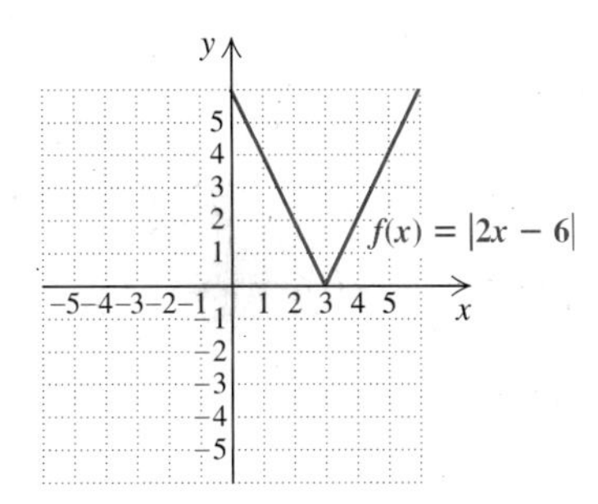

112. Describe a procedure that could be used to solve any equation of the form $g(x) < c$ graphically.

113. Use a grapher to check the solutions to Examples 3 and 5.

114. Use a grapher to check your answers to Exercises 1, 9, 15, 41, 53, 63, 71, and 95.

115. Isabel is using the following graph to solve $|x - 3| < 4$.

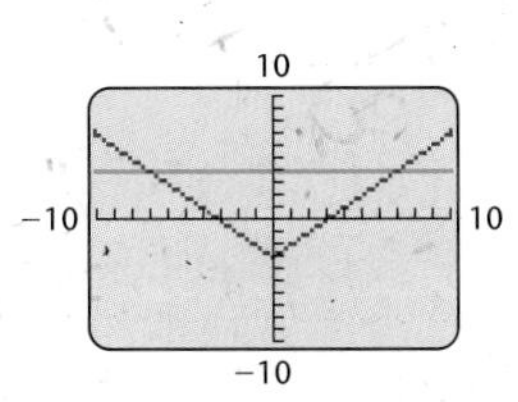

How can you tell that a mistake has been made?

Inequalities in Two Variables

4.4

Graphs of Linear Inequalities • Systems of Linear Inequalities

In Section 4.1, we graphed inequalities in one variable on a number line. Now we graph inequalities in two variables on a plane.

Graphs of Linear Inequalities

When the equals sign in a linear equation is replaced with an inequality sign, a **linear inequality** is formed. Solutions of linear inequalities are ordered pairs.

Example 1 Determine whether $(-3, 2)$ and $(6, -7)$ are solutions of the inequality $5x - 4y > 13$.

Solution Below, on the left, we replace x with -3 and y with 2. On the right, we replace x with 6 and y with -7.

$$\begin{array}{r|l} 5x - 4y & > 13 \\ \hline 5(-3) - 4 \cdot 2 & ?\ 13 \\ -15 - 8 & \\ -23 & 13 \quad \text{FALSE} \end{array}$$

Since $-23 > 13$ is false, $(-3, 2)$ is not a solution.

$$\begin{array}{r|l} 5x - 4y & > 13 \\ \hline 5(6) - 4(-7) & ?\ 13 \\ 30 + 28 & \\ 58 & 13 \quad \text{TRUE} \end{array}$$

Since $58 > 13$ is true, $(6, -7)$ is a solution.

The graph of a linear equation is a straight line. The graph of a linear inequality is a half-plane, bordered by the graph of the *related equation*. To find an inequality's related equation, we simply replace the inequality sign with an equals sign.

Example 2 Graph: $y \leq x$.

Solution We first graph the related equation $y = x$. Every solution of $y = x$ is an ordered pair, like $(3, 3)$, in which both coordinates are the same. The graph of $y = x$ is shown on the left below. Since the inequality symbol is $\leq$, the line is drawn solid and is part of the graph of $y \leq x$.

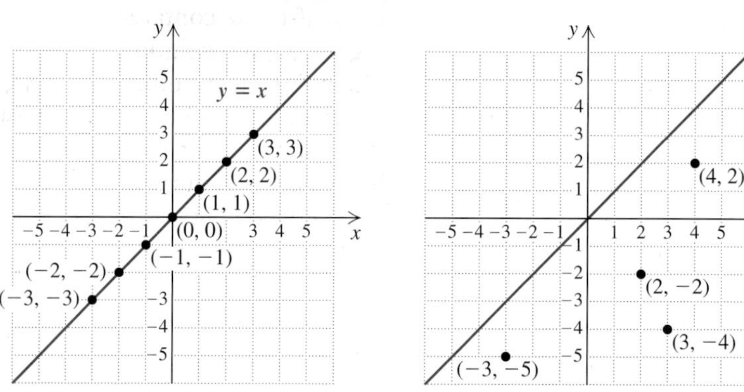

Study Tip

It is never too soon to begin reviewing for the final examination. Take a few minutes each week to review important problems, formulas, and properties. There is also at least one Connecting the Concepts in each chapter. Spend time reviewing the information in this special feature.

Note that in the graph on the right each ordered pair on the half-plane below $y = x$ contains a y-coordinate that is less than the x-coordinate. All these pairs represent solutions of $y \leq x$. We check one pair, $(4, 2)$, as follows:

$$\begin{array}{c|c} y & \leq x \\ \hline 2 & 4 \end{array} \quad \text{TRUE}$$

It turns out that *any* point on the same side of $y = x$ as $(4, 2)$ is also a solution. Thus, if one point in a half-plane is a solution, then *all* points in that half-plane are solutions. We complete the drawing of the solution set by shading the half-plane below $y = x$. The complete solution set consists of the shaded half-plane and the line. Note too that for any inequality of the form $y \leq f(x)$ or $y < f(x)$, we shade *below* the graph of $y = f(x)$.

For any point here, $y = x$.

$y \leq x$ (4, 2) (5, −2) (2, −4) (−3, −5)

For any point here, $y < x$.

Example 3 Graph: $8x + 3y > 24$.

Solution First, we sketch the line $8x + 3y = 24$. Since the inequality sign is $>$, points on this line do not represent solutions of the inequality, so the line is drawn dashed. Points representing solutions of $8x + 3y > 24$ are in either the half-plane above the line or the half-plane below the line. To determine which, we select a point that is not on the line and determine whether it is a solution of $8x + 3y > 24$. We try $(-3, 4)$ as a *test point*:

$$\begin{array}{r|l} 8x + 3y & > 24 \\ \hline 8(-3) + 3 \cdot 4 & ?\ 24 \\ -24 + 12 & \\ -12 & 24 \end{array} \quad \text{FALSE}$$

Since $-12 > 24$ is *false*, $(-3, 4)$ is not a solution. Thus no point in the half-plane containing $(-3, 4)$ is a solution. The points in the other half-plane *are* solutions, so we shade that half-plane and obtain the graph shown at right.

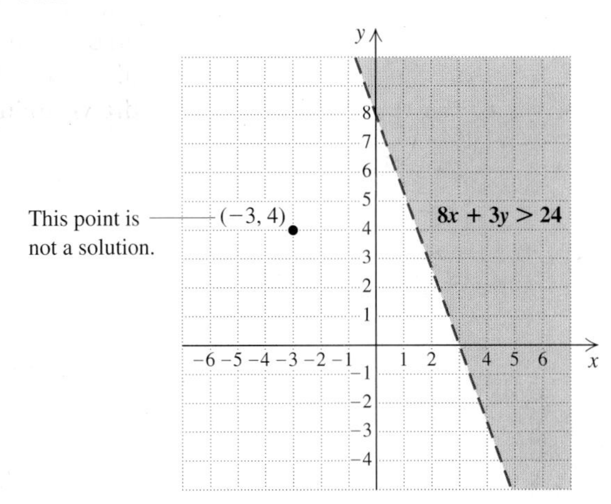

Steps for Graphing Linear Inequalities

1. Replace the inequality sign with an equals sign and graph this related equation. If the inequality symbol is $<$ or $>$, draw the line dashed. If the inequality symbol is $\leq$ or $\geq$, draw the line solid.
2. The graph consists of a half-plane on one side of the line and, if the line is solid, the line as well.

   a) If the inequality is of the form $y < f(x)$ or $y \leq f(x)$, shade *below* the line. If the inequality is of the form $y > f(x)$ or $y \geq f(x)$, shade *above* the line.

   b) If y is not isolated, either use algebra to isolate y and proceed as in part (a) or select a point not on the line. If the point represents a solution of the inequality, shade the half-plane containing the point. If it does not, shade the other half-plane.

Example 4

Graph: $6x - 2y < 12$.

Solution We first graph the related equation, $6x - 2y = 12$, as a dashed line. This line passes through the points $(2, 0)$, $(0, -6)$, and $(3, 3)$, and serves as the boundary of the solution set of the inequality. Since y is not isolated, we determine which half-plane to shade by testing a point *not* on the line. The pair $(0, 0)$ is easy to substitute:

$$\begin{array}{r|l} 6x - 2y & < 12 \\ \hline 6 \cdot 0 - 2 \cdot 0 & ?\ 12 \\ 0 - 0 & \\ 0 & 12 \quad \text{TRUE} \end{array}$$

Since the inequality $0 < 12$ is *true*, the point $(0, 0)$ is a solution, as are all points in the half-plane containing $(0, 0)$. The graph is shown below.

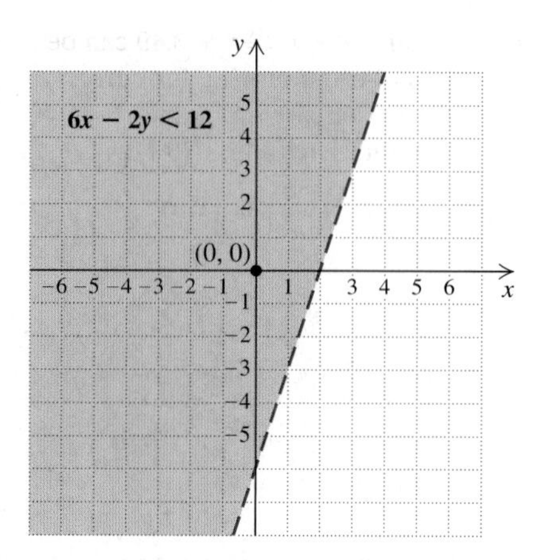

Example 5 Graph $x > -3$ on a plane.

Solution There is a missing variable in this inequality. If we graph the inequality on a line, its graph is as follows:

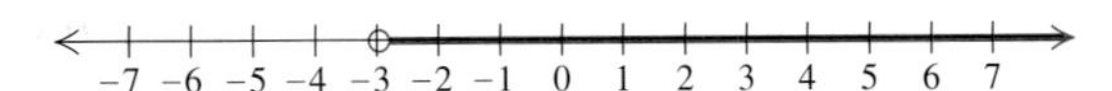

However, we can also write this inequality as $x + 0y > -3$ and graph it on a plane. We can use the same technique as in the examples above. First, we graph the related equation $x = -3$ in the plane, using a dashed line. Then we test some point, say, $(2, 5)$:

$$x + 0y > -3$$

$$2 + 0 \cdot 5 \ ? \ -3$$

$$2 \mid -3 \quad \text{TRUE}$$

Since $(2, 5)$ is a solution, all points in the half-plane containing $(2, 5)$ are solutions. We shade that half-plane. Another approach is to simply note that the solutions of $x > -3$ are all pairs with first coordinates greater than -3.

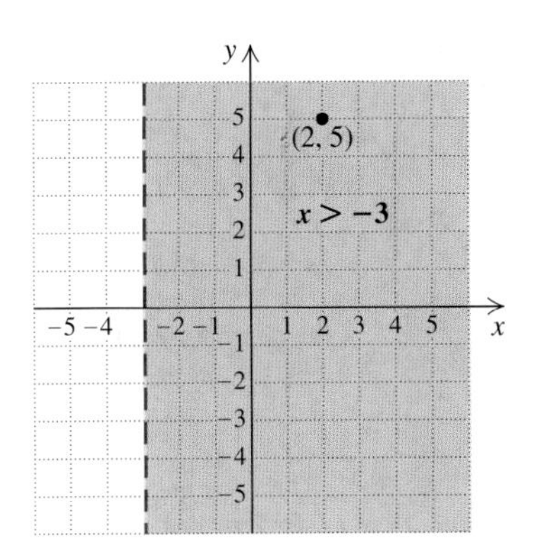

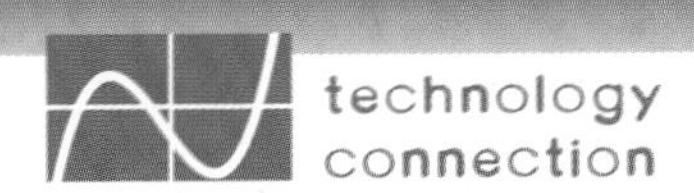

On many graphers, an inequality like $y < \frac{6}{5}x + 3.49$ can be drawn by entering $(6/5)x + 3.49$ as y_1, moving the cursor to the GraphStyle icon just to the left of y_1, pressing ENTER until ◣ appears, and then pressing GRAPH.

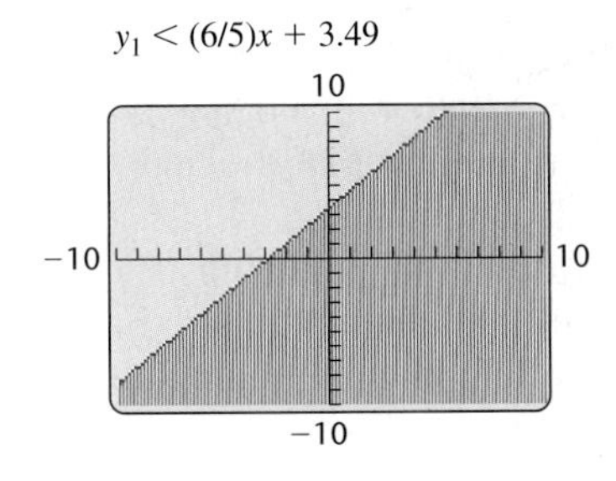

Had we wished to shade above the line, we would have selected ◥ as the GraphStyle. Note that on some graphers a SHADE option must be selected, and on nearly all graphers, the boundary line will always appear solid.

Graph each of the following. Solve for y first if necessary.

1. $y > x + 3.5$
2. $7y \le 2x + 5$
3. $8x - 2y < 11$
4. $11x + 13y + 4 \ge 0$

Example 6 Graph $y \leq 4$ on a plane.

Solution The inequality is of the form $y \leq f(x)$, so we shade below the solid line representing solutions of $y = f(x)$, or in this case, $y = 4$.

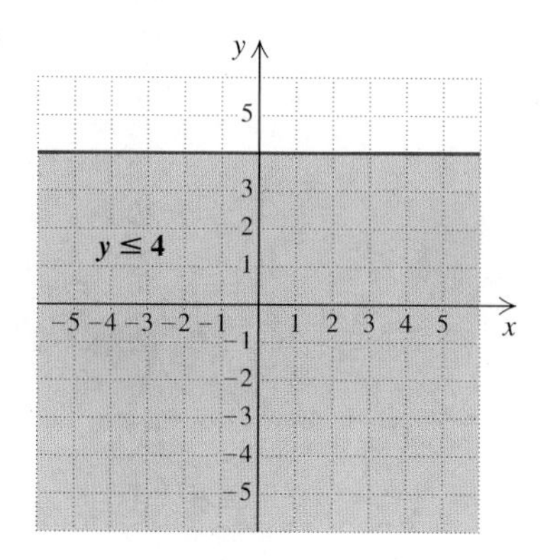

This inequality can also be graphed by drawing $y = 4$ and testing a point above or below the line. The student should check that this results in a graph identical to the one above.

Systems of Linear Inequalities

To graph a system of equations, we graph the individual equations and then find the intersection of the individual graphs. We do the same thing for a system of inequalities, that is, we graph each inequality and find the intersection of the individual graphs.

Example 7 Graph the system

$$x + y \leq 4,$$
$$x - y < 4.$$

Solution To graph $x + y \leq 4$, we graph $x + y = 4$ using a solid line. Since the test point $(0, 0)$ *is* a solution and $(0, 0)$ is below the line, we shade the half-plane below the graph red. The arrows near the ends of the line are another way of indicating the half-plane containing solutions.

Next, we graph $x - y < 4$. We graph $x - y = 4$ using a dashed line and consider $(0, 0)$ as a test point. Again, $(0, 0)$ is a solution, so we shade that side of the line blue. The solution set of the system is the region that is shaded purple (both red and blue) and part of the line $x + y = 4$.

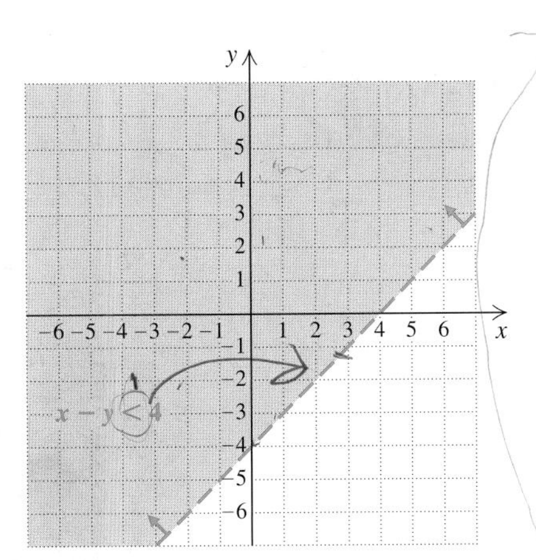

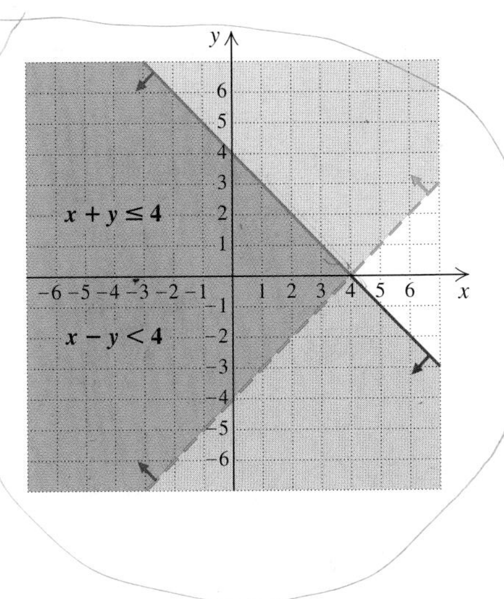

Example 8 Graph: $-2 < x \leq 3$.

Solution This is a system of inequalities:

$$-2 < x,$$
$$x \leq 3.$$

We graph the equation $-2 = x$, and see that the graph of the first inequality is the half-plane to the right of the line $-2 = x$. It is shaded red.

We graph the second inequality, starting with the line $x = 3$, and find that its graph is the line and also the half-plane to its left. It is shaded blue.

The solution set of the system is the region that is the intersection of the individual graphs. Since it is shaded both blue and red, it appears to be purple. All points in this region have x-coordinates that are greater than -2 but do not exceed 3.

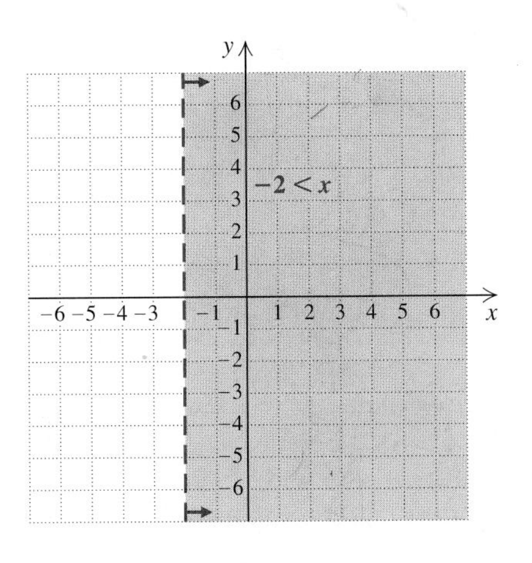

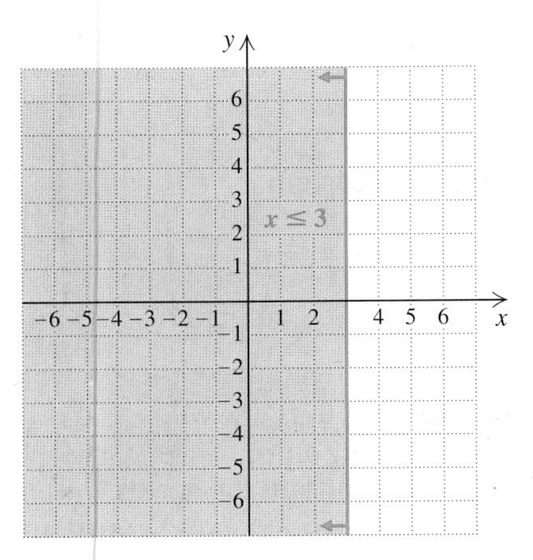

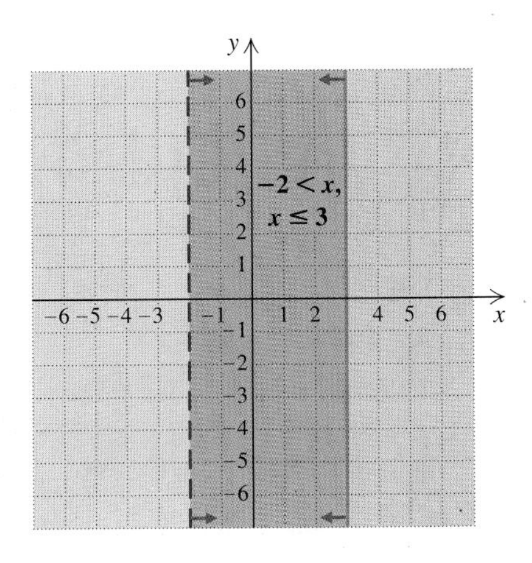

A system of inequalities may have a graph that consists of a polygon and its interior. In Section 4.5, we will have use for the corners, or *vertices* (singular, vertex), of such a graph.

Example 9 Graph the system of inequalities. Find the coordinates of any vertices formed.

$$6x - 2y \le 12, \quad (1)$$
$$y - 3 \le 0, \quad (2)$$
$$x + y \ge 0 \quad (3)$$

Solution We graph the lines

$$6x - 2y = 12,$$
$$y - 3 = 0,$$
and $$x + y = 0$$

using solid lines. The regions for each inequality are indicated by the arrows near the ends of the lines. We note where the regions overlap and shade the region of solutions purple.

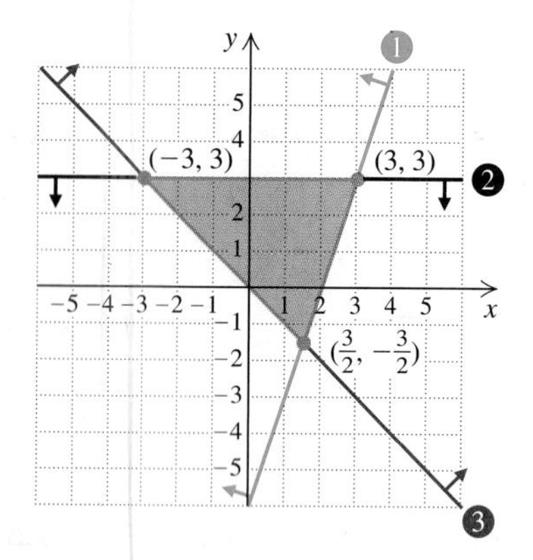

To find the vertices, we solve three different systems of equations. The system of related equations from inequalities (1) and (2) is

$$6x - 2y = 12,$$
$$y - 3 = 0.$$

Solving, we obtain the vertex $(3, 3)$.

The system of related equations from inequalities (1) and (3) is

$$6x - 2y = 12,$$
$$x + y = 0.$$

Solving, we obtain the vertex $(\frac{3}{2}, -\frac{3}{2})$.

The system of related equations from inequalities (2) and (3) is

$$y - 3 = 0,$$
$$x + y = 0.$$

Solving, we obtain the vertex $(-3, 3)$.

CONNECTING THE CONCEPTS

We have now solved a variety of equations, inequalities, systems of equations, and systems of inequalities. In each case, there are different ways to represent the solution. Below is a list of the different types of problems we have solved, along with illustrations of each type.

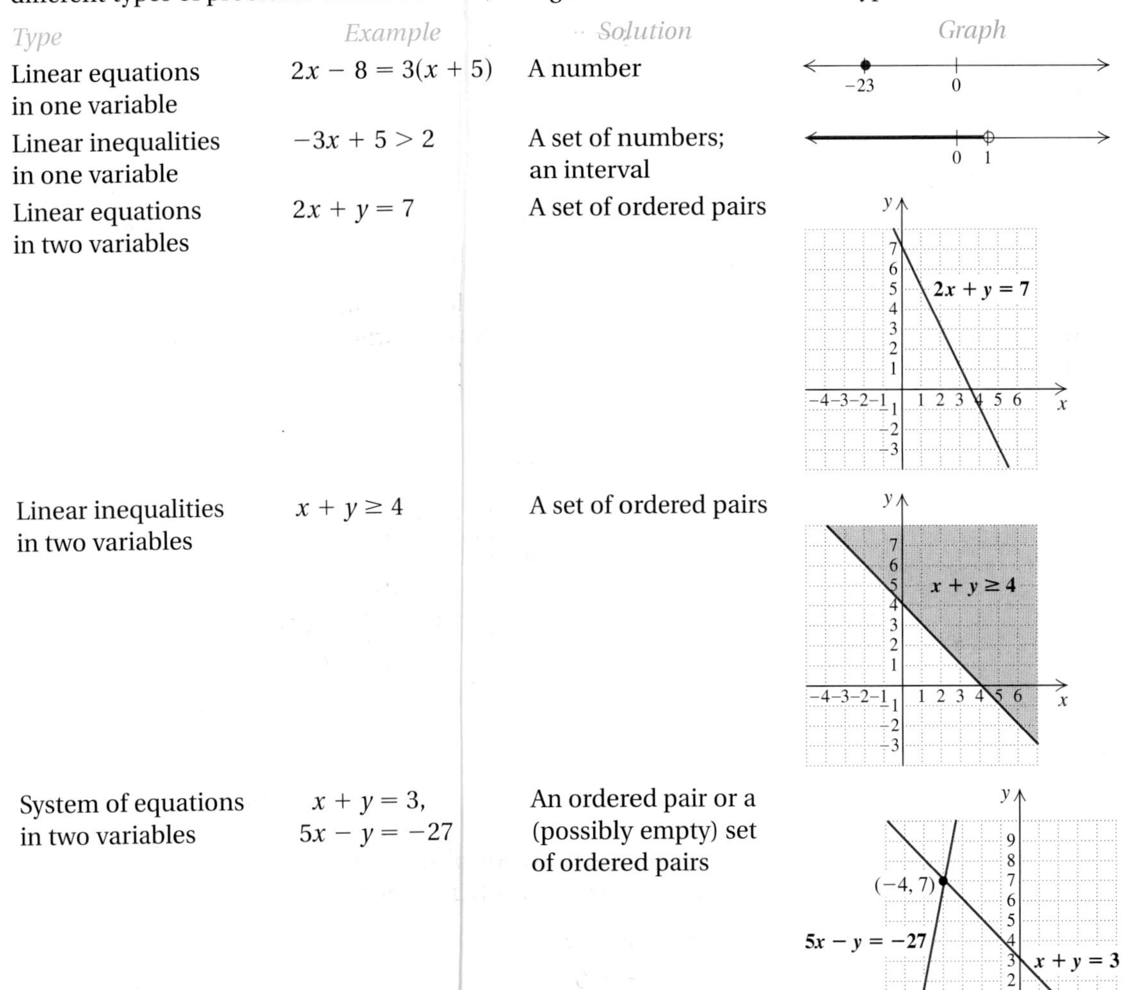

Type	*Example*	*Solution*	*Graph*
Linear equations in one variable	$2x - 8 = 3(x + 5)$	A number	
Linear inequalities in one variable	$-3x + 5 > 2$	A set of numbers; an interval	
Linear equations in two variables	$2x + y = 7$	A set of ordered pairs	
Linear inequalities in two variables	$x + y \geq 4$	A set of ordered pairs	
System of equations in two variables	$x + y = 3$, $5x - y = -27$	An ordered pair or a (possibly empty) set of ordered pairs	
System of inequalities in two variables	$6x - 2y \leq 12$, $y - 3 \leq 0$, $x + y \geq 0$	A set of ordered pairs	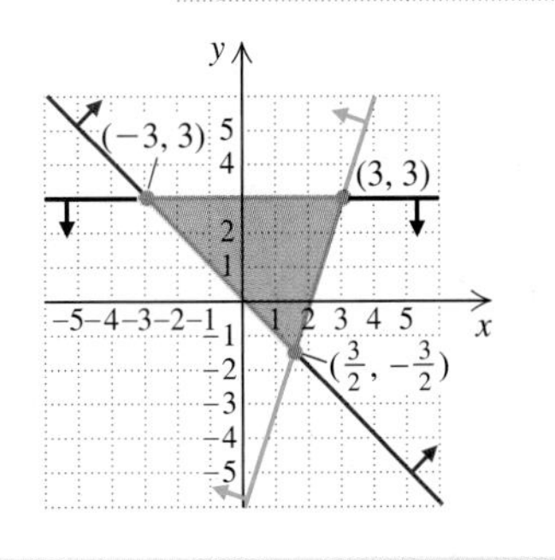

Keeping in mind how these solutions vry and what their graphs look like will help you as you prgress further in this book and in mathematics in general.

Exercise Set 4.4

FOR EXTRA HELP

Digital Video Tutor CD 3
Videotape 8

InterAct Math

Math Tutor Center

MathXL

MyMathLab.com

Determine whether each ordered pair is a solution of the given inequality.

1. $(-4, 2)$; $2x + 3y < -1$

2. $(3, -6)$; $4x + 2y \leq -2$

3. $(8, 14)$; $2y - 3x \geq 9$

4. $(7, 20)$; $3x - y > -1$

Graph on a plane.

5. $y > \frac{1}{2}x$

6. $y > 2x$

7. $y \geq x - 3$

8. $y < x + 3$

9. $y \leq x + 4$

10. $y > x - 2$

11. $x - y \leq 5$

12. $x + y < 4$

13. $2x + 3y < 6$

14. $3x + 4y \leq 12$

15. $2y - x \leq 4$

16. $2y - 3x > 6$

17. $2x - 2y \geq 8 + 2y$

18. $3x - 2 \leq 5x + y$

19. $y \geq 2$

20. $x < -5$

21. $x \leq 7$

22. $y > -3$

23. $-2 < y < 6$

24. $-4 < y < -1$

25. $-4 \leq x \leq 5$

26. $-3 \leq y \leq 4$

27. $0 \leq y \leq 3$

28. $0 \leq x \leq 6$

Graph each system.

29. $y > x$,
$y < -x + 2$

30. $y < x$,
$y > -x + 1$

31. $y \geq x$,
$y \geq 2x - 4$

32. $y \geq x$,
$y \leq -x + 4$

33. $y \leq -3$,
$x \geq -1$

34. $y \geq -3$,
$x \geq 1$

35. $x > -4$,
$y < -2x + 3$

36. $x < 3$,
$y > -3x + 2$

37. $y \leq 3$,
$y \geq -x + 2$

38. $y \geq -2$,
$y \geq x + 3$

39. $x + y \leq 6$,
$x - y \leq 4$

40. $x + y < 1$,
$x - y < 2$

41. $y + 3x > 0$,
$y + 3x < 2$

42. $y - 2x \geq 1$,
$y - 2x \leq 3$

Graph each system of inequalities. Find the coordinates of any vertices formed.

43. $y \leq 2x - 1$,
$y \geq -2x + 1$,
$x \leq 3$

44. $2y - x \leq 2$,
$y - 3x \geq -4$,
$y \geq -1$

45. $x + 2y \leq 12$,
$2x + y \leq 12$,
$x \geq 0$,
$y \geq 0$

46. $x - y \leq 2$,
$x + 2y \geq 8$,
$y \leq 4$

47. $8x + 5y \leq 40$,
$x + 2y \leq 8$,
$x \geq 0$,
$y \geq 0$

48. $4y - 3x \geq -12$,
$4y + 3x \geq -36$,
$y \leq 0$,
$x \leq 0$

49. $y - x \geq 1$,
$y - x \leq 3$,
$2 \leq x \leq 5$

50. $3x + 4y \geq 12$,
$5x + 6y \leq 30$,
$1 \leq x \leq 3$

51. In Example 7, is the point $(4, 0)$ part of the solution set? Why or why not?

52. When graphing linear inequalities, Ron makes a habit of always shading above the line when the symbol $\geq$ is used. Is this wise? Why or why not?

SKILL MAINTENANCE

53. *Catering.* Sandy's Catering needs to provide 10 lb of mixed nuts for a wedding reception. Peanuts cost $2.50 per pound and fancy nuts cost $7 per pound. If $40 has been allocated for nuts, how many pounds of each type should be mixed?

54. *Household waste.* The Hendersons generate two and a half times as much trash as their neighbors, the Savickis. Together, the two households produce 14 bags of trash each month. How much trash does each household produce?

55. *Paid admissions.* There were 203 tickets sold for a volleyball game. For activity-card holders the price was $1.25, and for noncard holders the price was $2. The total amount of money collected was $310. How many of each type of ticket were sold?

56. *Paid admissions.* There were 200 tickets sold for a women's basketball game. Tickets for students were \$2 each and for adults were \$3 each. The total amount collected was \$530. How many of each type of ticket were sold?

57. *Landscaping.* Grass seed is being spread on a triangular traffic island. If the grass seed can cover an area of 200 ft^2 and the island's base is 16 ft long, how tall a triangle can the seed fill?

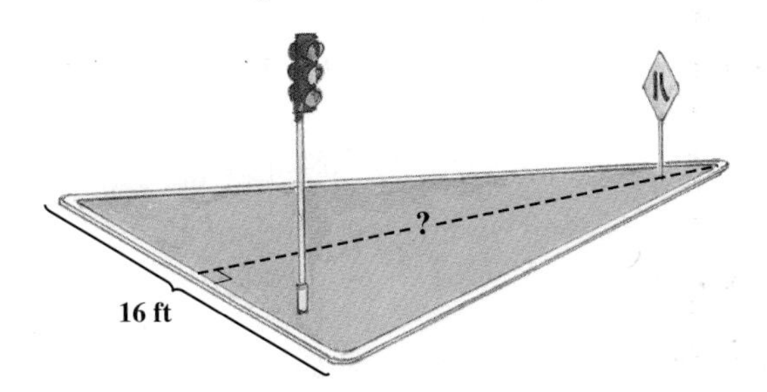

58. *Interest rate.* What rate of interest is required in order for a principal of \$320 to earn \$17.60 in half a year?

SYNTHESIS

59. Explain how a system of linear inequalities could have a solution set containing exactly one pair.

60. Do all systems of linear inequalities have solutions? Why or why not?

Graph.

61. $x + y > 8,$
$x + y \leq -2$

62. $x + y \geq 1,$
$-x + y \geq 2,$
$x \leq 4,$
$y \geq 0,$
$y \leq 4,$
$x \leq 2$

63. $x - 2y \leq 0,$
$-2x + y \leq 2,$
$x \leq 2,$
$y \leq 2,$
$x + y \leq 4$

64. Write four systems of four inequalities that describe a 2-unit by 2-unit square that has $(0, 0)$ as one of the vertices.

65. *Luggage size.* Unless an additional fee is paid, most major airlines will not check any luggage that is more than 62 in. long. The U.S. Postal Service will ship a package only if the sum of the package's length and girth (distance around its midsection) does not exceed 108 in. Concert Productions is ordering several 62-in. long trunks that will be both mailed and checked as luggage. Using w and h for width and height (in inches), respectively, write and graph an inequality that represents all acceptable combinations of width and height.

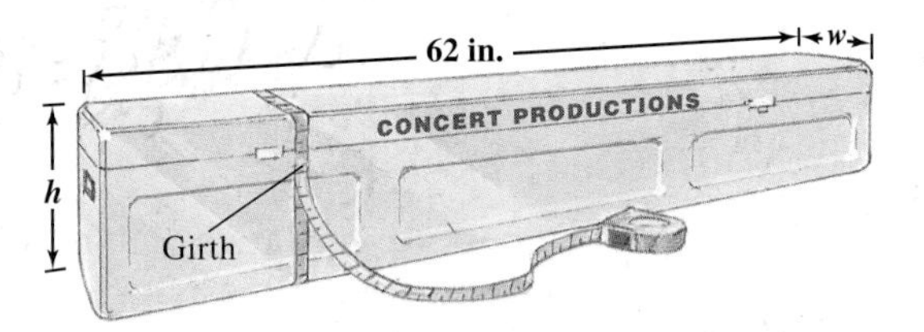

66. *Hockey wins and losses.* The Skating Stars figure that they need at least 60 points for the season in order to make the playoffs. A win is worth 2 points and a tie is worth 1 point. Graph a system of inequalities that describes the situation. (*Hint*: Let $w =$ the number of wins and $t =$ the number of ties.)

67. *Elevators.* Many elevators have a capacity of 1 metric ton (1000 kg). Suppose that c children, each weighing 35 kg, and a adults, each 75 kg, are on an elevator. Graph a system of inequalities that indicates when the elevator is overloaded.

68. *Widths of a basketball floor.* Sizes of basketball floors vary due to building sizes and other constraints such as cost. The length L is to be at most 94 ft and the width W is to be at most 50 ft. Graph a system of inequalities that describes the possible dimensions of a basketball floor.

69. Use a grapher to graph each inequality.

a) $3x + 6y > 2$ **b)** $x - 5y \leq 10$
c) $13x - 25y + 10 \leq 0$ **d)** $2x + 5y > 0$

70. Use a grapher to check your answers to Exercises 29–42. Then use INTERSECT to determine any point(s) of intersection.

COLLABORATIVE

CORNER

The Rule of 85

Focus: Linear inequalities

Time: 20–30 minutes

Group size: 3

Under a proposed "Rule of 85," full-time faculty in the California State Teachers Retirement System (kindergarten through community college) who are a years old with y years of service would have the option of retirement if $a + y \geq 85$.

ACTIVITY

1. Decide, as a group, the age range of full-time teachers. Express this age range as an inequality involving a.
2. Decide, as a group, the number of years someone could teach full-time before retiring. Express this answer as a compound inequality involving y.
3. Using the Rule of 85 and the answers to parts (1) and (2) above, write a system of inequalities. Then, using a scale of 5 yr per square, graph the system. To facilitate comparisons with graphs from other groups, plot a on the horizontal axis and y on the vertical axis.
4. Compare the graphs from all groups. Try to reach consensus on the graph that most clearly illustrates what the status would be of someone who would have the option of retirement under the Rule of 85.
5. If your instructor is agreeable to the idea, attempt to represent him or her with a point on your graph.

4.5 Applications Using Linear Programming

Objective Functions and Constraints • Linear Programming

There are many real-world situations in which we need to find a greatest value (a maximum) or a least value (a minimum). For example, most businesses would like to know how to make the *most* profit and how to make their expenses the *least* possible. Some such problems can be solved using systems of inequalities.

Objective Functions and Constraints

Often a quantity we wish to maximize depends on two or more other quantities. For example, a gardener's profits P might depend on the number of shrubs s and the number of trees t that are planted. If the gardener makes a \$5 profit

from each shrub and a \$9 profit from each tree, the total profit, in dollars, is given by the **objective function**

$$P = 5s + 9t.$$

Thus the gardener might be tempted to simply plant lots of trees since they yield the greater profit. This would be a good idea were it not for the fact that the number of trees and shrubs planted—and thus the total profit—is subject to the demands, or **constraints**, of the situation. For example, to improve drainage, the gardener might be required to plant at least 3 shrubs. Thus the objective function would be subject to the *constraint*

$$s \geq 3.$$

Because of the limited space, the gardener might also be required to plant no more than 10 plants. This would subject the objective function to a *second* constraint:

$$s + t \leq 10.$$

Finally, the gardener might be told to spend no more than \$350 on the plants. If the shrubs cost \$20 each and the trees cost \$50 each, the objective function is subject to a *third* constraint:

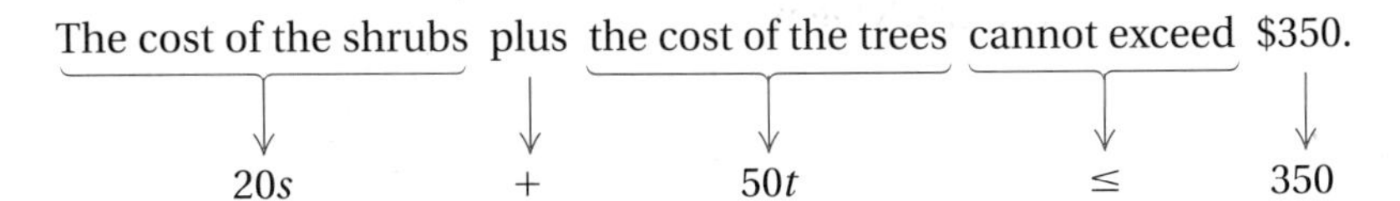
The cost of the shrubs
plus
the cost of the trees
cannot exceed
$350.
20s
+
50t
≤
350

In short, the gardener wishes to maximize the objective function

$$P = 5s + 9t$$

subject to the constraints

$$\begin{aligned} s + t &\leq 10, \\ s &\geq 3, \\ 20s + 50t &\leq 350, \\ \left.\begin{aligned} s &\geq 0, \\ t &\geq 0. \end{aligned}\right\} &\quad \text{Because the number of trees and shrubs cannot be negative} \end{aligned}$$

These constraints form a system of linear inequalities that can be graphed.

Linear Programming

The gardener's problem is "How many shrubs and trees should be planted, subject to the constraints listed, in order to maximize profit?" To solve such a problem, we use an important result from a branch of mathematics known as **linear programming**.

The Corner Principle

Suppose that an objective function $F = ax + by + c$ depends on x and y (with a, b, and c constant). Suppose also that F is subject to constraints on x and y, which form a system of linear inequalities. If F has a minimum or a maximum value, it can then be found as follows:

1. Graph the system of inequalities and find the vertices.
2. Find the value of the objective function at each vertex. The largest and the smallest of those values are the maximum and the minimum of the function, respectively.
3. The ordered pair at which the maximum or minimum occurs indicates the choice of (x, y) for which that maximum or minimum occurs.

This result was proven during World War II, when linear programming was developed to help with shipping troops and supplies to Europe.

Example 1 Solve the gardener's problem discussed above.

Solution We are asked to maximize $P = 5s + 9t$, subject to the constraints

$$\begin{aligned} s + t &\leq 10, \\ s &\geq 3, \\ 20s + 50t &\leq 350, \\ s &\geq 0, \\ t &\geq 0. \end{aligned}$$

We graph the system, using the techniques of Section 4.4. The portion of the graph that is shaded represents all pairs that satisfy the constraints. It is sometimes called the *feasible region*.

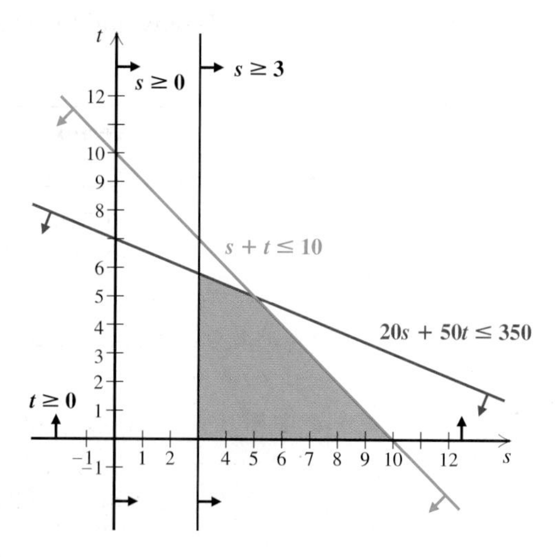

According to the corner principle, P is maximized at one of the vertices of the shaded region. To determine the coordinates of the vertices, we solve the following systems:

$$\left.\begin{aligned} 20s + 50t &= 350, \\ s &= 3; \end{aligned}\right\}$$ The student can verify that the solution of this system is (3, 5.8).

$$\left.\begin{aligned} s + t &= 10, \\ 20s + 50t &= 350; \end{aligned}\right\}$$ The student can verify that the solution of this system is (5, 5).

$$\left.\begin{aligned} s + t &= 10, \\ t &= 0; \end{aligned}\right\}$$ The solution of this system is (10, 0).

$$\left.\begin{aligned} t &= 0, \\ s &= 3. \end{aligned}\right\}$$ The solution of this system is (3, 0).

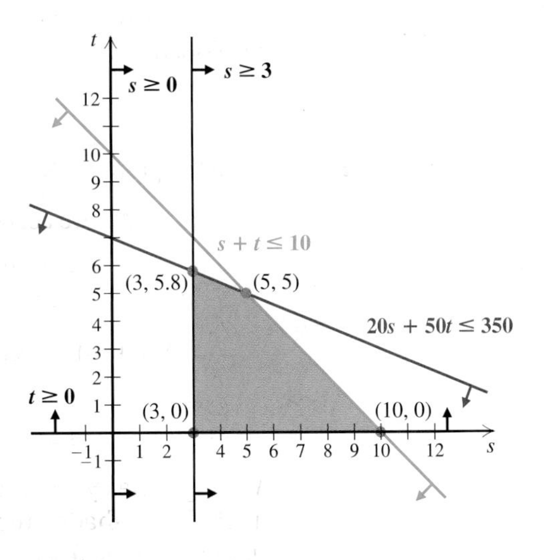

We now find the value of P at each vertex.

Vertex (s, t)	**Profit** $P = 5s + 9t$	
(3, 5.8)	$5(3) + 9(5.8) = 67.2$	
(5, 5)	$5(5) + 9(5) = 70$	← Maximum
(10, 0)	$5(10) + 9(0) = 50$	
(3, 0)	$5(3) + 9(0) = 15$	← Minimum

The largest value of P occurs at (5, 5). Thus profit is maximized at \$70 if the gardener plants 5 shrubs and 5 trees. Incidentally, we have also shown that profit is minimized at \$15 if 3 shrubs and 0 trees are planted.

Example 2

Test scores. Corinna is taking a test in which multiple-choice questions are worth 10 points each and short-answer questions are worth 15 points each. It takes her 3 min to answer each multiple-choice question and 6 min to answer each short-answer question. The total time allowed is 60 min, and no more than 16 questions can be answered. Assuming that all her answers are correct, how many items of each type should Corinna answer in order to get the best score?

Solution

1. **Familiarize.** Tabulating information will help us to see the picture.

Type	Number of Points for Each	Time Required for Each	Number Answered
Multiple-choice	10	3 min	x
Short-answer	15	6 min	y
Total time: 60 min			
Total number of items: 16 or fewer			

Note that we use x to represent the number of multiple-choice questions and y to represent the number of short-answer questions that are answered.

2. **Translate.** In this case, it helps to extend the table.

Type	Number of Points for Each	Time Required for Each	Number Answered	Total Time for Each Type	Total Points for Each Type
Multiple-choice	10	3 min	x	$3x$	$10x$
Short-answer	15	6 min	y	$6y$	$15y$
Total			$x + y \leq 16$	$3x + 6y \leq 60$	$10x + 15y$

Because no more than 16 items may be answered
Because the time cannot exceed 60 min
This is what we want to maximize: the total score on the test.

Suppose that the total score on the test is T. We write T as the objective function in terms of x and y:

$$T = 10x + 15y.$$

We wish to maximize T subject to the constraints on x and y listed above:

$$x + y \leq 16,$$
$$3x + 6y \leq 60,$$
$$\left.\begin{aligned} x &\geq 0, \\ y &\geq 0. \end{aligned}\right\} \quad \text{Because the number of questions answered cannot be negative}$$

Study Tip

On lengthy problems, like Example 2, it is helpful to carefully check each step of your work before proceeding to the next step.

3. **Carry out.** The mathematical manipulation consists of graphing the system and evaluating T at each vertex. The graph is as follows:

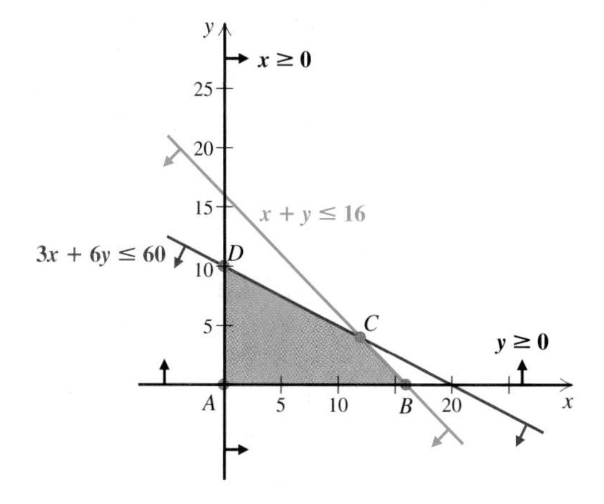

We can find the coordinates of each vertex by solving a system of two linear equations. The coordinates of point A are obviously $(0, 0)$. To find the coordinates of point C, we solve the system

$$x + y = 16, \quad (1)$$
$$3x + 6y = 60, \quad (2)$$

as follows:

$$-3x - 3y = -48 \quad \text{Multiplying both sides of equation (1) by } -3$$
$$3x + 6y = 60$$
$$3y = 12 \quad \text{Adding}$$
$$y = 4.$$

Then we find that $x = 12$. Thus the coordinates of vertex C are $(12, 4)$. Point B is the x-intercept of the line given by $x + y = 16$, so B is $(16, 0)$. Point D is the y-intercept of $3x + 6y = 60$, so D is $(0, 10)$. Computing the test score for each ordered pair, we obtain the following:

Vertex (x, y)	Score $T = 10x + 15y$
$A(0, 0)$	0
$B(16, 0)$	160
$C(12, 4)$	180
$D(0, 10)$	150

The greatest score in the table is 180, obtained when 12 multiple-choice and 4 short-answer questions are answered.

4. **Check.** We can check that $T \leq 180$ for any other pair in the shaded region. This is left to the student.

5. **State.** In order to maximize her score, Corinna should answer 12 multiple-choice questions and 4 short-answer questions.

Exercise Set 4.5

FOR EXTRA HELP

Digital Video Tutor CD 3
Videotape 8

InterAct Math

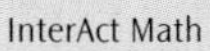

Math Tutor Center

MathXL

MyMathLab.com

Find the maximum and the minimum values of each objective function and the values of x and y at which they occur.

1. $F = 2x + 14y$,
subject to
$5x + 3y \leq 34$,
$3x + 5y \leq 30$,
$x \geq 0$,
$y \geq 0$

2. $G = 7x + 8y$,
subject to
$3x + 2y \leq 12$,
$2y - x \leq 4$,
$x \geq 0$,
$y \geq 0$

3. $P = 8x - y + 20$,
subject to
$6x + 8y \leq 48$,
$0 \leq y \leq 4$,
$0 \leq x \leq 7$

4. $Q = 24x - 3y + 52$,
subject to
$5x + 4y \leq 20$,
$0 \leq y \leq 4$,
$0 \leq x \leq 3$

5. $F = 2y - 3x$,
subject to
$y \leq 2x + 1$,
$y \geq -2x + 3$,
$x \leq 3$

6. $G = 5x + 2y + 4$,
subject to
$y \leq 2x + 1$,
$y \geq -x + 3$,
$x \leq 5$

Solve.

7. *Lunch-time profits.* Elrod's lunch cart sells burritos and chili. To stay in business, Elrod must sell at least 10 orders of chili and 30 burritos each day. Because of limited space, no more than 40 orders of chili or 70 burritos can be made. The total number of orders cannot exceed 90. If profit is \$1.65 per chili order and \$1.05 per burrito, how many of each item should Elrod sell in order to maximize profit?

8. *Milling.* Johnson Lumber can convert logs into either lumber or plywood. In a given week, the mill can turn out 400 units of production, of which 100 units of lumber and 150 units of plywood are required by regular customers. The profit on a unit of lumber is \$20 and on a unit of plywood is \$30. How many units of each type should the mill produce in order to maximize profit?

9. *Cycle production.* Yawaka manufactures motorcycles and bicycles. To stay in business, the number of bicycles made cannot exceed 3 times the number of motorcycles made. Yawaka lacks the facilities to produce more than 60 motorcycles or more than 120 bicycles. The total production of motorcycles and bicycles cannot exceed 160. The profit on a motorcycle is \$1340 and on a bicycle is \$200. Find the number of each that should be manufactured in order to maximize profit.

10. *Gas mileage.* Roschelle owns a car and a moped. She has at most 12 gal of gasoline to be used between the car and the moped. The car's tank holds at most 18 gal and the moped's 3 gal. The mileage for the car is 20 mpg and for the moped is 100 mpg. How many gallons of gasoline should each vehicle use if Roschelle wants to travel as far as possible? What is the maximum number of miles?

11. *Test scores.* Phil is about to take a test that contains matching questions worth 10 points each and essay questions worth 25 points each. He must do at least 3 matching questions, but time restricts doing more than 12. Phil must do at least 4 essays, but time restricts doing more than 15. If no more than 20 questions can be answered, how many of each type should Phil do in order to maximize his score? What is this maximum score?

12. *Test scores.* Edy is about to take a test that contains short-answer questions worth 4 points each and word problems worth 7 points each. Edy must do at least 5 short-answer questions, but time restricts doing more than 10. She must do at least 3 word problems, but time restricts doing more than 10. Edy can do no more than 18 questions in total. How many of each type of question must Edy do in order to maximize her score? What is this maximum score?

Aha! **13.** *Investing.* Rosa is planning to invest up to \$40,000 in corporate or municipal bonds, or both. She must invest from \$6000 to \$22,000 in corporate bonds, and she does not want to invest more than \$30,000 in municipal bonds. The interest on corporate bonds is 8% and on municipal bonds is $7\frac{1}{2}$%. This is simple interest for one year. How much should Rosa invest in each type of bond in order to earn the most interest? What is the maximum interest?

14. *Grape growing.* Auggie's vineyard consists of 240 acres upon which he wishes to plant Merlot and Cabernet grapes. Profit per acre of Merlot is \$400 and profit per acre of Cabernet is \$300. Furthermore, the total number of hours of labor available during the harvest season is 3200. Each acre of Merlot requires 20 hr of labor and each acre of Cabernet requires 10 hr of labor. Determine how the land should be divided between Merlot and Cabernet in order to maximize profit.

15. *Coffee blending.* The Coffee Peddler has 1440 lb of Sumatran coffee and 700 lb of Kona coffee. A batch of Hawaiian Blend requires 8 lb of Kona and 12 lb of Sumatran, and yields a profit of \$90. A batch of Classic Blend requires 4 lb of Kona and 16 lb of Sumatran, and yields a \$55 profit. How many batches of each kind should be made in order to maximize profit? What is the maximum profit? (*Hint*: Organize the information in a table.)

16. *Investing.* Jamaal is planning to invest up to \$22,000 in City Bank or the Southwick Credit Union, or both. He wants to invest at least \$2000 but no more than \$14,000 in City Bank. The Southwick Credit Union does not insure more than a \$15,000 investment, so he will invest no more than that in the Southwick Credit Union. The interest in City Bank is 6% and in the credit union is $6\frac{1}{2}$%. This is simple interest for one year. How much should he invest in each bank in order to earn the most interest? What is the maximum interest?

17. *Textile production.* It takes Cosmic Stitching 2 hr of cutting and 4 hr of sewing to make a knit suit. To make a worsted suit, it takes 4 hr of cutting and 2 hr of sewing. At most 20 hr per day are available for cutting and at most 16 hr per day are available for sewing. The profit on a knit suit is \$68 and on a worsted suit is \$62. How many of each kind of suit should be made in order to maximize profit?

18. *Biscuit production.* The Hockeypuck Biscuit Factory makes two types of biscuits, Biscuit Jumbos and Mitimite Biscuits. The oven can cook at most 200 biscuits per hour. Jumbos each require 2 oz of flour, Mitimites require 1 oz of flour, and there is at most 1440 oz of flour available. The income from Jumbos is \$1.00 and from Mitimites is \$0.80. How many of each type of biscuit should be made in order to maximize income? What is the maximum income?

19. Before a student begins work in this section, what three sections of the text would you suggest he or she study? Why?

20. What does the use of the word "constraint" in this section have in common with the use of the word in everyday speech?

SKILL MAINTENANCE

Evaluate.

21. $5x^3 - 4x^2 - 7x + 2$, for $x = -2$

22. $6t^3 - 3t^2 + 5t$, for $t = 2$

Simplify.

23. $3(2x - 5) + 4(x + 5)$

24. $4(5t - 7) + 6(t + 8)$

25. $6x - 3(x + 2)$

26. $8t - 2(3t - 1)$

SYNTHESIS

27. Explain how Exercises 16 and 18 can be answered by logical reasoning without linear programming.

28. Write a linear programming problem for a classmate to solve. Devise the problem so that profit must be maximized subject to at least two (nontrivial) constraints.

29. *Airplane production.* Alpha Tours has two types of airplanes, the T3 and the S5, and contracts requiring accommodations for a minimum of 2000 first-class, 1500 tourist-class, and 2400 economy-class passengers. The T3 costs \$30 per mile to operate and can accommodate 40 first-class, 40 tourist-class, and 120 economy-class passengers, whereas the S5 costs \$25 per mile to operate and can accommodate 80 first-class, 30 tourist-class, and 40 economy-class passengers. How many of each type of airplane should be used in order to minimize the operating cost?

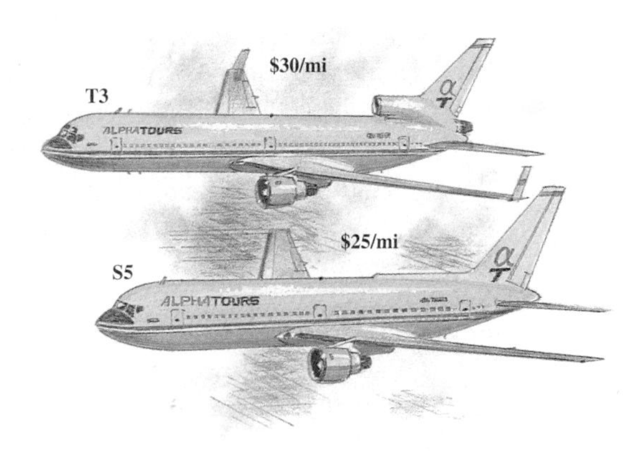

30. *Airplane production.* A new airplane, the T4, is now available, having an operating cost of $37.50 per mile and accommodating 40 first-class, 40 tourist-class, and 80 economy-class passengers. If the T3 of Exercise 29 were replaced with the T4, how many S5's and how many T4's would be needed in order to minimize the operating cost?

31. *Furniture production.* P. J. Edward Furniture Design produces chairs and sofas. The chairs require 20 ft of wood, 1 lb of foam rubber, and 2 sq yd of fabric. The sofas require 100 ft of wood, 50 lb of foam rubber, and 20 sq yd of fabric. The company has 1900 ft of wood, 500 lb of foam rubber, and 240 sq yd of fabric. The chairs can be sold for $80 each and the sofas for $1200 each. How many of each should be produced in order to maximize income?

Summary and Review 4

Key Terms

Inequality, p. 212
Solution, p. 212
Solution set, p. 212
Set-builder notation, p. 213
Interval notation, p. 213
Open interval, p. 213
Closed interval, p. 213
Half-open interval, p. 213
Compound inequality, p. 224
Intersection, p. 224
Conjunction, p. 224
Union, p. 227
Disjunction, p. 227
Absolute value, p. 233
Linear inequality, p. 242
Half-plane, p. 242
Related equation, p. 242
Test point, p. 243
Vertices (singular, vertex), p. 248
Objective function, p. 253
Constraint, p. 253
Linear programming, p. 253
Feasible region, p. 254

Important Properties and Formulas

The Addition Principle for Inequalities

For any real numbers a, b, and c:

$a < b$ is equivalent to $a + c < b + c$;
$a > b$ is equivalent to $a + c > b + c$.

Similar statements hold for $\leq$ and $\geq$.

The Multiplication Principle for Inequalities

For any real numbers a and b, and for any positive number c,

$a < b$ is equivalent to $ac < bc$;
$a > b$ is equivalent to $ac > bc$.

For any real numbers a and b, and for any *negative* number c,

$a < b$ is equivalent to $ac > bc$;
$a > b$ is equivalent to $ac < bc$.

Similar statements hold for $\leq$ and $\geq$.

Set intersection:

$A \cap B = \{x \mid x \text{ is in } A \text{ and } x \text{ is in } B\}$

Set union:

$A \cup B = \{x \mid x \text{ is in } A \text{ or in } B\text{, or both}\}$

Intersection corresponds to "and"; union corresponds to "or."

$|x| = x$ if $x \geq 0$; $|x| = -x$ if $x < 0$.

The Absolute-Value Principles for Equations and Inequalities

For any positive number p and any algebraic expression X:

a) The solutions of $|X| = p$ are those numbers that satisfy $X = -p$ or $X = p$.
b) The solutions of $|X| < p$ are those numbers that satisfy $-p < X < p$.
c) The solutions of $|X| > p$ are those numbers that satisfy $X < -p$ or $p < X$.

If $|X| = 0$, then $X = 0$. If p is negative, then $|X| = p$ and $|X| < p$ have no solution, and any value of X will satisfy $|X| > p$.

Steps for Graphing Linear Inequalities

1. Graph the related equation. Draw the line *dashed* if the inequality symbol is $<$ or $>$ and *solid* if the inequality symbol is $\leq$ or $\geq$.
2. The graph consists of a half-plane on one side of the line and, if the line is solid, the line as well.
   a) Shade *below* the line if the inequality is of the form $y < f(x)$ or $y \leq f(x)$. Shade *above* the line if the inequality is of the form $y > f(x)$ or $y \geq f(x)$.
   b) If y is not isolated, either isolate y and proceed as in part (a) or select a test point not on the line. If the point represents a solution of the inequality, shade the half-plane containing the point. If it does not, shade the other half-plane.

The Corner Principle

Suppose that an objective function $F = ax + by + c$ depends on x and y. Suppose also that F is subject to constraints on x and y, which form a system of linear inequalities. If F has a minimum or a maximum value, it can then be found as follows:

1. Graph the system of inequalities and find the vertices.
2. Find the value of the objective function at each vertex. The largest and the smallest of those values are the maximum and the minimum of the function, respectively.
3. The ordered pair at which the maximum or minimum occurs indicates the choice of (x, y) for which that maximum or minimum occurs.

Review Exercises

Graph each inequality and write the solution set using both set-builder and interval notation.

1. $x \leq -2$
2. $a + 7 \leq -14$
3. $y - 5 \geq -12$
4. $4y > -15$
5. $-0.3y < 9$
6. $-6x - 5 < 4$
7. $-\frac{1}{2}x - \frac{1}{4} > \frac{1}{2} - \frac{1}{4}x$
8. $0.3y - 7 < 2.6y + 15$
9. $-2(x - 5) \geq 6(x + 7) - 12$
10. Let $f(x) = 3x - 5$ and $g(x) = 11 - x$. Find all values of x for which $f(x) \leq g(x)$.

Solve.

11. Jessica can choose between two summer jobs. She can work as a checker in a discount store for \$8.40 an hour, or she can mow lawns for \$12.00 an hour. In order to mow lawns, she must buy a \$450 lawnmower. How many hours of labor will it take Jessica to make more money mowing lawns?

12. Clay is going to invest \$4500, part at 6% and the rest at 7%. What is the most he can invest at 6% and still be guaranteed \$300 in interest each year?

13. Find the intersection:

$$\{1, 2, 5, 6, 9\} \cap \{1, 3, 5, 9\}.$$

14. Find the union:

$$\{1, 2, 5, 6, 9\} \cup \{1, 3, 5, 9\}.$$

Graph and write interval notation.

15. $x \leq 3$ *and* $x > -5$

16. $x \leq 3$ *or* $x > -5$

Solve and graph each solution set.

17. $-4 < x + 3 \leq 5$

18. $-15 < -4x - 5 < 0$

19. $3x < -9$ *or* $-5x < -5$

20. $2x + 5 < -17$ *or* $-4x + 10 \leq 34$

21. $2x + 7 \leq -5$ *or* $x + 7 \geq 15$

22. $f(x) < -5$ *or* $f(x) > 5$, where $f(x) = 3 - 5x$

For $f(x)$ as given, use interval notation to write the domain of f.

23. $f(x) = \dfrac{x}{x - 3}$

24. $f(x) = \sqrt{x + 3}$

25. $f(x) = \sqrt{8 - 3x}$

Solve.

26. $|x| = 4$

27. $|t| \geq 3.5$

28. $|x - 2| = 7$

29. $|2x + 5| < 12$

30. $|3x - 4| \geq 15$

31. $|2x + 5| = |x - 9|$

32. $|5n + 6| = -8$

33. $\left|\dfrac{x + 4}{8}\right| \leq 1$

34. $2|x - 5| - 7 > 3$

35. Let $f(x) = |3x - 5|$. Find all x for which $f(x) < 0$.

36. Graph $x - 2y \geq 6$ on a plane.

Graph each system of inequalities. Find the coordinates of any vertices formed.

37. $x + 3y > -1,$
$x + 3y < 4$

38. $x - 3y \leq 3,$
$x + 3y \geq 9,$
$y \leq 6$

39. Find the maximum and the minimum values of

$$F = 3x + y + 4$$

subject to

$$y \leq 2x + 1,$$
$$x \leq 7,$$
$$y \geq 3.$$

40. Edsel Computers has two manufacturing plants. The Oregon plant cannot produce more than 60 computers a month, while the Ohio plant cannot produce more than 120 computers a month. The Electronics Outpost sells at least 160 Edsel computers each month. It costs \$40 to ship a computer to The Electronics Outpost from the Oregon plant and \$25 to ship from the Ohio plant. How many computers should be shipped from each plant in order to minimize cost?

SYNTHESIS

41. Explain in your own words why $|X| = p$ has two solutions when p is positive and no solution when p is negative.

42. Explain why the graph of the solution of a system of linear inequalities is the intersection, not the union, of the individual graphs.

43. Solve: $|2x + 5| \leq |x + 3|$.

44. Classify as true or false: If $x < 3$, then $x^2 < 9$. If false, give an example showing why.

45. Just-For-Fun manufactures marbles with a 1.1-cm diameter and a ± 0.03-cm manufacturing tolerance, or allowable variation in diameter. Write the tolerance as an inequality with absolute value.

Chapter Test 4

Graph each inequality and write the solution set using both set-builder and interval notation.

1. $x - 2 < 10$

2. $-0.6y < 30$

3. $-4y - 3 \geq 5$

4. $3a - 5 \leq -2a + 6$

5. $4(5 - x) < 2x + 5$

6. $-8(2x + 3) + 6(4 - 5x) \geq 2(1 - 7x) - 4(4 + 6x)$

7. Let $f(x) = -5x - 1$ and $g(x) = -9x + 3$. Find all values of x for which $f(x) > g(x)$.

8. Lia can rent a van for either $40 per day with unlimited mileage or $30 per day with 100 free miles and an extra charge of 15¢ for each mile over 100. For what numbers of miles traveled would the unlimited mileage plan save Lia money?

9. A refrigeration repair company charges $40 for the first half-hour of work and $30 for each additional hour. Blue Mountain Camp has budgeted $100 to repair its walk-in cooler. For what lengths of a service call will the budget not be exceeded?

10. Find the intersection:

$$\{1, 3, 5, 7, 9\} \cap \{3, 5, 11, 13\}.$$

11. Find the union:

$$\{1, 3, 5, 7, 9\} \cup \{3, 5, 11, 13\}.$$

12. Write the domain of f using interval notation if $f(x) = \sqrt{7 - x}$.

Solve and graph each solution set.

13. $-3 < x - 2 < 4$

14. $-11 \leq -5t - 2 < 0$

15. $3x - 2 < 7$ *or* $x - 2 > 4$

16. $-3x > 12$ *or* $4x > -10$

17. $-\frac{1}{3} \leq \frac{1}{6}x - 1 < \frac{1}{4}$

18. $|x| = 9$

19. $|a| > 3$

20. $|4x - 1| < 4.5$

21. $|-5t - 3| \geq 10$

22. $|2 - 5x| = -10$

23. $g(x) < -3$ *or* $g(x) > 3$, where $g(x) = 4 - 2x$

24. Let $f(x) = |x + 10|$ and $g(x) = |x - 12|$. Find all values of x for which $f(x) = g(x)$.

Graph the system of inequalities. Find the coordinates of any vertices formed.

25. $x + y \geq 3,$
$x - y \geq 5$

26. $2y - x \geq -7,$
$2y + 3x \leq 15,$
$y \leq 0,$
$x \leq 0$

27. Find the maximum and the minimum values of

$F = 5x + 3y$

subject to

$x + y \leq 15,$

$1 \leq x \leq 6,$

$0 \leq y \leq 12.$

28. Sassy Salon makes $12 on each manicure and $18 on each haircut. A manicure takes 30 minutes and a haircut takes 50 minutes, and there are 5 stylists who each work 6 hours a day. If the salon can schedule 50 appointments a day, how many should be manicures and how many haircuts in order to maximize profit? What is the maximum profit?

SYNTHESIS

Solve. Write the solution set using interval notation.

29. $|2x - 5| \leq 7$ *and* $|x - 2| \geq 2$

30. $7x < 8 - 3x < 6 + 7x$

31. Write an absolute-value inequality for which the interval shown is the solution.

5 Polynomials and Polynomial Functions

5.1 Introduction to Polynomials and Polynomial Functions

Connecting the Concepts

5.2 Multiplication of Polynomials

5.3 Common Factors and Factoring by Grouping

Connecting the Concepts

5.4 Factoring Trinomials

5.5 Factoring Perfect-Square Trinomials and Differences of Squares

5.6 Factoring Sums or Differences of Cubes

5.7 Factoring: A General Strategy

5.8 Applications of Polynomial Equations

Summary and Review

Test

AN APPLICATION

During intermission at sporting events, team mascots often use a powerful slingshot to launch tightly rolled tee shirts into the stands. The height $h(t)$, in feet, of a tee shirt t seconds after being launched can be approximated by

$$h(t) = -15t^2 + 75t + 10.$$

After peaking, a rolled-up tee shirt is caught by a fan 70 ft above ground level. How long was the tee shirt in the air?

This problem appears as Example 5 in Section 5.8.

Every design of a toy with moving parts, especially toy vehicles, requires an understanding of math and physics if it is ever to make it off the drawing board.

PETER HARRIS
Senior Industrial Designer
Fisher-Price
Aurora, NY

A polynomial is a type of algebraic expression that contains one or more terms. We define polynomials more specifically in this chapter and learn how to manipulate them so that we can use polynomials and polynomial functions in problem solving.

Introduction to Polynomials and Polynomial Functions

5.1

Algebraic Expressions and Polynomials • Polynomial Functions • Adding Polynomials • Opposites and Subtraction

CONNECTING THE CONCEPTS

Let's briefly summarize our work up to this point in the text: After a review of the basics of algebra and problem solving in Chapter 1, we turned our attention in Chapter 2 to equations in two variables for which graphs are used to represent the solution sets. Graphs also enabled us to visualize the solutions of systems of equations in Chapter 3. In Chapter 4, we continued finding solutions, but this time our work included absolute-value functions and inequalities.

Here in Chapter 5, we will take a break from solving equations and inequalities and concentrate on finding equivalent expressions. Our work with equivalent expressions will ultimately allow us to solve a new type of equation at the end of this chapter.

Chapters 6 and 7 will follow a similar pattern: After learning new ways of writing equivalent expressions, we will learn to solve new types of equations toward the end of each chapter.

In this section, we introduce a type of algebraic expression known as a *polynomial*. After developing some vocabulary, we study addition and subtraction of polynomials, and evaluate *polynomial functions*.

Algebraic Expressions and Polynomials

In Chapter 1, we introduced algebraic expressions like

$$5x^2, \quad \frac{3}{x^2+5}, \quad 9a^3b^4, \quad 3x^{-2}, \quad 6x^2+3x+1, \quad -9, \quad \text{and} \quad 5-2x.$$

Of the expressions listed, $5x^2, 9a^3b^4, 3x^{-2}$, and -9 are examples of *terms*. A **term** is simply a number or a product of a number and a variable or variables raised to a power.

When all variables in a term are raised to whole-number powers, the term is a **monomial** (pronounced mä-nō-mē-əl). Of the terms listed above, $5x^2$, $9a^3b^4$, and -9 are monomials. The **degree** of a monomial is the sum of the exponents of the variables. Thus, $5x^2$ has degree 2 and $9a^3b^4$ has degree 7. Nonzero constant terms, like -9, can be written $-9x^0$ and therefore have degree 0. The term 0 itself is said to have no degree.

The number 5 is said to be the **coefficient** of $5x^2$. Thus the coefficient of $9a^3b^4$ is 9 and the coefficient of $-2x$ is -2. The coefficient of a constant term is just that constant.

A **polynomial** is a monomial or a sum of monomials. Of the expressions listed, $5x^2$, $9a^3b^4$, $6x^2 + 3x + 1$, -9, and $5 - 2x$ are polynomials. In fact, with the exception of $9a^3b^4$, these are all polynomials *in one variable*. The expression $9a^3b^4$ is a *polynomial in two variables*. Note that $5 - 2x$ is the sum of 5 and $-2x$. Thus, 5 and $-2x$ are the terms in the polynomial $5 - 2x$.

The **leading term** of a polynomial is the term of highest degree. Its coefficient is called the **leading coefficient**. The **degree of a polynomial** is the same as the degree of its leading term.

Example 1

For each polynomial given, find the degree of each term, the degree of the polynomial, the leading term, and the leading coefficient.

a) $2x^3 + 8x^2 - 17x - 3$
b) $6x^2 + 8x^2y^3 - 17xy - 24xy^2z^4 + 2y + 3$

Solution

(a) $2x^3 + 8x^2 - 17x - 3$ **(b)** $6x^2 + 8x^2y^3 - 17xy - 24xy^2z^4 + 2y + 3$

Term	$2x^3$	$8x^2$	$-17x$	-3	$6x^2$	$8x^2y^3$	$-17xy$	$-24xy^2z^4$	$2y$	3
Degree	3	2	1	0	2	5	2	7	1	0
Leading Term	$2x^3$				$-24xy^2z^4$					
Leading Coefficient	2				-24					
Degree of Polynomial	3				7					

A polynomial of degree 0 or 1 is called **linear**. A polynomial in one variable is said to be **quadratic** if it is of degree 2 and **cubic** if it is of degree 3.

The following are some names for certain kinds of polynomials.

Type	**Definition**	**Examples**
Monomial	A polynomial of one term	$4,\ -3p,\ 5x^2,\ -7a^2b^3,\ 0,\ xyz$
Binomial	A polynomial of two terms	$2x + 7,\ a - 3b,\ 5x^2 + 7y^3$
Trinomial	A polynomial of three terms	$x^2 - 7x + 12,\ 4a^2 + 2ab + b^2$

We generally arrange polynomials in one variable so that the exponents *decrease* from left to right. This is called **descending order**. Some polynomials may be written with exponents *increasing* from left to right, which is **ascending order**. Generally, if an exercise is written in one kind of order, the answer is written in that same order.

Example 2 Arrange in ascending order: $12 + 2x^3 - 7x + x^2$.

Solution

$$12 + 2x^3 - 7x + x^2 = 12 - 7x + x^2 + 2x^3$$

Polynomials in several variables can be arranged with respect to the powers of one of the variables.

Example 3 Arrange in descending powers of x: $y^4 + 2 - 5x^2 + 3x^3y + 7xy^2$.

Solution

$$y^4 + 2 - 5x^2 + 3x^3y + 7xy^2 = 3x^3y - 5x^2 + 7xy^2 + y^4 + 2$$

Polynomial Functions

A *polynomial function* is a function in which ordered pairs are determined by evaluating a polynomial. For example, the function P given by

$$P(x) = 5x^7 + 3x^5 - 4x^2 - 5$$

is an example of a polynomial function. To evaluate a polynomial function, we substitute a number for the variable just as in Chapter 2. In this text, we limit ourselves to polynomial functions in one variable.

Example 4 For the polynomial function $P(x) = -x^2 + 4x - 1$, find the following: **(a)** $P(2)$; **(b)** $P(10)$; **(c)** $P(-10)$.

Solution

a) $P(2) = -2^2 + 4(2) - 1$ We square the input before taking its opposite.

$= -4 + 8 - 1 = 3$

b) $P(10) = -10^2 + 4(10) - 1$

$= -100 + 40 - 1 = -61$

c) $P(-10) = -(-10)^2 + 4(-10) - 1$

$= -100 - 40 - 1 = -141$

Example 5 ***Veterinary medicine.*** Gentamicin is an antibiotic frequently used by veterinarians. The concentration, in micrograms per milliliter (mcg/mL), of Gentamicin in a horse's bloodstream t hours after injection can be approximated by the polynomial function

$$C(t) = -0.005t^4 + 0.003t^3 + 0.35t^2 + 0.5t.$$

(*Source*: Michele Tulis, DVM, telephone interview)

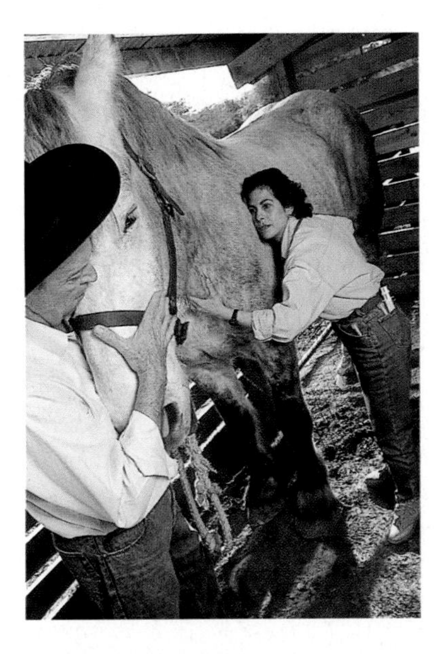

a) What is the concentration 2 hr after injection?
b) Use the graph below to estimate $C(4)$.

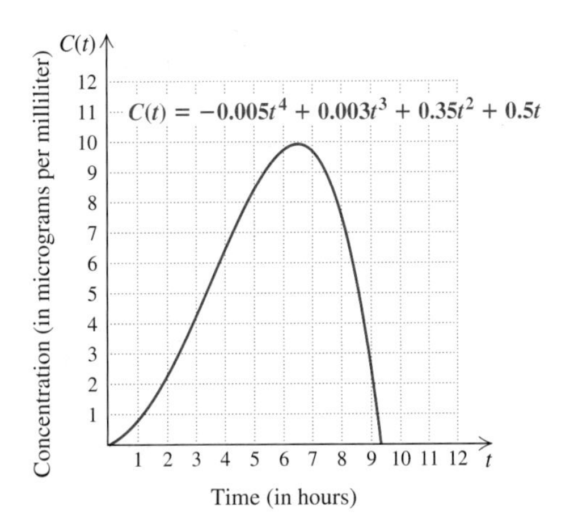

technology connection A

One way to evaluate a function is to enter and graph it as y_1 and then select TRACE. We can then enter any x-value that appears in that window and the corresponding y-value will appear. We use this approach below to check Example 5(a).

$y_1 = -0.005x^4 + 0.003x^3 + 0.35x^2 + 0.5x$

13, −1, 13, −1

$y_1 = -0.005x^4 + 0.003x^3 + 0.35x^2 + 0.5x$

13, −1, 13, −1

X = 2, Y = 2.344

Use this approach to check Examples 5(b) and 4.

Solution

a) We evaluate the function for $t = 2$:

$$\begin{aligned} C(2) &= -0.005(2)^4 + 0.003(2)^3 + 0.35(2)^2 + 0.5(2) \\ &= -0.005(16) + 0.003(8) + 0.35(4) + 0.5(2) \\ &= -0.08 + 0.024 + 1.4 + 1 \\ &= -0.08 + 2.424 \\ &= 2.344. \end{aligned}$$

We carry out the calculation using the rules for order of operations.

The concentration after 2 hr is 2.344 mcg/mL.

b) To estimate $C(4)$, the concentration after 4 hr, we locate 4 on the horizontal axis. From there we move vertically to the graph of the function and then horizontally to the $C(t)$-axis, as shown below. This locates a value of about 6.5. Thus,

$$C(4) \approx 6.5.$$

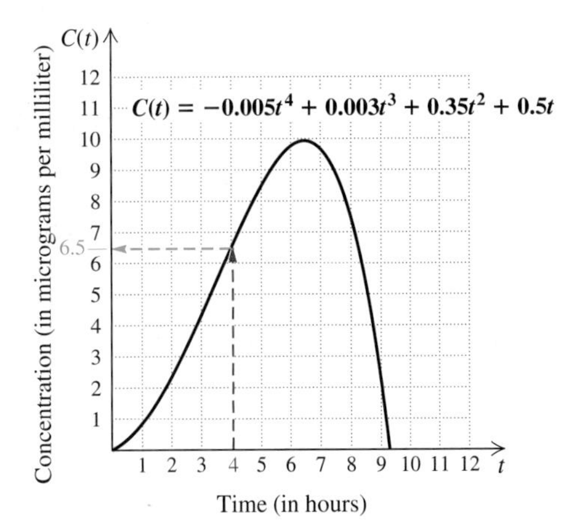

Adding Polynomials

Recall from Section 1.3 that when two terms have the same variable(s) raised to the same power(s), they are **similar**, or **like**, **terms** and can be "combined" or "collected."

Example 6

Combine like terms.

a) $3x^2 - 4y + 2x^2$
b) $9x^3 + 5x - 4x^2 - 2x^3 + 5x^2$
c) $3x^2y + 5xy^2 - 3x^2y - xy^2$

Solution

a) $3x^2 - 4y + 2x^2 = 3x^2 + 2x^2 - 4y$ Rearranging terms using the commutative law for addition

$= (3 + 2)x^2 - 4y$ Using the distributive law

$= 5x^2 - 4y$

b) $9x^3 + 5x - 4x^2 - 2x^3 + 5x^2 = 7x^3 + x^2 + 5x$ We usually perform the middle steps mentally and write just the answer.

c) $3x^2y + 5xy^2 - 3x^2y - xy^2 = 4xy^2$

The sum of two polynomials can be found by writing a plus sign between them and then combining like terms.

Example 7

Add: $(-3x^3 + 2x - 4) + (4x^3 + 3x^2 + 2)$.

Solution

$$(-3x^3 + 2x - 4) + (4x^3 + 3x^2 + 2) = x^3 + 3x^2 + 2x - 2$$

Using columns is sometimes helpful. To do so, we write the polynomials one under the other, listing like terms under one another and leaving spaces for missing terms.

Example 8

Add: $4ax^2 + 4bx - 5$ and $-6ax^2 + 8$.

Solution

$$\begin{array}{r} 4ax^2 + 4bx - 5 \\ \underline{-6ax^2 \qquad\quad + 8} \\ -2ax^2 + 4bx + 3 \end{array}$$

Combining like terms

Example 9 Add: $13x^3y + 3x^2y - 5y$ and $x^3y + 4x^2y - 3xy$.

Solution

$$(13x^3y + 3x^2y - 5y) + (x^3y + 4x^2y - 3xy) = 14x^3y + 7x^2y - 3xy - 5y$$

Opposites and Subtraction

If the sum of two polynomials is 0, the polynomials are *opposites*, or *additive inverses*, of each other. For example,

$$(3x^2 - 5x + 2) + (-3x^2 + 5x - 2) = 0,$$

so the opposite of $(3x^2 - 5x + 2)$ must be $(-3x^2 + 5x - 2)$. We can say the same thing using algebraic symbolism, as follows:

$$\underbrace{\text{The opposite of}}_{-}\quad \underbrace{(3x^2 - 5x + 2)}_{(3x^2 - 5x + 2)}\quad \underbrace{\text{is}}_{=}\quad \underbrace{(-3x^2 + 5x - 2).}_{-3x^2 + 5x - 2}$$

To form the opposite of a polynomial, we can think of distributing the "–" sign, or multiplying each term of the polynomial by -1, and removing the parentheses. The effect is to change the sign of each term in the polynomial.

> ***The Opposite of a Polynomial***
>
> The *opposite* of a polynomial P can be written as $-P$ or, equivalently, by replacing each term with its opposite.

Example 10 Write two equivalent expressions for the opposite of

$$7xy^2 - 6xy - 4y + 3.$$

Solution

a) The opposite of $7xy^2 - 6xy - 4y + 3$ can be written as

$$-(7xy^2 - 6xy - 4y + 3). \qquad \text{Writing the opposite of } P \text{ as } -P$$

b) The opposite of $7xy^2 - 6xy - 4y + 3$ can also be written as

$$-7xy^2 + 6xy + 4y - 3. \qquad \text{Multiplying each term by } -1 \text{ and removing parentheses}$$

To subtract a polynomial, we add its opposite.

Example 11 Subtract: $(-5x^2 + 4) - (2x^2 + 3x - 1)$.

Solution

$$(-5x^2 + 4) - (2x^2 + 3x - 1)$$
$$= (-5x^2 + 4) + (-2x^2 - 3x + 1) \quad \text{Adding the opposite of the polynomial being subtracted}$$
$$= -7x^2 - 3x + 5$$

With practice, you will find that you can skip more steps, by mentally taking the opposite of each term and then combining like terms. Eventually, all you will write is the answer.

To use columns for subtraction, we mentally change the signs of the terms being subtracted.

Example 12 Subtract:

$$(4x^2y - 6x^3y^2 + x^2y^2) - (4x^2y + x^3y^2 + 3x^2y^3 - 8x^2y^2).$$

Solution

Write: (Subtract)

$$\begin{array}{r} 4x^2y - 6x^3y^2 \qquad\qquad + x^2y^2 \\ -(4x^2y + x^3y^2 + 3x^2y^3 - 8x^2y^2) \\ \hline \end{array}$$

Think: (Add)

$$\begin{array}{r} 4x^2y - 6x^3y^2 \qquad\qquad + x^2y^2 \\ -4x^2y - x^3y^2 - 3x^2y^3 + 8x^2y^2 \\ \hline -7x^3y^2 - 3x^2y^3 + 9x^2y^2 \end{array}$$

Take the opposite of each term mentally and add.

One way to check problems like Example 11 is to note that if the subtraction is correct, then

$$(-5x^2 + 4) - (2x^2 + 3x - 1) = -7x^2 - 3x + 5$$

is an identity. Thus the graph of the left side,

$$y_1 = (-5x^2 + 4) - (2x^2 + 3x - 1),$$

and the graph of the right side,

$$y_2 = -7x^2 - 3x + 5,$$

should coincide. Be sure to use parentheses when entering y_1.

1. Graph y_1 and y_2 as described above, using the window $[-10, 10, -160, 30]$ with Xscl = 1 and Yscl = 8. Do the graphs coincide? How can you tell?

   Example 11 can also be checked by graphing $y_3 = y_2 - y_1$ (but not graphing y_1 or y_2), where the VARS key is used and y_2 and y_1 are as described above. If, as expected, the graphs of y_2 and y_1 are the same, then the graph of y_3 will be the x-axis.

2. Perform the check just described. What advantage(s) does this check have over the check used in (1) above?

Exercise Set 5.1

FOR EXTRA HELP

Digital Video Tutor CD 3
Videotape 9

InterAct Math

Math Tutor Center

MathXL

MyMathLab.com

Determine the degree of each term and the degree of the polynomial.

1. $-6x^5 - 8x^3 + x^2 + 3x - 4$
2. $t^3 - 5t^2 + 2t + 7$
3. $y^3 + 2y^7 + x^2y^4 - 8$
4. $-2u^2 + 3v^5 - u^3v^4 - 7$
5. $a^5 + 4a^2b^4 + 6ab + 4a - 3$
6. $8p^6 + 2p^4t^4 - 7p^3t + 5p^2 - 14$

Arrange in descending order. Then find the leading term and the leading coefficient.

7. $19 - 4y^3 + 7y - 6y^2$
8. $3 - 5y + 6y^2 + 11y^3 - 18y^4$
9. $5x^2 + 3x^7 - x + 12$
10. $9 - 3x - 10x^4 + 7x^2$
11. $a + 5a^3 - a^7 - 19a^2 + 8a^5$
12. $a^3 - 7 + 11a^4 + a^9 - 5a^2$

Arrange in ascending powers of x.

13. $6x - 9 + 3x^4 - 5x^2$
14. $-3x^4 + 4x - x^3 + 9$
15. $7x^3y + 3xy^3 + x^2y^2 - 5x^4$
16. $5x^2y^2 - 9xy + 8x^3y^2 - 5x^4$
17. $4ax - 7ab + 4x^6 - 7ax^2$
18. $5xy^8 - 3ax^5 + 4ax^3 - 12a + 5x^5$

Find the specified function values.

19. Find $P(4)$ and $P(0)$: $P(x) = 3x^2 - 2x + 7$.
20. Find $Q(3)$ and $Q(-1)$: $Q(x) = -4x^3 + 7x^2 - 6$.
21. Find $P(-2)$ and $P(\frac{1}{3})$: $P(y) = 8y^3 - 12y - 5$.
22. Find $Q(-3)$ and $Q(0)$:
$Q(y) = -8y^3 + 7y^2 - 4y - 9$.

Evaluate each polynomial for $x = 4$.

23. $-7x + 5$
24. $4x - 13$
25. $x^3 - 5x^2 + x$
26. $7 - x + 3x^2$

Evaluate each polynomial function for $x = -1$.

27. $f(x) = -5x^3 + 3x^2 - 4x - 3$
28. $g(x) = -4x^3 + 2x^2 + 5x - 7$

Electing officers. *For a club consisting of n people, the number of ways in which a president, vice president, and treasurer can be elected can be determined using the function given by*

$$p(n) = n^3 - 3n^2 + 2n.$$

29. The Southside Rugby Club has 20 members. In how many ways can they elect a president, vice president, and treasurer?
30. The Stage Right drama club has 12 members. In how many ways can a president, vice president, and treasurer be elected?

Falling distance. *The distance $s(t)$, in feet, traveled by a body falling freely from rest in t seconds is approximated by the function given by*

$$s(t) = 16t^2.$$

31. A paintbrush falls from a scaffold and takes 3 sec to hit the ground. How high is the scaffold?

32. A stone is dropped from the Briar Cliff lookout and takes 5 sec to hit the ground. How high is the cliff?

Total revenue. *An electronics firm is marketing a new kind of DVD player. The firm determines that when it sells x DVD players, its total revenue is*

$$R(x) = 280x - 0.4x^2 \text{ dollars.}$$

33. What is the total revenue from the sale of 75 DVD players?

34. What is the total revenue from the sale of 100 DVD players?

Total cost. *The electronics firm determines that the total cost, in dollars, of producing x DVD players is given by*

$$C(x) = 5000 + 0.6x^2.$$

35. What is the total cost of producing 75 DVD players?

36. What is the total cost of producing 100 DVD players?

Daily accidents. *The number of daily accidents (the average number of accidents per day) involving drivers of age a is approximated by the polynomial function*

$$P(a) = 0.4a^2 - 40a + 1039.$$

37. Find the number of daily accidents involving a 20-year-old driver.

38. Find the number of daily accidents involving a 25-year-old driver.

NASCAR attendance. *Attendance at NASCAR auto races has grown rapidly over the past 10 years. Attendance A, in millions, can be approximated by the polynomial function given by*

$$A(x) = 0.0024x^3 - 0.005x^2 + 0.31x + 3,$$

where x is the number of years since 1989. Use the following graph for Exercises 39–42.

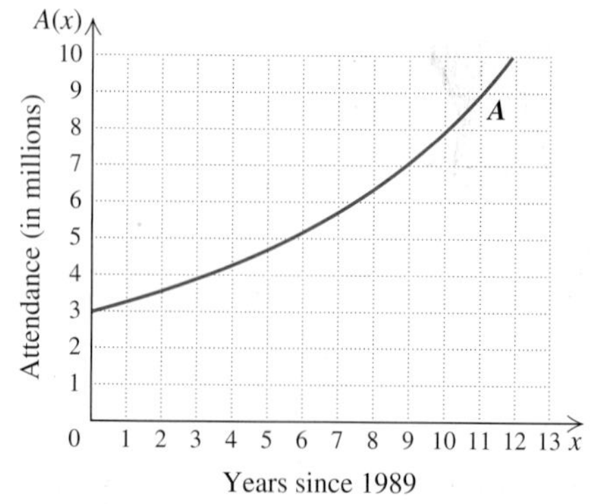

Source: NASCAR

39. Estimate the attendance at NASCAR races in 1995.

40. Estimate the attendance at NASCAR races in 1999.

41. Approximate $A(8)$.

42. Approximate $A(12)$.

Medicine. *Ibuprofen is a medication used to relieve pain. The polynomial function*

$$M(t) = 0.5t^4 + 3.45t^3 - 96.65t^2 + 347.7t,$$
$$0 \le t \le 6$$

can be used to estimate the number of milligrams of ibuprofen in the bloodstream t hours after 400 mg of the medication has been swallowed (Source: Based on data from Dr. P. Carey, Burlington, VT). Use the following graph for Exercises 43–46.

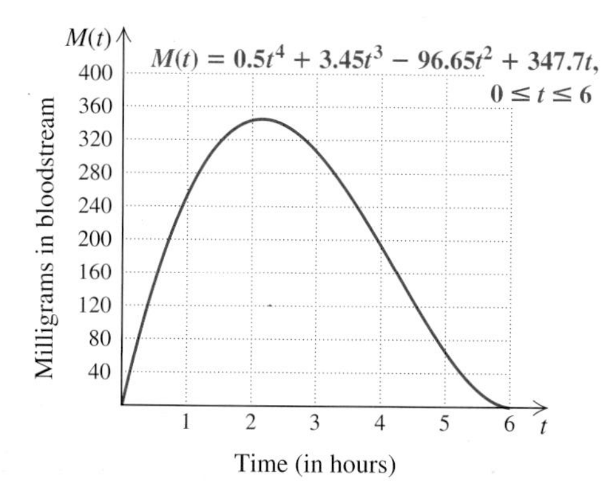

43. Use the graph above to estimate the number of milligrams of ibuprofen in the bloodstream 2 hr after 400 mg has been swallowed.

44. Use the graph above to estimate the number of milligrams of ibuprofen in the bloodstream 4 hr after 400 mg has been swallowed.

45. Approximate $M(5)$.

46. Approximate $M(3)$.

Surface area of a right circular cylinder. *The surface area of a right circular cylinder is given by the polynomial*

$$2\pi rh + 2\pi r^2,$$

where h is the height, r is the radius of the base, and h and r are given in the same units.

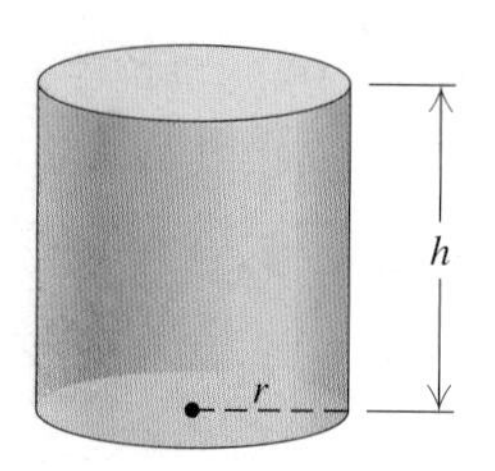

47. A 16-oz beverage can has height 6.3 in. and radius 1.2 in. Find the surface area of the can. (Use a calculator with a $\boxed{\pi}$ key or use 3.141592654 for π.)

48. A 12-oz beverage can has height 4.7 in. and radius 1.2 in. Find the surface area of the can. (Use a calculator with a $\boxed{\pi}$ key or use 3.141592654 for π.)

Combine like terms.

49. $5a + 6 - 4 + 2a^3 - 6a + 2$

50. $6x + 13 - 8 - 7x + 5x^2 + 10$

51. $3a^2b + 4b^2 - 9a^2b - 7b^2$

52. $5x^2y^2 + 4x^3 - 8x^2y^2 - 12x^3$

53. $9x^2 - 3xy + 12y^2 + x^2 - y^2 + 5xy + 4y^2$

54. $a^2 - 2ab + b^2 + 9a^2 + 5ab - 4b^2 + a^2$

Add.

55. $(8a + 6b - 3c) + (4a - 2b + 2c)$

56. $(7x - 5y + 3z) + (9x + 12y - 8z)$

57. $(a^2 - 3b^2 + 4c^2) + (-5a^2 + 2b^2 - c^2)$

58. $(x^2 - 5y^2 - 9z^2) + (-6x^2 + 9y^2 - 2z^2)$

59. $(x^2 + 2x - 3xy - 7) + (-3x^2 - x + 2xy + 6)$

60. $(3a^2 - 2b + ab + 6) + (-a^2 + 5b - 5ab - 2)$

61. $(8x^2y - 3xy^2 + 4xy) + (-2x^2y - xy^2 + xy)$

62. $(9ab - 3ac + 5bc) + (13ab - 15ac - 8bc)$

63. $(2r^2 + 12r - 11) + (6r^2 - 2r + 4) + (r^2 - r - 2)$

64. $(5x^2 + 19x - 23) + (-7x^2 - 11x + 12) + (-x^2 - 9x + 8)$

65. $\left(\frac{1}{8}xy - \frac{3}{5}x^3y^2 + 4.3y^3\right) + \left(-\frac{1}{3}xy - \frac{3}{4}x^3y^2 - 2.9y^3\right)$

66. $\left(\frac{2}{3}xy + \frac{5}{6}xy^2 + 5.1x^2y\right) + \left(-\frac{4}{5}xy + \frac{3}{4}xy^2 - 3.4x^2y\right)$

Write two equivalent expressions for the opposite, or additive inverse, of each polynomial.

67. $5x^3 - 7x^2 + 3x - 9$

68. $-8y^4 - 18y^3 + 4y - 7$

69. $-12y^5 + 4ay^4 - 7by^2$

70. $7ax^3y^2 - 8by^4 - 7abx - 12ay$

Subtract.

71. $(7x - 5) - (-3x + 4)$

72. $(8y + 2) - (-6y - 5)$

73. $(-3x^2 + 2x + 9) - (x^2 + 5x - 4)$

74. $(-9y^2 + 4y + 8) - (4y^2 + 2y - 3)$

75. $(6a - 2b + c) - (3a + 2b - 2c)$

76. $(7x - 4y + z) - (4x + 6y - 3z)$

77. $(3x^2 - 2x - x^3) - (5x^2 - 8x - x^3)$

78. $(8y^2 - 3y - 4y^3) - (3y^2 - 9y - 7y^3)$

79. $(5a^2 + 4ab - 3b^2) - (9a^2 - 4ab + 2b^2)$

80. $(7y^2 - 14yz - 8z^2) - (12y^2 - 8yz + 4z^2)$

81. $(6ab - 4a^2b + 6ab^2) - (3ab^2 - 10ab - 12a^2b)$

82. $(10xy - 4x^2y^2 - 3y^3) - (-9x^2y^2 + 4y^3 - 7xy)$

83. $\left(\frac{5}{8}x^4 - \frac{1}{4}x^2 - \frac{1}{2}\right) - \left(-\frac{3}{8}x^4 + \frac{3}{4}x^2 + \frac{1}{2}\right)$

84. $\left(\frac{5}{6}y^4 - \frac{1}{2}y^2 - 7.8y\right) - \left(-\frac{3}{8}y^4 + \frac{3}{4}y^2 + 3.4y\right)$

Total profit. Total profit is defined as total revenue minus total cost. In Exercises 85 and 86, $R(x)$ and $C(x)$ are the revenue and cost, respectively, from the sale of x futons.

85. If $R(x) = 280x - 0.4x^2$ and $C(x) = 5000 + 0.6x^2$, find the profit from the sale of 70 futons.

86. If $R(x) = 280x - 0.7x^2$ and $C(x) = 8000 + 0.5x^2$, find the profit from the sale of 100 futons.

87. Is the sum of two binomials always a binomial? Why or why not?

88. Ani claims that she can add any two polynomials but finds subtraction difficult. What advice would you offer her?

SKILL MAINTENANCE

Simplify.

89. $2(x + 3) + 5(x + 2)$

90. $7(a + 2) + 3(a + 15)$

91. $a(a - 1) + 4(a - 1)$

92. $x(x - 3) + 2(x - 1)$

93. $x^5 \cdot x^4$

94. $a^2 \cdot a^6$

SYNTHESIS

95. Write a problem in which revenue and cost functions are given and a profit function, $P(x)$, is required. Devise the problem so that $P(0) < 0$ and $P(100) > 0$.

96. Write a problem in which revenue and cost functions are given and a profit function, $P(x)$, is required. Devise the problem so that $P(10) < 0$ and $P(50) > 0$.

For P(x) and Q(x) as given, find the following.

$$P(x) = 13x^5 - 22x^4 - 36x^3 + 40x^2 - 16x + 75,$$
$$Q(x) = 42x^5 - 37x^4 + 50x^3 - 28x^2 + 34x + 100$$

97. $2[P(x)] + Q(x)$

98. $3[P(x)] - Q(x)$

99. $2[Q(x)] - 3[P(x)]$

100. $4[P(x)] + 3[Q(x)]$

101. *Volume of a display.* The number of spheres in a triangular pyramid with x layers is given by the function

$$N(x) = \frac{1}{6}x^3 + \frac{1}{2}x^2 + \frac{1}{3}x.$$

The volume of a sphere of radius r is given by the function

$$V(r) = \frac{4}{3}\pi r^3,$$

where π can be approximated as 3.14.

Chocolate Heaven has a window display of truffles piled in a triangular pyramid formation 5 layers deep. If the diameter of each truffle is 3 cm, find the volume of chocolate in the display.

102. If one large truffle were to have the same volume as the display of truffles in Exercise 101, what would be its diameter?

103. Find a polynomial function that gives the outside surface area of a box like this one, with an open top and dimensions as shown.

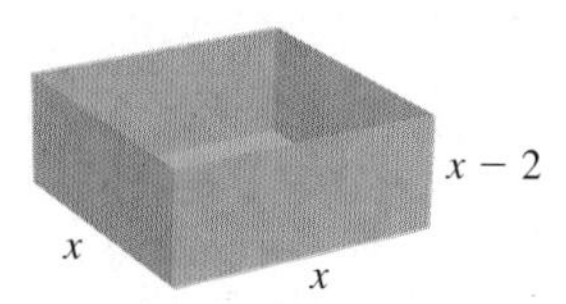

104. Develop a formula for the surface area of a right circular cylinder in which h is the height, in *centimeters*, and r is the radius, in *meters*. (See Exercises 47 and 48.)

Perform the indicated operation. Assume that the exponents are natural numbers.

105. $(2x^{2a} + 4x^a + 3) + (6x^{2a} + 3x^a + 4)$

106. $(3x^{6a} - 5x^{5a} + 4x^{3a} + 8) - (2x^{6a} + 4x^{4a} + 3x^{3a} + 2x^{2a})$

107. $(2x^{5b} + 4x^{4b} + 3x^{3b} + 8) - (x^{5b} + 2x^{3b} + 6x^{2b} + 9x^b + 8)$

108. Use a grapher to check your answers to Exercises 39, 41, and 43.

109. Use a grapher to check your answers to Exercises 20, 33, and 34.

110. A student who is trying to graph

$$p(x) = 0.05x^4 - x^2 + 5$$

gets the following screen.

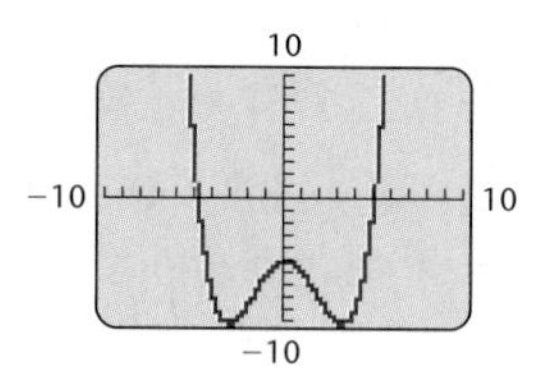

How can the student tell at a glance that a mistake has been made?

COLLABORATIVE

CORNER

How Many Handshakes?

Focus: Polynomial functions

Time: 20 minutes

Group size: 5

Activity

1. All group members should shake hands with each other. Without "double counting," determine how many handshakes occurred.
2. Complete the table in the next column.
3. Join another group to determine the number of handshakes for a group of size 10.
4. Try to find a function of the form $H(n) = an^2 + bn$, for which $H(n)$ is the number of different handshakes that are possible in a group of n people. Make sure that $H(n)$ produces all of the values in the table above. (*Hint*: Use the table to twice select n and $H(n)$. Then solve the resulting system of equations for a and b.)

Group Size	Number of Handshakes
1	
2	
3	
4	
5	

Multiplication of Polynomials

5.2

Multiplying Monomials • Multiplying Monomials and Binomials • Multiplying Any Two Polynomials • The Product of Two Binomials: FOIL • Squares of Binomials • Products of Sums and Differences • Function Notation

Just like numbers, polynomials can be multiplied. The product of two polynomials $P(x)$ and $Q(x)$ is a polynomial $R(x)$ that gives the same value as $P(x) \cdot Q(x)$ for any replacement of x.

Multiplying Monomials

To multiply monomials, we first multiply their coefficients. Then we multiply the variables using the rules for exponents and the commutative and associative laws. With practice, we can work mentally, writing only the answer.

Example 1 Multiply and simplify: **(a)** $(-8x^4y^7)(5x^3y^2)$; **(b)** $(-2x^2yz^5)(-6x^5y^{10}z^2)$.

Solution

a) $(-8x^4y^7)(5x^3y^2) = -8 \cdot 5 \cdot x^4 \cdot x^3 \cdot y^7 \cdot y^2$ Using the associative and commutative laws

$= -40x^{4+3}y^{7+2}$ Multiplying coefficients; adding exponents

$= -40x^7y^9$

b) $(-2x^2yz^5)(-6x^5y^{10}z^2) = (-2)(-6) \cdot x^2 \cdot x^5 \cdot y \cdot y^{10} \cdot z^5 \cdot z^2$

$= 12x^7y^{11}z^7$ Multiplying coefficients; adding exponents

Multiplying Monomials and Binomials

The distributive law is the basis for multiplying polynomials other than monomials. We first multiply a monomial and a binomial.

Example 2 Multiply: **(a)** $2x(3x - 5)$; **(b)** $3a^2b(a^2 - b^2)$.

Solution

a) $2x(3x - 5) = 2x \cdot 3x - 2x \cdot 5$ Using the distributive law

$= 6x^2 - 10x$ Multiplying monomials

b) $3a^2b(a^2 - b^2) = 3a^2b \cdot a^2 - 3a^2b \cdot b^2$ Using the distributive law

$= 3a^4b - 3a^2b^3$

The distributive law is also used when multiplying two binomials. In this case, however, we begin by distributing a *binomial* rather than a monomial. With practice, some of the following steps can be combined.

Example 3 Multiply: $(y^3 - 5)(2y^3 + 4)$.

Solution

$(y^3 - 5)(2y^3 + 4) = (y^3 - 5)2y^3 + (y^3 - 5)4$ "Distributing" the $y^3 - 5$

$= 2y^3(y^3 - 5) + 4(y^3 - 5)$ Using the commutative law for multiplication. Try to do this step mentally.

$= 2y^3 \cdot y^3 - 2y^3 \cdot 5 + 4 \cdot y^3 - 4 \cdot 5$ Using the distributive law (twice)

$= 2y^6 - 10y^3 + 4y^3 - 20$ Multiplying the monomials

$= 2y^6 - 6y^3 - 20$ Combining like terms

Multiplying Any Two Polynomials

Repeated use of the distributive law enables us to multiply *any* two polynomials, regardless of how many terms are in each.

Example 4 Multiply: $(p + 2)(p^4 - 2p^3 + 3)$.

Solution By the distributive law, we have

$$\begin{aligned}
&(p + 2)(p^4 - 2p^3 + 3) \\
&\quad = (p + 2)(p^4) - (p + 2)(2p^3) + (p + 2)(3) \\
&\quad = p^4(p + 2) - 2p^3(p + 2) + 3(p + 2) && \text{Using a commutative law} \\
&\quad = p^4 \cdot p + p^4 \cdot 2 - 2p^3 \cdot p - 2p^3 \cdot 2 + 3 \cdot p + 3 \cdot 2 \\
&\quad = p^5 + 2p^4 - 2p^4 - 4p^3 + 3p + 6 \\
&\quad = p^5 - 4p^3 + 3p + 6. && \text{Combining like terms}
\end{aligned}$$

The Product of Two Polynomials

The *product* of two polynomials $P(x)$ and $Q(x)$ is found by multiplying each term of $P(x)$ by every term of $Q(x)$ and then combining like terms.

It is also possible to stack the polynomials, multiplying each term at the top by every term below, keeping like terms in columns, and leaving spaces for missing terms. Then we add just as we do in long multiplication with numbers.

Example 5 Multiply: $(5x^3 + x - 4)(-2x^2 + 3x + 6)$.

Solution

$$\begin{array}{rrrrrrl}
 & & 5x^3 + & & x - & 4 & \\
 & & & -2x^2 + & 3x + & 6 & \\
\hline
 & & 30x^3 & & + 6x - & 24 & \text{Multiplying by } 6 \\
 & 15x^4 & & + 3x^2 - & 12x & & \text{Multiplying by } 3x \\
-10x^5 & & - 2x^3 + & 8x^2 & & & \text{Multiplying by } -2x^2 \\
\hline
-10x^5 + & 15x^4 + & 28x^3 + & 11x^2 - & 6x - & 24 & \text{Adding}
\end{array}$$

Example 6 Multiply $4x^4y - 7x^2y + 3y$ by $2y - 3x^2y$.

Solution

$$\begin{array}{rrrrl}
 & 4x^4y - & 7x^2y + & 3y & \\
 & & -3x^2y + & 2y & \text{Writing descending powers of } x \\
\hline
 & 8x^4y^2 - & 14x^2y^2 + & 6y^2 & \text{Multiplying by } 2y \\
-12x^6y^2 + & 21x^4y^2 - & 9x^2y^2 & & \text{Multiplying by } -3x^2y \\
\hline
-12x^6y^2 + & 29x^4y^2 - & 23x^2y^2 + & 6y^2 & \text{Adding}
\end{array}$$

The Product of Two Binomials: FOIL

We now consider what are called *special products*. These products of polynomials occur often and can be simplified using shortcuts that we now develop.

To find a faster special-product rule for the product of two binomials, consider $(x + 7)(x + 4)$. We multiply each term of $(x + 7)$ by each term of $(x + 4)$.

$$(x + 7)(x + 4) = x \cdot x + x \cdot 4 + 7 \cdot x + 7 \cdot 4.$$

A visualization of $(x + 7)(x + 4)$ using areas

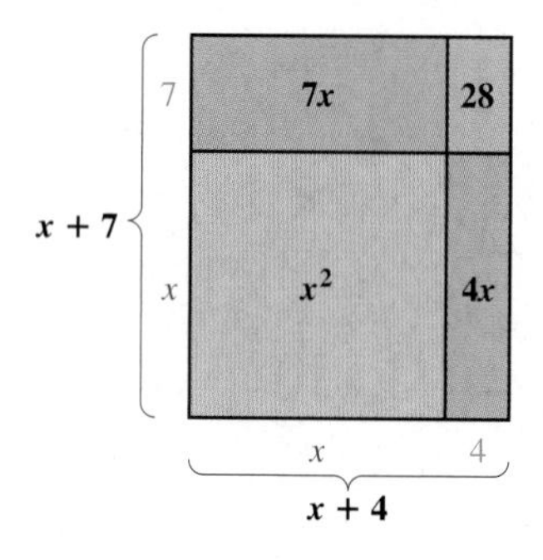

This multiplication illustrates a pattern that occurs whenever two binomials are multiplied:

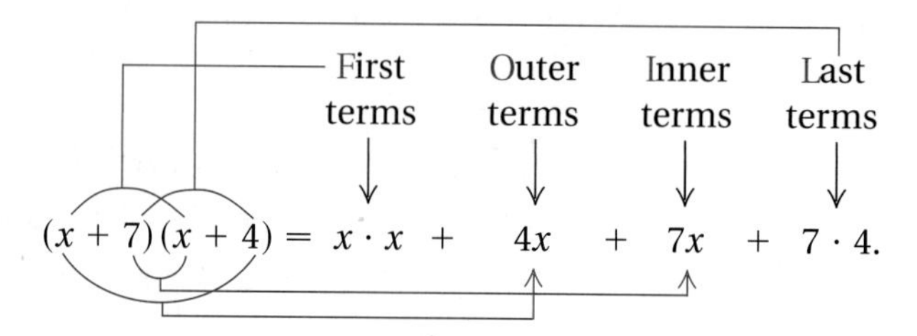

We use the mnemonic device FOIL to remember this method for multiplying.

The FOIL Method

To multiply two binomials $A + B$ and $C + D$, multiply the First terms AC, the Outer terms AD, the Inner terms BC, and then the Last terms BD. Then combine like terms, if possible.

$$(A + B)(C + D) = AC + AD + BC + BD$$

1. Multiply First terms: AC.
2. Multiply Outer terms: AD.
3. Multiply Inner terms: BC.
4. Multiply Last terms: BD.

↓
FOIL

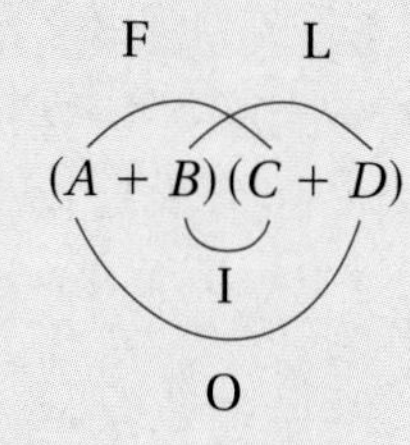

Example 7 Multiply.

a) $(x + 5)(x - 8)$ **b)** $(2x + 3y)(x - 4y)$ **c)** $(5xy + 2x)(x^2 + 2xy^2)$

Solution

F O I L

a) $(x + 5)(x - 8) = x^2 - 8x + 5x - 40$
$= x^2 - 3x - 40$ Combining like terms

b) $(2x + 3y)(x - 4y) = 2x^2 - 8xy + 3xy - 12y^2$ Using FOIL
$= 2x^2 - 5xy - 12y^2$ Combining like terms

c) $(5xy + 2x)(x^2 + 2xy^2) = 5x^3y + 10x^2y^3 + 2x^3 + 4x^2y^2$
There are no like terms to combine.

A visualization of $(A + B)^2$ using areas

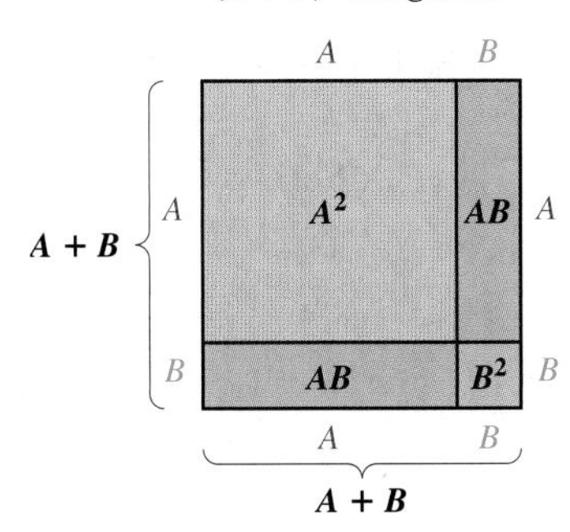

Squares of Binomials

A fast method for squaring binomials can be developed using FOIL:

$$
\begin{aligned}
(A + B)^2 &= (A + B)(A + B) \\
&= A^2 + AB + AB + B^2 \qquad \text{Note that } AB \text{ occurs twice.} \\
&= A^2 + 2AB + B^2; \\
(A - B)^2 &= (A - B)(A - B) \\
&= A^2 - AB - AB + B^2 \qquad \text{Note that } -AB \text{ occurs twice.} \\
&= A^2 - 2AB + B^2.
\end{aligned}
$$

Squaring a Binomial

$$(A + B)^2 = A^2 + 2AB + B^2;$$
$$(A - B)^2 = A^2 - 2AB + B^2$$

The square of a binomial is the square of the first term, plus twice the product of the two terms, plus the square of the last term.

It can help to remember the words of the rules and say them while multiplying.

Example 8 Multiply: **(a)** $(y - 5)^2$; **(b)** $(2x + 3y)^2$; **(c)** $(\frac{1}{2}x - 3y^4)^2$.

Solution

$$(A - B)^2 = A^2 - 2 \cdot A \cdot B + B^2$$
$$\downarrow \quad \downarrow \qquad \downarrow \qquad \downarrow \quad \downarrow \quad \downarrow$$

a) $(y - 5)^2 = y^2 - 2 \cdot y \cdot 5 + 5^2$ — Note that $-2 \cdot y \cdot 5$ is twice the product of y and -5.

$= y^2 - 10y + 25$

b) $(2x + 3y)^2 = (2x)^2 + 2 \cdot 2x \cdot 3y + (3y)^2$

$= 4x^2 + 12xy + 9y^2$ — Raising a product to a power

c) $(\frac{1}{2}x - 3y^4)^2 = (\frac{1}{2}x)^2 - 2 \cdot \frac{1}{2}x \cdot 3y^4 + (3y^4)^2$ — $2 \cdot \frac{1}{2}x \cdot (-3y^4) = -2 \cdot \frac{1}{2}x \cdot 3y^4$

$= \frac{1}{4}x^2 - 3xy^4 + 9y^8$ — Raising a product to a power; multiplying exponents

technology connection A

To verify that $(x + 3)^2 \neq x^2 + 9$, let $y_1 = (x + 3)^2$ and $y_2 = x^2 + 9$. Then compare y_1 and y_2 using a table of values or a graph. Don't forget to use parentheses.

Caution! Note that $(y - 5)^2 \neq y^2 - 5^2$. (To see this, replace y with 6 and note that $(6 - 5)^2 = 1^2 = 1$ and $6^2 - 5^2 = 36 - 25 = 11$.) More generally,

$$(A + B)^2 \neq A^2 + B^2 \quad \text{and} \quad (A - B)^2 \neq A^2 - B^2.$$

Products of Sums and Differences

Another pattern emerges when we are multiplying a sum and difference of the same two terms. Note the following:

$$\begin{aligned}(A+B)(A-B) &= \overset{F}{A^2} - \overset{O}{AB} + \overset{I}{AB} - \overset{L}{B^2}\\ &= A^2 - B^2. \qquad -AB + AB = 0\end{aligned}$$

The Product of a Sum and a Difference

$$(A+B)(A-B) = A^2 - B^2$$ This is called a *difference of two squares.*

The product of the sum and difference of the same two terms is the square of the first term minus the square of the second term.

Example 9 Multiply.

a) $(y+5)(y-5)$
b) $(2xy^2+3x)(2xy^2-3x)$
c) $(0.2t-1.4m)(0.2t+1.4m)$
d) $\left(\frac{2}{3}n-m^3\right)\left(\frac{2}{3}n+m^3\right)$

Solution

$(A+B)(A-B) = A^2 - B^2$

a) $(y+5)(y-5) = y^2 - 5^2$ Replacing A with y and B with 5
$= y^2 - 25$ Try to do problems like this mentally.

b) $(2xy^2+3x)(2xy^2-3x) = (2xy^2)^2 - (3x)^2$
$= 4x^2y^4 - 9x^2$ Raising a product to a power

c) $(0.2t-1.4m)(0.2t+1.4m) = (0.2t)^2 - (1.4m)^2$
$= 0.04t^2 - 1.96m^2$

d) $\left(\frac{2}{3}n-m^3\right)\left(\frac{2}{3}n+m^3\right) = \left(\frac{2}{3}n\right)^2 - (m^3)^2$
$= \frac{4}{9}n^2 - m^6$

Example 10 Multiply.

a) $(5y+4+3x)(5y+4-3x)$
b) $(3xy^2+4y)(-3xy^2+4y)$
c) $(a-5b)(a+5b)(a^2-25b^2)$

The check outlined in Technology Connection B in Section 5.1 can be used to check multiplication of polynomials in one variable. To check Example 3, using x in place of y, let
$y_1 = (x^3 - 5)(2x^3 + 4)$,
$y_2 = 2x^6 - 6x^3 - 20$, and
$y_3 = y_2 - y_1$. Since the multiplication is correct, $y_2 = y_1$ and $y_3 = 0$. Using the G-T mode, available on many graphers, and graphing with a wide line, we can split the window vertically to display both a graph and a table.

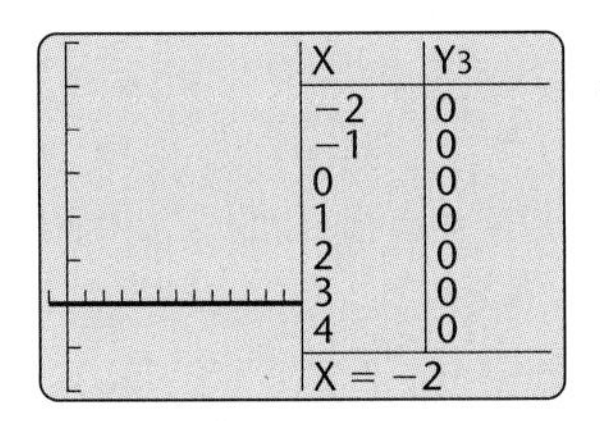

Had we found $y_3 \neq 0$, we would have known that a mistake had been made.

1. Use the procedure above to check Examples 4 and 5.

Solution

a) $(5y + 4 + 3x)(5y + 4 - 3x) = (5y + 4)^2 - (3x)^2$ Try to be alert for situations like this.

$$= 25y^2 + 40y + 16 - 9x^2$$

We can also multiply $(5y + 4 + 3x)(5y + 4 - 3x)$ using columns, but not as quickly.

b) $(3xy^2 + 4y)(-3xy^2 + 4y) = (4y + 3xy^2)(4y - 3xy^2)$ Rewriting

$$= (4y)^2 - (3xy^2)^2$$
$$= 16y^2 - 9x^2y^4$$

c) $(a - 5b)(a + 5b)(a^2 - 25b^2) = (a^2 - 25b^2)(a^2 - 25b^2)$

$$= (a^2 - 25b^2)^2$$
$$= (a^2)^2 - 2(a^2)(25b^2) + (25b^2)^2$$

Squaring a binomial

$$= a^4 - 50a^2b^2 + 625b^4$$

Function Notation

Let's stop for a moment and look back at what we have done in this section. We have shown, for example, that

$$(x - 2)(x + 2) = x^2 - 4,$$

that is, $x^2 - 4$ and $(x - 2)(x + 2)$ are equivalent expressions.

From the viewpoint of functions, if

$$f(x) = x^2 - 4$$

and

$$g(x) = (x - 2)(x + 2),$$

then for any given input x, the outputs $f(x)$ and $g(x)$ are identical. Thus the graphs of these functions are identical and we say that f and g represent the same function. Functions like these are graphed in detail in Chapter 8.

x	$f(x)$	$g(x)$
3	5	5
2	0	0
1	−3	−3
0	−4	−4
−1	−3	−3
−2	0	0
−3	5	5

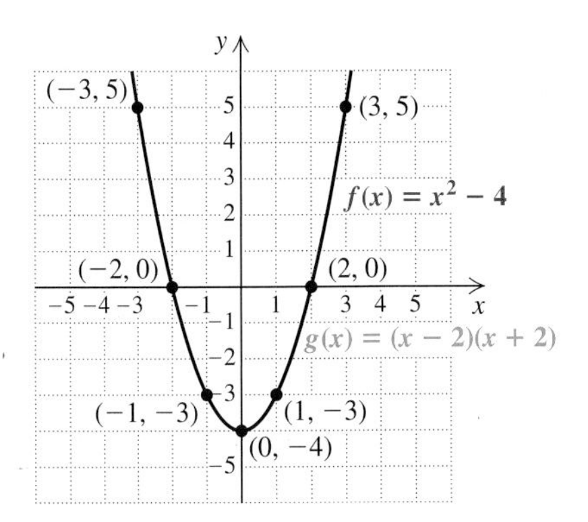

Our work with multiplying can be used when evaluating functions.

Example 11 Given $f(x) = x^2 - 4x + 5$, find and simplify each of the following.

a) $f(a + 3)$
b) $f(a + h) - f(a)$

Solution

a) To find $f(a + 3)$, we replace x with $a + 3$. Then we simplify:

$$\begin{aligned} f(a+3) &= (a+3)^2 - 4(a+3) + 5 \\ &= a^2 + 6a + 9 - 4a - 12 + 5 \\ &= a^2 + 2a + 2. \end{aligned}$$

b) To find $f(a + h)$ and $f(a)$, we replace x with $a + h$ and a, respectively.

$$\begin{aligned} f(a+h) - f(a) &= [(a+h)^2 - 4(a+h) + 5] - [a^2 - 4a + 5] \\ &= a^2 + 2ah + h^2 - 4a - 4h + 5 - a^2 + 4a - 5 \\ &= 2ah + h^2 - 4h \end{aligned}$$

Exercise Set 5.2

FOR EXTRA HELP

Digital Video Tutor CD 4
Videotape 9

InterAct Math

Math Tutor Center

MathXL

MyMathLab.com

Multiply.

1. $8a^2 \cdot 4a$

2. $-5x^3 \cdot 2x$

3. $5x(-4x^2y)$

4. $-3ab^2(2a^2b^2)$

5. $(2x^3y^2)(-5x^2y^4)$

6. $(7a^2bc^4)(-8ab^3c^2)$

7. $7x(3 - x)$

8. $3a(a^2 - 4a)$

9. $5cd(4c^2d - 5cd^2)$

10. $a^2(2a^2 - 5a^3)$

11. $(2x + 5)(3x - 4)$

12. $(2a + 3b)(4a - b)$

13. $(m + 2n)(m - 3n)$

14. $(m - 5)(m + 5)$

15. $(3y + 8x)(y - 7x)$

16. $(x + y)(x - 2y)$

17. $(a^2 - 2b^2)(a^2 - 3b^2)$

18. $(2m^2 - n^2)(3m^2 - 5n^2)$

19. $(x - 4)(x^2 + 4x + 16)$

20. $(y + 3)(y^2 - 3y + 9)$

21. $(x + y)(x^2 - xy + y^2)$

22. $(a - b)(a^2 + ab + b^2)$

23. $(a^2 + a - 1)(a^2 + 4a - 5)$

24. $(x^2 - 2x + 1)(x^2 + x + 2)$

25. $(3a^2b - 2ab + 3b^2)(ab - 2b + a)$

26. $(2x^2 + y^2 - 2xy)(x^2 - 2y^2 - xy)$

27. $\left(x - \frac{1}{2}\right)\left(x - \frac{1}{4}\right)$

28. $\left(b - \frac{1}{3}\right)\left(b - \frac{1}{3}\right)$

29. $(1.2x - 3y)(2.5x + 5y)$

30. $(40a - 0.24b)(0.3a + 10b)$

31. Let $P(x) = 3x^2 - 5$ and $Q(x) = 4x^2 - 7x + 1$. Find $P(x) \cdot Q(x)$.

32. Let $P(x) = x^2 - x + 1$ and $Q(x) = x^3 + x^2 + 5$. Find $P(x) \cdot Q(x)$.

Multiply.

33. $(a + 4)(a + 5)$

34. $(x + 3)(x + 2)$

35. $(y - 8)(y + 3)$

36. $(y - 1)(y + 5)$

37. $(x + 5)^2$

38. $(y - 7)^2$

39. $(x - 2y)^2$

40. $(2s + 3t)^2$

41. $(2x + 9)(x + 2)$

42. $(3b + 2)(2b - 5)$

43. $(10a - 0.12b)^2$

44. $(10p^2 + 2.3q)^2$

45. $(2x - 3y)(2x + y)$

46. $(2a - 3b)(2a - b)$

47. $(2x^3 - 3y^2)^2$

48. $(3s^2 + 4t^3)^2$

49. $(a^2b^2 + 1)^2$

50. $(x^2y - xy^2)^2$

51. Let $P(x) = 4x - 1$. Find $P(x) \cdot P(x)$.

52. Let $Q(x) = 3x^2 + 1$. Find $Q(x) \cdot Q(x)$.

53. Let $F(x) = 2x - \frac{1}{3}$. Find $[F(x)]^2$.

54. Let $G(x) = 5x - \frac{1}{2}$. Find $[G(x)]^2$.

Multiply.

55. $(c + 2)(c - 2)$

56. $(x - 3)(x + 3)$

57. $(4x + 1)(4x - 1)$

58. $(3 - 2x)(3 + 2x)$

59. $(3m - 2n)(3m + 2n)$

60. $(3x + 5y)(3x - 5y)$

61. $(x^3 + yz)(x^3 - yz)$

62. $(4a^3 + 5ab)(4a^3 - 5ab)$

63. $(-mn + m^2)(mn + m^2)$

64. $(-3b + a^2)(3b + a^2)$

65. $(x + 1)(x - 1)(x^2 + 1)$

66. $(y - 2)(y + 2)(y^2 + 4)$

67. $(a - b)(a + b)(a^2 - b^2)$

68. $(2x - y)(2x + y)(4x^2 - y^2)$

Aha! **69.** $(a + b + 1)(a + b - 1)$

70. $(m + n + 2)(m + n - 2)$

71. $(2x + 3y + 4)(2x + 3y - 4)$

72. $(3a - 2b + c)(3a - 2b - c)$

73. *Compounding interest.* Suppose that P dollars is invested in a savings account at interest rate i, compounded annually, for 2 yr. The amount A in the account after 2 yr is given by

$$A = P(1 + i)^2.$$

Find an equivalent expression for A.

74. *Compounding interest.* Suppose that P dollars is invested in a savings account at interest rate i, compounded semiannually, for 1 yr. The amount A in the account after 1 yr is given by

$$A = P\left(1 + \frac{i}{2}\right)^2.$$

Find an equivalent expression for A.

75. Given $f(x) = x^2 + 5$, find and simplify.

a) $f(t - 1)$
b) $f(a + h) - f(a)$
c) $f(a) - f(a - h)$

76. Given $f(x) = x^2 + 7$, find and simplify.

a) $f(p + 1)$
b) $f(a + h) - f(a)$
c) $f(a) - f(a - h)$

77. Find two binomials whose product is $x^2 - 25$ and explain how you decided on those two binomials.

78. Find two binomials whose product is $x^2 - 6x + 9$ and explain how you decided on those two binomials.

SKILL MAINTENANCE

Solve.

79. $ab + ac = d$, for a

80. $xy + yz = w$, for y

81. $mn + m = p$, for m

82. $rs + s = t$, for s

83. *Value of coins.* There are 50 dimes in a roll of dimes, 40 nickels in a roll of nickels, and 40 quarters in a roll of quarters. Kacie has 13 rolls of coins, which have a total value of \$89. There are three more rolls of dimes than nickels. How many of each type of roll does she have?

84. *Wages.* Takako worked a total of 17 days last month at her father's restaurant. She earned \$50 a day during the week and \$60 a day during the weekend. Last month Takako earned \$940. How many weekdays did she work?

SYNTHESIS

85. We have seen that $(a - b)(a + b) = a^2 - b^2$. Explain how this result can be used to develop a fast way of multiplying $95 \cdot 105$.

86. A student incorrectly claims that since $2x^2 \cdot 2x^2 = 4x^4$, it follows that $5x^5 \cdot 5x^5 = 25x^{25}$. What mistake is the student making?

Multiply. Assume that variables in exponents represent natural numbers.

87. $[(-x^ay^b)^4]^a$

88. $(z^{n^2})^{n^3}(z^{4n^3})^{n^2}$

89. $(a^xb^{2y})(\frac{1}{2}a^{3x}b)^2$

90. $(a^xb^y)^{w+z}$

91. $y^3z^n(y^{3n}z^3 - 4yz^{2n})$

92. $[(a + b)(a - b)][5 - (a + b)][5 + (a + b)]$

93. $(a - b + c - d)(a + b + c + d)$

94. $\left(\frac{2}{3}x + \frac{1}{3}y + 1\right)\left(\frac{2}{3}x - \frac{1}{3}y - 1\right)$

95. $(4x^2 + 2xy + y^2)(4x^2 - 2xy + y^2)$

96. $(x^2 - 3x + 5)(x^2 + 3x + 5)$

97. $(x^a + y^b)(x^a - y^b)(x^{2a} + y^{2b})$

98. $(x - 1)(x^2 + x + 1)(x^3 + 1)$

99. $(x^{a-b})^{a+b}$

100. $(M^{x+y})^{x+y}$

101. $(x - a)(x - b)(x - c) \cdots (x - z)$

102. Draw rectangles similar to those on p. 280 to show that $(x + 2)(x + 5) = x^2 + 7x + 10$.

103. Use a grapher to determine whether each of the following is an identity.

a) $(x - 1)^2 = x^2 - 1$
b) $(x - 2)(x + 3) = x^2 + x - 6$
c) $(x - 1)^3 = x^3 - 3x^2 + 3x - 1$
d) $(x + 1)^4 = x^4 + 1$
e) $(x + 1)^4 = x^4 + 4x^3 + 8x^2 + 4x + 1$

104. Use a grapher to check your answers to Exercises 23, 35, and 65.

COLLABORATIVE CORNER

Algebra and Number Tricks

Focus: Polynomial multiplication

Time: 15–20 minutes

Group size: 2

Consider the following dialogue:

Jinny: Cal, let me do a number trick with you. Think of a number between 1 and 7. I'll have you perform some manipulations to this number, you'll tell me the result, and I'll tell you your number.

Cal: Okay. I've thought of a number.

Jinny: Good. Write it down so I can't see it, double it, and then subtract x from the result.

Cal: Hey, this is algebra!

Jinny: I know. Now square your binomial and subtract x^2.

Cal: How did you know I had an x^2? I *thought* this was rigged!

Jinny: It is. Now, divide by 4 and tell me either your constant term or your x-term. I'll tell you the other term and the number you chose.

Cal: Okay. The constant term is 16.

Jinny: Then the other term is $-4x$ and the number you chose is 4.

Cal: You're right! How did you do it?

ACTIVITY

1. Each group member should follow Jinny's instructions. Then determine how Jinny determined Cal's number and the other term.
2. Suppose that, at the end, Cal told Jinny the x-term. How would Jinny have determined Cal's number and the other term?
3. Would Jinny's "trick" work with *any* real number? Why do you think she specified numbers between 1 and 7?
4. Each group member should create a new number "trick" and perform it on the other group member. Be sure to include a variable so that both members can gain practice with polynomials.

Common Factors and Factoring by Grouping

5.3

Terms with Common Factors • Factoring by Grouping

Factoring is the reverse of multiplication. To **factor** an expression means to write an equivalent expression that is a product. Skill at factoring will assist us when working with polynomial functions and solving polynomial equations later in this chapter.

CONNECTING THE CONCEPTS

Despite all the equals signs in Sections 5.1 and 5.2, we have not solved any equations since Chapter 4. We have concentrated instead on writing equivalent expressions by adding, subtracting, and multiplying polynomials. We found that we could not have multiplied polynomials had we not first learned how to add or subtract them. In a similar manner, we will find that our work with factoring polynomials in Sections 5.3–5.7 relies heavily on our ability to multiply polynomials. Not until Section 5.8 will we return to solving equations. At that point, however, we will study a new type of equation that cannot be solved without understanding how to factor. As we have seen before, mathematics consistently builds on learned concepts. The more mastery you develop in factoring, the better prepared you will be to solve the equations of Section 5.8.

Terms with Common Factors

When factoring, we look for factors common to every term in an expression and then use the distributive law.

Example 1 Factor out a common factor: $4y^2 - 8$.

Solution

$$4y^2 - 8 = 4 \cdot y^2 - 4 \cdot 2 \quad \text{Noting that 4 is a common factor}$$
$$= 4(y^2 - 2) \quad \text{Using the distributive law}$$

In some cases, there is more than one common factor. In $5x^4 + 20x^3$, for instance, 5 is a common factor, x^3 is a common factor, and $5x^3$ is a common factor. If there is more than one common factor, we factor out the *largest*, or *greatest, common factor*, that is, the common factor with the largest coefficient and the highest degree. In $5x^4 + 20x^3$, the largest common factor is $5x^3$.

Example 2

Factor out a common factor.

a) $5x^4 + 20x^3$ b) $12x^2y - 20x^3y$ c) $10p^6q^2 - 4p^5q^3 - 2p^4q^4$

Solution

a) $5x^4 + 20x^3 = 5x^3(x + 4)$ Try to write your answer directly. Multiply mentally to check your answer.

b) $12x^2y - 20x^3y = 4x^2y(3 - 5x)$

Check: $4x^2y \cdot 3 = 12x^2y$ and $4x^2y(-5x) = -20x^3y$, so $4x^2y(3 - 5x) = 12x^2y - 20x^3y$.

c) $10p^6q^2 - 4p^5q^3 - 2p^4q^4 = 2p^4q^2(5p^2 - 2pq - q^2)$

The check is left to the student.

The polynomials in Examples 1 and 2 cannot be factored further unless (in the cases of Examples 1 and 2c) square roots or (in the case of Example 2b) fractions are used. In both examples, we have **factored completely** over the set of integers. The factors used are said to be **prime polynomials** over the set of integers.

When a factor contains more than one term, it is usually desirable for the leading coefficient to be positive. To achieve this may require factoring out a common factor with a negative coefficient.

Example 3

Factor out a common factor with a negative coefficient.

a) $-4x - 24$ b) $-2x^3 + 6x^2 - 10x$

Solution

a) $-4x - 24 = -4(x + 6)$
b) $-2x^3 + 6x^2 - 10x = -2x(x^2 - 3x + 5)$

Example 4

Height of a thrown object. Suppose that a baseball is thrown upward with an initial velocity of 64 ft/sec. Its height in feet, $h(t)$, after t seconds is given by

$$h(t) = -16t^2 + 64t.$$

Find an equivalent expression for $h(t)$ by factoring out a common factor.

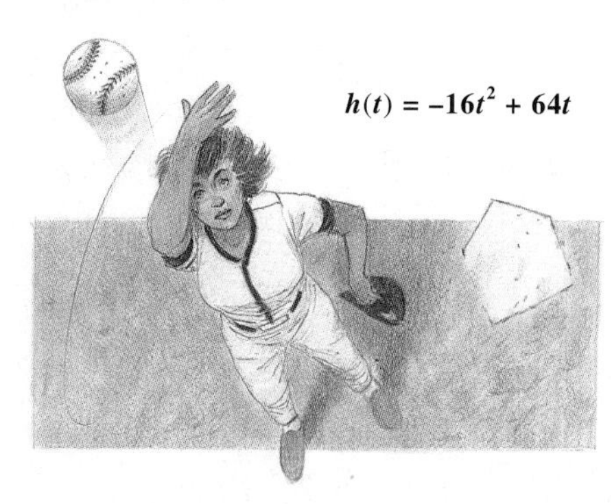

To check Example 4 with a table, let $y_1 = -16x^2 + 64x$ and $y_2 = -16x(x - 4)$. Then compare values of y_1 and y_2.

ΔTBL = 1

X	Y1	Y2
0	0	0
1	48	48
2	64	64
3	48	48
4	0	0
5	−80	−80
6	−192	−192

X = 0

1. How can $y_3 = y_2 - y_1$ and a table be used as a check?

Solution We factor out $-16t$ as follows:

$$h(t) = -16t^2 + 64t = -16t(t - 4).$$ *Check*: $-16t \cdot t = -16t^2$ and $-16t(-4) = 64t$.

Note that we can obtain function values using either expression for $h(t)$, since factoring forms equivalent expressions. For example,

$$h(1) = -16 \cdot 1^2 + 64 \cdot 1 = 48$$

and $$h(1) = -16 \cdot 1(1 - 4) = 48.$$ Using the factorization

In Example 4, we could have evaluated $-16t^2 + 64t$ and $-16t(t - 4)$ using any value for t. The results should always match. Thus a quick partial check of any factorization is to evaluate the factorization and the original polynomial for one or two convenient replacements. The check in Example 4 becomes foolproof if three replacements are used. In general, an nth-degree factorization is correct if it checks for $n + 1$ different replacements.

Factoring by Grouping

The largest common factor is sometimes a binomial.

Example 5 Factor: $(a - b)(x + 5) + (a - b)(x - y^2)$.

Solution Here the largest common factor is the binomial $a - b$:

$$\begin{aligned}(a - b)(x + 5) + (a - b)(x - y^2) &= (a - b)[(x + 5) + (x - y^2)] \\ &= (a - b)[2x + 5 - y^2].\end{aligned}$$

Often, in order to identify a common binomial factor, we must regroup into two groups of two terms each.

Example 6 Factor: **(a)** $y^3 + 3y^2 + 4y + 12$; **(b)** $4x^3 - 15 + 20x^2 - 3x$.

Solution

a) $y^3 + 3y^2 + 4y + 12 = (y^3 + 3y^2) + (4y + 12)$ Each grouping has a common factor.

$= y^2(y + 3) + 4(y + 3)$ Factoring out a common factor from each binomial

$= (y + 3)(y^2 + 4)$ Factoring out $y + 3$

b) When we try grouping $4x^3 - 15 + 20x^2 - 3x$ as

$$(4x^3 - 15) + (20x^2 - 3x),$$

we are unable to factor $4x^3 - 15$. When this happens, we can rearrange the polynomial and try a different grouping:

$$4x^3 - 15 + 20x^2 - 3x = 4x^3 + 20x^2 - 3x - 15$$ Using the commutative law to rearrange the terms

$$= 4x^2(x + 5) - 3(x + 5)$$
$$= (x + 5)(4x^2 - 3).$$

In Example 8 of Section 1.3 (see p. 25), we saw that

$$b - a, \quad -(a - b), \quad \text{and} \quad -1(a - b)$$

are equivalent. Remembering this can help anytime we wish to reverse subtraction (see the third step below).

Example 7 Factor: $ax - bx + by - ay$.

Solution We have

$ax - bx + by - ay = (ax - bx) + (by - ay)$	Grouping
$= x(a - b) + y(b - a)$	Factoring each binomial
$= x(a - b) + y(-1)(a - b)$	Factoring out -1 to reverse $b - a$
$= x(a - b) - y(a - b)$	Simplifying
$= (a - b)(x - y).$	Factoring out $a - b$

We can always check our factoring by multiplying:

Check: $(a - b)(x - y) = ax - ay - bx + by = ax - bx + by - ay.$

Some polynomials with four terms, like $x^3 + x^2 + 3x - 3$, are prime. Not only is there no common monomial factor, but no matter how we group terms, there is no common binomial factor:

$x^3 + x^2 + 3x - 3 = x^2(x + 1) + 3(x - 1);$	No common factor
$x^3 + 3x + x^2 - 3 = x(x^2 + 3) + (x^2 - 3);$	No common factor
$x^3 - 3 + x^2 + 3x = (x^3 - 3) + x(x + 3).$	No common factor

Exercise Set 5.3

FOR EXTRA HELP

Digital Video Tutor CD 4
Videotape 9

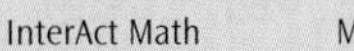
InterAct Math

Math Tutor Center

MathXL

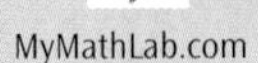
MyMathLab.com

Factor.

1. $2t^2 + 8t$
2. $3y^2 + 6y$
3. $y^2 - 5y$
4. $x^2 + 9x$
5. $y^3 + 9y^2$
6. $x^3 + 8x^2$
7. $15x^2 - 5x^4$
8. $8y^2 + 4y^4$
9. $4x^2y - 12xy^2$
10. $5x^2y^3 + 15x^3y^2$
11. $3y^2 - 3y - 9$
12. $5x^2 - 5x + 15$
13. $6ab - 4ad + 12ac$
14. $8xy + 10xz - 14xw$
15. $9x^3y^6z^2 - 12x^4y^4z^4 + 15x^2y^5z^3$
16. $14a^4b^3c^5 + 21a^3b^5c^4 - 35a^4b^4c^3$

Factor out a factor with a negative coefficient.

17. $-5x + 35$
18. $-5x - 40$
19. $-6y - 72$
20. $-8t + 72$
21. $-2x^2 + 4x - 12$
22. $-2x^2 + 12x + 40$
23. $3y - 24x$
24. $7x - 56y$
25. $7s - 14t$
26. $5r - 10s$
27. $-x^2 + 5x - 9$
28. $-p^3 - 4p^2 + 11$
29. $-a^4 + 2a^3 - 13a$
30. $-m^3 - m^2 + m - 2$

Factor.

31. $a(b - 5) + c(b - 5)$
32. $r(t - 3) - s(t - 3)$
33. $(x + 7)(x - 1) + (x + 7)(x - 2)$
34. $(a + 5)(a - 2) + (a + 5)(a + 1)$
35. $a^2(x - y) + 5(y - x)$
36. $5x^2(x - 6) + 2(6 - x)$
37. $ac + ad + bc + bd$
38. $xy + xz + wy + wz$
39. $b^3 - b^2 + 2b - 2$
40. $y^3 - y^2 + 3y - 3$
41. $a^3 - 3a^2 + 6 - 2a$
42. $t^3 + 6t^2 - 2t - 12$
43. $72x^3 - 36x^2 + 24x$
44. $12a^4 - 21a^3 - 9a^2$
45. $x^6 - x^5 - x^3 + x^4$
46. $y^4 - y^3 - y + y^2$
47. $2y^4 + 6y^2 + 5y^2 + 15$
48. $2xy - x^2y - 6 + 3x$

49. *Height of a baseball.* A baseball is popped up with an upward velocity of 72 ft/sec. Its height in feet, $h(t)$, after t seconds is given by

$$h(t) = -16t^2 + 72t.$$

a) Find an equivalent expression for $h(t)$ by factoring out a common factor with a negative coefficient.
b) Perform a partial check of part (a) by evaluating both expressions for $h(t)$ at $t = 1$.

50. *Height of a rocket.* A model rocket is launched upward with an initial velocity of 96 ft/sec. Its height in feet, $h(t)$, after t seconds is given by

$$h(t) = -16t^2 + 96t.$$

a) Find an equivalent expression for $h(t)$ by factoring out a common factor with a negative coefficient.
b) Check your factoring by evaluating both expressions for $h(t)$ at $t = 1$.

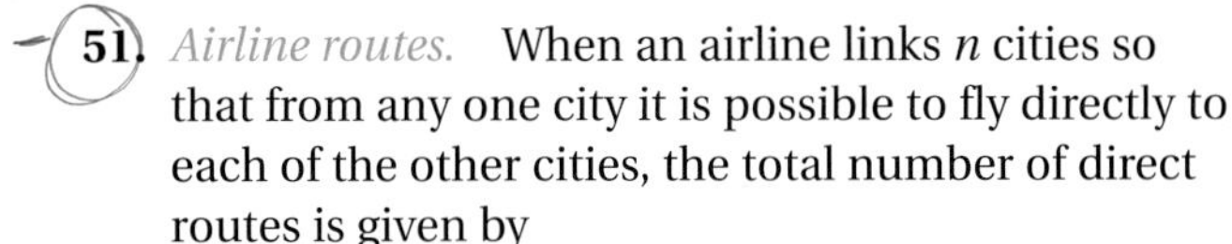

51. *Airline routes.* When an airline links n cities so that from any one city it is possible to fly directly to each of the other cities, the total number of direct routes is given by

$$R(n) = n^2 - n.$$

Find an equivalent expression for $R(n)$ by factoring out a common factor.

52. *Surface area of a silo.* A silo is a structure that is shaped like a right circular cylinder with a half sphere on top. The surface area of a silo of height h and radius r (including the area of the base) is given by the polynomial $2\pi rh + \pi r^2$. Find an equivalent expression by factoring out a common factor.

53. *Total profit.* When x hundred CD players are sold, Rolics Electronics collects a profit of $P(x)$, where

$$P(x) = x^2 - 3x,$$

and $P(x)$ is in thousands of dollars. Find an equivalent expression by factoring out a common factor.

54. *Total profit.* After t weeks of production, Claw Foot, Inc., is making a profit of $P(t) = t^2 - 5t$ from sales of their surfboards. Find an equivalent expression by factoring out a common factor.

55. *Total revenue.* Urban Sounds is marketing a new MP3 player. The firm determines that when it sells x units, the total revenue R is given by the polynomial function

$$R(x) = 280x - 0.4x^2 \text{ dollars.}$$

Find an equivalent expression for $R(x)$ by factoring out $0.4x$.

56. *Total cost.* Urban Sounds determines that the total cost C of producing x MP3 players is given by the polynomial function

$$C(x) = 0.18x + 0.6x^2.$$

Find an equivalent expression for $C(x)$ by factoring out $0.6x$.

57. *Counting spheres in a pile.* The number N of spheres in a triangular pile like the one shown here is a polynomial function given by

$$N(x) = \tfrac{1}{6}x^3 + \tfrac{1}{2}x^2 + \tfrac{1}{3}x,$$

where x is the number of layers and $N(x)$ is the number of spheres. Find an equivalent expression for $N(x)$ by factoring out $\frac{1}{6}$.

58. *Number of games in a league.* If there are n teams in a league and each team plays every other team once, we can find the total number of games played by using the polynomial function $f(n) = \frac{1}{2}n^2 - \frac{1}{2}n$. Find an equivalent expression by factoring out $\frac{1}{2}$.

59. *High-fives.* When a team of n players all give each other high-fives, a total of $H(n)$ hand slaps occurs, where

$$H(n) = \tfrac{1}{2}n^2 - \tfrac{1}{2}n.$$

Find an equivalent expression by factoring out $\frac{1}{2}n$.

60. *Number of diagonals.* The number of diagonals of a polygon having n sides is given by the polynomial function

$$P(n) = \tfrac{1}{2}n^2 - \tfrac{3}{2}n.$$

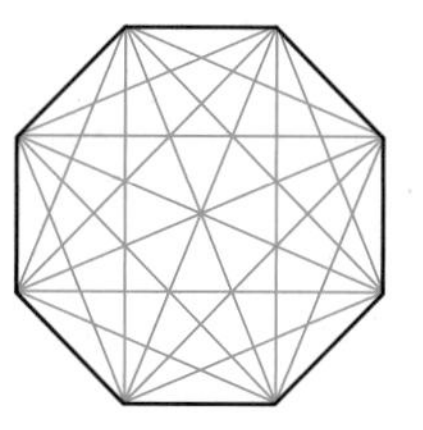

Find an equivalent expression for $P(n)$ by factoring out $\frac{1}{2}$.

61. Under what conditions would it be easier to evaluate a polynomial *after* it has been factored?

62. Explain in your own words why $-(a - b) = b - a$.

SKILL MAINTENANCE

Simplify.

63. $2(-3) + 4(-5)$

64. $-7(-2) + 5(-3)$

65. $4(-6) - 3(2)$

66. $5(-3) + 2(-2)$

67. *Geometry.* The perimeter of a triangle is 174. The lengths of the three sides are consecutive even numbers. What are the lengths of the sides of the triangle?

68. *Manufacturing.* In a factory, there are three machines A, B, and C. When all three are running, they produce 222 suitcases per day. If A and B work but C does not, they produce 159 suitcases per day. If B and C work but A does not, they produce 147 suitcases. What is the daily production of each machine?

SYNTHESIS

69. Is it true that if a polynomial's coefficients and exponents are all prime numbers, then the polynomial itself is prime? Why or why not?

70. Following Example 4, we stated that checking the factorization of a second-degree polynomial by making a single replacement is only a *partial* check. Write an *incorrect* factorization and explain how evaluating both the polynomial and the factorization might not catch the mistake.

Complete each of the following.

71. $x^5y^4 + \_\_\_\_ = x^3y(\_\_\_\_ + xy^5)$

72. $a^3b^7 - \_\_\_\_ = \_\_\_\_(ab^4 - c^2)$

Factor.

73. $rx^2 - rx + 5r + sx^2 - sx + 5s$

74. $3a^2 + 6a + 30 + 7a^2b + 14ab + 70b$

75. $a^4x^4 + a^4x^2 + 5a^4 + a^2x^4 + a^2x^2 + 5a^2 + 5x^4 + 5x^2 + 25$
(*Hint*: Use three groups of three.)

Factor out the smallest power of x in each of the following.

76. $x^{1/2} + 5x^{3/2}$

77. $x^{1/3} - 7x^{4/3}$

78. $x^{3/4} + x^{1/2} - x^{1/4}$

79. $x^{1/3} - 5x^{1/2} + 3x^{3/4}$

Factor. Assume that all exponents are natural numbers.

80. $2x^{3a} + 8x^a + 4x^{2a}$

81. $3a^{n+1} + 6a^n - 15a^{n+2}$

82. $4x^{a+b} + 7x^{a-b}$

83. $7y^{2a+b} - 5y^{a+b} + 3y^{a+2b}$

84. Use the TABLE feature of a grapher to check your answers to Exercises 17, 29, and 33.

85. Use a grapher to show that
$$(x^2 - 3x + 2)^4 = x^8 + 81x^4 + 16$$
is *not* an identity.

5.4 Factoring Trinomials

Factoring Trinomials of the Type $x^2 + bx + c$ • Factoring Trinomials of the Type $ax^2 + bx + c$, $a \neq 1$

Study Tip

Spending extra time studying this section will save you time when you work in Sections 5.5–5.8.

Our study of the factoring of trinomials begins with trinomials of the type $x^2 + bx + c$. We then move on to the type $ax^2 + bx + c$, where $a \neq 1$.

Factoring Trinomials of the Type $x^2 + bx + c$

When trying to factor trinomials of the type $x^2 + bx + c$, we can use a trial-and-error procedure.

Constant Term Positive

Recall the FOIL method of multiplying two binomials:

$$\begin{aligned}(x + 3)(x + 5) &= \overset{F}{x^2} + \underbrace{\overset{O}{5x} + \overset{I}{3x}} + \overset{L}{15}\\ &= x^2 + \quad 8x \quad + 15.\end{aligned}$$

Because the leading coefficient in each binomial is 1, the leading coefficient in the product is also 1. To factor $x^2 + 8x + 15$, we think of FOIL: The first term,

x^2, is the product of the First terms of two binomial factors, so the first term in each binomial must be x. The challenge is to find two numbers p and q such that

$$x^2 + 8x + 15 = (x + p)(x + q).$$
$$= x^2 + qx + px + pq.$$

Note that the Outer and Inner products, qx and px, can be written as $(p + q)x$. The Last product, pq, will be a constant. Thus the numbers p and q must be selected so that their product is 15 and their sum is 8. In this case, we know from above that these numbers are 3 and 5. The factorization is

$$(x + 3)(x + 5), \quad \text{or} \quad (x + 5)(x + 3). \qquad \text{Using a commutative law}$$

In general, to factor $x^2 + (p + q)x + pq$, we use FOIL in reverse:

$$x^2 + (p + q)x + pq = (x + p)(x + q).$$

Example 1 Factor: $x^2 + 9x + 8$.

Solution We think of FOIL in reverse. The first term of each factor is x. We are looking for numbers p and q such that

$$x^2 + 9x + 8 = (x + p)(x + q) = x^2 + (p + q)x + pq.$$

Thus we search for factors of 8 whose sum is 9.

Pair of Factors	Sum of Factors
2, 4	6
1, 8	9 ← The numbers we need are 1 and 8.

The factorization is thus $(x + 1)(x + 8)$. The student should check by multiplying to confirm that the product is the original trinomial.

When factoring trinomials with a leading coefficient of 1, it suffices to consider all pairs of factors along with their sums, as we did above. At times, however, you may be tempted to form factors without calculating any sums. It is essential that you check any attempt made in this manner! For example, if we attempt the factorization

$$x^2 + 9x + 8 \stackrel{?}{=} (x + 2)(x + 4),$$

a check reveals that

$$(x + 2)(x + 4) = x^2 + 6x + 8 \neq x^2 + 9x + 8.$$

This type of trial-and-error procedure becomes easier to use with time. As you gain experience, you will find that many trials can be performed mentally.

When the constant term of a trinomial is positive, the constant terms in both binomial factors must have the same sign. This ensures a positive product. The sign used is that of the trinomial's middle term.

Example 2

Factor: $y^2 - 9y + 20$.

Solution Since the constant term is positive and the coefficient of the middle term is negative, we look for a factorization of 20 in which both factors are negative. Their sum must be -9.

Pair of Factors	Sum of Factors
$-1, -20$	-21
$-2, -10$	-12
$-4, -5$	-9 ← The numbers we need are -4 and -5.

The factorization is $(y - 4)(y - 5)$.

Constant Term Negative

When the constant term of a trinomial is negative, we look for one negative factor and one positive factor. The sum of the factors must still be the coefficient of the middle term.

Example 3

Factor: $x^3 - x^2 - 30x$.

Solution *Always* look first for a common factor! This time there is one, x. We factor it out:

$$x^3 - x^2 - 30x = x(x^2 - x - 30).$$

Now we consider $x^2 - x - 30$. We need a factorization of -30 in which one factor is positive, the other factor is negative, and the sum of the factors is -1. Since the sum is to be negative, the negative factor must have the greater absolute value. Thus we need only consider the following pairs of factors.

Pair of Factors	Sum of Factors
$1, -30$	-29
$3, -10$	-7
$5, -6$	-1 ← The numbers we need are 5 and -6.

The factorization of $x^2 - x - 30$ is $(x + 5)(x - 6)$. *Don't forget to include the factor that was factored out earlier*! In this case, the factorization of the original trinomial is $x(x + 5)(x - 6)$.

Example 4 Factor: $2x^2 + 34x - 220$.

The method described in Technology Connection B in Section 5.1 can be used to check Example 4: Let
$y_1 = 2x^2 + 34x - 220$,
$y_2 = 2(x - 5)(x + 22)$, and
$y_3 = y_2 - y_1$.

1. How should the graphs of y_1 and y_2 compare?
2. What should the graph of y_3 look like?
3. Check Example 3 with a grapher.
4. Use graphs to show that $(2x + 5)(x - 3)$ is *not* a factorization of $2x^2 + x - 15$.

Solution *Always* look first for a common factor! This time we can factor out 2:

$$2x^2 + 34x - 220 = 2(x^2 + 17x - 110).$$

We next look for a factorization of -110 in which one factor is positive, the other factor is negative, and the sum of the factors is 17. Since the sum is to be positive, we examine only pairs of factors in which the positive term has the larger absolute value.

Pair of Factors	Sum of Factors
−1, 110	109
−2, 55	53
−5, 22	17

The numbers we need are −5 and 22.

The factorization of $x^2 + 17x - 110$ is $(x - 5)(x + 22)$. The factorization of the original trinomial, $2x^2 + 34x - 220$, is $2(x - 5)(x + 22)$.

Some polynomials are not factorable using integers.

Example 5 Factor: $x^2 - x - 7$.

Solution There are no factors of -7 whose sum is -1. This trinomial is *not* factorable into binomials with integer coefficients. Although $x^2 - x - 7$ can be factored using more advanced techniques, for our purposes the polynomial is *prime*.

Tips for Factoring $x^2 + bx + c$

1. If necessary, rewrite the trinomial in descending order. Search for factors of c that add up to b. Remember the following:
   - If c is positive, the signs of the factors are the same as the sign of b.
   - If c is negative, one factor is positive and the other is negative.
   - If the sum of the two factors is the opposite of b, changing the signs of both factors will give the desired factors whose sum is b.
2. Check the result by multiplying the binomials.

These tips still apply when a trinomial has more than one variable.

Example 6 Factor: $x^2 - 2xy - 48y^2$.

Solution We look for numbers p and q such that

$$x^2 - 2xy - 48y^2 = (x + py)(x + qy).$$

The x's and y's can be written in the binomials in advance.

Our thinking is much the same as if we were factoring $x^2 - 2x - 48$. We look for factors of -48 whose sum is -2. Those factors are 6 and -8. Thus,

$$x^2 - 2xy - 48y^2 = (x + 6y)(x - 8y).$$

The check is left to the student.

Factoring Trinomials of the Type $ax^2 + bx + c$, $a \neq 1$

Now we look at trinomials in which the leading coefficient is not 1. We consider two methods. Use what works best for you or what your instructor chooses for you.

Method 1: The FOIL Method

We first consider the **FOIL method** for factoring trinomials of the type

$$ax^2 + bx + c, \quad \text{where } a \neq 1.$$

Consider the following multiplication.

$$\begin{aligned}(3x + 2)(4x + 5) &= \overset{\text{F}}{12x^2} + \underbrace{\overset{\text{O}}{15x} + \overset{\text{I}}{8x}} + \overset{\text{L}}{10} \\ &= 12x^2 + \quad 23x \quad + 10\end{aligned}$$

To factor $12x^2 + 23x + 10$, we must reverse what we just did. We look for two binomials whose product is this trinomial. The product of the First terms must be $12x^2$. The product of the Outer terms plus the product of the Inner terms must be $23x$. The product of the Last terms must be 10. We know from the preceding discussion that the answer is

$$(3x + 2)(4x + 5).$$

In general, however, finding such an answer involves trial and error. We use the following method.

To Factor $ax^2 + bx + c$ Using FOIL

1. Factor out the largest common factor, if one exists. Here we assume none does.
2. Find two **F**irst terms whose product is ax^2:

$$(\square x + \quad)(\square x + \quad) = ax^2 + bx + c.$$

FOIL

3. Find two **L**ast terms whose product is c:

$$(\quad x + \square)(\quad x + \square) = ax^2 + bx + c.$$

FOIL

4. Repeat steps (2) and (3) until a combination is found for which the sum of the **O**uter and **I**nner products is bx:

$$(\square x + \square)(\square x + \square) = ax^2 + bx + c.$$

I O FOIL

Example 7

Factor: $3x^2 + 10x - 8$.

Solution

1. First, observe that there is no common factor (other than 1 or -1).
2. Next, factor the first term, $3x^2$. The only possibility for factors is $3x \cdot x$. Thus, if a factorization exists, it must be of the form

$$(3x + \square)(x + \square).$$

We need to find the right numbers for the blanks.

3. Note that the constant term, -8, can be factored as $(-8)(1)$, $8(-1)$, $(-2)4$, and $2(-4)$, as well as $(1)(-8)$, $(-1)8$, $4(-2)$, and $(-4)2$.
4. Find a pair of factors for which the sum of the products (the "outer" and "inner" parts of FOIL) is the middle term, $10x$. Each possibility should be checked by multiplying:

$$(3x - 8)(x + 1) = 3x^2 - 5x - 8. \qquad O + I = 3x + (-8x) = -5x$$

This gives a middle term with a negative coefficient. Since a positive coefficient is needed, a second possibility must be tried:

$$(3x + 8)(x - 1) = 3x^2 + 5x - 8. \qquad O + I = -3x + 8x = 5x$$

Note that changing the signs of the two constant terms changes only the sign of the middle term. We try again:

$$(3x - 2)(x + 4) = 3x^2 + 10x - 8. \qquad \text{This is what we wanted.}$$

Thus the desired factorization is $(3x - 2)(x + 4)$.

Example 8

Factor: $6x^6 - 19x^5 + 10x^4$.

Solution

1. First, factor out the common factor x^4:

$$x^4(6x^2 - 19x + 10).$$

2. Note that $6x^2 = 6x \cdot x$ and $6x^2 = 3x \cdot 2x$. Thus, $6x^2 - 19x + 10$ may factor into

$$(3x + \square)(2x + \square) \quad \text{or} \quad (6x + \square)(x + \square).$$

3. We factor the last term, 10. The possibilities are $10 \cdot 1$, $(-10)(-1)$, $5 \cdot 2$, and $(-5)(-2)$, as well as $1 \cdot 10$, $(-1)(-10)$, $2 \cdot 5$, and $(-2)(-5)$.
4. There are 8 possibilities for *each* factorization in step (2). We need factors for which the sum of the products (the "outer" and "inner" parts of FOIL) is the middle term, $-19x$. Since the x-coefficient is negative, we consider pairs of negative factors. Each possible factorization must be checked by multiplying:

$$(3x - 10)(2x - 1) = 6x^2 - 23x + 10.$$

We try again:

$$(3x - 5)(2x - 2) = 6x^2 - 16x + 10.$$

Actually this last attempt could have been rejected by simply noting that $2x - 2$ has a common factor, 2. Since the *largest* common factor was removed in step (1), no other common factors can exist. We try again, reversing the -5 and -2:

$$(3x - 2)(2x - 5) = 6x^2 - 19x + 10. \qquad \text{This is what we wanted.}$$

The factorization of $6x^2 - 19x + 10$ is $(3x - 2)(2x - 5)$. But do not forget the common factor! We must include it to get the complete factorization of the original trinomial:

$$6x^6 - 19x^5 + 10x^4 = x^4(3x - 2)(2x - 5).$$

Tips for Factoring with FOIL

1. If the largest common factor has been factored out of the original trinomial, then no binomial factor can have a common factor (other than 1 or -1).
2. If a and c are both positive, then the signs in the factors will be the same as the sign of b.
3. When a possible factoring produces the opposite of the desired middle term, reverse the signs of the constants in the factors.
4. Be systematic about your trials. Keep track of those possibilities that you have tried and those that you have not.

Keep in mind that this method of factoring involves trial and error. With practice, you will find yourself making fewer and better guesses.

Method 2: The Grouping Method

The second method for factoring trinomials of the type $ax^2 + bx + c, a \neq 1$, is known as the *grouping method*. It involves not only trial and error and FOIL but also factoring by grouping. We know that

$$\begin{aligned} x^2 + 7x + 10 &= x^2 + 2x + 5x + 10 \\ &= x(x + 2) + 5(x + 2) \\ &= (x + 2)(x + 5), \end{aligned}$$

but what if the leading coefficient is not 1? Consider $6x^2 + 23x + 20$. The method is similar to what we just did with $x^2 + 7x + 10$, but we need two more steps.* First, multiply the leading coefficient, 6, and the constant, 20, to get 120. Then find a factorization of 120 in which the sum of the factors is the coefficient of the middle term: 23. The middle term is then split into a sum or difference using these factors.

$6x^2 + 23x + 20$

(1) Multiply 6 and 20: $6 \cdot 20 = 120$.
(2) Factor 120: $120 = 8 \cdot 15$, and $8 + 15 = 23$.
(3) Split the middle term: $23x = 8x + 15x$.
(4) Factor by grouping.

We factor by grouping as follows:

$$\begin{aligned} 6x^2 + 23x + 20 &= 6x^2 + 8x + 15x + 20 \\ &= 2x(3x + 4) + 5(3x + 4) \\ &= (3x + 4)(2x + 5). \end{aligned}$$

Factoring by grouping

To Factor $ax^2 + bx + c$ Using Grouping

1. Make sure that any common factors have been factored out.
2. Multiply the leading coefficient a and the constant c.
3. Try to factor the product ac so that the sum of the factors is b. That is, find integers p and q so that $pq = ac$ and $p + q = b$.
4. Split the middle term. That is, write bx as $px + qx$.
5. Factor by grouping.

Example 9 Factor: $3x^2 + 10x - 8$.

Solution

1. First, look for a common factor. There is none (other than 1 or -1).
2. Multiply the leading coefficient and the constant, 3 and -8:

$$3(-8) = -24.$$

*The rationale behind these steps is outlined in Exercise 105.

3. Try to factor -24 so that the sum of the factors is 10:

$$-24 = 12(-2) \quad \text{and} \quad 12 + (-2) = 10.$$

4. Split $10x$ using the results of step (3):

$$10x = 12x - 2x.$$

5. Finally, factor by grouping:

$$\begin{aligned} 3x^2 + 10x - 8 &= 3x^2 + 12x - 2x - 8 \\ &= 3x(x + 4) - 2(x + 4) \\ &= (x + 4)(3x - 2). \end{aligned}$$ Factoring by grouping

Exercise Set 5.4

FOR EXTRA HELP

Digital Video Tutor CD 4 Videotape 9

InterAct Math

Math Tutor Center

MathXL

MyMathLab.com

Factor.

1. $x^2 + 8x + 12$
2. $x^2 + 6x + 5$
3. $t^2 + 8t + 15$
4. $y^2 + 12y + 27$
5. $x^2 - 27 - 6x$
6. $t^2 - 15 - 2t$
7. $2n^2 - 20n + 50$
8. $2a^2 - 16a + 32$
9. $a^3 - a^2 - 72a$
10. $x^3 + 3x^2 - 54x$
11. $14x + x^2 + 45$
12. $12y + y^2 + 32$
13. $y^2 + 2y - 63$
14. $p^2 - 3p - 40$
15. $t^2 - 14t + 45$
16. $a^2 - 11a + 28$
17. $3x + x^2 - 10$
18. $x + x^2 - 6$
19. $3x^2 + 15x + 18$
20. $5y^2 + 40y + 35$
21. $56 + x - x^2$
22. $32 + 4y - y^2$
23. $32y + 4y^2 - y^3$
24. $56x + x^2 - x^3$
25. $x^4 + 11x^3 - 80x^2$
26. $y^4 + 5y^3 - 84y^2$
27. $x^2 + 12x + 13$
28. $x^2 - 3x + 7$
29. $p^2 - 5pq - 24q^2$
30. $x^2 + 12xy + 27y^2$
31. $y^2 + 8yz + 16z^2$
32. $x^2 - 14xy + 49y^2$
33. $p^4 - 80p^3 + 79p^2$
34. $x^4 - 50x^3 + 49x^2$
35. $x^6 + 7x^5 - 18x^4$
36. $x^6 + 2x^5 - 63x^4$
37. $6x^2 - 5x - 25$
38. $3x^2 - 16x - 12$
39. $10y^3 - 12y - 7y^2$
40. $6x^3 - 15x - x^2$
41. $24a^2 - 14a + 2$
42. $3a^2 - 10a + 8$
43. $35y^2 + 34y + 8$
44. $9a^2 + 18a + 8$
45. $4t + 10t^2 - 6$
46. $8x + 30x^2 - 6$
47. $8x^2 - 16 - 28x$
48. $18x^2 - 24 - 6x$
49. $a^6 + a^5 - 6a^4$
50. $t^8 + 5t^7 - 14t^6$
51. $14x^4 - 19x^3 - 3x^2$
52. $70x^4 - 68x^3 + 16x^2$
53. $12a^2 - 4a - 16$
54. $12a^2 - 14a - 20$
55. $9x^2 + 15x + 4$
56. $6y^2 + 7y + 2$
57. (Aha!) $4x^2 + 15x + 9$
58. $2y^2 + 7y + 6$
59. $-8t^2 - 8t + 30$
60. $-36a^2 + 21a - 3$
61. $18xy^3 + 3xy^2 - 10xy$
62. $3x^3y^2 - 5x^2y^2 - 2xy^2$
63. $24x^2 - 2 - 47x$
64. $15y^2 - 10 - 47y$
65. $63x^3 + 111x^2 + 36x$
66. $50y^3 + 115y^2 + 60y$
67. $48x^4 + 4x^3 - 30x^2$
68. $40y^4 + 4y^3 - 12y^2$
69. $12a^2 - 17ab + 6b^2$
70. $20p^2 - 23pq + 6q^2$
71. $2x^2 + xy - 6y^2$
72. $8m^2 - 6mn - 9n^2$
73. $6x^2 - 29xy + 28y^2$
74. $10p^2 + 7pq - 12q^2$
75. $9x^2 - 30xy + 25y^2$
76. $4p^2 + 12pq + 9q^2$
77. $9x^2y^2 + 5xy - 4$
78. $7a^2b^2 + 13ab + 6$

79. How can one conclude that $x^2 + 5x + 200$ is a prime polynomial without performing any trials?

80. How can one conclude that $x^2 - 59x + 6$ is a prime polynomial without performing any trials?

SKILL MAINTENANCE

Factor.

81. $10x^3 - 35x^2 + 5x$

82. $12t^3 - 40t^2 - 8t$

Simplify.

83. $(5a^4)^3$

84. $(-2x^2)^5$

85. If $g(x) = -5x^2 - 7x$, find $g(-3)$.

86. *Height of a rocket.* A model rocket is launched upward with an initial velocity of 96 ft/sec from a height of 880 ft. Its height in feet, $h(t)$, after t seconds is given by

$$h(t) = -16t^2 + 96t + 880.$$

What is its height after 0 sec, 1 sec, 3 sec, 8 sec, and 10 sec?

SYNTHESIS

87. Describe in your own words an approach that can be used to factor any "nonprime" trinomial of the form $ax^2 + bx + c$.

88. Suppose $(rx + p)(sx - q) = ax^2 - bx + c$ is true. Explain how this can be used to factor $ax^2 + bx + c$.

Factor. Assume that variables in exponents represent positive integers.

89. $2a^4b^6 - 3a^2b^3 - 20ab^2$

90. $5x^8y^6 + 35x^4y^3 + 60$

91. $x^2 - \frac{4}{25} + \frac{3}{5}x$

92. $y^2 - \frac{8}{49} + \frac{2}{7}y$

93. $y^2 + 0.4y - 0.05$

94. $4x^{2a} - 4x^a - 3$

95. $x^{2a} + 5x^a - 24$

96. $x^2 + ax + bx + ab$

97. $bdx^2 + adx + bcx + ac$

98. $2ar^2 + 4asr + as^2 - asr$

99. $a^2p^{2a} + a^2p^a - 2a^2$

Aha! **100.** $(x + 3)^2 - 2(x + 3) - 35$

101. $6(x - 7)^2 + 13(x - 7) - 5$

102. Find all integers m for which $x^2 + mx + 75$ can be factored.

103. Find all integers q for which $x^2 + qx - 32$ can be factored.

104. One factor of $x^2 - 345x - 7300$ is $x + 20$. Find the other factor.

105. To better understand factoring $ax^2 + bx + c$ by grouping, suppose that

$$ax^2 + bx + c = (mx + r)(nx + s).$$

Show that if $P = ms$ and $Q = rn$, then $P + Q = b$ and $PQ = ac$.

106. Use the TABLE feature to check your answers to Exercises 11, 65, and 93.

107. Let $y_1 = 3x^2 + 10x - 8$, $y_2 = (x + 4)(3x - 2)$, and $y_3 = y_2 - y_1$ to check Example 9 graphically.

108. Explain how the following graph of

$$y = x^2 + 3x - 2 - (x - 2)(x + 1)$$

can be used to show that

$$x^2 + 3x - 2 \neq (x - 2)(x + 1).$$

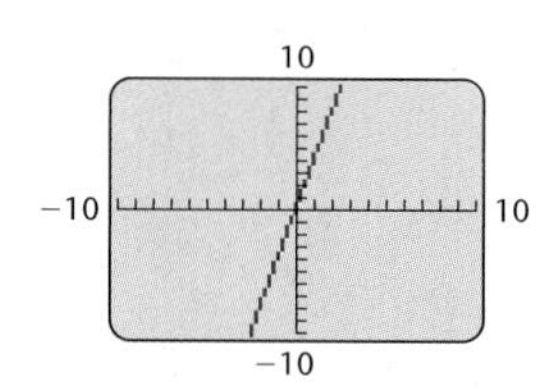

Factoring Perfect-Square Trinomials and Differences of Squares

5.5

Perfect-Square Trinomials • Differences of Squares • More Factoring by Grouping

We now introduce a faster way to factor trinomials that are squares of binomials. A method for factoring differences of squares is also developed.

Perfect-Square Trinomials

Consider the trinomial

$$x^2 + 6x + 9.$$

To factor it, we can proceed as in Section 5.4 and look for factors of 9 that add to 6. These factors are 3 and 3 and the factorization is

$$x^2 + 6x + 9 = (x + 3)(x + 3) = (x + 3)^2.$$

Note that the result is the square of a binomial. Because of this, we call $x^2 + 6x + 9$ a **perfect-square trinomial**. Although trial and error can be used to factor a perfect-square trinomial, a faster procedure can be used if we recognize when a trinomial is a perfect square.

To Recognize a Perfect-Square Trinomial

- Two of the terms must be squares, such as A^2 and B^2.
- There must be no minus sign before A^2 or B^2.
- The remaining term is twice the product of A and B, $2AB$, or its opposite, $-2AB$.

Example 1

Determine whether each polynomial is a perfect-square trinomial.

a) $x^2 + 10x + 25$
b) $4x + 16 + 3x^2$
c) $100y^2 + 81 - 180y$

Solution

a)
- Two of the terms in $x^2 + 10x + 25$ are squares: x^2 and 25.
- There is no minus sign before either x^2 or 25.
- The remaining term, $10x$, is twice the product of the square roots, x and 5.

Thus, $x^2 + 10x + 25$ *is* a perfect square.

b) In $4x + 16 + 3x^2$, only one term, 16, is a square ($3x^2$ is not a square because 3 is not a perfect-square integer and $4x$ is not a square because x is not a square).

Thus, $4x + 16 + 3x^2$ is not a perfect square.

c) It can help to first write the polynomial in descending order:

$$100y^2 - 180y + 81.$$

- Two of the terms, $100y^2$ and 81, are squares.
- There is no minus sign before either $100y^2$ or 81.
- If the product of the square roots, $10y$ and 9, is doubled, we get the opposite of the remaining term: $2(10y)(9) = 180y$ (the opposite of $-180y$).

Thus, $100y^2 + 81 - 180y$ *is* a perfect-square trinomial.

To factor a perfect-square trinomial, we reuse the patterns that we learned in Section 5.2.

Factoring a Perfect-Square Trinomial

$$A^2 + 2AB + B^2 = (A + B)^2;$$
$$A^2 - 2AB + B^2 = (A - B)^2$$

Example 2

Factor.

a) $x^2 - 10x + 25$
b) $16y^2 + 49 + 56y$
c) $-20xy + 4y^2 + 25x^2$

Solution

a) $x^2 - 10x + 25 = (x - 5)^2$ — We find the square terms and write the square roots with a minus sign between them.

Note the sign!

b) $16y^2 + 49 + 56y = 16y^2 + 56y + 49$ — Using a commutative law

$= (4y + 7)^2$ — We find the square terms and write the square roots with a plus sign between them.

c) $-20xy + 4y^2 + 25x^2 = 4y^2 - 20xy + 25x^2$ — Writing descending order with respect to y

$= (2y - 5x)^2$

This square can also be expressed as

$$25x^2 - 20xy + 4y^2 = (5x - 2y)^2.$$

As always, any factorization can be checked by multiplying:

$$(5x - 2y)^2 = (5x - 2y)(5x - 2y) = 25x^2 - 20xy + 4y^2.$$

When factoring, always look first for a factor common to all the terms.

Example 3 Factor: **(a)** $2x^2 - 12xy + 18y^2$; **(b)** $-4y^2 - 144y^8 + 48y^5$.

Solution

a) We first look for a common factor. This time, there is a common factor, 2.

$$2x^2 - 12xy + 18y^2 = 2(x^2 - 6xy + 9y^2) \quad \text{Factoring out the 2}$$
$$= 2(x - 3y)^2 \quad \text{Factoring the perfect-square trinomial}$$

b) $-4y^2 - 144y^8 + 48y^5 = -4y^2(1 + 36y^6 - 12y^3)$ Factoring out the common factor

$= -4y^2(36y^6 - 12y^3 + 1)$ Changing order. Note that $(y^3)^2 = y^6$.

$= -4y^2(6y^3 - 1)^2$ Factoring the perfect-square trinomial

Differences of Squares

When an expression like $x^2 - 9$ is recognized as a difference of two squares, we can reverse another pattern first seen in Section 5.2.

Factoring a Difference of Two Squares

$$A^2 - B^2 = (A + B)(A - B)$$

To factor a difference of two squares, write the product of the sum and the difference of the quantities being squared.

Example 4 Factor: **(a)** $x^2 - 9$; **(b)** $25y^6 - 49x^2$.

Solution

a) $x^2 - 9 = x^2 - 3^2 = (x + 3)(x - 3)$

$$A^2 - B^2 = (A + B)(A - B)$$

b) $25y^6 - 49x^2 = (5y^3)^2 - (7x)^2 = (5y^3 + 7x)(5y^3 - 7x)$

As always, the first step in factoring is to look for common factors.

Example 5 Factor: **(a)** $5 - 5x^2y^6$; **(b)** $16x^4y - 81y$.

Solution

a) $5 - 5x^2y^6 = 5(1 - x^2y^6)$ Factoring out the common factor

$= 5[1^2 - (xy^3)^2]$ Rewriting x^2y^6 as a quantity squared

$= 5(1 + xy^3)(1 - xy^3)$ Factoring the difference of squares

b) $16x^4y - 81y = y(16x^4 - 81)$ Factoring out the common factor

$= y[(4x^2)^2 - 9^2]$

$= y(4x^2 + 9)(4x^2 - 9)$ Factoring the difference of squares

$= y(4x^2 + 9)(2x + 3)(2x - 3)$ Factoring $4x^2 - 9$, which is *also* a difference of squares

In Example 5(b), it is tempting to try to factor $(4x^2 + 9)$. Note that it is a sum of two squares. Apart from possibly removing a common factor, it is impossible to factor a sum of squares using real numbers. Note also in Example 5(b) that $4x^2 - 9$ *could* be factored further. Whenever a factor itself can be factored, do so. We say that we have factored completely when none of the factors can be factored further.

More Factoring by Grouping

Sometimes, when factoring a polynomial with four terms, we may be able to factor further.

Example 6 Factor: $x^3 + 3x^2 - 4x - 12$.

Solution

$x^3 + 3x^2 - 4x - 12 = x^2(x + 3) - 4(x + 3)$ Factoring by grouping

$= (x + 3)(x^2 - 4)$ Factoring out $x + 3$

$= (x + 3)(x + 2)(x - 2)$ Factoring $x^2 - 4$

A difference of squares can have four or more terms. For example, one of the squares may be a trinomial. In this case, a type of grouping can be used.

Example 7 Factor: **(a)** $x^2 + 6x + 9 - y^2$; **(b)** $a^2 - b^2 + 8b - 16$.

Solution

a) $x^2 + 6x + 9 - y^2 = (x^2 + 6x + 9) - y^2$ Grouping as a perfect-square trinomial minus y^2 to show a difference of squares

$= (x + 3)^2 - y^2$

$= (x + 3 + y)(x + 3 - y)$

b) Grouping $a^2 - b^2 + 8b - 16$ into two groups of two terms does not yield a common binomial factor, so we look for a perfect-square trinomial. In this case, the perfect-square trinomial is being subtracted from a^2:

$$a^2 - b^2 + 8b - 16 = a^2 - (b^2 - 8b + 16)$$ Factoring out -1 and rewriting as subtraction

$$= a^2 - (b - 4)^2$$ Factoring the perfect-square trinomial

$$= (a + (b - 4))(a - (b - 4))$$ Factoring a difference of squares

$$= (a + b - 4)(a - b + 4)$$ Removing parentheses

Exercise Set 5.5

FOR EXTRA HELP

Digital Video Tutor CD 4 Videotape 10 | InterAct Math | Math Tutor Center | MathXL | MyMathLab.com

Factor completely.

1. $x^2 - 8x + 16$

2. $t^2 + 6t + 9$

3. $a^2 + 16a + 64$

4. $a^2 - 14a + 49$

5. $2a^2 + 8a + 8$

6. $4a^2 - 16a + 16$

7. $y^2 + 36 - 12y$

8. $y^2 + 36 + 12y$

9. $24a^2 + a^3 + 144a$

10. $-18y^2 + y^3 + 81y$

11. $32x^2 + 48x + 18$

12. $2x^2 - 40x + 200$

13. $64 + 25y^2 - 80y$

14. $1 - 8d + 16d^2$

15. $a^3 - 10a^2 + 25a$

16. $y^3 + 8y^2 + 16y$

17. $0.25x^2 + 0.30x + 0.09$

18. $0.04x^2 - 0.28x + 0.49$

19. $p^2 - 2pq + q^2$

20. $m^2 + 2mn + n^2$

21. $25a^2 + 30ab + 9b^2$

22. $49p^2 - 84pq + 36q^2$

23. $4t^2 - 8tr + 4r^2$

24. $5a^2 - 10ab + 5b^2$

25. $x^2 - 16$

26. $y^2 - 100$

27. $p^2 - 49$

28. $m^2 - 64$

29. $a^2b^2 - 81$

30. $p^2q^2 - 25$

31. $6x^2 - 6y^2$

32. $8x^2 - 8y^2$

33. $7xy^4 - 7xz^4$

34. $25ab^4 - 25az^4$

35. $4a^3 - 49a$

36. $9x^4 - 25x^2$

37. $3x^8 - 3y^8$

38. $9a^4 - a^2b^2$

39. $9a^4 - 25a^2b^4$

40. $16x^6 - 121x^2y^4$

41. $\frac{1}{25} - x^2$

42. $\frac{1}{16} - y^2$

43. $(a + b)^2 - 9$

44. $(p + q)^2 - 25$

45. $x^2 - 6x + 9 - y^2$

46. $a^2 - 8a + 16 - b^2$

47. $m^2 - 2mn + n^2 - 25$

48. $x^2 + 2xy + y^2 - 9$

49. $36 - (x + y)^2$

50. $49 - (a + b)^2$

51. $r^2 - 2r + 1 - 4s^2$

52. $c^2 + 4cd + 4d^2 - 9p^2$

Aha! **53.** $16 - a^2 - 2ab - b^2$

54. $9 - x^2 - 2xy - y^2$

55. $m^3 - 7m^2 - 4m + 28$

56. $x^3 + 8x^2 - x - 8$

57. $a^3 - ab^2 - 2a^2 + 2b^2$

58. $p^2q - 25q + 3p^2 - 75$

59. Are the product and power rules for exponents (see Section 1.6) important when factoring differences of squares? Why or why not?

60. Describe a procedure that could be used to find a polynomial with four terms that can be factored as a difference of two squares.

SKILL MAINTENANCE

Simplify.

61. $(2a^4b^5)^3$

62. $(5x^2y^4)^3$

63. $(x + y)^3$

64. $(a + 1)^3$

Solve.

65. $x - y + z = 6,$
$2x + y - z = 0,$
$x + 2y + z = 3$

66. $|5 - 7x| \geq 9$

67. $|5 - 7x| \leq 9$

68. $5 - 7x > -9 + 12x$

SYNTHESIS

69. Without finding the entire factorization, determine the number of factors of $x^{256} - 1$. Explain how you arrived at your answer.

70. Under what conditions can a sum of two squares be factored?

Factor completely. Assume that variables in exponents represent positive integers.

71. $-\frac{8}{27}r^2 - \frac{10}{9}rs - \frac{1}{6}s^2 + \frac{2}{3}rs$

72. $\frac{1}{36}x^8 + \frac{2}{9}x^4 + \frac{4}{9}$

73. $0.09x^8 + 0.48x^4 + 0.64$

74. $a^2 + 2ab + b^2 - c^2 + 6c - 9$

75. $r^2 - 8r - 25 - s^2 - 10s + 16$

76. $x^{2a} - y^2$

77. $x^{4a} - y^{2b}$

78. $4y^{4a} + 20y^{2a} + 20y^{2a} + 100$

79. $25y^{2a} - (x^{2b} - 2x^b + 1)$

80. $8(a - 3)^2 - 64(a - 3) + 128$

81. $3(x + 1)^2 + 12(x + 1) + 12$

82. $5c^{100} - 80d^{100}$

83. $9x^{2n} - 6x^n + 1$

84. $c^{2w+1} + 2c^{w+1} + c$

85. If $P(x) = x^2$, use factoring to simplify
$P(a + h) - P(a)$.

86. If $P(x) = x^4$, use factoring to simplify
$P(a + h) - P(a)$.

87. *Volume of carpeting.* The volume of a carpet that is rolled up can be estimated by the polynomial $\pi R^2h - \pi r^2h$.

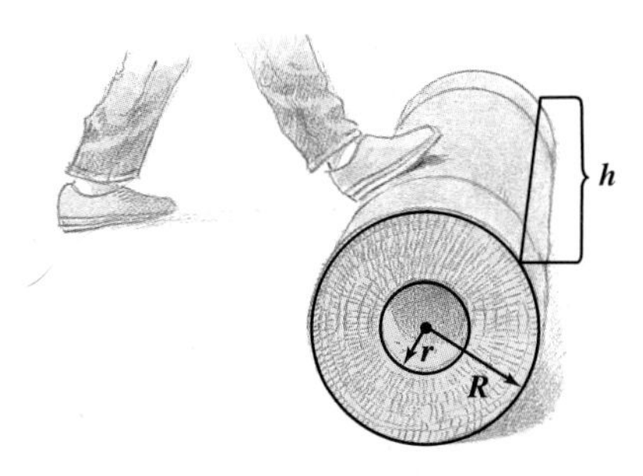

a) Factor the polynomial.

b) Use both the original and the factored forms to find the volume of a roll for which $R = 50$ cm, $r = 10$ cm, and $h = 4$ m. Use 3.14 for π.

88. Use a grapher to check your answers to Exercises 1, 35, and 55 graphically by examining $y_1 =$ the original polynomial, $y_2 =$ the factored polynomial, and $y_3 = y_2 - y_1$.

89. Check your answers to Exercises 1, 35, and 55 by using tables of values (see Exercise 88).

Factoring Sums or Differences of Cubes

5.6

Formulas for Factoring Sums or Differences of Cubes • Using the Formulas

Formulas for Factoring Sums or Differences of Cubes

We have seen that a difference of two squares can be factored but (unless a common factor exists) a *sum* of two squares cannot be factored. The situation is different with cubes: The difference *or sum* of two cubes can always be factored. To see this, consider the following products:

$$\begin{aligned}(A + B)(A^2 - AB + B^2) &= A(A^2 - AB + B^2) + B(A^2 - AB + B^2)\\ &= A^3 - A^2B + AB^2 + A^2B - AB^2 + B^3\\ &= A^3 + B^3 \qquad \text{Combining like terms}\end{aligned}$$

and

$$\begin{aligned}(A - B)(A^2 + AB + B^2) &= A(A^2 + AB + B^2) - B(A^2 + AB + B^2)\\ &= A^3 + A^2B + AB^2 - A^2B - AB^2 - B^3\\ &= A^3 - B^3. \qquad \text{Combining like terms}\end{aligned}$$

These products allow us to factor a sum or a difference of two cubes.

Factoring a Sum or Difference of Two Cubes

$$A^3 + B^3 = (A + B)(A^2 - AB + B^2);$$
$$A^3 - B^3 = (A - B)(A^2 + AB + B^2)$$

Using the Formulas

When factoring a sum or difference of cubes, it can be helpful to remember that $2^3 = 8$, $3^3 = 27$, $4^3 = 64$, $5^3 = 125$, $6^3 = 216$, and so on. We say that 2 is the *cube root* of 8, that 3 is the cube root of 27, and so on.

Example 1 Factor: $x^3 - 27$.

Solution We have

$$x^3 - 27 = x^3 - 3^3.$$

In one set of parentheses, we write the first cube root, x, minus the second cube root, 3:

$$(x - 3)(\qquad\qquad).$$

To get the other factor, we think of $x - 3$ and do the following:

Square the first term: x^2.
Multiply the terms and then change the sign: $3x$.
Square the second term: $(-3)^2$, or 9.

$$(x - 3)(x^2 + 3x + 9).$$

Thus, $x^3 - 27 = (x - 3)(x^2 + 3x + 9)$.

In Example 1, note that $x^2 + 3x + 9$ cannot be factored further. In general, unless A and B share a common factor, the trinomials $A^2 + AB + B^2$ and $A^2 - AB + B^2$ are prime.

Example 2 Factor.

a) $125x^3 + y^3$
b) $m^6 + 64$
c) $128y^7 - 250x^6y$
d) $r^6 - s^6$

Solution

a) We have

$$125x^3 + y^3 = (5x)^3 + y^3.$$

In one set of parentheses, we write the cube root of the first term, $5x$, plus the cube root of the second term, y:

$$(5x + y)(\qquad\qquad).$$

To get the other factor, we think of $5x + y$ and do the following:

Square the first term: $(5x)^2$, or $25x^2$.
Multiply the terms and then change the sign: $-5xy$.
Square the second term: y^2.

$$(5x + y)(25x^2 - 5xy + y^2).$$

Thus, $125x^3 + y^3 = (5x + y)(25x^2 - 5xy + y^2)$.

b) We have

$$m^6 + 64 = (m^2)^3 + 4^3. \qquad \text{Rewriting as quantities cubed}$$

Next, we reuse the pattern used in part (a) above:

$$A^3 + B^3 = (A + B)(A^2 - A \cdot B + B^2)$$

$$\begin{aligned}(m^2)^3 + 4^3 &= (m^2 + 4)((m^2)^2 - m^2 \cdot 4 + 4^2)\\ &= (m^2 + 4)(m^4 - 4m^2 + 16).\end{aligned}$$

c) We have

$$\begin{aligned}128y^7 - 250x^6y &= 2y(64y^6 - 125x^6) && \text{Remember: Always look for a common factor.}\\ &= 2y[(4y^2)^3 - (5x^2)^3] && \text{Rewriting as quantities cubed}\end{aligned}$$

To factor $(4y^2)^3 - (5x^2)^3$, it is essential to remember the pattern used in Example 1:

$$A^3 - B^3 = (A - B)(A^2 + A \cdot B + B^2)$$
$$\downarrow \quad \downarrow$$
$$\begin{aligned}(4y^2)^3 - (5x^2)^3 &= (4y^2 - 5x^2)((4y^2)^2 + 4y^2 \cdot 5x^2 + (5x^2)^2) \\ &= (4y^2 - 5x^2)(16y^4 + 20x^2y^2 + 25x^4).\end{aligned}$$

Thus,

$$128y^7 - 250x^6y = 2y(4y^2 - 5x^2)(16y^4 + 20x^2y^2 + 25x^4).$$

d) We have

$$\begin{aligned}r^6 - s^6 &= (r^3)^2 - (s^3)^2 \\ &= (r^3 + s^3)(r^3 - s^3) && \text{Factoring a difference of two } \textit{squares} \\ &= (r + s)(r^2 - rs + s^2)(r - s)(r^2 + rs + s^2). && \text{Factoring the sum and difference of two cubes}\end{aligned}$$

In Example 2(d), suppose we first factored $r^6 - s^6$ as a difference of two cubes:

$$\begin{aligned}(r^2)^3 - (s^2)^3 &= (r^2 - s^2)(r^4 + r^2s^2 + s^4) \\ &= (r + s)(r - s)(r^4 + r^2s^2 + s^4).\end{aligned}$$

In this case, we might have missed some factors; $r^4 + r^2s^2 + s^4$ can be factored as $(r^2 - rs + s^2)(r^2 + rs + s^2)$, but we probably would never have suspected that such a factorization exists. Given a choice, it is generally better to factor as a difference of squares before factoring as a sum or difference of cubes.

Try to remember the following:

Useful Factoring Facts

Sum of cubes: $A^3 + B^3 = (A + B)(A^2 - AB + B^2)$;

Difference of cubes: $A^3 - B^3 = (A - B)(A^2 + AB + B^2)$;

Difference of squares: $A^2 - B^2 = (A + B)(A - B)$;

Sum of squares: $A^2 + B^2$ cannot be factored using real numbers if the largest common factor has been removed.

Exercise Set 5.6

FOR EXTRA HELP

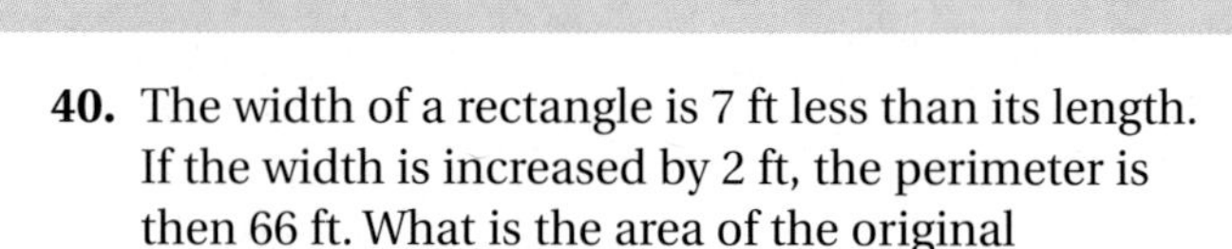

Digital Video Tutor CD 4 Videotape 10 | InterAct Math | Math Tutor Center | MathXL | MyMathLab.com

Factor completely.

1. $t^3 + 27$ **2.** $x^3 + 64$

3. $x^3 - 8$ **4.** $z^3 - 1$

5. $m^3 - 64$ **6.** $x^3 - 27$

7. $8a^3 + 1$ **8.** $27x^3 + 1$

9. $27 - 8t^3$ **10.** $64 - 125x^3$

11. $8x^3 + 27$ **12.** $27y^3 + 64$

13. $y^3 - z^3$ **14.** $x^3 - y^3$

15. $x^3 + \frac{1}{27}$ **16.** $a^3 + \frac{1}{8}$

17. $2y^3 - 128$ **18.** $8t^3 - 8$

19. $8a^3 + 1000$ **20.** $54x^3 + 2$

21. $rs^3 + 64r$ **22.** $ab^3 + 125a$

23. $2y^3 - 54z^3$ **24.** $5x^3 - 40z^3$

25. $y^3 + 0.125$ **26.** $x^3 + 0.001$

27. $125c^6 - 8d^6$ **28.** $64x^6 - 8t^6$

29. $3z^5 - 3z^2$ **30.** $2y^4 - 128y$

31. $t^6 + 1$ **32.** $z^6 - 1$

33. $p^6 - q^6$ **34.** $t^6 + 64y^6$

35. $a^9 + b^{12}c^{15}$ **36.** $x^{12} - y^3z^{12}$

37. How could you use factoring to convince someone that $x^3 + y^3 \neq (x + y)^3$?

38. Is the following statement true or false and why? If A^3 and B^3 have a common factor, then A and B have a common factor.

SKILL MAINTENANCE

39. *Height of a baseball.* A baseball is thrown upward with an initial velocity of 80 ft/sec from a 224-ft-high cliff. Its height in feet, $h(t)$, after t seconds is given by

$$h(t) = -16t^2 + 80t + 224.$$

What is the height of the ball after 0 sec, 1 sec, 3 sec, 4 sec, and 6 sec?

40. The width of a rectangle is 7 ft less than its length. If the width is increased by 2 ft, the perimeter is then 66 ft. What is the area of the original rectangle?

41. If $f(x) = 7 - x^2$, find $f(-3)$.

42. Find the slope and the y-intercept of the line given by $4x - 3y = 8$.

Solve.

43. $3x - 5 = 0$

44. $2x + 7 = 0$

SYNTHESIS

45. Explain how the geometric model below can be used to verify the formula for factoring $a^3 - b^3$.

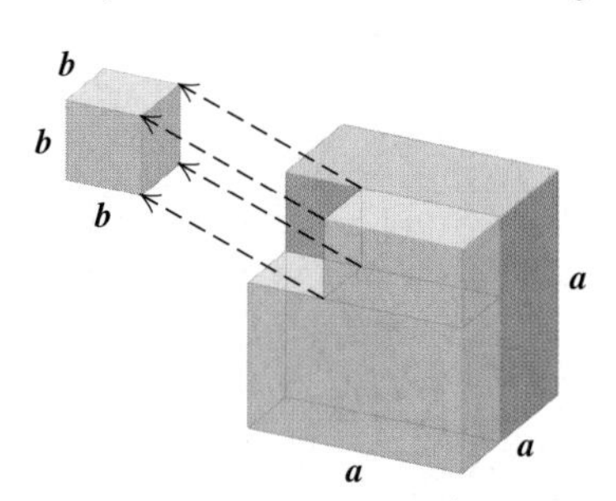

46. Explain how the formula for factoring a *difference* of two cubes can be used to factor $x^3 + 8$.

Factor.

47. $x^{6a} - y^{3b}$ **48.** $2x^{3a} + 16y^{3b}$

Aha! **49.** $(x + 5)^3 + (x - 5)^3$ **50.** $\frac{1}{16}x^{3a} + \frac{1}{2}y^{6a}z^{9b}$

51. $5x^3y^6 - \frac{5}{8}$ **52.** $x^3 - (x + y)^3$

53. $x^{6a} - (x^{2a} + 1)^3$ **54.** $(x^{2a} - 1)^3 - x^{6a}$

55. $t^4 - 8t^3 - t + 8$

56. If $P(x) = x^3$, use factoring to simplify

$$P(a + h) - P(a).$$

57. If $Q(x) = x^6$, use factoring to simplify

$$Q(a + h) - Q(a).$$

58. Using one set of axes, graph the following.

a) $f(x) = x^3$

b) $g(x) = x^3 - 8$

c) $h(x) = (x - 2)^3$

59. Use a grapher to check Example 1: Let $y_1 = x^3 - 27$, $y_2 = (x - 3)(x^2 + 3x + 9)$, and $y_3 = y_1 - y_2$.

60. Use the approach of Exercise 59 to check your answers to Exercises 7, 17, and 29.

Factoring: A General Strategy

5.7

Mixed Factoring Problems

Study Tip

Don't try studying when you are tired. A good night's sleep or a 10-minute "power nap" can often make a problem suddenly seem much easier to solve.

Factoring is an important algebraic skill. Once you recognize the kind of expression you have, factoring can be done without too much difficulty.

A Strategy for Factoring

A. Always factor out the largest common factor.

B. Once any common factor has been factored out, look at the number of terms.

Two terms: Try factoring as a difference of squares first. Next, try factoring as a sum or a difference of cubes. Do *not* try to factor a *sum* of squares.

Three terms: Try factoring as a perfect-square trinomial. Next, try trial and error, using the FOIL method or the grouping method.

Four or more terms: Try factoring by grouping and factoring out a common binomial factor. Next, try grouping into a difference of squares, one of which is a trinomial.

C. Always *factor completely*. If a factor with more than one term can itself be factored further, do so.

D. *Check* the factorization by multiplying.

Example 1 Factor: $10a^2x - 40b^2x$.

Solution

A. Look first for a common factor:

$$10a^2x - 40b^2x = 10x(a^2 - 4b^2).$$ Factoring out the largest common factor

B. The factor $a^2 - 4b^2$ has two terms. It is a difference of squares. We factor it, keeping the common factor:

$$10a^2x - 40b^2x = 10x(a + 2b)(a - 2b).$$

C. Have we factored completely? Yes, because no factor with more than one term can be factored further.
D. *Check:* $10x(a + 2b)(a - 2b) = 10x(a^2 - 4b^2) = 10a^2x - 40b^2x$.

Example 2

Factor: $x^6 - 64$.

Solution

A. Look for a common factor. There is none (other than 1 or -1).
B. There are two terms, a difference of squares: $(x^3)^2 - (8)^2$. We factor it:

$$x^6 - 64 = (x^3 + 8)(x^3 - 8). \qquad \text{Note that } x^6 = (x^3)^2.$$

C. One factor is a sum of two cubes, and the other factor is a difference of two cubes. We factor both:

$$x^6 - 64 = (x + 2)(x^2 - 2x + 4)(x - 2)(x^2 + 2x + 4).$$

The factorization is complete because no factor can be factored further.
D. The check is left to the student.

Example 3

Factor: $7x^6 + 35y^2$.

Solution

A. Factor out the largest common factor:

$$7x^6 + 35y^2 = 7(x^6 + 5y^2).$$

B. In the parentheses, the binomial that cannot be factored.
C. We cannot factor further.
D. *Check:* $7(x^6 + 5y^2) = 7x^6 + 35y^2$.

Example 4

Factor: $2x^2 + 50a^2 - 20ax$.

Solution

A. Factor out the largest common factor: $2(x^2 + 25a^2 - 10ax)$.
B. Next, we rearrange the trinomial in descending powers of x: $2(x^2 - 10ax + 25a^2)$. The trinomial is a perfect-square trinomial:

$$2x^2 + 50a^2 - 20ax = 2(x - 5a)^2.$$

C. No factor with more than one term can be factored further. Had we used descending powers of a, we would discover an equivalent factorization, $2(5a - x)^2$.
D. *Check:* $2(x - 5a)^2 = 2(x^2 - 10ax + 25a^2) = 2x^2 - 20ax + 50a^2 = 2x^2 + 50a^2 - 20ax$.

Example 5

Factor: $12x^2 - 40x - 32$.

Solution

A. Factor out the largest common factor: $4(3x^2 - 10x - 8)$.
B. The trinomial factor is not a square. We factor using trial and error:

$$12x^2 - 40x - 32 = 4(x - 4)(3x + 2).$$

C. We cannot factor further.
D. *Check:* $4(x - 4)(3x + 2) = 4(3x^2 + 2x - 12x - 8) = 4(3x^2 - 10x - 8) = 12x^2 - 40x - 32$.

Example 6

Factor: $3x + 12 + ax^2 + 4ax$.

Solution

A. There is no common factor (other than 1 or -1).
B. There are four terms. We try grouping to find a common binomial factor:

$$3x + 12 + ax^2 + 4ax = 3(x + 4) + ax(x + 4)$$ Factoring two grouped binomials

$$= (x + 4)(3 + ax).$$ Removing the common binomial factor

C. We cannot factor further.
D. *Check:* $(x + 4)(3 + ax) = 3x + ax^2 + 12 + 4ax = 3x + 12 + ax^2 + 4ax$.

Example 7

Factor: $y^2 - 9a^2 + 12y + 36$.

Solution

A. There is no common factor (other than 1 or -1).
B. There are four terms. We try grouping to remove a common binomial factor, but find none. Next, we try grouping as a difference of squares:

$$(y^2 + 12y + 36) - 9a^2$$ Grouping

$$= (y + 6)^2 - (3a)^2$$ Rewriting as a difference of squares

$$= (y + 6 + 3a)(y + 6 - 3a).$$ Factoring the difference of squares

C. No factor with more than one term can be factored further.
D. The check is left to the student.

Example 8

Factor: $x^3 - xy^2 + x^2y - y^3$.

Solution

A. There is no common factor (other than 1 or -1).
B. There are four terms. We try grouping to remove a common binomial factor:

$$x^3 - xy^2 + x^2y - y^3$$

$$= x(x^2 - y^2) + y(x^2 - y^2)$$ Factoring two grouped binomials

$$= (x^2 - y^2)(x + y).$$ Removing the common binomial factor

C. The factor $x^2 - y^2$ can be factored further:

$$x^3 - xy^2 + x^2y - y^3 = (x + y)(x - y)(x + y), \text{ or } (x + y)^2(x - y).$$

No factor can be factored further, so we have factored completely.

D. The check is left to the student.

Exercise Set 5.7

FOR EXTRA HELP

Digital Video Tutor CD 4
Videotape 10

InterAct Math

Math Tutor Center

MathXL

MyMathLab.com

Factor completely.

1. $5m^4 - 20$

2. $x^2 - 144$

3. $a^2 - 81$

4. $2a^2 - 11a + 12$

5. $8x^2 + 18x - 5$

6. $2xy^2 - 50x$

7. $a^2 + 25 - 10a$

8. $p^2 + 64 + 16p$

9. $3x^2 + 15x - 252$

10. $2y^2 + 10y - 132$

11. $9x^2 - 25y^2$

12. $16a^2 - 81b^2$

13. $t^6 + 1$

14. $64t^6 - 1$

15. $x^2 + 6x - y^2 + 9$

16. $t^2 + 10t - p^2 + 25$

17. $343x^3 + 27y^3$

18. $128a^3 + 250b^3$

19. $8m^3 + m^6 - 20$

20. $-37x^2 + x^4 + 36$

21. $ac + cd - ab - bd$

22. $xw - yw + xz - yz$

23. $4c^2 - 4cd + d^2$

24. $70b^2 - 3ab - a^2$

25. $24 + 9t^2 + 8t + 3t^3$

26. $4a - 14 + 2a^3 - 7a^2$

27. $2x^3 + 6x^2 - 8x - 24$

28. $3x^3 + 6x^2 - 27x - 54$

29. $54a^3 - 16b^3$

30. $54x^3 - 250y^3$

31. $36y^2 - 35 + 12y$

32. $2b - 28a^2b + 10ab$

33. $a^8 - b^8$

34. $2x^4 - 32$

35. $a^3b - 16ab^3$

36. $x^3y - 25xy^3$

Aha! **37.** $(a - 3)(a + 7) + (a - 3)(a - 1)$

38. $x^2(x + 3) - 4(x + 3)$

39. $7a^4 - 14a^3 + 21a^2 - 7a$

40. $a^3 - ab^2 + a^2b - b^3$

41. $42ab + 27a^2b^2 + 8$

42. $-23xy + 20x^2y^2 + 6$

43. $p - 64p^4$

44. $125a - 8a^4$

Aha! **45.** $a^2 - b^2 - 6b - 9$

46. $m^2 - n^2 - 8n - 16$

47. Emily has factored a polynomial as $(a - b)(x - y)$, while Jorge has factored the same polynomial as $(b - a)(y - x)$. Can they both be correct? Why or why not?

48. In your own words, outline a procedure that can be used to factor any polynomial.

SKILL MAINTENANCE

Solve.

49. $5x - 9 = 0$

50. $7x + 13 = 0$

Graph.

51. $g(x) = 3x - 7$

52. $f(x) = -2x + 8$

53. *Exam scores.* There are 75 questions on a college entrance examination. Two points are awarded for each correct answer, and one half point is deducted for each incorrect answer. Ralph scored 100 on the exam. How many correct and how many incorrect answers did Ralph have if all questions were answered?

54. *Perimeter.* A pentagon with all five sides the same size has the same perimeter as an octagon in which all eight sides are the same size. One side of the pentagon is 2 less than three times the length of one side of the octagon. Find the perimeters.

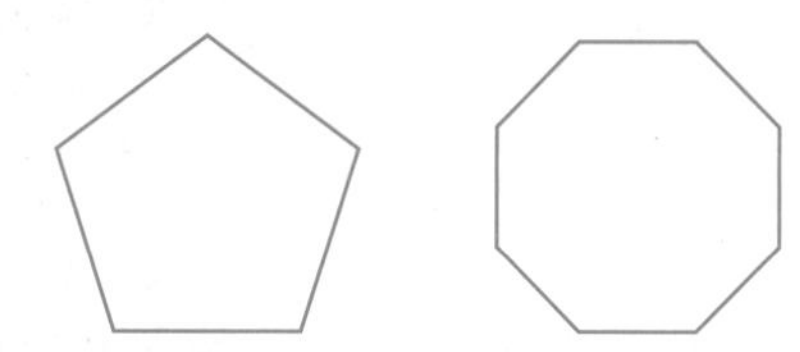

SYNTHESIS

55. Explain how one could construct a polynomial that is a difference of squares that contains a sum of two cubes and a difference of two cubes as factors.

56. Explain how one could construct a polynomial with four terms that can be factored by grouping three terms together.

Factor completely.

57. $60x^2 - 97xy^2 + 30y^4$

58. $28a^3 - 25a^2bc + 3ab^2c^2$

59. $-16 + 17(5 - y^2) - (5 - y^2)^2$

Aha! 60. $(x - p)^2 - p^2$

61. $a^4 - 50a^2b^2 + 49b^4$

62. $(y - 1)^4 - (y - 1)^2$

63. $27x^{6s} + 64y^{3t}$

64. $x^6 - 2x^5 + x^4 - x^2 + 2x - 1$

65. $4x^2 + 4xy + y^2 - r^2 + 6rs - 9s^2$

66. $(1 - x)^3 - (x - 1)^6$

67. $24t^{2a} - 6$

68. $a^{2w+1} + 2a^{w+1} + a$

69. $\dfrac{x^{27}}{1000} - 1$

70. $a - by^8 + b - ay^8$

71. $3(x + 1)^2 - 9(x + 1) - 12$

72. $3a^2 + 3b^2 - 3c^2 - 3d^2 + 6ab - 6cd$

73. $3(a + 2)^2 + 30(a + 2) + 75$

74. $(m - 1)^3 - (m + 1)^3$

75. If $\left(x + \dfrac{2}{x}\right)^2 = 6$, find $x^3 + \dfrac{8}{x^3}$.

Applications of Polynomial Equations

5.8

The Principle of Zero Products • Polynomial Functions and Graphs • Problem Solving

Whenever two polynomials are set equal to each other, we have a **polynomial equation**. Some examples of polynomial equations are

$$4x^3 + x^2 + 5x = 6x - 3,$$
$$x^2 - x = 6,$$

and

$$3y^4 + 2y^2 + 2 = 0.$$

The *degree of a polynomial equation* is the same as the highest degree of any term in the equation. Thus, from top to bottom, the degree of each equation listed above is 3, 2, and 4. A second-degree polynomial equation in one variable is usually called a **quadratic equation**. Of the equations listed above, only $x^2 - x = 6$ is a quadratic equation.

Polynomial equations, and quadratic equations in particular, occur frequently in applications, so the ability to solve them is an important skill. One way of solving certain polynomial equations involves factoring.

The Principle of Zero Products

When we multiply two or more numbers, the product is 0 if any one of those numbers (factors) is 0. Conversely, if a product is 0, then at least one of the factors must be 0. This property of 0 gives us a new principle for solving equations.

> ***The Principle of Zero Products***
>
> For any real numbers a and b:
>
> If $ab = 0$, then $a = 0$ or $b = 0$. If $a = 0$ or $b = 0$, then $ab = 0$.

To solve an equation using the principle of zero products, we first write it in *standard form*: with 0 on one side of the equation and the leading coefficient positive.

Example 1 Solve: $x^2 - x = 6$.

Solution To apply the principle of zero products, we need 0 on one side of the equation. Thus we subtract 6 from both sides:

$$x^2 - x - 6 = 0. \qquad \text{Getting 0 on one side}$$

In order to express the polynomial as a product, we factor:

$$(x - 3)(x + 2) = 0. \qquad \text{Factoring}$$

Since $(x - 3)(x + 2)$ is 0, the principle of zero products says that at least one factor is 0. Thus,

$$x - 3 = 0 \quad \text{or} \quad x + 2 = 0. \qquad \text{Using the principle of zero products}$$

Each of these linear equations is then solved separately:

$$x = 3 \quad \text{or} \quad x = -2.$$

We check as follows:

Check:

$x^2 - x$	$= 6$
$3^2 - 3$	? 6
$9 - 3$	
6	6 TRUE

$x^2 - x$	$= 6$
$(-2)^2 - (-2)$	? 6
$4 + 2$	
6	6 TRUE

Both 3 and -2 are solutions. The solution set is $\{3, -2\}$.

> ***To Use the Principle of Zero Products***
>
> 1. Obtain a 0 on one side of the equation using the addition principle.
> 2. Factor the nonzero side of the equation.
> 3. Set each factor that is not a constant equal to 0.
> 4. Solve the resulting equations.

Caution! When using the principle of zero products, we must make sure that there is a 0 on one side of the equation. If neither side of the equation is 0, the procedure will not work.

To see this, consider $x^2 - x = 6$ in Example 1 as

$$x(x - 1) = 6.$$

Knowing that the product of two numbers is 6 tells us nothing about either number. The numbers may be $2 \cdot 3$ or $6 \cdot 1$ or $12 \cdot \frac{1}{2}$ or $-\frac{3}{5} \cdot (-10)$ and so on.

Suppose we *incorrectly* set each factor equal to 6:

$$x = 6 \quad or \quad x - 1 = 6 \longleftarrow \text{This is wrong!}$$
$$x = 7.$$

Neither 6 nor 7 checks, as shown below:

$$\begin{array}{r|l} x^2 - x & = 6 \\ \hline 6^2 - 6 & ?\ 6 \\ 36 - 6 & \\ 30 & 6 \quad \text{FALSE} \end{array} \qquad \begin{array}{r|l} x^2 - x & = 6 \\ \hline 7^2 - 7 & ?\ 6 \\ 49 - 7 & \\ 42 & 6 \quad \text{FALSE} \end{array}$$

Example 2 Solve.

a) $5b^2 = 10b$
b) $x^2 - 6x + 9 = 0$
c) $3x^3 - 30x = 9x^2$

Solution

a) We have

$$5b^2 = 10b$$
$$5b^2 - 10b = 0 \qquad \text{Getting 0 on one side}$$
$$5b(b - 2) = 0 \qquad \text{Factoring}$$
$$5b = 0 \quad or \quad b - 2 = 0 \qquad \text{Using the principle of zero products}$$
$$b = 0 \quad or \quad b = 2. \qquad \text{The checks are left to the student.}$$

The solutions are 0 and 2. The solution set is $\{0, 2\}$.

b) We have

$$x^2 - 6x + 9 = 0$$
$$(x - 3)(x - 3) = 0 \qquad \text{Factoring}$$
$$x - 3 = 0 \quad or \quad x - 3 = 0 \qquad \text{Using the principle of zero products}$$
$$x = 3 \quad or \quad x = 3. \qquad \textit{Check:}\ 3^2 - 6 \cdot 3 + 9 = 9 - 18 + 9 = 0.$$

There is only one solution, 3. The solution set is $\{3\}$.

c) We have

$$3x^3 - 30x = 9x^2$$

$$3x^3 - 9x^2 - 30x = 0 \quad \text{Getting 0 on one side and writing in descending order}$$

$$3x(x^2 - 3x - 10) = 0 \quad \text{Factoring out a common factor}$$

$$3x(x + 2)(x - 5) = 0 \quad \text{Factoring the trinomial}$$

$$3x = 0 \quad or \quad x + 2 = 0 \quad or \quad x - 5 = 0 \quad \text{Using the principle of zero products}$$

$$x = 0 \quad or \quad x = -2 \quad or \quad x = 5.$$

Check:

$3x^3 - 30x$	$= 9x^2$	
$3 \cdot 0^3 - 30 \cdot 0$?	$9 \cdot 0^2$	
$0 - 0$	$9 \cdot 0$	
0	0	TRUE

$3x^3 - 30x$	$= 9x^2$	
$3(-2)^3 - 30(-2)$?	$9(-2)^2$	
$3(-8) + 60$	$9 \cdot 4$	
$-24 + 60$	36	
36	36	TRUE

$3x^3 - 30x$	$= 9x^2$	
$3 \cdot 5^3 - 30 \cdot 5$?	$9 \cdot 5^2$	
$3 \cdot 125 - 150$	$9 \cdot 25$	
$375 - 150$	225	
225	225	TRUE

The solutions are 0, -2, and 5. The solution set is $\{0, -2, 5\}$.

Example 3 Given that $f(x) = 3x^2 - 4x$, find all values of a for which $f(a) = 4$.

Solution We want all numbers a for which $f(a) = 4$. Since $f(a) = 3a^2 - 4a$, we must have

$$3a^2 - 4a = 4 \quad \text{Setting } f(a) \text{ equal to 4}$$

$$3a^2 - 4a - 4 = 0 \quad \text{Getting 0 on one side}$$

$$(3a + 2)(a - 2) = 0 \quad \text{Factoring}$$

$$3a + 2 = 0 \quad or \quad a - 2 = 0$$

$$a = -\tfrac{2}{3} \quad or \quad a = 2.$$

Check: $f(-\frac{2}{3}) = 3(-\frac{2}{3})^2 - 4(-\frac{2}{3}) = 3 \cdot \frac{4}{9} + \frac{8}{3} = \frac{4}{3} + \frac{8}{3} = \frac{12}{3} = 4.$
$f(2) = 3(2)^2 - 4(2) = 3 \cdot 4 - 8 = 12 - 8 = 4.$

To have $f(a) = 4$, we must have $a = -\frac{2}{3}$ or $a = 2$.

Example 4 Find the domain of F if $F(x) = \dfrac{x - 2}{x^2 + 2x - 15}$.

Solution The domain of F is the set of all values for which $F(x)$ is a real number. Since division by 0 is undefined, $F(x)$ cannot be calculated for any x-value for which the denominator, $x^2 + 2x - 15$, is 0. To make sure these values are *excluded*, we solve:

$$x^2 + 2x - 15 = 0 \quad \text{Setting the denominator equal to 0}$$
$$(x - 3)(x + 5) = 0 \quad \text{Factoring}$$
$$x - 3 = 0 \quad or \quad x + 5 = 0$$
$$x = 3 \quad or \quad x = -5. \quad \text{These are the values to } exclude.$$

The domain of F is $\{x \mid x \text{ is a real number } and\ x \neq -5 \text{ and } x \neq 3\}$.

technology connection

To use the INTERSECT option (see p. 113 in Section 2.4) to check Example 1, let $y_1 = x^2 - x$ and $y_2 = 6$. One intersection occurs at $(-2, 6)$. You should confirm that the other intersection occurs at $(3, 6)$.

$y_1 = x^2 - x,\ y_2 = 6$

10, −10, 10, −10

Intersection
X = −2, Y = 6

Another approach is to find where the graph of $y_1 = x^2 - x - 6$ crosses the x-axis. Using the ZERO option of the CALC menu, enter a number as a LEFT BOUND to the left of an x-intercept, a number as a RIGHT BOUND to the right of that x-intercept, and a GUESS in between the two bounds. The ZERO, or root, occurs at $x = -2$.

$y_1 = x^2 - x - 6$

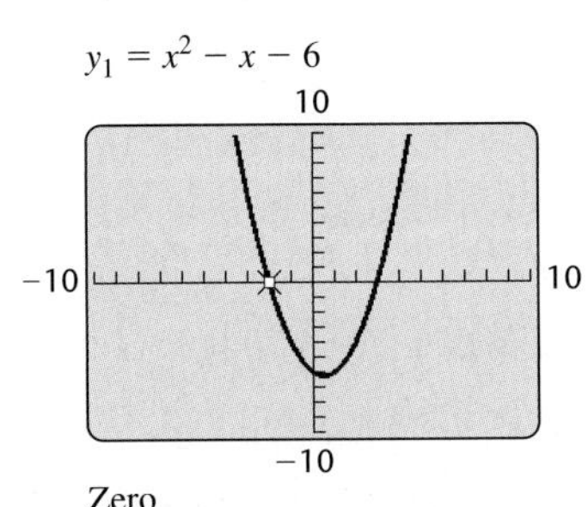

Zero
X = −2, Y = 0

Polynomial Functions and Graphs

Let's return for a moment to the equation in Example 1, $x^2 - x = 6$. One way to begin solving this equation is to either graph by hand or use computer graphics or a graphing calculator to draw the graph of the function given by $f(x) = x^2 - x$. We then check visually for an x-value that is paired with 6, as shown on the left below.

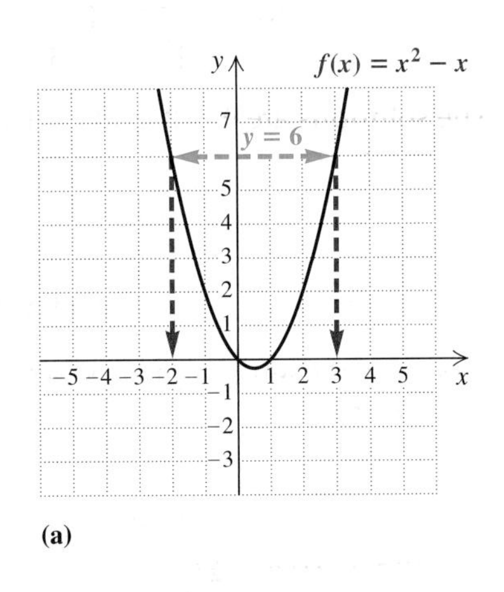

(a)

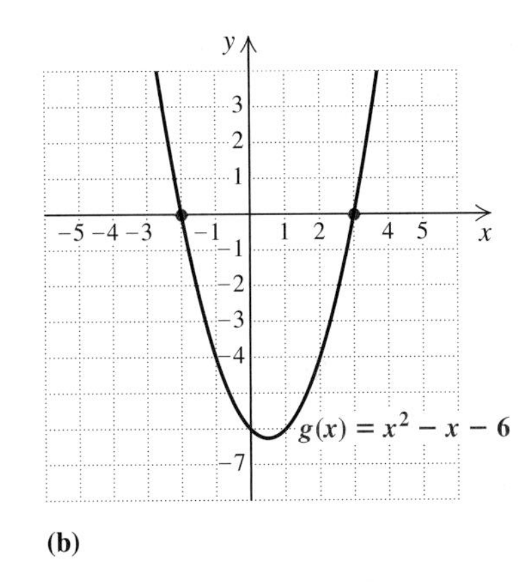

(b)

Equivalently, we could graph the function given by $g(x) = x^2 - x - 6$ and look for values of x for which $g(x) = 0$. See figure (b) above. Here you can visualize what we call the *roots*, or *zeros*, of a polynomial function.

It appears from the graph that $f(x) = 6$ and $g(x) = 0$ when $x \approx -2$ or $x \approx 3$. Although making a graph is not the fastest or most precise method of solving a polynomial equation, it gives us a visualization and is useful with problems that are more difficult to solve algebraically.

Problem Solving

Some problems can be translated to quadratic equations, which we can now solve. The problem-solving process is the same as for other kinds of problems.

Example 5

Prize tee shirts. During intermission at sporting events, it has become common for team mascots to use a powerful slingshot to launch tightly rolled tee shirts into the stands. The height $h(t)$, in feet, of an airborne tee shirt t seconds after being launched can be approximated by

$$h(t) = -15t^2 + 75t + 10.$$

After peaking, a rolled-up tee shirt is caught by a fan 70 ft above ground level. How long was the tee shirt in the air?

Solution

1. **Familiarize.** We make a drawing and label it, using the information provided (see the figure). If we wanted to, we could evaluate $h(t)$ for a few values of t. Note that t cannot be negative, since it represents time from launch.

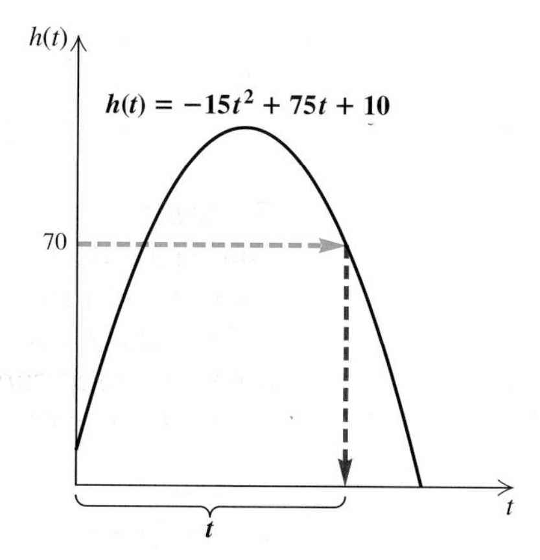

2. **Translate.** The relevant function has been provided. Since we are asked to determine how long it will take for the shirt to reach someone 70 ft above ground level, we are interested in the value of t for which $h(t) = 70$:

$$-15t^2 + 75t + 10 = 70.$$

3. **Carry out.** We solve the quadratic equation:

$$-15t^2 + 75t + 10 = 70$$

$$-15t^2 + 75t - 60 = 0 \quad \text{Subtracting 70 from both sides}$$

$$\left.\begin{aligned} -15(t^2 - 5t + 4) &= 0 \\ -15(t - 4)(t - 1) &= 0 \end{aligned}\right\} \quad \text{Factoring}$$

$$t - 4 = 0 \quad or \quad t - 1 = 0$$

$$t = 4 \quad or \quad t = 1.$$

The solutions appear to be 4 and 1.

4. **Check.** We have

$$h(4) = -15 \cdot 4^2 + 75 \cdot 4 + 10 = -240 + 300 + 10 = 70 \text{ ft};$$

$$h(1) = -15 \cdot 1^2 + 75 \cdot 1 + 10 = -15 + 75 + 10 = 70 \text{ ft}.$$

Both 4 and 1 check. However, the problem states that the tee shirt is caught after peaking. Thus we reject 1 since that would indicate when the height of the tee shirt was 70 ft on the way *up*.

5. **State.** The tee shirt was in the air for 4 sec before being caught 70 ft above ground level.

The following problem involves the **Pythagorean theorem**, which relates the lengths of the sides of a right triangle. A **right triangle** has a 90°, or right, angle, which is denoted in the triangle by the symbol ⌜ or ⌝. The longest side, opposite the 90° angle, is called the **hypotenuse**. The other sides, called **legs**, form the two sides of the right angle.

The Pythagorean Theorem

In any right triangle, if a and b are the lengths of the legs and c is the length of the hypotenuse, then

$$a^2 + b^2 = c^2.$$

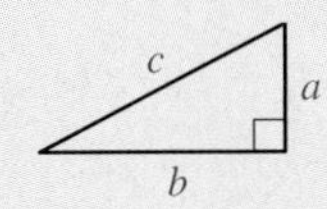

The symbol ⌜ denotes a 90° angle.

Example 6

Carpentry. In order to build a deck at a right angle to their house, Lucinda and Felipe decide to plant a stake in the ground a precise distance from the back wall of their house. This stake will combine with two marks on the house to form a right triangle. From a course in geometry, Lucinda remembers that there are three consecutive integers that can work as sides of a right triangle. Find the measurements of that triangle.

Solution

1. **Familiarize.** Recall that x, $x + 1$, and $x + 2$ can be used to represent three unknown consecutive integers. Since $x + 2$ is the largest number, it must represent the hypotenuse. The legs serve as the sides of the right angle, so one leg must be formed by the marks on the house. We make a drawing in which

$$x = \text{the distance between the marks on the house,}$$

$$x + 1 = \text{the length of the other leg,}$$

and

$$x + 2 = \text{the length of the hypotenuse.}$$

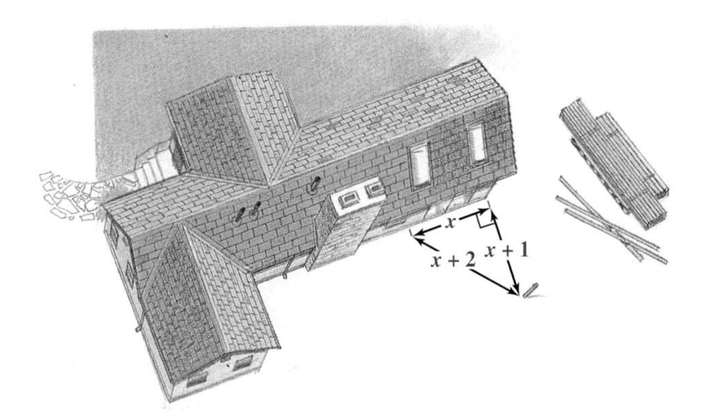

2. **Translate.** Applying the Pythagorean theorem, we translate as follows:

$$a^2 + b^2 = c^2$$
$$x^2 + (x + 1)^2 = (x + 2)^2.$$

3. **Carry out.** We solve the equation as follows:

$x^2 + (x^2 + 2x + 1) = x^2 + 4x + 4$	Squaring the binomials
$2x^2 + 2x + 1 = x^2 + 4x + 4$	Combining like terms
$x^2 - 2x - 3 = 0$	Subtracting $x^2 + 4x + 4$ from both sides
$(x - 3)(x + 1) = 0$	Factoring
$x - 3 = 0 \quad or \quad x + 1 = 0$	Using the principle of zero products
$x = 3 \quad or \quad x = -1.$	

4. **Check.** The integer -1 cannot be a length of a side because it is negative. For $x = 3$, we have $x + 1 = 4$, and $x + 2 = 5$. Since $3^2 + 4^2 = 5^2$, the lengths 3, 4, and 5 determine a right triangle. Thus, 3, 4, and 5 check.

5. **State.** Lucinda and Felipe should use a triangle with sides having a ratio of 3:4:5. Thus, if the marks on the house are 3 yd apart, they should locate the stake at the point in the yard that is precisely 4 yd from one mark and 5 yd from the other mark.

Example 7

Display of a sports card. A valuable sports card is 4 cm wide and 5 cm long. The card is to be sandwiched by two pieces of Lucite, each of which is $5\frac{1}{2}$ times the area of the card. Determine the dimensions of the Lucite that will ensure a uniform border around the card.

Solution

1. **Familiarize.** We make a drawing and label it, using x to represent the width of the border, in centimeters. Since the border extends uniformly around the entire card, the length of the Lucite must be $5 + 2x$ and the width must be $4 + 2x$.

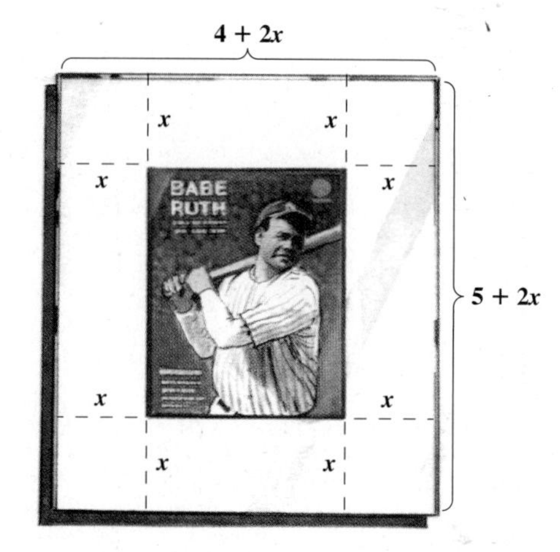

2. **Translate.** We rephrase the information given and translate as follows:

$$\underbrace{\text{Area of Lucite}}_{(5+2x)(4+2x)} \; \underbrace{\text{is}}_{=} \; \underbrace{5\tfrac{1}{2}\text{ times}}_{5\frac{1}{2}\,\cdot} \; \underbrace{\text{area of card.}}_{5\cdot 4}$$

3. **Carry out.** We solve the equation:

$$(5 + 2x)(4 + 2x) = 5\tfrac{1}{2} \cdot 5 \cdot 4$$

$$20 + 10x + 8x + 4x^2 = 110 \qquad \text{Multiplying}$$

$$4x^2 + 18x - 90 = 0 \qquad \text{Finding standard form}$$

$$\left.\begin{aligned} 2(2x^2 + 9x - 45) &= 0 \\ 2(2x + 15)(x - 3) &= 0 \end{aligned}\right\} \qquad \text{Factoring}$$

$$2x + 15 = 0 \quad \text{or} \quad x - 3 = 0 \qquad \text{Principle of zero products}$$

$$x = -7\tfrac{1}{2} \quad \text{or} \quad x = 3.$$

4. **Check.** We check 3 in the original problem. (Note that $-7\frac{1}{2}$ is not a solution because measurements cannot be negative.) If the border is 3 cm wide, the Lucite will have a length of $5 + 2 \cdot 3$, or 11 cm, and a width of $4 + 2 \cdot 3$, or 10 cm. The area of the Lucite is thus $11 \cdot 10$, or 110 cm^2. Since the area of the card is 20 cm^2 and 110 cm^2 is $5\frac{1}{2}$ times 20 cm^2, the number 3 checks.

5. **State.** Each piece of Lucite should be 11 cm long and 10 cm wide.

Exercise Set 5.8

FOR EXTRA HELP

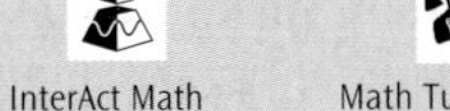

Digital Video Tutor CD 4 Videotape 10 | InterAct Math | Math Tutor Center | MathXL | MyMathLab.com

Solve.

1. $x^2 - 4x = 45$
2. $t^2 - 3t = 28$
3. $a^2 + 1 = 2a$
4. $r^2 + 16 = 8r$
5. $x^2 + 12x + 36 = 0$
6. $y^2 + 16y + 64 = 0$
7. $9x + x^2 + 20 = 0$
8. $8y + y^2 + 15 = 0$
9. $x^2 - 8x = 0$
10. $t^2 - 9t = 0$
11. $a^3 - 3a^2 = 40a$
12. $x^3 - 2x^2 = 63x$
13. (Aha!) $x^2 - 16 = 0$
14. $r^2 - 9 = 0$
15. $(t - 6)(t + 6) = 45$
16. $(a - 4)(a + 4) = 20$
17. $3x^2 - 8x + 4 = 0$
18. $9x^2 - 15x + 4 = 0$
19. $4t^3 + 11t^2 + 6t = 0$
20. $8y^3 + 10y^2 + 3y = 0$
21. $(y - 3)(y + 2) = 14$
22. $(z + 4)(z - 2) = -5$
23. $x(5 + 12x) = 28$
24. $a(1 + 21a) = 10$
25. $a^2 - \frac{1}{64} = 0$
26. $x^2 - \frac{1}{25} = 0$
27. $t^4 - 26t^2 + 25 = 0$
28. $t^4 - 13t^2 + 36 = 0$
29. Let $f(x) = x^2 + 12x + 40$. Find a such that $f(a) = 8$.
30. Let $f(x) = x^2 + 14x + 50$. Find a such that $f(a) = 5$.
31. Let $g(x) = 2x^2 + 5x$. Find a such that $g(a) = 12$.
32. Let $g(x) = 2x^2 - 15x$. Find a such that $g(a) = -7$.
33. Let $h(x) = 12x + x^2$. Find a such that $h(a) = -27$.
34. Let $h(x) = 4x - x^2$. Find a such that $h(a) = -32$.

Find the domain of the function f given by each of the following.

35. $f(x) = \dfrac{3}{x^2 - 4x - 5}$
36. $f(x) = \dfrac{2}{x^2 - 7x + 6}$

37. $f(x) = \dfrac{x}{6x^2 - 54}$

38. $f(x) = \dfrac{2x}{5x^2 - 20}$

39. $f(x) = \dfrac{x - 5}{9x - 18x^2}$

40. $f(x) = \dfrac{1 + x}{3x - 15x^2}$

41. $f(x) = \dfrac{7}{5x^3 - 35x^2 + 50x}$

42. $f(x) = \dfrac{3}{2x^3 - 2x^2 - 12x}$

Solve.

43. The square of a number plus the number is 132. What is the number?

44. The square of a number plus the number is 156. What is the number?

45. A photo is 5 cm longer than it is wide. Find the length and the width if the area is 84 cm^2.

46. An envelope is 4 cm longer than it is wide. The area is 96 cm^2. Find the length and the width.

47. *Geometry.* If each of the sides of a square is lengthened by 4 m, the area becomes 49 m^2. Find the length of a side of the original square.

48. *Geometry.* If each of the sides of a square is lengthened by 6 cm, the area becomes 144 cm^2. Find the length of a side of the original square.

49. *Framing a picture.* A picture frame measures 12 cm by 20 cm, and 84 cm^2 of picture shows. Find the width of the frame.

50. *Framing a picture.* A picture frame measures 14 cm by 20 cm, and 160 cm^2 of picture shows. Find the width of the frame.

51. *Landscaping.* A rectangular lawn measures 60 ft by 80 ft. Part of the lawn is torn up to install a sidewalk of uniform width around it. The area of the new lawn is 2400 ft^2. How wide is the sidewalk?

52. *Landscaping.* A rectangular garden is 30 ft by 40 ft. Part of the garden is removed in order to install a walkway of uniform width around it. The area of the new garden is one-half the area of the old garden. How wide is the walkway?

53. Three consecutive even integers are such that the square of the third is 76 more than the square of the second. Find the three integers.

54. Three consecutive even integers are such that the square of the first plus the square of the third is 136. Find the three integers.

55. *Tent design.* The triangular entrance to a tent is 2 ft taller than it is wide. The area of the entrance is 12 ft^2. Find the height and the base.

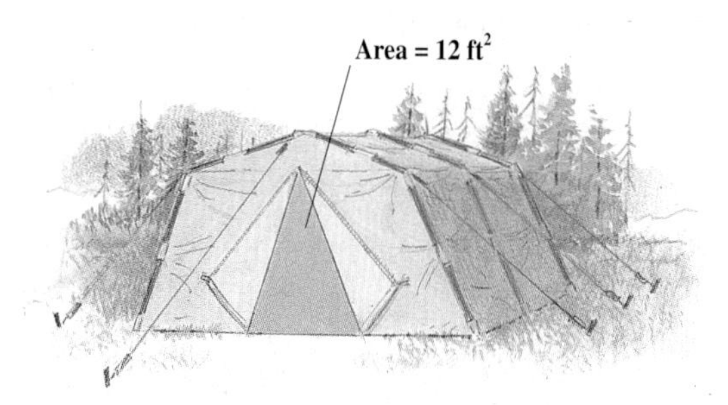

56. *Antenna wires.* A wire is stretched from the ground to the top of an antenna tower, as shown. The wire is 20 ft long. The height of the tower is 4 ft greater than the distance d from the tower's base to the bottom of the wire. Find the distance d and the height of the tower.

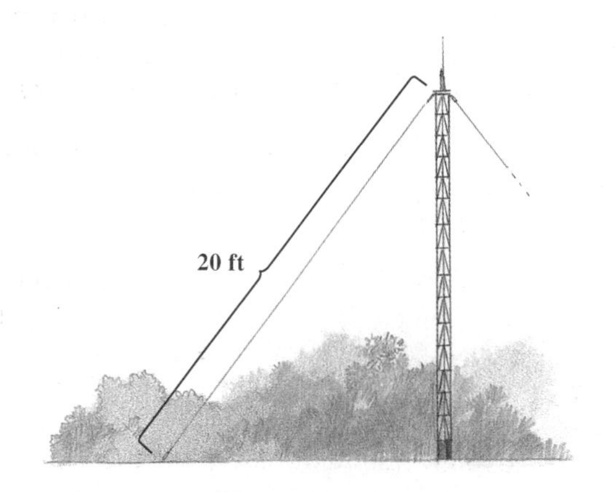

57. *Sailing.* A triangular sail is 9 m taller than it is wide. The area is 56 m^2. Find the height and the base of the sail.

58. *Parking lot design.* A rectangular parking lot is 50 ft longer than it is wide. Determine the dimensions of the parking lot if it measures 250 ft diagonally.

59. *Ladder location.* The foot of an extension ladder is 9 ft from a wall. The height that the ladder reaches on the wall and the length of the ladder are consecutive integers. How long is the ladder?

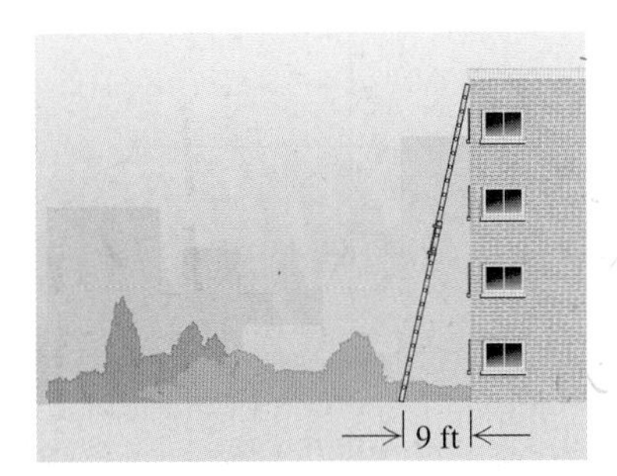

60. *Ladder location.* The foot of an extension ladder is 10 ft from a wall. The ladder is 2 ft longer than the height that it reaches on the wall. How far up the wall does the ladder reach?

61. *Garden design.* Ignacio is planning a garden that is 25 m longer than it is wide. The garden will have an area of 7500 m^2. What will its dimensions be?

62. *Garden design.* A flower bed is to be 3 m longer than it is wide. The flower bed will have an area of 108 m^2. What will its dimensions be?

63. *Cabinet making.* Dovetail Woodworking determines that the revenue R, in thousands of dollars, from the sale of x sets of cabinets is given by $R(x) = 2x^2 + x$. If the cost C, in thousands of dollars, of producing x sets of cabinets is given by $C(x) = x^2 - 2x + 10$, how many sets must be produced and sold in order for the company to break even?

64. *Camcorder production.* Suppose that the cost of making x video cameras is $C(x) = \frac{1}{9}x^2 + 2x + 1$, where $C(x)$ is in thousands of dollars. If the revenue from the sale of x video cameras is given by $R(x) = \frac{5}{36}x^2 + 2x$, where $R(x)$ is in thousands of dollars, how many cameras must be sold in order for the firm to break even?

65. *Prize tee shirts.* Using the model in Example 5, determine how long a tee shirt has been airborne if it is caught on the way *up* by a fan 100 ft above ground level.

66. *Prize tee shirts.* Using the model in Example 5, determine how long a tee shirt has been airborne if it is caught on the way *down* by a fan 10 ft above ground level.

67. *Fireworks displays.* Fireworks are typically launched from a mortar with an upward velocity (initial speed) of about 64 ft/sec. The height $h(t)$, in feet, of a "weeping willow" display, t seconds after having been launched from an 80-ft-high rooftop, is given by

$$h(t) = -16t^2 + 64t + 80.$$

After how long will the cardboard shell from the fireworks reach the ground?

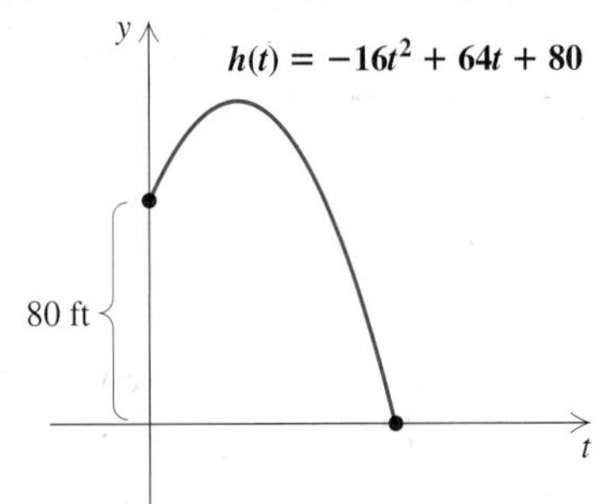

68. *Safety flares.* Suppose that a flare is launched upward with an initial velocity of 80 ft/sec from a height of 224 ft. Its height in feet, $h(t)$, after t seconds is given by

$$h(t) = -16t^2 + 80t + 224.$$

After how long will the flare reach the ground?

69. Suppose that you are given a detailed graph of $y = p(x)$, where $p(x)$ is some polynomial in x. How could the graph be used to help solve the equation $p(x) = 0$?

70. Can the number of solutions of a quadratic equation exceed two? Why or why not?

SKILL MAINTENANCE

Simplify.

71. $\dfrac{5 - 10 \cdot 3}{-4 + 11 \cdot 4}$

72. $\dfrac{2 \cdot 3 - 5 \cdot 2}{7 - 3^2}$

73. *Driving.* At noon, two cars start from the same location going in opposite directions at different speeds. After 7 hr, they are 651 mi apart. If one car is traveling 15 mph slower than the other car, what are their respective speeds?

74. *Television sales.* At the beginning of the month, J.C.'s Appliances had 150 televisions in stock. During the month, they sold 45% of their conventional televisions and 60% of their surround-sound televisions. If they sold a total of 78 televisions, how many of each type did they sell?

Solve.

75. $2x - 14 + 9x > -8x + 16 + 10x$

76. $x + y = 0,$
$z - y = -2,$
$x - z = 6$

SYNTHESIS

77. Explain how one could write a quadratic equation that has -3 and 5 as solutions.

78. If the graph of $f(x) = ax^2 + bx + c$ has no x-intercepts, what can you conclude about the equation $ax^2 + bx + c = 0$?

Solve.

79. $(8x + 11)(12x^2 - 5x - 2) = 0$

80. $(x + 1)^3 = (x - 1)^3 + 26$

81. $(x - 2)^3 = x^3 - 2$

82. Use the following graph of $g(x) = -x^2 - 2x + 3$ to solve $-x^2 - 2x + 3 = 0$ and to solve $-x^2 - 2x + 3 \geq -5$.

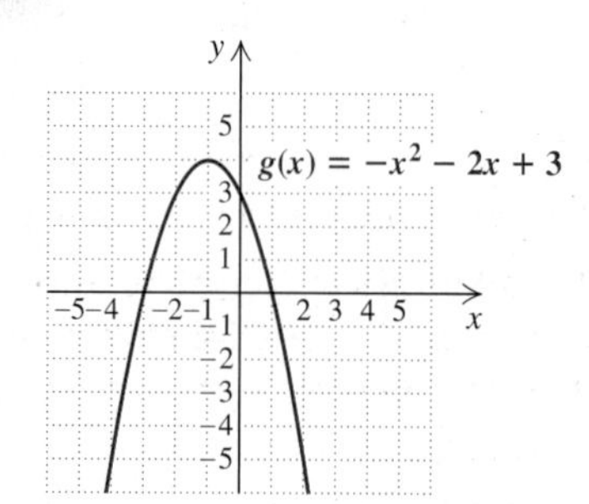

83. Find a polynomial function f for which $f(2) = 0$, $f(-1) = 0$, $f(3) = 0$, and $f(0) = 30$.

84. Find a polynomial function g for which $g(-3) = 0$, $g(1) = 0$, $g(5) = 0$, and $g(0) = 45$.

85. Use the following graph of $f(x) = x^2 - 2x - 3$ to solve $x^2 - 2x - 3 = 0$ and to solve $x^2 - 2x - 3 < 5$.

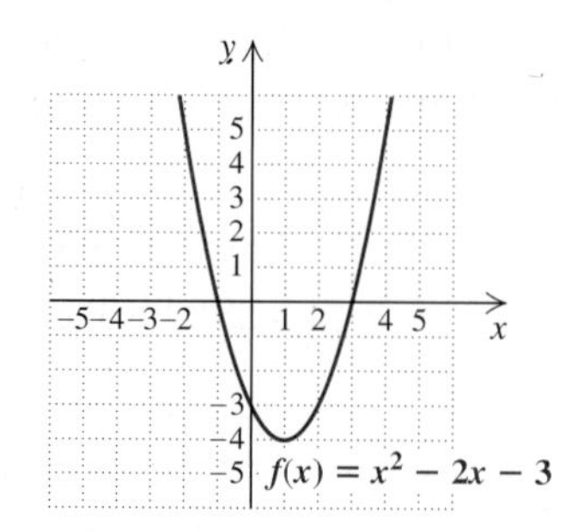

86. *Box construction.* A rectangular piece of tin is twice as long as it is wide. Squares 2 cm on a side are cut out of each corner, and the ends are turned up to make a box whose volume is 480 cm^3. What are the dimensions of the piece of tin?

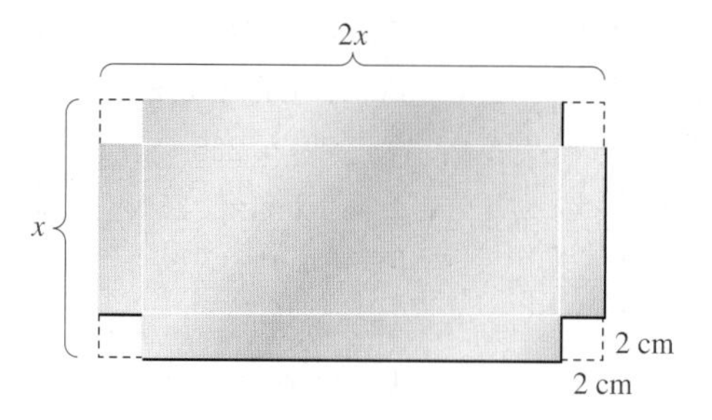

87. *Highway fatalities.* The function given by $n(t) = 0.015t^2 - 0.9t + 28$ can be used to estimate the number of deaths $n(t)$ per 100,000 drivers, for drivers t years above age 20 (*Source*: based on information in the *Statistical Abstract of the United States*, 2000). What age group of drivers has a fatality rate of 14.5 deaths per 100,000 drivers?

88. *Navigation.* A tugboat and a freighter leave the same port at the same time at right angles. The freighter travels 7 km/h slower than the tugboat. After 4 hr, they are 68 km apart. Find the speed of each boat.

89. *Skydiving.* During the first 13 sec of a jump, a skydiver falls approximately $11.12t^2$ feet in t seconds. A small heavy object (with less wind resistance) falls about $15.4t^2$ feet in t seconds. Suppose that a skydiver jumps from 30,000 ft, and 1 sec later a camera falls out of the airplane. How long will it take the camera to catch up to the skydiver?

90. Use the TABLE feature of a grapher to check that -5 and 3 are not in the domain of F as shown in Example 4.

91. Use the TABLE feature of a grapher to check your answers to Exercises 37, 39, and 41.

In Exercises 92–96, use a grapher to find any real-number solutions that exist accurate to two decimal places.

92. $x^2 - 2x - 8 = 0$ (Check by factoring.)

93. $x^2 + 3x - 4 = 0$ (Check by factoring.)

94. $-x^2 + 13.80x = 47.61$

95. $-x^2 + 3.63x + 34.34 = x^2$

96. $x^3 - 3.48x^2 + x = 3.48$

97. Mary Louise is attempting to solve $x^3 + 20x^2 + 4x + 80 = 0$ with a grapher. Unfortunately, when she graphs $y_1 = x^3 + 20x^2 + 4x + 80$ in a standard $[-10, 10, -10, 10]$ window, she sees no graph at all, let alone any x-intercept. Can this problem be solved graphically? If so, how? If not, why not?

Summary and Review 5

Key Terms

Term, p. 266
Monomial, p. 266
Degree, p. 266
Coefficient, p. 267
Polynomial, p. 267
Leading term, p. 267
Leading coefficient, p. 267
Linear, p. 267
Quadratic, p. 267
Cubic, p. 267
Binomial, p. 267
Trinomial, p. 267
Descending order, p. 268
Ascending order, p. 268
Polynomial function, p. 268
Similar, or like terms, p. 270
FOIL, p. 280
Square of a binomial, p. 281
Difference of two squares, p. 282
Factor, p. 287
Factored completely, p. 288
Prime polynomial, p. 288
Factoring by grouping, p. 289
Perfect-square trinomial, p. 303
Sum or difference of cubes, p. 309
Cube root, p. 309
Polynomial equation, p. 317
Quadratic equation, p. 317
Standard form, p. 318
Root, or zero, p. 321
Pythagorean theorem, p. 323
Right triangle, p. 323
Hypotenuse, p. 323
Leg, p. 323

Important Properties and Formulas

Factoring Formulas

$A^2 + 2AB + B^2 = (A + B)^2;$
$A^2 - 2AB + B^2 = (A - B)^2;$
$A^2 - B^2 = (A + B)(A - B);$
$A^3 + B^3 = (A + B)(A^2 - AB + B^2);$
$A^3 - B^3 = (A - B)(A^2 + AB + B^2)$

To Factor $ax^2 + bx + c$ Using FOIL

1. Factor out the largest common factor, if one exists. Here we assume none does.
2. Find two **F**irst terms whose product is ax^2:

$(\square x + \quad)(\square x + \quad) = ax^2 + bx + c.$

FOIL

3. Find two **L**ast terms whose product is c:

$$(\;\;x + \square)(\;\;x + \square) = ax^2 + bx + c.$$

FOIL

4. Repeat steps (2) and (3) until a combination is found for which the sum of the **O**uter and **I**nner products is bx:

$$(\square x + \square)(\square x + \square) = ax^2 + bx + c.$$

I O FOIL

To Factor $ax^2 + bx + c$ Using Grouping

1. Make sure that any common factors have been factored out.
2. Multiply the leading coefficient a and the constant c.
3. Try to factor the product ac so that the sum of the factors is b. That is, find integers p and q so that $pq = ac$ and $p + q = b$.
4. Split the middle term. That is, write bx as $px + qx$.
5. Factor by grouping.

To Factor a Polynomial

A. Always factor out the largest common factor.

B. Once any common factor has been factored out, look at the number of terms.

Two terms: Try factoring as a difference of squares first. Next, try factoring as a sum or a difference of cubes. Do *not* try to factor a *sum* of squares.

Three terms: Try factoring as a perfect-square trinomial. Next, try trial and error, using the FOIL method or the grouping method.

Four or more terms: Try factoring by grouping and factoring out a common binomial factor. Next, try grouping into a difference of squares, one of which is a trinomial.

C. Always *factor completely*. If a factor with more than one term can itself be factored further, do so.

D. Check the factorization by multiplying.

The Principle of Zero Products

For any real numbers a and b:

If $ab = 0$, then $a = 0$ or $b = 0$.
If $a = 0$ or $b = 0$, then $ab = 0$.

Review Exercises

1. Given the polynomial

$$2xy^6 - 7x^8y^3 + 2x^3 - 3,$$

determine the degree of each term and the degree of the polynomial.

2. Given the polynomial

$$4x - 5x^3 + 2x^2 - 7,$$

arrange in descending order and determine the leading term and the leading coefficient.

3. Arrange in ascending powers of x:

$$3x^6y - 7x^8y^3 + 2x^3 - 3x^2.$$

4. Find $P(0)$ and $P(-1)$:

$$P(x) = x^3 - x^2 + 4x.$$

5. Evaluate the polynomial function for $x = -2$:

$$P(x) = 4 - 2x - x^2.$$

Combine like terms.

6. $6 - 4a + a^2 - 2a^3 - 10 + a$

7. $4x^2y - 3xy^2 - 5x^2y + xy^2$

Add.

8. $(-6x^3 - 4x^2 + 3x + 1) + (5x^3 + 2x + 6x^2 + 1)$

9. $(3x^4 + 3x^3 - 8x + 9) + (-6x^4 + 4x + 7 + 3x)$

10. $(-9xy^2 - xy + 6x^2y) + (-5x^2y - xy + 4xy^2)$

Subtract.

11. $(3x - 5) - (-6x + 2)$

12. $(4a - b + 3c) - (6a - 7b - 4c)$

13. $(8x^2 - 4xy + y^2) - (2x^2 + 3xy - 2y^2)$

Multiply.

14. $(3x^2y)(-6xy^3)$

15. $(x^4 - 2x^2 + 3)(x^4 + x^2 - 1)$

16. $(4ab + 3c)(2ab - c)$

17. $(2x + 5y)(2x - 5y)$

18. $(2x - 5y)^2$

19. $(x + 3)(2x - 1)$

20. $(x^2 + 4y^3)^2$

21. $(x - 5)(x^2 + 5x + 25)$

22. $\left(x - \frac{1}{3}\right)\left(x - \frac{1}{6}\right)$

Factor.

23. $6x^2 + 5x$

24. $9y^4 - 3y^2$

25. $15x^4 - 18x^3 + 21x^2 - 9x$

26. $a^2 - 12a + 27$

27. $3m^2 + 14m + 8$

28. $25x^2 + 20x + 4$

29. $4y^2 - 16$

30. $5x^2 + x^3 - 14x$

31. $ax + 2bx - ay - 2by$

32. $3y^3 + 6y^2 - 5y - 10$

33. $a^4 - 81$

34. $4x^4 + 4x^2 + 20$

35. $27x^3 - 8$

36. $0.064b^3 - 0.125c^3$

37. $y^5 + y$

38. $2z^8 - 16z^6$

39. $54x^6y - 2y$

40. $36x^2 - 120x + 100$

41. $6t^2 + 17pt + 5p^2$

42. $x^3 + 2x^2 - 9x - 18$

43. $a^2 - 2ab + b^2 - 4t^2$

Solve.

44. $x^2 - 20x = -100$

45. $6b^2 - 13b + 6 = 0$

46. $8y^2 = 14y$

47. $r^2 = 16$

48. $a^3 = 4a^2 + 21a$

49. $(y - 1)(y - 4) = 10$

50. Let $f(x) = x^2 - 7x - 40$. Find a such that $f(a) = 4$.

51. Find the domain of the function f given by

$$f(x) = \frac{x - 3}{3x^2 + 19x - 14}.$$

52. The area of a square is 5 more than four times the length of a side. What is the length of a side of the square?

53. The sum of the squares of three consecutive odd numbers is 83. Find the numbers.

54. A photograph is 3 in. longer than it is wide. When a 2-in. border is placed around the photograph, the total area of the photograph and the border is 108 in^2. Find the dimensions of the photograph.

55. Tim is designing a rectangular garden with a width of 8 ft. The path that leads diagonally across the garden is 2 ft longer than the length of the garden. How long is the path?

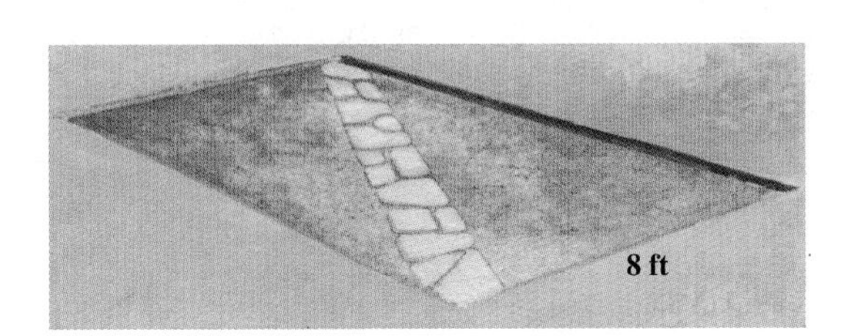

SYNTHESIS

56. Explain how to find the roots of a polynomial function from its graph.

57. Explain in your own words why there must be a 0 on one side of an equation before you can use the principle of zero products.

Factor.

58. $128x^6 - 2y^6$

59. $(x - 1)^3 - (x + 1)^3$

Multiply.

60. $[a - (b - 1)][(b - 1)^2 + a(b - 1) + a^2]$

61. $(z^{n^2})^{n^3}(z^{4n^3})^{n^2}$

62. Solve: $(x + 1)^3 = x^2(x + 1)$.

Chapter Test 5

Given the polynomial $3xy^3 - 4x^2y + 5x^5y^4 - 2x^4y$.

1. Determine the degree of the polynomial.

2. Arrange in descending powers of x.

3. Determine the leading term of the polynomial $8a - 2 + a^2 - 4a^3$.

4. Given $P(x) = 2x^3 + 3x^2 - x + 4$, find $P(0)$ and $P(-2)$.

5. Given $P(x) = x^2 - 5x$, find and simplify

$$P(a + h) - P(a).$$

6. Combine like terms:

$$5xy - 2xy^2 - 2xy + 5xy^2.$$

Add.

7. $(-6x^3 + 3x^2 - 4y) + (3x^3 - 2y - 7y^2)$

8. $(5m^3 - 4m^2n - 6mn^2 - 3n^3) + (9mn^2 - 4n^3 + 2m^3 + 6m^2n)$

Subtract.

9. $(9a - 4b) - (3a + 4b)$

10. $(6y^2 - 2y - 5y^3) - (4y^2 - 7y - 6y^3)$

Multiply.

11. $(-4x^2y)(-16xy^2)$

12. $(6a - 5b)(2a + b)$

13. $(x - y)(x^2 - xy - y^2)$

14. $(2x^3 + 5)^2$

15. $(4y - 9)^2$

16. $(x - 2y)(x + 2y)$

Factor.

17. $15x^2 - 5x^4$

18. $y^3 + 5y^2 - 4y - 20$

19. $p^2 - 12p - 28$

20. $12m^2 + 20m + 3$

21. $9y^2 - 25$

22. $3r^3 - 3$

23. $9x^2 + 25 - 30x$

24. $x^8 - y^8$

25. $y^2 + 8y + 16 - 100t^2$

26. $20a^2 - 5b^2$

27. $24x^2 - 46x + 10$

28. $16a^7b + 54ab^7$

29. $4y^4x + 36yx^2 + 8y^2x^3 - 16xy$

Solve.

30. $x^2 - 18 = 3x$

31. $5y^2 = 125$

32. $2x^3 + 21x = -17x^2$

33. $9x^2 + 3x = 0$

34. Let $f(x) = 3x^2 - 15x + 11$. Find a such that $f(a) = 11$.

35. Find the domain of the function f given by

$$f(x) = \frac{3 - x}{x^2 + 2x + 1}.$$

36. A photograph is 3 cm longer than it is wide. Its area is 40 cm^2. Find its length and its width.

37. To celebrate a town's centennial, fireworks are launched over a lake off a dam 36 ft above the water. The height of a display, t seconds after it has been launched, is given by

$$h(t) = -16t^2 + 64t + 36.$$

After how long will the shell from the fireworks reach the water?

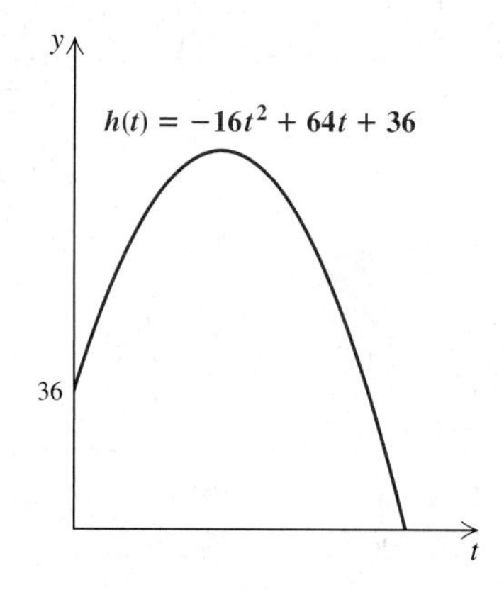

SYNTHESIS

38. a) Multiply: $(x^2 + x + 1)(x^3 - x^2 + 1)$.
b) Factor: $x^5 + x + 1$.

39. Factor: $6x^{2n} - 7x^n - 20$.

6 Rational Expressions, Equations, and Functions

6.1 Rational Expressions and Functions: Multiplying and Dividing

6.2 Rational Expressions and Functions: Adding and Subtracting

6.3 Complex Rational Expressions

6.4 Rational Equations

Connecting the Concepts

6.5 Solving Applications Using Rational Equations

6.6 Division of Polynomials

6.7 Synthetic Division

6.8 Formulas, Applications, and Variation

Summary and Review

Test

Cumulative Review

AN APPLICATION

The Honeywell HQ17 air cleaner takes twice as long as the EV25 to clean the same volume of air. Together the two machines can clean the air in a 24-ft by 24-ft banquet room in 10 min. How long would it take each machine, working alone, to clean the air in the room?

This problem appears as Exercise 15 in Section 6.5.

JEAN CHEN
Food and
Beverage Coordinator
East Boston, MA

In my work, I am always dealing with numbers. I use spreadsheets to track inventory and balance finances. I also use math to calculate the amount of food needed and cost per person for a banquet.

A rational expression is an expression, similar to the fractions in arithmetic, that indicates division. In this chapter, we add, subtract, multiply, and divide rational expressions, and use them in equations and functions. We then use rational expressions to solve problems that we could not have solved before.

6.1 Rational Expressions and Functions: Multiplying and Dividing

Rational Functions • Multiplying • Simplifying Rational Expressions • Dividing and Simplifying

An expression that consists of a polynomial divided by a nonzero polynomial is called a **rational expression**. The following are examples of rational expressions:

$$\frac{3}{4},\quad \frac{x}{y},\quad \frac{9}{a+b},\quad \frac{x^2+7xy-4}{x^3-y^3},\quad \frac{1+z^3}{1-z^6}.$$

Rational Functions

Like polynomials, certain rational expressions are used to describe functions. Such functions are called **rational functions**.

Example 1 The function given by

$$H(t) = \frac{t^2+5t}{2t+5}$$

gives the time, in hours, for two machines, working together, to complete a job that the first machine could do alone in t hours and the other machine could do in $t + 5$ hours. How long will the two machines, working together, require for the job if the first machine alone would take (**a**) 1 hour? (**b**) 5 hours?

Solution

a) $H(1) = \dfrac{1^2+5\cdot 1}{2\cdot 1+5} = \dfrac{1+5}{2+5} = \dfrac{6}{7}$ hr

b) $H(5) = \dfrac{5^2+5\cdot 5}{2\cdot 5+5} = \dfrac{25+25}{10+5} = \dfrac{50}{15} = \dfrac{10}{3}$ hr

In Section 5.8, we found that the domain of a rational function must exclude any numbers for which the denominator is 0. For a function like H above, the denominator is 0 when t is $-\frac{5}{2}$, so the domain of H is $\left(-\infty, -\frac{5}{2}\right) \cup \left(-\frac{5}{2}, \infty\right)$.

Although graphing rational functions is beyond the scope of this course, it is educational to examine a computer-generated graph of the above function.

Note that the graph consists of two unconnected "branches." Since $-\frac{5}{2}$ is not in the domain of H, a vertical line drawn at $-\frac{5}{2}$ does not touch the graph of H.

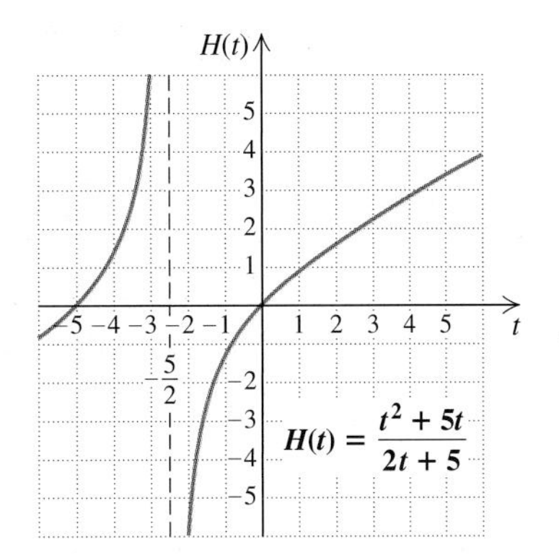

Multiplying

The calculations that are performed with rational expressions resemble those performed in arithmetic.

Products of Rational Expressions

To multiply two rational expressions, multiply numerators and multiply denominators:

$$\frac{A}{B} \cdot \frac{C}{D} = \frac{AC}{BD}, \quad \text{where } B \neq 0, D \neq 0.$$

Example 2 Multiply: $\dfrac{x+1}{y-3} \cdot \dfrac{x^2}{y+1}$.

Solution

$$\frac{x+1}{y-3} \cdot \frac{x^2}{y+1} = \frac{(x+1)x^2}{(y-3)(y+1)}$$ Multiplying the numerators and multiplying the denominators

Recall from arithmetic that multiplication by 1 can be used to find equivalent expressions:

$$\frac{3}{5} = \frac{3}{5} \cdot \frac{2}{2}$$ Multiplying by $\frac{2}{2}$, which is 1

$$= \frac{6}{10}.$$ $\frac{3}{5}$ and $\frac{6}{10}$ represent the same number.

Similarly, multiplication by 1 can be used to find equivalent rational expressions:

$$\left.\begin{aligned}\frac{x-5}{x+2} &= \frac{x-5}{x+2} \cdot \frac{x+3}{x+3} \\ &= \frac{(x-5)(x+3)}{(x+2)(x+3)}.\end{aligned}\right\}$$

Multiplying by $\frac{x+3}{x+3}$, which, provided $x \neq -3$, is 1

The expressions

$$\frac{x-5}{x+2} \quad \text{and} \quad \frac{(x-5)(x+3)}{(x+2)(x+3)}$$

are equivalent: So long as x is replaced with a number other than -2 or -3, both expressions represent the same number. For example, if $x = 4$, then

$$\frac{x-5}{x+2} = \frac{4-5}{4+2} = \frac{-1}{6}$$

and

$$\frac{(x-5)(x+3)}{(x+2)(x+3)} = \frac{(4-5)(4+3)}{(4+2)(4+3)} = \frac{-1 \cdot 7}{6 \cdot 7} = \frac{-7}{42} = \frac{-1}{6}.$$

Example 3 Multiply to find an equivalent expression:

$$\frac{4-x}{y-x} \cdot \frac{-1}{-1}.$$

Solution We have

$$\frac{4-x}{y-x} \cdot \frac{-1}{-1} = \frac{-4+x}{-y+x} = \frac{x-4}{x-y}.$$

Multiplying by $-1/-1$ is the same as multiplying by 1, so

$$\frac{4-x}{y-x} \quad \text{is equivalent to} \quad \frac{x-4}{x-y}.$$

Simplifying Rational Expressions

As in arithmetic, rational expressions are *simplified* by "removing" a factor equal to 1. This reverses the process shown above:

$$\frac{6}{10} = \frac{3 \cdot 2}{5 \cdot 2} = \frac{3}{5} \cdot \frac{2}{2} = \frac{3}{5}.$$

We "removed" the factor that equals 1: $\frac{2}{2} = 1$.

Similarly,

$$\frac{(x-5)(x+3)}{(x+2)(x+3)} = \frac{x-5}{x+2} \cdot \frac{x+3}{x+3} = \frac{x-5}{x+2}.$$

We "removed" the factor that equals 1: $\frac{x+3}{x+3} = 1$.

Example 4 Simplify: $\dfrac{7a^2 + 21a}{14a}$.

Solution We first factor the numerator and the denominator, looking for the largest factor common to both. Once the greatest common factor is found, we use it to write 1 and simplify:

$$\frac{7a^2 + 21a}{14a} = \frac{7a(a + 3)}{7 \cdot 2 \cdot a}$$ Factoring; the greatest common factor is $7a$.

$$= \frac{7a}{7a} \cdot \frac{a + 3}{2}$$ Rewriting as a product of two rational expressions

$$= 1 \cdot \frac{a + 3}{2}$$ $\dfrac{7a}{7a} = 1$; try to do this step mentally.

$$= \frac{a + 3}{2}.$$ Removing the factor 1

A rational expression is said to be **simplified** when no factors equal to 1 can be removed. Often, this may require two or more steps. For example, suppose we remove 7/7 instead of $(7a)/(7a)$ in Example 4. We would then have

$$\left.\begin{aligned}\frac{7a^2 + 21a}{14a} &= \frac{7(a^2 + 3a)}{7 \cdot 2a}\\ &= \frac{a^2 + 3a}{2a}.\end{aligned}\right\}$$ Removing a factor equal to 1: $\frac{7}{7} = 1$

Here, since 7 is not the *greatest* common factor, we need to simplify further:

$$\left.\begin{aligned}\frac{a^2 + 3a}{2a} &= \frac{a(a + 3)}{a \cdot 2}\\ &= \frac{a + 3}{2}.\end{aligned}\right\}$$ Removing another factor equal to 1: $a/a = 1$. The rational expression is now simplified.

Example 5 Simplify each of the following.

a) $\dfrac{x^2 - 4}{2x^2 - 3x - 2}$ **b)** $\dfrac{9x^2 + 6xy - 3y^2}{12x^2 - 12y^2}$

Solution

a) $$\frac{x^2 - 4}{2x^2 - 3x - 2} = \frac{(x - 2)(x + 2)}{(2x + 1)(x - 2)}$$ Factoring the numerator and the denominator

$$= \frac{x - 2}{x - 2} \cdot \frac{x + 2}{2x + 1}$$ Rewriting as a product of two rational expressions

$$= \frac{x + 2}{2x + 1}$$ Removing a factor equal to 1: $\dfrac{x - 2}{x - 2} = 1$

To check that the simplified form of an expression (in one variable) is equivalent to the original expression, we can let y_1 = the original expression, y_2 = the simplified expression, and $y_3 = y_1 - y_2$ (or $y_2 - y_1$). If y_1 and y_2 are indeed equivalent, a TABLE or the TRACE feature can be used to show that, except when y_1 or y_2 is undefined, we have $y_1 = y_2$ and $y_3 = 0$.

1. Use a grapher to check Example 4. Be careful to use parentheses as needed.
2. Use a grapher to show that $\dfrac{x+3}{x} \neq 3$. (See the Caution! box below.)

b) $$\frac{9x^2 + 6xy - 3y^2}{12x^2 - 12y^2} = \frac{3(x+y)(3x-y)}{12(x+y)(x-y)}$$ Factoring the numerator and the denominator

$$= \frac{3(x+y)}{3(x+y)} \cdot \frac{3x-y}{4(x-y)}$$ Rewriting as a product of two rational expressions

$$= \frac{3x-y}{4(x-y)}$$ Removing a factor equal to 1: $\dfrac{3(x+y)}{3(x+y)} = 1$

For purposes of later work, we usually do not multiply out the numerator and the denominator of the simplified expression.

Canceling

"Canceling" is a shortcut that you may have used for removing a factor equal to 1 when working with fractions. With some misgivings, we mention it here as a possible way to speed up your work. Canceling is one way to remove factors equal to 1 in products. It *cannot* be done in sums or when adding expressions together. If your instructor permits canceling (not all do), it is essential that it be done with care and understanding. Example 5(b) might have been done faster as follows:

$$\frac{9x^2 + 6xy - 3y^2}{12x^2 - 12y^2} = \frac{\cancel{3}\cancel{(x+y)}(3x-y)}{\cancel{3} \cdot 4\cancel{(x+y)}(x-y)}$$ When a factor that equals 1 is found, it is "canceled" as shown.

$$= \frac{3x-y}{4(x-y)}.$$ Removing a factor equal to 1: $\dfrac{3(x+y)}{3(x+y)} = 1$

Caution! Canceling is often performed incorrectly:

$$\frac{\cancel{x}+3}{\cancel{x}} = 3, \qquad \frac{\cancel{4}x+3}{\cancel{2}} = 2x+3, \qquad \frac{\cancel{5}}{\cancel{5}+x} = \frac{1}{x}$$

Incorrect! Incorrect! Incorrect!

To check that these are not equivalent, substitute a number for x.

In each of these situations, the expressions canceled are *not* factors that equal 1. Factors are parts of products. For example, in $x \cdot 3$, x and 3 are factors, but in $x + 3$, x and 3 are *not* factors, but terms. If you can't factor, you can't cancel! If in doubt, don't cancel!

Often, after multiplying two rational expressions, it is possible to simplify the result.

Example 6 Multiply. Then simplify by removing a factor equal to 1.

a) $\dfrac{x+2}{x-3} \cdot \dfrac{x^2-4}{x^2+x-2}$ **b)** $\dfrac{1-a^3}{a^2} \cdot \dfrac{a^5}{a^2-1}$

Solution

a) $\dfrac{x+2}{x-3} \cdot \dfrac{x^2-4}{x^2+x-2} = \dfrac{(x+2)(x^2-4)}{(x-3)(x^2+x-2)}$ Multiplying the numerators and also the denominators

$= \dfrac{(x+2)(x-2)(x+2)}{(x-3)(x+2)(x-1)}$ Factoring the numerator and the denominator and finding common factors

$= \dfrac{(x+2)\cancel{(x+2)}(x-2)}{(x-3)\cancel{(x+2)}(x-1)}$ Removing a factor equal to 1: $\dfrac{x+2}{x+2} = 1$

$= \dfrac{(x+2)(x-2)}{(x-3)(x-1)}$ Simplifying

b) $\dfrac{1-a^3}{a^2} \cdot \dfrac{a^5}{a^2-1}$

$= \dfrac{(1-a^3)a^5}{a^2(a^2-1)}$

$= \dfrac{(1-a)(1+a+a^2)a^5}{a^2(a-1)(a+1)}$ Factoring a difference of cubes and a difference of squares

$= \dfrac{-1(a-1)(1+a+a^2)a^5}{a^2(a-1)(a+1)}$ *Important!* Factoring out -1 reverses the subtraction.

$= \dfrac{\cancel{(a-1)a^2} \cdot a^3(-1)(1+a+a^2)}{\cancel{(a-1)a^2}(a+1)}$ Rewriting a^5 as $a^2 \cdot a^3$; removing a factor equal to 1: $\dfrac{(a-1)a^2}{(a-1)a^2} = 1$

$= \dfrac{-a^3(1+a+a^2)}{a+1}$ Simplifying

As in Example 5, there is no need for us to multiply out the numerator or the denominator of the final result.

Dividing and Simplifying

Two expressions are reciprocals of each other if their product is 1. As in arithmetic, to find the reciprocal of a rational expression, we interchange numerator and denominator.

The reciprocal of $\dfrac{x}{x^2+3}$ is $\dfrac{x^2+3}{x}$.

The reciprocal of $y-8$ is $\dfrac{1}{y-8}$.

Quotients of Rational Expressions

For any rational expressions A/B and C/D, with $B, C, D \neq 0$,

$$\frac{A}{B} \div \frac{C}{D} = \frac{A}{B} \cdot \frac{D}{C}.$$

(To divide two rational expressions, multiply by the reciprocal of the divisor. We often say that we "*invert* and multiply.")

Example 7

Divide. Simplify by removing a factor equal to 1 if possible.

a) $\dfrac{x-2}{x+1} \div \dfrac{x+5}{x-3}$ **b)** $\dfrac{a^2-1}{a-1} \div \dfrac{a^2-2a+1}{a+1}$

Solution

a) $\dfrac{x-2}{x+1} \div \dfrac{x+5}{x-3} = \dfrac{x-2}{x+1} \cdot \dfrac{x-3}{x+5}$ Multiplying by the reciprocal of the divisor

$= \dfrac{(x-2)(x-3)}{(x+1)(x+5)}$ Multiplying the numerators and the denominators

b) $\dfrac{a^2-1}{a-1} \div \dfrac{a^2-2a+1}{a+1} = \dfrac{a^2-1}{a-1} \cdot \dfrac{a+1}{a^2-2a+1}$ Multiplying by the reciprocal of the divisor

$= \dfrac{(a^2-1)(a+1)}{(a-1)(a^2-2a+1)}$ Multiplying the numerators and the denominators

$= \dfrac{(a+1)(a-1)(a+1)}{(a-1)(a-1)(a-1)}$ Factoring the numerator and the denominator

$= \dfrac{(a+1)\cancel{(a-1)}(a+1)}{(a-1)\cancel{(a-1)}(a-1)}$ Removing a factor equal to 1: $\dfrac{a-1}{a-1} = 1$

$= \dfrac{(a+1)(a+1)}{(a-1)(a-1)}$ Simplifying

Study Tip

The procedures covered in this chapter are by their nature rather long. As is often the case in mathematics, it may help to write out each step as you do the problems. If you have difficulty, consider starting over with a new sheet of paper. Don't squeeze your work into a small amount of space. When using lined paper, consider using two spaces at a time, writing the fraction bar on a line of the paper.

Exercise Set 6.1

FOR EXTRA HELP

Digital Video Tutor CD 4 Videotape 11

InterAct Math

Math Tutor Center

MathXL

MyMathLab.com

Photo developing. Rik usually takes 3 hr more than Pearl does to process a day's orders at Liberty Place Photo. If Pearl takes t hr to process a day's orders, the function given by

$$H(t) = \frac{t^2+3t}{2t+3}$$

can be used to determine how long it would take if they worked together.

1. How long will it take them, working together, to complete a day's orders if Pearl can process the orders alone in 5 hr?

2. How long will it take them, working together, to complete a day's orders if Pearl can process the orders alone in 7 hr?

For each rational function, find the function values indicated, provided the value exists.

3. $v(t) = \dfrac{4t^2 - 5t + 2}{t + 3}$; $v(0), v(-2), v(7)$

4. $f(x) = \dfrac{5x^2 + 4x - 12}{6 - x}$; $f(0), f(-1), f(3)$

5. $g(x) = \dfrac{2x^3 - 9}{x^2 - 4x + 4}$; $g(0), g(2), g(-1)$

6. $r(t) = \dfrac{t^2 - 5t + 4}{t^2 - 9}$; $r(1), r(2), r(-3)$

Multiply to obtain equivalent expressions. Do not simplify. Assume that all denominators are nonzero.

7. $\dfrac{4x}{4x} \cdot \dfrac{x - 3}{x + 2}$

8. $\dfrac{3 - a^2}{a - 7} \cdot \dfrac{-1}{-1}$

9. $\dfrac{t - 2}{t + 3} \cdot \dfrac{-1}{-1}$

10. $\dfrac{x - 4}{x + 5} \cdot \dfrac{x - 5}{x - 5}$

Simplify by removing a factor equal to 1.

11. $\dfrac{15x}{5x^2}$

12. $\dfrac{7a^3}{21a}$

13. $\dfrac{18t^3}{27t^7}$

14. $\dfrac{8y^5}{4y^9}$

15. $\dfrac{2a - 10}{2}$

16. $\dfrac{3a + 12}{3}$

17. $\dfrac{15}{25a - 30}$

18. $\dfrac{21}{6x - 9}$

19. $\dfrac{3x - 12}{3x + 15}$

20. $\dfrac{4y - 20}{4y + 12}$

21. $\dfrac{5x + 20}{x^2 + 4x}$

22. $\dfrac{3x + 21}{x^2 + 7x}$

23. $\dfrac{3a - 1}{2 - 6a}$

24. $\dfrac{6 - 5a}{10a - 12}$

25. $\dfrac{8t - 16}{t^2 - 4}$

26. $\dfrac{t^2 - 9}{5t + 15}$

27. $\dfrac{2t - 1}{1 - 4t^2}$

28. $\dfrac{3a - 2}{4 - 9a^2}$

29. $\dfrac{12 - 6x}{5x - 10}$

30. $\dfrac{21 - 7x}{3x - 9}$

31. $\dfrac{a^2 - 25}{a^2 + 10a + 25}$

32. $\dfrac{a^2 - 16}{a^2 - 8a + 16}$

33. $\dfrac{x^2 + 9x + 8}{x^2 - 3x - 4}$

34. $\dfrac{t^2 - 8t - 9}{t^2 + 5t + 4}$

35. $\dfrac{16 - t^2}{t^2 - 8t + 16}$

36. $\dfrac{25 - p^2}{p^2 + 10p + 25}$

Multiply and, if possible, simplify.

37. $\dfrac{5a^3}{3b} \cdot \dfrac{7b^3}{10a^7}$

38. $\dfrac{25a}{9b^8} \cdot \dfrac{3b^5}{5a^2}$

39. $\dfrac{8x - 16}{5x} \cdot \dfrac{x^3}{5x - 10}$

40. $\dfrac{5t^3}{4t - 8} \cdot \dfrac{6t - 12}{10t}$

41. $\dfrac{y^2 - 16}{4y + 12} \cdot \dfrac{y + 3}{y - 4}$

42. $\dfrac{m^2 - n^2}{4m + 4n} \cdot \dfrac{m + n}{m - n}$

43. $\dfrac{x^2 - 16}{x^2} \cdot \dfrac{x^2 - 4x}{x^2 - x - 12}$

44. $\dfrac{y^2 + 10y + 25}{y^2 - 9} \cdot \dfrac{y^2 + 3y}{y + 5}$

45. $\dfrac{7a - 14}{4 - a^2} \cdot \dfrac{5a^2 + 6a + 1}{35a + 7}$

46. $\dfrac{a^2 - 1}{2 - 5a} \cdot \dfrac{15a - 6}{a^2 + 5a - 6}$

Aha! 47. $\dfrac{t^3 - 4t}{t - t^4} \cdot \dfrac{t^4 - t}{4t - t^3}$

48. $\dfrac{x^2 - 6x + 9}{12 - 4x} \cdot \dfrac{x^6 - 9x^4}{x^3 - 3x^2}$

49. $\dfrac{x^2 - 2x - 35}{2x^3 - 3x^2} \cdot \dfrac{4x^3 - 9x}{7x - 49}$

50. $\dfrac{y^2 - 10y + 9}{y^2 - 1} \cdot \dfrac{1 - y^2}{y^2 - 5y - 36}$

51. $\dfrac{c^3 + 8}{c^5 - 4c^3} \cdot \dfrac{c^6 - 4c^5 + 4c^4}{c^2 - 2c + 4}$

52. $\dfrac{x^3 - 27}{x^4 - 9x^2} \cdot \dfrac{x^5 - 6x^4 + 9x^3}{x^2 + 3x + 9}$

53. $\dfrac{a^3 - b^3}{3a^2 + 9ab + 6b^2} \cdot \dfrac{a^2 + 2ab + b^2}{a^2 - b^2}$

54. $\dfrac{x^3 + y^3}{x^2 + 2xy - 3y^2} \cdot \dfrac{x^2 - y^2}{3x^2 + 6xy + 3y^2}$

55. $\dfrac{4x^2 - 9y^2}{8x^3 - 27y^3} \cdot \dfrac{4x^2 + 6xy + 9y^2}{4x^2 + 12xy + 9y^2}$

56. $\dfrac{3x^2 - 3y^2}{27x^3 - 8y^3} \cdot \dfrac{6x^2 + 5xy - 6y^2}{6x^2 + 12xy + 6y^2}$

Divide and, if possible, simplify.

57. $\dfrac{9x^5}{8y^2} \div \dfrac{3x}{16y^9}$

58. $\dfrac{16a^7}{3b^5} \div \dfrac{8a^3}{6b}$

59. $\dfrac{5x + 10}{x^8} \div \dfrac{x + 2}{x^3}$

60. $\dfrac{3y + 15}{y^7} \div \dfrac{y + 5}{y^2}$

61. $\dfrac{x^2 - 4}{x^3} \div \dfrac{x^5 - 2x^4}{x + 4}$

62. $\dfrac{y^2 - 9}{y^2} \div \dfrac{y^5 + 3y^4}{y + 2}$

63. $\dfrac{25x^2 - 4}{x^2 - 9} \div \dfrac{2 - 5x}{x + 3}$

64. $\dfrac{4a^2 - 1}{a^2 - 4} \div \dfrac{2a - 1}{2 - a}$

65. $\dfrac{5y - 5x}{15y^3} \div \dfrac{x^2 - y^2}{3x + 3y}$

66. $\dfrac{x^2 - y^2}{4x + 4y} \div \dfrac{3y - 3x}{12x^2}$

67. $\dfrac{x^2 - 16}{x^2 - 10x + 25} \div \dfrac{3x - 12}{x^2 - 3x - 10}$

68. $\dfrac{y^2 - 36}{y^2 - 8y + 16} \div \dfrac{3y - 18}{y^2 - y - 12}$

69. $\dfrac{y^3 + 3y}{y^2 - 9} \div \dfrac{y^2 + 5y - 14}{y^2 + 4y - 21}$

70. $\dfrac{a^3 + 4a}{a^2 - 16} \div \dfrac{a^2 + 8a + 15}{a^2 + a - 20}$

71. $\dfrac{x^3 - 64}{x^3 + 64} \div \dfrac{x^2 - 16}{x^2 - 4x + 16}$

72. $\dfrac{8y^3 - 27}{64y^3 - 1} \div \dfrac{4y^2 - 9}{16y^2 + 4y + 1}$

73. $\dfrac{8a^3 + b^3}{2a^2 + 3ab + b^2} \div \dfrac{8a^2 - 4ab + 2b^2}{4a^2 + 4ab + b^2}$

74. $\dfrac{x^3 + 8y^3}{2x^2 + 5xy + 2y^2} \div \dfrac{x^3 - 2x^2y + 4xy^2}{8x^2 - 2y^2}$

75. Is it possible to understand how to simplify rational expressions without first understanding how to multiply rational expressions? Why or why not?

76. Nancy *incorrectly* simplifies $\dfrac{x + 2}{x}$ as

$$\frac{x + 2}{x} = \frac{\not{x} + 2}{\not{x}} = 1 + 2 = 3.$$

She insists this is correct because it checks when x is replaced with 1. Explain her misconception.

SKILL MAINTENANCE

Simplify.

77. $\dfrac{3}{10} - \dfrac{8}{15}$

78. $\dfrac{3}{8} - \dfrac{7}{10}$

79. $\dfrac{2}{3} \cdot \dfrac{5}{7} - \dfrac{5}{7} \cdot \dfrac{1}{6}$

80. $\dfrac{4}{7} \cdot \dfrac{1}{5} - \dfrac{3}{10} \cdot \dfrac{2}{7}$

81. $(8x^3 - 5x^2 + 6x + 2) - (4x^3 + 2x^2 - 3x + 7)$

82. $(6t^4 + 9t^3 - t^2 + 4t) - (8t^4 - 2t^3 - 6t + 3)$

SYNTHESIS

83. Tony *incorrectly* argues that since

$$\frac{a^2 - 4}{a - 2} = \frac{a^2}{a} + \frac{-4}{-2} = a + 2,$$

is correct, it follows that

$$\frac{x^2 + 9}{x + 1} = \frac{x^2}{x} + \frac{9}{1} = x + 9.$$

Explain his misconception.

84. Explain why the graphs of $f(x) = 5x$ and $g(x) = \dfrac{5x^2}{x}$ differ.

85. Let

$$g(x) = \frac{2x + 3}{4x - 1}.$$

Determine each of the following.

a) $g(x + h)$
b) $g(2x - 2) \cdot g(x)$
c) $g\left(\frac{1}{2}x + 1\right) \cdot g(x)$

86. Graph the function given by

$$f(x) = \frac{x^2 - 9}{x - 3}.$$

(*Hint*: Determine the domain of f and simplify.)

Perform the indicated operations and simplify.

87. $\left[\dfrac{r^2 - 4s^2}{r + 2s} \div (r + 2s)\right] \cdot \dfrac{2s}{r - 2s}$

88. $\left[\dfrac{d^2 - d}{d^2 - 6d + 8} \cdot \dfrac{d - 2}{d^2 + 5d}\right] \div \dfrac{5d}{d^2 - 9d + 20}$

Aha! **89.** $\left[\dfrac{6t^2 - 26t + 30}{8t^2 - 15t - 21} \cdot \dfrac{5t^2 - 9t - 15}{6t^2 - 14t - 20}\right] \div \dfrac{5t^2 - 9t - 15}{6t^2 - 14t - 20}$

Simplify.

90. $\dfrac{x(x + 1) - 2(x + 3)}{(x + 3)(x + 1)(x + 2)}$

91. $\dfrac{m^2 - t^2}{m^2 + t^2 + m + t + 2mt}$

92. $\dfrac{a^3 - 2a^2 + 2a - 4}{a^3 - 2a^2 - 3a + 6}$

93. $\dfrac{x^3 + x^2 - y^3 - y^2}{x^2 - 2xy + y^2}$

94. $\dfrac{u^6 + v^6 + 2u^3v^3}{u^3 - v^3 + u^2v - uv^2}$

95. $\dfrac{x^5 - x^3 + x^2 - 1 - (x^3 - 1)(x + 1)^2}{(x^2 - 1)^2}$

96. Let

$$f(x) = \frac{4}{x^2 - 1} \quad \text{and} \quad g(x) = \frac{4x^2 + 8x + 4}{x^3 - 1}.$$

Find each of the following.

a) $(f \cdot g)(x)$
b) $(f/g)(x)$
c) $(g/f)(x)$

97. Use a grapher to check Example 6. Use the method described on p. 338.

98. Use a grapher to check your answers to Exercises 27, 45, and 71. Use the method described on p. 338.

99. Use a grapher to show that

$$\frac{x^2 - 16}{x + 2} \neq x - 8.$$

100. To check Example 4, Kara lets

$$y_1 = \frac{7x^2 + 21x}{14x} \quad \text{and} \quad y_2 = \frac{x + 3}{2}.$$

Since the graphs of y_1 and y_2 appear to be identical, Kara believes that the domains of the functions described by y_1 and y_2 are the same, $\mathbb{R}$. How could you convince Kara otherwise?

6.2 Rational Expressions and Functions: Adding and Subtracting

When Denominators Are the Same • When Denominators Are Different

Rational expressions are added in much the same way as the fractions of arithmetic.

When Denominators Are the Same

Addition and Subtraction with Like Denominators

To add or subtract when denominators are the same, add or subtract the numerators and keep the same denominator.

$$\frac{A}{C} + \frac{B}{C} = \frac{A + B}{C} \quad \text{and} \quad \frac{A}{C} - \frac{B}{C} = \frac{A - B}{C}, \quad \text{where } C \neq 0.$$

Example 1

Add: $\dfrac{3+x}{x}+\dfrac{4}{x}$.

Solution

$$\frac{3+x}{x}+\frac{4}{x}=\frac{3+x+4}{x}=\frac{x+7}{x}$$

Because x is not a factor of the numerator and the denominator, the result cannot be simplified.

Example 2

Add: $\dfrac{4x^2-5xy}{x^2-y^2}+\dfrac{2xy-y^2}{x^2-y^2}$.

Solution

$$\frac{4x^2-5xy}{x^2-y^2}+\frac{2xy-y^2}{x^2-y^2}=\frac{4x^2-3xy-y^2}{x^2-y^2}$$

Adding the numerators and combining like terms. The denominator is unchanged.

$$=\frac{(x-y)(4x+y)}{(x-y)(x+y)}$$

Factoring the numerator and the denominator and looking for common factors

$$=\frac{\cancel{(x-y)}(4x+y)}{\cancel{(x-y)}(x+y)}$$

Removing a factor equal to 1: $\dfrac{x-y}{x-y}=1$

$$=\frac{4x+y}{x+y}$$

Simplifying

Recall that a fraction bar is a grouping symbol. The next example shows that when a numerator is subtracted, care must be taken to subtract, or change the sign of, *each* term in that polynomial.

Example 3

If

$$f(x)=\frac{4x+5}{x+3}-\frac{x-2}{x+3},$$

find a simplified form of $f(x)$.

Solution

$$f(x)=\frac{4x+5}{x+3}-\frac{x-2}{x+3}$$

$$=\frac{4x+5-(x-2)}{x+3}$$

The parentheses remind us to subtract *both* terms.

$$=\frac{4x+5-x+2}{x+3}$$

$$=\frac{3x+7}{x+3}$$

Example 3 can be checked by comparing the graphs of

$$y_1=\frac{4x+5}{x+3}-\frac{x-2}{x+3}$$

and $$y_2=\frac{3x+7}{x+3}$$

on the same set of axes. Since the equations are equivalent, one curve (it has two branches) should appear. Equivalently, you can show that $y_3=y_2-y_1$ is 0 for all x not equal to -3. The TABLE or TRACE feature can assist in either type of check.

When Denominators Are Different

In order to add rational expressions such as

$$\frac{7}{12xy^2} + \frac{8}{15x^3y} \quad \text{or} \quad \frac{x}{x^2 - y^2} + \frac{y}{x^2 - 4xy + 3y^2},$$

we must first find common denominators. As in arithmetic, our work is easier when we use the *least common multiple* (LCM) of the denominators.

Least Common Multiple

To find the least common multiple (LCM) of two or more expressions, find the prime factorization of each expression and form a product that contains each factor the greatest number of times that it occurs in any one prime factorization.

Example 4

Find the least common multiple of each pair of polynomials.

a) $21x$ and $3x^2$

b) $x^2 + x - 12$ and $x^2 - 16$

Solution

a) We write the prime factorizations of $21x$ and $3x^2$:

$$21x = 3 \cdot 7 \cdot x \quad \text{and} \quad 3x^2 = 3 \cdot x \cdot x.$$

The factors 3, 7, and x must appear in the LCM if $21x$ is to be a factor of the LCM. The other polynomial, $3x^2$, is not a factor of $3 \cdot 7 \cdot x$ because the prime factors of $3x^2$—namely, 3, x, and x—do not all appear in $3 \cdot 7 \cdot x$. However, if $3 \cdot 7 \cdot x$ is multiplied by another factor of x, a product is formed that contains both $21x$ and $3x^2$ as factors:

$21x$ is a factor.

$$\text{LCM} = 3 \cdot 7 \cdot x \cdot x = 21x^2.$$

$3x^2$ is a factor.

Note that each factor (3, 7, and x) is used the greatest number of times that it occurs as a factor of either $21x$ or $3x^2$. The LCM is $3 \cdot 7 \cdot x \cdot x$, or $21x^2$.

b) We factor both expressions:

$$x^2 + x - 12 = (x - 3)(x + 4),$$
$$x^2 - 16 = (x + 4)(x - 4).$$

The LCM must contain each polynomial as a factor. By multiplying the factors of $x^2 + x - 12$ by $x - 4$, we form a product that contains both $x^2 + x - 12$ and $x^2 - 16$ as factors:

$x^2 + x - 12$ is a factor.

$$\text{LCM} = (x - 3)(x + 4)(x - 4).$$

There is no need to multiply this out.

$x^2 - 16$ is a factor.

Before adding or subtracting rational expressions with unlike denominators, we determine the *least common denominator*, or LCD, by finding the LCM of the denominators. Each rational expression is then multiplied by a form of 1, as needed, to form an equivalent expression that has the LCD.

Example 5

Add: $\frac{2}{21x} + \frac{5}{3x^2}$.

Solution In Example 4(a), we found that the LCD is $3 \cdot 7 \cdot x \cdot x$, or $21x^2$. We now multiply each rational expression by 1, using expressions for 1 that give us the LCD in each expression. To determine what to use, ask "$21x$ times what is $21x^2$?" and "$3x^2$ times what is $21x^2$?" The answers are x and 7, respectively, so we multiply by x/x and $7/7$:

$$\frac{2}{21x} \cdot \frac{x}{x} + \frac{5}{3x^2} \cdot \frac{7}{7} = \frac{2x}{21x^2} + \frac{35}{21x^2}$$ We now have a common denominator.

$$= \frac{2x + 35}{21x^2}.$$ This expression cannot be simplified.

Example 6

Add: $\frac{x^2}{x^2 + 2xy + y^2} + \frac{2x - 2y}{x^2 - y^2}$.

Solution Before we look for an LCD, each denominator is factored:

$$\frac{x^2}{x^2 + 2xy + y^2} + \frac{2x - 2y}{x^2 - y^2} = \frac{x^2}{(x + y)(x + y)} + \frac{2x - 2y}{(x + y)(x - y)}.$$

Although the numerators need not always be factored, we do so if it enables us to simplify. In this case, the rightmost rational expression can be simplified:

$$\frac{x^2}{x^2 + 2xy + y^2} + \frac{2x - 2y}{x^2 - y^2} = \frac{x^2}{(x + y)(x + y)} + \frac{2(x - y)}{(x + y)(x - y)}$$ Factoring

$$= \frac{x^2}{(x + y)(x + y)} + \frac{2}{x + y}.$$ Removing a factor equal to 1: $\frac{x - y}{x - y} = 1$

Note that the LCM of $(x + y)(x + y)$ and $(x + y)$ is $(x + y)(x + y)$. To get the LCD in the second expression, we multiply by 1, using $(x + y)/(x + y)$. Then we add and, if possible, simplify.

$$\frac{x^2}{(x + y)(x + y)} + \frac{2}{x + y} = \frac{x^2}{(x + y)(x + y)} + \frac{2}{x + y} \cdot \frac{x + y}{x + y}$$

$$= \frac{x^2}{(x + y)(x + y)} + \frac{2x + 2y}{(x + y)(x + y)}$$ We now have the LCD.

$$= \frac{x^2 + 2x + 2y}{(x + y)(x + y)}$$ Since the numerator cannot be factored, we cannot simplify further.

Example 7

Subtract: $\dfrac{2y+1}{y^2-7y+6} - \dfrac{y+3}{y^2-5y-6}$.

Solution

$$\frac{2y+1}{y^2-7y+6} - \frac{y+3}{y^2-5y-6}$$

$$= \frac{2y+1}{(y-6)(y-1)} - \frac{y+3}{(y-6)(y+1)}$$ The LCD is $(y-6)(y-1)(y+1)$.

$$= \frac{2y+1}{(y-6)(y-1)} \cdot \frac{y+1}{y+1} - \frac{y+3}{(y-6)(y+1)} \cdot \frac{y-1}{y-1}$$

Multiplying by 1 to get the LCD in each expression

$$= \frac{(2y+1)(y+1) - (y+3)(y-1)}{(y-6)(y-1)(y+1)}$$

$$= \frac{2y^2+3y+1-(y^2+2y-3)}{(y-6)(y-1)(y+1)}$$ The parentheses are important.

$$= \frac{2y^2+3y+1-y^2-2y+3}{(y-6)(y-1)(y+1)}$$

$$= \frac{y^2+y+4}{(y-6)(y-1)(y+1)}$$ We leave the denominator in factored form.

Example 8

Add: $\dfrac{3}{8a} + \dfrac{1}{-8a}$.

Solution

$$\frac{3}{8a} + \frac{1}{-8a} = \frac{3}{8a} + \frac{-1}{-1} \cdot \frac{1}{-8a}$$

When denominators are opposites, we multiply one rational expression by $-1/-1$ to get the LCD.

$$= \frac{3}{8a} + \frac{-1}{8a} = \frac{2}{8a}$$

$$= \frac{\not{2} \cdot 1}{\not{2} \cdot 4a} = \frac{1}{4a}$$ Simplifying by removing a factor equal to 1: $\dfrac{2}{2} = 1$

Example 9

Subtract: $\dfrac{5x}{x-2y} - \dfrac{3y-7}{2y-x}$.

Solution

$$\frac{5x}{x-2y} - \frac{3y-7}{2y-x} = \frac{5x}{x-2y} - \frac{-1}{-1} \cdot \frac{3y-7}{2y-x}$$ Note that $x - 2y$ and $2y - x$ are opposites.

$$= \frac{5x}{x-2y} - \frac{7-3y}{x-2y}$$ Performing the multiplication. *Note*: $-1(2y-x) = -2y + x = x - 2y$.

$$= \frac{5x-(7-3y)}{x-2y}$$

$$= \frac{5x-7+3y}{x-2y}$$

Subtracting. The parentheses are important.

In Example 9, you may have noticed that when $3y - 7$ is multiplied by -1 and subtracted, the result is $-7 + 3y$, which is equivalent to the original $3y - 7$. Thus, instead of multiplying the numerator by -1 and then subtracting, we could have simply *added* $3y - 7$ to $5x$, as in the following:

$$\frac{5x}{x-2y} - \frac{3y-7}{2y-x} = \frac{5x}{x-2y} + (-1)\cdot\frac{3y-7}{2y-x}$$ Rewriting subtraction as addition

$$= \frac{5x}{x-2y} + \frac{1}{-1}\cdot\frac{3y-7}{2y-x}$$ Writing -1 as $\frac{1}{-1}$

$$= \frac{5x}{x-2y} + \frac{3y-7}{x-2y}$$ The opposite of $2y - x$ is $x - 2y$.

$$= \frac{5x+3y-7}{x-2y}.$$ This checks with the answer to Example 9.

Example 10 Perform the indicated operations and simplify:

$$\frac{2x}{x^2-4} + \frac{5}{2-x} - \frac{1}{2+x}.$$

Solution We have

$$\frac{2x}{x^2-4} + \frac{5}{2-x} - \frac{1}{2+x} = \frac{2x}{(x-2)(x+2)} + \frac{5}{2-x} - \frac{1}{2+x}$$ Factoring

$$= \frac{2x}{(x-2)(x+2)} + \frac{-1}{-1}\cdot\frac{5}{(2-x)} - \frac{1}{x+2}$$ Multiplying by $\frac{-1}{-1}$ since $2 - x$ is the opposite of $x - 2$

$$= \frac{2x}{(x-2)(x+2)} + \frac{-5}{x-2} - \frac{1}{x+2}$$ The LCD is $(x - 2)(x + 2)$.

$$= \frac{2x}{(x-2)(x+2)} + \frac{-5}{x-2}\cdot\frac{x+2}{x+2} - \frac{1}{x+2}\cdot\frac{x-2}{x-2}$$ Multiplying by 1 to get the LCD

$$= \frac{2x-5(x+2)-(x-2)}{(x-2)(x+2)} = \frac{2x-5x-10-x+2}{(x-2)(x+2)}$$

$$= \frac{-4x-8}{(x-2)(x+2)} = \frac{-4(x+2)}{(x-2)(x+2)}$$

$$= \frac{-4\cancel{(x+2)}}{(x-2)\cancel{(x+2)}}$$ Removing a factor equal to 1: $\frac{x+2}{x+2} = 1$

$$= \frac{-4}{x-2}, \text{ or } -\frac{4}{x-2}.$$

Another correct answer is $\frac{4}{2-x}$. It is found by writing $-\frac{4}{x-2}$ as $\frac{4}{-(x-2)}$ and then using the distributive law to remove parentheses.

Study Tip

Because there are many steps to these problems, we recommend that you check each step as you go.

Our work in Example 10 indicates that if

$$f(x) = \frac{2x}{x^2 - 4} + \frac{5}{2 - x} - \frac{1}{2 + x}$$

and

$$g(x) = \frac{-4}{x - 2},$$

then, for $x \neq -2$ and $x \neq 2$, we have $f = g$. Note that whereas the domain of f includes all real numbers except -2 or 2, the domain of g excludes only 2. This is illustrated in the graphs below. Methods for drawing such graphs by hand are discussed in more advanced courses. The graphs are for visualization only.

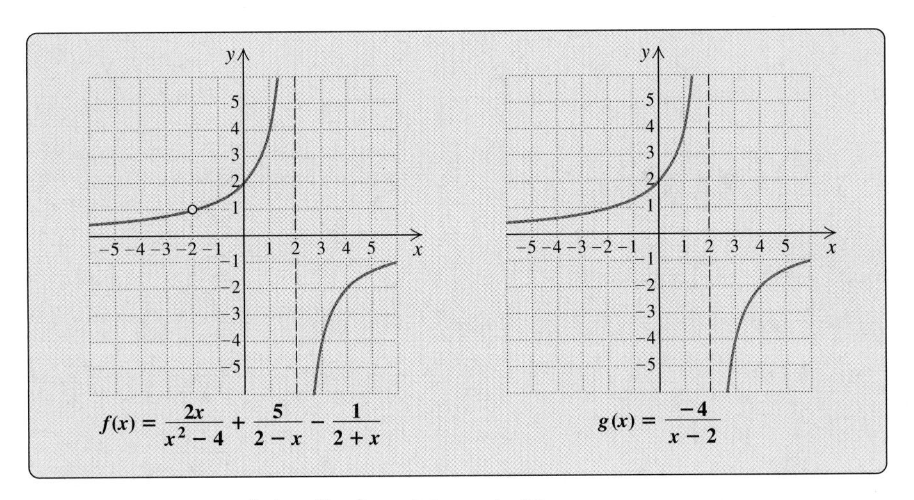

A computer-generated visualization of Example 10

Whenever rational expressions are simplified, a quick partial check is to evaluate both the original and the simplified expressions for a convenient choice of x. For instance, to check Example 10, if $x = 1$, we have

$$f(1) = \frac{2 \cdot 1}{1^2 - 4} + \frac{5}{2 - 1} - \frac{1}{2 + 1}$$

$$= \frac{2}{-3} + \frac{5}{1} - \frac{1}{3} = 5 - \frac{3}{3} = 4$$

and

$$g(1) = \frac{-4}{1 - 2} = \frac{-4}{-1} = 4.$$

Since both functions include the pair $(1, 4)$, our algebra was *probably* correct. Although this is only a partial check (on rare occasions, an incorrect answer might "check"), because it is so easy to perform, it is nonetheless very useful. Further evaluation provides a more definitive check.

Exercise Set 6.2

FOR EXTRA HELP

Digital Video Tutor CD 4
Videotape 11

InterAct Math

Math Tutor Center

MathXL

MyMathLab.com

Perform the indicated operations. Simplify when possible.

1. $\dfrac{5}{3a} + \dfrac{7}{3a}$

2. $\dfrac{3}{2y} + \dfrac{5}{2y}$

3. $\dfrac{1}{4a^2b} - \dfrac{5}{4a^2b}$

4. $\dfrac{5}{3m^2n^2} - \dfrac{4}{3m^2n^2}$

5. $\dfrac{a - 5b}{a + b} + \dfrac{a + 7b}{a + b}$

6. $\dfrac{x - 3y}{x + y} + \dfrac{x + 5y}{x + y}$

7. $\dfrac{4y + 2}{y - 2} - \dfrac{y - 3}{y - 2}$

8. $\dfrac{3t + 2}{t - 4} - \dfrac{t - 2}{t - 4}$

9. $\dfrac{3x - 4}{x^2 - 5x + 4} + \dfrac{3 - 2x}{x^2 - 5x + 4}$

10. $\dfrac{5x - 4}{x^2 - 6x - 7} + \dfrac{5 - 4x}{x^2 - 6x - 7}$

11. $\dfrac{3a - 2}{a^2 - 25} - \dfrac{4a - 7}{a^2 - 25}$

12. $\dfrac{2a - 5}{a^2 - 9} - \dfrac{3a - 8}{a^2 - 9}$

13. $\dfrac{a^2}{a - b} + \dfrac{b^2}{b - a}$

14. $\dfrac{s^2}{r - s} + \dfrac{r^2}{s - r}$

15. $\dfrac{7}{x} - \dfrac{8}{-x}$

16. $\dfrac{2}{a} - \dfrac{5}{-a}$

17. $\dfrac{x - 7}{x^2 - 16} - \dfrac{x - 1}{16 - x^2}$

18. $\dfrac{y - 4}{y^2 - 25} - \dfrac{9 - 2y}{25 - y^2}$

19. $\dfrac{t^2 + 3}{t^4 - 16} + \dfrac{7}{16 - t^4}$

20. $\dfrac{y^2 - 5}{y^4 - 81} + \dfrac{4}{81 - y^4}$

21. $\dfrac{m - 3n}{m^3 - n^3} - \dfrac{2n}{n^3 - m^3}$

22. $\dfrac{r - 6s}{r^3 - s^3} - \dfrac{5s}{s^3 - r^3}$

23. $\dfrac{a + 2}{a - 4} + \dfrac{a - 2}{a + 3}$

24. $\dfrac{a + 3}{a - 5} + \dfrac{a - 2}{a + 4}$

25. $4 + \dfrac{x - 3}{x + 1}$

26. $3 + \dfrac{y + 2}{y - 5}$

27. $\dfrac{4xy}{x^2 - y^2} + \dfrac{x - y}{x + y}$

28. $\dfrac{5ab}{a^2 - b^2} + \dfrac{a + b}{a - b}$

29. $\dfrac{8}{2x^2 - 7x + 5} + \dfrac{3x + 2}{2x^2 - x - 10}$

30. $\dfrac{7}{3y^2 + y - 4} + \dfrac{9y + 2}{3y^2 - 2y - 8}$

31. $\dfrac{4}{x + 1} + \dfrac{x + 2}{x^2 - 1} + \dfrac{3}{x - 1}$

32. $\dfrac{-2}{y + 2} + \dfrac{5}{y - 2} + \dfrac{y + 3}{y^2 - 4}$

33. $\dfrac{x + 6}{5x + 10} - \dfrac{x - 2}{4x + 8}$

34. $\dfrac{a + 3}{5a + 25} - \dfrac{a - 1}{3a + 15}$

35. $\dfrac{5ab}{a^2 - b^2} - \dfrac{a - b}{a + b}$

36. $\dfrac{6xy}{x^2 - y^2} - \dfrac{x + y}{x - y}$

37. $\dfrac{x}{x^2 + 9x + 20} - \dfrac{4}{x^2 + 7x + 12}$

38. $\dfrac{x}{x^2 + 11x + 30} - \dfrac{5}{x^2 + 9x + 20}$

39. $\dfrac{3y}{y^2 - 7y + 10} - \dfrac{2y}{y^2 - 8y + 15}$

40. $\dfrac{5x}{x^2 - 6x + 8} - \dfrac{3x}{x^2 - x - 12}$

41. $\dfrac{2x + 1}{x - y} + \dfrac{5x^2 - 5xy}{x^2 - 2xy + y^2}$

42. $\dfrac{2 - 3a}{a - b} + \dfrac{3a^2 + 3ab}{a^2 - b^2}$

43. $\dfrac{3y + 2}{y^2 + 5y - 24} + \dfrac{7}{y^2 + 4y - 32}$

44. $\dfrac{3x + 2}{x^2 - 7x + 10} + \dfrac{2x}{x^2 - 8x + 15}$

45. $\dfrac{a - 3}{a^2 - 16} - \dfrac{3a - 2}{a^2 + 2a - 24}$

46. $\dfrac{t + 4}{t^2 - 9} - \dfrac{3t - 1}{t^2 + 2t - 3}$

47. $\dfrac{2}{a^2 - 5a + 4} + \dfrac{-2}{a^2 - 4}$

48. $\dfrac{3}{a^2 - 7a + 6} + \dfrac{-3}{a^2 - 9}$

49. $5 + \dfrac{t}{t + 2} - \dfrac{8}{t^2 - 4}$

50. $2 + \dfrac{t}{t - 3} - \dfrac{18}{t^2 - 9}$

51. $\dfrac{2y - 6}{y^2 - 9} - \dfrac{y}{y - 1} + \dfrac{y^2 + 2}{y^2 + 2y - 3}$

52. $\dfrac{x - 1}{x^2 - 1} - \dfrac{x}{x - 2} + \dfrac{x^2 + 2}{x^2 - x - 2}$

Aha! **53.** $\dfrac{5y}{1 - 4y^2} - \dfrac{2y}{2y + 1} + \dfrac{5y}{4y^2 - 1}$

54. $\dfrac{4x}{x^2 - 1} + \dfrac{3x}{1 - x} - \dfrac{4}{x - 1}$

55. $\dfrac{2}{x^2 - 5x + 6} - \dfrac{4}{x^2 - 2x - 3} + \dfrac{2}{x^2 + 4x + 3}$

56. $\dfrac{1}{t^2 + 5t + 6} - \dfrac{2}{t^2 + 3t + 2} - \dfrac{1}{t^2 + 5t + 6}$

57. Janine found that the sum of two rational expressions was $(3 - x)/(x - 5)$. The answer given at the back of the book is $(x - 3)/(5 - x)$. Is Janine's answer incorrect? Why or why not?

58. When two rational expressions are added or subtracted, should the numerator of the result be factored? Why or why not?

SKILL MAINTENANCE

Simplify. Use only positive exponents in your answer.

59. $\dfrac{15x^{-7}y^{12}z^4}{35x^{-2}y^6z^{-3}}$

60. $\dfrac{21a^{-4}b^6c^8}{27a^{-2}b^{-5}c}$

61. $\dfrac{34s^9t^{-40}r^{30}}{10s^{-3}t^{20}r^{-10}}$

62. Find an equation for the line that passes through the point $(-2, 3)$ and is perpendicular to the line $f(x) = -\frac{4}{5}x + 7$.

63. *Value of coins.* There are 50 dimes in a roll of dimes, 40 nickels in a roll of nickels, and 40 quarters in a roll of quarters. Robert has a total of 12 rolls of coins with a total value of \$70.00. If he has 3 more rolls of nickels than dimes, how many of each roll of coins does he have?

64. *Audiotapes.* Anna wants to buy tapes for her work at the campus radio station. She needs some 30-min tapes and some 60-min tapes. If she buys 12 tapes with a total recording time of 10 hr, how many tapes of each length did she buy?

SYNTHESIS

65. Many students make the mistake of always multiplying denominators when looking for a common denominator. Use Example 7 to explain why this approach can yield results that are more difficult to simplify.

66. Is the sum of two rational expressions always a rational expression? Why or why not?

67. *Prescription drugs.* After visiting her doctor, Corinna went to the pharmacy for a two-week supply of Zyrtec®, a 20-day supply of Albuterol®, and a 30-day supply of Pepcid®. Corinna refills each prescription as soon as her supply runs out. How long will it be until she can refill all three prescriptions on the same day?

68. *Astronomy.* The earth, Jupiter, Saturn, and Uranus all revolve around the sun. The earth takes 1 yr, Jupiter 12 yr, Saturn 30 yr, and Uranus 84 yr. How frequently do these four planets line up with each other?

69. *Music.* To duplicate a common African polyrhythm, a drummer needs to play sextuplets (6 beats per measure) on a tom-tom while simultaneously playing quarter notes (4 beats per measure) on a bass drum. Into how many equally sized parts must a measure be divided, in order to precisely execute this rhythm?

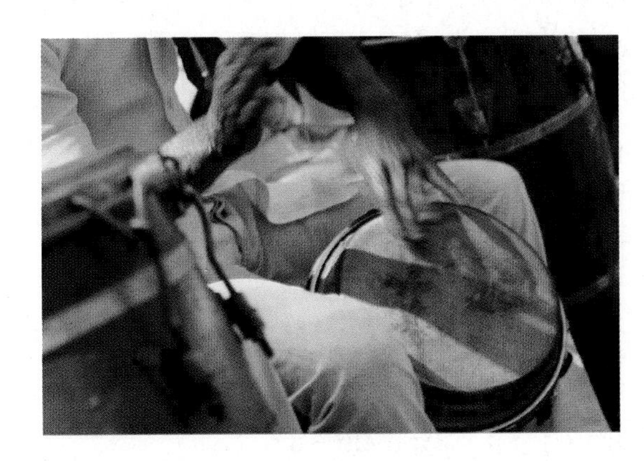

70. *Home appliances.* Refrigerators last an average of 20 yr, clothes washers about 14 yr, and dishwashers about 10 yr (*Source*: U.S. Department of Energy). In 1980, Westgate College bought new refrigerators for its dormitories. In 1990, the college bought new clothes washers and dishwashers. Predict the year in which the college will need to replace all three types of appliances at once.

Find the LCM.

71. $x^8 - x^4,\ x^5 - x^2,\ x^5 - x^3,\ x^5 + x^2$

72. $2a^3 + 2a^2b + 2ab^2,\ a^6 - b^6,$
$2b^2 + ab - 3a^2,\ 2a^2b + 4ab^2 + 2b^3$

73. The LCM of two expressions is $8a^4b^7$. One of the expressions is $2a^3b^7$. List all the possibilities for the other expression.

74. Determine the domain and the range of the function graphed below.

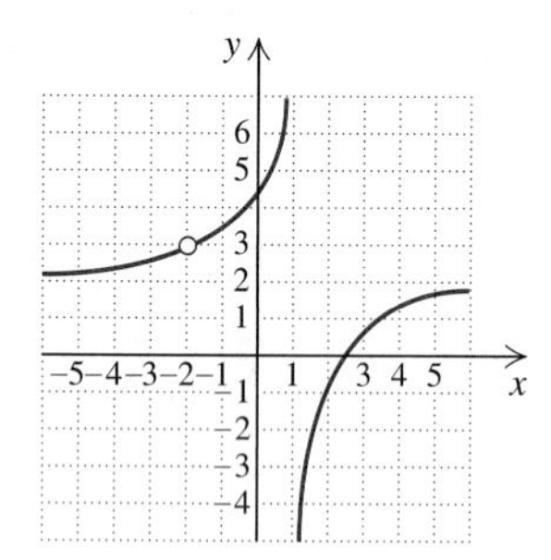

If

$$f(x) = \frac{x^3}{x^2 - 4} \quad \text{and} \quad g(x) = \frac{x^2}{x^2 + 3x - 10},$$

find each of the following.

75. $(f + g)(x)$

76. $(f - g)(x)$

77. $(f \cdot g)(x)$

78. $(f/g)(x)$

Perform the indicated operations and simplify.

79. $5(x - 3)^{-1} + 4(x + 3)^{-1} - 2(x + 3)^{-2}$

80. $4(y - 1)(2y - 5)^{-1} + 5(2y + 3)(5 - 2y)^{-1} + (y - 4)(2y - 5)^{-1}$

81. $\dfrac{x + 4}{6x^2 - 20x} \cdot \left(\dfrac{x}{x^2 - x - 20} + \dfrac{2}{x + 4}\right)$

82. $\dfrac{x^2 - 7x + 12}{x^2 - x - 29/3} \cdot \left(\dfrac{3x + 2}{x^2 + 5x - 24} + \dfrac{7}{x^2 + 4x - 32}\right)$

83. $\dfrac{8t^5}{2t^2 - 10t + 12} \div \left(\dfrac{2t}{t^2 - 8t + 15} - \dfrac{3t}{t^2 - 7t + 10}\right)$

84. $\dfrac{9t^3}{3t^3 - 12t^2 + 9t} \div \left(\dfrac{t + 4}{t^2 - 9} - \dfrac{3t - 1}{t^2 + 2t - 3}\right)$

85. Use a grapher to check your answers to Exercises 9, 25, and 40. Use the method discussed on p. 344.

86. Let

$$f(x) = 2 + \frac{x - 3}{x + 1}.$$

Use algebra, together with a grapher, to determine the domain and the range of f.

6.3 Complex Rational Expressions

Multiplying by 1 • Dividing Two Rational Expressions

A **complex rational expression** is a rational expression that contains rational expressions within its numerator and/or its denominator. Here are some examples:

$$\frac{x + \frac{5}{x}}{4x}, \quad \frac{\frac{x - y}{x + y}}{\frac{2x - y}{3x + y}}, \quad \frac{\frac{7x}{3} - \frac{4}{x}}{\frac{5x}{6} + \frac{8}{3}}, \quad \frac{\frac{r}{6} + \frac{r^2}{144}}{\frac{r}{12}}.$$

The rational expressions within each complex rational expression are red.

Complex rational expressions arise in a variety of real-world applications. For example, the complex rational expression on the far right at the bottom of p. 352 is used when calculating the size of certain loan payments.

Two methods are used to simplify complex rational expressions.

Method 1: Multiplying by 1

One method of simplifying a complex rational expression is to multiply the entire expression by 1. To write 1, we use the LCD of the rational expressions within the complex rational expression.

Example 1 Simplify:

$$\frac{\frac{1}{a^3b} + \frac{1}{b}}{\frac{1}{a^2b^2} - \frac{1}{b^2}}.$$

Solution The denominators within the complex rational expression are a^3b, b, a^2b^2, and b^2. Thus the LCD is a^3b^2. We multiply by 1, using $(a^3b^2)/(a^3b^2)$:

$$\frac{\frac{1}{a^3b} + \frac{1}{b}}{\frac{1}{a^2b^2} - \frac{1}{b^2}} = \frac{\frac{1}{a^3b} + \frac{1}{b}}{\frac{1}{a^2b^2} - \frac{1}{b^2}} \cdot \frac{a^3b^2}{a^3b^2}$$ Multiplying by 1, using the LCD

$$= \frac{\left(\frac{1}{a^3b} + \frac{1}{b}\right)a^3b^2}{\left(\frac{1}{a^2b^2} - \frac{1}{b^2}\right)a^3b^2}$$ Multiplying the numerator and the denominator. Remember to use parentheses.

$$= \frac{\frac{1}{a^3b} \cdot a^3b^2 + \frac{1}{b} \cdot a^3b^2}{\frac{1}{a^2b^2} \cdot a^3b^2 - \frac{1}{b^2} \cdot a^3b^2}$$ Using the distributive law to carry out the multiplications

$$= \frac{\frac{a^3b}{a^3b} \cdot b + \frac{b}{b} \cdot a^3b}{\frac{a^2b^2}{a^2b^2} \cdot a - \frac{b^2}{b^2} \cdot a^3}$$ Removing factors that equal 1. Study this carefully.

$$= \frac{b + a^3b}{a - a^3}$$ Simplifying

$$= \frac{b(1 + a^3)}{a(1 - a^2)}$$ Factoring

$$= \frac{b(1 + a)(1 - a + a^2)}{a(1 + a)(1 - a)}$$ Factoring further and identifying a factor that equals 1

$$= \frac{b(1 - a + a^2)}{a(1 - a)}.$$ Simplifying

Using Multiplication by 1 to Simplify a Complex Rational Expression

1. Find the LCD of all rational expressions *within* the complex rational expression.
2. Multiply the complex rational expression by 1, writing 1 as the LCD divided by itself.
3. Distribute and simplify so that the numerator and the denominator of the complex rational expression are polynomials.
4. Factor and, if possible, simplify.

Note that use of the LCD when multiplying by a form of 1 clears the numerator and the denominator of the complex rational expression of all rational expressions.

Example 2 Simplify:

$$\frac{\dfrac{3}{2x-2}-\dfrac{1}{x+1}}{\dfrac{1}{x-1}+\dfrac{x}{x^2-1}}.$$

Solution In this case, to find the LCD, we have to factor first:

$$\frac{\dfrac{3}{2x-2}-\dfrac{1}{x+1}}{\dfrac{1}{x-1}+\dfrac{x}{x^2-1}}=\frac{\dfrac{3}{2(x-1)}-\dfrac{1}{x+1}}{\dfrac{1}{x-1}+\dfrac{x}{(x-1)(x+1)}}$$

The LCD is $2(x-1)(x+1)$.

$$=\frac{\dfrac{3}{2(x-1)}-\dfrac{1}{x+1}}{\dfrac{1}{x-1}+\dfrac{x}{(x-1)(x+1)}}\cdot\frac{2(x-1)(x+1)}{2(x-1)(x+1)}$$

Multiplying by 1, using the LCD

$$=\frac{\dfrac{3}{2(x-1)}\cdot 2(x-1)(x+1)-\dfrac{1}{x+1}\cdot 2(x-1)(x+1)}{\dfrac{1}{x-1}\cdot 2(x-1)(x+1)+\dfrac{x}{(x-1)(x+1)}\cdot 2(x-1)(x+1)}$$

Using the distributive law

$$=\frac{\dfrac{2(x-1)}{2(x-1)}\cdot 3(x+1)-\dfrac{x+1}{x+1}\cdot 2(x-1)}{\dfrac{x-1}{x-1}\cdot 2(x+1)+\dfrac{(x-1)(x+1)}{(x-1)(x+1)}\cdot 2x}$$

Removing factors that equal 1

$$=\frac{3(x+1)-2(x-1)}{2(x+1)+2x}$$

Simplifying

$$= \frac{3x + 3 - 2x + 2}{2x + 2 + 2x}$$ Using the distributive law

$$= \frac{x + 5}{4x + 2}.$$ Combining like terms

Method 2: Dividing Two Rational Expressions

Another method for simplifying complex rational expressions involves first adding or subtracting, as necessary, to obtain one rational expression in the numerator and one rational expression in the denominator. The problem is thereby simplified to one involving the division of two rational expressions.

Example 3 Simplify:

$$\frac{\frac{3}{x} - \frac{2}{x^2}}{\frac{3}{x-2} + \frac{1}{x^2}}.$$

Solution

$$\frac{\frac{3}{x} - \frac{2}{x^2}}{\frac{3}{x-2} + \frac{1}{x^2}} = \frac{\frac{3}{x} \cdot \frac{x}{x} - \frac{2}{x^2}}{\frac{3}{x-2} \cdot \frac{x^2}{x^2} + \frac{1}{x^2} \cdot \frac{x-2}{x-2}}$$

Multiplying $3/x$ by 1 to obtain x^2 as a common denominator

Multiplying by 1, twice, to obtain $x^2(x - 2)$ as a common denominator

$$= \frac{\frac{3x}{x^2} - \frac{2}{x^2}}{\frac{3x^2}{(x-2)x^2} + \frac{x-2}{x^2(x-2)}}$$

There is now a common denominator in the numerator and a common denominator in the denominator of the complex rational expression.

$$= \frac{\frac{3x-2}{x^2}}{\frac{3x^2 + x - 2}{(x-2)x^2}}$$

Subtracting in the numerator and adding in the denominator. We now have one rational expression divided by another rational expression.

$$= \frac{3x-2}{x^2} \div \frac{3x^2 + x - 2}{(x-2)x^2}$$

Rewriting with a division symbol

$$= \frac{3x-2}{x^2} \cdot \frac{(x-2)x^2}{3x^2 + x - 2}$$

To divide, multiply by the reciprocal of the divisor.

$$= \frac{\cancel{(3x-2)}(x-2)\cancel{x^2}}{\cancel{x^2}\cancel{(3x-2)}(x+1)}$$

Factoring and removing a factor equal to 1: $\frac{x^2(3x - 2)}{x^2(3x - 2)} = 1$

$$= \frac{x-2}{x+1}$$

Using Division to Simplify a Complex Rational Expression

1. Add or subtract, as necessary, to get one rational expression in the numerator.
2. Add or subtract, as necessary, to get one rational expression in the denominator.
3. Perform the indicated division (invert the divisor and multiply).
4. Simplify, if possible, by removing any factors that equal 1.

Example 4 Simplify:

$$\frac{1 + \frac{2}{x}}{1 - \frac{4}{x^2}}.$$

Solution We have

$$\frac{1 + \frac{2}{x}}{1 - \frac{4}{x^2}} = \frac{\frac{x}{x} + \frac{2}{x}}{\frac{x^2}{x^2} - \frac{4}{x^2}} \quad \begin{matrix}\text{Finding a common denominator} \\ \text{Finding a common denominator}\end{matrix}$$

$$= \frac{\frac{x + 2}{x}}{\frac{x^2 - 4}{x^2}} \quad \begin{matrix}\text{Adding in the numerator} \\ \text{Subtracting in the denominator}\end{matrix}$$

$$= \frac{x + 2}{x} \cdot \frac{x^2}{x^2 - 4} \quad \text{Multiplying by the reciprocal of the divisor}$$

$$= \frac{(x + 2) \cdot x^2}{x(x + 2)(x - 2)} \quad \text{Factoring. Remember to simplify when possible.}$$

$$= \frac{\cancel{(x + 2)}\cancel{x} \cdot x}{\cancel{x}\cancel{(x + 2)}(x - 2)} \quad \text{Removing a factor equal to 1: } \frac{(x + 2)x}{(x + 2)x} = 1$$

$$= \frac{x}{x - 2}. \quad \text{Simplifying}$$

As a quick partial check, we select a convenient value for x—say, 1:

$$\frac{1 + \frac{2}{1}}{1 - \frac{4}{1^2}} = \frac{1 + 2}{1 - 4} = \frac{3}{-3} = -1 \quad \text{We evaluated the original expression for } x = 1.$$

and

$$\frac{1}{1 - 2} = \frac{1}{-1} = -1. \quad \text{We evaluated the simplified expression for } x = 1.$$

Since both expressions yield the same result, our simplification is probably correct. More evaluation would provide a more definitive check.

To check Example 4, we can show that the graphs of

$$y_1 = \frac{1 + \frac{2}{x}}{1 - \frac{4}{x^2}}$$

and

$$y_2 = \frac{x}{x - 2}$$

coincide, or we can show that (except for $x = -2$ or 0) their tables of values are identical. We could also check by showing that (except for $x = -2$ or 0 or 2) $y_2 - y_1 = 0$.

1. Use a grapher to check Example 3. What values, if any, can x *not* equal?

If negative exponents occur, we first find an equivalent expression using positive exponents and then proceed as in the preceding examples.

Example 5 Simplify:

$$\frac{a^{-1} + b^{-1}}{a^{-3} + b^{-3}}.$$

Solution

$$\frac{a^{-1} + b^{-1}}{a^{-3} + b^{-3}} = \frac{\dfrac{1}{a} + \dfrac{1}{b}}{\dfrac{1}{a^3} + \dfrac{1}{b^3}}$$ Rewriting with positive exponents. We continue, using method 2.

$$= \frac{\dfrac{1}{a} \cdot \dfrac{b}{b} + \dfrac{1}{b} \cdot \dfrac{a}{a}}{\dfrac{1}{a^3} \cdot \dfrac{b^3}{b^3} + \dfrac{1}{b^3} \cdot \dfrac{a^3}{a^3}}$$ Finding a common denominator; Finding a common denominator

$$= \frac{\dfrac{b}{ab} + \dfrac{a}{ab}}{\dfrac{b^3}{a^3b^3} + \dfrac{a^3}{a^3b^3}}$$

$$= \frac{\dfrac{b + a}{ab}}{\dfrac{b^3 + a^3}{a^3b^3}}$$ Adding in the numerator; Adding in the denominator

$$= \frac{b + a}{ab} \cdot \frac{a^3b^3}{b^3 + a^3}$$ Multiplying by the reciprocal of the divisor

$$= \frac{(b + a) \cdot ab \cdot a^2b^2}{ab(b + a)(b^2 - ab + a^2)}$$ Factoring and looking for common factors

$$= \frac{\cancel{(b + a)} \cdot \cancel{ab} \cdot a^2b^2}{\cancel{ab}\cancel{(b + a)}(b^2 - ab + a^2)}$$ Removing a factor equal to 1: $\frac{(b + a)ab}{(b + a)ab} = 1$

$$= \frac{a^2b^2}{b^2 - ab + a^2}$$

There is no one method that is best to use. For expressions like

$$\frac{\dfrac{3x + 1}{x - 5}}{\dfrac{2 - x}{x + 3}} \quad \text{or} \quad \frac{\dfrac{3}{x} - \dfrac{2}{x}}{\dfrac{1}{x + 1} + \dfrac{5}{x + 1}},$$

the second method is probably easier to use since it is little or no work to write the expression as a quotient of two rational expressions.

On the other hand, expressions like

$$\frac{\frac{3}{a^2b}-\frac{4}{bc^3}}{\frac{1}{b^3c}+\frac{2}{ac^4}} \quad \text{or} \quad \frac{\frac{5}{a^2-b^2}+\frac{2}{a^2+2ab+b^2}}{\frac{1}{a-b}+\frac{4}{a+b}}$$

require fewer steps if we use the first method. Either method can be used with any complex rational expression.

Exercise Set 6.3

FOR EXTRA HELP

Digital Video Tutor CD 5
Videotape 11

InterAct Math

Math Tutor Center

MathXL

MyMathLab.com

Simplify. If possible, use a second method or evaluation as a check.

1. $\dfrac{7+\frac{1}{a}}{\frac{1}{a}-3}$

2. $\dfrac{\frac{1}{y}+2}{\frac{1}{y}-3}$

3. $\dfrac{x-x^{-1}}{x+x^{-1}}$

4. $\dfrac{y+y^{-1}}{y-y^{-1}}$

5. $\dfrac{\frac{6}{x}+\frac{7}{y}}{\frac{7}{x}-\frac{6}{y}}$

6. $\dfrac{\frac{5}{z}+\frac{2}{y}}{\frac{4}{z}-\frac{1}{y}}$

7. $\dfrac{\frac{x^2-y^2}{xy}}{\frac{x-y}{y}}$

8. $\dfrac{\frac{a^2-b^2}{ab}}{\frac{a-b}{b}}$

9. $\dfrac{\frac{3x}{y}-x}{2y-\frac{y}{x}}$

10. $\dfrac{1-\frac{2}{3x}}{x-\frac{4}{9x}}$

11. $\dfrac{a^{-1}+b^{-1}}{\frac{a^2-b^2}{ab}}$

12. $\dfrac{x^{-1}+y^{-1}}{\frac{x^2-y^2}{xy}}$

13. $\dfrac{\frac{1}{x+h}-\frac{1}{x}}{h}$

14. $\dfrac{\frac{1}{a-h}-\frac{1}{a}}{h}$

15. $\dfrac{\frac{x^2-x-12}{x^2-2x-15}}{\frac{x^2+8x+12}{x^2-5x-14}}$

16. $\dfrac{\frac{a^2-4}{a^2+3a+2}}{\frac{a^2-5a-6}{a^2-6a-7}}$

17. $\dfrac{\frac{1}{x-2}+\frac{3}{x-1}}{\frac{2}{x-1}+\frac{5}{x-2}}$

18. $\dfrac{\frac{2}{y-3}+\frac{1}{y+1}}{\frac{3}{y+1}+\frac{4}{y-3}}$

19. $\dfrac{a(a+3)^{-1}-2(a-1)^{-1}}{a(a+3)^{-1}-(a-1)^{-1}}$

20. $\dfrac{a(a+2)^{-1}-3(a-3)^{-1}}{a(a+2)^{-1}-(a-3)^{-1}}$

21. $\dfrac{\frac{x}{x^2+3x-4}-\frac{1}{x^2+3x-4}}{\frac{x}{x^2+6x+8}+\frac{3}{x^2+6x+8}}$

22. $\dfrac{\frac{x}{x^2+5x-6}+\frac{6}{x^2+5x-6}}{\frac{x}{x^2-5x+4}-\frac{2}{x^2-5x+4}}$

23. $\dfrac{\frac{2}{a^2-1}+\frac{1}{a+1}}{\frac{3}{a^2-1}+\frac{2}{a-1}}$

24. $\dfrac{\frac{3}{a^2-9}+\frac{2}{a+3}}{\frac{4}{a^2-9}+\frac{1}{a+3}}$

25. $\dfrac{\frac{5}{x^2-4}-\frac{3}{x-2}}{\frac{4}{x^2-4}-\frac{2}{x+2}}$

26. $\dfrac{\frac{4}{x^2-1}-\frac{3}{x+1}}{\frac{5}{x^2-1}-\frac{2}{x-1}}$

27. $\dfrac{\frac{y}{y^2-4}+\frac{5}{4-y^2}}{\frac{y^2}{y^2-4}+\frac{25}{4-y^2}}$

28. $\dfrac{\frac{y}{y^2-1}+\frac{3}{1-y^2}}{\frac{y^2}{y^2-1}+\frac{9}{1-y^2}}$

29. $\dfrac{\dfrac{y^2}{y^2-9}-\dfrac{y}{y+3}}{\dfrac{y}{y^2-9}-\dfrac{1}{y-3}}$

30. $\dfrac{\dfrac{y^2}{y^2-25}-\dfrac{y}{y-5}}{\dfrac{y}{y^2-25}-\dfrac{1}{y+5}}$

31. $\dfrac{\dfrac{a}{a+3}+\dfrac{4}{5a}}{\dfrac{a}{2a+6}+\dfrac{3}{a}}$

32. $\dfrac{\dfrac{a}{a+2}+\dfrac{5}{a}}{\dfrac{a}{2a+4}+\dfrac{1}{3a}}$

33. $\dfrac{\dfrac{1}{x^2-3x+2}+\dfrac{1}{x^2-4}}{\dfrac{1}{x^2+4x+4}+\dfrac{1}{x^2-4}}$

34. $\dfrac{\dfrac{1}{x^2+3x+2}+\dfrac{1}{x^2-1}}{\dfrac{1}{x^2-1}+\dfrac{1}{x^2-4x+3}}$

35. $\dfrac{\dfrac{3}{a^2-4a+3}+\dfrac{3}{a^2-5a+6}}{\dfrac{3}{a^2-3a+2}+\dfrac{3}{a^2+3a-10}}$

36. $\dfrac{\dfrac{1}{a^2+7a+10}-\dfrac{2}{a^2-7a+12}}{\dfrac{2}{a^2-a-6}-\dfrac{1}{a^2+a-20}}$

Aha! **37.** $\dfrac{\dfrac{y}{y^2-4}-\dfrac{2y}{y^2+y-6}}{\dfrac{2y}{y^2+y-6}-\dfrac{y}{y^2-4}}$

38. $\dfrac{\dfrac{y}{y^2-1}-\dfrac{3y}{y^2+5y+4}}{\dfrac{3y}{y^2-1}-\dfrac{y}{y^2-4y+3}}$

39. $\dfrac{\dfrac{3}{x^2+2x-3}-\dfrac{1}{x^2-3x-10}}{\dfrac{3}{x^2-6x+5}-\dfrac{1}{x^2+5x+6}}$

40. $\dfrac{\dfrac{1}{a^2+7a+12}+\dfrac{1}{a^2+a-6}}{\dfrac{1}{a^2+2a-8}+\dfrac{1}{a^2+5a+4}}$

41. Michael *incorrectly* simplifies

$$\frac{a+b^{-1}}{a+c^{-1}} \quad \text{as} \quad \frac{a+c}{a+b}.$$

What mistake is he making and how could you convince him that this is incorrect?

42. To simplify a complex rational expression in which the sum of two fractions is divided by the difference of the same two fractions, which method is easier? Why?

SKILL MAINTENANCE

Solve.

43. $2(3x-1)+5(4x-3)=3(2x+1)$

44. $5(2x+3)-3(4x+1)=2(3x-5)$

45. Solve for y: $\dfrac{t}{s+y}=r$.

46. Solve: $|2x-3|=7$.

47. *Framing.* Andrea has two rectangular frames. The first frame is 3 cm shorter, and 4 cm narrower, than the second frame. If the perimeter of the second frame is 1 cm less than twice the perimeter of the first, what is the perimeter of each frame?

48. *Earnings.* Antonio received $28 in tips on Monday, $22 in tips on Tuesday, and $36 in tips on Wednesday. How much will Antonio need to receive in tips on Thursday if his average for the four days is to be $30?

SYNTHESIS

49. Lisa claims that she can simplify complex rational expressions without knowing how to add rational expressions. Is this possible? Why or why not?

50. In arithmetic, we are taught that

$$\frac{a}{b} \div \frac{c}{d} = \frac{a}{b} \cdot \frac{d}{c}$$

(to divide by a fraction, we invert and multiply). Use method 1 to explain *why* we do this.

Simplify.

51. $\dfrac{5x^{-2}+10x^{-1}y^{-1}+5y^{-2}}{3x^{-2}-3y^{-2}}$

52. $(a^2-ab+b^2)^{-1}(a^2b^{-1}+b^2a^{-1}) \times (a^{-2}-b^{-2})(a^{-2}+2a^{-1}b^{-1}+b^{-2})^{-1}$

53. *Astronomy.* When two galaxies are moving in opposite directions at velocities v_1 and v_2, an observer in one of the galaxies would see the other galaxy receding at speed

$$\frac{v_1 + v_2}{1 + \frac{v_1 v_2}{c^2}},$$

where c is the speed of light. Determine the observed speed if v_1 and v_2 are both one-fourth the speed of light.

Find and simplify

$$\frac{f(x+h) - f(x)}{h}$$

for each rational function f in Exercises 54–57.

54. $f(x) = \frac{2}{x^2}$

55. $f(x) = \frac{3}{x}$

56. $f(x) = \frac{x}{1 - x}$

57. $f(x) = \frac{2x}{1 + x}$

58. If

$$F(x) = \frac{3 + \frac{1}{x}}{2 - \frac{8}{x^2}},$$

find the domain of F.

59. If

$$G(x) = \frac{x - \frac{1}{x^2 - 1}}{\frac{1}{9} - \frac{1}{x^2 - 16}},$$

find the domain of G.

60. Find the reciprocal of y if

$$y = x^2 + x + 1 + \frac{1}{x} + \frac{1}{x^2}.$$

61. For $f(x) = \frac{2}{2 + x}$, find $f(f(a))$.

62. For $g(x) = \frac{x + 3}{x - 1}$, find $g(g(a))$.

63. Let

$$f(x) = \left[\frac{\frac{x+3}{x-3} + 1}{\frac{x+3}{x-3} - 1}\right]^4.$$

Find a simplified form of $f(x)$ and specify the domain of f.

64. Use a grapher to check your answers to Exercises 3, 17, 31, and 59.

65. Use a grapher to check your answers to Exercises 1, 10, 35, and 58.

66. Use algebra to determine the domain of the function given by

$$f(x) = \frac{\frac{1}{x - 2}}{\frac{x}{x - 2} - \frac{5}{x - 2}}.$$

Then explain how a grapher could be used to check your answer.

67. *Financial planning.* Alexis wishes to invest a portion of each month's pay in an account that pays 7.5% interest. If he wants to have \$30,000 in the account after 10 yr, the amount invested each month is given by

$$\frac{30{,}000 \cdot \frac{0.075}{12}}{\left(1 + \frac{0.075}{12}\right)^{120} - 1}.$$

Find the amount of Alexis' monthly investment.

COLLABORATIVE

CORNER

Which Method Is Best?

Focus: Complex rational expressions

Time: 10–15 minutes

Group size: 2–3

ACTIVITY

Consider the steps in Examples 2 and 3 for simplifying a complex rational expression by each of the two methods. Then, work as a group to simplify

$$\frac{\frac{5}{x+1}-\frac{1}{x}}{\frac{2}{x^2}+\frac{4}{x}}$$

subject to the following conditions.

1. The group should predict which method will more easily simplify this expression.
2. Using the method selected in part (1), one group member should perform the first step in the simplification and then pass the problem on to another member of the group. That person then checks the work, performs the next step, and passes the problem on to another group member. If a mistake is found, the problem should be passed to the person who made the mistake for repair. This process continues until, eventually, the simplification is complete.
3. At the same time that part (2) is being performed, another group member should perform the first step of the solution using the method not selected in part (1). He or she should then pass the problem to another group member and so on, just as in part (2).
4. What method *was* easier? Why? Compare your responses with those of other groups.

6.4 Rational Equations

Solving Rational Equations • Rational Equations and Graphs

CONNECTING THE CONCEPTS

As we mentioned in the Connecting the Concepts feature of Chapter 5, it is not unusual to learn a skill that enables us to write equivalent expressions and then to put that skill to work solving a new type of equation. That is precisely what we are now doing: Sections 6.1–6.3 have been devoted to writing equivalent expressions in which rational expressions appeared. There we used least common denominators to add or subtract. Here in Section 6.4, we return to the task of solving equations, but this time the equations contain rational expressions and the LCD is used as part of the multiplication principle.

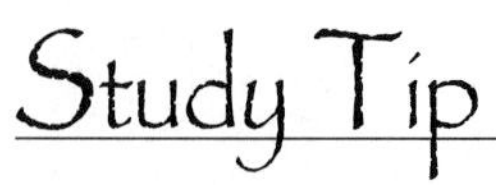

Make sure that all the examples make sense to you before attempting any exercises. This will prove to be a most efficient use of time.

Solving Rational Equations

In Sections 6.1–6.3, we learned how to *simplify expressions.* We now learn to *solve* a new type of *equation.* A **rational equation** is an equation that contains one or more rational expressions. Here are some examples:

$$\frac{2}{3} - \frac{5}{6} = \frac{1}{t}, \qquad \frac{a-1}{a-5} = \frac{4}{a^2-25}, \qquad x^3 + \frac{6}{x} = 5.$$

As you will see in Section 6.5, equations of this type occur frequently in applications. To solve rational equations, recall that one way to *clear fractions* from an equation is to multiply both sides of the equation by the LCD.

To Solve a Rational Equation

Multiply both sides of the equation by the LCD. This is called *clearing fractions* and produces an equation similar to those we have already solved.

Recall that division by 0 is undefined. Note, too, that variables usually appear in at least one denominator of a rational equation. Thus certain numbers can often be ruled out as possible solutions before we even attempt to solve a given rational equation.

Example 1 Solve: $\dfrac{x+4}{3x} + \dfrac{x+8}{5x} = 2.$

Solution Because the left side of this equation is undefined when x is 0, we state at the outset that $x \neq 0$. Next, we multiply both sides of the equation by the LCD, $3 \cdot 5 \cdot x$, or $15x$:

$$15x\left(\frac{x+4}{3x} + \frac{x+8}{5x}\right) = 15x \cdot 2$$ Multiplying by the LCD to clear fractions

$$15x \cdot \frac{x+4}{3x} + 15x \cdot \frac{x+8}{5x} = 15x \cdot 2$$ Using the distributive law

$$\frac{5 \cdot 3x \cdot (x+4)}{3x} + \frac{3 \cdot 5x \cdot (x+8)}{5x} = 30x$$ Locating factors equal to 1

$$5(x+4) + 3(x+8) = 30x$$ Removing factors equal to 1: $\frac{3x}{3x} = 1$; $\frac{5x}{5x} = 1$

$$5x + 20 + 3x + 24 = 30x$$ Using the distributive law

$$8x + 44 = 30x$$

$$44 = 22x$$

$$2 = x.$$ This should check since $x \neq 0$.

Check:

$$\begin{array}{c|c} \dfrac{x+4}{3x} + \dfrac{x+8}{5x} & 2 \\ \hline \dfrac{2+4}{3\cdot 2} + \dfrac{2+8}{5\cdot 2} \;?\; & 2 \\ \dfrac{6}{6} + \dfrac{10}{10} & \\ 2 & 2 \quad \text{TRUE} \end{array}$$

The number 2 is the solution.

Note that when we clear fractions, all denominators "disappear." This leaves an equation without rational expressions, which we know how to solve.

Example 2 Solve: $\dfrac{x-1}{x-5} = \dfrac{4}{x-5}$.

Solution To ensure that neither denominator is 0, we state at the outset the restriction that $x \neq 5$. Then we proceed as before, multiplying both sides by the LCD, $x - 5$:

$$(x-5)\cdot\frac{x-1}{x-5} = (x-5)\cdot\frac{4}{x-5}$$
$$x - 1 = 4$$
$$x = 5. \qquad \text{But recall that } x \neq 5.$$

In this case, it is important to remember that, because of the restriction above, 5 cannot be a solution. A check confirms the necessity of that restriction.

Check:

$$\begin{array}{c|c} \dfrac{x-1}{x-5} & \dfrac{4}{x-5} \\ \hline \dfrac{5-1}{5-5} \;?\; & \dfrac{4}{5-5} \\ \dfrac{4}{0} & \dfrac{4}{0} \end{array} \qquad \text{Division by 0 is undefined.}$$

This equation has no solution.

To see why 5 is not a solution of Example 2, note that the multiplication principle for equations requires that we multiply both sides by a *nonzero* number. When both sides of an equation are multiplied by an expression containing variables, it is possible that certain replacements will make that expression equal to 0. Thus it is safe to say that *if* a solution of

$$\frac{x-1}{x-5} = \frac{4}{x-5}$$

exists, then it is also a solution of $x - 1 = 4$. We *cannot* conclude that every solution of $x - 1 = 4$ is a solution of the original equation.

> ***Caution!*** When solving rational equations, do not forget to list any restrictions as part of the first step.

Example 3

Solve: $\dfrac{x^2}{x-3} = \dfrac{9}{x-3}$.

Solution Note that $x \neq 3$. Since the LCD is $x - 3$, we multiply both sides by $x - 3$:

$$(x-3) \cdot \frac{x^2}{x-3} = (x-3) \cdot \frac{9}{x-3}$$

$$x^2 = 9 \qquad \text{Simplifying}$$

$$x^2 - 9 = 0 \qquad \text{Getting 0 on one side}$$

$$(x-3)(x+3) = 0 \qquad \text{Factoring}$$

$$x = 3 \quad or \quad x = -3. \qquad \text{Using the principle of zero products}$$

Although 3 is a solution of $x^2 = 9$, it must be rejected as a solution of the rational equation. You should perform a check to confirm that -3 *is* a solution despite the fact that 3 is not.

Example 4

Solve: $\dfrac{2}{x+5} + \dfrac{1}{x-5} = \dfrac{16}{x^2-25}$.

Solution To find all restrictions and to assist in finding the LCD, we factor:

$$\frac{2}{x+5} + \frac{1}{x-5} = \frac{16}{(x+5)(x-5)}. \qquad \text{Factoring } x^2 - 25$$

Note that $x \neq -5$ and $x \neq 5$. We multiply by the LCD, $(x+5)(x-5)$, and then use the distributive law:

$$(x+5)(x-5)\left(\frac{2}{x+5} + \frac{1}{x-5}\right) = (x+5)(x-5) \cdot \frac{16}{(x+5)(x-5)}$$

$$(x+5)(x-5)\frac{2}{x+5} + (x+5)(x-5)\frac{1}{x-5}$$

$$= (x+5)(x-5) \cdot \frac{16}{(x+5)(x-5)}$$

$$2(x-5) + (x+5) = 16$$

$$2x - 10 + x + 5 = 16$$

$$3x - 5 = 16$$

$$3x = 21$$

$$x = 7.$$

A check will confirm that the solution is 7.

Rational equations often appear when we are working with functions.

Example 5 Let $f(x) = x + \frac{6}{x}$. Find all values of a for which $f(a) = 5$.

Solution Since $f(a) = a + \frac{6}{a}$, the problem asks that we find all values of a for which

$$a + \frac{6}{a} = 5.$$

First note that $a \neq 0$. To solve for a, we multiply both sides of the equation by the LCD, a:

$$a\left(a + \frac{6}{a}\right) = 5 \cdot a$$ Multiplying both sides by a. Parentheses are important.

$$a \cdot a + a \cdot \frac{6}{a} = 5a$$ Using the distributive law

$$a^2 + 6 = 5a$$ Simplifying

$$a^2 - 5a + 6 = 0$$ Getting 0 on one side

$$(a - 3)(a - 2) = 0$$ Factoring

$$a = 3 \quad or \quad a = 2.$$ Using the principle of zero products

Check: $f(3) = 3 + \frac{6}{3} = 3 + 2 = 5;$

$f(2) = 2 + \frac{6}{2} = 2 + 3 = 5.$

The solutions are 2 and 3. For $a = 2$ or $a = 3$, we have $f(a) = 5$.

technology connection

There are several ways in which Example 5 can be checked. One way is to confirm that the graphs of $y_1 = x + 6/x$ and $y_2 = 5$ intersect at $x = 2$ and $x = 3$. You can also use a table to check that $y_1 = y_2$ when x is 2 and again when x is 3.

Use a grapher to check Examples 1–3.

Rational Equations and Graphs

One way to visualize the solution to Example 5 is to make a graph. This can be done by graphing

$$f(x) = x + \frac{6}{x}$$

with a computer, with a calculator, or by hand. We then inspect the graph for any x-values that are paired with 5. (Note that no y-value is paired with 0, since 0 is not in the domain of f.) It appears from the graph that $f(x) = 5$ when $x \approx 2$ or $x \approx 3$. Although making a graph is not the fastest or most precise method of solving a rational equation, it provides visualization and is useful when problems are too difficult to solve algebraically.

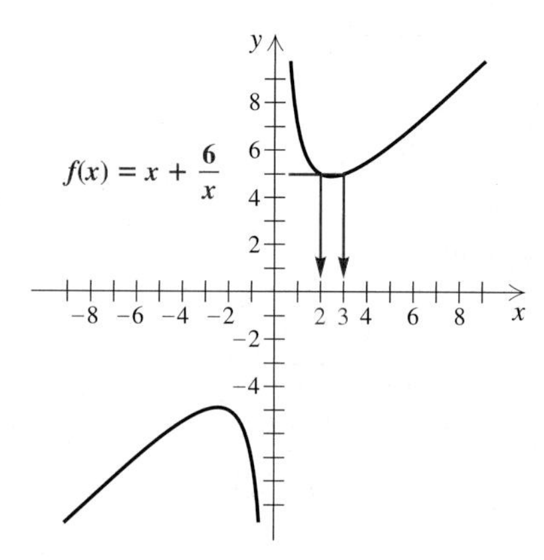

A computer-generated visualization of Example 5

Exercise Set 6.4

FOR EXTRA HELP

Digital Video Tutor CD 5 Videotape 11

InterAct Math

Math Tutor Center

MathXL

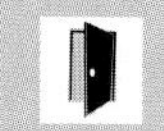
MyMathLab.com

Solve.

1. $\frac{4}{5} + \frac{1}{3} = \frac{x}{9}$
2. $\frac{7}{8} + \frac{2}{5} = \frac{x}{20}$
3. $\frac{x}{3} - \frac{x}{4} = 12$
4. $\frac{y}{5} - \frac{y}{3} = 15$
5. $\frac{1}{3} - \frac{1}{x} = \frac{5}{6}$
6. $\frac{5}{8} - \frac{1}{a} = \frac{2}{5}$
7. $\frac{1}{2} - \frac{2}{7} = \frac{3}{2x}$
8. $\frac{2}{3} - \frac{1}{5} = \frac{7}{3x}$
9. $\frac{12}{15} - \frac{1}{3x} = \frac{4}{5}$
10. $\frac{2}{6} + \frac{1}{2x} = \frac{1}{3}$
11. $\frac{4}{3y} - \frac{3}{y} = \frac{10}{3}$
12. $y + \frac{4}{y} = -5$
13. $\frac{x-2}{x-4} = \frac{2}{x-4}$
14. $\frac{y-1}{y-3} = \frac{2}{y-3}$
15. $\frac{5}{4t} = \frac{7}{5t-2}$
16. $\frac{3}{x-2} = \frac{5}{x+4}$
17. *Aha!* $\frac{x^2+4}{x-1} = \frac{5}{x-1}$
18. $\frac{x^2-1}{x+2} = \frac{3}{x+2}$
19. $\frac{6}{a+1} = \frac{a}{a-1}$
20. $\frac{4}{a-7} = \frac{-2a}{a+3}$
21. $\frac{60}{t-5} - \frac{18}{t} = \frac{40}{t}$
22. $\frac{50}{t-2} - \frac{16}{t} = \frac{30}{t}$
23. $\frac{3}{x} + \frac{x}{x+2} = \frac{4}{x^2+2x}$
24. $\frac{x}{x+1} + \frac{5}{x} = \frac{1}{x^2+x}$

In Exercises 25–30, a rational function f is given. Find all values of a for which f(a) is the indicated value.

25. $f(x) = 2x - \frac{15}{x}$; $f(a) = 7$
26. $f(x) = 2x - \frac{6}{x}$; $f(a) = 1$
27. $f(x) = \frac{x-5}{x+1}$; $f(a) = \frac{3}{5}$
28. $f(x) = \frac{x-3}{x+2}$; $f(a) = \frac{1}{5}$
29. $f(x) = \frac{12}{x} - \frac{12}{2x}$; $f(a) = 8$
30. $f(x) = \frac{6}{x} - \frac{6}{2x}$; $f(a) = 5$

Solve.

31. $\frac{5}{x+2} - \frac{3}{x-2} = \frac{2x}{4-x^2}$
32. $\frac{y+3}{y+2} - \frac{y}{y^2-4} = \frac{y}{y-2}$
33. $\frac{2}{a+4} + \frac{2a-1}{a^2+2a-8} = \frac{1}{a-2}$

34. $\dfrac{3}{x^2 - 6x + 9} + \dfrac{x - 2}{3x - 9} = \dfrac{x}{2x - 6}$

35. $\dfrac{2}{x + 3} - \dfrac{3x + 5}{x^2 + 4x + 3} = \dfrac{5}{x + 1}$

36. $\dfrac{3 - 2y}{y + 1} - \dfrac{10}{y^2 - 1} = \dfrac{2y + 3}{1 - y}$

37. $\dfrac{x - 1}{x^2 - 2x - 3} + \dfrac{x + 2}{x^2 - 9} = \dfrac{2x + 5}{x^2 + 4x + 3}$

38. $\dfrac{2x + 1}{x^2 - 3x - 10} + \dfrac{x - 1}{x^2 - 4} = \dfrac{3x - 1}{x^2 - 7x + 10}$

39. $\dfrac{3}{x^2 - x - 12} + \dfrac{1}{x^2 + x - 6} = \dfrac{4}{x^2 + 3x - 10}$

40. $\dfrac{3}{x^2 - 2x - 3} - \dfrac{1}{x^2 - 1} = \dfrac{2}{x^2 - 8x + 7}$

41. Explain how one can easily produce rational equations for which no solution exists. (*Hint*: Examine Example 2.)

42. Below are unlabeled graphs of $f(x) = x + 2$ and $g(x) = (x^2 - 4)/(x - 2)$. How could you determine which graph represents f and which graph represents g?

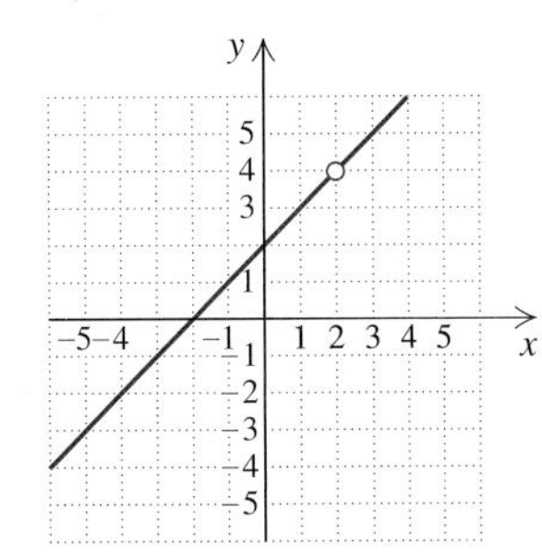

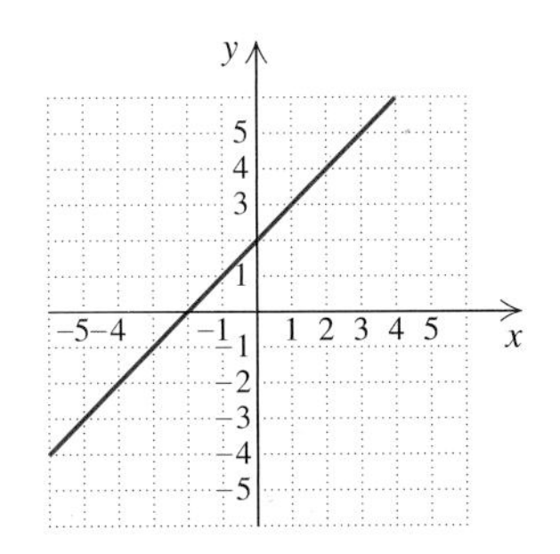

SKILL MAINTENANCE

43. *Test questions.* There are 70 questions on a test. The questions are either multiple-choice, true–false, or fill-in. There are twice as many true–false as fill-in and 5 fewer multiple-choice than true–false. How many of each type of question are there on the test?

44. *Dinner prices.* The Danville Volunteer Fire Department served 250 dinners. A child's dinner cost \$3 and an adult's dinner cost \$8. The total amount of money collected was \$1410. How many of each type of dinner was served?

45. *Agriculture.* The perimeter of a rectangular corn field is 628 m. The length of the field is 6 m greater than the width. Find the area of the field.

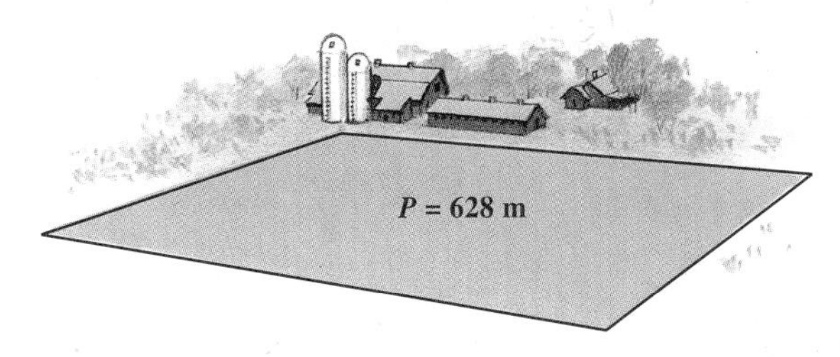

46. Find two consecutive positive even numbers whose product is 288.

47. Determine whether each of the following systems is consistent or inconsistent.

a) $2x - 3y = 4,$
$4x - 6y = 7$

b) $x + 3y = 2,$
$2x - 3y = 1$

48. Solve: $|x - 2| > 3$.

SYNTHESIS

49. Is the following statement true or false: "For any real numbers a, b, and c, if $ac = bc$, then $a = b$"? Explain why you answered as you did.

50. When checking a possible solution of a rational equation, is it sufficient to check that the "solution" does not make any denominator equal to 0? Why or why not?

For each pair of functions f and g, find all values of a for which $f(a) = g(a)$.

51. $f(x) = \dfrac{x - \frac{2}{3}}{x + \frac{1}{2}}, \quad g(x) = \dfrac{x + \frac{2}{3}}{x - \frac{3}{2}}$

52. $f(x) = \dfrac{2 - \frac{x}{4}}{2}, \quad g(x) = \dfrac{\frac{x}{4} - 2}{\frac{x}{2} + 2}$

53. $f(x) = \dfrac{x + 3}{x + 2} - \dfrac{x + 4}{x + 3}, \quad g(x) = \dfrac{x + 5}{x + 4} - \dfrac{x + 6}{x + 5}$

54. $f(x) = \dfrac{1}{1 + x} + \dfrac{x}{1 - x}, \quad g(x) = \dfrac{1}{1 - x} - \dfrac{x}{1 + x}$

55. $f(x) = \dfrac{0.793}{x} + 18.15, \quad g(x) = \dfrac{6.034}{x} - 43.17$

56. $f(x) = \frac{2.315}{x} - \frac{12.6}{17.4}$, $g(x) = \frac{6.71}{x} + 0.763$

Recall that identities are true for any possible replacement of the variable(s). Determine whether each of the following equations is an identity.

57. $\frac{x^2 + 6x - 16}{x - 2} = x + 8, \; x \neq 2$

58. $\frac{x^3 + 8}{x^2 - 4} = \frac{x^2 - 2x + 4}{x - 2}, \; x \neq -2, x \neq 2$

59. Use a grapher to check your answers to Exercises 3, 27, and 33.

60. Use a grapher to check your answers to Exercises 4, 18, and 28.

61. Use a grapher with a TABLE feature to show that 2 is not in the domain of f, if $f(x) = (x^2 - 4)/(x - 2)$. (See Exercise 42.)

62. Can Exercise 61 be answered on a grapher using only graphs? Why or why not?

Solving Applications Using Rational Equations

6.5

Problems Involving Work • Problems Involving Motion

Now that we are able to solve rational equations, it is possible to solve problems that we could not have handled before. The five problem-solving steps remain the same.

Problems Involving Work

Example 1

Sue can mow a lawn in 4 hr. Lenny can mow the same lawn in 5 hr. How long would it take both of them, working together, to mow the lawn?

Solution

1. **Familiarize.** We familiarize ourselves with the problem by considering two *incorrect* ways of translating the problem to mathematical language.

   a) One *incorrect* way to translate the problem is to add the two times:

   4 hr + 5 hr = 9 hr.

   Think about this. Sue can do the job *alone* in 4 hr. If Sue and Lenny work together, whatever time it takes them must be *less* than 4 hr.

   b) Another *incorrect* approach is to assume that each person mows half the lawn. Were this the case,

   Sue would mow $\frac{1}{2}$ the lawn in $\frac{1}{2}$(4 hr), or 2 hr

   and

   Lenny would mow $\frac{1}{2}$ the lawn in $\frac{1}{2}$(5 hr), or $2\frac{1}{2}$ hr.

   But time would be wasted since Sue would finish $\frac{1}{2}$ hr before Lenny. Were Sue to help Lenny after completing her half, the entire job would take between 2 and $2\frac{1}{2}$ hr. This information provides a partial check on any answer we get—the answer should be between 2 and $2\frac{1}{2}$ hr.

Let's consider how much of the job each person completes in 1 hr, 2 hr, 3 hr, and so on. Since Sue takes 4 hr to mow the entire lawn, in 1 hr she mows $\frac{1}{4}$ of the lawn. Since Lenny takes 5 hr to mow the entire lawn, in 1 hr he mows $\frac{1}{5}$ of the lawn. Thus Sue works at a rate of $\frac{1}{4}$ lawn per hr, and Lenny works at a rate of $\frac{1}{5}$ lawn per hr.

Working together, Sue and Lenny mow $\frac{1}{4} + \frac{1}{5}$ of the lawn in 1 hr, so their rate—as a team—is $\frac{5}{20} + \frac{4}{20} = \frac{9}{20}$ lawn per hr.

In 2 hr, Sue mows $\frac{1}{4} \cdot 2$ of the lawn and Lenny mows $\frac{1}{5} \cdot 2$ of the lawn. Working together, they mow

$\frac{1}{4} \cdot 2 + \frac{1}{5} \cdot 2$, or $\frac{9}{10}$ of the lawn in 2 hr. Note that $\frac{9}{20} \cdot 2 = \frac{9}{10}$.

Continuing this pattern, we can form a table.

Time	Fraction of the Lawn Mowed			
	By Sue	By Lenny	Together	
1 hr	$\frac{1}{4}$	$\frac{1}{5}$	$\frac{1}{4} + \frac{1}{5}$, or $\frac{9}{20}$	
2 hr	$\frac{1}{4} \cdot 2$	$\frac{1}{5} \cdot 2$	$(\frac{1}{4} + \frac{1}{5})2$, or $\frac{9}{20} \cdot 2$, or $\frac{9}{10}$	← This is too little.
3 hr	$\frac{1}{4} \cdot 3$	$\frac{1}{5} \cdot 3$	$(\frac{1}{4} + \frac{1}{5})3$, or $\frac{9}{20} \cdot 3$, or $\frac{27}{20}$	← This is too much.
t hr	$\frac{1}{4} \cdot t$	$\frac{1}{5} \cdot t$	$(\frac{1}{4} + \frac{1}{5})t$, or $\frac{9}{20} \cdot t$	

From the table, we note that the number of hours t required for Sue and Lenny to mow exactly one lawn is between 2 hr and 3 hr.

2. **Translate.** From the table, we see that t must be some number for which

Fraction of lawn done by Sue in t hr; Fraction of lawn done by Lenny in t hr

$$\frac{1}{4} \cdot t + \frac{1}{5} \cdot t = 1,$$

or

$$\frac{t}{4} + \frac{t}{5} = 1.$$

3. **Carry out.** We solve the equation:

$$\frac{t}{4} + \frac{t}{5} = 1$$

$$20\left(\frac{t}{4} + \frac{t}{5}\right) = 20 \cdot 1 \quad \text{Multiplying by the LCD}$$

$$\frac{20t}{4} + \frac{20t}{5} = 20 \quad \text{Distributing the 20}$$

$$5t + 4t = 20 \quad \text{Simplifying}$$

$$9t = 20$$

$$t = \frac{20}{9}, \text{ or } 2\tfrac{2}{9}.$$

4. **Check.** In $\frac{20}{9}$ hr, Sue mows $\frac{1}{4} \cdot \frac{20}{9}$, or $\frac{5}{9}$, of the lawn and Lenny mows $\frac{1}{5} \cdot \frac{20}{9}$, or $\frac{4}{9}$, of the lawn. Together, they mow $\frac{5}{9} + \frac{4}{9}$, or 1 lawn. The fact that our solution is between 2 and $2\frac{1}{2}$ hr (see step 1 above) is also a check.
5. **State.** It will take $2\frac{2}{9}$ hr for Sue and Lenny, working together, to mow the lawn.

Example 2

It takes Ruth 9 hr more than Willie to repaint a car. Working together, they can do the job in 20 hr. How long would it take, each working alone, to repaint a car?

Solution

1. **Familiarize.** Unlike Example 1, this problem does not provide us with the times required by the individuals to do the job alone. Let's have w = the number of hours it would take Willie working alone and $w + 9$ = the number of hours it would take Ruth working alone.

2. **Translate.** Using the same reasoning as in Example 1, we see that Willie completes $\dfrac{1}{w}$ of the job in 1 hr and Ruth completes $\dfrac{1}{w+9}$ of the job in 1 hr. In 2 hr, Willie completes $\dfrac{1}{w} \cdot 2$ of the job and Ruth completes $\dfrac{1}{w+9} \cdot 2$ of the job. We are told that, working together, Willie and Ruth can complete the entire job in 20 hr. This gives the following:

Fraction of job done by Willie in 20 hr; Fraction of job done by Ruth in 20 hr

$$\frac{1}{w} \cdot 20 + \frac{1}{w+9} \cdot 20 = 1$$

or

$$\frac{20}{w} + \frac{20}{w+9} = 1.$$

3. **Carry out.** We solve the equation:

$$\frac{20}{w} + \frac{20}{w+9} = 1$$

$$w(w+9)\left(\frac{20}{w} + \frac{20}{w+9}\right) = w(w+9)1 \quad \text{Multiplying by the LCD}$$

$$(w+9)20 + w \cdot 20 = w(w+9) \quad \text{Distributing and simplifying}$$

$$40w + 180 = w^2 + 9w$$

$$0 = w^2 - 31w - 180 \quad \text{Obtaining 0 on one side}$$

$$0 = (w-36)(w+5) \quad \text{Factoring}$$

$$w - 36 = 0 \quad or \quad w + 5 = 0 \quad \text{Principle of zero products}$$

$$w = 36 \quad or \quad w = -5.$$

4. **Check.** Since negative time has no meaning in the problem, -5 is not a solution to the original problem. The number 36 checks since, if Willie takes 36 hr alone and Ruth takes $36 + 9 = 45$ hr alone, in 20 hr they would have completed

$$\frac{20}{36} + \frac{20}{45} = \frac{5}{9} + \frac{4}{9} = 1 \text{ paint job.}$$

5. **State.** It would take Willie 36 hr to repaint a car alone, and Ruth 45 hr.

The equations used in Examples 1 and 2 can be generalized as follows.

Modeling Work Problems

If

$a =$ the time needed for A to complete the work alone,

$b =$ the time needed for B to complete the work alone, and

$t =$ the time needed for A and B to complete the work together,

then

$$\frac{t}{a} + \frac{t}{b} = 1.$$

The following are equivalent equations that can also be used:

$$\frac{1}{a} \cdot t + \frac{1}{b} \cdot t = 1 \quad \text{and} \quad \frac{1}{a} + \frac{1}{b} = \frac{1}{t}.$$

Problems Involving Motion

Problems dealing with distance, rate (or speed), and time are called **motion problems**. To translate them, we use either the basic motion formula, $d = rt$, or the formulas $r = d/t$ or $t = d/r$, which can be derived from $d = rt$.

Example 3 A racer is bicycling 15 km/h faster than a person on a mountain bike. In the time it takes the racer to travel 80 km, the person on the mountain bike has gone 50 km. Find the speed of each bicyclist.

Solution

1. **Familiarize.** Let's guess that the person on the mountain bike is going 10 km/h. The racer would then be traveling $10 + 15$, or 25 km/h. At 25 km/h, the racer will travel 80 km in $\frac{80}{25} = 3.2$ hr. Going 10 km/h, the mountain bike will cover 50 km in $\frac{50}{10} = 5$ hr. Since $3.2 \neq 5$, our guess was wrong, but we can see that if $r =$ the rate, in kilometers per hour, of the slower bike, then the rate of the racer $= r + 15$.

Making a drawing and constructing a table can be helpful.

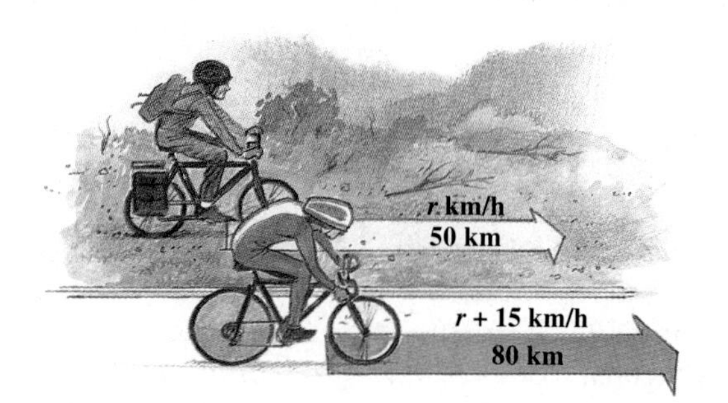

	Distance	Speed	Time
Mountain Bike	50	r	t
Racing Bike	80	$r + 15$	t

2. **Translate.** By looking at how we checked our guess, we see that in the **Time** column of the table, the t's can be replaced, using the formula *Time* = *Distance*/*Rate*, as follows.

	Distance	Speed	Time
Mountain Bike	50	r	$50/r$
Racing Bike	80	$r + 15$	$80/(r + 15)$

Since we are told that the times must be the same, we can write an equation:

$$\frac{50}{r} = \frac{80}{r + 15}.$$

3. **Carry out.** We solve the equation:

$$\frac{50}{r} = \frac{80}{r + 15}$$

$$r(r + 15)\frac{50}{r} = r(r + 15)\frac{80}{r + 15} \quad \text{Multiplying by the LCD}$$

$$50r + 750 = 80r \quad \text{Simplifying}$$

$$750 = 30r$$

$$25 = r.$$

4. **Check.** If our answer checks, the mountain bike is going 25 km/h and the racing bike is going $25 + 15 = 40$ km/h.

Traveling 80 km at 40 km/h, the racer is riding for $\frac{80}{40} = 2$ hr. Traveling 50 km at 25 km/h, the person on the mountain bike is riding for $\frac{50}{25} = 2$ hr. Our answer checks since the two times are the same.

5. **State.** The speed of the racer is 40 km/h, and the speed of the person on the mountain bike is 25 km/h.

In the following example, although the distance is the same in both directions, the key to the translation lies in an additional piece of given information.

Example 4 A Hudson River tugboat goes 10 mph in still water. It travels 24 mi upstream and 24 mi back in a total time of 5 hr. What is the speed of the current?

Solution

1. **Familiarize.** Let's guess that the speed of the current is 4 mph. The tugboat would then be moving $10 - 4 = 6$ mph upstream and $10 + 4 = 14$ mph downstream. The tugboat would require $\frac{24}{6} = 4$ hr to travel 24 mi upstream and $\frac{24}{14} = 1\frac{5}{7}$ hr to travel 24 mi downstream. Since the total time, $4 + 1\frac{5}{7} = 5\frac{5}{7}$ hr, is not the 5 hr mentioned in the problem, we know that our guess is wrong.

   Suppose that the current's speed $= c$ mph. The tugboat would then travel $10 - c$ mph when going upstream and $10 + c$ mph when going downstream.

   A sketch and table can help display the information.

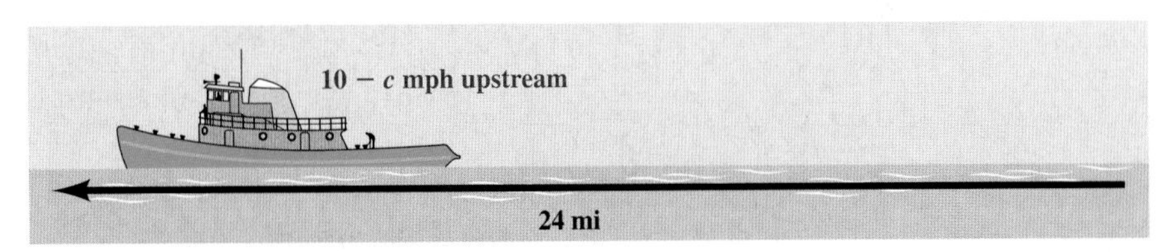

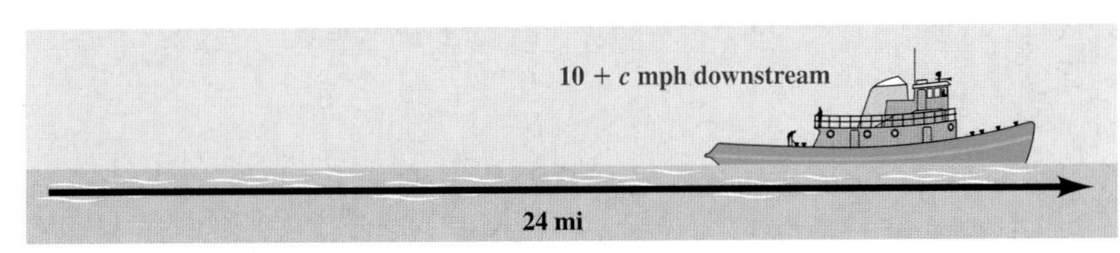

	Distance	Speed	Time
Upstream	24	$10 - c$	t_1
Downstream	24	$10 + c$	t_2

2. **Translate.** From examining our guess, we see that the time traveled can be represented using the formula *Time* = *Distance*/*Rate*:

	Distance	Speed	Time
Upstream	24	$10 - c$	$24/(10 - c)$
Downstream	24	$10 + c$	$24/(10 + c)$

Since the total time upstream and back is 5 hr, we use the last column of the table to form an equation:

$$\frac{24}{10 - c} + \frac{24}{10 + c} = 5.$$

3. **Carry out.** We solve the equation:

$$\frac{24}{10 - c} + \frac{24}{10 + c} = 5$$

$$(10 - c)(10 + c)\left[\frac{24}{10 - c} + \frac{24}{10 + c}\right] = (10 - c)(10 + c)5 \quad \text{Multiplying by the LCD}$$

$$24(10 + c) + 24(10 - c) = (100 - c^2)5$$

$$480 = 500 - 5c^2 \quad \text{Simplifying}$$

$$5c^2 - 20 = 0$$

$$5(c^2 - 4) = 0$$

$$5(c - 2)(c + 2) = 0$$

$$c = 2 \quad or \quad c = -2.$$

4. **Check.** Since speed cannot be negative in this problem, -2 cannot be a solution. You should confirm that 2 checks in the original problem.
5. **State.** The speed of the current is 2 mph.

Exercise Set 6.5

FOR EXTRA HELP

Digital Video Tutor CD 5
Videotape 12

InterAct Math

Math Tutor Center

MathXL

MyMathLab.com

Solve.

1. The reciprocal of 3, plus the reciprocal of 6, is the reciprocal of what number?
2. The reciprocal of 5, plus the reciprocal of 7, is the reciprocal of what number?
3. The sum of a number and 6 times its reciprocal is -5. Find the number.
4. The sum of a number and 21 times its reciprocal is -10. Find the number.
5. The reciprocal of the product of two consecutive integers is $\frac{1}{42}$. Find the two integers.
6. The reciprocal of the product of two consecutive integers is $\frac{1}{72}$. Find the two integers.
7. *Home restoration.* Cedric can refinish the floor of an apartment in 8 hr. Carolyn can refinish the floor in 6 hr. How long will it take them, working together, to refinish the floor?
8. *Mail order.* Zoe, an experienced shipping clerk, can fill a certain order in 5 hr. Willy, a new clerk, needs 9 hr to complete the same job. Working together, how long will it take them to fill the order?
9. *Filling a tank.* A community water tank can be filled in 18 hr by the town office well alone and in 22 hr by the high school well alone. How long will it take to fill the tank if both wells are working?
10. *Filling a pool.* A swimming pool can be filled in 12 hr if water enters through a pipe alone or in

30 hr if water enters through a hose alone. If water is entering through both the pipe and the hose, how long will it take to fill the pool?

11. *Hotel management.* The Honeywell HQ17 air cleaner can clean the air in a 12-ft by 14-ft conference room in 10 min. The HQ174 can clean the air in a room of the same size in 6 min. How long would it take the two machines together to clean the air in such a room?

12. *Printing.* Pronto Press can print an order of booklets in 4.5 hr. Red Dot Printers can do the same job in 5.5 hr. How long will it take if both presses are used?

13. *Cutting firewood.* Jake can cut and split a cord of firewood in 6 fewer hr than Skyler can. When they work together, it takes them 4 hr. How long would it take each of them to do the job alone?

14. *Wood cutting.* Damon can clear a lot in 5.5 hr. His partner, Tyron, can complete the same job in 7.5 hr. How long will it take them to clear the lot working together?

15. *Hotel management.* The Honeywell HQ17 air cleaner takes twice as long as the EV25 to clean the same volume of air. Together the two machines can clean the air in a 24-ft by 24-ft banquet room in 10 min. How long would it take each machine, working alone, to clean the air in the room?

16. *Computer printers.* The HP Office Jet G85 works twice as fast as the Laser Jet II. When the machines work together, a university can produce all its staff manuals in 15 hr. Find the time it would take each machine, working alone, to complete the same job.

17. *Painting.* Sara takes 3 hr longer to paint a floor than it takes Kate. When they work together, it takes them 2 hr. How long would each take to do the job alone?

18. *Painting.* Claudia can paint a neighbor's house 4 times as fast as Jan can. The year they worked together it took them 8 days. How long would it take each to paint the house alone?

19. *Waxing a car.* Rosita can wax her car in 2 hr. When she works together with Helga, they can wax the car in 45 min. How long would it take Helga, working by herself, to wax the car?

20. *Newspaper delivery.* Zsuzanna can deliver papers 3 times as fast as Stan can. If they work together, it takes them 1 hr. How long would it take each to deliver the papers alone?

21. *Sorting recyclables.* Together, it takes John and Deb 2 hr 55 min to sort recyclables. Alone, John would require 2 more hr than Deb. How long would it take Deb to do the job alone? (*Hint*: Convert minutes to hours or hours to minutes.)

22. *Paving.* Together, Larry and Mo require 4 hr 48 min to pave a driveway. Alone, Larry would require 4 hr more than Mo. How long would it take Mo to do the job alone? (*Hint*: Convert minutes to hours.)

23. *Kayaking.* The speed of the current in Catamount Creek is 3 mph. Zeno can kayak 4 mi upstream in the same time it takes him to kayak 10 mi downstream. What is the speed of Zeno's kayak in still water?

24. *Boating.* The current in the Lazy River moves at a rate of 4 mph. Monica's dinghy motors 6 mi upstream in the same time it takes to motor 12 mi downstream. What is the speed of the dinghy in still water?

25. *Moving sidewalks.* The moving sidewalk at O'Hare Airport in Chicago moves 1.8 ft/sec. Walking on the moving sidewalk, Camille travels 105 ft forward in the time it takes to travel 51 ft in the opposite direction. How fast would Camille be walking on a nonmoving sidewalk?

26. *Moving sidewalks.* Newark Airport's moving sidewalk moves at a speed of 1.7 ft/sec. Walking on the moving sidewalk, Benny can travel 120 ft forward in the same time it takes to travel 52 ft in the opposite direction. How fast would Benny be walking on a nonmoving sidewalk?

27. *Train speed.* The speed of the A&M freight train is 14 mph less than the speed of the A&M passenger train. The passenger train travels 400 mi in the same time that the freight train travels 330 mi. Find the speed of each train.

28. *Walking.* Rosanna walks 2 mph slower than Simone. In the time it takes Simone to walk 8 mi, Rosanna walks 5 mi. Find the speed of each person.

Aha! **29.** *Bus travel.* A local bus travels 7 mph slower than the express. The express travels 45 mi in the time it takes the local to travel 38 mi. Find the speed of each bus.

30. *Train speed.* The A train goes 12 mph slower than the E train. The A train travels 230 mi in the same time that the E train travels 290 mi. Find the speed of each train.

31. *Boating.* Audrey's paddleboat travels 2 km/h in still water. The boat is paddled 4 km downstream in the same time it takes to go 1 km upstream. What is the speed of the river?

32. *Boating.* Laverne's Mercruiser travels 15 km/h in still water. She motors 140 km downstream in the same time it takes to travel 35 km upstream. What is the speed of the river?

33. *Shipping.* A barge moves 7 km/h in still water. It travels 45 km upriver and 45 km downriver in a total time of 14 hr. What is the speed of the current?

34. *Moped speed.* Jaime's moped travels 8 km/h faster than Mara's. Jaime travels 69 km in the same time that Mara travels 45 km. Find the speed of each person's moped.

35. *Aviation.* A Citation II Jet travels 350 mph in still air and flies 487.5 mi into the wind and 487.5 mi with the wind in a total of 2.8 hr (*Source*: Eastern Air Charter). Find the wind speed.

36. *Canoeing.* Al paddles 55 m per minute in still water. He paddles 150 m upstream and 150 m downstream in a total time of 5.5 min. What is the speed of the current?

37. *Train travel.* A freight train covered 120 mi at a certain speed. Had the train been able to travel 10 mph faster, the trip would have been 2 hr shorter. How fast did the train go?

38. *Boating.* Julia's Boston Whaler cruised 45 mi upstream and 45 mi back in a total of 8 hr. The speed of the river is 3 mph. Find the speed of the boat in still water.

39. Two steamrollers are paving a parking lot. Working together, will the two steamrollers take less than half as long as the slower steamroller would working alone? Why or why not?

40. Two fuel lines are filling a freighter with oil. Will the faster fuel line take more or less than twice as long to fill the freighter by itself? Why?

SKILL MAINTENANCE

Simplify.

41. $\dfrac{35a^6b^8}{7a^2b^2}$

42. $\dfrac{20x^9y^6}{4x^3y^2}$

43. $\dfrac{36s^{15}t^{10}}{9s^5t^2}$

44. $6x^4 - 3x^2 + 9x - (8x^4 + 4x^2 - 2x)$

45. $2(x^3 + 4x^2 - 5x + 7) - 5(2x^3 - 4x^2 + 3x - 1)$

46. $9x^4 + 7x^3 + x^2 - 8 - (-2x^4 + 3x^2 + 4x + 2)$

SYNTHESIS

47. Write a work problem for a classmate to solve. Devise the problem so that the solution is "Liane and Michele will take 4 hr to complete the job, working together."

48. Write a work problem for a classmate to solve. Devise the problem so that the solution is "Jen takes 5 hr and Pablo takes 6 hr to complete the job alone."

49. *Filling a bog.* The Norwich cranberry bog can be filled in 9 hr and drained in 11 hr. How long will it take to fill the bog if the drainage gate is left open?

50. *Filling a tub.* Justine's hot tub can be filled in 10 min and drained in 8 min. How long will it take to empty a full tub if the water is left on?

51. Refer to Exercise 24. How long will it take Monica to motor 3 mi downstream?

52. Refer to Exercise 23. How long will it take Zeno to kayak 5 mi downstream?

53. *Escalators.* Together, a 100-cm-wide escalator and a 60-cm-wide escalator can empty a 1575-person auditorium in 14 min (*Source*: *McGraw-Hill Encyclopedia of Science and Technology*). The wider escalator moves twice as many people as the narrower one. How many people per hour does the 60-cm-wide escalator move?

54. *Aviation.* A Coast Guard plane has enough fuel to fly for 6 hr, and its speed in still air is 240 mph. The plane departs with a 40-mph tailwind and returns to the same airport flying into the same wind. How far can the plane travel under these conditions?

55. *Boating.* Shoreline Travel operates a 3-hr paddle-boat cruise on the Missouri River. If the speed of the boat in still water is 12 mph, how far upriver can the pilot travel against a 5-mph current before it is time to turn around?

56. *Boating.* The speed of a motor boat in still water is three times the speed of a river's current. A trip up the river and back takes 10 hr, and the total distance of the trip is 100 km. Find the speed of the current.

57. *Travel by car.* Melissa drives to work at 50 mph and arrives 1 min late. She drives to work at 60 mph and arrives 5 min early. How far does Melissa live from work?

58. At what time after 4:00 will the minute hand and the hour hand of a clock first be in the same position?

59. At what time after 10:30 will the hands of a clock first be perpendicular?

Average speed is defined as total distance divided by total time.

60. Lenore drove 200 km. For the first 100 km of the trip, she drove at a speed of 40 km/h. For the second half of the trip, she traveled at a speed of 60 km/h. What was the average speed of the entire trip? (It was *not* 50 km/h.)

61. For the first 50 mi of a 100-mi trip, Chip drove 40 mph. What speed would he have to travel for the last half of the trip so that the average speed for the entire trip would be 45 mph?

COLLABORATIVE CORNER

Does the Model Hold Water?

Focus: Testing a mathematical model

Time: 20–30 minutes

Group size: 2–3

Materials: An empty 1-gal plastic jug, a kitchen or laboratory sink, a stopwatch or a watch capable of measuring seconds, an inexpensive pen or pair of scissors or a nail or knife for poking holes in plastic.

Problems like Exercises 49 and 50 can be solved algebraically and then checked at home or in a laboratory.

ACTIVITY

1. While one group member fills the empty jug with water, the other group member(s) should record how many seconds this takes.
2. After carefully poking a few holes in the bottom of the jug, record how many seconds it takes the full jug to empty.
3. Using the information found in parts (1) and (2) above, use algebra to predict how long it will take to fill the punctured jug.
4. Test your prediction by timing how long it takes for the pierced jug to be filled. Be sure to run the water at the same rate as in part (1).
5. How accurate was your prediction? How might your prediction have been made more accurate?

6.6 Division of Polynomials

Divisor a Monomial • Divisor a Polynomial

A rational expression indicates division. Division of polynomials, like division of real numbers, relies on our multiplication and subtraction skills.

Divisor a Monomial

To divide a monomial by a monomial, we can subtract exponents when bases are the same (see Section 1.6). For example,

$$\frac{45x^{10}}{3x^4} = 15x^{10-4} = 15x^6, \qquad \frac{48a^2b^5}{-3ab^2} = \frac{48}{-3}a^{2-1}b^{5-2} = -16ab^3.$$

To divide a polynomial by a monomial, we regard the division as a sum of quotients of monomials. This uses the fact that since

$$\frac{A}{C} + \frac{B}{C} = \frac{A+B}{C}, \quad \text{we know that} \quad \frac{A+B}{C} = \frac{A}{C} + \frac{B}{C}.$$

Example 1 Divide $12x^3 + 8x^2 + x + 4$ by $4x$.

Solution

$$(12x^3 + 8x^2 + x + 4) \div (4x) = \frac{12x^3 + 8x^2 + x + 4}{4x}$$ Writing a rational expression

$$= \frac{12x^3}{4x} + \frac{8x^2}{4x} + \frac{x}{4x} + \frac{4}{4x}$$ Writing as a sum of quotients

$$= 3x^2 + 2x + \frac{1}{4} + \frac{1}{x}$$ Performing the four indicated divisions

Example 2 Divide: $(8x^4y^5 - 3x^3y^4 + 5x^2y^3) \div x^2y^3$.

Solution

$$\frac{8x^4y^5 - 3x^3y^4 + 5x^2y^3}{x^2y^3} = \frac{8x^4y^5}{x^2y^3} - \frac{3x^3y^4}{x^2y^3} + \frac{5x^2y^3}{x^2y^3}$$ Try to perform this step mentally.

$$= 8x^2y^2 - 3xy + 5$$

Division by a Monomial

To divide a polynomial by a monomial, divide each term of the polynomial by the monomial.

Divisor a Polynomial

When the divisor has more than one term, we use a procedure very similar to long division in arithmetic.

Example 3 Divide $2x^2 - 7x - 15$ by $x - 5$.

Solution We have

$$\begin{array}{r|l} 2x & \\ x - 5\overline{)2x^2 - 7x - 15} & \text{Divide } 2x^2 \text{ by } x\text{: } 2x^2/x = 2x. \\ \underline{-(2x^2 - 10x)} & \leftarrow \text{Multiply } x - 5 \text{ by } 2x. \\ 3x & \leftarrow \text{Subtract by mentally changing signs and adding: } -7x + 10x = 3x. \end{array}$$

We now "bring down" the other term in the dividend, -15.

$$\begin{array}{r|l} 2x + 3 & \\ x - 5\overline{)2x^2 - 7x - 15} & \\ \underline{2x^2 - 10x} & \\ 3x - 15 & \text{Divide } 3x \text{ by } x\text{: } 3x/x = 3. \\ \underline{-(3x - 15)} & \leftarrow \text{Multiply } x - 5 \text{ by } 3. \\ 0 & \leftarrow \text{Subtract.} \end{array}$$

Check: $(x - 5)(2x + 3) = 2x^2 - 7x - 15$. The answer checks.

The quotient is $2x + 3$.

To understand why we perform long division as we do, note that Example 3 amounts to "filling in" an unknown polynomial:

$$(x - 5)(\ ?\ ) = 2x^2 - 7x - 15.$$

We see that $2x$ must be in the unknown polynomial if we are to get the first term, $2x^2$, from the multiplication. To see what else is needed, note that

$$(x - 5)(2x \qquad) = 2x^2 - 10x \neq 2x^2 - 7x - 15.$$

The $2x$ can be regarded as a (poor) approximation of the quotient that we are after. To see how far off the approximation is, we subtract:

$$\left.\begin{array}{r} 2x^2 - 7x - 15 \\ \underline{-(2x^2 - 10x) } \\ 3x - 15 \end{array}\right\} \text{Note where this appeared in the long division above.}$$

To get the needed terms, $3x - 15$, we need another term in the unknown polynomial. We use 3 because $(x - 5) \cdot 3$ is $3x - 15$:

$$\begin{aligned} (x - 5)(2x + 3) &= 2x^2 - 10x + 3x - 15 \\ &= 2x^2 - 7x - 15. \end{aligned}$$

Now when we subtract the product $(x - 5)(2x + 3)$ from $2x^2 - 7x - 15$, the remainder is 0.

If a nonzero remainder occurs, when do we stop dividing? We continue until the degree of the remainder is less than the degree of the divisor.

Example 4 Divide $x^2 + 5x + 8$ by $x + 3$.

Solution We have

$$\begin{array}{r|l} x & \\ x + 3\overline{)x^2 + 5x + 8} & \text{Divide the first term of the dividend by the first term of the divisor: } x^2/x = x. \\ \underline{x^2 + 3x} \leftarrow & \text{Multiply } x \text{ above by } x + 3. \\ 2x \leftarrow & \text{Subtract.} \end{array}$$

The subtraction above is $(x^2 + 5x) - (x^2 + 3x)$. Remember: To subtract, add the opposite (change the sign of every term, then add).

We now "bring down" the next term of the dividend—in this case, 8—and repeat the process:

$$\begin{array}{r|l} x + 2 \leftarrow & \text{Divide the first term by the first term: } 2x/x = 2. \\ x + 3\overline{)x^2 + 5x + 8} & \\ \underline{x^2 + 3x} & \\ 2x + 8 \leftarrow & \text{The 8 has been "brought down."} \\ \underline{2x + 6} \leftarrow & \text{Multiply 2 by } x + 3. \\ 2 \leftarrow & \text{Subtract: } (2x + 8) - (2x + 6). \end{array}$$

The quotient is $x + 2$, with remainder 2. Note that the degree of the remainder is 0 and the degree of the divisor, $x + 3$, is 1. Since $0 < 1$, the process stops.

Check: $(x + 3)(x + 2) + 2 = x^2 + 5x + 6 + 2$ Add the remainder to the product.

$= x^2 + 5x + 8$

We write our answer as $x + 2$, R2, or as

$$\text{Quotient} + \frac{\text{Remainder}}{\text{Divisor}}$$

$$x + 2 + \frac{2}{x + 3}.$$

This is how answers are listed at the back of the book.

The last answer in Example 4 can also be checked by multiplying:

$$(x + 3)\left[(x + 2) + \frac{2}{x + 3}\right] = (x + 3)(x + 2) + (x + 3)\frac{2}{x + 3}$$

Using the distributive law

$$= x^2 + 5x + 6 + 2$$

$$= x^2 + 5x + 8.$$ This was the dividend in Example 4.

An equivalent, but quicker, check is to multiply the divisor by the quotient and then add the remainder. This is precisely what we did in the check in Example 4.

You may have noticed that it is helpful to have all polynomials written in descending order.

Tips for Dividing Polynomials

1. Arrange polynomials in descending order.
2. If there are missing terms in the dividend, either write them with 0 coefficients or leave space for them.
3. Continue the long division process until the degree of the remainder is less than the degree of the divisor.

Example 5 Divide: $(9a^2 + a^3 - 5) \div (a^2 - 1)$.

Solution We rewrite the problem in descending order:

$$(a^3 + 9a^2 - 5) \div (a^2 - 1).$$

Thus,

$$\begin{array}{r|l} & a + 9 \\ a^2 - 1 & a^3 + 9a^2 + 0a - 5 \\ & \underline{a^3 \qquad\quad - a} \\ & 9a^2 + a - 5 \\ & \underline{9a^2 \qquad - 9} \\ & a + 4 \end{array}$$

When there is a missing term, we can write it in, as in this example, or leave space, as in Example 6 below.

Subtracting:
$a^3 + 9a^2 - (a^3 - a) = 9a^2 + a$

The degree of the remainder is less than the degree of the divisor, so we are finished.

The answer is $a + 9 + \dfrac{a + 4}{a^2 - 1}$.

Example 6 Let $f(x) = 125x^3 - 8$ and $g(x) = 5x - 2$. If $F(x) = (f/g)(x)$, find a simplified expression for $F(x)$.

Solution Recall that $(f/g)(x) = f(x)/g(x)$. Thus,

$$F(x) = \frac{125x^3 - 8}{5x - 2}$$

and

$$\begin{array}{r|l} & 25x^2 + 10x + 4 \\ 5x - 2 & 125x^3 \qquad\qquad\quad - 8 \\ & \underline{125x^3 - 50x^2} \\ & 50x^2 \\ & \underline{50x^2 - 20x} \\ & 20x - 8 \\ & \underline{20x - 8} \\ & 0. \end{array}$$

Leaving space for the missing terms.
Subtracting:
$125x^3 - (125x^3 - 50x^2) = 50x^2$

Subtracting

Note that, because $F(x) = f(x)/g(x)$, $g(x)$ cannot be 0. Since $g(x)$ is 0 for $x = \frac{2}{5}$ (check this), we have

$$F(x) = 25x^2 + 10x + 4, \quad \text{provided } x \neq \tfrac{2}{5}.$$

Exercise Set 6.6

FOR EXTRA HELP

Digital Video Tutor CD 5
Videotape 12

InterAct Math

Math Tutor Center

MathXL

MyMathLab.com

Divide and check.

1. $\dfrac{34x^6 + 18x^5 - 28x^2}{6x^2}$
2. $\dfrac{30y^8 - 15y^6 + 40y^4}{5y^4}$
3. $\dfrac{21a^3 + 7a^2 - 3a - 14}{7a}$
4. $\dfrac{-25x^3 + 20x^2 - 3x + 7}{5x}$
5. $(14y^3 - 9y^2 - 8y) \div (2y^2)$
6. $(6a^4 + 9a^2 - 8) \div (2a)$
7. $(15x^7 - 21x^4 - 3x^2) \div (-3x^2)$
8. $(36y^6 - 18y^4 - 12y^2) \div (-6y)$
9. $(a^2b - a^3b^3 - a^5b^5) \div (a^2b)$
10. $(x^3y^2 - x^3y^3 - x^4y^2) \div (x^2y^2)$
11. $(6p^2q^2 - 9p^2q + 12pq^2) \div (-3pq)$
12. $(16y^4z^2 - 8y^6z^4 + 12y^8z^3) \div (4y^4z)$
13. (Aha!) $(x^2 + 10x + 21) \div (x + 7)$
14. $(y^2 - 8y + 16) \div (y - 4)$
15. $(a^2 - 8a - 16) \div (a + 4)$
16. $(y^2 - 10y - 25) \div (y - 5)$
17. $(x^2 - 9x + 21) \div (x - 5)$
18. $(x^2 - 11x + 23) \div (x - 7)$
19. $(y^2 - 25) \div (y + 5)$
20. $(a^2 - 81) \div (a - 9)$
21. $(y^3 - 4y^2 + 3y - 6) \div (y - 2)$
22. $(x^3 - 5x^2 + 4x - 7) \div (x - 3)$
23. $(2x^3 + 3x^2 - x - 3) \div (x + 2)$
24. $(3x^3 - 5x^2 - 3x - 2) \div (x - 2)$
25. $(a^3 - a + 10) \div (a - 4)$
26. $(x^3 - x + 6) \div (x + 2)$
27. $(10y^3 + 6y^2 - 9y + 10) \div (5y - 2)$
28. $(6x^3 - 11x^2 + 11x - 2) \div (2x - 3)$
29. $(2x^4 - x^3 - 5x^2 + x - 6) \div (x^2 + 2)$
30. $(3x^4 + 2x^3 - 11x^2 - 2x + 5) \div (x^2 - 2)$

For Exercises 31–38, $f(x)$ and $g(x)$ are as given. Find a simplified expression for $F(x)$ if $F(x) = (f/g)(x)$. (See Example 6.)

31. $f(x) = 8x^3 + 27,\ g(x) = 2x + 3$
32. $f(x) = 64x^3 - 8,\ g(x) = 4x - 2$
33. $f(x) = 6x^2 - 11x - 10,\ g(x) = 3x + 2$
34. $f(x) = 8x^2 - 22x - 21,\ g(x) = 2x - 7$
35. $f(x) = x^4 - 24x^2 - 25,\ g(x) = x^2 - 25$
36. $f(x) = x^4 - 3x^2 - 54,\ g(x) = x^2 - 9$
37. $f(x) = 8x^2 - 3x^4 - 2x^3 + 2x^5 - 5,\ g(x) = x^2 - 1$
38. $f(x) = 4x - x^3 - 10x^2 + 3x^4 - 8,\ g(x) = x^2 - 4$
39. Explain how factoring could be used to solve Example 6.
40. Explain how to construct a polynomial of degree 4 that has a remainder of 3 when divided by $x + 1$.

SKILL MAINTENANCE

Solve.

41. $ab - cd = k$, for c
42. $xy - wz = t$, for z

43. Find three consecutive positive integers such that the product of the first and second integers is 26 less than the product of the second and third integers.

44. If $f(x) = 2x^3$, find $f(-3a)$.

Solve.

45. $|2x - 3| > 7$

46. $|3x - 1| < 8$

SYNTHESIS

47. Explain how to construct a polynomial of degree 4 that has a remainder of 2 when divided by $x + c$.

48. Do addition, subtraction, and multiplication of polynomials always result in a polynomial? Does division? Why or why not?

Divide.

49. $(4a^3b + 5a^2b^2 + a^4 + 2ab^3) \div (a^2 + 2b^2 + 3ab)$

50. $(x^4 - x^3y + x^2y^2 + 2x^2y - 2xy^2 + 2y^3) \div (x^2 - xy + y^2)$

51. $(a^7 + b^7) \div (a + b)$

52. Find k such that when $x^3 - kx^2 + 3x + 7k$ is divided by $x + 2$, the remainder is 0.

53. When $x^2 - 3x + 2k$ is divided by $x + 2$, the remainder is 7. Find k.

54. Let

$$f(x) = \frac{3x + 7}{x + 2}.$$

a) Use division to find an expression equivalent to $f(x)$. Then graph f.

b) On the same set of axes, sketch both $g(x) = 1/(x + 2)$ and $h(x) = 1/x$.

c) How do the graphs of f, g, and h compare?

55. Jamaladeen incorrectly states that

$$(x^3 + 9x^2 - 6) \div (x^2 - 1) = x + 9 + \frac{x + 4}{x^2 - 1}.$$

Without performing any long division, how could you show Jamaladeen that his division cannot possibly be correct?

56. Check Example 3 by setting $y_1 = (2x^2 - 7x - 15)/(x - 5)$ and $y_2 = 2x + 3$. Then use either the TRACE feature (after selecting the ZOOM Z INTEGER option) or the TABLE feature (with TblMin = 0 and $\triangle$Tbl = 1) to show that $y_1 \neq y_2$ for $x = 5$.

57. Use a grapher to check Example 5. Perform the check using $y_1 = (9x^2 + x^3 - 5)/(x^2 - 1)$, $y_2 = x + 9 + (x + 4)/(x^2 - 1)$, and $y_3 = y_2 - y_1$.

Synthetic Division

6.7

Streamlining Long Division • The Remainder Theorem

Streamlining Long Division

To divide a polynomial by a binomial of the type $x - a$, we can streamline the usual procedure to develop a process called *synthetic division*.

Compare the following. In each stage, we attempt to write a bit less than in the previous stage, while retaining enough essentials to solve the problem. At the end, we will return to the usual polynomial notation.

Stage 1

When a polynomial is written in descending order, the coefficients provide the essential information:

$$\begin{array}{r} 4x^2 + 5x + 11 \\ x - 2\overline{)4x^3 - 3x^2 + x + 7} \\ \underline{4x^3 - 8x^2} \\ 5x^2 + x \\ \underline{5x^2 - 10x} \\ 11x + 7 \\ \underline{11x - 22} \\ 29 \end{array} \qquad \begin{array}{r} 4 + 5 + 11 \\ 1 - 2\overline{)4 - 3 + 1 + 7} \\ \underline{4 - 8} \\ 5 + 1 \\ \underline{5 - 10} \\ 11 + 7 \\ \underline{11 - 22} \\ 29 \end{array}$$

Because the leading coefficient in the divisor is 1, each time we multiply the divisor by a term in the answer, the leading coefficient of that product duplicates a coefficient in the answer. In the next stage, we don't bother to duplicate these numbers. We also show where -2 is used and drop the 1 from the divisor.

Stage 2

$$\begin{array}{r} 4x^2 + 5x + 11 \\ x - 2\overline{)4x^3 - 3x^2 + x + 7} \\ \underline{4x^3 - 8x^2} \\ 5x^2 + x \\ \underline{5x^2 - 10x} \\ 11x + 7 \\ \underline{11x - 22} \\ 29 \end{array} \qquad \begin{array}{rl} 4 + 5 + 11 & \\ -2\overline{)4 - 3 + 1 + 7} & \\ \underline{-8} & \leftarrow \text{Multiply: } -2 \cdot 4 = -8. \\ 5 + 1 & \text{Subtract: } -3 - (-8) = 5. \\ \underline{-10} & \leftarrow \text{Multiply: } -2 \cdot 5 = -10. \\ 11 + 7 & \text{Subtract: } 1 - (-10) = 11. \\ \underline{-22} & \leftarrow \text{Multiply: } -2 \cdot 11 = -22. \\ 29 & \leftarrow \text{Subtract: } 7 - (-22) = 29. \end{array}$$

To simplify further, we now reverse the sign of the -2 in the divisor and, in exchange, *add* at each step in the long division.

Stage 3

$$\begin{array}{r} 4x^2 + 5x + 11 \\ x - 2\overline{)4x^3 - 3x^2 + x + 7} \\ \underline{4x^3 - 8x^2} \\ 5x^2 + x \\ \underline{5x^2 - 10x} \\ 11x + 7 \\ \underline{11x - 22} \\ 29 \end{array} \qquad \begin{array}{rl} 4 + 5 + 11 & \\ 2\overline{)4 - 3 + 1 + 7} & \text{Replace the } -2 \text{ with } 2. \\ \underline{8} & \leftarrow \text{Multiply: } 2 \cdot 4 = 8. \\ 5 + 1 & \text{Add: } -3 + 8 = 5. \\ \underline{10} & \leftarrow \text{Multiply: } 2 \cdot 5 = 10. \\ 11 + 7 & \text{Add: } 1 + 10 = 11. \\ \underline{22} & \leftarrow \text{Multiply: } 2 \cdot 11 = 22. \\ 29 & \leftarrow \text{Add: } 7 + 22 = 29. \end{array}$$

The blue numbers can be eliminated if we look at the red numbers instead.

Stage 4

$$\begin{array}{r} 4x^2 + 5x + 11 \\ x - 2\overline{)4x^3 - 3x^2 + x + 7} \\ \underline{4x^3 - 8x^2} \\ 5x^2 + x \\ \underline{5x^2 - 10x} \\ 11x + 7 \\ \underline{11x - 22} \\ 29 \end{array} \qquad \begin{array}{r} 4 \quad 5 \quad 11 \\ 2\overline{)4 \quad -3 \quad 1 \quad 7} \\ \underline{8} \quad \underline{10} \quad \underline{22} \\ 5 \quad 11 \quad 29 \end{array}$$

Don't lose sight of how the products 8, 10, and 22 are found. Also, note that the 5 and 11 preceding the remainder 29 coincide with the 5 and 11 following the 4 on the top line. By writing a 4 to the left of 5 on the bottom line, we can eliminate the top line in stage 4 and read our answer from the bottom line. This final stage is commonly called **synthetic division**.

Stage 5

$$\begin{array}{r} 4 \quad 5 \quad 11 \\ 2\overline{)4 \quad -3 \quad 1 \quad 7} \\ \underline{8} \quad \underline{10} \quad \underline{22} \\ 5 \quad 11 \quad 29 \end{array} \qquad \begin{array}{r|rrrr} 2 & 4 & -3 & 1 & 7 \\ & & 8 & 10 & 22 \\ \hline & 4 & 5 & 11 & 29 \end{array}$$

29 ← This is the remainder.
11 ← This is the zero-degree coefficient.
5 ← This is the first-degree coefficient.
4 ← This is the second-degree coefficient.

The quotient is $4x^2 + 5x + 11$. The remainder is 29.

> Remember that in order for this method to work, the divisor must be of the form $x - a$, that is, a variable minus a constant. The coefficient of the variable must be 1.

Example 1 Use synthetic division to divide: $(x^3 + 6x^2 - x - 30) \div (x - 2)$.

Solution

$$\begin{array}{r|rrrr} 2 & 1 & 6 & -1 & -30 \\ & & & & \\ \hline & 1 & & & \end{array}$$

Write the 2 of $x - 2$ and the coefficients of the dividend.

Bring down the first coefficient.

$$\begin{array}{r|rrrr} 2 & 1 & 6 & -1 & -30 \\ & & 2 & & \\ \hline & 1 & 8 & & \end{array}$$

Multiply 1 by 2 to get 2.

Add 6 and 2.

$$\begin{array}{r|rrrr} 2 & 1 & 6 & -1 & -30 \\ & & 2 & 16 & \\ \hline & 1 & 8 & 15 & \end{array}$$

Multiply 8 by 2.

Add -1 and 16.

2	1	6	−1	−30	
		2	16	30	Multiply 15 by 2 and add.
	1	8	15	0	

The answer is $x^2 + 8x + 15$ with R0, or just $x^2 + 8x + 15$.

Example 2

Use synthetic division to divide.

a) $(2x^3 + 7x^2 - 5) \div (x + 3)$
b) $(10x^2 - 13x + 3x^3 - 20) \div (4 + x)$

Solution

a) $(2x^3 + 7x^2 - 5) \div (x + 3)$

The dividend has no x-term, so we need to write 0 for its coefficient of x. Note that $x + 3 = x - (-3)$, so we write -3 inside the ⌋.

−3	2	7	0	−5
		−6	−3	9
	2	1	−3	4

The answer is $2x^2 + x - 3$, with R4, or $2x^2 + x - 3 + \frac{4}{x + 3}$.

b) We first rewrite $(10x^2 - 13x + 3x^3 - 20) \div (4 + x)$ in descending order:

$$(3x^3 + 10x^2 - 13x - 20) \div (x + 4).$$

Next, we use synthetic division. Note that $x + 4 = x - (-4)$.

−4	3	10	−13	−20
		−12	8	20
	3	−2	−5	0

The answer is $3x^2 - 2x - 5$.

In Example 1, the division by $x - 2$ gave a remainder of 0. The remainder theorem tells us that this means that when $x = 2$, the value of $x^3 + 6x^2 - x - 30$ is 0. Check this both graphically and algebraically (by substitution). Then perform a similar check for Example 2(b).

The Remainder Theorem

Because the remainder is 0, Example 1 shows that $x - 2$ is a factor of $x^3 + 6x^2 - x - 30$ and that we can write $x^3 + 6x^2 - x - 30$ as $(x - 2)(x^2 + 8x + 15)$. Using this result and the principle of zero products, we know that if $f(x) = x^3 + 6x^2 - x - 30$, then $f(2) = 0$ (since $x - 2$ is a factor of $f(x)$). Similarly, from Example 2(b), we know that $x + 4$ is a factor of $g(x) = 10x^2 - 13x + 3x^3 - 20$. This tells us that $g(-4) = 0$. In both examples, the remainder from the division, 0, can serve as a function value. Remarkably, this pattern extends to nonzero remainders. To see this, note that the remainder in Example 2(a) is 4, and if $f(x) = 2x^3 + 7x^2 - 5$, then $f(-3)$ is also 4 (you should check this). The fact that the remainder and the function value coincide is predicted by the remainder theorem, which follows.

The Remainder Theorem

The remainder obtained by dividing $P(x)$ by $x - r$ is $P(r)$.

A proof of this result is outlined in Exercise 31.

Example 3 Let $f(x) = 8x^5 - 6x^3 + x - 8$. Use synthetic division to find $f(2)$.

Solution The remainder theorem tells us that $f(2)$ is the remainder when $f(x)$ is divided by $x - 2$. We use synthetic division to find that remainder:

2	8	0	−6	0	1	−8
		16	32	52	104	210
	8	16	26	52	105	202

Although the bottom line can be used to find the quotient for the division $(8x^5 - 6x^3 + x - 8) \div (x - 2)$, what we are really interested in is the remainder. It tells us that $f(2) = 202$.

Exercise Set 6.7

FOR EXTRA HELP

Digital Video Tutor CD 5
Videotape 12

InterAct Math

Math Tutor Center

MathXL

MyMathLab.com

Use synthetic division to divide.

1. $(x^3 - 2x^2 + 2x - 7) \div (x + 1)$
2. $(x^3 - 2x^2 + 2x - 7) \div (x - 1)$
3. $(a^2 + 8a + 11) \div (a + 3)$
4. $(a^2 + 8a + 11) \div (a + 5)$
5. $(x^3 - 7x^2 - 13x + 3) \div (x - 2)$
6. $(x^3 - 7x^2 - 13x + 3) \div (x + 2)$
7. $(3x^3 + 7x^2 - 4x + 3) \div (x + 3)$
8. $(3x^3 + 7x^2 - 4x + 3) \div (x - 3)$
9. $(y^3 - 3y + 10) \div (y - 2)$
10. $(x^3 - 2x^2 + 8) \div (x + 2)$
11. $(x^5 - 32) \div (x - 2)$
12. $(y^5 - 1) \div (y - 1)$
13. $(3x^3 + 1 - x + 7x^2) \div \left(x + \frac{1}{3}\right)$
14. $(8x^3 - 1 + 7x - 6x^2) \div \left(x - \frac{1}{2}\right)$

Use synthetic division to find the indicated function value.

15. $f(x) = 5x^4 + 12x^3 + 28x + 9;\ f(-3)$
16. $g(x) = 3x^4 - 25x^2 - 18;\ g(3)$
17. $P(x) = 6x^4 - x^3 - 7x^2 + x + 2;\ P(-1)$
18. $F(x) = 3x^4 + 8x^3 + 2x^2 - 7x - 4;\ F(-2)$
19. $f(x) = x^4 - x^3 - 19x^2 + 49x - 30;\ f(4)$
20. $p(x) = x^4 + 7x^3 + 11x^2 - 7x - 12;\ p(2)$
21. Why is it that we *add* when performing synthetic division, but *subtract* when performing long division?
22. Explain how synthetic division could be useful when factoring a polynomial.

SKILL MAINTENANCE

Solve.

23. $9 + cb = a - b$, for b
24. $8 + ac = bd + ab$, for a

Find the domain of f.

25. $f(x) = \dfrac{5}{3x^2 - 75}$

26. $f(x) = \dfrac{7}{2x^2 + 7x - 9}$

Graph.

27. $y - 2 = \frac{3}{4}(x + 1)$

28. $y = -\frac{4}{3}x + 2$

SYNTHESIS

29. Let $Q(x)$ be a polynomial function with $p(x)$ a factor of $Q(x)$. If $p(3) = 0$, does it follow that $Q(3) = 0$? Why or why not? If $Q(3) = 0$, does it follow that $p(3) = 0$? Why or why not?

30. What adjustments must be made if synthetic division is to be used to divide a polynomial by a binomial of the form $ax + b$, with $a > 1$?

31. To prove the remainder theorem, note that any polynomial $P(x)$ can be rewritten as $(x - r) \cdot Q(x) + R$, where $Q(x)$ is the quotient polynomial that arises when $P(x)$ is divided by $x - r$, and R is some constant (the remainder).

a) How do we know that R must be a constant?

b) Show that $P(r) = R$ (this says that $P(r)$ is the remainder when $P(x)$ is divided by $x - r$).

32. Let $f(x) = 4x^3 + 16x^2 - 3x - 45$. Find $f(-3)$ and then solve the equation $f(x) = 0$.

33. Let $f(x) = 6x^3 - 13x^2 - 79x + 140$. Find $f(4)$ and then solve the equation $f(x) = 0$.

34. Use the TRACE feature on a grapher to check your answer to Exercise 32.

35. Use the TRACE feature on a grapher to check your answer to Exercise 33.

Nested evaluation. One way to evaluate a polynomial function like $P(x) = 3x^4 - 5x^3 + 4x^2 - 1$ *is to successively factor out x as shown:*

$$P(x) = x(x(x(3x - 5) + 4) + 0) - 1.$$

Computations are then performed using this "nested" form of P(x).

36. Use nested evaluation to find $f(-3)$ in Exercise 32. Note the similarities to the calculations performed with synthetic division.

37. Use nested evaluation to find $f(4)$ in Exercise 33. Note the similarities to the calculations performed with synthetic division.

Formulas, Applications, and Variation

6.8

Formulas • Direct Variation • Inverse Variation • Joint and Combined Variation

Formulas

Formulas occur frequently as mathematical models. Many formulas contain rational expressions, and to solve such formulas for a specified letter, we proceed as when solving rational equations.

Example 1

Optics. The formula $f = L/d$ tells how to calculate a camera's "f-stop." In this formula, f is the f-stop, L is the focal length (approximately the distance from the lens to the film), and d is the diameter of the lens. Solve for d.

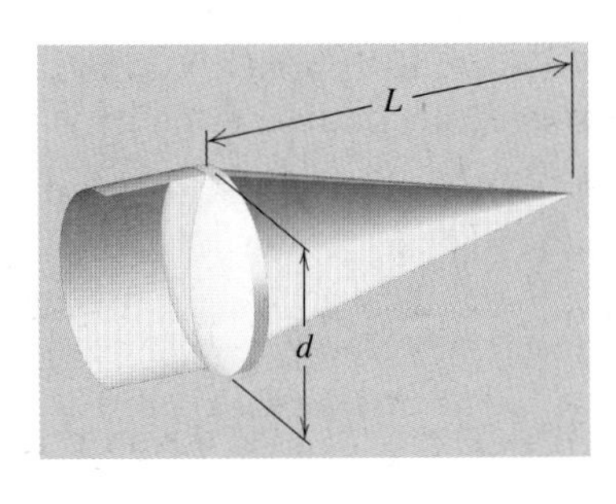

Solution We solve this equation as we did those in Section 6.4:

$$f = \frac{L}{d}$$

$$d \cdot f = d \cdot \frac{L}{d} \qquad \text{Multiplying both sides by the LCD to clear fractions}$$

$$df = L$$

$$df \cdot \frac{1}{f} = L \cdot \frac{1}{f} \qquad \text{Multiplying both sides by } \frac{1}{f} \text{ or dividing by } f$$

$$d = \frac{L}{f}. \qquad \text{Simplifying and removing a factor equal to 1: } \frac{f}{f} = 1$$

The formula $d = L/f$ can now be used to determine the diameter of a lens if we know the focal length and the f-stop.

Example 2 ***Astronomy.*** The formula

$$\frac{V^2}{R^2} = \frac{2g}{R + h}$$

is used to find a satellite's *escape velocity* V, where R is a planet's radius, h is the satellite's height above the planet, and g is the planet's gravitational constant. Solve for h.

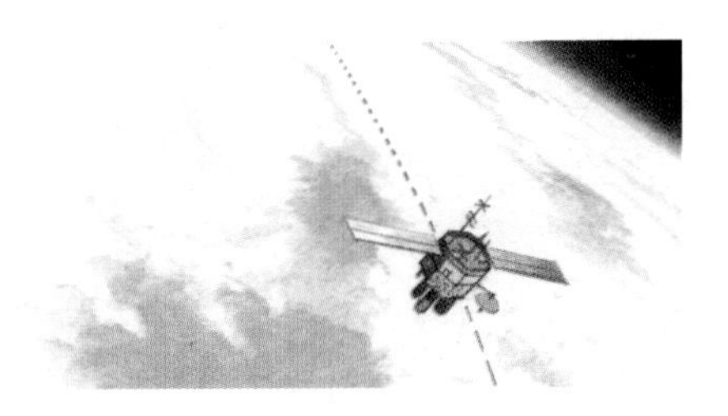

Solution We first clear fractions by multiplying by the LCD, which is $R^2(R + h)$:

$$\frac{V^2}{R^2} = \frac{2g}{R + h}$$

$$R^2(R + h)\frac{V^2}{R^2} = R^2(R + h)\frac{2g}{R + h}$$

$$\frac{R^2(R + h)V^2}{R^2} = \frac{R^2(R + h)2g}{R + h}$$

$$(R + h)V^2 = R^2 \cdot 2g. \qquad \text{Removing factors equal to 1: } \frac{R^2}{R^2} = 1 \text{ and } \frac{R + h}{R + h} = 1$$

Remember: We are solving for h. Although we *could* distribute V^2, since h appears only within the factor $R + h$, it is easier to divide both sides by V^2:

$$\frac{(R+h)V^2}{V^2} = \frac{2R^2g}{V^2} \quad \text{Dividing both sides by } V^2$$

$$R + h = \frac{2R^2g}{V^2} \quad \text{Removing a factor equal to 1: } \frac{V^2}{V^2} = 1$$

$$h = \frac{2R^2g}{V^2} - R. \quad \text{Subtracting } R \text{ from both sides}$$

The last equation can be used to determine the height of a satellite above a planet when the planet's radius and gravitational constant, along with the satellite's escape velocity, are known.

Example 3 ***Acoustics (the Doppler Effect).*** The formula

$$f = \frac{sg}{s+v}$$

is used to determine the frequency f of a sound that is moving at velocity v toward a listener who hears the sound as frequency g. Here s is the speed of sound in a particular medium. Solve for s.

Solution We first clear fractions by multiplying by the LCD, $s + v$:

$$f \cdot (s+v) = \frac{sg}{s+v}(s+v)$$

$$fs + fv = sg. \quad \text{The variable for which we are solving appears on both sides, forcing us to distribute on the left side.}$$

Next, we must get all terms containing s on one side:

$$fv = sg - fs \quad \text{Subtracting } fs \text{ from both sides}$$

$$fv = s(g - f) \quad \text{Factoring out } s$$

$$\frac{fv}{g-f} = s. \quad \text{Dividing both sides by } g - f$$

Since s is isolated on one side, we have solved for s. This last equation can be used to determine the speed of sound whenever f, v, and g are known.

To Solve a Rational Equation for a Specified Unknown

1. If necessary, multiply both sides by the LCD to clear fractions.
2. Multiply, as needed, to remove parentheses.
3. Get all terms with the specified unknown alone on one side.
4. Factor out the specified unknown if it is in more than one term.
5. Multiply or divide on both sides to isolate the specified unknown.

Variation

To extend our study of formulas and functions, we now examine three real-world situations: direct variation, inverse variation, and combined variation.

Direct Variation

A hair stylist earns $18 per hour. In 1 hr, $18 is earned. In 2 hr, $36 is earned. In 3 hr, $54 is earned, and so on. This gives rise to a set of ordered pairs:

$$(1, 18), (2, 36), (3, 54), (4, 72), \quad \text{and so on.}$$

Note that the ratio of earnings E to time t is $\frac{18}{1}$ in every case.

If a situation gives rise to pairs of numbers in which the ratio is constant, we say that there is **direct variation**. Here earnings *vary directly* as the time:

We have $\dfrac{E}{t} = 18$, so $E = 18t$ or, using function notation, $E(t) = 18t$.

Direct Variation

When a situation gives rise to a linear function of the form $f(x) = kx$, or $y = kx$, where k is a nonzero constant, we say that there is *direct variation,* that *y varies directly* as x, or that *y is proportional to x*. The number k is called the *variation constant,* or *constant of proportionality.*

Note that for $k > 0$, any equation of the form $y = kx$ indicates that as x increases, y increases as well.

Example 4 Find the variation constant and an equation of variation if y varies directly as x, and $y = 32$ when $x = 2$.

Solution We know that $(2, 32)$ is a solution of $y = kx$. Therefore,

$$32 = k \cdot 2 \qquad \text{Substituting}$$

$$\frac{32}{2} = k, \quad \text{or} \quad k = 16. \qquad \text{Solving for } k$$

The variation constant is 16. The equation of variation is $y = 16x$. The notation $y(x) = 16x$ or $f(x) = 16x$ is also used.

Example 5

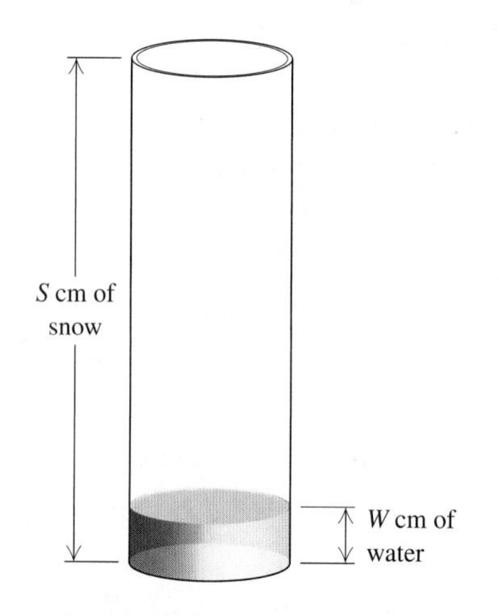

Water from melting snow. The number of centimeters W of water produced from melting snow varies directly as the number of centimeters S of snow. Meteorologists know that under certain conditions, 150 cm of snow will melt to 16.8 cm of water. The average annual snowfall in Alta, Utah, is 500 in. Assuming the above conditions, how much water will replace the 500 in. of snow?

Solution

1. **Familiarize.** Because of the phrase "W... varies directly as ...S," we express the amount of water as a function of the amount of snow. Thus, $W(S) = kS$, where k is the variation constant. Knowing that 150 cm of snow becomes 16.8 cm of water, we have $W(150) = 16.8$. Because we are using ratios, it does not matter whether we work in inches or centimeters, provided the same units are used for W and S.

2. **Translate.** We find the variation constant using the data and then find the equation of variation:

$$W(S) = kS$$

$$W(150) = k \cdot 150 \qquad \text{Replacing } S \text{ with } 150$$

$$16.8 = k \cdot 150 \qquad \text{Replacing } W(150) \text{ with } 16.8$$

$$\frac{16.8}{150} = k \qquad \text{Solving for } k$$

$$0.112 = k. \qquad \text{This is the variation constant.}$$

The equation of variation is $W(S) = 0.112S$. This is the translation.

3. **Carry out.** To find how much water 500 in. of snow will become, we compute $W(500)$:

$$W(S) = 0.112S$$

$$W(500) = 0.112(500) \qquad \text{Substituting 500 for } S$$

$$W = 56.$$

4. **Check.** To check, we could reexamine all our calculations. Note that our answer seems reasonable since 500/56 and 150/16.8 are equal.

5. **State.** Alta's 500 in. of snow will be replaced with 56 in. of water.

Inverse Variation

To see what we mean by inverse variation, suppose a bus is traveling 20 mi. At 20 mph, the trip will take 1 hr. At 40 mph, it will take $\frac{1}{2}$ hr. At 60 mph, it will take $\frac{1}{3}$ hr, and so on. This gives rise to pairs of numbers, all having the same product:

$$(20, 1), \left(40, \tfrac{1}{2}\right), \left(60, \tfrac{1}{3}\right), \left(80, \tfrac{1}{4}\right), \quad \text{and so on.}$$

Note that the product of each pair of numbers is 20. Whenever a situation gives rise to pairs of numbers for which the product is constant, we say that there is **inverse variation**. Since $r \cdot t = 20$, the time t, in hours, required for the bus to travel 20 mi at r mph is given by

$$t = \frac{20}{r} \quad \text{or, using function notation,} \quad t(r) = \frac{20}{r}.$$

Inverse Variation

When a situation gives rise to a rational function of the form $f(x) = k/x$, or $y = k/x$, where k is a nonzero constant, we say that there is *inverse variation,* that *y varies inversely as x,* or that *y is inversely proportional to x.* The number k is called the *variation constant,* or *constant of proportionality.*

Note that for $k > 0$, any equation of the form $y = k/x$ indicates that as x increases, y decreases.

Example 6 Find the variation constant and an equation of variation if y varies inversely as x, and $y = 32$ when $x = 0.2$.

Solution We know that $(0.2, 32)$ is a solution of

$$y = \frac{k}{x}.$$

Therefore,

$$32 = \frac{k}{0.2} \qquad \text{Substituting}$$

$$(0.2)32 = k$$

$$6.4 = k. \qquad \text{Solving for } k$$

The variation constant is 6.4. The equation of variation is

$$y = \frac{6.4}{x}.$$

There are many real-life problems that translate to an equation of inverse variation.

Example 7 ***Ultraviolet index.*** The ultraviolet, or UV, index is a measure issued daily by the National Weather Service that indicates the strength of the sun's rays in a particular locale. For those people whose skin is quite sensitive, a UV rating of 7 will cause sunburn after 10 min (*Source*: *Los Angeles Times*, 3/24/98). Given that the number of minutes it takes to burn, t, varies inversely as the UV rating u, how long will it take a highly-sensitive person to burn on a day with a UV rating of 2?

Solution

1. **Familiarize.** Because of the phrase "...varies inversely as the UV index," we express the amount of time needed to burn as a function of the UV rating: $t(u) = k/u$.
2. **Translate.** We use the given information to solve for k. Then we find the equation of variation.

$$t(u) = \frac{k}{u} \quad \text{Using function notation}$$

$$t(7) = \frac{k}{7} \quad \text{Replacing } u \text{ with 7}$$

$$10 = \frac{k}{7} \quad \text{Replacing } t(7) \text{ with 10}$$

$$70 = k \quad \text{Solving for } k\text{, the variation constant}$$

The equation of variation is $t(u) = 70/u$. This is the translation.

3. **Carry out.** To find how long it would take a highly-sensitive person to burn on a day with a UV index of 2, we calculate $t(2)$:

$$t(2) = \frac{70}{2} = 35. \qquad t = 35 \text{ when } u = 2.$$

4. **Check.** We could now recheck each step. Note that, as expected, as the UV rating goes *down*, the time it takes to burn goes *up*.
5. **State.** On a day with a UV rating of 2, a highly-sensitive person will begin to burn after 35 min of exposure.

Joint and Combined Variation

When a variable varies directly with more than one other variable, we say that there is *joint variation*. For example, in the formula for the volume of a right circular cylinder, $V = \pi r^2 h$, we say that V varies *jointly* as h and the square of r.

Joint Variation

y varies *jointly* as x and z if, for some nonzero constant k, $y = kxz$.

Example 8

Find an equation of variation if y varies jointly as x and z, and $y = 30$ when $x = 2$ and $z = 3$.

Solution We have

$$y = kxz,$$

so

$$30 = k \cdot 2 \cdot 3$$

$$k = 5. \qquad \text{The variation constant is 5.}$$

The equation of variation is $y = 5xz$.

Joint variation is one form of *combined variation*. In general, when a variable varies directly and/or inversely, at the same time, with more than one other variable, there is **combined variation**. Examples 8 and 9 are both examples of combined variation.

Example 9 Find an equation of variation if y varies jointly as x and z and inversely as the square of w, and $y = 105$ when $x = 3$, $z = 20$, and $w = 2$.

Solution The equation of variation is of the form

$$y = k \cdot \frac{xz}{w^2},$$

so, substituting, we have

$$105 = k \cdot \frac{3 \cdot 20}{2^2}$$

$$105 = k \cdot 15$$

$$k = 7.$$

Thus,

$$y = 7 \cdot \frac{xz}{w^2}.$$

Exercise Set 6.8

FOR EXTRA HELP

Digital Video Tutor CD 5
Videotape 12

InterAct Math

Math Tutor Center

MathXL

MyMathLab.com

Solve the formula for the specified letter.

1. $\frac{W_1}{W_2} = \frac{d_1}{d_2}$; d_1

2. $\frac{W_1}{W_2} = \frac{d_1}{d_2}$; W_1

3. $s = \frac{(v_1 + v_2)t}{2}$; v_1

4. $s = \frac{(v_1 + v_2)t}{2}$; t

5. $\frac{1}{R} = \frac{1}{r_1} + \frac{1}{r_2}$; r_1

6. $\frac{1}{R} = \frac{1}{r_1} + \frac{1}{r_2}$; R

7. $I = \frac{2V}{R + 2r}$; R

8. $I = \frac{2V}{R + 2r}$; r

9. $R = \frac{gs}{g + s}$; g

10. $K = \frac{rt}{r - t}$; t

11. $I = \frac{nE}{R + nr}$; n

12. $I = \frac{nE}{R + nr}$; r

13. $\frac{1}{p} + \frac{1}{q} = \frac{1}{f}$; q

14. $\frac{1}{p} + \frac{1}{q} = \frac{1}{f}$; p

15. $S = \frac{H}{m(t_1 - t_2)}$; t_1

16. $S = \frac{H}{m(t_1 - t_2)}$; H

17. $\frac{E}{e} = \frac{R + r}{r}$; r

18. $\frac{E}{e} = \frac{R + r}{R}$; R

19. $S = \frac{a}{1 - r}$; r

20. $S = \frac{a - ar^n}{1 - r}$; a

Aha! **21.** $c = \frac{f}{(a + b)c}$; $a + b$

22. $d = \frac{g}{d(c + f)}$; $c + f$

23. *Taxable interest.* The formula

$$I_t = \frac{I_f}{1 - T}$$

gives the *taxable interest rate* I_t equivalent to the *tax-free interest rate* I_f for a person in the $(100 \cdot T)\%$ tax bracket. Solve for T.

24. *Interest.* The formula

$$P = \frac{A}{1 + r}$$

is used to determine what principal P should be invested for one year at $(100 \cdot r)\%$ simple interest in order to have A dollars after a year. Solve for r.

25. *Electricity.* Electricians regularly use the formula

$$\frac{1}{R} = \frac{1}{r_1} + \frac{1}{r_2}$$

to determine the resistance R that corresponds to two resistors r_1 and r_2 connected in parallel. Solve for r_2.

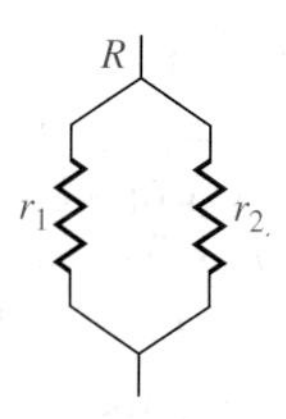

26. *Work rate.* The formula

$$\frac{1}{t} = \frac{1}{a} + \frac{1}{b}$$

gives the total time t required for two workers to complete a job, if the workers' individual times are a and b. Solve for t.

27. *Average acceleration.* The formula

$$a = \frac{v_2 - v_1}{t_2 - t_1}$$

gives a vehicle's *average acceleration* when its velocity changes from v_1 at time t_1 to v_2 at time t_2. Solve for t_1.

28. *Average speed.* The formula

$$v = \frac{d_2 - d_1}{t_2 - t_1}$$

gives an object's average speed v when that object has traveled d_1 miles in t_1 hours and d_2 miles in t_2 hours. Solve for t_2.

29. *Semester average.* The formula

$$A = \frac{2Tt + Qq}{2T + Q}$$

gives a student's average A after T tests and Q quizzes, where each test counts as 2 quizzes, t is the test average, and q is the quiz average. Solve for Q.

30. *Astronomy.* The formula

$$L = \frac{dR}{D - d},$$

where D is the diameter of the sun, d is the diameter of the earth, R is the earth's distance from the sun, and L is some fixed distance, is used in calculating when lunar eclipses occur. Solve for D.

Find the variation constant and an equation of variation if y varies directly as x and the following conditions apply.

31. $y = 28$ when $x = 4$

32. $y = 5$ when $x = 12$

33. $y = 3.4$ when $x = 2$

34. $y = 2$ when $x = 5$

35. $y = 2$ when $x = \frac{1}{3}$

36. $y = 0.9$ when $x = 0.5$

37. *Hooke's law.* Hooke's law states that the distance d that a spring is stretched by a hanging object varies directly as the mass m of the object. If the distance is 20 cm when the mass is 3 kg, what is the distance when the mass is 5 kg?

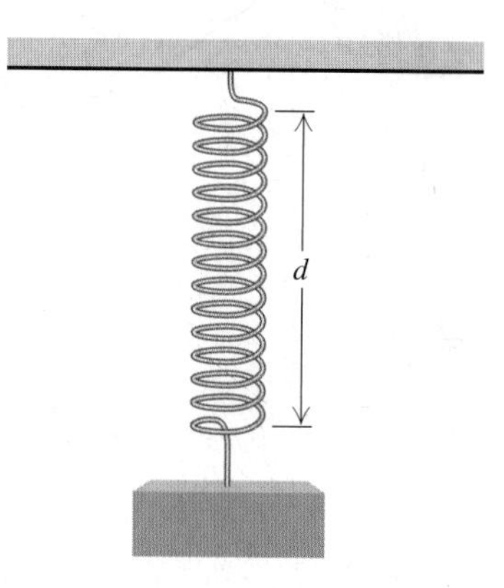

38. *Ohm's law.* The electric current I, in amperes, in a circuit varies directly as the voltage V. When 15 volts are applied, the current is 5 amperes. What is the current when 18 volts are applied?

39. *Use of aluminum cans.* The number N of aluminum cans used each year varies directly as the number of people using the cans. If 250 people use 60,000 cans in one year, how many cans are used each year in Dallas, which has a population of 1,008,000?

40. *Weekly allowance.* According to Fidelity Investments *Investment Vision Magazine,* the average weekly allowance A of children varies directly as their grade level, G. In a recent year, the average allowance of a 9th-grade student was \$9.66 per week. What was the average weekly allowance of a 4th-grade student?

Aha! **41.** *Mass of water in a human.* The number of kilograms W of water in a human body varies directly as the mass of the body. A 96-kg person contains 64 kg of water. How many kilograms of water are in a 48-kg person?

42. *Weight on Mars.* The weight M of an object on Mars varies directly as its weight E on Earth. A person who weighs 95 lb on Earth weighs 38 lb on Mars. How much would a 100-lb person weigh on Mars?

43. *Relative aperture.* The relative aperture, or f-stop, of a 23.5-mm lens is directly proportional to the focal length F of the lens. If a lens with a 150-mm focal length has an f-stop of 6.3, find the f-stop of a 23.5-mm lens with a focal length of 80 mm.

44. *Lead pollution.* The average U.S. community of population 12,500 released about 385 tons of lead into the environment in a recent year.* How many tons were released nationally? Use 250,000,000 as the U.S. population.

Find the variation constant and an equation of variation in which y varies inversely as x, and the following conditions exist.

45. $y = 3$ when $x = 20$

46. $y = 16$ when $x = 4$

47. $y = 28$ when $x = 4$

48. $y = 9$ when $x = 5$

49. $y = 27$ when $x = \frac{1}{3}$

50. $y = 81$ when $x = \frac{1}{9}$

Solve.

51. *Ultraviolet index.* At an ultraviolet, or UV, rating of 4, those people who are moderately sensitive to the sun will burn in 70 min (*Source*: *Los Angeles Times,* 3/24/98). Given that the number of minutes it takes to burn, t, varies inversely with the UV rating, u, how long will it take moderately-sensitive people to burn when the UV rating is 14?

52. *Current and resistance.* The current I in an electrical conductor varies inversely as the resistance R of the conductor. If the current is $\frac{1}{2}$ ampere when the resistance is 240 ohms, what is the current when the resistance is 540 ohms?

53. *Volume and pressure.* The volume V of a gas varies inversely as the pressure P upon it. The volume of a gas is 200 cm^3 under a pressure of 32 kg/cm^2. What will be its volume under a pressure of 40 kg/cm^2?

54. *Pumping rate.* The time t required to empty a tank varies inversely as the rate r of pumping. If a Briggs and Stratton pump can empty a tank in 45 min at the rate of 600 kL/min, how long will it take the pump to empty the tank at 1000 kL/min?

55. *Work rate.* The time T required to do a job varies inversely as the number of people P working. It takes 5 hr for 7 volunteers to pick up rubbish from 1 mi of roadway. How long would it take 10 volunteers to complete the job?

56. *Wavelength and frequency.* The wavelength W of a radio wave varies inversely as its frequency F. A wave with a frequency of 1200 kilohertz has a length of 300 meters. What is the length of a wave with a frequency of 800 kilohertz?

Find an equation of variation in which:

57. y varies directly as the square of x, and $y = 6$ when $x = 3$.

58. y varies directly as the square of x, and $y = 0.15$ when $x = 0.1$.

59. y varies inversely as the square of x, and $y = 6$ when $x = 3$.

60. y varies inversely as the square of x, and $y = 0.15$ when $x = 0.1$.

61. y varies jointly as x and the square of z, and $y = 105$ when $x = 14$ and $z = 5$.

62. y varies jointly as x and z and inversely as w, and $y = \frac{3}{2}$ when $x = 2$, $z = 3$, and $w = 4$.

63. y varies jointly as w and the square of x and inversely as z, and $y = 49$ when $w = 3$, $x = 7$, and $z = 12$.

**Conservation Matters,* Autumn 1995 issue. (Boston: Conservation Law Foundation), p. 30.

64. y varies directly as x and inversely as w and the square of z, and $y = 4.5$ when $x = 15$, $w = 5$, and $z = 2$.

Solve.

65. *Intensity of light.* The intensity I of light from a light bulb varies inversely as the square of the distance d from the bulb. Suppose I is 90 W/m^2 (watts per square meter) when the distance is 5 m. What would the intensity be 7.5 m from the bulb?

66. *Stopping distance of a car.* The stopping distance d of a car after the brakes have been applied varies directly as the square of the speed r. If a car traveling 60 mph can stop in 200 ft, what stopping distance corresponds to a speed of 36 mph?

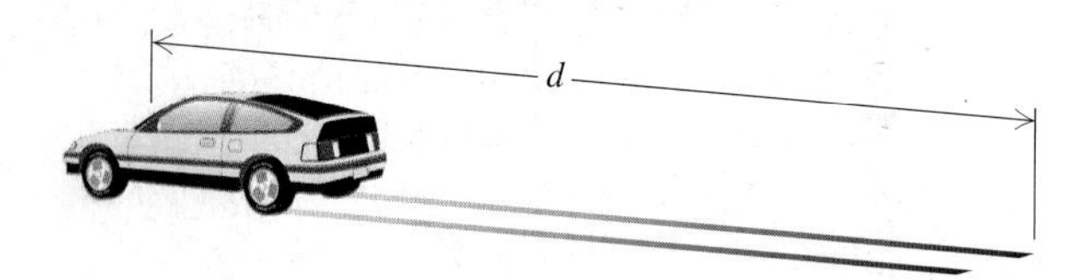

67. *Volume of a gas.* The volume V of a given mass of a gas varies directly as the temperature T and inversely as the pressure P. If $V = 231$ cm^3 when $T = 42°$ and $P = 20$ kg/cm^2, what is the volume when $T = 30°$ and $P = 15$ kg/cm^2?

68. *Intensity of a signal.* The intensity I of a television signal varies inversely as the square of the distance d from the transmitter. If the intensity is 25 W/m^2 at a distance of 2 km, what is the intensity 6.25 km from the transmitter?

69. *Atmospheric drag.* Wind resistance, or atmospheric drag, tends to slow down moving objects. Atmospheric drag W varies jointly as an object's surface area A and velocity v. If a car traveling at a speed of 40 mph with a surface area of 37.8 ft^2 experiences a drag of 222 N (Newtons), how fast must a car with 51 ft^2 of surface area travel in order to experience a drag force of 430 N?

70. *Drag force.* The drag force F on a boat varies jointly as the wetted surface area A and the square of the velocity of the boat. If a boat going 6.5 mph experiences a drag force of 86 N when the wetted surface area is 41.2 ft^2, find the wetted surface area of a boat traveling 8.2 mph with a drag force of 94 N.

71. Which exercise did you find easier to work: Exercise 7 or Exercise 11? Why?

72. If y varies directly as x, does doubling x cause y to be doubled as well? Why or why not?

SKILL MAINTENANCE

Find the domain of f.

73. $f(x) = \dfrac{2x - 1}{x^2 + 1}$

74. $f(x) = |2x - 1|$

75. Graph on a plane: $6x - y < 6$.

76. If $f(x) = x^3 - x$, find $f(2a)$.

77. Factor: $t^3 + 8b^3$.

78. Solve: $6x^2 = 11x + 35$.

SYNTHESIS

79. Suppose that the number of customer complaints is inversely proportional to the number of employees hired. Will a firm reduce the number of complaints more by expanding from 5 to 10 employees, or from 20 to 25? Explain. Consider using a graph to help justify your answer.

80. Why do you think subscripts are used in Exercises 3 and 15 but not in Exercises 17 and 18?

81. *Escape velocity.* A satellite's escape velocity is 6.5 mi/sec, the radius of the earth is 3960 mi, and the earth's gravitational constant is 32.2 ft/sec^2. How far is the satellite from the surface of the earth? (See Example 2.)

82. The *harmonic mean* of two numbers a and b is a number M such that the reciprocal of M is the average of the reciprocals of a and b. Find a formula for the harmonic mean.

83. *Health-care.* Young's rule for determining the size of a particular child's medicine dosage c is

$$c = \frac{a}{a + 12} \cdot d,$$

where a is the child's age and d is the typical adult dosage (*Source*: Olsen, June Looby, Leon J. Ablon, and Anthony Patrick Giangrasso, *Medical Dosage Calculations,* 6th ed.). If a child's age is doubled, the dosage increases. Find the ratio of the larger dosage to the smaller dosage. By what percent does the dosage increase?

84. Solve for x:

$$x^2\left(1 - \frac{2pq}{x}\right) = \frac{2p^2q^3 - pq^2x}{-q}.$$

85. *Average acceleration.* The formula

$$a = \frac{\dfrac{d_4 - d_3}{t_4 - t_3} - \dfrac{d_2 - d_1}{t_2 - t_1}}{t_4 - t_2}$$

can be used to approximate average acceleration, where the d's are distances and the t's are the corresponding times. Solve for t_1.

86. If y varies inversely as the cube of x and x is multiplied by 0.5, what is the effect on y?

Describe, in words, the variation given by the equation. Assume k is a constant.

87. $Q = \dfrac{kp^2}{q^3}$

88. $W = \dfrac{km_1M_1}{d^2}$

89. *Tension of a musical string.* The tension T on a string in a musical instrument varies jointly as the string's mass per unit length m, the square of its length l, and the square of its fundamental frequency f. A 2-m–long string of mass 5 gm/m with a fundamental frequency of 80 has a tension of 100 N. How long should the same string be if its tension is going to be changed to 72 N?

90. *Volume and cost.* A peanut butter jar in the shape of a right circular cylinder is 4 in. high and 3 in. in diameter and sells for \$1.20. If we assume that cost is proportional to volume, how much should a jar 6 in. high and 6 in. in diameter cost?

91. *Golf distance finder.* A device used in golf to estimate the distance d to a hole measures the size s that the 7-ft pin *appears* to be in a viewfinder. The viewfinder uses the principle, diagrammed here, that s gets bigger when d gets smaller. If $s = 0.56$ in. when $d = 50$ yd, find an equation of variation that expresses d as a function of s. What is d when $s = 0.40$ in.?

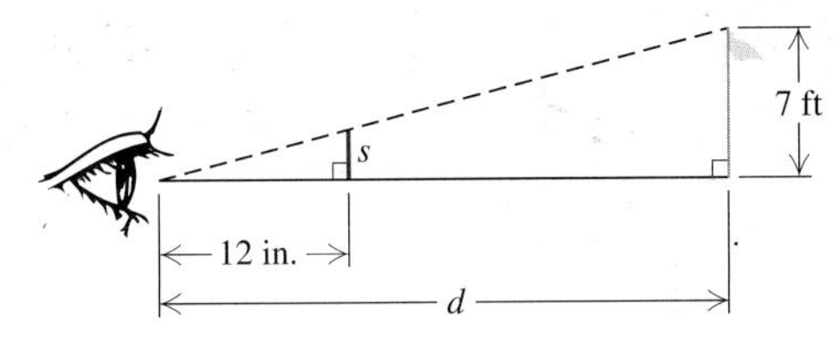

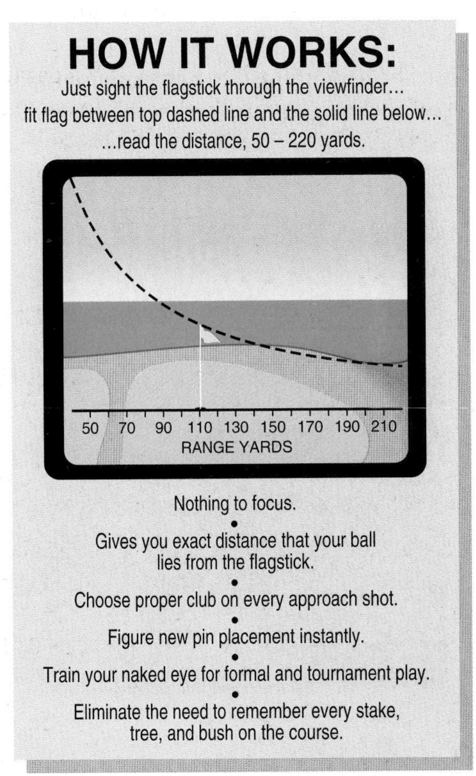

HOW IT WORKS:
Just sight the flagstick through the viewfinder...
fit flag between top dashed line and the solid line below...
...read the distance, 50 – 220 yards.
50 70 90 110 130 150 170 190 210
RANGE YARDS
Nothing to focus.
Gives you exact distance that your ball lies from the flagstick.
Choose proper club on every approach shot.
Figure new pin placement instantly.
Train your naked eye for formal and tournament play.
Eliminate the need to remember every stake, tree, and bush on the course.

COLLABORATIVE

CORNER

How Many Is a Million?

Focus: Direct variation and estimation

Time: 15 minutes

Group size: 2 or 3 and entire class

The National Park Service's estimates of crowd sizes for static (stationary) mass demonstrations vary directly as the area covered by the crowd. Park Service officials have found that at basic "shoulder-to-shoulder" demonstrations, 1 acre of land (about 45,000 ft^2) holds about 9000 people. Using aerial photographs, officials impose a grid to estimate the total area covered by the demonstrators. Once this has been accomplished, estimates of crowd size can be prepared.

ACTIVITY

1. In the grid imposed on the photograph below, each square represents 10,000 ft^2. Estimate the size of the crowd photographed. Then compare your group's estimate with those of other groups. What might explain discrepancies between estimates? List ways in which your group's estimate could be made more accurate.
2. Park Service officials use an "acceptable margin of error" of no more than 20%. Using all estimates from part (1) above and allowing for error, find a range of values within which you feel certain that the actual crowd size lies.
3. The Million Man March of 1995 was not a static demonstration because of a periodic turnover of people in attendance (many people stayed for only part of the day's festivities). How might you change your methodology to compensate for this complication?

Summary and Review 6

Key Terms

Rational expression, p. 334
Rational function, p. 334
Simplified, p. 337
Least common multiple, LCM, p. 345
Least common denominator, LCD, p. 346
Complex rational expression, p. 352
Rational equation, p. 362
Clear fractions, p. 362
Motion problem, p. 371
Synthetic division, p. 385
Direct variation, p. 391
Variation constant, p. 391
Constant of proportionality, p. 391
Inverse variation, p. 392
Joint variation, p. 394
Combined variation, p. 395

Important Properties and Formulas

Addition: $\frac{A}{C} + \frac{B}{C} = \frac{A + B}{C}$

Subtraction: $\frac{A}{C} - \frac{B}{C} = \frac{A - B}{C}$

Multiplication: $\frac{A}{B} \cdot \frac{C}{D} = \frac{AC}{BD}$

Division: $\frac{A}{B} \div \frac{C}{D} = \frac{A}{B} \cdot \frac{D}{C}$

To find the least common multiple, LCM, use each factor the greatest number of times that it occurs in any one prime factorization.

Simplifying Complex Rational Expressions

I: By using multiplication by 1

1. Find the LCD of all rational expressions *within* the complex rational expression.
2. Multiply the complex rational expression by 1, writing 1 as the LCD divided by itself.
3. Distribute and simplify so that the numerator and the denominator of the complex rational expression are polynomials.
4. Factor and, if possible, simplify.

II. By using division

1. Add or subtract, as necessary, to get one rational expression in the numerator.
2. Add or subtract, as necessary, to get one rational expression in the denominator.
3. Perform the indicated division (invert the divisor and multiply).
4. Simplify, if possible, by removing any factors equal to 1.

Modeling Work Problems

If

a = the time needed for A to complete the work alone,
b = the time needed for B to complete the work alone, and
t = the time needed for A and B to complete the work together,

then

$\frac{t}{a} + \frac{t}{b} = 1$ and $\frac{1}{a} \cdot t + \frac{1}{b} \cdot t = 1$

and $\frac{1}{a} + \frac{1}{b} = \frac{1}{t}$.

Motion Formula

$d = rt$ or $r = d/t$ or $t = d/r$

The Remainder Theorem

The remainder obtained by dividing $P(x)$ by $x - r$ is $P(r)$.

Variation

y varies directly as x if there is some nonzero constant k such that $y = kx$.

y varies inversely as x if there is some nonzero constant k such that $y = k/x$.

y varies jointly as x and z if there is some nonzero constant k such that $y = kxz$.

Review Exercises

1. If

$$f(t) = \frac{t^2 - 3t + 2}{t^2 - 9},$$

find the following function values.

a) $f(0)$ **b)** $f(-1)$ **c)** $f(2)$

Find the LCD.

2. $\frac{7}{6x^3}, \frac{y}{16x^2}$

3. $\frac{x+8}{x^2+x-20}, \frac{x}{x^2+3x-10}$

Perform the indicated operations and, if possible, simplify.

4. $\frac{x^2}{x-3} - \frac{9}{x-3}$

5. $\frac{4x-2}{x^2-5x+4} - \frac{3x+2}{x^2-5x+4}$

6. $\frac{3a^2b^3}{5c^3d^2} \cdot \frac{15c^9d^4}{9a^7b}$

7. $\frac{5}{6m^2n^3p} + \frac{7}{9mn^4p^2}$

8. $\frac{y^2-64}{2y+10} \cdot \frac{y+5}{y+8}$

9. $\frac{x^3-8}{x^2-25} \cdot \frac{x^2+10x+25}{x^2+2x+4}$

10. $\frac{9a^2-1}{a^2-9} \div \frac{3a+1}{a+3}$

11. $\frac{x^3-64}{x^2-16} \div \frac{x^2+5x+6}{x^2-3x-18}$

12. $\frac{x}{x^2+5x+6} - \frac{2}{x^2+3x+2}$

13. $\frac{-4xy}{x^2-y^2} + \frac{x+y}{x-y}$

14. $\frac{2x^2}{x-y} + \frac{2y^2}{y-x}$

15. $\frac{3}{y+4} - \frac{y}{y-1} + \frac{y^2+3}{y^2+3y-4}$

Simplify.

16. $\dfrac{\frac{5}{x} - 5}{\frac{7}{x} - 7}$

17. $\dfrac{\frac{2}{a} + \frac{2}{b}}{\frac{4}{a^3} + \frac{4}{b^3}}$

18. $\dfrac{\frac{y^2+4y-77}{y^2-10y+25}}{\frac{y^2-5y-14}{y^2-25}}$

19. $\dfrac{\frac{5}{x^2-9} - \frac{3}{x+3}}{\frac{4}{x^2+6x+9} + \frac{2}{x-3}}$

Solve.

20. $\frac{6}{x} + \frac{4}{x} = 5$

21. $\frac{5}{3x+2} = \frac{3}{2x}$

22. $\frac{4x}{x+1} + \frac{4}{x} + 9 = \frac{4}{x^2+x}$

23. $\dfrac{x+6}{x^2+x-6}+\dfrac{x}{x^2+4x+3}=\dfrac{x+2}{x^2-x-2}$

24. If

$$f(x)=\frac{2}{x-1}+\frac{2}{x+2},$$

find all a for which $f(a)=1$.

Solve.

25. Kim can set up for a banquet in 12 hr. Kelly can set up for the same banquet in 9 hr. How long would it take them, working together, to set up for the banquet?

26. A research company uses personal computers to process data while the owner is not using the computer. A Pentium III 850 megahertz processor can process a megabyte of data in 15 sec less time than a Celeron 700 megahertz processor. Working together, the computers can process a megabyte of data in 18 sec. How long does it take each computer to process one megabyte of data?

27. The Gold River's current is 6 mph. A boat travels 50 mi downstream in the same time that it takes to travel 30 mi upstream. What is the speed of the boat in still water?

28. A car and a motorcycle leave a rest area at the same time, with the car traveling 8 mph faster than the motorcycle. The car then travels 105 mi in the time it takes the motorcycle to travel 93 mi. Find the speed of each vehicle.

Divide.

29. $(20r^2s^3+15r^2s^2-10r^3s^3)\div(5r^2s)$

30. $(y^3+125)\div(y+5)$

31. $(4x^3+3x^2-5x-2)\div(x^2+1)$

32. Divide using synthetic division:

$(x^3+3x^2+2x-6)\div(x-3)$.

33. If $f(x)=4x^3-6x^2-9$, use synthetic division to find $f(5)$.

Solve.

34. $R=\dfrac{gs}{g+s}$, for s

35. $S=\dfrac{H}{m(t_1-t_2)}$, for m

36. $\dfrac{1}{ac}=\dfrac{2}{ab}-\dfrac{3}{bc}$, for c

37. $T=\dfrac{A}{v(t_2-t_1)}$, for t_1

38. The amount of waste generated by a restaurant varies directly as the number of customers served. A typical McDonalds that serves 2000 customers per day generates 238 lb of waste daily (*Source*: Environmental Defense Fund Study, November 1990). How many pounds of waste would be generated daily by a McDonalds that serves 1700 customers a day?

39. A warning dye is used by people in lifeboats to aid search planes. The volume V of the dye used varies directly as the square of the diameter d of the circular patch of water formed by the dye. If 4 L of dye is required for a 10-m wide circle, how much dye is needed for a 40-m wide circle?

40. Find an equation of variation in which y varies inversely as x, and $y=3$ when $x=\frac{1}{4}$.

SYNTHESIS

41. Discuss at least three different uses of the LCD studied in this chapter.

42. Explain the difference between a rational expression and a rational equation.

Solve.

43. $\dfrac{5}{x-13}-\dfrac{5}{x}=\dfrac{65}{x^2-13x}$

44. $\dfrac{\dfrac{x}{x^2-25}+\dfrac{2}{x-5}}{\dfrac{3}{x-5}-\dfrac{4}{x^2-10x+25}}=1$

45. A Pentium 4 1.5-gigahertz processor can process a megabyte of data in 20 sec. How long would it take a Pentium 4 working together with the Pentium III and Celeron processors (see Exercise 26) to process a megabyte of data?

Chapter Test 6

Simplify.

1. $\dfrac{t-1}{t+3} \cdot \dfrac{3t+9}{4t^2-4}$

2. $\dfrac{x^3+27}{x^2-16} \div \dfrac{x^2+8x+15}{x^2+x-20}$

3. Find the LCD:
$$\frac{3x}{x^2+8x-33}, \quad \frac{x+1}{x^2-12x+27}.$$

Perform the indicated operation and simplify when possible.

4. $\dfrac{25x}{x+5} + \dfrac{x^3}{x+5}$

5. $\dfrac{3a^2}{a-b} - \dfrac{3b^2-6ab}{b-a}$

6. $\dfrac{4ab}{a^2-b^2} + \dfrac{a^2+b^2}{a+b}$

7. $\dfrac{6}{x^3-64} - \dfrac{4}{x^2-16}$

8. $\dfrac{4}{y+3} - \dfrac{y}{y-2} + \dfrac{y^2+4}{y^2+y-6}$

Simplify.

9. $\dfrac{\dfrac{2}{a}+\dfrac{3}{b}}{\dfrac{5}{ab}+\dfrac{1}{a^2}}$

10. $\dfrac{\dfrac{x^2-5x-36}{x^2-36}}{\dfrac{x^2+x-12}{x^2-12x+36}}$

11. $\dfrac{\dfrac{4}{x+3}-\dfrac{2}{x^2-3x+2}}{\dfrac{3}{x-2}+\dfrac{1}{x^2+2x-3}}$

Solve.

12. $\dfrac{4}{2x-5} = \dfrac{6}{5x+3}$

13. $\dfrac{t+11}{t^2-t-12} + \dfrac{1}{t-4} = \dfrac{4}{t+3}$

For Exercises 14 and 15, let $f(x) = \dfrac{x+3}{x-1}$.

14. Find $f(2)$ and $f(-3)$.

15. Find all a for which $f(a) = 7$.

16. Kyla can lay vinyl in a kitchen in 3.5 hr. Brock can lay the same vinyl in 4.5 hr. How long will it take them, working together, to lay the vinyl?

Divide.

17. $(16ab^3c - 10ab^2c^2 + 12a^2b^2c) \div (4a^2b)$

18. $(y^2 - 20y + 64) \div (y - 6)$

19. $(6x^4 + 3x^2 + 5x + 4) \div (x^2 + 2)$

20. Divide using synthetic division:
$$(x^3 + 5x^2 + 4x - 7) \div (x - 4).$$

21. If $f(x) = 3x^4 - 5x^3 + 2x - 7$, use synthetic division to find $f(4)$.

22. Solve $A = \dfrac{h(b_1 + b_2)}{2}$ for b_1.

23. The product of the reciprocals of two consecutive integers is $\frac{1}{30}$. Find the integers.

24. Georgia bicycles 12 mph with no wind. Against the wind, she bikes 8 mi in the same time that it takes to bike 14 mi with the wind. What is the speed of the wind?

25. The number of workers n needed to clean a stadium after a game varies inversely as the amount of time t allowed for the cleanup. If it takes 25 workers to clean the stadium when there are 6 hr allowed for the job, how many workers are needed if the stadium must be cleaned in 5 hr?

26. The surface area of a balloon varies directly as the square of its radius. The area is 325 in^2 when the radius is 5 in. What is the area when the radius is 7 in.?

SYNTHESIS

27. Let
$$f(x) = \frac{1}{x+3} + \frac{5}{x-2}.$$
Find all a for which $f(a) = f(a+5)$.

28. Solve: $\dfrac{6}{x-15} - \dfrac{6}{x} = \dfrac{90}{x^2-15x}$.

29. Find the x- and y-intercepts for the function given by
$$f(x) = \frac{\dfrac{5}{x+4} - \dfrac{3}{x-2}}{\dfrac{2}{x-3} + \dfrac{1}{x+4}}.$$

30. One summer, Hans mowed 4 lawns for every 3 lawns mowed by his brother Franz. Together, they mowed 98 lawns. How many lawns did each mow?

Cumulative Review 1–6

1. Evaluate

$$\frac{2x - y^2}{x + y}$$

for $x = 3$ and $y = -4$.

2. Convert to scientific notation: 5,760,000,000.

3. Determine the slope and the y-intercept for the line given by $7x - 4y = 12$.

4. Find an equation for the line that passes through the points $(-1, 7)$ and $(2, -3)$.

5. Solve the system

$$5x - 2y = -23,$$
$$3x + 4y = 7.$$

6. Solve the system

$$-3x + 4y + z = -5,$$
$$x - 3y - z = 6,$$
$$2x + 3y + 5z = -8.$$

7. Briar Creek Elementary School sold 45 pizzas for a fundraiser. Small pizzas sold for \$7.00 each and large pizzas for \$10.00 each. The total amount of funds received from the sale was \$402. How many of each size pizza were sold?

8. The sum of three numbers is 20. The first number is 3 less than twice the third number. The second number minus the third number is -7. What are the numbers?

9. Trex Company makes decking material from waste wood fibers and reclaimed polyethylene. Its sales rose from \$3.5 million in 1993 to \$74.3 million in 1999 (*Source*: *Business Week*, May 29, 2000). Calculate the rate at which sales were rising.

10. In 1989, the average length of a visit to a physician in an HMO was 15.4 min; and in 1998, it was 17.9 min (*Sources*: Rutgers University Study and National Center for Health Statistics). Let V represent the average length of a visit t years after 1989.

a) Find a linear function $V(t)$ that fits the data.

b) Use the function of part (a) to predict the average length of a visit in 2005.

11. If

$$f(x) = \frac{x - 2}{x - 5},$$

find **(a)** $f(3)$ and **(b)** the domain of f.

Solve.

12. $8x = 1 + 16x^2$

13. $625 = 49y^2$

14. $20 > 2 - 6x$

15. $\frac{1}{3}x - \frac{1}{5} \geq \frac{1}{5}x - \frac{1}{3}$

16. $-8 < x + 2 < 15$

17. $3x - 2 < -6$ *or* $x + 3 > 9$

18. $|x| > 6.4$

19. $|4x - 1| \leq 14$

20. $\frac{2}{n} - \frac{7}{n} = 3$

21. $\frac{6}{x - 5} = \frac{2}{2x}$

22. $\frac{3x}{x - 2} - \frac{6}{x + 2} = \frac{24}{x^2 - 4}$

23. $\frac{3x^2}{x + 2} + \frac{5x - 22}{x - 2} = \frac{-48}{x^2 - 4}$

24. Let $f(x) = |3x - 5|$. Find all values of x for which $f(x) = 2$.

25. Write the domain of f using interval notation if $f(x) = \sqrt{x - 7}$.

Solve.

26. $5m - 3n = 4m + 12$, for n

27. $P = \frac{3a}{a + b}$, for a

Graph on a plane.

28. $4x \geq 5y + 20$

29. $y = \frac{1}{3}x - 2$

Perform the indicated operations and simplify.

30. $(2x^2 - 3x + 1) + (6x - 3x^3 + 7x^2 - 4)$

31. $(5x^3y^2)(-3xy^2)$

32. $(3a + b - 2c) - (-4b + 3c - 2a)$

33. $(5x^2 - 2x + 1)(3x^2 + x - 2)$

34. $(2x^2 - y)^2$

35. $(2x^2 - y)(2x^2 + y)$

36. $(-5m^3n^2 - 3mn^3) + (-4m^2n^2 + 4m^3n^2) - (2mn^3 - 3m^2n^2)$

37. $\frac{y^2 - 36}{2y + 8} \cdot \frac{y + 4}{y + 6}$

38. $\frac{x^4 - 1}{x^2 - x - 2} \div \frac{x^2 + 1}{x - 2}$

39. $\frac{5ab}{a^2 - b^2} + \frac{a + b}{a - b}$

40. $\frac{2}{m + 1} + \frac{3}{m - 5} - \frac{m^2 - 1}{m^2 - 4m - 5}$

41. $y - \frac{2}{3y}$

42. Simplify: $\frac{\frac{1}{x} - \frac{1}{y}}{x + y}$.

43. Divide: $(9x^3 + 5x^2 + 2) \div (x + 2)$.

Factor.

44. $4x^3 + 18x^2$

45. $x^2 + 8x - 84$

46. $16y^2 - 81$

47. $64x^3 + 8$

48. $t^2 - 16t + 64$

49. $x^6 - x^2$

50. $0.027b^3 - 0.008c^3$

51. $20x^2 + 7x - 3$

52. $3x^2 - 17x - 28$

53. $x^5 - x^3y + x^2y - y^2$

54. If $f(x) = x^2 - 4$ and $g(x) = x^2 - 7x + 10$, find the domain of f/g.

55. A digital data circuit can transmit a particular set of data in 4 sec. An analog phone circuit can transmit the same data in 20 sec. How long would it take, working together, for both circuits to transmit the data?

56. The floor area of a rental trailer is rectangular. The length is 3 ft more than the width. A rug of area 54 ft^2 exactly fills the floor of the trailer. Find the perimeter of the trailer.

57. The sum of the squares of three consecutive even integers is equal to 8 more than three times the square of the second number. Find the integers.

58. *Logging.* The volume of wood V in a tree trunk varies jointly as the height h and the square of the girth g (girth is distance around). If the volume is 35 ft^3 when the height is 20 ft and the girth is 5 ft, what is the height when the volume is 85.75 ft^3 and the girth is 7 ft?

SYNTHESIS

59. Multiply: $(x - 4)^3$.

60. Find all roots for $f(x) = x^4 - 34x^2 + 225$.

Solve.

61. $4 \le |3 - x| \le 6$

62. $\frac{18}{x - 9} + \frac{10}{x + 5} = \frac{28x}{x^2 - 4x - 45}$

63. $16x^3 = x$

7 Exponents and Radicals

7.1 Radical Expressions and Functions
7.2 Rational Numbers as Exponents
7.3 Multiplying Radical Expressions
7.4 Dividing Radical Expressions
7.5 Expressions Containing Several Radical Terms
7.6 Solving Radical Equations
Connecting the Concepts
7.7 Geometric Applications
7.8 The Complex Numbers
Summary and Review
Test

AN APPLICATION

In steel production, the temperature of the molten metal is so great that conventional thermometers melt. Instead, sound is transmitted across the surface of the metal to a receiver on the far side and the speed of sound is measured. The formula

$$S(t) = 1087.7\sqrt{\frac{9t + 2617}{2457}}$$

gives the speed of sound $S(t)$, in feet per second, at a temperature of t degrees Celsius. Find the temperature of a blast furnace where sound travels 1502.3 ft/sec.

This problem appears as Exercise 59 in Section 7.6.

Mathematics has transformed the way we produce steel from using the skills of an experienced operator, to an accurate science using highly automated computer control systems.

DANIEL GOLDSTEIN
Research Engineer
Bethlehem Steel
Bethlehem, PA

In this chapter, we learn about square roots, cube roots, fourth roots, and so on. These roots are studied in connection with the manipulation of radical expressions and the solution of real-world applications. Fractional exponents are also studied and are used to ease some of our work with radicals. The chapter closes with an examination of the complex-number system.

7.1 Radical Expressions and Functions

Square Roots and Square Root Functions • Expressions of the Form $\sqrt{a^2}$ • Cube Roots • Odd and Even *n*th Roots

Study Tip

The best way to prepare for a final exam is to do so over a period of at least two weeks. First review each chapter, studying the formulas, problems, properties, and procedures in the chapter Summary and Review. Then retake your quizzes and tests. If you miss any questions, spend extra time reviewing the corresponding topics. Watch the videotapes that accompany the text or use the InterAct Math Tutorial Software. Also consider participating in a study group or attending a tutoring or review session.

In this section, we consider roots, such as square roots and cube roots. We look at the symbolism that is used and ways in which symbols can be manipulated to get equivalent expressions. All of this will be important in problem solving.

Square Roots and Square Root Functions

When a number is raised to the second power, the number is squared. Often we need to know what number was squared in order to produce some value a. If such a number can be found, we call that number a *square root* of a.

Square Root

The number c is a *square root* of a if $c^2 = a$.

For example,

9 has -3 and 3 as square roots because $(-3)^2 = 9$ and $3^2 = 9$.

25 has -5 and 5 as square roots because $(-5)^2 = 25$ and $5^2 = 25$.

-4 does not have a real-number square root because there is no real number c such that $c^2 = -4$.

Note that every positive number has two square roots, whereas 0 has only itself as a square root. Negative numbers do not have real-number square roots, although later in this chapter we will work with a number system in which such square roots do exist.

Example 1 Find the two square roots of 64.

Solution The square roots are 8 and -8, because $8^2 = 64$ and $(-8)^2 = 64$.

Whenever we refer to *the* square root of a number, we mean the nonnegative square root of that number. This is often referred to as the *principal square root* of the number.

Principal Square Root

The *principal square root* of a nonnegative number is its nonnegative square root. The symbol $\sqrt{}$ is called a *radical sign* and is used to indicate the principal square root of the number over which it appears.

Example 2 Simplify each of the following.

a) $\sqrt{25}$ **b)** $\sqrt{\frac{25}{64}}$

c) $-\sqrt{64}$ **d)** $\sqrt{0.0049}$

Solution

a) $\sqrt{25} = 5$ — $\sqrt{}$ indicates the principal square root. Note that $\sqrt{25} \neq -5$.

b) $\sqrt{\frac{25}{64}} = \frac{5}{8}$ — Since $\left(\frac{5}{8}\right)^2 = \frac{25}{64}$

c) $-\sqrt{64} = -8$ — Since $\sqrt{64} = 8$, $-\sqrt{64} = -8$.

d) $\sqrt{0.0049} = 0.07$ — $(0.07)(0.07) = 0.0049$

In addition to being read as "the principal square root of a," $\sqrt{a}$ is also read as "the square root of a," or simply "root a." Any expression in which a radical sign appears is called a *radical expression.* The following are radical expressions:

$$\sqrt{5}, \quad \sqrt{a}, \quad -\sqrt{3x}, \quad \sqrt{\frac{y^2 + 7}{y}}.$$

The expression under the radical sign is called the **radicand**. In the expressions above, the radicands are 5, a, $3x$, and $(y^2 + 7)/y$.

All but the most basic calculators give values for square roots. These values are, for the most part, approximations. For example, on many calculators, if you enter 5 and then press $\boxed{\sqrt{}}$, a number like

2.23606798

appears, depending on how the calculator rounds. (On some calculators, the $\boxed{\sqrt{}}$ key is pressed first.) The exact value of $\sqrt{5}$ is not given by any repeating or terminating decimal. The same is true for the square root of any whole number that is not a perfect square. We discussed such *irrational numbers* in Chapter 1.

The square-root function, given by

$$f(x) = \sqrt{x},$$

has the interval $[0, \infty)$ as its domain. We can draw its graph by selecting convenient values for x and calculating the corresponding outputs. Once these ordered pairs have been graphed, a smooth curve can be drawn.

$f(x) = \sqrt{x}$

x	$\sqrt{x}$	$(x, f(x))$
0	0	(0, 0)
1	1	(1, 1)
4	2	(4, 2)
9	3	(9, 3)

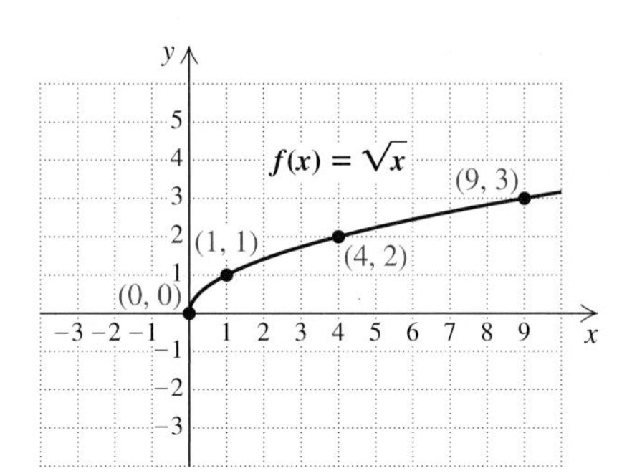

Example 3 For each function, find the indicated function value.

a) $f(x) = \sqrt{3x - 2}$; $f(1)$ **b)** $g(z) = -\sqrt{6z + 4}$; $g(3)$

Solution

a) $f(1) = \sqrt{3 \cdot 1 - 2}$ Substituting

$= \sqrt{1} = 1$ Simplifying

b) $g(3) = -\sqrt{6 \cdot 3 + 4}$ Substituting

$= -\sqrt{22}$ Simplifying

≈ -4.69041576 Using a calculator to approximate $\sqrt{22}$

Expressions of the Form $\sqrt{a^2}$

It is tempting to write $\sqrt{a^2} = a$, but the next example shows that, as a rule, this is untrue.

Example 4 Evaluate $\sqrt{x^2}$ for the following values: **(a)** 5; **(b)** 0; **(c)** -5.

Solution

a) $\sqrt{5^2} = \sqrt{25} = 5$ — Same

b) $\sqrt{0^2} = \sqrt{0} = 0$ — Same

c) $\sqrt{(-5)^2} = \sqrt{25} = 5$ — Opposites Note that $\sqrt{(-5)^2} \neq -5$.

You may have noticed that evaluating $\sqrt{a^2}$ is just like evaluating $|a|$.

Simplifying $\sqrt{a^2}$

For any real number a,

$$\sqrt{a^2} = |a|.$$

(The principal square root of a^2 is the absolute value of a.)

When a radicand is the square of a variable expression, like $(x + 5)^2$ or $36t^2$, absolute-value signs are needed when simplifying. We use absolute-value signs unless we know that the expression being squared is nonnegative. This assures that our result is never negative.

Example 5

Simplify each expression. Assume that the variable can represent any real number.

a) $\sqrt{(x + 1)^2}$ **b)** $\sqrt{x^2 - 8x + 16}$

c) $\sqrt{a^8}$ **d)** $\sqrt{t^6}$

Solution

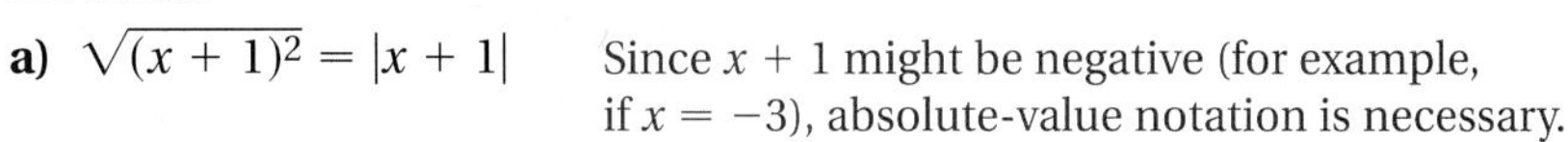

a) $\sqrt{(x + 1)^2} = |x + 1|$ Since $x + 1$ might be negative (for example, if $x = -3$), absolute-value notation is necessary.

b) $\sqrt{x^2 - 8x + 16} = \sqrt{(x - 4)^2} = |x - 4|$ Since $x - 4$ might be negative, absolute-value notation is necessary.

c) Note that $(a^4)^2 = a^8$ and that a^4 is never negative. Thus,

$\sqrt{a^8} = a^4.$ Absolute-value notation is unnecessary here.

d) Note that $(t^3)^2 = t^6$. Thus,

$\sqrt{t^6} = |t^3|.$ Since t^3 might be negative, absolute-value notation is necessary.

To see the necessity of the absolute-value signs, let $y_1 = \sqrt{x^2}$, $y_2 = x$, and $y_3 = \text{ABS}(x)$. Then use a graph or table to show that $y_1 \neq y_2$ and $y_3 \neq y_2$, but $y_1 = y_3$.

Example 6

Simplify each expression. Assume that no radicands were formed by raising negative quantities to even powers.

a) $\sqrt{y^2}$ **b)** $\sqrt{a^{10}}$ **c)** $\sqrt{9x^2 - 6x + 1}$

Solution

a) $\sqrt{y^2} = y$ We are assuming that y is nonnegative, so no absolute-value notation is necessary. When y *is* negative, $\sqrt{y^2} \neq y$.

b) $\sqrt{a^{10}} = a^5$ Assuming that a^5 is nonnegative. Note that $(a^5)^2 = a^{10}$.

c) $\sqrt{9x^2 - 6x + 1} = \sqrt{(3x - 1)^2} = 3x - 1$ Assuming that $3x - 1$ is nonnegative

Cube Roots

We often need to know what number was cubed in order to produce a certain value. When such a number is found, we say that we have found a *cube root*. For example,

$$2 \text{ is the cube root of } 8 \text{ because } 2^3 = 2 \cdot 2 \cdot 2 = 8;$$
$$-4 \text{ is the cube root of } -64 \text{ because } (-4)^3 = (-4)(-4)(-4) = -64.$$

Cube Root

The number c is the *cube root* of a if $c^3 = a$. In symbols, we write $\sqrt[3]{a}$ to denote the cube root of a.

The cube-root function, given by

$$f(x) = \sqrt[3]{x},$$

has $\mathbb{R}$ as its domain. We can draw its graph by selecting convenient values for x and calculating the corresponding outputs. Once these ordered pairs have been graphed, a smooth curve can be drawn.

$f(x) = \sqrt[3]{x}$

x	$\sqrt[3]{x}$	$(x, f(x))$
0	0	(0, 0)
1	1	(1, 1)
8	2	(8, 2)
−1	−1	(−1, −1)
−8	−2	(−8, −2)

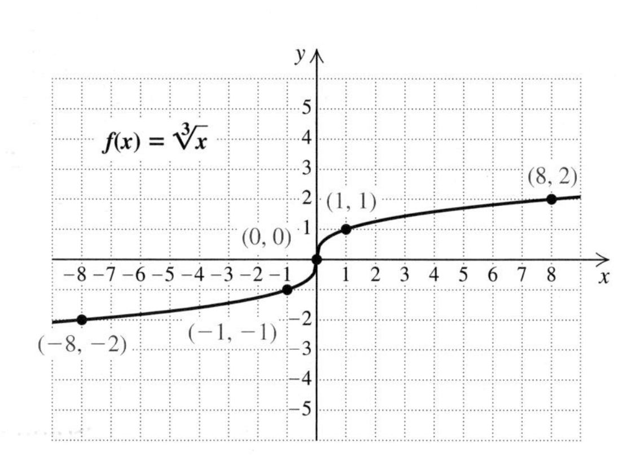

In the real-number system, every number has exactly one cube root. The cube root of a positive number is positive, and the cube root of a negative number is negative. Absolute-value signs are not used when finding cube roots.

Example 7 For each function, find the indicated function value.

a) $f(y) = \sqrt[3]{y}$; $f(125)$ **b)** $g(x) = \sqrt[3]{x - 3}$; $g(-24)$

Solution

a) $f(125) = \sqrt[3]{125} = 5$ Since $5 \cdot 5 \cdot 5 = 125$

b) $g(-24) = \sqrt[3]{-24 - 3}$
$= \sqrt[3]{-27}$
$= -3$ Since $(-3)(-3)(-3) = -27$

Example 8

Simplify: $\sqrt[3]{-8y^3}$.

Solution

$$\sqrt[3]{-8y^3} = -2y \qquad \text{Since } (-2y)(-2y)(-2y) = -8y^3$$

Odd and Even nth Roots

The fifth root of a number a is the number c for which $c^5 = a$. There are also 6th roots, 7th roots, and so on. We write $\sqrt[n]{a}$ for the nth root. The number n is called the *index* (plural, *indices*). When the index is 2, we do not write it.

Example 9

Find each of the following.

a) $\sqrt[5]{32}$

b) $\sqrt[5]{-32}$

c) $-\sqrt[5]{32}$

d) $-\sqrt[5]{-32}$

Solution

a) $\sqrt[5]{32} = 2$ — Since $2^5 = 32$

b) $\sqrt[5]{-32} = -2$ — Since $(-2)^5 = -32$

c) $-\sqrt[5]{32} = -2$ — Taking the opposite of $\sqrt[5]{32}$

d) $-\sqrt[5]{-32} = -(-2) = 2$ — Taking the opposite of $\sqrt[5]{-32}$

Note that every number has just one real root when n is odd. Odd roots of positive numbers are positive and odd roots of negative numbers are negative. Absolute-value signs are not used when finding odd roots.

Example 10

Find each of the following.

a) $\sqrt[7]{x^7}$

b) $\sqrt[9]{(x-1)^9}$

Solution

a) $\sqrt[7]{x^7} = x$

b) $\sqrt[9]{(x-1)^9} = x - 1$

When the index n is even, we say that we are taking an *even root.* Every positive real number has two real nth roots when n is even. One root is positive and one is negative. Negative numbers do not have real nth roots when n is even.

When n is even, the notation $\sqrt[n]{a}$ indicates the nonnegative nth root. Thus, when we are finding even nth roots, absolute-value signs are often necessary.

Example 11

Simplify each expression, if possible. Assume that variables can represent any real number.

a) $\sqrt[4]{16}$

b) $-\sqrt[4]{16}$

c) $\sqrt[4]{-16}$

d) $\sqrt[4]{81x^4}$

e) $\sqrt[6]{(y+7)^6}$

Solution

a) $\sqrt[4]{16} = 2$ — Since $2^4 = 16$

b) $-\sqrt[4]{16} = -2$ — Taking the opposite of $\sqrt[4]{16}$

c) $\sqrt[4]{-16}$ cannot be simplified. — $\sqrt[4]{-16}$ is not a real number.

d) $\sqrt[4]{81x^4} = 3|x|$ — Use absolute-value notation since x could represent a negative number.

e) $\sqrt[6]{(y+7)^6} = |y+7|$ — Use absolute-value notation since $y + 7$ could be negative.

We summarize as follows.

Simplifying *n*th Roots

n	a	$\sqrt[n]{a}$	$\sqrt[n]{a^n}$
Even	Positive	Positive	$\|a\|$ (or a)
	Negative	Not a real number	$\|a\|$ (or $-a$)
Odd	Positive	Positive	a
	Negative	Negative	a

Example 12

Determine the domain of $g(x) = \sqrt[6]{7-3x}$.

Solution Since the index is even, the radicand, $7 - 3x$, must be nonnegative. We solve the inequality:

$7 - 3x \geq 0$ — We cannot find the 6th root of a negative number.

$-3x \geq -7$

$x \leq \frac{7}{3}$. — Multiplying both sides by $-\frac{1}{3}$ and reversing the inequality

Thus,

$$\text{Domain of } g = \left\{x \mid x \leq \tfrac{7}{3}\right\} = \left(-\infty, \tfrac{7}{3}\right].$$

Use a grapher to draw the graph of $f(x) = \sqrt{x+2}$. Then check graphically that the domain of f is $[-2, \infty)$.

Exercise Set 7.1

FOR EXTRA HELP

Digital Video Tutor CD 5 Videotape 13 | InterAct Math | Math Tutor Center | MathXL | MyMathLab.com

For each number, find the square roots.

1. 16 **2.** 49

3. 144 **4.** 9

5. 81 **6.** 400

7. 900 **8.** 225

Simplify.

9. $-\sqrt{\frac{49}{36}}$ **10.** $-\sqrt{\frac{361}{9}}$ **11.** $\sqrt{441}$

12. $\sqrt{196}$ **13.** $-\sqrt{\frac{16}{81}}$ **14.** $-\sqrt{\frac{81}{144}}$

15. $\sqrt{0.09}$ **16.** $\sqrt{0.36}$

17. $-\sqrt{0.0049}$ **18.** $\sqrt{0.0144}$

Identify the radicand and the index for each expression.

19. $5\sqrt{p^2 + 4}$ **20.** $-7\sqrt{y^2 - 8}$

21. $x^2y^3\sqrt[3]{\frac{x}{y+4}}$ **22.** $a^2b^3\sqrt[3]{\frac{a}{a^2-b}}$

For each function, find the specified function value, if it exists.

23. $f(t) = \sqrt{5t - 10}$; $f(6), f(2), f(1), f(-1)$

24. $g(x) = \sqrt{x^2 - 25}$; $g(-6), g(3), g(6), g(13)$

25. $t(x) = -\sqrt{2x + 1}$; $t(4), t(0), t(-1), t\left(-\frac{1}{2}\right)$

26. $p(z) = \sqrt{2z^2 - 20}$; $p(4), p(3), p(-5), p(0)$

27. $f(t) = \sqrt{t^2 + 1}$; $f(0), f(-1), f(-10)$

28. $g(x) = -\sqrt{(x + 1)^2}$; $g(-3), g(4), g(-5)$

29. $g(x) = \sqrt{x^3 + 9}$; $g(-2), g(-3), g(3)$

30. $f(t) = \sqrt{t^3 - 10}$; $f(2), f(3), f(4)$

Simplify. Remember to use absolute-value notation when necessary.

31. $\sqrt{36x^2}$ **32.** $\sqrt{25t^2}$

33. $\sqrt{(-6b)^2}$ **34.** $\sqrt{(-7c)^2}$

35. $\sqrt{(7 - t)^2}$ **36.** $\sqrt{(a + 1)^2}$

37. $\sqrt{y^2 + 16y + 64}$ **38.** $\sqrt{x^2 - 4x + 4}$

39. $\sqrt{9x^2 - 30x + 25}$ **40.** $\sqrt{4x^2 + 28x + 49}$

41. $-\sqrt[4]{625}$ **42.** $\sqrt[4]{256}$

43. $-\sqrt[5]{3^5}$ **44.** $\sqrt[5]{-1}$

45. $\sqrt[5]{-\frac{1}{32}}$ **46.** $\sqrt[5]{-\frac{32}{243}}$

47. $\sqrt[8]{y^8}$ **48.** $\sqrt[6]{x^6}$

49. $\sqrt[4]{(7b)^4}$ **50.** $\sqrt[4]{(5a)^4}$

51. $\sqrt[12]{(-10)^{12}}$ **52.** $\sqrt[10]{(-6)^{10}}$

53. $\sqrt[1976]{(2a + b)^{1976}}$ **54.** $\sqrt[414]{(a + b)^{414}}$

55. $\sqrt{x^{10}}$ **56.** $\sqrt{a^{22}}$

57. $\sqrt{a^{14}}$ **58.** $\sqrt{x^{16}}$

Simplify. Assume that no radicands were formed by raising negative quantities to even powers.

59. $\sqrt{25t^2}$ **60.** $\sqrt{16x^2}$

61. $\sqrt{(7c)^2}$ **62.** $\sqrt{(6b)^2}$

63. $\sqrt{(5 + b)^2}$ **64.** $\sqrt{(a + 1)^2}$

65. $\sqrt{9x^2 + 36x + 36}$ **66.** $\sqrt{4x^2 + 8x + 4}$

67. $\sqrt{25t^2 - 20t + 4}$ **68.** $\sqrt{9t^2 - 12t + 4}$

69. $-\sqrt[3]{64}$ **70.** $\sqrt[3]{27}$

71. $\sqrt[4]{81x^4}$ **72.** $\sqrt[4]{16x^4}$

73. $-\sqrt[5]{-100{,}000}$ **74.** $\sqrt[3]{-216}$

75. $-\sqrt[3]{-64x^3}$ **76.** $-\sqrt[3]{-125y^3}$

77. $\sqrt{a^{14}}$ **78.** $\sqrt{a^{22}}$

79. $\sqrt{(x + 3)^{10}}$ **80.** $\sqrt{(x - 2)^8}$

For each function, find the specified function value, if it exists.

81. $f(x) = \sqrt[3]{x + 1}$; $f(7), f(26), f(-9), f(-65)$

82. $g(x) = -\sqrt[3]{2x - 1}$; $g(0), g(-62), g(-13), g(63)$

83. $g(t) = \sqrt[4]{t - 3}$; $g(19), g(-13), g(1), g(84)$

84. $f(t) = \sqrt[4]{t + 1}$; $f(0), f(15), f(-82), f(80)$

Determine the domain of each function described.

85. $f(x) = \sqrt{x - 5}$ **86.** $g(x) = \sqrt{x + 8}$

87. $g(t) = \sqrt[4]{t+3}$

88. $f(x) = \sqrt[4]{x-7}$

89. $g(x) = \sqrt[4]{5-x}$

90. $g(t) = \sqrt[3]{2t-5}$

91. $f(t) = \sqrt[5]{2t+9}$

92. $f(t) = \sqrt[6]{2t+5}$

93. $h(z) = -\sqrt[6]{5z+3}$

94. $d(x) = -\sqrt[4]{7x-5}$

Aha! 95. $f(t) = 7 + \sqrt[8]{t^8}$

96. $g(t) = 9 + \sqrt[6]{t^6}$

97. Explain how to write the negative square root of a number using radical notation.

98. Does the square root of a number's absolute value always exist? Why or why not?

SKILL MAINTENANCE

Simplify. Do not use negative exponents in your answer.

99. $(a^3b^2c^5)^3$

100. $(5a^7b^8)(2a^3b)$

101. $(2a^{-2}b^3c^{-4})^{-3}$

102. $(5x^{-3}y^{-1}z^2)^{-2}$

103. $\dfrac{8x^{-2}y^5}{4x^{-6}z^{-2}}$

104. $\dfrac{10a^{-6}b^{-7}}{2a^{-2}c^{-3}}$

105. Under what conditions does the nth root of x^3 exist? Explain your reasoning.

106. Under what conditions does the nth root of x^2 exist? Explain your reasoning.

107. *Spaces in a parking lot.* A parking lot has attendants to park the cars. The number N of stalls needed for waiting cars before attendants can get to them is given by the formula $N = 2.5\sqrt{A}$, where A is the number of arrivals in peak hours. Find the number of spaces needed for the given number of arrivals in peak hours: **(a)** 25; **(b)** 36; **(c)** 49; **(d)** 64.

Determine the domain of each function described. Then draw the graph of each function.

108. $f(x) = \sqrt{x+5}$

109. $g(x) = \sqrt{x} + 5$

110. $g(x) = \sqrt{x} - 2$

111. $f(x) = \sqrt{x-2}$

112. Find the domain of f if

$$f(x) = \frac{\sqrt{x+3}}{\sqrt[4]{2-x}}.$$

113. Find the domain of g if

$$g(x) = \frac{\sqrt[4]{5-x}}{\sqrt[6]{x+4}}.$$

114. Use a grapher to check your answers to Exercises 31, 39, and 49. On some graphers, a MATH key is needed to enter higher roots.

115. Use a grapher to check your answers to Exercises 112 and 113. (See Exercise 114.)

7.2 Rational Numbers as Exponents

Rational Exponents • Negative Rational Exponents • Laws of Exponents • Simplifying Radical Expressions

In Section 1.1, we considered the natural numbers as exponents. Our discussion of exponents was expanded to include all integers in Section 1.6. In this section, we expand the study still further—to include all rational numbers. This will give meaning to expressions like $a^{1/3}$, $7^{-1/2}$, and $(3x)^{4/5}$. Such notation will help us simplify certain radical expressions.

Rational Exponents

Consider $a^{1/2} \cdot a^{1/2}$. If we still want to add exponents when multiplying, it must follow that $a^{1/2} \cdot a^{1/2} = a^{1/2+1/2}$, or a^1. This suggests that $a^{1/2}$ is a square root of a. Similarly, $a^{1/3} \cdot a^{1/3} \cdot a^{1/3} = a^{1/3+1/3+1/3}$, or a^1, so $a^{1/3}$ should mean $\sqrt[3]{a}$.

$a^{1/n} = \sqrt[n]{a}$

$a^{1/n}$ means $\sqrt[n]{a}$. When a is nonnegative, n can be any natural number greater than 1. When a is negative, n must be odd.

Note that the denominator of the exponent becomes the index and the base becomes the radicand.

Example 1

Write an equivalent expression using radical notation.

a) $x^{1/2}$ b) $(-8)^{1/3}$ c) $(abc)^{1/5}$

Solution

a) $x^{1/2} = \sqrt{x}$

b) $(-8)^{1/3} = \sqrt[3]{-8} = -2$

c) $(abc)^{1/5} = \sqrt[5]{abc}$

The denominator of the exponent becomes the index. The base becomes the radicand.

Example 2

Write an equivalent expression using exponential notation.

a) $\sqrt[5]{9xy}$ b) $\sqrt[7]{\dfrac{x^3y}{4}}$ c) $\sqrt{5x}$

Solution Parentheses are required to indicate the base.

a) $\sqrt[5]{9xy} = (9xy)^{1/5}$

b) $\sqrt[7]{\dfrac{x^3y}{4}} = \left(\dfrac{x^3y}{4}\right)^{1/7}$

The index becomes the denominator of the exponent. The radicand becomes the base.

c) $\sqrt{5x} = (5x)^{1/2}$ For square roots, the index 2 is understood without being written.

How shall we define $a^{2/3}$? If the property for multiplying exponents is to hold, we must have $a^{2/3} = (a^{1/3})^2$ and $a^{2/3} = (a^2)^{1/3}$. This would suggest that $a^{2/3} = (\sqrt[3]{a})^2$ and $a^{2/3} = \sqrt[3]{a^2}$. We make our definition accordingly.

Positive Rational Exponents

For any natural numbers m and n ($n \neq 1$) and any real number a for which $\sqrt[n]{a}$ exists,

$$a^{m/n} \text{ means } (\sqrt[n]{a})^m, \text{ or } \sqrt[n]{a^m}.$$

Example 3 Write an equivalent expression using radical notation and simplify.

a) $27^{2/3}$ b) $25^{3/2}$

Solution

a) $27^{2/3} = \sqrt[3]{27^2}$, or $(\sqrt[3]{27})^2$ It is easier to simplify using $(\sqrt[3]{27})^2$.
$= 3^2$, or 9

b) $25^{3/2} = \sqrt[2]{25^3}$, or $(\sqrt[2]{25})^3$ We normally omit the index 2.
$= 5^3$, or 125 Taking the square root and cubing

Example 4 Write an equivalent expression using exponential notation.

a) $\sqrt[3]{9^4}$ b) $(\sqrt[4]{7xy})^5$

Solution

a) $\sqrt[3]{9^4} = 9^{4/3}$
b) $(\sqrt[4]{7xy})^5 = (7xy)^{5/4}$
} The index becomes the denominator of the fractional exponent.

To approximate $7^{2/3}$, we enter 7 [^] (2/3).

1. Why are the parentheses needed above?
2. Compare the graphs of $y_1 = x^{1/2}$, $y_2 = x$, and $y_3 = x^{3/2}$.

Negative Rational Exponents

Recall that $x^{-2} = 1/x^2$. Negative rational exponents behave similarly.

Negative Rational Exponents

For any rational number m/n and any nonzero real number a for which $a^{m/n}$ exists,

$$a^{-m/n} \text{ means } \frac{1}{a^{m/n}}.$$

Caution! A negative exponent does not indicate that the expression in which it appears is negative.

Example 5 Write an equivalent expression with positive exponents and, if possible, simplify.

a) $9^{-1/2}$ b) $(5xy)^{-4/5}$ c) $64^{-2/3}$

d) $4x^{-2/3}y^{1/5}$ e) $\left(\dfrac{3r}{7s}\right)^{-5/2}$

Solution

a) $9^{-1/2} = \dfrac{1}{9^{1/2}}$ $9^{-1/2}$ is the reciprocal of $9^{1/2}$.

Since $9^{1/2} = \sqrt{9} = 3$, the answer simplifies to $\dfrac{1}{3}$.

b) $(5xy)^{-4/5} = \dfrac{1}{(5xy)^{4/5}}$ $(5xy)^{-4/5}$ is the reciprocal of $(5xy)^{4/5}$.

c) $64^{-2/3} = \frac{1}{64^{2/3}}$ $\quad 64^{-2/3}$ is the reciprocal of $64^{2/3}$.

Since $64^{2/3} = (\sqrt[3]{64})^2 = 4^2 = 16$, the answer simplifies to $\frac{1}{16}$.

d) $4x^{-2/3}y^{1/5} = 4 \cdot \frac{1}{x^{2/3}} \cdot y^{1/5} = \frac{4y^{1/5}}{x^{2/3}}$

e) In Section 1.6, we found that $(a/b)^{-n} = (b/a)^n$. This property holds for *any* negative exponent:

$$\left(\frac{3r}{7s}\right)^{-5/2} = \left(\frac{7s}{3r}\right)^{5/2}.$$

Writing the reciprocal of the base and changing the sign of the exponent

Laws of Exponents

The same laws hold for rational exponents as for integer exponents.

Laws of Exponents

For any real numbers a and b and any rational exponents m and n for which a^m, a^n, and b^m are defined:

1. $a^m \cdot a^n = a^{m+n}$ — In multiplying, add exponents if the bases are the same.

2. $\frac{a^m}{a^n} = a^{m-n}$ — In dividing, subtract exponents if the bases are the same. (Assume $a \neq 0$.)

3. $(a^m)^n = a^{m \cdot n}$ — To raise a power to a power, multiply the exponents.

4. $(ab)^m = a^m b^m$ — To raise a product to a power, raise each factor to the power and multiply.

Example 6 Use the laws of exponents to simplify.

a) $3^{1/5} \cdot 3^{3/5}$ **b)** $a^{1/4}/a^{1/2}$

c) $(7.2^{2/3})^{3/4}$ **d)** $(a^{-1/3}b^{2/5})^{1/2}$

Solution

a) $3^{1/5} \cdot 3^{3/5} = 3^{1/5+3/5} = 3^{4/5}$ — Adding exponents

b) $\frac{a^{1/4}}{a^{1/2}} = a^{1/4-1/2} = a^{1/4-2/4}$ — Subtracting exponents after finding a common denominator

$= a^{-1/4}$, or $\frac{1}{a^{1/4}}$ — $a^{-1/4}$ is the reciprocal of $a^{1/4}$.

c) $(7.2^{2/3})^{3/4} = 7.2^{2/3 \cdot 3/4} = 7.2^{6/12}$ — Multiplying exponents

$= 7.2^{1/2}$ — Using arithmetic to simplify the exponent

d) $(a^{-1/3}b^{2/5})^{1/2} = a^{-1/3 \cdot 1/2} \cdot b^{2/5 \cdot 1/2}$ — Raising a product to a power and multiplying exponents

$= a^{-1/6}b^{1/5}$, or $\frac{b^{1/5}}{a^{1/6}}$

Simplifying Radical Expressions

Many radical expressions can be simplified using rational exponents.

> ***To Simplify Radical Expressions***
>
> **1.** Convert radical expressions to exponential expressions.
> **2.** Use arithmetic and the laws of exponents to simplify.
> **3.** Convert back to radical notation as needed.

Example 7 Use rational exponents to simplify. Do not use fractional exponents in the final answer.

a) $\sqrt[6]{(5x)^3}$ **b)** $\sqrt[5]{t^{20}}$

c) $(\sqrt[3]{ab^2c})^{12}$ **d)** $\sqrt{\sqrt[3]{x}}$

Solution

a) $\sqrt[6]{(5x)^3} = (5x)^{3/6}$ Converting to exponential notation
$= (5x)^{1/2}$ Simplifying the exponent
$= \sqrt{5x}$ Returning to radical notation

b) $\sqrt[5]{t^{20}} = t^{20/5}$ Converting to exponential notation
$= t^4$ Simplifying the exponent

c) $(\sqrt[3]{ab^2c})^{12} = (ab^2c)^{12/3}$ Converting to exponential notation
$= (ab^2c)^4$ Simplifying the exponent
$= a^4b^8c^4$ Using the laws of exponents

d) $\sqrt{\sqrt[3]{x}} = \sqrt{x^{1/3}}$ Converting the radicand to exponential notation
$= (x^{1/3})^{1/2}$ Try to go directly to this step.
$= x^{1/6}$ Using the laws of exponents
$= \sqrt[6]{x}$ Returning to radical notation

> **technology connection B**
>
> One way to check Example 7(a) is to let $y_1 = (5x)^{3/6}$ and $y_2 = \sqrt{5x}$. Then see if the graphs of y_1 and y_2 coincide. An alternative is to let $y_3 = y_2 - y_1$ and see if $y_3 = 0$. Check Example 7(a) using one of the checks just described.
>
> **1.** Why are rational exponents especially useful when working on a grapher?

Exercise Set 7.2

FOR EXTRA HELP

Digital Video Tutor CD 5
Videotape 13

InterAct Math

Math Tutor Center

MathXL

MyMathLab.com

Note: Assume for all exercises that even roots are of nonnegative quantities and that all denominators are nonzero.

Write an equivalent expression using radical notation and, if possible, simplify.

1. $x^{1/4}$ **2.** $y^{1/5}$ **3.** $16^{1/2}$

4. $8^{1/3}$ **5.** $81^{1/4}$ **6.** $64^{1/6}$

7. $9^{1/2}$ **8.** $25^{1/2}$ **9.** $(xyz)^{1/3}$

10. $(ab)^{1/4}$ **11.** $(a^2b^2)^{1/5}$ **12.** $(x^3y^3)^{1/4}$

13. $a^{2/3}$ **14.** $b^{3/2}$ **15.** $16^{3/4}$

16. $4^{7/2}$ **17.** $49^{3/2}$ **18.** $27^{4/3}$

19. $9^{5/2}$ **20.** $81^{3/2}$ **21.** $(81x)^{3/4}$

22. $(125a)^{2/3}$ **23.** $(25x^4)^{3/2}$ **24.** $(9y^6)^{3/2}$

Write an equivalent expression using exponential notation.

25. $\sqrt[3]{20}$ **26.** $\sqrt[3]{19}$ **27.** $\sqrt{17}$

28. $\sqrt{6}$ **29.** $\sqrt{x^3}$ **30.** $\sqrt{a^5}$

31. $\sqrt[5]{m^2}$ **32.** $\sqrt[5]{n^4}$ **33.** $\sqrt[4]{cd}$

34. $\sqrt[5]{xy}$ **35.** $\sqrt[5]{xy^2z}$ **36.** $\sqrt[7]{x^3y^2z^2}$

37. $(\sqrt{3mn})^3$ **38.** $(\sqrt[3]{7xy})^4$ **39.** $(\sqrt[7]{8x^2y})^5$

40. $(\sqrt[6]{2a^5b})^7$ **41.** $\dfrac{2x}{\sqrt[3]{z^2}}$ **42.** $\dfrac{3a}{\sqrt[5]{c^2}}$

Write an equivalent expression with positive exponents and, if possible, simplify.

43. $x^{-1/3}$ **44.** $y^{-1/4}$

45. $(2rs)^{-3/4}$ **46.** $(5xy)^{-5/6}$

47. $\left(\dfrac{1}{8}\right)^{-2/3}$ **48.** $\left(\dfrac{1}{16}\right)^{-3/4}$

49. $\dfrac{1}{a^{-5/7}}$ **50.** $\dfrac{1}{a^{-3/5}}$

51. $2a^{3/4}b^{-1/2}c^{2/3}$ **52.** $5x^{-2/3}y^{4/5}z$

53. $2^{-1/3}x^4y^{-2/7}$ **54.** $3^{-5/2}a^3b^{-7/3}$

55. $\left(\dfrac{7x}{8yz}\right)^{-3/5}$ **56.** $\left(\dfrac{2ab}{3c}\right)^{-5/6}$

57. $\dfrac{7x}{\sqrt[3]{z}}$ **58.** $\dfrac{6a}{\sqrt[4]{b}}$

59. $\dfrac{5a}{3c^{-1/2}}$ **60.** $\dfrac{2z}{5x^{-1/3}}$

Use the laws of exponents to simplify. Do not use negative exponents in any answers.

61. $5^{3/4} \cdot 5^{1/8}$ **62.** $11^{2/3} \cdot 11^{1/2}$

63. $\dfrac{3^{5/8}}{3^{-1/8}}$ **64.** $\dfrac{8^{7/11}}{8^{-2/11}}$

65. $\dfrac{4.1^{-1/6}}{4.1^{-2/3}}$ **66.** $\dfrac{2.3^{-3/10}}{2.3^{-1/5}}$

67. $(10^{3/5})^{2/5}$ **68.** $(5^{5/4})^{3/7}$

69. $a^{2/3} \cdot a^{5/4}$ **70.** $x^{3/4} \cdot x^{2/3}$

Aha! **71.** $(64^{3/4})^{4/3}$ **72.** $(27^{-2/3})^{3/2}$

73. $(m^{2/3}n^{-1/4})^{1/2}$ **74.** $(x^{-1/3}y^{2/5})^{1/4}$

Use rational exponents to simplify. Do not use fractional exponents in the final answer.

75. $\sqrt[6]{a^2}$ **76.** $\sqrt[6]{t^4}$

77. $\sqrt[3]{x^{15}}$ **78.** $\sqrt[4]{a^{12}}$

79. $\sqrt[6]{x^{18}}$ **80.** $\sqrt[5]{a^{10}}$

81. $(\sqrt[3]{ab})^{15}$ **82.** $(\sqrt[7]{xy})^{14}$

83. $\sqrt[8]{(3x)^2}$ **84.** $\sqrt[4]{(7a)^2}$

85. $(\sqrt[10]{3a})^5$ **86.** $(\sqrt[8]{2x})^6$

87. $\sqrt[4]{\sqrt{x}}$ **88.** $\sqrt[3]{\sqrt[6]{m}}$

89. $\sqrt{(ab)^6}$ **90.** $\sqrt[4]{(xy)^{12}}$

91. $(\sqrt[3]{x^2y^5})^{12}$ **92.** $(\sqrt[5]{a^2b^4})^{15}$

93. $\sqrt[3]{\sqrt[4]{xy}}$ **94.** $\sqrt[5]{\sqrt{2a}}$

95. If $f(x) = (x + 5)^{1/2}(x + 7)^{-1/2}$, find the domain of f. Explain how you found your answer.

96. Explain why $\sqrt[3]{x^6} = x^2$ for any value of x, whereas $\sqrt[2]{x^6} = x^3$ only when $x \geq 0$.

SKILL MAINTENANCE

Simplify.

97. $3x(x^3 - 2x^2) + 4x^2(2x^2 + 5x)$

98. $5t^3(2t^2 - 4t) - 3t^4(t^2 - 6t)$

99. $(3a - 4b)(5a + 3b)$

100. $(7x - y)^2$

101. *Real estate taxes.* For homes under \$100,000, the real-estate transfer tax in Vermont is 0.5% of the selling price. Find the selling price of a home that had a transfer tax of \$467.50.

102. What numbers are their own squares?

SYNTHESIS

103. Let $f(x) = 5x^{-1/3}$. Under what condition will we have $f(x) > 0$? Why?

104. If $g(x) = x^{3/n}$, in what way does the domain of g depend on whether n is odd or even?

Use rational exponents to simplify.

105. $\sqrt[5]{x^2y\sqrt{xy}}$

106. $\sqrt{x\sqrt[3]{x^2}}$

107. $\sqrt[4]{\sqrt[3]{8x^3y^6}}$

108. $\sqrt[12]{p^2 + 2pq + q^2}$

*Music. The function $f(x) = k2^{x/12}$ can be used to determine the frequency, in cycles per second, of a musical note that is x half-steps above a note with frequency k.**

109. The frequency of middle C on a piano is 262 cycles per second. Find the frequency of the C that is one octave (12 half-steps) higher.

110. The frequency of concert A for a trumpet is 440 cycles per second. Find the frequency of the A that is two octaves (24 half-steps) above concert A (few trumpeters can reach this note.)

111. Show that the G that is 7 half-steps (a "perfect fifth") above middle C (see Exercise 109) has a frequency that is about 1.5 times that of middle C.

112. Show that the C sharp that is 4 half-steps (a "major third") above concert A (see Exercise 110) has a frequency that is about 25% greater than that of concert A.

113. *Road pavement messages.* In a psychological study, it was determined that the proper length L of the letters of a word printed on pavement is given by

$$L = \frac{0.000169d^{2.27}}{h},$$

where d is the distance of a car from the lettering and h is the height of the eye above the surface of the road. All units are in meters. This formula says that if a person is h meters above the surface of the road and is to be able to recognize a message d meters away, that message will be the most recognizable if the length of the letters is L. Find L to the nearest tenth of a meter, given d and h.

a) $h = 1$ m, $d = 60$ m
b) $h = 0.9906$ m, $d = 75$ m
c) $h = 2.4$ m, $d = 80$ m
d) $h = 1.1$ m, $d = 100$ m

114. *Dating fossils.* The function $r(t) = 10^{-12}2^{-t/5700}$ expresses the ratio of carbon isotopes to carbon atoms in a fossil that is t years old. What ratio of carbon isotopes to carbon atoms would be present in a 1900-year-old bone?

115. *Physics.* The equation $m = m_0(1 - v^2c^{-2})^{-1/2}$, developed by Albert Einstein, is used to determine the mass m of an object that is moving v meters per second and has mass m_0 before the motion begins. The constant c is the speed of light, approximately 3×10^8 m/sec. Suppose that a particle with mass 8 mg is accelerated to a speed of $\frac{9}{5} \times 10^8$ m/sec. Without using a calculator, find the new mass of the particle.

116. Use a grapher in the SIMULTANEOUS mode with the LABEL OFF format to graph

$$y_1 = x^{1/2}, \quad y_2 = 3x^{2/5},$$
$$y_3 = x^{4/7}, \quad \text{and} \quad y_4 = \tfrac{1}{5}x^{3/4}.$$

Then, looking only at coordinates, match each graph with its equation.

*This application was inspired by information provided by Dr. Homer B. Tilton of Pima Community College East.

COLLABORATIVE

CORNER

Are Equivalent Fractions Equivalent Exponents?

Focus: Functions and rational exponents

Time: 10–20 minutes

Group size: 3

Materials: Graph paper

In arithmetic, we have seen that $\frac{1}{3}$, $\frac{1}{6} \cdot 2$, and $2 \cdot \frac{1}{6}$ all represent the same number. Interestingly,

$$f(x) = x^{1/3},$$
$$g(x) = (x^{1/6})^2, \quad \text{and}$$
$$h(x) = (x^2)^{1/6}$$

represent three *different* functions.

ACTIVITY

1. Selecting a variety of values for x and using the definition of positive rational exponents, one group member should graph f, a second group member should graph g, and a third group member should graph h. Be sure to check whether negative x-values are in the domain of the function.
2. Compare the three graphs and check each other's work. How and why do the graphs differ?
3. Decide as a group which graph, if any, would best represent the graph of $k(x) = x^{2/6}$. Then be prepared to explain your reasoning to the entire class. (*Hint*: Study the definition of $a^{m/n}$ on p. 417 carefully.)

Multiplying Radical Expressions

7.3

Multiplying Radical Expressions • Simplifying by Factoring • Multiplying and Simplifying

Multiplying Radical Expressions

Note that $\sqrt{4}\,\sqrt{25} = 2 \cdot 5 = 10$. Also $\sqrt{4 \cdot 25} = \sqrt{100} = 10$. Likewise,

$$\sqrt[3]{27}\,\sqrt[3]{8} = 3 \cdot 2 = 6 \quad \text{and} \quad \sqrt[3]{27 \cdot 8} = \sqrt[3]{216} = 6.$$

These examples suggest the following.

The Product Rule for Radicals

For any real numbers $\sqrt[n]{a}$ and $\sqrt[n]{b}$,

$$\sqrt[n]{a} \cdot \sqrt[n]{b} = \sqrt[n]{a \cdot b}.$$

(To multiply, when the indices match, multiply the radicands.)

Fractional exponents can be used to derive this rule:

$$\sqrt[n]{a} \cdot \sqrt[n]{b} = a^{1/n} \cdot b^{1/n} = (a \cdot b)^{1/n} = \sqrt[n]{a \cdot b}.$$

Example 1 Multiply.

a) $\sqrt{3} \cdot \sqrt{5}$

b) $\sqrt{x+3}\,\sqrt{x-3}$

c) $\sqrt[3]{4} \cdot \sqrt[3]{5}$

d) $\sqrt[4]{\dfrac{y}{5}} \cdot \sqrt[4]{\dfrac{7}{x}}$

Solution

a) When no index is written, roots are understood to be square roots with an unwritten index of two. We apply the product rule:

$$\sqrt{3} \cdot \sqrt{5} = \sqrt{3 \cdot 5}$$
$$= \sqrt{15}.$$

b) $\sqrt{x+3}\,\sqrt{x-3} = \sqrt{(x+3)(x-3)}$
$= \sqrt{x^2 - 9}$

The product of two square roots is the square root of the product.

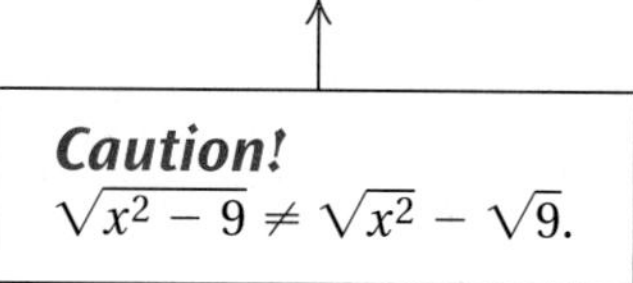

Caution!
$\sqrt{x^2 - 9} \neq \sqrt{x^2} - \sqrt{9}.$

c) Both $\sqrt[3]{4}$ and $\sqrt[3]{5}$ have indices of three, so to multiply we can use the product rule:

$$\sqrt[3]{4} \cdot \sqrt[3]{5} = \sqrt[3]{4 \cdot 5} = \sqrt[3]{20}.$$

d) $\sqrt[4]{\dfrac{y}{5}} \cdot \sqrt[4]{\dfrac{7}{x}} = \sqrt[4]{\dfrac{y}{5} \cdot \dfrac{7}{x}} = \sqrt[4]{\dfrac{7y}{5x}}$

In Section 7.4, we discuss other ways to write answers like this.

technology connection A

To check Example 1(b), let $y_1 = \sqrt{x+3}\,\sqrt{x-3}$ and $y_2 = \sqrt{x^2 - 9}$ and compare.

1. What should the graph of

$$y = \sqrt{x^2 - 9} - \sqrt{x+3}\,\sqrt{x-3}$$

look like?

Important: The product rule for radicals applies only when radicals have the same index.

Simplifying by Factoring

An integer p is a *perfect square* if there exists a rational number q for which $q^2 = p$. We say that p is a *perfect cube* if $q^3 = p$ for some rational number q. In general, p is a *perfect nth power* if $q^n = p$ for some rational number q. The product rule allows us to simplify $\sqrt[n]{ab}$ when a or b is a perfect nth power.

Using the Product Rule to Simplify

$$\sqrt[n]{ab} = \sqrt[n]{a} \cdot \sqrt[n]{b}.$$

To illustrate, suppose we wish to simplify $\sqrt{20}$. Since this is a *square* root, we check to see if there is a factor of 20 that is a perfect square. There is one, 4, so we express 20 as $4 \cdot 5$ and use the product rule:

$$\begin{aligned} \sqrt{20} &= \sqrt{4 \cdot 5} && \text{Factoring the radicand (4 is a perfect square)} \\ &= \sqrt{4} \cdot \sqrt{5} && \text{Factoring into two radicals} \\ &= 2\sqrt{5}. && \text{Taking the square root of 4} \end{aligned}$$

To Simplify a Radical Expression with Index n by Factoring

1. Express the radicand as a product in which one factor is the largest perfect nth power possible.
2. Take the nth root of each factor.
3. Simplification is complete when no radicand has a factor that is a perfect nth power.

Example 2

Simplify by factoring: **(a)** $\sqrt{200}$; **(b)** $\sqrt[3]{32}$; **(c)** $\sqrt[4]{48}$; **(d)** $\sqrt{18x^2y}$.

Solution

a) $\sqrt{200} = \sqrt{100 \cdot 2} = \sqrt{100} \cdot \sqrt{2} = 10\sqrt{2}$

This is the largest perfect-square factor of 200. (refers to 100)

b) $\sqrt[3]{32} = \sqrt[3]{8 \cdot 4} = \sqrt[3]{8} \cdot \sqrt[3]{4} = 2\sqrt[3]{4}$

This is the largest perfect-cube (third-power) factor of 32. (refers to 8)

c) $\sqrt[4]{48} = \sqrt[4]{16 \cdot 3} = \sqrt[4]{16} \cdot \sqrt[4]{3} = 2\sqrt[4]{3}$

This is the largest fourth-power factor of 48. (refers to 16)

d)
$$\begin{aligned} \sqrt{18x^2y} &= \sqrt{9x^2 \cdot 2y} && 9x^2 \text{ is a perfect square.} \\ &= \sqrt{9x^2} \cdot \sqrt{2y} && \text{Factoring into two radicals} \\ &= |3x|\sqrt{2y}, \text{ or } 3|x|\sqrt{2y} && \text{Taking the square root of } 9x^2 \end{aligned}$$

technology connection B

To use a grapher to check Example 3, let
$y_1 = \sqrt{(3x^2 - 6x + 3)}$,
$y_2 = \text{abs}(x - 1)\sqrt{3}$, and
$y_3 = (x - 1)\sqrt{3}$. Do the graphs all coincide? Why or why not?

Example 3

If $f(x) = \sqrt{3x^2 - 6x + 3}$, find a simplified form for $f(x)$.

Solution

$$\begin{aligned} f(x) &= \sqrt{3x^2 - 6x + 3} \\ &= \sqrt{3(x^2 - 2x + 1)} && \text{Factoring the radicand; } x^2 - 2x + 1 \\ &= \sqrt{(x-1)^2 \cdot 3} && \text{is a perfect square.} \\ &= \sqrt{(x-1)^2} \cdot \sqrt{3} && \text{Factoring into two radicals} \\ &= |x - 1|\sqrt{3} && \text{Taking the square root of } (x-1)^2 \end{aligned}$$

In many situations that do not involve functions, it is safe to assume that no radicands were formed by raising negative quantities to even powers. We now make this assumption and thus discontinue the use of absolute-value notation when taking even roots unless functions are involved.

Example 4 Simplify: **(a)** $\sqrt{x^7y^{11}z^9}$; **(b)** $\sqrt[3]{16a^7b^{14}}$.

Solution

a) There are many ways to factor $x^7y^{11}z^9$. Because of the square root (index of 2), we identify the largest exponents that are multiples of 2:

$$\sqrt{x^7y^{11}z^9} = \sqrt{x^6 \cdot x \cdot y^{10} \cdot y \cdot z^8 \cdot z}$$ Using the largest even powers of x, y, and z

$$= \sqrt{x^6}\,\sqrt{y^{10}}\,\sqrt{z^8}\,\sqrt{xyz}$$ Factoring into several radicals

$$= x^{6/2}\;y^{10/2}\;z^{8/2}\sqrt{xyz}$$ Converting to fractional exponents

$$= x^3y^5z^4\sqrt{xyz}.$$

Check: $(x^3y^5z^4\sqrt{xyz})^2 = (x^3)^2(y^5)^2(z^4)^2(\sqrt{xyz})^2$
$= x^6 \cdot y^{10} \cdot z^8 \cdot xyz = x^7y^{11}z^9$

Our check shows that $x^3y^5z^4\sqrt{xyz}$ is the square root of $x^7y^{11}z^9$.

b) There are many ways to factor $16a^7b^{14}$. Because of the cube root (index of 3), we identify factors with the largest exponents that are multiples of 3:

$$\sqrt[3]{16a^7b^{14}} = \sqrt[3]{8 \cdot 2 \cdot a^6 \cdot a \cdot b^{12} \cdot b^2}$$ Using the largest perfect-cube factors

$$= \sqrt[3]{8}\,\sqrt[3]{a^6}\,\sqrt[3]{b^{12}}\,\sqrt[3]{2ab^2}$$ Factoring into several radicals

$$= 2\;a^{6/3}\;b^{12/3}\sqrt[3]{2ab^2}$$ Converting to fractional exponents

$$= 2a^2b^4\sqrt[3]{2ab^2}$$

Check: $(2a^2b^4\sqrt[3]{2ab^2})^3 = 2^3(a^2)^3(b^4)^3(\sqrt[3]{2ab^2})^3$
$= 8 \cdot a^6 \cdot b^{12} \cdot 2ab^2 = 16a^7b^{14}$

We see that $2a^2b^4\sqrt[3]{2ab^2}$ is the cube root of $16a^7b^{14}$.

Example 4 demonstrates the following.

> To simplify an nth root, identify factors in the radicand with exponents that are multiples of n.

Multiplying and Simplifying

We have used the product rule for radicals to find products and also to simplify radical expressions. For some radical expressions, it is possible to do both: First find a product and then simplify.

Example 5 Multiply and simplify.

a) $\sqrt{15}\,\sqrt{6}$ b) $3\sqrt[3]{25} \cdot 2\sqrt[3]{5}$ c) $\sqrt[4]{8x^3y^5}\,\sqrt[4]{4x^2y^3}$

Solution

a) $\sqrt{15}\,\sqrt{6} = \sqrt{15 \cdot 6}$ Multiplying radicands

$= \sqrt{90} = \sqrt{9 \cdot 10}$ 9 is a perfect square.

$= 3\sqrt{10}$

b) $3\sqrt[3]{25} \cdot 2\sqrt[3]{5} = 3 \cdot 2 \cdot \sqrt[3]{25 \cdot 5}$ Using a commutative law; multiplying radicands

$= 6 \cdot \sqrt[3]{125}$ 125 is a perfect cube.

$= 6 \cdot 5$, or 30

c) $\sqrt[4]{8x^3y^5}\,\sqrt[4]{4x^2y^3} = \sqrt[4]{32x^5y^8}$ Multiplying radicands

$= \sqrt[4]{16x^4y^8 \cdot 2x}$ Identifying perfect fourth-power factors

$= \sqrt[4]{16}\,\sqrt[4]{x^4}\,\sqrt[4]{y^8}\,\sqrt[4]{2x}$ Factoring into radicals

$= 2xy^2\sqrt[4]{2x}$ Finding the fourth roots; assume $x \geq 0$.

The checks are left to the student.

Exercise Set 7.3

FOR EXTRA HELP

Digital Video Tutor CD 5 Videotape 13

InterAct Math

Math Tutor Center

MathXL

MyMathLab.com

Multiply.

1. $\sqrt{10}\,\sqrt{7}$
2. $\sqrt{5}\,\sqrt{7}$
3. $\sqrt[3]{2}\,\sqrt[3]{5}$
4. $\sqrt[3]{7}\,\sqrt[3]{2}$
5. $\sqrt[4]{8}\,\sqrt[4]{9}$
6. $\sqrt[4]{6}\,\sqrt[4]{3}$
7. $\sqrt{5a}\,\sqrt{6b}$
8. $\sqrt{2x}\,\sqrt{13y}$
9. $\sqrt[5]{9t^2}\,\sqrt[5]{2t}$
10. $\sqrt[5]{8y^3}\,\sqrt[5]{10y}$
11. $\sqrt{x-a}\,\sqrt{x+a}$
12. $\sqrt{y-b}\,\sqrt{y+b}$
13. $\sqrt[3]{0.5x}\,\sqrt[3]{0.2x}$
14. $\sqrt[3]{0.7y}\,\sqrt[3]{0.3y}$
15. $\sqrt[4]{x-1}\,\sqrt[4]{x^2+x+1}$
16. $\sqrt[5]{x-2}\,\sqrt[5]{(x-2)^2}$
17. $\sqrt{\dfrac{x}{6}}\,\sqrt{\dfrac{7}{y}}$
18. $\sqrt{\dfrac{7}{t}}\,\sqrt{\dfrac{s}{11}}$
19. $\sqrt[7]{\dfrac{x-3}{4}}\,\sqrt[7]{\dfrac{5}{x+2}}$
20. $\sqrt[6]{\dfrac{a}{b-2}}\,\sqrt[6]{\dfrac{3}{b+2}}$

Simplify by factoring.

21. $\sqrt{50}$
22. $\sqrt{27}$
23. $\sqrt{28}$
24. $\sqrt{45}$
25. $\sqrt{8}$
26. $\sqrt{18}$
27. $\sqrt{198}$
28. $\sqrt{325}$
29. $\sqrt{36a^4b}$
30. $\sqrt{175y^8}$
31. $\sqrt[3]{8x^3y^2}$
32. $\sqrt[3]{27ab^6}$
33. $\sqrt[3]{-16x^6}$
34. $\sqrt[3]{-32a^6}$

Find a simplified form of $f(x)$. Assume that x can be any real number.

35. $f(x) = \sqrt[3]{125x^5}$
36. $f(x) = \sqrt[3]{16x^6}$
37. $f(x) = \sqrt{49(x-3)^2}$
38. $f(x) = \sqrt{81(x-1)^2}$
39. $f(x) = \sqrt{5x^2 - 10x + 5}$
40. $f(x) = \sqrt{2x^2 + 8x + 8}$

Simplify. Assume that no radicands were formed by raising negative numbers to even powers.

41. $\sqrt{a^3b^4}$
42. $\sqrt{x^6y^9}$
43. $\sqrt[3]{x^5y^6z^{10}}$
44. $\sqrt[3]{a^6b^7c^{13}}$
45. $\sqrt[5]{-32a^7b^{11}}$
46. $\sqrt[4]{16x^5y^{11}}$

47. $\sqrt[5]{a^6b^8c^9}$

48. $\sqrt[5]{x^{13}y^8z^{17}}$

49. $\sqrt[4]{810x^9}$

50. $\sqrt[3]{-80a^{14}}$

Multiply and simplify.

51. $\sqrt{15}\sqrt{5}$

52. $\sqrt{6}\sqrt{3}$

53. $\sqrt{10}\sqrt{14}$

54. $\sqrt{15}\sqrt{21}$

55. $\sqrt[3]{2}\sqrt[3]{4}$

56. $\sqrt[3]{9}\sqrt[3]{3}$

Aha! **57.** $\sqrt{18a^3}\sqrt{18a^3}$

58. $\sqrt{75x^7}\sqrt{75x^7}$

59. $\sqrt[3]{5a^2}\sqrt[3]{2a}$

60. $\sqrt[3]{7x}\sqrt[3]{3x^2}$

61. $\sqrt{3x^5}\sqrt{15x^2}$

62. $\sqrt{5a^7}\sqrt{15a^3}$

63. $\sqrt[3]{s^2t^4}\sqrt[3]{s^4t^6}$

64. $\sqrt[3]{x^2y^4}\sqrt[3]{x^2y^6}$

65. $\sqrt[3]{(x+5)^2}\sqrt[3]{(x+5)^4}$

66. $\sqrt[3]{(a-b)^5}\sqrt[3]{(a-b)^7}$

67. $\sqrt[4]{12a^3b^7}\sqrt[4]{4a^2b^5}$

68. $\sqrt[4]{9x^7y^2}\sqrt[4]{9x^2y^9}$

69. $\sqrt[5]{x^3(y+z)^4}\sqrt[5]{x^3(y+z)^6}$

70. $\sqrt[5]{a^3(b-c)^4}\sqrt[5]{a^7(b-c)^4}$

71. Why do we need to know how to multiply radical expressions before learning how to simplify radical expressions?

72. Why is it incorrect to say that, in general, $\sqrt{x^2} = x$?

SKILL MAINTENANCE

Perform the indicated operation and, if possible, simplify.

73. $\dfrac{3x}{16y} + \dfrac{5y}{64x}$

74. $\dfrac{2}{a^3b^4} + \dfrac{6}{a^4b}$

75. $\dfrac{4}{x^2-9} - \dfrac{7}{2x-6}$

76. $\dfrac{8}{x^2-25} - \dfrac{3}{2x-10}$

Simplify.

77. $\dfrac{9a^4b^7}{3a^2b^5}$

78. $\dfrac{12a^2b^7}{4ab^2}$

SYNTHESIS

79. Explain why it is true that

$$\sqrt[n]{ab} = \sqrt[n]{a} \cdot \sqrt[n]{b}.$$

80. Is the equation $\sqrt{(2x+3)^8} = (2x+3)^4$ always, sometimes, or never true? Why?

81. *Speed of a skidding car.* Police can estimate the speed at which a car was traveling by measuring its skid marks. The function

$$r(L) = 2\sqrt{5L}$$

can be used, where L is the length of a skid mark, in feet, and $r(L)$ is the speed, in miles per hour. Find the exact speed and an estimate (to the nearest tenth mile per hour) for the speed of a car that left skid marks **(a)** 20 ft long; **(b)** 70 ft long; **(c)** 90 ft long.

82. *Wind chill temperature.* When the temperature is T degrees Celsius and the wind speed is v meters per second, the *wind chill temperature,* T_W, is the temperature (with no wind) that it feels like. Here is a formula for finding wind chill temperature:

$$T_W = 33 - \frac{(10.45 + 10\sqrt{v} - v)(33 - T)}{22}.$$

Estimate the wind chill temperature (to the nearest tenth of a degree) for the given actual temperatures and wind speeds.

a) $T = 7°C$, $v = 8$ m/sec
b) $T = 0°C$, $v = 12$ m/sec
c) $T = -5°C$, $v = 14$ m/sec
d) $T = -23°C$, $v = 15$ m/sec

Simplify. Assume that all variables are nonnegative.

83. $(\sqrt{r^3t})^7$

84. $(\sqrt[3]{25x^4})^4$

85. $(\sqrt[3]{a^2b^4})^5$

86. $(\sqrt{a^3b^5})^7$

Draw and compare the graphs of each group of equations.

87. $f(x) = \sqrt{x^2 - 2x + 1}$,
$g(x) = x - 1$,
$h(x) = |x - 1|$

88. $f(x) = \sqrt{x^2 + 2x + 1}$,
$g(x) = x + 1$,
$h(x) = |x + 1|$

89. If $f(t) = \sqrt{t^2 - 3t - 4}$, what is the domain of f?

90. What is the domain of g, if $g(x) = \sqrt{x^2 - 6x + 8}$?

Solve.

91. $\sqrt[3]{5x^{k+1}}\sqrt[3]{25x^k} = 5x^7$, for k

92. $\sqrt[5]{4a^{3k+2}}\sqrt[5]{8a^{6-k}} = 2a^4$, for k

93. Use a grapher to check your answers to Exercises 15, 35, and 59.

94. Rony is puzzled. When he uses a grapher to graph $y = \sqrt{x} \cdot \sqrt{x}$, he gets the following screen. Explain why Rony did not get the complete line $y = x$.

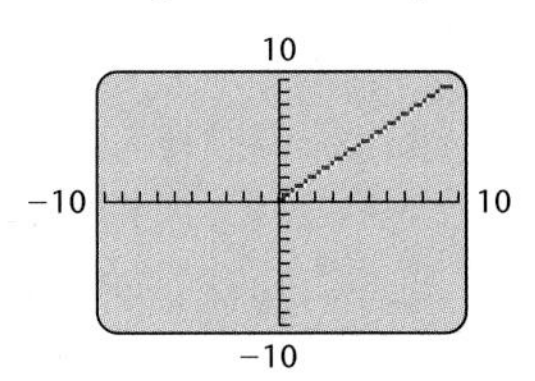

7.4 Dividing Radical Expressions

Dividing and Simplifying • Rationalizing Denominators and Numerators

> **Study Tip**
>
> It is always best to study for a final exam over a period of at least two weeks. If you have only one or two days of study time, however, begin by studying the formulas, problems, properties, and procedures in each chapter Summary and Review. Then do the exercises in the Cumulative Reviews. Make sure to attend a review session if one is offered.

Dividing and Simplifying

Just as the root of a product can be expressed as the product of two roots, the root of a quotient can be expressed as the quotient of two roots. For example,

$$\sqrt[3]{\frac{27}{8}} = \frac{3}{2} \quad \text{and} \quad \frac{\sqrt[3]{27}}{\sqrt[3]{8}} = \frac{3}{2}.$$

This example suggests the following.

> ***The Quotient Rule for Radicals***
>
> For any real numbers $\sqrt[n]{a}$ and $\sqrt[n]{b}$, $b \neq 0$,
>
> $$\sqrt[n]{\frac{a}{b}} = \frac{\sqrt[n]{a}}{\sqrt[n]{b}}.$$

Remember that an nth root is simplified when its radicand has no factors that are perfect nth powers. Recall too that we assume that no radicands represent negative quantities raised to an even power.

Example 1 Simplify by taking the roots of the numerator and the denominator.

a) $\sqrt[3]{\dfrac{27}{125}}$ **b)** $\sqrt{\dfrac{25}{y^2}}$

Solution

a) $\sqrt[3]{\dfrac{27}{125}} = \dfrac{\sqrt[3]{27}}{\sqrt[3]{125}} = \dfrac{3}{5}$ Taking the cube roots of the numerator and the denominator

b) $\sqrt{\dfrac{25}{y^2}} = \dfrac{\sqrt{25}}{\sqrt{y^2}} = \dfrac{5}{y}$ Taking the square roots of the numerator and the denominator. Assume $y > 0$.

As in Section 7.3, any radical expressions appearing in the answers should be simplified as much as possible.

Example 2 Simplify: **(a)** $\sqrt{\dfrac{16x^3}{y^8}}$; **(b)** $\sqrt[3]{\dfrac{27y^{14}}{8x^3}}$.

Solution

a) $\sqrt{\dfrac{16x^3}{y^8}} = \dfrac{\sqrt{16x^3}}{\sqrt{y^8}}$

$= \dfrac{\sqrt{16x^2 \cdot x}}{\sqrt{y^8}} = \dfrac{4x\sqrt{x}}{y^4}$ Simplifying the numerator and the denominator

b) $\sqrt[3]{\dfrac{27y^{14}}{8x^3}} = \dfrac{\sqrt[3]{27y^{14}}}{\sqrt[3]{8x^3}}$

$= \dfrac{\sqrt[3]{27y^{12}y^2}}{\sqrt[3]{8x^3}} = \dfrac{\sqrt[3]{27y^{12}}\,\sqrt[3]{y^2}}{\sqrt[3]{8x^3}} = \dfrac{3y^4\sqrt[3]{y^2}}{2x}$ Simplifying the numerator and the denominator

If we read from right to left, the quotient rule tells us that to divide two radical expressions that have the same index, we can divide the radicands.

Example 3 Divide and, if possible, simplify.

a) $\dfrac{\sqrt{80}}{\sqrt{5}}$ **b)** $\dfrac{5\sqrt[3]{32}}{\sqrt[3]{2}}$ **c)** $\dfrac{\sqrt{72xy}}{2\sqrt{2}}$ **d)** $\dfrac{\sqrt[4]{18a^9b^5}}{\sqrt[4]{3b}}$

Solution

a) $\dfrac{\sqrt{80}}{\sqrt{5}} = \sqrt{\dfrac{80}{5}} = \sqrt{16} = 4$ Because the indices match, we can divide the radicands.

b) $\dfrac{5\sqrt[3]{32}}{\sqrt[3]{2}} = 5\sqrt[3]{\dfrac{32}{2}} = 5\sqrt[3]{16}$

$= 5\sqrt[3]{8 \cdot 2}$

$= 5\sqrt[3]{8}\,\sqrt[3]{2} = 5 \cdot 2\sqrt[3]{2}$

$= 10\sqrt[3]{2}$

c) $\dfrac{\sqrt{72xy}}{2\sqrt{2}} = \dfrac{1}{2}\dfrac{\sqrt{72xy}}{\sqrt{2}} = \dfrac{1}{2}\sqrt{\dfrac{72xy}{2}} = \dfrac{1}{2}\sqrt{36xy}$ Because the indices match, we can divide the radicands.

$= \dfrac{1}{2}\sqrt{36}\sqrt{xy} = \dfrac{1}{2}\cdot 6\sqrt{xy}$

$= 3\sqrt{xy}$

d) $\dfrac{\sqrt[4]{18a^9b^5}}{\sqrt[4]{3b}} = \sqrt[4]{\dfrac{18a^9b^5}{3b}}$

$= \sqrt[4]{6a^9b^4} = \sqrt[4]{a^8b^4}\sqrt[4]{6a}$ Note that 8 is the largest power less than 9 that is a multiple of the index 4.

$= a^2b\sqrt[4]{6a}$ *Partial check*: $(a^2b)^4 = a^8b^4$

Rationalizing Denominators and Numerators

When a radical expression appears in a denominator, it can be useful to find an equivalent expression in which the denominator no longer contains a radical.* The procedure for finding such an expression is called **rationalizing the denominator**. We carry this out by multiplying by 1 in either of two ways.

One way is to multiply by 1 *under* the radical to make the denominator of the radicand a perfect power.

Example 4 Rationalize each denominator.

a) $\sqrt{\dfrac{7}{3}}$ b) $\sqrt[3]{\dfrac{5}{16}}$

Solution

a) We multiply by 1 under the radical, using $\frac{3}{3}$. We do this so that the denominator of the radicand will be a perfect square:

$\sqrt{\dfrac{7}{3}} = \sqrt{\dfrac{7}{3}\cdot\dfrac{3}{3}}$ Multiplying by 1 under the radical

$= \sqrt{\dfrac{21}{9}}$ The denominator, 9, is now a perfect square.

$= \dfrac{\sqrt{21}}{\sqrt{9}}$ Using the quotient rule for radicals

$= \dfrac{\sqrt{21}}{3}.$

*See Exercise 65 on p. 434.

b) Note that $16 = 4^2$. Thus, to make the denominator a perfect cube, we multiply under the radical by $\frac{4}{4}$:

$$\sqrt[3]{\frac{5}{16}} = \sqrt[3]{\frac{5}{4 \cdot 4} \cdot \frac{4}{4}}$$ Since the index is 3, we need 3 identical factors in the denominator.

$$= \sqrt[3]{\frac{20}{4^3}}$$ The denominator is now a perfect cube.

$$= \frac{\sqrt[3]{20}}{\sqrt[3]{4^3}}$$

$$= \frac{\sqrt[3]{20}}{4}.$$

Another way to rationalize a denominator is to multiply by 1 *outside* the radical.

Example 5 Rationalize each denominator.

a) $\sqrt{\frac{4}{5b}}$ **b)** $\frac{\sqrt[3]{a}}{\sqrt[3]{9x}}$ **c)** $\frac{3x}{\sqrt[5]{2x^2y^3}}$

Solution

a) We rewrite the expression as a quotient of two radicals. Then we simplify and multiply by 1:

$$\sqrt{\frac{4}{5b}} = \frac{\sqrt{4}}{\sqrt{5b}} = \frac{2}{\sqrt{5b}}$$ We assume $b > 0$.

$$= \frac{2}{\sqrt{5b}} \cdot \frac{\sqrt{5b}}{\sqrt{5b}}$$ Multiplying by 1

$$= \frac{2\sqrt{5b}}{(\sqrt{5b})^2}$$ Try to do this step mentally.

$$= \frac{2\sqrt{5b}}{5b}.$$

b) To rationalize the denominator $\sqrt[3]{9x}$, note that $9x$ is $3 \cdot 3 \cdot x$. In order for this radicand to be a cube, we need another factor of 3 and two more factors of x. Thus we multiply by 1, using $\sqrt[3]{3x^2}/\sqrt[3]{3x^2}$:

$$\frac{\sqrt[3]{a}}{\sqrt[3]{9x}} = \frac{\sqrt[3]{a}}{\sqrt[3]{9x}} \cdot \frac{\sqrt[3]{3x^2}}{\sqrt[3]{3x^2}}$$ Multiplying by 1

$$= \frac{\sqrt[3]{3ax^2}}{\sqrt[3]{27x^3}}$$ ← This radicand is now a perfect cube.

$$= \frac{\sqrt[3]{3ax^2}}{3x}.$$

c) To change the radicand $2x^2y^3$ into a perfect fifth power, we need four more factors of 2, three more factors of x, and two more factors of y. Thus we multiply by 1, using $\sqrt[5]{2^4x^3y^2}/\sqrt[5]{2^4x^3y^2}$, or $\sqrt[5]{16x^3y^2}/\sqrt[5]{16x^3y^2}$:

$$\frac{3x}{\sqrt[5]{2x^2y^3}} = \frac{3x}{\sqrt[5]{2x^2y^3}} \cdot \frac{\sqrt[5]{16x^3y^2}}{\sqrt[5]{16x^3y^2}}$$ Multiplying by 1

$$= \frac{3x\sqrt[5]{16x^3y^2}}{\sqrt[5]{32x^5y^5}}$$ This radicand is now a perfect fifth power.

$$= \frac{3x\sqrt[5]{16x^3y^2}}{2xy} = \frac{3\sqrt[5]{16x^3y^2}}{2y}.$$ Always simplify if possible.

Sometimes in calculus it is necessary to rationalize a numerator. To do so, we multiply by 1 to make the radicand in the *numerator* a perfect power.

Example 6 Rationalize each numerator: **(a)** $\sqrt{\frac{7}{5}}$; **(b)** $\frac{\sqrt[3]{4a^2}}{\sqrt[3]{5b}}$.

Solution

a) $$\sqrt{\frac{7}{5}} = \sqrt{\frac{7}{5} \cdot \frac{7}{7}}$$ Multiplying by 1 under the radical. We also could have multiplied by $\sqrt{7}/\sqrt{7}$ outside the radical.

$$= \sqrt{\frac{49}{35}}$$ The numerator is now a perfect square.

$$= \frac{\sqrt{49}}{\sqrt{35}}$$ Using the quotient rule for radicals

$$= \frac{7}{\sqrt{35}}$$

b) $$\frac{\sqrt[3]{4a^2}}{\sqrt[3]{5b}} = \frac{\sqrt[3]{4a^2}}{\sqrt[3]{5b}} \cdot \frac{\sqrt[3]{2a}}{\sqrt[3]{2a}}$$ Multiplying by 1

$$= \frac{\sqrt[3]{8a^3}}{\sqrt[3]{10ba}}$$ This radicand is now a perfect cube.

$$= \frac{2a}{\sqrt[3]{10ab}}$$

In Section 7.5, we will discuss rationalizing denominators and numerators in which two terms appear.

Exercise Set 7.4

FOR EXTRA HELP

Digital Video Tutor CD 5
Videotape 13

InterAct Math

Math Tutor Center

MathXL

MyMathLab.com

Simplify by taking the roots of the numerator and the denominator. Assume all variables represent positive numbers.

1. $\sqrt{\dfrac{25}{36}}$
2. $\sqrt{\dfrac{100}{81}}$
3. $\sqrt[3]{\dfrac{64}{27}}$
4. $\sqrt[3]{\dfrac{343}{1000}}$
5. $\sqrt{\dfrac{49}{y^2}}$
6. $\sqrt{\dfrac{121}{x^2}}$
7. $\sqrt{\dfrac{25y^3}{x^4}}$
8. $\sqrt{\dfrac{36a^5}{b^6}}$
9. $\sqrt[3]{\dfrac{27a^4}{8b^3}}$
10. $\sqrt[3]{\dfrac{64x^7}{216y^6}}$
11. $\sqrt[4]{\dfrac{16a^4}{b^4c^8}}$
12. $\sqrt[4]{\dfrac{81x^4}{y^8z^4}}$
13. $\sqrt[4]{\dfrac{a^5b^8}{c^{10}}}$
14. $\sqrt[4]{\dfrac{x^9y^{12}}{z^6}}$
15. $\sqrt[5]{\dfrac{32x^6}{y^{11}}}$
16. $\sqrt[5]{\dfrac{243a^9}{b^{13}}}$
17. $\sqrt[6]{\dfrac{x^6y^8}{z^{15}}}$
18. $\sqrt[6]{\dfrac{a^9b^{12}}{c^{13}}}$

Divide and, if possible, simplify. Assume all variables represent positive numbers.

19. $\dfrac{\sqrt{35x}}{\sqrt{7x}}$
20. $\dfrac{\sqrt{28y}}{\sqrt{4y}}$
21. $\dfrac{\sqrt[3]{270}}{\sqrt[3]{10}}$
22. $\dfrac{\sqrt[3]{40}}{\sqrt[3]{5}}$
23. $\dfrac{\sqrt{40xy^3}}{\sqrt{8x}}$
24. $\dfrac{\sqrt{56ab^3}}{\sqrt{7a}}$
25. $\dfrac{\sqrt[3]{96a^4b^2}}{\sqrt[3]{12a^2b}}$
26. $\dfrac{\sqrt[3]{189x^5y^7}}{\sqrt[3]{7x^2y^2}}$
27. $\dfrac{\sqrt{100ab}}{5\sqrt{2}}$
28. $\dfrac{\sqrt{75ab}}{3\sqrt{3}}$
29. $\dfrac{\sqrt[4]{48x^9y^{13}}}{\sqrt[4]{3xy^{-2}}}$
30. $\dfrac{\sqrt[5]{64a^{11}b^{28}}}{\sqrt[5]{2ab^{-2}}}$
31. $\dfrac{\sqrt[3]{x^3-y^3}}{\sqrt[3]{x-y}}$
32. $\dfrac{\sqrt[3]{r^3+s^3}}{\sqrt[3]{r+s}}$

Hint: Factor and then simplify.

Rationalize each denominator. Assume all variables represent positive numbers.

33. $\sqrt{\dfrac{5}{7}}$
34. $\sqrt{\dfrac{11}{6}}$
35. $\dfrac{6\sqrt{5}}{5\sqrt{3}}$
36. $\dfrac{4\sqrt{5}}{3\sqrt{2}}$
37. $\sqrt[3]{\dfrac{16}{9}}$
38. $\sqrt[3]{\dfrac{2}{9}}$
39. $\dfrac{\sqrt[3]{3a}}{\sqrt[3]{5c}}$
40. $\dfrac{\sqrt[3]{7x}}{\sqrt[3]{3y}}$
41. $\dfrac{\sqrt[3]{5y^4}}{\sqrt[3]{6x^4}}$
42. $\dfrac{\sqrt[3]{3a^4}}{\sqrt[3]{7b^2}}$
43. $\sqrt[3]{\dfrac{2}{x^2y}}$
44. $\sqrt[3]{\dfrac{5}{ab^2}}$
45. $\sqrt{\dfrac{7a}{18}}$
46. $\sqrt{\dfrac{3x}{10}}$
47. $\sqrt{\dfrac{9}{20x^2y}}$
48. $\sqrt{\dfrac{7}{32a^2b}}$
49. Aha! $\sqrt{\dfrac{10ab^2}{72a^3b}}$
50. $\sqrt{\dfrac{21x^2y}{75xy^5}}$

Rationalize each numerator. Assume all variables represent positive numbers.

51. $\dfrac{\sqrt{5}}{\sqrt{7x}}$
52. $\dfrac{\sqrt{10}}{\sqrt{3x}}$
53. $\sqrt{\dfrac{14}{21}}$
54. $\sqrt{\dfrac{12}{15}}$
55. $\dfrac{4\sqrt{13}}{3\sqrt{7}}$
56. $\dfrac{5\sqrt{21}}{2\sqrt{5}}$
57. $\dfrac{\sqrt[3]{7}}{\sqrt[3]{2}}$
58. $\dfrac{\sqrt[3]{5}}{\sqrt[3]{4}}$
59. $\sqrt{\dfrac{7x}{3y}}$
60. $\sqrt{\dfrac{6a}{5b}}$
61. $\sqrt[3]{\dfrac{2a^5}{5b}}$
62. $\sqrt[3]{\dfrac{2a^4}{7b}}$
63. $\sqrt{\dfrac{x^3y}{2}}$
64. $\sqrt{\dfrac{ab^5}{3}}$

65. Explain why it is easier to approximate
$$\frac{\sqrt{2}}{2} \text{ than } \frac{1}{\sqrt{2}}$$
if no calculator is available and $\sqrt{2} \approx 1.414213562$.

66. A student *incorrectly* claims that
$$\frac{5+\sqrt{2}}{\sqrt{18}} = \frac{5+\sqrt{1}}{\sqrt{9}} = \frac{5+1}{3}.$$
How could you convince the student that a mistake has been made? How would you explain the correct way of rationalizing the denominator?

SKILL MAINTENANCE

Multiply.

67. $\dfrac{3}{x-5} \cdot \dfrac{x-1}{x+5}$
68. $\dfrac{7}{x+4} \cdot \dfrac{x-2}{x-4}$

Simplify.

69. $\dfrac{a^2 - 8a + 7}{a^2 - 49}$

70. $\dfrac{t^2 + 9t - 22}{t^2 - 4}$

71. $(5a^3b^4)^3$

72. $(3x^4)^2(5xy^3)^2$

SYNTHESIS

73. Is it possible to understand how to rationalize a denominator without knowing how to multiply rational expressions? Why or why not?

74. Is the quotient of two irrational numbers always an irrational number? Why or why not?

75. *Pendulums.* The *period* of a pendulum is the time it takes to complete one cycle, swinging to and fro. For a pendulum that is L centimeters long, the period T is given by the formula

$$T = 2\pi\sqrt{\frac{L}{980}},$$

where T is in seconds. Find, to the nearest hundredth of a second, the period of a pendulum of length **(a)** 65 cm; **(b)** 98 cm; **(c)** 120 cm. Use a calculator's $\boxed{\pi}$ key if possible.

Perform the indicated operations.

76. $\dfrac{7\sqrt{a^2b}\sqrt{25xy}}{5\sqrt{a^{-4}b^{-1}}\sqrt{49x^{-1}y^{-3}}}$

77. $\dfrac{(\sqrt[3]{81mn^2})^2}{(\sqrt[3]{mn})^2}$

78. $\dfrac{\sqrt{44x^2y^9z}\sqrt{22y^9z^6}}{(\sqrt{11xy^8z^2})^2}$

79. $\sqrt{a^2 - 3} - \dfrac{a^2}{\sqrt{a^2 - 3}}$

80. $5\sqrt{\dfrac{x}{y}} + 4\sqrt{\dfrac{y}{x}} - \dfrac{3}{\sqrt{xy}}$

81. Provide a reason for each step in the following derivation of the quotient rule:

$$\sqrt[n]{\frac{a}{b}} = \left(\frac{a}{b}\right)^{1/n} \quad \underline{\qquad\qquad}$$

$$= \frac{a^{1/n}}{b^{1/n}} \quad \underline{\qquad\qquad}$$

$$= \frac{\sqrt[n]{a}}{\sqrt[n]{b}} \quad \underline{\qquad\qquad}.$$

82. Show that $\dfrac{\sqrt[n]{a}}{\sqrt[n]{b}}$ is the nth root of $\dfrac{a}{b}$ by raising it to the nth power and simplifying.

83. Let $f(x) = \sqrt{18x^3}$ and $g(x) = \sqrt{2x}$. Find $(f/g)(x)$ and specify the domain of f/g.

84. Let $f(t) = \sqrt{2t}$ and $g(t) = \sqrt{50t^3}$. Find $(f/g)(t)$ and specify the domain of f/g.

85. Let $f(x) = \sqrt{x^2 - 9}$ and $g(x) = \sqrt{x - 3}$. Find $(f/g)(x)$ and specify the domain of f/g.

Expressions Containing Several Radical Terms

7.5

Adding and Subtracting Radical Expressions • Products and Quotients of Two or More Radical Terms • Terms with Differing Indices

Radical expressions like $6\sqrt{7} + 4\sqrt{7}$ or $(\sqrt{a} + \sqrt{b})(\sqrt{a} - \sqrt{b})$ contain more than one *radical term* and can sometimes be simplified.

Adding and Subtracting Radical Expressions

When two radical expressions have the same indices and radicands, they are said to be **like radicals**. Like radicals can be combined (added or subtracted) in much the same way that we combined like terms earlier in this text.

Example 1

Simplify by combining like radical terms.

a) $6\sqrt{7} + 4\sqrt{7}$

b) $\sqrt[3]{2} - 7x\sqrt[3]{2} + 5\sqrt[3]{2}$

c) $6\sqrt[5]{4x} + 3\sqrt[5]{4x} - \sqrt[3]{4x}$

Solution

a) $6\sqrt{7} + 4\sqrt{7} = (6 + 4)\sqrt{7}$ Using the distributive law (factoring out $\sqrt{7}$)

$= 10\sqrt{7}$ You can think: 6 square roots of 7 plus 4 square roots of 7 results in 10 square roots of 7.

b) $\sqrt[3]{2} - 7x\sqrt[3]{2} + 5\sqrt[3]{2} = (1 - 7x + 5)\sqrt[3]{2}$ Factoring out $\sqrt[3]{2}$

$= (6 - 7x)\sqrt[3]{2}$ These parentheses are important!

c) $6\sqrt[5]{4x} + 3\sqrt[5]{4x} - \sqrt[3]{4x} = (6 + 3)\sqrt[5]{4x} - \sqrt[3]{4x}$ Try to do this step mentally.

$= 9\sqrt[5]{4x} - \sqrt[3]{4x}$ Because the indices differ, we are done.

Our ability to simplify radical expressions can help us to find like radicals even when, at first, it may appear that none exists.

Example 2

Simplify by combining like radical terms, if possible.

a) $3\sqrt{8} - 5\sqrt{2}$

b) $9\sqrt{5} - 4\sqrt{3}$

c) $\sqrt[3]{2x^6y^4} + 7\sqrt[3]{2y}$

Solution

a)
$$\left.\begin{aligned} 3\sqrt{8} - 5\sqrt{2} &= 3\sqrt{4 \cdot 2} - 5\sqrt{2} \\ &= 3\sqrt{4} \cdot \sqrt{2} - 5\sqrt{2} \\ &= 3 \cdot 2 \cdot \sqrt{2} - 5\sqrt{2} \end{aligned}\right\} \text{Simplifying } \sqrt{8}$$

$= 6\sqrt{2} - 5\sqrt{2}$

$= \sqrt{2}$ Combining like radicals

b) $9\sqrt{5} - 4\sqrt{3}$ cannot be simplified.

c)
$$\left.\begin{aligned} \sqrt[3]{2x^6y^4} + 7\sqrt[3]{2y} &= \sqrt[3]{x^6y^3 \cdot 2y} + 7\sqrt[3]{2y} \\ &= \sqrt[3]{x^6y^3} \cdot \sqrt[3]{2y} + 7\sqrt[3]{2y} \\ &= x^2y \cdot \sqrt[3]{2y} + 7\sqrt[3]{2y} \end{aligned}\right\} \text{Simplifying } \sqrt[3]{2x^6y^4}$$

$= (x^2y + 7)\sqrt[3]{2y}$ Factoring to combine like radical terms

Products and Quotients of Two or More Radical Terms

Radical expressions often contain factors that have more than one term. The procedure for multiplying out such expressions is similar to finding products of polynomials. Some products will yield like radical terms, which we can now combine.

Example 3 Multiply.

a) $\sqrt{3}(x - \sqrt{5})$

b) $\sqrt[3]{y}(\sqrt[3]{y^2} + \sqrt[3]{2})$

c) $(4\sqrt{3} + \sqrt{2})(\sqrt{3} - 5\sqrt{2})$

d) $(\sqrt{a} + \sqrt{b})(\sqrt{a} - \sqrt{b})$

Solution

a) $\sqrt{3}(x - \sqrt{5}) = \sqrt{3} \cdot x - \sqrt{3} \cdot \sqrt{5}$ Using the distributive law

$= x\sqrt{3} - \sqrt{15}$ Multiplying radicals

b) $\sqrt[3]{y}(\sqrt[3]{y^2} + \sqrt[3]{2}) = \sqrt[3]{y} \cdot \sqrt[3]{y^2} + \sqrt[3]{y} \cdot \sqrt[3]{2}$ Using the distributive law

$= \sqrt[3]{y^3} + \sqrt[3]{2y}$ Multiplying radicals

$= y + \sqrt[3]{2y}$ Simplifying $\sqrt[3]{y^3}$

c) $(4\sqrt{3} + \sqrt{2})(\sqrt{3} - 5\sqrt{2}) = \overset{F}{4(\sqrt{3})^2} - \overset{O}{20\sqrt{3} \cdot \sqrt{2}} + \overset{I}{\sqrt{2} \cdot \sqrt{3}} - \overset{L}{5(\sqrt{2})^2}$

$= 4 \cdot 3 - 20\sqrt{6} + \sqrt{6} - 5 \cdot 2$ Multiplying radicals

$= 12 - 20\sqrt{6} + \sqrt{6} - 10$

$= 2 - 19\sqrt{6}$ Combining like terms

d) $(\sqrt{a} + \sqrt{b})(\sqrt{a} - \sqrt{b}) = (\sqrt{a})^2 - \sqrt{a}\sqrt{b} + \sqrt{a}\sqrt{b} - (\sqrt{b})^2$ Using FOIL

$= a - b$ Combining like terms

In Example 3(d) above, you may have noticed that since the outer and inner products in FOIL are opposites, the result, $a - b$, is not itself a radical expression. Pairs of radical terms, like $\sqrt{a} + \sqrt{b}$ and $\sqrt{a} - \sqrt{b}$, are called **conjugates**. The use of conjugates allows us to rationalize denominators or numerators with two terms.

Example 4 Rationalize each denominator: **(a)** $\dfrac{4}{\sqrt{3} + x}$; **(b)** $\dfrac{4 + \sqrt{2}}{\sqrt{5} - \sqrt{2}}$.

Solution

a) $\dfrac{4}{\sqrt{3}+x} = \dfrac{4}{\sqrt{3}+x} \cdot \dfrac{\sqrt{3}-x}{\sqrt{3}-x}$ Multiplying by 1, using the conjugate of $\sqrt{3}+x$, which is $\sqrt{3}-x$

$= \dfrac{4(\sqrt{3}-x)}{(\sqrt{3}+x)(\sqrt{3}-x)}$ Multiplying numerators and denominators

$= \dfrac{4(\sqrt{3}-x)}{(\sqrt{3})^2-x^2}$ Using FOIL in the denominator

$= \dfrac{4\sqrt{3}-4x}{3-x^2}$ Simplifying

b) $\dfrac{4+\sqrt{2}}{\sqrt{5}-\sqrt{2}} = \dfrac{4+\sqrt{2}}{\sqrt{5}-\sqrt{2}} \cdot \dfrac{\sqrt{5}+\sqrt{2}}{\sqrt{5}+\sqrt{2}}$ Multiplying by 1, using the conjugate of $\sqrt{5}-\sqrt{2}$, which is $\sqrt{5}+\sqrt{2}$

$= \dfrac{(4+\sqrt{2})(\sqrt{5}+\sqrt{2})}{(\sqrt{5}-\sqrt{2})(\sqrt{5}+\sqrt{2})}$ Multiplying numerators and denominators

$= \dfrac{4\sqrt{5}+4\sqrt{2}+\sqrt{2}\,\sqrt{5}+(\sqrt{2})^2}{(\sqrt{5})^2-(\sqrt{2})^2}$ Using FOIL

$= \dfrac{4\sqrt{5}+4\sqrt{2}+\sqrt{10}+2}{5-2}$ Squaring in the denominator and the numerator

$= \dfrac{4\sqrt{5}+4\sqrt{2}+\sqrt{10}+2}{3}$

To rationalize a numerator with more than one term, we use the conjugate of the numerator.

Example 5 Rationalize the numerator: $\dfrac{4+\sqrt{2}}{\sqrt{5}-\sqrt{2}}$.

Solution

$\dfrac{4+\sqrt{2}}{\sqrt{5}-\sqrt{2}} = \dfrac{4+\sqrt{2}}{\sqrt{5}-\sqrt{2}} \cdot \dfrac{4-\sqrt{2}}{4-\sqrt{2}}$ Multiplying by 1, using the conjugate of $4+\sqrt{2}$, which is $4-\sqrt{2}$

$= \dfrac{16-(\sqrt{2})^2}{4\sqrt{5}-\sqrt{5}\,\sqrt{2}-4\sqrt{2}+(\sqrt{2})^2}$

$= \dfrac{14}{4\sqrt{5}-\sqrt{10}-4\sqrt{2}+2}$

Terms with Differing Indices

Sometimes it is necessary to determine products or quotients involving radical terms with indices that differ from each other. When this occurs, we can convert to exponential notation, use the rules for exponents, and then convert back to radical notation.

Example 6 Divide and, if possible, simplify: $\dfrac{\sqrt[4]{(x+y)^3}}{\sqrt{x+y}}$.

Solution

$$\frac{\sqrt[4]{(x+y)^3}}{\sqrt{x+y}} = \frac{(x+y)^{3/4}}{(x+y)^{1/2}} \quad \text{Converting to exponential notation}$$

$$= (x+y)^{3/4-1/2} \quad \text{Since the bases are identical, we can subtract exponents: } \tfrac{3}{4}-\tfrac{1}{2}=\tfrac{3}{4}-\tfrac{2}{4}=\tfrac{1}{4}.$$

$$\left.\begin{aligned} &= (x+y)^{1/4} \\ &= \sqrt[4]{x+y} \end{aligned}\right\} \quad \text{Converting back to radical notation}$$

The steps used in Example 6 can be used in a variety of situations.

To Simplify Products or Quotients with Differing Indices

1. Convert all radical expressions to exponential notation.
2. When the bases are identical, subtract exponents to divide and add exponents to multiply. This may require finding a common denominator.
3. Convert back to radical notation and, if possible, simplify.

Example 7 Multiply and simplify: $\sqrt{x^3}\,\sqrt[3]{x}$.

Solution

$$\sqrt{x^3}\,\sqrt[3]{x} = x^{3/2} \cdot x^{1/3} \quad \text{Converting to exponential notation}$$

$$= x^{11/6} \quad \text{Adding exponents: } \tfrac{3}{2}+\tfrac{1}{3}=\tfrac{9}{6}+\tfrac{2}{6}$$

$$= \sqrt[6]{x^{11}} \quad \text{Converting back to radical notation}$$

$$\left.\begin{aligned} &= \sqrt[6]{x^6}\,\sqrt[6]{x^5} \\ &= x\sqrt[6]{x^5} \end{aligned}\right\} \quad \text{Simplifying}$$

Example 8 If $f(x) = \sqrt[3]{x^2}$ and $g(x) = \sqrt{x} + \sqrt[4]{x}$, find $(f \cdot g)(x)$.

Solution Recall from Section 2.6 that $(f \cdot g)(x) = f(x) \cdot g(x)$. Thus,

$$(f \cdot g)(x) = \sqrt[3]{x^2}\left(\sqrt{x} + \sqrt[4]{x}\right) \quad x \text{ is assumed to be nonnegative.}$$

$$= x^{2/3}(x^{1/2} + x^{1/4}) \quad \text{Converting to exponential notation}$$

$$= x^{2/3} \cdot x^{1/2} + x^{2/3} \cdot x^{1/4} \quad \text{Using the distributive law}$$

$$= x^{2/3+1/2} + x^{2/3+1/4} \quad \text{Adding exponents}$$

$$= x^{7/6} + x^{11/12} \quad \tfrac{2}{3}+\tfrac{1}{2}=\tfrac{4}{6}+\tfrac{3}{6};\ \tfrac{2}{3}+\tfrac{1}{4}=\tfrac{8}{12}+\tfrac{3}{12}$$

$$= \sqrt[6]{x^7} + \sqrt[12]{x^{11}} \quad \text{Converting back to radical notation}$$

$$\left.\begin{aligned} &= \sqrt[6]{x^6}\,\sqrt[6]{x} + \sqrt[12]{x^{11}} \\ &= x\sqrt[6]{x} + \sqrt[12]{x^{11}} \end{aligned}\right\} \quad \text{Simplifying}$$

If factors are raised to powers that share a common denominator, we can write the final result as a single radical expression.

Example 9 Divide and, if possible, simplify: $\dfrac{\sqrt[3]{a^2b^4}}{\sqrt{ab}}$.

Solution

$$\frac{\sqrt[3]{a^2b^4}}{\sqrt{ab}} = \frac{(a^2b^4)^{1/3}}{(ab)^{1/2}} \quad \text{Converting to exponential notation}$$
$$= \frac{a^{2/3}b^{4/3}}{a^{1/2}b^{1/2}} \quad \text{Using the product and power rules}$$
$$= a^{2/3-1/2}b^{4/3-1/2} \quad \text{Subtracting exponents}$$
$$= a^{1/6}b^{5/6}$$
$$= \sqrt[6]{a}\,\sqrt[6]{b^5} \quad \text{Converting to radical notation}$$
$$= \sqrt[6]{ab^5} \quad \text{Using the product rule for radicals}$$

Exercise Set 7.5

FOR EXTRA HELP

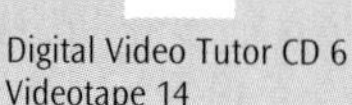
Digital Video Tutor CD 6
Videotape 14

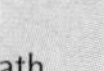
InterAct Math

Math Tutor Center

MathXL

MyMathLab.com

Add or subtract. Simplify by combining like radical terms, if possible. Assume that all variables and radicands represent nonnegative real numbers.

1. $3\sqrt{7} + 2\sqrt{7}$
2. $8\sqrt{5} + 9\sqrt{5}$
3. $9\sqrt[3]{5} - 6\sqrt[3]{5}$
4. $14\sqrt[5]{2} - 6\sqrt[5]{2}$
5. $4\sqrt[3]{y} + 9\sqrt[3]{y}$
6. $9\sqrt[4]{t} - 3\sqrt[4]{t}$
7. $8\sqrt{2} - 6\sqrt{2} + 5\sqrt{2}$
8. $2\sqrt{6} + 8\sqrt{6} - 3\sqrt{6}$
9. $9\sqrt[3]{7} - \sqrt{3} + 4\sqrt[3]{7} + 2\sqrt{3}$
10. $5\sqrt{7} - 8\sqrt[4]{11} + \sqrt{7} + 9\sqrt[4]{11}$
11. $8\sqrt{27} - 3\sqrt{3}$
12. $9\sqrt{50} - 4\sqrt{2}$
13. $3\sqrt{45} + 7\sqrt{20}$
14. $5\sqrt{12} + 16\sqrt{27}$
15. $3\sqrt[3]{16} + \sqrt[3]{54}$
16. $\sqrt[3]{27} - 5\sqrt[3]{8}$
17. $\sqrt{5a} + 2\sqrt{45a^3}$
18. $4\sqrt{3x^3} - \sqrt{12x}$
19. $\sqrt[3]{6x^4} + \sqrt[3]{48x}$
20. $\sqrt[3]{54x} - \sqrt[3]{2x^4}$
21. $\sqrt{4a - 4} + \sqrt{a - 1}$
22. $\sqrt{9y + 27} + \sqrt{y + 3}$
23. $\sqrt{x^3 - x^2} + \sqrt{9x - 9}$
24. $\sqrt{4x - 4} - \sqrt{x^3 - x^2}$

Multiply. Assume all variables represent nonnegative real numbers.

25. $\sqrt{7}(3 - \sqrt{7})$
26. $\sqrt{3}(4 + \sqrt{3})$
27. $4\sqrt{2}(\sqrt{3} - \sqrt{5})$
28. $3\sqrt{5}(\sqrt{5} - \sqrt{2})$
29. $\sqrt{3}(2\sqrt{5} - 3\sqrt{4})$
30. $\sqrt{2}(3\sqrt{10} - 2\sqrt{2})$
31. $\sqrt[3]{2}(\sqrt[3]{4} - 2\sqrt[3]{32})$
32. $\sqrt[3]{3}(\sqrt[3]{9} - 4\sqrt[3]{21})$

33. $\sqrt[3]{a}(\sqrt[3]{a^2} + \sqrt[3]{24a^2})$

34. $\sqrt[3]{x}(\sqrt[3]{3x^2} - \sqrt[3]{81x^2})$

35. $(5 + \sqrt{6})(5 - \sqrt{6})$

36. $(2 - \sqrt{5})(2 + \sqrt{5})$

37. $(3 - 2\sqrt{7})(3 + 2\sqrt{7})$

38. $(4 + 3\sqrt{2})(4 - 3\sqrt{2})$

39. $(3 + \sqrt{5})^2$

40. $(7 + \sqrt{3})^2$

41. $(2\sqrt{7} - 4\sqrt{2})(3\sqrt{7} + 6\sqrt{2})$

42. $(4\sqrt{5} + 3\sqrt{3})(3\sqrt{5} - 4\sqrt{3})$

43. $(2\sqrt[3]{3} - \sqrt[3]{2})(\sqrt[3]{3} + 2\sqrt[3]{2})$

44. $(3\sqrt[4]{7} + \sqrt[4]{6})(2\sqrt[4]{9} - 3\sqrt[4]{6})$

45. $(\sqrt{3x} + \sqrt{y})^2$

46. $(\sqrt{t} - \sqrt{2r})^2$

Rationalize each denominator.

47. $\frac{2}{3 + \sqrt{5}}$

48. $\frac{3}{4 - \sqrt{7}}$

49. $\frac{2 + \sqrt{5}}{6 - \sqrt{3}}$

50. $\frac{1 + \sqrt{2}}{3 + \sqrt{5}}$

51. $\frac{\sqrt{a}}{\sqrt{a} + \sqrt{b}}$

52. $\frac{\sqrt{z}}{\sqrt{x} - \sqrt{z}}$

Aha! **53.** $\frac{\sqrt{7} - \sqrt{3}}{\sqrt{3} - \sqrt{7}}$

54. $\frac{\sqrt{7} + \sqrt{5}}{\sqrt{5} + \sqrt{2}}$

55. $\frac{3\sqrt{2} - \sqrt{7}}{4\sqrt{2} + \sqrt{5}}$

56. $\frac{5\sqrt{3} - \sqrt{11}}{2\sqrt{3} - 5\sqrt{2}}$

57. $\frac{5\sqrt{3} - 3\sqrt{2}}{3\sqrt{2} - 2\sqrt{3}}$

58. $\frac{7\sqrt{2} + 4\sqrt{3}}{4\sqrt{3} - 3\sqrt{2}}$

Rationalize each numerator.

59. $\frac{\sqrt{7} + 2}{5}$

60. $\frac{\sqrt{3} + 1}{4}$

61. $\frac{\sqrt{6} - 2}{\sqrt{3} + 7}$

62. $\frac{\sqrt{10} + 4}{\sqrt{2} - 3}$

63. $\frac{\sqrt{x} - \sqrt{y}}{\sqrt{x} + \sqrt{y}}$

64. $\frac{\sqrt{a} + \sqrt{b}}{\sqrt{a} - \sqrt{b}}$

Perform the indicated operation and simplify. Assume all variables represent nonnegative real numbers.

65. $\sqrt{a}\sqrt[4]{a^3}$

66. $\sqrt[3]{x^2}\sqrt[6]{x^5}$

67. $\sqrt[5]{b^2}\sqrt{b^3}$

68. $\sqrt[4]{a^3}\sqrt[3]{a^2}$

69. $\sqrt{xy^3}\sqrt[3]{x^2y}$

70. $\sqrt[5]{a^3b}\sqrt{ab}$

71. $\sqrt[4]{9ab^3}\sqrt{3a^4b}$

72. $\sqrt{2x^3y^3}\sqrt[3]{4xy^2}$

73. $\sqrt[3]{xy^2z}\sqrt{x^3yz^2}$

74. $\sqrt{a^4b^3c^4}\sqrt[3]{ab^2c}$

75. $\frac{\sqrt[3]{x^2}}{\sqrt[5]{x}}$

76. $\frac{\sqrt[3]{a^2}}{\sqrt[4]{a}}$

77. $\frac{\sqrt[5]{a^4b}}{\sqrt[3]{ab}}$

78. $\frac{\sqrt[4]{x^2y^3}}{\sqrt[3]{xy}}$

79. $\frac{\sqrt[5]{x^3y^4}}{\sqrt{xy}}$

80. $\frac{\sqrt{ab^3}}{\sqrt[5]{a^2b^3}}$

81. $\frac{\sqrt[3]{(2 + 5x)^2}}{\sqrt[4]{2 + 5x}}$

82. $\frac{\sqrt[4]{(3x - 1)^3}}{\sqrt[5]{(3x - 1)^3}}$

83. $\frac{\sqrt[4]{(5 + 3x)^3}}{\sqrt[3]{(5 + 3x)^2}}$

84. $\frac{\sqrt[3]{(2x + 1)^2}}{\sqrt[5]{(2x + 1)^2}}$

85. $\sqrt[3]{x^2y}(\sqrt{xy} - \sqrt[5]{xy^3})$

86. $\sqrt[4]{a^2b}(\sqrt[3]{a^2b} - \sqrt[5]{a^2b^2})$

87. $(m + \sqrt[3]{n^2})(2m + \sqrt[4]{n})$

88. $(r - \sqrt[4]{s^3})(3r - \sqrt[5]{s})$

In Exercises 89–92, $f(x)$ and $g(x)$ are as given. Find $(f \cdot g)(x)$. Assume all variables represent nonnegative real numbers.

89. $f(x) = \sqrt[4]{x}$, $g(x) = \sqrt[4]{2x} - \sqrt[4]{x^{11}}$

90. $f(x) = \sqrt[4]{x^7} + \sqrt[4]{3x^2}$, $g(x) = \sqrt[4]{x}$

91. $f(x) = x + \sqrt{7}$, $g(x) = x - \sqrt{7}$

92. $f(x) = x - \sqrt{2}$, $g(x) = x + \sqrt{6}$

Let $f(x) = x^2$. Find each of the following.

93. $f(5 - \sqrt{2})$

94. $f(7 + \sqrt{3})$

95. $f(\sqrt{3} + \sqrt{5})$

96. $f(\sqrt{6} - \sqrt{3})$

97. Why do we need to know how to multiply radical expressions before learning how to add them?

98. In what way(s) is combining like radical terms the same as combining like terms that are monomials?

SKILL MAINTENANCE

Solve.

99. $\frac{12x}{x - 4} - \frac{3x^2}{x + 4} = \frac{384}{x^2 - 16}$

100. $\frac{2}{3} + \frac{1}{t} = \frac{4}{5}$

101. The width of a rectangle is one-fourth the length. The area is twice the perimeter. Find the dimensions of the rectangle.

102. The sum of a number and its square is 20. Find the number.

103. $5x^2 - 6x + 1 = 0$

104. $7t^2 - 8t + 1 = 0$

SYNTHESIS

105. Ramon *incorrectly* writes

$$\sqrt[5]{x^2} \cdot \sqrt{x^3} = x^{2/5} \cdot x^{3/2} = \sqrt[5]{x^3}.$$

What mistake do you suspect he is making?

106. After examining the expression $\sqrt[4]{25xy^3}\sqrt{5x^4y}$ Dyan (correctly) concludes that x and y are both nonnegative. Explain how she could reach this conclusion.

For Exercises 107–110, fill in the blanks by selecting from the following words:

radicands, indices, bases, denominators.

Words can be used more than once.

107. To add radical expressions, the __________ and the __________ must be the same.

108. To multiply radical expressions, the __________ must be the same.

109. To add rational expressions, the __________ must be the same.

110. To find a product by adding exponents, the __________ must be the same.

Find a simplified form for $f(x)$. Assume $x \geq 0$.

111. $f(x) = \sqrt{20x^2 + 4x^3} - 3x\sqrt{45 + 9x} + \sqrt{5x^2 + x^3}$

112. $f(x) = \sqrt{x^3 - x^2} + \sqrt{9x^3 - 9x^2} - \sqrt{4x^3 - 4x^2}$

113. $f(x) = \sqrt[4]{x^5 - x^4} + 3\sqrt[4]{x^9 - x^8}$

114. $f(x) = \sqrt[4]{16x^4 + 16x^5} - 2\sqrt[4]{x^8 + x^9}$

Simplify.

115. $\frac{1}{2}\sqrt{36a^5bc^4} - \frac{1}{2}\sqrt[3]{64a^4bc^6} + \frac{1}{6}\sqrt{144a^3bc^6}$

116. $7x\sqrt{(x + y)^3} - 5xy\sqrt{x + y} - 2y\sqrt{(x + y)^3}$

117. $\sqrt{27a^5(b + 1)}\sqrt[3]{81a(b + 1)^4}$

118. $\sqrt{8x(y + z)^5}\sqrt[3]{4x^2(y + z)^2}$

119. $\dfrac{\dfrac{1}{\sqrt{w}} - \sqrt{w}}{\dfrac{\sqrt{w} + 1}{\sqrt{w}}}$

120. $\dfrac{1}{4 + \sqrt{3}} + \dfrac{1}{\sqrt{3}} + \dfrac{1}{\sqrt{3} - 4}$

Express each of the following as the product of two radical expressions.

121. $x - 5$

122. $y - 7$

123. $x - a$

Multiply.

124. $\sqrt{9 + 3\sqrt{5}}\sqrt{9 - 3\sqrt{5}}$

125. $(\sqrt{x + 2} - \sqrt{x - 2})^2$

For Exercises 126–129, assume that all radicands are positive and that no denominator is 0.

Rationalize each denominator.

126. $\dfrac{a - \sqrt{a + b}}{\sqrt{a + b} - b}$

127. $\dfrac{b + \sqrt{b}}{1 + b + \sqrt{b}}$

Rationalize each numerator.

128. $\dfrac{\sqrt{y + 18} - \sqrt{y}}{18}$

129. $\dfrac{\sqrt{x + 6} - 5}{\sqrt{x + 6} + 5}$

130. Use a grapher to check your answers to Exercises 19, 33, and 75.

7.6 Solving Radical Equations

The Principle of Powers • Equations with Two Radical Terms

CONNECTING THE CONCEPTS

In Sections 7.1–7.5, we learned how to manipulate radical expressions as well as expressions containing rational exponents. We performed this work to find *equivalent expressions*.

Now that we know how to work with radicals and rational exponents, we can learn how to solve a new type of equation. As in our earlier work with equations, finding *equivalent equations* will be part of our strategy. What is different, however, is that now we will use a step that does not always produce equivalent equations. Checking solutions will therefore be more important than ever.

The Principle of Powers

A **radical equation** is an equation in which the variable appears in a radicand. Examples are

$$\sqrt[3]{2x} + 1 = 5, \quad \sqrt{a} + \sqrt{a - 2} = 7, \quad \text{and} \quad 4 - \sqrt{3x + 1} = \sqrt{6 - x}.$$

To solve such equations, we need a new principle. Suppose an equation $a = b$ is true. If we square both sides, we get another true equation: $a^2 = b^2$. This can be generalized.

The Principle of Powers

If $a = b$, then $a^n = b^n$ for any exponent n.

Note that the principle of powers is an "if–then" statement. The statement obtained by interchanging the two parts of the sentence—"if $a^n = b^n$ for some exponent n, then $a = b$"—*is not always true.* For example, $3^2 = (-3)^2$ *is* true, but $3 = -3$ *is not* true. More generally, $3^n = (-3)^n$ is true for any even number n, whereas $3 = -3$ is false. For this reason, when we raise both sides of an equation to an even power, it will be essential for us to check the answer in the original equation.

Example 1 Solve: $\sqrt{x} - 3 = 4$.

Solution

$$\sqrt{x} - 3 = 4$$

$\sqrt{x} = 7$ — Adding 3 to both sides to isolate the radical

$(\sqrt{x})^2 = 7^2$ — Using the principle of powers

$$x = 49$$

Check:

$$\begin{array}{r|l} \sqrt{x} - 3 & = 4 \\ \hline \sqrt{49} - 3 & ?\ 4 \\ 7 - 3 & \\ 4 & 4 \quad \text{TRUE} \end{array}$$

The solution is 49.

Example 2 Solve: $\sqrt{x} - 5 = -7$.

Solution

$$\sqrt{x} - 5 = -7$$

$\sqrt{x} = -2$ — Adding 5 to both sides to isolate the radical

The equation $\sqrt{x} = -2$ has no solution because the principal square root of a number is never negative. We continue as in Example 1 for comparison.

$(\sqrt{x})^2 = (-2)^2$ — Using the principle of powers

$$x = 4$$

Check:

$$\begin{array}{r|l} \sqrt{x} - 5 & = -7 \\ \hline \sqrt{4} - 5 & ?\ -7 \\ 2 - 5 & \\ -3 & -7 \quad \text{FALSE} \end{array}$$

The number 4 does not check. Thus the equation $\sqrt{x} - 5 = -7$ has no real-number solution.

> ***Caution!*** Raising both sides of an equation to an even power may not produce an equivalent equation. In this case, a check is essential.

Note in Example 2 that $x = 4$ has solution 4, but that $\sqrt{x} - 5 = -7$ has *no* solution. Thus the equations $x = 4$ and $\sqrt{x} - 5 = -7$ are *not* equivalent.

To Solve an Equation with a Radical Term

1. Isolate the radical term on one side of the equation.
2. Use the principle of powers and solve the resulting equation.
3. Check any possible solution in the original equation.

Example 3

Solve: $x = \sqrt{x + 7} + 5$.

Solution

$$x = \sqrt{x + 7} + 5$$

$$x - 5 = \sqrt{x + 7}$$ Subtracting 5 from both sides. This isolates the radical term.

$$(x - 5)^2 = (\sqrt{x + 7})^2$$
$$x^2 - 10x + 25 = x + 7$$ Using the principle of powers; squaring both sides

$$x^2 - 11x + 18 = 0$$ Adding $-x - 7$ to both sides to write the quadratic equation in standard form

$$(x - 9)(x - 2) = 0$$ Factoring

$$x = 9 \quad or \quad x = 2$$ Using the principle of zero products

The possible solutions are 9 and 2. Let's check.

Check: For 9:

$$x = \sqrt{x + 7} + 5$$
$$9 \; ? \; \sqrt{9 + 7} + 5$$
$$9 \mid 9$$ TRUE

For 2:

$$x = \sqrt{x + 7} + 5$$
$$2 \; ? \; \sqrt{2 + 7} + 5$$
$$2 \mid 8$$ FALSE

Since 9 checks but 2 does not, the solution is 9.

To solve Example 3 with a grapher, graph the curves $y_1 = x$ and $y_2 = (x + 7)^{1/2} + 5$ on the same set of axes.

$y_1 = x, \; y_2 = (x + 7)^{1/2} + 5$

10, −10, 10, −10; y_1, y_2

Using the INTERSECT option of the CALC menu, determine the point of intersection. The intersection should appear to occur when $x = 9$. Note that there is no intersection when $x = 2$, as predicted in the check of Example 3.

1. Use a grapher to solve Examples 1, 2, 4, 5, and 6. Compare your answers with those found using the algebraic methods shown.

It is important to isolate a radical term before using the principle of powers. Suppose in Example 3 that both sides of the equation were squared *before* isolating the radical. We then would have had the expression $(\sqrt{x + 7} + 5)^2$ or $x + 7 + 10\sqrt{x + 7} + 25$ on the right side, and the radical would have remained in the problem.

Example 4

Solve: $(2x + 1)^{1/3} + 5 = 0$.

Solution We need not use radical notation to solve:

$$(2x + 1)^{1/3} + 5 = 0$$

$$(2x + 1)^{1/3} = -5$$ Subtracting 5 from both sides

$$[(2x + 1)^{1/3}]^3 = (-5)^3$$ Cubing both sides

$$(2x + 1)^1 = (-5)^3$$ Multiplying exponents. Try to do this mentally.

$$2x + 1 = -125$$

$$2x = -126$$ Subtracting 1 from both sides

$$x = -63.$$

Because both sides were raised to an *odd* power, it is not essential that we check the answer. The student can show that -63 checks and is the solution.

Equations with Two Radical Terms

A strategy for solving equations with two or more radical terms is as follows.

> ***To Solve an Equation with Two or More Radical Terms***
>
> 1. Isolate one of the radical terms.
> 2. Use the principle of powers.
> 3. If a radical remains, perform steps (1) and (2) again.
> 4. Solve the resulting equation.
> 5. Check possible solutions in the original equation.

Example 5 Solve: $\sqrt{2x-5} = 1 + \sqrt{x-3}$.

Solution

$$\sqrt{2x-5} = 1 + \sqrt{x-3}$$

$$(\sqrt{2x-5})^2 = (1 + \sqrt{x-3})^2$$ One radical is already isolated. We square both sides.

This is like squaring a binomial. We square 1, then find twice the product of 1 and $\sqrt{x-3}$ and then the square of $\sqrt{x-3}$.

$$2x - 5 = 1 + 2\sqrt{x-3} + (\sqrt{x-3})^2$$

$$2x - 5 = 1 + 2\sqrt{x-3} + (x-3)$$

$$x - 3 = 2\sqrt{x-3}$$ Isolating the remaining radical term

$$(x-3)^2 = (2\sqrt{x-3})^2$$ Squaring both sides

$$x^2 - 6x + 9 = 4(x-3)$$ Remember to square both the 2 and the $\sqrt{x-3}$ on the right side.

$$x^2 - 6x + 9 = 4x - 12$$

$$x^2 - 10x + 21 = 0$$

$$(x-7)(x-3) = 0$$ Factoring

$$x = 7 \quad or \quad x = 3$$ Using the principle of zero products

We leave it to the student to show that 7 and 3 both check and are the solutions.

Caution! A common error in solving equations like

$$\sqrt{2x - 5} = 1 + \sqrt{x - 3}$$

is to obtain $1 + (x - 3)$ as the square of the right side. This is wrong because $(A + B)^2 \neq A^2 + B^2$. For example,

$$(1 + 2)^2 \neq 1^2 + 2^2$$
$$3^2 \neq 1 + 4$$
$$9 \neq 5.$$

Example 6 Let $f(x) = \sqrt{x + 5} - \sqrt{x - 7}$. Find all x-values for which $f(x) = 2$.

Solution We must have $f(x) = 2$, or

$$\sqrt{x + 5} - \sqrt{x - 7} = 2. \qquad \text{Substituting for } f(x)$$

To solve, we isolate one radical term and square both sides:

$\sqrt{x + 5} = 2 + \sqrt{x - 7}$	Adding $\sqrt{x - 7}$ to both sides. This isolates one of the radical terms.
$\left(\sqrt{x + 5}\right)^2 = \left(2 + \sqrt{x - 7}\right)^2$	Using the principle of powers (squaring both sides)
$x + 5 = 4 + 4\sqrt{x - 7} + (x - 7)$	Using $(A + B)^2 = A^2 + 2AB + B^2$
$5 = 4\sqrt{x - 7} - 3$	Adding $-x$ to both sides and combining like terms
$8 = 4\sqrt{x - 7}$	Isolating the remaining radical term
$2 = \sqrt{x - 7}$	
$2^2 = \left(\sqrt{x - 7}\right)^2$	Squaring both sides
$4 = x - 7$	
$11 = x.$	

Check:

$$\begin{aligned} f(11) &= \sqrt{11 + 5} - \sqrt{11 - 7} \\ &= \sqrt{16} - \sqrt{4} \\ &= 4 - 2 = 2. \end{aligned}$$

We will have $f(x) = 2$ when $x = 11$.

Exercise Set 7.6

FOR EXTRA HELP

Digital Video Tutor CD 6 Videotape 14

InterAct Math

Math Tutor Center

MathXL

MyMathLab.com

Solve.

1. $\sqrt{x+3} = 5$

2. $\sqrt{5x+1} = 8$

3. $\sqrt{2x} - 1 = 2$

4. $\sqrt{3x} + 1 = 6$

5. $\sqrt{x-2} - 7 = -4$

6. $\sqrt{y+1} - 5 = 8$

7. $\sqrt{y+4} + 6 = 7$

8. $\sqrt{x-7} + 3 = 10$

9. $\sqrt[3]{x-2} = 3$

10. $\sqrt[3]{x+5} = 2$

11. $\sqrt[4]{x+3} = 2$

12. $\sqrt[4]{y-1} = 3$

13. $8\sqrt{y} = y$

14. $3\sqrt{x} = x$

15. $3x^{1/2} + 12 = 9$

16. $2y^{1/2} - 7 = 9$

17. $\sqrt[3]{y} = -4$

18. $\sqrt[3]{x} = -3$

19. $x^{1/4} - 2 = 1$

20. $t^{1/3} - 2 = 3$

Aha! **21.** $(y-3)^{1/2} = -2$

22. $(x+2)^{1/2} = -4$

23. $\sqrt[4]{3x+1} - 4 = -1$

24. $\sqrt[4]{2x+3} - 5 = -2$

25. $(x+7)^{1/3} = 4$

26. $(y-7)^{1/4} = 3$

27. $\sqrt[3]{3y+6} + 7 = 8$

28. $\sqrt[3]{6x+9} + 5 = 2$

29. $\sqrt{3t+4} = \sqrt{4t+3}$

30. $\sqrt{2t-7} = \sqrt{3t-12}$

31. $3(4-t)^{1/4} = 6^{1/4}$

32. $2(1-x)^{1/3} = 4^{1/3}$

33. $3 + \sqrt{5-x} = x$

34. $x = \sqrt{x-1} + 3$

35. $\sqrt{4x-3} = 2 + \sqrt{2x-5}$

36. $3 + \sqrt{z-6} = \sqrt{z+9}$

37. $\sqrt{20-x} + 8 = \sqrt{9-x} + 11$

38. $4 + \sqrt{10-x} = 6 + \sqrt{4-x}$

39. $\sqrt{x+2} + \sqrt{3x+4} = 2$

40. $\sqrt{6x+7} - \sqrt{3x+3} = 1$

41. If $f(x) = \sqrt{x} + \sqrt{x-9}$, find x such that $f(x) = 1$.

42. If $g(x) = \sqrt{x} + \sqrt{x-5}$, find x such that $g(x) = 5$.

43. If $f(x) = \sqrt{x-2} - \sqrt{4x+1}$, find a such that $f(a) = -3$.

44. If $g(x) = \sqrt{2x+7} - \sqrt{x+15}$, find a such that $g(a) = -1$.

45. If $f(x) = \sqrt{2x-3}$ and $g(x) = \sqrt{x+7} - 2$, find x such that $f(x) = g(x)$.

46. If $f(x) = 2\sqrt{3x+6}$ and $g(x) = 5 + \sqrt{4x+9}$, find x such that $f(x) = g(x)$.

47. If $f(t) = 4 - \sqrt{t-3}$ and $g(t) = (t+5)^{1/2}$, find a such that $f(a) = g(a)$.

48. If $f(t) = 7 + \sqrt{2t-5}$ and $g(t) = 3(t+1)^{1/2}$, find a such that $f(a) = g(a)$.

49. Explain in your own words why it is important to check your answers when using the principle of powers.

50. Describe a procedure that could be used to write radical equations that have no solution.

SKILL MAINTENANCE

51. The base of a triangle is 2 in. longer than the height. The area is $31\frac{1}{2}$ in². Find the height and the base.

52. During a one-hour television show, there were 12 commercials. Some of the commercials were 30 sec long and the others were 60 sec long. If the number of 30-sec commercials was 6 less than the total number of minutes of commercial time during the show, how many 60-sec commercials were used?

53. Elaine can sew a quilt in 6 fewer hours than Gonzalo can. When they work together, it takes them 4 hr. How long would it take each of them alone to sew the quilt?

54. Jackie can paint an apartment in 5.5 hr. Her partner, Grant, can paint the same apartment in 7.5 hr. How long would it take the two of them, working together, to paint the apartment?

Graph.

55. $y > 3x + 5$

56. $f(x) = \frac{2}{3}x - 7$

SYNTHESIS

57. The principle of powers is an "if–then" statement that becomes false when the sentence parts are interchanged. Give an example of another such if–then statement.

58. Is checking essential when the principle of powers is used with an odd power n? Why or why not?

Steel manufacturing. *In the production of steel and other metals, the temperature of the molten metal is so great that conventional thermometers melt. Instead, sound is transmitted across the surface of the metal to a receiver on the far side and the speed of the sound is measured. The formula*

$$S(t) = 1087.7\sqrt{\frac{9t + 2617}{2457}}$$

gives the speed of sound $S(t)$, in feet per second, at a temperature of t degrees Celsius.

59. Find the temperature of a blast furnace where sound travels 1502.3 ft/sec.

60. Find the temperature of a blast furnace where sound travels 1880 ft/sec.

61. Solve the above equation for t.

Automotive repair. *For an engine with a displacement of* 2.8 L, *the function given by*

$$d(n) = 0.75\sqrt{2.8n}$$

can be used to determine the diameter size of the carburetor's opening, in millimeters. Here n is the number of rpm's at which the engine achieves peak performance. (Source: macdizzy.com)

62. If a carburetor's opening is 81 mm, for what number of rpm's will the engine produce peak power?

63. If a carburetor's opening is 84 mm, for what number of rpm's will the engine produce peak power?

Escape velocity. *A formula for the escape velocity v of a satellite is*

$$v = \sqrt{2gr}\sqrt{\frac{h}{r + h}},$$

where g is the force of gravity, r is the planet or star's radius, and h is the height of the satellite above the planet or star's surface.

64. Solve for h.

65. Solve for r.

Sighting to the horizon. *The function $D(h) = 1.2\sqrt{h}$ can be used to approximate the distance D, in miles, that a person can see to the horizon from a height h, in feet.*

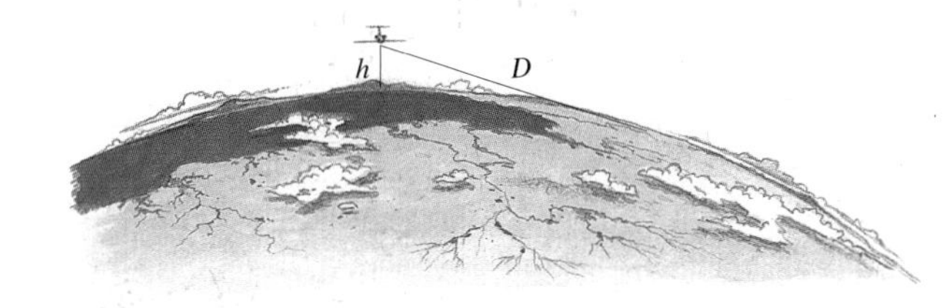

66. How far above sea level must a pilot fly in order to see a horizon that is 180 mi away?

67. How high above sea level must a sailor climb in order to see 10.2 mi out to sea?

Solve.

68. $\dfrac{x + \sqrt{x + 1}}{x - \sqrt{x + 1}} = \dfrac{5}{11}$

69. $\left(\dfrac{z}{4} - 5\right)^{2/3} = \dfrac{1}{25}$

70. $(z^2 + 17)^{3/4} = 27$

71. $\sqrt{\sqrt{y} + 49} = 7$

72. $x^2 - 5x - \sqrt{x^2 - 5x - 2} = 4$
(*Hint*: Let $u = x^2 - 5x - 2$.)

73. $\sqrt{8 - b} = b\sqrt{8 - b}$

Without graphing, determine the x-intercepts of the graphs given by each of the following.

74. $f(x) = \sqrt{x - 2} - \sqrt{x + 2} + 2$

75. $g(x) = 6x^{1/2} + 6x^{-1/2} - 37$

76. $f(x) = (x^2 + 30x)^{1/2} - x - (5x)^{1/2}$

77. Use a grapher to check your answers to Exercises 4, 10, and 26.

78. Saul is trying to solve Exercise 67 using a grapher. Without resorting to trial and error, how can he determine a suitable viewing window for finding the solution?

79. Use a grapher to check your answers to Exercises 21, 29, and 35.

COLLABORATIVE

CORNER

Tailgater Alert

Focus: Radical equations and problem solving

Time: 15–25 minutes

Group size: 2–3

Materials: Calculators or square-root tables

The faster a car is traveling, the more distance it needs to stop. Thus it is important for drivers to allow sufficient space between their vehicle and the vehicle in front of them. Police recommend that for each 10 mph of speed, a driver allow 1 car length. Thus a driver going 30 mph should have at least 3 car lengths between his or her vehicle and the one in front.

In Exercise Set 7.3, the function $r(L) = 2\sqrt{5L}$ was used to find the speed, in miles per hour, that a car was traveling when it left skid marks L feet long.

ACTIVITY

1. Each group member should estimate the length of a car in which he or she frequently travels. (Each should use a different length, if possible.)
2. Using a calculator as needed, each group member should complete the table below.

Column 1 gives a car's speed s, column 2 lists the minimum amount of space between cars traveling s miles per hour, as recommended by police. Column 3 is the speed that a vehicle *could* travel were it forced to stop in the distance listed in column 2, using the above function.

Column 1 s (in miles per hour)	Column 2 $L(s)$ (in feet)	Column 3 $r(L)$ (in miles per hour)
20		
30		
40		
50		
60		
70		

3. Determine whether there are any speeds at which the "1 car length per 10 mph" guideline might not suffice. On what reasoning do you base your answer? Compare tables to determine how car length affects the results. What recommendations would your group make to a new driver?

7.7 Geometric Applications

Using the Pythagorean Theorem • Two Special Triangles

Using the Pythagorean Theorem

There are many kinds of problems that involve powers and roots. Many also involve right triangles and the Pythagorean theorem, which we studied in Section 5.8 and restate here.

*The Pythagorean Theorem**

In any right triangle, if a and b are the lengths of the legs and c is the length of the hypotenuse, then

$$a^2 + b^2 = c^2.$$

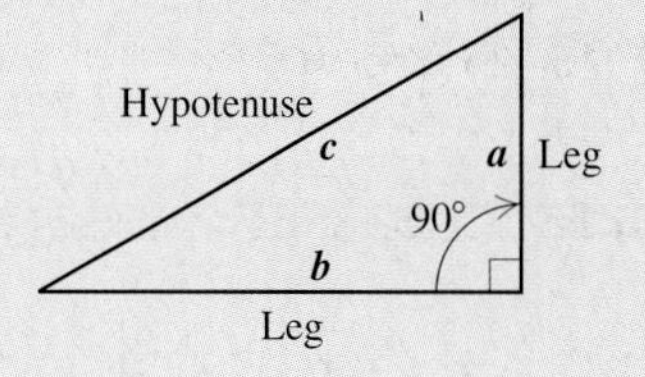

In using the Pythagorean theorem, we often make use of the following principle.

The Principle of Square Roots

For any nonnegative real number n,

If $x^2 = n$, then $x = \sqrt{n}$ or $x = -\sqrt{n}$.

Example 1

Baseball. A baseball diamond is actually a square 90 ft on a side. Suppose a catcher fields a ball along the third-base line 10 ft from home plate. How far would the catcher's throw to first base be? Give an exact answer and an approximation to three decimal places.

Solution We first make a drawing and let $d =$ the distance, in feet, to first base. Note that a right triangle is formed in which the length of the leg from home to first base is 90 ft. The length of the leg from home to where the catcher fields the ball is 10 ft.

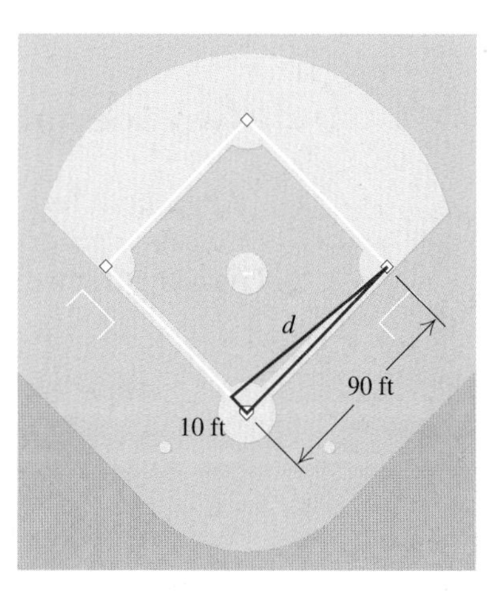

*The converse of the Pythagorean theorem also holds. That is, if a, b, and c are the lengths of the sides of a triangle and $a^2 + b^2 = c^2$, then the triangle is a right triangle.

We substitute these values into the Pythagorean theorem to find d:

$$d^2 = 90^2 + 10^2$$
$$d^2 = 8100 + 100$$
$$d^2 = 8200.$$

We now use the principle of square roots: If $d^2 = 8200$, then $d = \sqrt{8200}$ or $d = -\sqrt{8200}$. In this case, since d is a length, it follows that d is the positive square root of 8200:

$d = \sqrt{8200}$ ft This is an exact answer.

$d \approx 90.6$ ft. Using a calculator for an approximation

Example 2 **Guy wires.** The base of a 40-ft-long guy wire is located 15 ft from the telephone pole that it is anchoring. How high up the pole does the guy wire reach? Give an exact answer and an approximation to three decimal places.

Solution We make a drawing and let h = the height on the pole that the guy wire reaches. A right triangle is formed in which the length of one leg is 15 ft and the length of the hypotenuse is 40 ft. Using the Pythagorean theorem, we have

$$h^2 + 15^2 = 40^2$$
$$h^2 + 225 = 1600$$
$$h^2 = 1375$$
$$h = \sqrt{1375}.$$

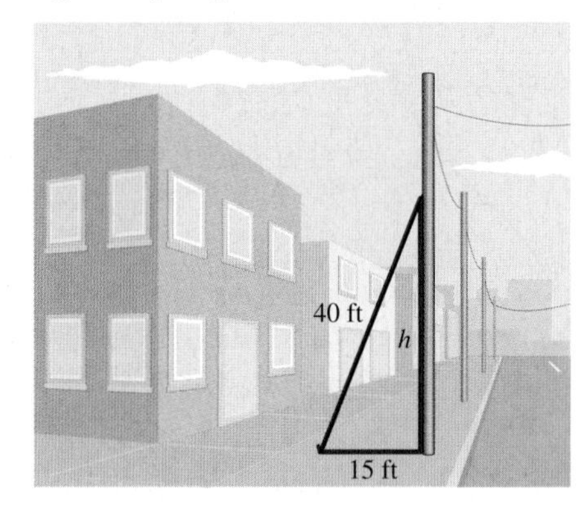

Exact answer:

$h = \sqrt{1375}$ ft

Approximation:

$h \approx 37.081$ ft Using a calculator

Two Special Triangles

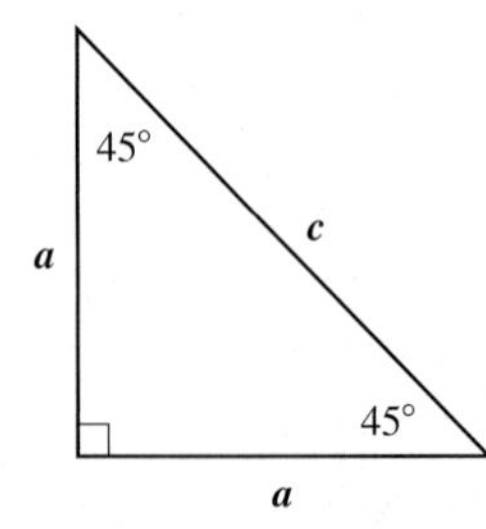

When both legs of a right triangle are the same size, we call the triangle an *isosceles right triangle,* as shown at left. If one leg of an isosceles right triangle has length a, we can find a formula for the length of the hypotenuse as follows:

$c^2 = a^2 + b^2$

$c^2 = a^2 + a^2$ Because the triangle is isosceles, both legs are the same size: $a = b$.

$c^2 = 2a^2.$ Combining like terms

Next, we use the principle of square roots. Because a, b, and c are lengths, there is no need to consider negative square roots or absolute values. Thus,

$c = \sqrt{2a^2}$ Using the principle of square roots

$c = \sqrt{a^2 \cdot 2} = a\sqrt{2}.$

Example 3 One leg of an isosceles right triangle measures 7 cm. Find the length of the hypotenuse. Give an exact answer and an approximation to three decimal places.

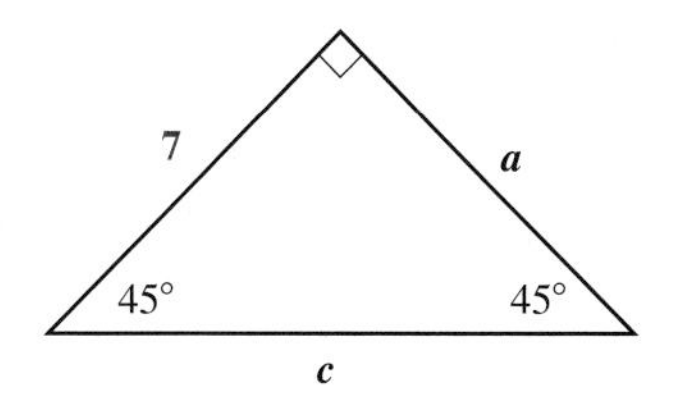

Solution We substitute:

$$c = a\sqrt{2} \qquad \text{This equation is worth memorizing.}$$
$$c = 7\sqrt{2}.$$

Exact answer: $c = 7\sqrt{2}$ cm

Approximation: $c \approx 9.899$ cm Using a calculator

When the hypotenuse of an isosceles right triangle is known, the lengths of the legs can be found.

Example 4 The hypotenuse of an isosceles right triangle is 5 ft long. Find the length of a leg. Give an exact answer and an approximation to three decimal places.

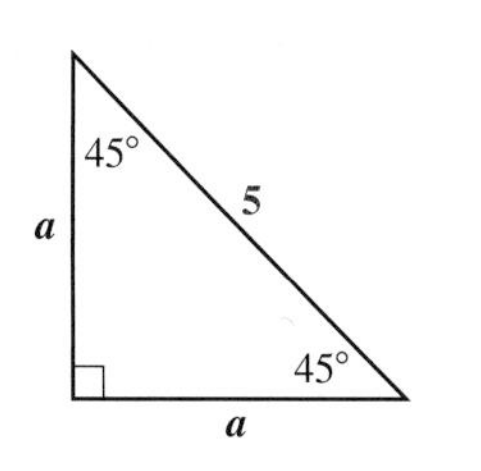

Solution We replace c with 5 and solve for a:

$$5 = a\sqrt{2} \qquad \text{Substituting 5 for } c \text{ in } c = a\sqrt{2}$$
$$\frac{5}{\sqrt{2}} = a \qquad \text{Dividing both sides by } \sqrt{2}$$
$$\frac{5\sqrt{2}}{2} = a. \qquad \text{Rationalize the denominator if desired.}$$

Exact answer: $a = \dfrac{5}{\sqrt{2}}$ ft, or $\dfrac{5\sqrt{2}}{2}$ ft

Approximation: $a \approx 3.536$ ft Using a calculator

A second special triangle is known as a 30°–60°–90° right triangle, so named because of the measures of its angles. Note that in an equilateral triangle, all sides have the same length and all angles are 60°. An altitude, drawn dashed in the figure, bisects, or splits in half, one angle and one side. Two 30°–60°–90° right triangles are thus formed.

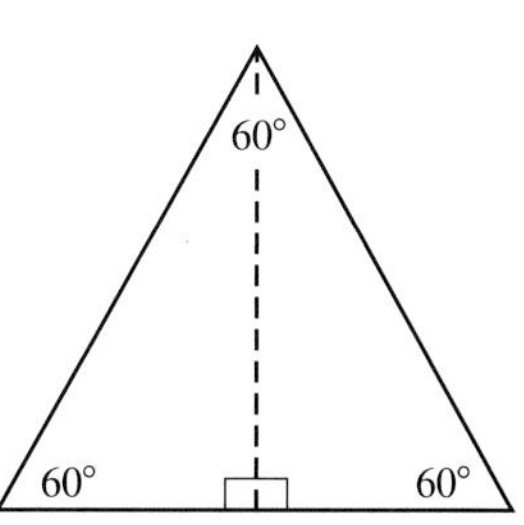

Because of the way in which the altitude is drawn, if a represents the length of the shorter leg in a 30°–60°–90° right triangle, then $2a$ represents the length of the hypotenuse. We have

$$a^2 + b^2 = (2a)^2 \qquad \text{Using the Pythagorean theorem}$$
$$a^2 + b^2 = 4a^2$$
$$b^2 = 3a^2 \qquad \text{Adding } -a^2 \text{ to both sides}$$
$$b = \sqrt{3a^2}$$
$$= \sqrt{a^2 \cdot 3} = a\sqrt{3}.$$

Example 5 The shorter leg of a 30°–60°–90° right triangle measures 8 in. Find the lengths of the other sides. Give exact answers and, where appropriate, an approximation to three decimal places.

Solution The hypotenuse is twice as long as the shorter leg, so we have

$$\begin{aligned} c &= 2a && \text{This relationship is worth memorizing.} \\ &= 2 \cdot 8 = 16 \text{ in.} \end{aligned}$$

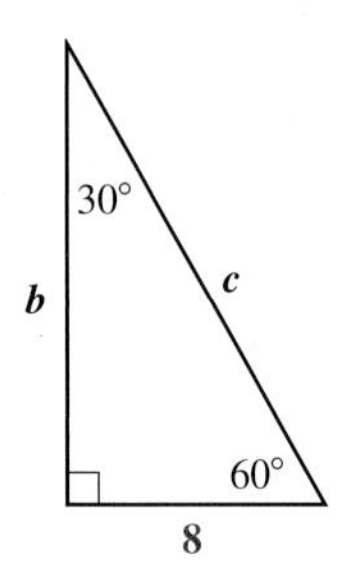

The length of the longer leg is the length of the shorter leg times $\sqrt{3}$. This gives us

$$\begin{aligned} b &= a\sqrt{3} && \text{This is also worth memorizing.} \\ &= 8\sqrt{3} \text{ in.} \end{aligned}$$

Exact answer: $c = 16$ in., $b = 8\sqrt{3}$ in.

Approximation: $b \approx 13.856$ in.

Example 6 The length of the longer leg of a 30°–60°–90° right triangle is 14 cm. Find the length of the hypotenuse. Give an exact answer and an approximation to three decimal places.

Solution The length of the hypotenuse is twice the length of the shorter leg. We first find a, the length of the shorter leg, by using the length of the longer leg:

$$\begin{aligned} 14 &= a\sqrt{3} && \text{Substituting 14 for } b \text{ in } b = a\sqrt{3} \\ \frac{14}{\sqrt{3}} &= a. && \text{Dividing by } \sqrt{3} \end{aligned}$$

Since the hypotenuse is twice as long as the shorter leg, we have

$$\begin{aligned} c &= 2a \\ &= 2 \cdot \frac{14}{\sqrt{3}} && \text{Substituting} \\ &= \frac{28}{\sqrt{3}} \text{ cm.} \end{aligned}$$

Exact answer: $c = \dfrac{28}{\sqrt{3}}$ cm, or $\dfrac{28\sqrt{3}}{3}$ cm if the denominator is rationalized.

Approximation: $c \approx 16.166$ cm

Lengths Within Isosceles and 30°–60°–90° Right Triangles

The length of the hypotenuse in an isosceles right triangle is the length of a leg times $\sqrt{2}$.

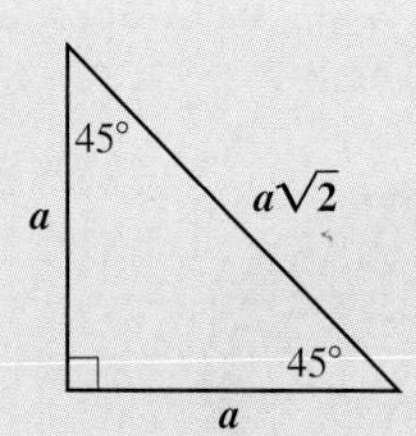

The length of the longer leg in a 30°–60°–90° right triangle is the length of the shorter leg times $\sqrt{3}$. The hypotenuse is twice as long as the shorter leg.

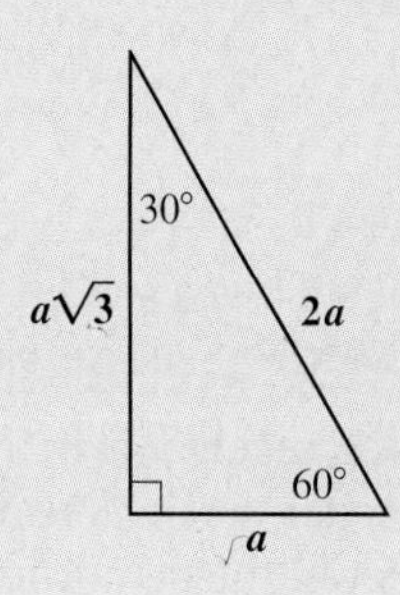

Exercise Set 7.7

FOR EXTRA HELP

Digital Video Tutor CD 6
Videotape 14

InterAct Math

Math Tutor Center

MathXL

MyMathLab.com

In a right triangle, find the length of the side not given. Give an exact answer and, where appropriate, an approximation to three decimal places.

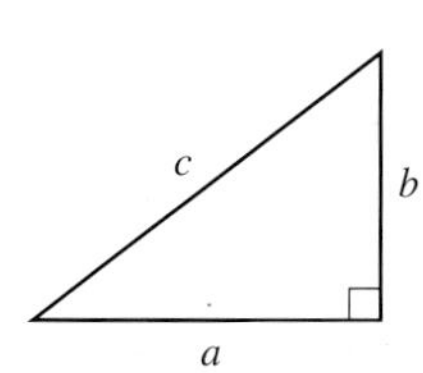

1. $a = 5, b = 3$

2. $a = 8, b = 10$

Aha! **3.** $a = 9, b = 9$

4. $a = 10, b = 10$

5. $b = 12, c = 13$

6. $a = 5, c = 12$

7. $c = 6, a = \sqrt{5}$

8. $c = 8, a = 4\sqrt{3}$

9. $b = 2, c = \sqrt{15}$

10. $a = 1, c = \sqrt{20}$

Aha! **11.** $a = 1, c = \sqrt{2}$

12. $c = 2, a = 1$

In Exercises 13–20, give an exact answer and, where appropriate, an approximation to three decimal places.

13. *Guy wire.* How long is a guy wire if it reaches from the top of a 15-ft pole to a point on the ground 10 ft from the pole?

14. *Softball.* A slow-pitch softball diamond is actually a square 65 ft on a side. How far is it from home to second base?

15. *Baseball.* Suppose the catcher in Example 1 makes a throw to second base from the same location. How far is that throw?

16. *Television sets.* What does it mean to refer to a 20-in. TV set or a 25-in. TV set? Such units refer to the diagonal of the screen. A 20-in. TV set has a width of 16 in. What is its height?

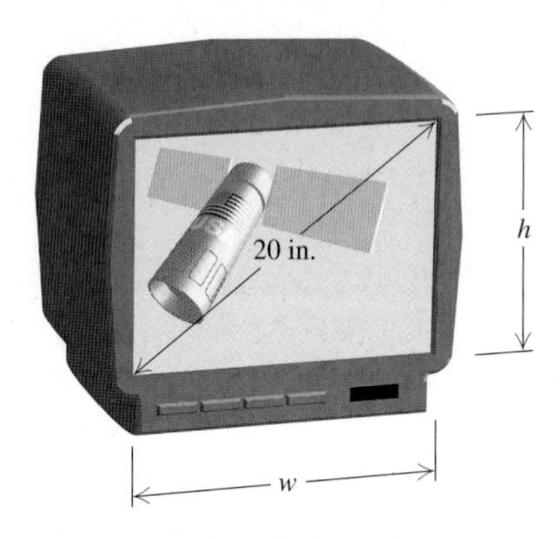

17. *Television sets.* A 25-in. TV set has a screen with a height of 15 in. What is its width? (See Exercise 16.)

18. *Speaker placement.* A stereo receiver is in a corner of a 12-ft by 14-ft room. Speaker wire will run under a rug, diagonally, to a speaker in the far corner. If 4 ft of slack is required on each end, how long a piece of wire should be purchased?

19. *Distance over water.* To determine the width of a pond, a surveyor locates two stakes at either end of the pond and uses instrumentation to place a third stake so that the distance across the pond is the length of a hypotenuse. If the third stake is 90 m from one stake and 70 m from the other, how wide is the pond?

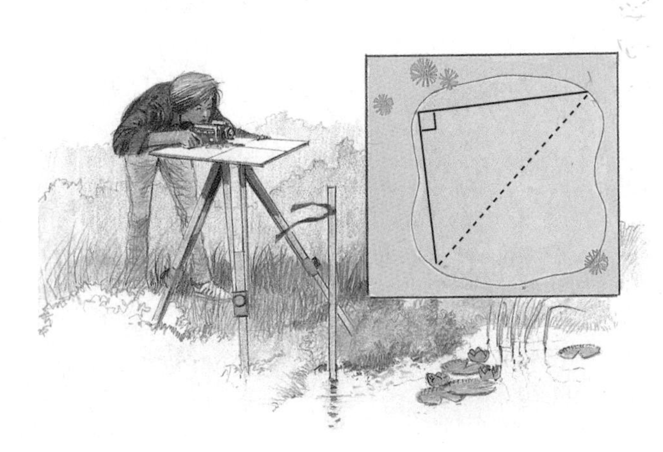

20. *Vegetable garden.* Benito and Dominique are planting a 30-ft by 40-ft vegetable garden and are laying it out using string. They would like to know the length of a diagonal to make sure that right angles are formed. Find the length of a diagonal.

For each triangle, find the missing length(s). Give an exact answer and, where appropriate, an approximation to three decimal places.

21.

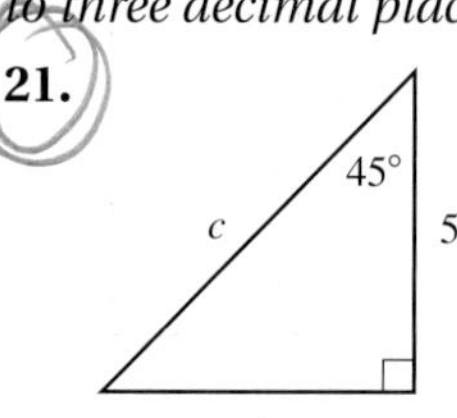

22.

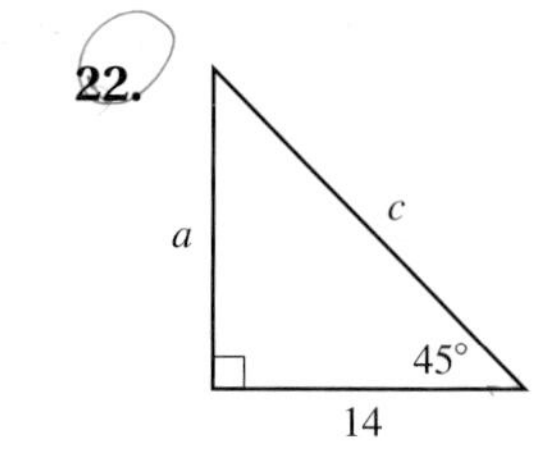

23.

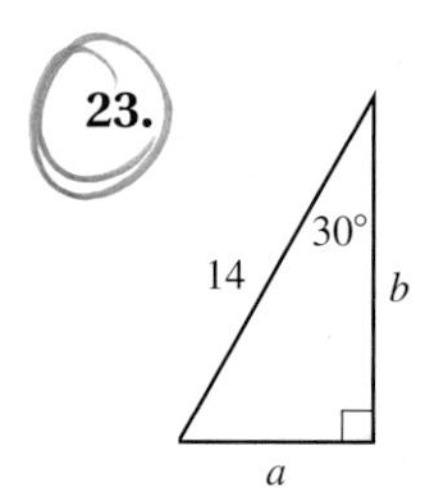

24.

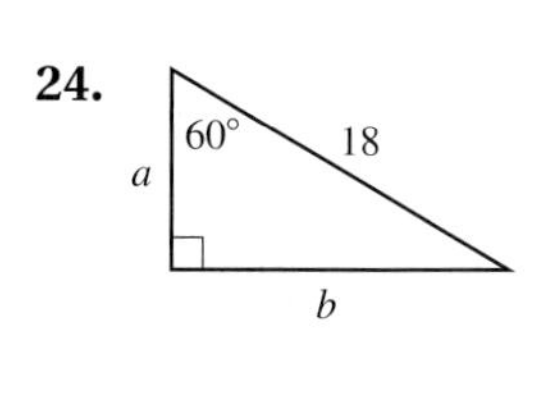

25.

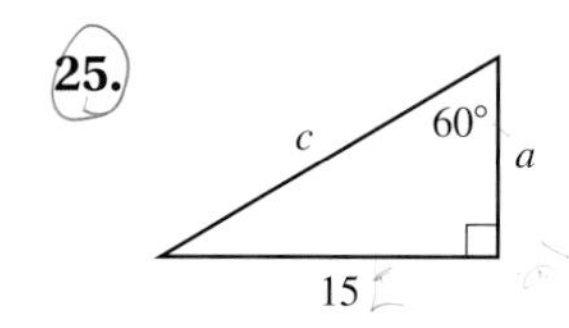

26.

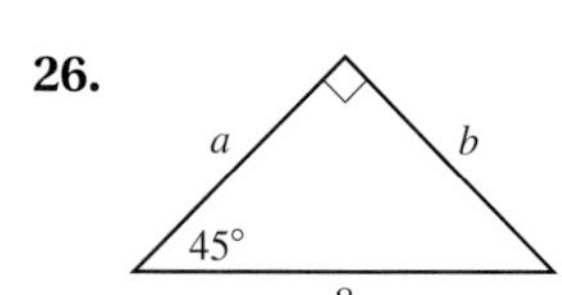

27.

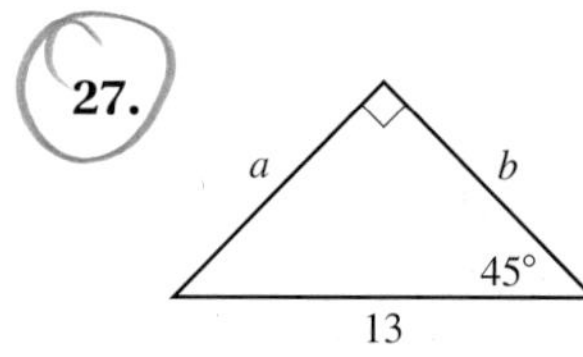

28.

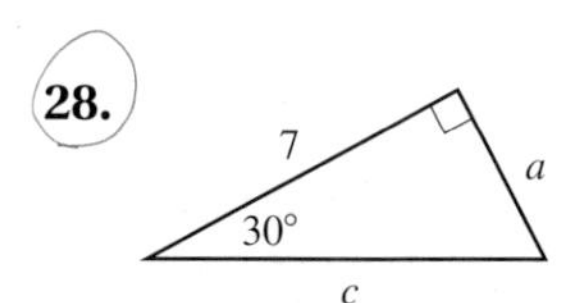

29.

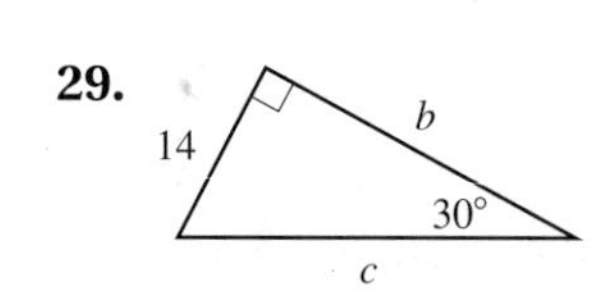

30.

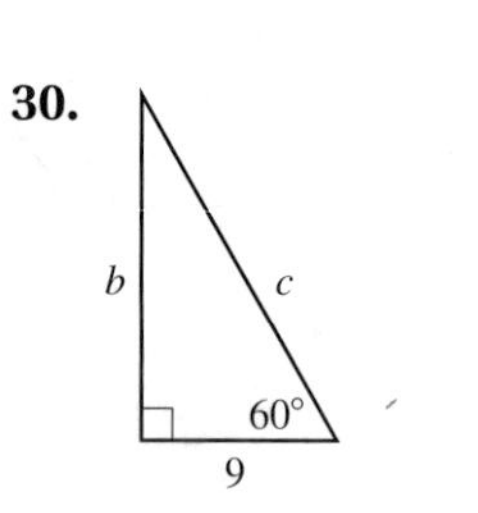

31.

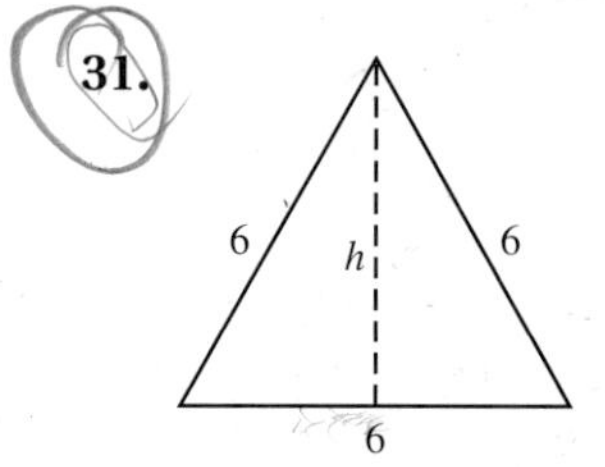

32.

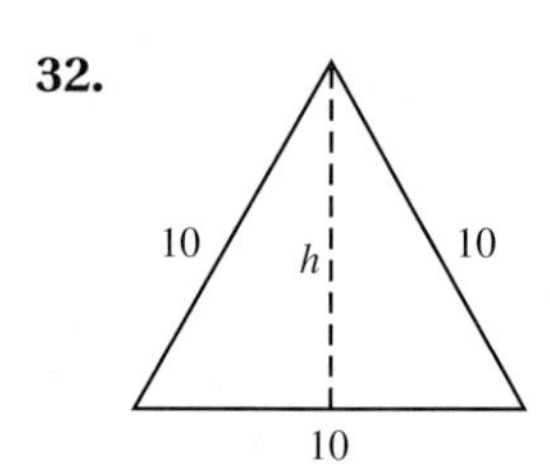

33.

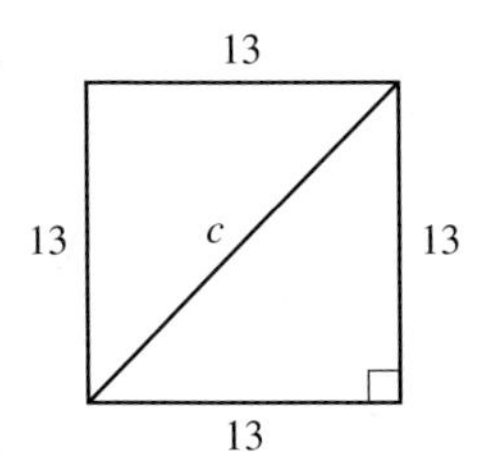

34.

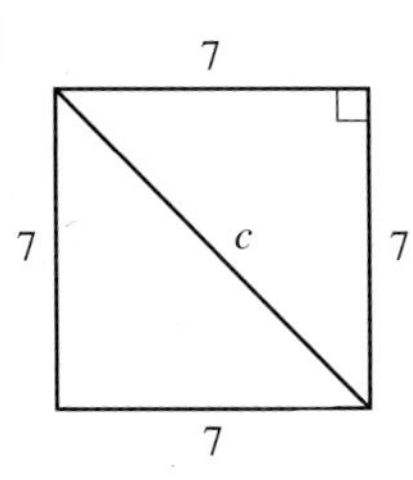

35.

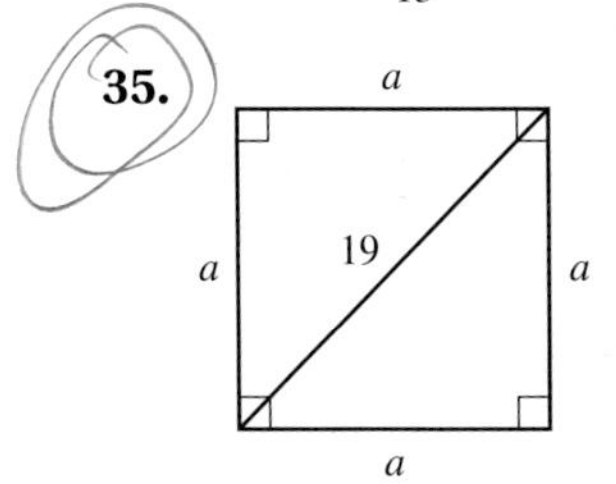

36.

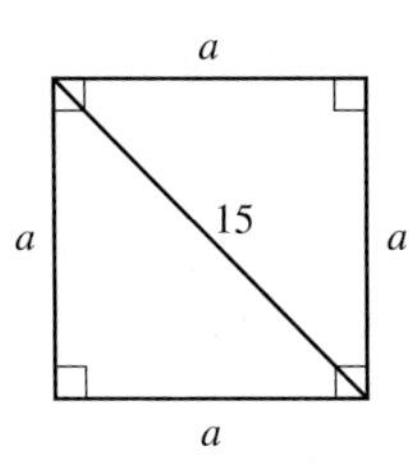

In Exercises 37–42, give an exact answer and, where appropriate, an approximation to three decimal places.

37. *Bridge expansion.* During the summer heat, a 2-mi bridge expands 2 ft in length. If we assume that the bulge occurs straight up the middle, how high is the bulge? (The answer may surprise you. Most bridges have expansion spaces to avoid such buckling.)

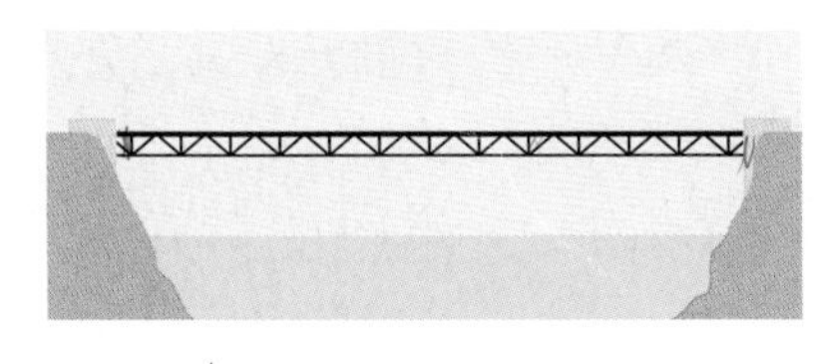

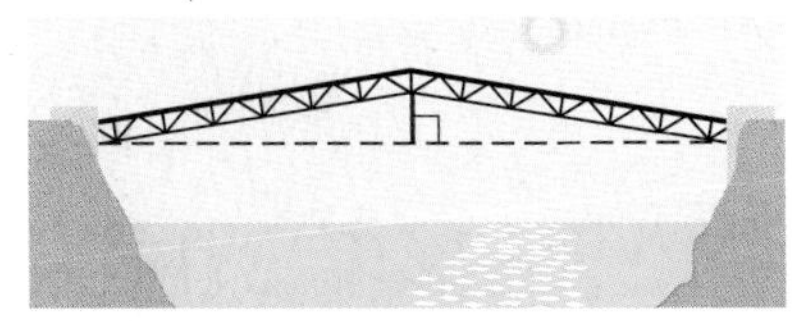

38. Triangle ABC has sides of lengths 25 ft, 25 ft, and 30 ft. Triangle PQR has sides of lengths 25 ft, 25 ft, and 40 ft. Which triangle has the greater area and by how much?

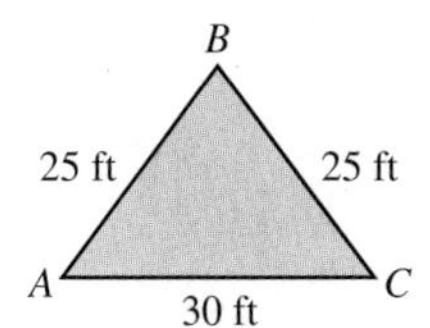

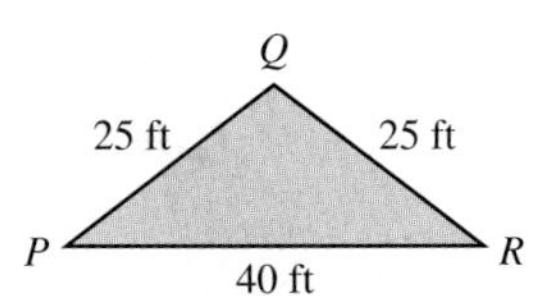

39. *Camping tent.* The entrance to a pup tent is the shape of an equilateral triangle. If the base of the tent is 4 ft wide, how tall is the tent?

40. Each side of a regular octagon has length s. Find a formula for the distance d between the parallel sides of the octagon.

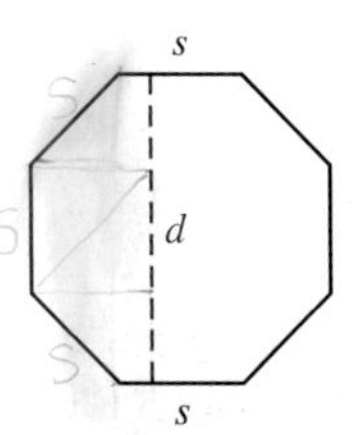

41. The diagonal of a square has length $8\sqrt{2}$ ft. Find the length of a side of the square.

42. The length and the width of a rectangle are given by consecutive integers. The area of the rectangle is 90 cm². Find the length of a diagonal of the rectangle.

43. Find all points on the y-axis of a Cartesian coordinate system that are 5 units from the point (3, 0).

44. Find all points on the x-axis of a Cartesian coordinate system that are 5 units from the point (0, 4).

45. Write a problem for a classmate to solve in which the solution is: "The height of the tepee is $5\sqrt{3}$ yd."

46. Write a problem for a classmate to solve in which the solution is: "The height of the window is $15\sqrt{3}$ ft."

SKILL MAINTENANCE

Simplify.

47. $47(-1)^{19}$

48. $(-5)(-1)^{13}$

Factor.

49. $x^3 - 9x$

50. $7a^3 - 28a$

Solve.

51. $|3x - 5| = 7$

52. $|2x - 3| = |x + 7|$

SYNTHESIS

53. Are there any right triangles, other than those with sides measuring 3, 4, and 5, that have consecutive numbers for the lengths of the sides? Why or why not?

54. If a 30°–60°–90° triangle and an isosceles right triangle have the same perimeter, which will have the greater area? Why?

55. A cube measures 5 cm on each side. How long is the diagonal that connects two opposite corners of the cube? Give an exact answer.

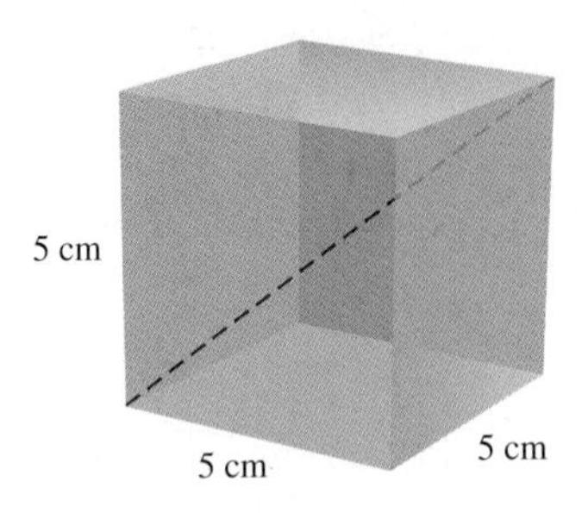

56. *Roofing.* Kit's cottage, which is 24 ft wide and 32 ft long, needs a new roof. By counting clapboards that are 4 in. apart, Kit determines that the peak of the roof is 6 ft higher than the sides. If one packet of shingles covers 100 square feet, how many packets will the job require?

57. *Painting.* (Refer to Exercise 56.) A gallon of paint covers about 275 square feet. If Kit's first floor is 10 ft high, how many gallons of paint should be bought to paint the house? What assumption(s) is made in your answer?

58. *Contracting.* Oxford Builders has an extension cord on their generator that permits them to work, with electricity, anywhere in a circular area of 3850 ft^2. Find the dimensions of the largest square room they could work on without having to relocate the generator to reach each corner of the floor plan.

59. *Contracting.* Cerrelli Construction has an extension cord on their generator that permits them to work, with electricity, anywhere in a circular area of 6160 ft^2. Find the dimensions of the largest cube-shaped room they could work on without having to relocate the generator to reach the corners of the ceiling. Assume that the generator sits on the floor.

7.8 The Complex Numbers

Imaginary and Complex Numbers • Addition and Subtraction • Multiplication • Conjugates and Division • Powers of *i*

Imaginary and Complex Numbers

Negative numbers do not have square roots in the real-number system. However, a larger number system that contains the real-number system is designed so that negative numbers *do* have square roots. That system is called the **complex-number system**, and it makes use of a number that is a square root of -1. We call this new number i.

The Number i

We define the number i such that $i = \sqrt{-1}$ and $i^2 = -1$.

To express roots of negative numbers in terms of i, we can use the fact that in the complex numbers, $\sqrt{-p} = \sqrt{-1}\sqrt{p} = i\sqrt{p}$ or $\sqrt{p}i$, for any positive number p.

Example 1 Express in terms of i: **(a)** $\sqrt{-7}$; **(b)** $\sqrt{-16}$; **(c)** $-\sqrt{-13}$; **(d)** $-\sqrt{-50}$.

Solution

a) $\sqrt{-7} = \sqrt{-1 \cdot 7} = \sqrt{-1} \cdot \sqrt{7} = i\sqrt{7}$, or $\sqrt{7}i$ — i is *not* under the radical.

b) $\sqrt{-16} = \sqrt{-1 \cdot 16} = \sqrt{-1} \cdot \sqrt{16} = i \cdot 4 = 4i$

c) $-\sqrt{-13} = -\sqrt{-1 \cdot 13} = -\sqrt{-1} \cdot \sqrt{13} = -i\sqrt{13}$, or $-\sqrt{13}i$

d) $-\sqrt{-50} = -\sqrt{-1} \cdot \sqrt{25} \cdot \sqrt{2} = -i \cdot 5 \cdot \sqrt{2} = -5i\sqrt{2}$, or $-5\sqrt{2}i$

Imaginary Numbers

An *imaginary number* is a number that can be written in the form $a + bi$, where a and b are real numbers and $b \neq 0$.

Don't let the name "imaginary" fool you. Imaginary numbers appear in fields such as engineering and the physical sciences. The following are examples of imaginary numbers:

$5 + 4i$, Here $a = 5$, $b = 4$.

$\sqrt{5} - \pi i$, Here $a = \sqrt{5}$, $b = -\pi$.

$17i$. Here $a = 0$, $b = 17$.

When a and b are real numbers and b is allowed to be 0, the number $a + bi$ is said to be **complex**.

Complex Numbers

A *complex number* is any number that can be written in the form $a + bi$, where a and b are real numbers. (Note that a and b both can be 0.)

The following are examples of complex numbers:

$7 + 3i$ (here $a \neq 0$, $b \neq 0$); $4i$ (here $a = 0$, $b \neq 0$);

8 (here $a \neq 0$, $b = 0$); 0 (here $a = 0$, $b = 0$).

Complex numbers like $17i$ or $4i$, in which $a = 0$ and $b \neq 0$, are imaginary numbers with no real part. Such numbers are called *pure imaginary numbers.*

Note that when $b = 0$, we have $a + 0i = a$, so every real number is a complex number. The relationships among various real and complex numbers are shown below.

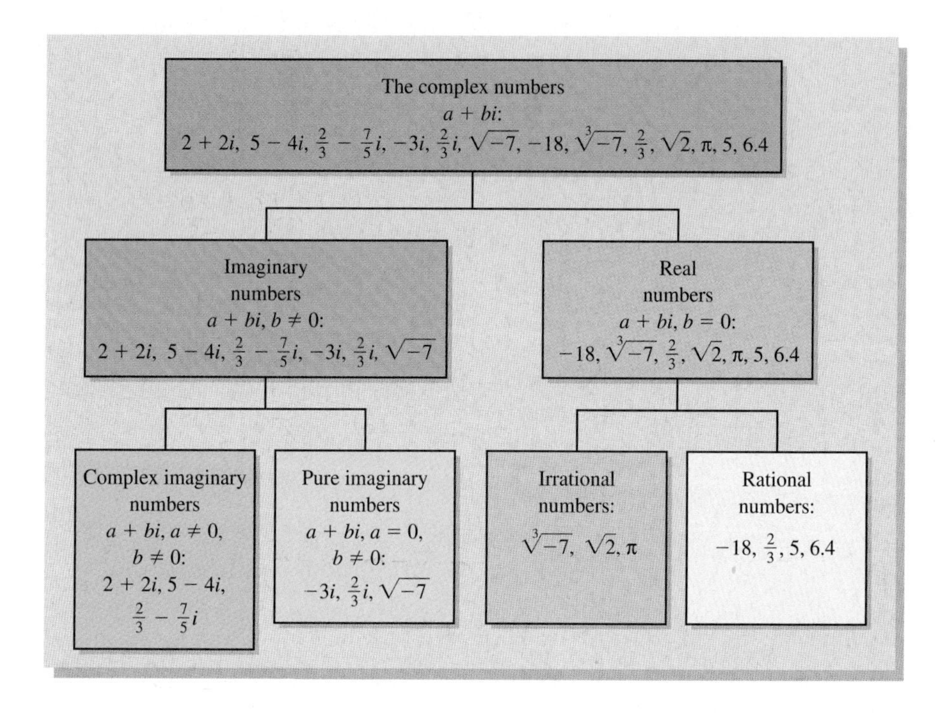

Note that although $\sqrt{-7}$ and $\sqrt[3]{-7}$ are both complex numbers, $\sqrt{-7}$ is imaginary whereas $\sqrt[3]{-7}$ is real.

Addition and Subtraction

The complex numbers obey the commutative, associative, and distributive laws. Thus we can add and subtract them as we do binomials.

Example 2 Add or subtract and simplify.

a) $(8 + 6i) + (3 + 2i)$ **b)** $(4 + 5i) - (6 - 3i)$

Solution

a) $(8 + 6i) + (3 + 2i) = (8 + 3) + (6i + 2i)$ Combining the real parts and the imaginary parts

$= 11 + (6 + 2)i = 11 + 8i$

b) $(4 + 5i) - (6 - 3i) = (4 - 6) + [5i - (-3i)]$ Note that the 6 and the $-3i$ are both being subtracted.

$= -2 + 8i$

Multiplication

For complex numbers, the property $\sqrt{a}\sqrt{b} = \sqrt{ab}$ does *not* hold in general, but it does hold when $a = -1$ and b is nonnegative. To multiply square roots of negative real numbers, we first express them in terms of i. For example,

$$\begin{aligned}\sqrt{-2} \cdot \sqrt{-5} &= \sqrt{-1} \cdot \sqrt{2} \cdot \sqrt{-1} \cdot \sqrt{5} \\ &= i \cdot \sqrt{2} \cdot i \cdot \sqrt{5} \\ &= i^2\sqrt{10} \\ &= -1\sqrt{10} = -\sqrt{10} \text{ is correct!}\end{aligned}$$

> ***Caution!*** With complex numbers, simply multiplying radicands is *incorrect*: $\sqrt{-2} \cdot \sqrt{-5} \neq \sqrt{10}$.

With this in mind, we can now multiply complex numbers.

Example 3 Multiply and simplify. When possible, write answers in the form $a + bi$.

a) $\sqrt{-16} \cdot \sqrt{-25}$ **b)** $\sqrt{-5} \cdot \sqrt{-7}$ **c)** $-3i \cdot 8i$

d) $-4i(3 - 5i)$ **e)** $(1 + 2i)(1 + 3i)$

Solution

a)
$$\begin{aligned}\sqrt{-16} \cdot \sqrt{-25} &= \sqrt{-1} \cdot \sqrt{16} \cdot \sqrt{-1} \cdot \sqrt{25} \\ &= i \cdot 4 \cdot i \cdot 5 \\ &= i^2 \cdot 20 \\ &= -1 \cdot 20 && i^2 = -1 \\ &= -20\end{aligned}$$

b)
$$\begin{aligned}\sqrt{-5} \cdot \sqrt{-7} &= \sqrt{-1} \cdot \sqrt{5} \cdot \sqrt{-1} \cdot \sqrt{7} && \text{Try to do this step mentally.} \\ &= i \cdot \sqrt{5} \cdot i \cdot \sqrt{7} \\ &= i^2 \cdot \sqrt{35} \\ &= -1 \cdot \sqrt{35} && i^2 = -1 \\ &= -\sqrt{35}\end{aligned}$$

c)
$$\begin{aligned}-3i \cdot 8i &= -24 \cdot i^2 \\ &= -24 \cdot (-1) && i^2 = -1 \\ &= 24\end{aligned}$$

d)
$$\begin{aligned}-4i(3 - 5i) &= -4i \cdot 3 + (-4i)(-5i) && \text{Using the distributive law} \\ &= -12i + 20i^2 \\ &= -12i - 20 && i^2 = -1 \\ &= -20 - 12i && \text{Writing in the form } a + bi\end{aligned}$$

e)
$$\begin{aligned}(1 + 2i)(1 + 3i) &= 1 + 3i + 2i + 6i^2 && \text{Multiplying every term of one number by every term of the other (FOIL)} \\ &= 1 + 3i + 2i - 6 && i^2 = -1 \\ &= -5 + 5i && \text{Combining like terms}\end{aligned}$$

Conjugates and Division

Conjugates of complex numbers are defined as follows.

Conjugate of a Complex Number

The *conjugate* of a complex number $a + bi$ is $a - bi$, and the *conjugate* of $a - bi$ is $a + bi$.

Example 4

Find the conjugate.

a) $-3 + 7i$

b) $14 - 5i$

c) $4i$

Solution

a) $-3 + 7i$ The conjugate is $-3 - 7i$.

b) $14 - 5i$ The conjugate is $14 + 5i$.

c) $4i$ The conjugate is $-4i$. Note that $4i = 0 + 4i$.

The product of a complex number and its conjugate is a real number.

Example 5

Multiply: $(5 + 7i)(5 - 7i)$.

Solution

$$\begin{aligned}(5 + 7i)(5 - 7i) &= 5^2 - (7i)^2 && \text{Using } (A + B)(A - B) = A^2 - B^2\\ &= 25 - 49i^2 \\ &= 25 - 49(-1) && i^2 = -1\\ &= 25 + 49 = 74\end{aligned}$$

Conjugates are used when dividing complex numbers. The procedure is much like that used to rationalize denominators in Section 7.5.

Example 6

Divide and simplify to the form $a + bi$.

a) $\dfrac{-5 + 9i}{1 - 2i}$

b) $\dfrac{7 + 3i}{5i}$

Solution

a) To divide and simplify $(-5 + 9i)/(1 - 2i)$, we multiply by 1, using the conjugate of the denominator to form 1:

$$\frac{-5 + 9i}{1 - 2i} = \frac{-5 + 9i}{1 - 2i} \cdot \frac{1 + 2i}{1 + 2i}$$ Multiplying by 1 using the conjugate of the denominator in the symbol for 1

$$= \frac{(-5 + 9i)(1 + 2i)}{(1 - 2i)(1 + 2i)}$$ Multiplying numerators; multiplying denominators

$$= \frac{-5 - 10i + 9i + 18i^2}{1^2 - 4i^2}$$ Using FOIL

$$= \frac{-5 - i - 18}{1 - 4(-1)}$$ $i^2 = -1$

$$= \frac{-23 - i}{5}$$

$$= -\frac{23}{5} - \frac{1}{5}i$$ Writing in the form $a + bi$

b) When the denominator is a pure imaginary number, it is easiest if we multiply by i/i:

$$\frac{7 + 3i}{5i} = \frac{7 + 3i}{5i} \cdot \frac{i}{i}$$ Multiplying by 1 using i/i. We can also use the conjugate of $5i$ to write $-5i/(-5i)$.

$$= \frac{7i + 3i^2}{5i^2}$$ Multiplying

$$= \frac{7i + 3(-1)}{5(-1)}$$ $i^2 = -1$

$$= \frac{7i - 3}{-5}$$

$$= \frac{-3}{-5} + \frac{7}{-5}i, \text{ or } \frac{3}{5} - \frac{7}{5}i.$$

Powers of i

Answers to problems involving complex numbers are generally written in the form $a + bi$. In the following discussion, we show why there is no need to use powers of i (other than 1) when writing answers.

Recall that -1 raised to an *even* power is 1, and -1 raised to an *odd* power is -1. Simplifying powers of i can then be done by using the fact that $i^2 = -1$ and expressing the given power of i in terms of i^2. Consider the following:

$$i, \text{ or } \sqrt{-1},$$
$$i^2 = -1,$$
$$i^3 = i^2 \cdot i = (-1)i = -i,$$
$$i^4 = (i^2)^2 = (-1)^2 = 1,$$
$$\left.\begin{array}{l} i^5 = i^4 \cdot i = (i^2)^2 \cdot i = (-1)^2 \cdot i = i, \\ i^6 = (i^2)^3 = (-1)^3 = -1. \end{array}\right\}$$ The pattern is now repeating.

Note that the powers of i cycle themselves through the values i, -1, $-i$, and 1 and that even powers of i are -1 or 1 whereas odd powers of i are i or $-i$.

Example 7 Simplify: **(a)** i^{18}; **(b)** i^{24}; **(c)** i^{29}; **(d)** i^{75}.

Solution

a) $i^{18} = (i^2)^9$ Using the power rule

$= (-1)^9 = -1$ -1 to an odd power is -1

b) $i^{24} = (i^2)^{12}$ Using the power rule

$= (-1)^{12} = 1$ -1 to an even power is 1

c) $i^{29} = i^{28}i^1$ Using the product rule. This is a key step when i is raised to an odd power.

$= (i^2)^{14}i$ Using the power rule

$= (-1)^{14}i$

$= 1 \cdot i = i$

d) $i^{75} = i^{74}i^1$ Using the product rule

$= (i^2)^{37}i$ Using the power rule

$= (-1)^{37}i$

$= -1 \cdot i = -i$

Exercise Set 7.8

FOR EXTRA HELP

Digital Video Tutor CD 6 Videotape 14

InterAct Math

Math Tutor Center

MathXL

MyMathLab.com

Express in terms of i.

1. $\sqrt{-25}$ **2.** $\sqrt{-36}$ **3.** $\sqrt{-13}$

4. $\sqrt{-19}$ **5.** $\sqrt{-18}$ **6.** $\sqrt{-98}$

7. $\sqrt{-3}$ **8.** $\sqrt{-4}$ **9.** $\sqrt{-81}$

10. $\sqrt{-27}$ **11.** $\sqrt{-300}$ **12.** $-\sqrt{-75}$

13. $-\sqrt{-49}$ **14.** $-\sqrt{-125}$

15. $4 - \sqrt{-60}$ **16.** $6 - \sqrt{-84}$

17. $\sqrt{-4} + \sqrt{-12}$ **18.** $-\sqrt{-76} + \sqrt{-125}$

19. $\sqrt{-72} - \sqrt{-25}$ **20.** $\sqrt{-18} - \sqrt{-100}$

Perform the indicated operation and simplify. Write each answer in the form a + bi.

21. $(7 + 8i) + (5 + 3i)$ **22.** $(4 - 5i) + (3 + 9i)$

23. $(9 + 8i) - (5 + 3i)$ **24.** $(9 + 7i) - (2 + 4i)$

25. $(5 - 3i) - (9 + 2i)$ **26.** $(7 - 4i) - (5 - 3i)$

27. $(-2 + 6i) - (-7 + i)$ **28.** $(-5 - i) - (7 + 4i)$

29. $6i \cdot 9i$ **30.** $7i \cdot 6i$

31. $7i \cdot (-8i)$ **32.** $(-4i)(-6i)$

33. $\sqrt{-49}\sqrt{-25}$ **34.** $\sqrt{-36}\sqrt{-9}$

35. $\sqrt{-6}\sqrt{-7}$ **36.** $\sqrt{-5}\sqrt{-2}$

37. $\sqrt{-15}\sqrt{-10}$ **38.** $\sqrt{-6}\sqrt{-21}$

39. $2i(7 + 3i)$ **40.** $5i(2 + 6i)$

41. $-4i(6 - 5i)$ **42.** $-7i(3 - 4i)$

43. $(1 + 5i)(4 + 3i)$ **44.** $(1 + i)(3 + 2i)$

45. $(5 - 6i)(2 + 5i)$ **46.** $(6 - 5i)(3 + 4i)$

47. $(-4 + 5i)(3 - 4i)$ **48.** $(7 - 2i)(2 - 6i)$

49. $(7 - 3i)(4 - 7i)$ **50.** $(5 - 3i)(4 - 5i)$

51. $(-3 + 6i)(-3 + 4i)$

52. $(-2 + 3i)(-2 + 5i)$

53. $(2 + 9i)(-3 - 5i)$

54. $(-5 - 4i)(3 + 7i)$

55. $(1 - 2i)^2$

56. $(4 - 2i)^2$

57. $(3 + 2i)^2$

58. $(2 + 3i)^2$

59. $(-5 - 2i)^2$

60. $(-2 + 3i)^2$

61. $\dfrac{3}{2 - i}$

62. $\dfrac{4}{3 + i}$

63. $\dfrac{3i}{5 + 2i}$

64. $\dfrac{4i}{5 - 3i}$

65. $\dfrac{7}{9i}$

66. $\dfrac{5}{8i}$

67. $\dfrac{5 - 3i}{4i}$

68. $\dfrac{2 + 7i}{5i}$

Aha! **69.** $\dfrac{7i + 14}{7i}$

70. $\dfrac{6i + 3}{3i}$

71. $\dfrac{4 + 5i}{3 - 7i}$

72. $\dfrac{5 + 3i}{7 - 4i}$

73. $\dfrac{3 - 2i}{4 + 3i}$

74. $\dfrac{5 - 2i}{3 + 6i}$

Simplify.

75. i^7

76. i^{11}

77. i^{24}

78. i^{35}

79. i^{42}

80. i^{64}

81. i^9

82. $(-i)^{71}$

83. $(-i)^6$

84. $(-i)^4$

85. $(5i)^3$

86. $(-3i)^5$

87. $i^2 + i^4$

88. $5i^5 + 4i^3$

89. Is the product of two imaginary numbers always an imaginary number? Why or why not?

90. In what way(s) are conjugates of complex numbers similar to the conjugates used in Section 7.5?

SKILL MAINTENANCE

For Exercises 91–94, let

$$f(x) = x^2 - 3x \quad \text{and} \quad g(x) = 2x - 5.$$

91. Find $(f + g)(-2)$.

92. Find $(f - g)(4)$.

93. Find $(f \cdot g)(5)$.

94. Find $(f/g)(3)$.

Solve.

95. $28 = 3x^2 - 17x$

96. $|3x + 7| < 22$

SYNTHESIS

97. Is the set of real numbers a subset of the complex numbers? Why or why not?

98. Is the union of the set of imaginary numbers and the set of real numbers the set of complex numbers? Why or why not?

A function g is given by

$$g(z) = \frac{z^4 - z^2}{z - 1}.$$

99. Find $g(3i)$.

100. Find $g(1 + i)$.

101. Find $g(5i - 1)$.

102. Find $g(2 - 3i)$.

103. Evaluate

$$\frac{1}{w - w^2} \quad \text{for} \quad w = \frac{1 - i}{10}.$$

Simplify.

104. $\dfrac{i^5 + i^6 + i^7 + i^8}{(1 - i)^4}$

105. $(1 - i)^3(1 + i)^3$

106. $\dfrac{5 - \sqrt{5}i}{\sqrt{5}i}$

107. $\dfrac{6}{1 + \dfrac{3}{i}}$

108. $\left(\dfrac{1}{2} - \dfrac{1}{3}i\right)^2 - \left(\dfrac{1}{2} + \dfrac{1}{3}i\right)^2$

109. $\dfrac{i - i^{38}}{1 + i}$

Summary and Review 7

Key Terms

Square root, p. 408
Principal square root, p. 409
Radical sign, p. 409
Radical expression, p. 409
Radicand, p. 409
Square-root function, p. 410
Cube root, p. 412
*n*th root, p. 413
Index (plural, indices), p. 413
Even root, p. 413
Rational exponent, p. 416
Perfect square, p. 424
Perfect cube, p. 424
Perfect *n*th power, p. 424
Rationalizing, p. 431
Radical term, p. 435
Like radicals, p. 436
Conjugates, p. 437
Radical equation, p. 443
Isosceles right triangle, p. 452
30°–60°–90° right triangle, p. 453
Complex-number system, p. 458
Imaginary number, p. 459
Complex number, p. 459
Pure imaginary number, p. 460
Conjugate of a complex number, p. 462

Important Properties and Formulas

The number c is a square root of a if $c^2 = a$.

The number c is the cube root of a if $c^3 = a$.

For any real number a:

a) $\sqrt[n]{a^n} = |a|$ when n is even. Unless a is known to be nonnegative, absolute-value notation is needed when n is even.

b) $\sqrt[n]{a^n} = a$ when n is odd. Absolute-value notation is not used when n is odd.

$a^{1/n}$ means $\sqrt[n]{a}$. When a is nonnegative, n can be any natural number greater than 1. When a is negative, n must be odd.

For any natural numbers m and n $(n \neq 1)$, and any real number a,

$$a^{m/n} \quad \text{means} \quad \left(\sqrt[n]{a}\right)^m \quad \text{or} \quad \sqrt[n]{a^m}.$$

When a is negative, n must be odd.

For any rational number m/n and any nonzero real number a for which $a^{m/n}$ exists,

$$a^{-m/n} \quad \text{means} \quad \frac{1}{a^{m/n}}.$$

For any real numbers a and b and any rational exponents m and n for which a^m, a^n, and b^m are defined:

1. $a^m \cdot a^n = a^{m+n}$ — In multiplying, add exponents if the bases are the same.

2. $\dfrac{a^m}{a^n} = a^{m-n}$ — In dividing, subtract exponents if the bases are the same. (Assume $a \neq 0$.)

3. $(a^m)^n = a^{m \cdot n}$ — To raise a power to a power, multiply the exponents.

4. $(ab)^m = a^m b^m$ — To raise a product to a power, raise each factor to the power and multiply.

The Product Rule for Radicals

For any real numbers $\sqrt[n]{a}$ and $\sqrt[n]{b}$,

$\sqrt[n]{a}\sqrt[n]{b} = \sqrt[n]{a \cdot b}$.

The Quotient Rule for Radicals

For any real numbers $\sqrt[n]{a}$ and $\sqrt[n]{b}$, $b \neq 0$,

$$\sqrt[n]{\frac{a}{b}} = \frac{\sqrt[n]{a}}{\sqrt[n]{b}}.$$

Some Ways to Simplify Radical Expressions

1. *Simplifying by factoring.* Factor the radicand and look for factors raised to powers that are divisible by the index.

   Example: $\sqrt[3]{a^6b} = \sqrt[3]{a^6}\,\sqrt[3]{b} = a^2\sqrt[3]{b}$

2. *Using rational exponents to simplify.* Convert to exponential notation and then use arithmetic and the laws of exponents to simplify the exponents. Then convert back to radical notation as needed.

   Example: $\sqrt[3]{p} \cdot \sqrt[4]{q^3} = p^{1/3} \cdot q^{3/4}$
   $= p^{4/12} \cdot q^{9/12}$
   $= \sqrt[12]{p^4q^9}$

3. *Combining like radical terms.*

   Example:
   $\sqrt{8} + 3\sqrt{2} = \sqrt{4} \cdot \sqrt{2} + 3\sqrt{2}$
   $= 2\sqrt{2} + 3\sqrt{2} = 5\sqrt{2}$

The Principle of Powers

If $a = b$, then $a^n = b^n$ for any exponent n.

To solve an equation with a radical term:

1. Isolate the radical term on one side of the equation.
2. Use the principle of powers and solve the resulting equation.
3. Check any possible solution in the original equation.

To solve an equation with two or more radical terms:

1. Isolate one of the radical terms.
2. Use the principle of powers.
3. If a radical remains, repeat steps (1) and (2).
4. Solve the resulting equation.
5. Check possible solutions in the original equation.

The Pythagorean Theorem

$a^2 + b^2 = c^2$

The Principle of Square Roots

If $x^2 = n$, then $x = \sqrt{n}$ or $x = -\sqrt{n}$.

Special Triangles

The length of the hypotenuse in an isosceles right triangle is the length of a leg times $\sqrt{2}$.

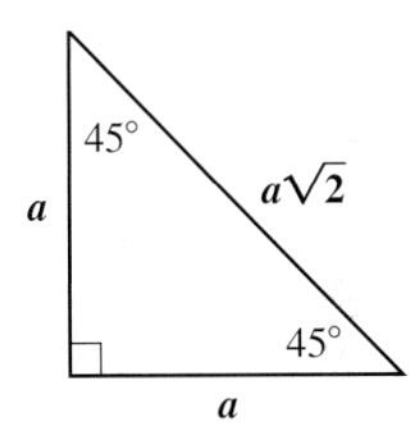

The length of the longer leg in a 30°–60°–90° right triangle is the length of the shorter leg times $\sqrt{3}$. The hypotenuse is twice as long as the shorter leg.

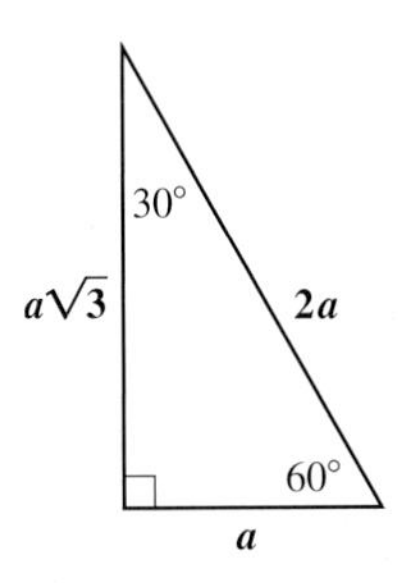

A complex number is any number that can be written in the form $a + bi$, where a and b are real numbers and $i = \sqrt{-1}$.

Review Exercises

Simplify.

1. $\sqrt{\dfrac{49}{36}}$

2. $-\sqrt{0.25}$

Let $f(x) = \sqrt{2x - 7}$. Find the following.

3. $f(16)$

4. The domain of f

Simplify. Assume that each variable can represent any real number.

5. $\sqrt{49a^2}$

6. $\sqrt{(c + 8)^2}$

7. $\sqrt{x^2 - 6x + 9}$

8. $\sqrt{4x^2 + 4x + 1}$

9. $\sqrt[5]{-32}$

10. $\sqrt[3]{-\dfrac{64x^6}{27}}$

11. $\sqrt[4]{x^{12}y^8}$

12. $\sqrt[6]{64x^{12}}$

13. Write an equivalent expression using exponential notation: $(\sqrt[3]{5ab})^4$.

14. Write an equivalent expression using radical notation: $(16a^6)^{3/4}$.

Use rational exponents to simplify. Assume $x, y \geq 0$.

15. $\sqrt{x^6y^{10}}$

16. $(\sqrt[6]{x^2y})^2$

Simplify. Do not use negative exponents in the answers.

17. $(x^{-2/3})^{3/5}$

18. $\dfrac{7^{-1/3}}{7^{-1/2}}$

19. If $f(x) = \sqrt{25(x - 3)^2}$, find a simplified form for $f(x)$.

Perform the indicated operation and, if possible, simplify. Write all answers using radical notation.

20. $\sqrt{5x}\sqrt{3y}$

21. $\sqrt[3]{a^5b}\sqrt[3]{27b}$

22. $\sqrt[3]{-24x^{10}y^8}\sqrt[3]{18x^7y^4}$

23. $\dfrac{\sqrt[3]{60xy^3}}{\sqrt[3]{10x}}$

24. $\dfrac{\sqrt{75x}}{2\sqrt{3}}$

25. $\sqrt[4]{\dfrac{48a^{11}}{c^8}}$

26. $5\sqrt[3]{x} + 2\sqrt[3]{x}$

27. $2\sqrt{75} - 7\sqrt{3}$

28. $\sqrt[3]{8x^4} + \sqrt[3]{xy^6}$

29. $\sqrt{50} + 2\sqrt{18} + \sqrt{32}$

30. $(\sqrt{5} - 3\sqrt{8})(\sqrt{5} + 2\sqrt{8})$

31. $\sqrt[4]{x}\sqrt{x}$

32. $\dfrac{\sqrt[3]{x^2}}{\sqrt[4]{x}}$

33. If $f(x) = x^2$, find $f(a - \sqrt{2})$.

34. Rationalize the denominator:
$$\frac{2\sqrt{3}}{\sqrt{2} + \sqrt{3}}.$$

35. Rationalize the numerator of the expression in Exercise 34.

Solve.

36. $\sqrt{y + 4} - 2 = 3$

37. $(x + 1)^{1/3} = -5$

38. $1 + \sqrt{x} = \sqrt{3x - 3}$

39. If $f(x) = \sqrt[4]{x + 2}$, find a such that $f(a) = 2$.

Solve. Give an exact answer and, where appropriate, an approximation to three decimal places.

40. The diagonal of a square has length 10 cm. Find the length of a side of the square.

41. A bookcase is 5 ft tall and has a 7-ft diagonal brace, as shown. How wide is the bookcase?

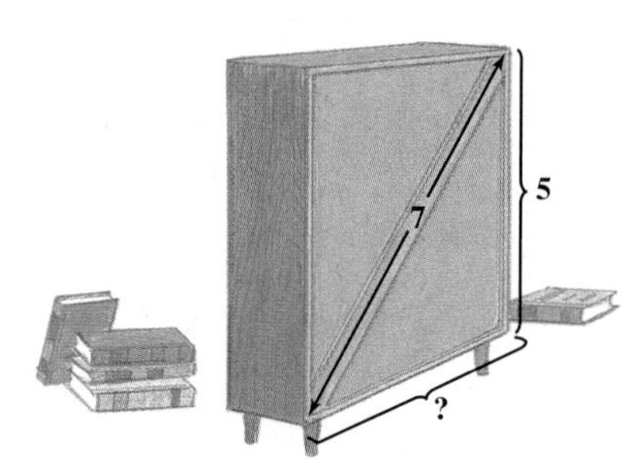

42. Find the missing lengths. Give exact answers and, where appropriate, an approximation to three decimal places.

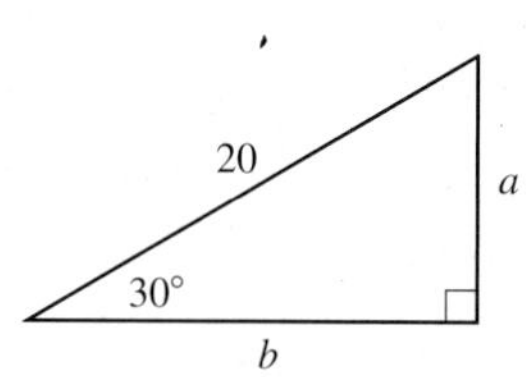

43. Express in terms of i and simplify: $-\sqrt{-8}$.

44. Add: $(-4 + 3i) + (2 - 12i)$.

45. Subtract: $(4 - 7i) - (3 - 8i)$.

Multiply.

46. $(2 + 5i)(2 - 5i)$

47. i^{13}

48. $(6 - 3i)(2 - i)$

49. Divide and simplify to the form $a + bi$:

$$\frac{7 - 2i}{3 + 4i}.$$

SYNTHESIS

50. Explain why $\sqrt[n]{x^n} = |x|$ when n is even, but $\sqrt[n]{x^n} = x$ when n is odd.

51. What is the difference between real numbers and complex numbers?

52. Solve:

$$\sqrt{11x + \sqrt{6 + x}} = 6.$$

53. Simplify:

$$\frac{2}{1 - 3i} - \frac{3}{4 + 2i}.$$

Chapter Test 7

Simplify. Assume that variables can represent any real number.

1. $\sqrt{75}$

2. $\sqrt[3]{-\frac{8}{x^6}}$

3. $\sqrt{100a^2}$

4. $\sqrt{x^2 - 8x + 16}$

5. $\sqrt[5]{x^{12}y^8}$

6. $\sqrt{\frac{25x^2}{36y^4}}$

7. $\sqrt[3]{2x}\,\sqrt[3]{5y^2}$

8. $\frac{\sqrt[5]{x^3y^4}}{\sqrt[5]{xy^2}}$

9. $\sqrt[4]{x^3y^2}\,\sqrt{xy}$

10. $\frac{\sqrt[5]{a^2}}{\sqrt[4]{a}}$

11. $7\sqrt{2} - 2\sqrt{2}$

12. $\sqrt{x^4y} + \sqrt{9y^3}$

13. $(7 + \sqrt{x})(2 - 3\sqrt{x})$

14. Write an equivalent expression using radical notation: $(2a^3b)^{5/6}$.

15. Write an equivalent expression using exponential notation: $\sqrt{7xy}$.

16. If $f(x) = \sqrt{8 - 4x}$, determine the domain of f.

17. If $f(x) = x^2$, find $f(5 + \sqrt{2})$.

18. Rationalize the denominator:

$$\frac{\sqrt{3}}{1 + \sqrt{2}}.$$

Solve.

19. $x = \sqrt{2x - 5} + 4$

20. $\sqrt{x} = \sqrt{x + 1} - 5$

Solve. Give exact answers and, where appropriate, approximations to three decimal places.

21. One leg of an isosceles right triangle is 7 cm long. Find the lengths of the other sides.

22. A referee jogs diagonally from one corner of a 50-ft by 90-ft basketball court to the far corner. How far does she jog? Give an exact answer and an approximation to three decimal places.

23. Express in terms of i and simplify: $\sqrt{-50}$.

24. Subtract: $(7 + 8i) - (-3 + 6i)$.

25. Multiply: $\sqrt{-16}\,\sqrt{-36}$.

26. Multiply. Write the answer in the form $a + bi$.

$$(4 - i)^2$$

27. Divide and simplify to the form $a + bi$:

$$\frac{-3 + i}{2 - 7i}.$$

28. Simplify: i^{37}.

SYNTHESIS

29. Solve:

$$\sqrt{2x - 2} + \sqrt{7x + 4} = \sqrt{13x + 10}.$$

30. Simplify:

$$\frac{1 - 4i}{4i(1 + 4i)^{-1}}.$$

8 Quadratic Functions and Equations

8.1 Quadratic Equations
8.2 The Quadratic Formula
8.3 Applications Involving Quadratic Equations
8.4 Studying Solutions of Quadratic Equations
8.5 Equations Reducible to Quadratic
8.6 Quadratic Functions and Their Graphs
Connecting the Concepts
8.7 More About Graphing Quadratic Functions
8.8 Problem Solving and Quadratic Functions
8.9 Polynomial and Rational Inequalities
Summary and Review
Test

AN APPLICATION

The number of pounds of milk per day recommended for a calf that is x weeks old can be approximated by $p(x)$, where $p(x) = -0.2x^2 + 1.3x + 6.2$ (*Source*: C. Chaloux, University of Vermont, 1998). When is the milk consumption of a calf greatest and how much milk does the calf consume at that time?

This problem appears as Example 1 in Section 8.8.

Determining an animal's nutritional requirements and adjusting dosages of medicine require quick calculations and an understanding of the appropriate formulas.

KAREN ANDERSON
Doctor of Veterinary Medicine
Waitsfield, VT

In translating problem situations to mathematics, we often obtain a function or equation containing a second-degree polynomial in one variable. Such functions or equations are said to be quadratic. *In this chapter, we will study a variety of equations, inequalities, and applications for which we will need to solve quadratic equations or graph quadratic functions.*

8.1 Quadratic Equations

The Principle of Square Roots • Completing the Square

In Section 5.8, we solved quadratic equations like $3x^2 = 2 - x$ by factoring. Let's review that procedure.

Example 1 Solve: $3x^2 = 2 - x$.

Solution To use the principle of zero products, we must first have zero on one side of the equation. We then factor:

$$3x^2 = 2 - x$$

$$3x^2 + x - 2 = 0 \quad \text{Adding } -2 + x \text{ to both sides to obtain standard form}$$

$$(3x - 2)(x + 1) = 0 \quad \text{Factoring}$$

$$3x - 2 = 0 \quad or \quad x + 1 = 0 \quad \text{Using the principle of zero products}$$

$$3x = 2 \quad or \quad x = -1$$

$$x = \tfrac{2}{3} \quad or \quad x = -1.$$

Check: For $\frac{2}{3}$:

$$\begin{array}{c|l} 3x^2 & = 2 - x \\ \hline 3\left(\frac{2}{3}\right)^2 & ?\ 2 - \frac{2}{3} \\ 3 \cdot \frac{4}{9} & \frac{6}{3} - \frac{2}{3} \\ \frac{4}{3} & \frac{4}{3} \quad \text{TRUE} \end{array}$$

For -1:

$$\begin{array}{c|l} 3x^2 & = 2 - x \\ \hline 3(-1)^2 & ?\ 2 - (-1) \\ 3 \cdot 1 & 2 + 1 \\ 3 & 3 \quad \text{TRUE} \end{array}$$

The solutions are -1 and $\frac{2}{3}$.

Example 2 Solve: $x^2 = 25$.

Solution We have

$$x^2 = 25$$

$$x^2 - 25 = 0 \quad \text{Writing in standard form}$$

$$(x - 5)(x + 5) = 0 \quad \text{Factoring}$$

$$x - 5 = 0 \quad or \quad x + 5 = 0 \quad \text{Using the principle of zero products}$$

$$x = 5 \quad or \quad x = -5.$$

The solutions are 5 and -5. The checks are left to the student.

The Principle of Square Roots

Consider the equation $x^2 = 25$ again. We know from Chapter 7 that the number 25 has two real-number square roots, namely, 5 and -5. Note that these are the solutions of the equation in Example 2. Thus square roots can provide a quick method for solving equations of the type $x^2 = k$.

The Principle of Square Roots

For any real number k, if $x^2 = k$, then $x = \sqrt{k}$ or $x = -\sqrt{k}$.

Example 3

Solve: $3x^2 = 6$.

Solution We have

$$3x^2 = 6$$
$$x^2 = 2 \qquad \text{Multiplying by } \tfrac{1}{3}$$
$$x = \sqrt{2} \quad or \quad x = -\sqrt{2}. \qquad \text{Using the principle of square roots}$$

We often use the symbol $\pm\sqrt{2}$ to represent the two numbers $\sqrt{2}$ and $-\sqrt{2}$. We check as follows.

Check: For $\sqrt{2}$:

$$\begin{array}{c|c} 3x^2 & 6 \\ \hline 3(\sqrt{2})^2 & 6 \\ 3 \cdot 2 & \\ 6 & 6 \quad \text{TRUE} \end{array}$$

For $-\sqrt{2}$:

$$\begin{array}{c|c} 3x^2 & 6 \\ \hline 3(-\sqrt{2})^2 & 6 \\ 3 \cdot 2 & \\ 6 & 6 \quad \text{TRUE} \end{array}$$

The solutions are $\sqrt{2}$ and $-\sqrt{2}$, or $\pm\sqrt{2}$.

Sometimes we rationalize denominators to simplify answers, although this is not as common as it once was.

Example 4

Solve: $-5x^2 + 2 = 0$.

Solution We have

$$-5x^2 + 2 = 0$$
$$x^2 = \frac{2}{5} \qquad \text{Isolating } x^2$$
$$x = \sqrt{\frac{2}{5}} \quad or \quad x = -\sqrt{\frac{2}{5}}. \qquad \text{Using the principle of square roots}$$

The solutions are $\sqrt{\frac{2}{5}}$ and $-\sqrt{\frac{2}{5}}$. This can also be written as $\pm\sqrt{\frac{2}{5}}$, or, if we rationalize the denominator, $\pm\frac{\sqrt{10}}{5}$. The checks are left to the student.

Sometimes we get solutions that are imaginary numbers.

Example 5

Solve: $4x^2 + 9 = 0$.

Solution We have

$$4x^2 + 9 = 0$$
$$x^2 = -\tfrac{9}{4} \qquad \text{Isolating } x^2$$
$$x = \sqrt{-\tfrac{9}{4}} \quad or \quad x = -\sqrt{-\tfrac{9}{4}} \qquad \text{Using the principle of square roots}$$
$$x = \sqrt{\tfrac{9}{4}}\sqrt{-1} \quad or \quad x = -\sqrt{\tfrac{9}{4}}\sqrt{-1}$$
$$x = \tfrac{3}{2}i \quad or \quad x = -\tfrac{3}{2}i.$$

Check: Since the solutions are opposites and the equation has an x^2-term and no x-term, we can check both solutions at once.

$$\begin{array}{r|l} 4x^2 + 9 & 0 \\ \hline 4(\pm\tfrac{3}{2}i)^2 + 9 \ ? & 0 \\ 4 \cdot \tfrac{9}{4} \cdot i^2 + 9 & \\ 9(-1) + 9 & \\ 0 & 0 \ \text{TRUE} \end{array}$$

The solutions are $\tfrac{3}{2}i$ and $-\tfrac{3}{2}i$, or $\pm\tfrac{3}{2}i$.

The principle of square roots can be restated in a more general form that pertains to more complicated algebraic expressions than just x.

> ***The Principle of Square Roots (Generalized Form)***
>
> For any real number k and any algebraic expression X,
>
> If $X^2 = k$, then $X = \sqrt{k}$ or $X = -\sqrt{k}$.

Example 6

Let $f(x) = (x - 2)^2$. Find all x-values for which $f(x) = 7$.

Solution We are asked to find all x-values for which

$$f(x) = 7,$$

or

$$(x - 2)^2 = 7. \qquad \text{Substituting } (x-2)^2 \text{ for } f(x)$$

The generalized principle of square roots gives us

$$x - 2 = \sqrt{7} \quad or \quad x - 2 = -\sqrt{7} \qquad \text{Replacing } X \text{ with } x - 2$$
$$x = 2 + \sqrt{7} \quad or \quad x = 2 - \sqrt{7}.$$

Check: $f(2 + \sqrt{7}) = (2 + \sqrt{7} - 2)^2 = (\sqrt{7})^2 = 7.$

Similarly,

$$f(2 - \sqrt{7}) = (2 - \sqrt{7} - 2)^2 = (-\sqrt{7})^2 = 7.$$

The solutions are $2 + \sqrt{7}$ and $2 - \sqrt{7}$, or $2 \pm \sqrt{7}$.

In Example 6, one side of the equation is the square of a binomial and the other side is a constant. Sometimes an equation must be factored in order to appear in this form.

Example 7 Solve: $x^2 + 6x + 9 = 2$.

Solution We have

$x^2 + 6x + 9 = 2$ The left side is the square of a binomial.

$(x + 3)^2 = 2$ Factoring

$x + 3 = \sqrt{2}$ *or* $x + 3 = -\sqrt{2}$ Using the principle of square roots

$x = -3 + \sqrt{2}$ *or* $x = -3 - \sqrt{2}$. Adding -3 to both sides

The solutions are $-3 + \sqrt{2}$ and $-3 - \sqrt{2}$, or $-3 \pm \sqrt{2}$. The checks are left to the student.

Completing the Square

By using a method called *completing the square,* we can use the principle of square roots to solve *any* quadratic equation.

Example 8 Solve: $x^2 + 6x + 4 = 0$.

Solution We have

$x^2 + 6x + 4 = 0$

$x^2 + 6x = -4$ Subtracting 4 from both sides

$x^2 + 6x + 9 = -4 + 9$ Adding 9 to both sides. We explain this shortly.

$(x + 3)^2 = 5$ Factoring the perfect-square trinomial

$x + 3 = \pm\sqrt{5}$ Using the principle of square roots. Remember that $\pm\sqrt{5}$ represents two numbers.

$x = -3 \pm \sqrt{5}$. Adding -3 to both sides

technology connection A

One way to check Example 8 is to store $-3 + \sqrt{5}$ as x using the STO→ key. We can then evaluate $x^2 + 6x + 4$ by entering $x^2 + 6x + 4$ and ENTER.

1. Check Example 8 using the method described above.

Check: For $-3 + \sqrt{5}$:

$$
\begin{array}{c|l}
x^2 + 6x + 4 & = 0 \\
\hline
(-3 + \sqrt{5})^2 + 6(-3 + \sqrt{5}) + 4 & ?\ 0 \\
9 - 6\sqrt{5} + 5 - 18 + 6\sqrt{5} + 4 & \\
9 + 5 - 18 + 4 - 6\sqrt{5} + 6\sqrt{5} & \\
0 & 0 \quad \text{TRUE}
\end{array}
$$

For $-3 - \sqrt{5}$:

$$
\begin{array}{c|l}
x^2 + 6x + 4 & = 0 \\
\hline
(-3 - \sqrt{5})^2 + 6(-3 - \sqrt{5}) + 4 & ?\ 0 \\
9 + 6\sqrt{5} + 5 - 18 - 6\sqrt{5} + 4 & \\
9 + 5 - 18 + 4 + 6\sqrt{5} - 6\sqrt{5} & \\
0 & 0 \quad \text{TRUE}
\end{array}
$$

The solutions are $-3 + \sqrt{5}$ and $-3 - \sqrt{5}$, or $-3 \pm \sqrt{5}$.

Let's examine how the above solutions were found. The decision to add 9 to both sides in Example 8 was made because it creates a perfect-square trinomial on the left side. The 9 was determined by taking half of the coefficient of x and squaring it—that is,

$$\left(\tfrac{1}{2} \cdot 6\right)^2 = 3^2, \quad \text{or} \quad 9.$$

To help see why this procedure works, examine the following drawings.

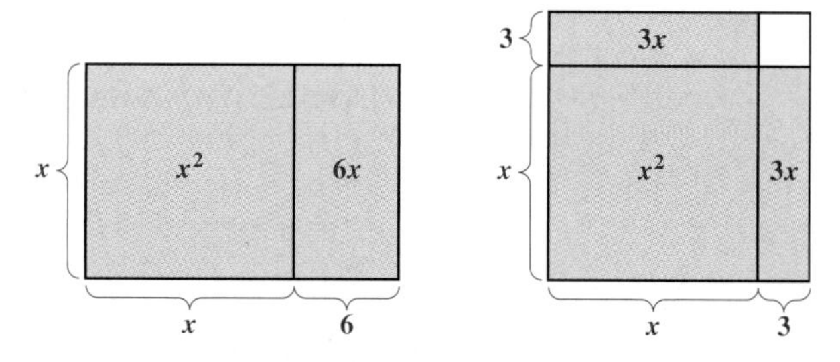

Note that the shaded areas in both figures represent the same area, $x^2 + 6x$. However, only the figure on the right, in which the $6x$ is halved, can be converted into a square with the addition of a constant term. The constant term, 9, can be interpreted as the "missing" piece of the diagram on the right. It *completes* the square.

To complete the square for $x^2 + bx$, we add $(b/2)^2$.

Example 9, which follows, is one of the few examples in this text for which we are neither solving an equation nor writing an equivalent expression. Instead we are simply gaining practice in finding numbers that complete the square. The trinomial that we create is *not* equivalent to the original binomial.

Example 9

Complete the square. Then write the trinomial in factored form.

a) $x^2 + 14x$
b) $x^2 - 5x$
c) $x^2 + \frac{3}{4}x$

Solution

a) We take half of the coefficient of x and square it.

$$x^2 + 14x \quad \longrightarrow \text{Half of 14 is 7, and } 7^2 = 49. \text{ We add 49.}$$

Thus, $x^2 + 14x + 49$ is a perfect-square trinomial. It is equivalent to $(x + 7)^2$. We must add 49 in order for $x^2 + 14x$ to become a perfect-square trinomial.

b) We take half of the coefficient of x and square it:

$$x^2 - 5x \quad \longrightarrow \frac{1}{2} \cdot (-5) = -\frac{5}{2}, \quad \text{and} \quad \left(-\frac{5}{2}\right)^2 = \frac{25}{4}.$$

Thus, $x^2 - 5x + \frac{25}{4}$ is a perfect-square trinomial. It is equivalent to $\left(x - \frac{5}{2}\right)^2$. Note that for purposes of factoring, it is best *not* to convert $\frac{25}{4}$ to decimal notation.

c) We take half of the coefficient of x and square it:

$$x^2 + \frac{3}{4}x \quad \longrightarrow \frac{1}{2} \cdot \frac{3}{4} = \frac{3}{8}, \quad \text{and} \quad \left(\frac{3}{8}\right)^2 = \frac{9}{64}.$$

Thus, $x^2 + \frac{3}{4}x + \frac{9}{64}$ is a perfect-square trinomial. It is equivalent to $\left(x + \frac{3}{8}\right)^2$.

We can now use the method of completing the square to solve equations similar to Example 8.

Example 10

Solve: **(a)** $x^2 - 8x - 7 = 0$; **(b)** $x^2 + 5x - 3 = 0$.

Solution

a)

$x^2 - 8x - 7 = 0$	
$x^2 - 8x = 7$	Adding 7 to both sides. We can now complete the square on the left side.
$x^2 - 8x + 16 = 7 + 16$	Adding 16 to both sides to complete the square: $\frac{1}{2}(-8) = -4$, and $(-4)^2 = 16$
$(x - 4)^2 = 23$	Factoring
$x - 4 = \pm\sqrt{23}$	Using the principle of square roots
$x = 4 \pm \sqrt{23}$	Adding 4 to both sides

The solutions are $4 - \sqrt{23}$ and $4 + \sqrt{23}$, or $4 \pm \sqrt{23}$. The checks are left to the student.

b) $x^2 + 5x - 3 = 0$

$x^2 + 5x = 3$ Adding 3 to both sides

$x^2 + 5x + \frac{25}{4} = 3 + \frac{25}{4}$ Completing the square: $\frac{1}{2} \cdot 5 = \frac{5}{2}$, and $\left(\frac{5}{2}\right)^2 = \frac{25}{4}$

$\left(x + \frac{5}{2}\right)^2 = \frac{37}{4}$ Factoring and simplifying

$x + \frac{5}{2} = \pm\frac{\sqrt{37}}{2}$ Using the principle of square roots and the quotient rule for radicals

$x = \frac{-5 \pm \sqrt{37}}{2}$ Adding $-\frac{5}{2}$ to both sides

The checks are left to the student. The solutions are $(-5 - \sqrt{37})/2$ and $(-5 + \sqrt{37})/2$, or $(-5 \pm \sqrt{37})/2$.

Before we can complete the square, the coefficient of x^2 must be 1. When it is not 1, we divide both sides of the equation by whatever that coefficient may be.

Example 11 Solve: $3x^2 + 7x - 2 = 0$.

Solution We have

$3x^2 + 7x - 2 = 0$

$3x^2 + 7x = 2$ Adding 2 to both sides

$x^2 + \frac{7}{3}x = \frac{2}{3}$ Dividing both sides by 3

$x^2 + \frac{7}{3}x + \frac{49}{36} = \frac{2}{3} + \frac{49}{36}$ Completing the square: $\left(\frac{1}{2} \cdot \frac{7}{3}\right)^2 = \frac{49}{36}$

$\left(x + \frac{7}{6}\right)^2 = \frac{73}{36}$ Factoring and simplifying

$x + \frac{7}{6} = \pm\frac{\sqrt{73}}{6}$ Using the principle of square roots and the quotient rule for radicals

$x = \frac{-7 \pm \sqrt{73}}{6}$. Adding $-\frac{7}{6}$ to both sides

The solutions are

$$\frac{-7 - \sqrt{73}}{6} \text{ and } \frac{-7 + \sqrt{73}}{6}, \text{ or } \frac{-7 \pm \sqrt{73}}{6}.$$

The checks are left to the student.

The procedure used in Example 11 is important because it can be used to solve *any* quadratic equation.

To Solve a Quadratic Equation in x by Completing the Square

1. Isolate the terms with variables on one side of the equation, and arrange them in descending order.
2. Divide both sides by the coefficient of x^2 if that coefficient is not 1.
3. Complete the square by taking half of the coefficient of x and adding its square to both sides.
4. Express one side as the square of a binomial and simplify the other side.
5. Use the principle of square roots.
6. Solve for x by adding or subtracting on both sides.

Problem Solving

If you put money in a savings account, the bank will pay you interest. As interest is paid into your account, the bank will start paying you interest on both the original amount and the interest already earned. This is called **compounding interest**. If interest is paid yearly, we say that it is **compounded annually**.

The Compound-Interest Formula

If an amount of money P is invested at interest rate r, compounded annually, then in t years, it will grow to the amount A given by

$$A = P(1 + r)^t.$$

We can use quadratic equations to solve certain interest problems.

Example 12

Investment growth. Rosa invested \$4000 at interest rate r, compounded annually. In 2 yr, it grew to \$4410. What was the interest rate?

Solution

1. **Familiarize.** We are already familiar with the compound-interest formula. If we were not, we would need to consult an outside source.
2. **Translate.** The translation consists of substituting into the formula:

$$A = P(1 + r)^t$$
$$4410 = 4000(1 + r)^2. \quad \text{Substituting}$$

3. **Carry out.** We solve for r:

$$4410 = 4000(1 + r)^2$$

$$\frac{4410}{4000} = (1 + r)^2 \quad \text{Dividing both sides by 4000}$$

$$\frac{441}{400} = (1 + r)^2 \quad \text{Simplifying}$$

$$\pm\sqrt{\frac{441}{400}} = 1 + r \quad \text{Using the principle of square roots}$$

$$\pm\frac{21}{20} = 1 + r \quad \text{Simplifying}$$

$$-\frac{20}{20} \pm \frac{21}{20} = r \quad \text{Adding } -1\text{, or } -\frac{20}{20}\text{, to both sides}$$

$$\frac{1}{20} = r \quad or \quad -\frac{41}{20} = r.$$

4. **Check.** Since the interest rate cannot be negative, we need only check $\frac{1}{20}$, or 5%. If \$4000 were invested at 5% interest, compounded annually, then in 2 yr it would grow to $4000(1.05)^2$, or \$4410. The number 5% checks.

5. **State.** The interest rate was 5%.

Example 13

Free-falling objects. The formula $s = 16t^2$ is used to approximate the distance s, in feet, that an object falls freely from rest in t seconds. The RCA Building in New York City is 850 ft tall. How long will it take an object to fall from the top?

Solution

1. **Familiarize.** We make a drawing to help visualize the problem.

2. **Translate.** We substitute into the formula:

$$s = 16t^2$$

$$850 = 16t^2.$$

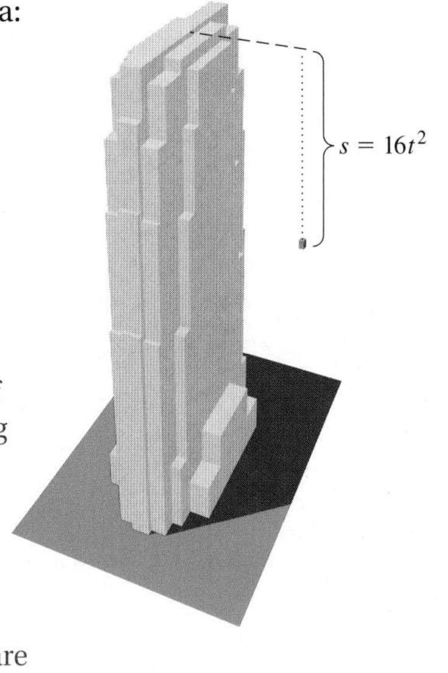

3. **Carry out.** We solve for t:

$$850 = 16t^2$$

$$\frac{850}{16} = t^2$$

$$53.125 = t^2$$

$$\sqrt{53.125} = t \quad \text{Using the principle of square roots; rejecting the negative square root since } t \text{ cannot be negative in this problem}$$

$$7.3 \approx t. \quad \text{Using a calculator to approximate the square root and rounding to the nearest tenth}$$

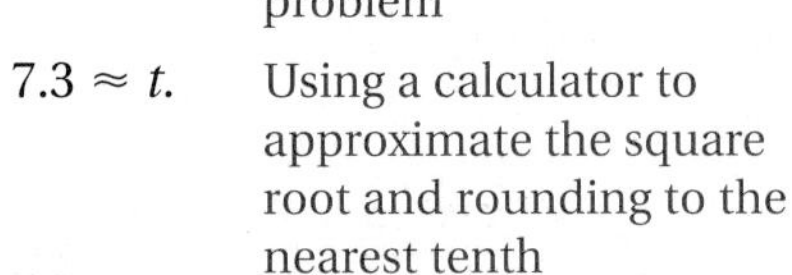

4. **Check.** Since $16(7.3)^2 = 852.64 \approx 850$, our answer checks.

5. **State.** It takes about 7.3 sec for an object to fall freely from the top of the RCA Building.

As we saw in Section 5.8, a grapher can be used to find approximate solutions of any quadratic equation that has real-number solutions.

To check Example 10(a), we graph $y = x^2 - 8x - 7$ and use the ZERO or ROOT option of the CALC menu. When asked for a Left and Right Bound, we enter cursor positions to the left of and to the right of the root. A Guess between the bounds is entered and a value for the root then appears.

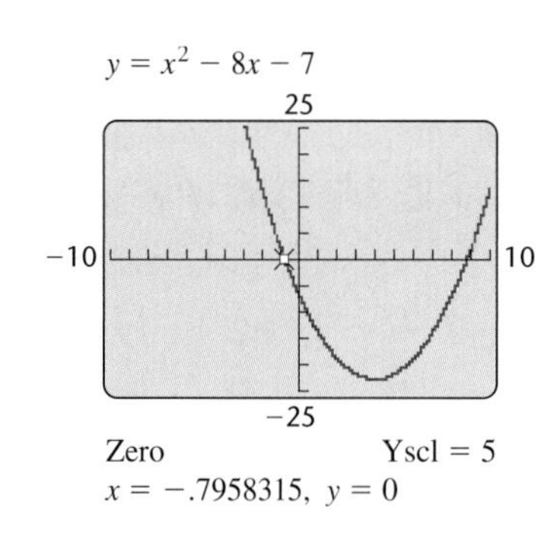

1. Use a grapher to check the second solution of Example 10(a).
2. Use a grapher to confirm the solutions in Examples 8 and 10(b).
3. Can a grapher be used to find *exact* solutions in Example 11? Why or why not?
4. Use a grapher to confirm that there are no real-number solutions in Example 5.

Exercise Set 8.1

FOR EXTRA HELP

Digital Video Tutor CD 6
Videotape 15

InterAct Math

Math Tutor Center

MathXL

MyMathLab.com

Solve.

1. $7x^2 = 21$

2. $4x^2 = 20$

3. $25x^2 + 4 = 0$

4. $9x^2 + 16 = 0$

5. $3t^2 - 2 = 0$

6. $5t^2 - 7 = 0$

7. $(x + 2)^2 = 25$

8. $(x - 1)^2 = 49$

9. $(a + 5)^2 = 8$

10. $(a - 13)^2 = 18$

11. $(x - 1)^2 = -49$

12. $(x + 1)^2 = -9$

13. $\left(t + \frac{3}{2}\right)^2 = \frac{7}{2}$

14. $\left(y + \frac{3}{4}\right)^2 = \frac{17}{16}$

15. $x^2 - 6x + 9 = 100$

16. $x^2 - 10x + 25 = 64$

17. Let $f(x) = (x - 5)^2$. Find x such that $f(x) = 16$.

18. Let $g(x) = (x - 2)^2$. Find x such that $g(x) = 25$.

19. Let $F(t) = (t + 4)^2$. Find t such that $F(t) = 13$.

20. Let $f(t) = (t + 6)^2$. Find t such that $f(t) = 15$.

Aha! **21.** Let $g(x) = x^2 + 14x + 49$. Find x such that $g(x) = 49$.

22. Let $F(x) = x^2 + 8x + 16$. Find x such that $F(x) = 9$.

Complete the square. Then write the perfect-square trinomial in factored form.

23. $x^2 + 8x$

24. $x^2 + 16x$

25. $x^2 - 6x$

26. $x^2 - 10x$

27. $x^2 - 24x$

28. $x^2 - 18x$

29. $t^2 + 9t$

30. $t^2 + 3t$

31. $x^2 - 3x$

32. $x^2 - 7x$

33. $x^2 + \frac{2}{3}x$

34. $x^2 + \frac{2}{5}x$

35. $t^2 - \frac{5}{3}t$

36. $t^2 - \frac{5}{6}t$

37. $x^2 + \frac{9}{5}x$

38. $x^2 + \frac{9}{4}x$

Solve by completing the square. Show your work.

39. $x^2 + 6x = 7$
40. $x^2 + 8x = 9$
41. $x^2 - 10x = 22$
42. $x^2 - 4x = -9$
43. $x^2 + 8x + 7 = 0$
44. $x^2 + 10x + 9 = 0$
45. $x^2 - 10x + 21 = 0$
46. $x^2 - 10x + 24 = 0$
47. $t^2 + 5t + 3 = 0$
48. $t^2 + 6t + 7 = 0$
49. $x^2 + 10 = 6x$
50. $x^2 + 23 = 10x$
51. $s^2 + 4s + 13 = 0$
52. $t^2 + 12t + 25 = 0$

Solve by completing the square. Remember to first divide, as in Example 11, to make sure that the coefficient of x^2 is 1.

53. $2x^2 - 5x - 3 = 0$
54. $3x^2 + 5x - 2 = 0$
55. $4x^2 + 8x + 3 = 0$
56. $9x^2 + 18x + 8 = 0$
57. $6x^2 - x = 15$
58. $6x^2 - x = 2$
59. $2x^2 + 4x + 1 = 0$
60. $2x^2 + 5x + 2 = 0$
61. $3x^2 - 5x - 3 = 0$
62. $4x^2 - 6x - 1 = 0$

Interest. *Use $A = P(1 + r)^t$ to find the interest rate in Exercises 63–68. Refer to Example 12.*

63. \$2000 grows to \$2420 in 2 yr

64. \$2560 grows to \$2890 in 2 yr

65. \$1280 grows to \$1805 in 2 yr

66. \$1000 grows to \$1440 in 2 yr

67. \$6250 grows to \$6760 in 2 yr

68. \$6250 grows to \$7290 in 2 yr

Free-falling objects. *Use $s = 16t^2$ for Exercises 69–72. Refer to Example 13.*

69. The CN Tower in Toronto, at 1815 ft, is the world's tallest self-supporting tower (no guy wires) (*Source*: *The Guinness Book of Records*). How long would it take an object to fall freely from the top?

70. Reaching 745 ft above the water, the towers of California's Golden Gate Bridge are the world's tallest bridge towers (*Source*: *The Guinness Book of Records*). How long would it take an object to fall freely from the top?

71. The Gateway Arch in St. Louis is 640 ft high. How long would it take an object to fall freely from the top?

72. The Sears Tower in Chicago is 1454 ft tall. How long would it take an object to fall freely from the top?

73. Explain in your own words a sequence of steps that can be used to solve any quadratic equation in the quickest way.

74. Write an interest-rate problem for a classmate to solve. Devise the problem so that the solution is "The loan was made at 7% interest."

SKILL MAINTENANCE

Evaluate.

75. $at^2 - bt$, for $a = 3$, $b = 5$, and $t = 4$

76. $mn^2 - mp$, for $m = -2$, $n = 7$, and $p = 3$

Simplify.

77. $\sqrt[3]{270}$
78. $\sqrt{80}$

Let $f(x) = \sqrt{3x - 5}$.

79. Find $f(10)$.
80. Find $f(18)$.

SYNTHESIS

81. What would be better: to receive 3% interest every 6 months, or to receive 6% interest every 12 months? Why?

82. Write a problem involving a free-falling object for a classmate to solve (see Example 13). Devise the problem so that the solution is "The object takes about 4.5 sec to fall freely from the top of the structure."

Find b such that each trinomial is a square.

83. $x^2 + bx + 81$
84. $x^2 + bx + 49$

85. If $f(x) = 2x^5 - 9x^4 - 66x^3 + 45x^2 + 280x$ and $x^2 - 5$ is a factor of $f(x)$, find all a for which $f(a) = 0$.

86. If $f(x) = \left(x - \frac{1}{3}\right)(x^2 + 6)$ and $g(x) = \left(x - \frac{1}{3}\right)\left(x^2 - \frac{2}{3}\right)$, find all a for which $(f + g)(a) = 0$.

87. *Boating.* A barge and a fishing boat leave a dock at the same time, traveling at a right angle to each other. The barge travels 7 km/h slower than the fishing boat. After 4 hr, the boats are 68 km apart. Find the speed of each boat.

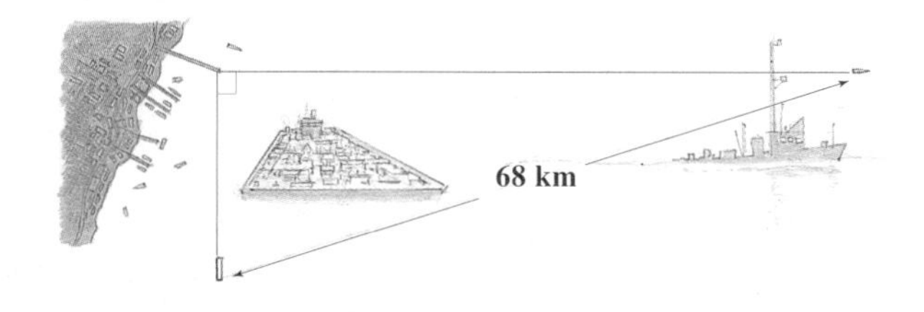

88. Find three consecutive integers such that the square of the first plus the product of the other two is 67.

89. Exercises 17, 21, and 41 can be solved on a grapher without first rewriting in standard form. Simply let y_1 represent the left side of the equation and y_2 the right side. Then use a grapher to determine the x-coordinate of any point of intersection. Use a grapher to solve Exercises 17, 21, and 41 in this manner.

90. Use a grapher to check your answers to Exercises 5, 45, 59, and 61.

91. Example 12 can be solved with a grapher by graphing each side of

$$4410 = 4000(1 + r)^2.$$

How could you determine, from a reading of the problem, a suitable viewing window? What might that window be?

The Quadratic Formula

8.2

Solving Using the Quadratic Formula • Approximating Solutions

There are at least two reasons for learning to complete the square. One is to enhance your ability to graph certain equations that appear later in this and other chapters. Another is to develop a general formula for solving quadratic equations.

Solving Using the Quadratic Formula

Each time you solve by completing the square, the procedure is the same. In mathematics, when a procedure is repeated many times, a formula is often developed to speed up our work.

We begin with a quadratic equation in standard form,

$$ax^2 + bx + c = 0,$$

with $a > 0$. For $a < 0$, a slightly different derivation is needed (see Exercise 52), but the result is the same. Let's solve by completing the square. As the steps are performed, compare them with Example 11 on p. 478.

$$ax^2 + bx = -c \qquad \text{Adding } -c \text{ to both sides}$$

$$x^2 + \frac{b}{a}x = -\frac{c}{a} \qquad \text{Dividing both sides by } a$$

Half of $\frac{b}{a}$ is $\frac{b}{2a}$ and $\left(\frac{b}{2a}\right)^2$ is $\frac{b^2}{4a^2}$. We add $\frac{b^2}{4a^2}$ to both sides:

$$x^2 + \frac{b}{a}x + \frac{b^2}{4a^2} = -\frac{c}{a} + \frac{b^2}{4a^2}$$ Adding $\frac{b^2}{4a^2}$ to complete the square

$$\left(x + \frac{b}{2a}\right)^2 = -\frac{4ac}{4a^2} + \frac{b^2}{4a^2}$$

$$\left(x + \frac{b}{2a}\right)^2 = \frac{b^2 - 4ac}{4a^2}$$ Factoring on the left side; finding a common denominator on the right side

$$x + \frac{b}{2a} = \pm\frac{\sqrt{b^2 - 4ac}}{2a}$$ Using the principle of square roots and the quotient rule for radicals; since $a > 0$, $\sqrt{4a^2} = 2a$

$$x = \frac{-b \pm \sqrt{b^2 - 4ac}}{2a}.$$ Adding $-\frac{b}{2a}$ to both sides

It is important that you remember the quadratic formula and know how to use it.

The Quadratic Formula

The solutions of $ax^2 + bx + c = 0$, $a \neq 0$, are given by

$$x = \frac{-b \pm \sqrt{b^2 - 4ac}}{2a}.$$

Example 1 Solve $5x^2 + 8x = -3$ using the quadratic formula.

Solution We first find standard form and determine a, b, and c:

$$5x^2 + 8x + 3 = 0;$$ Adding 3 to both sides to get 0 on one side

$$a = 5, \quad b = 8, \quad c = 3.$$

Next, we use the quadratic formula:

$$x = \frac{-b \pm \sqrt{b^2 - 4ac}}{2a}$$

$$x = \frac{-8 \pm \sqrt{8^2 - 4 \cdot 5 \cdot 3}}{2 \cdot 5}$$ Substituting

Be sure to write the fraction bar all the way across.

$$x = \frac{-8 \pm \sqrt{64 - 60}}{10}$$

$$x = \frac{-8 \pm \sqrt{4}}{10} = \frac{-8 \pm 2}{10}$$

$$x = \frac{-8 + 2}{10} \quad or \quad x = \frac{-8 - 2}{10}$$

$$x = \frac{-6}{10} \quad or \quad x = \frac{-10}{10}$$

$$x = -\frac{3}{5} \quad or \quad x = -1.$$

The solutions are $-\frac{3}{5}$ and -1. The checks are left to the student.

Because $5x^2 + 8x + 3$ can be factored as $(5x + 3)(x + 1)$, the quadratic formula may not have been the fastest way of solving Example 1. However, because the quadratic formula works for *any* quadratic equation, we need not spend too much time struggling to solve a quadratic equation by factoring.

To Solve a Quadratic Equation

1. If the equation can be easily written in the form $ax^2 = p$ or $(x + k)^2 = d$, use the principle of square roots as in Section 8.1.
2. If step (1) does not apply, write the equation in the form $ax^2 + bx + c = 0$.
3. Try factoring and using the principle of zero products.
4. If factoring seems to be difficult or impossible, use the quadratic formula.

The solutions of a quadratic equation can always be found using the quadratic formula. They cannot always be found by factoring.

Recall that a second-degree polynomial in one variable is said to be quadratic. Similarly, a second-degree polynomial function in one variable is said to be a **quadratic function**.

Example 2 For the quadratic function given by $f(x) = 5x^2 - 8x - 3$, find all x for which $f(x) = 0$.

Solution We substitute and solve for x:

$$f(x) = 0$$

$$5x^2 - 8x - 3 = 0 \qquad \text{Substituting. This cannot be solved by factoring.}$$

$$a = 5, \quad b = -8, \quad c = -3.$$

We then substitute into the quadratic formula:

$$x = \frac{-(-8) \pm \sqrt{(-8)^2 - 4 \cdot 5 \cdot (-3)}}{2 \cdot 5}$$

$$= \frac{8 \pm \sqrt{64 + 60}}{10}$$

$$= \frac{8 \pm \sqrt{124}}{10}. \qquad \text{Note that 4 is a perfect-square factor of 124.}$$

A

On many graphers, it is possible to check Example 2 by graphing $y_1 = 5x^2 - 8x - 3$, pressing **TRACE**, and entering $(4 + \sqrt{(31)})/5$. A rational approximation and the y-value 0 appear.

$y = 5x^2 - 8x - 3$

10

−5 5

−10

$x = 1.9135529, \ y = 0$

Use this approach to check the other solution of Example 2.

Thus,

$$x = \frac{8 \pm \sqrt{4 \cdot 31}}{10} \qquad 124 = 4 \cdot 31$$

$$= \frac{8 \pm 2\sqrt{31}}{10} \qquad \sqrt{4} = 2$$

$$= \frac{2(4 \pm \sqrt{31})}{2 \cdot 5} = \frac{4 \pm \sqrt{31}}{5}. \qquad \text{Removing a factor equal to 1: } \frac{2}{2} = 1$$

Caution! To avoid a common error, *factor the numerator and the denominator* when removing a factor equal to 1.

The solutions are

$$\frac{4 + \sqrt{31}}{5} \quad \text{and} \quad \frac{4 - \sqrt{31}}{5}.$$

The checks are left to the student.

Some quadratic equations have solutions that are imaginary numbers.

Example 3

Solve: $x^2 + 2 = -x$.

Solution We first find standard form:

$$x^2 + x + 2 = 0. \qquad \text{Adding } x \text{ to both sides}$$

Since we cannot factor $x^2 + x + 2$, we use the quadratic formula with $a = 1$, $b = 1$, and $c = 2$:

$$x = \frac{-1 \pm \sqrt{1^2 - 4 \cdot 1 \cdot 2}}{2 \cdot 1} \qquad \text{Substituting}$$

$$= \frac{-1 \pm \sqrt{1 - 8}}{2}$$

$$= \frac{-1 \pm \sqrt{-7}}{2}$$

$$= \frac{-1 \pm i\sqrt{7}}{2}, \text{ or } -\frac{1}{2} \pm \frac{\sqrt{7}}{2}i.$$

The solutions are $-\frac{1}{2} - \frac{\sqrt{7}}{2}i$ and $-\frac{1}{2} + \frac{\sqrt{7}}{2}i$. The checks are left to the student.

The quadratic formula is sometimes used to solve equations that do not originally appear to be quadratic.

Example 4 If $f(x) = 2 + \frac{7}{x}$ and $g(x) = \frac{4}{x^2}$, find all x for which $f(x) = g(x)$.

We saw in Sections 5.8 and 8.1 how graphers can solve quadratic equations. To determine whether quadratic equations are solved more quickly on a grapher or by using the quadratic formula, solve Examples 2 and 4 both ways. Which method is faster? Which method is more precise? Why?

Solution We set $f(x)$ equal to $g(x)$ and solve:

$$f(x) = g(x)$$

$$2 + \frac{7}{x} = \frac{4}{x^2}. \quad \text{Substituting. Note that } x \neq 0.$$

This is a rational equation similar to those in Section 6.4. To solve, we multiply both sides by the LCD, x^2:

$$x^2\left(2 + \frac{7}{x}\right) = x^2 \cdot \frac{4}{x^2}$$

$$2x^2 + 7x = 4 \quad \text{Simplifying}$$

$$2x^2 + 7x - 4 = 0. \quad \text{Subtracting 4 from both sides}$$

We have

$$a = 2, \quad b = 7, \quad \text{and} \quad c = -4.$$

Substituting then gives us

$$x = \frac{-7 \pm \sqrt{7^2 - 4 \cdot 2 \cdot (-4)}}{2 \cdot 2}$$

$$= \frac{-7 \pm \sqrt{49 + 32}}{4}$$

$$= \frac{-7 \pm \sqrt{81}}{4}$$

$$= \frac{-7 \pm 9}{4}$$

$$x = \frac{-7 + 9}{4} = \frac{1}{2} \quad \text{or} \quad x = \frac{-7 - 9}{4} = -4. \quad \text{Both answers should check since } x \neq 0.$$

You can confirm that $f(\frac{1}{2}) = g(\frac{1}{2})$ and $f(-4) = g(-4)$. The solutions are $\frac{1}{2}$ and -4.

Checking the solutions of Examples 2 and 3 can be cumbersome. Fortunately, when the quadratic formula is used to solve a quadratic equation, the results will always check in that equation provided the formula has been properly used. Thus checking for computational errors is usually sufficient.

Approximating Solutions

When the solution of an equation is irrational, a rational-number approximation is often useful. This is often the case in real-world applications similar to those found in Section 8.3.

Example 5 Use a calculator to approximate the solutions of Example 2.

Solution On most calculators, one of the following sequences of keystrokes can be used to approximate $(4+\sqrt{31})/5$:

(4 + √ 31)) ÷ 5 ENTER ; or

31 √ + 4 = ÷ 5 = .

Similar keystrokes can be used to approximate $(4-\sqrt{31})/5$.

The solutions of Example 2 are approximately 1.913552873 and −0.3135528726.

Exercise Set 8.2

FOR EXTRA HELP

MyMathLab.com

Solve.

1. $x^2+7x-3=0$
2. $x^2-7x+4=0$
3. $3p^2=18p-6$
4. $3u^2=8u-5$
5. $x^2+x+2=0$
6. $x^2+x+1=0$
7. $x^2+13=4x$
8. $x^2+13=6x$
9. $h^2+4=6h$
10. $r^2+3r=8$
11. $3+\frac{8}{x}=\frac{1}{x^2}$
12. $2+\frac{5}{x^2}=\frac{9}{x}$
13. $3x+x(x-2)=4$
14. $4x+x(x-3)=5$
15. $12t^2+9t=1$
16. $15t^2+7t=2$
17. $25x^2-20x+4=0$
18. $36x^2+84x+49=0$
19. $7x(x+2)+5=3x(x+1)$
20. $5x(x-1)-7=4x(x-2)$
21. $14(x-4)-(x+2)=(x+2)(x-4)$
22. $11(x-2)+(x-5)=(x+2)(x-6)$
23. $5x^2=13x+17$
24. $25x=3x^2+28$
25. $x^2+9=4x$
26. $x^2+7=3x$
27. $x^3-8=0$ (*Hint*: Factor the difference of cubes. Then use the quadratic formula.)
28. $x^3+1=0$
29. Let $f(x)=3x^2-5x-1$. Find x such that $f(x)=0$.
30. Let $g(x)=4x^2-2x-3$. Find x such that $g(x)=0$.
31. Let
$$f(x)=\frac{7}{x}+\frac{7}{x+4}.$$
Find all x for which $f(x)=1$.
32. Let
$$g(x)=\frac{2}{x}+\frac{2}{x+3}.$$
Find all x for which $g(x)=1$.
33. Let
$$F(x)=\frac{x+3}{x} \quad\text{and}\quad G(x)=\frac{x-4}{3}.$$
Find all x for which $F(x)=G(x)$.
34. Let
$$f(x)=\frac{3-x}{4} \quad\text{and}\quad g(x)=\frac{1}{4x}.$$
Find all x for which $f(x)=g(x)$.
35. Let
$$f(x)=\frac{15-2x}{6} \quad\text{and}\quad g(x)=\frac{3}{x}.$$
Find all x for which $f(x)=g(x)$.

36. Let

$$f(x) = x + 5 \quad \text{and} \quad g(x) = \frac{3}{x - 5}.$$

Find all x for which $f(x) = g(x)$.

Solve. Use a calculator to approximate solutions as rational numbers.

37. $x^2 + 4x - 7 = 0$

38. $x^2 + 6x + 4 = 0$

39. $x^2 - 6x + 4 = 0$

40. $x^2 - 4x + 1 = 0$

41. $2x^2 - 3x - 7 = 0$

42. $3x^2 - 3x - 2 = 0$

43. Are there any equations that can be solved by the quadratic formula but not by completing the square? Why or why not?

44. The list on p. 485 does not mention completing the square as a method of solving quadratic equations. Why not?

SKILL MAINTENANCE

45. *Coffee beans.* Twin Cities Roasters has Kenyan coffee for which they pay \$6.75 a pound and Peruvian coffee for which they pay \$11.25 a pound. How much of each kind should be mixed in order to obtain a 50-lb mixture that is worth \$8.55 a pound?

46. *Donuts.* South Street Bakers charges \$1.10 for a cream-filled donut and 85¢ for a glazed donut. On a recent Sunday, a total of 90 glazed and cream-filled donuts were sold for \$88.00. How many of each type were sold?

Simplify.

47. $\sqrt{27a^2b^5} \cdot \sqrt{6a^3b}$

48. $\sqrt{8a^3b} \cdot \sqrt{12ab^5}$

49. $$\frac{\frac{3}{x-1}}{\frac{1}{x+1} + \frac{2}{x-1}}$$

50. $$\frac{\frac{4}{a^2b}}{\frac{3}{a} - \frac{4}{b^2}}$$

SYNTHESIS

51. Suppose you had a large number of quadratic equations to solve and none of the equations had a constant term. Would you use factoring or the quadratic formula to solve these equations? Why?

52. If $a < 0$ and $ax^2 + bx + c = 0$, then $-a$ is positive and the equivalent equation, $-ax^2 - bx - c = 0$, can be solved using the quadratic formula.

a) Find this solution, replacing a, b, and c in the formula with $-a$, $-b$, and $-c$ from the equation.

b) How does the result of part (a) indicate that the quadratic formula "works" regardless of the sign of a?

For Exercises 53–55, let

$$f(x) = \frac{x^2}{x - 2} + 1 \quad \text{and} \quad g(x) = \frac{4x - 2}{x - 2} + \frac{x + 4}{2}.$$

53. Find the x-intercepts of the graph of f.

54. Find the x-intercepts of the graph of g.

55. Find all x for which $f(x) = g(x)$.

Solve.

56. $x^2 - 0.75x - 0.5 = 0$

57. $z^2 + 0.84z - 0.4 = 0$

58. $(1 + \sqrt{3})x^2 - (3 + 2\sqrt{3})x + 3 = 0$

59. $\sqrt{2}x^2 + 5x + \sqrt{2} = 0$

60. $ix^2 - 2x + 1 = 0$

61. One solution of $kx^2 + 3x - k = 0$ is -2. Find the other.

62. Use a grapher to solve Exercises 3, 17, and 37.

63. Use a grapher to solve Exercises 9, 25, and 33. Use the method of graphing each side of the equation.

64. Can a grapher be used to solve *any* quadratic equation? Why or why not?

Applications Involving Quadratic Equations

8.3

Solving Problems • Solving Formulas

Solving Problems

As we found in Section 6.5, some problems translate to rational equations. The solution of such rational equations can involve quadratic equations.

Example 1

Motorcycle travel. Makita rode her motorcycle 300 mi at a certain average speed. Had she averaged 10 mph more, the trip would have taken 1 hr less. Find the average speed of the motorcycle.

Solution

1. **Familiarize.** We make a drawing, labeling it with the known and unknown information. As in Section 6.5, we can organize the information in a table. We let r represent the rate, in miles per hour, and t the time, in hours, for Makita's trip.

300 miles

Time t Speed r

300 miles

Time $t - 1$ Speed $r + 10$

Distance	Speed	Time	
300	r	t	$\longrightarrow r = \dfrac{300}{t}$
300	$r + 10$	$t - 1$	$\longrightarrow r + 10 = \dfrac{300}{t - 1}$

Recall that the definition of speed, $r = d/t$, relates the three quantities.

2. **Translate.** From the first two lines of the table, we obtain

$$r = \frac{300}{t} \quad \text{and} \quad r + 10 = \frac{300}{t - 1}.$$

3. **Carry out.** A system of equations has been formed. We substitute for r from the first equation into the second and solve the resulting equation:

$$\frac{300}{t} + 10 = \frac{300}{t-1}$$ Substituting $300/t$ for r

$$t(t-1) \cdot \left[\frac{300}{t} + 10\right] = t(t-1) \cdot \frac{300}{t-1}$$ Multiplying by the LCD

$$\cancel{t}(t-1) \cdot \frac{300}{\cancel{t}} + t(t-1) \cdot 10 = t\cancel{(t-1)} \cdot \frac{300}{\cancel{t-1}}$$ Using the distributive law and removing factors that equal 1: $\frac{t}{t} = 1$; $\frac{t-1}{t-1} = 1$

$$\left.\begin{aligned} 300(t-1) + 10(t^2 - t) &= 300t \\ 300t - 300 + 10t^2 - 10t &= 300t \\ 10t^2 - 10t - 300 &= 0 \end{aligned}\right\}$$ Rewriting in standard form

$$t^2 - t - 30 = 0$$ Multiplying by $\frac{1}{10}$ or dividing by 10

$$(t-6)(t+5) = 0$$ Factoring

$$t = 6 \quad or \quad t = -5.$$ Principle of zero products

4. **Check.** Note that we have solved for t, not r as required. Since negative time has no meaning here, we disregard the -5 and use 6 hr to find r:

$$r = \frac{300 \text{ mi}}{6 \text{ hr}} = 50 \text{ mph}.$$

> ***Caution!*** Always make sure that you find the quantity asked for in the problem.

To see if 50 mph checks, we increase the speed 10 mph to 60 mph and see how long the trip would have taken at that speed:

$$t = \frac{d}{r} = \frac{300 \text{ mi}}{60 \text{ mph}} = 5 \text{ hr}.$$ Note that $\text{mi/mph} = \text{mi} \div \frac{\text{mi}}{\text{hr}} = \cancel{\text{mi}} \cdot \frac{\text{hr}}{\cancel{\text{mi}}} = \text{hr}.$

This is 1 hr less than the trip actually took, so the answer checks.

5. **State.** Makita's motorcycle traveled at an average speed of 50 mph.

Solving Formulas

Recall that to solve a formula for a certain letter, we use the principles for solving equations to get that letter alone on one side.

Example 2 **Period of a pendulum.** The time T required for a pendulum of length l to swing back and forth (complete one period) is given by the formula $T = 2\pi\sqrt{l/g}$, where g is the earth's gravitational constant. Solve for l.

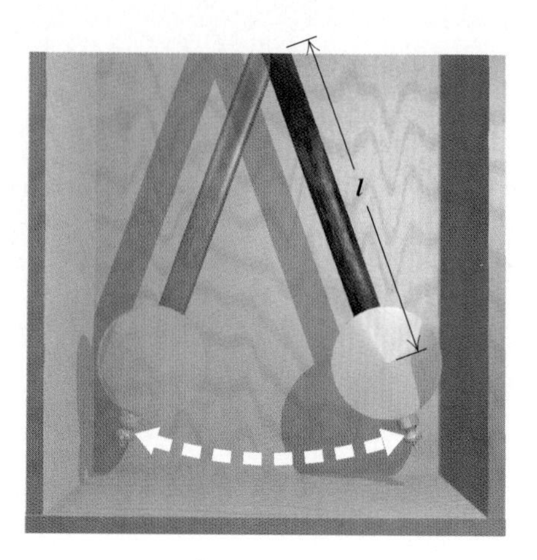

Solution We have

$$T = 2\pi\sqrt{\frac{l}{g}}$$ This is a radical equation (see Section 7.6).

$$T^2 = \left(2\pi\sqrt{\frac{l}{g}}\right)^2$$ Principle of powers (squaring both sides)

$$T^2 = 2^2\pi^2\frac{l}{g}$$

$$gT^2 = 4\pi^2 l$$ Multiplying both sides by g to clear fractions

$$\frac{gT^2}{4\pi^2} = l.$$ Dividing both sides by $4\pi^2$

We now have l alone on one side and l does not appear on the other side, so the formula is solved for l.

In most formulas, variables represent nonnegative numbers, so we do not need to use absolute-value signs when taking square roots.

Example 3 **Hang time.*** An athlete's *hang time* is the amount of time that the athlete can remain airborne when jumping. A formula relating an athlete's vertical leap V, in inches, to hang time T, in seconds, is $V = 48T^2$. Solve for T.

*This formula is taken from an article by Peter Brancazio, "The Mechanics of a Slam Dunk," *Popular Mechanics,* November 1991. Courtesy of Professor Peter Brancazio, Brooklyn College.

Solution

$$48T^2 = V$$

$$T^2 = \frac{V}{48} \qquad \text{Dividing by 48 to get } T^2 \text{ alone}$$

$$T = \frac{\sqrt{V}}{\sqrt{48}} \qquad \text{Using the principle of square roots and the quotient rule for radicals}$$

$$= \frac{\sqrt{V}}{\sqrt{16}\sqrt{3}} = \frac{\sqrt{V}}{4\sqrt{3}}$$

$$= \frac{\sqrt{V}}{4\sqrt{3}} \cdot \frac{\sqrt{3}}{\sqrt{3}} = \frac{\sqrt{3V}}{12}. \qquad \text{Rationalizing the denominator}$$

Example 4

Falling distance. An object tossed downward with an initial speed (velocity) of v_0 will travel a distance of s meters, where $s = 4.9t^2 + v_0 t$ and t is measured in seconds. Solve for t.

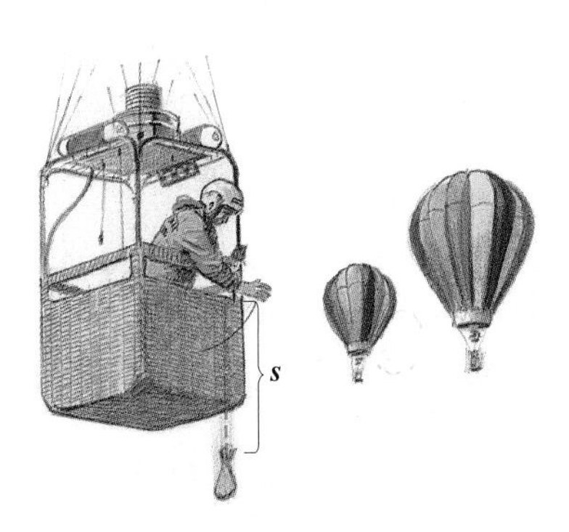

Solution Since t is squared in one term and raised to the first power in the other term, the equation is quadratic in t.

$$4.9t^2 + v_0 t = s$$

$$4.9t^2 + v_0 t - s = 0 \qquad \text{Writing standard form}$$

$$a = 4.9, \quad b = v_0, \quad c = -s$$

$$t = \frac{-v_0 \pm \sqrt{v_0^2 - 4(4.9)(-s)}}{2(4.9)} \qquad \text{Using the quadratic formula}$$

Since the negative square root would yield a negative value for t, we use only the positive root:

$$t = \frac{-v_0 + \sqrt{v_0^2 + 19.6s}}{9.8}.$$

The following list of steps should help you when solving formulas for a given letter. Try to remember that when solving a formula, you use the same approach that you would to solve an equation.

To Solve a Formula for a Letter—Say, b

1. Clear fractions and use the principle of powers, as needed. (In some cases, you may clear the fractions first, and in some cases, you may use the principle of powers first.) Perform these steps until radicals containing b are gone and b is not in any denominator.
2. Combine all terms with b^2 in them. Also combine all terms with b in them.
3. If b^2 does not appear, you can solve by using just the addition and multiplication principles as in Sections 1.5 and 6.8.
4. If b^2 appears but b does not, solve the equation for b^2. Then use the principle of square roots to solve for b.
5. If there are terms containing both b and b^2, put the equation in standard form and use the quadratic formula.

Exercise Set 8.3

FOR EXTRA HELP

Digital Video Tutor CD 6
Videotape 15

InterAct Math

Math Tutor Center

MathXL

MyMathLab.com

Solve.

1. *Canoeing.* During the first part of a canoe trip, Tim covered 60 km at a certain speed. He then traveled 24 km at a speed that was 4 km/h slower. If the total time for the trip was 8 hr, what was the speed on each part of the trip?

2. *Car trips.* During the first part of a trip, Meira's Honda traveled 120 mi at a certain speed. Meira then drove another 100 mi at a speed that was 10 mph slower. If Meira's total trip time was 4 hr, what was her speed on each part of the trip?

3. *Car trips.* Sandi's Subaru travels 280 mi averaging a certain speed. If the car had gone 5 mph faster, the trip would have taken 1 hr less. Find Sandi's average speed.

4. *Car trips.* Petra's Plymouth travels 200 mi averaging a certain speed. If the car had gone 10 mph faster, the trip would have taken 1 hr less. Find Petra's average speed.

5. *Air travel.* A Cessna flies 600 mi at a certain speed. A Beechcraft flies 1000 mi at a speed that is 50 mph faster, but takes 1 hr longer. Find the speed of each plane.

6. *Air travel.* A turbo-jet flies 50 mph faster than a super-prop plane. If a turbo-jet goes 2000 mi in 3 hr less time than it takes the super-prop to go 2800 mi, find the speed of each plane.

7. *Bicycling.* Naoki bikes the 40 mi to Hillsboro averaging a certain speed. The return trip is made at a speed that is 6 mph slower. Total time for the round trip is 14 hr. Find Naoki's average speed on each part of the trip.

8. *Car speed.* On a sales trip, Gail drives the 600 mi to Richmond averaging a certain speed. The return trip is made at an average speed that is 10 mph slower. Total time for the round trip is 22 hr. Find Gail's average speed on each part of the trip.

9. *Navigation.* The current in a typical Mississippi River shipping route flows at a rate of 4 mph. In order for a barge to travel 24 mi upriver and then return in a total of 5 hr, approximately how fast must the barge be able to travel in still water?

10. *Navigation.* The Hudson River flows at a rate of 3 mph. A patrol boat travels 60 mi upriver and returns in a total time of 9 hr. What is the speed of the boat in still water?

11. *Filling a pool.* Two wells are used to fill a swimming pool. Working together, they can fill the pool in 4 hr. One well, working alone, can fill the pool in 6 hr less time than the other. How long would the smaller one take, working alone, to fill the pool?

12. *Filling a tank.* Two pipes are connected to the same tank. Working together, they can fill the tank in 2 hr. The larger pipe, working alone, can fill the tank in 3 hr less time than the smaller one. How long would the smaller one take, working alone, to fill the tank?

13. *Paddleboats.* Ellen paddles 1 mi upstream and 1 mi back in a total time of 1 hr. The speed of the river is 2 mph. Find the speed of Ellen's paddleboat in still water.

14. *Rowing.* Dan rows 10 km upstream and 10 km back in a total time of 3 hr. The speed of the river is 5 km/h. Find Dan's speed in still water.

Solve each formula for the indicated letter. Assume that all variables represent nonnegative numbers.

15. $A = 4\pi r^2$, for r
(Surface area of a sphere)

16. $A = 6s^2$, for s
(Surface area of a cube)

17. $A = 2\pi r^2 + 2\pi rh$, for r
(Surface area of a right cylindrical solid)

18. $F = \dfrac{Gm_1m_2}{r^2}$, for r
(Law of gravity)

19. $N = \dfrac{kQ_1Q_2}{s^2}$, for s
(Number of phone calls between two cities)

20. $A = \pi r^2$, for r
(Area of a circle)

21. $T = 2\pi\sqrt{\dfrac{l}{g}}$, for g
(A pendulum formula)

22. $a^2 + b^2 = c^2$, for b
(Pythagorean formula in two dimensions)

23. $a^2 + b^2 + c^2 = d^2$, for c
(Pythagorean formula in three dimensions)

24. $N = \dfrac{k^2 - 3k}{2}$, for k
(Number of diagonals of a polygon)

25. $s = v_0t + \dfrac{gt^2}{2}$, for t
(A motion formula)

26. $A = \pi r^2 + \pi rs$, for r
(Surface area of a cone)

27. $N = \frac{1}{2}(n^2 - n)$, for n
(Number of games if n teams play each other once)

28. $A = A_0(1 - r)^2$, for r
(A business formula)

29. $V = 3.5\sqrt{h}$, for h
(Distance to horizon from a height)

30. $W = \sqrt{\dfrac{1}{LC}}$, for L
(An electricity formula)

Aha! **31.** $at^2 + bt + c = 0$, for t
(An algebraic formula)

32. $A = P_1(1 + r)^2 + P_2(1 + r)$, for r
(An investment formula)

Solve. Refer to Exercises 15–32 and Examples 2–4 for the appropriate formula.

33. *Falling distance.*

a) An object is dropped 500 m from an airplane. How long does it take the object to reach the ground?

b) An object is thrown downward 500 m from the plane at an initial velocity of 30 m/sec. How long does it take the object to reach the ground?

c) How far will an object fall in 5 sec, when thrown downward at an initial velocity of 30 m/sec?

34. *Falling distance.*

a) An object is dropped 75 m from an airplane. How long does it take the object to reach the ground?

b) An object is thrown downward with an initial velocity of 30 m/sec from a plane 75 m above the ground. How long does it take the object to reach the ground?

c) How far will an object fall in 2 sec, if thrown downward at an initial velocity of 30 m/sec?

35. *Bungee jumping.* Jesse is tied to one end of a 40-m elasticized (bungee) cord. The other end of the cord is tied to the middle of a train trestle. If Jesse jumps off the bridge, for how long will he fall before the cord begins to stretch? (See Example 4 and let $v_0 = 0$.)

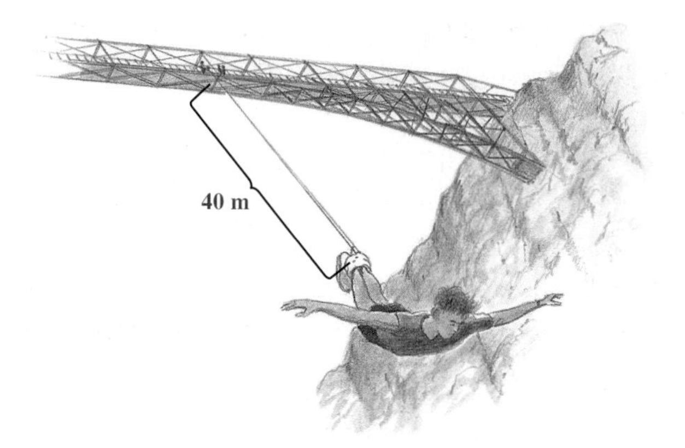

36. *Bungee jumping.* Sheila is tied to a bungee cord (see Exercise 35) and falls for 2.5 sec before her cord begins to stretch. How long is the bungee cord?

37. *Hang time.* The NBA's Vince Carter has a vertical leap of about 36 in. What is his hang time?

38. *League schedules.* In a volleyball league, each team plays each of the other teams once. If a total of 66 games is played, how many teams are in the league?

39. *Downward speed.* An object thrown downward from a 100-m cliff travels 51.6 m in 3 sec. What was the initial velocity of the object?

40. *Downward speed.* An object thrown downward from a 200-m cliff travels 91.2 m in 4 sec. What was the initial velocity of the object?

41. *Compound interest.* A firm invests \$3000 in a savings account for 2 yr. At the beginning of the second year, an additional \$1700 is invested. If a total of \$5253.70 is in the account at the end of the second year, what is the annual interest rate? (*Hint*: See Exercise 32.)

42. *Compound interest.* A business invests \$10,000 in a savings account for 2 yr. At the beginning of the second year, an additional \$3500 is invested. If a total of \$15,569.75 is in the account at the end of the second year, what is the annual interest rate? (*Hint*: See Exercise 32.)

43. Marti is tied to a bungee cord that is twice as long as the cord tied to Pedro. Will Marti's fall take twice as long as Pedro's before their cords begin to stretch? Why or why not? (See Exercises 35 and 36.)

44. Under what circumstances would a negative value for t, time, have meaning?

SKILL MAINTENANCE

Evaluate.

45. $b^2 - 4ac$, for $a = 5$, $b = 6$, and $c = 7$

46. $\sqrt{b^2 - 4ac}$, for $a = 3$, $b = 4$, and $c = 5$

Simplify.

47. $\dfrac{x^2 + xy}{2x}$

48. $\dfrac{a^3 - ab^2}{ab}$

49. $\dfrac{3 + \sqrt{45}}{6}$

50. $\dfrac{2 - \sqrt{28}}{10}$

SYNTHESIS

51. In what ways do the motion problems of this section (like Example 1) differ from the motion problems in Chapter 6 (see p. 373)?

52. Write a problem for a classmate to solve. Devise the problem so that **(a)** the solution is found after solving a rational equation and **(b)** the solution is "The express train travels 90 mph."

53. *Biochemistry.* The equation

$$A = 6.5 - \frac{20.4t}{t^2 + 36}$$

is used to calculate the acid level A in a person's blood t minutes after sugar is consumed. Solve for t.

54. *Special relativity.* Einstein found that an object of mass m_0, traveling velocity v, has its mass become

$$m = \frac{m_0}{\sqrt{1 - \frac{v_2}{c^2}}},$$

where c is the speed of light. Solve the formula for c.

55. *The Golden Rectangle.* For over 2000 yr, the proportions of a "golden" rectangle have been considered visually appealing. A rectangle of width w and length l is considered "golden" if

$$\frac{w}{l} = \frac{l}{w + l}.$$

Solve for l.

56. *Diagonal of a cube.* Find a formula that expresses the length of the three-dimensional diagonal of a cube as a function of the cube's surface area.

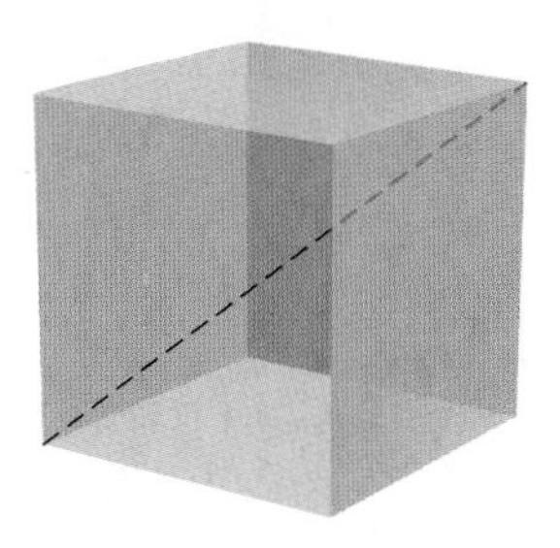

57. Find a number for which the reciprocal of 1 less than the number is the same as 1 more than the number.

58. *Purchasing.* A discount store bought a quantity of beach towels for \$250 and sold all but 15 at a profit of \$3.50 per towel. With the total amount received, the manager could buy 4 more than twice as many as were bought before. Find the cost per towel.

59. Solve for n:

$$mn^4 - r^2pm^3 - r^2n^2 + p = 0.$$

60. *Surface area.* Find a formula that expresses the diameter of a right cylindrical solid as a function of its surface area and its height.

61. A sphere is inscribed in a cube as shown in the figure below. Express the surface area of the sphere as a function of the surface area S of the cube.

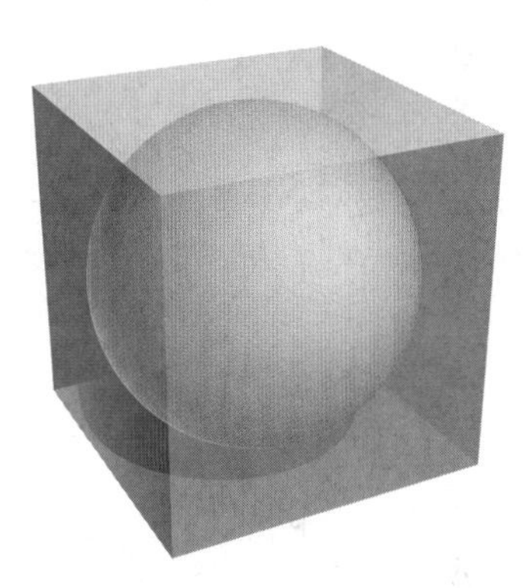

62. Explain how Exercises 1–14 can be solved without factoring, completing the square, or using the quadratic formula.

Studying Solutions of Quadratic Equations

8.4

The Discriminant • Writing Equations from Solutions

The Discriminant

Sometimes in mathematics it is enough to know what *type* of number a solution will be, without actually solving the equation. To illustrate, suppose we want to know if the equation $4x^2 + 7x - 15 = 0$ has rational solutions (and thus can be solved by factoring). Using the quadratic formula, we would have

$$x = \frac{-b \pm \sqrt{b^2 - 4ac}}{2a} = \frac{-7 \pm \sqrt{7^2 - 4 \cdot 4 \cdot (-15)}}{2 \cdot 4}.$$

Note that the radicand, $7^2 - 4 \cdot 4 \cdot (-15)$, determines what type of number the solutions will be. Since $7^2 - 4 \cdot 4 \cdot (-15) = 49 - 16(-15) = 289$, and since 289 is a perfect square $(\sqrt{289} = 17)$, we know that the solutions of the equation will be two rational numbers. This means that $4x^2 + 7x - 15 = 0$ *can* be solved by factoring.

It is the expression $b^2 - 4ac$, known as the **discriminant**, that determines what type of number the solutions of a quadratic equation will be:

- When $b^2 - 4ac$ simplifies to 0, it doesn't matter if we use $+\sqrt{b^2 - 4ac}$ or $-\sqrt{b^2 - 4ac}$; we get the same solution twice. Thus, when the discriminant is 0, there is one *repeated* solution and it will be rational.
- When the discriminant is positive, there are two different real-number solutions. As we saw above, when $b^2 - 4ac$ is a perfect square, these solutions are rational numbers. When $b^2 - 4ac$ is positive but not a perfect square, there are two irrational solutions and they will be conjugates of each other (see p. 437).
- When the discriminant is negative, there are two imaginary-number solutions and they will be complex conjugates of each other.

Discriminant $b^2 - 4ac$	Nature of Solutions
0	One solution; a rational number.
Positive Perfect square Not a perfect square	Two different real-number solutions Solutions are rational. Solutions are irrational conjugates.
Negative	Two different imaginary-number solutions (complex conjugates)

Example 1 For each equation, determine what type of number the solutions will be and how many solutions exist.

a) $9x^2 - 12x + 4 = 0$ **b)** $x^2 + 5x + 8 = 0$ **c)** $2x^2 + 7x - 3 = 0$

Recall that the real-number solutions of $ax^2 + bx + c = 0$ are the x-intercepts of the graph of $y = ax^2 + bx + c$. Use a grapher to confirm that part (a) of Example 1 has one real solution, part (b) has no real solution, and part (c) has two real solutions.

Solution

a) For $9x^2 - 12x + 4 = 0$, we have

$$a = 9, \quad b = -12, \quad c = 4.$$

We substitute and compute the discriminant:

$$\begin{aligned} b^2 - 4ac &= (-12)^2 - 4 \cdot 9 \cdot 4 \\ &= 144 - 144 = 0. \end{aligned}$$

There is just one solution, and it is rational. This tells us that $9x^2 - 12x + 4 = 0$ can be solved by factoring.

b) For $x^2 + 5x + 8 = 0$, we have

$$a = 1, \quad b = 5, \quad c = 8.$$

We substitute and compute the discriminant:

$$\begin{aligned} b^2 - 4ac &= 5^2 - 4 \cdot 1 \cdot 8 \\ &= 25 - 32 = -7. \end{aligned}$$

Since the discriminant is negative, there are two imaginary-number solutions that are complex conjugates of each other.

c) For $2x^2 + 7x - 3 = 0$, we have

$$\begin{aligned} a = 2, \quad b = 7, \quad c &= -3; \\ b^2 - 4ac &= 7^2 - 4 \cdot 2(-3) \\ &= 49 - (-24) = 73. \end{aligned}$$

The discriminant is a positive number that is not a perfect square. Thus there are two irrational solutions that are conjugates of each other.

Discriminants can also be used to determine the number of real-number solutions of $ax^2 + bx + c = 0$. This can be used as an aid in graphing.

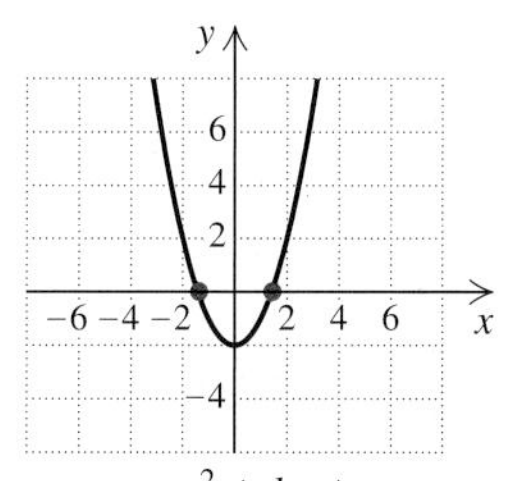

$y = ax^2 + bx + c$
$b^2 - 4ac > 0$
Two real solutions
of $ax^2 + bx + c = 0$
Two x-intercepts

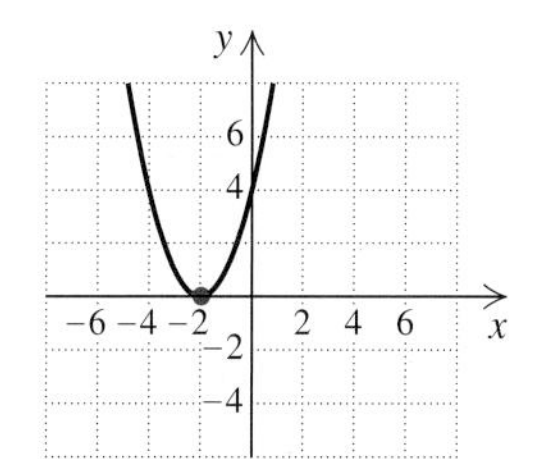

$y = ax^2 + bx + c$
$b^2 - 4ac = 0$
One real solution
of $ax^2 + bx + c = 0$
One x-intercept

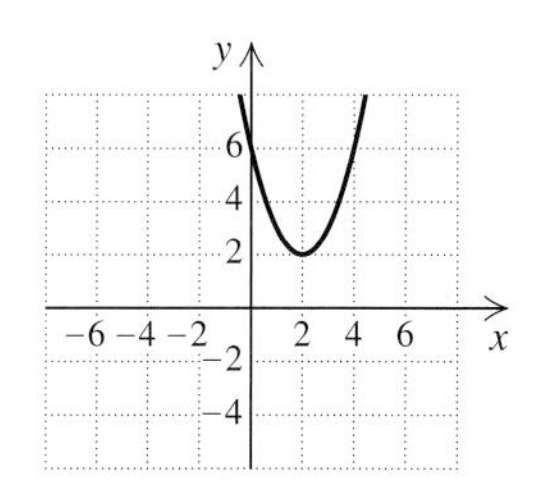

$y = ax^2 + bx + c$
$b^2 - 4ac < 0$
No real solutions
of $ax^2 + bx + c = 0$
No x-intercept

Writing Equations from Solutions

We know by the principle of zero products that $(x - 2)(x + 3) = 0$ has solutions 2 and -3. If we know the solutions of an equation, we can write an equation, using the principle in reverse.

Example 2 Find an equation for which the given numbers are solutions.

a) 3 and $-\frac{2}{5}$

b) $2i$ and $-2i$

c) $5\sqrt{7}$ and $-5\sqrt{7}$

d) -4, 0, and 1

Solution

a)

$$x = 3 \quad or \quad x = -\tfrac{2}{5}$$

$$x - 3 = 0 \quad or \quad x + \tfrac{2}{5} = 0$$ Getting 0's on one side

$$(x - 3)\left(x + \tfrac{2}{5}\right) = 0$$ Using the principle of zero products (multiplying)

$$x^2 + \tfrac{2}{5}x - 3x - 3 \cdot \tfrac{2}{5} = 0$$ Multiplying

$$x^2 - \tfrac{13}{5}x - \tfrac{6}{5} = 0$$ Combining like terms

$$5x^2 - 13x - 6 = 0$$ Multiplying both sides by 5 to clear fractions

Note that multiplying both sides by the LCD, 5, clears the equation of fractions. Had we preferred, we could have multiplied $x + \frac{2}{5} = 0$ by 5, thus clearing fractions *before* using the principle of zero products.

b)

$$x = 2i \quad or \quad x = -2i$$

$$x - 2i = 0 \quad or \quad x + 2i = 0$$ Getting 0's on one side

$$(x - 2i)(x + 2i) = 0$$ Using the principle of zero products (multiplying)

$$x^2 - (2i)^2 = 0$$ Finding the product of a sum and difference

$$x^2 - 4i^2 = 0$$

$$x^2 + 4 = 0$$ $i^2 = -1$

c)

$$x = 5\sqrt{7} \quad or \quad x = -5\sqrt{7}$$

$$x - 5\sqrt{7} = 0 \quad or \quad x + 5\sqrt{7} = 0$$ Getting 0's on one side

$$\left(x - 5\sqrt{7}\right)\left(x + 5\sqrt{7}\right) = 0$$ Using the principle of zero products

$$x^2 - \left(5\sqrt{7}\right)^2 = 0$$ Finding the product of a sum and difference

$$x^2 - 25 \cdot 7 = 0$$

$$x^2 - 175 = 0$$

d)

$$x = -4 \quad or \quad x = 0 \quad or \quad x = 1$$

$$x + 4 = 0 \quad or \quad x = 0 \quad or \quad x - 1 = 0$$ Getting 0's on one side

$$(x + 4)x(x - 1) = 0$$ Using the principle of zero products

$$x(x^2 + 3x - 4) = 0$$ Multiplying

$$x^3 + 3x^2 - 4x = 0$$

To check any of these equations, we can simply substitute one or more of the given solutions. For example, in Example 2(d) above,

$$(-4)^3 + 3(-4)^2 - 4(-4) = -64 + 3 \cdot 16 + 16$$
$$= -64 + 48 + 16 = 0.$$

The other checks are left to the student.

Exercise Set 8.4

FOR EXTRA HELP

Digital Video Tutor CD 6
Videotape 15

InterAct Math

Math Tutor Center

MathXL

MyMathLab.com

For each equation, determine what type of number the solutions are and how many solutions exist.

1. $x^2 - 5x + 3 = 0$

2. $x^2 - 7x + 5 = 0$

3. $x^2 + 5 = 0$

4. $x^2 + 3 = 0$

5. $x^2 - 3 = 0$

6. $x^2 - 5 = 0$

7. $4x^2 - 12x + 9 = 0$

8. $4x^2 + 8x - 5 = 0$

9. $x^2 - 2x + 4 = 0$

10. $x^2 + 4x + 6 = 0$

11. $6t^2 - 19t - 20 = 0$

12. $9t^2 - 48t + 64 = 0$

13. $6x^2 + 5x - 4 = 0$

14. $10x^2 - x - 2 = 0$

Aha! **15.** $9t^2 - 3t = 0$

16. $4m^2 + 7m = 0$

17. $x^2 + 4x = 8$

18. $x^2 + 5x = 9$

19. $2a^2 - 3a = -5$

20. $3a^2 + 5 = 7a$

21. $y^2 + \frac{9}{4} = 4y$

22. $x^2 = \frac{1}{2}x - \frac{3}{5}$

Write a quadratic equation having the given numbers as solutions.

23. $-7, 3$

24. $-6, 4$

25. 3, only solution (*Hint*: It must be a repeated solution.)

26. -5, only solution

27. $-2, -5$

28. $-1, -3$

29. $4, \frac{2}{3}$

30. $5, \frac{3}{4}$

31. $\frac{1}{2}, \frac{1}{3}$

32. $-\frac{1}{4}, -\frac{1}{2}$

33. $-0.6, 1.4$

34. $2.4, -0.4$

35. $-\sqrt{7}, \sqrt{7}$

36. $-\sqrt{3}, \sqrt{3}$

37. $3\sqrt{2}, -3\sqrt{2}$

38. $2\sqrt{5}, -2\sqrt{5}$

39. $3i, -3i$

40. $4i, -4i$

41. $5 - 2i, 5 + 2i$

42. $2 - 7i, 2 + 7i$

43. $2 - \sqrt{10}, 2 + \sqrt{10}$

44. $3 - \sqrt{14}, 3 + \sqrt{14}$

Write a third-degree equation having the given numbers as solutions.

45. $-2, 1, 5$

46. $-5, 0, 2$

47. $-1, 0, 3$

48. $-2, 2, 3$

49. Under what condition(s) is the discriminant *not* the fastest way to determine how many and what type of solutions exist?

50. Describe a procedure that could be used to write an equation having the first 7 natural numbers as solutions.

SKILL MAINTENANCE

Simplify.

51. $(3a^2)^4$

52. $(4x^3)^2$

Find the x-intercepts of the graph of f.

53. $f(x) = x^2 - 7x - 8$

54. $f(x) = x^2 - 6x + 8$

55. During a one-hour television show, there were 12 commercials. Some of the commercials were 30 sec long and the others were 60 sec long. The amount of time for 30-sec commercials was 6 min less than the total number of minutes of commercial time during the show. How many 30-sec commercials were used?

56. Graph: $y = -\frac{3}{7}x + 4$.

SYNTHESIS

57. If we assume that a quadratic equation has integers for coefficients, will the product of the solutions always be a real number? Why or why not?

58. Can a fourth-degree equation have three irrational solutions? Why or why not?

59. The graph of an equation of the form

$$y = ax^2 + bx + c$$

is a curve similar to the one shown below. Determine a, b, and c from the information given.

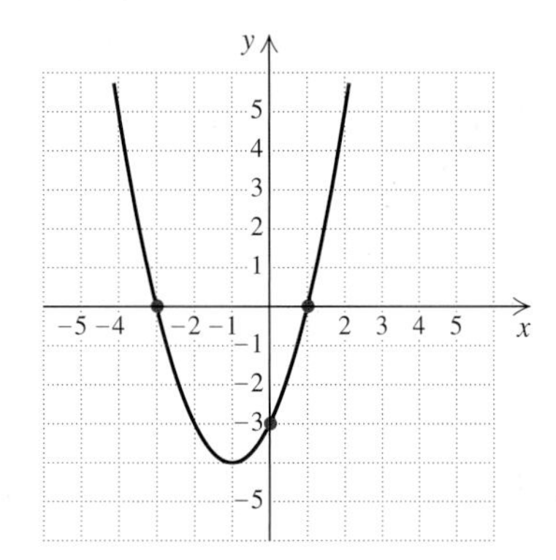

60. Show that the product of the solutions of $ax^2 + bx + c = 0$ is c/a.

For each equation under the given condition, **(a)** *find k and* **(b)** *find the other solution.*

61. $kx^2 - 2x + k = 0$; one solution is -3

62. $x^2 - kx + 2 = 0$; one solution is $1 + i$

63. $x^2 - (6 + 3i)x + k = 0$; one solution is 3

64. Show that the sum of the solutions of $ax^2 + bx + c = 0$ is $-b/a$.

65. Show that whenever there is just one solution of $ax^2 + bx + c = 0$, that solution is of the form $-b/2a$.

66. Find h and k, where $3x^2 - hx + 4k = 0$, the sum of the solutions is -12, and the product of the solutions is 20. (*Hint*: See Exercises 60 and 64.)

67. Suppose that $f(x) = ax^2 + bx + c$, with $f(-3) = 0$, $f\left(\frac{1}{2}\right) = 0$, and $f(0) = -12$. Find a, b, and c.

68. Find an equation for which $2 - \sqrt{3}$, $2 + \sqrt{3}$, $5 - 2i$, and $5 + 2i$ are solutions.

Aha! 69. Find an equation for which $1 - \sqrt{5}$ and $3 + 2i$ are two of the solutions.

70. A discriminant that is a perfect square indicates that factoring can be used to solve the quadratic equation. Why?

71. While solving a quadratic equation of the form $ax^2 + bx + c = 0$ with a grapher, Shawn-Marie gets the following screen.

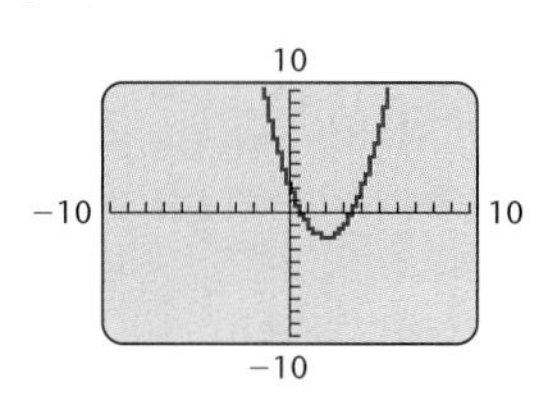

How could the sign of the discriminant help her check the graph?

8.5 Equations Reducible to Quadratic

Recognizing Equations in Quadratic Form • Radical and Rational Equations

Recognizing Equations in Quadratic Form

Certain equations that are not really quadratic can be thought of in such a way that they can be solved as quadratic. For example, because the square of x^2 is

x^4, the equation $x^4 - 9x^2 + 8 = 0$ is said to be "quadratic in x^2":

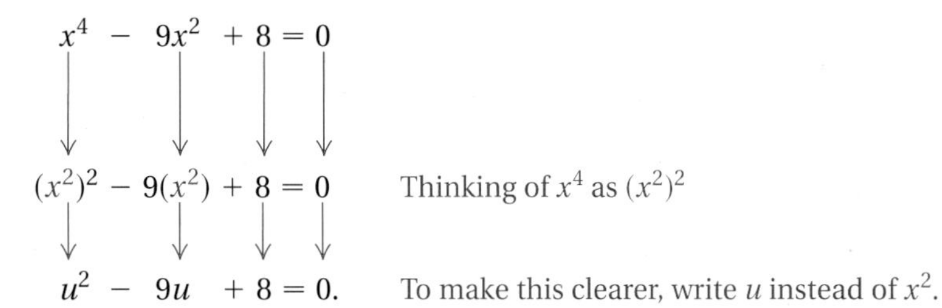

$x^4 - 9x^2 + 8 = 0$
$(x^2)^2 - 9(x^2) + 8 = 0$
Thinking of x^4 as $(x^2)^2$
$u^2 - 9u + 8 = 0.$
To make this clearer, write u instead of x^2.

The equation $u^2 - 9u + 8 = 0$ can be solved by factoring or by the quadratic formula. Then, remembering that $u = x^2$, we can solve for x. Equations that can be solved like this are said to be *reducible to quadratic*, or *in quadratic form*.

Example 1

Solve: $x^4 - 9x^2 + 8 = 0$.

Solution Let $u = x^2$. Then we solve by substituting u for x^2 and u^2 for x^4:

$$u^2 - 9u + 8 = 0$$
$$(u-8)(u-1) = 0 \qquad \text{Factoring}$$
$$u - 8 = 0 \quad \text{or} \quad u - 1 = 0 \qquad \text{Principle of zero products}$$
$$u = 8 \quad \text{or} \quad u = 1.$$

> **Caution!** A common error is to solve for u and then forget to solve for x. Remember that you must find values for the *original* variable!

We replace u with x^2 and solve these equations:

$$x^2 = 8 \quad \text{or} \quad x^2 = 1$$
$$x = \pm\sqrt{8} \quad \text{or} \quad x = \pm 1$$
$$x = \pm 2\sqrt{2} \quad \text{or} \quad x = \pm 1.$$

To check, note that for $x = 2\sqrt{2}$, we have $x^2 = 8$ and $x^4 = 64$. Similarly, for $x = -2\sqrt{2}$, we have $x^2 = 8$ and $x^4 = 64$. When $x = 1$, we have $x^2 = 1$ and $x^4 = 1$, and when $x = -1$, we have $x^2 = 1$ and $x^4 = 1$. Thus instead of making four checks, we need make only two.

Check:

For $\pm 2\sqrt{2}$:

$$\begin{array}{r|l} x^4 - 9x^2 + 8 = & 0 \\ \hline (\pm 2\sqrt{2})^4 - 9(\pm 2\sqrt{2})^2 + 8 \ ? & 0 \\ 64 - 9 \cdot 8 + 8 & \\ 0 & 0 \quad \text{TRUE} \end{array}$$

For ± 1:

$$\begin{array}{r|l} x^4 - 9x^2 + 8 = & 0 \\ \hline (\pm 1)^4 - 9(\pm 1)^2 + 8 \ ? & 0 \\ 1 - 9 + 8 & \\ 0 & 0 \quad \text{TRUE} \end{array}$$

The solutions are $1, -1, 2\sqrt{2}$, and $-2\sqrt{2}$.

Example 1 can be solved directly by factoring:

$$x^4 - 9x^2 + 8 = 0$$
$$(x^2 - 1)(x^2 - 8) = 0$$
$$x^2 - 1 = 0 \quad or \quad x^2 - 8 = 0$$
$$x^2 = 1 \quad or \quad x^2 = 8$$
$$x = \pm 1 \quad or \quad x = \pm 2\sqrt{2}.$$

There is nothing wrong with this approach. However, in the examples that follow, you will note that it becomes increasingly difficult to solve the equation without first making a substitution.

Radical and Rational Equations

Sometimes rational equations, radical equations, or equations containing fractional exponents are reducible to quadratic. It is especially important that answers to these equations be checked in the original equation.

Example 2 Solve: $x - 3\sqrt{x} - 4 = 0$.

Solution This radical equation could be solved using the method discussed in Section 7.6. However, if we note that the square of $\sqrt{x}$ is x, we can regard the equation as "quadratic in $\sqrt{x}$."

We let $u = \sqrt{x}$ and consequently $u^2 = x$:

$$x - 3\sqrt{x} - 4 = 0$$
$$u^2 - 3u - 4 = 0 \qquad \text{Substituting}$$
$$(u - 4)(u + 1) = 0$$
$$u = 4 \quad or \quad u = -1. \qquad \text{Using the principle of zero products}$$

Next, we replace u with $\sqrt{x}$ and solve these equations:

$$\sqrt{x} = 4 \quad or \quad \sqrt{x} = -1.$$

Squaring gives us $x = 16$ or $x = 1$ and also makes checking essential.

Check: For 16:

$$\begin{array}{c|l} x - 3\sqrt{x} - 4 & = 0 \\ \hline 16 - 3\sqrt{16} - 4 & ?\ 0 \\ 16 - 3 \cdot 4 - 4 & \\ 0 & 0 \quad \text{TRUE} \end{array}$$

For 1:

$$\begin{array}{c|l} x - 3\sqrt{x} - 4 & = 0 \\ \hline 1 - 3\sqrt{1} - 4 & ?\ 0 \\ 1 - 3 \cdot 1 - 4 & \\ -6 & 0 \quad \text{FALSE} \end{array}$$

The number 16 checks, but 1 does not. Had we noticed that $\sqrt{x} = -1$ has no solution (since principal roots are never negative), we could have solved only the equation $\sqrt{x} = 4$. The solution is 16.

Example 3 Find the x-intercepts of the graph of $f(x) = (x^2 - 1)^2 - (x^2 - 1) - 2$.

Check Example 3 with a grapher. Use the ZERO, ROOT, or INTERSECT option, if possible.

Solution The x-intercepts occur where $f(x) = 0$ so we must have

$$(x^2 - 1)^2 - (x^2 - 1) - 2 = 0. \qquad \text{Setting } f(x) \text{ equal to } 0$$

This equation is quadratic in $x^2 - 1$, so we let $u = x^2 - 1$ and $u^2 = (x^2 - 1)^2$:

$$u^2 - u - 2 = 0 \qquad \text{Substituting in } (x^2 - 1)^2 - (x^2 - 1) - 2 = 0$$
$$(u - 2)(u + 1) = 0$$
$$u = 2 \quad or \quad u = -1. \qquad \text{Using the principle of zero products}$$

Next, we replace u with $x^2 - 1$ and solve these equations:

$$x^2 - 1 = 2 \quad or \quad x^2 - 1 = -1$$
$$x^2 = 3 \quad or \quad x^2 = 0 \qquad \text{Adding 1 to both sides}$$
$$x = \pm\sqrt{3} \quad or \quad x = 0. \qquad \text{Using the principle of square roots}$$

The x-intercepts occur at $(-\sqrt{3}, 0)$, $(0, 0)$, and $(\sqrt{3}, 0)$.

Sometimes great care must be taken in deciding what substitution to make.

Example 4 Solve: $m^{-2} - 6m^{-1} + 4 = 0$.

Solution Note that the square of m^{-1} is $(m^{-1})^2$, or m^{-2}. This allows us to regard the equation as quadratic in m^{-1}.

We let $u = m^{-1}$ and $u^2 = m^{-2}$:

$$u^2 - 6u + 4 = 0 \qquad \text{Substituting}$$
$$u = \frac{-(-6) \pm \sqrt{(-6)^2 - 4 \cdot 1 \cdot 4}}{2 \cdot 1} \qquad \text{Using the quadratic formula}$$
$$\left.\begin{aligned} u &= \frac{6 \pm \sqrt{20}}{2} = \frac{2 \cdot 3 \pm 2\sqrt{5}}{2} \\ u &= 3 \pm \sqrt{5}. \end{aligned}\right\} \qquad \text{Simplifying}$$

Next, we replace u with m^{-1} and solve:

$$m^{-1} = 3 \pm \sqrt{5}$$
$$\frac{1}{m} = 3 \pm \sqrt{5} \qquad \text{Recall that } m^{-1} = \frac{1}{m}.$$
$$1 = m(3 \pm \sqrt{5}) \qquad \text{Multiplying both sides by } m$$
$$\frac{1}{3 \pm \sqrt{5}} = m. \qquad \text{Dividing both sides by } 3 \pm \sqrt{5}$$

We can check both solutions as follows.

Check:

For $1/(3-\sqrt{5})$:

$$\begin{array}{c|c} m^{-2} - 6m^{-1} + 4 = 0 & \\ \hline \left(\frac{1}{3-\sqrt{5}}\right)^{-2} - 6\left(\frac{1}{3-\sqrt{5}}\right)^{-1} + 4 & ?\ 0 \\ (3-\sqrt{5})^2 - 6(3-\sqrt{5}) + 4 & \\ 9 - 6\sqrt{5} + 5 - 18 + 6\sqrt{5} + 4 & \\ 0 & 0 \quad \text{TRUE} \end{array}$$

For $1/(3+\sqrt{5})$:

$$\begin{array}{c|c} m^{-2} - 6m^{-1} + 4 = 0 & \\ \hline \left(\frac{1}{3+\sqrt{5}}\right)^{-2} - 6\left(\frac{1}{3+\sqrt{5}}\right)^{-1} + 4 & ?\ 0 \\ (3+\sqrt{5})^2 - 6(3+\sqrt{5}) + 4 & \\ 9 + 6\sqrt{5} + 5 - 18 - 6\sqrt{5} + 4 & \\ 0 & 0 \quad \text{TRUE} \end{array}$$

Both numbers check. The solutions are $1/(3-\sqrt{5})$ and $1/(3+\sqrt{5})$, or approximately 1.309016994 and 0.1909830056.

Example 5 Solve: $t^{2/5} - t^{1/5} - 2 = 0$.

Solution Note that the square of $t^{1/5}$ is $(t^{1/5})^2$, or $t^{2/5}$. The equation is therefore quadratic in $t^{1/5}$, so we let $u = t^{1/5}$ and $u^2 = t^{2/5}$:

$$u^2 - u - 2 = 0 \qquad \text{Substituting}$$
$$(u-2)(u+1) = 0$$
$$u = 2 \quad or \quad u = -1. \qquad \text{Using the principle of zero products}$$

Now we replace u with $t^{1/5}$ and solve:

$$t^{1/5} = 2 \quad or \quad t^{1/5} = -1$$
$$t = 32 \quad or \quad t = -1. \qquad \text{Principle of powers; raising to the 5th power}$$

Check:

For 32:

$$\begin{array}{c|c} t^{2/5} - t^{1/5} - 2 = 0 & \\ \hline 32^{2/5} - 32^{1/5} - 2 & ?\ 0 \\ (32^{1/5})^2 - 32^{1/5} - 2 & \\ 2^2 - 2 - 2 & \\ 0 & 0 \quad \text{TRUE} \end{array}$$

For -1:

$$\begin{array}{c|c} t^{2/5} - t^{1/5} - 2 = 0 & \\ \hline (-1)^{2/5} - (-1)^{1/5} - 2 & ?\ 0 \\ [(-1)^{1/5}]^2 - (-1)^{1/5} - 2 & \\ (-1)^2 - (-1) - 2 & \\ 0 & 0 \quad \text{TRUE} \end{array}$$

Both numbers check. The solutions are 32 and -1.

The following tips may prove useful.

To Solve an Equation That Is Reducible to Quadratic

1. The equation is quadratic in form if the variable factor in one term is the square of the variable factor in the other variable term.
2. Write down any substitutions that you are making.
3. Whenever you make a substitution, be sure to solve for the variable that is used in the original equation.
4. Check possible answers in the original equation.

18

Exercise Set 8.5

FOR EXTRA HELP

Digital Video Tutor CD 7
Videotape 15

InterAct Math

Math Tutor Center

MathXL

MyMathLab.com

Solve.

1. $x^4 - 10x^2 + 9 = 0$
2. $x^4 - 5x^2 + 4 = 0$
3. $x^4 - 12x^2 + 27 = 0$
4. $x^4 - 9x^2 + 20 = 0$
5. $9x^4 - 14x^2 + 5 = 0$
6. $4x^4 - 19x^2 + 12 = 0$
7. $x - 4\sqrt{x} - 1 = 0$
8. $x - 2\sqrt{x} - 6 = 0$
9. $(x^2 - 7)^2 - 3(x^2 - 7) + 2 = 0$
10. $(x^2 - 1)^2 - 5(x^2 - 1) + 6 = 0$
11. $(1 + \sqrt{x})^2 + 5(1 + \sqrt{x}) + 6 = 0$
12. $(3 + \sqrt{x})^2 + 3(3 + \sqrt{x}) - 10 = 0$
13. $x^{-2} - x^{-1} - 6 = 0$
14. $2x^{-2} - x^{-1} - 1 = 0$
15. $4x^{-2} + x^{-1} - 5 = 0$
16. $m^{-2} + 9m^{-1} - 10 = 0$
17. $t^{2/3} + t^{1/3} - 6 = 0$
18. $w^{2/3} - 2w^{1/3} - 8 = 0$
19. $y^{1/3} - y^{1/6} - 6 = 0$
20. $t^{1/2} + 3t^{1/4} + 2 = 0$
21. $t^{1/3} + 2t^{1/6} = 3$
22. $m^{1/2} + 6 = 5m^{1/4}$
23. $(3 - \sqrt{x})^2 - 10(3 - \sqrt{x}) + 23 = 0$
24. $(5 + \sqrt{x})^2 - 12(5 + \sqrt{x}) + 33 = 0$
25. $16\left(\frac{x-1}{x-8}\right)^2 + 8\left(\frac{x-1}{x-8}\right) + 1 = 0$
26. $9\left(\frac{x+2}{x+3}\right)^2 - 6\left(\frac{x+2}{x+3}\right) + 1 = 0$

Find all x-intercepts of the given function f. If none exist, state this.

27. $f(x) = 5x + 13\sqrt{x} - 6$
28. $f(x) = 3x + 10\sqrt{x} - 8$
29. $f(x) = (x^2 - 3x)^2 - 10(x^2 - 3x) + 24$
30. $f(x) = (x^2 - 6x)^2 - 2(x^2 - 6x) - 35$
31. $f(x) = x^{2/5} + x^{1/5} - 6$
32. $f(x) = x^{1/2} - x^{1/4} - 6$
33. (Aha!) $f(x) = \left(\frac{x^2+2}{x}\right)^4 + 7\left(\frac{x^2+2}{x}\right)^2 + 5$
34. $f(x) = \left(\frac{x^2+1}{x}\right)^4 + 4\left(\frac{x^2+1}{x}\right)^2 + 12$
35. To solve $25x^6 - 10x^3 + 1 = 0$, Don lets $u = 5x^3$ and Robin lets $u = x^3$. Can they both be correct? Why or why not?
36. Can the examples and exercises of this section be understood without knowing the rules for exponents? Why or why not?

SKILL MAINTENANCE

Graph.

37. $f(x) = \frac{3}{2}x$
38. $f(x) = -\frac{2}{3}x$
39. $f(x) = \frac{2}{x}$
40. $f(x) = \frac{3}{x}$
41. Solution A is 18% alcohol and solution B is 45% alcohol. How much of each should be mixed together to get 12 L of a solution that is 36% alcohol?
42. If $g(x) = x^2 - x$, find $g(a + 1)$.

SYNTHESIS

43. Describe a procedure that could be used to solve any equation of the form $ax^4 + bx^2 + c = 0$.
44. Describe a procedure that could be used to write an equation that is quadratic in $3x^2 - 1$. Then explain how the procedure could be adjusted to write equations that are quadratic in $3x^2 - 1$ and have no real-number solution.

Solve.

45. $5x^4 - 7x^2 + 1 = 0$

46. $3x^4 + 5x^2 - 1 = 0$

47. $(x^2 - 4x - 2)^2 - 13(x^2 - 4x - 2) + 30 = 0$

48. $(x^2 - 5x - 1)^2 - 18(x^2 - 5x - 1) + 65 = 0$

49. $\frac{x}{x-1} - 6\sqrt{\frac{x}{x-1}} - 40 = 0$

50. $\left(\sqrt{\frac{x}{x-3}}\right)^2 - 24 = 10\sqrt{\frac{x}{x-3}}$

51. $a^5(a^2 - 25) + 13a^3(25 - a^2) + 36a(a^2 - 25) = 0$

52. $a^3 - 26a^{3/2} - 27 = 0$

53. $x^6 - 28x^3 + 27 = 0$

54. $x^6 + 7x^3 - 8 = 0$

55. Use a grapher to check your answers to Exercises 1, 3, 29, and 47.

56. Use a grapher to solve

$$x^4 - x^3 - 13x^2 + x + 12 = 0.$$

57. While trying to solve $0.05x^4 - 0.8 = 0$ with a grapher, Murray gets the following screen.

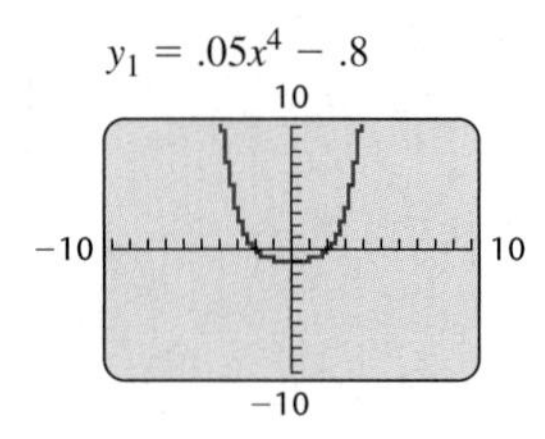

Can Murray solve this equation with a grapher? Why or why not?

Quadratic Functions and Their Graphs

8.6

The Graph of $f(x) = ax^2$ • The Graph of $f(x) = a(x - h)^2$ • The Graph of $f(x) = a(x - h)^2 + k$

We have already used quadratic functions when we solved equations earlier in this chapter. In this section and the next, we learn to graph such functions.

The Graph of $f(x) = ax^2$

The most basic quadratic function is $f(x) = x^2$.

Example 1 Graph: $f(x) = x^2$.

Solution We choose some values for x and compute $f(x)$ for each. Then we plot the ordered pairs and connect them with a smooth curve.

To examine the effect of a when graphing $f(x) = ax^2$, first graph $y_1 = x^2$ in a $[-5, 5, -10, 10]$ window. Without erasing this curve, graph $y_2 = 3x^2$. How do the graphs compare? Now include the graph of $y_3 = \frac{1}{3}x^2$. Describe the effect of multiplying x^2 by a, for $a > 1$ or $0 < a < 1$.

Clear the display and graph $y_1 = x^2$ again. Now include the graph of $y_2 = -x^2$. Note how it differs from the graph of $y_1 = x^2$. Next, graph $y_3 = \frac{2}{3}x^2$ and $y_4 = -\frac{2}{3}x^2$ and compare them. Describe the effect of multiplying x^2 by a, for $a < -1$ or $-1 < a < 0$.

x	$f(x) = x^2$	$(x, f(x))$
-3	9	$(-3, 9)$
-2	4	$(-2, 4)$
-1	1	$(-1, 1)$
0	0	$(0, 0)$
1	1	$(1, 1)$
2	4	$(2, 4)$
3	9	$(3, 9)$

All quadratic functions have graphs similar to the one in Example 1. Such curves are called *parabolas*. They are cup-shaped and symmetric with respect to a vertical line known as the parabola's *axis of symmetry*. For the graph of $f(x) = x^2$, the y-axis (or the line $x = 0$) is the axis of symmetry. Were the paper folded on this line, the two halves of the curve would match. The point $(0, 0)$ is known as the *vertex* of this parabola.

By plotting points, we can compare the graphs of $g(x) = \frac{1}{2}x^2$ and $h(x) = 2x^2$ with the graph of $f(x) = x^2$.

x	$h(x) = 2x^2$
-3	18
-2	8
-1	2
0	0
1	2
2	8
3	18

x	$g(x) = \frac{1}{2}x^2$
-3	$\frac{9}{2}$
-2	2
-1	$\frac{1}{2}$
0	0
1	$\frac{1}{2}$
2	2
3	$\frac{9}{2}$

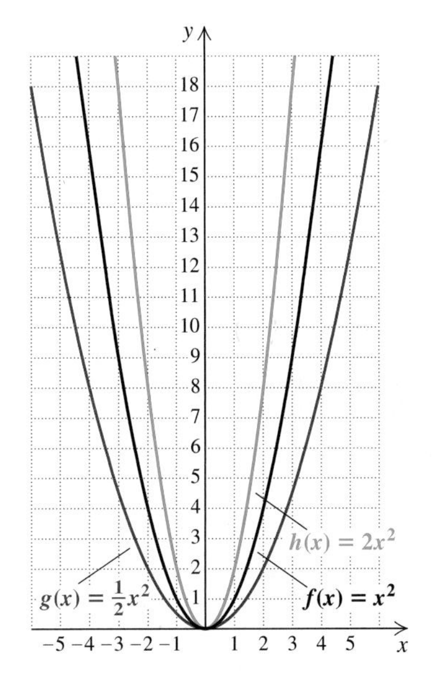

Note that the graph of $g(x) = \frac{1}{2}x^2$ is a wider parabola than the graph of $f(x) = x^2$, and the graph of $h(x) = 2x^2$ is narrower. The vertex and the axis of symmetry, however, remain $(0, 0)$ and $x = 0$, respectively.

When we consider the graph of $k(x) = -\frac{1}{2}x^2$, we see that the parabola is the same shape as the graph of $g(x) = \frac{1}{2}x^2$, but opens downward. We say that the graphs of k and g are *reflections* of each other across the x-axis.

x	$k(x) = -\frac{1}{2}x^2$
-3	$-\frac{9}{2}$
-2	-2
-1	$-\frac{1}{2}$
0	0
1	$-\frac{1}{2}$
2	-2
3	$-\frac{9}{2}$

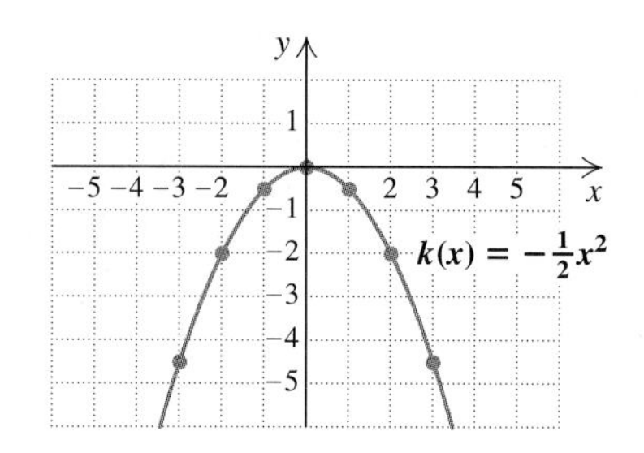

To investigate the effect of h on the graph of $f(x) = a(x - h)^2$, let $y_1 = 7x^2$ and $y_2 = 7(x - 1)^2$. Graph both y_1 and y_2 in the window $[-5, 5, -5, 5]$ and compare. Using the TABLE feature, compare y-values, beginning at $x = 1$ and increasing x by one unit at a time. On many graphers, the G-T or HORIZ modes can be used to view a split screen showing both the graph and the table.

Next, let $y_3 = 7(x - 2)^2$ and compare its graph and y-values with those of y_1 and y_2.

Finally, replace y_2 and y_3 with $y_2 = 7(x + 1)^2$ and $y_3 = 7(x + 2)^2$. Compare graphs and y-values and describe the effect of h on the graph of $f(x) = a(x - h)^2$.

Graphing $f(x) = ax^2$

The graph of $f(x) = ax^2$ is a parabola with $x = 0$ as its axis of symmetry. Its vertex is the origin.

For $a > 0$, the parabola opens upward. For $a < 0$, the parabola opens downward.

If $|a|$ is greater than 1, the parabola is narrower than $y = x^2$.

If $|a|$ is between 0 and 1, the parabola is wider than $y = x^2$.

The Graph of $f(x) = a(x - h)^2$

Why not now consider graphs of

$$f(x) = ax^2 + bx + c,$$

where b and c are not both 0? In effect, we will do that, but in a disguised form. It turns out to be convenient to first graph $f(x) = a(x - h)^2$, where h is some constant. This allows us to observe similarities to the graphs drawn above.

Example 2 Graph: $f(x) = (x - 3)^2$.

Solution We choose some values for x and compute $f(x)$. Note that when an input here is 3 more than an input for Example 1, the outputs match. We plot the points and draw the curve.

x	$f(x) = (x-3)^2$
−1	16
0	9
1	4
2	1
3	0 ← Vertex
4	1
5	4
6	9

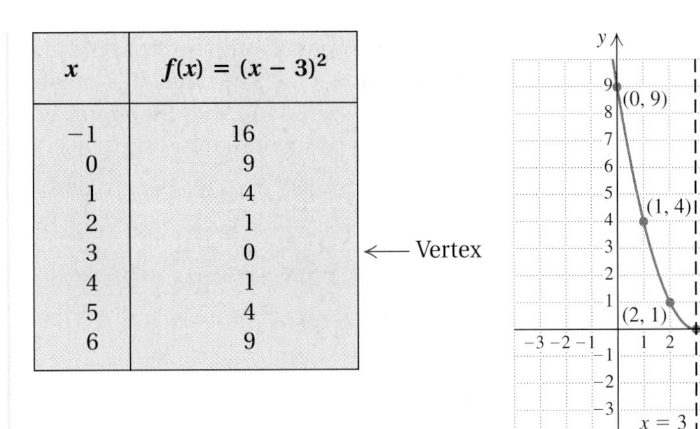

Note that $f(x)$ is smallest when $x - 3$ is 0, that is, for $x = 3$. Thus the line $x = 3$ is now the axis of symmetry and the point $(3, 0)$ is the vertex. Had we recognized earlier that $x = 3$ is the axis of symmetry, we could have computed some values on one side, such as $(4, 1)$, $(5, 4)$, and $(6, 9)$, and then used symmetry to get their mirror images $(2, 1)$, $(1, 4)$, and $(0, 9)$ without further computation.

Example 3 Graph: $g(x) = -2(x+3)^2$.

Solution We choose some values for x and compute $g(x)$. Note that $g(x)$ is greatest when $x + 3$ is 0, that is, for $x = -3$. Thus the line given by $x = -3$ is the axis of symmetry and the point $(-3, 0)$ is the vertex. We plot some points and draw the curve.

x	$g(x) = -2(x+3)^2$
−5	−8
−4	−2
−3	0
−2	−2
−1	−8

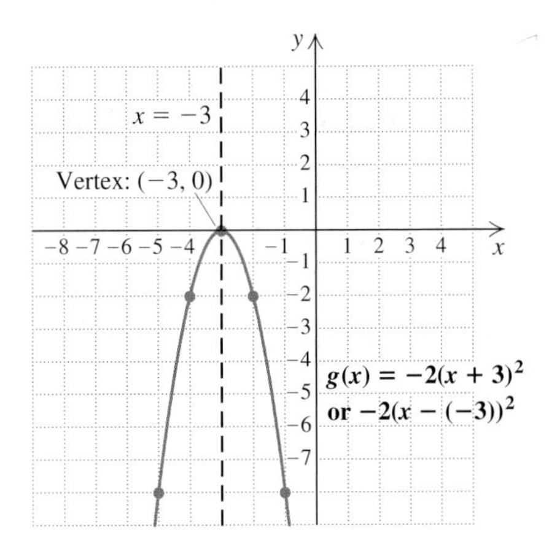

In Example 2, we found that the graph of $f(x) = (x - 3)^2$ looks just like the graph of $y = x^2$, except that it is moved, or *translated*, 3 units to the right. In Example 3, we found that the graph of $g(x) = -2(x + 3)^2$ looks like the graph of $y = -2x^2$, except that it is shifted 3 units to the left. These results can be generalized as follows.

> ***Graphing $f(x) = a(x - h)^2$***
>
> The graph of $f(x) = a(x - h)^2$ has the same shape as the graph of $y = ax^2$.
>
> If h is positive, the graph of $y = ax^2$ is shifted h units to the right.
>
> If h is negative, the graph of $y = ax^2$ is shifted $|h|$ units to the left.
>
> The vertex is $(h, 0)$ and the axis of symmetry is $x = h$.

The Graph of $f(x) = a(x - h)^2 + k$

Given a graph of $f(x) = a(x - h)^2$, what happens if we add a constant k? Suppose that we add 2. This increases $f(x)$ by 2, so the curve is moved up. If k is negative, the curve is moved down. The axis of symmetry for the parabola remains $x = h$, but the vertex will be at (h, k), or, equivalently, $(h, f(h))$.

Note that if a parabola opens upward ($a > 0$), the function value, or y-value, at the vertex is a least, or *minimum*, value. That is, it is less than the y-value at any other point on the graph. If the parabola opens downward ($a < 0$), the function value at the vertex is a greatest, or *maximum*, value.

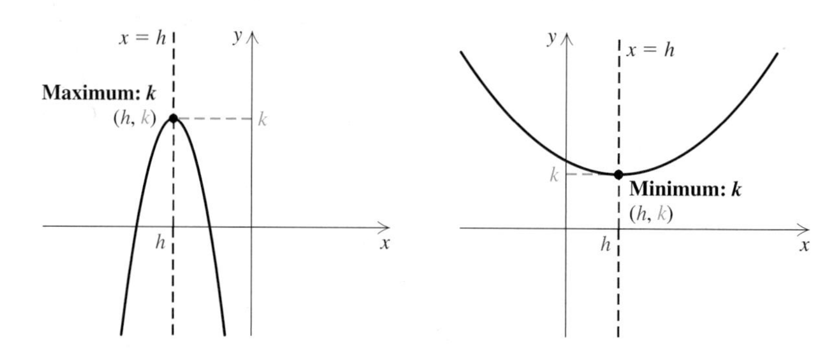

> ***Graphing $f(x) = a(x - h)^2 + k$***
>
> The graph of $f(x) = a(x - h)^2 + k$ has the same shape as the graph of $y = a(x - h)^2$.
>
> If k is positive, the graph of $y = a(x - h)^2$ is shifted k units up.
>
> If k is negative, the graph of $y = a(x - h)^2$ is shifted $|k|$ units down.
>
> The vertex is (h, k), and the axis of symmetry is $x = h$.
>
> For $a > 0$, k is the minimum function value. For $a < 0$, k is the maximum function value.

technology connection

To study the effect of k on the graph of $f(x) = a(x - h)^2 + k$, let $y_1 = 7(x - 1)^2$ and $y_2 = 7(x - 1)^2 + 2$. Graph both y_1 and y_2 in the window $[-5, 5, -5, 5]$ and use TRACE or a TABLE to compare the y-values for any given x-value.

Next, let $y_3 = 7(x - 1)^2 - 4$ and compare its graph and y-values with those of y_1 and y_2. Try other values of k, including decimals and fractions. Describe the effect of adding k to the right side of $f(x) = a(x - h)^2$.

Example 4 Graph $g(x) = (x - 3)^2 - 5$, and find the minimum function value.

Solution The graph will look like that of $f(x) = (x - 3)^2$ (see Example 2) but shifted 5 units down. You can confirm this by plotting some points. For instance, $g(4) = (4 - 3)^2 - 5 = -4$, whereas in Example 2, $f(4) = (4 - 3)^2 = 1$.

The vertex is now $(3, -5)$, and the minimum function value is -5.

x	$g(x) = (x - 3)^2 - 5$
0	4
1	−1
2	−4
3	−5 ← Vertex
4	−4
5	−1
6	4

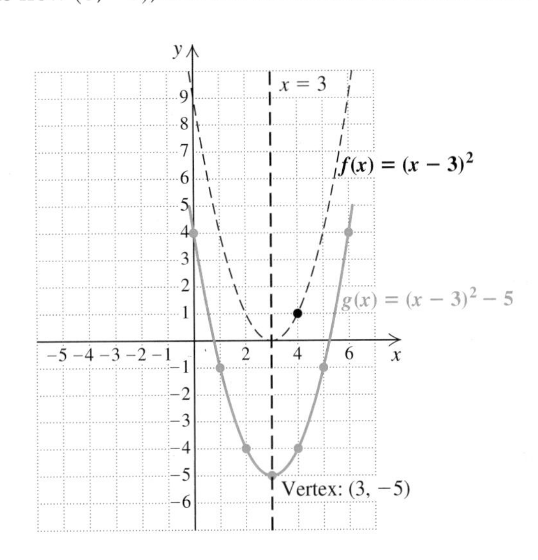

Example 5 Graph $h(x) = \frac{1}{2}(x - 3)^2 + 5$, and find the minimum function value.

Solution The graph looks just like that of $f(x) = \frac{1}{2}x^2$ but moved 3 units to the right and 5 units up. The vertex is $(3, 5)$, and the axis of symmetry is $x = 3$. We draw $f(x) = \frac{1}{2}x^2$ and then shift the curve over and up. The minimum function value is 5. By plotting some points, we have a check.

x	$h(x) = \frac{1}{2}(x - 3)^2 + 5$
0	$9\frac{1}{2}$
1	7
3	5 ← Vertex
5	7
6	$9\frac{1}{2}$

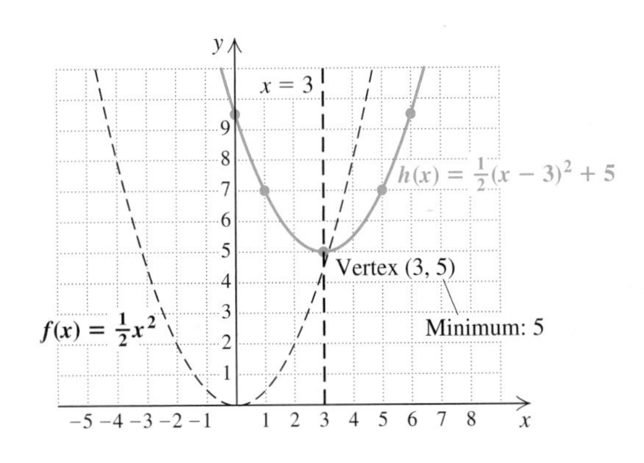

Example 6 Graph $y = -2(x + 3)^2 + 5$. Find the vertex, the axis of symmetry, and the maximum or minimum value.

Solution We first express the equation in the equivalent form

$$y = -2[x - (-3)]^2 + 5.$$

The graph looks like that of $y = -2x^2$ translated 3 units to the left and 5 units up. The vertex is $(-3, 5)$, and the axis of symmetry is $x = -3$. Since -2 is negative, we know that 5, the second coordinate of the vertex, is the maximum y-value.

We compute a few points as needed, selecting convenient x-values on either side of the vertex. The graph is shown here.

x	$y = -2(x + 3)^2 + 5$	
-4	3	
-3	5	← Vertex
-2	3	

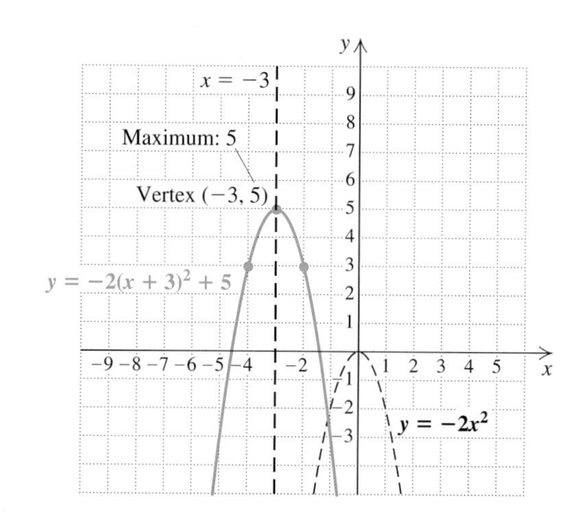

CONNECTING THE CONCEPTS

The ability to graph a function is an important skill. Later in this chapter, as well as in future courses, you will find that graphs of polynomial functions can be used as a tool for solving equations, inequalities, and real-world applications. In the process of learning how to graph quadratic functions, we have developed the ability to reflect or shift (translate) a graph. This skill will prove useful not only in future courses, but in Chapters 9 and 10 as well.

Exercise Set 8.6

FOR EXTRA HELP

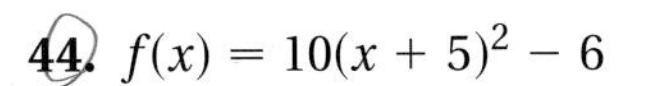

Digital Video Tutor CD 7 Videotape 16 | InterAct Math | Math Tutor Center | MathXL | MyMathLab.com

Graph.

1. $f(x) = x^2$

2. $f(x) = -x^2$

3. $f(x) = -2x^2$

4. $f(x) = -3x^2$

5. $g(x) = \frac{1}{3}x^2$

6. $g(x) = \frac{1}{4}x^2$

Aha! **7.** $h(x) = -\frac{1}{3}x^2$

8. $h(x) = -\frac{1}{4}x^2$

9. $f(x) = \frac{5}{2}x^2$

10. $f(x) = \frac{3}{2}x^2$

For each of the following, graph the function, label the vertex, and draw the axis of symmetry.

11. $g(x) = (x + 4)^2$

12. $g(x) = (x + 1)^2$

13. $f(x) = (x - 1)^2$

14. $f(x) = (x - 2)^2$

15. $h(x) = (x - 3)^2$

16. $h(x) = (x - 4)^2$

17. $f(x) = -(x + 4)^2$

18. $f(x) = -(x - 2)^2$

19. $g(x) = -(x - 1)^2$

20. $g(x) = -(x + 1)^2$

21. $f(x) = 2(x + 1)^2$

22. $f(x) = 2(x + 4)^2$

23. $h(x) = -\frac{1}{2}(x - 4)^2$

24. $h(x) = -\frac{3}{2}(x - 2)^2$

25. $f(x) = \frac{1}{2}(x - 1)^2$

26. $f(x) = \frac{1}{3}(x + 2)^2$

27. $f(x) = -2(x + 5)^2$

28. $f(x) = 2(x + 7)^2$

29. $h(x) = -3\left(x - \frac{1}{2}\right)^2$

30. $h(x) = -2\left(x + \frac{1}{2}\right)^2$

For each of the following, graph the function and find the vertex, the axis of symmetry, and the maximum value or the minimum value.

31. $f(x) = (x - 5)^2 + 2$

32. $f(x) = (x + 3)^2 - 2$

33. $f(x) = (x + 1)^2 - 3$

34. $f(x) = (x - 1)^2 + 2$

35. $g(x) = (x + 4)^2 + 1$

36. $g(x) = -(x - 2)^2 - 4$

37. $h(x) = -2(x - 1)^2 - 3$

38. $h(x) = -2(x + 1)^2 + 4$

39. $f(x) = 2(x + 4)^2 + 1$

40. $f(x) = 2(x - 5)^2 - 3$

41. $g(x) = -\frac{3}{2}(x - 1)^2 + 4$

42. $g(x) = \frac{3}{2}(x + 2)^2 - 3$

Without graphing, find the vertex, the axis of symmetry, and the maximum value or the minimum value.

43. $f(x) = 8(x - 9)^2 + 7$

44. $f(x) = 10(x + 5)^2 - 6$

45. $h(x) = -\frac{2}{7}(x + 6)^2 + 11$

46. $h(x) = -\frac{3}{11}(x - 7)^2 - 9$

47. $f(x) = 5\left(x + \frac{1}{4}\right)^2 - 13$

48. $f(x) = 6\left(x - \frac{1}{4}\right)^2 + 15$

49. $f(x) = \sqrt{2}(x + 4.58)^2 + 65\pi$

50. $f(x) = 4\pi(x - 38.2)^2 - \sqrt{34}$

51. Explain, without plotting points, why the graph of $y = x^2 - 4$ looks like the graph of $y = x^2$ translated 4 units down.

52. Explain, without plotting points, why the graph of $y = (x + 2)^2$ looks like the graph of $y = x^2$ translated 2 units to the left.

SKILL MAINTENANCE

Graph using intercepts.

53. $2x - 7y = 28$

54. $6x - 3y = 36$

Solve each system.

55. $3x + 4y = -19,$
$7x - 6y = -29$

56. $5x + 7y = 9,$
$3x - 4y = -11$

Complete the square.

57. $x^2 + 5x +$ ____

58. $x^2 - 9x +$ ____

SYNTHESIS

59. Before graphing a quadratic function, Sophie always plots five points. First, she calculates and plots the coordinates of the vertex. Then she plots *four* more points after calculating *two* more ordered pairs. How is this possible?

60. If the graphs of $f(x) = a_1(x - h_1)^2 + k_1$ and $g(x) = a_2(x - h_2)^2 + k_2$ have the same shape, what, if anything, can you conclude about the a's, the h's, and the k's? Why?

Write an equation for a function having a graph with the same shape as the graph of $f(x) = \frac{3}{5}x^2$, but with the given point as the vertex.

61. (4, 1) **62.** (2, 6) **63.** (3, −1)

64. (5, −6) **65.** (−2, −5) **66.** (−4, −2)

For each of the following, write the equation of the parabola that has the shape of $f(x) = 2x^2$ or $g(x) = -2x^2$ and has a maximum or minimum value at the specified point.

67. Maximum: (5, 0) **68.** Minimum: (2, 0)

69. Minimum: (−4, 0) **70.** Maximum: (0, 3)

71. Maximum: (3, 8) **72.** Minimum: (−2, 3)

Find an equation for a quadratic function F that satisfies the following conditions.

73. The graph of F is the same shape as the graph of f, where $f(x) = 3(x + 2)^2 + 7$, and $F(x)$ is a minimum at the same point that $g(x) = -2(x - 5)^2 + 1$ is a maximum.

74. The graph of F is the same shape as the graph of f, where $f(x) = -\frac{1}{3}(x - 2)^2 + 7$, and $F(x)$ is a maximum at the same point that $g(x) = 2(x + 4)^2 - 6$ is a minimum.

Functions other than parabolas can be translated. When calculating $f(x)$, if we replace x with $x - h$, where h is a constant, the graph will be moved horizontally. If we replace $f(x)$ with $f(x) + k$, the graph will be moved vertically.

Use the graph below for Exercises 75–80.

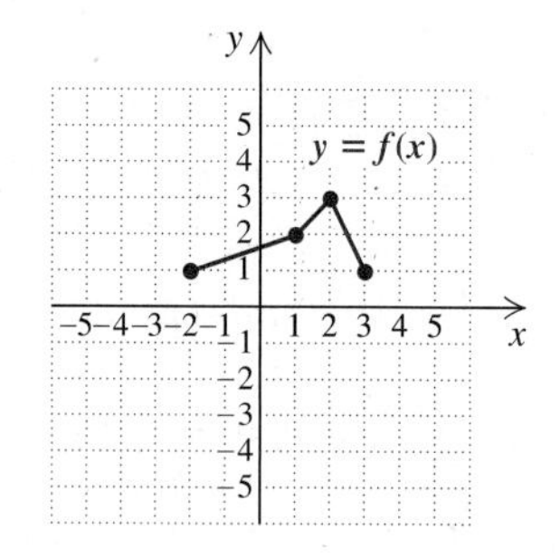

Draw a graph of each of the following.

75. $y = f(x - 1)$

76. $y = f(x + 2)$

77. $y = f(x) + 2$

78. $y = f(x) - 3$

79. $y = f(x + 3) - 2$

80. $y = f(x - 3) + 1$

81. Use the TRACE and/or TABLE features of a grapher to confirm the maximum and minimum values given as answers to Exercises 43, 45, and 47. Be sure to adjust the window appropriately. On some graphers, a maximum or minimum option may be available by using a CALC key.

82. Use a grapher to check your graphs for Exercises 10, 20, and 40.

83. While trying to graph $y = -\frac{1}{2}x^2 + 3x + 1$, Omar gets the following screen.

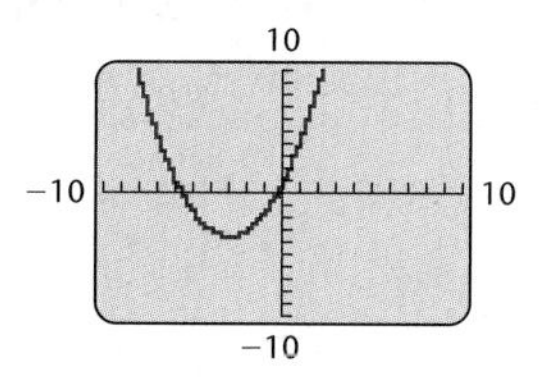

How can Omar tell at a glance that a mistake has been made?

COLLABORATIVE

CORNER

Match the Graph

Focus: Graphing quadratic functions

Time: 15–20 minutes

Group size: 6

Materials: Index cards

ACTIVITY

1. On each of six index cards, write one of the following equations:

$$y = \tfrac{1}{2}(x - 3)^2 + 1; \qquad y = \tfrac{1}{2}(x - 1)^2 + 3;$$
$$y = \tfrac{1}{2}(x + 1)^2 - 3; \qquad y = \tfrac{1}{2}(x + 3)^2 + 1;$$
$$y = \tfrac{1}{2}(x + 3)^2 - 1; \qquad y = \tfrac{1}{2}(x + 1)^2 + 3.$$

2. Fold each index card and mix up the six cards in a hat or bag. Then, one by one, each group member should select one of the equations. Do not let anyone see your equation.

3. Each group member should carefully graph the equation selected. Make the graph large enough so that when it is finished, it can be easily viewed by the rest of the group. Be sure to scale the axes and label the vertex, but **do not label the graph with the equation used.**

4. When all group members have drawn a graph, place the graphs in a pile. The group should then match and agree on the correct equation for each graph *with no help from the person who drew the graph.* If a mistake has been made and a graph has no match, determine what its equation *should* be.

5. Compare your group's labeled graphs with those of other groups to reach consensus within the class on the correct label for each graph.

More About Graphing Quadratic Functions

8.7

Completing the Square • Finding Intercepts

Completing the Square

By *completing the square* (see Section 8.1), we can rewrite any polynomial $ax^2 + bx + c$ in the form $a(x - h)^2 + k$. Once that has been done, the procedures discussed in Section 8.6 will enable us to graph any quadratic function.

Example 1 Graph: $g(x) = x^2 - 6x + 4$.

Solution We have

$$\begin{aligned} g(x) &= x^2 - 6x + 4 \\ &= (x^2 - 6x) + 4. \end{aligned}$$

To complete the square inside the parentheses, we take half the x-coefficient, $\frac{1}{2} \cdot (-6) = -3$, and square it to get $(-3)^2 = 9$. Then we add $9 - 9$ inside the parentheses:

$$g(x) = (x^2 - 6x + 9 - 9) + 4 \quad \text{The effect is of adding 0.}$$
$$= (x^2 - 6x + 9) + (-9 + 4) \quad \text{Using the associative law of addition to regroup}$$
$$= (x - 3)^2 - 5. \quad \text{Factoring and simplifying}$$

This equation was graphed in Example 4 of Section 8.6. The graph is that of $f(x) = x^2$ translated right 3 units and down 5 units. The vertex is $(3, -5)$.

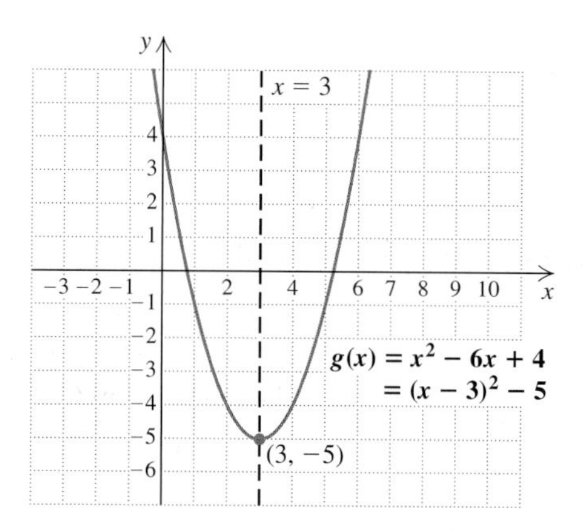

When the leading coefficient is not 1, we factor out that number from the first two terms. Then we complete the square.

Example 2 Graph: $f(x) = 3x^2 + 12x + 13$.

Solution Since the coefficient of x^2 is not 1, we need to factor out that number—in this case, 3—from the first two terms. Remember that we want the form $f(x) = a(x - h)^2 + k$:

$$f(x) = 3x^2 + 12x + 13$$
$$= 3(x^2 + 4x) + 13.$$

Now we complete the square as before. We take half of the x-coefficient, $\frac{1}{2} \cdot 4 = 2$, and square it: $2^2 = 4$. Then we add $4 - 4$ inside the parentheses:

$$f(x) = 3(x^2 + 4x + 4 - 4) + 13. \quad \text{Adding } 4 - 4\text{, or 0, inside the parentheses}$$

The distributive law allows us to separate the -4 from the perfect-square trinomial so long as it is multiplied by 3:

$$f(x) = 3(x^2 + 4x + 4) + 3(-4) + 13 \quad \text{This leaves a perfect-square trinomial inside the parentheses.}$$
$$= 3(x + 2)^2 + 1. \quad \text{Factoring and simplifying}$$

The vertex is $(-2, 1)$, and the axis of symmetry is $x = -2$. The coefficient of x^2 is 3, so the graph is narrow and opens upward. We choose a few x-values on either side of the vertex, compute y-values, and then graph the parabola.

x	$f(x) = 3(x + 2)^2 + 1$	
-2	1	← Vertex
-3	4	
-1	4	

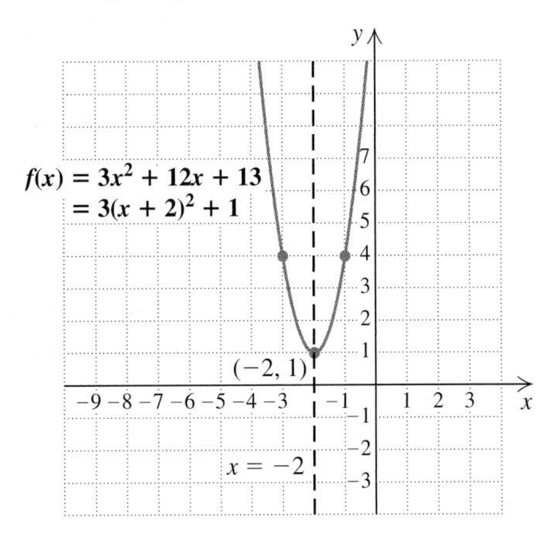

Example 3 Graph: $f(x) = -2x^2 + 10x - 7$.

Solution We first find the vertex by completing the square. To do so, we factor out -2 from the first two terms of the expression. This makes the coefficient of x^2 inside the parentheses 1:

$$f(x) = -2x^2 + 10x - 7$$
$$= -2(x^2 - 5x) - 7.$$

Now we complete the square as before. We take half of the x-coefficient and square it to get $\frac{25}{4}$. Then we add $\frac{25}{4} - \frac{25}{4}$ inside the parentheses:

$$f(x) = -2\left(x^2 - 5x + \tfrac{25}{4} - \tfrac{25}{4}\right) - 7$$
$$= -2\left(x^2 - 5x + \tfrac{25}{4}\right) + (-2)\left(-\tfrac{25}{4}\right) - 7$$ Multiplying by -2, using the distributive law, and regrouping
$$= -2\left(x - \tfrac{5}{2}\right)^2 + \tfrac{11}{2}.$$ Factoring and simplifying

The vertex is $\left(\frac{5}{2}, \frac{11}{2}\right)$, and the axis of symmetry is $x = \frac{5}{2}$. The coefficient of x^2, -2, is negative, so the graph opens downward. We plot a few points on either side of the vertex, including the y-intercept, $f(0)$, and graph the parabola.

x	$f(x)$	
$\frac{5}{2}$	$\frac{11}{2}$	← Vertex
0	-7	← y-intercept
1	1	
4	1	

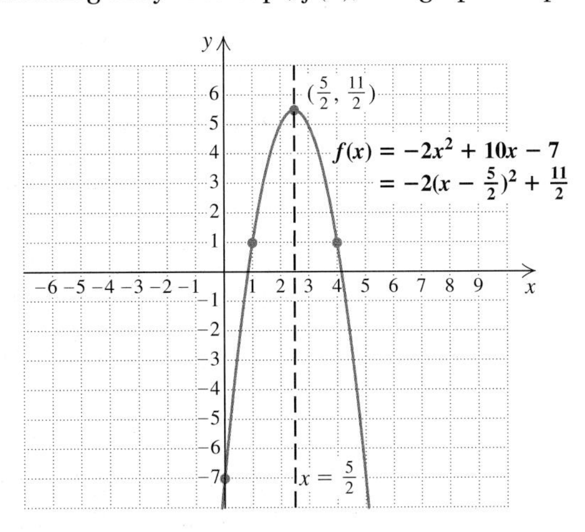

The method used in Examples 1–3 can be generalized to find a formula for locating the vertex. We complete the square as follows:

$$f(x) = ax^2 + bx + c$$

$$= a\left(x^2 + \frac{b}{a}x\right) + c. \qquad \text{Factoring } a \text{ out of the first two terms. Check by multiplying.}$$

Half of the x-coefficient, $\frac{b}{a}$, is $\frac{b}{2a}$. We square it to get $\frac{b^2}{4a^2}$ and add $\frac{b^2}{4a^2} - \frac{b^2}{4a^2}$ inside the parentheses. Then we distribute the a and regroup terms:

$$f(x) = a\left(x^2 + \frac{b}{a}x + \frac{b^2}{4a^2} - \frac{b^2}{4a^2}\right) + c$$

$$= a\left(x^2 + \frac{b}{a}x + \frac{b^2}{4a^2}\right) + a\left(-\frac{b^2}{4a^2}\right) + c \qquad \text{Using the distributive law}$$

$$= a\left(x + \frac{b}{2a}\right)^2 + \frac{-b^2}{4a} + \frac{4ac}{4a} \qquad \text{Factoring and finding a common denominator}$$

$$= a\left[x - \left(-\frac{b}{2a}\right)\right]^2 + \frac{4ac - b^2}{4a}.$$

Thus we have the following.

The Vertex of a Parabola

The vertex of the parabola given by $f(x) = ax^2 + bx + c$ is

$$\left(-\frac{b}{2a}, f\left(-\frac{b}{2a}\right)\right) \quad \text{or} \quad \left(-\frac{b}{2a}, \frac{4ac - b^2}{4a}\right).$$

The x-coordinate of the vertex is $-b/(2a)$. The axis of symmetry is $x = -b/(2a)$. The second coordinate of the vertex is most commonly found by computing $f\left(-\frac{b}{2a}\right)$.

Let's reexamine Example 3 to see how we could have found the vertex directly. From the formula above,

$$\text{the } x\text{-coordinate of the vertex is } -\frac{b}{2a} = -\frac{10}{2(-2)} = \frac{5}{2}.$$

Substituting $\frac{5}{2}$ into $f(x) = -2x^2 + 10x - 7$, we find the second coordinate of the vertex:

$$f\left(\tfrac{5}{2}\right) = -2\left(\tfrac{5}{2}\right)^2 + 10\left(\tfrac{5}{2}\right) - 7$$

$$= -2\left(\tfrac{25}{4}\right) + 25 - 7$$

$$= -\tfrac{25}{2} + 18$$

$$= -\tfrac{25}{2} + \tfrac{36}{2} = \tfrac{11}{2}.$$

The vertex is $\left(\frac{5}{2}, \frac{11}{2}\right)$. The axis of symmetry is $x = \frac{5}{2}$.

We have actually developed two methods for finding the vertex. One is by completing the square and the other is by using a formula. You should check with your instructor about which method to use.

Finding Intercepts

The points at which a graph crosses an axis are called intercepts. We saw in Chapter 2 and again in Example 3 that the y-intercept occurs at $f(0)$. For $f(x) = ax^2 + bx + c$, the y-intercept is simply $(0, c)$. To find x-intercepts, we look for points where $y = 0$ or $f(x) = 0$. To find the x-intercepts of the quadratic function given by $f(x) = ax^2 + bx + c$, we solve the equation

$$0 = ax^2 + bx + c.$$

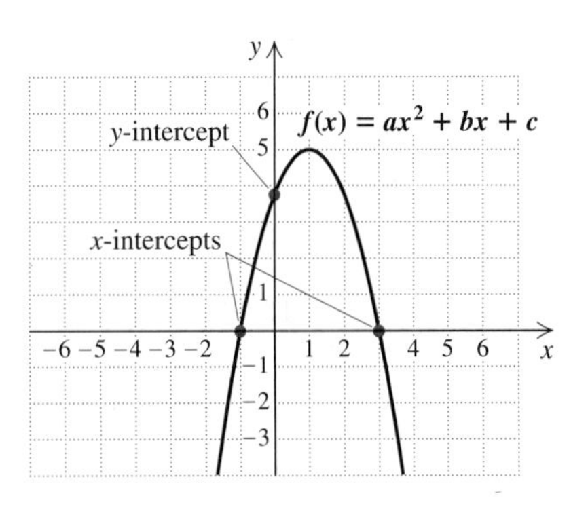

Example 4 Find the x- and y-intercepts of the graph of $f(x) = x^2 - 2x - 2$.

Solution The y-intercept is simply $(0, f(0))$, or $(0, -2)$. To find the x-intercepts, we solve the equation

$$0 = x^2 - 2x - 2.$$

We are unable to factor $x^2 - 2x - 2$, so we use the quadratic formula and get $x = 1 \pm \sqrt{3}$. Thus the x-intercepts are $\left(1 - \sqrt{3}, 0\right)$ and $\left(1 + \sqrt{3}, 0\right)$.

If graphing, we would approximate, to get $(-0.7, 0)$ and $(2.7, 0)$.

Exercise Set 8.7

FOR EXTRA HELP

Digital Video Tutor CD 7
Videotape 16

InterAct Math

Math Tutor Center

MathXL

MyMathLab.com

For each quadratic function, **(a)** *find the vertex and the axis of symmetry and* **(b)** *graph the function.*

1. $f(x) = x^2 + 4x + 5$
2. $f(x) = x^2 + 2x - 5$
3. $g(x) = x^2 - 6x + 13$
4. $g(x) = x^2 - 4x + 5$
5. $f(x) = x^2 + 8x + 20$
6. $f(x) = x^2 - 10x + 21$
7. $h(x) = 2x^2 - 16x + 25$
8. $h(x) = 2x^2 + 16x + 23$
9. $f(x) = -x^2 + 2x + 5$
10. $f(x) = -x^2 - 2x + 7$
11. $g(x) = x^2 + 3x - 10$
12. $g(x) = x^2 + 5x + 4$
13. $f(x) = 3x^2 - 24x + 50$
14. $f(x) = 4x^2 + 8x - 3$
15. $h(x) = x^2 + 7x$
16. $h(x) = x^2 - 5x$
17. $f(x) = -2x^2 - 4x - 6$
18. $f(x) = -3x^2 + 6x + 2$
19. $g(x) = 2x^2 - 8x + 3$
20. $g(x) = 2x^2 + 5x - 1$
21. $f(x) = -3x^2 + 5x - 2$
22. $f(x) = -3x^2 - 7x + 2$
23. $h(x) = \frac{1}{2}x^2 + 4x + \frac{19}{3}$
24. $h(x) = \frac{1}{2}x^2 - 3x + 2$

Find the x- and y-intercepts. If no x-intercepts exist, state this.

25. $f(x) = x^2 - 6x + 3$
26. $f(x) = x^2 + 5x + 2$
27. $g(x) = -x^2 + 2x + 3$
28. $g(x) = x^2 - 6x + 9$
29. Aha! $f(x) = x^2 - 9x$
30. $f(x) = x^2 - 7x$
31. $h(x) = -x^2 + 4x - 4$
32. $h(x) = 4x^2 - 12x + 3$
33. $f(x) = 2x^2 - 4x + 6$
34. $f(x) = x^2 - x + 2$
35. Does the graph of every quadratic function have a y-intercept? Why or why not?
36. Is it possible for the graph of a quadratic function to have only one x-intercept if the vertex is off the x-axis? Why or why not?

SKILL MAINTENANCE

Solve each system.

37. $5x - 3y = 16,$
 $4x + 2y = 4$
38. $2x - 5y = 9,$
 $5x - 15y = 20$
39. $4a - 5b + c = 3,$
 $3a - 4b + 2c = 3,$
 $a + b - 7c = -2$
40. $2a - 7b + c = 25,$
 $a + 5b - 2c = -18,$
 $3a - b + 4c = 14$

Solve.

41. $\sqrt{4x - 4} = \sqrt{x + 4} + 1$
42. $\sqrt{5x - 4} + \sqrt{13 - x} = 7$

SYNTHESIS

43. If the graphs of two quadratic functions have the same x-intercepts, will they also have the same vertex? Why or why not?
44. Suppose that the graph of $f(x) = ax^2 + bx + c$ has $(x_1, 0)$ and $(x_2, 0)$ as x-intercepts. Explain why the graph of $g(x) = -ax^2 - bx - c$ will also have $(x_1, 0)$ and $(x_2, 0)$ as x-intercepts.

For each quadratic function, find **(a)** *the maximum or minimum value and* **(b)** *the x- and y-intercepts.*

45. $f(x) = 2.31x^2 - 3.135x - 5.89$
46. $f(x) = -18.8x^2 + 7.92x + 6.18$
47. Graph the function

$$f(x) = x^2 - x - 6.$$

Then use the graph to approximate solutions to each of the following equations.

a) $x^2 - x - 6 = 2$
b) $x^2 - x - 6 = -3$

48. Graph the function

$$f(x) = \frac{x^2}{2} + x - \frac{3}{2}.$$

Then use the graph to approximate solutions to each of the following equations.

a) $\frac{x^2}{2} + x - \frac{3}{2} = 0$

b) $\frac{x^2}{2} + x - \frac{3}{2} = 1$

c) $\frac{x^2}{2} + x - \frac{3}{2} = 2$

Find an equivalent equation of the type

$$f(x) = a(x - h)^2 + k.$$

49. $f(x) = mx^2 - nx + p$

50. $f(x) = 3x^2 + mx + m^2$

51. A quadratic function has $(-1, 0)$ as one of its intercepts and $(3, -5)$ as its vertex. Find an equation for the function.

52. A quadratic function has $(4, 0)$ as one of its intercepts and $(-1, 7)$ as its vertex. Find an equation for the function.

Graph.

53. $f(x) = |x^2 - 1|$

54. $f(x) = |x^2 - 3x - 4|$

55. $f(x) = |2(x - 3)^2 - 5|$

56. Use a grapher to check your answers to Exercises 9, 23, 33, 45, and 47.

8.8 Problem Solving and Quadratic Functions

Maximum and Minimum Problems • Fitting Quadratic Functions to Data

Let's look now at some of the many situations in which quadratic functions are used for problem solving.

Maximum and Minimum Problems

We have seen that for any quadratic function f, the value of $f(x)$ at the vertex is either a maximum or a minimum. Thus problems in which a quantity must be maximized or minimized can often be solved by finding the coordinates of a vertex. This assumes that the problem can be modeled with a quadratic function.

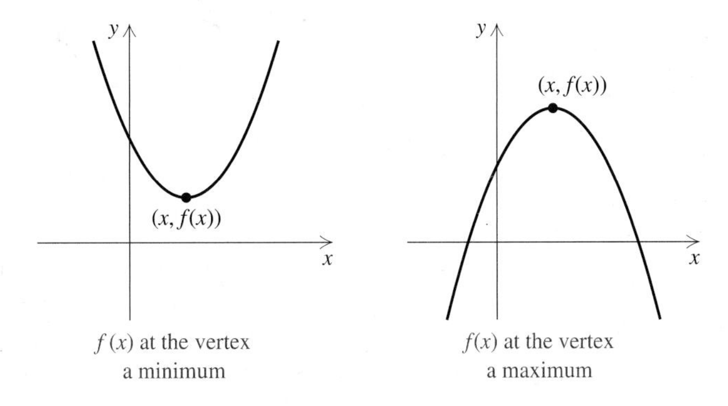

$f(x)$ at the vertex a minimum

$f(x)$ at the vertex a maximum

Example 1

Newborn calves. The number of pounds of milk per day recommended for a calf that is x weeks old can be approximated by $p(x)$, where $p(x) = -0.2x^2 + 1.3x + 6.2$ (*Source*: C. Chaloux, University of Vermont, 1998). When is a calf's milk consumption greatest and how much milk does it consume at that time?

Solution

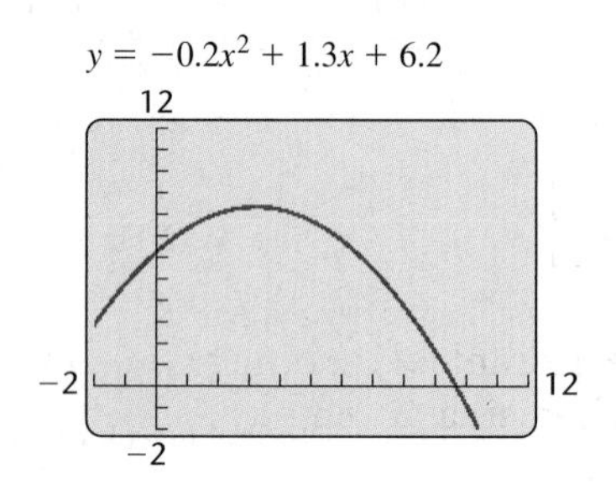

A visualization for Example 1

1., 2. Familiarize and **Translate.** We are given the function for milk consumption by a calf. Note that it is a quadratic function of x, the calf's age in weeks. If it is not difficult to do so, we can generate a graph of the function. The graph (shown at left) indicates that the calf's consumption increases and then decreases.

3. Carry out. We can either complete the square,

$$\begin{aligned} p(x) &= -0.2x^2 + 1.3x + 6.2 \\ &= -0.2(x^2 - 6.5x) + 6.2 \\ &= -0.2(x^2 - 6.5x + 3.25^2 - 3.25^2) + 6.2 \quad \text{Completing the square; } \frac{6.5}{2} = 3.25 \\ &= -0.2(x^2 - 6.5x + 3.25^2) + (-0.2)(-3.25^2) + 6.2 \\ &= -0.2(x - 3.25)^2 + 8.3125 \quad \text{Factoring and simplifying} \end{aligned}$$

or we can use $-b/(2a) = -1.3/(-0.4) = 3.25$. Using a calculator, we find that

$$p(3.25) = -0.2(3.25)^2 + 1.3(3.25) + 6.2 = 8.3125.$$

4. Check. Both of the approaches in step (3) indicate that a maximum occurs when $x = 3.25$, or $3\frac{1}{4}$. The graph also serves as a check.

5. State. A calf's milk consumption is greatest when the calf is $3\frac{1}{4}$ weeks old. At that time, it drinks about 8.3 lb of milk per day.

Example 2

Fenced-in land. What are the dimensions of the largest rectangular pen that a farmer can enclose with 64 m of electric fence?

Solution

1. Familiarize. We make a drawing and label it. Recall these important formulas:

Perimeter: $2w + 2l$;

Area: $l \cdot w$.

To get a better feel for the problem, we can look at some possible dimensions for a rectangular pen that can be enclosed with 64 m of fence. All possibilities are chosen so that $2w + 2l = 64$.

l	w	Perimeter	Area
22 m	10 m	64 m	220 m^2
20 m	12 m	64 m	240 m^2
18 m	14 m	64 m	252 m^2
⋮	⋮		⋮

What choice of l and w will maximize A?

To generate a table of values on your grapher, let x represent the width of the pen, in meters. If l represents the length, in meters, we must have $64 = 2x + 2l$. Next, solve for l and use that expression for y_1. Then let $y_2 = x$ and $y_3 = y_1 \cdot y_2$ and create a table.

2. **Translate.** We have two equations: One guarantees that the perimeter is 64 m; the other expresses area in terms of length and width.

$$2w + 2l = 64$$
$$A = l \cdot w$$

3. **Carry out.** We need to express A as a function of l or w but not both. To do so, we solve for l in the first equation to obtain $l = 32 - w$. Substituting for l in the second equation, we get a quadratic function:

$$A = (32 - w)w \qquad \text{Substituting for } l$$
$$= -w^2 + 32w. \qquad \text{This is a parabola opening downward, so a maximum exists.}$$

Completing the square, we get

$$A = -(w^2 - 32w + 256 - 256)$$
$$= -(w - 16)^2 + 256.$$

The maximum function value, 256 m^2, occurs when $w = 16$ m and $l = 32 - 16$, or 16 m.

4. **Check.** Note that 256 m^2 is greater than any of the values for A found in the *Familiarize* step. To be more certain, we could check values other than those used in that step. For example, if $w = 15$ m, then $l = 32 - 15 = 17$ m, and $A = 15 \cdot 17 = 255$ m^2. The same area results if $w = 17$ m and $l = 15$ m. Since 256 m^2 is greater than 255 m^2, it looks as though we have a maximum.

5. **State.** The largest rectangular pen that can be enclosed is 16 m by 16 m.

Fitting Quadratic Functions to Data

Whenever a certain quadratic function fits a situation, that function can be determined if three inputs and their outputs are known. Each of the given ordered pairs is called a *data point*.

Example 3 ***Hydrology.*** The drawing below shows the cross section of a river. Typically rivers are deepest in the middle, with the depth decreasing to 0 at the edges. A hydrologist measures the depths D, in feet, of a river at distances x, in feet, from one bank. The results are listed in the table below.

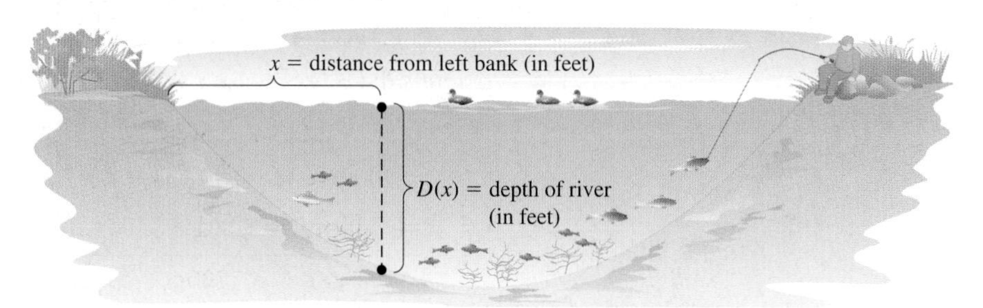

Distance x, from the Left Bank (in feet)	Depth, D, of the River (in feet)
0	0
15	10.2
25	17
50	20
90	7.2
100	0

a) Plot the data and decide whether the data seem to fit a quadratic function.
b) Use the data points $(0, 0)$, $(50, 20)$, and $(100, 0)$ to find a quadratic function that fits the data.
c) Use the function to estimate the depth of the river 70 ft from the left bank.

Solution

a) We plot the data points as follows.

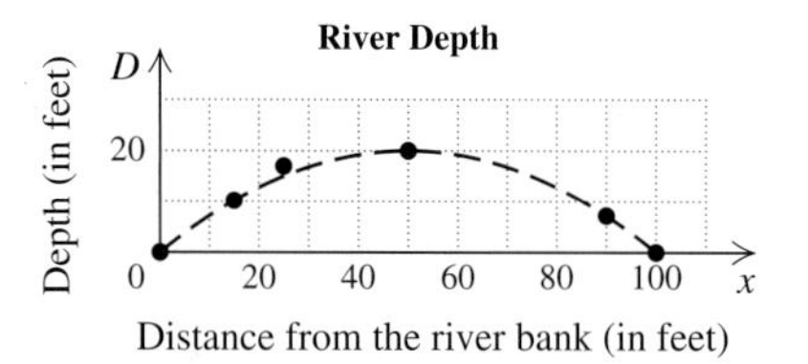

The data seem to rise and fall in a manner similar to a quadratic function. The dashed curve represents a quadratic function that closely fits the data. Note that it need not pass through each point.

b) We are looking for a quadratic function of the form

$$D(x) = ax^2 + bx + c.$$

To determine the constants a, b, and c, we use the points $(0, 0)$, $(50, 20)$, and $(100, 0)$ and substitute:

$$0 = a \cdot 0^2 + b \cdot 0 + c,$$
$$20 = a \cdot 50^2 + b \cdot 50 + c,$$
$$0 = a \cdot 100^2 + b \cdot 100 + c.$$

Enter the data from the table in Example 3 using the EDIT option of the STAT menu. Then use the CALC option of the STAT menu to find the quadratic function that most closely fits the data. Finally, graph the function just developed.

After simplifying, we see that we need to solve the system

$$\begin{aligned} 0 &= c, \\ 20 &= 2{,}500a + 50b + c, \\ 0 &= 10{,}000a + 100b + c. \end{aligned}$$

Since $c = 0$, the system reduces to a system of two equations in two variables:

$$\begin{aligned} 20 &= 2{,}500a + 50b, && (1) \\ 0 &= 10{,}000a + 100b. && (2) \end{aligned}$$

We multiply both sides of equation (1) by -2, add, and solve for a:

$$\begin{aligned} -40 &= -5{,}000a - 100b, \\ 0 &= 10{,}000a + 100b \\ \hline -40 &= 5000a && \text{Adding} \\ \frac{-40}{5000} &= a && \text{Solving for } a \\ -0.008 &= a. \end{aligned}$$

Next, we substitute -0.008 for a in equation (2) and solve for b:

$$\begin{aligned} 0 &= 10{,}000(-0.008) + 100b \\ 0 &= -80 + 100b \\ 80 &= 100b \\ 0.8 &= b. \end{aligned}$$

We can now rewrite $D(x) = ax^2 + bx + c$ as

$$D(x) = -0.008x^2 + 0.8x.$$

c) To find the depth 70 ft from the riverbank, we substitute:

$$D(70) = -0.008(70)^2 + 0.8(70) = 16.8.$$

At a distance of 70 ft from the riverbank, the river is 16.8 ft deep.

Exercise Set 8.8

FOR EXTRA HELP

Digital Video Tutor CD 7
Videotape 16

InterAct Math

Math Tutor Center

MathXL

MyMathLab.com

Solve.

1. *Stock prices.* The value of a share of R. P. Mugahti can be represented by $V(x) = x^2 - 6x + 13$, where x is the number of months after January 2001. What is the lowest value $V(x)$ will reach, and when will that occur?

2. *Minimizing cost.* Aki's Bicycle Designs has determined that when x hundred bicycles are built, the average cost per bicycle is given by

$$C(x) = 0.1x^2 - 0.7x + 2.425,$$

where $C(x)$ is in hundreds of dollars. What is the minimum average cost per bicycle and how many bicycles should be built to achieve that minimum?

3. *Ticket sales.* The number of tickets sold each day for an upcoming performance of Handel's Messiah is given by

$$N(x) = -0.4x^2 + 9x + 11,$$

where x is the number of days since the concert was first announced. When will daily ticket sales peak and how many tickets will be sold that day?

4. *Maximizing profit.* Recall that total profit P is the difference between total revenue R and total cost C. Given $R(x) = 1000x - x^2$ and $C(x) = 3000 + 20x$, find the total profit, the maximum value of the total profit, and the value of x at which it occurs.

5. *Architecture.* An architect is designing an atrium for a hotel. The atrium is to be rectangular with a perimeter of 720 ft of brass piping. What dimensions will maximize the area of the atrium?

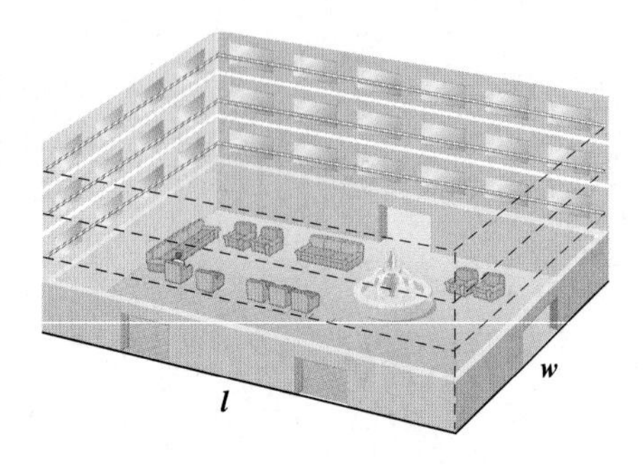

6. *Stained-glass window design.* An artist is designing a rectangular stained-glass window with a perimeter of 84 in. What dimensions will yield the maximum area?

7. *Garden design.* A farmer decides to enclose a rectangular garden, using the side of a barn as one side of the rectangle. What is the maximum area that the farmer can enclose with 40 ft of fence? What should the dimensions of the garden be in order to yield this area?

8. *Patio design.* A stone mason has enough stones to enclose a rectangular patio with 60 ft of perimeter, assuming that the attached house forms one side of the rectangle. What is the maximum area that the mason can enclose? What should the dimensions of the patio be in order to yield this area?

9. *Molding plastics.* Economite Plastics plans to produce a one-compartment vertical file by bending the long side of an 8-in. by 14-in. sheet of plastic along two lines to form a U shape. How tall should the file be in order to maximize the volume that the file can hold?

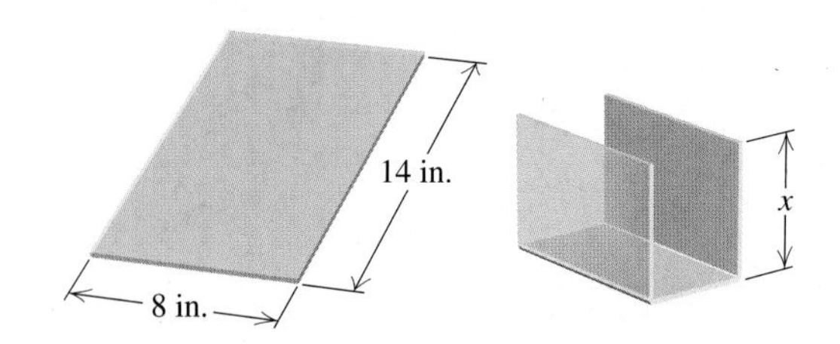

10. *Composting.* A rectangular compost container is to be formed in a corner of a fenced yard, with 8 ft

of chicken wire completing the other two sides of the rectangle. If the chicken wire is 3 ft high, what dimensions of the base will maximize the container's volume?

11. What is the maximum product of two numbers that add to 18? What numbers yield this product?

12. What is the maximum product of two numbers that add to 26? What numbers yield this product?

13. What is the minimum product of two numbers that differ by 8? What are the numbers?

14. What is the minimum product of two numbers that differ by 7? What are the numbers?

Aha! **15.** What is the maximum product of two numbers that add to -10? What numbers yield this product?

16. What is the maximum product of two numbers that add to -12? What numbers yield this product?

For Exercises 17–24, state whether the graph appears to represent a quadratic function.

17.

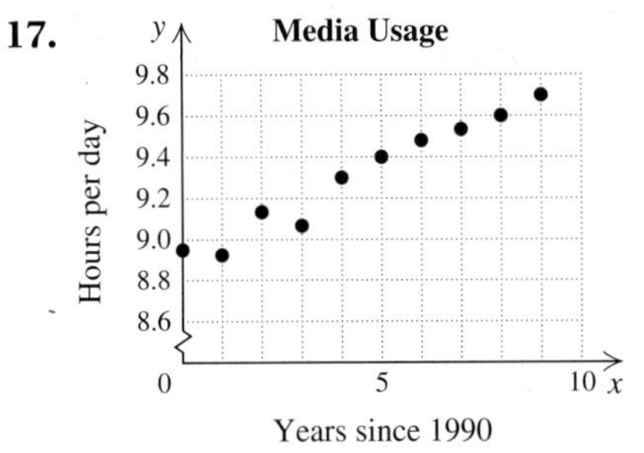

18.

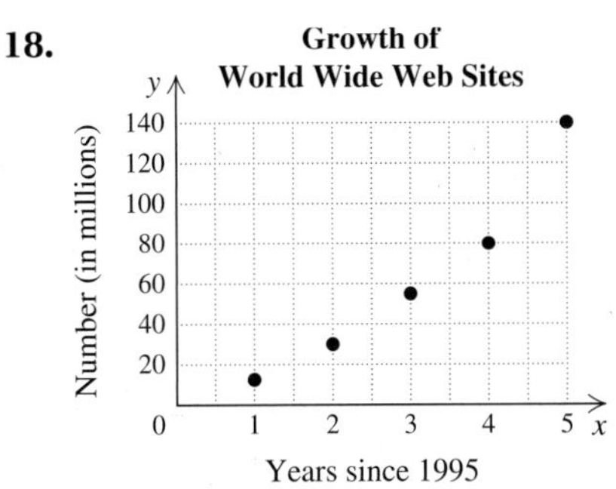

19.

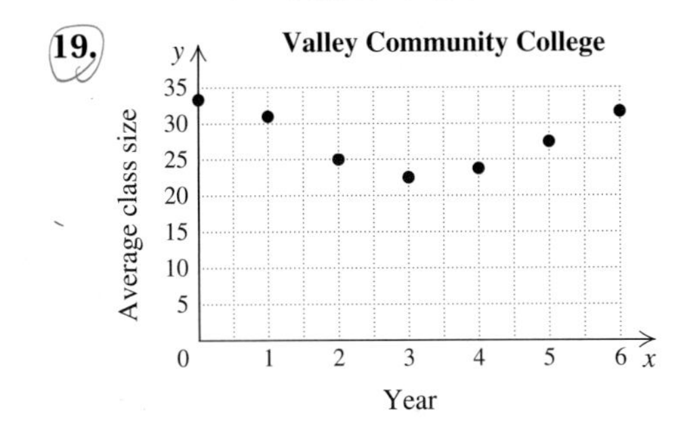

20.

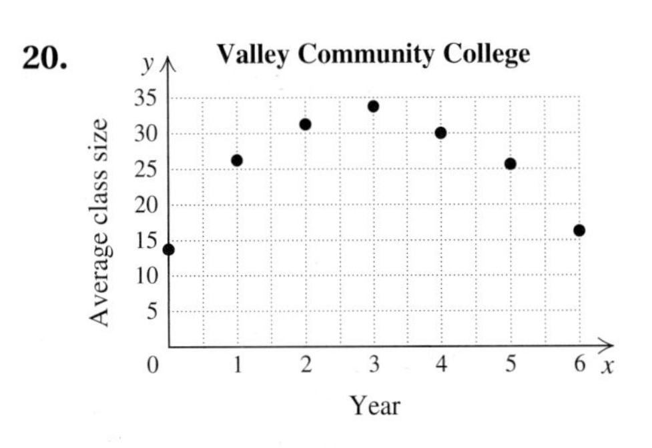

21.

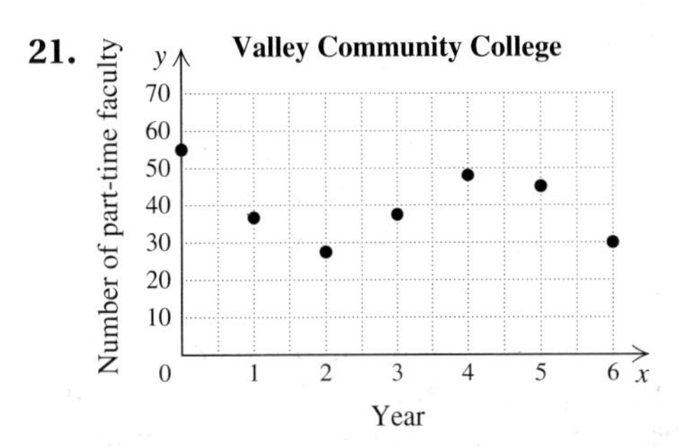

22.

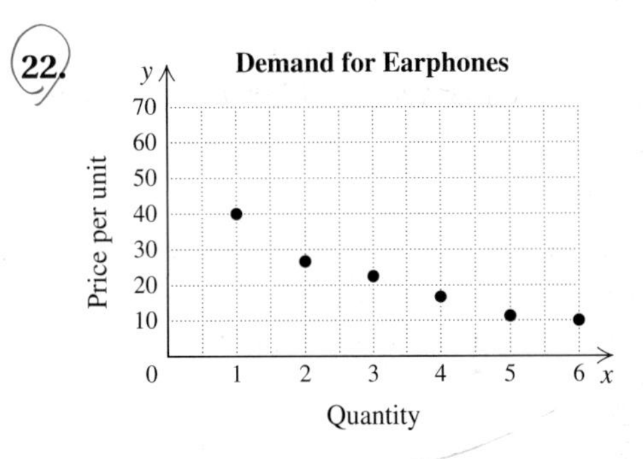

23.

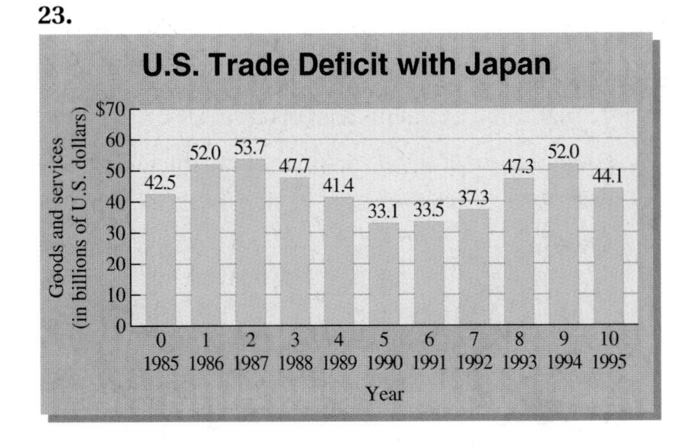

24.

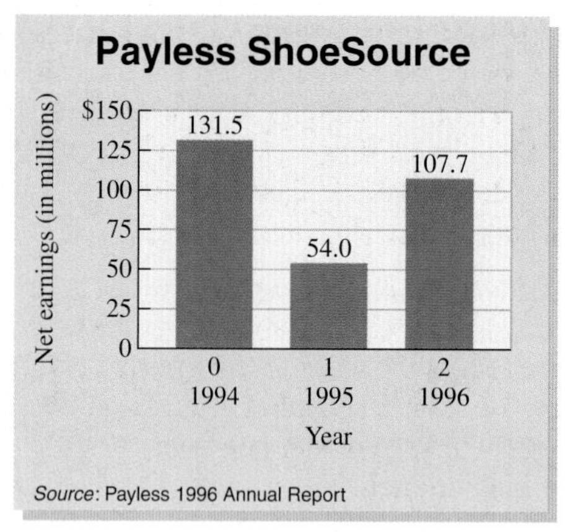

Find a quadratic function that fits the set of data points.

25. $(1, 4), (-1, -2), (2, 13)$

26. $(1, 4), (-1, 6), (-2, 16)$

27. $(2, 0), (4, 3), (12, -5)$

28. $(-3, -30), (3, 0), (6, 6)$

29. a) Find a quadratic function that fits the following data.

Travel Speed (in kilometers per hour)	Number of Nighttime Accidents (for every 200 million kilometers driven)
60	400
80	250
100	250

b) Use the function to estimate the number of nighttime accidents that occur at 50 km/h.

30. a) Find a quadratic function that fits the following data.

Travel Speed (in kilometers per hour)	Number of Daytime Accidents (for every 200 million kilometers driven)
60	100
80	130
100	200

b) Use the function to estimate the number of daytime accidents that occur at 50 km/h.

31. *Archery.* The Olympic flame tower at the 1992 Summer Olympics was lit at a height of about 27 m by a flaming arrow that was launched about 63 m from the base of the tower. If the arrow landed about 63 m beyond the tower, find a quadratic function that expresses the height h of the arrow as a function of the distance d that it traveled horizontally.

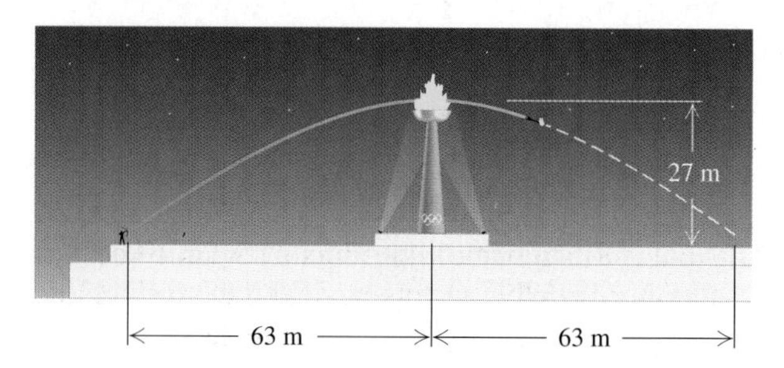

32. *Pizza prices.* Pizza Unlimited has the following prices for pizzas.

Diameter	Price
8 in.	$ 6.00
12 in.	$ 8.50
16 in.	$11.50

Is price a quadratic function of diameter? It probably should be, because the price should be proportional to the area, and the area is a quadratic function of the diameter. (The area of a circular region is given by $A = \pi r^2$ or $(\pi/4) \cdot d^2$.)

a) Express price as a quadratic function of diameter using the data points $(8, 6)$, $(12, 8.50)$, and $(16, 11.50)$.

b) Use the function to find the price of a 14-in. pizza.

33. Does every nonlinear function have a minimum or maximum value? Why or why not?

34. Explain how the leading coefficient of a quadratic function can be used to determine if a maximum or a minimum function value exists.

SKILL MAINTENANCE

Simplify.

35. $\dfrac{x}{x^2 + 17x + 72} - \dfrac{8}{x^2 + 15x + 56}$

36. $\dfrac{x^2 - 9}{x^2 - 8x + 7} \div \dfrac{x^2 + 6x + 9}{x^2 - 1}$

37. $\dfrac{t^2 - 4}{t^2 - 7t - 8} \cdot \dfrac{t^2 - 64}{t^2 - 5t + 6}$

38. $\dfrac{t}{t^2 - 10t + 21} + \dfrac{t}{t^2 - 49}$

Solve.

39. $5x - 9 < 31$

40. $3x - 8 \geq 22$

SYNTHESIS

41. Write a problem for a classmate to solve. Design the problem so that its solution requires finding the minimum or maximum value.

42. Explain what restrictions should be placed on the quadratic functions developed in Exercises 29 and 32 and why such restrictions are needed.

43. *Norman window.* A *Norman window* is a rectangle with a semicircle on top. Big Sky Windows is designing a Norman window that will require 24 ft of trim. What dimensions will allow the maximum amount of light to enter a house?

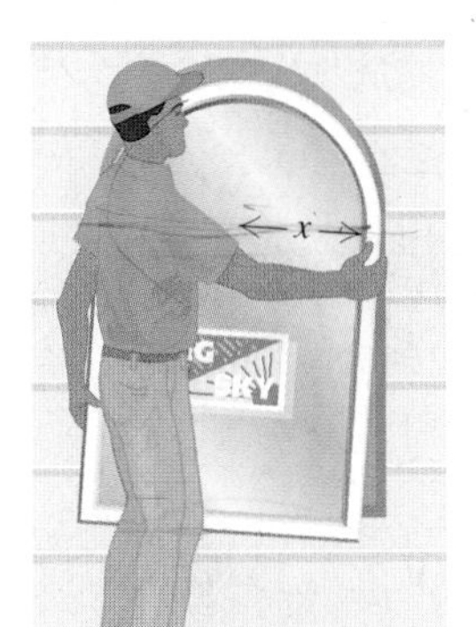

44. *Minimizing area.* A 36-in. piece of string is cut into two pieces. One piece is used to form a circle while the other is used to form a square. How should the string be cut so that the sum of the areas is a minimum?

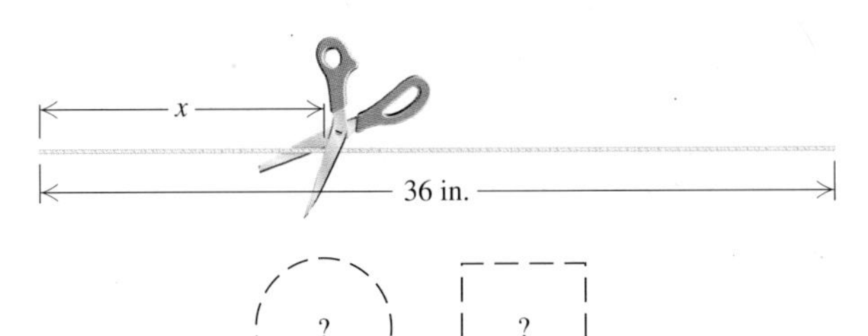

45. *Crop yield.* An orange grower finds that she gets an average yield of 40 bushels (bu) per tree when she plants 20 trees on an acre of ground. Each time she adds a tree to an acre, the yield per tree decreases by 1 bu, due to congestion. How many trees per acre should she plant for maximum yield?

46. *Cover charges.* When the owner of Sweet Sounds charges a \$10 cover charge, an average of 80 people will attend a show. For each 25¢ increase in admission price, the average number attending decreases by 1. What should the owner charge in order to make the most money?

47. *Trajectory of a launched object.* The height above the ground of a launched object is a quadratic function of the time that it is in the air. Suppose that a flare is launched from a cliff 64 ft above sea level. If 3 sec after being launched the flare is again level with the cliff, and if 2 sec after that it lands in the sea, what is the maximum height that the flare will reach?

48. *Bridge design.* The cables supporting a straight-line suspension bridge are nearly parabolic in shape. Suppose that a suspension bridge is being designed with concrete supports 160 ft apart and with vertical cables 30 ft above road level at the midpoint of the bridge and 80 ft above road level at a point 50 ft from the midpoint of the bridge. How long are the longest vertical cables?

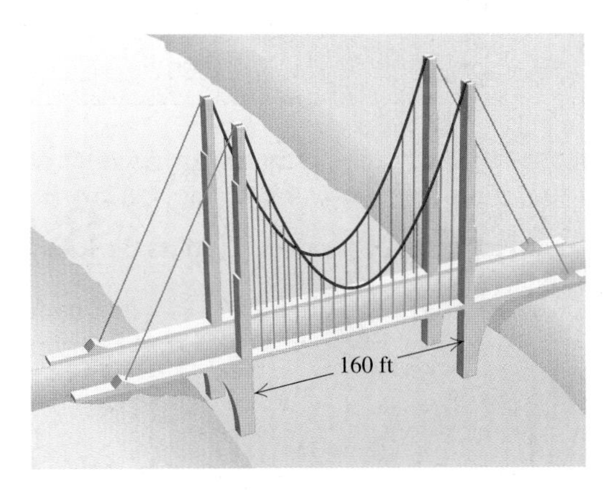

49. Many graphers have a QUADREG option as part of the CALC option of the STAT menu. Such a feature can quickly fit a quadratic function to a LIST of ordered pairs. Use the QUADREG feature to check your answers to Exercises 25, 27, and 29.

COLLABORATIVE CORNER

Quadratic Counter Settings

Focus: Modeling quadratic functions
Time: 20–30 minutes
Group size: 3 or 4
Materials: Graphers are optional.

The Panasonic Portable Stereo System RX-DT680® has a counter for finding locations on an audio cassette. When a fully wound cassette with 45 min of music on a side begins to play, the counter is at 0. After 15 min of music has played, the counter reads 250 and after 35 min, it reads 487. When the 45-min side is finished playing, the counter reads 590.

ACTIVITY

1. The paragraph above describes four ordered pairs of the form (counter number, minutes played). Three pairs are enough to find a function of the form

$$T(n) = an^2 + bn + c,$$

where $T(n)$ represents the time, in minutes, that the tape has run at counter reading n hundred. Each group member should select a different set of three points from the four given and then fit a quadratic function to the data.
2. Of the 3 or 4 functions found in part (1) above, which fits the data "best"? One way to answer this is to see how well each function predicts other pairs. The same counter used above reads 432 after a 45-min tape has played for 30 min. Which function comes closest to predicting this?
3. If a grapher is available with a QUADREG option (see Exercise 49), what function does it fit to the four pairs originally listed?
4. If a class member has access to a Panasonic System RX-DT680, see how well the functions developed above predict the counter readings for a tape that has played for 5 or 10 min.

8.9 Polynomial and Rational Inequalities

Quadratic and Other Polynomial Inequalities • Rational Inequalities

Quadratic and Other Polynomial Inequalities

Inequalities like the following are called *polynomial inequalities*:

$$x^3 - 5x > x^2 + 7, \qquad 4x - 3 < 9, \qquad 5x^2 - 3x + 2 \geq 0.$$

Second-degree polynomial inequalities in one variable are called *quadratic inequalities.* To solve polynomial inequalities, we often focus attention on where the outputs of a polynomial function are positive and where they are negative.

Example 1

Solve: $x^2 + 3x - 10 > 0$.

Solution Consider the "related" function $f(x) = x^2 + 3x - 10$ and its graph. Its graph opens upward since the leading coefficient is positive.

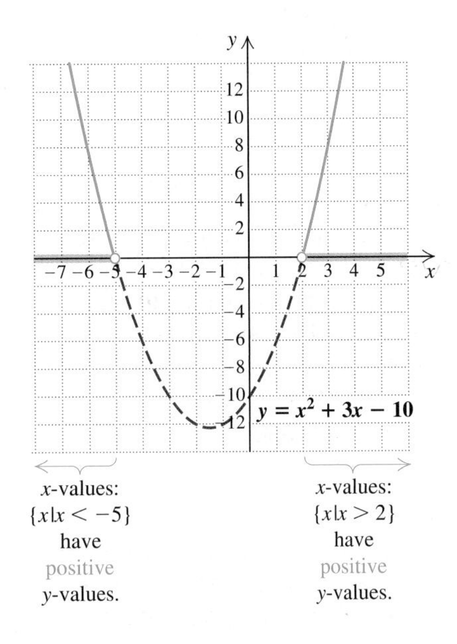
y
x
$y = x^2 + 3x - 10$
x-values:
{x|x < −5}
have
positive
y-values.
x-values:
{x|x > 2}
have
positive
y-values.

Values of y will be positive to the left and right of the x-intercepts, as shown. To find the intercepts, we set the polynomial equal to 0 and solve:

$$x^2 + 3x - 10 = 0$$
$$(x + 5)(x - 2) = 0$$
$$x + 5 = 0 \quad or \quad x - 2 = 0$$
$$x = -5 \quad or \quad x = 2.$$

Thus the solution set of the inequality is

$$\{x \mid x < -5 \; or \; x > 2\}, \quad \text{or} \quad (-\infty, -5) \cup (2, \infty).$$

Any inequality with 0 on one side can be solved by considering a graph of the related function and finding intercepts as in Example 1. Sometimes the quadratic formula is needed to find the intercepts.

Example 2

Solve: $x^2 - 2x \leq 2$.

Solution We first find standard form with 0 on one side:

$$x^2 - 2x - 2 \leq 0.$$ This is equivalent to the original inequality.

The graph of $f(x) = x^2 - 2x - 2$ is a parabola opening upward. Values of $f(x)$ are negative for x-values between the x-intercepts. We find the x-intercepts by solving $f(x) = 0$:

$$x = \frac{-b \pm \sqrt{b^2 - 4ac}}{2a}$$

$$= \frac{-(-2) \pm \sqrt{(-2)^2 - 4 \cdot 1(-2)}}{2 \cdot 1}$$

$$= \frac{2 \pm \sqrt{12}}{2} = \frac{2 \pm 2\sqrt{3}}{2} = 1 \pm \sqrt{3}.$$

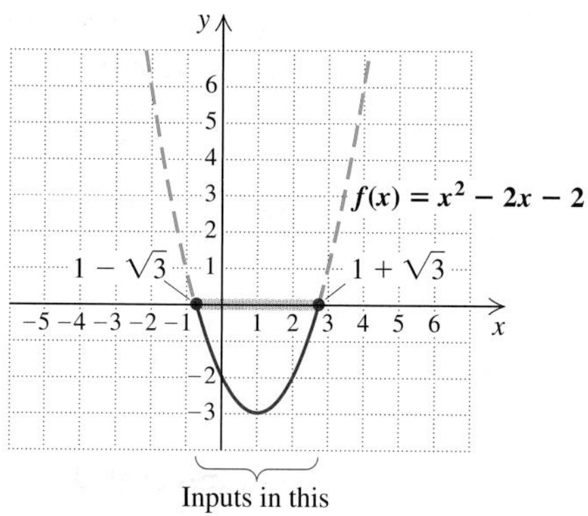

At the x-intercepts, $1 - \sqrt{3}$ and $1 + \sqrt{3}$, the value of $f(x)$ is 0. Thus the solution set of the inequality is

$$\left[1 - \sqrt{3}, 1 + \sqrt{3}\right], \quad \text{or} \quad \left\{x \middle| 1 - \sqrt{3} \le x \le 1 + \sqrt{3}\right\}.$$

Note that in Example 2 it was not essential to actually draw the graph. The important information came from the location of the x-intercepts and the sign of $f(x)$ on each side of those intercepts.

In the next example, we solve a third-degree polynomial inequality, without graphing, by locating the x-intercepts, or **zeros**, of f and then using *test points* to determine the sign of $f(x)$ over each interval of the x-axis.

Example 3 Solve: $5x^3 + 10x^2 - 15x > 0$.

Solution We first solve the related equation:

$$5x^3 + 10x^2 - 15x = 0$$

$$5x(x^2 + 2x - 3) = 0$$

$$5x(x + 3)(x - 1) = 0$$

$$5x = 0 \quad or \quad x + 3 = 0 \quad or \quad x - 1 = 0$$

$$x = 0 \quad or \quad x = -3 \quad or \quad x = 1.$$

We see that if $f(x) = 5x^3 + 10x^2 - 15x$, then the zeros of f are $-3, 0$, and 1. These zeros divide the number line, or x-axis, into four intervals: A, B, C, and D.

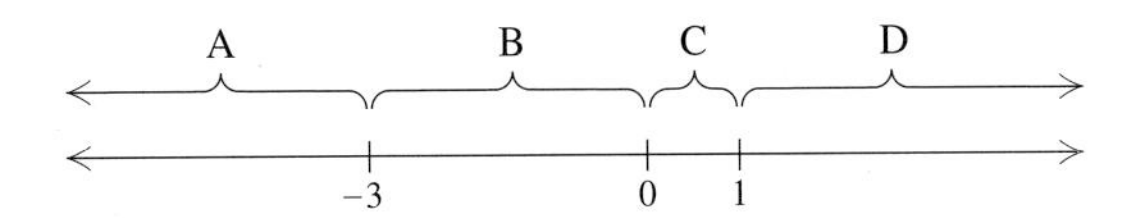

Next, using one convenient test value from each interval, we determine the sign of $f(x)$ for that interval. We know that, within each interval, the sign of $f(x)$ cannot change. If it did, there would need to be another zero in that interval. Using the factored form of $f(x)$ eases the computations:

$$f(x) = 5x(x+3)(x-1).$$

For interval A,

$$f(-4) = 5(-4)((-4)+3)((-4)-1)$$ -4 is a convenient value in interval A.

$$= -20(-1)(-5)$$

$$= -100.$$ $f(-4)$ is negative.

For interval B,

$$f(-1) = 5(-1)((-1)+3)((-1)-1)$$ -1 is a convenient value in interval B.

$$= -5(2)(-2)$$

$$= 20.$$ $f(-1)$ is positive.

For interval C,

$$f\left(\tfrac{1}{2}\right) = \underbrace{\underbrace{5 \cdot \tfrac{1}{2}}_{\text{Positive}} \cdot \underbrace{\left(\tfrac{1}{2}+3\right)}_{\text{Positive}} \cdot \underbrace{\left(\tfrac{1}{2}-1\right)}_{\text{Negative}}}_{\text{Negative}}.$$

$\frac{1}{2}$ is a convenient value in interval C. Only the sign is important. The product is negative, so $f\left(\frac{1}{2}\right)$ is negative.

For interval D,

$$f(2) = \underbrace{5 \cdot 2}_{\text{Positive}} \cdot \underbrace{(2+3)}_{\text{Positive}} \cdot \underbrace{(2-1)}_{\text{Positive}}.$$

2 is a convenient value in interval D. $f(2)$ is positive.

Recall that we are looking for all x for which $5x^3 + 10x^2 - 15x > 0$. The calculations above indicate that $f(x)$ is positive for any number in intervals B and D. The solution set of the original inequality is

$$(-3, 0) \cup (1, \infty), \quad \text{or} \quad \{x \mid -3 < x < 0 \text{ or } x > 1\}.$$

Note that the calculations in Example 3 were made simpler by using the factored form of the polynomial. The process was simplified further when, for intervals C and D, we concentrated on only the *sign* of $f(x)$. In the next example, we determine the sign of a polynomial function over each interval by tracking the sign of each factor. By looking at how many positive or negative factors are being multiplied, we will be able to determine the sign of the polynomial function.

Example 4

Solve: $4x^3 - 4x \leq 0$.

Solution We first solve the related equation:

$$4x^3 - 4x = 0$$
$$4x(x^2 - 1) = 0$$
$$4x(x + 1)(x - 1) = 0$$
$$4x = 0 \quad or \quad x + 1 = 0 \quad or \quad x - 1 = 0$$
$$x = 0 \quad or \quad x = -1 \quad or \quad x = 1.$$

The function $f(x) = 4x^3 - 4x$ has zeros at $-1, 0$, and 1. We could now use test values, as in Example 3. Instead, let's use the factorization $f(x) = 4x(x + 1)(x - 1)$. The product $4x(x + 1)(x - 1)$ is positive or negative, depending on the signs of the factors $4x$, $x + 1$, and $x - 1$. This is easily determined using a chart.

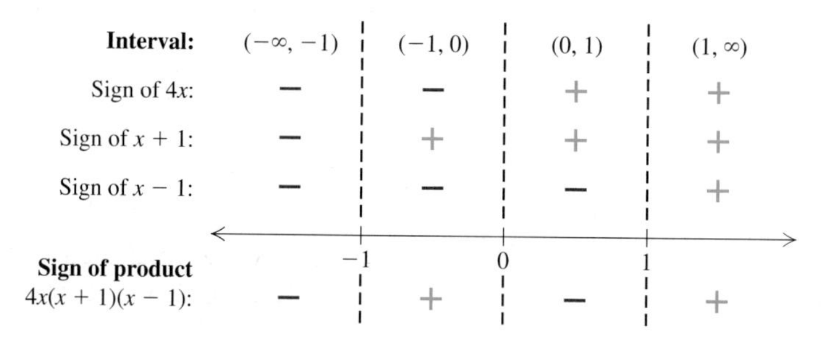

Interval: (−∞, −1) (−1, 0) (0, 1) (1, ∞)
Sign of 4x: − − + +
Sign of x + 1: − + + +
Sign of x − 1: − − − +
−1 0 1
Sign of product 4x(x + 1)(x − 1): − + − +

A product is negative when it has an odd number of negative factors. Since the $\leq$ sign allows for equality, the endpoints $-1, 0$, and 1 are solutions. From the chart, we see that the solution set is

$$(-\infty, -1] \cup [0, 1], \quad \text{or} \quad \{x \mid x \leq -1 \text{ or } 0 \leq x \leq 1\}.$$

To Solve a Polynomial Inequality Using Factors

1. Get 0 on one side and solve the related polynomial equation by factoring.
2. Use the numbers found in step (1) to divide the number line into intervals.
3. Using a test value from each interval, determine the sign of each factor over that interval.
4. Determine the sign of the product of the factors over each interval. Remember that the product of an odd number of negative numbers is negative.
5. Select the interval(s) for which the inequality is satisfied and write set-builder notation or interval notation for the solution set. Include the endpoints of the intervals when $\leq$ or $\geq$ is used.

To solve $2.3x^2 \leq 9.11 - 2.94x$, we first rewrite the inequality in the form $2.3x^2 + 2.94x - 9.11 \leq 0$ and graph the function $f(x) = 2.3x^2 + 2.94x - 9.11$.

$y = 2.3x^2 + 2.94x - 9.11$

10, −10, 10, −10

To find the values of x for which $f(x) \leq 0$, we focus on the region in which the graph lies *on or below* the x-axis. From this graph, it appears that this region begins somewhere between -3 and -2, and continues to somewhere between 1 and 2. Using the ZERO or ROOT option of CALC, we can find the endpoints of this region. To two decimal places, the endpoints are -2.73 and 1.45. The solution set is approximately $\{x \mid -2.73 \leq x \leq 1.45\}$.

Had the inequality been $2.3x^2 > 9.11 - 2.94x$, we would look for portions of the graph that lie *above* the x-axis. An approximate solution set of such an inequality would be $\{x \mid x < -2.73 \text{ or } x > 1.45\}$.

Use a grapher to solve each inequality. Round the values of the endpoints to the nearest hundredth.

1. $4.32x^2 - 3.54x - 5.34 \leq 0$
2. $7.34x^2 - 16.55x - 3.89 \geq 0$
3. $10.85x^2 + 4.28x + 4.44 > 7.91x^2 + 7.43x + 13.03$
4. $5.79x^3 - 5.68x^2 + 10.68x > 2.11x^3 + 16.90x - 11.69$

Rational Inequalities

Inequalities involving rational expressions are called **rational inequalities.** Like polynomial inequalities, rational inequalities can be solved using test values. Unlike polynomials, however, rational expressions often have values for which the expression is undefined.

Example 5 Solve: $\frac{x-3}{x+4} \geq 2$.

Solution We write the related equation by changing the $\geq$ symbol to $=$:

$$\frac{x-3}{x+4} = 2.$$

Next, we solve this related equation:

$$(x+4) \cdot \frac{x-3}{x+4} = (x+4) \cdot 2 \quad \text{Multiplying both sides by the LCD, } x+4$$

$$x - 3 = 2x + 8$$

$$-11 = x. \quad \text{Solving for } x$$

In the case of rational inequalities, we also need to find any values that make the denominator 0. We set the denominator equal to 0 and solve:

$$\left.\begin{aligned} x + 4 &= 0 \\ x &= -4. \end{aligned}\right\} \quad \text{This tells us that } -4 \text{ is not in the domain of } f \text{ if } f(x) = \frac{x-3}{x+4}.$$

Now we use -11 and -4 to divide the number line into intervals:

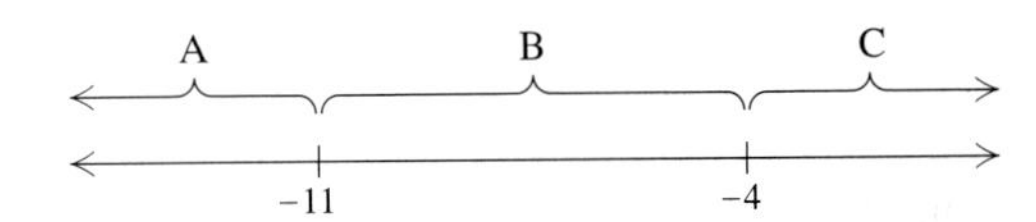

We test a number in each interval to see where the original inequality is satisfied:

$$\frac{x-3}{x+4} \geq 2.$$

A: Test -15, $\frac{-15-3}{-15+4} = \frac{-18}{-11}$

$= \frac{18}{11} \not\geq 2$ $\quad$ -15 *is not* a solution, so interval A is not part of the solution set.

B: Test -8, $\frac{-8-3}{-8+4} = \frac{-11}{-4}$

$= \frac{11}{4} \geq 2$ $\quad$ -8 *is* a solution, so interval B is part of the solution set.

C: Test 1, $\dfrac{1-3}{1+4} = \dfrac{-2}{5}$

$= -\dfrac{2}{5} \not\geq 2$ 1 *is not* a solution, so interval C is not part of the solution set.

The solution set includes the interval B. The endpoint -11 is included because the inequality symbol is $\geq$ and -11 is a solution of the related equation. The number -4 is *not* included because $(x-3)/(x+4)$ is undefined for $x = -4$. Thus the solution set of the original inequality is

$$[-11, -4), \quad \text{or} \quad \{x \mid -11 \leq x < -4\}.$$

There is an interesting visual interpretation of Example 5. If we graph the function $f(x) = (x-3)/(x+4)$, we see that the solutions of the inequality $(x-3)/(x+4) \geq 2$ can be found by inspection. We simply sketch the line $y = 2$ and locate all x-values for which $f(x) \geq 2$.

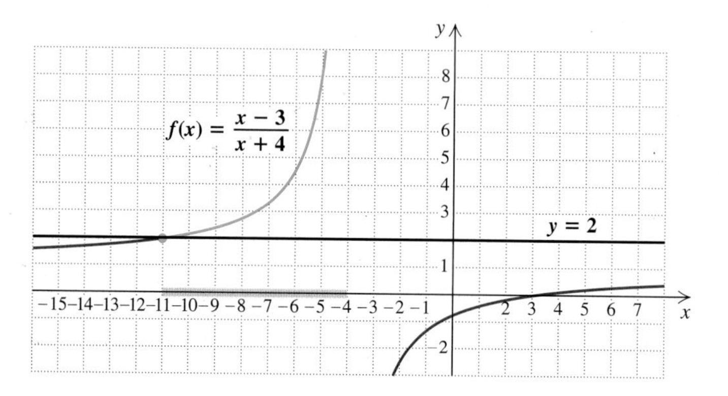

Because graphing rational functions can be very time-consuming, we generally just use test values.

To Solve a Rational Inequality

1. Change the inequality symbol to an equals sign and solve the related equation.
2. Find any replacements for which the rational expression is undefined.
3. Use the numbers found in steps (1) and (2) to divide the number line into intervals.
4. Substitute a test value from each interval into the inequality. If the number is a solution, then the interval to which it belongs is part of the solution set.
5. Select the interval(s) and any endpoints for which the inequality is satisfied and write set-builder or interval notation for the solution set. If the inequality symbol is $\leq$ or $\geq$, then the solutions from step (1) are also included in the solution set. Those numbers found in step (2) should be excluded from the solution set, even if they are solutions from step (1).

Exercise Set 8.9

FOR EXTRA HELP

Digital Video Tutor CD 7
Videotape 16

InterAct Math

Math Tutor Center

MathXL

MyMathLab.com

Solve.

1. $(x + 4)(x - 3) < 0$

2. $(x - 5)(x + 2) > 0$

3. $(x + 7)(x - 2) \geq 0$

4. $(x - 1)(x + 4) \leq 0$

5. $x^2 - x - 2 < 0$

6. $x^2 + x - 2 < 0$

7. $25 - x^2 \geq 0$

8. $4 - x^2 \geq 0$

Aha! **9.** $x^2 + 4x + 4 < 0$

10. $x^2 + 6x + 9 < 0$

11. $x^2 - 4x < 12$

12. $x^2 + 6x > -8$

13. $3x(x + 2)(x - 2) < 0$

14. $5x(x + 1)(x - 1) > 0$

15. $(x + 3)(x - 2)(x + 1) > 0$

16. $(x - 1)(x + 2)(x - 4) < 0$

17. $(x + 3)(x + 2)(x - 1) < 0$

18. $(x - 2)(x - 3)(x + 1) < 0$

19. $\dfrac{1}{x + 3} < 0$

20. $\dfrac{1}{x + 4} > 0$

21. $\dfrac{x + 1}{x - 5} \geq 0$

22. $\dfrac{x - 2}{x + 5} \leq 0$

23. $\dfrac{3x + 2}{2x - 4} \leq 0$

24. $\dfrac{5 - 2x}{4x + 3} \leq 0$

25. $\dfrac{x + 1}{x + 6} > 1$

26. $\dfrac{x - 1}{x - 2} < 1$

27. $\dfrac{(x - 2)(x + 1)}{x - 5} \leq 0$

28. $\dfrac{(x + 4)(x - 1)}{x + 3} \geq 0$

29. $\dfrac{x}{x + 3} \geq 0$

30. $\dfrac{x - 2}{x} \leq 0$

31. $\dfrac{x - 5}{x} < 1$

32. $\dfrac{x}{x - 1} > 2$

33. $\dfrac{x - 1}{(x - 3)(x + 4)} \leq 0$

34. $\dfrac{x + 2}{(x - 2)(x + 7)} \geq 0$

35. $4 < \dfrac{1}{x}$

36. $\dfrac{1}{x} \leq 5$

37. Explain how any quadratic inequality can be solved by examining a parabola.

38. Describe a method for creating a quadratic inequality for which there is no solution.

SKILL MAINTENANCE

Simplify.

39. $(2a^3b^2c^4)^3$

40. $(5a^4b^7)^2$

41. 2^{-5}

42. 3^{-4}

43. If $f(x) = 3x^2$, find $f(a + 1)$.

44. If $g(x) = 5x - 3$, find $g(a + 2)$.

SYNTHESIS

45. Step (5) on p. 538 states that even when the inequality symbol is $\leq$ or $\geq$, the solutions from step (1) are not always part of the solution set. Why?

46. Describe a method that could be used to create quadratic inequalities that have $(-\infty, a] \cup [b, \infty)$ as the solution set.

Find each solution set.

47. $x^2 + 2x < 5$

48. $x^4 + 2x^2 \geq 0$

49. $x^4 + 3x^2 \leq 0$

50. $\left|\dfrac{x + 2}{x - 1}\right| \leq 3$

51. *Total profit.* Derex, Inc., determines that its total-profit function is given by

$$P(x) = -3x^2 + 630x - 6000.$$

a) Find all values of x for which Derex makes a profit.

b) Find all values of x for which Derex loses money.

52. *Height of a thrown object.* The function

$$S(t) = -16t^2 + 32t + 1920$$

gives the height S, in feet, of an object thrown from a cliff that is 1920 ft high. Here t is the time, in seconds, that the object is in the air.

a) For what times does the height exceed 1920 ft?

b) For what times is the height less than 640 ft?

53. *Number of handshakes.* There are n people in a room. The number N of possible handshakes by the people is given by the function

$$N(n) = \frac{n(n-1)}{2}.$$

For what number of people n is $66 \le N \le 300$?

54. *Number of diagonals.* A polygon with n sides has D diagonals, where D is given by the function

$$D(n) = \frac{n(n-3)}{2}.$$

Find the number of sides n if

$$27 \le D \le 230.$$

Use a grapher to graph each function and find solutions of $f(x) = 0$. Then solve the inequalities $f(x) < 0$ and $f(x) > 0$.

55. $f(x) = x^3 - 2x^2 - 5x + 6$

56. $f(x) = \frac{1}{3}x^3 - x + \frac{2}{3}$

57. $f(x) = x + \frac{1}{x}$

58. $f(x) = x - \sqrt{x}, x \ge 0$

59. $f(x) = \frac{x^3 - x^2 - 2x}{x^2 + x - 6}$

60. $f(x) = x^4 - 4x^3 - x^2 + 16x - 12$

61. Use a grapher to solve Exercises 11, 25, and 35 by drawing two curves, one for each side of the inequality.

Summary and Review 8

Key Terms

Quadratic equation, p. 472
Principle of square roots, p. 473
Completing the square, p. 475
Compounding interest annually, p. 479
The quadratic formula, p. 483
Quadratic function, p. 485
Discriminant, p. 498
Reducible to quadratic, p. 503
Parabola, p. 509
Axis of symmetry, p. 509
Vertex, p. 509
Reflection, p. 510
Translated, p. 511
Minimum value, p. 512
Maximum value, p. 512
Data point, p. 525
Polynomial inequality, p. 532
Quadratic inequality, p. 532
Zero, p. 534
Rational inequality, p. 537

Important Properties and Formulas

The Principle of Square Roots

For any real number k, if $x^2 = k$, then $x = \sqrt{k}$ or $x = -\sqrt{k}$.

For any real number k and any algebraic expression X, if $X^2 = k$, then $X = \sqrt{k}$ or $X = -\sqrt{k}$.

To complete the square for $x^2 + bx$, add $(b/2)^2$.

To solve a quadratic equation in x by completing the square:

1. Isolate the terms with variables on one side of the equation, and arrange them in descending order.
2. Divide both sides by the coefficient of x^2 if that coefficient is not 1.
3. Complete the square by taking half of the coefficient of x and adding its square to both sides.

4. Express one side as the square of a binomial and simplify the other side.
5. Use the principle of square roots.
6. Solve for x by adding or subtracting on both sides.

The Quadratic Formula

The solutions of $ax^2 + bx + c = 0, a \neq 0$, are given by

$$x = \frac{-b \pm \sqrt{b^2 - 4ac}}{2a}.$$

To solve a quadratic equation:

1. If the equation can be easily written in the form $ax^2 = p$ or $(x + k)^2 = d$, use the principle of square roots.
2. If step (1) does not apply, write the equation in $ax^2 + bx + c = 0$ form.
3. Try factoring and using the principle of zero products.
4. If factoring seems to be difficult or impossible, use the quadratic formula.

The solutions of a quadratic equation can always be found using the quadratic formula. They cannot always be found by factoring.

To solve a formula for a letter—say, b:

1. Clear fractions and use the principle of powers, as needed. (In some cases you may clear the fractions first, and in some cases you may use the principle of powers first.) Perform these steps until radicals containing b are gone and b is not in any denominator.
2. Combine all terms with b^2 in them. Also combine all terms with b in them.
3. If b^2 does not appear, you can solve by using just the addition and multiplication principles as in Sections 1.5 and 6.8.
4. If b^2 appears but b does not, solve the equation for b^2. Then use the principle of square roots to solve for b.
5. If there are terms containing both b and b^2, put the equation in standard form and use the quadratic formula.

Discriminant $b^2 - 4ac$	**Nature of Solutions**
0	One solution; a rational number
Positive	Two different real-number solutions
Perfect square	Solutions are rational.
Not a perfect square	Solutions are irrational conjugates.
Negative	Two different imaginary-number solutions (complex conjugates)

The graph of $g(x) = ax^2$ is a parabola with $x = 0$ as its axis of symmetry; its vertex is the origin.

For $a > 0$, the parabola opens upward. For $a < 0$, the parabola opens downward.

If $|a|$ is greater than 1, the parabola is narrower than $f(x) = x^2$.

If $|a|$ is between 0 and 1, the parabola is wider than $f(x) = x^2$.

The graph of $f(x) = a(x - h)^2$ has the same shape as the graph of $y = ax^2$.

If h is positive, the graph of $y = ax^2$ is shifted h units to the right.

If h is negative, the graph of $y = ax^2$ is shifted $|h|$ units to the left.

The vertex is $(h, 0)$, and the axis of symmetry is $x = h$.

The graph of $f(x) = a(x - h)^2 + k$ has the same shape as the graph of $y = a(x - h)^2$.

If k is positive, the graph of $y = a(x - h)^2$ is shifted k units up.

If k is negative, the graph of $y = a(x - h)^2$ is shifted $|k|$ units down.

The vertex is (h, k), and the axis of symmetry is $x = h$.

For $a > 0$, k is the minimum function value. For $a < 0$, k is the maximum function value.

Formulas

Compound interest: $A = P(1 + r)^t$
Free-fall distance, in feet: $s = 16t^2$

The vertex of the parabola given by $f(x) = ax^2 + bx + c$ is

$$\left(-\frac{b}{2a}, f\left(-\frac{b}{2a}\right)\right),$$

or

$$\left(-\frac{b}{2a}, \frac{4ac - b^2}{4a}\right).$$

The x-coordinate of the vertex is $-b/(2a)$. The axis of symmetry is $x = -b/(2a)$.

To solve a polynomial inequality using factors:

1. Get 0 on one side and solve the related polynomial equation by factoring.
2. Use the numbers found in step (1) to divide the number line into intervals.
3. Using a test value from each interval, determine the sign of each factor over that interval.
4. Determine the sign of the product of the factors over each interval. Remember that the product of an odd number of negative numbers is negative.
5. Select the interval(s) for which the inequality is satisfied and write set-builder or interval notation for the solution set. Include the endpoints of the intervals when $\leq$ or $\geq$ is used.

To solve a rational inequality:

1. Change the inequality symbol to an equals sign and solve the related equation.
2. Find any replacements for which the rational expression is undefined.
3. Use the numbers found in steps (1) and (2) to divide the number line into intervals.
4. Substitute a test value from each interval into the inequality. If the number is a solution, then the interval to which it belongs is part of the solution set.
5. Select the interval(s) and any endpoints for which the inequality is satisfied and write set-builder or interval notation for the solution set. If the inequality symbol is $\leq$ or $\geq$, then the solutions to step (1) are also included in the solution set. Those numbers found in step (2) should be excluded from the solution set, even if they are solutions from step (1).

Review Exercises

Solve.

1. $2x^2 - 7 = 0$
2. $14x^2 + 5x = 0$
3. $x^2 - 12x + 36 = 9$
4. $x^2 - 5x + 9 = 0$
5. $x(3x + 4) = 4x(x - 1) + 15$
6. $x^2 + 9x = 1$
7. $x^2 - 5x - 2 = 0$. Use a calculator to approximate the solutions with rational numbers.
8. Let $f(x) = 4x^2 - 3x - 1$. Find x such that $f(x) = 0$.

Complete the square. Then write the perfect-square trinomial in factored form.

9. $x^2 - 12x$
10. $x^2 + \frac{3}{5}x$
11. Solve by completing the square. Show your work.
$$x^2 - 6x + 1 = 0$$
12. \$2500 grows to \$3025 in 2 yr. Use the formula $A = P(1 + r)^t$ to find the interest rate.
13. The Peachtree Center Plaza in Atlanta, Georgia, is 723 ft tall. Use the formula $s = 16t^2$ to approximate how long it would take an object to fall from the top.

Solve.

14. A corporate pilot must fly from company headquarters to a manufacturing plant and back in 4 hr. The distance between headquarters and the plant is 300 mi. If there is a 20-mph headwind going and a 20-mph tailwind returning, how fast must the plane be able to travel in still air?
15. Working together, Erica and Shawna can answer a day's worth of technical support questions in 4 hr. Working alone, Erica takes 6 hr longer than Shawna. How long would it take Shawna to answer the questions alone?

For each equation, determine what type of number the solutions will be.

16. $x^2 + 3x - 6 = 0$
17. $x^2 + 2x + 5 = 0$
18. Write a quadratic equation having the solutions $\sqrt{5}$ and $-\sqrt{5}$.
19. Write a quadratic equation having -4 as its only solution.
20. Find all x-intercepts of the graph of
$$f(x) = x^4 - 13x^2 + 36.$$

Solve.

21. $15x^{-2} - 2x^{-1} - 1 = 0$
22. $(x^2 - 4)^2 - (x^2 - 4) - 6 = 0$
23. **a)** Graph: $f(x) = -3(x + 2)^2 + 4$.
    b) Label the vertex.
    c) Draw the axis of symmetry.
    d) Find the maximum or the minimum value.
24. For the function $f(x) = 2x^2 - 12x + 23$:
    a) find the vertex and the axis of symmetry;
    b) graph the function.
25. Find the x- and y-intercepts of
$$f(x) = x^2 - 9x + 14.$$
26. Solve $N = 3\pi\sqrt{1/p}$ for p.
27. Solve $2A + T = 3T^2$ for T.

State whether the graph appears to represent a quadratic function.

28.

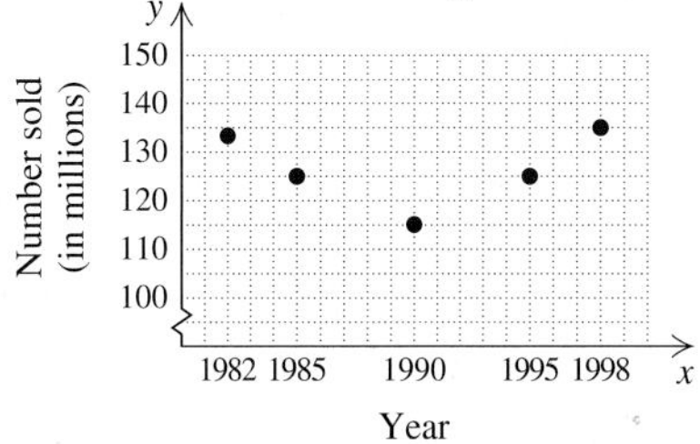

Source: *Statistical Abstract of the United States*, 2000

29.

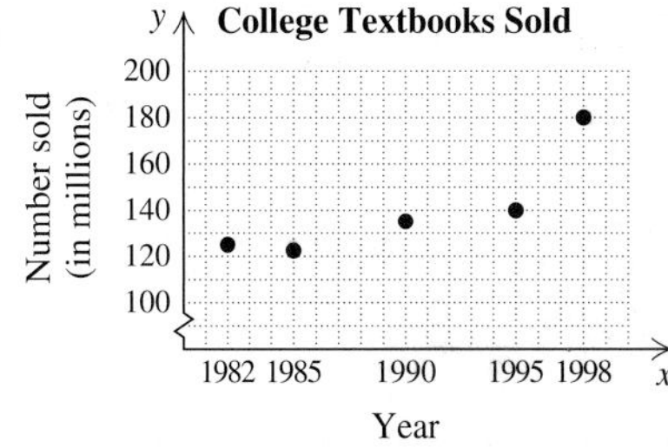

Source: *Statistical Abstract of the United States*, 2000

30. Eastgate Consignments wants to build an area for children to play in while their parents shop. They have 30 ft of low fencing. What is the maximum area they can enclose? What dimensions will yield this area?

31. The following table lists the number of books (in millions) sold through bookclubs x years after 1980. (See Exercise 28.)

Years Since 1980	Number of Books Sold Through Bookclubs (in millions)
5	130
10	108
15	123

a) Find the quadratic function that fits the data.
b) Use the function to estimate the number of books sold through bookclubs in 2000.

Solve.

32. $x^3 - 3x > 2x^2$

33. $\dfrac{x-5}{x+3} \leq 0$

SYNTHESIS

34. Explain how the x-intercepts of a quadratic function can be used to help find the maximum or minimum value of the function.

35. Suppose that the quadratic formula is used to solve a quadratic equation. If the discriminant is a perfect square, could factoring have been used to solve the equation? Why or why not?

36. What is the greatest number of solutions that an equation of the form $ax^4 + bx + c = 0$ can have? Why?

37. Discuss two ways in which completing the square was used in this chapter.

38. A quadratic function has x-intercepts at -3 and 5. If the y-intercept is at -7, find an equation for the function.

39. Find h and k if, for $3x^2 - hx + 4k = 0$, the sum of the solutions is 20 and the product is 80.

40. The average of two positive integers is 171. One of the numbers is the square root of the other. Find the integers.

Chapter Test 8

Solve.

1. $3x^2 - 16 = 0$

2. $4x(x-2) - 3x(x+1) = -18$

3. $x^2 + x + 1 = 0$

4. $2x + 5 = x^2$

5. $x^{-2} - x^{-1} = \frac{3}{4}$

6. $x^2 + 3x = 5$. Use a calculator to approximate the solutions with rational numbers.

7. Let $f(x) = 12x^2 - 19x - 21$. Find x such that $f(x) = 0$.

Complete the square. Then write the perfect-square trinomial in factored form.

8. $x^2 + 14x$

9. $x^2 - \frac{2}{7}x$

10. Solve by completing the square. Show your work.

$$x^2 + 10x + 15 = 0$$

Solve.

11. The Connecticut River flows at a rate of 4 km/h for the length of a popular scenic route. In order for a cruiser to travel 60 km upriver and then return in a total of 8 hr, how fast must the boat be able to travel in still water?

12. Brock and Ian can assemble a swing set in $1\frac{1}{2}$ hr. Working alone, it takes Ian 4 hr longer than Brock to assemble the swing set. How long would it take Brock, working alone, to assemble the swing set?

13. Determine the type of number that the solutions of $x^2 + 5x + 17 = 0$ will be.

14. Write a quadratic equation having solutions -2 and $\frac{1}{3}$.

15. Find all x-intercepts of the graph of

$$f(x) = (x^2 + 4x)^2 + 2(x^2 + 4x) - 3.$$

16. **a)** Graph: $f(x) = 4(x - 3)^2 + 5$.
b) Label the vertex.
c) Draw the axis of symmetry.
d) Find the maximum or the minimum function value.

17. For the function $f(x) = 2x^2 + 4x - 6$:
a) find the vertex and the axis of symmetry;
b) graph the function.

18. Find the x- and y-intercepts of

$$f(x) = x^2 - x - 6.$$

19. Solve $V = \frac{1}{3}\pi(R^2 + r^2)$ for r.

20. State whether the graph appears to represent a quadratic function.

21. Jay's Custom Pickups has determined that when x hundred truck caps are built, the average cost per cap is given by

$$C(x) = 0.2x^2 - 1.3x + 3.4025,$$

where $C(x)$ is in hundreds of dollars. What is the minimum cost per truck cap and how many caps should be built to achieve that minimum?

22. Find the quadratic function that fits the data points $(0, 0)$, $(3, 0)$, and $(5, 2)$.

Solve.

23. $x^2 + 5x \leq 6$

24. $x - \dfrac{1}{x} > 0$

SYNTHESIS

25. One solution of $kx^2 + 3x - k = 0$ is -2. Find the other solution.

26. Find a fourth-degree polynomial equation, with integer coefficients, for which $2 - \sqrt{3}$ and $5 - i$ are solutions.

27. Find a polynomial equation, with integer coefficients, for which 5 is a repeated root and $\sqrt{2}$ and $\sqrt{3}$ are solutions.

9 Exponential and Logarithmic Functions

9.1 Composite and Inverse Functions
9.2 Exponential Functions
Connecting the Concepts
9.3 Logarithmic Functions
9.4 Properties of Logarithmic Functions
9.5 Common and Natural Logarithms
9.6 Solving Exponential and Logarithmic Equations
9.7 Applications of Exponential and Logarithmic Functions
Summary and Review
Test
Cumulative Review

AN APPLICATION

The number of computers infected by a virus t days after it first appears usually increases exponentially. In 2000, the "Love Bug" virus spread from 100 computers to about 1,000,000 computers in 2 hr (120 min). Assuming exponential growth, estimate how long it took the Love Bug virus to infect 80,000 computers.

This problem appears as Example 5 in Section 9.7.

I use algebra to calculate bandwidth availability and network capacity and to create security profiles. I also use matrices in data storage.

CHRISTOPHER KENLY
Chief Consultant
TOS Consulting
Dallas, TX

The functions that we consider in this chapter are interesting not only from a purely intellectual point of view, but also for their rich applications to many fields. We will look at applications such as compound interest and population growth, to name just two.

The basis of the theory centers on functions having variable exponents (exponential functions). *Results follow from those functions and their properties.*

9.1 Composite and Inverse Functions

Composite Functions • Inverses and One-to-One Functions • Finding Formulas for Inverses • Graphing Functions and Their Inverses • Inverse Functions and Composition

Composite Functions

In the real world, functions frequently occur in which some quantity depends on a variable that, in turn, depends on another variable. For instance, a firm's profits may depend on the number of items the firm produces, which may in turn depend on the number of employees hired. Functions like this are called **composite functions**.

For example, the function g that gives a correspondence between women's shoe sizes in the United States and those in Italy is given by $g(x) = 2x + 24$, where x is the U.S. size and $g(x)$ is the Italian size. Thus a U.S. size 4 corresponds to a shoe size of $g(4) = 2 \cdot 4 + 24$, or 32, in Italy.

There is also a function that gives a correspondence between women's shoe sizes in Italy and those in Britain. This particular function is given by $f(x) = \frac{1}{2}x - 14$, where x is the Italian size and $f(x)$ is the corresponding British size. Thus an Italian size 32 corresponds to a British size $f(32) = \frac{1}{2} \cdot 32 - 14$, or 2.

It seems reasonable to conclude that a shoe size of 4 in the United States corresponds to a size of 2 in Britain and that some function h describes this correspondence. Can we find a formula for h? If we look at the following tables, we might guess that such a formula is $h(x) = x - 2$, and that is indeed correct. But, for more complicated formulas, we would need to use algebra.

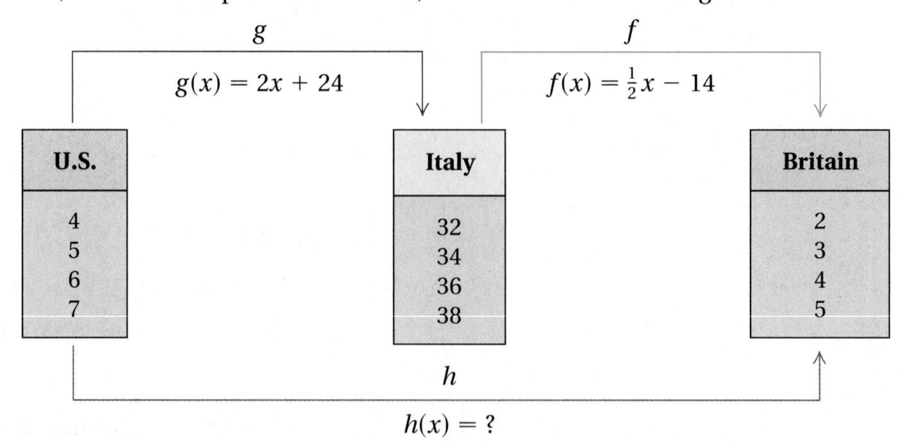

Size x shoes in the United States correspond to size $g(x)$ shoes in Italy, where

$$g(x) = 2x + 24.$$

Size n shoes in Italy correspond to size $f(n)$ shoes in Britain. Thus size $g(x)$ shoes in Italy correspond to size $f(g(x))$ shoes in Britain. Since the x in the expression $f(g(x))$ represents a U.S. shoe size, we can find the British shoe size that corresponds to a U.S. size x as follows:

$$f(g(x)) = f(2x + 24) = \frac{1}{2} \cdot (2x + 24) - 14 \qquad \text{Using } g(x) \text{ as an input}$$
$$= x + 12 - 14 = x - 2.$$

This gives a formula for h: $h(x) = x - 2$. Thus a shoe size of 4 in the United States corresponds to a shoe size of $h(4) = 4 - 2$, or 2, in Britain. The function h is called the *composition* of f and g and is denoted $f \circ g$ (read "the composition of f and g," "f composed with g," or "f circle g").

Composition of Functions

The *composite function* $f \circ g$, the *composition* of f and g, is defined as

$$(f \circ g)(x) = f(g(x)).$$

We can visualize the composition of functions as follows.

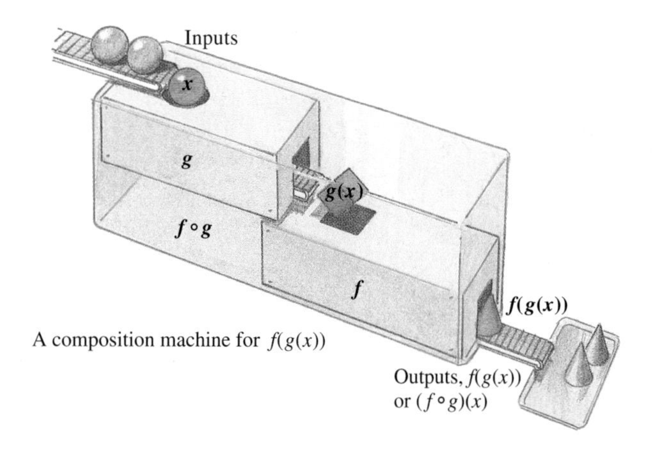

A composition machine for $f(g(x))$

Example 1 Given $f(x) = 3x$ and $g(x) = 1 + x^2$:

a) Find $(f \circ g)(5)$ and $(g \circ f)(5)$.
b) Find $(f \circ g)(x)$ and $(g \circ f)(x)$.

Solution Consider each function separately:

$f(x) = 3x$ This function multiplies each input by 3.

and

$g(x) = 1 + x^2.$ This function adds 1 to the square of each input.

In Example 2, we will see that if $g(x) = x - 1$ and $f(x) = \sqrt{x}$, then $f(g(x)) = \sqrt{x-1}$. One way to show this on a grapher is to let $y_1 = x - 1$ and $y_2 = \sqrt{y_1}$ (this is accomplished by using the Y-VARS option of the VARS key's menu to enter y_1). We then let $y_3 = \sqrt{x-1}$ and use graphs or a table to show that $y_2 = y_3$.

Another approach is to let $y_1 = x - 1$ and $y_2 = \sqrt{x}$ and have $y_4 = y_2(y_1)$. If we again let $y_3 = \sqrt{x-1}$, we complete the check by showing that $y_3 = y_4$.

1. Check Example 3 by using one of the above approaches.

a) To find $(f \circ g)(5)$, we first find $g(5)$ by substituting in the formula for g: Square 5 and add 1, to get 26. We then use 26 as an input for f:

$$(f \circ g)(5) = f(g(5)) = f(1 + 5^2) \qquad \text{Using } g(x) = 1 + x^2$$
$$= f(26) = 3 \cdot 26 = 78. \qquad \text{Using } f(x) = 3x$$

To find $(g \circ f)(5)$, we first find $f(5)$ by substituting into the formula for f: Multiply 5 by 3, to get 15. We then use 15 as an input for g:

$$(g \circ f)(5) = g(f(5)) = g(3 \cdot 5) \qquad \text{Note that } f(5) = 3 \cdot 5 = 15.$$
$$= g(15) = 1 + 15^2 = 1 + 225 = 226.$$

b) We find $(f \circ g)(x)$ by substituting $g(x)$ for x in the equation for $f(x)$:

$$(f \circ g)(x) = f(g(x)) = f(1 + x^2) \qquad \text{Using } g(x) = 1 + x^2$$
$$= 3(1 + x^2) = 3 + 3x^2. \qquad \text{Using } f(x) = 3x. \textit{ These} \text{ parentheses indicate multiplication.}$$

To find $(g \circ f)(x)$, we substitute $f(x)$ for x in the equation for $g(x)$:

$$(g \circ f)(x) = g(f(x)) = g(3x) \qquad \text{Substituting } 3x \text{ for } f(x)$$
$$= 1 + (3x)^2 = 1 + 9x^2.$$

As a check, note that $(g \circ f)(5) = 1 + 9 \cdot 5^2 = 1 + 9 \cdot 25 = 226$, as expected from part (a) above.

Example 1 shows that, in general, $(f \circ g)(5) \neq (g \circ f)(5)$ and $(f \circ g)(x) \neq (g \circ f)(x)$.

Example 2

Given $f(x) = \sqrt{x}$ and $g(x) = x - 1$, find $(f \circ g)(x)$ and $(g \circ f)(x)$.

Solution

$$(f \circ g)(x) = f(g(x)) = f(x - 1) = \sqrt{x-1}$$
$$(g \circ f)(x) = g(f(x)) = g(\sqrt{x}) = \sqrt{x} - 1$$

In fields ranging from chemistry to geology and economics, one needs to recognize how a function can be regarded as the composition of two "simpler" functions. This is sometimes called *de*composition.

Example 3

If $h(x) = (7x + 3)^2$, find $f(x)$ and $g(x)$ such that $h(x) = (f \circ g)(x)$.

Solution To find $h(x)$, we can think of first forming $7x + 3$ and then squaring. This suggests that $g(x) = 7x + 3$ and $f(x) = x^2$. We check by forming the composition:

$$h(x) = (f \circ g)(x) = f(g(x))$$
$$= f(7x + 3) = (7x + 3)^2.$$

This is probably the most "obvious" answer to the question. There can be other less obvious answers. For example, if

$$f(x) = (x - 1)^2$$

and

$$g(x) = 7x + 4,$$

then

$$h(x) = (f \circ g)(x) = f(g(x)) = f(7x + 4)$$
$$= (7x + 4 - 1)^2 = (7x + 3)^2.$$

Inverses and One-to-One Functions

Let's consider the following two functions. We think of them as relations, or correspondences.

Professions and Their Median Yearly Salary in 1998 Dollars*

Domain (Set of Inputs)		Range (Set of Outputs)
Teacher	→	\$39,300
Registered nurse	→	\$40,690
Emergency medical technician	→	\$20,290
Computer programmer	→	\$47,550
Veterinarian	→	\$50,950
Accountant	→	\$37,860

U.S. Senators and Their States

Domain (Set of Inputs)		Range (Set of Outputs)
Clinton	→	New York
Schumer	→	New York
McCain	→	Arizona
Kyl	→	Arizona
Feinstein	→	California
Boxer	→	California

Suppose we reverse the arrows. We obtain what is called the **inverse relation**. Are these inverse relations functions?

Professions and Their Median Yearly Salary in 1998 Dollars

Range (Set of Outputs)		Domain (Set of Inputs)
Teacher	←	\$39,300
Registered nurse	←	\$40,690
Emergency medical technician	←	\$20,290
Computer programmer	←	\$47,550
Veterinarian	←	\$50,950
Accountant	←	\$37,860

U.S. Senators and Their States

Range (Set of Outputs)		Domain (Set of Inputs)
Clinton	←	New York
Schumer	←	New York
McCain	←	Arizona
Kyl	←	Arizona
Feinstein	←	California
Boxer	←	California

Recall that for each input, a function provides exactly one output. However, a function can have the same output for two or more different inputs. Thus it is possible for different inputs to correspond to the same output. Only when this

*Source: U.S. Bureau of Labor Statistics, 2001

possibility is *excluded* will the inverse be a function. For the functions listed above, this means the inverse of the "Profession" correspondence is a function, but the inverse of the "U.S. Senator" correspondence is not.

In the Profession function, different inputs have different outputs. It is an example of a **one-to-one function**. In the U.S. Senator function, *Clinton* and *Schumer* are both paired with *New York*. Thus the U.S. Senator function is not one-to-one.

One-To-One Function

A function f is *one-to-one* if different inputs have different outputs. That is, if for any $a \neq b$, we have $f(a) \neq f(b)$, the function f is one-to-one. If a function is one-to-one, then its inverse correspondence is also a function.

How can we tell graphically whether a function is one-to-one?

Example 4 Shown here is the graph of a function similar to those we will study in Section 9.2. Determine whether the function is one-to-one and thus has an inverse that is a function.

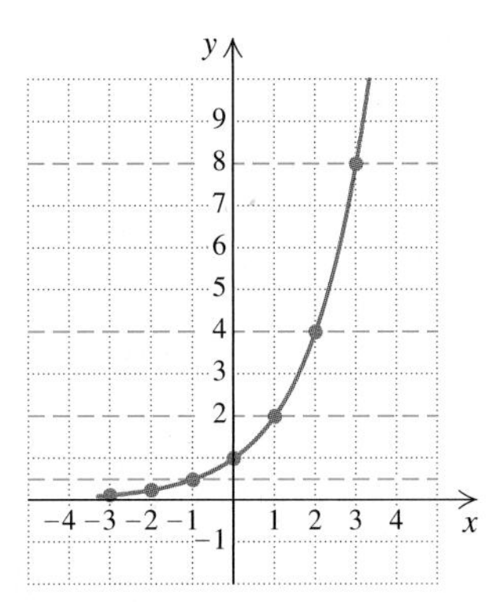

Solution A function is one-to-one if different inputs have different outputs—that is, if no two x-values have the same y-value. For this function, we cannot find two x-values that have the same y-value. Note that this means that no horizontal line can be drawn so that it crosses the graph more than once. The function is one-to-one so its inverse is a function.

The graph of every function must pass the vertical-line test. In order for a function to have an inverse that is a function, it must pass the *horizontal-line test* as well.

The Horizontal-Line Test

A function is one-to-one, and thus has an inverse that is a function, if it is impossible to draw a horizontal line that intersects its graph more than once.

Example 5 Determine whether the function $f(x) = x^2$ is one-to-one and thus has an inverse that is a function.

Solution The graph of $f(x) = x^2$ is shown here. Many horizontal lines cross the graph more than once—in particular, the line $y = 4$. Note that where the line crosses, the first coordinates are -2 and 2. Although these are different inputs, they have the same output. That is, $-2 \neq 2$, but

$$f(-2) = (-2)^2 = 4 = 2^2 = f(2).$$

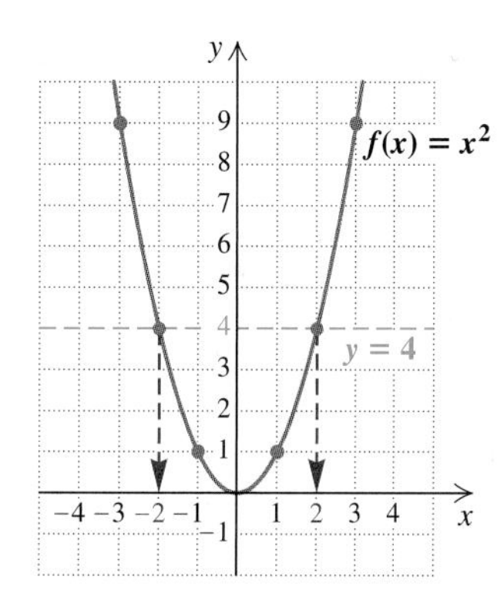

Thus the function is not one-to-one and no inverse function exists.

Finding Formulas for Inverses

When the inverse of f is also a function, it is denoted f^{-1} (read "f-inverse").

> ***Caution!*** The -1 in f^{-1} is *not* an exponent!

Suppose a function is described by a formula. If it has an inverse that is a function, how do we find a formula for the inverse? For any equation in two variables, if we interchange the variables, we obtain an equation of the inverse correspondence. If it is a function, we proceed as follows to find a formula for f^{-1}.

To Find a Formula for f^{-1}

First check a graph to make sure that f is one-to-one. Then:

1. Replace $f(x)$ with y.
2. Interchange x and y. (This gives the inverse function.)
3. Solve for y.
4. Replace y with $f^{-1}(x)$. (This is inverse function notation.)

Example 6

Determine if each function is one-to-one and if it is, find a formula for $f^{-1}(x)$.

a) $f(x) = x + 2$ **b)** $f(x) = 2x - 3$

Solution

a) The graph of $f(x) = x + 2$ is shown below. It passes the horizontal-line test, so it is one-to-one. Thus its inverse is a function.

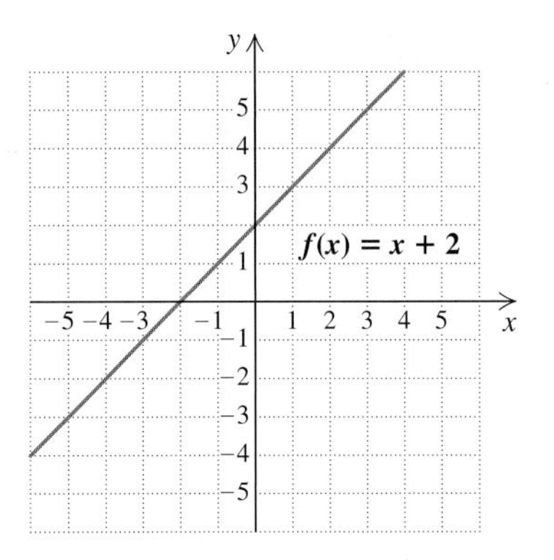

1. Replace $f(x)$ with y: $y = x + 2$.
2. Interchange x and y: $x = y + 2$. This gives the inverse function.
3. Solve for y: $x - 2 = y$.
4. Replace y with $f^{-1}(x)$: $f^{-1}(x) = x - 2$. We also "reversed" the equation.

In this case, the function f added 2 to all inputs. Thus, to "undo" f, the function f^{-1} must subtract 2 from its inputs.

b) The function $f(x) = 2x - 3$ is also linear. Any linear function that is not constant will pass the horizontal-line test. Thus, f is one-to-one.

1. Replace $f(x)$ with y: $y = 2x - 3$.
2. Interchange x and y: $x = 2y - 3$.
3. Solve for y: $x + 3 = 2y$

$$\frac{x+3}{2} = y.$$

4. Replace y with $f^{-1}(x)$: $f^{-1}(x) = \dfrac{x+3}{2}$.

Graphing Functions and Their Inverses

How do the graphs of a function and its inverse compare?

Example 7

Graph $f(x) = 2x - 3$ and $f^{-1}(x) = (x + 3)/2$ on the same set of axes. Then compare.

Solution The graph of each function follows. Note that the graph of f^{-1} can be drawn by reflecting the graph of f across the line $y = x$. That is, if we graph $f(x) = 2x - 3$ in wet ink and fold the paper along the line $y = x$, the graph of $f^{-1}(x) = (x + 3)/2$ will appear as the impression made by f.

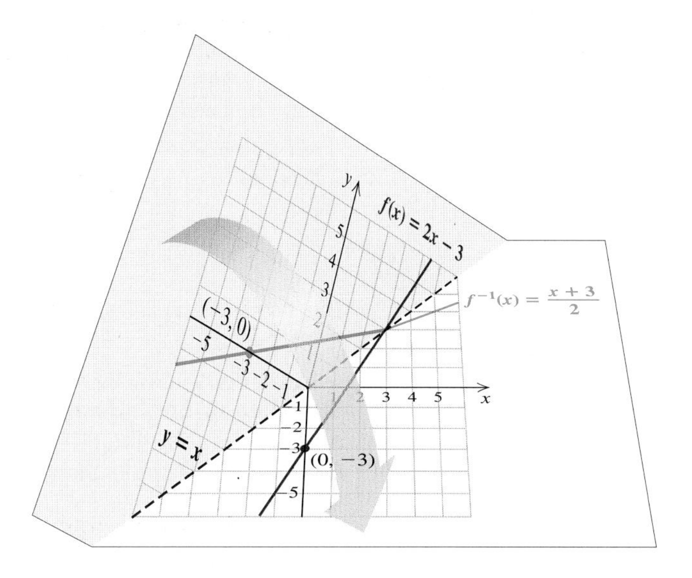

When x and y are interchanged to find a formula for the inverse, we are, in effect, reflecting or flipping the graph of $f(x) = 2x - 3$ across the line $y = x$. For example, when the coordinates of the y-intercept of the graph of f, $(0, -3)$, are reversed, we get the x-intercept of the graph of f^{-1}, $(-3, 0)$.

Visualizing Inverses

The graph of f^{-1} is a reflection of the graph of f across the line $y = x$.

Example 8 Consider $g(x) = x^3 + 2$.

a) Determine whether the function is one-to-one.

b) If it is one-to-one, find a formula for its inverse.

c) Graph the inverse, if it exists.

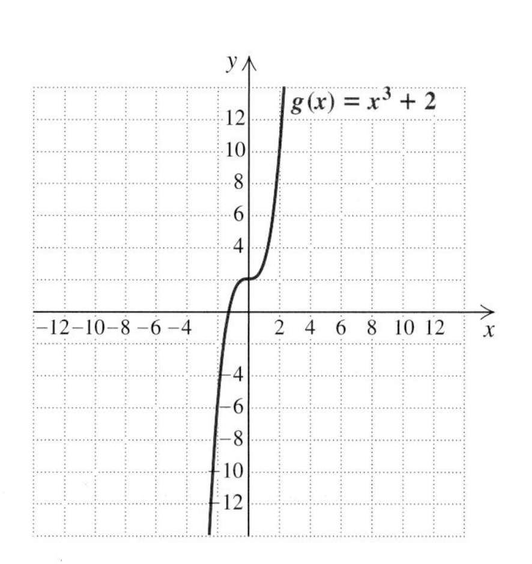

Solution

a) The graph of $g(x) = x^3 + 2$ is shown at right. It passes the horizontal-line test and thus has an inverse.

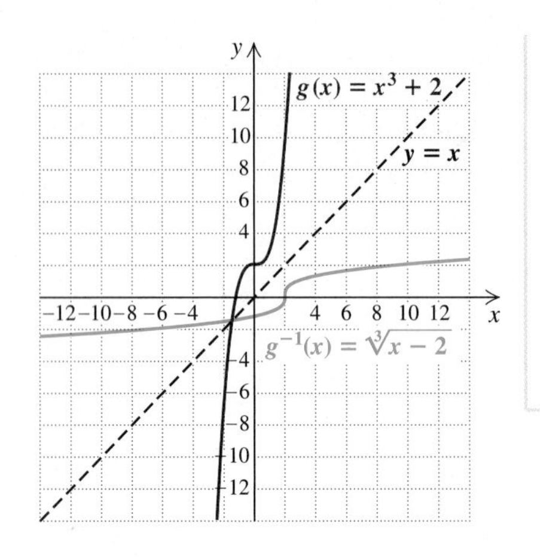

b) **1.** Replace $g(x)$ with y: $y = x^3 + 2$. Using $g(x) = x^3 + 2$

2. Interchange x and y: $x = y^3 + 2$.

3. Solve for y: $x - 2 = y^3$

$\sqrt[3]{x-2} = y$. Since a number has only one cube root, we can solve for y.

4. Replace y with $g^{-1}(x)$: $g^{-1}(x) = \sqrt[3]{x-2}$.

c) To find the graph, we reflect the graph of $g(x) = x^3 + 2$ across the line $y = x$, as we did in Example 7. We can also substitute into $g^{-1}(x) = \sqrt[3]{x-2}$ and plot points. The graphs of g and g^{-1} are shown together at left

Inverse Functions and Composition

Let's consider inverses of functions in terms of a function machine. Suppose that a one-to-one function f is programmed into a machine. If the machine has a reverse switch, when the switch is thrown, the machine performs the inverse function f^{-1}. Inputs then enter at the opposite end, and the entire process is reversed.

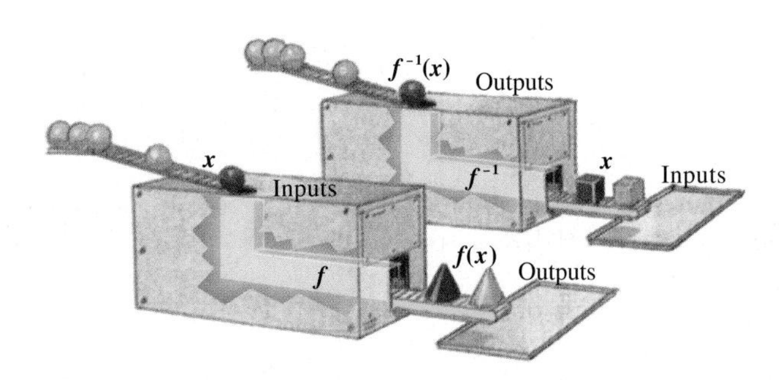

Consider $g(x) = x^3 + 2$ and $g^{-1}(x) = \sqrt[3]{x-2}$ from Example 8. For the input 3,

$$g(3) = 3^3 + 2 = 27 + 2 = 29.$$

The output is 29. Now we use 29 for the input in the inverse:

$$g^{-1}(29) = \sqrt[3]{29-2} = \sqrt[3]{27} = 3.$$

The function g takes 3 to 29. The inverse function g^{-1} takes the number 29 back to 3.

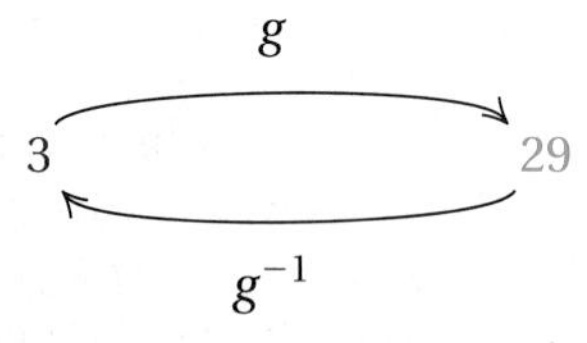

In general, for any output $f(x)$, the function f^{-1} takes that output back to x. Similarly, for any output $f^{-1}(x)$, the function f takes that output back to x.

Composition and Inverses

If a function f is one-to-one, then f^{-1} is the unique function for which

$$(f^{-1} \circ f)(x) = x \quad \text{and} \quad (f \circ f^{-1})(x) = x.$$

Example 9 Let $f(x) = 2x + 1$. Show that

$$f^{-1}(x) = \frac{x - 1}{2}.$$

Solution We find $(f^{-1} \circ f)(x)$ and $(f \circ f^{-1})(x)$ and check to see that each is x.

$$\begin{aligned}(f^{-1} \circ f)(x) &= f^{-1}(f(x)) = f^{-1}(2x + 1) \\ &= \frac{(2x + 1) - 1}{2} \\ &= \frac{2x}{2} = x\end{aligned}$$

$$\begin{aligned}(f \circ f^{-1})(x) &= f(f^{-1}(x)) = f\left(\frac{x - 1}{2}\right) \\ &= 2 \cdot \frac{x - 1}{2} + 1 \\ &= x - 1 + 1 = x\end{aligned}$$

technology connection B

To determine whether $y_1 = 2x + 6$ and $y_2 = \frac{1}{2}x - 3$ might be inverses of each other, we have drawn both functions, along with the line $y = x$, on a "squared" set of axes. It *appears* that y_1 and y_2 are inverses of each other. For further verification, we can examine a table of values in which $y_1 = 2x + 6$ and $y_2 = \frac{1}{2} \cdot y_1 - 3$. Note that y_2 "undoes" what y_1 "does."

$y_1 = 2x + 6,\ y_2 = \frac{1}{2}x - 3$

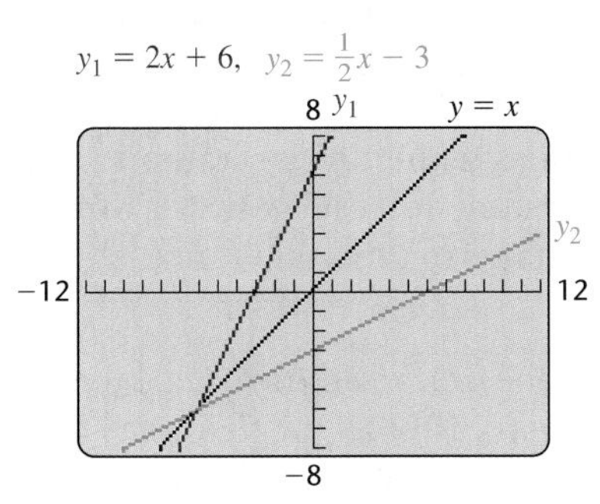

TBL MIN = −3 ΔTBL = 1 $y_2 = \frac{1}{2}y_1 - 3$

X	Y1	Y2
−3	0	−3
−2	2	−2
−1	4	−1
0	6	0
1	8	1
2	10	2
3	12	3

X = 3

A final, visual, check can be made by graphing

$$y_1 = 2x + 6 \quad \text{and} \quad y_2 = \tfrac{1}{2}x - 3$$

and then pressing DRAW and selecting the DRAWINV option. Once this has been selected, we use VARS to enter y_1. The resulting graph of the inverse of y_1 should coincide with y_2.

1. Use a grapher to check Examples 7, 8, and 9.
2. Will DRAWINV work for *any* choice of y_1? Why or why not?

Exercise Set 9.1

FOR EXTRA HELP

Digital Video Tutor CD 7 Videotape 17

InterAct Math

Math Tutor Center

MathXL

MyMathLab.com

Find $(f \circ g)(1)$, $(g \circ f)(1)$, $(f \circ g)(x)$, *and* $(g \circ f)(x)$.

1. $f(x) = x^2 + 3;\ g(x) = 2x + 1$

2. $f(x) = 2x + 1;\ g(x) = x^2 - 5$

3. $f(x) = 3x - 1;\ g(x) = 5x^2 + 2$

4. $f(x) = 3x^2 + 4;\ g(x) = 4x - 1$

5. $f(x) = x + 7;\ g(x) = 1/x^2$

6. $f(x) = 1/x^2;\ g(x) = x + 2$

Find f(x) and g(x) such that $h(x) = (f \circ g)(x)$. *Answers may vary.*

7. $h(x) = (7 + 5x)^2$

8. $h(x) = (3x - 1)^2$

9. $h(x) = \sqrt{2x + 7}$

10. $h(x) = \sqrt{5x + 2}$

11. $h(x) = \dfrac{2}{x - 3}$

12. $h(x) = \dfrac{3}{x} + 4$

13. $h(x) = \dfrac{1}{\sqrt{7x + 2}}$

14. $h(x) = \sqrt{x - 7} - 3$

15. $h(x) = \dfrac{1}{\sqrt{3x}} + \sqrt{3x}$

16. $h(x) = \dfrac{1}{\sqrt{2x}} - \sqrt{2x}$

Determine whether each function is one-to-one.

17. $f(x) = x - 5$

18. $f(x) = 5 - 2x$

Aha! **19.** $f(x) = x^2 + 1$

20. $f(x) = 1 - x^2$

21. $g(x) = x^3$

22. $g(x) = \sqrt{x} + 1$

23. $g(x) = |x|$

24. $h(x) = |x| - 1$

For each function, **(a)** *determine whether it is one-to-one;* **(b)** *if it is one-to-one, find a formula for the inverse.*

25. $f(x) = x - 4$

26. $f(x) = x - 2$

27. $f(x) = 3 + x$

28. $f(x) = 9 + x$

29. $g(x) = x + 5$

30. $g(x) = x + 8$

31. $f(x) = 4x$

32. $f(x) = 7x$

33. $g(x) = 4x - 1$

34. $g(x) = 4x - 6$

35. $h(x) = 5$

36. $h(x) = -2$

Aha! **37.** $f(x) = \dfrac{1}{x}$

38. $f(x) = \dfrac{3}{x}$

39. $f(x) = \dfrac{2x + 1}{3}$

40. $f(x) = \dfrac{3x + 2}{5}$

41. $f(x) = x^3 - 5$

42. $f(x) = x^3 + 2$

43. $g(x) = (x - 2)^3$

44. $g(x) = (x + 7)^3$

45. $f(x) = \sqrt{x}$

46. $f(x) = \sqrt{x - 1}$

47. $f(x) = 2x^2 + 1, x \geq 0$

48. $f(x) = 3x^2 - 2, x \geq 0$

Graph each function and its inverse using the same set of axes.

49. $f(x) = \frac{1}{3}x - 2$

50. $g(x) = x + 4$

51. $f(x) = x^3$

52. $f(x) = x^3 - 1$

53. $g(x) = -2x + 3$

54. $g(x) = \sqrt{x}$

55. $F(x) = -\sqrt{x}$

56. $f(x) = -\frac{1}{2}x + 1$

57. $f(x) = 3 - x^2, x \geq 0$

58. $f(x) = x^2 - 1, x \leq 0$

59. Let $f(x) = \frac{4}{5}x$. Show that $f^{-1}(x) = \frac{5}{4}x$.

60. Let $f(x) = (x + 7)/3$. Show that $f^{-1}(x) = 3x - 7$.

61. Let $f(x) = (1 - x)/x$. Show that

$$f^{-1}(x) = \frac{1}{x + 1}.$$

62. Let $f(x) = x^3 - 5$. Show that $f^{-1}(x) = \sqrt[3]{x + 5}$.

63. *Dress sizes in the United States and France.* A size-6 dress in the United States is size 38 in France. A function that converts dress sizes in the United States to those in France is

$$f(x) = x + 32.$$

a) Find the dress sizes in France that correspond to sizes 8, 10, 14, and 18 in the United States.

b) Determine whether this function has an inverse that is a function. If so, find a formula for the inverse.

c) Use the inverse function to find dress sizes in the United States that correspond to sizes 40, 42, 46, and 50 in France.

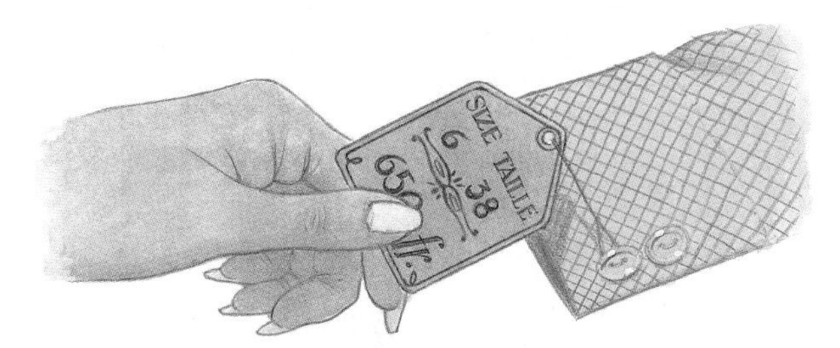

64. *Dress sizes in the United States and Italy.* A size-6 dress in the United States is size 36 in Italy. A function that converts dress sizes in the United States to those in Italy is

$$f(x) = 2(x + 12).$$

a) Find the dress sizes in Italy that correspond to sizes 8, 10, 14, and 18 in the United States.

b) Determine whether this function has an inverse that is a function. If so, find a formula for the inverse.

c) Use the inverse function to find dress sizes in the United States that correspond to sizes 40, 44, 52, and 60 in Italy.

65. Is there a one-to-one relationship between the numbers and letters on the keypad of a telephone? Why or why not?

66. Mathematicians usually try to select "logical" words when forming definitions. Does the term "one-to-one" seem logical? Why or why not?

SKILL MAINTENANCE

Simplify.

67. $(a^5b^4)^2(a^3b^5)$

68. $(x^3y^5)^2(x^4y^2)$

69. $27^{4/3}$

70. $25^{3/2}$

Solve.

71. $x = \frac{2}{3}y - 7$, for y

72. $x = 10 - 3y$, for y

SYNTHESIS

73. The function $V(t) = 750(1.2)^t$ is used to predict the value, $V(t)$, of a certain rare stamp t years from 2001. Do not calculate $V^{-1}(t)$, but explain how V^{-1} could be used.

74. An organization determines that the cost per person of chartering a bus is given by the function

$$C(x) = \frac{100 + 5x}{x},$$

where x is the number of people in the group and $C(x)$ is in dollars. Determine $C^{-1}(x)$ and explain how this inverse function could be used.

For Exercises 75 and 76, graph the inverse of f.

75.

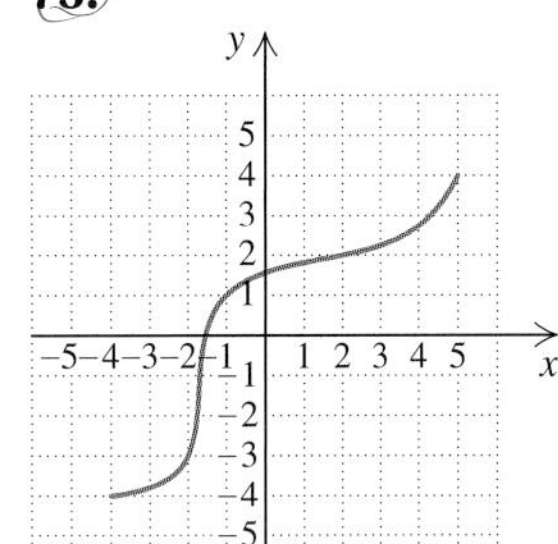

76.

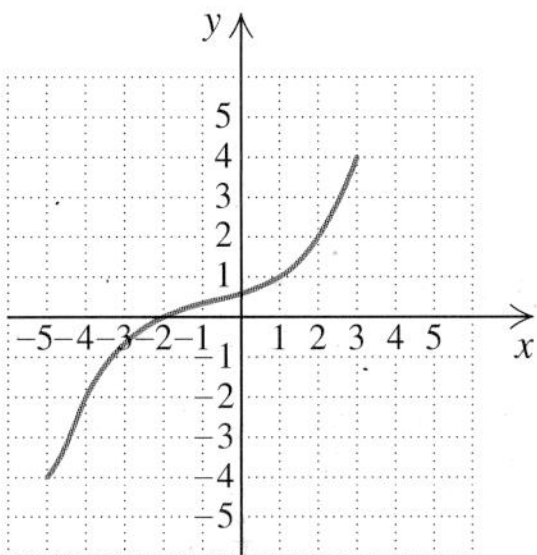

77. *Dress sizes in France and Italy.* Use the information in Exercises 63 and 64 to find a function for the French dress size that corresponds to a size x dress in Italy.

78. *Dress sizes in Italy and France.* Use the information in Exercises 63 and 64 to find a function for the Italian dress size that corresponds to a size x dress in France.

79. What relationship exists between the answers to Exercises 77 and 78? Explain how you determined this.

80. Show that function composition is associative by showing that $((f \circ g) \circ h)(x) = (f \circ (g \circ h))(x)$.

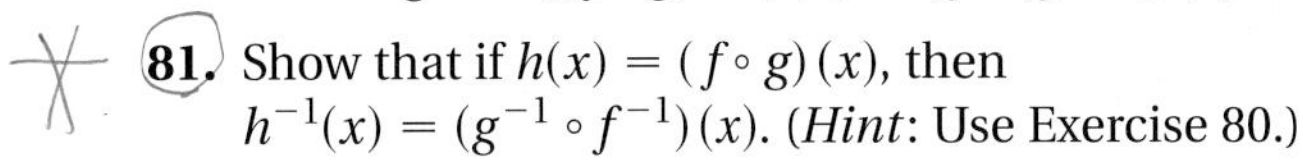

81. Show that if $h(x) = (f \circ g)(x)$, then $h^{-1}(x) = (g^{-1} \circ f^{-1})(x)$. (*Hint*: Use Exercise 80.)

Determine whether or not the given functions are inverses of each other.

82. $f(x) = 0.75x^2 + 2$; $g(x) = \sqrt{\dfrac{4(x-2)}{3}}$

83. $f(x) = 1.4x^3 + 3.2$; $g(x) = \sqrt[3]{\dfrac{x - 3.2}{1.4}}$

84. $f(x) = \sqrt{2.5x + 9.25}$;
$g(x) = 0.4x^2 - 3.7, x \geq 0$

85. $f(x) = 0.8x^{1/2} + 5.23$;
$g(x) = 1.25(x^2 - 5.23), x \geq 0$

86. $f(x) = 2.5(x^3 - 7.1)$;
$g(x) = \sqrt[3]{0.4x + 7.1}$

87. Match each function in Column A with its inverse from Column B.

Column A	*Column B*
(1) $y = 5x^3 + 10$	**A.** $y = \dfrac{\sqrt[3]{x} - 10}{5}$
(2) $y = (5x + 10)^3$	**B.** $y = \sqrt[3]{\dfrac{x}{5}} - 10$
(3) $y = 5(x + 10)^3$	**C.** $y = \sqrt[3]{\dfrac{x - 10}{5}}$
(4) $y = (5x)^3 + 10$	**D.** $y = \dfrac{\sqrt[3]{x - 10}}{5}$

88. How could a grapher be used to determine whether a function is one-to-one?

89. Examine the following table. Does it appear that *f* and *g* could be inverses of each other? Why or why not?

x	$f(x)$	$g(x)$
6	6	6
7	6.5	8
8	7	10
9	7.5	12
10	8	14
11	8.5	16
12	9	18

90. Assume in Exercise 89 that *f* and *g* are both linear functions. Find equations for $f(x)$ and $g(x)$. Are *f* and *g* inverses of each other?

9.2 Exponential Functions

Graphing Exponential Functions • Equations with x and y Interchanged • Applications of Exponential Functions

CONNECTING THE CONCEPTS

Composite and inverse functions, as shown in Section 9.1, are very useful in and of themselves. The reason they are included in this chapter, however, is because they are needed in order to understand the logarithmic functions that appear in Section 9.3. Here in Section 9.2, we make no reference to composite or inverse functions. Instead, we introduce a new type of function, the *exponential function,* so that we can study both it and its inverse in Sections 9.3–9.7.

Consider the graph below. The rapidly rising curve approximates the graph of an *exponential function*. We now consider such functions and some of their applications.

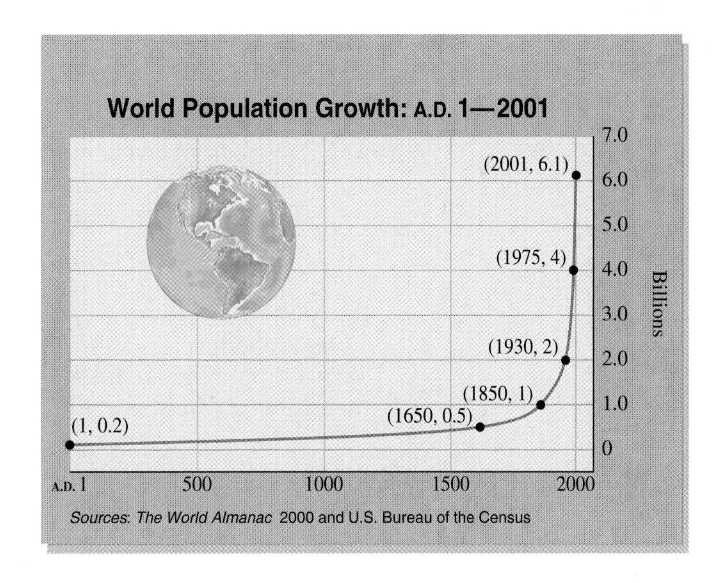

Graphing Exponential Functions

In Chapter 7, we studied exponential expressions with rational-number exponents, such as

$$5^{1/4}, \quad 3^{-3/4}, \quad 7^{2.34}, \quad 5^{1.73}.$$

For example, $5^{1.73}$, or $5^{173/100}$, represents the 100th root of 5 raised to the 173rd power. What about expressions with irrational exponents, such as $5^{\sqrt{3}}$ or $7^{-\pi}$? To attach meaning to $5^{\sqrt{3}}$, consider a rational approximation, r, of $\sqrt{3}$. As r gets closer to $\sqrt{3}$, the value of 5^r gets closer to some real number p.

r closes in on $\sqrt{3}$.	5^r closes in on some real number p.
$1.7 < r < 1.8$	$15.426 \approx 5^{1.7} < p < 5^{1.8} \approx 18.119$
$1.73 < r < 1.74$	$16.189 \approx 5^{1.73} < p < 5^{1.74} \approx 16.452$
$1.732 < r < 1.733$	$16.241 \approx 5^{1.732} < p < 5^{1.733} \approx 16.267$

We define $5^{\sqrt{3}}$ to be the number p. To eight decimal places,

$$5^{\sqrt{3}} \approx 16.24245082.$$

Any positive irrational exponent can be defined in a similar way. Negative irrational exponents are then defined using reciprocals. Thus, so long as a is positive, a^x has meaning for *any* real number x. All of the laws of exponents still hold, but we will not prove that here. We now define an *exponential function.*

Exponential Function

The function $f(x) = a^x$, where a is a positive constant, $a \neq 1$, is called the *exponential function,* base a.

We require the base a to be positive to avoid imaginary numbers that would result from taking even roots of negative numbers. The restriction $a \neq 1$ is made to exclude the constant function $f(x) = 1^x$, or $f(x) = 1$.

The following are examples of exponential functions:

$$f(x) = 2^x, \qquad f(x) = \left(\tfrac{1}{3}\right)^x, \qquad f(x) = 5^{-3x}. \qquad \text{Note that } 5^{-3x} = (5^{-3})^x.$$

Like polynomial functions, the domain of an exponential function is the set of all real numbers. In contrast to polynomial functions, exponential functions have a variable exponent. Because of this, graphs of exponential functions either rise or fall dramatically.

Example 1 Graph the exponential function $y = f(x) = 2^x$.

Solution We compute some function values, thinking of y as $f(x)$, and list the results in a table. It is a good idea to start by letting $x = 0$.

$$f(0) = 2^0 = 1; \qquad f(-1) = 2^{-1} = \frac{1}{2^1} = \frac{1}{2};$$
$$f(1) = 2^1 = 2;$$
$$f(2) = 2^2 = 4; \qquad f(-2) = 2^{-2} = \frac{1}{2^2} = \frac{1}{4};$$
$$f(3) = 2^3 = 8; \qquad f(-3) = 2^{-3} = \frac{1}{2^3} = \frac{1}{8}$$

x	y, or $f(x)$
0	1
1	2
2	4
3	8
-1	$\frac{1}{2}$
-2	$\frac{1}{4}$
-3	$\frac{1}{8}$

Next, we plot these points and connect them with a smooth curve.

The curve comes very close to the x-axis, but does not touch or cross it.

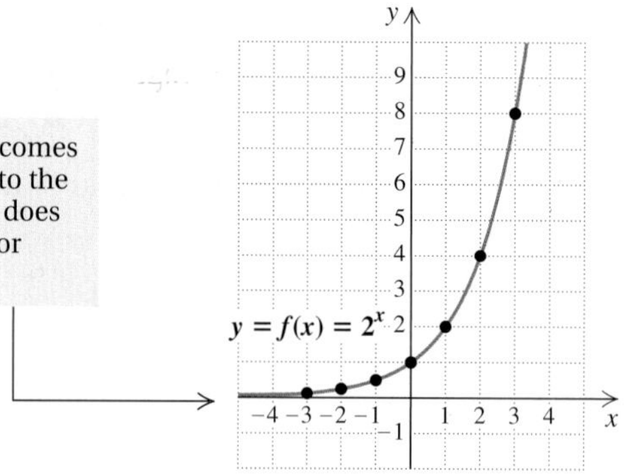

$y = f(x) = 2^x$

Be sure to plot enough points to determine how steeply the curve rises.

Note that as x increases, the function values increase without bound. As x decreases, the function values decrease, getting very close to 0. The x-axis, or the line $y = 0$, is a horizontal *asymptote,* meaning that the curve gets closer and closer to this line the further we move to the left.

Example 2 Graph the exponential function $y = f(x) = \left(\frac{1}{2}\right)^x$.

Solution We compute some function values, thinking of y as $f(x)$, and list the results in a table. Before we do this, note that

$$y = f(x) = \left(\tfrac{1}{2}\right)^x = (2^{-1})^x = 2^{-x}.$$

Then we have

$$f(0) = 2^{-0} = 1;$$

$$f(1) = 2^{-1} = \frac{1}{2^1} = \frac{1}{2};$$

$$f(2) = 2^{-2} = \frac{1}{2^2} = \frac{1}{4};$$

$$f(3) = 2^{-3} = \frac{1}{2^3} = \frac{1}{8};$$

$$f(-1) = 2^{-(-1)} = 2^1 = 2;$$

$$f(-2) = 2^{-(-2)} = 2^2 = 4;$$

$$f(-3) = 2^{-(-3)} = 2^3 = 8.$$

x	y, or $f(x)$
0	1
1	$\frac{1}{2}$
2	$\frac{1}{4}$
3	$\frac{1}{8}$
−1	2
−2	4
−3	8

Next, we plot these points and connect them with a smooth curve. Note that this curve is a mirror image, or *reflection*, of the above graph of $y = 2^x$ across the y-axis. The line $y = 0$ is again an asymptote.

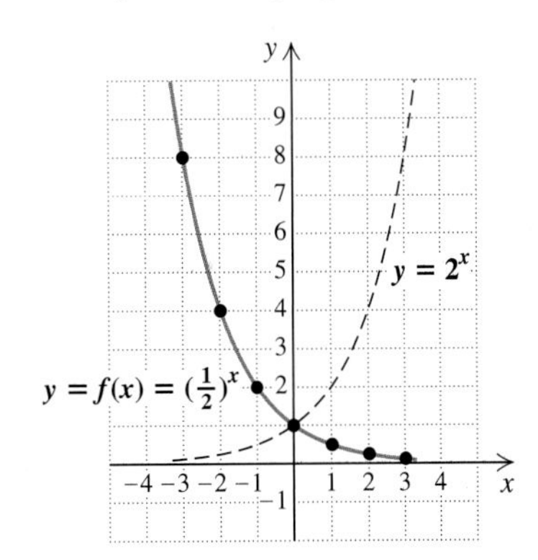

Graphers are especially helpful when bases are not whole numbers and when function values become large. For example, consider the graph of $y = 5000(1.075)^x$. Because y-values will be positive and will increase rapidly, an appropriate window might be $[-10, 10, 0, 15000]$, where the *scale* of the y-axis is 1000.

$y = 5000(1.075)^x$

15000

−10 10

0

Yscl = 1000

Graph each pair of functions. Select an appropriate window and scale.

1. $y_1 = \left(\frac{5}{2}\right)^x$ and $y_2 = \left(\frac{2}{5}\right)^x$
2. $y_1 = 3.2^x$ and $y_2 = 3.2^{-x}$
3. $y_1 = \left(\frac{3}{7}\right)^x$ and $y_2 = \left(\frac{7}{3}\right)^x$
4. $y_1 = 5000(1.08)^x$ and $y_2 = 5000(1.08)^{x-3}$

From Examples 1 and 2, we can make the following observations.

A. For $a > 1$, the graph of $f(x) = a^x$ increases from left to right. The greater the value of a, the steeper the curve. (See the figure below.)

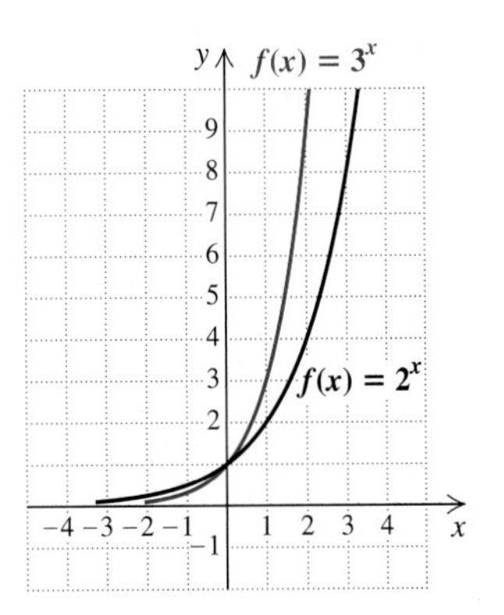

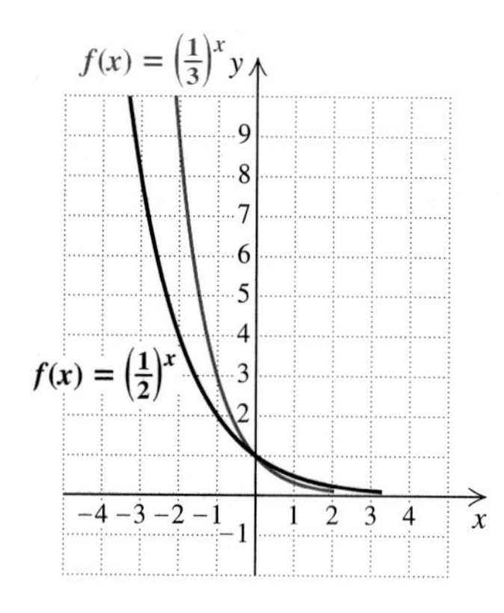

B. For $0 < a < 1$, the graph of $f(x) = a^x$ decreases from left to right. For smaller values of a, the curve becomes steeper. (See the figure at left.)

C. All graphs of $f(x) = a^x$ go through the y-intercept $(0, 1)$.

D. If $f(x) = a^x$, with $a > 0$, $a \neq 1$, the domain of f is all real numbers, and the range of f is all positive real numbers.

E. For $a > 0$, $a \neq 1$, the function given by $f(x) = a^x$ is one-to-one. Its graph passes the horizontal-line test.

Example 3 Graph: $y = f(x) = 2^{x-2}$.

Solution We construct a table of values. Then we plot the points and connect them with a smooth curve. Here $x - 2$ is the *exponent.*

$$f(0) = 2^{0-2} = 2^{-2} = \frac{1}{4};$$
$$f(1) = 2^{1-2} = 2^{-1} = \frac{1}{2};$$
$$f(2) = 2^{2-2} = 2^0 = 1;$$
$$f(3) = 2^{3-2} = 2^1 = 2;$$
$$f(4) = 2^{4-2} = 2^2 = 4;$$
$$f(-1) = 2^{-1-2} = 2^{-3} = \frac{1}{8};$$
$$f(-2) = 2^{-2-2} = 2^{-4} = \frac{1}{16}$$

x	y, or $f(x)$
0	$\frac{1}{4}$
1	$\frac{1}{2}$
2	1
3	2
4	4
−1	$\frac{1}{8}$
−2	$\frac{1}{16}$

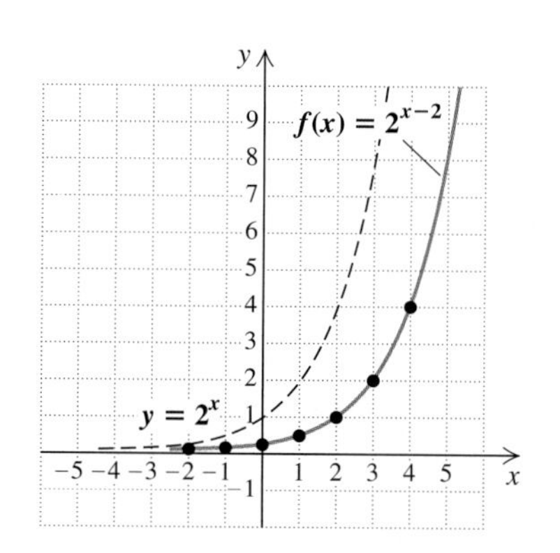

The graph looks just like the graph of $y = 2^x$, but it is translated 2 units to the right. The y-intercept of $y = 2^x$ is $(0, 1)$. The y-intercept of $y = 2^{x-2}$ is $\left(0, \frac{1}{4}\right)$. The line $y = 0$ is again the asymptote.

Equations with x and y Interchanged

It will be helpful in later work to be able to graph an equation in which the x and the y in $y = a^x$ are interchanged.

Example 4 Graph: $x = 2^y$.

Solution Note that x is alone on one side of the equation. To find ordered pairs that are solutions, we choose values for y and then compute values for x:

For $y = 0$, $x = 2^0 = 1$.
For $y = 1$, $x = 2^1 = 2$.
For $y = 2$, $x = 2^2 = 4$.
For $y = 3$, $x = 2^3 = 8$.
For $y = -1$, $x = 2^{-1} = \frac{1}{2}$.
For $y = -2$, $x = 2^{-2} = \frac{1}{4}$.
For $y = -3$, $x = 2^{-3} = \frac{1}{8}$.

x	y
1	0
2	1
4	2
8	3
$\frac{1}{2}$	-1
$\frac{1}{4}$	-2
$\frac{1}{8}$	-3

(1) Choose values for y.
(2) Compute values for x.

We plot the points and connect them with a smooth curve.

This curve does not touch or cross the y-axis, which serves as a vertical asymptote.

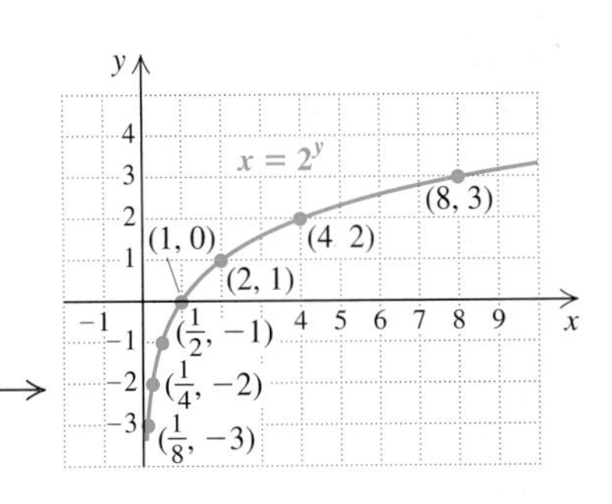

Note too that this curve looks just like the graph of $y = 2^x$, except that it is reflected across the line $y = x$, as shown here.

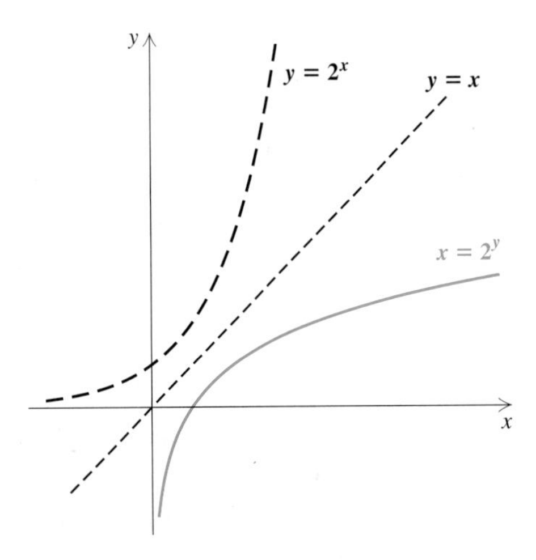

Applications of Exponential Functions

Example 5 ***Interest compounded annually.*** The amount of money A that a principal P will be worth after t years at interest rate i, compounded annually, is given by the formula

$$A = P(1 + i)^t. \qquad \text{You might review Example 12 in Section 8.1.}$$

Suppose that \$100,000 is invested at 8% interest, compounded annually.

a) Find a function for the amount in the account after t years.
b) Find the amount of money in the account at $t = 0$, $t = 4$, $t = 8$, and $t = 10$.
c) Graph the function.

Solution

a) If $P = \$100{,}000$ and $i = 8\% = 0.08$, we can substitute these values and form the following function:

$$\begin{aligned} A(t) &= \$100{,}000(1 + 0.08)^t \qquad \text{Using } A = P(1 + i)^t \\ &= \$100{,}000(1.08)^t. \end{aligned}$$

technology connection B

Graphers can quickly find many function values at the touch of a few keys. To see this, let $y_1 = 100{,}000(1.08)^x$. Then use the TABLE feature to check Example 5(b).

b) To find the function values, a calculator with a power key is helpful.

$$\begin{aligned} A(0) &= \$100{,}000(1.08)^0 \\ &= \$100{,}000(1) \\ &= \$100{,}000 \end{aligned} \qquad \begin{aligned} A(8) &= \$100{,}000(1.08)^8 \\ &\approx \$100{,}000(1.85093021) \\ &\approx \$185{,}093.02 \end{aligned}$$

$$\begin{aligned} A(4) &= \$100{,}000(1.08)^4 \\ &= \$100{,}000(1.36048896) \\ &\approx \$136{,}048.90 \end{aligned} \qquad \begin{aligned} A(10) &= \$100{,}000(1.08)^{10} \\ &\approx \$100{,}000(2.158924997) \\ &\approx \$215{,}892.50 \end{aligned}$$

c) We use the function values computed in part (b), and others if we wish, to draw the graph as follows. Note that the axes are scaled differently because of the large numbers.

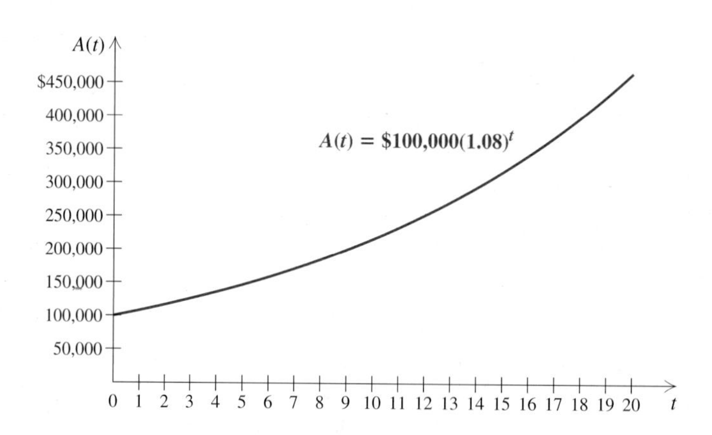

Exercise Set 9.2

FOR EXTRA HELP

Digital Video Tutor CD 7
Videotape 17

InterAct Math

Math Tutor Center

MathXL

MyMathLab.com

Graph.

1. $y = f(x) = 2^x$

2. $y = f(x) = 3^x$

3. $y = 5^x$

4. $y = 6^x$

5. $y = 2^x + 3$

6. $y = 2^x + 1$

7. $y = 3^x - 1$

8. $y = 3^x - 2$

9. $y = 2^x - 4$

10. $y = 2^x - 5$

11. $y = 2^{x-1}$

12. $y = 2^{x-2}$

13. $y = 2^{x+3}$

14. $y = 2^{x+1}$

15. $y = \left(\frac{1}{5}\right)^x$

16. $y = \left(\frac{1}{4}\right)^x$

17. $y = \left(\frac{1}{2}\right)^x$

18. $y = \left(\frac{1}{3}\right)^x$

19. $y = 2^{x-3} - 1$

20. $y = 2^{x+1} - 3$

21. $x = 3^y$

22. $x = 6^y$

23. $x = 2^{-y}$

24. $x = 3^{-y}$

25. $x = 5^y$

26. $x = 4^y$

27. $x = \left(\frac{3}{2}\right)^y$

28. $x = \left(\frac{4}{3}\right)^y$

Graph both equations using the same set of axes.

29. $y = 3^x,\ x = 3^y$

30. $y = 2^x,\ x = 2^y$

31. $y = \left(\frac{1}{2}\right)^x,\ x = \left(\frac{1}{2}\right)^y$

32. $y = \left(\frac{1}{4}\right)^x,\ x = \left(\frac{1}{4}\right)^y$

Solve.

33. *Population growth.* The world population $P(t)$, in billions, t years after 1975 can be approximated by

$$P(t) = 4(1.0164)^t$$

(*Sources*: Data from *The World Almanac* 2000 and U.S. Bureau of the Census).

a) What will the world population be in 2004? in 2008? in 2012?

b) Graph the function.

34. *Growth of bacteria.* The bacteria *Escherichi coli* are commonly found in the human bladder. Suppose that 3000 of the bacteria are present at time $t = 0$. Then t minutes later, the number of bacteria present will be

$$N(t) = 3000(2)^{t/20}.$$

a) How many bacteria will be present after 10 min? 20 min? 30 min? 40 min? 60 min?

b) Graph the function.

35. *Marine biology.* Due to excessive whaling prior to the mid 1970s, the humpback whale is considered an endangered species. The worldwide population of humpbacks, $P(t)$, in thousands, t years after 1900 ($t < 70$) can be approximated by*

$$P(t) = 150(0.960)^t.$$

a) How many humpback whales were alive in 1930? in 1960?

b) Graph the function.

36. *Marine biology.* As a result of preservation efforts in most countries in which whaling was common, the humpback whale population has grown since the 1970s. The worldwide population of humpbacks, $P(t)$, in thousands, t years after 1982 can be approximated by*

$$P(t) = 5.5(1.047)^t.$$

a) How many humpback whales were alive in 1992? in 2001?

b) Graph the function.

37. *Recycling aluminum cans.* It is estimated that $\frac{2}{3}$ of all aluminum cans distributed will be recycled each year. A beverage company distributes 250,000 cans. The number still in use after time t, in years, is given by the exponential function

$$N(t) = 250{,}000\left(\tfrac{2}{3}\right)^t.$$

a) How many cans are still in use after 0 yr? 1 yr? 4 yr? 10 yr?

b) Graph the function.

38. *Salvage value.* A photocopier is purchased for \$5200. Its value each year is about 80% of the value of the preceding year. Its value, in dollars, after t years is given by the exponential function

$$V(t) = 5200(0.8)^t.$$

a) Find the value of the machine after 0 yr, 1 yr, 2 yr, 5 yr, and 10 yr.

b) Graph the function.

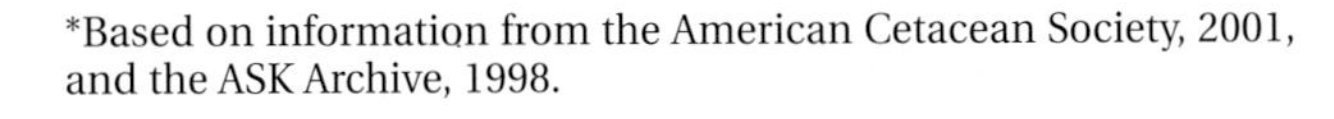

*Based on information from the American Cetacean Society, 2001, and the ASK Archive, 1998.

39. *Cellular phones.* The number of cellular phones in use in the United States is increasing exponentially. The number N, in millions, in use is given by the exponential function

$$N(t) = 0.3(1.4477)^t,$$

where t is the number of years after 1985 (*Source*: Cellular Telecommunications and Internet Association).

a) Find the number of cellular phones in use in 1985, 1995, 2005, and 2010.

b) Graph the function.

40. *Spread of zebra mussels.* Beginning in 1988, infestations of zebra mussels started spreading throughout North American waters.* These mussels spread with such speed that water treatment facilities, power plants, and entire ecosystems can become threatened. The function

$$A(t) = 10 \cdot 34^t$$

can be used to estimate the number of square centimeters of lake bottom that will be covered with mussels t years after an infestation covering 10 cm^2 first occurs.

a) How many square centimeters of lake bottom will be covered with mussels 5 years after an infestation covering 10 cm^2 first appears? 7 years after the infestation first appears?

b) Graph the function.

41. Without using a calculator, explain why 2^π must be greater than 8 but less than 16.

42. Suppose that \$1000 is invested for 5 yr at 7% interest, compounded annually. In what year will the most interest be earned? Why?

SKILL MAINTENANCE

43. 5^{-2}

44. 2^{-5}

45. $1000^{2/3}$

46. $25^{-3/2}$

47. $\dfrac{10a^8b^7}{2a^2b^4}$

48. $\dfrac{24x^6y^4}{4x^2y^3}$

SYNTHESIS

49. Examine Exercise 39. Do you believe that the equation for the number of cellular phones in use in the United States will be accurate 20 yr from now? Why or why not?

50. Why was it necessary to discuss irrational exponents before graphing exponential functions?

Determine which of the two numbers is larger. Do not use a calculator.

51. $\pi^{1.3}$ or $\pi^{2.4}$

52. $\sqrt{8^3}$ or $8^{\sqrt{3}}$

Graph.

53. $f(x) = 3.8^x$

54. $f(x) = 2.3^x$

55. $y = 2^x + 2^{-x}$

56. $y = \left|\left(\frac{1}{2}\right)^x - 1\right|$

57. $y = |2^x - 2|$

58. $y = 2^{-(x-1)^2}$

59. $y = |2^{x^2} - 1|$

60. $y = 3^x + 3^{-x}$

Graph both equations using the same set of axes.

61. $y = 3^{-(x-1)}, \; x = 3^{-(y-1)}$

62. $y = 1^x, \; x = 1^y$

63. *Sales of DVD players.* As prices of DVD players continue to drop, sales have grown from \$171 million in 1997 to \$421 million in 1998 and \$1099 million in 1999 (*Source*: *Statistical Abstract of the United States*, 2000). Use the REGRESSION feature in the STAT CALC menu to find an exponential function that models the total sales of DVD players t years after 1997. Then use that function to predict the total sales in 2005.

64. *Spread of AIDS.* In 1985, a total of 8249 cases of AIDS was reported in the United States; in 1988, a total of 31,001 cases; in 1989, a total of 33,722 cases; and in 1990, a total of 41,595 cases. Use the STAT REGRESSION feature of a grapher to find a model for $N(t)$, the number of AIDS cases in the United States t years after 1985. Then estimate the number of cases reported in 1987 and in 1997.

65. *Keyboarding speed.* Ali is studying keyboarding. After he has studied for t hours, Ali's speed, in words per minute, is given by the exponential function

$$S(t) = 200[1 - (0.99)^t].$$

Use a graph and/or table of values to predict Ali's speed after studying for 10 hr, 40 hr, and 80 hr.

66. Consider any exponential function of the form $f(x) = a^x$ with $a > 1$. Will it always follow that $f(3) - f(2) > f(2) - f(1)$, and, in general, $f(n + 2) - f(n + 1) > f(n + 1) - f(n)$? Why or why not? (*Hint*: Think graphically.)

*Many thanks to Dr. Gerald Mackie of the Department of Zoology at the University of Guelph in Ontario for the background information for this exercise.

COLLABORATIVE

CORNER

The True Cost of a New Car

Focus: Car loans and exponential functions

Time: 30 minutes

Group size: 2

Materials: Calculators with exponentiation keys

The formula

$$M = \frac{Pr}{1 - (1 + r)^{-n}}$$

is used to determine the payment size, M, when a loan of P dollars is to be repaid in n equally sized monthly payments. Here r represents the monthly interest rate. Loans repaid in this fashion are said to be *amortized* (spread out equally) over a period of n months.

ACTIVITY

1. Suppose one group member is selling the other a car for $2600, financed at 1% interest per month for 24 months. What should be the size of each monthly payment?

2. Suppose both group members are shopping for the same model new car. To save time, each group member visits a different dealer. One dealer offers the car for $13,000 at 10.5% interest (0.00875 monthly interest) for 60 months (no down payment). The other dealer offers the same car for $12,000, but at 12% interest (0.01 monthly interest) for 48 months (no down payment).

a) Determine the monthly payment size for each offer (remember to use the *monthly* interest rates). Then determine the total amount paid for the car under each offer. How much of each total is interest?

b) Work together to find the annual interest rate for which the total cost of 60 monthly payments for the $13,000 car would equal the total amount paid for the $12,000 car (as found in part a above).

9.3 Logarithmic Functions

Graphs of Logarithmic Functions • Converting Exponential and Logarithmic Equations • Solving Certain Logarithmic Equations

We are now ready to study inverses of exponential functions. These functions have many applications and are referred to as *logarithm,* or *logarithmic, functions.*

Graphs of Logarithmic Functions

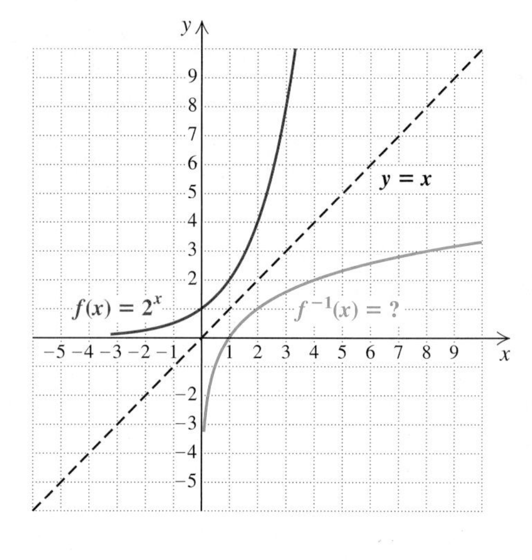

Consider the exponential function $f(x) = 2^x$. Like all exponential functions, f is one-to-one. Can a formula for f^{-1} be found? To answer this, we use the method of Section 9.1:

1. Replace $f(x)$ with y:

$$y = 2^x.$$

2. Interchange x and y:

$$x = 2^y.$$

3. Solve for y: $y =$ the power to which we raise 2 to get x.
4. Replace y with $f^{-1}(x)$: $f^{-1}(x) =$ the power to which we raise 2 to get x.

We now define a new symbol to replace the words "the power to which we raise 2 to get x":

> **$\log_2 x$, read "the logarithm, base 2, of x", or "log, base 2, of x," means "the power to which we raise 2 to get x."**

Thus if $f(x) = 2^x$, then $f^{-1}(x) = \log_2 x$. Note that $f^{-1}(8) = \log_2 8 = 3$, because 3 is *the power to which we raise* 2 *to get* 8.

Example 1 Simplify: **(a)** $\log_2 32$; **(b)** $\log_2 1$; **(c)** $\log_2 \frac{1}{8}$.

Solution

a) Think of the meaning of $\log_2 32$. It is the exponent to which we raise 2 to get 32. That exponent is 5. Therefore, $\log_2 32 = 5$.

b) We ask ourselves: "To what power do we raise 2 in order to get 1?" That power is 0 (recall that $2^0 = 1$). Thus, $\log_2 1 = 0$.

c) To what power do we raise 2 in order to get $\frac{1}{8}$? Since $2^{-3} = \frac{1}{8}$, we have $\log_2 \frac{1}{8} = -3$.

Although expressions like $\log_2 13$ can only be approximated, we must remember that $\log_2 13$ represents *the power to which we raise* 2 *to get* 13. That is, $2^{\log_2 13} = 13$. A calculator can be used to show that $\log_2 13 \approx 3.7$ and $2^{3.7} \approx 13$. Later in this chapter, we will discuss how a calculator can be used to find such approximations.

For any exponential function $f(x) = a^x$, the inverse is called a **logarithmic function, base *a*.** The graph of the inverse can, of course, be drawn by reflecting the graph of $f(x) = a^x$ across the line $y = x$. It will be helpful to remember that the inverse of $f(x) = a^x$ is given by $f^{-1}(x) = \log_a x$.

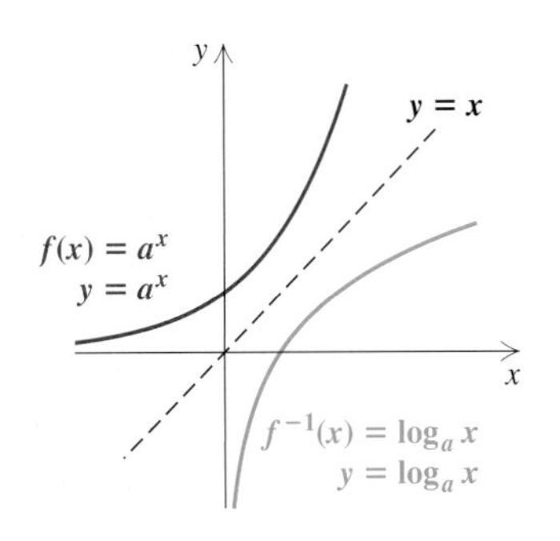

The Meaning of $\log_a x$

For $x > 0$ and a a positive constant other than 1, $\log_a x$ is the power to which a must be raised in order to get x. Thus,

$$a^{\log_a x} = x \qquad \text{or equivalently,} \qquad \text{if } y = \log_a x, \text{ then } a^y = x.$$

It is important to remember that *a logarithm is an exponent.* It might help to repeat several times: "The logarithm, base a, of a number x is the power to which a must be raised in order to get x."

Example 2 Simplify: $7^{\log_7 85}$.

Solution Remember that $\log_7 85$ is the power to which 7 is raised to get 85. Raising 7 to that power, we have

$$7^{\log_7 85} = 85.$$

Because logarithmic and exponential functions are inverses of each other, the result in Example 2 should come as no surprise: If $f(x) = \log_7 x$, then

$$\text{for} \quad f(x) = \log_7 x, \text{ we have } f^{-1}(x) = 7^x$$

$$\text{and} \quad f^{-1}(f(x)) = f^{-1}(\log_7 x) = 7^{\log_7 x} = x.$$

Thus, $f^{-1}(f(85)) = 7^{\log_7 85} = 85$.

The following is a comparison of exponential and logarithmic functions.

Exponential Function	Logarithmic Function
$y = a^x$	$x = a^y$
$f(x) = a^x$	$f(x) = \log_a x$
$a > 0, a \neq 1$	$a > 0, a \neq 1$
The domain is $\mathbb{R}$.	The range is $\mathbb{R}$.
$y > 0$ (Outputs are positive.)	$x > 0$ (Inputs are positive.)
$f^{-1}(x) = \log_a x$	$f^{-1}(x) = a^x$

To see that $f(x) = 10^x$ and $g(x) = \log_{10} x$ are inverses of each other, let $y_1 = 10^x$ and $y_2 = \log_{10} x = \log x$. Then, using a squared window, compare both graphs. Finally, let $y_3 = y_1(y_2)$ and $y_4 = y_2(y_1)$ to show, using a table or graphs, that $y_3 = y_4 = x$.

Example 3 Graph: $y = f(x) = \log_5 x$.

Solution If $y = \log_5 x$, then $5^y = x$. We can find ordered pairs that are solutions by choosing values for y and computing the x-values.

For $y = 0$, $x = 5^0 = 1$.

For $y = 1$, $x = 5^1 = 5$. (1) Select y.

For $y = 2$, $x = 5^2 = 25$. (2) Compute x.

For $y = -1$, $x = 5^{-1} = \frac{1}{5}$.

For $y = -2$, $x = 5^{-2} = \frac{1}{25}$.

x, or 5^y	y
1	0
5	1
25	2
$\frac{1}{5}$	-1
$\frac{1}{25}$	-2

This table shows the following:

$$\left.\begin{aligned} \log_5 1 &= 0; \\ \log_5 5 &= 1; \\ \log_5 25 &= 2; \\ \log_5 \tfrac{1}{5} &= -1; \\ \log_5 \tfrac{1}{25} &= -2. \end{aligned}\right\}$$

These can all be checked using the equations above.

We plot the set of ordered pairs and connect the points with a smooth curve. The graph of $y = 5^x$ is shown only for reference.

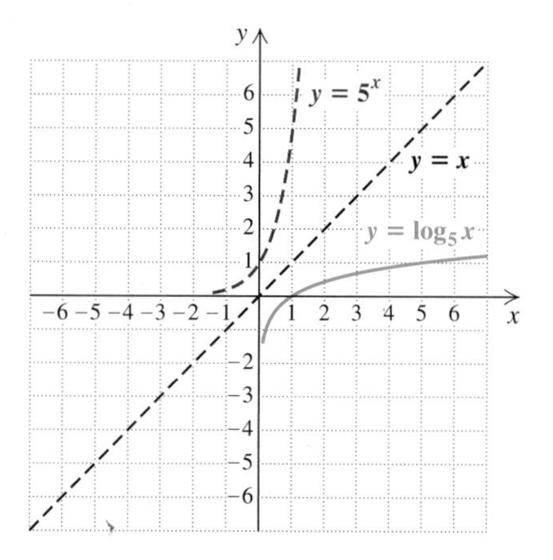

Converting Exponential and Logarithmic Equations

We use the definition of logarithm to convert from *exponential equations* to *logarithmic equations*:

$$\mathbf{y = \log_a x \quad \text{is equivalent to} \quad a^y = x.}$$

Caution! **Do not forget this relationship!** It is probably the most important definition in the chapter. Many times this definition will be used to justify a property we are considering.

Example 4

Convert each to a logarithmic equation: **(a)** $8 = 2^x$; **(b)** $y^{-1} = 4$; **(c)** $a^b = c$.

Solution

a) $8 = 2^x$ is equivalent to $x = \log_2 8$ — The exponent is the logarithm. The base remains the same.

b) $y^{-1} = 4$ is equivalent to $-1 = \log_y 4$

c) $a^b = c$ is equivalent to $b = \log_a c$

We also use the definition of logarithm to convert from logarithmic equations to exponential equations.

Example 5

Convert each to an exponential equation: **(a)** $y = \log_3 5$; **(b)** $-2 = \log_a 7$; **(c)** $a = \log_b d$.

Solution

a) $y = \log_3 5$ is equivalent to $3^y = 5$ — The logarithm is the exponent. The base remains the same.

b) $-2 = \log_a 7$ is equivalent to $a^{-2} = 7$

c) $a = \log_b d$ is equivalent to $b^a = d$

Solving Certain Logarithmic Equations

Some logarithmic equations can be solved by converting to exponential equations.

Example 6

Solve: **(a)** $\log_2 x = -3$; **(b)** $\log_x 16 = 2$.

Solution

a) $\log_2 x = -3$

$2^{-3} = x$ — Converting to an exponential equation

$\frac{1}{8} = x$ — Computing 2^{-3}

Check: $\log_2 \frac{1}{8}$ is the power to which 2 is raised to get $\frac{1}{8}$. Since that power is -3, we have a check. The solution is $\frac{1}{8}$.

b) $\log_x 16 = 2$

$x^2 = 16$ — Converting to an exponential equation

$x = 4$ *or* $x = -4$ — Principle of square roots

Check: $\log_4 16 = 2$ because $4^2 = 16$. Thus, 4 is a solution of $\log_x 16 = 2$. Because all logarithmic bases must be positive, -4 cannot be a solution. Logarithmic bases must be positive because logarithms are defined using exponential functions that require positive bases. The solution is 4.

One method for solving certain logarithmic and exponential equations relies on the following property, which results from the fact that exponential functions are one-to-one.

The Principle of Exponential Equality

For any real number b, where $b \neq -1, 0$, or 1,

$$b^x = b^y \quad \text{is equivalent to} \quad x = y.$$

(Powers of the same base are equal if and only if the exponents are equal.)

Example 7

Solve: **(a)** $\log_{10} 1000 = x$; **(b)** $\log_4 1 = t$.

Solution

a) We convert $\log_{10} 1000 = x$ to exponential form and solve:

$10^x = 1000$	Converting to an exponential equation
$10^x = 10^3$	Writing 1000 as a power of 10
$x = 3.$	Equating exponents

Check: This equation can also be solved directly by determining the power to which we raise 10 in order to get 1000. In both cases we find that $\log_{10} 1000 = 3$, so we have a check. The solution is 3.

b) We convert $\log_4 1 = t$ to exponential form and solve:

$4^t = 1$	Converting to an exponential equation
$4^t = 4^0$	Writing 1 as a power of 4. This can be done mentally.
$t = 0.$	Equating exponents

Check: As in part (a), this equation can be solved directly by determining the power to which we raise 4 in order to get 1. In both cases we find that $\log_4 1 = 0$, so we have a check. The solution is 0.

Example 7 illustrates an important property of logarithms.

$\log_a 1$

The logarithm, base a, of 1 is always 0: $\log_a 1 = 0$.

This follows from the fact that $a^0 = 1$ is equivalent to the logarithmic equation $\log_a 1 = 0$. Thus, $\log_{10} 1 = 0$, $\log_7 1 = 0$, and so on.

Another property results from the fact that $a^1 = a$. This is equivalent to the equation $\log_a a = 1$.

$\log_a a$

The logarithm, base a, of a is always 1: $\log_a a = 1$.

Thus, $\log_{10} 10 = 1$, $\log_8 8 = 1$, and so on.

Exercise Set 9.3

FOR EXTRA HELP

Digital Video Tutor CD 7
Videotape 17

InterAct Math

Math Tutor Center

MathXL

MyMathLab.com

Simplify.

1. $\log_{10} 100$
2. $\log_{10} 1000$
3. $\log_2 8$
4. $\log_2 16$
5. $\log_3 81$
6. $\log_3 27$
7. $\log_4 \frac{1}{16}$
8. $\log_4 \frac{1}{4}$
9. $\log_7 \frac{1}{7}$
10. $\log_7 \frac{1}{49}$
11. $\log_5 625$
12. $\log_5 125$
13. $\log_6 6$
14. $\log_7 1$
15. $\log_8 1$
16. $\log_8 8$
17. Aha! $\log_9 9^7$
18. $\log_9 9^{10}$
19. $\log_{10} 0.1$
20. $\log_{10} 0.01$
21. $\log_9 3$
22. $\log_{16} 4$
23. $\log_9 27$
24. $\log_{16} 64$
25. $\log_{1000} 100$
26. $\log_{27} 9$
27. $5^{\log_5 7}$
28. $6^{\log_6 13}$

Graph.

29. $y = \log_{10} x$
30. $y = \log_2 x$
31. $y = \log_3 x$
32. $y = \log_7 x$
33. $f(x) = \log_6 x$
34. $f(x) = \log_4 x$
35. $f(x) = \log_{2.5} x$
36. $f(x) = \log_{1/2} x$

Graph both functions using the same set of axes.

37. $f(x) = 3^x$, $f^{-1}(x) = \log_3 x$
38. $f(x) = 4^x$, $f^{-1}(x) = \log_4 x$

Convert to logarithmic equations.

39. $10^2 = 100$
40. $10^4 = 10{,}000$
41. $4^{-5} = \frac{1}{1024}$
42. $5^{-3} = \frac{1}{125}$
43. $16^{3/4} = 8$
44. $8^{1/3} = 2$
45. $10^{0.4771} = 3$
46. $10^{0.3010} = 2$
47. $p^k = 3$
48. $m^n = r$
49. $p^m = V$
50. $Q^t = x$
51. $e^3 = 20.0855$
52. $e^2 = 7.3891$
53. $e^{-4} = 0.0183$
54. $e^{-2} = 0.1353$

Convert to exponential equations.

55. $t = \log_3 8$
56. $h = \log_7 10$
57. $\log_5 25 = 2$
58. $\log_6 6 = 1$
59. $\log_{10} 0.1 = -1$
60. $\log_{10} 0.01 = -2$
61. $\log_{10} 7 = 0.845$
62. $\log_{10} 3 = 0.4771$
63. $\log_c m = 8$
64. $\log_b n = 23$
65. $\log_t Q = r$
66. $\log_m P = a$
67. $\log_e 0.25 = -1.3863$

68. $\log_e 0.989 = -0.0111$

69. $\log_r T = -x$

70. $\log_c M = -w$

Solve.

71. $\log_3 x = 2$

72. $\log_4 x = 3$

73. $\log_x 64 = 3$

74. $\log_x 125 = 3$

75. $\log_5 25 = x$

76. $\log_2 16 = x$

77. $\log_4 16 = x$

78. $\log_3 27 = x$

79. $\log_x 7 = 1$

80. $\log_x 8 = 1$

81. $\log_9 x = 1$

82. $\log_6 x = 0$

83. $\log_3 x = -2$

84. $\log_2 x = -1$

85. $\log_{32} x = \frac{2}{5}$

86. $\log_8 x = \frac{2}{3}$

87. Express in words what number is represented by $\log_b c$.

88. Is it true that $2 = b^{\log_b 2}$? Why or why not?

SKILL MAINTENANCE

Simplify.

89. $\dfrac{x^{12}}{x^4}$

90. $\dfrac{a^{15}}{a^3}$

91. $(a^4b^6)(a^3b^2)$

92. $(x^3y^5)(x^2y^7)$

93. $\dfrac{\dfrac{3}{x} - \dfrac{2}{xy}}{\dfrac{2}{x^2} + \dfrac{1}{xy}}$

94. $\dfrac{\dfrac{4 + x}{x^2 + 2x + 1}}{\dfrac{3}{x + 1} - \dfrac{2}{x + 2}}$

SYNTHESIS

95. Would a manufacturer be pleased or unhappy if sales of a product grew logarithmically? Why?

96. Explain why the number $\log_2 13$ must be between 3 and 4.

97. Graph both equations using the same set of axes:

$$y = \left(\tfrac{3}{2}\right)^x, \qquad y = \log_{3/2} x.$$

Graph.

98. $y = \log_2 (x - 1)$

99. $y = \log_3 |x + 1|$

Solve.

100. $|\log_3 x| = 2$

101. $\log_{125} x = \frac{2}{3}$

102. $\log_4 (3x - 2) = 2$

103. $\log_8 (2x + 1) = -1$

104. $\log_{10} (x^2 + 21x) = 2$

Simplify.

105. $\log_{1/4} \frac{1}{64}$

106. $\log_{1/5} 25$

107. $\log_{81} 3 \cdot \log_3 81$

108. $\log_{10} (\log_4 (\log_3 81))$

109. $\log_2 (\log_2 (\log_4 256))$

110. Show that $b^x = b^y$ is *not* equivalent to $x = y$ for $b = 0$ or $b = 1$.

111. If $\log_b a = x$, does it follow that $\log_a b = 1/x$? Why or why not?

Properties of Logarithmic Functions

9.4

Logarithms of Products • Logarithms of Powers • Logarithms of Quotients • Using the Properties Together

Logarithmic functions are important in many applications and in more advanced mathematics. We now establish some basic properties that are useful in manipulating expressions involving logarithms. As their proofs reveal, the properties of logarithms are related to the properties of exponents.

Logarithms of Products

The first property we discuss is related to the product rule for exponents: $a^m \cdot a^n = a^{m+n}$.

The Product Rule for Logarithms

For any positive numbers M, N, and a ($a \neq 1$),

$$\log_a MN = \log_a M + \log_a N.$$

(The logarithm of a product is the sum of the logarithms of the factors.)

Example 1

Express as a sum of logarithms: $\log_2 (4 \cdot 16)$.

Solution We have

$$\log_2 (4 \cdot 16) = \log_2 4 + \log_2 16. \qquad \text{Using the product rule for logarithms}$$

As a check, note that

$$\log_2 (4 \cdot 16) = \log_2 64 = 6 \qquad 2^6 = 64$$

and that

$$\log_2 4 + \log_2 16 = 2 + 4 = 6. \qquad 2^2 = 4 \text{ and } 2^4 = 16$$

Example 2

Express as a single logarithm: $\log_b 7 + \log_b 5$.

Solution We have

$$\begin{aligned} \log_b 7 + \log_b 5 &= \log_b (7 \cdot 5) \qquad \text{Using the product rule for logarithms} \\ &= \log_b 35. \end{aligned}$$

The check is left to the student.

A Proof of the Product Rule. Let $\log_a M = x$ and $\log_a N = y$. Converting to exponential equations, we have $a^x = M$ and $a^y = N$.

Now we multiply the last two equations, to obtain

$$MN = a^x \cdot a^y, \quad \text{or} \quad MN = a^{x+y}.$$

Converting back to a logarithmic equation, we get

$$\log_a MN = x + y.$$

Recalling what x and y represent, we get

$$\log_a MN = \log_a M + \log_a N.$$

Logarithms of Powers

The second basic property is related to the power rule for exponents: $(a^m)^n = a^{mn}$.

The Power Rule for Logarithms

For any positive numbers M and a ($a \neq 1$), and any real number p,

$$\log_a M^p = p \cdot \log_a M.$$

(The logarithm of a power of M is the exponent times the logarithm of M.)

To better understand the power rule, note that

$$\log_a M^3 = \log_a (M \cdot M \cdot M) = \log_a M + \log_a M + \log_a M = 3 \log_a M.$$

Example 3 Express as a product: **(a)** $\log_a 9^{-5}$; **(b)** $\log_7 \sqrt[3]{x}$.

Solution

a) $\log_a 9^{-5} = -5 \log_a 9$ Using the power rule for logarithms

b) $\log_7 \sqrt[3]{x} = \log_7 x^{1/3}$ Writing exponential notation

$= \frac{1}{3} \log_7 x$ Using the power rule for logarithms

A Proof of the Power Rule. Let $x = \log_a M$. We then write the equivalent exponential equation, $a^x = M$. Raising both sides to the pth power, we get

$$(a^x)^p = M^p, \quad \text{or} \quad a^{xp} = M^p. \qquad \text{Multiplying exponents}$$

Converting back to a logarithmic equation gives us

$$\log_a M^p = xp.$$

But $x = \log_a M$, so substituting, we have

$$\log_a M^p = (\log_a M)p = p \cdot \log_a M.$$

Logarithms of Quotients

The third property that we study is similar to the quotient rule for exponents: $a^m/a^n = a^{m-n}$.

The Quotient Rule for Logarithms

For any positive numbers M, N, and a ($a \neq 1$),

$$\log_a \frac{M}{N} = \log_a M - \log_a N.$$

(The logarithm of a quotient is the logarithm of the dividend minus the logarithm of the divisor.)

To better understand the quotient rule, note that

$$\log_a\left(\frac{b^5}{b^3}\right) = \log_a b^2 = 2\log_a b = 5\log_a b - 3\log_a b$$
$$= \log_a b^5 - \log_a b^3.$$

Example 4

Express as a difference of logarithms: $\log_t (6/U)$.

Solution

$$\log_t \frac{6}{U} = \log_t 6 - \log_t U$$ Using the quotient rule for logarithms

Example 5

Express as a single logarithm: $\log_b 17 - \log_b 27$.

Solution

$$\log_b 17 - \log_b 27 = \log_b \frac{17}{27}$$ Using the quotient rule for logarithms "in reverse"

A Proof of the Quotient Rule. Our proof uses both the product and power rules:

$$\log_a \frac{M}{N} = \log_a MN^{-1}$$ Rewriting $\frac{M}{N}$ with a negative exponent

$$= \log_a M + \log_a N^{-1}$$ Using the product rule for logarithms

$$= \log_a M + (-1)\log_a N$$ Using the power rule for logarithms

$$= \log_a M - \log_a N.$$

Using the Properties Together

Example 6

Express in terms of the individual logarithms of x, y, and z.

a) $\log_b \dfrac{x^3}{yz}$ **b)** $\log_a \sqrt[4]{\dfrac{xy}{z^3}}$

Solution

a) $\log_b \dfrac{x^3}{yz} = \log_b x^3 - \log_b yz$ Using the quotient rule for logarithms

$= 3\log_b x - \log_b yz$ Using the power rule for logarithms

$= 3\log_b x - (\log_b y + \log_b z)$ Using the product rule for logarithms. Because of the subtraction, parentheses are essential.

$= 3\log_b x - \log_b y - \log_b z$ Using the distributive law

b) $\log_a \sqrt[4]{\frac{xy}{z^3}} = \log_a\left(\frac{xy}{z^3}\right)^{1/4}$ — Writing exponential notation

$= \frac{1}{4} \cdot \log_a \frac{xy}{z^3}$ — Using the power rule for logarithms

$= \frac{1}{4}(\log_a xy - \log_a z^3)$ — Using the quotient rule for logarithms. Parentheses are important.

$= \frac{1}{4}(\log_a x + \log_a y - 3\log_a z)$ — Using the product and power rules for logarithms

Caution! When subtraction or multiplication precedes use of the product or quotient rule, parentheses are needed, as in Example 6.

Example 7

Express as a single logarithm.

a) $\frac{1}{2}\log_a x - 7\log_a y + \log_a z$

b) $\log_a \frac{b}{\sqrt{x}} + \log_a \sqrt{bx}$

Solution

a) $\frac{1}{2}\log_a x - 7\log_a y + \log_a z$

$= \log_a x^{1/2} - \log_a y^7 + \log_a z$ — Using the power rule for logarithms

$= (\log_a \sqrt{x} - \log_a y^7) + \log_a z$ — Using parentheses to emphasize the order of operations; $x^{1/2} = \sqrt{x}$

$= \log_a \frac{\sqrt{x}}{y^7} + \log_a z$ — Using the quotient rule for logarithms

$= \log_a \frac{z\sqrt{x}}{y^7}$ — Using the product rule for logarithms

b) $\log_a \frac{b}{\sqrt{x}} + \log_a\sqrt{bx} = \log_a \frac{b \cdot \sqrt{bx}}{\sqrt{x}}$ — Using the product rule for logarithms

$= \log_a b\sqrt{b}$ — Removing a factor equal to 1: $\frac{\sqrt{x}}{\sqrt{x}} = 1$

$= \log_a b^{3/2}$, or $\frac{3}{2}\log_a b$ — Since $b\sqrt{b} = b^1 \cdot b^{1/2}$

If we know the logarithms of two different numbers (to the same base), the properties allow us to calculate other logarithms.

Example 8

Given $\log_a 2 = 0.431$ and $\log_a 3 = 0.683$, find each of the following.

a) $\log_a 6$ **b)** $\log_a \frac{2}{3}$ **c)** $\log_a 81$
d) $\log_a \frac{1}{3}$ **e)** $\log_a 2a$ **f)** $\log_a 5$

Solution

a) $\log_a 6 = \log_a (2 \cdot 3) = \log_a 2 + \log_a 3$ Using the product rule for logarithms
$= 0.431 + 0.683 = 1.114$

Check: $a^{1.114} = a^{0.431} \cdot a^{0.683} = 2 \cdot 3 = 6$

b) $\log_a \frac{2}{3} = \log_a 2 - \log_a 3$ Using the quotient rule for logarithms
$= 0.431 - 0.683 = -0.252$

c) $\log_a 81 = \log_a 3^4 = 4 \log_a 3$ Using the power rule for logarithms
$= 4(0.683) = 2.732$

d) $\log_a \frac{1}{3} = \log_a 1 - \log_a 3$ Using the quotient rule for logarithms
$= 0 - 0.683 = -0.683$

e) $\log_a 2a = \log_a 2 + \log_a a$ Using the product rule for logarithms
$= 0.431 + 1 = 1.431$

f) $\log_a 5$ *cannot be found using these properties.* $(\log_a 5 \neq \log_a 2 + \log_a 3)$

A final property follows from the product rule: Since $\log_a a^k = k \log_a a$, and $\log_a a = 1$, we have $\log_a a^k = k$.

The Logarithm of the Base to a Power

For any base a,

$$\log_a a^k = k.$$

(The logarithm, base a, of a to a power is the power.)

This property also follows from the definition of logarithm: k is the power to which you raise a in order to get a^k.

Example 9 Simplify: **(a)** $\log_3 3^7$; **(b)** $\log_{10} 10^{-5.2}$.

Solution

a) $\log_3 3^7 = 7$ 7 is the power to which you raise 3 in order to get 3^7.
b) $\log_{10} 10^{-5.2} = -5.2$

We summarize the properties covered in this section as follows.

For any positive numbers M, N, and a $(a \neq 1)$:

$\log_a MN = \log_a M + \log_a N$; $\log_a M^p = p \cdot \log_a M$;

$\log_a \dfrac{M}{N} = \log_a M - \log_a N$; $\log_a a^k = k$.

Caution! Keep in mind that, in general,

$\log_a (M + N) \neq \log_a M + \log_a N$, $\quad \log_a MN \neq (\log_a M)(\log_a N)$,

$\log_a (M - N) \neq \log_a M - \log_a N$, $\quad \log_a \dfrac{M}{N} \neq \dfrac{\log_a M}{\log_a N}$.

Exercise Set 9.4

FOR EXTRA HELP

Digital Video Tutor CD 7
Videotape 17

InterAct Math

Math Tutor Center

MathXL

MyMathLab.com

Express as a sum of logarithms.

1. $\log_3 (81 \cdot 27)$

2. $\log_2 (16 \cdot 32)$

3. $\log_4 (64 \cdot 16)$

4. $\log_5 (25 \cdot 125)$

5. $\log_c rst$

6. $\log_t 3ab$

Express as a single logarithm.

7. $\log_a 5 + \log_a 14$

8. $\log_b 65 + \log_b 2$

9. $\log_c t + \log_c y$

10. $\log_t H + \log_t M$

Express as a product.

11. $\log_a r^8$

12. $\log_b t^5$

13. $\log_c y^6$

14. $\log_{10} y^7$

15. $\log_b C^{-3}$

16. $\log_c M^{-5}$

Express as a difference of logarithms.

17. $\log_2 \frac{53}{17}$

18. $\log_3 \frac{23}{9}$

19. $\log_b \dfrac{m}{n}$

20. $\log_a \dfrac{y}{x}$

Express as a single logarithm.

21. $\log_a 15 - \log_a 3$

22. $\log_b 42 - \log_b 7$

23. $\log_b 36 - \log_b 4$

24. $\log_a 26 - \log_a 2$

25. $\log_a 7 - \log_a 18$

26. $\log_b 5 - \log_b 13$

Express in terms of the individual logarithms of w, x, y, and z.

27. $\log_a x^5y^7z^6$

28. $\log_a xy^4z^3$

29. $\log_b \dfrac{xy^2}{z^3}$

30. $\log_b \dfrac{x^2y^5}{w^4z^7}$

31. $\log_a \dfrac{x^4}{y^3z}$

32. $\log_a \dfrac{x^4}{yz^2}$

33. $\log_b \dfrac{xy^2}{wz^3}$

34. $\log_b \dfrac{w^2x}{y^3z}$

35. $\log_a \sqrt{\dfrac{x^7}{y^5z^8}}$

36. $\log_c \sqrt[3]{\dfrac{x^4}{y^3z^2}}$

37. $\log_a \sqrt[3]{\dfrac{x^6y^3}{a^2z^7}}$

38. $\log_a \sqrt[4]{\dfrac{x^8y^{12}}{a^3z^5}}$

Express as a single logarithm and, if possible, simplify.

39. $7 \log_a x + 3 \log_a z$

40. $2 \log_b m + \frac{1}{2} \log_b n$

41. $\log_a x^2 - 2 \log_a \sqrt{x}$

42. $\log_a \dfrac{a}{\sqrt{x}} - \log_a \sqrt{ax}$

43. $\frac{1}{2} \log_a x + 5 \log_a y - 2 \log_a x$

44. $\log_a 2x + 3(\log_a x - \log_a y)$

45. $\log_a (x^2 - 4) - \log_a (x + 2)$

46. $\log_a (2x + 10) - \log_a (x^2 - 25)$

Given $\log_b 3 = 0.792$ *and* $\log_b 5 = 1.161$. *If possible, find each of the following.*

47. $\log_b 15$

48. $\log_b \frac{5}{3}$

49. $\log_b \frac{3}{5}$

50. $\log_b \frac{1}{3}$

51. $\log_b \frac{1}{5}$

52. $\log_b \sqrt{b}$

53. $\log_b \sqrt{b^3}$

54. $\log_b 3b$

55. $\log_b 8$

56. $\log_b 45$

57. $\log_b 75$

58. $\log_b 20$

Simplify.

Aha! **59.** $\log_t t^9$

60. $\log_p p^4$

61. $\log_e e^m$

62. $\log_Q Q^{-2}$

63. A student *incorrectly* reasons that

$$\log_b \frac{1}{x} = \log_b \frac{x}{xx}$$
$$= \log_b x - \log_b x + \log_b x = \log_b x.$$

What mistake has the student made?

64. How could you convince someone that

$$\log_a c \neq \log_c a?$$

SKILL MAINTENANCE

Graph.

65. $f(x) = \sqrt{x} - 3$

66. $g(x) = \sqrt{x} + 2$

67. $g(x) = \sqrt[3]{x} + 1$

68. $f(x) = \sqrt[3]{x} - 1$

Simplify.

69. $(a^3b^2)^5(a^2b^7)$

70. $(x^5y^3z^2)(x^2yz^2)^3$

SYNTHESIS

71. Is it possible to express $\log_b \frac{x}{5}$ as a difference of two logarithms without using the quotient rule? Why or why not?

72. Is it true that $\log_a x + \log_b x = \log_{ab} x$? Why or why not?

Express as a single logarithm and, if possible, simplify.

73. $\log_a (x^8 - y^8) - \log_a (x^2 + y^2)$

74. $\log_a (x + y) + \log_a (x^2 - xy + y^2)$

Express as a sum or difference of logarithms.

75. $\log_a \sqrt{1 - s^2}$

76. $\log_a \frac{c - d}{\sqrt{c^2 - d^2}}$

77. If $\log_a x = 2$, $\log_a y = 3$, and $\log_a z = 4$, what is

$$\log_a \frac{\sqrt[3]{x^2z}}{\sqrt[3]{y^2z^{-2}}}?$$

78. If $\log_a x = 2$, what is $\log_a (1/x)$?

79. If $\log_a x = 2$, what is $\log_{1/a} x$?

Classify each of the following as true or false. Assume a, x, P, and Q > 0.

80. $\log_a \left(\frac{P}{Q}\right)^x = x \log_a P - \log_a Q$

81. $\log_a (Q + Q^2) = \log_a Q + \log_a (Q + 1)$

82. Use graphs to show that

$$\log x^2 \neq \log x \cdot \log x.$$ (*Note*: log means $\log_{10}$.)

9.5 Common and Natural Logarithms

Common Logarithms on a Calculator • The Base *e* and Natural Logarithms on a Calculator • Changing Logarithmic Bases • Graphs of Exponential and Logarithmic Functions, Base *e*

Any positive number other than 1 can serve as the base of a logarithmic function. However, some numbers are easier to use than others, and there are logarithmic bases that fit into certain applications more naturally than others.

Base-10 logarithms, called **common logarithms**, are useful because they have the same base as our "commonly" used decimal system. Before calculators became so widely available, common logarithms were helpful when performing tedious calculations. In fact, that is why logarithms were invented.

Another logarithmic base widely used today is an irrational number named *e*. We will consider *e* and base *e*, or *natural*, logarithms later in this section. First we examine common logarithms.

Common Logarithms on a Calculator

Before the advent of calculators, tables were developed to list common logarithms. Today we find common logarithms using calculators.

Here, and in most books, the abbreviation **log**, with no base written, is understood to mean logarithm base 10, or a common logarithm. Thus,

$\log 17$ means $\log_{10} 17$. It is important to remember this abbreviation.

On scientific calculators, the key for common logarithms is usually marked [log]. To find the common logarithm of a number, we enter that number and press the [log] key. On most graphing calculators, we press [log], the number, and then [ENTER].

Example 1 Use a calculator to find each number: **(a)** $\log 53{,}128$; **(b)** $\dfrac{\log 6500}{\log 0.007}$.

Solution

a) We enter 53,128 and then press [log]. We find that

$\log 53{,}128 \approx 4.7253.$ Rounded to four decimal places

b) We enter 6500 and then press [log]. Next, we press [÷], enter 0.007, press [log] [=]. Be careful not to round until the end:

$\dfrac{\log 6500}{\log 0.007} \approx -1.7694.$ Rounded to four decimal places

The inverse of a logarithmic function is an exponential function. Because of this, on many calculators the [log] key doubles as the [10^x] key after a [2nd] or [SHIFT] key is pressed.

Example 2 Use a calculator to find $10^{3.417}$.

Solution We enter 3.417 and then press [10^x]. On some calculators, [10^x] is pressed first, followed by 3.417 and [ENTER]. Since $10^{3.417}$ is irrational, our answer is approximate:

$10^{3.417} \approx 2612.161354.$

On calculators without a [10^x] key, an exponential key, labeled [x^y], [a^x], or [∧] may be available. Such a key can raise any positive real number to any real-numbered power.

The Base e and Natural Logarithms on a Calculator

When interest is compounded n times a year, the compound interest formula is

$$A = P\left(1 + \frac{r}{n}\right)^{nt},$$

where A is the amount that an initial investment P will be worth after t years at interest rate r. Suppose that \$1 is invested at 100% interest for 1 year (no bank would pay this). The preceding formula becomes a function A defined in terms of the number of compounding periods n:

$$A(n) = \left(1 + \frac{1}{n}\right)^{n}.$$

Let's find some function values. We round to six decimal places, using a calculator.

n	$A(n) = \left(1 + \frac{1}{n}\right)^{n}$
1 (compounded annually)	\$2.00
2 (compounded semiannually)	\$2.25
3	\$2.370370
4 (compounded quarterly)	\$2.441406
5	\$2.488320
100	\$2.704814
365 (compounded daily)	\$2.714567
8760 (compounded hourly)	\$2.718127

The numbers in this table approach a very important number in mathematics, called e. Because e is irrational, its decimal representation does not terminate or repeat.

The Number e

$e \approx 2.7182818284\ldots$

Logarithms base e are called **natural logarithms**, or **Napierian logarithms**, in honor of John Napier (1550–1617), who first "discovered" logarithms.

The abbreviation "ln" is generally used with natural logarithms. Thus,

$\ln 53$ means $\log_e 53$. It is important to remember this abbreviation.

On most scientific calculators, to find the natural logarithm of a number, we enter that number and press [ln]. On most graphing calculators, we press [ln], the number, and then [ENTER].

Example 3 Use a calculator to find ln 4568.

Solution We enter 4568 and then press [ln]. We find that

$$\ln 4568 \approx 8.4268. \quad \text{Rounded to four decimal places}$$

On many calculators, the [ln] key doubles as the [e^x] key after a [2nd] or [SHIFT] key has been pressed.

Example 4 Use a calculator to find $e^{-1.524}$.

Solution We enter -1.524 and then press [e^x]. On some calculators, [e^x] is pressed first, followed by -1.524 and [ENTER]. Since $e^{-1.524}$ is irrational, our answer is approximate:

$$e^{-1.524} \approx 0.2178387868.$$

Changing Logarithmic Bases

Most calculators can find both common logarithms and natural logarithms. To find a logarithm with some other base, a conversion formula is needed.

The Change-of-Base Formula

For any logarithmic bases a and b, and any positive number M,

$$\log_b M = \frac{\log_a M}{\log_a b}.$$

(To find the log, base b, of some number M, find the log of M using another base—usually 10 or e—and divide by the log of b to that same base.)

Proof. Let $x = \log_b M$. Then,

$$b^x = M \quad \text{Rewriting } x = \log_b M \text{ in exponential form}$$

$$\log_a b^x = \log_a M \quad \text{Taking the logarithm, base } a\text{, on both sides}$$

$$x \log_a b = \log_a M \quad \text{Using the power rule for logarithms}$$

$$x = \frac{\log_a M}{\log_a b}. \quad \text{Dividing both sides by } \log_a b$$

But at the outset we stated that $x = \log_b M$. Thus, by substitution, we have

$$\log_b M = \frac{\log_a M}{\log_a b},$$

which is the change-of-base formula.

Example 5

Find $\log_5 8$ using the change-of-base formula.

Solution We use the change-of-base formula with $a = 10$, $b = 5$, and $M = 8$:

$$\log_5 8 = \frac{\log_{10} 8}{\log_{10} 5}$$ Substituting into $\log_b M = \frac{\log_a M}{\log_a b}$

$$\approx \frac{0.903089987}{0.6989700043}$$ Using `log` twice

$$\approx 1.2920.$$ When using a calculator, it is best not to round before dividing.

To check, note that $\ln 8/\ln 5 \approx 1.2920$. We can also use a calculator to verify that $5^{1.2920} \approx 8$.

Example 6

Find $\log_4 31$.

Solution As shown in the check of Example 5, base e can also be used.

$$\log_4 31 = \frac{\log_e 31}{\log_e 4}$$ Substituting into $\log_b M = \frac{\log_a M}{\log_a b}$

$$= \frac{\ln 31}{\ln 4} \approx \frac{3.433987204}{1.386294361}$$ Using `ln` twice

$$\approx 2.4771.$$ *Check*: $4^{2.4771} \approx 31$

Graphs of Exponential and Logarithmic Functions, Base *e*

Example 7

Graph $f(x) = e^x$ and $g(x) = e^{-x}$ and state the domain and the range of f and g.

Solution We use a calculator with an `e^x` key to find approximate values of e^x and e^{-x}. Using these values, we can graph the functions.

x	e^x	e^{-x}
0	1	1
1	2.7	0.4
2	7.4	0.1
−1	0.4	2.7
−2	0.1	7.4

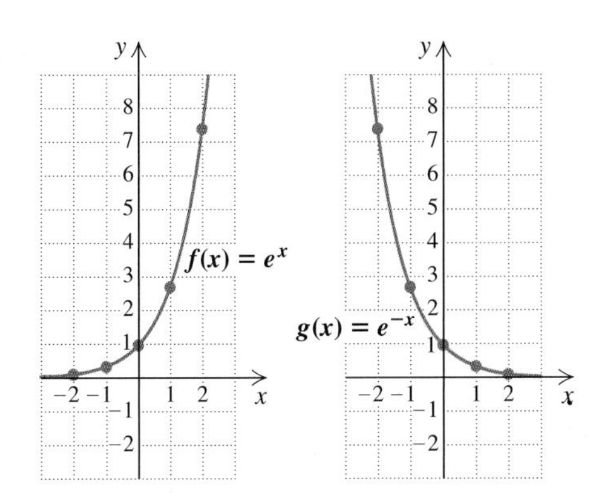

The domain of each function is $\mathbb{R}$ and the range of each function is $(0, \infty)$.

Example 8

Graph $f(x) = e^{-0.5x} + 1$ and state the domain and the range of f.

Solution We find some solutions with a calculator, plot them, and then draw the graph. For example, $f(2) = e^{-0.5(2)} + 1 = e^{-1} + 1 \approx 1.4$.

x	$e^{-0.5x} + 1$
0	2
1	1.6
2	1.4
3	1.2
−1	2.6
−2	3.7
−3	5.5

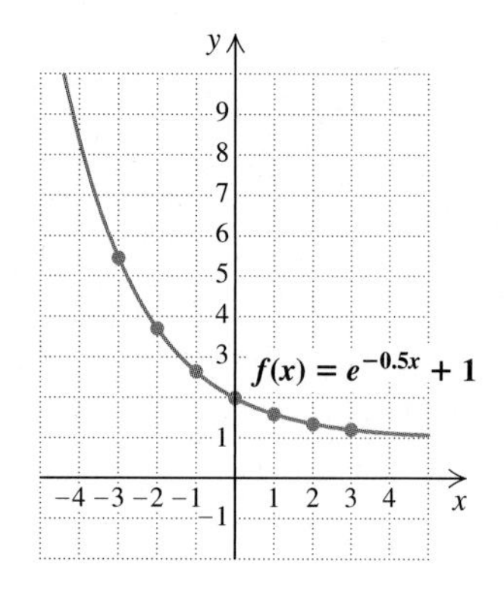

The domain of f is $\mathbb{R}$ and the range is $(1, \infty)$.

Example 9

Graph and state the domain and the range of each function.

a) $g(x) = \ln x$ **b)** $f(x) = \ln(x + 3)$

Solution

a) We find some solutions with a calculator and then draw the graph. As expected, the graph is a reflection across the line $y = x$ of the graph of $y = e^x$.

x	$\ln x$
1	0
4	1.4
7	1.9
0.5	−0.7

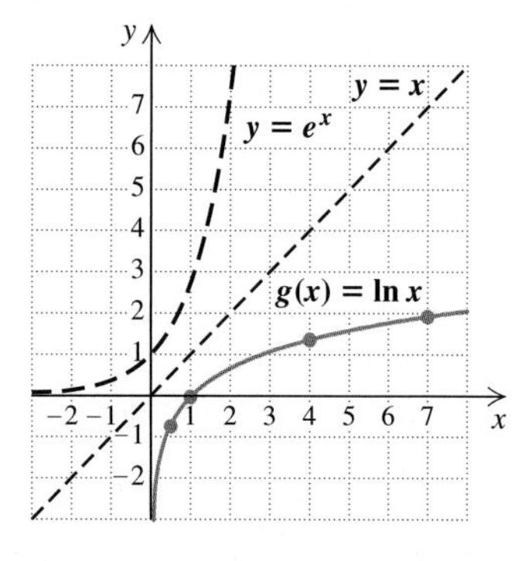

The domain of g is $(0, \infty)$ and the range is $\mathbb{R}$.

technology connection

Logarithmic functions with bases other than 10 or e can be easily drawn on a grapher, provided the change-of-base formula is used.

1. Graph $y = \log_5 x$.
2. Graph $y = \log_7 x$.
3. Graph $y = \log_5 (x + 2)$.
4. Graph $y = \log_7 x + 2$.

b) We find some solutions with a calculator, plot them, and draw the graph.

x	$\ln(x + 3)$
0	1.1
1	1.4
2	1.6
3	1.8
4	1.9
−1	0.7
−2	0
−2.5	−0.7

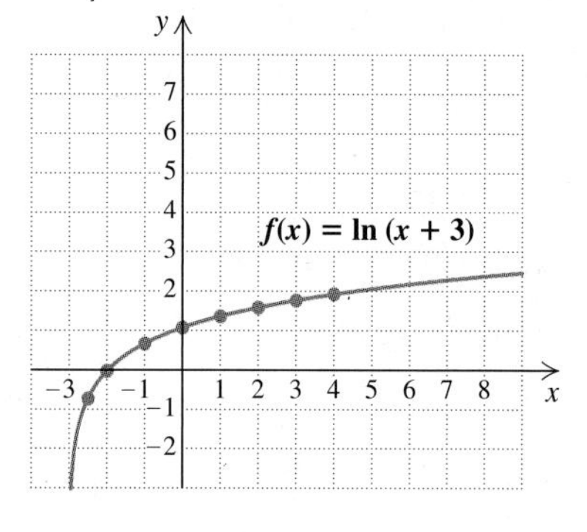

The graph of $y = \ln(x + 3)$ is the graph of $y = \ln x$ translated 3 units to the left. The domain is $(-3, \infty)$ and the range is $\mathbb{R}$.

Exercise Set 9.5

FOR EXTRA HELP

Digital Video Tutor CD 8
Videotape 18

InterAct Math

Math Tutor Center

MathXL

MyMathLab.com

Use a calculator to find each of the following to the nearest ten-thousandth.

1. $\log 6$ **2.** $\log 5$ **3.** $\log 72.8$

4. $\log 73.9$ Aha! **5.** $\log 1000$ **6.** $\log 100$

7. $\log 0.527$ **8.** $\log 0.493$ **9.** $\dfrac{\log 8200}{\log 150}$

10. $\dfrac{\log 5700}{\log 90}$ **11.** $10^{2.3}$ **12.** $10^{3.4}$

13. $10^{0.173}$ **14.** $10^{0.247}$ **15.** $10^{-2.9523}$

16. $10^{4.8982}$ **17.** $\ln 5$ **18.** $\ln 2$

19. $\ln 57$ **20.** $\ln 30$ **21.** $\ln 0.0062$

22. $\ln 0.00073$ **23.** $\dfrac{\ln 2300}{0.08}$ **24.** $\dfrac{\ln 1900}{0.07}$

25. $e^{2.71}$ **26.** $e^{3.06}$ **27.** $e^{-3.49}$

28. $e^{-2.64}$ **29.** $e^{4.7}$ **30.** $e^{1.23}$

Find each of the following logarithms using the change-of-base formula. Round answers to the nearest ten-thousandth.

31. $\log_6 92$ **32.** $\log_3 78$ **33.** $\log_2 100$

34. $\log_7 100$ **35.** $\log_7 65$ **36.** $\log_5 42$

37. $\log_{0.5} 5$ **38.** $\log_{0.1} 3$ **39.** $\log_2 0.2$

40. $\log_2 0.08$ **41.** $\log_\pi 58$ **42.** $\log_\pi 200$

Graph and state the domain and the range of each function.

43. $f(x) = e^x$ **44.** $f(x) = e^{0.5x}$

45. $f(x) = e^{-0.4x}$ **46.** $f(x) = e^{-x}$

47. $f(x) = e^x + 1$ **48.** $f(x) = e^x + 2$

49. $f(x) = e^x - 2$ **50.** $f(x) = e^x - 3$

51. $f(x) = 0.5e^x$ **52.** $f(x) = 2e^x$

53. $f(x) = 2e^{-0.5x}$ **54.** $f(x) = 0.5e^{2x}$

55. $f(x) = e^{x-2}$ **56.** $f(x) = e^{x-3}$

57. $f(x) = e^{x+3}$ **58.** $f(x) = e^{x+2}$

59. $g(x) = 2 \ln x$ **60.** $g(x) = 3 \ln x$

61. $g(x) = 0.5 \ln x$ **62.** $g(x) = 0.4 \ln x$

63. $g(x) = \ln x + 3$ **64.** $g(x) = \ln x + 2$

65. $g(x) = \ln x - 2$ **66.** $g(x) = \ln x - 3$

67. $g(x) = \ln(x + 1)$ **68.** $g(x) = \ln(x + 2)$

69. $g(x) = \ln(x - 3)$ **70.** $g(x) = \ln(x - 1)$

71. Using a calculator, Zeno *incorrectly* says that log 79 is between 4 and 5. How could you convince him, without using a calculator, that he is mistaken?

72. Examine Exercise 71. What mistake do you believe Zeno made?

SKILL MAINTENANCE

Solve.

73. $4x^2 - 25 = 0$ **74.** $5x^2 - 7x = 0$

75. $17x - 15 = 0$ **76.** $9 - 13x = 0$

77. $x^{1/2} - 6x^{1/4} + 8 = 0$ **78.** $2y - 7\sqrt{y} + 3 = 0$

SYNTHESIS

79. Explain how the graph of $f(x) = e^x$ could be used to graph the function given by $g(x) = 1 + \ln x$.

80. Explain how the graph of $f(x) = \ln x$ could be used to graph the function given by $g(x) = e^{x-1}$.

Given that $\log 2 \approx 0.301$ *and* $\log 3 \approx 0.477$, *find each of the following.*

81. $\log_6 81$ **82.** $\log_9 16$ **83.** $\log_{12} 36$

84. Find a formula for converting common logarithms to natural logarithms.

85. Find a formula for converting natural logarithms to common logarithms.

Solve for x.

86. $\log(275x^2) = 38$

87. $\log(492x) = 5.728$

88. $\dfrac{3.01}{\ln x} = \dfrac{28}{4.31}$

89. $\log 692 + \log x = \log 3450$

For each function given below, **(a)** *determine the domain and the range,* **(b)** *set an appropriate window, and* **(c)** *draw the graph. Graphs may vary, depending on the scale used.*

90. $f(x) = 7.4e^x \ln x$

91. $f(x) = 3.4 \ln x - 0.25e^x$

92. $f(x) = x \ln(x - 2.1)$

93. $f(x) = 2x^3 \ln x$

94. Use a grapher to check your answers to Exercises 45, 53, and 59.

95. Use a grapher to check your answers to Exercises 44, 52, and 60.

96. In an attempt to solve $\ln x = 1.5$, Emma gets the following graph.

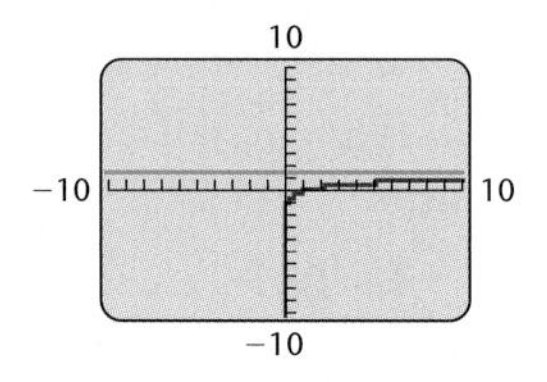

How can Emma tell at a glance that she has made a mistake?

9.6 Solving Exponential and Logarithmic Equations

Solving Exponential Equations • Solving Logarithmic Equations

Solving Exponential Equations

Equations with variables in exponents, such as $5^x = 12$ and $2^{7x} = 64$, are called **exponential equations**. In Section 9.3, we solved certain exponential equations by using the principle of exponential equality. We restate that principle below.

The Principle of Exponential Equality

For any real number b, where $b \neq -1, 0$, or 1,

$$b^x = b^y \quad \text{is equivalent to} \quad x = y.$$

(Powers of the same base are equal if and only if the exponents are equal.)

Example 1 Solve: $4^{3x-5} = 16$.

Solution Note that $16 = 4^2$. Thus we can write each side as a power of the same number:

$$4^{3x-5} = 4^2.$$

Since the base is the same, 4, the exponents must be the same. Thus,

$$3x - 5 = 2 \qquad \text{Equating exponents}$$
$$3x = 7$$
$$x = \tfrac{7}{3}.$$

Check:

4^{3x-5}	$= 16$	
$4^{3 \cdot 7/3 - 5}$	? 16	
4^{7-5}		
4^2	16	TRUE

The solution is $\frac{7}{3}$.

When it does not seem possible to write both sides of an equation as powers of the same base, we can use the following principle along with the properties developed in Section 9.4.

The Principle of Logarithmic Equality

For any logarithmic base a, and for $x, y > 0$,

$$x = y \quad \text{is equivalent to} \quad \log_a x = \log_a y.$$

(Two expressions are equal if and only if the logarithms of those expressions are equal.)

Because calculators can generally find only common or natural logarithms (without resorting to the change-of-base formula), we usually take the common or natural logarithm on both sides of the equation.

The principle of logarithmic equality is useful anytime a variable appears as an exponent.

Example 2 Solve: $5^x = 12$.

Solution We have

$$5^x = 12$$

$\log 5^x = \log 12$ Using the principle of logarithmic equality to take the common logarithm on both sides. Natural logarithms also would work.

$x \log 5 = \log 12$ Using the power rule for logarithms

$x = \dfrac{\log 12}{\log 5}$ ← **Caution!** This is *not* $\log 12 - \log 5$.

$\approx 1.544.$ Using a calculator and rounding to three decimal places

Since $5^{1.544} \approx 12$, we have a check. The solution is $\log 12/\log 5$, or approximately 1.544.

Example 3 Solve: $e^{0.06t} = 1500$.

Solution Since one side is a power of e, we take the *natural logarithm* on both sides:

$\ln e^{0.06t} = \ln 1500$ Taking the natural logarithm on both sides

$0.06t = \ln 1500$ Finding the logarithm of the base to a power: $\log_a a^k = k$

$t = \frac{\ln 1500}{0.06}$ Dividing both sides by 0.06

$\approx 121.887.$ Using a calculator and rounding to three decimal places

Solving Logarithmic Equations

Equations containing logarithmic expressions are called **logarithmic equations**. We saw in Section 9.3 that certain logarithmic equations can be solved by writing an equivalent exponential equation.

Example 4 Solve: $\log_4 (8x - 6) = 3$.

Solution We write an equivalent exponential equation:

$4^3 = 8x - 6$ Remember: $\log_a X = y$ is equivalent to $a^y = X$.

$64 = 8x - 6$

$70 = 8x$ Adding 6 to both sides

$x = \frac{70}{8}$, or $\frac{35}{4}$.

The check is left to the student. The solution is $\frac{35}{4}$.

Often the properties for logarithms are needed. The goal is to first write an equivalent equation in which the variable appears in just one logarithmic expression. We then isolate that term and solve as in Example 4.

Example 5 Solve.

a) $\log x + \log (x - 3) = 1$
b) $\log_2 (x + 7) - \log_2 (x - 7) = 3$
c) $\log_7 (x + 1) + \log_7 (x - 1) = \log_7 8$

Solution

a) As an aid in solving, we write in the base, 10.

$$\log_{10} x + \log_{10}(x-3) = 1$$

$$\log_{10}[x(x-3)] = 1$$ Using the product rule for logarithms to obtain a single logarithm

$$x(x-3) = 10^1$$ Writing an equivalent exponential equation

$$x^2 - 3x = 10$$

$$x^2 - 3x - 10 = 0$$

$$(x+2)(x-5) = 0$$ Factoring

$$x + 2 = 0 \quad or \quad x - 5 = 0$$ Using the principle of zero products

$$x = -2 \quad or \quad x = 5$$

Check:

For -2:

$$\log x + \log(x-3) = 1$$

$$\log(-2) + \log(-2-3) \;?\; 1 \quad \text{FALSE}$$

For 5:

$$\log x + \log(x-3) = 1$$

$$\log 5 + \log(5-3) \;?\; 1$$

$$\log 5 + \log 2$$

$$\log 10$$

$$1 \;|\; 1 \quad \text{TRUE}$$

The number -2 *does not check* because the logarithm of a negative number is undefined. The solution is 5.

b) We have

$$\log_2(x+7) - \log_2(x-7) = 3$$

$$\log_2 \frac{x+7}{x-7} = 3$$ Using the quotient rule for logarithms to obtain a single logarithm

$$\frac{x+7}{x-7} = 2^3$$ Writing an equivalent exponential equation

$$\frac{x+7}{x-7} = 8$$

$$x + 7 = 8(x-7)$$ Multiplying by the LCD, $x - 7$

$$x + 7 = 8x - 56$$ Using the distributive law

$$63 = 7x$$

$$9 = x.$$ Dividing by 7

Check:

$$\log_2(x+7) - \log_2(x-7) = 3$$

$$\log_2(9+7) - \log_2(9-7) \;?\; 3$$

$$\log_2 16 - \log_2 2$$

$$4 - 1$$

$$3 \;|\; 3 \quad \text{TRUE}$$

The solution is 9.

c) We have

$$\log_7 (x + 1) + \log_7 (x - 1) = \log_7 8$$

$$\log_7 [(x + 1)(x - 1)] = \log_7 8 \quad \text{Using the product rule for logarithms}$$

$$\log_7 (x^2 - 1) = \log_7 8 \quad \text{Multiplying}$$

$$x^2 - 1 = 8 \quad \text{Using the principle of logarithmic equality. Study this step carefully.}$$

$$x^2 - 9 = 0$$

$$(x - 3)(x + 3) = 0 \quad \text{Solving the quadratic equation}$$

$$x = 3 \quad \text{or} \quad x = -3.$$

We leave it to the student to show that 3 checks but -3 does not. The solution is 3.

technology connection

Exponential and logarithmic equations can be solved by graphing each side of the equation and using INTERSECT to determine the x-coordinate at each intersection.

For example, to solve $e^{0.5x} - 7 = 2x + 6$, we graph $y_1 = e^{0.5x} - 7$ and $y_2 = 2x + 6$ as shown at right. We then use the INTERSECT option of the CALC menu. The x-coordinates at the intersections are approximately -6.48 and 6.52.

$y_1 = e^{0.5x} - 7,\ y_2 = 2x + 6$

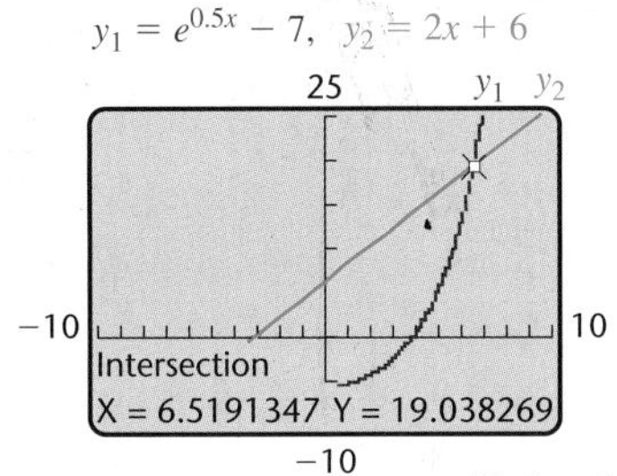

Use a grapher to find solutions, accurate to the nearest hundredth, for each of the following equations.

1. $e^{7x} = 14$

2. $8e^{0.5x} = 3$

3. $xe^{3x-1} = 5$

4. $4 \ln (x + 3.4) = 2.5$

5. $\ln 3x = 0.5x - 1$

6. $\ln x^2 = -x^2$

Exercise Set 9.6

FOR EXTRA HELP

Digital Video Tutor CD 8
Videotape 18

InterAct Math

Math Tutor Center

MathXL

MyMathLab.com

Solve. Where appropriate, include approximations to the nearest thousandth.

1. $2^x = 16$

2. $2^x = 8$

3. $3^x = 27$

4. $5^x = 125$

5. $2^{x+3} = 32$

6. $4^{x-2} = 64$

7. $5^{3x} = 625$

8. $3^{2x} = 27$

Aha! **9.** $7^{4x} = 1$

10. $8^{5x} = 1$

11. $4^{2x-1} = 64$

12. $5^{2x-3} = 25$

13. $3^{x^2} \cdot 3^{3x} = 81$

14. $3^{4x} \cdot 3^{x^2} = \frac{1}{27}$

15. $2^x = 15$

16. $2^x = 19$

17. $4^{x+1} = 13$

18. $8^{x-1} = 17$

19. $e^t = 100$

20. $e^t = 1000$

21. $e^{-0.07t} + 3 = 3.08$

22. $e^{0.03t} + 2 = 7$

23. $2^x = 3^{x-1}$

24. $5^x = 3^{x+1}$

25. $4^{x+1} = 5^x$

26. $2^{x+3} = 7^x$

27. $7.2^x - 65 = 0$

28. $4.9^x - 87 = 0$

29. $\log_5 x = 3$

30. $\log_3 x = 4$

31. $\log_4 x = \frac{1}{2}$

32. $\log_2 x = -3$

33. $\log x = 3$

34. $\log x = 1$

35. $2 \log x = -8$

36. $4 \log x = -16$

Aha! **37.** $\ln x = 1$

38. $\ln x = 2$

39. $5 \ln x = -15$

40. $3 \ln x = -3$

41. $\log_2 (8 - 6x) = 5$

42. $\log_5 (2x - 7) = 3$

43. $\log (x - 9) + \log x = 1$

44. $\log (x + 9) + \log x = 1$

45. $\log x - \log (x + 3) = 1$

46. $\log x - \log (x + 7) = -1$

47. $\log_4 (x + 3) - \log_4 (x - 5) = 2$

48. $\log_2 (x + 3) + \log_2 (x - 3) = 4$

49. $\log_7 (x + 1) + \log_7 (x + 2) = \log_7 6$

50. $\log_6 (x + 3) + \log_6 (x + 2) = \log_6 20$

51. $\log_3 (x + 4) + \log_3 (x - 4) = 2$

52. $\log_{14} (x + 3) + \log_{14} (x - 2) = 1$

53. $\log_{12} (x + 5) - \log_{12} (x - 4) = \log_{12} 3$

54. $\log_6 (x + 7) - \log_6 (x - 2) = \log_6 5$

55. $\log_2 (x - 2) + \log_2 x = 3$

56. $\log_4 (x + 6) - \log_4 x = 2$

57. Could Example 2 have been solved by taking the natural logarithm on both sides? Why or why not?

58. Christina finds that the solution of $\log_3 (x + 4) = 1$ is -1, but rejects -1 as an answer. What mistake is she making?

SKILL MAINTENANCE

59. Find an equation of variation if y varies directly as x, and $y = 7.2$ when $x = 0.8$.

60. Find an equation of variation if y varies inversely as x, and $y = 3.5$ when $x = 6.1$.

Solve.

61. $T = 2\pi\sqrt{L/32}$, for L

62. $E = mc^2$, for c
(Assume $E, m, c > 0$.)

63. Joni can key in a musical score in 2 hr. Miles takes 3 hr to key in the same score. How long would it take them, working together, to key in the score?

64. The side exit at the Flynn Theater can empty a capacity crowd in 25 min. The main exit can empty a capacity crowd in 15 min. How long will it take to empty a capacity crowd when both exits are in use?

SYNTHESIS

65. Can the principle of logarithmic equality be expanded to include all functions? That is, is the statement "$m = n$ is equivalent to $f(m) = f(n)$" true for any function f? Why or why not?

66. Explain how Exercises 33–36 could be solved using the graph of $f(x) = \log x$.

Solve.

67. $100^{3x} = 1000^{2x+1}$

68. $27^x = 81^{2x-3}$

69. $8^x = 16^{3x+9}$

70. $\log_x (\log_3 27) = 3$

71. $\log_6 (\log_2 x) = 0$

72. $x \log \frac{1}{8} = \log 8$

73. $\log_5 \sqrt{x^2 - 9} = 1$

74. $2^{x^2+4x} = \frac{1}{8}$

75. $\log (\log x) = 5$

76. $\log_5 |x| = 4$

77. $\log x^2 = (\log x)^2$

78. $\log \sqrt{2x} = \sqrt{\log 2x}$

79. $\log x^{\log x} = 25$

80. $3^{2x} - 8 \cdot 3^x + 15 = 0$

81. $(81^{x-2})(27^{x+1}) = 9^{2x-3}$

82. $3^{2x} - 3^{2x-1} = 18$

83. Given that $2^y = 16^{x-3}$ and $3^{y+2} = 27^x$, find the value of $x + y$.

84. If $x = (\log_{125} 5)^{\log_5 125}$, what is the value of $\log_3 x$?

85. Find the value of x for which the natural logarithm is the same as the common logarithm.

86. Use a grapher to check your answers to Exercises 3, 21, 25, 37, and 77.

9.7 Applications of Exponential and Logarithmic Functions

Applications of Logarithmic Functions • Applications of Exponential Functions

We now consider applications of exponential and logarithmic functions.

Applications of Logarithmic Functions

Example 1

Sound levels. To measure the volume, or "loudness," of a sound, the *decibel* scale is used. The loudness L, in decibels (dB), of a sound is given by

$$L = 10 \cdot \log \frac{I}{I_0},$$

where I is the intensity of the sound, in watts per square meter (W/m^2), and $I_0 = 10^{-12}\,W/m^2$. (I_0 is approximately the intensity of the softest sound that can be heard by the human ear.)

a) It is common for the intensity of sound at live performances of rock music to reach $10^{-1}\ \text{W/m}^2$ (even higher close to the stage). How loud, in decibels, is the sound level?

b) The Occupational Safety and Health Administration (OSHA) considers sound levels of 85 db and above unsafe. What is the intensity of such sounds?

Solution

a) To find the loudness, in decibels, we use the above formula:

$$\begin{aligned} L &= 10 \cdot \log \frac{I}{I_0} && \\ &= 10 \cdot \log \frac{10^{-1}}{10^{-12}} && \text{Substituting} \\ &= 10 \cdot \log 10^{11} && \text{Subtracting exponents} \\ &= 10 \cdot 11 && \log 10^a = a \\ &= 110. && \end{aligned}$$

The volume of the music is 110 decibels.

b) We substitute and solve for I:

$$\begin{aligned} L &= 10 \cdot \log \frac{I}{I_0} && \\ 85 &= 10 \cdot \log \frac{I}{10^{-12}} && \text{Substituting} \\ 8.5 &= \log \frac{I}{10^{-12}} && \text{Dividing both sides by 10} \\ 8.5 &= \log I - \log 10^{-12} && \text{Using the quotient rule for logarithms} \\ 8.5 &= \log I - (-12) && \log 10^a = a \\ -3.5 &= \log I && \text{Adding } -12 \text{ to both sides} \\ 10^{-3.5} &= I. && \text{Converting to an exponential equation} \end{aligned}$$

Earplugs would be recommended for sounds with intensities exceeding $10^{-3.5}\ \text{W/m}^2$.

Example 2

Chemistry: pH of liquids. In chemistry, the pH of a liquid is a measure of its acidity. We calculate pH as follows:

$$\text{pH} = -\log[\text{H}^+],$$

where $[\text{H}^+]$ is the hydrogen ion concentration in moles per liter.

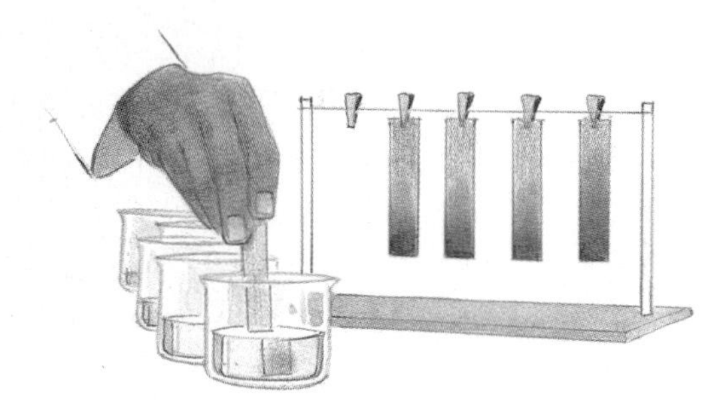

a) The hydrogen ion concentration of human blood is normally about 3.98×10^{-8} moles per liter. Find the pH.
b) The pH of seawater is about 8.3. Find the hydrogen ion concentration.

Solution

a) To find the pH of blood, we use the above formula:

$$\begin{aligned} \text{pH} &= -\log[\text{H}^+] \\ &= -\log[3.98 \times 10^{-8}] \\ &\approx -(-7.400117) && \text{Using a calculator} \\ &\approx 7.4. \end{aligned}$$

The pH of human blood is normally about 7.4.

b) We substitute and solve for $[\text{H}^+]$:

$$\begin{aligned} 8.3 &= -\log[\text{H}^+] && \text{Using pH} = -\log[\text{H}^+] \\ -8.3 &= \log[\text{H}^+] && \text{Dividing both sides by } -1 \\ 10^{-8.3} &= [\text{H}^+] && \text{Converting to an exponential equation} \\ 5.01 \times 10^{-9} &\approx [\text{H}^+]. && \text{Using a calculator; writing scientific notation} \end{aligned}$$

The hydrogen ion concentration of seawater is about 5.01×10^{-9} moles per liter.

Applications of Exponential Functions

Example 3

Interest compounded annually. Suppose that \$30,000 is invested at 8% interest, compounded annually. In t years, it will grow to the amount A given by the function

$$A(t) = 30{,}000(1.08)^t.$$

(See Example 5 in Section 9.2.)

a) How long will it take to accumulate \$150,000 in the account?
b) Find the amount of time it takes for the \$30,000 to double itself.

Solution

a) We set $A(t) = 150{,}000$ and solve for t:

$$150{,}000 = 30{,}000(1.08)^t$$

$$\frac{150{,}000}{30{,}000} = 1.08^t \qquad \text{Dividing both sides by 30,000}$$

$$5 = 1.08^t$$

$$\log 5 = \log 1.08^t \qquad \text{Taking the common logarithm on both sides}$$

$$\log 5 = t \log 1.08 \qquad \text{Using the power rule for logarithms}$$

$$\frac{\log 5}{\log 1.08} = t \qquad \text{Dividing both sides by log 1.08}$$

$$20.9 \approx t. \qquad \text{Using a calculator}$$

Remember that when doing a calculation like this on a calculator, it is best to wait until the end to round off. At an interest rate of 8% per year, it will take about 20.9 yr for \$30,000 to grow to \$150,000.

b) To find the *doubling time*, we replace $A(t)$ with 60,000 and solve for t:

$$60{,}000 = 30{,}000(1.08)^t$$

$$2 = (1.08)^t \qquad \text{Dividing both sides by 30,000}$$

$$\log 2 = \log (1.08)^t \qquad \text{Taking the common logarithm on both sides}$$

$$\log 2 = t \log 1.08 \qquad \text{Using the power rule for logarithms}$$

$$t = \frac{\log 2}{\log 1.08} \approx 9.0. \qquad \text{Dividing both sides by log 1.08 and using a calculator}$$

At an interest rate of 8% per year, the doubling time is about 9 yr.

Like investments, populations often grow exponentially.

Exponential Growth

An **exponential growth model** is a function of the form

$$P(t) = P_0 e^{kt}, \quad k > 0,$$

where P_0 is the population at time 0, $P(t)$ is the population at time t, and k is the **exponential growth rate** for the situation. The **doubling time** is the amount of time necessary for the population to double in size.

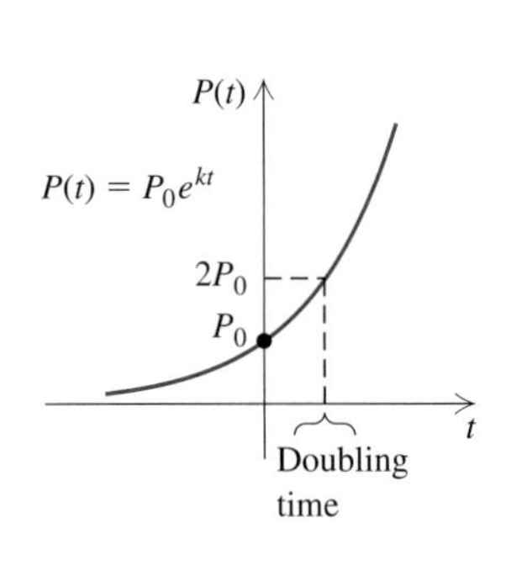

The exponential growth rate is the rate of growth of a population at any *instant* in time. Since the population is continually growing, the percent of total growth after one year will exceed the exponential growth rate.

Example 4

Growth of zebra mussel populations. Zebra mussels, inadvertently imported from Europe, began fouling North American waters in 1988. These mussels are so prolific that lake and river bottoms, as well as water intake pipes, can become blanketed with them, altering an entire ecosystem. In 2000, a portion of the Hudson River contained an average of 10 zebra mussels per square mile. The exponential growth rate was 340% per year.

a) Find the exponential growth model.
b) Predict the number of mussels per square mile in 2003.

Solution

a) In 2000, at $t = 0$, the population was $10/\text{mi}^2$. We substitute 10 for P_0 and 340%, or 3.4, for k. This gives the exponential growth function

$$P(t) = 10e^{3.4t}.$$

b) In 2003, we have $t = 3$ (since 3 yr have passed since 2000). To find the population in 2003, we compute $P(3)$:

$$\begin{aligned} P(3) &= 10e^{3.4(3)} && \text{Using } P(t) = 10e^{3.4t} \text{ from part (a)} \\ &= 10e^{10.2} \\ &\approx 269{,}000. && \text{Using a calculator} \end{aligned}$$

The population of zebra mussels in the specified portion of the Hudson River will reach approximately 269,000 per square mile in 2003.

Example 5 ***Spread of a computer virus.*** The number of computers infected by a virus t days after it first appears usually increases exponentially. In 2000, the "Love Bug" virus spread from 100 computers to about 1,000,000 computers in 2 hr (120 min).

a) Find the exponential growth rate and the exponential growth function.

b) Assuming exponential growth, estimate how long it took the Love Bug virus to infect 80,000 computers.

Solution

a) We use $N(t) = N_0e^{kt}$, where t is the number of minutes since the first 100 computers were infected. Substituting 100 for N_0 gives

$$N(t) = 100e^{kt}.$$

To find the exponential growth rate, k, note that after 120 min, 1,000,000 computers were infected:

$$\begin{aligned} \left.\begin{aligned} N(120) &= 100e^{k\cdot 120} \\ 1{,}000{,}000 &= 100e^{120k} \end{aligned}\right\} && \text{Substituting} \\ 10{,}000 = e^{120k} && \text{Dividing both sides by 100} \\ \ln 10{,}000 = \ln e^{120k} && \text{Taking the natural logarithm on both sides} \\ \ln 10{,}000 = 120k && \ln e^a = a \\ \frac{\ln 10{,}000}{120} = k && \text{Dividing both sides by 120} \\ 0.077 \approx k. && \text{Using a calculator and rounding} \end{aligned}$$

The exponential growth function is given by $N(t) = 100e^{0.077t}$.

b) To estimate how long it took for 80,000 computers to be infected, we replace $N(t)$ with 80,000 and solve for t:

$$80{,}000 = 100e^{0.077t}$$

$$800 = e^{0.077t} \qquad \text{Dividing both sides by 100}$$

$$\ln 800 = \ln e^{0.077t} \qquad \text{Taking the natural logarithm on both sides}$$

$$\ln 800 = 0.077t \qquad \ln e^a = a$$

$$\frac{\ln 800}{0.077} = t \qquad \text{Dividing both sides by 0.077}$$

$$86.8 \approx t. \qquad \text{Using a calculator}$$

Rounding up to 87, we see that, according to this model, it took about 87 min, or 1 hr 27 min, for 80,000 computers to be infected.

Example 6

Interest compounded continuously. Suppose that an amount of money P_0 is invested in a savings account at interest rate k, compounded continuously. That is, suppose that interest is computed every "instant" and added to the amount in the account. The balance $P(t)$, after t years, is given by the exponential growth model

$$P(t) = P_0e^{kt}.$$

a) Suppose that \$30,000 is invested and grows to \$44,754.75 in 5 yr. Find the exponential growth function.

b) What is the doubling time?

Solution

a) We have $P(0) = 30{,}000$. Thus the exponential growth function is

$$P(t) = 30{,}000e^{kt}, \quad \text{where } k \text{ must still be determined.}$$

Knowing that for $t = 5$ we have $P(5) = 44{,}754.75$, it is possible to solve for k:

$$44{,}754.75 = 30{,}000e^{k(5)} = 30{,}000e^{5k}$$

$$\frac{44{,}754.75}{30{,}000} = e^{5k} \qquad \text{Dividing both sides by 30,000}$$

$$1.491825 = e^{5k}$$

$$\ln 1.491825 = \ln e^{5k} \qquad \text{Taking the natural logarithm on both sides}$$

$$\ln 1.491825 = 5k \qquad \ln e^a = a$$

$$\frac{\ln 1.491825}{5} = k \qquad \text{Dividing both sides by 5}$$

$$0.08 \approx k. \qquad \text{Using a calculator and rounding}$$

The interest rate is about 0.08, or 8%, compounded continuously. Note that since interest is being compounded continuously, the interest earned each year is more than 8%. The exponential growth function is

$$P(t) = 30{,}000e^{0.08t}.$$

b) To find the doubling time T, we replace $P(T)$ with 60,000 and solve for T:

$$60{,}000 = 30{,}000e^{0.08T}$$

$$2 = e^{0.08T} \quad \text{Dividing both sides by 30,000}$$

$$\ln 2 = \ln e^{0.08T} \quad \text{Taking the natural logarithm on both sides}$$

$$\ln 2 = 0.08T \quad \ln e^a = a$$

$$\frac{\ln 2}{0.08} = T \quad \text{Dividing both sides by 0.08}$$

$$8.7 \approx T. \quad \text{Using a calculator and rounding}$$

Thus the original investment of \$30,000 will double in about 8.7 yr.

As the results of Examples 3(b) and 6(b) imply, for any specified interest rate, continuous compounding gives the highest yield and the shortest doubling time.

In some real-life situations, a quantity or population is *decreasing* or *decaying* exponentially.

Exponential Decay

An **exponential decay model** is a function of the form

$$P(t) = P_0e^{-kt}, \quad k > 0,$$

where P_0 is the quantity present at time 0, $P(t)$ is the amount present at time t, and k is the **decay rate.** The **half-life** is the amount of time necessary for half of the quantity to decay.

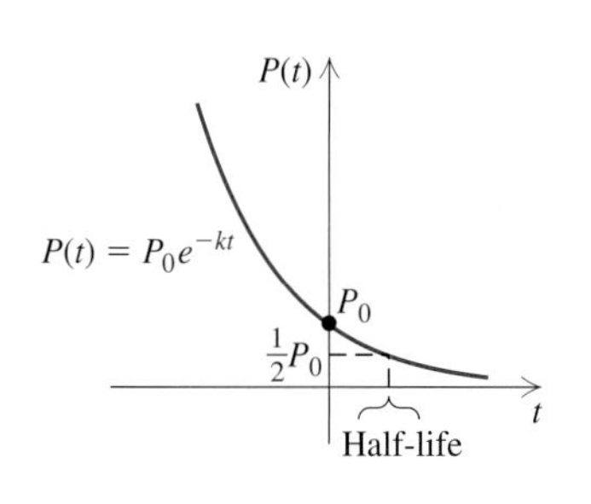

Example 7 ***Carbon dating.*** The radioactive element carbon-14 has a half-life of 5750 yr. The percentage of carbon-14 present in the remains of organic matter can be used to determine the age of that organic matter. Recently, while digging in Chaco Canyon, New Mexico, archaeologists found corn pollen that had lost 38.1% of its carbon-14. The age of this corn pollen was evidence that Indians had been cultivating crops in the Southwest centuries earlier than scientists had thought. What was the age of the pollen? (*Source*: *American Anthropologist*)

Solution We first find k. To do so, we use the concept of half-life. When $t = 5750$ (the half-life), $P(t)$ will be half of P_0. Then

$$0.5P_0 = P_0e^{-k(5750)} \quad \text{Substituting in } P(t) = P_0e^{-kt}$$
$$0.5 = e^{-5750k} \quad \text{Dividing both sides by } P_0$$
$$\ln 0.5 = \ln e^{-5750k} \quad \text{Taking the natural logarithm on both sides}$$
$$\ln 0.5 = -5750k \quad \ln e^a = a$$
$$\frac{\ln 0.5}{-5750} = k \quad \text{Dividing}$$
$$0.00012 \approx k. \quad \text{Using a calculator and rounding}$$

Now we have a function for the decay of carbon-14:

$$P(t) = P_0e^{-0.00012t}. \quad \text{This completes the first part of our solution.}$$

(*Note*: This equation can be used for any subsequent carbon-dating problem.) If the corn pollen has lost 38.1% of its carbon-14 from an initial amount P_0, then 100% − 38.1%, or 61.9%, of P_0 is still present. To find the age t of the pollen, we solve this equation for t:

$$0.619P_0 = P_0e^{-0.00012t} \quad \text{We want to find } t \text{ for which } P(t) = 0.619P_0.$$
$$0.619 = e^{-0.00012t} \quad \text{Dividing both sides by } P_0$$
$$\ln 0.619 = \ln e^{-0.00012t} \quad \text{Taking the natural logarithm on both sides}$$
$$\ln 0.619 = -0.00012t \quad \ln e^a = a$$
$$\frac{\ln 0.619}{-0.00012} = t \quad \text{Dividing}$$
$$4000 \approx t. \quad \text{Using a calculator}$$

The pollen is about 4000 yr old.

Exercise Set 9.7

FOR EXTRA HELP

Digital Video Tutor CD 8
Videotape 18

InterAct Math

Math Tutor Center

MathXL

MyMathLab.com

Solve.

1. *Cellular phones.* The number of cellular phones in use in the United States t years after May 2001 can be predicted by

$$N(t) = 112(1.37)^t,$$

where $N(t)$ is the number of cellular phones in use, in millions.

a) Determine the year in which 200 million cellular phones would be in use.

b) What is the doubling time for the number of cellular phones in use?

2. *Compact discs.* The number of compact discs N purchased each year, in millions, can be approximated by

$$N(t) = 384(1.13)^t,$$

where t is the number of years after 1991.*

a) After what amount of time will two billion compact discs be sold in a year?

b) What is the doubling time for the number of compact discs sold in one year?

*Based on data from the Recording Industry Association, Washington, DC.

3. *Student loan repayment.* A college loan of \$29,000 is made at 8% interest, compounded annually. After t years, the amount due, A, is given by the function

$$A(t) = 29{,}000(1.08)^t.$$

a) After what amount of time will the amount due reach \$40,000?

b) Find the doubling time.

4. *Spread of a rumor.* The number of people who have heard a rumor increases exponentially. If all who hear a rumor repeat it to two people a day, and if 20 people start the rumor, the number of people N who have heard the rumor after t days is given by

$$N(t) = 20(3)^t.$$

a) After what amount of time will 1000 people have heard the rumor?

b) What is the doubling time for the number of people who have heard the rumor?

5. *Skateboarding.* The number of skateboarders of age x (with $x \geq 16$), in thousands, can be approximated by

$$N(x) = 600(0.873)^{x-16}$$

(*Sources*: Based on figures from the U.S. Bureau of the Census and *Statistical Abstract of the United States,* 2000).

a) Estimate the number of 41-yr-old skateboarders.

b) At what age are there only 2000 skateboarders?

6. *Recycling aluminum cans.* Approximately two thirds of all aluminum cans distributed will be recycled each year. A beverage company distributes 250,000 cans. The number still in use after t years is given by the function

$$N(t) = 250{,}000\left(\tfrac{2}{3}\right)^t.$$

a) After how many years will 60,000 cans still be in use?

b) After what amount of time will only 1000 cans still be in use?

Use the pH formula given in Example 2 for Exercises 7–10.

7. *Chemistry.* The hydrogen ion concentration of fresh-brewed coffee is about 1.3×10^{-5} moles per liter. Find the pH.

8. *Chemistry.* The hydrogen ion concentration of milk is about 1.6×10^{-7} moles per liter. Find the pH.

9. *Medicine.* When the pH of a patient's blood drops below 7.4, a condition called *acidosis* sets in. Acidosis can be deadly when the patient's pH reaches 7.0. What would the hydrogen ion concentration of the patient's blood be at that point?

10. *Medicine.* When the pH of a patient's blood rises above 7.4, a condition called *alkalosis* sets in. Alkalosis can be deadly when the patient's pH reaches 7.8. What would the hydrogen ion concentration of the patient's blood be at that point?

Use the decibel formula given in Example 1 for Exercises 11–14.

11. *Audiology.* The intensity of sound in normal conversation is about 3.2×10^{-6} W/m^2. How loud in decibels is this sound level?

12. *Audiology.* The intensity of a riveter at work is about 3.2×10^{-3} W/m^2. How loud in decibels is this sound level?

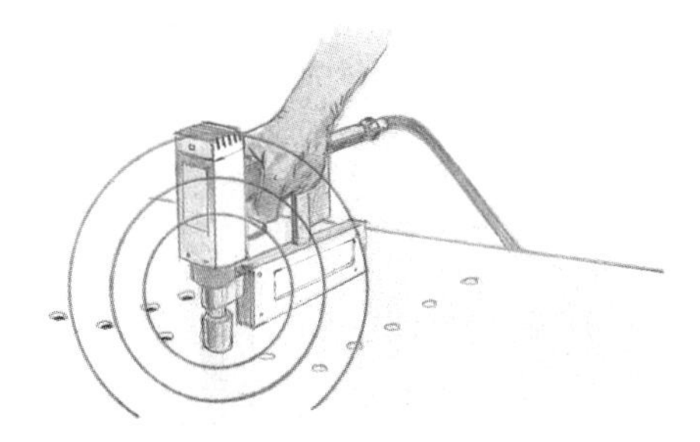

13. *Music.* The band U2 recently performed and sound measurements of 105 dB were recorded. What is the intensity of such sounds?

14. *Music.* The band Strange Folk recently performed in Burlington, VT, and reached sound levels of 111 dB (*Source*: Melissa Garrido, *Burlington Free Press*). What is the intensity of such sounds?

Use the compound-interest formula in Example 6 for Exercises 15 and 16.

15. *Interest compounded continuously.* Suppose that P_0 is invested in a savings account where interest is compounded continuously at 6% per year.

a) Express $P(t)$ in terms of P_0 and 0.06.
b) Suppose that $5000 is invested. What is the balance after 1 yr? after 2 yr?
c) When will an investment of $5000 double itself?

16. *Interest compounded continuously.* Suppose that P_0 is invested in a savings account where interest is compounded continuously at 5% per year.

a) Express $P(t)$ in terms of P_0 and 0.05.
b) Suppose that $1000 is invested. What is the balance after 1 yr? after 2 yr?
c) When will an investment of $1000 double itself?

17. *Population growth.* In 2000, the population of the United States was 283.75 million and the exponential growth rate was 1.3% per year (*Source*: U.S. Bureau of the Census).

a) Find the exponential growth function.
b) Predict the U.S. population in 2005.
c) When will the U.S. population reach 325 million?

18. *World population growth.* In 2001, the world population was 6.1 billion and the exponential growth rate was 1.4% per year (*Source*: U.S. Bureau of the Census).

a) Find the exponential growth function.
b) Predict the world population in 2005.
c) When will the world population be 8.0 billion?

19. *Growth of bacteria.* The bacteria *Escherichi coli* are commonly found in the human bladder. Suppose that 3000 of the bacteria are present at time $t = 0$. Then t minutes later, the number of bacteria present is

$$N(t) = 3000(2)^{t/20}.$$

a) After what amount of time will there be 60,000 bacteria?
b) If 100,000,000 bacteria accumulate, a bladder infection can occur. What amount of time would have to pass in order for a possible infection to occur?
c) What is the doubling time?

20. *Population growth.* The exponential growth rate of the population of Central America is 3.5% per year (one of the highest in the world). What is the doubling time?

21. *Advertising.* A model for advertising response is given by

$$N(a) = 2000 + 500 \log a, \quad a \geq 1,$$

where $N(a)$ is the number of units sold and a is the amount spent on advertising, in thousands of dollars.

a) How many units were sold after spending $1000 ($a = 1$) on advertising?
b) How many units were sold after spending $8000?
c) Graph the function.
d) How much would have to be spent in order to sell 5000 units?

22. *Forgetting.* Students in an English class took a final exam. They took equivalent forms of the exam at monthly intervals thereafter. The average score $S(t)$, in percent, after t months was found to be given by

$$S(t) = 68 - 20 \log (t + 1), \quad t \geq 0.$$

a) What was the average score when they initially took the test, $t = 0$?
b) What was the average score after 4 months? after 24 months?
c) Graph the function.
d) After what time t was the average score 50?

23. *Public health.* In 1995, an outbreak of Herpes infected 17 people in a large community. By 1996, the number of those infected had grown to 29.

a) Find an exponential growth function that fits the data.
b) Predict the number of people who will be infected in 2001.

24. *Heart transplants.* In 1967, Dr. Christiaan Barnard of South Africa stunned the world by performing the first heart transplant. There were 1418 heart transplants in 1987, and 2185 such transplants in 1999.

a) Find an exponential growth function that fits the data from 1987 and 1999.
b) Use the function to predict the number of heart transplants in 2012.

25. *Oil demand.* The exponential growth rate of the demand for oil in the United States is 10% per year. In what year will the demand be double that of May 1995?

26. *Coal demand.* The exponential growth rate of the demand for coal in the world is 4% per year. When will the demand be double that of 1995?

27. *Decline of discarded yard waste.* The amount of discarded yard waste has declined considerably in recent years because of increased recycling and composting. In 1996, 17.5 million tons were discarded, but by 1998 the figure dropped to 14.5 million tons (*Source*: *Statistical Abstract of the United States*, 2000). Assume the amount of discarded yard waste is decreasing according to the exponential decay model.

a) Find the value k, and write an exponential function that describes the amount of yard waste discarded t years after 1996.

b) Estimate the amount of discarded yard waste in 2006.

c) In what year (theoretically) will only 1 ton of yard waste be discarded?

28. *Decline in cases of mumps.* The number of cases of mumps has dropped exponentially from 5300 in 1990 to 800 in 1996 (*Source*: *Statistical Abstract of the United States*, 2000).

a) Find the value k, and write an exponential function that can be used to estimate the number of cases t years after 1990.

b) Estimate the number of cases of mumps in 2004.

c) In what year (theoretically) will there be only 1 case of mumps?

29. *Archaeology.* When archaeologists found the Dead Sea scrolls, they determined that the linen wrapping had lost 22.3% of its carbon-14. How old is the linen wrapping? (See Example 7.)

30. *Archaeology.* In 1996, researchers found an ivory tusk that had lost 18% of its carbon-14. How old was the tusk? (See Example 7.)

31. *Chemistry.* The exponential decay rate of iodine-131 is 9.6% per day. What is its half-life?

32. *Chemistry.* The decay rate of krypton-85 is 6.3% per year. What is its half-life?

33. *Home construction.* The chemical urea formaldehyde was found in some insulation used in houses built during the mid to late 60s. Unknown at the time was the fact that urea formaldehyde emitted toxic fumes as it decayed. The half-life of urea formaldehyde is 1 yr. What is its decay rate?

34. *Plumbing.* Lead pipes and solder are often found in older buildings. Unfortunately, as lead decays, toxic chemicals can get in the water resting in the pipes. The half-life of lead is 22 yr. What is its decay rate?

35. *Value of a sports card.* Legend has it that because he objected to smoking, and because his first baseball card was issued in cigarette packs, the great shortstop Honus Wagner halted production of his card before many were produced. One of these cards was sold in 1996 for $640,500 and again in 2000 for $1.1 million. For the following questions, assume that the card's value increases exponentially, as it has for many years.

a) Find the exponential growth rate k, and determine an exponential function V that can be used to estimate the dollar value, $V(t)$, of the card t years after 1996.

b) Predict the value of the card in 2006.

c) What is the doubling time for the value of the card?

d) In what year will the value of the card first exceed $2,000,000?

36. *Portrait of Dr. Gachet.* As of May 2001, the most ever paid for a painting is $82.5 million, paid in 1990 for Vincent Van Gogh's *Portrait of Dr. Gachet.* The same painting sold for $58 million in 1987. Assume that the growth in the value V of the painting is exponential.

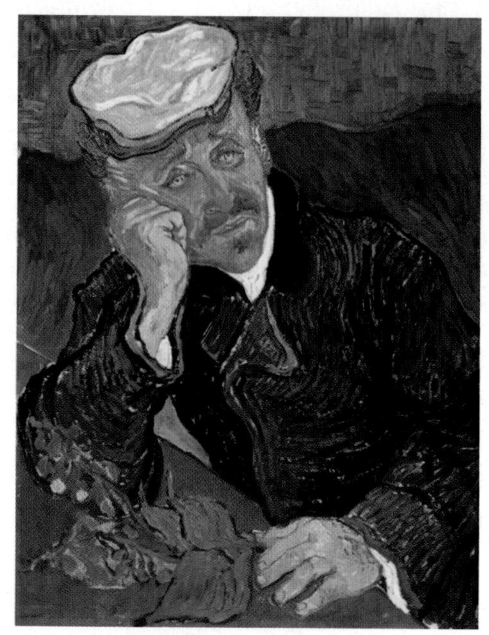

Van Gogh's *Portrait of Dr. Gachet*, oil on canvas.

a) Find the exponential growth rate k, and determine the exponential growth function V, for which $V(t)$ is the painting's value, in millions of dollars, t years after 1987.

b) Estimate the value of the painting in 2007.

c) What is the doubling time for the value of the painting?

d) How long after 1987 will the value of the painting be $1 billion?

37. Write a problem for a classmate to solve in which information is provided and the classmate is asked to find an exponential growth function. Make the problem as realistic as possible.

38. Examine the restriction on t in Exercise 22.

a) What upper limit might be placed on t?

b) In practice, would this upper limit ever be enforced? Why or why not?

SKILL MAINTENANCE

Graph.

39. $y = x^2 - 8x$

40. $y = x^2 - 5x - 6$

41. $f(x) = 3x^2 - 5x - 1$

42. $g(x) = 2x^2 - 6x + 3$

Solve by completing the square.

43. $x^2 - 8x = 7$

44. $x^2 + 10x = 6$

SYNTHESIS

45. Will the model used to predict the number of cellular phones in Exercise 1 still be realistic in 2020? Why or why not?

46. *Atmospheric pressure.* Atmospheric pressure P at altitude a is given by

$$P = P_0 e^{-0.00005a},$$

where P_0 is the pressure at sea level $\approx 14.7\ \text{lb/in}^2$ (pounds per square inch). Explain how a barometer, or some other device for measuring atmospheric pressure, can be used to find the height of a skyscraper.

47. *Sports salaries.* In 2001, Derek Jeter of the New York Yankees signed a $189 million 10-yr contract that will pay him $21 million in 2010. How much would Yankee owner George Steinbrenner need to invest in 2001 at 5% interest compounded continuously, in order to have the $21 million for Jeter in 2010? (This is much like finding what $21 million in 2010 will be worth in 2001 dollars.)

48. *Supply and demand.* The supply and demand for the sale of stereos by Sound Ideas are given by

$$S(x) = e^x \quad \text{and} \quad D(x) = 162{,}755e^{-x},$$

where $S(x)$ is the price at which the company is willing to supply x stereos and $D(x)$ is the demand price for a quantity of x stereos. Find the equilibrium point. (For reference, see Section 3.8.)

49. Use Exercises 1 and 17 to form a model for the percentage of U.S. residents owning a cellular phone t years after 2001.

50. Use the model developed in Exercise 49 to predict the percentage of U.S. residents who will own cellular phones in 2010. Does your prediction seem plausible? Why or why not?

51. *Nuclear energy.* Plutonium-239 (Pu-239) is used in nuclear energy plants. The half-life of Pu-239 is 24,360 yr (*Source*: *Microsoft Encarta 97 Encyclopedia*). How long will it take for a fuel rod of Pu-239 to lose 90% of its radioactivity?

COLLABORATIVE

CORNER

Investments in Collectibles

Focus: Exponential-growth models

Time: 30 minutes

Group size: 6

Prepaid calling cards have become a big business, not only as a convenient way to place a telephone call, but also as collectibles. Some telephone cards are worth far more than their original cost, particularly if they are unused.

Collectors often estimate the future value of their collections by examining the value's growth in the past. The value of some collectibles grows exponentially.

ACTIVITY

1. Suppose that in 2001, each group member needed to invest $1200 in the telephone cards listed in the following table. Looking only at the approximate value in 2001, each student should select $1200 worth of cards to buy. More than one card of each type can be selected. The card or cards chosen will become that student's portfolio.
2. Each group member should be assigned one of the cards. That person should then form an exponential growth model for the value of that card using the year of issue and the original cost of the card.
3. Using the models developed above, each group member should predict the value of his or her portfolio in 2005. Compare the values. Why, when buying a collector's item, is it important to consider its previous worth?
4. Look at the predicted value of each card in the year 2008. Does an exponential growth model seem appropriate? If possible, find the current value of some of the cards and compare those values with the predicted values.

Card	Approximate Value in 2001	Year of Issue	Original Cost
World Rowing Championships, Spec Set	$200	1994	$16
AT&T, America's Cup	$1200	1992	$50
McDonald's Hamburgers, Proof	$60	1996	$2
Sprint, Coca-Cola collection	$150	1995	$17
New York Telephone, 1992 Democratic National Convention	$350	1992	$1
McDonald's Extra-Value Meal Promotion	$75	1995	$3

Source: Moneycard.com, June 2001

Summary and Review 9

Key Terms

Composite functions, p. 548
Inverse relation, p. 551
One-to-one function, p. 552
Horizontal-line test, p. 552
Exponential function, p. 561
Asymptote, p. 562
Logarithmic function, p. 570
Common logarithm, p. 583
Natural logarithm, p. 585
Exponential equation, p. 590
Logarithmic equation, p. 592
Exponential growth model, p. 600
Exponential growth rate, p. 600
Doubling time, p. 600
Exponential decay model, p. 603
Exponential decay rate, p. 603
Half-life, p. 603

Important Properties and Formulas

To Find a Formula for the Inverse of a Function

First check a graph to make sure that the function f is one-to-one. Then:

1. Replace $f(x)$ with y.
2. Interchange x and y.
3. Solve for y.
4. Replace y with $f^{-1}(x)$.

Composition of f and g:	$(f \circ g)(x) = f(g(x))$
Composition and inverses:	$(f^{-1} \circ f)(x) = (f \circ f^{-1})(x) = x$
Exponential function:	$f(x) = a^x, \quad a > 0, \quad a \neq 1$
Interest compounded annually:	$A = P(1 + i)^t$

For $x > 0$ and a a positive constant other than 1, the number $\log_a x$ is the power to which a must be raised in order to get x. Thus, $a^{\log_a x} = x$, or equivalently, if $y = \log_a x$, then $a^y = x$. *A logarithm is an exponent.*

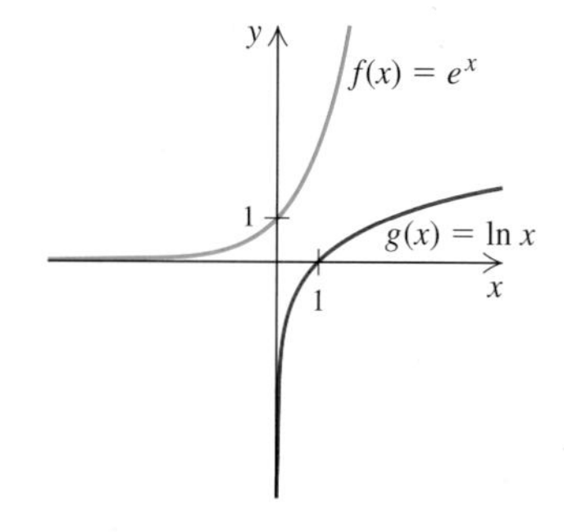

The Principle of Exponential Equality

For any real number b, $b \neq -1, 0$, or 1:

$b^x = b^y$ is equivalent to $x = y$.

The Principle of Logarithmic Equality

For any logarithmic base a, and for $x, y > 0$;

$x = y$ is equivalent to $\log_a x = \log_a y$.

Properties of Logarithms

$\log_a MN = \log_a M + \log_a N,$

$\log_a \dfrac{M}{N} = \log_a M - \log_a N,$

$\log_a M^p = p \cdot \log_a M,$

$\log_a 1 = 0,$

$\log_a a = 1,$

$\log_a a^k = k,$

$\log M = \log_{10} M,$

$\ln M = \log_e M,$

$\log_b M = \dfrac{\log_a M}{\log_a b}$

$e \approx 2.7182818284\ldots$

Loudness of sound:	$L = 10 \cdot \log \dfrac{I}{I_0}$
pH:	$\text{pH} = -\log[H^+]$
Exponential growth:	$P(t) = P_0e^{kt}, k > 0$
Exponential decay:	$P(t) = P_0e^{-kt}, k > 0$
Interest compounded continuously:	$P(t) = P_0e^{kt}$, where P_0 is the principal invested for t years at interest rate k
Carbon dating:	$P(t) = P_0e^{-0.00012t}$

Review Exercises

1. Find $(f \circ g)(x)$ and $(g \circ f)(x)$ if $f(x) = x^2 + 1$ and $g(x) = 2x - 3$.

2. If $h(x) = \sqrt{3 - x}$, find $f(x)$ and $g(x)$ such that $h(x) = (f \circ g)(x)$. Answers may vary.

3. Determine whether $f(x) = 4 - x^2$ is one-to-one.

Find a formula for the inverse of each function.

4. $f(x) = x - 3$

5. $g(x) = \dfrac{3x + 1}{2}$

6. $f(x) = 27x^3$

Graph.

7. $f(x) = 3^x + 1$

8. $x = \left(\frac{1}{4}\right)^y$

9. $y = \log_5 x$

Simplify.

10. $\log_3 9$

11. $\log_{10} \frac{1}{10}$

12. $\log_{25} 5$

13. $\log_3 3^{12}$

Convert to logarithmic equations.

14. $10^{-2} = \frac{1}{100}$

15. $25^{1/2} = 5$

Convert to exponential equations.

16. $\log_4 16 = x$

17. $\log_8 1 = 0$

Express in terms of logarithms of x, y, and z.

18. $\log_a x^4 y^2 z^3$

19. $\log_a \dfrac{x^3}{yz^2}$

20. $\log \sqrt[4]{\dfrac{z^2}{x^3 y}}$

Express as a single logarithm and, if possible, simplify.

21. $\log_a 8 + \log_a 15$

22. $\log_a 72 - \log_a 12$

23. $\frac{1}{2} \log a - \log b - 2 \log c$

24. $\frac{1}{3}[\log_a x - 2 \log_a y]$

Simplify.

25. $\log_m m$

26. $\log_m 1$

27. $\log_m m^{17}$

Given $\log_a 2 = 1.8301$ *and* $\log_a 7 = 5.0999$, *find each of the following.*

28. $\log_a 14$

29. $\log_a \frac{2}{7}$

30. $\log_a 28$

31. $\log_a 3.5$

32. $\log_a \sqrt{7}$

33. $\log_a \frac{1}{4}$

Use a calculator to find each of the following to the nearest ten-thousandth.

34. $\log 82$

35. $10^{1.789}$

36. $\ln 0.05$

37. $e^{-0.98}$

Find each of the following logarithms using the change-of-base formula. Round answers to the nearest ten-thousandth.

38. $\log_5 2$

39. $\log_{12} 70$

Graph and state the domain and the range of each function.

40. $f(x) = e^x - 1$

41. $g(x) = 0.6 \ln x$

Solve. Where appropriate, include approximations to the nearest ten-thousandth.

42. $2^x = 32$

43. $3^x = \frac{1}{9}$

44. $\log_3 x = -2$

45. $\log_x 32 = 5$

46. $\log x = -4$

47. $3 \ln x = -6$

48. $4^{2x-5} = 16$

49. $2^{x^2} \cdot 2^{4x} = 32$

50. $4^x = 8.3$

51. $e^{-0.1t} = 0.03$

52. $2 \ln x = -6$

53. $\log_3 (2x - 5) = 1$

54. $\log_4 x + \log_4 (x - 6) = 2$

55. $\log x + \log (x - 15) = 2$

56. $\log_3 (x - 4) = 3 - \log_3 (x + 4)$

57. In a business class, students were tested at the end of the course with a final exam. They were tested again after 6 months. The forgetting formula was determined to be

$$S(t) = 62 - 18 \log (t + 1),$$

where t is the time, in months, after taking the first test.

a) Determine the average score when they first took the test (when $t = 0$).

b) What was the average score after 6 months?

c) After what time was the average score 34?

58. A color photocopier is purchased for \$5200. Its value each year is about 80% of its value in the preceding year. Its value in dollars after t years is given by the exponential function

$$V(t) = 5200(0.8)^t.$$

a) After what amount of time will the salvage value be \$1200?

b) After what amount of time will the salvage value be half the original value?

59. The number of customer service complaints against U.S. airlines has grown exponentially from 667 in 1995 to 3664 in 1999 (*Sources*: U.S. Bureau of the Census and *Statistical Abstract of the United States*, 2000).

a) Find the value k, and write an exponential function that describes the number of customer service complaints t years after 1995.

b) Predict the number of complaints in 2005.

c) In what year will there be 10,000 customer service complaints?

60. The value of Jose's stock market portfolio doubled in 3 yr. What was the exponential growth rate?

61. How long will it take \$7600 to double itself if it is invested at 8.4%, compounded continuously?

62. How old is a skeleton that has lost 34% of its carbon-14? (Use $P(t) = P_0 e^{-0.00012t}$.)

63. What is the pH of a substance if its hydrogen ion concentration is 2.3×10^{-7} moles per liter? (Use $\text{pH} = -\log [\text{H}^+]$.)

64. The intensity of the sound of water at the foot of the Niagara Falls is about 10^{-3} W/m^2.* How loud in decibels is this sound level?

$$\left(\text{Use } L = 10 \cdot \log \frac{I}{10^{-12}}.\right)$$

SYNTHESIS

65. Explain why negative numbers do not have logarithms.

66. Explain why taking the natural or common logarithm on each side of an equation produces an equivalent equation.

Solve.

67. $\ln (\ln x) = 3$

68. $2^{x^2+4x} = \frac{1}{8}$

69. $5^{x+y} = 25,$
$2^{2x-y} = 64$

**Sound and Hearing*, Life Science Library. (New York: Time Incorporated, 1965), p. 173.

Chapter Test 9

1. Find $(f \circ g)(x)$ and $(g \circ f)(x)$ if $f(x) = x + x^2$ and $g(x) = 2x + 1$.

2. If

$$h(x) = \frac{1}{2x^2 + 1},$$

find $f(x)$ and $g(x)$ such that $h(x) = (f \circ g)(x)$. Answers may vary.

3. Determine whether $f(x) = |x + 1|$ is one-to-one.

Find a formula for the inverse of each function.

4. $f(x) = 4x - 3$

5. $g(x) = (x + 1)^3$

Graph.

6. $f(x) = 2^x - 3$

7. $g(x) = \log_7 x$

Simplify.

8. $\log_5 125$

9. $\log_{100} 10$

10. $3^{\log_3 18}$

Convert to logarithmic equations.

11. $4^{-3} = \frac{1}{64}$

12. $256^{1/2} = 16$

Convert to exponential equations.

13. $m = \log_7 49$

14. $\log_3 81 = 4$

15. Express in terms of logarithms of a, b, and c:

$$\log \frac{a^3 b^{1/2}}{c^2}.$$

16. Express as a single logarithm:

$$\tfrac{1}{3} \log_a x + 2 \log_a z.$$

Simplify.

17. $\log_p p$

18. $\log_t t^{23}$

19. $\log_c 1$

Given $\log_a 2 = 0.301$, $\log_a 6 = 0.778$, *and* $\log_a 7 = 0.845$, *find each of the following.*

20. $\log_a \frac{2}{7}$

21. $\log_a 12$

22. $\log_a 16$

Use a calculator to find each of the following to the nearest ten-thousandth.

23. $\log 12.3$

24. $10^{-0.8}$

25. $\ln 0.035$

26. $e^{4.8}$

27. Find $\log_3 10$ using the change-of-base formula. Round to the nearest ten-thousandth.

Graph and state the domain and the range of each function.

28. $f(x) = e^x + 3$

29. $g(x) = \ln(x - 4)$

Solve. Where appropriate, include approximations to the nearest ten-thousandth.

30. $2^x = \frac{1}{32}$

31. $\log_x 25 = 2$

32. $\log_4 x = \frac{1}{2}$

33. $\log x = 4$

34. $5^{4-3x} = 125$

35. $7^x = 1.2$

36. $\ln x = \frac{1}{4}$

37. $\log(x - 3) + \log(x + 1) = \log 5$

38. The average walking speed R of people living in a city of population P, in thousands, is given by $R = 0.37 \ln P + 0.05$, where R is in feet per second.

a) The population of Albuquerque, New Mexico, is 679,000. Find the average walking speed.

b) A city has an average walking speed of 2.6 ft/sec. Find the population.

39. The population of Kenya was 30 million in 2000, and the exponential growth rate was 1.5% per year.

a) Write an exponential function describing the population of Kenya.

b) What will the population be in 2003? in 2010?

c) When will the population be 50 million?

d) What is the doubling time?

40. The U.S. Consumer Price Index, a method of comparing prices of basic items, grew exponentially from 60.0 in 1920 to 511.5 in 2000 (*Sources*: U.S. Bureau of Labor Statistics and U.S. Department of Labor).

a) Find the value k, and write an exponential function that approximates the Consumer Price Index t years after 1920.

b) Predict the Consumer Price Index in 2010.

c) In what year will the Consumer Price Index be 1000?

41. An investment with interest compounded continuously doubled itself in 15 yr. What is the interest rate?

42. How old is an animal bone that has lost 43% of its carbon-14? (Use $P(t) = P_0 e^{-0.00012t}$.)

43. The sound of traffic at a busy intersection averages 75 dB. What is the intensity of such a sound? $\left(\text{Use } L = 10 \cdot \log \frac{I}{I_0}.\right)$

44. The hydrogen ion concentration of water is 1.0×10^{-7} moles per liter. What is the pH? (Use $\text{pH} = -\log[\text{H}^+]$.)

SYNTHESIS

45. Solve: $\log_5 |2x - 7| = 4$.

46. If $\log_a x = 2$, $\log_a y = 3$, and $\log_a z = 4$, find

$$\log_a \frac{\sqrt[3]{x^2 z}}{\sqrt[3]{y^2 z^{-1}}}.$$

Cumulative Review 1–9

1. Evaluate $\frac{x^0 + y}{-z}$ for $x = 6$, $y = 9$, and $z = -5$.

Simplify.

2. $\left|-\frac{5}{2} + \left(-\frac{7}{2}\right)\right|$

3. $(-2x^2y^{-3})^{-4}$

4. $(-5x^4y^{-3}z^2)(-4x^2y^2)$

5. $\frac{3x^4y^6z^{-2}}{-9x^4y^2z^3}$

6. $2x - 3 - 2[5 - 3(2 - x)]$

7. $3^3 + 2^2 - (32 \div 4 - 16 \div 8)$

Solve.

8. $8(2x - 3) = 6 - 4(2 - 3x)$

9. $4x - 3y = 15,$
$3x + 5y = 4$

10. $x + y - 3z = -1,$
$2x - y + z = 4,$
$-x - y + z = 1$

11. $x(x - 3) = 10$

12. $\frac{7}{x^2 - 5x} - \frac{2}{x - 5} = \frac{4}{x}$

13. $\frac{8}{x + 1} + \frac{11}{x^2 - x + 1} = \frac{24}{x^3 + 1}$

14. $\sqrt{4 - 5x} = 2x - 1$

15. $\sqrt[3]{2x} = 1$

16. $3x^2 + 75 = 0$

17. $x - 8\sqrt{x} + 15 = 0$

18. $x^4 - 13x^2 + 36 = 0$

19. $\log_8 x = 1$

20. $\log_x 49 = 2$

21. $9^x = 27$

22. $3^{5x} = 7$

23. $\log x - \log (x - 8) = 1$

24. $x^2 + 4x > 5$

25. If $f(x) = x^2 + 6x$, find a such that $f(a) = 11$.

26. If $f(x) = |2x - 3|$, find all x for which $f(x) \geq 9$.

Solve.

27. $D = \frac{ab}{b + a}$, for a

28. $\frac{1}{p} + \frac{1}{q} = \frac{1}{f}$, for q

29. $M = \frac{2}{3}(A + B)$, for B

Evaluate.

30. $\begin{vmatrix} 6 & -5 \\ 4 & -3 \end{vmatrix}$

31. $\begin{vmatrix} 7 & -6 & 0 \\ -2 & 1 & 2 \\ -1 & 1 & -1 \end{vmatrix}$

32. Find the domain of the function f given by

$$f(x) = \frac{-4}{3x^2 - 5x - 2}.$$

Solve.

33. The number of Americans filing taxes on the Internet (e-filers) in January and February increased from 2.4 million in 2000 to 3.4 million in 2001 (*Source*: *USA Today*, April 2, 2001).

a) At what rate was the number of January and February e-filers increasing?

b) Find a linear function $E(t)$ that fits the data. Let t represent the number of years since 2000.

c) Use the function of part (b) to predict the number of e-filers in January and February of 2005.

34. The perimeter of a rectangular garden is 112 m. The length is 16 m more than the width. Find the length and the width.

35. In triangle ABC, the measure of angle B is three times the measure of angle A. The measure of angle C is 105° greater than the measure of angle A. Find the angle measures.

36. Good's Candies makes all their chocolates by hand. It takes Anne 10 min to coat a tray of candies in chocolate. It takes Clay 12 min to coat a tray of candies. How long would it take Anne and Clay, working together, to coat the candies?

37. Joe's Thick and Tasty salad dressing gets 45% of its calories from fat. The Light and Lean dressing gets 20% of its calories from fat. How many ounces of each should be mixed in order to get 15 oz of dressing that gets 30% of its calories from fat?

38. A fishing boat with a trolling motor can move at a speed of 5 km/h in still water. The boat travels 42 km downstream in the same time that it takes to travel 12 km upstream. What is the speed of the stream?

39. What is the minimum product of two numbers whose difference is 14? What are the numbers that yield this product?

Students in a biology class just took a final exam. A formula for determining what the average exam grade will be t months later is

$$S(t) = 78 - 15 \log (t + 1).$$

40. The average score when the students first took the test occurs when $t = 0$. Find the students' average score on the final exam.

41. What would the average score be on a retest after 4 months?

The population of Mozambique was 19 million in 2000, and the exponential growth rate was 1.5% per year.

42. Write an exponential function describing the growth of the population of Mozambique.

43. Predict what the population will be in 2005 and in 2012.

44. What is the doubling time of the population?

45. y varies directly as the square of x and inversely as z, and $y = 2$ when $x = 5$ and $z = 100$. What is y when $x = 3$ and $z = 4$?

Perform the indicated operations and simplify.

46. $(5p^2q^3 + 6pq - p^2 + p) + (2p^2q^3 + p^2 - 5pq - 9)$

47. $(11x^2 - 6x - 3) - (3x^2 + 5x - 2)$

48. $(3x^2 - 2y)^2$

49. $(5a + 3b)(2a - 3b)$

50. $\dfrac{x^2 + 8x + 16}{2x + 6} \div \dfrac{x^2 + 3x - 4}{x^2 - 9}$

51. $\dfrac{1 + \dfrac{3}{x}}{x - 1 - \dfrac{12}{x}}$

52. $\dfrac{a^2 - a - 6}{a^3 - 27} \cdot \dfrac{a^2 + 3a + 9}{6}$

53. $\dfrac{3}{x + 6} - \dfrac{2}{x^2 - 36} + \dfrac{4}{x - 6}$

Factor.

54. $xy - 2xz + xw$

55. $1 - 125x^3$

56. $6x^2 + 8xy - 8y^2$

57. $x^4 - 4x^3 + 7x - 28$

58. $2m^2 + 12mn + 18n^2$

59. $x^4 - 16y^4$

60. For the function described by

$$h(x) = -3x^2 + 4x + 8,$$

find $h(-2)$.

61. Divide: $(x^4 - 5x^3 + 2x^2 - 6) \div (x - 3)$.

62. Multiply $(5.2 \times 10^4)(3.5 \times 10^{-6})$. Write scientific notation for the answer.

For the radical expressions that follow, assume that all variables represent positive numbers.

63. Divide and simplify:

$$\frac{\sqrt[3]{40xy^8}}{\sqrt[3]{5xy}}.$$

64. Multiply and simplify: $\sqrt{7xy^3} \cdot \sqrt{28x^2y}$.

65. Rewrite without rational exponents: $(27a^6b)^{4/3}$.

66. Rationalize the denominator:

$$\frac{3 - \sqrt{y}}{2 - \sqrt{y}}.$$

67. Divide and simplify:

$$\frac{\sqrt{x + 5}}{\sqrt[5]{x + 5}}.$$

68. Multiply these complex numbers:

$$\left(1 + i\sqrt{3}\right)\left(6 - 2i\sqrt{3}\right).$$

69. Add: $(3 - 2i) + (5 + 3i)$.

70. Find the inverse of f if $f(x) = 7 - 2x$.

71. Find a linear function with a graph that contains the points $(0, -3)$ and $(-1, 2)$.

72. Find an equation of the line whose graph has a y-intercept of $(0, 7)$ and is perpendicular to the line given by $2x + y = 6$.

Graph.

73. $5x = 15 + 3y$

74. $y = 2x^2 - 4x - 1$

75. $y = \log_3 x$

76. $y = 3^x$

77. $-2x - 3y \leq 6$

78. Graph: $f(x) = 2(x + 3)^2 + 1$.

a) Label the vertex.
b) Draw the axis of symmetry.
c) Find the maximum or minimum value.

79. Graph $f(x) = 2e^x$ and determine the domain and the range.

80. Express in terms of logarithms of a, b, and c:

$$\log\left(\frac{a^2c^3}{b}\right).$$

81. Express as a single logarithm:

$$3 \log x - \frac{1}{2} \log y - 2 \log z.$$

82. Convert to an exponential equation: $\log_a 5 = x$.

83. Convert to a logarithmic equation: $x^3 = t$.

Find each of the following using a calculator. Round to the nearest ten-thousandth.

84. $\log 0.05566$

85. $10^{2.89}$

86. $\ln 12.78$

87. $e^{-1.4}$

SYNTHESIS

Solve.

88. $\dfrac{5}{3x - 3} + \dfrac{10}{3x + 6} = \dfrac{5x}{x^2 + x - 2}$

89. $\log \sqrt{3x} = \sqrt{\log 3x}$

90. A train travels 280 mi at a certain speed. If the speed had been increased by 5 mph, the trip could have been made in 1 hr less time. Find the actual speed.

10 Conic Sections

10.1 Conic Sections: Parabolas and Circles
10.2 Conic Sections: Ellipses
10.3 Conic Sections: Hyperbolas
Connecting the Concepts
10.4 Nonlinear Systems of Equations
Summary and Review
Test

AN APPLICATION

Each side edge of the Burton® Twin 53 snowboard is an arc of a circle with a "running length" of 1150 mm and a "sidecut depth" of 19.5 mm. What radius is used to manufacture the edge of this board?

This problem, along with a diagram, appears as Exercise 97 in Section 10.1.

A snowboard's shape is designed by complex 3-D geometry, from the side-cut to the running length to the transition zones. By using these geometrical tools, we can design boards to perform like no others.

PETER BERGENDAHL
Board Design Engineer
Burlington, VT

The arcs described in the chapter opening are parts of a circle. A circle is one example of a conic section, *meaning that it can be regarded as a cross section of a cone. This chapter presents a variety of equations with graphs that are conic sections. We have already worked with two conic sections,* lines *and* parabolas, *in Chapters 2 and 8. There are many applications involving conics, and we will consider some in this chapter.*

10.1 Conic Sections: Parabolas and Circles

Parabolas • The Distance and Midpoint Formulas • Circles

This section and the next two examine curves formed by cross sections of cones. These curves are graphs of second-degree equations in two variables. Some are shown below:

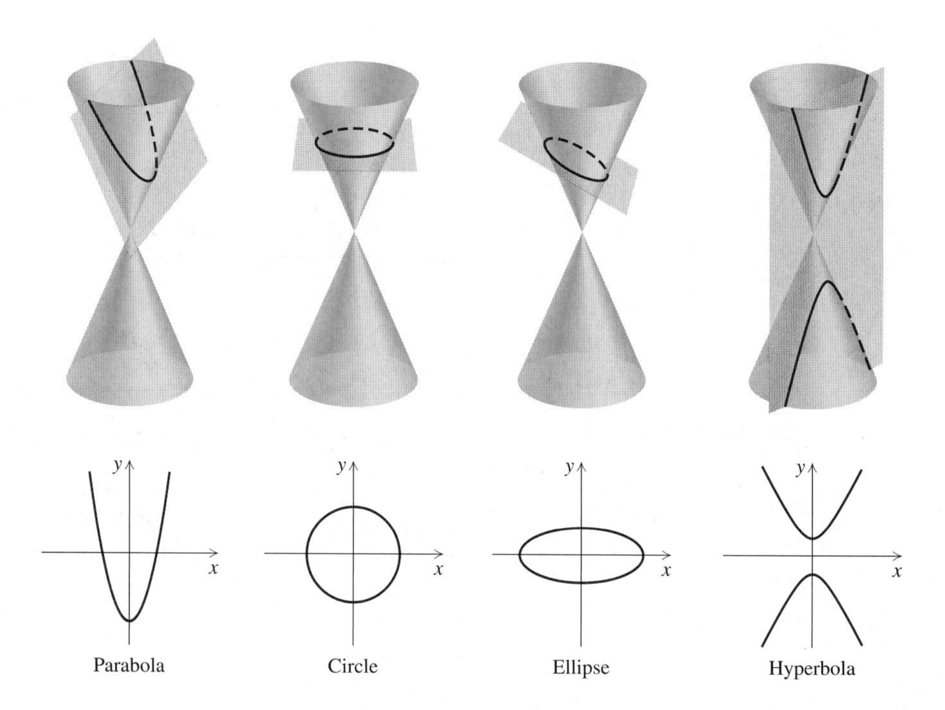

Parabolas

When a cone is cut as shown in the first figure above, the conic section formed is a **parabola**. Parabolas have many applications in electricity, mechanics, and optics. A cross section of a contact lens or satellite dish is a parabola, and arches that support certain bridges are parabolas.

Equation of a Parabola

A parabola with a vertical axis of symmetry opens upward or downward and has an equation that can be written in the form

$$y = ax^2 + bx + c.$$

A parabola with a horizontal axis of symmetry opens to the right or left and has an equation that can be written in the form

$$x = ay^2 + by + c.$$

Parabolas with equations of the form $f(x) = ax^2 + bx + c$ were graphed in Chapter 8.

Example 1 Graph: $y = x^2 - 4x + 9$.

Solution To locate the vertex, we can use either of two approaches. One way is to complete the square:

$$\begin{aligned} y &= (x^2 - 4x) + 9 && \text{Note that half of } -4 \text{ is } -2, \text{ and } (-2)^2 = 4. \\ &= (x^2 - 4x + 4 - 4) + 9 && \text{Adding and subtracting 4} \\ &= (x^2 - 4x + 4) + (-4 + 9) && \text{Regrouping} \\ &= (x - 2)^2 + 5. && \text{Factoring and simplifying} \end{aligned}$$

The vertex is $(2, 5)$.

A second way to find the vertex is to recall that the x-coordinate of the vertex of the parabola given by $y = ax^2 + bx + c$ is $-b/(2a)$:

$$x = -\frac{b}{2a} = -\frac{-4}{2(1)} = 2.$$

To find the y-coordinate of the vertex, we substitute 2 for x:

$$y = x^2 - 4x + 9 = 2^2 - 4(2) + 9 = 5.$$

Either way, the vertex is $(2, 5)$. Next, we calculate and plot some points on each side of the vertex. Since the x^2-coefficient, 1, is positive, the graph opens upward.

x	y	
2	5	← Vertex
0	9	← y-intercept
1	6	
3	6	
4	9	

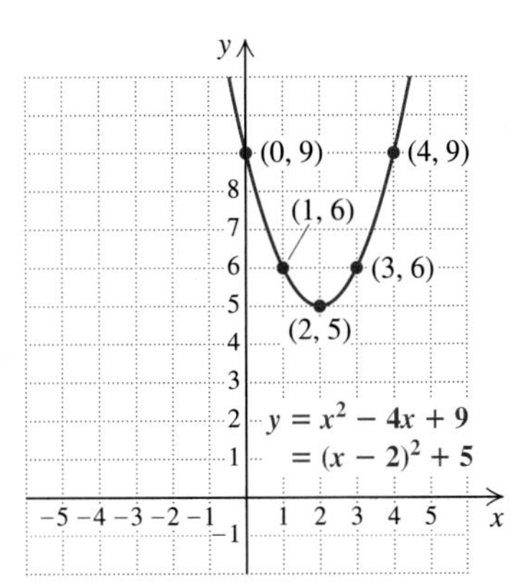

To Graph an Equation of the Form $y = ax^2 + bx + c$

1. Find the vertex (h, k) either by completing the square to find an equivalent equation $y = a(x - h)^2 + k$, or by using $-b/(2a)$ to find the x-coordinate and substituting to find the y-coordinate.
2. Choose other values for x on each side of the vertex, and compute the corresponding y-values.
3. The graph opens upward for $a > 0$ and downward for $a < 0$.

Equations of the form $x = ay^2 + by + c$ represent horizontal parabolas. These parabolas open to the right for $a > 0$, open to the left for $a < 0$, and have axes of symmetry parallel to the x-axis.

Example 2 Graph: $x = y^2 - 4y + 9$.

Solution This equation is like that in Example 1 except that x and y are interchanged. The vertex is $(5, 2)$ instead of $(2, 5)$. To find ordered pairs, we choose values for y on each side of the vertex. Then we compute values for x. Note that the x- and y-values of the table in Example 1 are now switched. You should confirm that, by completing the square, we get $x = (y - 2)^2 + 5$.

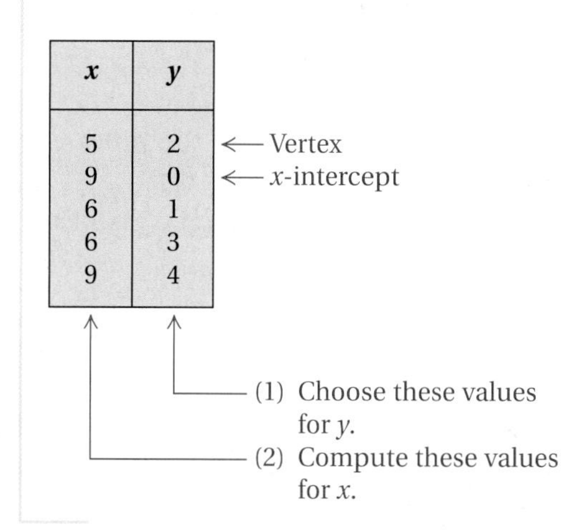

x	y
5	2
9	0
6	1
6	3
9	4

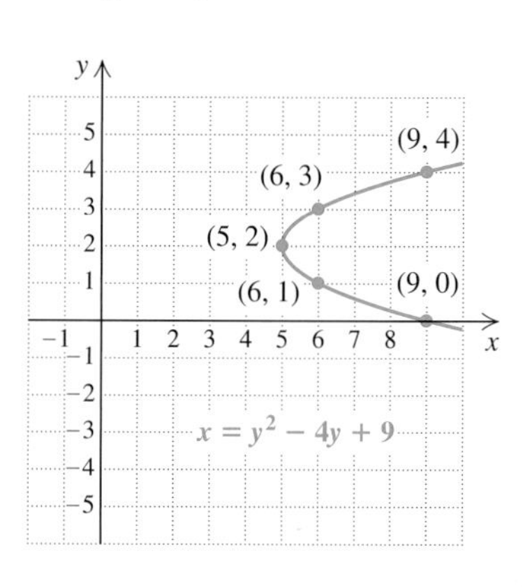

To Graph an Equation of the Form $x = ay^2 + by + c$

1. Find the vertex (h, k) either by completing the square to find an equivalent equation

$$x = a(y - k)^2 + h,$$

   or by using $-b/(2a)$ to find the y-coordinate and substituting to find the x-coordinate.
2. Choose other values for y that are above and below the vertex, and compute the corresponding x-values.
3. The graph opens to the right if $a > 0$ and to the left if $a < 0$.

Example 3 Graph: $x = -2y^2 + 10y - 7$.

Solution We use the method of completing the square:

$$\begin{aligned} x &= -2y^2 + 10y - 7 \\ &= -2(y^2 - 5y \qquad) - 7 \\ &= -2\left(y^2 - 5y + \tfrac{25}{4}\right) - 7 - (-2)\tfrac{25}{4} && \tfrac{1}{2}(-5) = \tfrac{-5}{2};\ \left(\tfrac{-5}{2}\right)^2 = \tfrac{25}{4}; \text{ we add and subtract } (-2)\tfrac{25}{4}. \\ &= -2\left(y - \tfrac{5}{2}\right)^2 + \tfrac{11}{2}. && \text{Factoring and simplifying} \end{aligned}$$

The vertex is $\left(\frac{11}{2}, \frac{5}{2}\right)$.

For practice, we also find the vertex by first computing its y-coordinate, $-b/(2a)$, and then substituting to find the x-coordinate:

$$y = -\frac{b}{2a} = -\frac{10}{2(-2)} = \frac{5}{2}$$

$$x = -2y^2 + 10y - 7 = -2\left(\tfrac{5}{2}\right)^2 + 10\left(\tfrac{5}{2}\right) - 7$$

$$= \tfrac{11}{2}.$$

To find ordered pairs, we choose values for y on each side of the vertex and then compute values for x. A table is shown below, together with the graph. The graph opens to the left because the y^2-coefficient, -2, is negative.

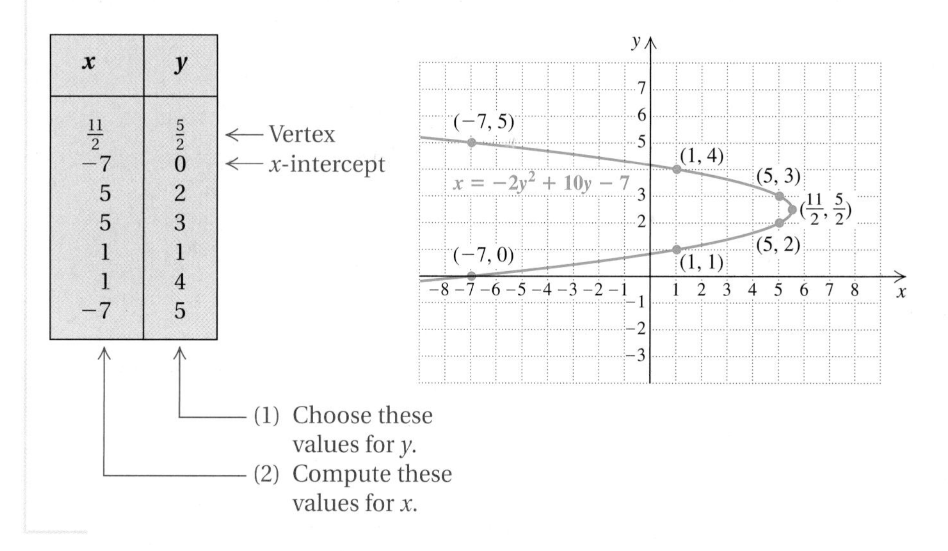

x	y	
$\frac{11}{2}$	$\frac{5}{2}$	← Vertex
-7	0	← x-intercept
5	2	
5	3	
1	1	
1	4	
-7	5	

The Distance and Midpoint Formulas

Suppose that two points are on a horizontal line, and thus have the same second coordinate. We can find the distance between them by subtracting their first coordinates. This difference may be negative, depending on the order in which we subtract. So, to make sure we get a positive number, we take the absolute value of this difference. The distance between the points (x_1, y_1) and (x_2, y_1) on a horizontal line is thus $|x_2 - x_1|$. Similarly, the distance between the points (x_2, y_1) and (x_2, y_2) on a vertical line is $|y_2 - y_1|$.

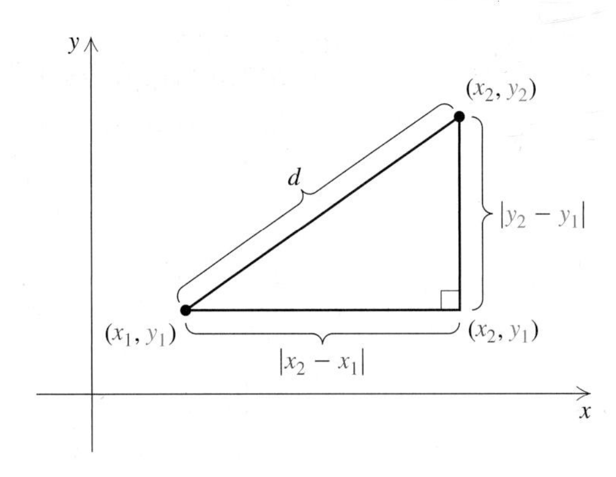

Now consider *any* two points (x_1, y_1) and (x_2, y_2). If $x_1 \neq x_2$ and $y_1 \neq y_2$, these points, along with the point (x_2, y_1), describe a right triangle. The lengths of the legs are $|x_2 - x_1|$ and $|y_2 - y_1|$. We find d, the length of the hypotenuse, by using the Pythagorean theorem:

$$d^2 = |x_2 - x_1|^2 + |y_2 - y_1|^2.$$

Since the square of a number is the same as the square of its opposite, we don't really need these absolute-value signs. Thus,

$$d^2 = (x_2 - x_1)^2 + (y_2 - y_1)^2.$$

Taking the principal square root, we obtain the distance between two points.

The Distance Formula

The distance d between any two points (x_1, y_1) and (x_2, y_2) is given by

$$d = \sqrt{(x_2 - x_1)^2 + (y_2 - y_1)^2}.$$

Example 4 Find the distance between $(5, -1)$ and $(-4, 6)$. Find an exact answer and an approximation to three decimal places.

Solution We substitute into the distance formula:

$$\begin{aligned} d &= \sqrt{(-4 - 5)^2 + [6 - (-1)]^2} && \text{Substituting} \\ &= \sqrt{(-9)^2 + 7^2} \\ &= \sqrt{130} && \text{This is exact.} \\ &\approx 11.402. && \text{Using a calculator for an approximation} \end{aligned}$$

The distance formula is needed to develop the formula for a circle, which follows, and to verify certain properties of conic sections. It is also needed to verify a formula for the coordinates of the *midpoint* of a segment connecting two points. We state the midpoint formula and leave its proof to the exercises. Note that although the distance formula involves both subtraction and addition, the midpoint formula uses only addition.

The Midpoint Formula

If the endpoints of a segment are (x_1, y_1) and (x_2, y_2), then the coordinates of the midpoint are

$$\left(\frac{x_1 + x_2}{2}, \frac{y_1 + y_2}{2}\right).$$

(To locate the midpoint, average the x-coordinates and average the y-coordinates.)

Example 5 Find the midpoint of the segment with endpoints $(-2, 3)$ and $(4, -6)$.

Solution Using the midpoint formula, we obtain

$$\left(\frac{-2 + 4}{2}, \frac{3 + (-6)}{2}\right), \quad \text{or} \quad \left(\frac{2}{2}, \frac{-3}{2}\right), \quad \text{or} \quad \left(1, -\frac{3}{2}\right).$$

Circles

One conic section, the **circle**, is a set of points in a plane that are a fixed distance r, called the **radius** (plural, **radii**), from a fixed point (h, k), called the **center**. Note that the word radius can mean either any segment connecting a point on a circle and the center of the circle or the length of such a segment. If a point (x, y) is on the circle, then by the definition of a circle and the distance formula, it follows that

$$r = \sqrt{(x - h)^2 + (y - k)^2}.$$

Squaring both sides gives the equation of a circle in standard form.

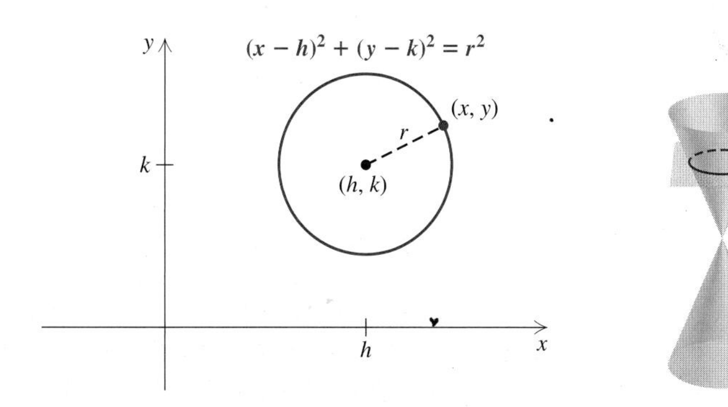

Equation of a Circle

The equation of a circle, centered at (h, k), with radius r, is given by

$$(x - h)^2 + (y - k)^2 = r^2.$$

Note that when $h = 0$ and $k = 0$, the circle is centered at the origin. Otherwise, the circle is translated $|h|$ units horizontally and $|k|$ units vertically.

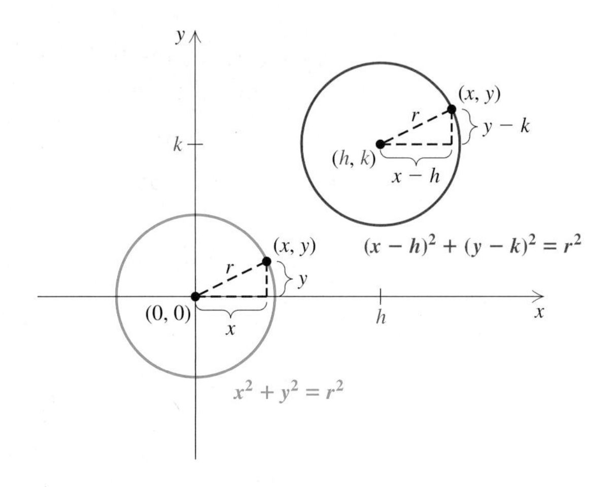

Example 6 Find an equation of the circle having center $(4, 5)$ and radius 6.

Solution Using the standard form, we obtain

$$(x - 4)^2 + (y - 5)^2 = 6^2, \qquad \text{Using } (x - h)^2 + (y - k)^2 = r^2$$

or

$$(x - 4)^2 + (y - 5)^2 = 36.$$

Example 7 Find the center and the radius and then graph each circle.

a) $(x - 2)^2 + (y + 3)^2 = 4^2$
b) $x^2 + y^2 + 8x - 2y + 15 = 0$

Solution

a) We write standard form:

$$(x - 2)^2 + [y - (-3)]^2 = 4^2.$$

The center is $(2, -3)$ and the radius is 4. To graph, we can use a compass or plot the points $(2, 1)$, $(6, -3)$, $(2, -7)$, and $(-2, -3)$, which are 4 units above, below, left, and right of $(2, -3)$, respectively, and then sketch a circle by hand.

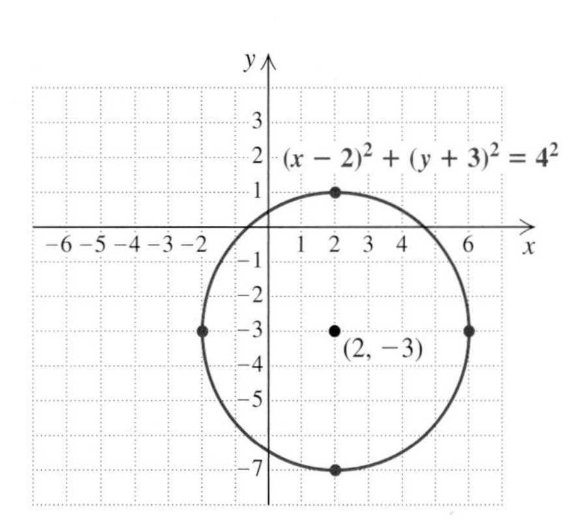

b) To write the equation $x^2 + y^2 + 8x - 2y + 15 = 0$ in standard form, we complete the square twice, once with $x^2 + 8x$ and once with $y^2 - 2y$:

$$x^2 + y^2 + 8x - 2y + 15 = 0$$

$$x^2 + 8x \quad + y^2 - 2y \quad = -15$$ Grouping the x-terms and the y-terms; adding -15 to both sides

$$x^2 + 8x + 16 + y^2 - 2y + 1 = -15 + 16 + 1$$ Adding $\left(\frac{8}{2}\right)^2$, or 16, and $\left(-\frac{2}{2}\right)^2$, or 1, to both sides

$$(x + 4)^2 + (y - 1)^2 = 2$$ Factoring

$$[x - (-4)]^2 + (y - 1)^2 = \left(\sqrt{2}\right)^2.$$ Writing standard form

The center is $(-4, 1)$ and the radius is $\sqrt{2}$.

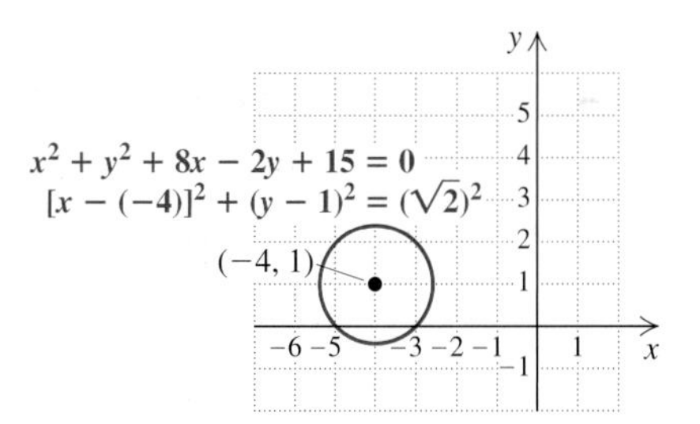

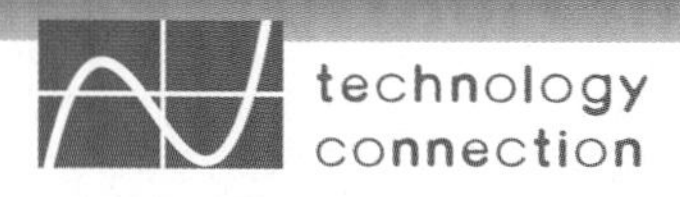

Because most graphers can graph only functions, graphing the equation of a circle usually requires two steps:

1. Solve the equation for y. The result will include a $\pm$ sign in front of a radical.
2. Graph two functions, one for the $+$ sign and the other for the $-$ sign, on the same set of axes.

For example, to graph $(x-3)^2 + (y+1)^2 = 16$, solve for $y+1$ and then y:

$$(y+1)^2 = 16 - (x-3)^2$$
$$y + 1 = \pm\sqrt{16-(x-3)^2}$$
$$y = -1 \pm \sqrt{16-(x-3)^2},$$

or $\quad y_1 = -1 + \sqrt{16-(x-3)^2}$

and $\quad y_2 = -1 - \sqrt{16-(x-3)^2}.$

When both functions are graphed (in a "squared" window to eliminate distortion), the result is as follows.

$y_1 = -1 + \sqrt{16-(x-3)^2}$,
$y_2 = -1 - \sqrt{16-(x-3)^2}$

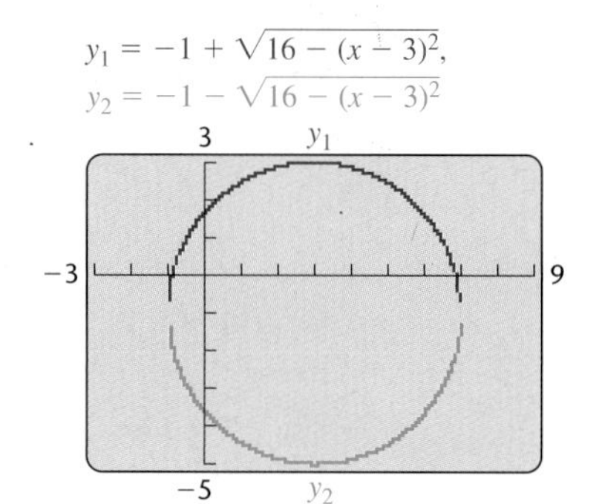

Circles can also be drawn using the CIRCLE option of the DRAW menu, but such graphs cannot be traced.

Use a grapher to graph each of the following equations.

1. $x^2 + y^2 - 16 = 0$
2. $4x^2 + 4y^2 = 100$
3. $x^2 + y^2 + 14x - 16y + 54 = 0$
4. $x^2 + y^2 - 10x - 11 = 0$

Exercise Set 10.1

FOR EXTRA HELP

Digital Video Tutor CD 8
Videotape 19

InterAct Math

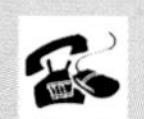
Math Tutor Center

MathXL

MyMathLab.com

Graph. Be sure to label each vertex.

1. $y = -x^2$
2. $y = 2x^2$
3. $y = -x^2 + 4x - 5$
4. $x = 4 - 3y - y^2$
5. $x = y^2 - 4y + 1$
6. $y = x^2 + 2x + 3$
7. $x = y^2 + 1$
8. $x = 2y^2$
9. $x = -\frac{1}{2}y^2$
10. $x = y^2 - 1$
11. $x = -y^2 - 4y$
12. $x = y^2 + y - 6$
13. $x = 8 - y - y^2$
14. $y = x^2 + 2x + 1$
15. $y = x^2 - 2x + 1$
16. $y = -\frac{1}{2}x^2$
17. $x = -y^2 + 2y - 1$
18. $x = -y^2 - 2y + 3$
19. $x = -2y^2 - 4y + 1$
20. $x = 2y^2 + 4y - 1$

Find the distance between each pair of points. Where appropriate, find an approximation to three decimal places.

21. $(1, 6)$ and $(5, 9)$
22. $(1, 10)$ and $(7, 2)$
23. $(0, -7)$ and $(3, -4)$
24. $(6, 2)$ and $(6, -8)$
25. $(-4, 4)$ and $(6, -6)$
26. $(5, 21)$ and $(-3, 1)$
27. $(8.6, -3.4)$ and $(-9.2, -3.4)$
28. $(5.9, 2)$ and $(3.7, -7.7)$
29. $\left(\frac{5}{7}, \frac{1}{14}\right)$ and $\left(\frac{1}{7}, \frac{11}{14}\right)$

30. $(0, \sqrt{7})$ and $(\sqrt{6}, 0)$

31. $(-\sqrt{6}, \sqrt{2})$ and $(0, 0)$

32. $(\sqrt{5}, -\sqrt{3})$ and $(0, 0)$

33. $(\sqrt{2}, -\sqrt{3})$ and $(-\sqrt{7}, \sqrt{5})$

34. $(\sqrt{8}, \sqrt{3})$ and $(-\sqrt{5}, -\sqrt{6})$

35. $(0, 0)$ and (s, t)

36. (p, q) and $(0, 0)$

Find the midpoint of each segment with the given endpoints.

37. $(-7, 6)$ and $(9, 2)$

38. $(6, 7)$ and $(7, -9)$

39. $(2, -1)$ and $(5, 8)$

40. $(-1, 2)$ and $(1, -3)$

41. $(-8, -5)$ and $(6, -1)$

42. $(8, -2)$ and $(-3, 4)$

43. $(-3.4, 8.1)$ and $(2.9, -8.7)$

44. $(4.1, 6.9)$ and $(5.2, -6.9)$

45. $\left(\frac{1}{6}, -\frac{3}{4}\right)$ and $\left(-\frac{1}{3}, \frac{5}{6}\right)$

46. $\left(-\frac{4}{5}, -\frac{2}{3}\right)$ and $\left(\frac{1}{8}, \frac{3}{4}\right)$

47. $(\sqrt{2}, -1)$ and $(\sqrt{3}, 4)$

48. $(9, 2\sqrt{3})$ and $(-4, 5\sqrt{3})$

Find an equation of the circle satisfying the given conditions.

49. Center $(0, 0)$, radius 6

50. Center $(0, 0)$, radius 5

51. Center $(7, 3)$, radius $\sqrt{5}$

52. Center $(5, 6)$, radius $\sqrt{2}$

53. Center $(-4, 3)$, radius $4\sqrt{3}$

54. Center $(-2, 7)$, radius $2\sqrt{5}$

55. Center $(-7, -2)$, radius $5\sqrt{2}$

56. Center $(-5, -8)$, radius $3\sqrt{2}$

57. Center $(0, 0)$, passing through $(-3, 4)$

58. Center $(3, -2)$, passing through $(11, -2)$

59. Center $(-4, 1)$, passing through $(-2, 5)$

60. Center $(-1, -3)$, passing through $(-4, 2)$

Find the center and the radius of each circle. Then graph the circle.

61. $x^2 + y^2 = 49$

62. $x^2 + y^2 = 36$

63. $(x + 1)^2 + (y + 3)^2 = 4$

64. $(x - 2)^2 + (y + 3)^2 = 1$

65. $(x - 4)^2 + (y + 3)^2 = 10$

66. $(x + 5)^2 + (y - 1)^2 = 15$

67. $x^2 + y^2 = 7$

68. $x^2 + y^2 = 8$

69. $(x - 5)^2 + y^2 = \frac{1}{4}$

70. $x^2 + (y - 1)^2 = \frac{1}{25}$

71. $x^2 + y^2 + 8x - 6y - 15 = 0$

72. $x^2 + y^2 + 6x - 4y - 15 = 0$

73. $x^2 + y^2 - 8x + 2y + 13 = 0$

74. $x^2 + y^2 + 6x + 4y + 12 = 0$

75. $x^2 + y^2 + 10y - 75 = 0$

76. $x^2 + y^2 - 8x - 84 = 0$

77. $x^2 + y^2 + 7x - 3y - 10 = 0$

78. $x^2 + y^2 - 21x - 33y + 17 = 0$

79. $36x^2 + 36y^2 = 1$

80. $4x^2 + 4y^2 = 1$

81. Describe a procedure that would use the distance formula to determine whether three points, (x_1, y_1), (x_2, y_2), and (x_3, y_3), are vertices of a right triangle.

82. Does the graph of an equation of a circle include the point that is the center? Why or why not?

SKILL MAINTENANCE

Solve.

83. $\frac{x}{4} + \frac{5}{6} = \frac{2}{3}$

84. $\frac{t}{6} - \frac{1}{9} = \frac{7}{12}$

85. A rectangle 10 in. long and 6 in. wide is bordered by a strip of uniform width. If the perimeter of the larger rectangle is twice that of the smaller rectangle, what is the width of the border?

86. One airplane flies 60 mph faster than another. To fly a certain distance, the faster plane takes 4 hr and the slower plane takes 4 hr and 24 min. What is the distance?

Solve each system.

87. $3x - 8y = 5,$
$2x + 6y = 5$

88. $4x - 5y = 9,$
$12x - 10y = 18$

SYNTHESIS

89. Outline a procedure that would use the distance formula to determine whether three points, (x_1, y_1), (x_2, y_2), and (x_3, y_3), are collinear (lie on the same line).

90. Why does the discussion of the distance formula precede the discussion of circles?

Find an equation of a circle satisfying the given conditions.

91. Center $(3, -5)$ and tangent to (touching at one point) the y-axis

92. Center $(-7, -4)$ and tangent to the x-axis

93. The endpoints of a diameter are $(7, 3)$ and $(-1, -3)$.

94. Center $(-3, 5)$ with a circumference of 8π units

95. Find the point on the y-axis that is equidistant from $(2, 10)$ and $(6, 2)$.

96. Find the point on the x-axis that is equidistant from $(-1, 3)$ and $(-8, -4)$.

97. *Snowboarding.* Each side edge of the Burton® Twin 53 snowboard is an arc of a circle with a "running length" of 1150 mm and a "sidecut depth" of 19.5 mm (see the figure below).

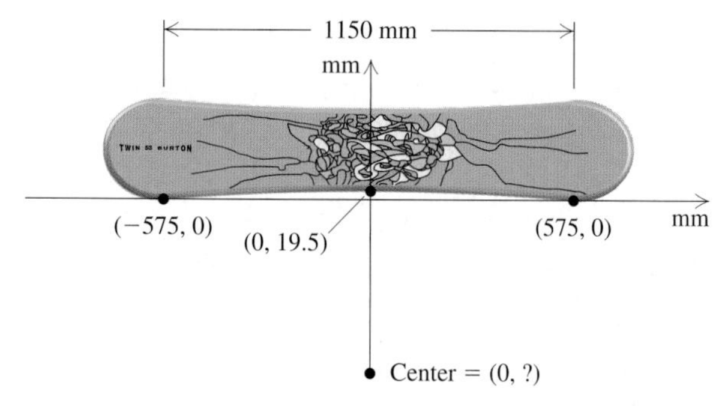

a) Using the coordinates shown, locate the center of the circle. (*Hint*: Equate distances.)
b) What radius is used for the edge of the board?

98. *Snowboarding.* The Burton® Twin 44 snowboard has a running length of 1070 mm and a sidecut depth of 17.5 mm (see Exercise 97). What radius is used for the edge of this snowboard?

99. *Skiing.* The Rossignol® Cut 10.4 ski, when lying flat and viewed from above, has edges that are arcs of a circle. (Actually, each edge is made of two arcs of slightly different radii. The arc for the rear half of the ski edge has a slightly larger radius.)

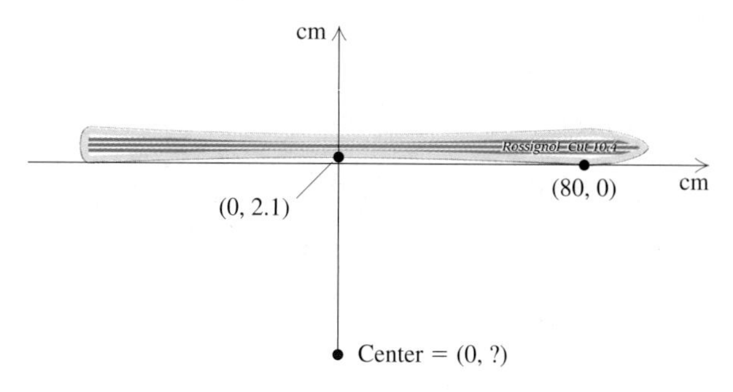

a) Using the coordinates shown, locate the center of the circle. (*Hint*: Equate distances.)
b) What radius is used for the arc passing through (0, 2.1) and (80, 0)?

100. *Doorway construction.* Ace Carpentry needs to cut an arch for the top of an entranceway. The arch needs to be 8 ft wide and 2 ft high. To draw the arch, the carpenters will use a stretched string with chalk attached at an end as a compass.

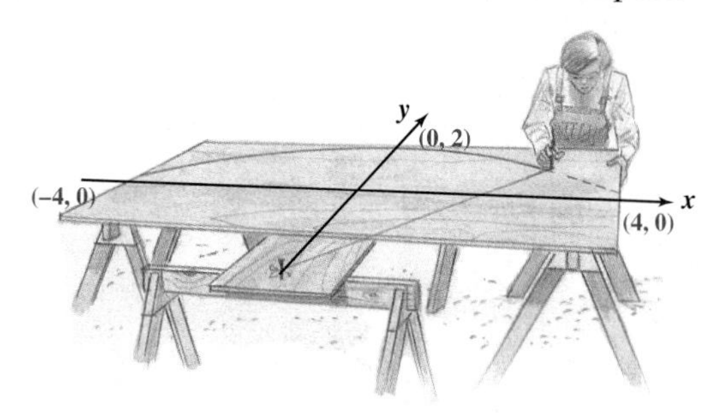

a) Using a coordinate system, locate the center of the circle.
b) What radius should the carpenters use to draw the arch?

101. *Archaeology.* During an archaeological dig, Martina finds the bowl fragment shown below. What was the original diameter of the bowl?

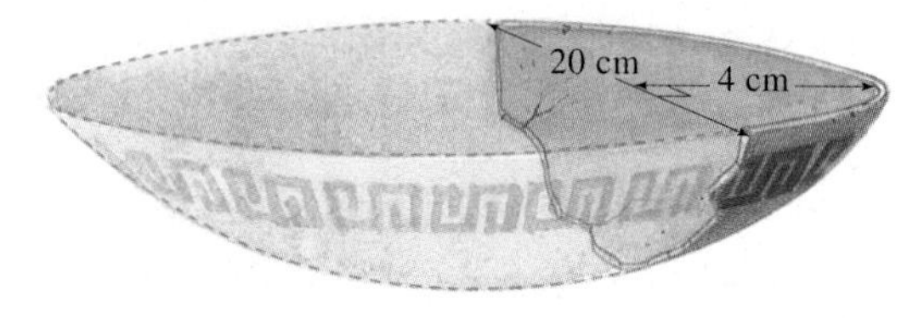

102. *Ferris wheel design.* A ferris wheel has a radius of 24.3 ft. Assuming that the center is 30.6 ft off the ground and that the origin is below the center, as in the following figure, find an equation of the circle.

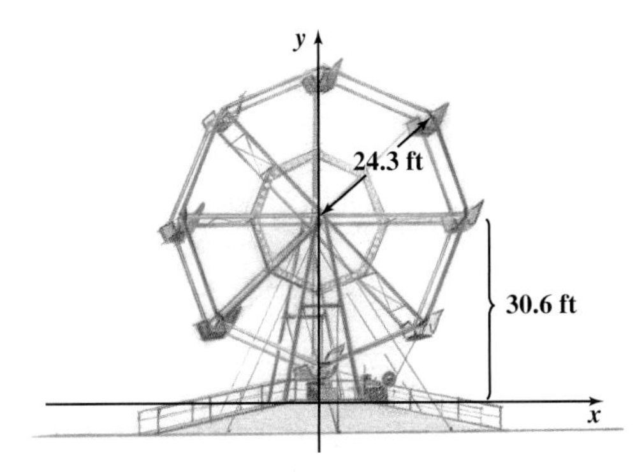

103. Use a graph of the equation $x = y^2 - y - 6$ to approximate to the nearest tenth the solutions of each of the following equations.

a) $y^2 - y - 6 = 2$ (*Hint*: Graph $x = 2$ on the same set of axes as the graph of $x = y^2 - y - 6$.)

b) $y^2 - y - 6 = -3$

104. *Power of a motor.* The horsepower of a certain kind of engine is given by the formula

$$H = \frac{D^2N}{2.5},$$

where N is the number of cylinders and D is the diameter, in inches, of each piston. Graph this equation, assuming that $N = 6$ (a six-cylinder engine). Let D run from 2.5 to 8.

105. Prove the midpoint formula by showing that

i) the distance from (x_1, y_1) to

$$\left(\frac{x_1 + x_2}{2}, \frac{y_1 + y_2}{2}\right)$$

equals the distance from (x_2, y_2) to

$$\left(\frac{x_1 + x_2}{2}, \frac{y_1 + y_2}{2}\right);$$

and

ii) the points

$$(x_1, y_1), \left(\frac{x_1 + x_2}{2}, \frac{y_1 + y_2}{2}\right),$$

and

(x_2, y_2)

lie on the same line (see Exercise 89).

106. If the equation $x^2 + y^2 - 6x + 2y - 6 = 0$ is written as $y^2 + 2y + (x^2 - 6x - 6) = 0$, it can be regarded as quadratic in y.

a) Use the quadratic formula to solve for y.

b) Show that the graph of your answer to part (a) coincides with the graph in the Technology Connection on p. 628.

107. How could a grapher best be used to help you sketch the graph of an equation of the form $x = ay^2 + by + c$?

108. Why should a grapher's window be "squared" before graphing a circle?

Conic Sections: Ellipses

10.2

Ellipses Centered at (0, 0) • Ellipses Centered at (*h*, *k*)

When a cone is cut at an angle, as shown on the following page, the conic section formed is an *ellipse*. To draw an ellipse, stick two tacks in a piece of cardboard. Then tie a string to the tacks, place a pencil as shown, and draw an oval by moving the pencil while keeping the string taut.

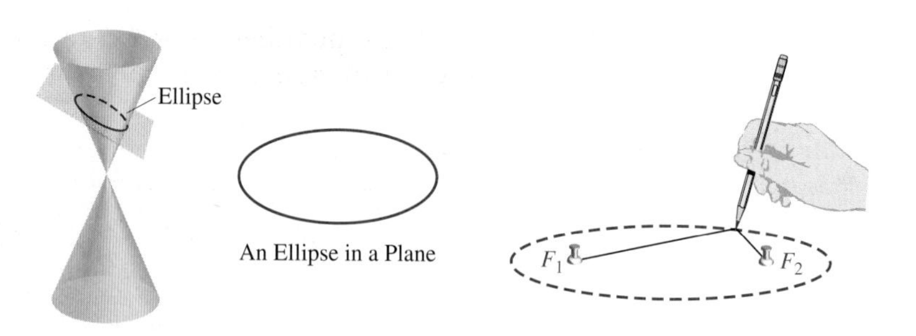

Ellipses Centered at (0, 0)

An **ellipse** is defined as the set of all points in a plane for which the *sum* of the distances from two fixed points F_1 and F_2 is constant. The points F_1 and F_2 are called **foci** (pronounced fō-sī), the plural of focus. In the figure above, the tacks are at the foci and the length of the string is the constant sum of the distances. The midpoint of the segment F_1F_2 is the **center**. The equation of an ellipse is as follows. Its derivation is left to the exercises.

Equation of an Ellipse Centered at the Origin

The equation of an ellipse centered at the origin and symmetric with respect to both axes is

$$\frac{x^2}{a^2} + \frac{y^2}{b^2} = 1, \quad a, b > 0. \qquad \text{(Standard form)}$$

To graph an ellipse centered at the origin, it helps to first find the intercepts. If we replace x with 0, we can find the y-intercepts:

$$\frac{0^2}{a^2} + \frac{y^2}{b^2} = 1$$

$$\frac{y^2}{b^2} = 1$$

$$y^2 = b^2 \quad \text{or} \quad y = \pm b.$$

Thus the y-intercepts are $(0, b)$ and $(0, -b)$. Similarly, the x-intercepts are $(a, 0)$ and $(-a, 0)$. If $a > b$, the ellipse is said to be horizontal and $(-a, 0)$ and $(a, 0)$ are referred to as the **vertices** (singular, **vertex**). If $b > a$, the ellipse is said to be vertical and $(0, -b)$ and $(0, b)$ are then the vertices.

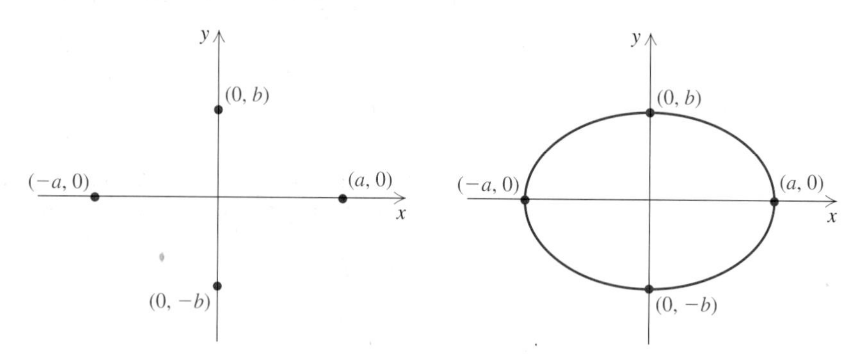

Plotting these four points and drawing an oval-shaped curve, we get a graph of the ellipse. If a more precise graph is desired, we can plot more points.

Using a and b to Graph an Ellipse

For the ellipse

$$\frac{x^2}{a^2} + \frac{y^2}{b^2} = 1,$$

the x-intercepts are $(-a, 0)$ and $(a, 0)$. The y-intercepts are $(0, -b)$ and $(0, b)$.

Example 1 Graph the ellipse

$$\frac{x^2}{4} + \frac{y^2}{9} = 1.$$

Solution Note that

$$\frac{x^2}{4} + \frac{y^2}{9} = \frac{x^2}{2^2} + \frac{y^2}{3^2}.$$

Identifying a and b. Since $b > a$, the ellipse is vertical.

Thus the x-intercepts are $(-2, 0)$ and $(2, 0)$, and the y-intercepts are $(0, -3)$ and $(0, 3)$. We plot these points and connect them with an oval-shaped curve. To plot some other points, we let $x = 1$ and solve for y:

$$\frac{1^2}{4} + \frac{y^2}{9} = 1$$

$$36\left(\frac{1}{4} + \frac{y^2}{9}\right) = 36 \cdot 1$$

$$36 \cdot \frac{1}{4} + 36 \cdot \frac{y^2}{9} = 36$$

$$9 + 4y^2 = 36$$

$$4y^2 = 27$$

$$y^2 = \frac{27}{4}$$

$$y = \pm\sqrt{\frac{27}{4}}$$

$$y \approx \pm 2.6.$$

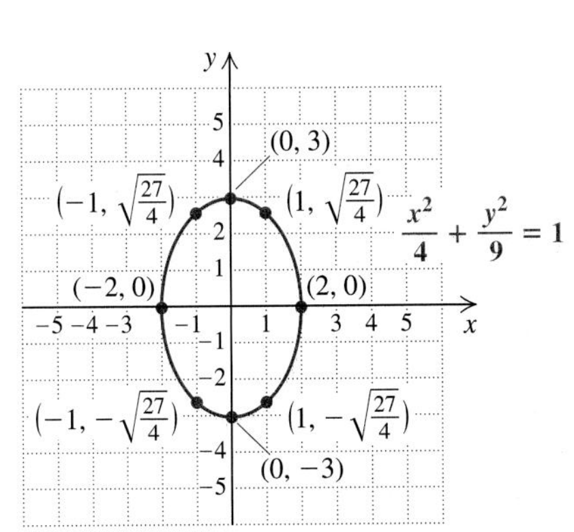

Thus, $(1, 2.6)$ and $(1, -2.6)$ can also be used to draw the graph. Similarly, the points $(-1, 2.6)$ and $(-1, -2.6)$ can also be computed and plotted.

Example 2

Graph: $4x^2 + 25y^2 = 100$.

Solution To write the equation in standard form, we divide both sides by 100 to get 1 on the right side:

$$\frac{4x^2 + 25y^2}{100} = \frac{100}{100} \qquad \text{Dividing by 100 to get 1 on the right side}$$

$$\left.\begin{aligned} \frac{4x^2}{100} + \frac{25y^2}{100} &= 1 \\ \frac{x^2}{25} + \frac{y^2}{4} &= 1 \end{aligned}\right\} \qquad \text{Simplifying}$$

$$\frac{x^2}{5^2} + \frac{y^2}{2^2} = 1. \qquad a = 5,\ b = 2$$

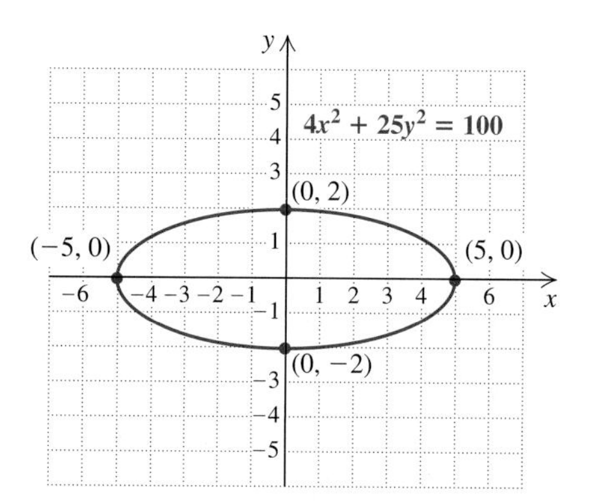

The x-intercepts are $(-5, 0)$ and $(5, 0)$, and the y-intercepts are $(0, -2)$ and $(0, 2)$. We plot the intercepts and connect them with an oval-shaped curve. Other points can also be computed and plotted.

Graphing an ellipse on a grapher is much like graphing a circle: We graph it in two pieces after solving for y. To illustrate, let's check Example 2:

$$\begin{aligned} 4x^2 + 25y^2 &= 100 \\ 25y^2 &= 100 - 4x^2 \\ y^2 &= 4 - \tfrac{4}{25}x^2 \\ y &= \pm\sqrt{4 - \tfrac{4}{25}x^2}. \end{aligned}$$

Using a squared window, we have our check:

$y_1 = -\sqrt{4 - \frac{4}{25}x^2}$, $y_2 = \sqrt{4 - \frac{4}{25}x^2}$

6

−9 9

−6

Ellipses Centered at (h, k)

Horizontal and vertical translations, similar to those used in Chapter 8, can be used to graph ellipses that are not centered at the origin.

Equation of an Ellipse Centered at (h, k)

The standard form of a horizontal or vertical ellipse centered at (h, k) is

$$\frac{(x-h)^2}{a^2} + \frac{(y-k)^2}{b^2} = 1.$$

The vertices are $(h + a, k)$ and $(h - a, k)$ if horizontal; $(h, k + b)$ and $(h, k - b)$ if vertical.

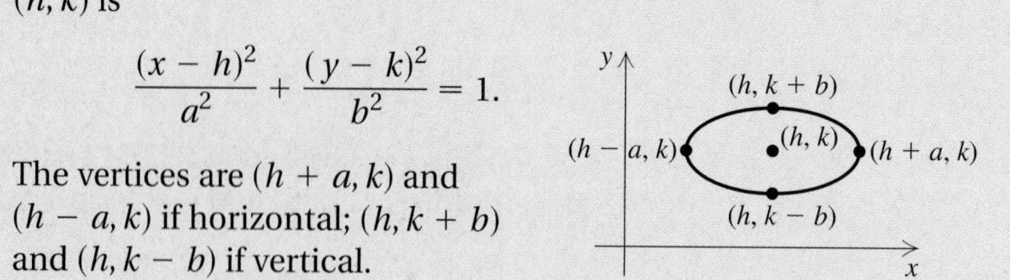

Example 3 Graph the ellipse

$$\frac{(x-1)^2}{4} + \frac{(y+5)^2}{9} = 1.$$

Solution Note that

$$\frac{(x-1)^2}{4} + \frac{(y+5)^2}{9} = \frac{(x-1)^2}{2^2} + \frac{(y+5)^2}{3^2}.$$

Thus, $a = 2$ and $b = 3$. To determine the center of the ellipse, (h, k), note that

$$\frac{(x-1)^2}{2^2} + \frac{(y+5)^2}{3^2} = \frac{(x-1)^2}{2^2} + \frac{(y-(-5))^2}{3^2}.$$

Thus the center is $(1, -5)$. We plot the points 2 units to the left and right of center, as well as the points 3 units above and below center. These are the points $(3, -5)$, $(-1, -5)$, $(1, -2)$, and $(1, -8)$.

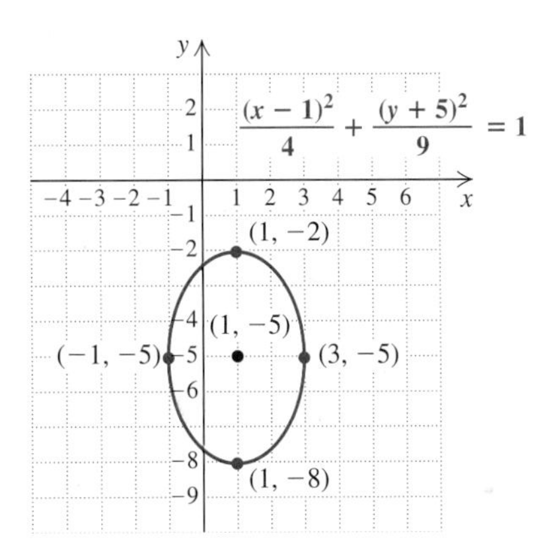

Note that this ellipse is the same as the ellipse in Example 1 but translated 1 unit to the right and 5 units down.

Ellipses have many applications. Communications satellites move in elliptical orbits with the earth as a focus while the earth itself follows an elliptical path around the sun. A medical instrument, the lithotripter, uses shock waves originating at one focus to crush a kidney stone located at the other focus.

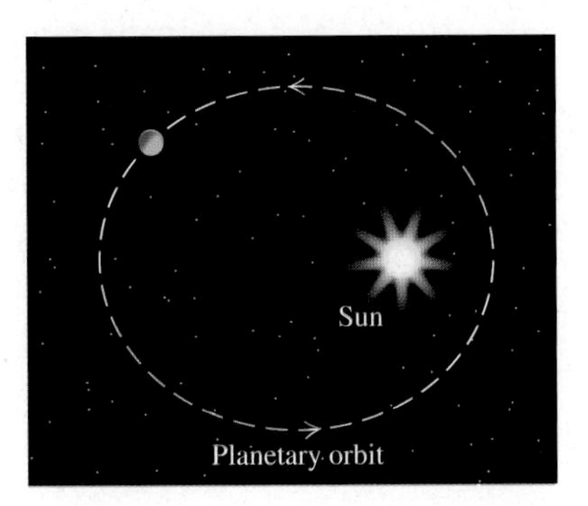

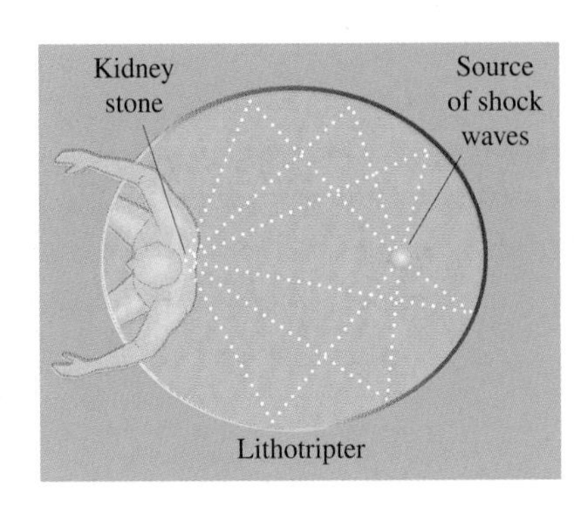

In some buildings, an ellipsoidal ceiling creates a "whispering gallery" in which a person at one focus can whisper and still be heard clearly at the other focus. This happens because sound waves coming from one focus are all reflected to the other focus. Similarly, light waves bouncing off an ellipsoidal mirror are used in a dentist's or surgeon's reflector light. The light source is located at one focus while the patient's mouth is at the other.

Exercise Set 10.2

FOR EXTRA HELP

Digital Video Tutor CD 8
Videotape 19

InterAct Math

Math Tutor Center

MathXL

MyMathLab.com

Graph each of the following equations.

1. $\dfrac{x^2}{1} + \dfrac{y^2}{4} = 1$

2. $\dfrac{x^2}{4} + \dfrac{y^2}{1} = 1$

3. $\dfrac{x^2}{25} + \dfrac{y^2}{9} = 1$

4. $\dfrac{x^2}{16} + \dfrac{y^2}{25} = 1$

5. $4x^2 + 9y^2 = 36$

6. $9x^2 + 4y^2 = 36$

7. $16x^2 + 9y^2 = 144$

8. $9x^2 + 16y^2 = 144$

9. $2x^2 + 3y^2 = 6$

10. $5x^2 + 7y^2 = 35$

Aha! 11. $5x^2 + 5y^2 = 125$

12. $8x^2 + 5y^2 = 80$

13. $3x^2 + 7y^2 - 63 = 0$

14. $3x^2 + 8y^2 - 72 = 0$

15. $8x^2 = 96 - 3y^2$

16. $6y^2 = 24 - 8x^2$

17. $16x^2 + 25y^2 = 1$

18. $9x^2 + 4y^2 = 1$

19. $\dfrac{(x-2)^2}{9} + \dfrac{(y-1)^2}{25} = 1$

20. $\dfrac{(x-3)^2}{25} + \dfrac{(y-4)^2}{9} = 1$

21. $\dfrac{(x+4)^2}{16} + \dfrac{(y-3)^2}{49} = 1$

22. $\dfrac{(x+5)^2}{4} + \dfrac{(y-2)^2}{36} = 1$

23. $12(x-1)^2 + 3(y+4)^2 = 48$
(*Hint*: Divide both sides by 48.)

24. $4(x-6)^2 + 9(y+2)^2 = 36$

Aha! 25. $4(x+3)^2 + 4(y+1)^2 - 10 = 90$

26. $9(x+6)^2 + (y+2)^2 - 20 = 61$

27. Is the center of an ellipse part of the ellipse itself? Why or why not?

28. Can an ellipse ever be the graph of a function? Why or why not?

SKILL MAINTENANCE

Solve.

29. $\dfrac{3}{x-2} - \dfrac{5}{x-2} = 9$

30. $\dfrac{7}{x+3} - \dfrac{2}{x+3} = 8$

31. $\dfrac{x}{x-4} - \dfrac{3}{x-5} = \dfrac{2}{x-4}$

32. $\dfrac{7}{x-3} - \dfrac{x}{x-2} = \dfrac{4}{x-2}$

33. $9 - \sqrt{2x+1} = 7$

34. $5 - \sqrt{x+3} = 9$

SYNTHESIS

35. An eccentric person builds a pool table in the shape of an ellipse with a hole at one focus and a tiny dot at the other. Guests are amazed at how many bank shots the owner of the pool table makes. Explain why this occurs.

36. Can a circle be considered a special type of ellipse? Why or why not?

Find an equation of an ellipse that contains the following points.

37. $(-9, 0)$, $(9, 0)$, $(0, -11)$, and $(0, 11)$

38. $(-7, 0)$, $(7, 0)$, $(0, -5)$, and $(0, 5)$

39. $(-2, -1)$, $(6, -1)$, $(2, -4)$, and $(2, 2)$

40. $(-6, 3)$, $(4, 3)$, $(-1, 7)$, and $(-1, -1)$

41. *Astronomy.* The maximum distance of the planet Mars from the sun is 2.48×10^8 mi. The minimum distance is 3.46×10^7 mi. The sun is at one focus of the elliptical orbit. Find the distance from the sun to the other focus.

42. Let $(-c, 0)$ and $(c, 0)$ be the foci of an ellipse. Any point $P(x, y)$ is on the ellipse if the sum of the distances from the foci to P is some constant. Use $2a$ to represent this constant.

a) Show that an equation for the ellipse is given by

$$\frac{x^2}{a^2} + \frac{y^2}{a^2 - c^2} = 1.$$

b) Substitute b^2 for $a^2 - c^2$ to get standard form.

43. *President's office.* The Oval Office of the President of the United States is an ellipse 31 ft wide and 38 ft long. Show in a sketch precisely where the President and an adviser could sit to best hear each other using the room's acoustics. (*Hint*: See Exercise 42(b) and the discussion following Example 3.)

44. *Dentistry.* The light source in a dental lamp shines against a reflector that is shaped like a portion of an ellipse in which the light source is one focus of the ellipse. Reflected light enters a patient's mouth at the other focus of the ellipse. If the ellipse from which the reflector was formed is 2 ft wide and 6 ft long, how far should the patient's mouth be from the light source? (*Hint*: See Exercise 42(b).)

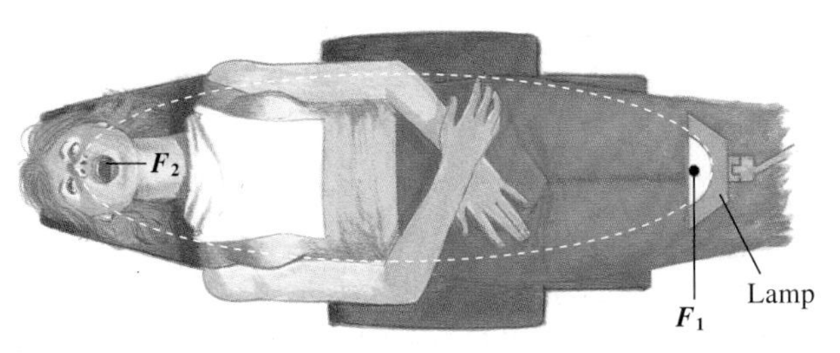

45. *Firefighting.* The size and shape of certain forest fires can be approximated as the union of two "half-ellipses." For the blaze modeled below, the equation of the smaller ellipse—the part of the fire moving *into* the wind—is

$$\frac{x^2}{40{,}000} + \frac{y^2}{10{,}000} = 1.$$

The equation of the other ellipse—the part moving *with* the wind—is

$$\frac{x^2}{250{,}000} + \frac{y^2}{10{,}000} = 1.$$

(*Source for figure*: "Predicting Wind-Driven Wild Land Fire Size and Shape," Hal E. Anderson, Research Paper INT-305, U.S. Department of Agriculture, Forest Service, February 1983).

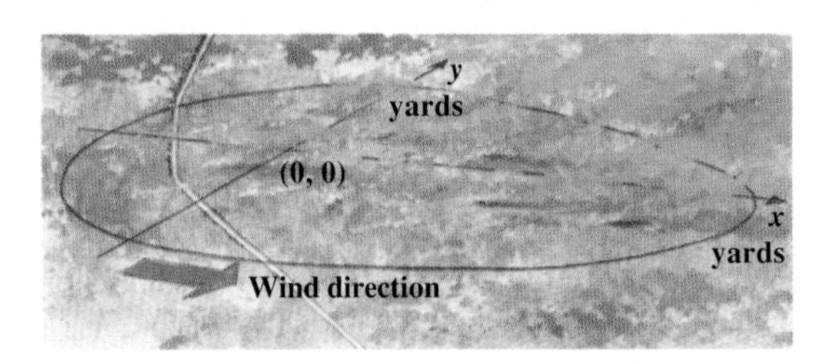

Determine the width and the length of the fire.

For each of the following equations, complete the square as needed and find an equivalent equation in standard form. Then graph the ellipse.

46. $x^2 - 4x + 4y^2 + 8y - 8 = 0$

47. $4x^2 + 24x + y^2 - 2y - 63 = 0$

 **48.** Use a grapher to check your answers to Exercises 3, 17, 21, and 25.

COLLABORATIVE

CORNER

A Cosmic Path

Focus: Ellipses

Time: 20–30 minutes

Group size: 2

Materials: Scientific calculators

In March 1996, the comet Hyakutake came within 21 million mi of the sun, and closer to Earth than any comet in over 500 yr (*Source*: Associated Press newspaper story, 3/20/96). Hyakutake is traveling in an elliptical orbit with the sun at one focus. The comet's average speed is about 100,000 mph (it actually goes much faster near its foci and slower as it gets further from the foci) and one orbit takes about 15,000 yr. (Astronomers estimate the time at 10,000–20,000 yr.)

ACTIVITY

1. The elliptical orbit of Hyakutake is so elongated that the distance traveled in one orbit can be estimated by $4a$ (see the following figure). Use the information above to estimate the distance, in millions of miles, traveled in one orbit. Then determine a.

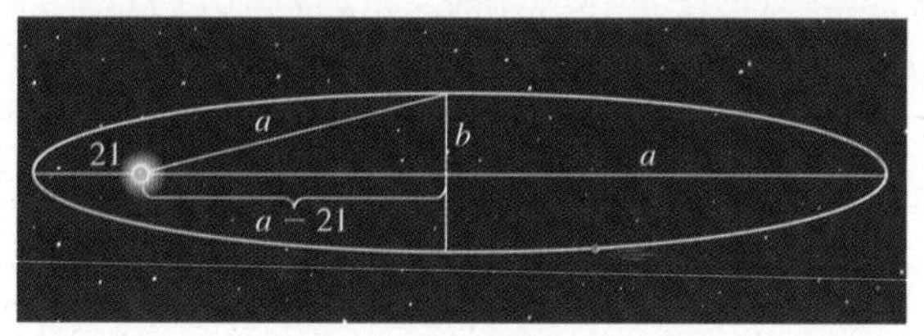

Units are millions of miles.

2. Using the figure above, express b^2 as a function of a. Then solve for b using the value found for a in part (1).

3. Approximately how far will Hyakutake be from the sun at the most distant part of its orbit?

4. Repeat parts (1)–(3), with one group member using the lower estimate of orbit time (10,000 yr) and the other using the upper estimate of orbit time (20,000 yr). By how much do the three answers to part (3) vary?

Conic Sections: Hyperbolas

10.3

Hyperbolas • Hyperbolas (Nonstandard Form) • Classifying Graphs of Equations

Hyperbolas

A **hyperbola** looks like a pair of parabolas, but the shapes are actually different. A hyperbola has two **vertices** and the line through the vertices is known as an **axis**. The point halfway between the vertices is called the **center**. The two curves that comprise a hyperbola are called **branches**.

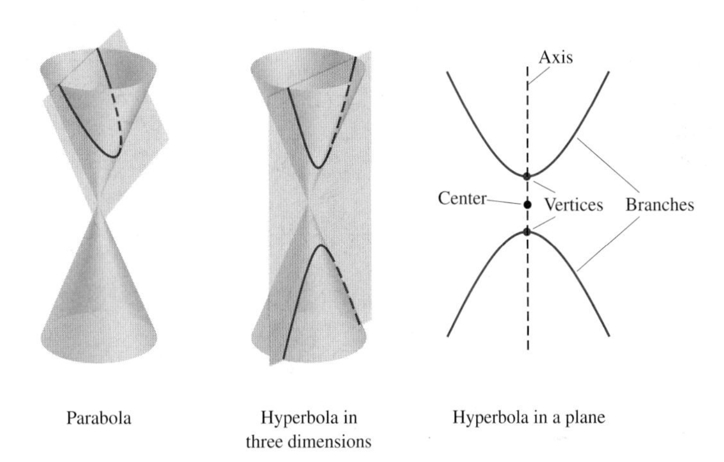

Parabola

Hyperbola in three dimensions

Hyperbola in a plane

Equation of a Hyperbola Centered at the Origin

Hyperbolas with their centers at the origin* have equations as follows:

$$\frac{x^2}{a^2} - \frac{y^2}{b^2} = 1 \qquad \text{(Axis horizontal);}$$

$$\frac{y^2}{b^2} - \frac{x^2}{a^2} = 1 \qquad \text{(Axis vertical).}$$

Note that both equations have a 1 on the right-hand side and a subtraction symbol between the terms. For the discussion that follows, we assume $a, b > 0$.

To graph a hyperbola, it helps to begin by graphing two lines called **asymptotes**. Although the asymptotes themselves are not part of the graph, they serve as guidelines for an accurate sketch.

*Hyperbolas with horizontal or vertical axes and centers *not* at the origin are discussed in Exercises 51–56.

Asymptotes of a Hyperbola

For hyperbolas with equations as shown below, the asymptotes are the lines

$$y = \frac{b}{a}x \quad \text{and} \quad y = -\frac{b}{a}x.$$

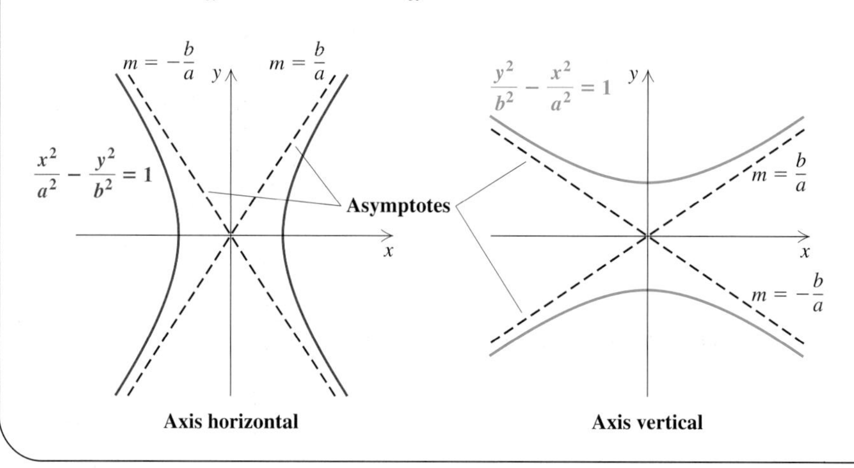

As a hyperbola gets farther away from the origin, it gets closer and closer to its asymptotes. The larger $|x|$ gets, the closer the graph gets to an asymptote. The asymptotes act to "constrain" the graph of a hyperbola. Parabolas are *not* constrained by any asymptotes.

In Section 10.2, we found that a and b can be used to determine the width and the length of an ellipse. For hyperbolas, a and b can be used to determine the base and the height of a rectangle that can be used as an aid in sketching asymptotes and locating vertices. This is illustrated in the following example.

Example 1 Graph: $\frac{x^2}{4} - \frac{y^2}{9} = 1.$

Solution Note that

$$\frac{x^2}{4} - \frac{y^2}{9} = \frac{x^2}{2^2} - \frac{y^2}{3^2}, \qquad \text{Identifying } a \text{ and } b$$

so $a = 2$ and $b = 3$. The asymptotes are thus

$$y = \frac{3}{2}x \quad \text{and} \quad y = -\frac{3}{2}x.$$

To help us sketch asymptotes and locate vertices, we use a and b—in this case, 2 and 3—to form the pairs $(-2, 3)$, $(2, 3)$, $(2, -3)$, and $(-2, -3)$. We plot these pairs and lightly sketch a rectangle. The asymptotes pass through the corners and, since this is a horizontal hyperbola, the vertices are where the rectangle intersects the x-axis. Finally, we draw the hyperbola, as shown on the next page.

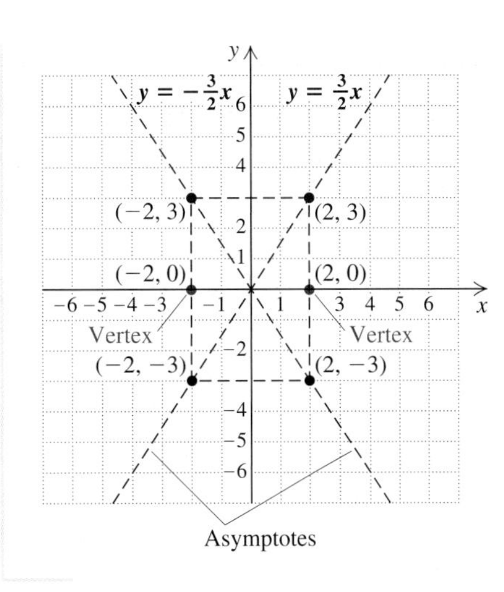

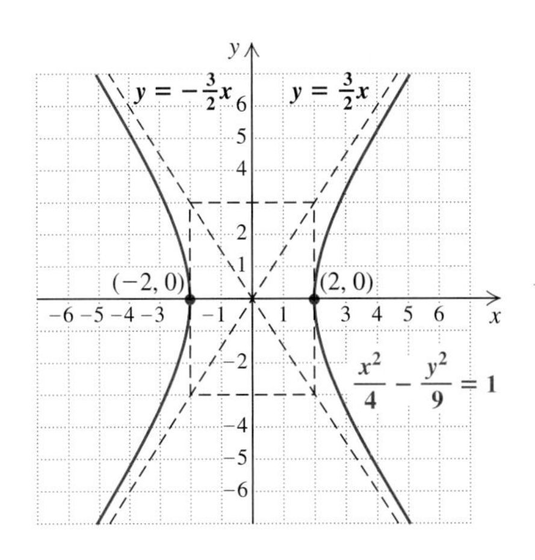

Example 2 Graph: $\frac{y^2}{36} - \frac{x^2}{4} = 1.$

Solution Note that

$$\frac{y^2}{36} - \frac{x^2}{4} = \frac{y^2}{6^2} - \frac{x^2}{2^2} = 1.$$

Whether the hyperbola is horizontal or vertical is determined by the nonnegative term. Here there is a y in this term, so the hyperbola is vertical.

Using ± 2 as x-coordinates and ± 6 as y-coordinates, we plot $(2, 6)$, $(2, -6)$, $(-2, 6)$, and $(-2, -6)$, and lightly sketch a rectangle through them. The asymptotes pass through the corners (see the figure on the left below). Since the hyperbola is vertical, its vertices are $(0, 6)$ and $(0, -6)$. Finally, we draw curves through the vertices toward the asymptotes, as shown below.

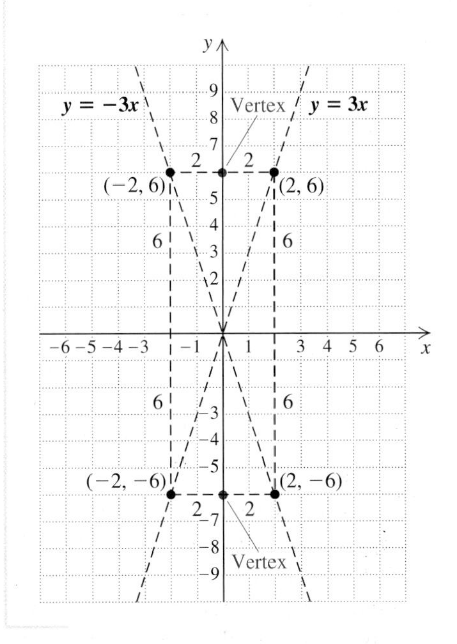

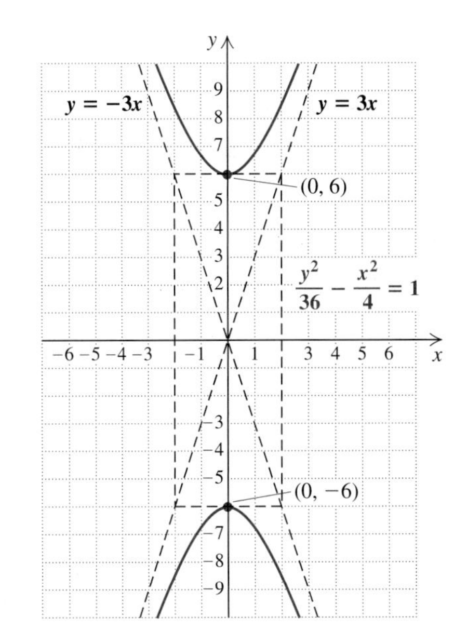

Hyperbolas (Nonstandard Form)

The equations for hyperbolas just examined are the standard ones, but there are other hyperbolas. We consider some of them.

Equation of a Hyperbola in Nonstandard Form

Hyperbolas having the x- and y-axes as asymptotes have equations as follows:

$$xy = c, \quad \text{where } c \text{ is a nonzero constant.}$$

Example 3 Graph: $xy = -8$.

Solution We first solve for y:

$$y = -\frac{8}{x}. \quad \text{Dividing both sides by } x. \text{ Note that } x \neq 0.$$

Next, we find some solutions, keeping the results in a table. Note that x cannot be 0 and that for large values of $|x|$, y will be close to 0. Thus the x- and y-axes serve as asymptotes. We plot the points and draw two curves.

x	y
2	−4
−2	4
4	−2
−4	2
1	−8
−1	8
8	−1
−8	1

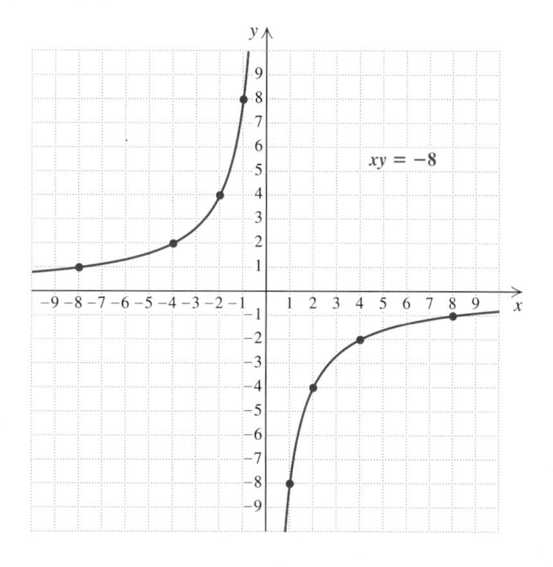

Hyperbolas have many applications. A jet breaking the sound barrier creates a sonic boom with a wave front the shape of a cone. The intersection of the cone with the ground is one branch of a hyperbola. Some comets travel in hyperbolic orbits, and a cross section of many lenses is hyperbolic in shape.

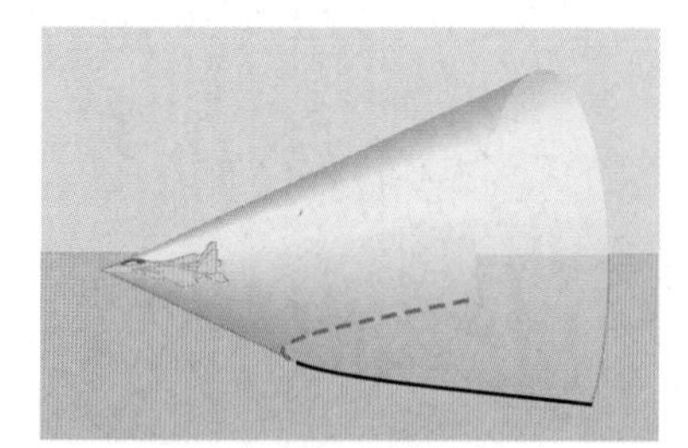

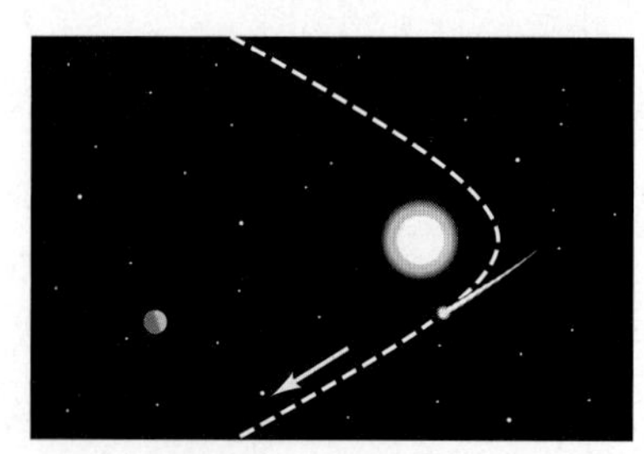

Classifying Graphs of Equations

CONNECTING THE CONCEPTS

Recall that the vertical-line test tells us that circles, ellipses, and hyperbolas in standard form do not represent functions. Of the graphs examined in this chapter, only vertical parabolas and hyperbolas similar to the one in Example 3 can represent functions. Because functions are so important, circles and ellipses generally appear in applications of a purely geometric nature, whereas vertical parabolas and nonstandard hyperbolas can be used in applications involving geometry or functions.

In Section 10.4, we return to the challenge of solving real-world problems that translate to a system of equations. There we will find that knowing the general shape of the graph of an equation can help us determine how many solutions, if any, may exist.

The procedure used to graph a hyperbola in standard form on a grapher is similar to that used to draw a circle or ellipse. Consider the graph of the hyperbola given by the equation.

$$\frac{x^2}{25} - \frac{y^2}{49} = 1.$$

The student should confirm that solving for y yields

$$y_1 = \frac{\sqrt{49x^2 - 1225}}{5}$$

$$= \frac{7}{5}\sqrt{x^2 - 25}$$

and

$$y_2 = \frac{-\sqrt{49x^2 - 1225}}{5}$$

$$= -\frac{7}{5}\sqrt{x^2 - 25},$$

or

$$y_2 = -y_1.$$

When the two pieces are drawn on the same squared window, the result is as shown. Note the problem that the grapher has at points where the graph is nearly vertical.

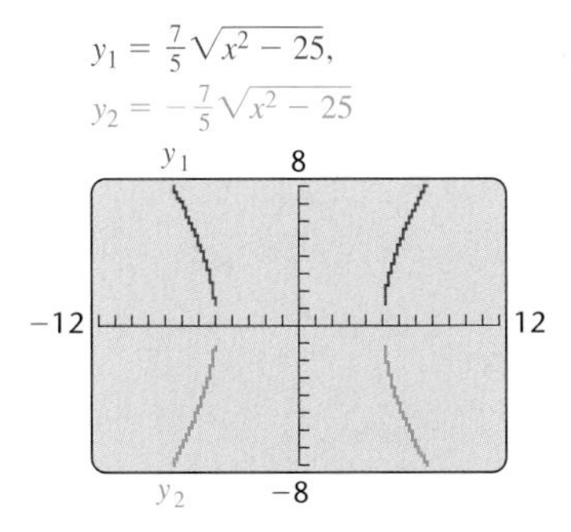

Use a grapher to draw the graph of each hyperbola. Use a squared window so that the shapes are not distorted.

1. $\frac{x^2}{16} - \frac{y^2}{60} = 1$

2. $16x^2 - 3y^2 = 64$

3. $\frac{y^2}{20} - \frac{x^2}{64} = 1$

4. $45y^2 - 9x^2 = 441$

We summarize the equations and the graphs of the conic sections studied. We resume the examples with Example 4 on p. 646.

Parabola

$$y = ax^2 + bx + c, \quad a > 0$$
$$= a(x - h)^2 + k$$

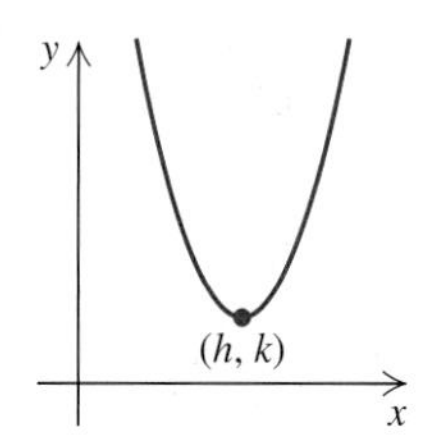

$$y = ax^2 + bx + c, \quad a < 0$$
$$= a(x - h)^2 + k$$

$$x = ay^2 + by + c, \quad a > 0$$
$$= a(y - k)^2 + h$$

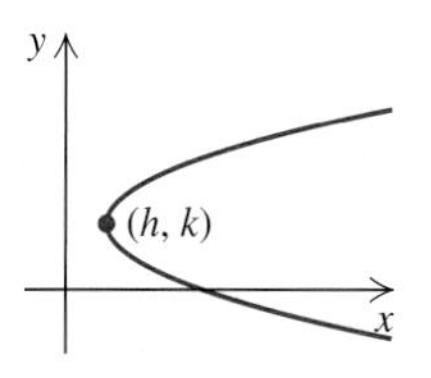

$$x = ay^2 + by + c, \quad a < 0$$
$$= a(y - k)^2 + h$$

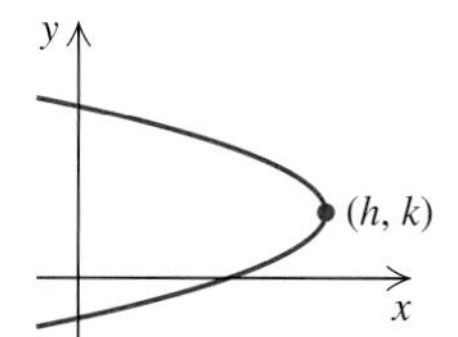

Circle

Center at the origin:

$$x^2 + y^2 = r^2$$

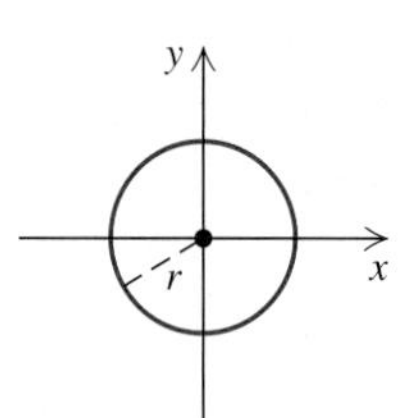

Center at (h, k):

$$(x - h)^2 + (y - k)^2 = r^2$$

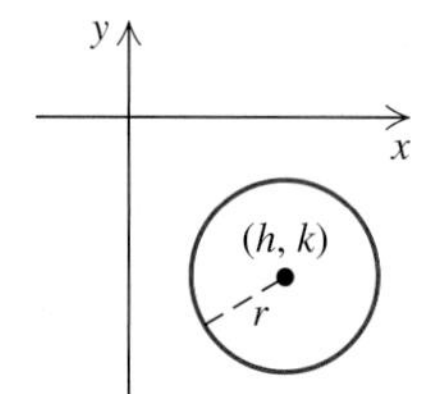

Hyperbola

Center at the origin:

$$\frac{x^2}{a^2} - \frac{y^2}{b^2} = 1$$

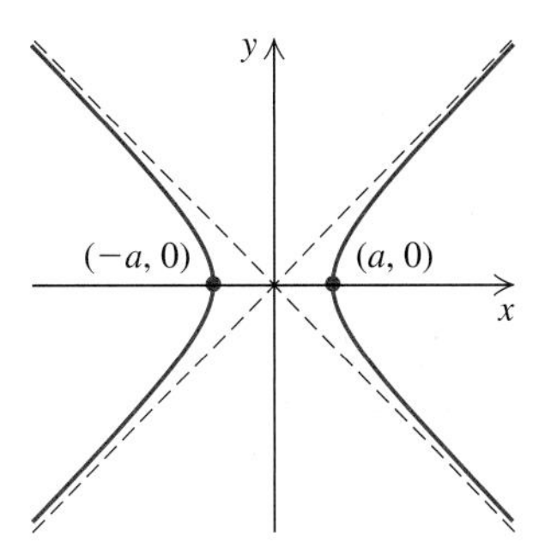

$$\frac{y^2}{b^2} - \frac{x^2}{a^2} = 1$$

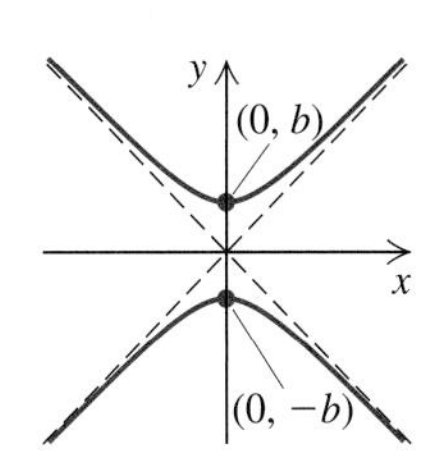

$xy = c, \quad c > 0$

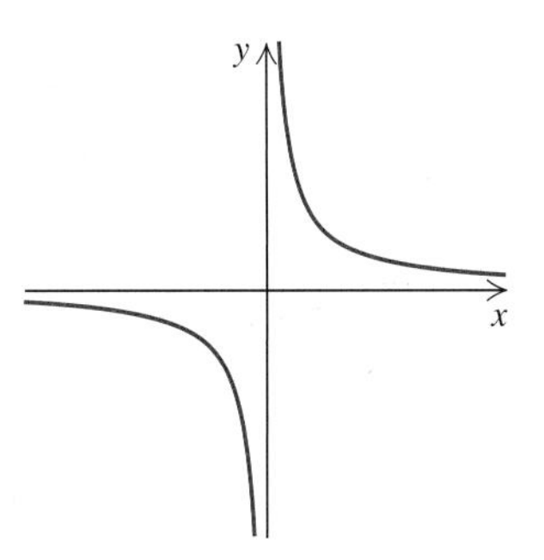

$xy = c, \quad c < 0$

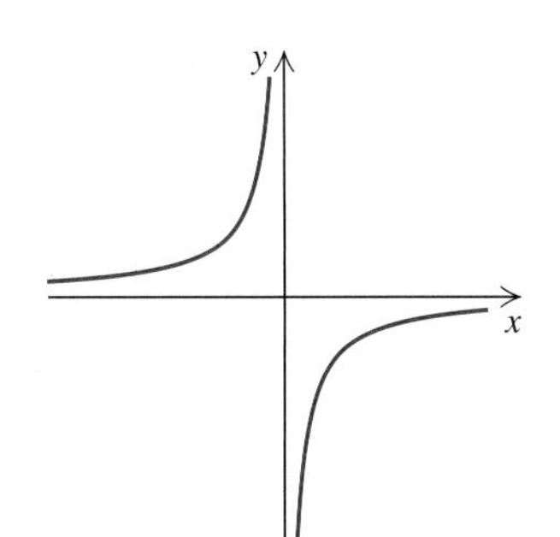

Center at (h, k)*:

$$\frac{(x-h)^2}{a^2} - \frac{(y-k)^2}{b^2} = 1$$

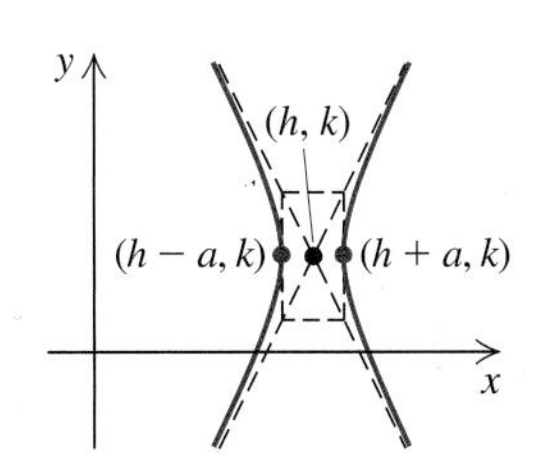

$$\frac{(y-k)^2}{b^2} - \frac{(x-h)^2}{a^2} = 1$$

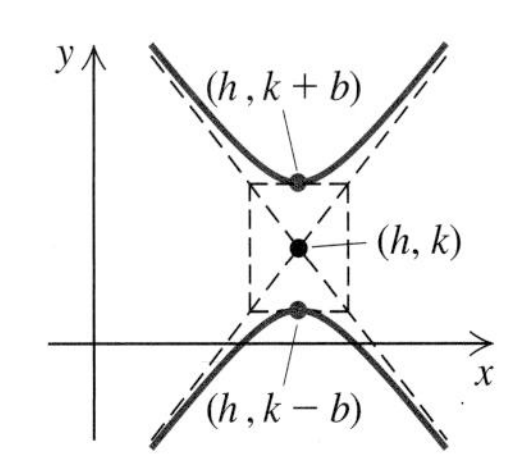

*See Exercises 51–56.

Ellipse

Center at the origin:

$$\frac{x^2}{a^2} + \frac{y^2}{b^2} = 1$$

Center at (h, k):

$$\frac{(x - h)^2}{a^2} + \frac{(y - k)^2}{b^2} = 1$$

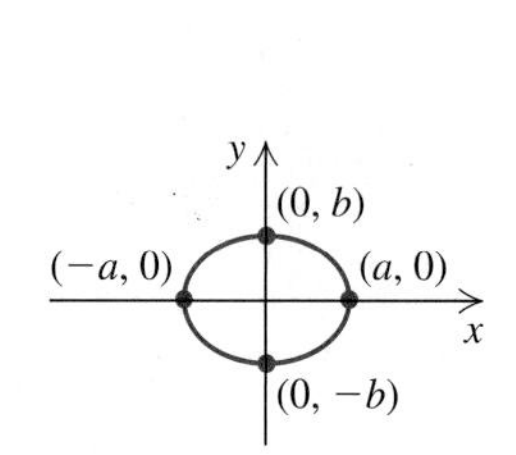

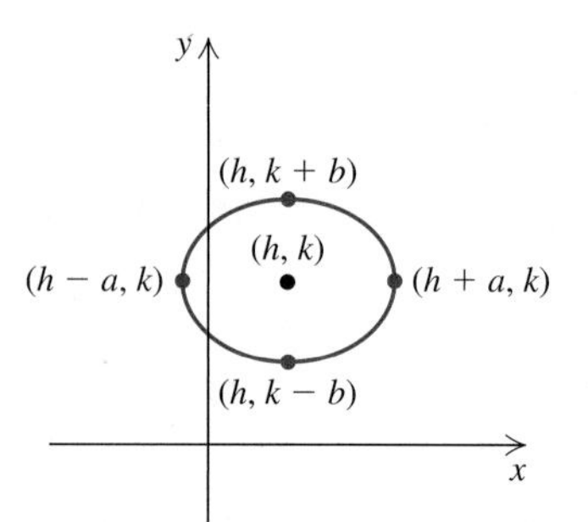

Algebraic manipulations may be needed to express an equation in one of the preceding forms.

Example 4 Classify the graph of each equation as a circle, an ellipse, a parabola, or a hyperbola.

a) $5x^2 = 20 - 5y^2$
b) $x + 3 + 8y = y^2$
c) $x^2 = y^2 + 4$
d) $x^2 = 16 - 4y^2$

Solution

a) We get the terms with variables on one side by adding $5y^2$:

$$5x^2 + 5y^2 = 20.$$

Since x and y are *both* squared, we do not have a parabola. The fact that the squared terms are *added* tells us that we do not have a hyperbola. Do we have a circle? To find out, we need to get $x^2 + y^2$ by itself. We can do that by factoring the 5 out of both terms on the left and then dividing by 5:

$$5(x^2 + y^2) = 20 \quad \text{Factoring out 5}$$
$$x^2 + y^2 = 4 \quad \text{Dividing both sides by 5}$$
$$x^2 + y^2 = 2^2. \quad \text{This is an equation for a circle.}$$

We can see that the graph is a circle with center at the origin and radius 2.

b) The equation $x + 3 + 8y = y^2$ has only one variable squared, so we solve for the other variable:

$$x = y^2 - 8y - 3. \quad \text{This is an equation for a parabola.}$$

The graph is a horizontal parabola that opens to the right.

c) In $x^2 = y^2 + 4$, both variables are squared, so the graph is not a parabola. We subtract y^2 on both sides and divide by 4 to obtain

$$\frac{x^2}{2^2} - \frac{y^2}{2^2} = 1. \quad \text{This is an equation for a hyperbola.}$$

The minus sign here indicates that the graph of this equation is a hyperbola. Because it is the x^2-term that is nonnegative, the hyperbola is horizontal.

d) In $x^2 = 16 - 4y^2$, both variables are squared, so the graph cannot be a parabola. We obtain the following equivalent equation:

$$x^2 + 4y^2 = 16.$$

If the coefficients of the terms were the same, we would have the graph of a circle, as in part (a), but they are not. Dividing both sides by 16 yields

$$\frac{x^2}{16} + \frac{y^2}{4} = 1. \qquad \text{This is an equation for an ellipse.}$$

The graph of this equation is a horizontal ellipse.

Exercise Set 10.3

FOR EXTRA HELP

Digital Video Tutor CD 8
Videotape 19

InterAct Math

Math Tutor Center

MathXL

MyMathLab.com

Graph each hyperbola. Label all vertices and sketch all asymptotes.

1. $\frac{y^2}{9} - \frac{x^2}{9} = 1$

2. $\frac{x^2}{16} - \frac{y^2}{16} = 1$

3. $\frac{x^2}{4} - \frac{y^2}{25} = 1$

4. $\frac{y^2}{16} - \frac{x^2}{9} = 1$

5. $\frac{y^2}{36} - \frac{x^2}{9} = 1$

6. $\frac{x^2}{25} - \frac{y^2}{36} = 1$

7. $y^2 - x^2 = 25$

8. $x^2 - y^2 = 4$

9. $25x^2 - 16y^2 = 400$

10. $4y^2 - 9x^2 = 36$

Graph.

11. $xy = -6$

12. $xy = 6$

13. $xy = 4$

14. $xy = -9$

15. $xy = -2$

16. $xy = -1$

17. $xy = 1$

18. $xy = 2$

Classify each of the following as the equation of a circle, an ellipse, a parabola, or a hyperbola.

19. $x^2 + y^2 - 10x + 8y - 40 = 0$

20. $y + 7 = 3x^2$

21. $9x^2 + 4y^2 - 36 = 0$

22. $1 + 3y = 2y^2 - x$

23. $4x^2 - 9y^2 - 72 = 0$

24. $y^2 + x^2 = 8$

25. $x^2 + y^2 = 2x + 4y + 4$

26. $2y + 13 + x^2 = 8x - y^2$

27. $4x^2 = 64 - y^2$

28. $y = \frac{2}{x}$

29. $x - \frac{3}{y} = 0$

30. $x - 4 = y^2 - 3y$

31. $y + 6x = x^2 + 5$

32. $x^2 = 16 + y^2$

33. $9y^2 = 36 + 4x^2$

34. $3x^2 + 5y^2 + x^2 = y^2 + 49$

35. $3x^2 + y^2 - x = 2x^2 - 9x + 10y + 40$

36. $4y^2 + 20x^2 + 1 = 8y - 5x^2$

37. $16x^2 + 5y^2 - 12x^2 + 8y^2 - 3x + 4y = 568$

38. $56x^2 - 17y^2 = 234 - 13x^2 - 38y^2$

39. What does graphing hyperbolas have in common with graphing ellipses?

40. Is it possible for a hyperbola to represent the graph of a function? Why or why not?

SKILL MAINTENANCE

Solve each system.

41. $5x + 6y = -12,$
$3x + 9y = 15$

42. $2x + 6y = -6,$
$3x + 5y = 7$

Solve.

43. $y^2 - 3 = 6$

44. $x^2 + 3 = 4$

45. The price of a radio, including 5% sales tax, is \$36.75. Find the price of the radio before the tax was added.

46. A basketball team increases its score by 7 points in each of three consecutive games. If the team scored a total of 228 points in all three games, what was its score in the first game?

SYNTHESIS

47. What is it in the equation of a hyperbola that controls how wide open the branches are? Explain your reasoning.

48. If, in

$$\frac{x^2}{a^2} - \frac{y^2}{b^2} = 1,$$

$a = b$, what are the asymptotes of the graph? Why?

Find an equation of a hyperbola satisfying the given conditions.

49. Having intercepts $(0, 6)$ and $(0, -6)$ and asymptotes $y = 3x$ and $y = -3x$

50. Having intercepts $(8, 0)$ and $(-8, 0)$ and asymptotes $y = 4x$ and $y = -4x$

The standard equations for horizontal or vertical hyperbolas centered at (h, k) are as follows:

$$\frac{(x-h)^2}{a^2} - \frac{(y-k)^2}{b^2} = 1$$

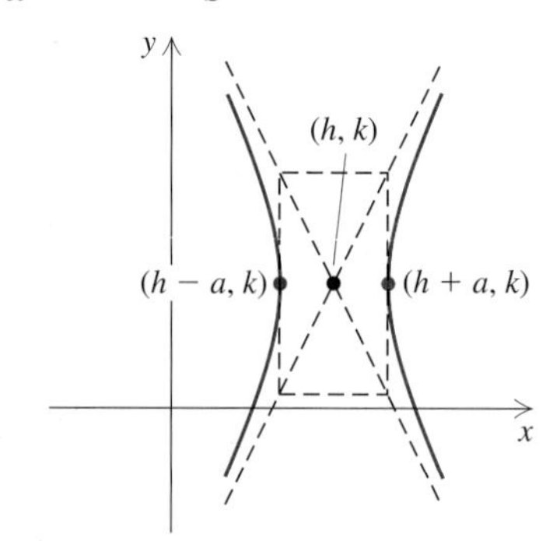

$$\frac{(y-k)^2}{b^2} - \frac{(x-h)^2}{a^2} = 1$$

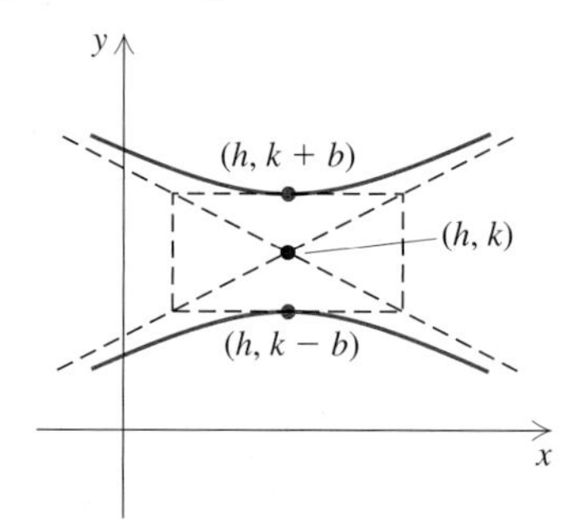

The vertices are as labeled and the asymptotes are

$$y - k = \frac{b}{a}(x - h) \quad \textit{and} \quad y - k = -\frac{b}{a}(x - h).$$

For each of the following equations of hyperbolas, complete the square, if necessary, and write in standard form. Find the center, the vertices, and the asymptotes. Then graph the hyperbola.

51. $\dfrac{(x-5)^2}{36} - \dfrac{(y-2)^2}{25} = 1$

52. $\dfrac{(x-2)^2}{9} - \dfrac{(y-1)^2}{4} = 1$

53. $8(y + 3)^2 - 2(x - 4)^2 = 32$

54. $25(x - 4)^2 - 4(y + 5)^2 = 100$

55. $4x^2 - y^2 + 24x + 4y + 28 = 0$

56. $4y^2 - 25x^2 - 8y - 100x - 196 = 0$

57. Use a grapher to check your answers to Exercises 5, 17, 23, and 51.

Nonlinear Systems of Equations

10.4

Systems Involving One Nonlinear Equation • Systems of Two Nonlinear Equations • Problem Solving

The equations appearing in systems of two equations have thus far all been linear. We now consider systems of two equations in which at least one equation is nonlinear.

Systems Involving One Nonlinear Equation

Suppose that a system consists of an equation of a circle and an equation of a line. In what ways can the circle and the line intersect? The figures below represent three ways in which the situation can occur. We see that such a system will have 0, 1, or 2 real solutions.

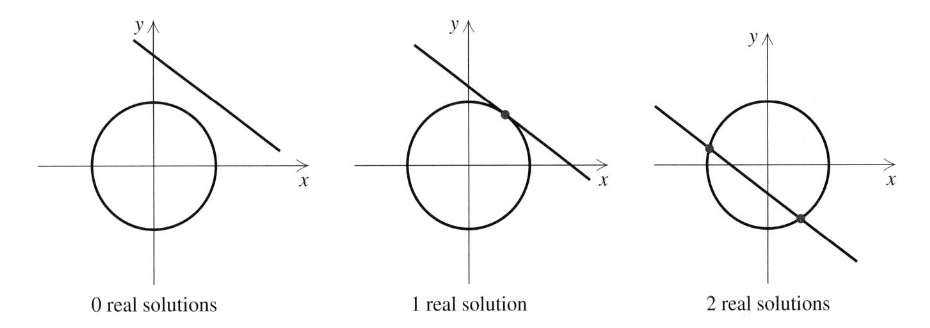

Recall that, in addition to graphing, we used both *elimination* and *substitution* to solve systems of linear equations. When solving systems in which one equation is of first degree and one is of second degree, it is preferable to use the *substitution* method.

Example 1 Solve the system

$$x^2 + y^2 = 25, \quad (1) \quad \text{(The graph is a circle.)}$$
$$3x - 4y = 0. \quad (2) \quad \text{(The graph is a line.)}$$

Solution First, we solve the linear equation, (2), for x:

$$x = \tfrac{4}{3}y. \quad (3) \quad \text{We could have solved for } y \text{ instead.}$$

Then we substitute $\frac{4}{3}y$ for x in equation (1) and solve for y:

$$\left(\tfrac{4}{3}y\right)^2 + y^2 = 25$$
$$\tfrac{16}{9}y^2 + y^2 = 25$$
$$\tfrac{25}{9}y^2 = 25$$
$$y^2 = 9 \quad \text{Multiplying both sides by } \tfrac{9}{25}$$
$$y = \pm 3. \quad \text{Using the principle of square roots}$$

Now we substitute these numbers for y in equation (3) and solve for x:

$$\text{for } y = 3, \quad x = \tfrac{4}{3}(3) = 4;$$

$$\text{for } y = -3, \quad x = \tfrac{4}{3}(-3) = -4.$$

Check: For $(4, 3)$:

$$\begin{array}{c|l} x^2 + y^2 & = 25 \\ \hline 4^2 + 3^2 & ?\ 25 \\ 16 + 9 & \\ 25 & 25 \quad \text{TRUE} \end{array} \qquad \begin{array}{c|l} 3x - 4y & = 0 \\ \hline 3(4) - 4(3) & ?\ 0 \\ 12 - 12 & \\ 0 & 0 \quad \text{TRUE} \end{array}$$

It is left to the student to confirm that $(-4, -3)$ also checks in both equations.

The pairs $(4, 3)$ and $(-4, -3)$ check, so they are solutions. We can see the solutions in the graph. Intersections occur at $(4, 3)$ and $(-4, -3)$.

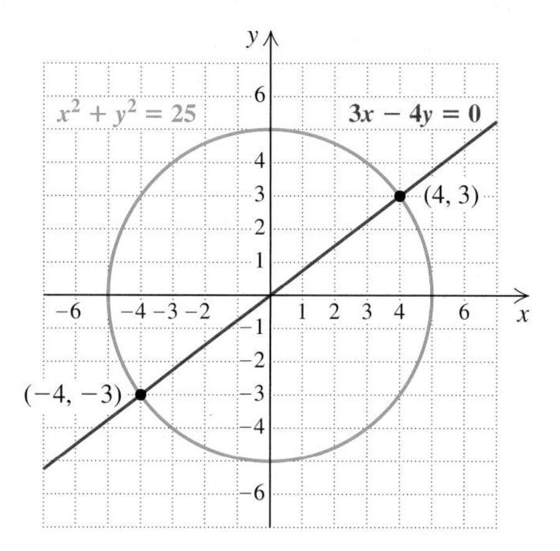

Although we may not know what the graph of each equation in a system looks like, the algebraic approach of Example 1 can still be used.

Example 2 Solve the system

$$y + 3 = 2x, \qquad (1)$$

$$x^2 + 2xy = -1. \qquad (2)$$

Solution First, we solve the linear equation (1) for y:

$$y = 2x - 3. \quad (3)$$

Then we substitute $2x - 3$ for y in equation (2) and solve for x:

$$x^2 + 2x(2x - 3) = -1$$

$$x^2 + 4x^2 - 6x = -1$$

$$5x^2 - 6x + 1 = 0$$

$$(5x - 1)(x - 1) = 0 \qquad \text{Factoring}$$

$$5x - 1 = 0 \quad \text{or} \quad x - 1 = 0 \qquad \text{Using the principle of zero products}$$

$$x = \tfrac{1}{5} \quad \text{or} \quad x = 1.$$

Now we substitute these numbers for x in equation (3) and solve for y:

$$\text{for } x = \tfrac{1}{5}, \quad y = 2\left(\tfrac{1}{5}\right) - 3 = -\tfrac{13}{5};$$
$$\text{for } x = 1, \quad y = 2(1) - 3 = -1.$$

You can confirm that $\left(\frac{1}{5}, -\frac{13}{5}\right)$ and $(1, -1)$ check, so they are both solutions.

Example 3

Systems of equations offer a fine opportunity to use the INTERSECT feature of a grapher, although most graphers will restrict solutions to real numbers.

To solve Example 2,

$$y + 3 = 2x,$$
$$x^2 + 2xy = -1,$$

we solve each equation for y and then graph:

$$\left.\begin{aligned} y_1 &= 2x - 3, \\ y_2 &= \frac{-1 - x^2}{2x}. \end{aligned}\right\} \quad \text{Note that } x, y \neq 0.$$

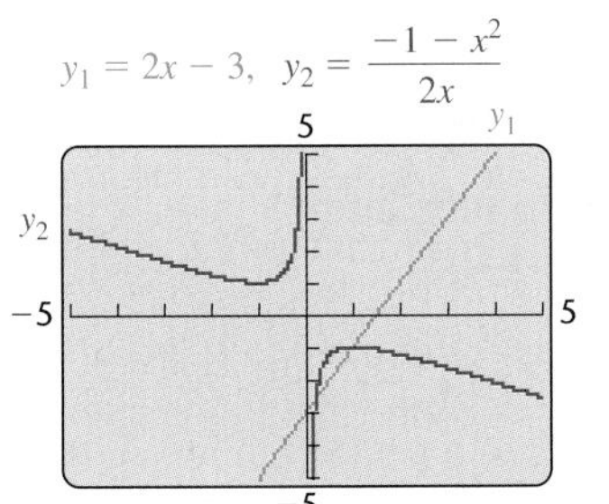

Using INTERSECT, we find the solutions to be $(0.2, -2.6)$ and $(1, -1)$.

Use a grapher to solve each system. Round all values to two decimal places.

1. $4xy - 7 = 0,$
   $x - 3y - 2 = 0$
2. $x^2 + y^2 = 14,$
   $16x + 7y^2 = 0$

Solve the system

$$x + y = 5, \quad (1) \quad \text{(The graph is a line.)}$$
$$y = 3 - x^2. \quad (2) \quad \text{(The graph is a parabola.)}$$

Solution We substitute $3 - x^2$ for y in the first equation:

$$x + 3 - x^2 = 5$$
$$-x^2 + x - 2 = 0 \quad \text{Adding } -5 \text{ to both sides and rearranging}$$
$$x^2 - x + 2 = 0. \quad \text{Multiplying both sides by } -1$$

Since $x^2 - x + 2$ does not factor, we need the quadratic formula:

$$x = \frac{-b \pm \sqrt{b^2 - 4ac}}{2a}$$
$$= \frac{-(-1) \pm \sqrt{(-1)^2 - 4 \cdot 1 \cdot 2}}{2(1)} \quad \text{Substituting}$$
$$= \frac{1 \pm \sqrt{1 - 8}}{2} = \frac{1 \pm \sqrt{-7}}{2} = \frac{1}{2} \pm \frac{\sqrt{7}}{2}i.$$

Solving equation (1) for y gives us $y = 5 - x$. Substituting values for x gives

$$y = 5 - \left(\frac{1}{2} + \frac{\sqrt{7}}{2}i\right) = \frac{9}{2} - \frac{\sqrt{7}}{2}i \quad \text{and}$$
$$y = 5 - \left(\frac{1}{2} - \frac{\sqrt{7}}{2}i\right) = \frac{9}{2} + \frac{\sqrt{7}}{2}i.$$

The solutions are

$$\left(\frac{1}{2} + \frac{\sqrt{7}}{2}i, \frac{9}{2} - \frac{\sqrt{7}}{2}i\right) \quad \text{and} \quad \left(\frac{1}{2} - \frac{\sqrt{7}}{2}i, \frac{9}{2} + \frac{\sqrt{7}}{2}i\right).$$

There are no real-number solutions. Note in the figure at right that the graphs do not intersect. Getting only nonreal solutions tells us that the graphs do not intersect.

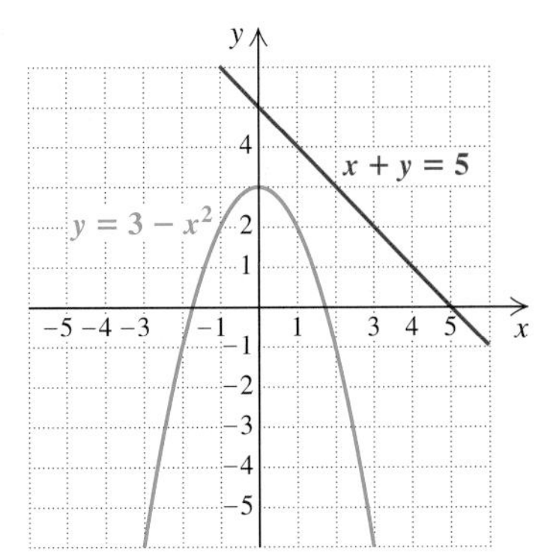

Systems of Two Nonlinear Equations

We now consider systems of two second-degree equations. Graphs of such systems can involve any two conic sections. The following figure shows some ways in which a circle and a hyperbola can intersect.

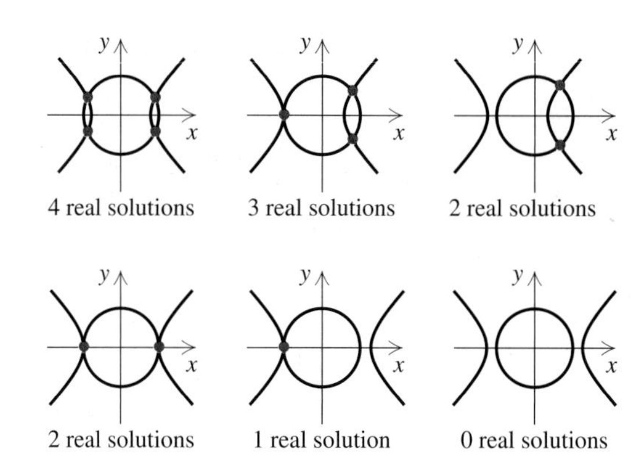

To solve systems of two second-degree equations, we can use either substitution or elimination. The elimination method is generally better when both equations are of the form $Ax^2 + By^2 = C$. Then we can eliminate an x^2- or y^2-term in a manner similar to the procedure used in Chapter 3.

Example 4 Solve the system

$$2x^2 + 5y^2 = 22, \quad (1)$$
$$3x^2 - y^2 = -1. \quad (2)$$

Solution Here we multiply equation (2) by 5 and then add:

$$2x^2 + 5y^2 = 22$$
$$15x^2 - 5y^2 = -5 \quad \text{Multiplying both sides of equation (2) by 5}$$
$$17x^2 = 17 \quad \text{Adding}$$
$$x^2 = 1$$
$$x = \pm 1.$$

There is no x-term, and whether x is -1 or 1, we have $x^2 = 1$. Thus we can simultaneously substitute 1 and -1 for x in equation (2), and we have

$$\left.\begin{aligned} 3 \cdot (\pm 1)^2 - y^2 &= -1 \\ 3 - y^2 &= -1 \\ -y^2 &= -4 \end{aligned}\right\} \quad \text{Since } (-1)^2 = 1^2 \text{, we can evaluate for } x = -1 \text{ and } x = 1 \text{ simultaneously.}$$
$$y^2 = 4 \quad \text{or} \quad y = \pm 2.$$

Thus, if $x = 1$, then $y = 2$ or $y = -2$; and if $x = -1$, then $y = 2$ or $y = -2$. The four possible solutions are $(1, 2)$, $(1, -2)$, $(-1, 2)$, and $(-1, -2)$.

Check: Since $(2)^2 = (-2)^2$ and $(1)^2 = (-1)^2$, we can check all four pairs at once.

$$\begin{array}{r|l} \multicolumn{2}{c}{2x^2 + 5y^2 = 22} \\ \hline 2(\pm 1)^2 + 5(\pm 2)^2 & ?\ 22 \\ 2 + 20 & \\ 22 & 22 \quad \text{TRUE} \end{array} \qquad \begin{array}{r|l} \multicolumn{2}{c}{3x^2 - y^2 = -1} \\ \hline 3(\pm 1)^2 - (\pm 2)^2 & ?\ -1 \\ 3 - 4 & \\ -1 & -1 \quad \text{TRUE} \end{array}$$

The solutions are $(1, 2)$, $(1, -2)$, $(-1, 2)$, and $(-1, -2)$.

When a product of variables is in one equation and the other equation is of the form $Ax^2 + By^2 = C$, we often solve for a variable in the equation with the product and then use substitution.

Example 5

Solve the system

$$x^2 + 4y^2 = 20, \quad (1)$$
$$xy = 4. \quad (2)$$

Solution First, we solve equation (2) for y:

$$y = \frac{4}{x}. \qquad \text{Dividing both sides by } x. \text{ Note that } x \neq 0.$$

Then we substitute $4/x$ for y in equation (1) and solve for x:

$$x^2 + 4\left(\frac{4}{x}\right)^2 = 20$$

$$x^2 + \frac{64}{x^2} = 20$$

$$x^4 + 64 = 20x^2 \qquad \text{Multiplying by } x^2$$

$$x^4 - 20x^2 + 64 = 0 \qquad \text{Obtaining standard form. This equation is reducible to quadratic.}$$

$$(x^2 - 4)(x^2 - 16) = 0 \qquad \text{Factoring. If you prefer, let } u = x^2 \text{ and substitute.}$$

$$(x - 2)(x + 2)(x - 4)(x + 4) = 0 \qquad \text{Factoring}$$

$$x = 2 \quad or \quad x = -2 \quad or \quad x = 4 \quad or \quad x = -4. \qquad \text{Using the principle of zero products}$$

Since $y = 4/x$, for $x = 2$, we have $y = 4/2$, or 2. Thus, $(2, 2)$ is a solution. Similarly, $(-2, -2)$, $(4, 1)$, and $(-4, -1)$ are solutions. You can show that all four pairs check.

Before Example 4 can be checked with a grapher, each equation must first be solved for y. When this has been done, we have $y_1 = \sqrt{(22 - 2x^2)/5}$ and $y_2 = -\sqrt{(22 - 2x^2)/5}$ for equation (1) and $y_3 = \sqrt{3x^2 + 1}$ and $y_4 = -\sqrt{3x^2 + 1}$ for equation (2). The graph verifies the solutions found algebraically.

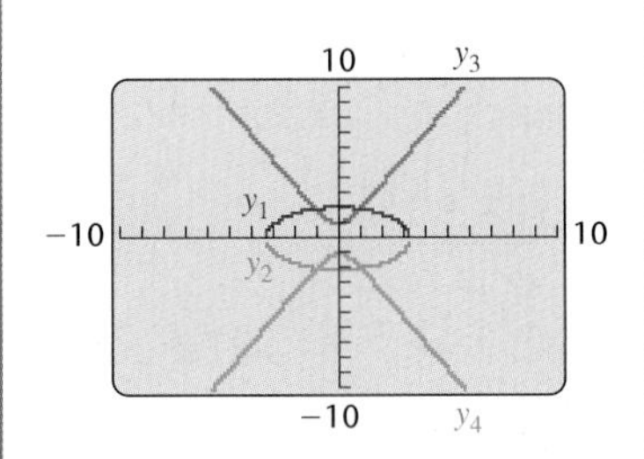

1. Use a grapher to provide a visual check for Example 5.

Problem Solving

We now consider applications that can be modeled by a system of equations in which at least one equation is not linear.

Example 6 ***Architecture.*** For a college gymnasium, an architect wants to lay out a rectangular piece of land that has a perimeter of 204 m and an area of 2565 m^2. Find the dimensions of the piece of land.

Solution

1. **Familiarize.** We draw and label a sketch, letting l = the length and w = the width, both in meters.

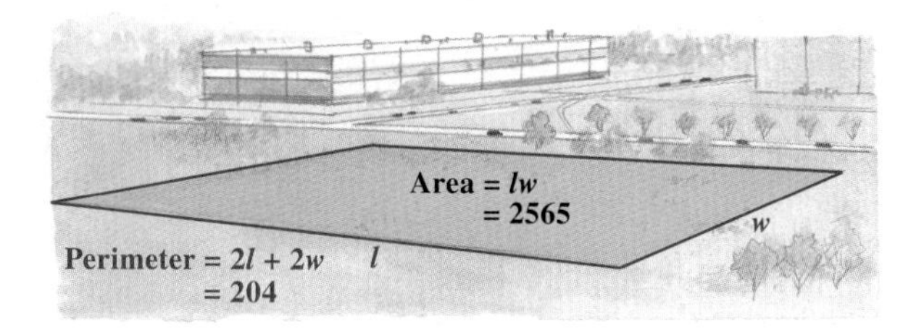

2. **Translate.** We then have the following translation:

$$\text{Perimeter: } 2w + 2l = 204;$$
$$\text{Area: } lw = 2565.$$

3. **Carry out.** We solve the system

$$2w + 2l = 204,$$
$$lw = 2565.$$

Solving the second equation for l gives us $l = 2565/w$. Then we substitute $2565/w$ for l in the first equation and solve for w:

$$2w + 2\left(\frac{2565}{w}\right) = 204$$

$$2w^2 + 2(2565) = 204w \quad \text{Multiplying both sides by } w$$

$$2w^2 - 204w + 2(2565) = 0 \quad \text{Standard form}$$

$$w^2 - 102w + 2565 = 0 \quad \text{Multiplying by } \tfrac{1}{2}$$

Factoring could be used instead of the quadratic formula, but the numbers are quite large.

$$w = \frac{-(-102) \pm \sqrt{(-102)^2 - 4 \cdot 1 \cdot 2565}}{2 \cdot 1}$$

$$w = \frac{102 \pm \sqrt{144}}{2} = \frac{102 \pm 12}{2}$$

$$w = 57 \quad or \quad w = 45.$$

If $w = 57$, then $l = 2565/w = 2565/57 = 45$. If $w = 45$, then $l = 2565/w = 2565/45 = 57$. Since length is usually considered to be longer than width, we have the solution $l = 57$ and $w = 45$, or $(57, 45)$.

4. **Check.** If $l = 57$ and $w = 45$, the perimeter is $2 \cdot 57 + 2 \cdot 45$, or 204. The area is $57 \cdot 45$, or 2565. The numbers check.

5. **State.** The length is 57 m and the width is 45 m.

Example 7

HDTV dimensions. High-definition television (HDTV) offers greater clarity than conventional television. The Kaplans' new HDTV screen has an area of 1296 in^2 and has a $\sqrt{3033}$-in. (about 55-in.) diagonal screen. Find the width and the length of the screen.

Solution

1. **Familiarize.** We make a drawing and label it. Note that there is a right triangle in the figure. We let l = the length and w = the width, both in inches.

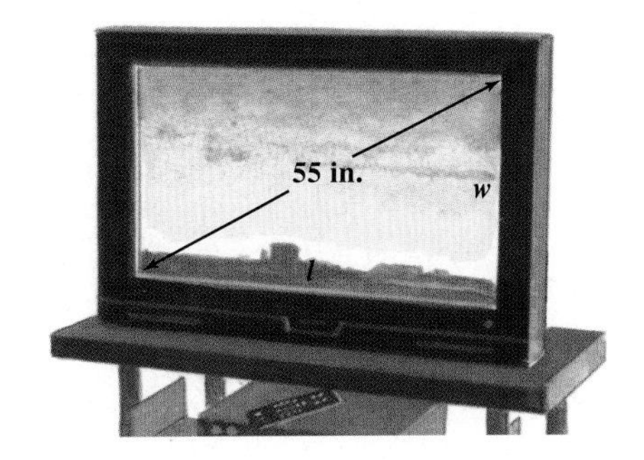

2. **Translate.** We translate to a system of equations:

$$l^2 + w^2 = \sqrt{3033}^2 \qquad \text{Using the Pythagorean theorem}$$
$$lw = 1296. \qquad \text{Using the formula for the area of a rectangle}$$

3. **Carry out.** We solve the system

$$\left.\begin{aligned} l^2 + w^2 &= 3033, \\ lw &= 1296 \end{aligned}\right\} \quad \text{You should complete the solution of this system.}$$

to get $(48, 27)$, $(27, 48)$, $(-48, -27)$, and $(-27, -48)$.

4. **Check.** Measurements cannot be negative and length is usually greater than width, so we check only $(48, 27)$. In the right triangle, $48^2 + 27^2 = 2304 + 729 = 3033$ or $\sqrt{3033}^2$. The area is $48 \cdot 27 = 1296$, so our answer checks.

5. **State.** The length is 48 in. and the width is 27 in.

Exercise Set 10.4

FOR EXTRA HELP

Digital Video Tutor CD 8
Videotape 19

InterAct Math

Math Tutor Center

MathXL

MyMathLab.com

Solve. Remember that graphs can be used to confirm all real solutions.

1. $x^2 + y^2 = 25,$
$y - x = 1$

2. $x^2 + y^2 = 100,$
$y - x = 2$

3. $9x^2 + 4y^2 = 36,$
$3x + 2y = 6$

4. $4x^2 + 9y^2 = 36,$
$3y + 2x = 6$

5. $y = x^2,$
$3x = y + 2$

6. $y^2 = x + 3,$
$2y = x + 4$

7. $2y^2 + xy + x^2 = 7,$
$x - 2y = 5$

8. $x^2 - xy + 3y^2 = 27,$
$x - y = 2$

9. $x^2 - y^2 = 16,$
$x - 2y = 1$

10. $x^2 + 4y^2 = 25,$
$x + 2y = 7$

11. $m^2 + 3n^2 = 10,$
$m - n = 2$

12. $x^2 - xy + 3y^2 = 5,$
$x - y = 2$

13. $2y^2 + xy = 5,$
$4y + x = 7$

14. $3x + y = 7,$
$4x^2 + 5y = 24$

15. $p + q = -6,$
$pq = -7$

16. $a + b = 7,$
$ab = 4$

17. $4x^2 + 9y^2 = 36,$
$x + 3y = 3$

18. $2a + b = 1,$
$b = 4 - a^2$

19. $xy = 4,$
$x + y = 5$

20. $a^2 + b^2 = 89,$
$a - b = 3$

Aha! **21.** $y = x^2,$
$x = y^2$

22. $x^2 + y^2 = 25,$
$y^2 = x + 5$

23. $x^2 + y^2 = 9,$
$x^2 - y^2 = 9$

24. $y^2 - 4x^2 = 4,$
$4x^2 + y^2 = 4$

25. $x^2 + y^2 = 25,$
$xy = 12$

26. $x^2 - y^2 = 16,$
$x + y^2 = 4$

27. $x^2 + y^2 = 4,$
$9x^2 + 16y^2 = 144$

28. $x^2 + y^2 = 9,$
$25x^2 + 16y^2 = 400$

29. $x^2 + y^2 = 16,$
$y^2 - 2x^2 = 10$

30. $x^2 + y^2 = 14,$
$x^2 - y^2 = 4$

31. $x^2 + y^2 = 5,$
$xy = 2$

32. $x^2 + y^2 = 20,$
$xy = 8$

33. $x^2 + y^2 = 13,$
$xy = 6$

34. $x^2 + 4y^2 = 20,$
$xy = 4$

35. $3xy + x^2 = 34,$
$2xy - 3x^2 = 8$

36. $2xy + 3y^2 = 7,$
$3xy - 2y^2 = 4$

37. $xy - y^2 = 2,$
$2xy - 3y^2 = 0$

38. $4a^2 - 25b^2 = 0,$
$2a^2 - 10b^2 = 3b + 4$

39. $x^2 - y = 5,$
$x^2 + y^2 = 25$

40. $ab - b^2 = -4,$
$ab - 2b^2 = -6$

Solve.

41. *Computer parts.* Dataport Electronics needs a rectangular memory board that has a perimeter of 28 cm and a diagonal of length 10 cm. What should the dimensions of the board be?

42. *Geometry.* A rectangle has an area of 2 yd^2 and a perimeter of 6 yd. Find its dimensions.

43. *Geometry.* A rectangle has an area of 20 in^2 and a perimeter of 18 in. Find its dimensions.

44. *Tile design.* The New World tile company wants to make a new rectangular tile that has a perimeter of 6 in. and a diagonal of length $\sqrt{5}$ in. What should the dimensions of the tile be?

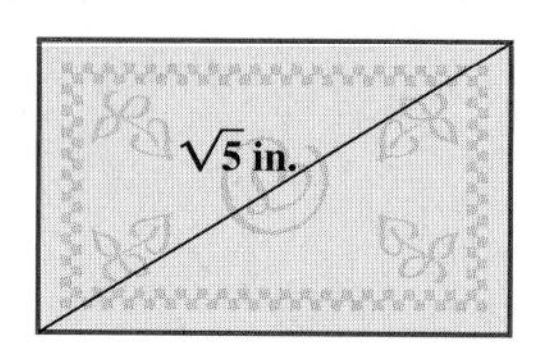

45. *Design of a van.* The cargo area of a delivery van must be 60 ft^2, and the length of a diagonal must accommodate a 13-ft board. Find the dimensions of the cargo area.

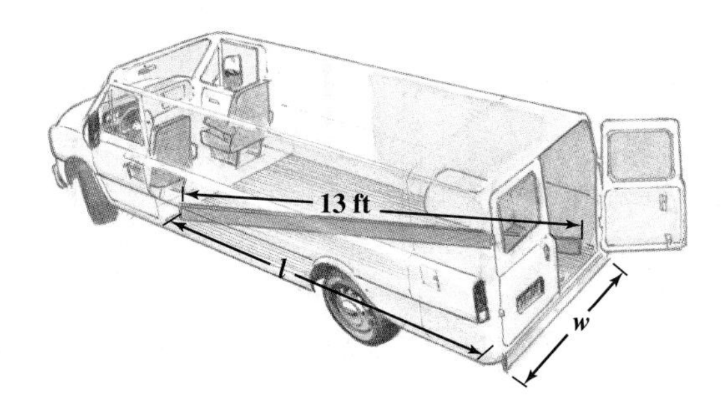

46. *Dimensions of a rug.* The diagonal of a Persian rug is 25 ft. The area of the rug is 300 ft^2. Find the length and the width of the rug.

47. The product of the lengths of the legs of a right triangle is 156. The hypotenuse has length $\sqrt{313}$. Find the lengths of the legs.

48. The product of two numbers is 60. The sum of their squares is 136. Find the numbers.

49. *Investments.* A certain amount of money saved for 1 yr at a certain interest rate yielded \$225 in interest. If \$750 more had been invested and the rate had been 1% less, the interest would have been the same. Find the principal and the rate.

50. *Garden design.* A garden contains two square peanut beds. Find the length of each bed if the sum of their areas is 832 ft^2 and the difference of their areas is 320 ft^2.

51. The area of a rectangle is $\sqrt{3}$ m^2, and the length of a diagonal is 2 m. Find the dimensions.

52. The area of a rectangle is $\sqrt{2}$ m^2, and the length of a diagonal is $\sqrt{3}$ m. Find the dimensions.

53. How can an understanding of conic sections be helpful when a system of nonlinear equations is being solved algebraically?

54. Suppose a system of equations is comprised of one linear and one nonlinear equation. Is it possible for such a system to have three solutions? Why or why not?

SKILL MAINTENANCE

Simplify.

55. $(-1)^9(-2)^4$

56. $(-1)^{10}(-2)^5$

Evaluate each of the following.

57. $\dfrac{(-1)^k}{k-5}$, for $k = 6$

58. $\dfrac{(-1)^k}{k-5}$, for $k = 9$

59. $\dfrac{n}{2}(3 + n)$, for $n = 8$

60. $\dfrac{7(1 - r^2)}{1 - r}$, for $r = 3$

SYNTHESIS

61. Write a problem that translates to a system of two equations. Design the problem so that at least one equation is nonlinear and so that no real solution exists.

62. Write a problem for a classmate to solve. Devise the problem so that a system of two nonlinear equations with exactly one real solution is solved.

63. A piece of wire 100 cm long is to be cut into two pieces and those pieces are each to be bent to make a square. The area of one square is to be 144 cm^2 greater than that of the other. How should the wire be cut?

64. Find the equation of a circle that passes through $(-2, 3)$ and $(-4, 1)$ and whose center is on the line $5x + 8y = -2$.

65. Find the equation of an ellipse centered at the origin that passes through the points $(2, -3)$ and $(1, \sqrt{13})$.

Solve.

66. $p^2 + q^2 = 13$,

$\dfrac{1}{pq} = -\dfrac{1}{6}$

67. $a + b = \dfrac{5}{6}$,

$\dfrac{a}{b} + \dfrac{b}{a} = \dfrac{13}{6}$

68. *Box design.* Four squares with sides 5 in. long are cut from the corners of a rectangular metal sheet that has an area of 340 in^2. The edges are bent up to form an open box with a volume of 350 in^3. Find the dimensions of the box.

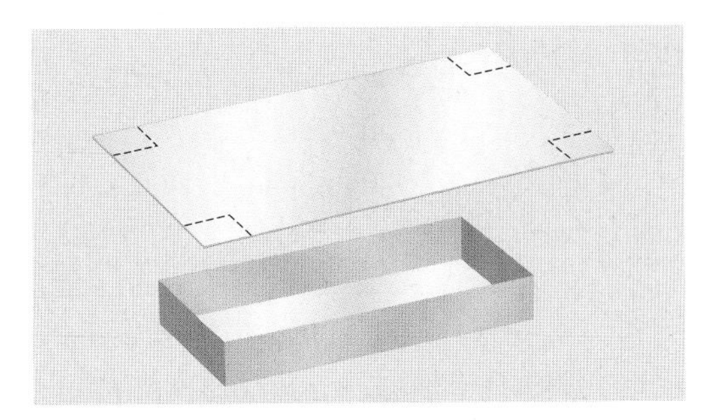

69. *Computer screens.* The ratio of the length to the height of the screen on a computer monitor is 4 to 3. An IBM Thinkpad laptop has a 31-cm diagonal screen. Find the dimensions of the screen.

70. *HDTV screens.* The ratio of the length to the height of an HDTV screen (see Example 7) is 16 to 9. The Remton Lounge has an HDTV screen with a $\sqrt{4901}$-in. (about 70-in.) diagonal screen. Find the dimensions of the screen.

71. *Railing sales.* Fireside Castings finds that the total revenue R from the sale of x units of railing is given by

$$R = 100x + x^2.$$

Fireside also finds that the total cost C of producing x units of the same product is given by

$$C = 80x + 1500.$$

A break-even point is a value of x for which total revenue is the same as total cost; that is, $R = C$. How many units must be sold to break even?

72. Use a grapher to check your answers to Exercises 5, 17, and 39.

Summary and Review 10

Key Terms

Conic section, p. 620
Parabola, p. 620
Midpoint, p. 624
Circle, p. 625
Radius (plural, radii), p. 625
Center, p. 625
Ellipse, p. 631
Foci (singular, focus), p. 632
Vertices (singular, vertex), p. 632
Hyperbola, p. 639
Axis, p. 639
Branches, p. 639
Asymptote, p. 639

Important Properties and Formulas

The Distance Formula

The distance d between any two points (x_1, y_1) and (x_2, y_2) is given by

$d = \sqrt{(x_2 - x_1)^2 + (y_2 - y_1)^2}$.

The Midpoint Formula

If the endpoints of a segment are (x_1, y_1) and (x_2, y_2), then the coordinates of the midpoint are

$\left(\frac{x_1 + x_2}{2}, \frac{y_1 + y_2}{2}\right)$.

(See the summary of graphs on pp. 644–646.)

Parabola

Vertical with vertex at(h, k):

$$y = ax^2 + bx + c$$
$$= a(x - h)^2 + k$$

Horizontal with vertex at (h, k):

$$x = ay^2 + by + c$$
$$= a(y - k)^2 + h$$

Circle

Center at the origin:

$$x^2 + y^2 = r^2$$

Center at (h, k):

$$(x - h)^2 + (y - k)^2 = r^2$$

Ellipse

Center at the origin:

$$\frac{x^2}{a^2} + \frac{y^2}{b^2} = 1$$

Center at (h, k):

$$\frac{(x - h)^2}{a^2} + \frac{(y - k)^2}{b^2} = 1$$

Hyperbola:

Center at the origin

Axis horizontal: $\frac{x^2}{a^2} - \frac{y^2}{b^2} = 1$ Axis vertical: $\frac{y^2}{b^2} - \frac{x^2}{a^2} = 1$

With x- and y-axes as asymptotes: $xy = c$

Review Exercises

Find the distance between each pair of points. Where appropriate, find an approximation to three decimal places.

1. $(2, 6)$ and $(6, 6)$

2. $(-1, 1)$ and $(-5, 4)$

3. $(1.4, 3.6)$ and $(4.7, -5.3)$

4. $(2, 3a)$ and $(-1, a)$

Find the midpoint of the segment with the given endpoints.

5. $(1, 6)$ and $(7, 6)$

6. $(-1, 1)$ and $(-5, 4)$

7. $(1, \sqrt{3})$ and $\left(\frac{1}{2}, -\sqrt{2}\right)$

8. $(2, 3a)$ and $(-1, a)$

Find the center and the radius of each circle.

9. $(x + 2)^2 + (y - 3)^2 = 2$

10. $(x - 5)^2 + y^2 = 49$

11. $x^2 + y^2 - 6x - 2y + 1 = 0$

12. $x^2 + y^2 + 8x - 6y = 10$

13. Find an equation of the circle with center $(-4, 3)$ and radius $4\sqrt{3}$.

14. Find an equation of the circle with center $(7, -2)$ and radius $2\sqrt{5}$.

Classify each equation as a circle, an ellipse, a parabola, or a hyperbola. Then graph.

15. $4x^2 + 4y^2 = 100$

16. $9x^2 + 2y^2 = 18$

17. $y = -x^2 + 2x - 3$

18. $\dfrac{y^2}{9} - \dfrac{x^2}{4} = 1$

19. $xy = 9$

20. $x = y^2 + 2y - 2$

21. $\dfrac{(x + 1)^2}{3} + (y - 3)^2 = 1$

22. $x^2 + y^2 + 6x - 8y - 39 = 0$

Solve.

23. $x^2 - y^2 = 33,$
$x + y = 11$

24. $x^2 - 2x + 2y^2 = 8,$
$2x + y = 6$

25. $x^2 - y = 3,$
$2x - y = 3$

26. $x^2 + y^2 = 25,$
$x^2 - y^2 = 7$

27. $x^2 - y^2 = 3,$
$y = x^2 - 3$

28. $x^2 + y^2 = 18,$
$2x + y = 3$

29. $x^2 + y^2 = 100,$
$2x^2 - 3y^2 = -120$

30. $x^2 + 2y^2 = 12,$
$xy = 4$

31. A rectangular garden has a perimeter of 38 m and an area of 84 m^2. What are the dimensions of the garden?

32. One type of carton used by table products.com exactly fits both a folded napkin of area 108 in^2 and a candle of length 15 in., laid diagonally on the bottom of the carton. What are the dimensions of the carton?

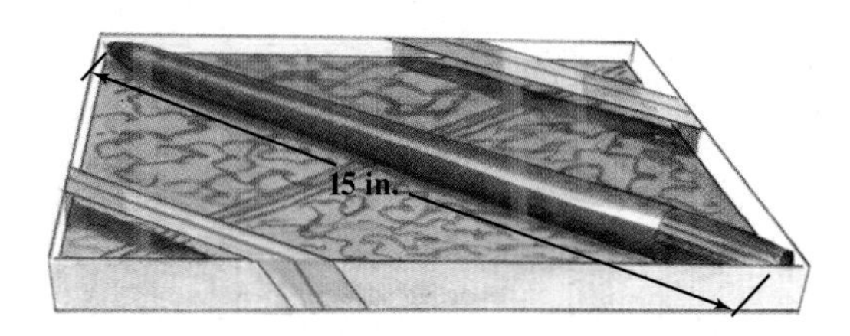

33. The perimeter of a square is 12 cm more than the perimeter of another square. Its area exceeds the area of the other by 39 cm^2. Find the perimeter of each square.

34. The sum of the areas of two circles is 130π ft^2. The difference of the circumferences is 16π ft. Find the radius of each circle.

SYNTHESIS

35. How does the graph of a hyperbola differ from the graph of a parabola?

36. Explain why function notation is not used in this chapter, and list the graphs discussed for which function notation could be used.

37. Solve:
$$4x^2 - x - 3y^2 = 9,$$
$$-x^2 + x + y^2 = 2.$$

38. Find the points whose distance from $(8, 0)$ and from $(-8, 0)$ is 10.

39. Find an equation of the circle that passes through $(-2, -4)$, $(5, -5)$, and $(6, 2)$.

40. Find an equation of the ellipse with the following intercepts: $(-7, 0)$, $(7, 0)$, $(0, -3)$, and $(0, 3)$.

41. Find the point on the x-axis that is equidistant from $(-3, 4)$ and $(5, 6)$.

Chapter Test 10

Find the distance between each pair of points. Where appropriate, find an approximation to three decimal places.

1. $(4, -1)$ and $(-5, 8)$

2. $(3, -a)$ and $(-3, a)$

Find the midpoint of the segment with the given endpoints.

3. $(4, -1)$ and $(-5, 8)$

4. $(3, -a)$ and $(-3, a)$

Find the center and the radius of each circle.

5. $(x + 2)^2 + (y - 3)^2 = 64$

6. $x^2 + y^2 + 4x - 6y + 4 = 0$

Classify the equation as a circle, an ellipse, a parabola, or a hyperbola. Then graph.

7. $y = x^2 - 4x - 1$

8. $x^2 + y^2 + 2x + 6y + 6 = 0$

9. $\dfrac{x^2}{9} - \dfrac{y^2}{4} = 1$

10. $16x^2 + 4y^2 = 64$

11. $xy = -5$

12. $x = -y^2 + 4y$

Solve.

13. $\dfrac{x^2}{16} + \dfrac{y^2}{9} = 1,$
$3x + 4y = 12$

14. $x^2 + y^2 = 16,$
$\dfrac{x^2}{16} - \dfrac{y^2}{9} = 1$

15. $x^2 - 2y^2 = 1.$
$xy = 6$

16. $x^2 + y^2 = 10,$
$x^2 = y^2 + 2$

17. A rectangle with diagonal of length $5\sqrt{5}$ has an area of 22. Find the dimensions of the rectangle.

18. Two squares are such that the sum of their areas is $8\ \text{m}^2$ and the difference of their areas is $2\ \text{m}^2$. Find the length of a side of each square.

19. A rectangle has a diagonal of length 20 ft and a perimeter of 56 ft. Find the dimensions of the rectangle.

20. Nikki invested a certain amount of money for 1 yr and earned $72 in interest. Erin invested $240 more than Nikki at an interest rate that was 83% of the rate given to Nikki, but she earned the same amount of interest. Find the principal and interest rate of Nikki's investment.

SYNTHESIS

21. Find an equation of the ellipse passing through $(6, 0)$ and $(6, 6)$ with vertices at $(1, 3)$ and $(11, 3)$.

22. Find the point on the y-axis that is equidistant from $(-3, -5)$ and $(4, -7)$.

23. The sum of two numbers is 36, and the product is 4. Find the sum of the reciprocals of the numbers.

11 Sequences, Series, and the Binomial Theorem

11.1 Sequences and Series
11.2 Arithmetic Sequences and Series
11.3 Geometric Sequences and Series
11.4 The Binomial Theorem
Connecting the Concepts
Summary and Review
Test
Cumulative Review

AN APPLICATION

At one point in a recent season, Derek Jeter of the New York Yankees had a batting average of .325. At that time, if someone were to randomly select five of his "at-bats," the probability of his getting exactly 3 hits would be the 3rd term of the binomial expansion of $(0.325 + 0.675)^5$. Find that term and use a calculator to estimate the probability.

This problem appears as Exercise 54 in Section 11.4.

I use math daily, especially to figure percentages such as batting averages, earned run averages, slugging percentages, and on-base percentages. Understanding math also helps me to apply the percentages with meaning.

CHARLIE SCOGGINS
Baseball Writer
The Lowell Sun
Lowell, MA

The first three sections of this chapter are devoted to sequences *and* series. *A sequence is simply an ordered list. For example, when a baseball coach writes a batting order, a sequence is being formed. When the members of a sequence are numbers, we can discuss their sum. Such a sum is called a* series.

Section 11.4 presents the binomial theorem, *which is used to expand expressions of the form* $(a + b)^n$. *Such an expansion is itself a series.*

Sequences and Series

11.1

Sequences • Finding the General Term • Sums and Series • Sigma Notation

Sequences

Suppose that \$1000 is invested at 8%, compounded annually. The amounts to which the money grows after 1 year, 2 years, 3 years, and so on, are as follows:

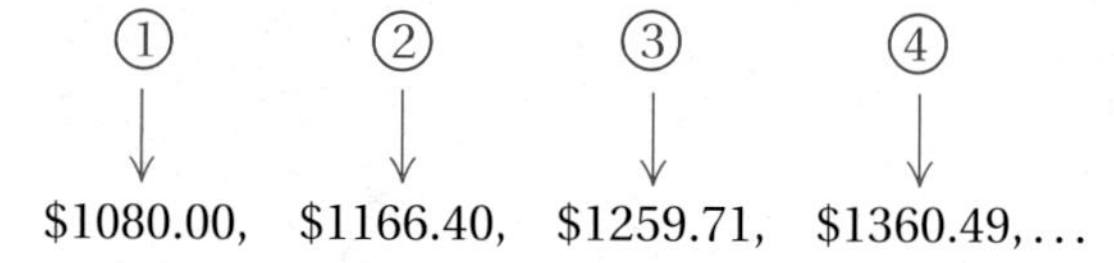

\$1080.00, \$1166.40, \$1259.71, \$1360.49, ...

We can regard this as a function that pairs 1 with \$1080.00, 2 with \$1166.40, 3 with \$1259.71, and so on. A **sequence** (or **progression**) is thus a function, where the domain is a set of consecutive positive integers beginning with 1, and the range varies from sequence to sequence.

If we continue computing the amounts in the account forever, we obtain an **infinite sequence**, with function values

\$1080.00, \$1166.40, \$1259.71, \$1360.49, \$1469.33, \$1586.87, ...

The three dots at the end indicate that the sequence goes on without stopping. If we stop after a certain number of years, we obtain a **finite sequence:**

\$1080.00, \$1166.40, \$1259.71, \$1360.49

Sequences

An *infinite sequence* is a function having for its domain the set of natural numbers: $\{1, 2, 3, 4, 5, \ldots\}$.

A *finite sequence* is a function having for its domain a set of natural numbers: $\{1, 2, 3, 4, 5, \ldots, n\}$, for some natural number n.

As another example, consider the sequence given by

$$a(n) = 2^n, \quad \text{or} \quad a_n = 2^n.$$

The notation a_n means the same as $a(n)$ but is used more commonly with sequences. Some function values (also called *terms* of the sequence) follow:

$$a_1 = 2^1 = 2,$$
$$a_2 = 2^2 = 4,$$
$$a_3 = 2^3 = 8,$$
$$a_6 = 2^6 = 64.$$

The first term of the sequence is a_1, the fifth term is a_5, and the nth term, or **general term**, is a_n. This sequence can also be denoted in the following ways:

$2, 4, 8, \ldots;$

or $2, 4, 8, \ldots, 2^n, \ldots.$ The 2^n emphasizes that the nth term of this sequence is found by raising 2 to the nth power.

Example 1

Find the first four terms and the 57th term of the sequence for which the general term is given by $a_n = (-1)^n/(n + 1)$.

Solution We have

$$a_1 = \frac{(-1)^1}{1+1} = -\frac{1}{2},$$
$$a_2 = \frac{(-1)^2}{2+1} = \frac{1}{3},$$
$$a_3 = \frac{(-1)^3}{3+1} = -\frac{1}{4},$$
$$a_4 = \frac{(-1)^4}{4+1} = \frac{1}{5},$$
$$a_{57} = \frac{(-1)^{57}}{57+1} = -\frac{1}{58}.$$

Note that the expression $(-1)^n$ causes the signs of the terms to alternate between positive and negative, depending on whether n is even or odd.

technology connection

Sequences are entered and graphed much like functions. The difference is that the SEQUENCE MODE must be selected. You can then enter U_n or V_n using n as the variable. Use this approach to check Example 1 with a table of values for the sequence.

Finding the General Term

When only the first few terms of a sequence are known, it is impossible to be certain what the general term is, but a prediction can be made by looking for a pattern.

Example 2

For each sequence, predict the general term.

a) $1, 4, 9, 16, 25, \ldots$

b) $-1, 2, -4, 8, -16, \ldots$

c) $2, 4, 8, \ldots$

Solution

a) $1, 4, 9, 16, 25, \ldots$

These are squares of consecutive positive integers, so the general term could be n^2.

b) $-1, 2, -4, 8, -16, \ldots$

These are powers of 2 with alternating signs, so the general term may be $(-1)^n[2^{n-1}]$. To check, note that 8 is the fourth term, and

$$\begin{aligned}(-1)^4[2^{4-1}] &= 1 \cdot 2^3 \\ &= 8.\end{aligned}$$

c) $2, 4, 8, \ldots$

We regard the pattern as powers of 2, in which case 16 would be the next term and 2^n the general term. The sequence could then be written with more terms as

$$2, 4, 8, 16, 32, 64, 128, \ldots$$

In part (c) above, suppose that the second term is found by adding 2, the third term by adding 4, the next term by adding 6, and so on. In this case, 14 would be the next term and the sequence would be

$$2, 4, 8, 14, 22, 32, 44, 58, \ldots$$

This illustrates that the fewer terms we are given, the greater the uncertainty about the nth term.

Sums and Series

Series

Given the infinite sequence

$$a_1,\ a_2,\ a_3,\ a_4,\ \ldots,\ a_n, \ldots,$$

the sum of the terms

$$a_1 + a_2 + a_3 + \cdots + a_n + \cdots$$

is called an *infinite series.* A *partial sum* is the sum of the first n terms:

$$a_1 + a_2 + a_3 + \cdots + a_n.$$

A partial sum is also called a *finite series* and is denoted S_n.

Example 3 For the sequence -2, 4, -6, 8, -10, 12, -14, find: **(a)** S_2; **(b)** S_3; **(c)** S_7.

Solution

a) $S_2 = -2 + 4 = 2$ This is the sum of the first 2 terms.

b) $S_3 = -2 + 4 + (-6) = -4$ This is the sum of the first 3 terms.

c) $S_7 = -2 + 4 + (-6) + 8 + (-10) + 12 + (-14) = -8$ This is the sum of the first 7 terms.

Sigma Notation

When the general term of a sequence is known, the Greek letter Σ (capital sigma) can be used to write a series. For example, the sum of the first four terms of the sequence 3, 5, 7, 9, 11, ..., $2k + 1$, ... can be named as follows, using *sigma notation*, or *summation notation*:

$$\sum_{k=1}^{4} (2k + 1).$$

This represents
$(2 \cdot 1 + 1) + (2 \cdot 2 + 1) + (2 \cdot 3 + 1) + (2 \cdot 4 + 1)$.

This is read "the sum as k goes from 1 to 4 of $(2k + 1)$." The letter k is called the *index of summation*. The index of summation need not start at 1.

Example 4 Write out and evaluate each sum.

a) $\sum_{k=1}^{5} k^2$ **b)** $\sum_{k=4}^{6} (-1)^k(2k)$ **c)** $\sum_{k=0}^{3} (2^k + 5)$

Solution

a) $\sum_{k=1}^{5} k^2 = 1^2 + 2^2 + 3^2 + 4^2 + 5^2 = 1 + 4 + 9 + 16 + 25 = 55$

Evaluate k^2 for all integers from 1 through 5. Then add.

b) $\sum_{k=4}^{6} (-1)^k(2k) = (-1)^4(2 \cdot 4) + (-1)^5(2 \cdot 5) + (-1)^6(2 \cdot 6)$

$= 8 - 10 + 12 = 10$

c) $\sum_{k=0}^{3} (2^k + 5) = (2^0 + 5) + (2^1 + 5) + (2^2 + 5) + (2^3 + 5)$

$= 6 + 7 + 9 + 13 = 35$

Example 5 Write sigma notation for each sum.

a) $1 + 4 + 9 + 16 + 25$ **b)** $-1 + 3 - 5 + 7$

c) $3 + 9 + 27 + 81 + \cdots$

Solution

a) $1 + 4 + 9 + 16 + 25$

Note that this is a sum of squares, $1^2 + 2^2 + 3^2 + 4^2 + 5^2$, so the general term is k^2. Sigma notation is

$$\sum_{k=1}^{5} k^2.$$

The sum starts with 1^2 and ends with 5^2.

Answers may vary here. For example, another—perhaps less obvious—way of writing $1 + 4 + 9 + 16 + 25$ is

$$\sum_{k=2}^{6} (k - 1)^2.$$

b) $-1 + 3 - 5 + 7$

Except for the alternating signs, this is the sum of the first four positive odd numbers. Note that $2k - 1$ is a formula for the kth positive odd number. It is also important to note that since $(-1)^k = 1$ when k is even and $(-1)^k = -1$ when k is odd, the factor $(-1)^k$ can be used to create the alternating signs. The general term is thus $(-1)^k(2k - 1)$, beginning with $k = 1$. Sigma notation is

$$\sum_{k=1}^{4} (-1)^k(2k - 1).$$

To check, we can evaluate $(-1)^k(2k - 1)$ using 1, 2, 3, and 4. Then we can write the sum of the four terms. We leave this to the student.

c) $3 + 9 + 27 + 81 + \cdots$

This is a sum of powers of 3, and it is also an infinite series. We use the symbol ∞ for infinity and write the series using sigma notation:

$$\sum_{k=1}^{\infty} 3^k.$$

Exercise Set 11.1

FOR EXTRA HELP

Math Tutor Center

MathXL

In each of the following, the nth term of a sequence is given. In each case, find the first 4 terms; the 10th term, a_{10}; and the 15th term, a_{15}.

1. $a_n = 5n - 2$

2. $a_n = 2n + 3$

3. $a_n = \dfrac{n}{n + 1}$

4. $a_n = n^2 + 2$

5. $a_n = n^2 - 2n$

6. $a_n = \dfrac{n^2 - 1}{n^2 + 1}$

7. $a_n = n + \dfrac{1}{n}$

8. $a_n = \left(-\dfrac{1}{2}\right)^{n-1}$

9. $a_n = (-1)^n n^2$

10. $a_n = (-1)^n(n + 3)$

11. $a_n = (-1)^{n+1}(3n - 5)$

12. $a_n = (-1)^n(n^3 - 1)$

Find the indicated term of each sequence.

13. $a_n = 2n - 5;\ a_7$

14. $a_n = 3n + 2;\ a_8$

15. $a_n = (3n + 1)(2n - 5);\ a_9$

16. $a_n = (3n + 2)^2;\ a_6$

17. $a_n = (-1)^{n-1}(3.4n - 17.3);\ a_{12}$

18. $a_n = (-2)^{n-2}(45.68 - 1.2n);\ a_{23}$

19. $a_n = 3n^2(9n - 100);\ a_{11}$

20. $a_n = 4n^2(2n - 39);\ a_{22}$

21. $a_n = \left(1 + \dfrac{1}{n}\right)^2;\ a_{20}$

22. $a_n = \left(1 - \dfrac{1}{n}\right)^3;\ a_{15}$

Look for a pattern and then predict the general term, or nth term, a_n, of each sequence. Answers may vary.

23. $1, 3, 5, 7, 9, \ldots$

24. $2, 4, 6, 8, \ldots$

25. $1, -1, 1, -1, \ldots$

26. $-1, 1, -1, 1, \ldots$

27. $-1, 2, -3, 4, \ldots$

28. $1, -2, 3, -4, \ldots$

29. $-2, 6, -18, 54, \ldots$

30. $-2, 3, 8, 13, 18, \ldots$

31. $\frac{1}{2}, \frac{2}{3}, \frac{3}{4}, \frac{4}{5}, \frac{5}{6}, \ldots$

32. $1 \cdot 2, 2 \cdot 3, 3 \cdot 4, 4 \cdot 5, \ldots$

33. $5, 25, 125, 625, \ldots$

34. $4, 16, 64, 256, \ldots$

35. $-1, 4, -9, 16, \ldots$

36. $1, -4, 9, -16, \ldots$

Find the indicated partial sum for each sequence.

37. $1, -2, 3, -4, 5, -6, \ldots; \; S_7$

38. $1, -3, 5, -7, 9, -11, \ldots; \; S_8$

39. $2, 4, 6, 8, \ldots; \; S_5$

40. $1, \frac{1}{4}, \frac{1}{9}, \frac{1}{16}, \frac{1}{25}, \ldots; \; S_5$

Write out and evaluate each sum.

41. $\sum_{k=1}^{5} \frac{1}{2k}$

42. $\sum_{k=1}^{6} \frac{1}{2k-1}$

43. $\sum_{k=0}^{4} 3^k$

44. $\sum_{k=4}^{7} \sqrt{2k+1}$

45. $\sum_{k=1}^{8} \frac{k}{k+1}$

46. $\sum_{k=1}^{4} \frac{k-2}{k+3}$

47. $\sum_{k=1}^{8} (-1)^{k+1} 2^k$

48. $\sum_{k=1}^{7} (-1)^k 4^{k+1}$

49. $\sum_{k=0}^{5} (k^2 - 2k + 3)$

50. $\sum_{k=0}^{5} (k^2 - 3k + 4)$

51. $\sum_{k=3}^{5} \frac{(-1)^k}{k(k+1)}$

52. $\sum_{k=3}^{7} \frac{k}{2^k}$

Rewrite each sum using sigma notation. Answers may vary.

53. $\frac{2}{3} + \frac{3}{4} + \frac{4}{5} + \frac{5}{6} + \frac{6}{7}$

54. $3 + 6 + 9 + 12 + 15$

55. $1 + 4 + 9 + 16 + 25 + 36$

56. $\frac{1}{1^2} + \frac{1}{2^2} + \frac{1}{3^2} + \frac{1}{4^2} + \frac{1}{5^2}$

57. $4 - 9 + 16 - 25 + \cdots + (-1)^n n^2$

58. $9 - 16 + 25 + \cdots + (-1)^{n+1} n^2$

59. $5 + 10 + 15 + 20 + 25 + \cdots$

60. $7 + 14 + 21 + 28 + 35 + \cdots$

61. $\frac{1}{1 \cdot 2} + \frac{1}{2 \cdot 3} + \frac{1}{3 \cdot 4} + \frac{1}{4 \cdot 5} + \cdots$

62. $\frac{1}{1 \cdot 2^2} + \frac{1}{2 \cdot 3^2} + \frac{1}{3 \cdot 4^2} + \frac{1}{4 \cdot 5^2} + \cdots$

63. The sequence $1, 4, 9, 16, \ldots$ can be written as $f(x) = x^2$ with the domain the set of all positive integers. Explain how the graph of f would compare with the graph of $y = x^2$.

64. Eric says he expects he will prefer sequences to functions because he dislikes fractions. Will his expectations prove correct? Why or why not?

SKILL MAINTENANCE

Evaluate.

65. $\frac{7}{2}(a_1 + a_7)$, for $a_1 = 8$ and $a_7 = 14$

66. $a_1 + (n-1)d$, for $a_1 = 3$, $n = 6$, and $d = 4$

Multiply.

67. $(x + y)^3$

68. $(a - b)^3$

69. $(2a - b)^3$

70. $(2x + y)^3$

SYNTHESIS

71. Explain why the equation

$$\sum_{k=1}^{n} (a_k + b_k) = \sum_{k=1}^{n} a_k + \sum_{k=1}^{n} b_k$$

is true for any positive integer n. What laws are used to justify this result?

72. Consider the sums

$$\sum_{k=1}^{5} 3k^2 \quad \text{and} \quad 3\sum_{k=1}^{5} k^2.$$

a) Which is easier to evaluate and why?

b) Is it true that

$$\sum_{k=1}^{n} ca_k = c\sum_{k=1}^{n} a_k?$$

Why or why not?

Some sequences are given by a recursive *definition. The value of the first term, a_1, is given, and then we are told how to find any subsequent term from the term preceding it. Find the first six terms of each of the following recursively defined sequences.*

73. $a_1 = 1, a_{n+1} = 5a_n - 2$

74. $a_1 = 0, a_{n+1} = a_n^2 + 3$

75. *Cell biology.* A single cell of bacterium divides into two every 15 min. Suppose that the same rate of division is maintained for 4 hr. Give a sequence that lists the number of cells after successive 15-min periods.

76. *Value of a copier.* The value of a color photocopier is \$5200. Its scrap value each year is 75% of its value the year before. Give a sequence that lists the scrap value of the machine at the start of each year for a 10-yr period.

77. Find S_{100} and S_{101} for the sequence in which $a_n = (-1)^n$.

Find the first five terms of each sequence; then find S_5.

78. $a_n = \frac{1}{2^n}\log 1000^n$

79. $a_n = i^n, i = \sqrt{-1}$

80. Find all values for x that solve the following:

$$\sum_{k=1}^{x} i^k = -1.$$

81. The nth term of a sequence is given by

$$a_n = n^5 - 14n^4 + 6n^3 + 416n^2 - 655n - 1050.$$

Use a grapher with a TABLE feature to determine what term in the sequence is 6144.

82. To define a sequence recursively on a grapher (see Exercises 73 and 74), the SEQ MODE is used. The general term U_n or V_n can often be expressed in terms of U_{n-1} or V_{n-1} by pressing [2nd] [7] or [2nd] [8]. The starting values of U_n, V_n, and n are set as one of the WINDOW variables.

Use recursion to determine how many handshakes will occur if a group of 50 people shake hands with one another. To develop the recursion formula, begin with a group of 2 and determine how many additional handshakes occur with the arrival of each new group member. (See the Collaborative Corner following Exercise Set 5.1 on p. 277.)

Arithmetic Sequences and Series

11.2

Arithmetic Sequences • Sum of the First n Terms of an Arithmetic Sequence • Problem Solving

In this section, we concentrate on sequences and series that are said to be arithmetic (pronounced ar-ith-MET-ik).

Arithmetic Sequences

In an **arithmetic sequence** (or **progression**), any term (other than the first) can be found by adding the same number to its preceding term. For example, the sequence 2, 5, 8, 11, 14, 17, ... is arithmetic because adding 3 to any term produces the next term.

Arithmetic Sequence

A sequence is *arithmetic* if there exists a number d, called the *common difference,* such that $a_{n+1} = a_n + d$ for any integer $n \geq 1$.

Example 1 For each arithmetic sequence, identify the first term, a_1, and the common difference, d.

a) 4, 9, 14, 19, 24, . . . **b)** 27, 20, 13, 6, −1, −8, . . .

Solution To find a_1, we simply use the first term listed. To find d, we choose any term beyond the first and subtract the preceding term from it.

	Sequence	*First Term, a_1*	*Common Difference, d*
a)	4, 9, 14, 19, 24, . . .	4	$5 \longleftarrow 9 - 4 = 5$
b)	27, 20, 13, 6, −1, −8, . . .	27	$-7 \longleftarrow 20 - 27 = -7$

To find the common difference, we subtracted a_1 from a_2. Had we subtracted a_2 from a_3 or a_3 from a_4, we would have found the same values for d.

Check: As a check, note that when d is added to each term, the result is the next term in the sequence.

a) $4 + 5 = 9, \quad 9 + 5 = 14, \quad 14 + 5 = 19, \quad 19 + 5 = 24$
b) $27 + (-7) = 20, \quad 20 + (-7) = 13, \quad 13 + (-7) = 6, \quad 6 + (-7) = -1,$
$-1 + (-7) = -8$

To find a formula for the general, or nth, term of any arithmetic sequence, we denote the common difference by d and write out the first few terms:

$$a_1,$$
$$a_2 = a_1 + d,$$
$$a_3 = a_2 + d = (a_1 + d) + d = a_1 + 2d, \qquad \text{Substituting } a_1 + d \text{ for } a_2$$
$$a_4 = a_3 + d = (a_1 + 2d) + d = a_1 + 3d. \qquad \text{Substituting } a_1 + 2d \text{ for } a_3$$

Note that the coefficient of d in each case is 1 less than the subscript.

Generalizing, we obtain the following formula.

To Find a_n for an Arithmetic Sequence

The nth term of an arithmetic sequence with common difference d is

$$a_n = a_1 + (n - 1)d, \quad \text{for any integer } n \geq 1.$$

Example 2

Find the 14th term of the arithmetic sequence 6, 9, 12, 15,...

Solution First we note that $a_1 = 6$, $d = 3$, and $n = 14$. Using the formula for the nth term of an arithmetic sequence, we have

$$a_n = a_1 + (n - 1)d$$
$$a_{14} = 6 + (14 - 1) \cdot 3 = 6 + 13 \cdot 3 = 6 + 39 = 45.$$

The 14th term is 45.

Example 3

For the sequence in Example 2, which term is 300? That is, find n if $a_n = 300$.

Solution We substitute into the formula for the nth term of an arithmetic sequence and solve for n:

$$a_n = a_1 + (n - 1)d$$
$$300 = 6 + (n - 1) \cdot 3$$
$$300 = 6 + 3n - 3$$
$$297 = 3n$$
$$99 = n.$$

The term 300 is the 99th term of the sequence.

Given two terms and their places in an arithmetic sequence, we can construct the sequence.

Example 4

The 3rd term of an arithmetic sequence is 14, and the 16th term is 79. Find a_1 and d and construct the sequence.

Solution We know that $a_3 = 14$ and $a_{16} = 79$. Thus we would have to add d 13 times to get from 14 to 79. That is,

$$14 + 13d = 79. \qquad a_3 \text{ and } a_{16} \text{ are 13 terms apart; } 16 - 3 = 13$$

Solving $14 + 13d = 79$, we obtain

$$13d = 65 \qquad \text{Subtracting 14 from both sides}$$
$$d = 5. \qquad \text{Dividing both sides by 13}$$

We subtract d twice from a_3 to get to a_1. Thus,

$$a_1 = 14 - 2 \cdot 5 = 4. \qquad a_1 \text{ and } a_3 \text{ are 2 terms apart; } 3 - 1 = 2$$

The sequence is 4, 9, 14, 19, ... Note that we could have subtracted d 15 times from a_{16} in order to find a_1.

In general, d should be subtracted $(n - 1)$ times from a_n in order to find a_1.

Sum of the First n Terms of an Arithmetic Sequence

When the terms of an arithmetic sequence are added, an **arithmetic series** is formed. To find a formula for computing S_n when the series is arithmetic, we denote the first n terms as follows:

This is the next-to-last term. If you add d to this term, the result is a_n.

$$a_1, (a_1 + d), (a_1 + 2d), \ldots, (a_n - 2d), (a_n - d), a_n$$

This term is two terms back from the end. If you add d to this term, you get the next-to-last term, $a_n - d$.

Thus, S_n is given by

$$S_n = a_1 + (a_1 + d) + (a_1 + 2d) + \cdots + (a_n - 2d) + (a_n - d) + a_n.$$

Using a commutative law, we have a second equation:

$$S_n = a_n + (a_n - d) + (a_n - 2d) + \cdots + (a_1 + 2d) + (a_1 + d) + a_1.$$

Adding corresponding terms on each side of the above equations, we get

$$\begin{aligned} 2S_n = {} & [a_1 + a_n] + [(a_1 + d) + (a_n - d)] + [(a_1 + 2d) + (a_n - 2d)] \\ & + \cdots + [(a_n - 2d) + (a_1 + 2d)] + [(a_n - d) + (a_1 + d)] \\ & + [a_n + a_1]. \end{aligned}$$

This simplifies to

$$\begin{aligned} 2S_n = {} & [a_1 + a_n] + [a_1 + a_n] + [a_1 + a_n] \\ & + \cdots + [a_n + a_1] + [a_n + a_1] + [a_n + a_1]. \end{aligned}$$

There are n bracketed sums.

Since $[a_1 + a_n]$ is being added n times, it follows that

$$2S_n = n[a_1 + a_n].$$

Dividing both sides by 2 leads to the following formula.

To Find S_n for an Arithmetic Sequence

The sum of the first n terms of an arithmetic sequence is given by

$$S_n = \frac{n}{2}(a_1 + a_n).$$

Example 5 Find the sum of the first 100 positive even numbers.

Solution The sum is

$$2 + 4 + 6 + \cdots + 198 + 200.$$

This is the sum of the first 100 terms of the arithmetic sequence for which

$$a_1 = 2, \qquad n = 100, \quad \text{and} \quad a_n = 200.$$

Substituting in the formula

$$S_n = \frac{n}{2}(a_1 + a_n),$$

we get

$$S_{100} = \frac{100}{2}(2 + 200)$$
$$= 50(202) = 10{,}100.$$

The above formula is useful when we know the first and last terms, a_1 and a_n. To find S_n when a_n is unknown, but a_1, n, and d are known, we can use the formula $a_n = a_1 + (n - 1)d$ to calculate a_n and then proceed as in Example 5.

Example 6 Find the sum of the first 15 terms of the arithmetic sequence 4, 7, 10, 13, ...

Solution Note that

$$a_1 = 4, \qquad n = 15, \quad \text{and} \quad d = 3.$$

Before using the formula for S_n, we find a_{15}:

$$a_{15} = 4 + (15 - 1)3 \qquad \text{Substituting into the formula for } a_n$$
$$= 4 + 14 \cdot 3 = 46.$$

Thus, knowing that $a_{15} = 46$, we have

$$S_{15} = \tfrac{15}{2}(4 + 46) \qquad \text{Using the formula for } S_n$$
$$= \tfrac{15}{2}(50) = 375.$$

Problem Solving

For some problem-solving situations, the translation may involve sequences or series. In Examples 7 and 8, the calculations and translations can be done in a number of ways. There is often a variety of ways in which a problem can be solved. You should use the one that is best or easiest for you. In this chapter, however, we will try to emphasize sequences and series and their related formulas.

Example 7

Hourly wages. Chris accepts a job managing a CD shop, starting with an hourly wage of \$14.25, and is promised a raise of 15¢ per hour every 2 months for 5 years. After 5 years of work, what will be Chris's hourly wage?

Solution

1. **Familiarize.** It helps to write down the hourly wage for several two-month time periods.

   Beginning: 14.25,
   After two months: 14.40,
   After four months: 14.55,
   and so on.

   What appears is a sequence of numbers: 14.25, 14.40, 14.55, ... Since the same amount is added each time, the sequence is arithmetic.

   We list what we know about arithmetic sequences. The pertinent formulas are

   $$a_n = a_1 + (n - 1)d$$

   and

   $$S_n = \frac{n}{2}(a_1 + a_n).$$

   In this case, we are not looking for a sum, so it is probably the first formula that will give us our answer. We want to determine the last term in a sequence. To do so, we need to know a_1, n, and d. From our list above, we see that

   $$a_1 = 14.25 \quad \text{and} \quad d = 0.15.$$

   What is n? That is, how many terms are in the sequence? After 1 year, there have been 6 raises, since Chris gets a raise every 2 months. There are 5 years, so the total number of raises will be $5 \cdot 6$, or 30. Altogether, there will be 31 terms: the original wage and 30 increased rates.

2. **Translate.** We want to find a_n for the arithmetic sequence in which $a_1 = 14.25$, $n = 31$, and $d = 0.15$.

3. **Carry out.** Substituting in the formula for a_n gives us

   $$\begin{aligned} a_{31} &= 14.25 + (31 - 1) \cdot 0.15 \\ &= 18.75. \end{aligned}$$

4. **Check.** We can check by redoing the calculations or we can calculate in a slightly different way for another check. For example, at the end of a year, there will be 6 raises, for a total raise of \$0.90. At the end of 5 years, the total raise will be $5 \times \$0.90$, or \$4.50. If we add that to the original wage of \$14.25, we obtain \$18.75. The answer checks.

5. **State.** After 5 years, Chris's hourly wage will be \$18.75.

Example 8 ***Telephone pole storage.*** A stack of telephone poles has 30 poles in the bottom row. There are 29 poles in the second row, 28 in the next row, and so on. How many poles are in the stack if there are 5 poles in the top row?

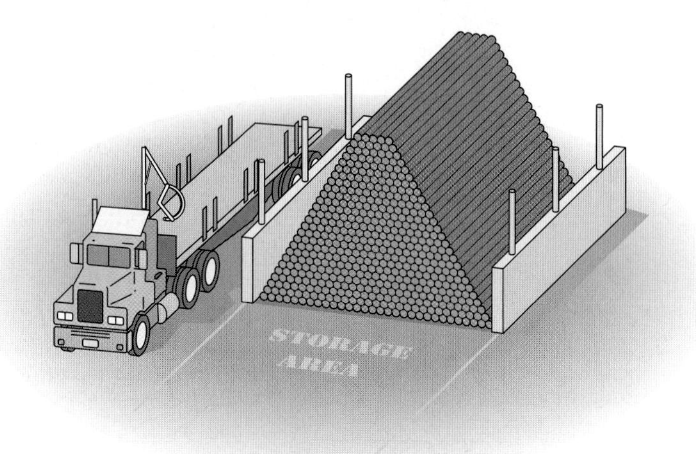

Solution

1. **Familiarize.** A picture will help in this case. The following figure shows the ends of the poles and the way in which they stack. There are 30 poles on the bottom, and we see that there will be one fewer in each succeeding row. How many rows will there be?

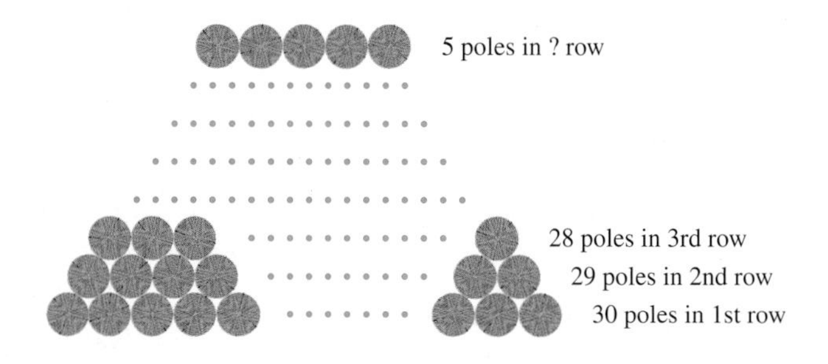

Note that there are $30 - 1 = 29$ poles in the 2nd row, $30 - 2 = 28$ poles in the 3rd row, $30 - 3 = 27$ poles in the 4th row, and so on. The pattern leads to $30 - 25 = 5$ poles in the 26th row.

The situation is represented by the equation

$30 + 29 + 28 + \cdots + 5.$ There are 26 terms in this series.

Thus we have an arithmetic series. We recall the formula

$$S_n = \frac{n}{2}(a_1 + a_n).$$

2. **Translate.** We want to find the sum of the first 26 terms of an arithmetic sequence in which $a_1 = 30$ and $a_{26} = 5$.

3. **Carry out.** Substituting into the above formula gives us

$$S_{26} = \tfrac{26}{2}(30 + 5)$$
$$= 13 \cdot 35 = 455.$$

4. **Check.** In this case, we can check the calculations by doing them again. A longer, harder way would be to do the entire addition:

$$30 + 29 + 28 + \cdots + 5.$$

5. **State.** There are 455 poles in the stack.

Exercise Set 11.2

FOR EXTRA HELP

Digital Video Tutor CD 9
Videotape 20

InterAct Math

Math Tutor Center

MathXL

MyMathLab.com

Find the first term and the common difference.

1. $2, 6, 10, 14, \ldots$
2. $1.06, 1.12, 1.18, 1.24, \ldots$
3. $6, 2, -2, -6, \ldots$
4. $-9, -6, -3, 0, \ldots$
5. $\frac{3}{2}, \frac{9}{4}, 3, \frac{15}{4}, \ldots$
6. $\frac{3}{5}, \frac{1}{10}, -\frac{2}{5}, \ldots$
7. $\$5.12, \$5.24, \$5.36, \$5.48, \ldots$
8. $\$214, \$211, \$208, \$205, \ldots$
9. Find the 12th term of the arithmetic sequence $3, 7, 11, \ldots$.
10. Find the 11th term of the arithmetic sequence $0.07, 0.12, 0.17, \ldots$.
11. Find the 17th term of the arithmetic sequence $7, 4, 1, \ldots$.
12. Find the 14th term of the arithmetic sequence $3, \frac{7}{3}, \frac{5}{3}, \ldots$.
13. Find the 13th term of the arithmetic sequence $\$1200, \$964.32, \$728.64, \ldots$.
14. Find the 10th term of the arithmetic sequence $\$2345.78, \$2967.54, \$3589.30, \ldots$.
15. In the sequence of Exercise 9, what term is 107?
16. In the sequence of Exercise 10, what term is 1.67?
17. In the sequence of Exercise 11, what term is -296?
18. In the sequence of Exercise 12, what term is -27?
19. Find a_{17} when $a_1 = 2$ and $d = 5$.
20. Find a_{20} when $a_1 = 14$ and $d = -3$.
21. Find a_1 when $d = 4$ and $a_8 = 33$.
22. Find a_1 when $d = 8$ and $a_{11} = 26$.
23. Find n when $a_1 = 5$, $d = -3$, and $a_n = -76$.
24. Find n when $a_1 = 25$, $d = -14$, and $a_n = -507$.
25. For an arithmetic sequence in which $a_{17} = -40$ and $a_{28} = -73$, find a_1 and d. Write the first five terms of the sequence.
26. In an arithmetic sequence, $a_{17} = \frac{25}{3}$ and $a_{32} = \frac{95}{6}$. Find a_1 and d. Write the first five terms of the sequence.
27. *Aha!* Find a_1 and d if $a_{13} = 13$ and $a_{54} = 54$.
28. Find a_1 and d if $a_{12} = 24$ and $a_{25} = 50$.
29. Find the sum of the first 20 terms of the arithmetic series $1 + 5 + 9 + 13 + \cdots$.
30. Find the sum of the first 14 terms of the arithmetic series $11 + 7 + 3 + \cdots$.
31. Find the sum of the first 250 natural numbers.
32. Find the sum of the first 400 natural numbers.
33. Find the sum of the even numbers from 2 to 100, inclusive.
34. Find the sum of the odd numbers from 1 to 99, inclusive.
35. Find the sum of all multiples of 6 from 6 to 102, inclusive.

36. Find the sum of all multiples of 4 that are between 15 and 521.

37. An arithmetic series has $a_1 = 4$ and $d = 5$. Find S_{20}.

38. An arithmetic series has $a_1 = 9$ and $d = -3$. Find S_{32}.

Solve.

39. *Band formations.* The Duxbury marching band has 14 marchers in the front row, 16 in the second row, 18 in the third row, and so on, for 15 rows. How many marchers are in the last row? How many marchers are there altogether?

40. *Gardening.* A gardener is planting bulbs near an entrance to a college. She has 39 plants in the front row, 35 in the second row, 31 in the third row, and so on. If the pattern is consistent, how many plants will be in the last row? How many plants will there be altogether?

41. *Telephone pole piles.* How many poles will be in a pile of telephone poles if there are 50 in the first layer, 49 in the second, and so on, until there are 6 in the last layer?

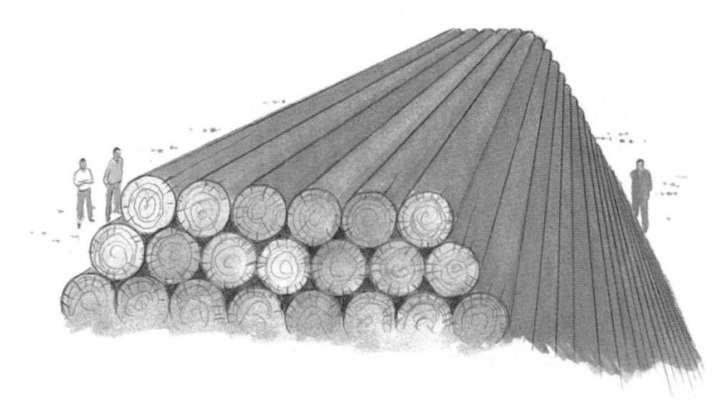

42. *Accumulated savings.* If 10¢ is saved on October 1, another 20¢ on October 2, another 30¢ on October 3, and so, how much is saved during October? (October has 31 days.)

43. *Accumulated savings.* Renata saves money in an arithmetic sequence: $600 for the first year, another $700 the second, and so on, for 20 yr. How much does she save in all (disregarding interest)?

44. *Spending.* Jacob spent $30 on August 1, $50 on August 2, $70 on August 3, and so on. How much did Jacob spend in August? (August has 31 days.)

45. *Auditorium design.* Theaters are often built with more seats per row as the rows move toward the back. The Sanders Amphitheater has 20 seats in the first row, 22 in the second, 24 in the third, and so on, for 19 rows. How many seats are in the amphitheater?

46. *Accumulated savings.* Shirley sets up an investment such that it will return $5000 the first year, $6125 the second year, $7250 the third year, and so on, for 25 yr. How much in all is received from the investment?

47. It is said that as a young child, the mathematician Karl F. Gauss (1777–1855) was able to compute the sum $1 + 2 + 3 + \cdots + 100$ very quickly in his head. Explain how Gauss might have done this and present a formula for the sum of the first n natural numbers. (*Hint*: $1 + 99 = 100$.)

48. If every number in a sequence is doubled and then added, is the result the same as if the numbers were first added and the sum then doubled? Why or why not?

SKILL MAINTENANCE

Simplify.

49. $\frac{3}{10x} + \frac{2}{15x}$

50. $\frac{2}{9t} + \frac{5}{12t}$

Convert to an exponential equation.

51. $\log_a P = k$

52. $\ln t = a$

Find an equation of the circle satisfying the given conditions.

53. Center $(0, 0)$, radius 9

54. Center $(-2, 5)$, radius $3\sqrt{2}$

SYNTHESIS

55. Write a problem for a classmate to solve. Devise the problem so that its solution requires computing S_{17} for an arithmetic sequence.

56. The sum of the first n terms of an arithmetic sequence is also given by

$$S_n = \frac{n}{2}[2a_1 + (n - 1)d].$$

Use the earlier formulas for a_n and S_n to explain how this equation was developed.

57. Find a formula for the sum of the first n consecutive odd numbers starting with 1:

$$1 + 3 + 5 + \cdots + (2n - 1).$$

58. Find three numbers in an arithmetic sequence for which the sum of the first and third is 10 and the product of the first and second is 15.

59. In an arithmetic sequence, $a_1 = \$8760$ and $d = -\$798.23$. Find the first 10 terms of the sequence.

60. Find the sum of the first 10 terms of the sequence given in Exercise 59.

61. Prove that if p, m, and q are consecutive terms in an arithmetic sequence, then

$$m = \frac{p + q}{2}.$$

62. *Straight-line depreciation.* A company buys a color copier for \$5200 on January 1 of a given year. The machine is expected to last for 8 yr, at the end of which time its *trade-in,* or *salvage, value* will be \$1100. If the company figures the decline in value to be the same each year, then the trade-in values, after t years, $0 \leq t \leq 8$, form an arithmetic sequence given by

$$a_t = C - t\left(\frac{C - S}{N}\right),$$

where C is the original cost of the item, N the years of expected life, and S the salvage value.

a) Find the formula for a_t for the straight-line depreciation of the copier.

b) Find the salvage value after 0 yr, 1 yr, 2 yr, 3 yr, 4 yr, 7 yr, and 8 yr.

c) Find a formula that expresses a_t recursively.

63. Use your answer to Exercise 31 to find the sum of all integers from 501 through 750.

Geometric Sequences and Series

11.3

Geometric Sequences • Sum of the First n Terms of a Geometric Sequence • Infinite Geometric Series • Problem Solving

In an arithmetic sequence, a certain number is added to each term to get the next term. When each term in a sequence is *multiplied* by a certain number to get the next term, the sequence is **geometric**. In this section, we examine geometric sequences (or progressions) and *geometric series.*

Geometric Sequences

Consider the sequence

$$2, 6, 18, 54, 162, \ldots.$$

If we multiply each term by 3, we obtain the next term. The multiplier is called the *common ratio* because it is found by dividing any term by the preceding term.

Geometric Sequence

A sequence is *geometric* if there exists a number r, called the *common ratio*, for which

$$\frac{a_{n+1}}{a_n} = r, \quad \text{or} \quad a_{n+1} = a_n \cdot r \quad \text{for any integer } n \geq 1.$$

Example 1 For each geometric sequence, find the common ratio.

a) 3, 6, 12, 24, 48, . . . **b)** 3, −6, 12, −24, 48, −96, . . .
c) \$5200, \$3900, \$2925, \$2193.75, . . .

Solution

Sequence	*Common Ratio*	
a) 3, 6, 12, 24, 48, . . .	2	$\frac{6}{3} = 2, \frac{12}{6} = 2$, and so on
b) 3, −6, 12, −24, 48, −96, . . .	−2	$\frac{-6}{3} = -2, \frac{12}{-6} = -2$ and so on
c) \$5200, \$3900, \$2925, \$2193.75, . . .	0.75	$\frac{\$3900}{\$5200} = 0.75, \frac{\$2925}{\$3900} = 0.75$

To develop a formula for the general, or nth, term of a geometric sequence, let a_1 be the first term and let r be the common ratio. We write out the first few terms as follows:

$$a_1,$$
$$a_2 = a_1 r,$$
$$a_3 = a_2 r = (a_1 r)r = a_1 r^2, \quad \text{Substituting } a_1 r \text{ for } a_2$$
$$a_4 = a_3 r = (a_1 r^2)r = a_1 r^3. \quad \text{Substituting } a_1 r^2 \text{ for } a_3$$

Note that the exponent is 1 less than the subscript.

Generalizing, we obtain the following.

To Find a_n for a Geometric Sequence

The nth term of a geometric sequence with common ratio r is given by

$$a_n = a_1 r^{n-1}, \quad \text{for any integer } n \geq 1.$$

Example 2 Find the 7th term of the geometric sequence 4, 20, 100, ...

Solution First we note that

$$a_1 = 4 \quad \text{and} \quad n = 7.$$

To find the common ratio, we can divide any term (other than the first) by the term preceding it. Since the second term is 20 and the first is 4,

$$r = \frac{20}{4}, \quad \text{or } 5.$$

The formula

$$a_n = a_1 r^{n-1}$$

gives us

$$a_7 = 4 \cdot 5^{7-1} = 4 \cdot 5^6 = 4 \cdot 15{,}625 = 62{,}500.$$

Example 3 Find the 10th term of the geometric sequence

$$64, -32, 16, -8, \ldots.$$

Solution First, we note that

$$a_1 = 64, \qquad n = 10, \quad \text{and} \quad r = \frac{-32}{64} = -\frac{1}{2}.$$

Then, using the formula for the nth term of a geometric series, we have

$$a_{10} = 64 \cdot \left(-\frac{1}{2}\right)^{10-1} = 64 \cdot \left(-\frac{1}{2}\right)^9 = 2^6 \cdot \left(-\frac{1}{2^9}\right) = -\frac{1}{2^3} = -\frac{1}{8}.$$

The 10th term is $-\frac{1}{8}$.

Sum of the First n Terms of a Geometric Sequence

We next develop a formula for S_n when a sequence is geometric:

$$a_1, a_1r, a_1r^2, a_1r^3, \ldots, a_1r^{n-1}, \ldots.$$

The **geometric series** S_n is given by

$$S_n = a_1 + a_1r + a_1r^2 + \cdots + a_1r^{n-2} + a_1r^{n-1}. \tag{1}$$

Multiplying both sides by r gives us

$$rS_n = a_1r + a_1r^2 + a_1r^3 + \cdots + a_1r^{n-1} + a_1r^n. \tag{2}$$

When we subtract corresponding sides of equation (2) from equation (1), the color terms drop out, leaving

$$S_n - rS_n = a_1 - a_1r^n$$

$$S_n(1 - r) = a_1(1 - r^n), \qquad \text{Factoring}$$

or

$$S_n = \frac{a_1(1 - r^n)}{1 - r}. \qquad \text{Dividing both sides by } 1 - r$$

To Find S_n for a Geometric Sequence

The sum of the first n terms of a geometric sequence with common ratio r is given by

$$S_n = \frac{a_1(1 - r^n)}{1 - r}, \quad \text{for any } r \neq 1.$$

Example 4 Find the sum of the first 7 terms of the geometric sequence 3, 15, 75, 375, ...

Solution First, we note that

$$a_1 = 3, \qquad n = 7, \quad \text{and} \quad r = \frac{15}{3} = 5.$$

Then, substituting in the formula $S_n = \dfrac{a_1(1 - r^n)}{1 - r}$, we have

$$\begin{aligned} S_7 &= \frac{3(1 - 5^7)}{1 - 5} = \frac{3(1 - 78{,}125)}{-4} \\ &= \frac{3(-78{,}124)}{-4} \\ &= 58{,}593. \end{aligned}$$

Infinite Geometric Series

Suppose we consider the sum of the terms of an infinite geometric sequence, such as 2, 4, 8, 16, 32, ... We get what is called an **infinite geometric series**:

$$2 + 4 + 8 + 16 + 32 + \cdots.$$

Here, as n grows larger and larger, the sum of the first n terms, S_n, becomes larger and larger without bound. There are also infinite series that get closer and closer to some specific number. Here is an example:

$$\frac{1}{2} + \frac{1}{4} + \frac{1}{8} + \frac{1}{16} + \cdots + \frac{1}{2^n} + \cdots.$$

Let's consider S_n for the first four values of n:

$$\begin{aligned} S_1 &= \tfrac{1}{2} &&= \tfrac{1}{2} = 0.5, \\ S_2 &= \tfrac{1}{2} + \tfrac{1}{4} &&= \tfrac{3}{4} = 0.75, \\ S_3 &= \tfrac{1}{2} + \tfrac{1}{4} + \tfrac{1}{8} &&= \tfrac{7}{8} = 0.875, \\ S_4 &= \tfrac{1}{2} + \tfrac{1}{4} + \tfrac{1}{8} + \tfrac{1}{16} &&= \tfrac{15}{16} = 0.9375. \end{aligned}$$

↑

The denominator of the sum is 2^n, where n is the subscript of S. The numerator is $2^n - 1$.

Thus, for this particular series, we have

$$S_n = \frac{2^n - 1}{2^n} = \frac{2^n}{2^n} - \frac{1}{2^n} = 1 - \frac{1}{2^n}.$$

Note that the value of S_n is less than 1 for any value of n, but as n gets larger and larger, the values of $1/2^n$ get closer to 0 and the values of S_n get closer to 1. We say that 1 is the *limit* of S_n and that 1 is the sum of this infinite geometric sequence. An infinite geometric series is denoted S_∞. It can be shown (but we will not do it here) that the sum of the terms of an infinite geometric sequence exists if and only if $|r| < 1$ (that is, the absolute value of the common ratio is less than 1).

To find a formula for the sum of an infinite geometric sequence, we first consider the sum of the first n terms:

$$S_n = \frac{a_1(1 - r^n)}{1 - r} = \frac{a_1 - a_1 r^n}{1 - r}. \qquad \text{Using the distributive law}$$

For $|r| < 1$, it follows that values of r^n get closer to 0 as n gets larger. (Check this by selecting a number between -1 and 1 and finding larger and larger powers on a calculator.) As r^n gets closer to 0, so does $a_1 r^n$. Thus, S_n gets closer to $a_1/(1 - r)$.

The Limit of an Infinite Geometric Series

When $|r| < 1$, the limit of an infinite geometric series is given by

$$S_\infty = \frac{a_1}{1 - r}. \qquad \text{(For } |r| \geq 1 \text{, no limit exists.)}$$

Example 5

Determine whether each series has a limit. If a limit exists, find it.

a) $1 + 3 + 9 + 27 + \cdots$

b) $-2 + 1 - \frac{1}{2} + \frac{1}{4} - \frac{1}{8} + \cdots$

Solution

a) Here $r = 3$, so $|r| = |3| = 3$. Since $|r| \not< 1$, the series does *not* have a limit.

b) Here $r = -\frac{1}{2}$, so $|r| = \left|-\frac{1}{2}\right| = \frac{1}{2}$. Since $|r| < 1$, the series *does* have a limit. We find the limit by substituting into the formula for S_∞:

$$S_\infty = \frac{-2}{1 - \left(-\frac{1}{2}\right)} = \frac{-2}{\frac{3}{2}} = -2 \cdot \tfrac{2}{3} = -\frac{4}{3}.$$

Example 6

Find fractional notation for 0.63636363....

Solution We can express this as

$$0.63 + 0.0063 + 0.000063 + \cdots.$$

This is an infinite geometric series, where $a_1 = 0.63$ and $r = 0.01$. Since $|r| < 1$, this series has a limit:

$$S_\infty = \frac{a_1}{1 - r} = \frac{0.63}{1 - 0.01} = \frac{0.63}{0.99} = \frac{63}{99}.$$

Thus fractional notation for 0.63636363... is $\frac{63}{99}$, or $\frac{7}{11}$.

Problem Solving

For some problem-solving situations, the translation may involve geometric sequences or series.

Example 7

Daily wages. Suppose someone offered you a job for the month of September (30 days) under the following conditions. You will be paid \$0.01 for the first day, \$0.02 for the second, \$0.04 for the third, and so on, doubling your previous day's salary each day. How much would you earn? (Would you take the job? Make a guess before reading further.)

Solution

1. **Familiarize.** You earn \$0.01 the first day, \$0.01(2) the second day, \$0.01(2)(2) the third day, and so on. Since each day's wages are a constant multiple of the previous day's wages, a geometric sequence is formed.
2. **Translate.** The amount earned is the geometric series

$$\$0.01 + \$0.01(2) + \$0.01(2^2) + \$0.01(2^3) + \cdots + \$0.01(2^{29}),$$

where

$$a_1 = \$0.01, \quad n = 30, \quad \text{and} \quad r = 2.$$

3. **Carry out.** Using the formula

$$S_n = \frac{a_1(1 - r^n)}{1 - r},$$

we have

$$\begin{aligned} S_{30} &= \frac{\$0.01(1 - 2^{30})}{1 - 2} \\ &= \frac{\$0.01(-1{,}073{,}741{,}823)}{-1} \qquad \text{Using a calculator} \\ &= \$10{,}737{,}418.23. \end{aligned}$$

4. **Check.** The calculations can be repeated as a check.
5. **State.** The pay exceeds \$10.7 million for the month. Most people would probably take the job!

Example 8

Loan repayment. Francine's student loan is in the amount of \$6000. Interest is to be 9% compounded annually, and the entire amount is to be paid after 10 yr. How much is to be paid back?

Solution

1. **Familiarize.** Suppose we let P represent any principal amount. At the end of one year, the amount owed will be $P + 0.09P$, or $1.09P$. That amount will be the principal for the second year. The amount owed at the end of the

second year will be $1.09 \times$ New principal $= 1.09(1.09P)$, or 1.09^2P. Thus the amount owed at the beginning of successive years is as follows:

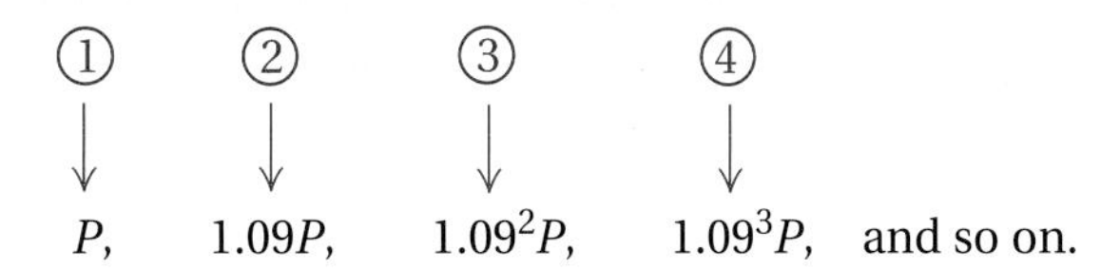

①
②
③
④
P, $1.09P$, 1.09^2P, 1.09^3P, and so on.

We have a geometric sequence. The amount owed at the beginning of the 11th year will be the amount owed at the end of the 10th year.

2. **Translate.** We have a geometric sequence with $a_1 = 6000$, $r = 1.09$, and $n = 11$. The appropriate formula is

$$a_n = a_1 r^{n-1}.$$

3. **Carry out.** We substitute and calculate:

$$a_{11} = \$6000(1.09)^{11-1} = \$6000(1.09)^{10}$$
$$\approx \$14{,}204.18. \quad \text{Using a calculator and rounding to the nearest hundredth}$$

4. **Check.** A check, by repeating the calculations, is left to the student.
5. **State.** Francine will owe \$14,204.18 at the end of 10 yr.

Example 9

Bungee jumping. A bungee jumper rebounds 60% of the height jumped. A bungee jump is made using a cord that stretches to 200 ft.

a) After jumping and then rebounding 9 times, how far has a bungee jumper traveled upward (the total rebound distance)?
b) Approximately how far will a jumper have traveled upward (bounced) before coming to rest?

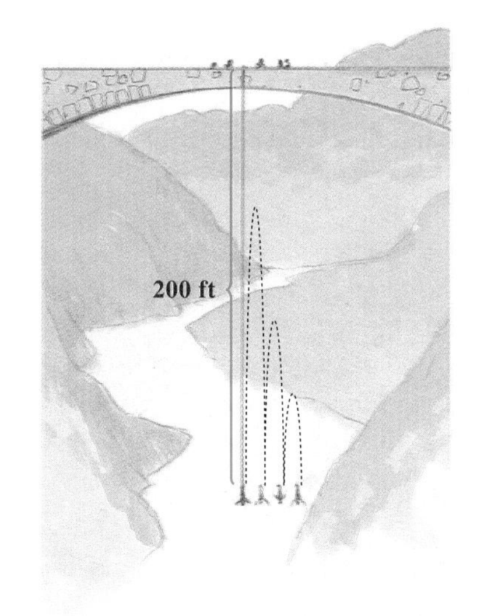

Solution

1. **Familiarize.** Let's do some calculations and look for a pattern.

First fall:	200 ft
First rebound:	0.6×200, or 120 ft
Second fall:	120 ft, or 0.6×200
Second rebound:	0.6×120, or $0.6(0.6 \times 200)$, which is 72 ft
Third fall:	72 ft, or $0.6(0.6 \times 200)$
Third rebound:	0.6×72, or $0.6(0.6(0.6 \times 200))$, which is 43.2 ft

The rebound distances form a geometric sequence:

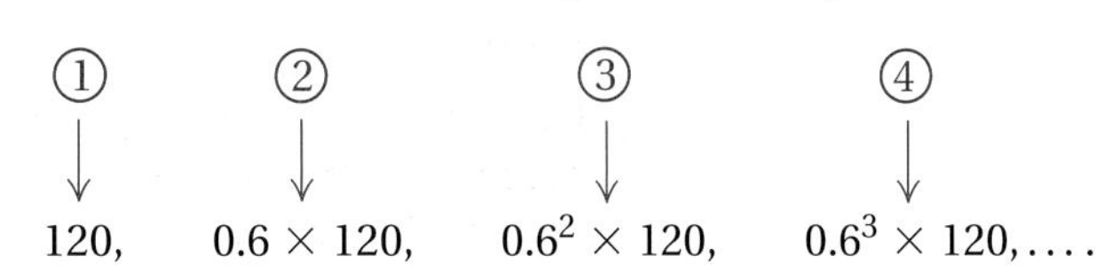

①
②
③
④
120, 0.6×120, $0.6^2 \times 120$, $0.6^3 \times 120, \ldots$.

2. Translate.

a) The total rebound distance after 9 bounces is the sum of a geometric sequence. The first term is 120 and the common ratio is 0.6. There will be 9 terms, so we can use the formula

$$S_n = \frac{a_1(1 - r^n)}{1 - r}.$$

b) Theoretically, the jumper will never stop bouncing. Realistically, the bouncing will eventually stop. To approximate the actual distance bounced, we consider an infinite number of bounces and use the formula

$$S_\infty = \frac{a_1}{1 - r}.$$ Since $r = 0.6$ and $|0.6| < 1$, we know that S_∞ exists.

3. Carry out.

a) We substitute into the formula and calculate:

$$S_9 = \frac{120[1 - (0.6)^9]}{1 - 0.6} \approx 297.$$ Using a calculator

b) We substitute and calculate:

$$S_\infty = \frac{120}{1 - 0.6} = 300.$$

4. Check. We can do the calculations again.

5. State.

a) In 9 bounces, the bungee jumper will have traveled upward a total distance of about 297 ft.

b) The jumper will have traveled upward a total of about 300 ft before coming to rest.

Exercise Set 11.3

FOR EXTRA HELP

Digital Video Tutor CD 9
Videotape 20

InterAct Math

Math Tutor Center

MathXL

MyMathLab.com

Find the common ratio for each geometric sequence.

1. $7, 14, 28, 56, \ldots$

2. $2, 6, 18, 54, \ldots$

3. $5, -5, 5, -5, \ldots$

4. $-5, -0.5, -0.05, -0.005, \ldots$

5. $\frac{1}{2}, -\frac{1}{4}, \frac{1}{8}, -\frac{1}{16}, \ldots$

6. $\frac{2}{3}, -\frac{4}{3}, \frac{8}{3}, -\frac{16}{3}, \ldots$

7. $75, 15, 3, \frac{3}{5}, \ldots$

8. $12, -4, \frac{4}{3}, -\frac{4}{9}, \ldots$

9. $\frac{1}{m}, \frac{3}{m^2}, \frac{9}{m^3}, \frac{27}{m^4}, \ldots$

10. $4, \frac{4m}{5}, \frac{4m^2}{25}, \frac{4m^3}{125}, \ldots$

Find the indicated term for each geometric sequence.

11. $3, 6, 12, \ldots$; the 7th term

12. $2, 8, 32, \ldots$; the 9th term

13. $5, 5\sqrt{2}, 10, \ldots$; the 9th term

14. $4, 4\sqrt{3}, 12, \ldots$; the 8th term

15. $-\frac{8}{243}, \frac{8}{81}, -\frac{8}{27}, \ldots$; the 10th term

16. $\frac{7}{625}, \frac{-7}{125}, \frac{7}{25}, \ldots$; the 13th term

17. \$1000, \$1080, \$1166.40, ...; the 12th term

18. \$1000, \$1070, \$1144.90, ...; the 11th term

Find the nth, or general, term for each geometric sequence.

19. $1, 3, 9, \ldots$

20. $25, 5, 1, \ldots$

21. $1, -1, 1, -1, \ldots$

22. $2, 4, 8, \ldots$

23. $\frac{1}{x}, \frac{1}{x^2}, \frac{1}{x^3}, \ldots$

24. $5, \frac{5m}{2}, \frac{5m^2}{4}, \ldots$

For Exercises 25–32, use the formula for S_n to find the indicated sum.

25. S_7 for the geometric series $6 + 12 + 24 + \cdots$

26. S_6 for the geometric series $16 - 8 + 4 - \cdots$

27. S_7 for the geometric series $\frac{1}{18} - \frac{1}{6} + \frac{1}{2} - \cdots$

Aha! **28.** S_5 for the geometric series $7 + 0.7 + 0.07 + \cdots$

29. S_8 for the series $1 + x + x^2 + x^3 + \cdots$

30. S_{10} for the series $1 + x^2 + x^4 + x^6 + \cdots$

31. S_{16} for the geometric sequence
\$200, \$200(1.06), \$200(1.06)^2, ...

32. S_{23} for the geometric sequence
\$1000, \$1000(1.08), \$1000(1.08)^2, ...

Determine whether each infinite geometric series has a limit. If a limit exists, find it.

33. $16 + 4 + 1 + \cdots$

34. $8 + 4 + 2 + \cdots$

35. $7 + 3 + \frac{9}{7} + \cdots$

36. $12 + 9 + \frac{27}{4} + \cdots$

37. $3 + 15 + 75 + \cdots$

38. $2 + 3 + \frac{9}{2} + \cdots$

39. $4 - 6 + 9 - \frac{27}{2} + \cdots$

40. $-6 + 3 - \frac{3}{2} + \frac{3}{4} - \cdots$

41. $0.43 + 0.0043 + 0.000043 + \cdots$

42. $0.37 + 0.0037 + 0.000037 + \cdots$

43. $\$500(1.02)^{-1} + \$500(1.02)^{-2} + \$500(1.02)^{-3} + \cdots$

44. $\$1000(1.08)^{-1} + \$1000(1.08)^{-2} + \$1000(1.08)^{-3} + \cdots$

Find fractional notation for each infinite sum. (These are geometric series.)

45. 0.7777...

46. 0.2222...

47. 8.3838...

48. 7.4747...

49. 0.15151515...

50. 0.12121212...

Solve. Use a calculator as needed for evaluating formulas.

51. *Rebound distance.* A ping-pong ball is dropped from a height of 20 ft and always rebounds one fourth of the distance fallen. How high does it rebound the 6th time?

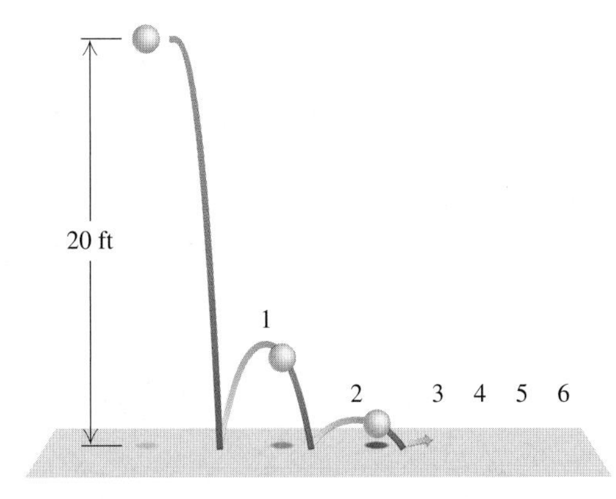

52. *Rebound distance.* Approximate the total of the rebound heights of the ball in Exercise 51.

53. *Population growth.* Yorktown has a current population of 100,000, and the population is increasing by 3% each year. What will the population be in 15 yr?

54. *Doubling time.* How long will it take for the population of Yorktown to double? (See Exercise 53.)

55. *Amount owed.* Gilberto borrows \$15,000. The loan is to be repaid in 13 yr at 8.5% interest, compounded annually. How much will be repaid at the end of 13 yr?

56. *Shrinking population.* A population of 5000 fruit flies is dying off at a rate of 4% per minute. How many flies will be alive after 15 min?

57. *Shrinking population.* For the population of fruit flies in Exercise 56, how long will it take for only 1800 fruit flies to remain alive? (See Exercise 56 and use logarithms.) Round to the nearest minute.

58. *Investing.* Leslie is saving money in a retirement account. At the beginning of each year, she invests \$1000 at 7%, compounded annually. How much will be in the retirement fund at the end of 40 yr?

59. *Rebound distance.* A superball dropped from the top of the Washington Monument (556 ft high) rebounds three fourths of the distance fallen. How far (up and down) will the ball have traveled when it hits the ground for the 6th time?

60. *Rebound distance.* Approximate the total distance that the ball of Exercise 59 will have traveled when it comes to rest.

61. *Stacking paper.* Construction paper is about 0.02 in. thick. Beginning with just one piece, a stack is doubled again and again 10 times. Find the height of the final stack.

62. *Monthly earnings.* Suppose you accepted a job for the month of February (28 days) under the following conditions. You will be paid \$0.01 the first day, \$0.02 the second, \$0.04 the third, and so on, doubling your previous day's salary each day. How much would you earn?

Aha! **63.** Under what circumstances is it possible for the 5th term of a geometric sequence to be greater than the 4th term but less than the 7th term?

64. When r is negative, a series is said to be *alternating*. Why do you suppose this terminology is used?

SKILL MAINTENANCE

Multiply.

65. $(x + y)(x^2 + 2xy + y^2)$

66. $(a - b)(a^2 - 2ab + b^2)$

Solve the system.

67. $5x - 2y = -3,$
$2x + 5y = -24$

68. $x - 2y + 3z = 4,$
$2x - y + z = -1,$
$4x + y + z = 1$

SYNTHESIS

69. Write a problem for a classmate to solve. Devise the problem so that a geometric series is involved and the solution is "The total amount in the bank is $\$900(1.08)^{40}$, or about \$19,550."

70. The infinite series

$$S_\infty = 2 + \frac{1}{2} + \frac{1}{2 \cdot 3} + \frac{1}{2 \cdot 3 \cdot 4} + \frac{1}{2 \cdot 3 \cdot 4 \cdot 5} + \frac{1}{2 \cdot 3 \cdot 4 \cdot 5 \cdot 6} + \cdots$$

is not geometric, but it does have a sum. Using S_1, S_2, S_3, S_4, S_5, and S_6, make a conjecture about the value of S_∞ and explain your reasoning.

71. Find the sum of the first n terms of

$$x^2 - x^3 + x^4 - x^5 + \cdots.$$

72. Find the sum of the first n terms of

$$1 + x + x^2 + x^3 + \cdots.$$

73. The sides of a square are each 16 cm long. A second square is inscribed by joining the midpoints of the sides, successively. In the second square we repeat the process, inscribing a third square. If this process is continued indefinitely, what is the sum of all of the areas of all the squares? (*Hint*: Use an infinite geometric series.)

74. Show that 0.999... is 1.

75. Using Example 5 and Exercises 33–44, explain how the graph of a geometric sequence can be used to determine whether a geometric series has a limit.

76. To compare the *graphs* of an arithmetic and a geometric sequence, we plot n on the horizontal axis and a_n on the vertical axis. Graph Example 1(a) of Section 11.2 and Example 1(a) of Section 11.3 on the same set of axes. How do the graphs of geometric sequences differ from the graphs of arithmetic sequences?

COLLABORATIVE CORNER

Bargaining for a Used Car

Focus: Geometric series

Time: 30 minutes

Group size: 2

Materials: Graphing calculators are optional.

*ACTIVITY**

1. One group member ("the seller") has a car for sale and is asking \$3500. The second ("the buyer") offers \$1500. The seller splits the difference (\$2000 ÷ 2 = \$1000) and lowers the price to \$2500. The buyer then splits the difference again (\$1000 ÷ 2 = \$500) and counters with \$2000. Continue in this manner and stop when you are able to agree on the car's selling price to the nearest penny.

2. What should the buyer's initial offer be in order to achieve a purchase price of \$2000? (Check several guesses to find the appropriate initial offer.)

3. The seller's price in the bargaining above can be modeled recursively (see Exercises 73, 74, and 82 in Section 11.1) by the sequence

$$a_1 = 3500, \qquad a_n = a_{n-1} - \frac{d}{2^{2n-3}},$$

where d is the difference between the initial price and the first offer. Use this recursively defined sequence to solve parts (1) and (2) above either manually or by using the SEQ MODE and the TABLE feature of a grapher.

4. The first four terms in the sequence in part (3) can be written as

$$a_1, \quad a_1 - \frac{d}{2}, \quad a_1 - \frac{d}{2} - \frac{d}{8},$$
$$a_1 - \frac{d}{2} - \frac{d}{8} - \frac{d}{32}.$$

Use the formula for the limit of an infinite geometric series to find a simple algebraic formula for the eventual sale price, P, when the bargaining process from above is followed. Verify the formula by using it to solve parts (1) and (2) above.

*This activity is based on the article, "Bargaining Theory, or Zeno's Used Cars," by James C. Kirby, *The College Mathematics Journal,* **27**(4), September 1996.

The Binomial Theorem

11.4

Binomial Expansion Using Pascal's Triangle • Binomial Expansion Using Factorial Notation

CONNECTING THE CONCEPTS

Sequences and series occur in many settings, some of which were mentioned in Sections 11.1–11.3. Although you may not have viewed it this way before, the expression $(x + y)^2$ can be regarded as a series: $x^2 + 2xy + y^2$.

In Chapter 5, we found that the expansion of $(x + y)^n$, for powers greater than 2, can be quite time-consuming. The reason for this extends all the way back to Chapter 1 and the rules for the order of operations and the properties of exponents: $(x + y)^n \neq x^n + y^n$. Since the terms in the expansion of $(x + y)^n$ have many uses, we devote this section to two methods that streamline the expansion of this important algebraic expression.

Binomial Expansion Using Pascal's Triangle

Consider the following expanded powers of $(a + b)^n$:

$$\begin{aligned}
(a + b)^0 &= 1\\
(a + b)^1 &= a + b\\
(a + b)^2 &= a^2 + 2a^1b^1 + b^2\\
(a + b)^3 &= a^3 + 3a^2b^1 + 3a^1b^2 + b^3\\
(a + b)^4 &= a^4 + 4a^3b^1 + 6a^2b^2 + 4a^1b^3 + b^4\\
(a + b)^5 &= a^5 + 5a^4b^1 + 10a^3b^2 + 10a^2b^3 + 5a^1b^4 + b^5.
\end{aligned}$$

Each expansion is a polynomial. There are some patterns to be noted:

1. There is one more term than the power of the binomial, n. That is, there are $n + 1$ terms in the expansion of $(a + b)^n$.
2. In each term, the sum of the exponents is the power to which the binomial is raised.
3. The exponents of a start with n, the power of the binomial, and decrease to 0 (since $a^0 = 1$, the last term has no factor of a). The first term has no factor of b, so powers of b start with 0 and increase to n.
4. The coefficients start at 1, increase through certain values, and then decrease through these same values back to 1. Let's study the coefficients further.

Suppose we wish to expand $(a + b)^8$. The patterns we noticed above indicate 9 terms in the expansion:

$$a^8 + c_1a^7b + c_2a^6b^2 + c_3a^5b^3 + c_4a^4b^4 + c_5a^3b^5 + c_6a^2b^6 + c_7ab^7 + b^8.$$

How can we determine the values for the c's? One method seems very simple, but it has some drawbacks. It involves writing down the coefficients in a triangular array as follows. We form what is known as **Pascal's triangle**:

$$
\begin{array}{lccccccccccc}
(a+b)^0: & & & & & & 1 & & & & & \\
(a+b)^1: & & & & & 1 & & 1 & & & & \\
(a+b)^2: & & & & 1 & & 2 & & 1 & & & \\
(a+b)^3: & & & 1 & & 3 & & 3 & & 1 & & \\
(a+b)^4: & & 1 & & 4 & & 6 & & 4 & & 1 & \\
(a+b)^5: & 1 & & 5 & & 10 & & 10 & & 5 & & 1
\end{array}
$$

There are many patterns in the triangle. Find as many as you can.

Perhaps you discovered a way to write the next row of numbers, given the numbers in the row above it. There are always 1's on the outside. Each remaining number is the sum of the two numbers above:

$$
\begin{array}{ccccccccccccc}
 & & & & & & 1 & & & & & & \\
 & & & & & 1 & & 1 & & & & & \\
 & & & & 1 & & 2 & & 1 & & & & \\
 & & & 1 & & 3 & & 3 & & 1 & & & \\
 & & 1 & & 4 & & 6 & & 4 & & 1 & & \\
 & 1 & & 5 & & 10 & & 10 & & 5 & & 1 & \\
1 & & 6 & & 15 & & 20 & & 15 & & 6 & & 1
\end{array}
$$

We see that in the bottom (seventh) row

the 1st and last numbers are 1;
the 2nd number is $1 + 5$, or 6:
the 3rd number is $5 + 10$, or 15;
the 4th number is $10 + 10$, or 20;
the 5th number is $10 + 5$, or 15; and
the 6th number is $5 + 1$, or 6.

Thus the expansion of $(a + b)^6$ is

$$(a+b)^6 = 1a^6 + 6a^5b + 15a^4b^2 + 20a^3b^3 + 15a^2b^4 + 6ab^5 + 1b^6.$$

To expand $(a + b)^8$, we complete two more rows of Pascal's triangle:

$$
\begin{array}{ccccccccccccccccc}
 & & & & & & & & 1 & & & & & & & & \\
 & & & & & & & 1 & & 1 & & & & & & & \\
 & & & & & & 1 & & 2 & & 1 & & & & & & \\
 & & & & & 1 & & 3 & & 3 & & 1 & & & & & \\
 & & & & 1 & & 4 & & 6 & & 4 & & 1 & & & & \\
 & & & 1 & & 5 & & 10 & & 10 & & 5 & & 1 & & & \\
 & & 1 & & 6 & & 15 & & 20 & & 15 & & 6 & & 1 & & \\
 & 1 & & 7 & & 21 & & 35 & & 35 & & 21 & & 7 & & 1 & \\
1 & & 8 & & 28 & & 56 & & 70 & & 56 & & 28 & & 8 & & 1
\end{array}
$$

Thus the expansion of $(a + b)^8$ has coefficients found in the 9th row above:

$$(a + b)^8 = 1a^8 + 8a^7b + 28a^6b^2 + 56a^5b^3 + 70a^4b^4 + 56a^3b^5 + 28a^2b^6 + 8ab^7 + 1b^8.$$

We can generalize our results as follows:

The Binomial Theorem (Form 1)

For any binomial $a + b$ and any natural number n,

$$(a + b)^n = c_0a^nb^0 + c_1a^{n-1}b^1 + c_2a^{n-2}b^2 + \cdots + c_{n-1}a^1b^{n-1} + c_na^0b^n,$$

where the numbers $c_0, c_1, c_2, \ldots, c_n$ are from the $(n + 1)$st row of Pascal's triangle.

Example 1 Expand: $(u - v)^5$.

Solution Using the binomial theorem, we have $a = u$, $b = \quad v$, and $n = 5$. We use the 6th row of Pascal's triangle: 1 5 10 10 5 1. Thus,

$$\begin{aligned}(u - v)^5 &= [u + (-v)]^5 \qquad \text{Rewriting } u - v \text{ as a sum}\\ &= 1(u)^5 + 5(u)^4(-v)^1 + 10(u)^3(-v)^2 + 10(u)^2(-v)^3\\ &\quad + 5(u)^1(-v)^4 + 1(-v)^5\\ &= u^5 - 5u^4v + 10u^3v^2 - 10u^2v^3 + 5uv^4 - v^5.\end{aligned}$$

Note that the signs of the terms alternate between + and −. When $-v$ is raised to an odd power, the sign is −.

Example 2 Expand: $\left(2t + \dfrac{3}{t}\right)^6$.

Solution Note that $a = 2t$, $b = 3/t$, and $n = 6$. We use the 7th row of Pascal's triangle: 1 6 15 20 15 6 1. Thus,

$$\begin{aligned}\left(2t + \frac{3}{t}\right)^6 &= 1(2t)^6 + 6(2t)^5\left(\frac{3}{t}\right)^1 + 15(2t)^4\left(\frac{3}{t}\right)^2 + 20(2t)^3\left(\frac{3}{t}\right)^3\\ &\quad + 15(2t)^2\left(\frac{3}{t}\right)^4 + 6(2t)^1\left(\frac{3}{t}\right)^5 + 1\left(\frac{3}{t}\right)^6\\ &= 64t^6 + 6(32t^5)\left(\frac{3}{t}\right) + 15(16t^4)\left(\frac{9}{t^2}\right) + 20(8t^3)\left(\frac{27}{t^3}\right)\\ &\quad + 15(4t^2)\left(\frac{81}{t^4}\right) + 6(2t)\left(\frac{243}{t^5}\right) + \frac{729}{t^6}\\ &= 64t^6 + 576t^4 + 2160t^2 + 4320 + 4860t^{-2} + 2916t^{-4}\\ &\quad + 729t^{-6}.\end{aligned}$$

Binomial Expansion Using Factorial Notation

The drawback to using Pascal's triangle is that we must compute all the preceding rows in the table to obtain the row needed for the expansion in which we are interested. The following method avoids this difficulty. It will also enable us to find a specific term— say, the 8th term—without computing all the other terms in the expansion. This method is useful in such courses as finite mathematics, calculus, and statistics.

To develop the method, we need some new notation. Products of successive natural numbers, such as $6 \cdot 5 \cdot 4 \cdot 3 \cdot 2 \cdot 1$ and $8 \cdot 7 \cdot 6 \cdot 5 \cdot 4 \cdot 3 \cdot 2 \cdot 1$, have a special notation. For the product $6 \cdot 5 \cdot 4 \cdot 3 \cdot 2 \cdot 1$, we write $6!$, read "6 factorial."

Factorial Notation

For any natural number n,

$$n! = n(n-1)(n-2) \cdots (3)(2)(1).$$

Here are some examples:

$$\begin{aligned}
6! &= 6 \cdot 5 \cdot 4 \cdot 3 \cdot 2 \cdot 1 = 720,\\
5! &= 5 \cdot 4 \cdot 3 \cdot 2 \cdot 1 = 120,\\
4! &= 4 \cdot 3 \cdot 2 \cdot 1 = 24,\\
3! &= 3 \cdot 2 \cdot 1 = 6,\\
2! &= 2 \cdot 1 = 2,\\
1! &= 1 = 1.
\end{aligned}$$

We also define $0!$ to be 1 for reasons explained shortly.

To simplify expressions like

$$\frac{8!}{5!\,3!},$$

note that

$$8! = 8 \cdot 7 \cdot 6 \cdot 5 \cdot 4 \cdot 3 \cdot 2 \cdot 1 = 8 \cdot 7! = 8 \cdot 7 \cdot 6! = 8 \cdot 7 \cdot 6 \cdot 5!,$$

and so on.

Caution! $\dfrac{6!}{3!} \neq 2!$ To see this, note that

$$\frac{6!}{3!} = \frac{6 \cdot 5 \cdot 4 \cdot \cancel{3} \cdot \cancel{2} \cdot \cancel{1}}{\cancel{3} \cdot \cancel{2} \cdot \cancel{1}} = 6 \cdot 5 \cdot 4.$$

Example 3 Simplify: $\dfrac{8!}{5!\,3!}$.

Solution

$$\frac{8!}{5!\,3!} = \frac{8 \cdot 7 \cdot 6 \cdot 5!}{5! \cdot 3 \cdot 2 \cdot 1} = 8 \cdot 7 \qquad \text{Removing a factor equal to 1: } \frac{6 \cdot 5!}{5! \cdot 3 \cdot 2} = 1$$

$$= 56.$$

The following notation is used in our second formulation of the binomial theorem.

$\binom{n}{r}$ Notation

For n, r nonnegative integers with $n \geq r$,

$$\binom{n}{r}, \quad \text{read "}n\text{ choose }r\text{,"} \quad \text{means} \quad \frac{n!}{(n-r)!\,r!}.^{*}$$

Example 4 Simplify: **(a)** $\binom{7}{2}$; **(b)** $\binom{9}{6}$; **(c)** $\binom{6}{6}$.

Solution

a) $$\binom{7}{2} = \frac{7!}{(7-2)!\,2!}$$

$$= \frac{7!}{5!\,2!} = \frac{7 \cdot 6 \cdot 5!}{5! \cdot 2 \cdot 1} = \frac{7 \cdot 6}{2}$$

$$= 7 \cdot 3$$

$$= 21$$

b) $$\binom{9}{6} = \frac{9!}{3!\,6!}$$

$$= \frac{9 \cdot 8 \cdot 7 \cdot 6!}{3 \cdot 2 \cdot 1 \cdot 6!} = \frac{9 \cdot 8 \cdot 7}{3 \cdot 2}$$

$$= 3 \cdot 4 \cdot 7$$

$$= 84$$

c) $$\binom{6}{6} = \frac{6!}{0!\,6!} = \frac{6!}{1 \cdot 6!} \qquad \text{Since } 0! = 1$$

$$= \frac{6!}{6!}$$

$$= 1$$

The PRB option of the MATH menu provides access to both factorial calculations and nCr. In both cases, a number must be entered first. To find $\binom{7}{2}$, we press [7] [MATH], select PRB and nCr, and press [2] [ENTER].

```
7 nCr 2
                21
```

Use a grapher to check Example 4(b).

*In many books and for many calculators, the notation ${}_nC_r$ is used instead of $\binom{n}{r}$.

Now we can restate the binomial theorem using our new notation.

The Binomial Theorem (Form 2)

For any binomial $a + b$ and any natural number n,

$$(a + b)^n = \binom{n}{0}a^n + \binom{n}{1}a^{n-1}b + \binom{n}{2}a^{n-2}b^2 + \cdots + \binom{n}{n}b^n.$$

Example 5

Expand: $(3x + y)^4$.

Solution We use the binomial theorem (Form 2) with $a = 3x$, $b = y$, and $n = 4$:

$$\begin{aligned}(3x + y)^4 &= \binom{4}{0}(3x)^4 + \binom{4}{1}(3x)^3y + \binom{4}{2}(3x)^2y^2 + \binom{4}{3}(3x)y^3 + \binom{4}{4}y^4 \\ &= \frac{4!}{4!\,0!}3^4x^4 + \frac{4!}{3!\,1!}3^3x^3y + \frac{4!}{2!\,2!}3^2x^2y^2 + \frac{4!}{1!\,3!}3xy^3 + \frac{4!}{0!\,4!}y^4 \\ &= 81x^4 + 108x^3y + 54x^2y^2 + 12xy^3 + y^4. \quad \text{Simplifying}\end{aligned}$$

Example 6

Expand: $(x^2 - 2y)^5$.

Solution In this case, $a = x^2$, $b = -2y$, and $n = 5$:

$$\begin{aligned}(x^2 - 2y)^5 &= \binom{5}{0}(x^2)^5 + \binom{5}{1}(x^2)^4(-2y) + \binom{5}{2}(x^2)^3(-2y)^2 \\ &\quad + \binom{5}{3}(x^2)^2(-2y)^3 + \binom{5}{4}(x^2)(-2y)^4 + \binom{5}{5}(-2y)^5 \\ &= \frac{5!}{5!\,0!}x^{10} + \frac{5!}{4!\,1!}x^8(-2y) + \frac{5!}{3!\,2!}x^6(-2y)^2 + \frac{5!}{2!\,3!}x^4(-2y)^3 \\ &\quad + \frac{5!}{1!\,4!}x^2(-2y)^4 + \frac{5!}{0!\,5!}(-2y)^5 \\ &= x^{10} - 10x^8y + 40x^6y^2 - 80x^4y^3 + 80x^2y^4 - 32y^5.\end{aligned}$$

Note that in the binomial theorem (Form 2), $\binom{n}{0}a^nb^0$ gives us the first term, $\binom{n}{1}a^{n-1}b^1$ gives us the second term, $\binom{n}{2}a^{n-2}b^2$ gives us the third term, and so on. This can be generalized to give a method for finding a specific term without writing the entire expansion.

Finding a Specific Term

The $(r + 1)$st term of $(a + b)^n$ is

$$\binom{n}{r} a^{n-r} b^r.$$

Example 7 Find the 5th term in the expansion of $(2x - 3y)^7$.

Solution First, we note that $5 = 4 + 1$. Thus, $r = 4$, $a = 2x$, $b = -3y$, and $n = 7$. Then the 5th term of the expansion is

$$\binom{7}{4}(2x)^{7-4}(-3y)^4, \quad \text{or} \quad \frac{7!}{3!\,4!}(2x)^3(-3y)^4, \quad \text{or} \quad 22{,}680x^3y^4.$$

It is because of the binomial theorem that $\binom{n}{r}$ is called a *binomial coefficient.* We can now explain why 0! is defined to be 1. In the binomial expansion, we want $\binom{n}{0}$ to equal 1 and we also want the definition

$$\binom{n}{r} = \frac{n!}{(n-r)!\,r!}$$

to hold for all whole numbers n and r. Thus we must have

$$\binom{n}{0} = \frac{n!}{(n-0)!\,0!} = \frac{n!}{n!\,0!} = 1.$$

This is satisfied only if 0! is defined to be 1.

Exercise Set 11.4

FOR EXTRA HELP

Digital Video Tutor CD 9
Videotape 20

InterAct Math

Math Tutor Center

MathXL

MyMathLab.com

Simplify.

1. $8!$

2. $9!$

3. $10!$

4. $11!$

5. $\dfrac{7!}{4!}$

6. $\dfrac{8!}{6!}$

7. $\dfrac{10!}{7!}$

8. $\dfrac{9!}{5!}$

9. $\binom{8}{2}$

10. $\binom{7}{4}$

11. $\binom{10}{6}$

12. $\binom{9}{5}$

13. $\binom{20}{18}$

14. $\binom{30}{3}$

15. $\binom{35}{2}$

16. $\binom{40}{38}$

Expand. Use both of the methods shown in this section.

17. $(m + n)^5$

18. $(a - b)^4$

19. $(x - y)^6$

20. $(p + q)^7$

21. $(x^2 - 3y)^5$

22. $(3c - d)^7$

23. $(3c - d)^6$

24. $(t^{-2} + 2)^6$

25. $(x - y)^3$

26. $(x - y)^5$

27. $\left(x + \frac{2}{y}\right)^9$

28. $\left(3s + \frac{1}{t}\right)^9$

29. $(a^2 - b^3)^5$

30. $(x^3 - 2y)^5$

31. $(\sqrt{3} - t)^4$

32. $(\sqrt{5} + t)^6$

33. $(x^{-2} + x^2)^4$

34. $\left(\frac{1}{\sqrt{x}} - \sqrt{x}\right)^6$

Find the indicated term for each binomial expression.

35. 3rd, $(a + b)^6$

36. 6th, $(x + y)^7$

37. 12th, $(a - 3)^{14}$

38. 11th, $(x - 2)^{12}$

39. 5th, $(2x^3 + \sqrt{y})^8$

40. 4th, $\left(\frac{1}{b^2} + c\right)^7$

41. Middle, $(2u - 3v^2)^{10}$

42. Middle two, $(\sqrt{x} + \sqrt{3})^5$

Aha! **43.** 9th, $(x - y)^8$

44. 10th, $(a - b)^9$

45. Maya claims that she can calculate mentally the first two and the last two terms of the expansion of $(a + b)^n$ for any whole number n. How do you think she does this?

46. Without performing any calculations, how can you tell if the expansions of $(x - y)^8$ and $(y - x)^8$ are equal?

SKILL MAINTENANCE

Solve.

47. $\log_2 x + \log_2(x - 2) = 3$

48. $\log_3(x + 2) - \log_3(x - 2) = 2$

49. $e^t = 280$

50. $\log_5 x^2 = 2$

SYNTHESIS

51. Explain how someone can determine the x^2-term of the expansion of $\left(x - \frac{3}{x}\right)^{10}$ without calculating any other terms.

52. Devise two problems requiring the use of the binomial theorem. Design the problems so that one is solved more easily using Form 1 and the other is solved more easily using Form 2. Then explain what makes one form easier to use than the other in each case.

53. Show that there are exactly $\binom{5}{3}$ ways of forming a subset of size 3 from a set of 5 elements.

54. *Baseball.* At one point in a recent season. Derek Jeter of the New York Yankees had a batting average of .325. At that time, if someone were to randomly select 5 of his "at-bats," the probability of his getting exactly 3 hits would be the 3rd term of the binomial expansion of $(0.325 + 0.675)^5$. Find that term and use a calculator to estimate the probability.

55. *Widows or divorcees.* The probability that a woman will be either widowed or divorced is 85%. If 8 women are randomly selected, the probability that exactly 5 of them will be either widowed or divorced is the 6th term of the binomial expansion of $(0.15 + 0.85)^8$. Use a calculator to estimate that probability.

56. *Baseball.* In reference to Exercise 54, the probability that Jeter will get *at most* 3 hits is found by adding the last 4 terms of the binomial expansion of $(0.325 + 0.675)^5$. Find these terms and use a calculator to estimate the probability.

57. *Widows or divorcees.* In reference to Exercise 55, the probability that *at least* 6 of the women will be widowed or divorced is found by adding the last three terms of the binomial expansion of $(0.15 + 0.85)^8$. Find these terms and use a calculator to estimate the probability.

58. Prove that

$$\binom{n}{r} = \binom{n}{n-r}$$

for any whole numbers n and r. Assume $r \leq n$.

59. Find the term of

$$\left(\frac{3x^2}{2} - \frac{1}{3x}\right)^{12}$$

that does not contain x.

60. Find the middle term of $(x^2 - 6y^{3/2})^6$.

61. Find the ratio of the 4th term of

$$\left(p^2 - \frac{1}{2}p\sqrt[3]{q}\right)^5$$

to the 3rd term.

62. Find the term containing $\frac{1}{x^{1/6}}$ of

$$\left(\sqrt[3]{x} - \frac{1}{\sqrt{x}}\right)^7.$$

63. What is the degree of $(x^2 + 3)^4$?

Aha! **64.** Multiply: $(x^2 + 2xy + y^2)(x^2 + 2xy + y^2)^2(x + y)$.

Summary and Review 11

Key Terms

Sequence, p. 662
Progression, p. 662
Infinite sequence, p. 662
Finite sequence, p. 662
General term, p. 663
Series, p. 664
Infinite series, p. 664
Partial sum, p. 664
Finite series, p. 664
Sigma notation, p. 665
Summation notation, p. 665
Index of summation, p. 665
Arithmetic sequence, p. 668
Arithmetic progression, p. 668
Common difference, p. 669
Arithmetic series, p. 671
Geometric sequence, p. 678
Geometric progression, p. 678
Common ratio, p. 678
Geometric series, p. 679
Infinite geometric series, p. 680
Pascal's triangle, p. 689
Binomial theorem, p. 690
Factorial, p. 691
Binomial coefficient, p. 694

Important Properties and Formulas

Arithmetic sequence: $a_{n+1} = a_n + d$

nth term of an arithmetic sequence: $a_n = a_1 + (n - 1)d$

Sum of the first n terms of an arithmetic sequence: $S_n = \frac{n}{2}(a_1 + a_n)$

Geometric sequence: $a_{n+1} = a_n \cdot r$

nth term of a geometric sequence: $a_n = a_1 r^{n-1}$

Sum of the first n terms of a geometric sequence: $S_n = \frac{a_1(1 - r^n)}{1 - r}$

Limit of an infinite geometric series: $S_\infty = \frac{a_1}{1 - r}, \quad |r| < 1$

Factorial notation: $n! = n(n - 1)(n - 2) \cdots 3 \cdot 2 \cdot 1$

Binomial coefficient: $\binom{n}{r} = \frac{n!}{(n - r)!\, r!}$

Binomial theorem:

$$(a + b)^n = \binom{n}{0}a^n + \binom{n}{1}a^{n-1}b + \binom{n}{2}a^{n-2}b^2 + \cdots + \binom{n}{n}b^n$$

$(r + 1)$st term of $(a + b)^n$: $\binom{n}{r}a^{n-r}b^r$

Review Exercises

Find the first four terms; the 8th term, a_8; and the 12th term, a_{12}.

1. $a_n = 4n - 3$

2. $a_n = \dfrac{n-1}{n^2+1}$

Predict the general term. Answers may vary.

3. $-2, -4, -6, -8, -10, \ldots$

4. $-1, 3, -5, 7, -9, \ldots$

Write out and evaluate each sum.

5. $\sum_{k=1}^{5} (-2)^k$

6. $\sum_{k=2}^{7} (1 - 2k)$

Rewrite using sigma notation.

7. $4 + 8 + 12 + 16 + 20$

8. $\dfrac{-1}{2} + \dfrac{1}{4} + \dfrac{-1}{8} + \dfrac{1}{16} + \dfrac{-1}{32}$

9. Find the 14th term of the arithmetic sequence $-6, 1, 8, \ldots$.

10. Find d when $a_1 = 11$ and $a_{10} = 35$. Assume an arithmetic sequence.

11. Find a_1 and d when $a_{12} = 25$ and $a_{24} = 40$. Assume an arithmetic sequence.

12. Find the sum of the first 17 terms of the arithmetic series $-8 + (-11) + (-14) + \cdots$.

13. Find the sum of all the multiples of 6 from 12 to 318, inclusive.

14. Find the 20th term of the geometric sequence $2, 2\sqrt{2}, 4, \ldots$.

15. Find the common ratio of the geometric sequence $2, \frac{4}{3}, \frac{8}{9}, \ldots$.

16. Find the nth term of the geometric sequence $-2, 2, -2, \ldots$.

17. Find the nth term of the geometric sequence $3, \frac{3}{4}x, \frac{3}{16}x^2, \ldots$.

18. Find S_6 for the geometric series

$$3 + 12 + 48 + \cdots.$$

19. Find S_{12} for the geometric series

$$3x - 6x + 12x - \cdots.$$

Determine whether each infinite geometric series has a limit. If a limit exists, find it.

20. $6 + 3 + 1.5 + 0.75 + \cdots$

21. $7 - 4 + \frac{16}{7} - \cdots$

22. $2 + (-2) + 2 + (-2) + \cdots$

23. $0.04 + 0.08 + 0.16 + 0.32 + \cdots$

24. $\$2000 + \$1900 + \$1805 + \$1714.75 + \cdots$

25. Find fractional notation for $0.555555\ldots$.

26. Find fractional notation for $1.39393939\ldots$.

Solve.

27. Adam took a telemarketing job, starting with an hourly wage of \$11.40. He was promised a raise of 20¢ per hour every 3 mos for 8 yr. At the end of 8 yr, what will be his hourly wage?

28. A stack of poles has 42 poles in the bottom row. There are 41 poles in the second row, 40 poles in the third row, and so on, ending with 1 pole in the top row. How many poles are in the stack?

29. A student loan is in the amount of \$10,000. Interest is 7%, compounded annually, and the amount is to be paid off in 12 yr. How much is to be paid back?

30. Find the total rebound distance of a ball, given that it is dropped from a height of 12 m and each rebound is one-third of the preceding one.

Simplify.

31. $7!$

32. $\dbinom{8}{3}$

33. Find the 3rd term of $(a + b)^{20}$.

34. Expand: $(x - 2y)^4$.

SYNTHESIS

35. What happens to the terms of a geometric sequence with $|r| < 1$ as n gets larger? Why?

36. Compare the two forms of the binomial theorem given in the text. Under what circumstances would one be more useful than the other?

37. Find the sum of the first n terms of the geometric series $1 - x + x^2 - x^3 + \cdots$.

38. Expand: $(x^{-3} + x^3)^5$.

Chapter Test 11

1. Find the first five terms and the 16th term of a sequence with general term $a_n = 6n - 5$.

2. Predict the general term of the sequence
$\frac{4}{3}, \frac{4}{9}, \frac{4}{27}, \ldots$.

3. Write out and evaluate:

$$\sum_{k=1}^{5} (3 - 2^k).$$

4. Rewrite using sigma notation:

$$1 + (-8) + 27 + (-64) + 125.$$

5. Find the 12th term, a_{12}, of the arithmetic sequence $9, 4, -1, \ldots$.

Assume arithmetic sequences for Questions 6 and 7.

6. Find the common difference d when $a_1 = 9$ and $a_7 = 11\frac{1}{4}$.

7. Find a_1 and d when $a_5 = 16$ and $a_{10} = -3$.

8. Find the sum of all the multiples of 12 from 24 to 240, inclusive.

9. Find the 6th term of the geometric sequence $72, 18, 4\frac{1}{2}, \ldots$.

10. Find the common ratio of the geometric sequence $22\frac{1}{2}, 15, 10, \ldots$.

11. Find the nth term of the geometric sequence $3, -9, 27, \ldots$.

12. Find the sum of the first nine terms of the geometric series

$$(1 + x) + (2 + 2x) + (4 + 4x) + \cdots.$$

Determine whether each infinite geometric series has a limit. If a limit exists, find it.

13. $0.5 + 0.25 + 0.125 + \cdots$

14. $0.5 + 1 + 2 + 4 + \cdots$

15. $\$1000 + \$80 + \$6.40 + \cdots$

16. Find fraction notation for $0.85858585\ldots$.

17. An auditorium has 31 seats in the first row, 33 seats in the second row, 35 seats in the third row, and so on, for 18 rows. How many seats are in the 17th row?

18. Lindsay's Uncle Ken gave her \$100 for her first birthday, \$200 for her second birthday, and so on, until her eighteenth birthday. How much did he give her in all?

19. Each week the price of a \$15,000 boat will be reduced 5% of the previous week's price. If we assume that it is not sold, what will be the price after 10 weeks?

20. Find the total rebound distance of a ball that is dropped from a height of 18 m, with each rebound two thirds of the preceding one.

21. Simplify: $\binom{13}{11}$.

22. Expand: $(x^2 - 3y)^5$.

23. Find the 4th term in the expansion of $(a + x)^{12}$.

SYNTHESIS

24. Find a formula for the sum of the first n even natural numbers:

$$2 + 4 + 6 + \cdots + 2n.$$

25. Find the sum of the first n terms of

$$1 + \frac{1}{x} + \frac{1}{x^2} + \frac{1}{x^3} + \cdots.$$

Cumulative Review 1–11

Simplify.

1. $(-9x^2y^3)(5x^4y^{-7})$

2. $|-3.5 + 9.8|$

3. $2y - [3 - 4(5 - 2y) - 3y]$

4. $(10 \cdot 8 - 9 \cdot 7)^2 - 54 \div 9 - 3$

5. Evaluate

$$\frac{ab - ac}{bc}$$

for $a = -2$, $b = 3$, and $c = -4$.

Perform the indicated operations and simplify.

6. $(5a^2 - 3ab - 7b^2) - (2a^2 + 5ab + 8b^2)$

7. $(-3x^2 + 4x^3 - 5x - 1) + (9x^3 - 4x^2 + 7 - x)$

8. $(2a - 1)(3a + 5)$

9. $(3a^2 - 5y)^2$

10. $\dfrac{1}{x - 2} - \dfrac{4}{x^2 - 4} + \dfrac{3}{x + 2}$

11. $\dfrac{x^2 - 6x + 8}{3x + 9} \cdot \dfrac{x + 3}{x^2 - 4}$

12. $\dfrac{3x + 3y}{5x - 5y} \div \dfrac{3x^2 + 3y^2}{5x^3 - 5y^3}$

13. $\dfrac{x - \dfrac{a^2}{x}}{1 + \dfrac{a}{x}}$

Factor.

14. $4x^2 - 12x + 9$

15. $27a^3 - 8$

16. $a^3 + 3a^2 - ab - 3b$

17. $15y^4 + 33y^2 - 36$

18. For the function described by

$$f(x) = 3x^2 - 4x,$$

find $f(-2)$.

19. Divide:

$$(7x^4 - 5x^3 + x^2 - 4) \div (x - 2).$$

Solve.

20. $9(x - 1) - 3(x - 2) = 1$

21. $\dfrac{6}{x} + \dfrac{6}{x + 2} = \dfrac{5}{2}$

22. $2x + 1 > 5$ *or* $x - 7 \le 3$

23. $5x + 3y = 2,$
$3x + 5y = -2$

24. $x + y - z = 0,$
$3x + y + z = 6,$
$x - y + 2z = 5$

25. $3\sqrt{x - 1} = 5 - x$

26. $x^4 - 29x^2 + 100 = 0$

27. $x^2 + y^2 = 8,$
$x^2 - y^2 = 2$

28. $5^x = 8$

29. $\log(x^2 - 25) - \log(x + 5) = 3$

30. $\log_4 x = -2$

31. $7^{2x+3} = 49$

32. $|2x - 1| \le 5$

33. $7x^2 + 14 = 0$

34. $x^2 + 4x = 3$

35. $y^2 + 3y > 10$

36. Let $f(x) = x^2 - 2x$. Find a such that $f(a) = 48$.

37. If $f(x) = \sqrt{x + 1}$ and $g(x) = \sqrt{x - 2} + 3$, find a such that $f(a) = g(a)$.

Solve.

38. The perimeter of a rectangle is 34 ft. The length of a diagonal is 13 ft. Find the dimensions of the rectangle.

39. A music club offers two types of membership. Limited members pay a fee of \$10 a year and can buy CDs for \$10 each. Preferred members pay \$20 a year and can buy CDs for \$7.50 each. For what numbers of annual CD purchases would it be less expensive to be a preferred member?

40. Find three consecutive integers whose sum is 198.

41. A pentagon with all five sides the same size has a perimeter equal to that of an octagon in which all eight sides are the same size. One side of the pentagon is 2 less than three times one side of the octagon. What is the perimeter of each figure?

42. Mark's Natural Foods mixes herbs that cost \$2.68 an ounce with herbs that cost \$4.60 an ounce to create a seasoning that costs \$3.80 an ounce. How many ounces of each herb should be mixed together to make 24 oz of the seasoning?

43. An airplane can fly 190 mi with the wind in the same time it takes to fly 160 mi against the wind. The speed of the wind is 30 mph. How fast can the plane fly in still air?

44. Bianca can tap the sugar maple trees in Southway Park in 21 hr. Delia can tap the trees in 14 hr. How long would it take them, working together, to tap the trees?

45. The centripetal force F of an object moving in a circle varies directly as the square of the velocity v and inversely as the radius r of the circle. If $F = 8$ when $v = 1$ and $r = 10$, what is F when $v = 2$ and $r = 16$?

46. A farmer wants to fence in a rectangular area next to a river. (Note that no fence will be needed along the river.) What is the area of the largest region that can be fenced in with 100 ft of fencing?

Graph.

47. $3x - y = 6$

48. $\frac{x^2}{25} + \frac{y^2}{4} = 1$

49. $y = \log_2 x$

50. $2x - 3y < -6$

51. Graph: $f(x) = -2(x - 3)^2 + 1$.

a) Label the vertex.

b) Draw the axis of symmetry.

c) Find the maximum or minimum value.

52. Solve $V = P - Prt$ for r.

53. Solve $I = \frac{R}{R + r}$ for R.

54. Find a linear equation whose graph has a y-intercept of $(0, -3)$ and is parallel to the line whose equation is $3x - y = 6$.

Find the domain of each function.

55. $f(x) = \sqrt{5 - 3x}$

56. $g(x) = \frac{x - 4}{x^2 - 2x + 1}$

57. Multiply $(8.9 \times 10^{-17})(7.6 \times 10^4)$. Write scientific notation for the answer.

58. Multiply and simplify: $\sqrt{8x}\sqrt{8x^3y}$.

59. Simplify: $(25x^{4/3}y^{1/2})^{3/2}$.

60. Divide and simplify:

$$\frac{\sqrt[3]{15x}}{\sqrt[3]{3y^2}}.$$

61. Rationalize the denominator:

$$\frac{1 - \sqrt{x}}{1 + \sqrt{x}}.$$

62. Multiply these complex numbers:

$$(3 + 2i)(4 - 7i).$$

63. Write a quadratic equation whose solutions are $5\sqrt{2}$ and $-5\sqrt{2}$.

64. Find the center and the radius of the circle

$$x^2 + y^2 - 4x + 6y - 23 = 0.$$

65. Express as a single logarithm:

$$\frac{2}{3}\log_a x - \frac{1}{2}\log_a y + 5\log_a z.$$

66. Convert to an exponential equation: $\log_a c = 5$.

Find each of the following using a calculator.

67. $\log 5677.2$

68. $10^{-3.587}$

69. $\ln 5677.2$

70. $e^{-3.587}$

71. The number of personal computers in Mexico has grown exponentially from 0.12 million in 1985 to 6.0 million in 2000.

a) Find the exponential growth rate, k, to three decimal places and write an exponential function describing the number of personal computers in Mexico t years after 1985.

b) Predict the number of personal computers in Mexico in 2008.

72. Find the distance between the points $(-1, -5)$ and $(2, -1)$.

73. Find the 21st term of the arithmetic sequence $19, 12, 5, \ldots$.

74. Find the sum of the first 25 terms of the arithmetic series $-1 + 2 + 5 + \cdots$.

75. Find the general term of the geometric sequence $16, 4, 1, \ldots$.

76. Find the 7th term of $(a - 2b)^{10}$.

77. Find the sum of the first nine terms of the geometric series $x + 1.5x + 2.25x + \cdots$.

78. On Elyse's 9th birthday, her grandmother opened a savings account for her with \$100. The account draws 6% interest, compounded annually. If Elyse neither adds to nor withdraws any money from the bank, how much will be in the account on her 18th birthday?

SYNTHESIS

Solve.

79. $\dfrac{9}{x} - \dfrac{9}{x+12} = \dfrac{108}{x^2 + 12x}$

80. $\log_2 (\log_3 x) = 2$

81. y varies directly as the cube of x and x is multiplied by 0.5. What is the effect on y?

82. Divide these complex numbers:

$$\frac{2\sqrt{6} + 4\sqrt{5}i}{2\sqrt{6} - 4\sqrt{5}i}.$$

83. Diaphantos, a famous mathematician, spent $\frac{1}{6}$ of his life as a child, $\frac{1}{12}$ as an adolescent, and $\frac{1}{7}$ as a bachelor. Five years after he was married, he had a son who died 4 years before his father at half his father's final age. How long did Diaphantos live?

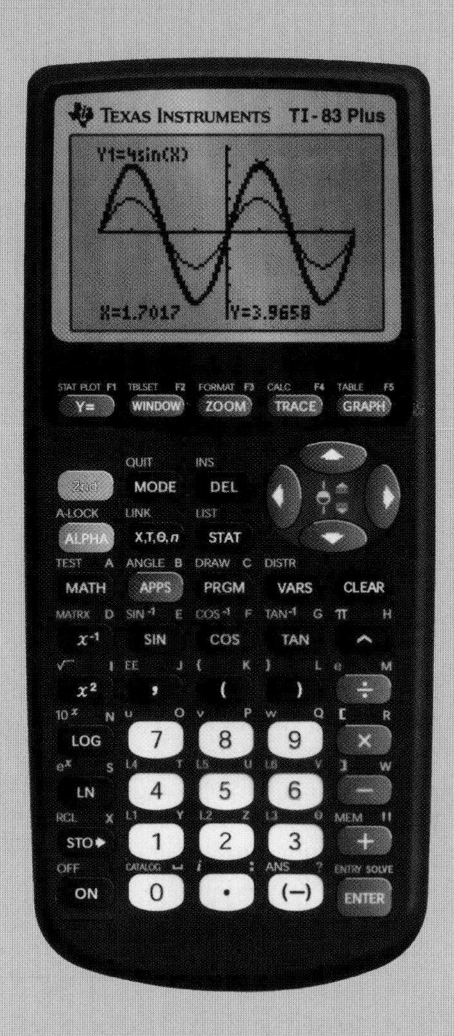

APPENDIX

The Graphing Calculator

I.1 Introduction to the Graphing Calculator

I.2 Variables and Functions

2.1 Graphs

2.5 Other Equations of Lines

4.4 Inequalities in Two Variables

8.8 Problem Solving and Quadratic Functions

Graphing calculators and computer graphing software, often referred to collectively as graphers, *make possible the quick graphing of equations. This appendix discusses the graphing calculator. Sections I.1 and I.2 of this appendix introduce the calculator, while the other sections are each linked to a corresponding section in the text.*

Different calculators may require different keystrokes to use the same feature. In this appendix, exact keystrokes referenced and screens shown apply to the Texas Instruments TI-83® Plus calculator. If your calculator does not have a key by the name shown, consult the user's manual that came with your calculator to learn how to perform the procedure. Often it is helpful to simply experiment by trying different combinations of keystrokes.

I.1 Introduction to the Graphing Calculator

A graphing calculator does not look exactly like the scientific calculator to which you may be accustomed. The screen is much larger: A typical screen holds up to 8 lines of text and is 16 columns wide. Screens are composed of *pixels*, or dots. After the calculator has been turned on, a blinking rectangle or line appears on the screen. This rectangle or line is called the *cursor*. It indicates your current position on the screen. In the graphing mode, the cursor may appear as a cross. The four arrow keys are used to move the cursor.

The *contrast* controls how dark the characters appear on the screen. Once the grapher is on, you can adjust the contrast by pressing **2nd** and either the up or down arrow key.

Entering and Editing Expressions

The large screen allows room for graphs and lets you view an expression in its entirety before it is executed. Pressing the operation keys ($+$, $-$, $\times$, $\div$) does not automatically perform the operation; you must also press an Execute or ENTER key.

Example 1 Calculate: $4 + 3(7 - 10)$.

Solution We first enter the expression as it is written and check our typing. If it is correct, pressing **ENTER** will perform all the operations in the correct order. The answer will appear on the right side of the screen, as shown below.

```
4+3(7−10)
               −5
```

Thus, $4 + 3(7 - 10) = -5$.

If, before pressing ENTER, you see that your typing is not correct, use the arrow keys to move through the expression and correct any errors. Pressing the DEL key will delete the character under the cursor. To replace the character under the cursor with another character, simply press the desired key. If you left out a character, press 2nd INS. The cursor will change to a line, and characters can then be inserted to the left of the cursor.

After pressing ENTER, you can still edit, or change, the previous expression. To do so, press 2nd ENTRY to copy the previous expression on the screen. You can then change it as described above.

A graphing calculator has many more features than those written on the keys. Written above each key are color-coded secondary features, which are performed by first pressing 2nd and then the key below the desired feature. Above many keys, there is also an *alpha character*. These are accessed by first pressing ALPHA and then the key below the character. The ALPHA key may be locked on by pressing 2nd A-LOCK. Press ALPHA again to unlock the key.

Other features are available from *menus*. When certain keys are pressed, a menu, or list of further options, appears on the screen. You can select an item from the menu by pressing the number of the item or by using the cursor keys to highlight the item and then pressing ENTER. Sometimes a menu has a *secondary menu* from which to choose. For example, pressing MATH gives you the menu shown below. The secondary menu is listed across the top of the screen.

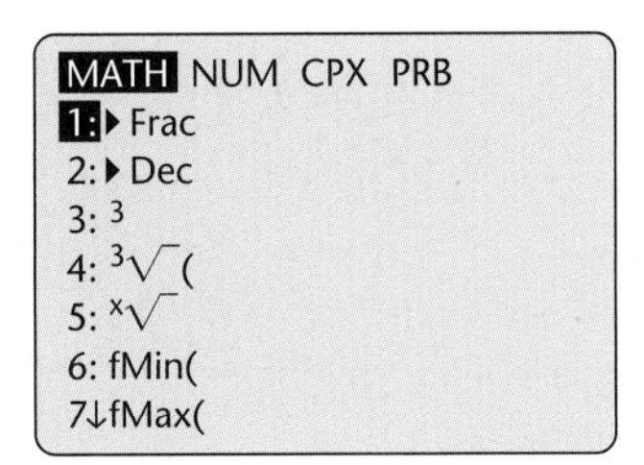

The screen in which you perform calculations is called the *home screen*. To return to the home screen, simply press 2nd QUIT.

Functions and Grouping Symbols

On a graphing calculator, most operations can be entered in the order in which you would write them on paper, except that you may need to add parentheses. For example, on most graphers, the square root is found by first pressing the square root key, then entering the number, and then pressing ENTER. A fraction bar is entered as a division symbol, /, and does not serve as a grouping symbol, so numerators and denominators may need to be enclosed in parentheses. It is important to remember that the subtraction symbol and the negative symbol are two different keys. To enter a negative number, we press (−) and to subtract, we press −.

Example 2 Calculate each of the following:

$$\sqrt{3 \cdot 5 - 3(6 - 4)} \quad \text{and} \quad \frac{4 + 1.5}{3.2 - 1}.$$

```
√(3*5−3(6−4))
                    3
(4+1.5)/(3.2−1)
                  2.5
```

Solution The keystrokes and the results of the calculations can be seen in the screen at left. Note that the entire radicand, as well as the numerator and the denominator of the rational expression, must be enclosed in parentheses.

Other functions, like absolute value, abs, work in a similar way. On some graphers, the left parenthesis is supplied automatically.

Any number can be squared by typing the number, pressing the x^2 key, and then pressing ENTER. In general, for exponents other than 2, use the $\wedge$ key, followed by the exponent. If the exponent is itself an expression, enclose it within parentheses. An exponent that is an integer or decimal does not need to be enclosed in parentheses. But an exponent written as a fraction must be in parentheses. (For more on rational exponents, see Chapter 7.)

Example 3 Calculate: $(3 + 5)^{-2}$, 7^{8-1}, and $8^{2/3}$.

Solution The results of the calculations are shown in the screen below.

```
(3 + 5)^ − 2
              .015625
7^(8 − 1)
               823543
8^(2/3)
                    4
```

Exercise Set I.1

FOR EXTRA HELP

Digital Video Tutor CD: None
Videotape: None

InterAct Math

Math Tutor Center

MathXL

MyMathLab.com

Short exercise sets appear throughout this appendix. These provide an opportunity to practice the skills just discussed. Answers can be found at the back of the book.

Use a grapher to evaluate each expression.

1. $\sqrt{-7 + 2(10 - 2)}$

2. $|-9|$

3. $|9 - (5 + (-3(6.4 - 19)))|$

4. $\dfrac{-4 - 3|13 - 5.3| + \sqrt{2.25}}{5.6 \div 7 - 55 \div 10}$

5. $(3.4 - 5.6(7.3 - 8.79))^2$

6. $2.6^{2(3-1)}$

7. $\left(\dfrac{3}{4}\right)^{-2}$

8. $16^{3/4}$

I.2 Variables and Functions

Graphers have memory available in which values of variables, equations of functions, and programs can be stored. (A *program* is a series of instructions that can be performed repeatedly. Consult your user's manual for information about storing and running programs.)

To store a numerical value for later use, type the number, press [STO▶], and then type the name of the variable using an alpha character. To recall that value, press [2nd] [RCL] and the variable name or press [ALPHA] followed by the variable name to recall its value. This can be done in the middle of a calculation.

To store the equation of a function, press [Y=]. We enter an equation the same way in which we would write it on paper. Always let the variable be x. To enter an x, we use the [X, T, θ, n] key.

Example 1 Enter the equations

$$y_1 = x^2 - x - 6, \qquad y_2 = 2x - 6, \quad \text{and} \quad y_3 = 2x^4 - 5x^3 + x^2 - 10.$$

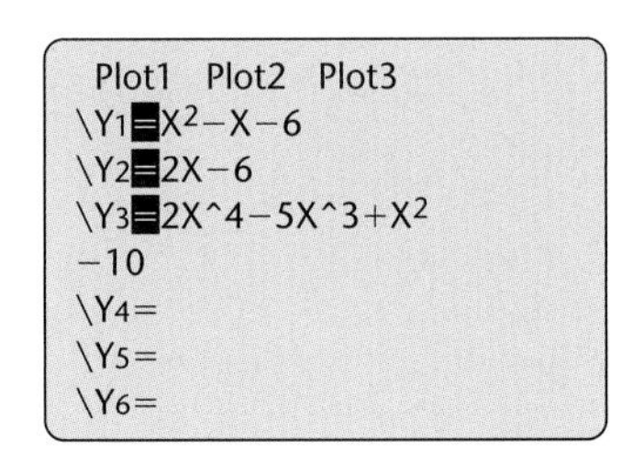

Solution Press [Y=] [X, T, θ, n] [x²] [−] [X, T, θ, n] [−] [6] [ENTER] to enter the first equation. You will now be on the line for Y2 and can press [2] [X, T, θ, n] [−] [6] [ENTER]. Enter y_3 in a similar way. Your screen should look like the one at left.

Remember: After you have worked on a screen other than the home screen, press [2nd] [QUIT] to return to the home screen.

Once entered, a function can be graphed or used in equations. Graphing functions and solving equations are discussed throughout the text and in the remainder of this appendix.

To evaluate a function for a particular value of x, you can use function notation or store the value to the variable X. Shown below are the keystrokes for each method of finding $f(2)$, where $f(x) = y_1 = x^2 - x - 6$.

Function Notation

[VARS] [▷] [1] [1] [(] [2] [)] [ENTER]

Storing the Value to X

[2] [STO▶] [X, T, θ, n] [ENTER]

[VARS] [▷] [1] [1] [ENTER]

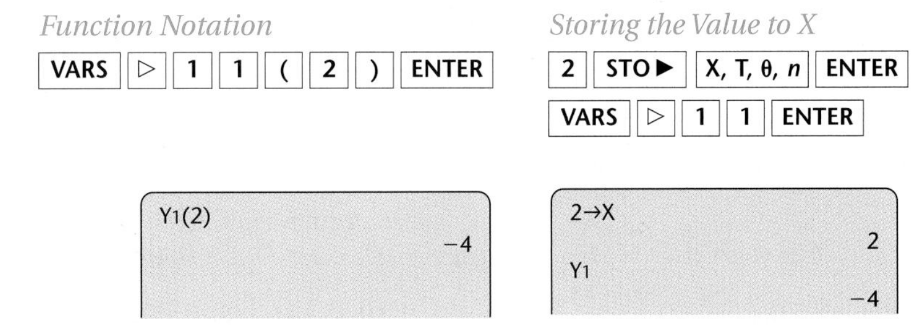

Exercise Set I.2

FOR EXTRA HELP

Digital Video Tutor CD: None
Videotape: None

InterAct Math

Math Tutor Center

MathXL

MyMathLab.com

1. Find the circumference and the area of a circle with radius 10.3452 cm using the following steps.

a) Store the value 10.3452 to the variable R.

b) Calculate the circumference using the formula $C = 2\pi R$. Use the π key and the stored value for R. (You can type simply $2\pi R$ and press ENTER.) Round to four decimal places.

c) Calculate the area using the formula $A = \pi R^2$. Round to four decimal places.

2. The formula $I = PRT$ gives the amount of simple interest earned on a principal P at an interest rate R (in decimal form) for a length of time T (in years). You have $1452.39 to invest for 7 months. Store the values for P and T and use them to calculate the interest earned when the interest rate is each of the following.

a) 10% **b)** 6%
c) 4.375% **d)** 8.65%

Evaluate each polynomial for the given values.

3. $y = 3x^4 - 2x^3 + 9x - 17$, for $x = 4, -1, 36$, and 98.

4. $y = 1.5x^6 - 2.8x^4 + 0.1x^2$, for $x = 1.23, -1.23, 107$, and -107.

2.1 Graphs

When we draw a graph with a grapher, the part of the graph shown is in a *window*. The window is described by four numbers: the minimum x-value shown (Xmin), the maximum x-value (Xmax), the minimum y-value (Ymin), and the maximum y-value (Ymax). These dimensions are often written in the form [L, R, B, T], or Left and Right endpoints of the x-axis and Bottom and Top endpoints of the y-axis. The number of markings, or tick marks, on each axis is determined by setting the x-scale (Xscl) and the y-scale (Yscl). The *scale* is the distance between the tick marks. The graphs in this appendix are labeled to indicate the size of the viewing window used. If the Xscl or Yscl is not 1, we indicate that as well.

Choosing appropriate dimensions for a viewing window is important. The window dimensions should be large enough to include important features of the graph. However, if the dimensions are too large, it may be difficult to see the curvature of the graph.

Example 1 Graph $y = x^3 + 3x^2 - x + 1$ using the windows

$$[0, 10, 0, 10], \quad [-100, 100, -100, 100], \quad \text{and} \quad [-10, 10, -10, 10].$$

Solution First enter the equation using Y= and then set the window dimensions using WINDOW. A good choice of scale might be 1 for the first and last graphs and 10 for the second graph. (Scales need not be the same for both axes.) Graph the equation by pressing GRAPH after setting each window.

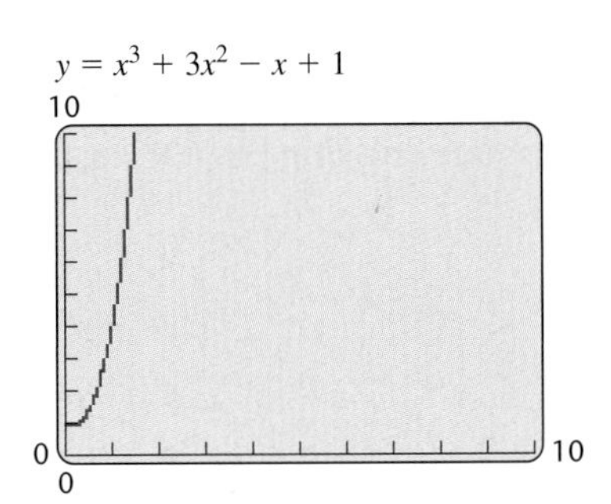

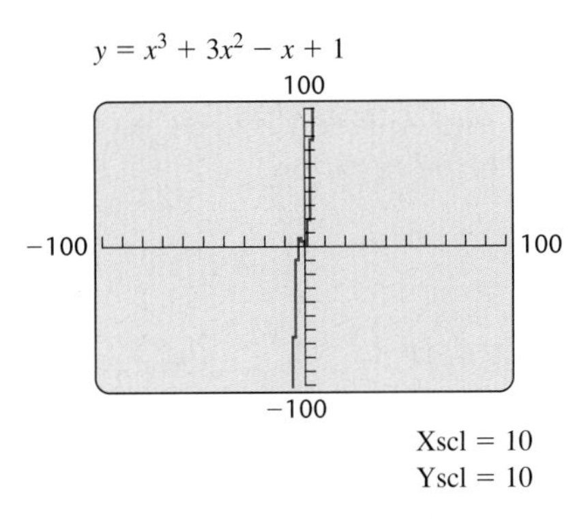

Xscl = 10
Yscl = 10

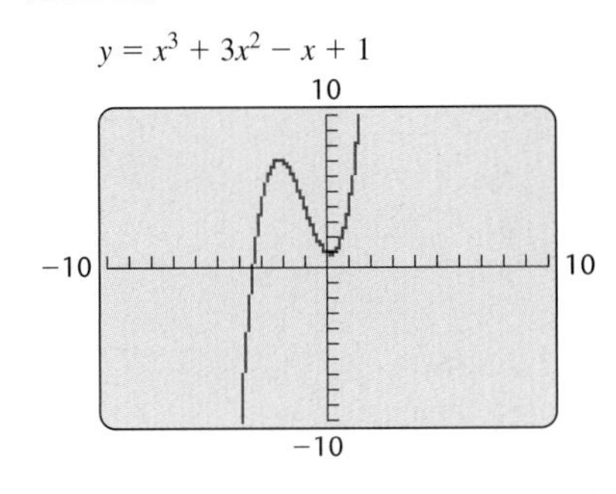

Which viewing window is the best? The answer depends on what we need to see from the graph. In Example 1, the third window would probably be the best choice since it shows where the x-axis is crossed, as well as the overall curvature.

Some window choices can be set quickly using the ZOOM option. The **standard viewing window** $[-10, 10, -10, 10]$ can be set up directly by pressing ZOOM 6. The ZOOM menu can also be used to magnify a portion of the graph (Zoom In) and to set up an appropriate window for a set of data (ZoomStat).

The VALUE option in the CALC menu gives a function value for a given value of x. Alternatively, pressing TRACE displays the coordinates of various points on the graph. The points traced by the cursor depend on the window. The Zdecimal option in the ZOOM menu sets up a window that traces by tenths. If we are interested in evaluating a function for a particular x-value, most graphers allow us to simply enter that x-value.

Example 2 Graph $y = x^2 - 4x + 2$ using the standard viewing window. Then use TRACE to find the value of y when x is 1.1.

Solution We enter the equation and then press ZOOM 6 to set the window and graph the equation. (See the graph on the left below.) Next, we press ZOOM 4. The equation is graphed again in a smaller window. Using TRACE and the arrow keys, we find that y is -1.19 when x is 1.1.

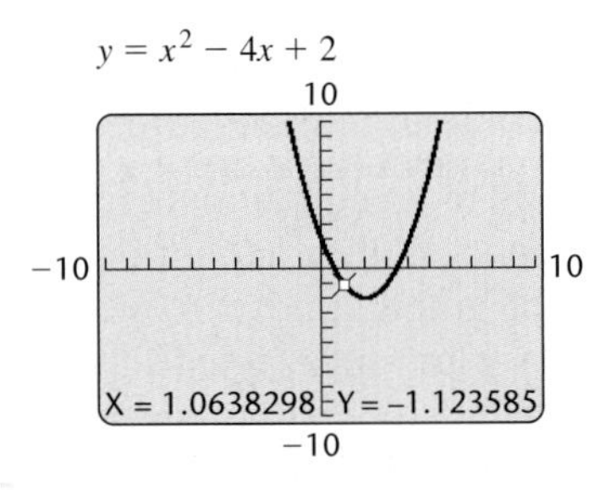

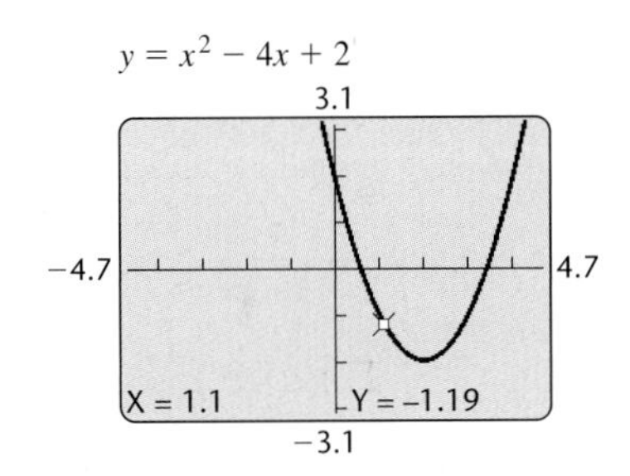

On most graphers, two or more equations can be graphed one after the other (*sequentially*) or at the same time (*simultaneously*). If equations are graphed simultaneously, all y-values are calculated and plotted for one x-value, then all y-values are calculated for the next x-value, and so on. Often a grapher can be set to the SEQUENTIAL or SIMULTANEOUS mode using the MODE key. To *select* the equation(s) that is graphed, locate the cursor on the equals sign of the equation you wish to graph, and press ENTER. When the equals sign is highlighted, the equation is selected and can be graphed. Pressing ENTER again deselects the equation.

Exercise Set 2.1

Graph each equation or pair of equations using the given viewing windows. Use an Xscl and a Yscl of 1 *unless otherwise indicated.*

1. $y = 2x - 5$

a) $[-10, 10, -10, 10]$
b) $[0, 10, 0, 10]$

2. $y = 2/x$

a) $[-100, 100, -100, 100]$
Xscl = 10 Yscl = 10
b) $[-10, 10, -10, 10]$

3. $y_1 = 3x - 2, y_2 = x^2 + x + 1$
$[-10, 10, -10, 10]$

a) Graph sequentially.
b) Graph simultaneously.

4. Use the graph of $y = 3x^2 - x - 1$ to find the value of y when $x = -1.2$.

Other Equations of Lines 2.5

The graph of an equation can be used to visually check what type of equation it represents. If the graph is obviously not a straight line, the equation is not linear. If the graph appears to be a straight line, the equation is *probably* linear. However, a line appearing to be linear may have a different shape when viewed using a different window. Algebraic checks for linearity are more accurate than graphing.

Example 1

Determine whether $y = 3x^2 + 5x - 1$ is linear.

Solution For the sake of illustration, the graph of the equation is shown using two different viewing windows. Note that in the figure on the left, the graph appears to be a straight line.

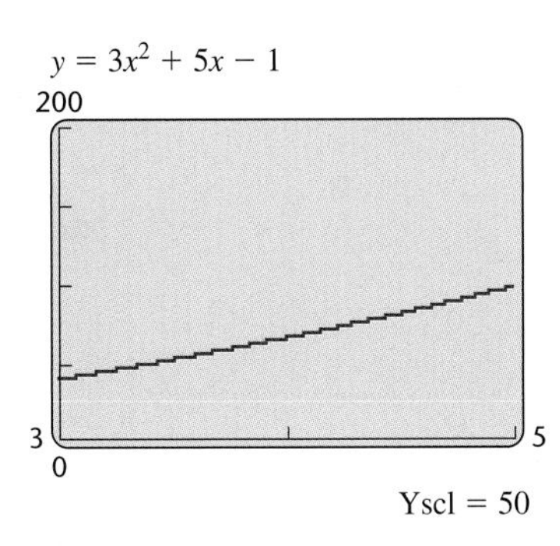

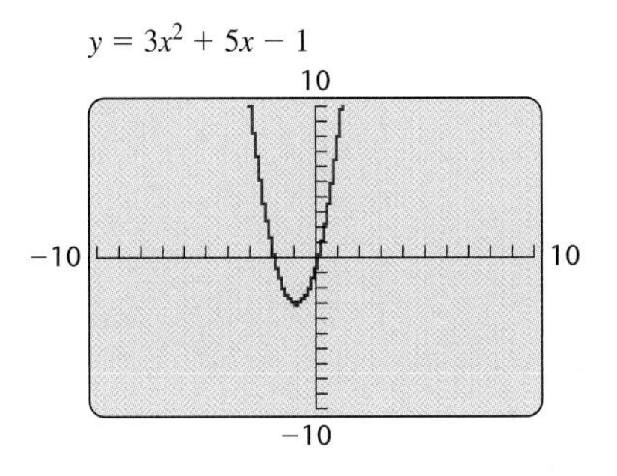

From the figure on the right above, we can see that the equation is not linear. Algebraically, it is not linear because it cannot be written in the standard form of a linear equation.

If a set of data is graphed and seems to lie on a straight line, we say that the data appear to be linear. We can find the equation of a line through two points. If more than two data points are known, the line that *best* describes the data may not actually go through any of the given points. The most commonly used method for fitting an equation to data is called *regression*. Most graphing calculators offer regression as a way of fitting a line or curve to a set of data. If the equation is a line, the method is called *linear regression*.

Example 2

The number of shopping centers in the United States has grown in recent years, as shown in the following table (*Source*: International Council of Shopping Centers).

Years After 1990	Number of Centers (in thousands)
0	36.515
2	38.966
5	41.235
6	42.130
8	43.661
9	44.426

a) Plot the data and determine whether they appear to be linear.

b) If the data appear to be linear, fit a linear equation to the data and graph the line.

c) Use the equation found in part (b) to predict the number of shopping centers in the United States in 2005.

Solution

a) We enter the data as lists of numbers using the EDIT menu. To do so, we press STAT and choose the EDIT option. Then we clear any data already in the lists by moving the cursor to the list name (L_1, L_2) and pressing CLEAR ENTER.

Next, we enter the years (the first column in the table) as L1 and the number of shopping centers as L2, and type the data in the correct row and column and press ENTER.

L1	L2	L3
0	36.515	---------
2	38.966	
5	41.235	
6	42.13	
8	43.661	
9	44.426	
---------	---------	

L1(1) = 0

To determine the dimensions of the viewing window, we look at the numbers in the table. The years range from 0 to 9, so we set Xmin = 0 and Xmax = 20. Since the number of shopping centers, in thousands, ranges from 36.515 to 44.426, we can set Ymin = 0, Ymax = 75, and Yscl = 5.

To graph the points, we turn the STAT PLOT on and press 2nd STAT PLOT ENTER. With the cursor on ON, we press ENTER again. Using the arrow keys and ENTER, we choose the first graph Type. If necessary, we set Xlist to L_1 and Ylist to L_2 using the functions above the 1 and 2 keys on the keypad.

Finally, we clear or deselect any equations stored in the grapher and plot the points. The data appear to be linear—that is, they appear to lie in a straight line.

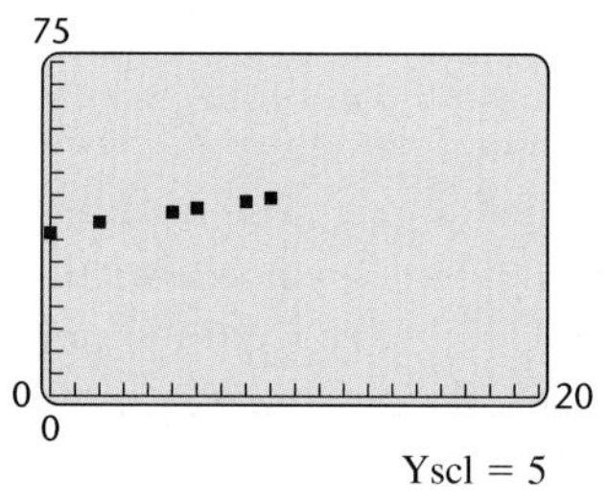

b) To calculate the linear regression line, we press STAT. Next, we choose the CALC submenu from the top of the screen and choose option **4: LinReg (ax + b)**. In order to be able to graph the equation, we copy the regression equation as Y1 by pressing VARS ► 1 1 and then ENTER.

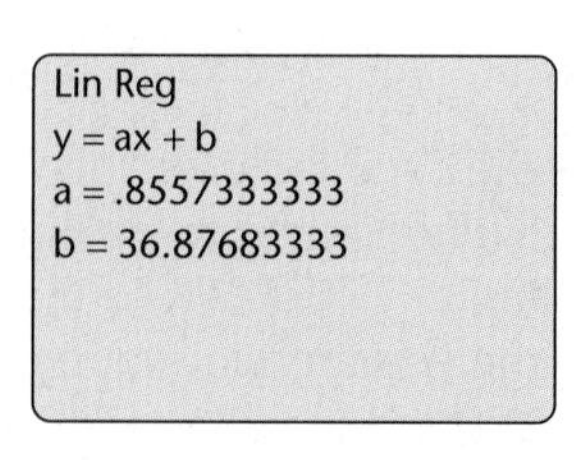

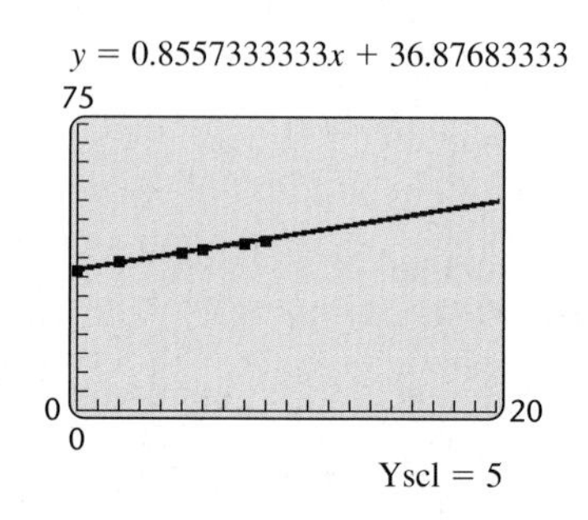

The equation is given in slope–intercept form $y = ax + b$. (The grapher uses a instead of m.) Substituting the given values of a and b, we have the equation $y = 0.8557333333x + 36.87683333$. We press GRAPH to see the graph of the equation, shown at the bottom of p. 712.

c) To predict the number of shopping centers in the United States in 2005, we evaluate the linear equation for $x = 15$. When $x = 15$, $y = 49.712833$, so we predict that the number of shopping centers in 2005 will be about 49,713.

Exercise Set 2.5

FOR EXTRA HELP

Digital Video Tutor CD: None
Videotape: None

InterAct Math

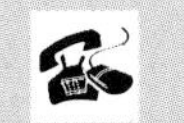
Math Tutor Center

MathXL

MyMathLab.com

Determine whether each equation is linear.

1. $y = 2.98x + 1.307$

2. $y = |3x - 6.5|$

3. $y = 0.002x^3$

4. The following table shows the life expectancy of males born in the United States in selected years (*Source*: *Statistical Abstract of the United States*, 2000).

Years Since 1970	Life Expectancy, y (in years)
0	67.1
10	70.0
20	71.8
30	74.2

a) Use linear regression to find a linear function that can be used to predict a man's life expectancy as a function of the year in which he was born, where x represents the number of years since 1970.

b) Predict the life expectancy in 2010.

5. The following table shows the estimated annual expenditures on a child in 1999 by a family with an annual income between \$36,800 and \$61,900 (*Source*: *Statistical Abstract of the United States*, 2000).

Age of Child	Annual Expenditures
1	\$8450
4	8660
7	8700
13	9390
16	9530

a) Use linear regression to find a linear function that can be used to predict the annual expenditure on a child as a function of the age of the child.

b) Estimate the annual expenditure on a 10-year-old child in 1999.

6. The following table gives the total waste generated in the United States in millions of tons per year (*Source*: Characterization of MSW in the US, US EPA, Washington, DC).

Years Since 1960	Total Waste Generated (in millions of tons)
0	88.1
10	121.1
20	151.6
30	196.9
40	221.7

a) Use linear regression to find a linear function that can be used to predict the total waste generated as a function of the number of years since 1960.

b) Estimate the total waste generated in the United States in 2010.

Inequalities in Two Variables 4.4

We can graph a linear inequality on most graphers by selecting the appropriate graph style in the Y= editing screen. The graph style icon immediately precedes the Y-variable. If we position the cursor on the icon and press ENTER, the icon will rotate through seven styles. Two of the styles shade above or below the graph of the corresponding equation.

Example 1 Graph: $y < 4x - 1$.

Solution We enter $4x - 1$ as y_1 and then move the cursor to the icon directly to the left of y_1. The region below the line $y = 4x - 1$ is the graph of the inequality. We press ENTER until the icon showing shading below a line appears, as shown on the left below.

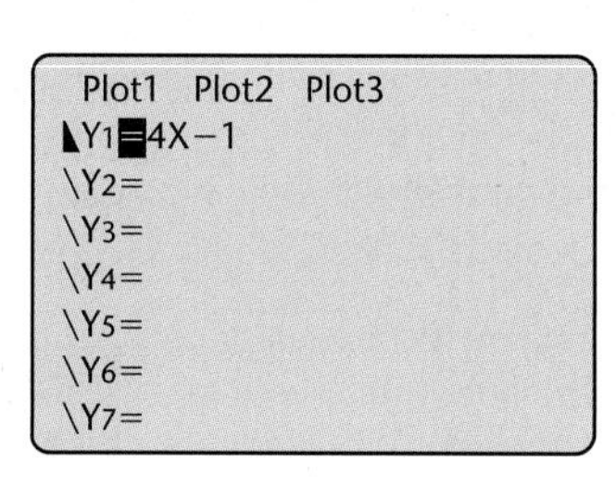

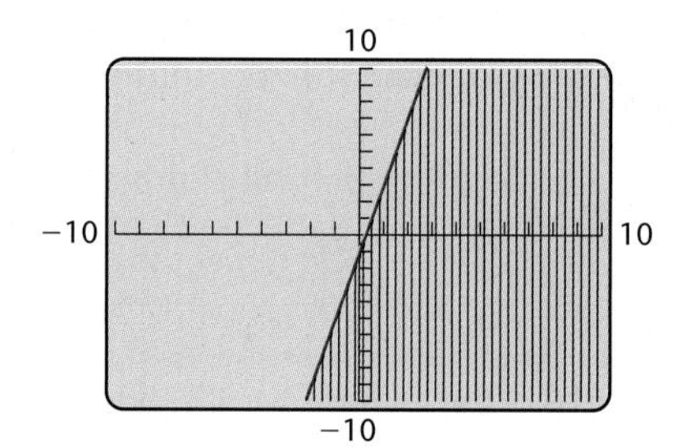

Now we press GRAPH. The graph of the inequality is shown on the right above.

Because most graphers use a variety of shading patterns, they can be used to graph systems of inequalities.

Example 2 Graph the system

$$2.65x - 1.3y < 7.3,$$
$$4.8x + 2.9y < 3.78.$$

Solution Solving for y, we can write the system as

$$y_1 > (2.65/1.3)x - (7.3/1.3)$$

and

$$y_2 < (-4.8/2.9)x + (3.78/2.9).$$

We enter the related equations, and shade above y_1 and below y_2. The grapher will use vertical lines for the first shading and horizontal lines for the second shading. The graph of the system is the intersection of the individual solution sets, or the region that is shaded by both vertical and horizontal lines.

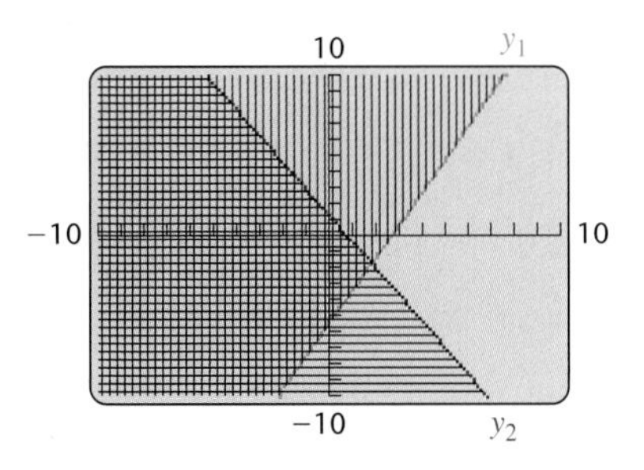

Exercise Set 4.4

FOR EXTRA HELP

Digital Video Tutor CD: None
Videotape : None

InterAct Math

Math Tutor Center

MathXL

MyMathLab.com

Graph each inequality.

1. $y > 2.7x + 4$
2. $y \leq 4 - x$
3. $2.8x - 4.6 \geq 0.5y$
4. Graph the system
$$7.4x + 3.8y > 1.9,$$
$$4.3x - 1.5y < -2.8.$$

8.8 Problem Solving and Quadratic Functions

The point at which a maximum or minimum value of a quadratic f ...ion occurs can often be seen directly from its graph. To help identify th many graphers have a feature that calculates the maximum or minim of a function over a specified interval.

Example 1 Estimate the minimum value of the function $g(x) = 1.65x^2 - 3.7x - 2.95$.

Solution We can see from the graph that $g(x)$ has a minimum value somewhere near $x = 1$.

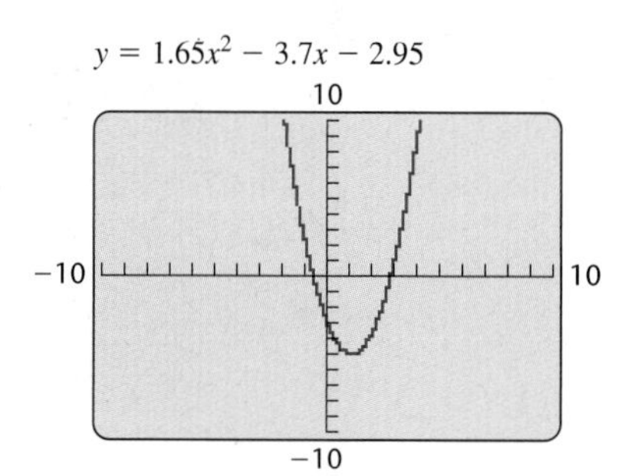

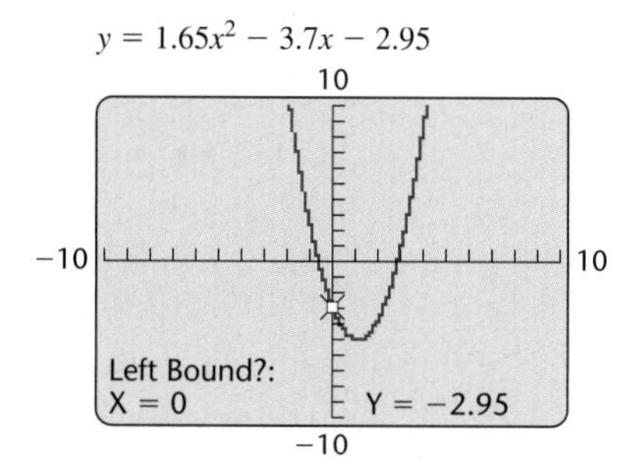

The CALC menu, which is the 2nd function of the TRACE key, lists some information that can be calculated from the graph. When we select option **3: minimum**, the graph is displayed, along with the question "Left Bound?" as shown in the figure on the right above. We position the cursor on the graph somewhere to the left of the minimum and press ENTER. We can also simply enter a value to the left of the minimum.

Next, we are asked for a right bound, as shown on the left below. We position the cursor to the *right* of the minimum and press ENTER or simply enter a value that is to the right of the minimum.

Finally, the grapher asks for a guess, as shown below in the center. We move the cursor closer to the minimum and press ENTER.

At the bottom of the graph on the right below, we can now read the coordinates of the minimum point on the graph. To the nearest hundredth, the minimum value of the function is -5.02.

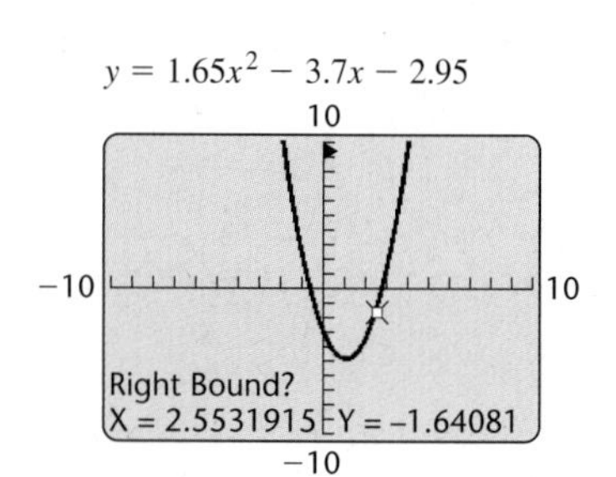

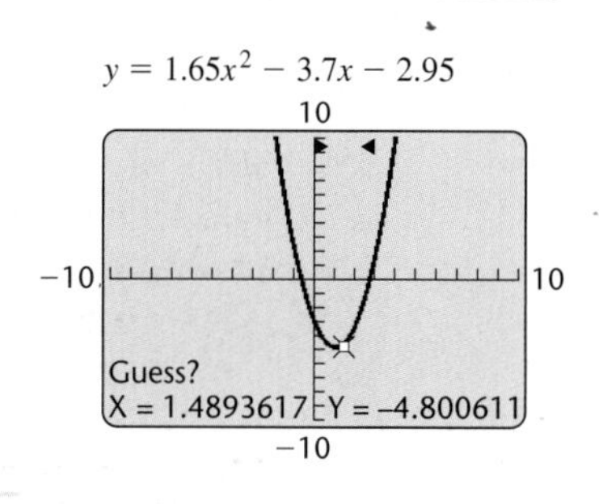

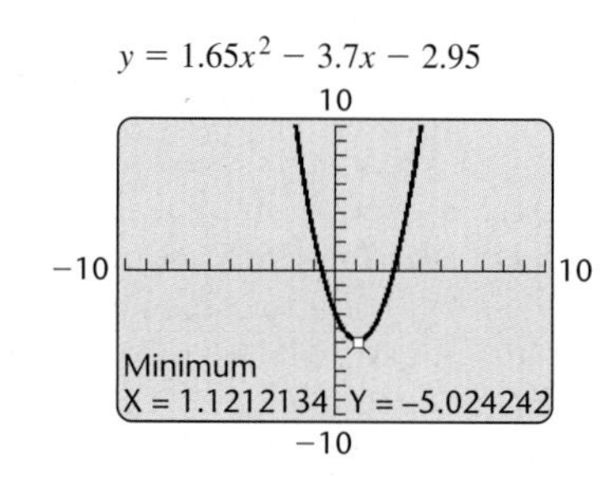

Exercise Set 8.8

FOR EXTRA HELP

Digital Video Tutor CD: None
Videotape: None

InterAct Math

Math Tutor Center

MathXL

MyMathLab.com

Find the maximum or minimum value of each function and the value of x at which it occurs. Round to the nearest hundredth.

1. $f(x) = -2.15x^2 + 3.8x - 6.9$

2. $f(x) = 8.3x^2 + 6.7x + 3.9$

3. $g(x) = 0.016 + 2.3x + 0.003x^2$

Answers

CHAPTER 1

Exercise Set 1.1, pp. 9–11

1. Let n represent the number; $n - 3$ **3.** Let t represent the number; $12t$ **5.** Let x represent the number; $0.65x$, or $\frac{65}{100}x$ **7.** Let y represent the number; $2y + 10$
9. Let s represent the number; $0.1s + 8$, or $\frac{10}{100}s + 8$
11. Let m and n represent the numbers; $m - n - 1$
13. $90 \div 4$, or $\frac{90}{4}$ **15.** 25 **17.** 16 **19.** 27 **21.** 0
23. 5 **25.** 25 **27.** 109 **29.** 17.5 sq ft
31. 11.2 sq m **33.** {a, e, i, o, u} or {a, e, i, o, u, y}
35. $\{1, 3, 5, 7, \ldots\}$ **37.** $\{5, 10, 15, 20, \ldots\}$
39. $\{x \mid x$ is an odd number between 10 and 20$\}$
41. $\{x \mid x$ is a whole number less than 5$\}$
43. $\{n \mid n$ is a multiple of 5 between 7 and 79$\}$ **45.** True
47. True **49.** False **51.** True **53.** True **55.** True
57. **59.** **61.** Let a and b represent the numbers; $\dfrac{a+b}{a-b}$ **63.** Let r and s represent the numbers; $\frac{1}{2}(r^2 - s^2)$, or $\dfrac{r^2 - s^2}{2}$ **65.** $\{0\}$ **67.** $\{5, 10, 15, 20, \ldots\}$
69. $\{1, 3, 5, 7, \ldots\}$ **71.**

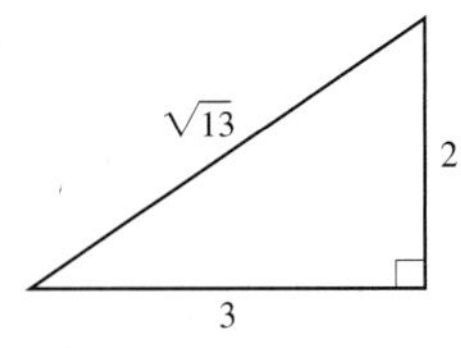

Exercise Set 1.2, pp. 20–21

1. 9 **3.** 6 **5.** 6.2 **7.** 0 **9.** $1\frac{7}{8}$ **11.** 4.21 **13.** -6 is less than or equal to -2; true **15.** -9 is greater than 1; false **17.** 3 is greater than or equal to -5; true **19.** -8 is less than -3; true **21.** -4 is greater than or equal to -4; true **23.** -5 is less than -5; false **25.** 11
27. -11 **29.** -1.2 **31.** $-\frac{11}{35}$ **33.** -8.5 **35.** $\frac{5}{9}$
37. -4.5 **39.** 0 **41.** -6.4 **43.** -3.14 **45.** $4\frac{1}{3}$
47. 0 **49.** -7 **51.** 2.7 **53.** -1.79 **55.** 0 **57.** 7
59. -7 **61.** 4 **63.** -19 **65.** -3.1 **67.** $-\frac{11}{10}$
69. 2.9 **71.** 7.9 **73.** -30 **75.** 24 **77.** -21
79. $-\frac{3}{7}$ **81.** 0 **83.** 5.44 **85.** 5 **87.** -5 **89.** -73
91. 0 **93.** $\frac{1}{4}$ **95.** $-\frac{1}{9}$ **97.** $\frac{3}{2}$ **99.** $-\frac{11}{3}$ **101.** $\frac{5}{6}$
103. $-\frac{6}{5}$ **105.** $\frac{1}{36}$ **107.** 1 **109.** 23 **111.** $-\frac{6}{11}$
113. Undefined **115.** $\frac{11}{43}$ **117.** 28 **119.** -7
121. 10 **123.** $7b + 4a$; $a4 + b7$; $4a + b7$ **125.** $y(7x)$; $(x7)y$ **127.** $3(xy)$ **129.** $(x + 2y) + 5$ **131.** $3a + 21$
133. $4x - 4y$ **135.** $-10a - 15b$
137. $9ab - 9ac + 9ad$ **139.** $5(x + 5)$ **141.** $3(p - 3)$
143. $7(x - 3y + 2z)$ **145.** $17(15 - 2b)$ **147.**
149. [1.1] 16; 16 **150.** [1.1] 11; 11 **151.**
153. $(8 - 5)^3 + 9 = 36$ **155.** $5 \cdot 2^3 \div (3 - 4)^4 = 40$
157. -6.2 **159.**

Exercise Set 1.3, pp. 28–29

1. Equivalent **3.** Equivalent **5.** Not equivalent
7. Not equivalent **9.** 16.3 **11.** 9 **13.** 18 **15.** 8
17. 24 **19.** $10x$ **21.** $-2rt$ **23.** $10t^2$ **25.** $11a$
27. $-7n$ **29.** $10x$ **31.** $7x - 2x^2$ **33.** $5a + 11a^2$
35. $22x + 18$ **37.** $-5t^2 + 2t + 4t^3$ **39.** $5a - 5$
41. $-2m + 1$ **43.** $5d - 12$ **45.** $-7x + 14$
47. $-10x + 21$ **49.** $44a - 22$ **51.** $-100a - 90$

53. $-12y - 145$ **55.** 7 **57.** 21 **59.** 4 **61.** 2 **63.** 3 **65.** 7 **67.** 5 **69.** 2 **71.** 2 **73.** $\frac{49}{9}$ **75.** $\frac{4}{5}$ **77.** $\frac{23}{8}$ **79.** $-\frac{1}{2}$ **81.** $\varnothing$; contradiction **83.** $\{0\}$; conditional **85.** $\varnothing$; contradiction **87.** $\mathbb{R}$; identity **89.** **91.** [1.1] Let n represent the number; $2n + 9$, or $9 + 2n$ **92.** [1.1] Let n represent the number; $0.42\left(\frac{n}{2}\right)$ **93.** **95.** -4.176190476 **97.** 8 **99.** $\frac{224}{29}$ **101.**

Exercise Set 1.4, pp. 36–38

1. Let x and $x + 7$ represent the numbers; $x + (x + 7) = 65$ **3.** Let t represent the time, in hours, that it will take the Memphis Queen to cruise 2 mi upstream; $4t = 2$ **5.** Let t represent the time, in seconds, that it takes Alida to walk the length of the sidewalk; $9t = 300$ **7.** Let x, $x + 1$, and $x + 2$ be the angle measures; $x + (x + 1) + (x + 2) = 180$ **9.** Let c represent the original cost of the order, in dollars; $c - 0.1c = 279$ **11.** Let t represent the number of minutes spent climbing; $3500t = 29{,}000 - 8000$ **13.** Let x represent the measure of the second angle, in degrees; $3x + x + (2x - 12) = 180$ **15.** Let x represent the first even number; $2x + 3(x + 2) = 76$ **17.** Let s represent the length, in centimeters, of a side of the smaller triangle; $3s + 3 \cdot 2s = 90$ **19.** Let x represent the score on the next test; $\frac{93 + 89 + 72 + 80 + 96 + x}{6} = 88$ **21.** 50.3 **23.** 320 **25.** 63 **27.** Length: 45 cm; width: 15 cm **29.** Length: 52 m; width: 13 m **31.** $\frac{2}{3}$ hr **33.** 100°, 25°, 55° **35.** \$150 **37.** \$1.16 **39.** **41.** [1.3] $\frac{9}{2}$ **42.** [1.3] 36 **43.** [1.3] 19 **44.** [1.3] $\frac{4}{3}$ **45.** **47.** 10 **49.** \$110,000

Exercise Set 1.5, pp. 44–46

1. $r = \frac{d}{t}$ **3.** $a = \frac{F}{m}$ **5.** $I = \frac{W}{E}$ **7.** $h = \frac{V}{lw}$ **9.** $k = Ld^2$ **11.** $n = \frac{G - w}{150}$ **13.** $l = p - 2w - 2h$ **15.** $y = \frac{8 - 5x}{2}$, or $y = 4 - \frac{5}{2}x$ **17.** $y = \frac{C - Ax}{B}$ **19.** $F = \frac{9}{5}C + 32$ **21.** $r^3 = \frac{3V}{4\pi}$ **23.** $b_2 = \frac{2A}{h} - b_1$ **25.** $n = \frac{q_1 + q_2 + q_3}{A}$ **27.** $t = \frac{d_2 - d_1}{v}$ **29.** $d_1 = d_2 - vt$ **31.** $m = \frac{r}{1 + np}$ **33.** $a = \frac{y}{b - c^2}$ **35.** 12% **37.** 6 cm **39.** About 22 g more **41.** About 8.5 cm **43.** 9 ft **45.** 1 yr **47.** 34 appointments **49.** **51.** [1.2] $(7 \cdot 3)(a \cdot a)$; $a \cdot 3 \cdot a \cdot 7$ **52.** [1.2] $4 \cdot x \cdot y \cdot y$; $(y \cdot 4)(x \cdot y)$ **53.** **55.** About 10.9 g **57.** About 610 cm **59.** $l = \frac{A - w^2}{4w}$ **61.** $T_1 = \frac{P_1V_1T_2}{P_2V_2}$ **63.** $d = \frac{me^2}{f}$ **65.** $t = \frac{1}{s}$

Technology Connection, p. 51

1. Answers may vary; 2 x^y 5 – =, 2 ^ (–) 5 ENTER, 2 x^y $-x$ 5 ENTER **2.** Compute $\frac{1}{2^5}$.

Exercise Set 1.6, pp. 54–56

1. 2^{11} **3.** 5^9 **5.** t^8 **7.** $18x^7$ **9.** $21m^{13}$ **11.** $x^{10}y^{10}$ **13.** a^6 **15.** $3t^5$ **17.** m^5n^4 **19.** $9x^6y^4$ **21.** $-4x^8y^6z^6$ **23.** -1 **25.** 1 **27.** 81 **29.** -81 **31.** $\frac{1}{16}$ **33.** $-\frac{1}{16}$ **35.** $-\frac{1}{16}$ **37.** $-\frac{1}{64}$ **39.** $\frac{1}{a^3}$ **41.** 125 **43.** $\frac{4}{x^3}$ **45.** $\frac{2a^3}{b^6}$ **47.** $\frac{1}{3x^5z^4}$ **49.** $\frac{y^7z^4}{x^2}$ **51.** 3^{-4} **53.** $(-16)^{-2}$ **55.** $\frac{1}{x^{-5}}$ **57.** $\frac{6}{x^{-2}}$ **59.** $(5y)^{-3}$ **61.** $\frac{y^{-4}}{3}$ **63.** 8^{-6}, or $\frac{1}{8^6}$ **65.** b^{-3}, or $\frac{1}{b^3}$ **67.** a^3 **69.** $-18m^4n^5$ **71.** $-14x^{-11}$, or $-\frac{14}{x^{11}}$ **73.** $10a^{-6}b^{-2}$, or $\frac{10}{a^6b^2}$ **75.** 10^{-9}, or $\frac{1}{10^9}$ **77.** 2^{-2}, or $\frac{1}{2^2}$, or $\frac{1}{4}$ **79.** y^9 **81.** $-3ab^2$ **83.** $-\frac{7}{4}a^{-4}b^2$, or $-\frac{7b^2}{4a^4}$ **85.** $-\frac{1}{6}x^3y^{-2}z^{10}$, or $-\frac{x^3z^{10}}{6y^2}$ **87.** x^{12} **89.** 9^{-12}, or $\frac{1}{9^{12}}$ **91.** t^{40} **93.** $36x^2y^2$ **95.** $a^{12}b^4$ **97.** $5x^{-14}y^{-14}$, or $\frac{5}{x^{14}y^{14}}$ **99.** $a^{-7}b^4$, or $\frac{b^4}{a^7}$ **101.** 1 **103.** $\frac{9x^8y^9}{2}$ **105.** $\frac{625}{256}x^{-20}y^{24}$, or $\frac{625y^{24}}{256x^{20}}$ **107.** $9a^{-4}b^{-10}$, or $\frac{9}{a^4b^{10}}$ **109.** 1 **111.** **113.** [1.1], [1.2] 35.1 **114.** [1.1], [1.2] 44 **115.** **117.** $4a^{-x-4}$ **119.** y^4 **121.** 3^{a^2+2a} **123.** $2x^{a+2}y^{b-2}$ **125.** $2^{-2a-2b+ab}$ **127.** $\frac{2}{27}$

Exercise Set 1.7, pp. 61–63

1. 8.3×10^{10} **3.** 8.63×10^{17} **5.** 1.6×10^{-8} **7.** 7×10^{-11} **9.** 8.03×10^{11} **11.** 9.04×10^{-7} **13.** 4.317×10^{11} **15.** 0.0005 **17.** 973,000,000 **19.** 0.0000000004923 **21.** 90,300,000,000 **23.** 0.00000004037 **25.** 7,010,000,000,000

27. 9.7×10^{-5} **29.** 1.3×10^{-11} **31.** 1.4×10^{11}
33. 4.6×10^{3} **35.** 6.0 **37.** 1.5×10^{3}
39. 3.0×10^{-5} **41.** 4.0×10^{-16} **43.** 3.00×10^{-22}
45. 2.00×10^{26} **47.** 1×10^{11} **49.** 8.3×10^{23}
51. 6.79×10^{8} km **53.** 2.2×10^{-3} lb **55.** 8.00 light years **57.** 3.08×10^{26} Å **59.** 1×10^{22} cu Å or $1 \times 10^{-8}\ \text{m}^3$ **61.** 7.90×10^{7} bacteria
63. 4.49×10^{4} km/h **65.** **67.** [1.1] 8
68. [1.1], [1.2] 32 **69.** **71.** 2.5×10^{-10} oz to 9.5×10^{-10} oz **73.** $8 \cdot 10^{-90}$ is larger by 7.1×10^{-90}.
75. 8 **77.** 8×10^{18} grains

Review Exercises: Chapter 1, pp. 66–67

1. [1.1] Let x and y represent the numbers; $\dfrac{x}{y} - 3$
2. [1.1] 22 **3.** [1.1] $\{2, 4, 6, 8, 10, 12\}$; $\{x \mid x$ is an even number between 1 and 13$\}$
4. [1.1] 1750 sq cm **5.** [1.2] 7.3 **6.** [1.2] 4.09
7. [1.2] 0 **8.** [1.2] -10.2 **9.** [1.2] $-\frac{23}{35}$ **10.** [1.2] $\frac{7}{15}$
11. [1.2] -11.5 **12.** [1.2] $-\frac{1}{6}$ **13.** [1.2] -5.4
14. [1.2] 12.6 **15.** [1.2] $-\frac{5}{12}$ **16.** [1.2] -4.8
17. [1.2] 6 **18.** [1.2] -9.1 **19.** [1.2] $-\frac{21}{4}$
20. [1.2] 4.01 **21.** [1.2] $a + 7$ **22.** [1.2] $y7$
23. [1.2] $x5 + y$, or $y + 5x$ **24.** [1.2] $4 + (a + b)$
25. [1.2] $x(y7)$ **26.** [1.2] $7m(n + 2)$
27. [1.3] $6x^3 - 8x^2 + 2$ **28.** [1.3] $47x - 60$
29. [1.3] 6.6 **30.** [1.3] $\frac{27}{2}$ **31.** [1.3] $-\frac{4}{11}$
32. [1.3] $\mathbb{R}$; identity **33.** [1.3] $\varnothing$; contradiction
34. [1.4] Let x represent the number; $2x + 13 = 21$
35. [1.4] 49 **36.** [1.4] 90°, 30°, 60° **37.** [1.5] $m = PS$
38. [1.5] $x = \dfrac{c}{m - r}$ **39.** [1.5] 4 cm **40.** [1.6] $-10a^5b^8$
41. [1.6] $4xy^6$ **42.** [1.6] 1, 28.09, -28.09
43. [1.6] 3^3, or 27 **44.** [1.6] 5^3a^6, or $125a^6$
45. [1.6] $-\dfrac{a^9}{8b^6}$ **46.** [1.6] $\dfrac{z^8}{x^4y^6}$ **47.** [1.6] $\dfrac{b^{16}}{16a^{20}}$
48. [1.2] $\frac{3}{7}$ **49.** [1.2] 0 **50.** [1.7] 1.03×10^{-7}
51. [1.7] 3.086×10^{13} **52.** [1.7] 3.7×10^{7}
53. [1.7] 2.0×10^{-6}
54. [1.7] $1.4 \times 10^{4}\ \text{mm}^3$, or $1.4 \times 10^{-5}\ \text{m}^3$
55. [1.3] To write an equation that has no solution, begin with a simple equation that is false for any value of x, such as $x = x + 1$. Then add or multiply by the same quantities on both sides of the equation to construct a more complicated equation with no solution.
56. [1.3] Use the distributive law to rewrite a sum of like terms as a single term by first writing the sum as a product. For example, $2a + 5a = (2 + 5)a = 7a$.
57. [1.7] 0.0000003% **58.** [1.1], [1.6] $-\frac{23}{24}$
59. [1.4], [1.5] The 17-in. pizza is a better deal. It costs about 5¢ per square inch; the 13-in. pizza costs about 6¢ per square inch.
60. [1.5] 729 cm^3 **61.** [1.5] $z = y - \dfrac{x}{m}$
62. [1.6] $3^{-2a+2b-8ab}$ **63.** [1.4] $88.\overline{3}$ **64.** [1.3] -39
65. [1.3] $-40x$
66. [1.2] $a2 + cb + cd + ad = ad + a2 + cb + cd = a(d + 2) + c(b + d)$ **67.** [1.1] $\sqrt{5}/4$; answers may vary

Test: Chapter 1, p. 68

1. [1.1] Let m and n represent the numbers; $mn + 3$
2. [1.1], [1.2] -47 **3.** [1.1] 3.75 sq cm **4.** [1.2] -31
5. [1.2] -3.7 **6.** [1.2] -5.11 **7.** [1.2] -14.2
8. [1.2] -43.2 **9.** [1.2] -33.92 **10.** [1.2] $-\frac{19}{12}$
11. [1.2] $\frac{5}{49}$ **12.** [1.2] 6 **13.** [1.2] $-\frac{4}{3}$ **14.** [1.2] $-\frac{5}{2}$
15. [1.2] $y + 7x$; $x7 + y$; answers may vary **16.** [1.3] $10y$
17. [1.3] $a^2b - 4ab^2 + 2$ **18.** [1.3] $3x + 8$
19. [1.3] -2 **20.** [1.3] $\mathbb{R}$; identity
21. [1.5] $P_2 = \dfrac{P_1V_1T_2}{T_1V_2}$ **22.** [1.4] 94 **23.** [1.4] 17, 19, 21
24. [1.3] $-8x - 1$ **25.** [1.3] $24b - 9$ **26.** [1.6] $-\dfrac{42}{x^{10}y^6}$
27. [1.6] $-\dfrac{1}{3^2}$, or $-\dfrac{1}{9}$ **28.** [1.6] $\dfrac{y^8}{36x^4}$ **29.** [1.6] $\dfrac{x^6}{4y^8}$
30. [1.6] 1 **31.** [1.7] 2.01×10^{-7} **32.** [1.7] 2.0×10^{10}
33. [1.7] 3.8×10^{2} **34.** [1.7] 4.2×10^{8} mi
35. [1.6] $16^c x^{6ac} y^{2bc+2c}$ **36.** [1.6] $-9a^3$ **37.** [1.6] $\dfrac{4}{7y^2}$

CHAPTER 2

Technology Connection, p. 71

1. $y = -4x + 3$

10, −10, 10, −10; X = 1 Y = −1

Technology Connection, p. 75

1. $y = 5x - 3$

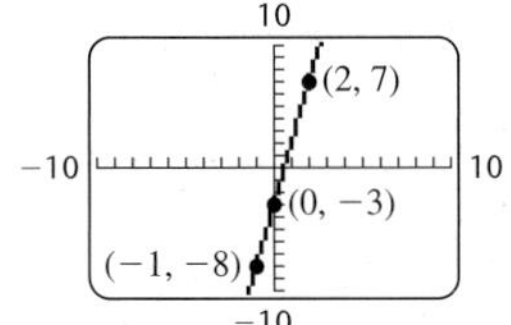

X	Y1
−1	−8
−.9	−7.5
−.8	−7
−.7	−6.5
−.6	−6
−.5	−5.5
−.4	−5

X = −1

2. $y = x^2 - 4x + 3$

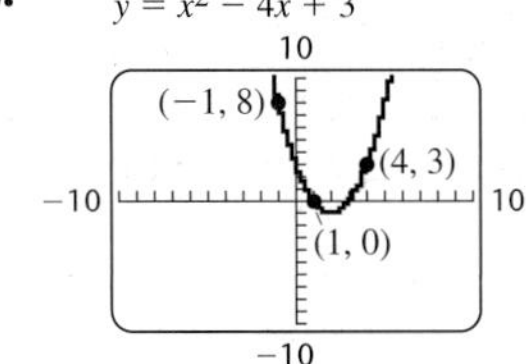

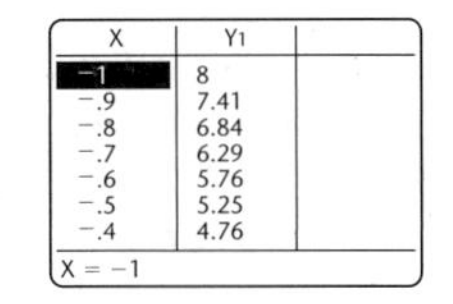

X	Y1
−1	8
−.9	7.41
−.8	6.84
−.7	6.29
−.6	5.76
−.5	5.25
−.4	4.76

X = −1

3. $y = (x + 4)^2$

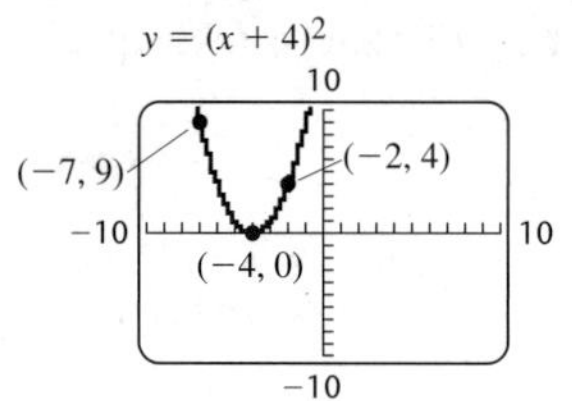

X	Y1
−1	9
−.9	9.61
−.8	10.24
−.7	10.89
−.6	11.56
−.5	12.25
−.4	12.96
X = −1	

4. $y = \sqrt{x + 2}$

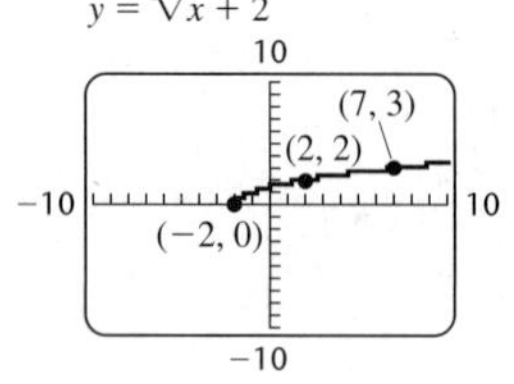

X	Y1
−1	1
−.9	1.0488
−.8	1.0954
−.7	1.1402
−.6	1.1832
−.5	1.2247
−.4	1.2649
X = −1	

5. $y = |x + 2|$

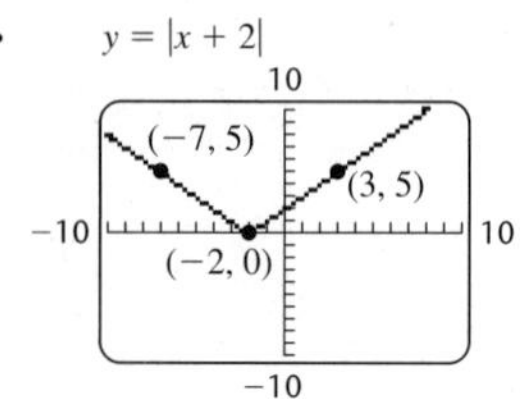

X	Y1
−1	1
−.9	1.1
−.8	1.2
−.7	1.3
−.6	1.4
−.5	1.5
−.4	1.6
X = −1	

Exercise Set 2.1, pp. 76–78

1. (5, 3), (−4, 3), (0, 2), (−2, −3), (4, −2), and (−5, 0)

3.

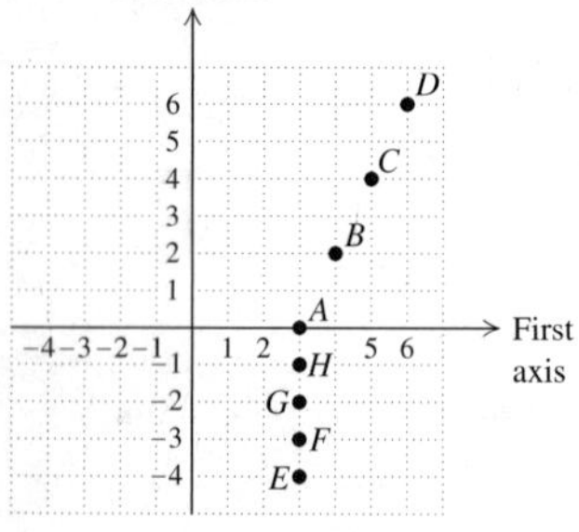

5.

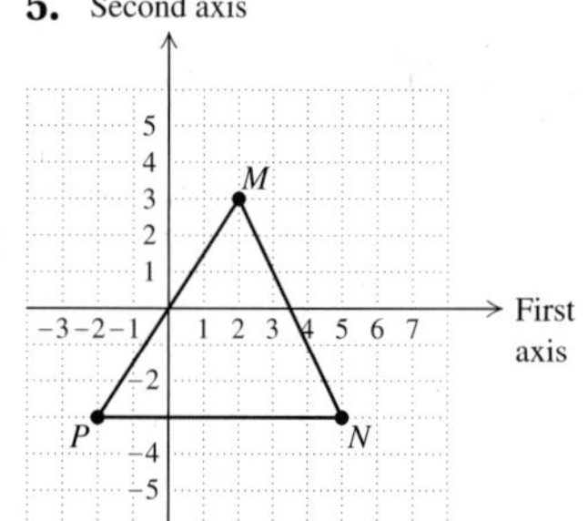

Triangle, 21 units^2

7. III **9.** II **11.** I **13.** IV **15.** Yes **17.** No **19.** Yes **21.** Yes **23.** Yes **25.** Yes **27.** Yes **29.** No

31.

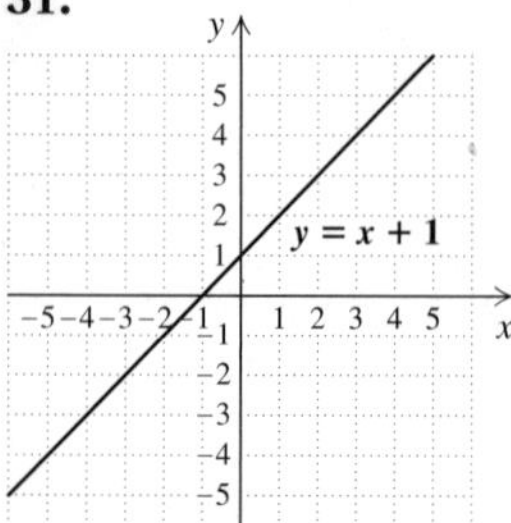

33.

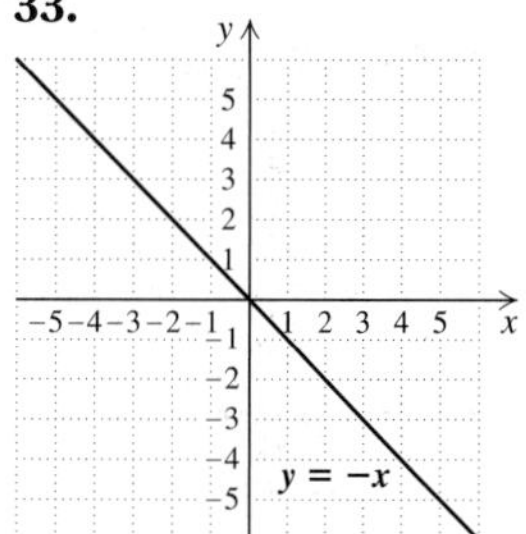

35.

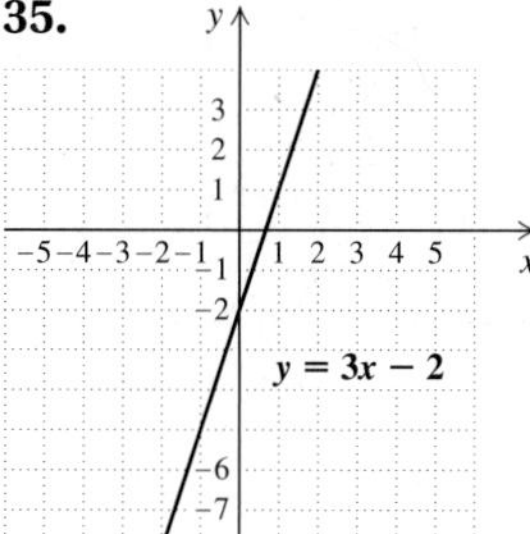

37.

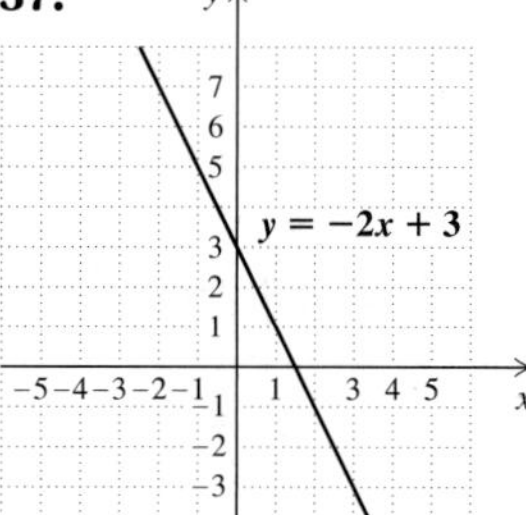

39.

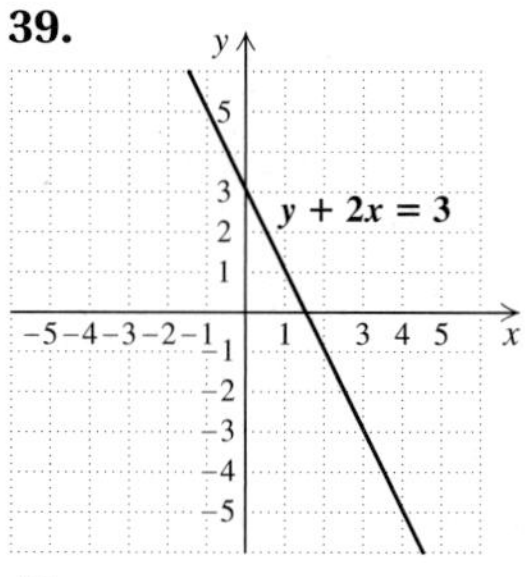

41.

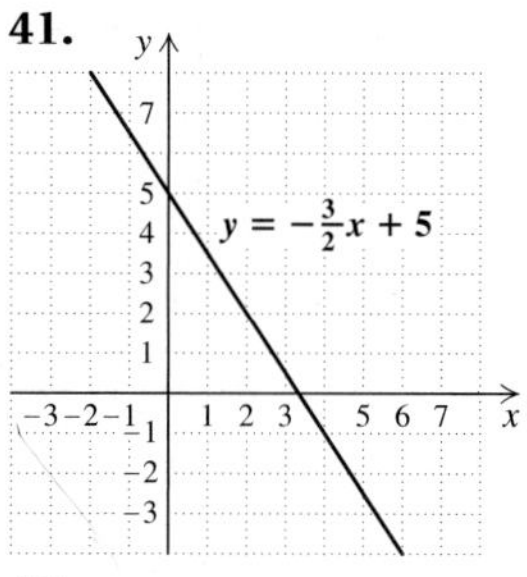

43.

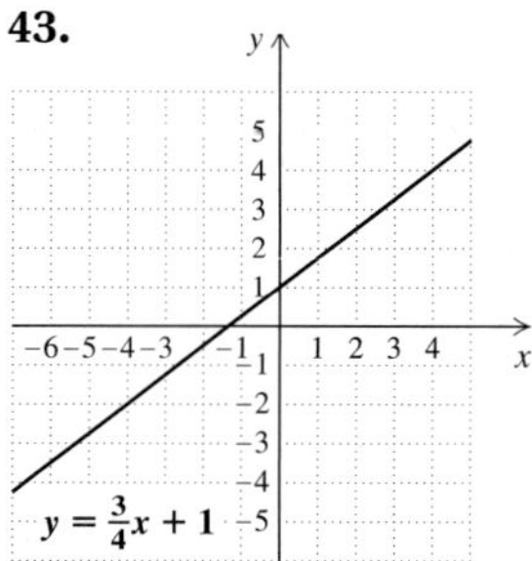

45.

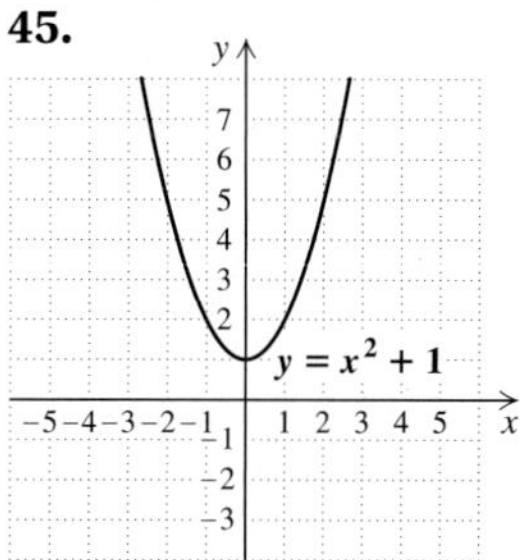

47.

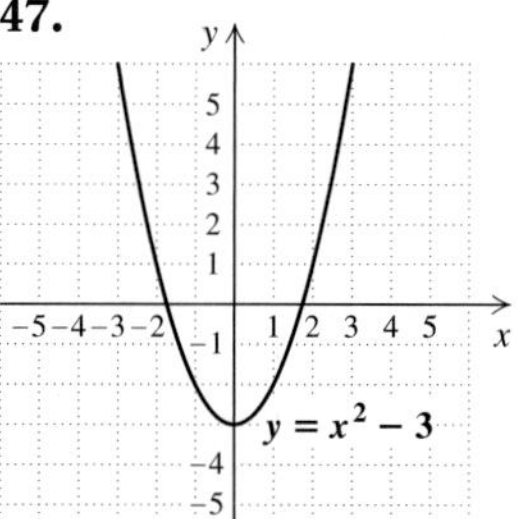

49.

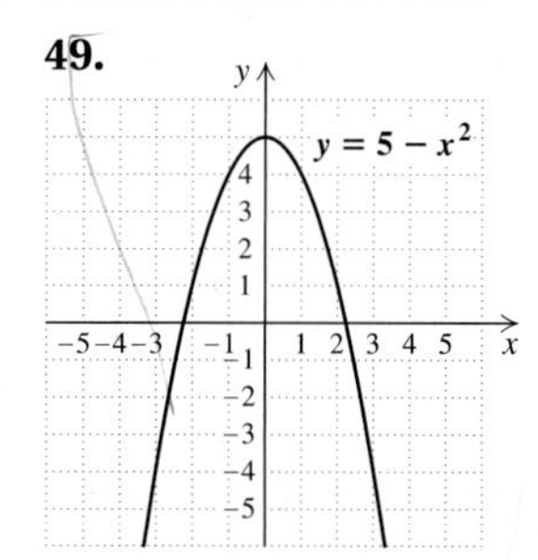

51.

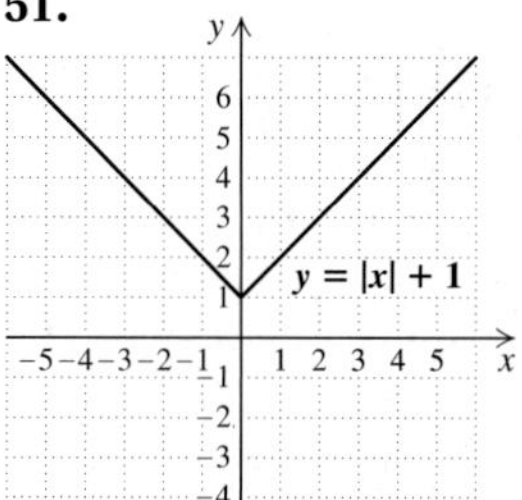

53.

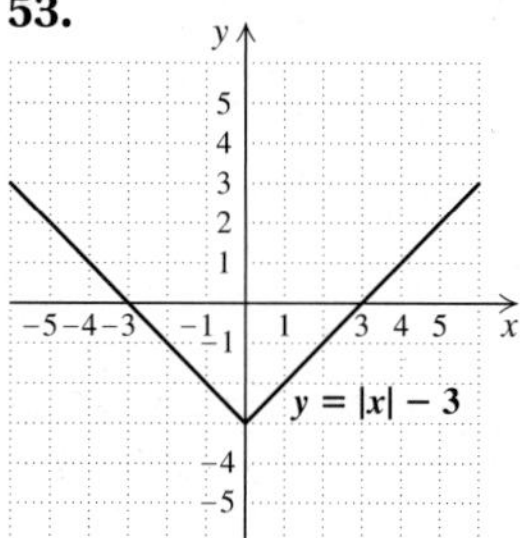

55. ▯ **57.** [1.2] −2 **58.** [1.1] 13 **59.** [1.1] 100 **60.** [1.1] 49 **61.** [1.2] 0 **62.** [1.2] −3 **63.** ▯ **65.** ▯ **67.** **(a)** III; **(b)** II; **(c)** I; **(d)** IV **69.** (a), (d)

71.

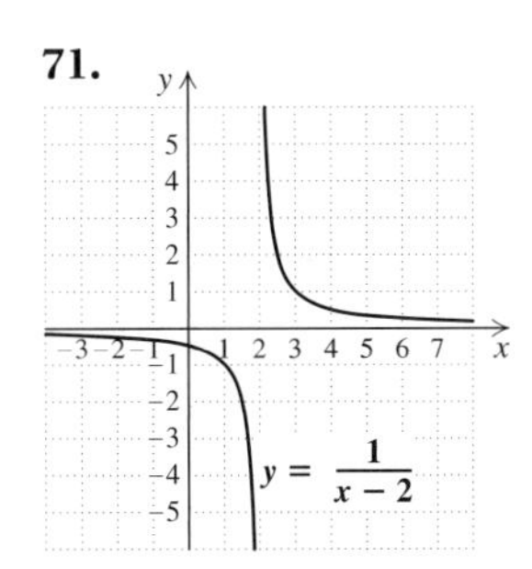

73.

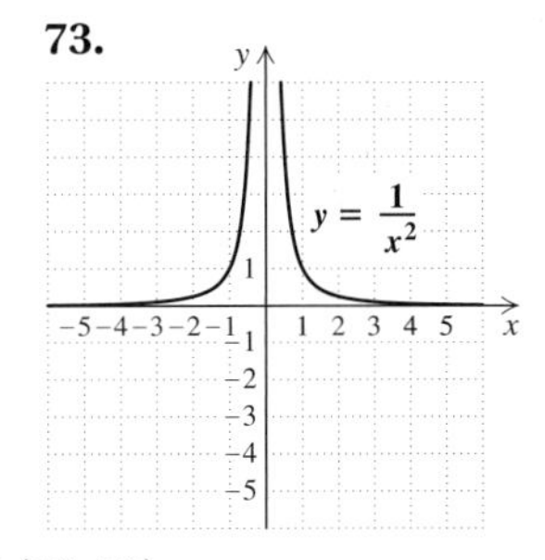

75. $(-1, -2)$, $(-19, -2)$, and $(13, 10)$

77. **(a)** $y = 2.3x^4 + 3.4x^2 + 1.2x - 4$

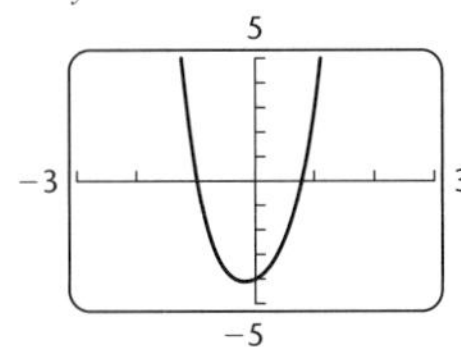

X	Y1
0	−4
.1	−3.846
.2	−3.62
.3	−3.315
.4	−2.917
.5	−2.406
.6	−1.758

X = 0

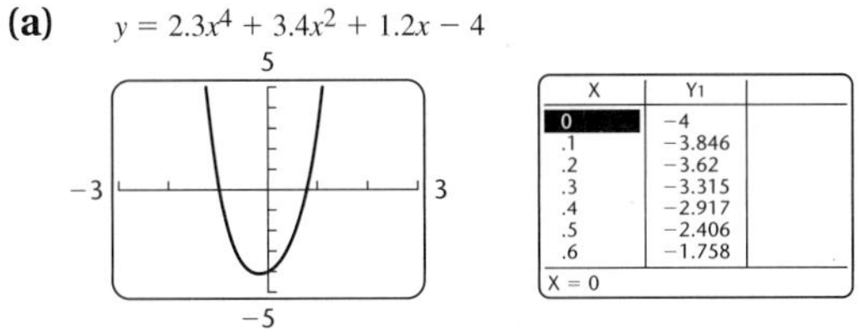

(b) $y = 12.3x - 3.5$

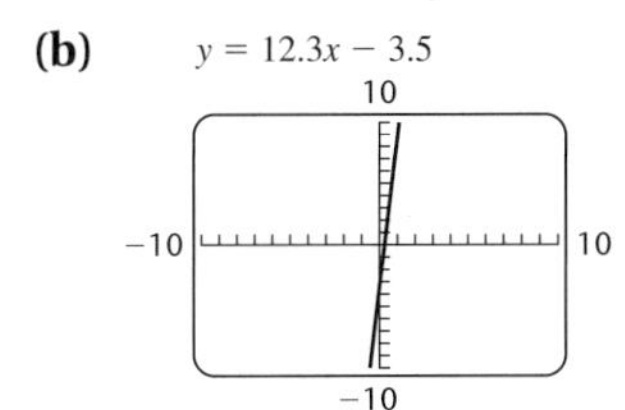

X	Y1
0	−3.5
.1	−2.27
.2	−1.04
.3	.19
.4	1.42
.5	2.65
.6	3.88

X = 0

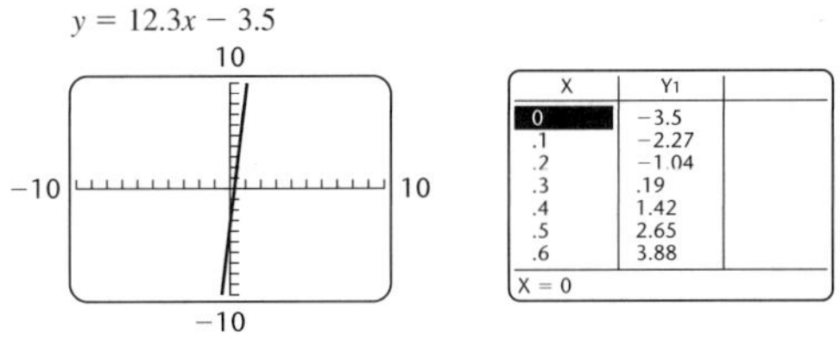

(c) $y = 3(x + 2.3)^2 + 2.3$

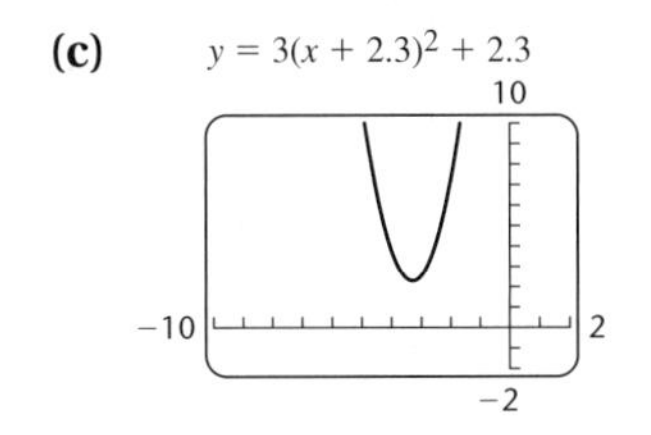

X	Y1
0	18.17
.1	19.58
.2	21.05
.3	22.58
.4	24.17
.5	25.82
.6	27.53

X = 0

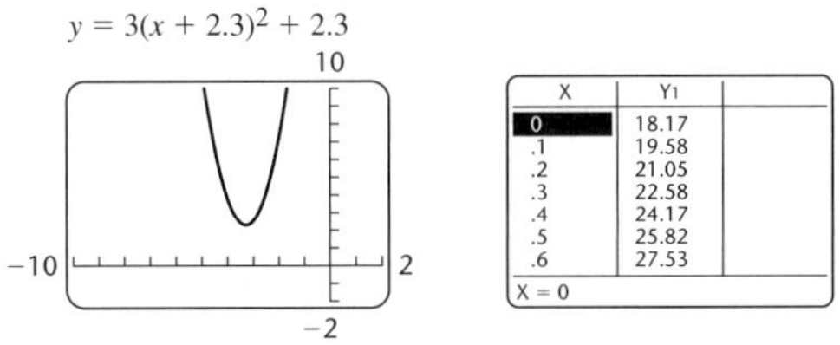

Exercise Set 2.2, pp. 87–91

1. No **3.** Yes **5.** Yes **7.** Function **9.** A relation but not a function **11.** Function
13. **(a)** -2; **(b)** $\{x \mid -2 \le x \le 5\}$; **(c)** 4; **(d)** $\{y \mid -3 \le y \le 4\}$
15. **(a)** 3; **(b)** $\{x \mid -1 \le x \le 4\}$; **(c)** 3; **(d)** $\{y \mid 1 \le y \le 4\}$
17. **(a)** -2; **(b)** $\{x \mid -4 \le x \le 2\}$; **(c)** -2; **(d)** $\{y \mid -3 \le y \le 3\}$
19. **(a)** 3; **(b)** $\{x \mid -4 \le x \le 3\}$; **(c)** -3; **(d)** $\{y \mid -2 \le y \le 5\}$
21. **(a)** 1; **(b)** $\{-3, -1, 1, 3, 5\}$; **(c)** 3; **(d)** $\{-1, 0, 1, 2, 3\}$
23. **(a)** 4; **(b)** $\{x \mid -3 \le x \le 4\}$; **(c)** $-1, 3$; **(d)** $\{y \mid -4 \le y \le 5\}$
25. **(a)** 1; **(b)** $\{x \mid -4 < x \le 5\}$; **(c)** $\{x \mid 2 < x \le 5\}$; **(d)** $\{-1, 1, 2\}$ **27.** Yes **29.** Yes **31.** No **33.** No
35. **(a)** 3; **(b)** -1; **(c)** -4; **(d)** 11; **(e)** $a + 5$
37. **(a)** 0; **(b)** 1; **(c)** 57; **(d)** $5t^2 + 4t$; **(e)** $20a^2 + 8a$
39. **(a)** $\frac{3}{5}$; **(b)** $\frac{1}{3}$; **(c)** $\frac{4}{7}$; **(d)** 0; **(e)** $\dfrac{x-1}{2x-1}$
41. **(a)** $\{x \mid x \text{ is a real number } \textit{and } x \ne 3\}$; **(b)** $\{x \mid x \text{ is a real number } \textit{and } x \ne 6\}$; **(c)** $\mathbb{R}$; **(d)** $\mathbb{R}$; **(e)** $\left\{x \mid x \text{ is a real number } \textit{and } x \ne \frac{5}{2}\right\}$; **(f)** $\mathbb{R}$
43. $4\sqrt{3}\ \text{cm}^2 \approx 6.93\ \text{cm}^2$ **45.** $36\pi\ \text{in}^2 \approx 113.10\ \text{in}^2$
47. 14°F **49.** 159.48 cm **51.** 75 heart attacks per 10,000 men **53.** 56%
55. 3.5 drinks

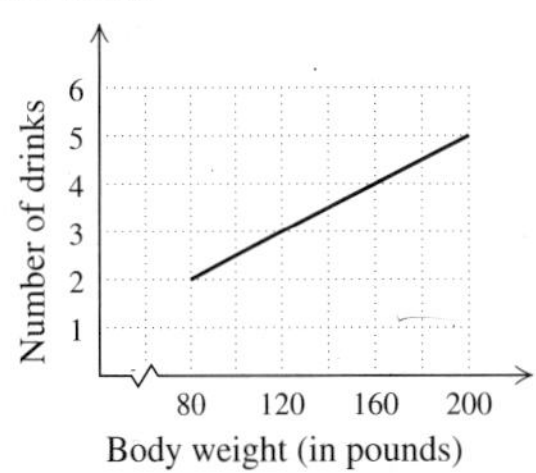

57. About 21,000 cases

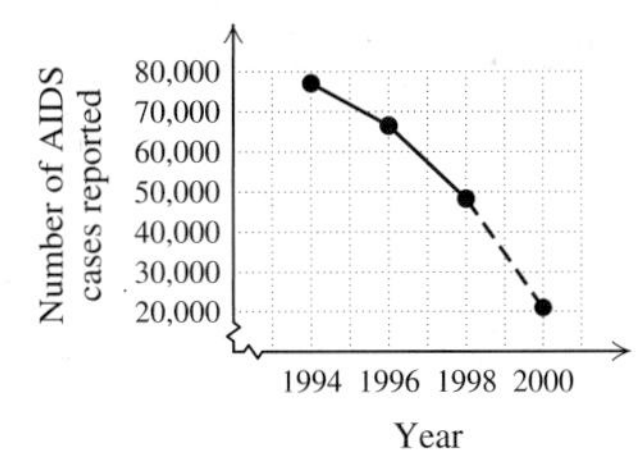

59. About 65,000

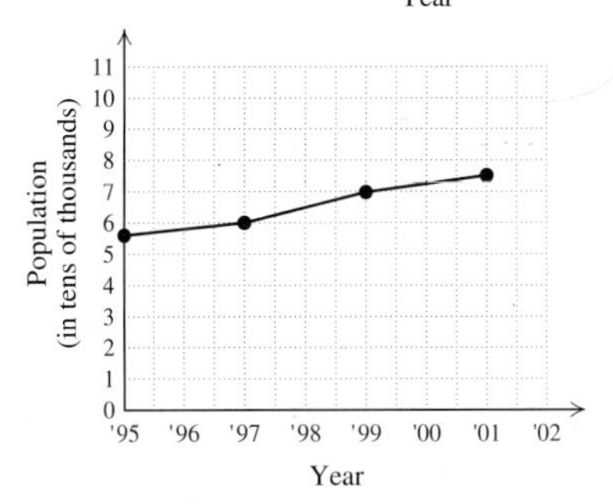

61. About $313,000

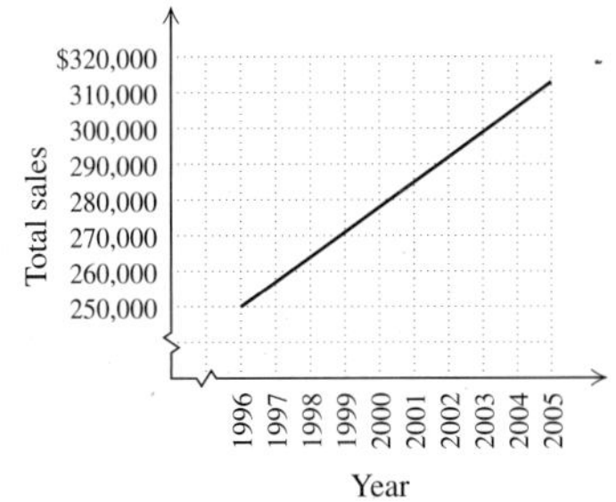

63. **65.** [1.2] $\frac{1}{3}$ **66.** [1.2] -1
67. [1.5] $l = \dfrac{S - 2wh}{2h + 2w}$ **68.** [1.5] $w = \dfrac{S - 2lh}{2l + 2h}$
69. [1.5] $y = -\frac{2}{3}x + 2$ **70.** [1.5] $y = \frac{5}{4}x - 2$ **71.**
73. 26; 99 **75.** About 22 mm **77.**
79.

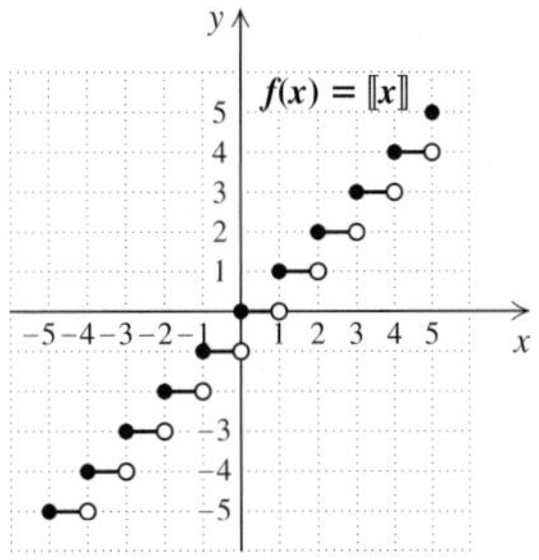

81. Bicycling 14 mph for 1 hr

Exercise Set 2.3, pp. 101–105

1.

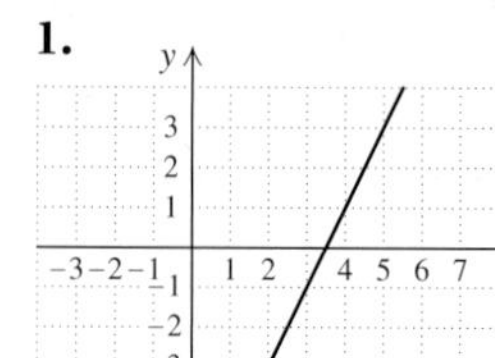

3.

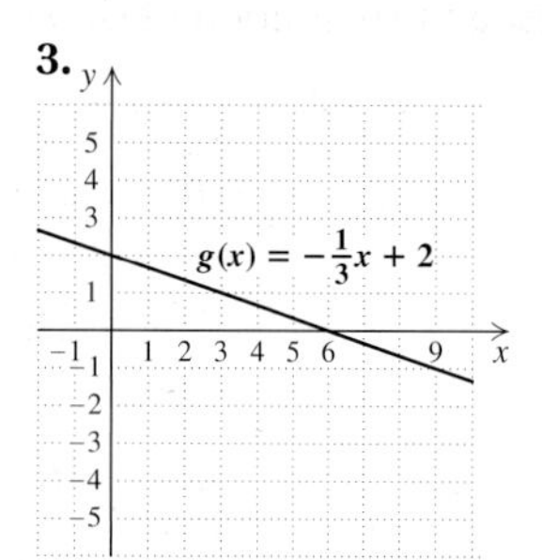

5.

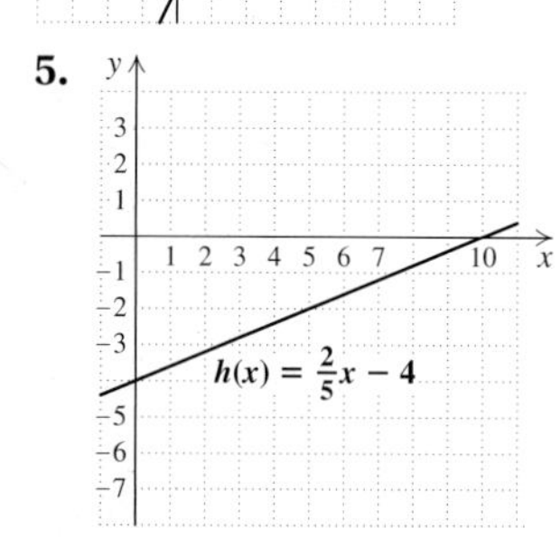

7. $(0, 7)$ **9.** $(0, -6)$ **11.** $(0, -4.5)$ **13.** $(0, -9)$
15. $(0, 204)$ **17.** 2 **19.** -2 **21.** $-\frac{1}{3}$ **23.** 0
25. Slope: $\frac{5}{2}$; y-intercept: $(0, 3)$

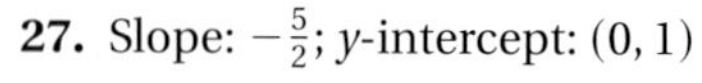

27. Slope: $-\frac{5}{2}$; y-intercept: $(0, 1)$

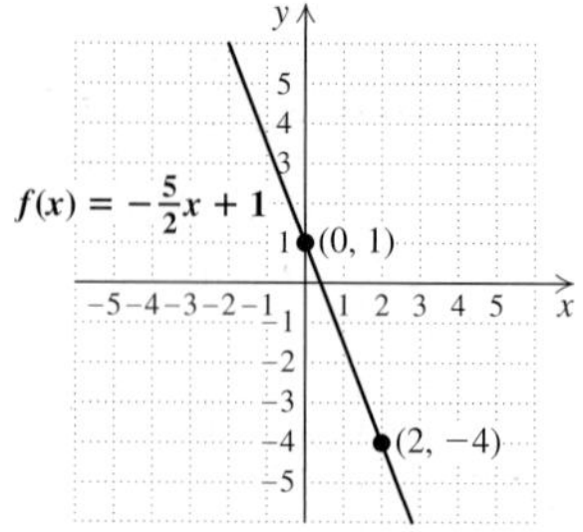

29. Slope: 2; y-intercept: $(0, -5)$

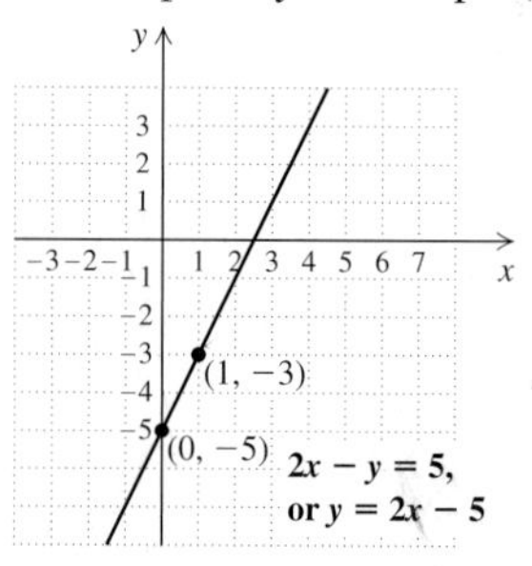

31. Slope: $\frac{1}{3}$; y-intercept: $(0, 2)$

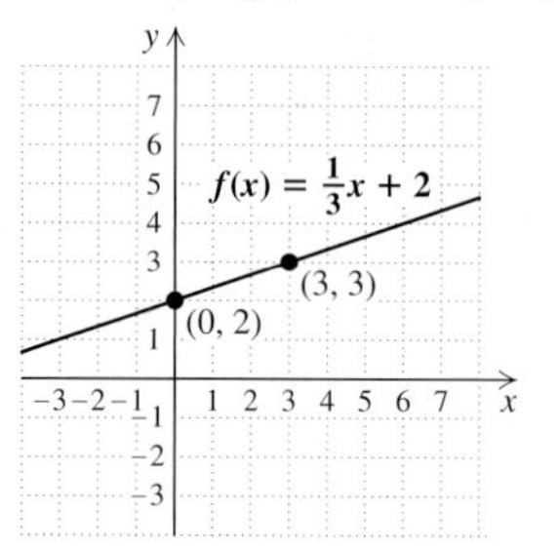

33. Slope: $-\frac{2}{7}$; y-intercept: $(0, 1)$

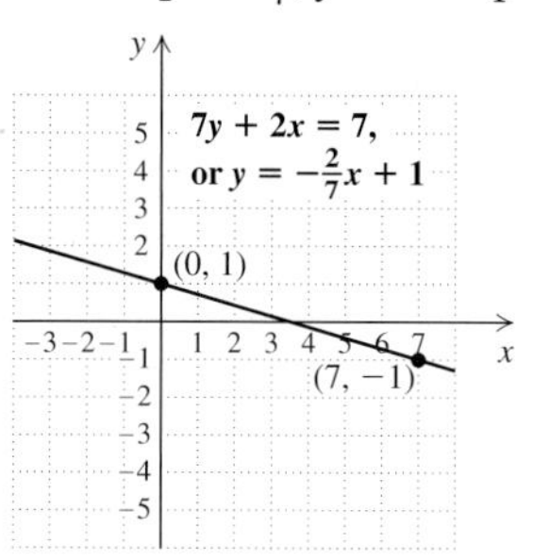

35. Slope: -0.25; y-intercept: $(0, 0)$

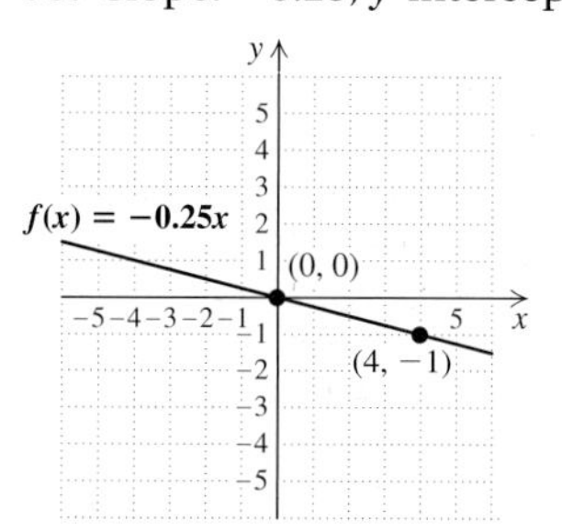

37. Slope: $\frac{4}{5}$; y-intercept: $(0, -2)$

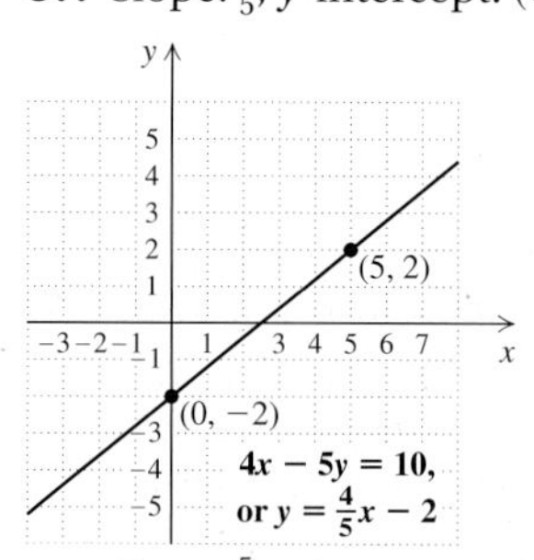

39. Slope: $\frac{5}{4}$; y-intercept: $(0, -2)$

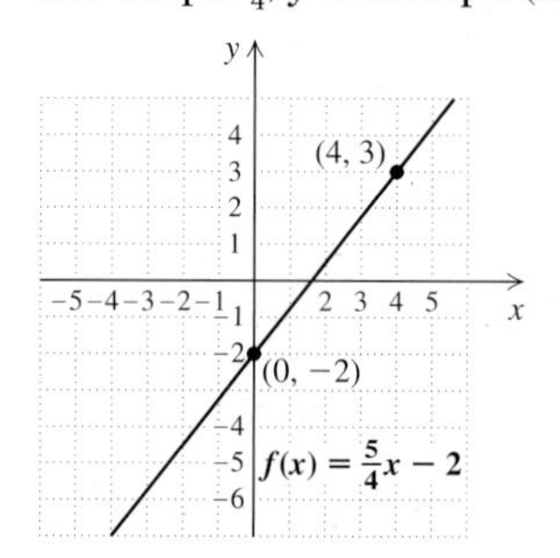

41. Slope: $-\frac{3}{4}$; y-intercept: $(0, 3)$

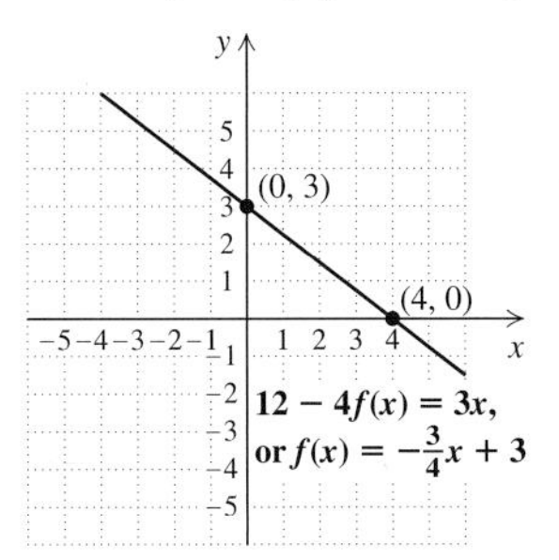

43. Slope: 0; y-intercept: $(0, 2.5)$

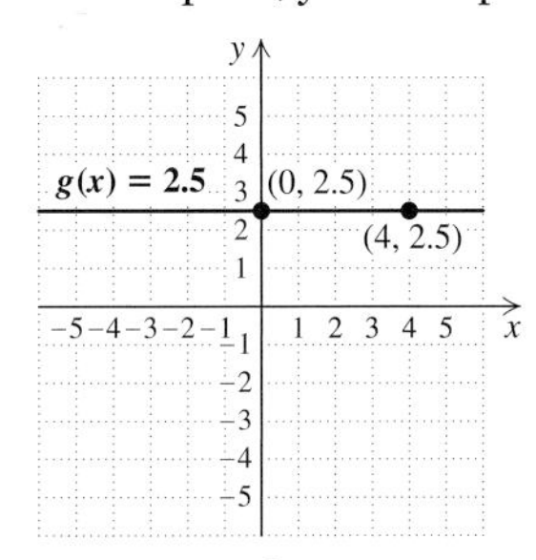

45. $f(x) = \frac{2}{3}x - 9$ **47.** $f(x) = -5x + 2$
49. $f(x) = -\frac{7}{9}x + 5$ **51.** $f(x) = 5x + \frac{1}{2}$
53. The value is decreasing at a rate of \$900 per year.
55. The distance is increasing at a rate of 3 miles per hour.
57. The number of pages read is increasing at a rate of 75 pages per day. **59.** The average SAT math score is increasing at a rate of 1 point per thousand dollars of family income. **61.** 12 km/h **63.** 175,000 hits/yr
65. 300 ft/min **67.** **(a)** II; **(b)** IV; **(c)** I; **(d)** III
69. 25 signifies that the cost per person is \$25; 75 signifies that the setup cost for the party is \$75.
71. $\frac{1}{2}$ signifies that Tina's hair grows $\frac{1}{2}$ inch per month; 1 signifies that her hair is 1 inch long when cut.
73. $\frac{3}{20}$ signifies that the life expectancy of American women increases $\frac{3}{20}$ year per year, for years after 1950; 72 signifies that the life expectancy in 1950 was 72 years.
75. 2.6 signifies that sales increase \$2.6 billion per year, for years after 1975; 17.8 signifies that sales in 1975 were \$17.8 billion. **77.** 0.75 signifies that the cost per mile of a taxi ride is \$0.75; 2 signifies that the minimum cost of a taxi ride is \$2. **79.** **(a)** -5000 represents a depreciation of \$5000 per year; 90,000 represents the original value of the truck, \$90,000; **(b)** 18 yr; **(c)** $\{t \mid 0 \le t \le 18\}$
81. **(a)** -150 signifies that the depreciation is \$150 per winter of use; 900 signifies that the original value of the snowblower was \$900; **(b)** after 4 winters of use; **(c)** $\{0, 1, 2, 3, 4, 5, 6\}$ **83.** **85.** [1.3] $-\frac{8}{5}$
86. [1.3] -5 **87.** [1.3] -25 **88.** [1.3] 3 **89.** [1.3] $-\frac{9}{2}$
90. [1.3] $\frac{3}{4}$ **91.** **93.** Slope: $-\dfrac{r}{p}$; y-intercept: $\left(0, \dfrac{s}{p}\right)$
95. Since (x_1, y_1) and (x_2, y_2) are two points on the graph of $y = mx + b$, then $y_1 = mx_1 + b$ and $y_2 = mx_2 + b$. Using the definition of slope, we have

$$\begin{aligned}\text{Slope} &= \frac{y_2 - y_1}{x_2 - x_1} \\ &= \frac{(mx_2 + b) - (mx_1 + b)}{x_2 - x_1} \\ &= \frac{m(x_2 - x_1)}{x_2 - x_1} \\ &= m.\end{aligned}$$

97. False **99.** False **101.** **(a)** III; **(b)** IV; **(c)** I; **(d)** II
103.

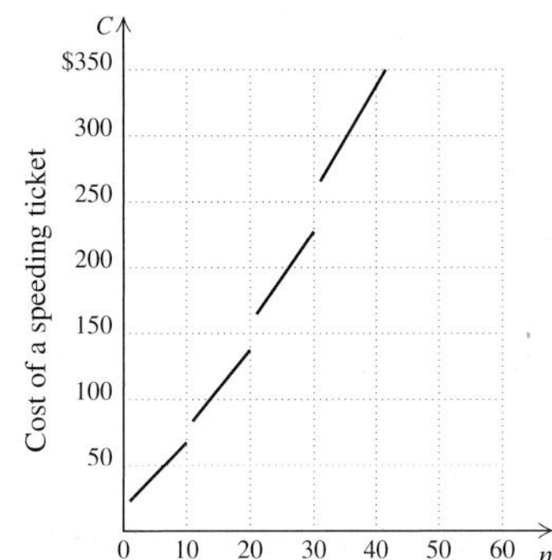

105.

Exercise Set 2.4, pp. 115–117

1. 0 **3.** Undefined **5.** 0 **7.** Undefined
9. Undefined **11.** 0 **13.** Undefined **15.** Undefined
17. $-\frac{2}{3}$

19.

y = 4

21.

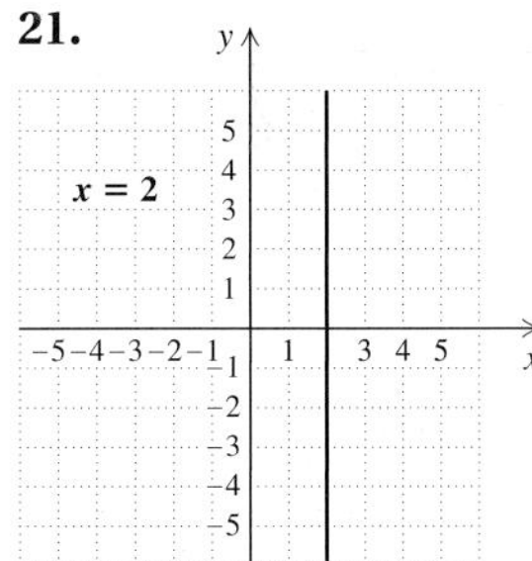

23.

4 · f(x) = 20

25.

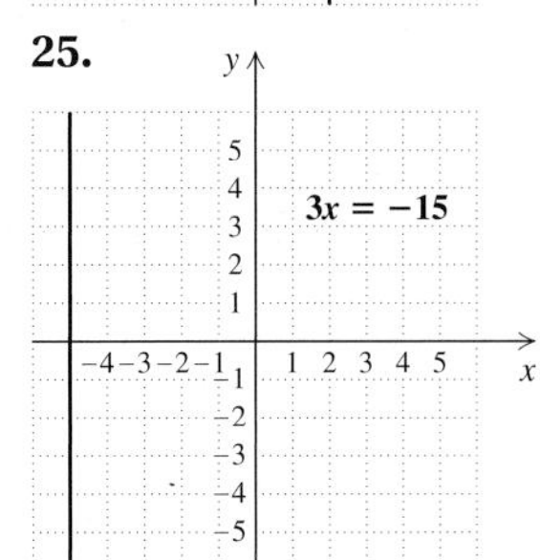

27.

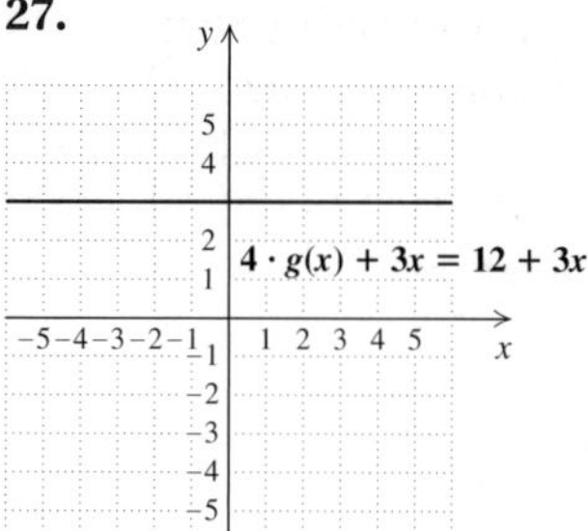

29.

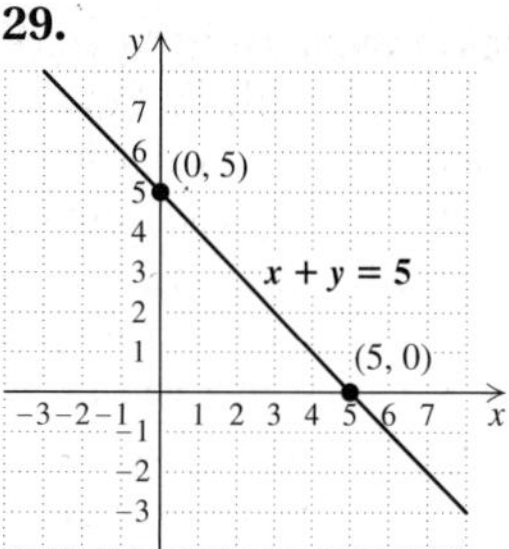

31.

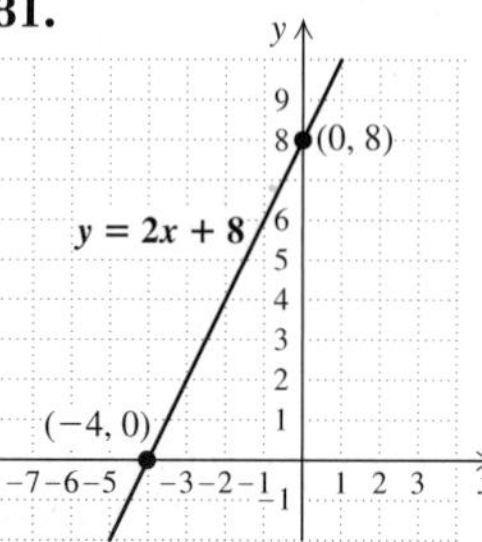

33.

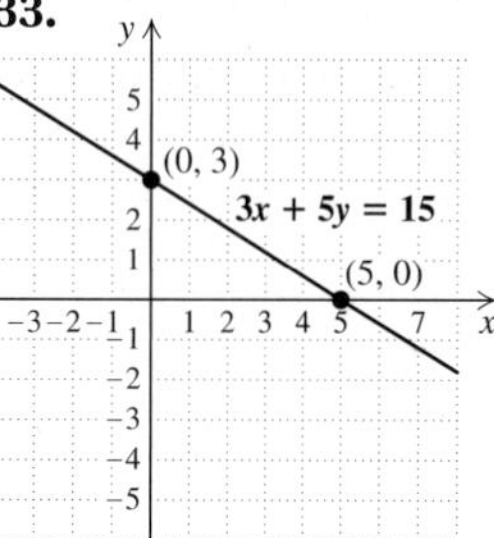

35.

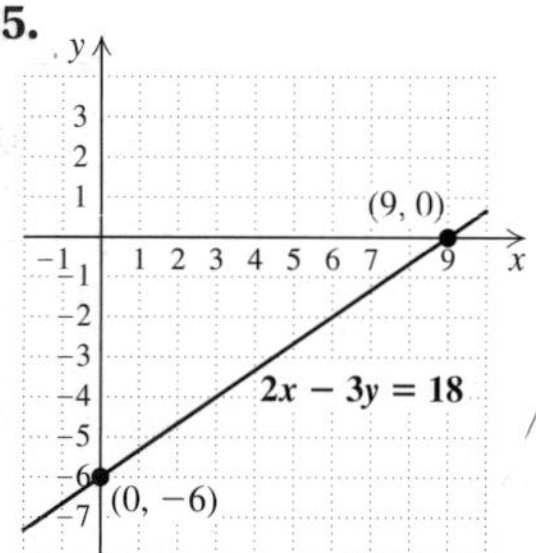

37.

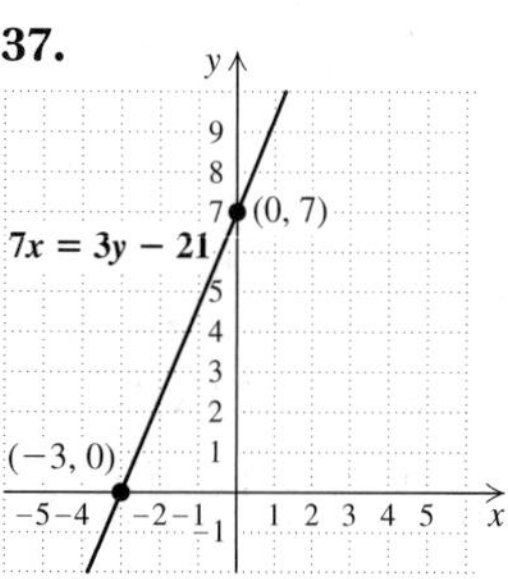

39.

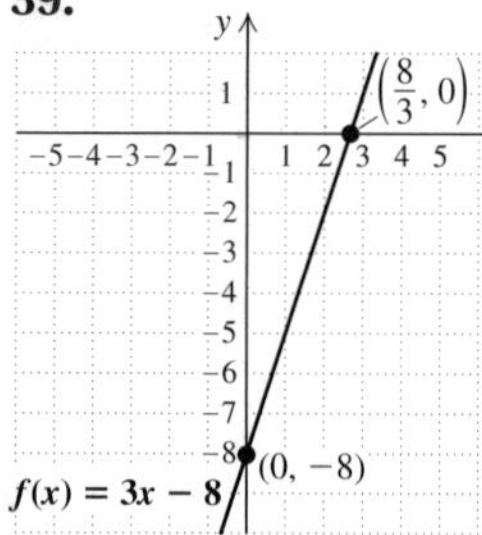

41.

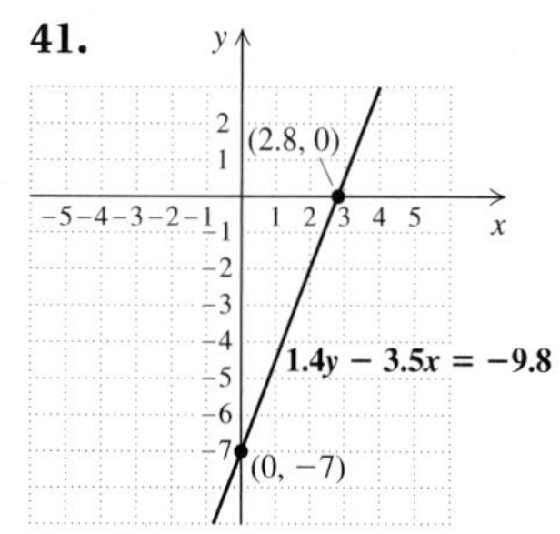

43.

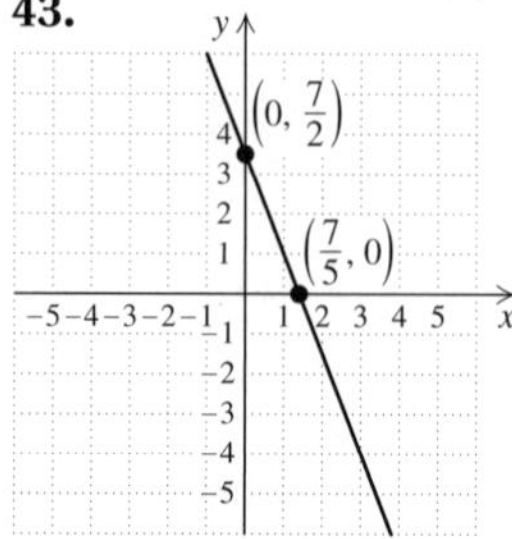

45. 7 **47.** $\frac{5}{3}$ **49.** -8 **51.** -1 **53.** 4 **55.** 2
57. 4 months **59.** 2 hr, 15 min **61.** 13 lb
63. Linear; $\frac{5}{3}$ **65.** Linear; 0 **67.** Not linear

69. Linear; $\frac{14}{3}$ **71.** Not linear **73.** Not linear **75.**
77. [1.2] -1 **78.** [1.2] -1 **79.** [1.3] $-5x - 15$
80. [1.3] $-2x - 8$ **81.** [1.3] $\frac{2}{3}x - \frac{2}{3}$ **82.** [1.3] $-\frac{3}{2}x - \frac{12}{5}$
83. **85.** $4x - 5y = 20$ **87.** Linear **89.** Linear
91. The slope of equation B is $\frac{1}{2}$ the slope of equation A.
93. $a = 7, b = -3$ **95.**

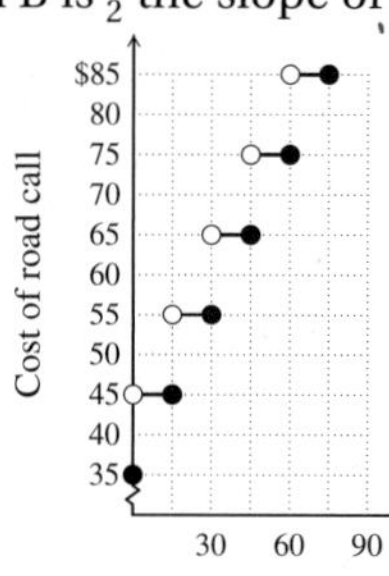

97. $0.\overline{6}$ **99.** 2.6 **101.** 249

Technology Connection, p. 120

1. $y_1 = \frac{3}{4}x + 2;\ y_2 = -\frac{4}{3}x - 1$

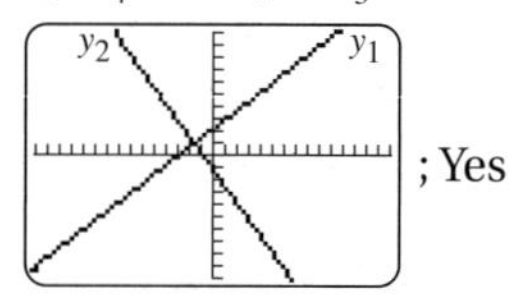

; Yes

2. $y_1 = -\frac{2}{5}x - 4;\ y_2 = \frac{5}{2}x + 3$

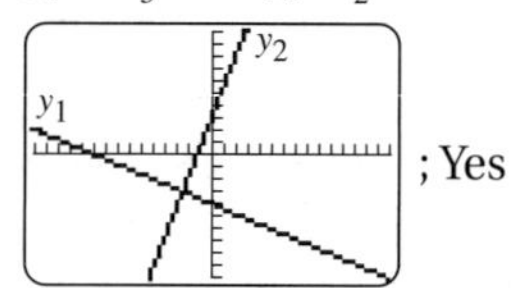

; Yes

3. $y_1 = \frac{31}{40}x + 2;\ y_2 = -\frac{40}{30}x - 1$

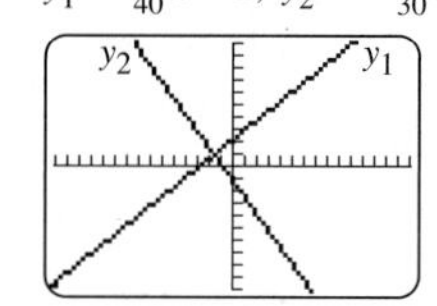

; No: $-\frac{40}{30} \neq -\frac{1}{\frac{31}{40}}$

Although the lines appear to be perpendicular, they are not, because the product of their slopes is not -1:

$$\frac{31}{40}\left(-\frac{40}{30}\right) = -\frac{1240}{1200} \neq -1.$$

Exercise Set 2.5, pp. 122–125

1.

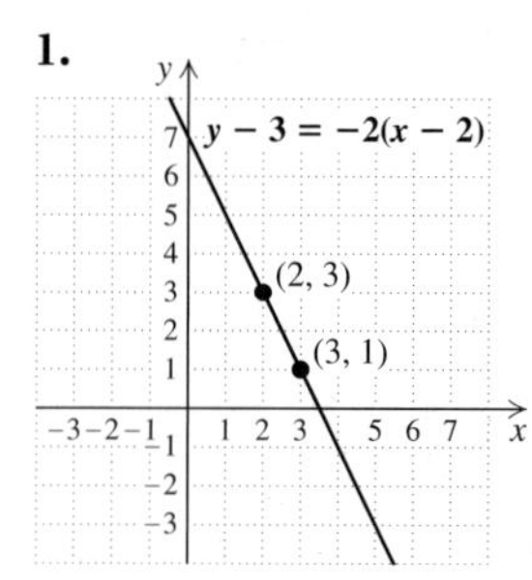

3.

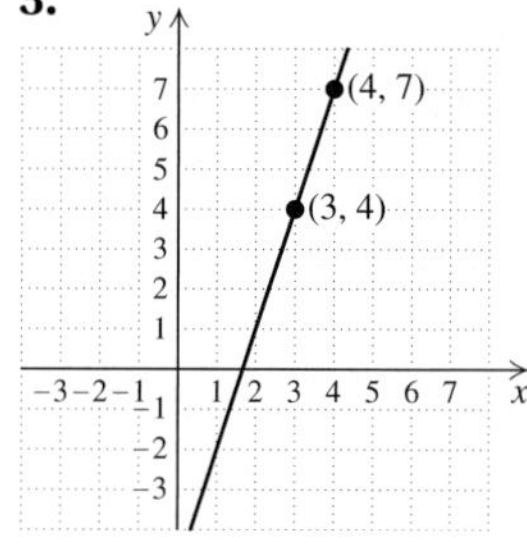

5.

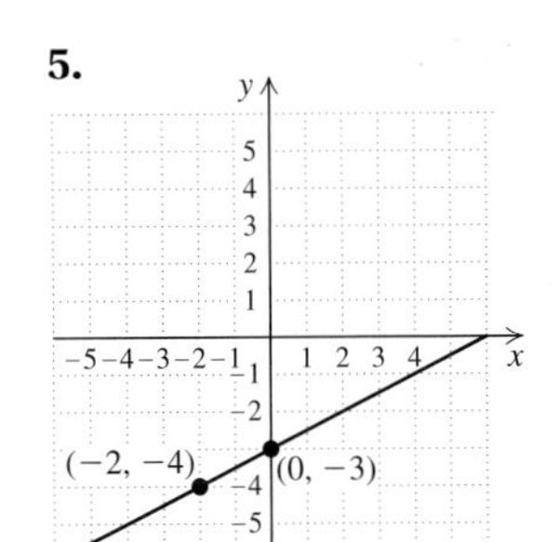

$y - (-4) = \frac{1}{2}(x - (-2))$, or

$y + 4 = \frac{1}{2}(x + 2)$

7.

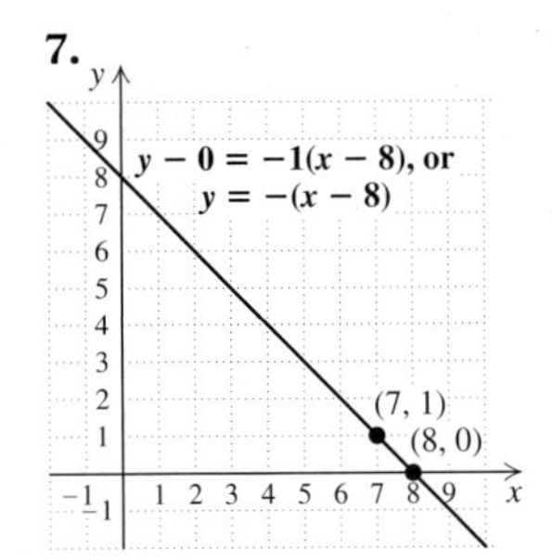

9.

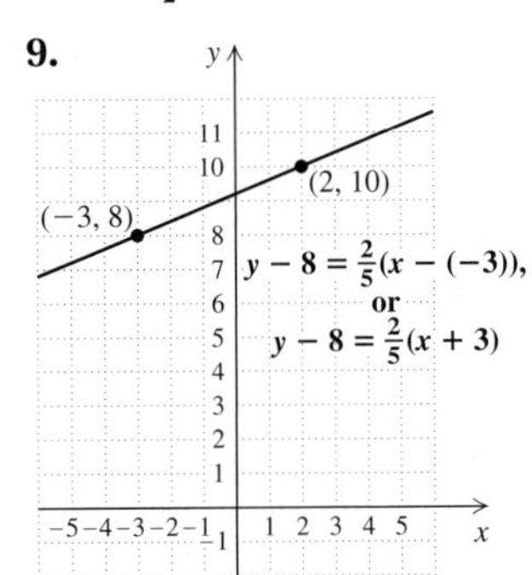

11. $\frac{2}{7}$; $(1, 4)$ **13.** -5; $(7, -2)$ **15.** $-\frac{5}{3}$; $(-2, 1)$
17. $\frac{4}{7}$; $(0, 0)$ **19.** $f(x) = 4x - 11$ **21.** $f(x) = -\frac{3}{5}x - \frac{23}{5}$
23. $f(x) = -0.6x - 5.8$ **25.** $f(x) = \frac{2}{7}x - 5$
27. $f(x) = \frac{1}{2}x + \frac{7}{2}$ **29.** $f(x) = 1.5x - 6.75$
31. $f(x) = 5x - 2$ **33.** $f(x) = \frac{3}{2}x$
35. **(a)** $R(t) = -0.075t + 46.8$; **(b)** 41.325 sec; 41.1 sec; **(c)** 2021 **37.** **(a)** $A(t) = 8.0625t + 178.6$; **(b)** \$291,475,000 **39.** **(a)** $N(t) = 4.25t + 43.8$; **(b)** 94.8 million tons **41.** **(a)** $E(t) = \frac{2}{35}t + 78.8$; **(b)** 79.8 yr **43.** **(a)** $A(t) = \frac{13}{15}t + 74.9$; **(b)** 85.3 million acres **45.** Yes **47.** Yes **49.** No **51.** $y = -\frac{1}{2}x + \frac{17}{2}$
53. $y = \frac{5}{7}x - \frac{17}{7}$ **55.** $y = \frac{1}{3}x + 4$ **57.** $y = -\frac{3}{2}x - \frac{13}{2}$
59. $y = 2x - 7$ **61.** Yes **63.** No **65.** $y = \frac{1}{2}x + 4$
67. $y = \frac{4}{3}x - 6$ **69.** $y = \frac{5}{2}x + 9$ **71.** $y = -\frac{5}{3}x - \frac{41}{3}$
73. $y = -\frac{1}{2}x + 6$ **75.** 📓 **77.** [1.3] $3x^2 + 7x - 4$
78. [1.3] $5t^2 - 6t - 3$ **79.** [1.1] 0 **80.** [1.2] 0
81. [1.2] 11 **82.** [1.2] -34 **83.** 📓 **85.** \$1350
87. \$30 **89.** 82.8% **91.** $\{p \mid 0 < p \leq 10.6\}$
93. **(a)** $g(x) = x - 8$; **(b)** -10; **(c)** 83 **95.** 7 **97.** 📓

Exercise Set 2.6, pp. 131–134

1. 1 **3.** -41 **5.** 12 **7.** $\frac{13}{18}$ **9.** 5 **11.** $x^2 - 3x + 3$
13. $x^2 - x + 3$ **15.** 23 **17.** 5 **19.** 56
21. $\dfrac{x^2 - 2}{5 - x}$, $x \neq 5$ **23.** $\frac{2}{7}$ **25.** $0.75 + 2.5 = 3.25$
27. Women under 30 **29.** About 50 million; the number of passengers using Newark and LaGuardia in 1998
31. About 8 million; how many more passengers used Kennedy than LaGuardia in 1994 **33.** About 89 million; the number of passengers using the three airports in 1999
35. $\mathbb{R}$ **37.** $\{x \mid x \text{ is a real number } and\ x \neq 3\}$
39. $\{x \mid x \text{ is a real number } and\ x \neq 0\}$
41. $\{x \mid x \text{ is a real number } and\ x \neq 1\}$
43. $\{x \mid x \text{ is a real number } and\ x \neq 2\ and\ x \neq 4\}$
45. $\{x \mid x \text{ is a real number } and\ x \neq 3\}$
47. $\{x \mid x \text{ is a real number } and\ x \neq 4\}$
49. $\{x \mid x \text{ is a real number } and\ x \neq 4\ and\ x \neq 5\}$
51. $\{x \mid x \text{ is a real number } and\ x \neq -1\ and\ x \neq -\frac{5}{2}\}$
53. 4; 3 **55.** 5; -1 **57.** $\{x \mid 0 \leq x \leq 9\}$; $\{x \mid 3 \leq x \leq 10\}$; $\{x \mid 3 \leq x \leq 9\}$; $\{x \mid 3 \leq x \leq 9\}$

59.

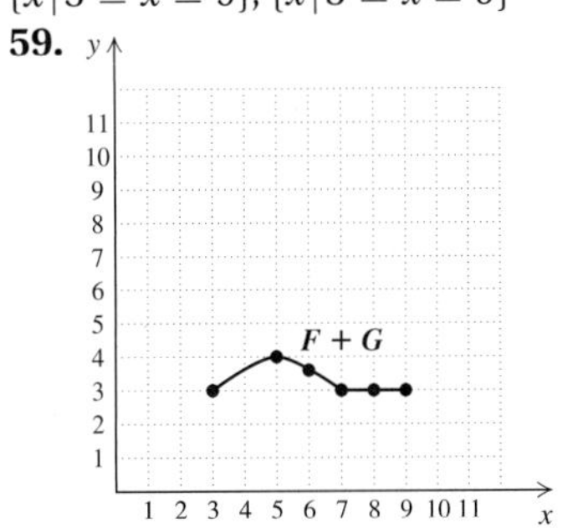

61. 📓

63. [1.5] $x = \frac{7}{4}y + 2$ **64.** [1.5] $y = \frac{3}{8}x - \frac{5}{8}$
65. [1.5] $y = -\frac{5}{2}x - \frac{3}{2}$ **66.** [1.5] $x = -\frac{5}{6}y - \frac{1}{3}$
67. [1.4] Let n represent the number; $2n + 5 = 49$
68. [1.4] Let x represent the number; $\frac{1}{2}x - 3 = 57$
69. [1.4] Let x represent the first integer; $x + (x + 1) = 145$
70. [1.4] Let n represent the number; $n - (-n) = 20$
71. 📓
73. $\{x \mid x \text{ is a real number } and\ x \neq -\frac{5}{2}\ and\ x \neq -3\ and\ x \neq 1\ and\ x \neq -1\}$
75. Answers may vary.

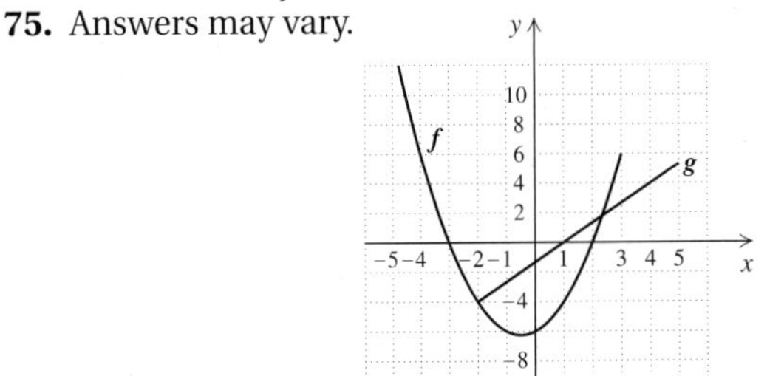

77. Domain of $f + g$ = Domain of $f - g$ = Domain of $f \cdot g = \{-2, -1, 0, 1\}$; Domain of $f/g = \{-2, 0, 1\}$
79. Answers may vary. $f(x) = \dfrac{1}{x + 2}$, $g(x) = \dfrac{1}{x - 5}$
81. Left to the student

Review Exercises: Chapter 2, pp. 136–137

1. [2.1] Yes **2.** [2.1] No **3.** [2.1] Yes **4.** [2.1] No

5. [2.1], [2.3]

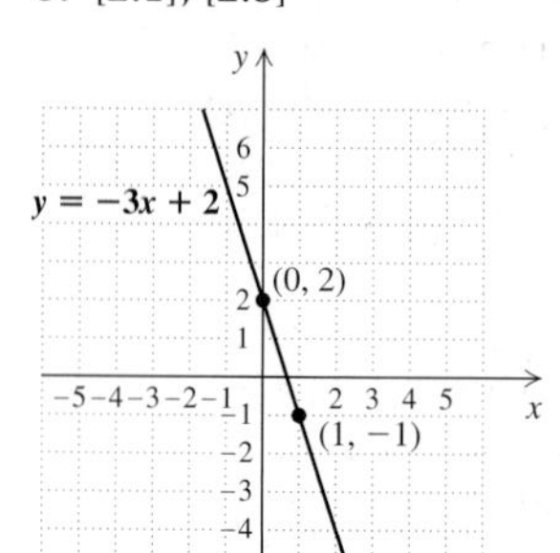

6. [2.1]

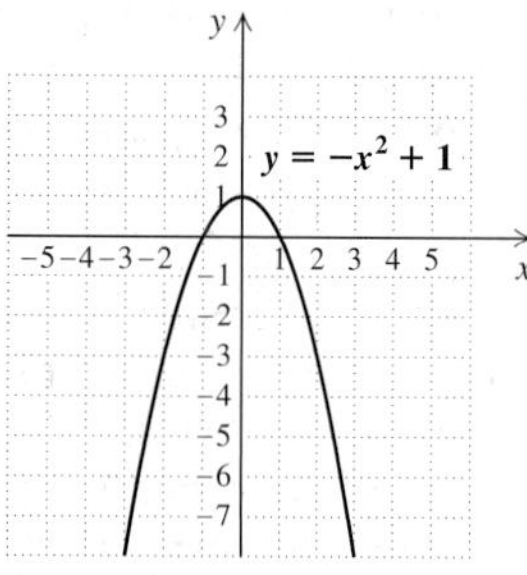

7. [2.4]

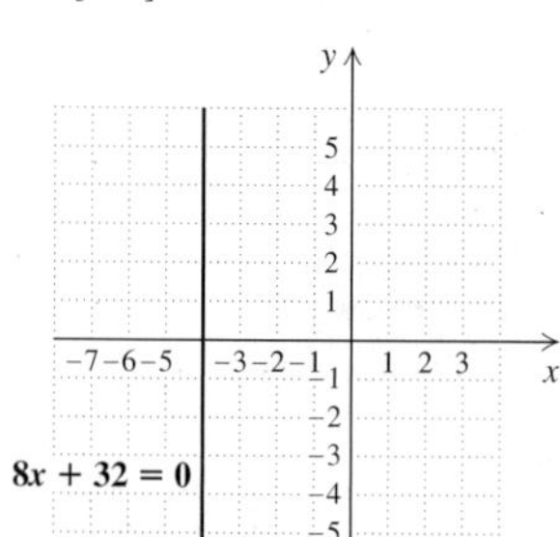

8. [2.4]

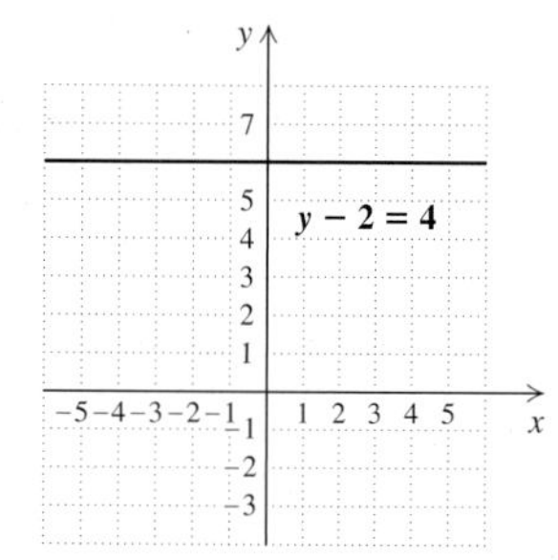

9. [2.2] **(a)** 3; **(b)** $\{x \mid -2 \leq x \leq 4\}$; **(c)** -1; **(d)** $\{y \mid 1 \leq y \leq 5\}$
10. [2.2] 10.53 yr
11. [2.3] Slope: -4; y-intercept: $(0, -9)$
12. [2.3] Slope: $\frac{1}{3}$; y-intercept: $\left(0, -\frac{7}{6}\right)$
13. [2.2] About 510 cans per person **14.** [2.2] About 670 cans per person **15.** [2.3] $\frac{4}{7}$ **16.** [2.4] Undefined
17. [2.3] \$2500 of personal income per year since high school **18.** [2.3] 132,500 homes per month
19. [2.3] 645 signifies that tuition is increasing at a rate of \$645 a year; 9800 represents tuition costs in 1997
20. [2.3] $f(x) = \frac{2}{7}x - 6$
21. [2.4]

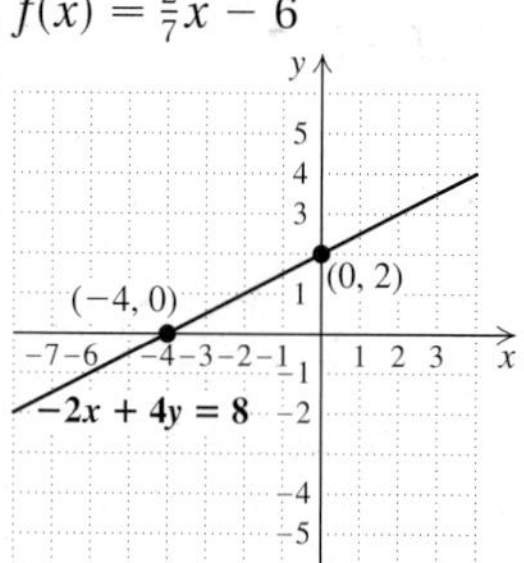

22. [2.4] -1 **23.** [2.4] 7 months **24.** [2.4] Yes
25. [2.4] Yes **26.** [2.4] No **27.** [2.4] No
28. [2.5] $y - 4 = -2(x - (-3))$, or $y - 4 = -2(x + 3)$
29. [2.5] $f(x) = \frac{4}{3}x + \frac{7}{3}$ **30.** [2.5] Perpendicular
31. [2.5] Parallel **32.** [2.5] **(a)** $W(t) = \frac{11}{105}t + 0.75$; **(b)** \$5.99 **33.** [2.5] $y = \frac{3}{5}x - \frac{31}{5}$ **34.** [2.5] $y = -\frac{5}{3}x - \frac{5}{3}$
35. [2.2] -6 **36.** [2.2] 26 **37.** [2.6] 102
38. [2.6] -17 **39.** [2.6] $-\frac{9}{2}$ **40.** [2.2] $3a + 3b - 6$
41. [2.6] $\mathbb{R}$ **42.** [2.6] $\{x \mid x \text{ is a real number } and\ x \neq 2\}$
43. [2.2] For a function, every member of the domain corresponds to *exactly one* member of the range. Thus, for any function, each member of the domain corresponds to *at least one* member of the range. Therefore, a function is a relation. In a relation, every member of the domain corresponds to *at least one,* but not necessarily *exactly one,* member of the range. Therefore, a relation may or may not be a function. **44.** [2.4] The slope of a line is the rise between two points on the line divided by the run between those points. For a vertical line, there is no run between any two points, and division by 0 is undefined; therefore, the slope is undefined. For a horizontal line, there is no rise between any two points, so the slope is 0/run, or 0. **45.** [1.6], [2.4] -9 **46.** [2.5] $-\frac{9}{2}$
47. [2.5] $f(x) = 3.09x + 3.75$
48. [2.3] **(a)** III; **(b)** IV; **(c)** I; **(d)** II

Test: Chapter 2, pp. 138–139

1. [2.1] Yes **2.** [2.1] No
3. [2.1], [2.3]

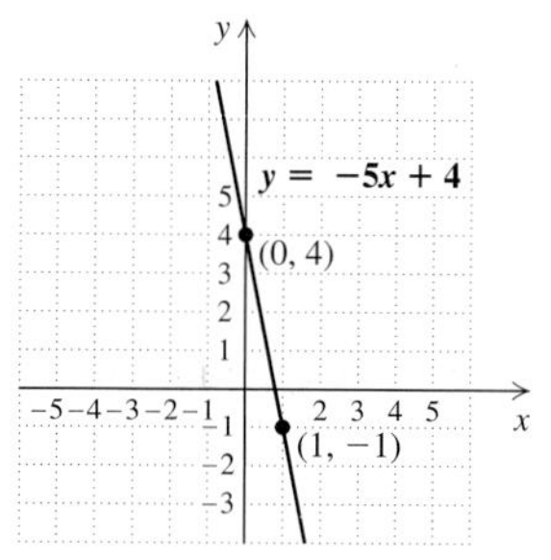

4. [2.1]

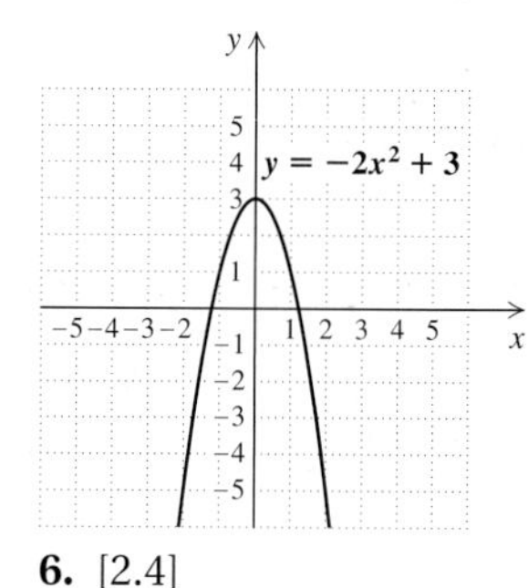

5. [2.4]

$f(x) = 5$

6. [2.4]

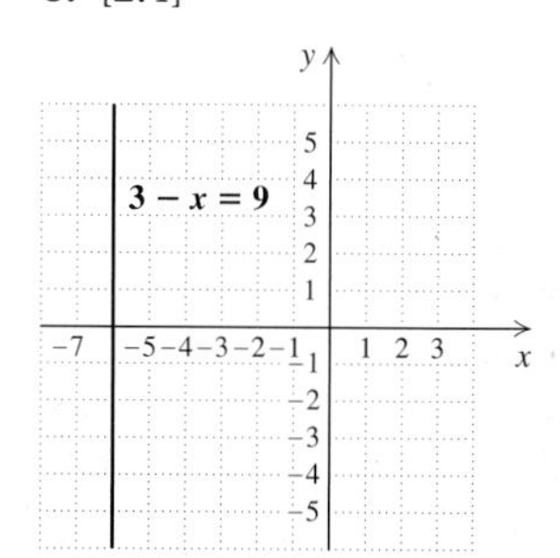

7. [2.2] **(a)** 1; **(b)** $\{x \mid -3 \leq x \leq 4\}$; **(c)** 3; **(d)** $\{y \mid -1 \leq y \leq 2\}$
8. (a) [2.2] \$40.6 billion; **(b)** [2.3] 1.2 signifies that the rate of increase in U.S. book sales was \$1.2 billion per year; 21.4 signifies that U.S. book sales were \$21.4 billion in 1992 **9.** [2.2] 46 million
10. [2.3] Slope: $-\frac{3}{5}$; y-intercept: $(0, 12)$
11. [2.3] Slope: $-\frac{2}{5}$; y-intercept: $\left(0, -\frac{7}{5}\right)$ **12.** [2.3] $\frac{5}{8}$
13. [2.3] 0 **14.** [2.3] 75 calories per 30 minutes, or 2.5 calories per minute **15.** [2.3] $f(x) = -5x - 1$

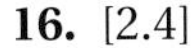

16. [2.4]

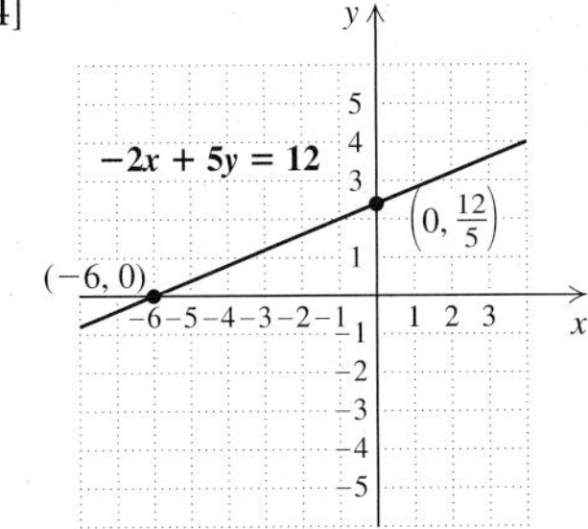

17. [2.4] 3 **18.** [2.4] (a), (c)
19. [2.5] $y - (-4) = 4(x - (-2)), y + 4 = 4(x + 2)$
20. [2.5] $f(x) = -x + 2$ **21.** [2.5] Parallel
22. [2.5] Perpendicular **23.** [2.5] $y = \frac{2}{5}x + \frac{16}{5}$
24. [2.5] $y = -\frac{5}{2}x - \frac{11}{2}$ **25.** [2.6] **(a)** 5; **(b)** -130;
(c) $\{x \mid x \text{ is a real number } and\ x \neq -\frac{4}{3}\}$
26. [2.5] **(a)** $C(m) = 0.3m + 25$; **(b)** \$175
27. [2.2], [2.3] **(a)** 30 mi; **(b)** 15 mph
28. [2.5] $s = -\dfrac{3}{2}r + \dfrac{27}{2}$, or $\dfrac{27 - 3r}{2}$
29. [2.6] $h(x) = 7x - 2$

CHAPTER 3

Technology Connection, p. 147

1. (1.53, 2.58) **2.** (−0.26, 57.06) **3.** (2.23, 1.14)
4. (0.87, −0.32)

Exercise Set 3.1, pp. 148–150

1. Yes **3.** No **5.** Yes **7.** Yes **9.** (4, 1)
11. (2, −1) **13.** (4, 3) **15.** (−3, −2) **17.** (−3, 2)
19. (3, −7) **21.** (7, 2) **23.** (4, 0) **25.** No solution
27. $\{(x, y) \mid y = 3 - x\}$ **29.** All except 25 **31.** 27
33. Let x represent the larger number and y the smaller number; $x - y = 11, 3x + 2y = 123$ **35.** Let x represent the number of \$8.50 brushes sold and y the number of \$9.75 brushes sold; $x + y = 45, 8.50x + 9.75y = 398.75$
37. Let x and y represent the angles; $x + y = 180$, $x = 2y - 3$ **39.** Let x represent the number of two-point shots and y the number of free throws; $x + y = 64$, $2x + y = 100$ **41.** Let h represent the number of vials of Humulin sold and n the number of vials of Novolin; $h + n = 50, 21.95h + 20.95n = 1077.50$ **43.** Let l represent the length, in feet, and w the width, in feet; $2l + 2w = 228, w = l - 42$ **45.** Let w represent the number of wins and t the number of ties; $2w + t = 60$, $w = t + 9$ **47.** Let x represent the number of ounces of lemon juice and y the number of ounces of linseed oil; $y = 2x, x + y = 32$ **49.** Let x represent the number of general-interest films rented and y the number of children's films rented; $x + y = 77, 3x + 1.5y = 213$ **51.**
53. [1.3] 15 **54.** [1.3] $\frac{19}{12}$ **55.** [1.3] $\frac{9}{20}$ **56.** [1.3] $\frac{13}{3}$
57. [1.5] $y = -\frac{3}{4}x + \frac{7}{4}$ **58.** [1.5] $y = \frac{2}{5}x - \frac{9}{5}$ **59.**
61. 1994 **63.** Answers may vary. **(a)** $x + y = 6$, $x - y = 4$; **(b)** $x + y = 1, 2x + 2y = 3$; **(c)** $x + y = 1$, $2x + 2y = 2$ **65.** $A = -\frac{17}{4}, B = -\frac{12}{5}$
67. Let x and y represent the number of years that Lou and Juanita have taught at the university, respectively; $x + y = 46, x - 2 = 2.5(y - 2)$ **69.** Let s and v represent the number of ounces of baking soda and vinegar needed, respectively; $s = 4v, s + v = 16$
71. (0, 0), (1, 1) **73.** (0.07, −7.95) **75.** (0.01, 1.25)

Exercise Set 3.2, pp. 157–158

1. (2, −3) **3.** $\left(\frac{21}{5}, \frac{12}{5}\right)$ **5.** (2, −2)
7. $\{(x, y) \mid 2x - 3 = y\}$ **9.** (−2, 1) **11.** $\left(\frac{1}{2}, \frac{1}{2}\right)$
13. $\left(\frac{19}{8}, \frac{1}{8}\right)$ **15.** No solution **17.** (1, 2) **19.** (3, 0)
21. (−1, 2) **23.** $\left(\frac{128}{31}, -\frac{17}{31}\right)$ **25.** (6, 2)
27. No solution **29.** $\left(\frac{110}{19}, -\frac{12}{19}\right)$ **31.** (3, −1)
33. $\{(x, y) \mid -4x + 2y = 5\}$ **35.** $\left(\frac{140}{13}, -\frac{50}{13}\right)$ **37.** (−2, −9)
39. (30, 6) **41.** $\{(x, y) \mid x = 2 + 3y\}$ **43.** No solution
45. (140, 60) **47.** $\left(\frac{1}{3}, -\frac{2}{3}\right)$ **49.** **51.** [1.4] 4 mi
52. [1.4] 86 **53.** [1.4] Bathrooms: \$11$\frac{2}{3}$ billion; kitchens: \$23$\frac{1}{3}$ billion **54.** [1.4] 30 m, 90 m, 360 m
55. [1.4] 450.5 mi **56.** [1.4] 460 mi **57.**
59. $m = -\frac{1}{2}, b = \frac{5}{2}$ **61.** $a = 5, b = 2$ **63.** $\left(-\frac{32}{17}, \frac{38}{17}\right)$
65. $\left(-\frac{1}{5}, \frac{1}{10}\right)$ **67.**

Exercise Set 3.3, pp. 169–171

1. 29, 18 **3.** \$8.50 brushes: 32; \$9.75 brushes: 13
5. 119°, 61° **7.** Two-point shots: 36; foul shots: 28
9. Humulin: 30 vials; Novolin: 20 vials
11. Width: 36 ft; length: 78 ft **13.** Wins: 23; ties: 14
15. Lemon juice: 10$\frac{2}{3}$ oz; linseed oil: 21$\frac{1}{3}$ oz
17. General-interest: 65; children's: 12 **19.** Boxes: 13; four-packs: 27 **21.** Kenyan: 8 lb; Sumatran: 12 lb
23. 5 lb of each **25.** 25%-acid: 4 L; 50%-acid: 6 L
27. \$7500 at 6%; \$4500 at 9% **29.** Arctic Antifreeze: 12.5 L; Frost No-More: 7.5 L **31.** Length: 76 m; width: 19 m **33.** Quarters: 17; fifty-cent pieces: 13
35. 375 km **37.** 14 km/h **39.** About 1489 mi **41.**
43. [1.1] 16 **44.** [1.1] 11 **45.** [1.1], [1.2] −28
46. [1.1], [1.2] −10 **47.** [1.1] $\frac{49}{12}$ **48.** [1.1] $\frac{13}{10}$ **49.**
51. Burl: 40; son: 20 **53.** Length: $\frac{288}{5}$ in.; width: $\frac{102}{5}$ in.
55. $\frac{120}{7}$ lb **57.** 4 km **59.** 82 **61.** Brown: 0.8 gal; neutral: 0.2 gal **63.** 45 L **65.** $P(x) = \dfrac{0.1 + x}{1.5}$ (This expresses the percent as a decimal quantity.)

Exercise Set 3.4, pp. 179–180

1. No **3.** $(1, 2, 3)$ **5.** $(-1, 5, -2)$ **7.** $(3, 1, 2)$ **9.** $(-3, -4, 2)$ **11.** $(2, 4, 1)$ **13.** $(-3, 0, 4)$ **15.** The equations are dependent. **17.** $(3, -5, 8)$ **19.** $\left(\frac{3}{5}, \frac{2}{3}, -3\right)$ **21.** $\left(4, \frac{1}{2}, -\frac{1}{2}\right)$ **23.** $(17, 9, 79)$ **25.** $\left(\frac{1}{4}, -\frac{1}{2}, -\frac{1}{4}\right)$ **27.** $(20, 62, 100)$ **29.** No solution **31.** The equations are dependent. **33.** **35.** [1.1] Let x and y represent the numbers; $x = 2y$ **36.** [1.1] Let x and y represent the numbers; $x + y = 3x$ **37.** [1.1] Let x represent the first number; $x + (x + 1) + (x + 2) = 45$ **38.** [1.1] Let x and y represent the numbers; $x + 2y = 17$ **39.** [1.1] Let x, y, and z represent the numbers; $x + y = 5z$ **40.** [1.1] Let x and y represent the numbers; $xy = 2(x + y)$ **41.** **43.** $(1, -1, 2)$ **45.** $(-3, -1, 0, 4)$ **47.** $\left(-\frac{1}{2}, -1, -\frac{1}{3}\right)$ **49.** 14 **51.** $z = 8 - 2x - 4y$

Exercise Set 3.5, pp. 184–186

1. 16, 19, 22 **3.** 8, 21, -3 **5.** 32°, 96°, 52° **7.** Automatic transmission: $865; power door locks: $520; air conditioning: $375 **9.** Elrod: 20; Dot: 24; Wendy: 30 **11.** 10-oz cups: 11; 14-oz cups: 15; 20-oz cups: 8 **13.** First fund: $45,000; second fund: $10,000; third fund: $25,000 **15.** Roast beef: 2; baked potato: 1; broccoli: 2 **17.** Man: 3.6; woman: 18.1; one-year-old child: 50 **19.** Two-point field goals: 32; three-point field goals: 5; foul shots: 13 **21.** **23.** [1.2] -8 **24.** [1.2] 33 **25.** [1.2] -55 **26.** [1.2] -71 **27.** [1.2] $-14x + 21y - 35z$ **28.** [1.2] $-24a - 42b + 54c$ **29.** [1.2] $-5a$ **30.** [1.2] $11x$ **31.** **33.** 464 **35.** Adults: 5; students: 1; children: 94 **37.** 180°

Exercise Set 3.6, pp. 190–191

1. $\left(-\frac{1}{3}, -4\right)$ **3.** $(-4, 3)$ **5.** $\left(\frac{3}{2}, \frac{5}{2}\right)$ **7.** $\left(2, \frac{1}{2}, -2\right)$ **9.** $(2, -2, 1)$ **11.** $\left(4, \frac{1}{2}, -\frac{1}{2}\right)$ **13.** $(1, -3, -2, -1)$ **15.** Dimes: 4; nickels: 30 **17.** $4.05-per-pound granola: 5 lb; $2.70-per-pound granola: 10 lb **19.** $400 at 7%; $500 at 8%; $1600 at 9% **21.** **23.** [1.2] 13 **24.** [1.2] -22 **25.** [1.2] 37 **26.** [1.2] 422 **27.** **29.** 1324

Exercise Set 3.7, pp. 196–197

1. 18 **3.** 36 **5.** 27 **7.** -3 **9.** -5 **11.** $(-3, 2)$ **13.** $\left(\frac{9}{19}, \frac{51}{38}\right)$ **15.** $\left(-1, -\frac{6}{7}, \frac{11}{7}\right)$ **17.** $(2, -1, 4)$ **19.** $(1, 2, 3)$ **21.** **23.** [1.3] $\frac{333}{245}$ **24.** [1.3] -12 **25.** [1.4] One piece: 20.8 ft; other piece: 12 ft **26.** [3.3] Scientific calculators: 18; graphing calculators: 27 **27.** [3.3] Mazzas: 28 rolls; Kranepools: 8 rolls **28.** [3.3] Buckets: 17; dinners: 11 **29.** **31.** 12 **33.** 10

Exercise Set 3.8, pp. 201–203

1. (a) $P(x) = 20x - 300{,}000$; **(b)** 15,000 units **3. (a)** $P(x) = 50x - 120{,}000$; **(b)** 2400 units **5. (a)** $P(x) = 45x - 22{,}500$; **(b)** 500 units **7. (a)** $P(x) = 18x - 16{,}000$; **(b)** 889 units **9. (a)** $P(x) = 50x - 100{,}000$; **(b)** 2000 units **11.** ($70, 300) **13.** ($22, 474) **15.** ($50, 6250) **17.** ($10, 1070) **19. (a)** $C(x) = 125{,}300 + 450x$; **(b)** $R(x) = 800x$; **(c)** $P(x) = 350x - 125{,}300$; **(d)** $90,300 loss, $14,700 profit; **(e)** (358 computers, $286,400) **21. (a)** $C(x) = 16{,}404 + 6x$; **(b)** $R(x) = 18x$; **(c)** $P(x) = 12x - 16{,}404$; **(d)** $19,596 profit, $4404 loss; **(e)** (1367 dozen caps, $24,606) **23.** **25.** [1.3] 12 **26.** [1.3] 15 **27.** [1.3] $\frac{8}{3}$ **28.** [1.3] 4 **29.** [1.3] $\frac{9}{2}$ **30.** [1.3] $\frac{1}{3}$ **31.** **33.** ($5, 300 yo-yo's) **35. (a)** $8.74; **(b)** 24,509 units

Review Exercises: Chapter 3, pp. 205–206

1. [3.1] $(-2, 1)$ **2.** [3.1] $(3, 2)$ **3.** [3.2] $\left(-\frac{11}{15}, -\frac{43}{30}\right)$ **4.** [3.2] No solution **5.** [3.2] $\left(-\frac{4}{5}, \frac{2}{5}\right)$ **6.** [3.2] $\left(\frac{37}{19}, \frac{53}{19}\right)$ **7.** [3.2] $\left(\frac{76}{17}, -\frac{2}{119}\right)$ **8.** [3.2] $(2, 2)$ **9.** [3.2] $\{(x, y) \mid 3x + 4y = 6\}$ **10.** [3.3] DVD: $29; videocassette: $14 **11.** [3.3] 4 hr **12.** [3.3] 8% juice: 10 L; 15% juice: 4 L **13.** [3.4] $(4, -8, 10)$ **14.** [3.4] The equations are dependent. **15.** [3.4] $(2, 0, 4)$ **16.** [3.2] No solution **17.** [3.4] $\left(\frac{8}{9}, -\frac{2}{3}, \frac{10}{9}\right)$ **18.** [3.5] A: 90°; B: 67.5°; C: 22.5° **19.** [3.5] 641 **20.** [3.5] $20 bills: 7; $5 bills: 3; $1 bills: 4 **21.** [3.6] $\left(55, -\frac{89}{2}\right)$ **22.** [3.6] $(-1, 1, 3)$ **23.** [3.7] 2 **24.** [3.7] 9 **25.** [3.7] $(6, -2)$ **26.** [3.7] $(-3, 0, 4)$ **27.** [3.8] ($3, 81) **28.** [3.8] **(a)** $C(x) = 1.5x + 18{,}000$; **(b)** $R(x) = 6x$; **(c)** $P(x) = 4.5x - 18{,}000$; **(d)** $11,250 loss, $4500 profit; **(e)** (4000 pints of honey, $24,000) **29.** [3.5] To solve a problem involving four variables, go through the *Familiarize* and *Translate* steps as usual. The resulting system of equations can be solved using the elimination method just as for three variables but likely with more steps. **30.** [3.4] A system of equations can be both dependent and inconsistent if it is equivalent to a system with fewer equations that has no solution. An example is a system of three equations in three unknowns in which two of the equations represent the same plane, and the third represents a parallel plane. **31.** [3.8] 10,000 pints **32.** [3.1] $(0, 2)$, $(1, 3)$ **33.** [3.5] $a = -\frac{2}{3}$, $b = -\frac{4}{3}$, $c = 3$; $f(x) = -\frac{2}{3}x^2 - \frac{4}{3}x + 3$

Test: Chapter 3, pp. 206–207

1. [3.1] $(2, 4)$ **2.** [3.2] $\left(3, -\frac{11}{3}\right)$ **3.** [3.2] $\left(\frac{15}{7}, -\frac{18}{7}\right)$ **4.** [3.2] $\left(-\frac{3}{2}, -\frac{3}{2}\right)$ **5.** [3.2] No solution **6.** [3.3] Length: 30 units; width: 18 units **7.** [3.5] Mortgage: $74,000;

car loan: \$600; credit-card bill: \$700
8. [3.4] The equations are dependent.
9. [3.4] $\left(2, -\frac{1}{2}, -1\right)$ **10.** [3.4] No solution
11. [3.4] $(0, 1, 0)$ **12.** [3.6] $\left(\frac{34}{107}, -\frac{104}{107}\right)$
13. [3.6] $(3, 1, -2)$ **14.** [3.7] 34 **15.** [3.7] 133
16. [3.7] $\left(\frac{13}{18}, \frac{7}{27}\right)$ **17.** [3.5] 3.5 hr **18.** [3.8] $(\$3, 55)$
19. [3.8] **(a)** $C(x) = 25x + 40{,}000$; **(b)** $R(x) = 70x$;
(c) $P(x) = 45x - 40{,}000$; **(d)** \$26,500 loss, \$500 profit;
(e) (889 radios, \$62,230) **20.** [2.3], [3.3] $m = 7, b = 10$
21. [3.5] Adult: 1346; senior citizen: 335; child: 1651

Cumulative Review: Chapters 1–3, pp. 207–209

1. [1.3] 14.87 **2.** [1.3] -22 **3.** [1.3] -42.9
4. [1.3] 20 **5.** [1.3] $-\frac{21}{4}$ **6.** [1.3] 6 **7.** [1.3] -5
8. [1.3] $\frac{10}{9}$ **9.** [1.3] $-\frac{32}{5}$ **10.** [1.3] $\frac{18}{17}$ **11.** [1.6] x^{11}
12. [1.6] $-\dfrac{40x}{y^5}$ **13.** [1.6] $-288x^4y^{18}$ **14.** [1.6] y^{10}
15. [1.6] $-\dfrac{2a^{11}}{5b^{33}}$ **16.** [1.6] $\dfrac{81x^{36}}{256y^8}$ **17.** [1.7] 1.12×10^6
18. [1.7] 4.00×10^6 **19.** [1.5] $b = \dfrac{2A}{h} - t$, or $\dfrac{2A - ht}{h}$
20. [2.1] Yes
21. [2.3] **22.** [2.1]

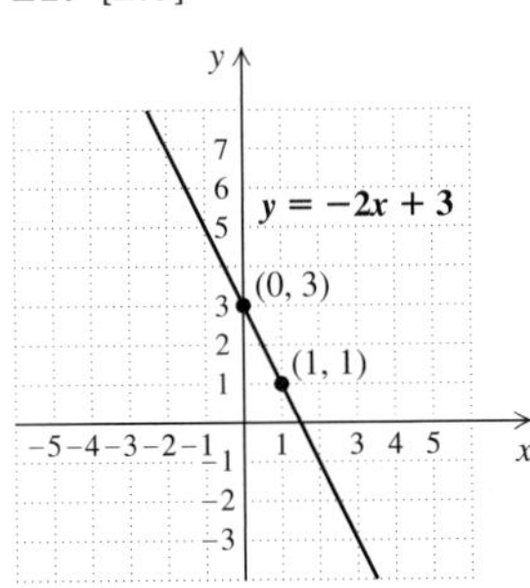

23. [2.4] **24.** [2.3]

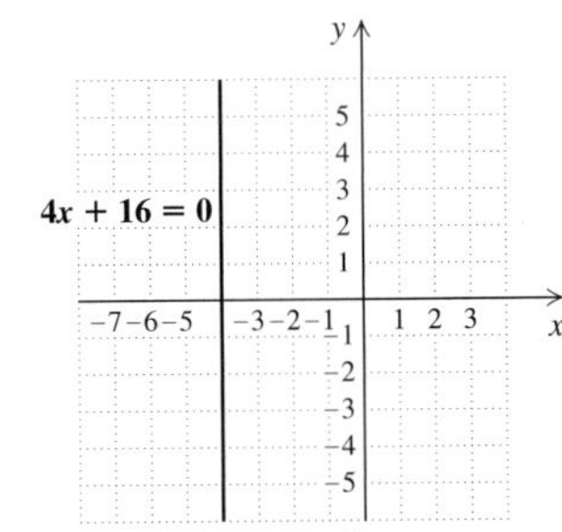

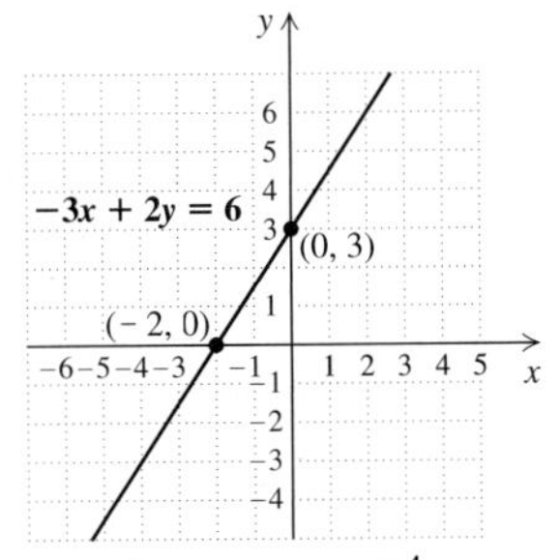

25. [2.3] Slope: $\frac{9}{4}$; y-intercept: $(0, -3)$ **26.** [2.3] $\frac{4}{3}$
27. [2.5] $y = -3x - 5$ **28.** [2.5] $y = -\frac{1}{10}x + \frac{12}{5}$
29. [2.5] Parallel **30.** [2.5] $y = -2x + 5$
31. [2.2] $\{-5, -3, -1, 1, 3\}$; $\{-3, -2, 1, 4, 5\}$; -2; 3
32. [2.2] $\left\{x \mid x \text{ is a real number } and\ x \neq \frac{1}{2}\right\}$
33. [2.2] -31 **34.** [2.2] 3 **35.** [2.6] 7
36. [2.6] $8a^2 + 4a - 4$ **37.** [3.2] $(1, 1)$
38. [3.2] $(-2, 3)$ **39.** [3.2] $\left(-3, \frac{2}{5}\right)$ **40.** [3.4] $(-3, 2, -4)$
41. [3.4] $(0, -1, 2)$ **42.** [3.7] 14 **43.** [3.7] 0
44. [3.3] 8, 18 **45.** [2.3] $90\frac{1}{3}$ recovery plans per year
46. [2.3] 9 signifies that the number of U.S. pleasure trips is increasing by 9 million each year; 616 signifies that there were 616 million U.S. pleasure trips in 1994.
47. [2.5] **(a)** $A(t) = 0.2t + 6.6$; **(b)** 10.8 million
48. [3.3] Soakem: $48\frac{8}{9}$ oz; Rinsem: $71\frac{1}{9}$ oz **49.** [1.4] 3, 5, 7
50. [1.4] 90 **51.** [3.3] Length: 10 cm; width: 6 cm
52. [3.3] Nickels: 19; dimes: 15 **53.** [3.5] \$120
54. [3.5] Wins: 23; losses: 33; ties: 8
55. [3.5] Cookie: 90; banana: 80; yogurt: 165
56. [1.6] $-12x^{2a}y^{b+y+3}$ **57.** [2.5] \$151,000
58. [2.5], [3.3] $m = -\frac{5}{9}, b = -\frac{2}{9}$

CHAPTER 4

Exercise Set 4.1, pp. 220–223

1. No, no, yes, yes **3.** No, yes, yes, no
5. $\{y \mid y < 6\}$, $(-\infty, 6)$
7. $\{x \mid x \geq -4\}$, $[-4, \infty)$
9. $\{t \mid t > -3\}$, $(-3, \infty)$
11. $\{x \mid x \leq -7\}$, $(-\infty, -7]$
13. $\{x \mid x > -5\}$, or $(-5, \infty)$
15. $\{a \mid a \leq -20\}$, or $(-\infty, -20]$
17. $\{x \mid x \leq 12\}$, or $(-\infty, 12]$
19. $\{y \mid y > -9\}$, or $(-9, \infty)$
21. $\{y \mid y \leq 14\}$, or $(-\infty, 14]$
23. $\{t \mid t < -9\}$, or $(-\infty, -9)$
25. $\{x \mid x < 50\}$, or $(-\infty, 50)$
27. $\{y \mid y \geq -0.4\}$, or $[-0.4, \infty)$
29. $\left\{y \mid y \geq \frac{9}{10}\right\}$, or $\left[\frac{9}{10}, \infty\right)$
31. $\{y \mid y > 3\}$, or $(3, \infty)$
33. $\{x \mid x \leq 4\}$, or $(-\infty, 4]$
35. $\left\{x \mid x < -\frac{2}{5}\right\}$, or $\left(-\infty, -\frac{2}{5}\right)$
37. $\{x \mid x \geq 11.25\}$, or $[11.25, \infty)$
39. $\left\{y \mid y \leq -\frac{53}{6}\right\}$, or $\left(-\infty, -\frac{53}{6}\right]$ **41.** $\left\{t \mid t < \frac{29}{5}\right\}$, or $\left(-\infty, \frac{29}{5}\right)$
43. $\left\{m \mid m > \frac{7}{3}\right\}$, or $\left(\frac{7}{3}, \infty\right)$ **45.** $\{x \mid x \geq 2\}$, or $[2, \infty)$

47. $\{y \mid y < 5\}$, or $(-\infty, 5)$ **49.** $\left\{x \mid x \leq \frac{4}{7}\right\}$, or $\left(-\infty, \frac{4}{7}\right]$
51. Mileages less than or equal to 150 mi
53. \$11,500 or more **55.** For 1175 min or more
57. More than 25 **59.** Gross sales greater than \$7000
61. More than \$1850 **63.** At least 625 people
65. **(a)** $\left\{x \mid x < 8181\frac{9}{11}\right\}$, or $\{x \mid x \leq 8181\}$ **(b)** $\left\{x \mid x > 8181\frac{9}{11}\right\}$, or $\{x \mid x \geq 8182\}$ **67.**
69. [2.2] $\{x \mid x$ is a real number *and* $x \neq 2\}$
70. [2.2] $\{x \mid x$ is a real number *and* $x \neq -3\}$
71. [2.2] $\left\{x \mid x \text{ is a real number and } x \neq \frac{7}{2}\right\}$
72. [2.2] $\left\{x \mid x \text{ is a real number and } x \neq \frac{9}{4}\right\}$
73. [1.2] $7x + 10$ **74.** [1.2] $22x - 7$ **75.**
77. $\left\{x \mid x \leq \dfrac{2}{a-1}\right\}$ **79.** $\left\{y \mid y \geq \dfrac{2a+5b}{b(a-2)}\right\}$
81. $\left\{x \mid x > \dfrac{4m-2c}{d-(5c+2m)}\right\}$ **83.** False; $2 < 3$ and $4 < 5$, but $2 - 4 = 3 - 5$. **85.**
87. $\mathbb{R}$
89. $\{x \mid x$ is a real number *and* $x \neq 0\}$ **91.**

Technology Connection, p. 230

1. Domain of $y_1 = \{x \mid x \leq 3\}$, or $(-\infty, 3]$; domain of $y_2 = \{x \mid x \geq -1\}$, or $[-1, \infty)$
2. Domain of $y_1 + y_2$ = domain of $y_1 - y_2$ = domain of $y_2 - y_1$ = domain of $y_1 \cdot y_2 = \{x \mid -1 \leq x \leq 3\}$, or $[-1, 3]$

Exercise Set 4.2, pp. 230–232

1. $\{9, 11\}$ **3.** $\{1, 5, 10, 15, 20\}$ **5.** $\{b, d, f\}$
7. $\{r, s, t, u, v\}$ **9.** $\varnothing$ **11.** $\{3, 5, 7\}$
13. $(3, 8)$
15. $[-6, -2]$
17. $(-\infty, -2) \cup (3, \infty)$
19. $(-\infty, -1] \cup (5, \infty)$
21. $(-2, 4]$
23. $(-2, 4)$
25. $(-\infty, 5) \cup (7, \infty)$
27. $(-\infty, -4] \cup [5, \infty)$
29. $[-6, 4)$
31. $[3, 7)$
33. $(-\infty, 5)$
35. $(-\infty, \infty)$
37. $(7, \infty)$
39. $\{t \mid -3 < t < 5\}$, or $(-3, 5)$
41. $\{x \mid -1 < x \leq 4\}$, or $(-1, 4]$
43. $\{a \mid -2 \leq a < 2\}$, or $[-2, 2)$
45. $\mathbb{R}$, or $(-\infty, \infty)$
47. $\{x \mid 1 \leq x \leq 3\}$, or $[1, 3]$
49. $\left\{x \mid -\frac{7}{2} < x \leq 7\right\}$, or $\left(-\frac{7}{2}, 7\right]$
51. $\{x \mid x \leq 1 \text{ or } x \geq 3\}$, or $(-\infty, 1] \cup [3, \infty)$
53. $\{x \mid x < 3 \text{ or } x > 4\}$, or $(-\infty, 3) \cup (4, \infty)$
55. $\left\{a \mid a < \frac{7}{2}\right\}$, or $\left(-\infty, \frac{7}{2}\right)$
57. $\{a \mid a < -5\}$, or $(-\infty, -5)$
59. $\mathbb{R}$, or $(-\infty, \infty)$
61. $\{t \mid t \leq 6\}$, or $(-\infty, 6]$
63. $(-\infty, -7) \cup (-7, \infty)$ **65.** $[6, \infty)$
67. $\left(-\infty, \frac{5}{2}\right) \cup \left(\frac{5}{2}, \infty\right)$ **69.** $[-4, \infty)$ **71.** $(-\infty, 4]$
73.
75. [2.4]

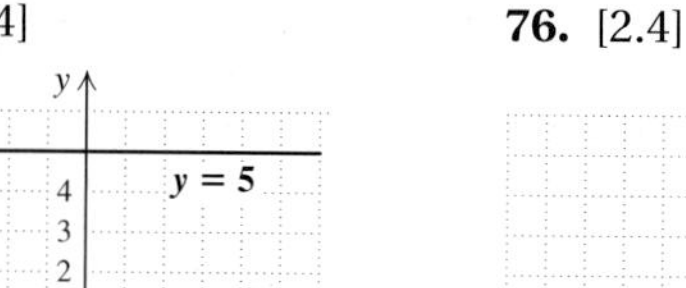

76. [2.4]

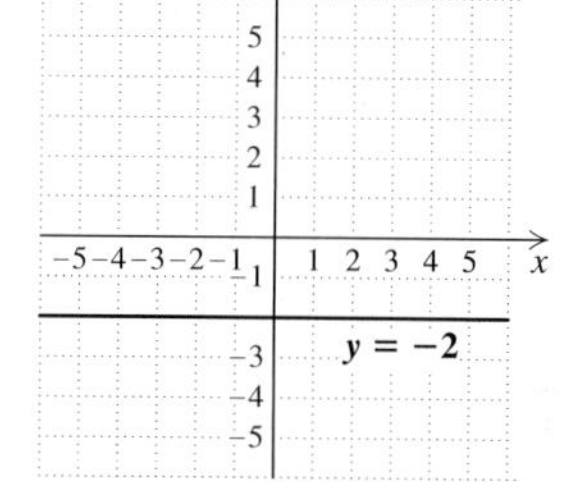

77. [2.2]

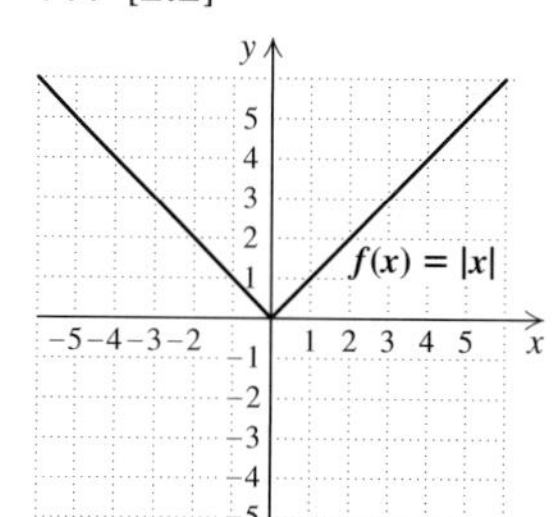

78. [2.3]

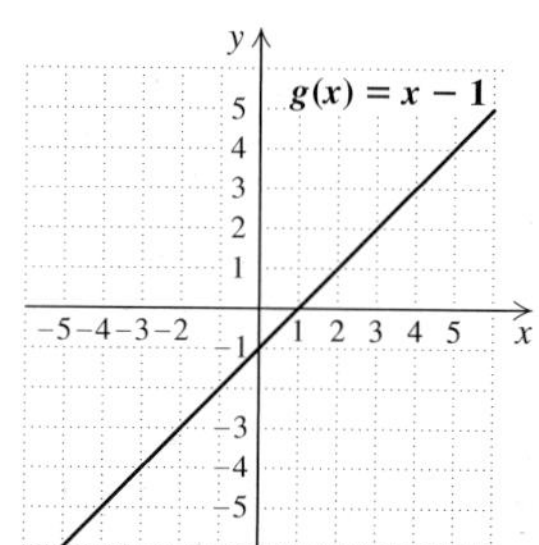

79. [3.1] $(8, 5)$ **80.** [3.1] $(-5, -3)$ **81.** **83.** $(-1, 6)$
85. Between 12 and 240 trips **87.** Sizes between 6 and 13
89. **(a)** $1945.4° \leq F < 4820°$; **(b)** $1761.44° \leq F < 3956°$
91. $\left\{a \mid -\frac{3}{2} \leq a \leq 1\right\}$, or $\left[-\frac{3}{2}, 1\right]$;

93. $\{x \mid -4 < x \leq 1\}$, or $(-4, 1]$;

95. True **97.** False **99.** $\left[-\frac{5}{2}, 1\right) \cup (1, \infty)$ **101.**
103.

Technology Connection, p. 235

1. The graphs of $y_1 = \text{abs}(4 - 7x)$ and $y_2 = -8$ do not intersect.

Exercise Set 4.3, pp. 240–241

1. $\{-4, 4\}$ **3.** $\varnothing$ **5.** $\{-7.3, 7.3\}$ **7.** $\{0\}$ **9.** $\left\{-\frac{9}{5}, 1\right\}$
11. $\varnothing$ **13.** $\{-5, 11\}$ **15.** $\{5, 7\}$ **17.** $\{-1, 9\}$
19. $\{-9, 9\}$ **21.** $\{-2, 2\}$ **23.** $\left\{-\frac{14}{5}, \frac{22}{5}\right\}$ **25.** $\{4, 10\}$
27. $\left\{\frac{3}{2}, \frac{7}{2}\right\}$ **29.** $\left\{-\frac{4}{3}, 4\right\}$ **31.** $\{-8.3, 8.3\}$ **33.** $\left\{-\frac{8}{3}, 4\right\}$
35. $\{1, 11\}$ **37.** $\left\{\frac{3}{2}\right\}$ **39.** $\left\{-4, -\frac{10}{9}\right\}$ **41.** $\mathbb{R}$ **43.** $\{1\}$
45. $\left\{32, \frac{8}{3}\right\}$
47. $\{a \mid -7 \leq a \leq 7\}$, or $[-7, 7]$
49. $\{x \mid x < -8 \text{ or } x > 8\}$, or $(-\infty, -8) \cup (8, \infty)$
51. $\{t \mid t < 0 \text{ or } t > 0\}$, or $(-\infty, 0) \cup (0, \infty)$
53. $\{x \mid -2 < x < 8\}$, or $(-2, 8)$
55. $\{x \mid -8 \leq x \leq 4\}$, or $[-8, 4]$
57. $\{x \mid x < -2 \text{ or } x > 8\}$, or $(-\infty, -2) \cup (8, \infty)$
59. $\mathbb{R}$, or $(-\infty, \infty)$
61. $\left\{a \mid a \leq -\frac{2}{3} \text{ or } a \geq \frac{10}{3}\right\}$, or $\left(-\infty, -\frac{2}{3}\right] \cup \left[\frac{10}{3}, \infty\right)$
63. $\{y \mid -9 < y < 15\}$, or $(-9, 15)$;
65. $\{x \mid x \leq -8 \text{ or } x \geq 0\}$, or $(-\infty, -8] \cup [0, \infty)$
67. $\left\{y \mid y < -\frac{4}{3} \text{ or } y > 4\right\}$, or $\left(-\infty, -\frac{4}{3}\right) \cup (4, \infty)$;
69. $\varnothing$
71. $\left\{x \mid x \leq -\frac{2}{15} \text{ or } x \geq \frac{14}{15}\right\}$, or $\left(-\infty, -\frac{2}{15}\right] \cup \left[\frac{14}{15}, \infty\right)$;
73. $\{m \mid -12 \leq m \leq 2\}$, or $[-12, 2]$;
75. $\{a \mid -6 < a < 0\}$, or $(-6, 0)$
77. $\left\{x \mid -\frac{1}{2} \leq x \leq \frac{7}{2}\right\}$, or $\left[-\frac{1}{2}, \frac{7}{2}\right]$
79. $\left\{x \mid x \leq -\frac{7}{3} \text{ or } x \geq 5\right\}$, or $\left(-\infty, -\frac{7}{3}\right] \cup [5, \infty)$
81. $\{x \mid -4 < x < 5\}$, or $(-4, 5)$
83. **85.** [3.2] $\left(-\frac{16}{13}, -\frac{41}{13}\right)$ **86.** [3.2] $(-2, -3)$
87. [3.2] $(10, 4)$ **88.** [3.2] $(-1, 7)$ **89.** [3.1] $(1, 4)$
90. [3.1] $(24, -41)$ **91.** **93.** $\left\{t \mid t \geq \frac{5}{3}\right\}$, or $\left[\frac{5}{3}, \infty\right)$
95. $\{x \mid -4 \leq x \leq -1 \text{ or } 3 \leq x \leq 6\}$, or $[-4, -1] \cup [3, 6]$
97. $\left\{x \mid x \leq \frac{5}{2}\right\}$, or $\left(-\infty, \frac{5}{2}\right]$ **99.** $|y| \leq 5$ **101.** $|x| > 4$
103. $|x + 2| < 3$ **105.** $|x - 5| < 1$, or $|5 - x| < 1$
107. $|x - 2| < 6$ **109.** $|x - 7| \leq 5$ **111.** $\{x \mid 1 \leq x \leq 5\}$, or $[1, 5]$ **113.** **115.**

Technology Connection, p. 245

1. $y > x + 3.5$

2. $7y \leq 2x + 5$

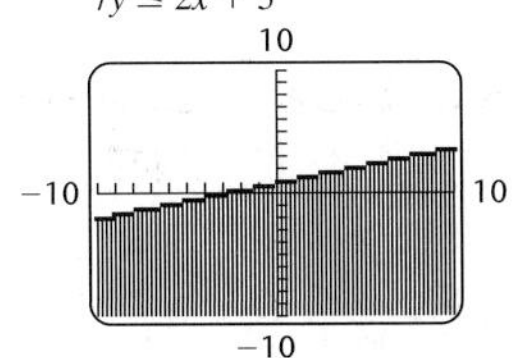

3. $8x - 2y < 11$

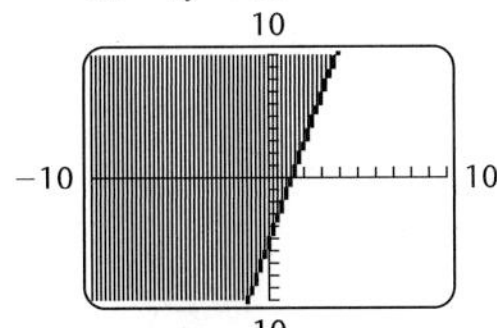

4. $11x + 13y + 4 \geq 0$

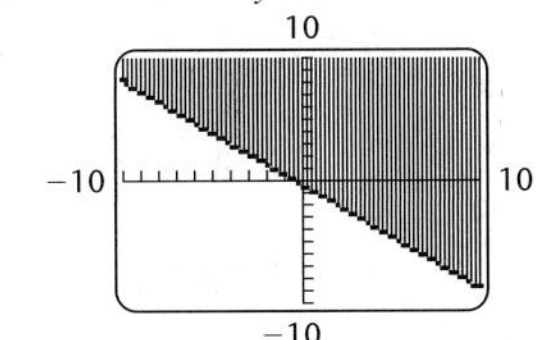

Exercise Set 4.4, pp. 250–251

1. Yes **3.** No

5.

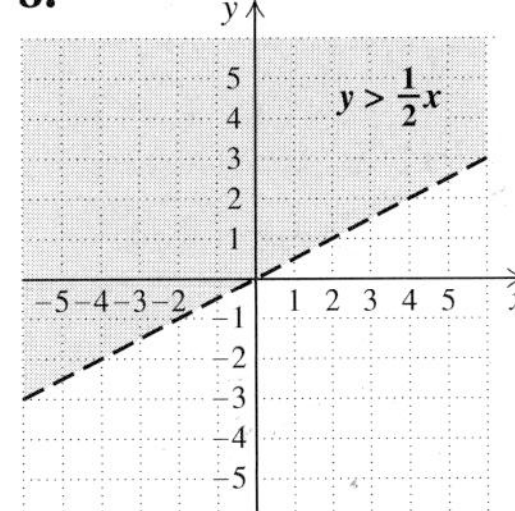

7.

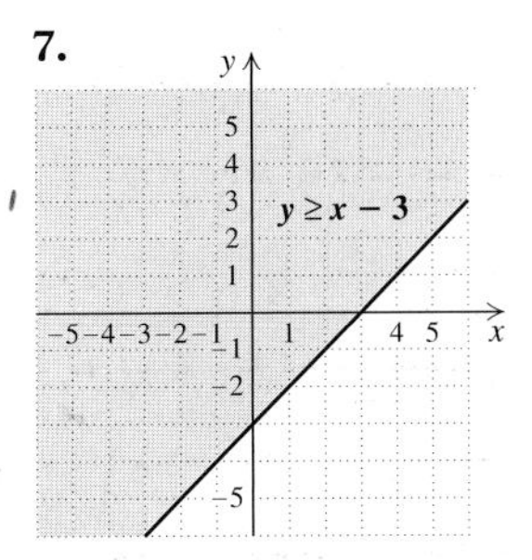

9.

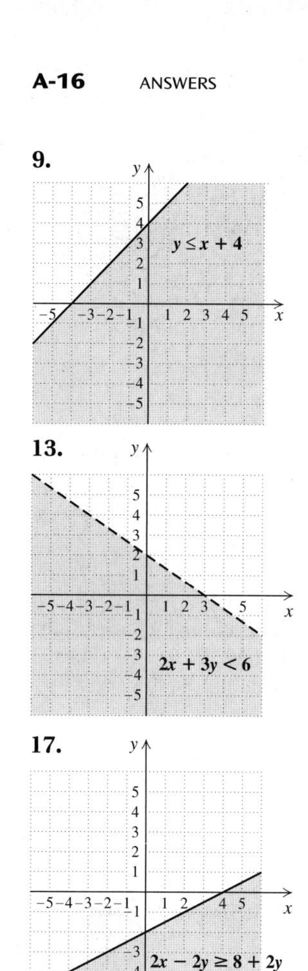

11.

$x - y \le 5$

13.

15.

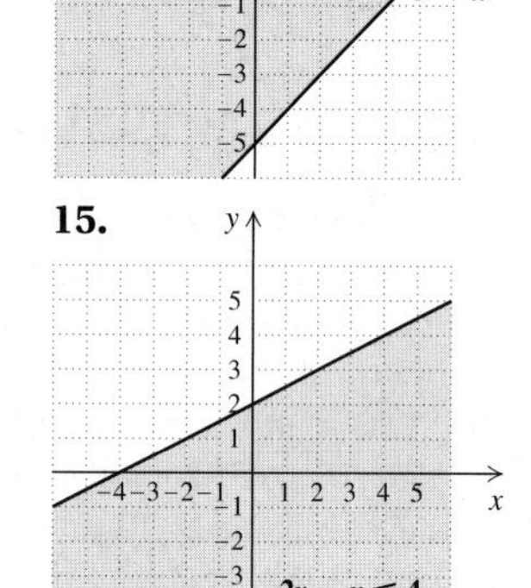

17.

19.

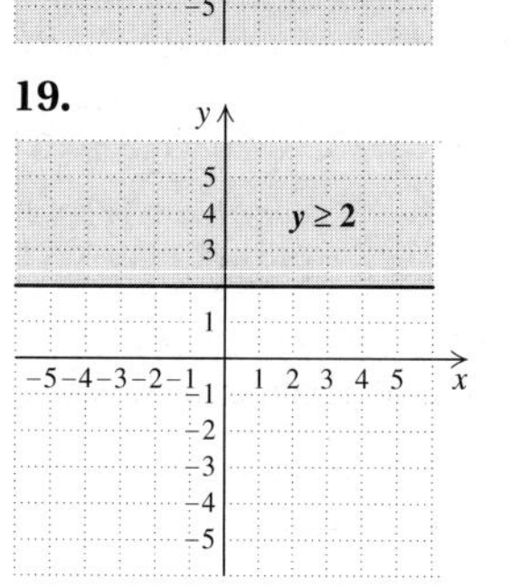

21.

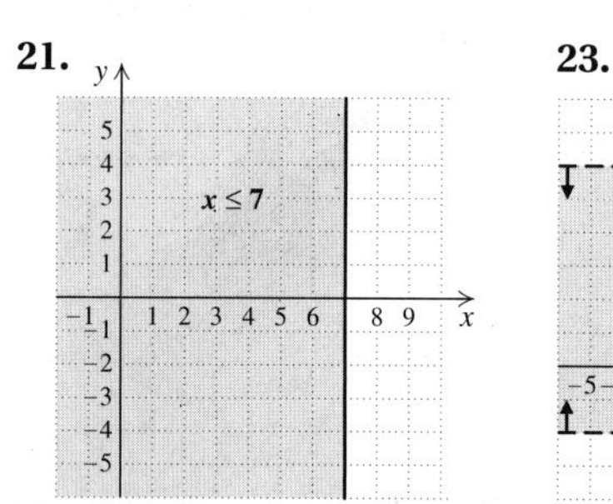

23.

$-2 < y < 6$

25.

27.

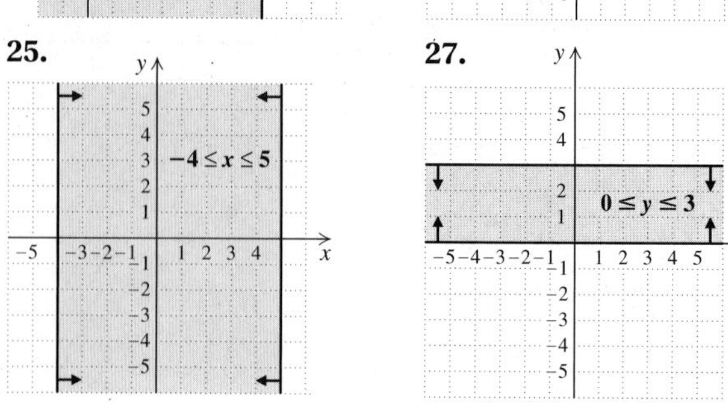

29.

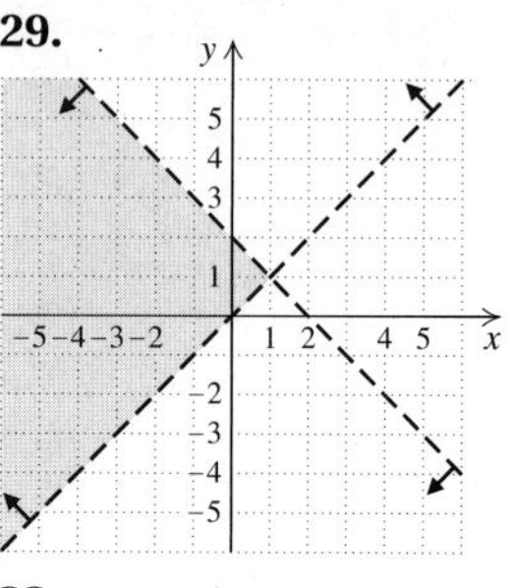

31.

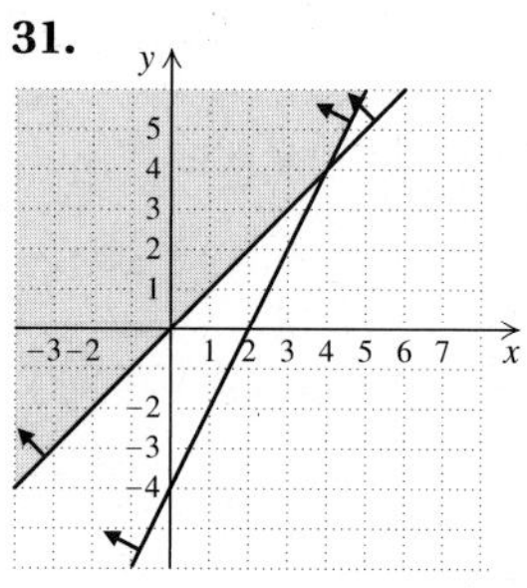

33.

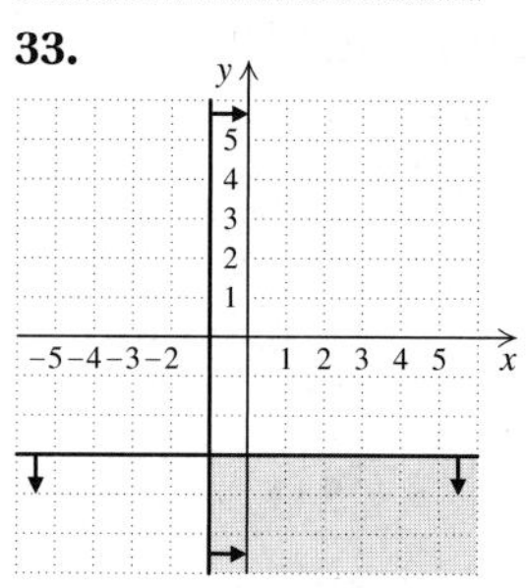

35.

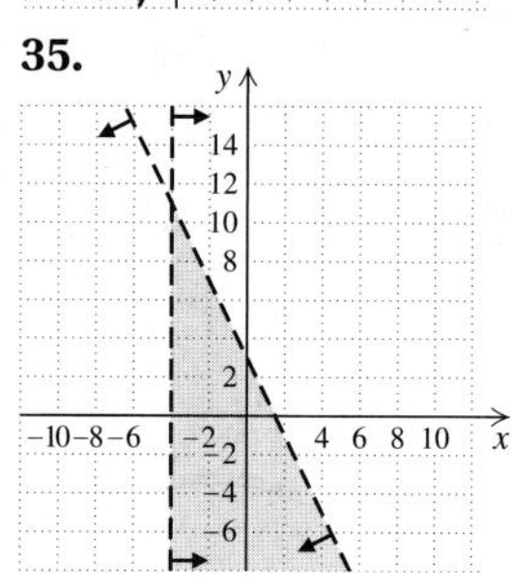

37.

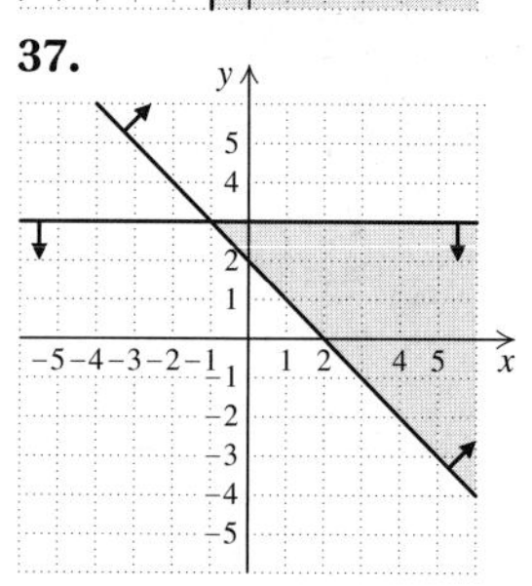

39.

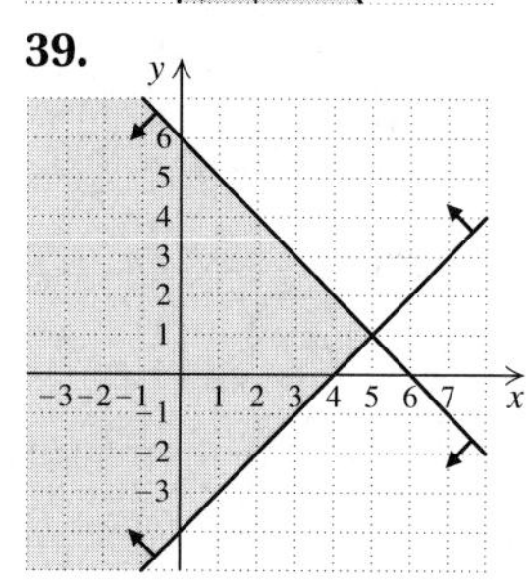

41.

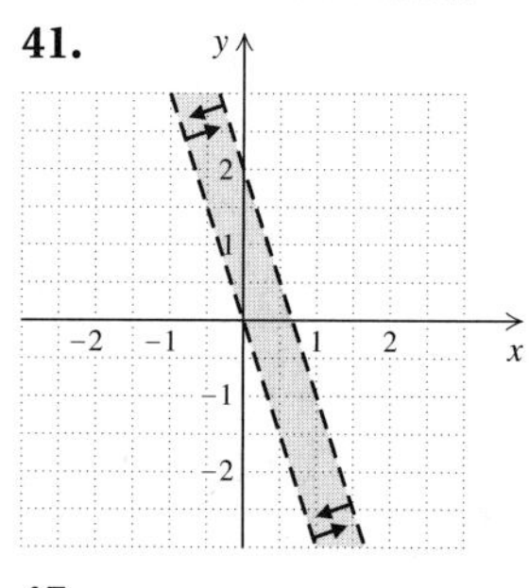

43.

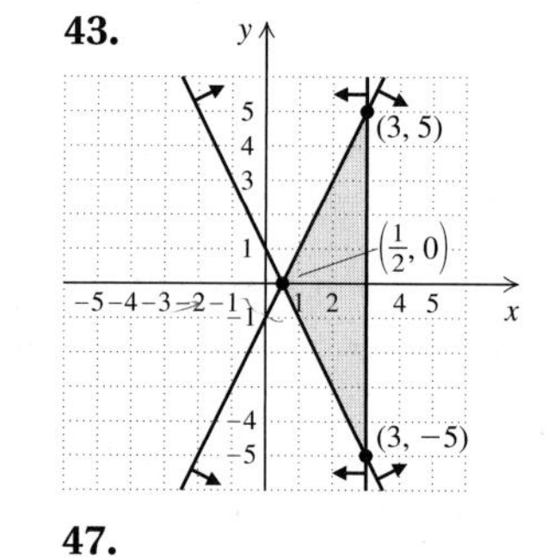

45.

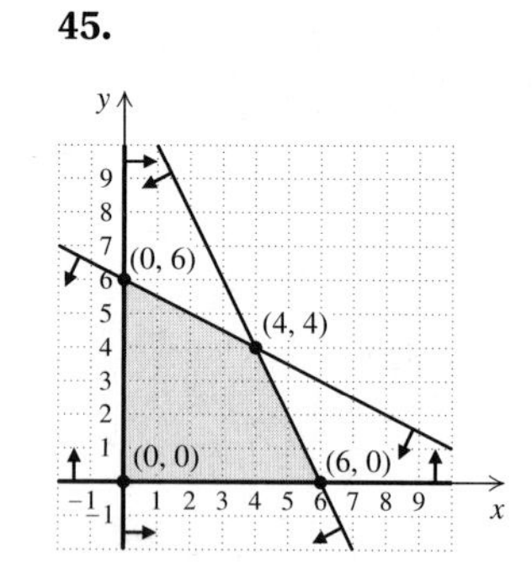

47.

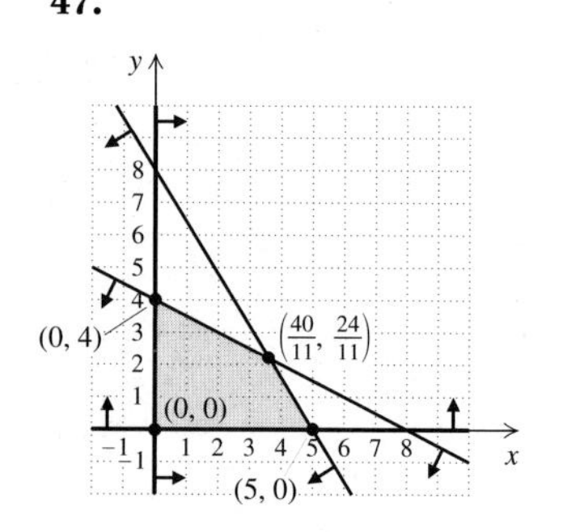

49.

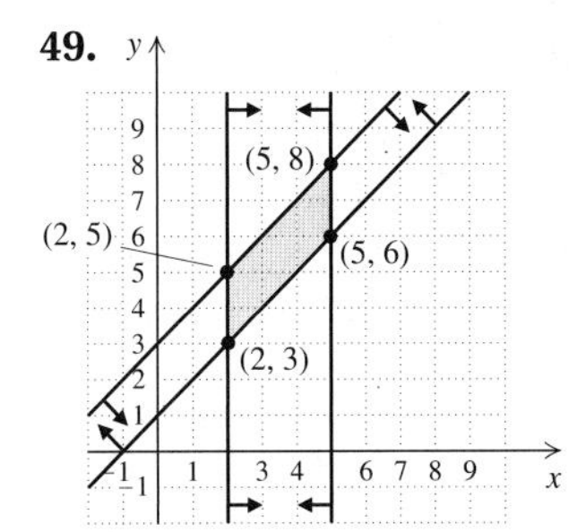

51. ▯ **53.** [3.3] Peanuts: $6\frac{2}{3}$ lb; fancy nuts: $3\frac{1}{3}$ lb
54. [3.3] Hendersons: 10 bags; Savickis: 4 bags
55. [3.3] Activity-card holders: 128; noncard holders: 75
56. [3.3] Students: 70; adults: 130 **57.** [1.4] 25 ft
58. [1.4] 11% **59.** ▯

61.

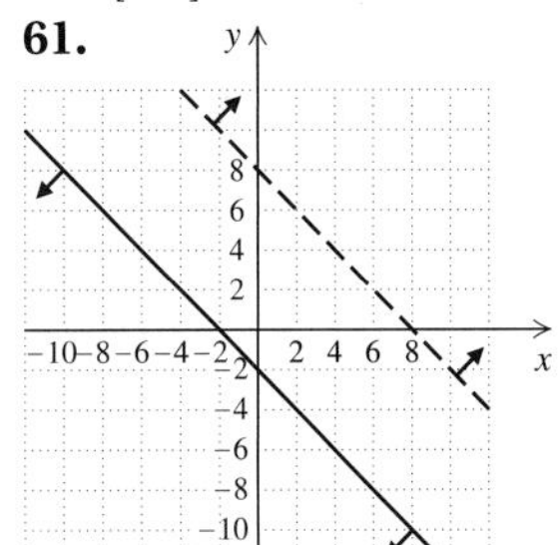

63.

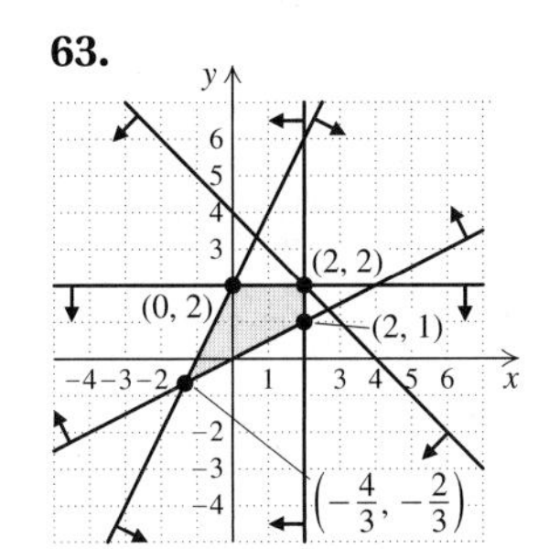

65. $0 < w \leq 62$,
$0 < h \leq 62$,
$62 + 2w + 2h \leq 108$,
or $w + h \leq 23$

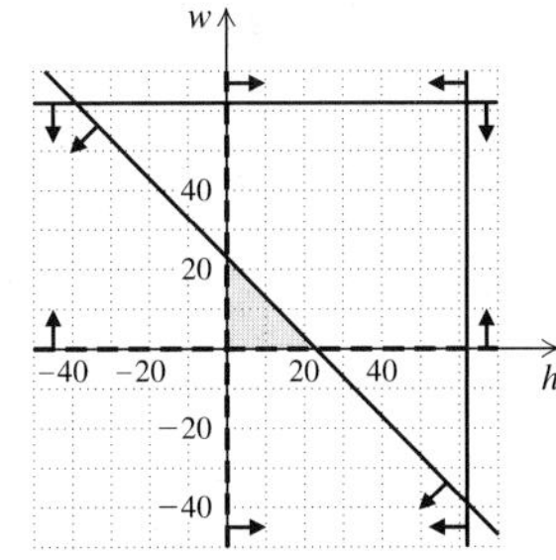

67. $35c + 75a < 1000$,
$c \geq 0$,
$a \geq 0$

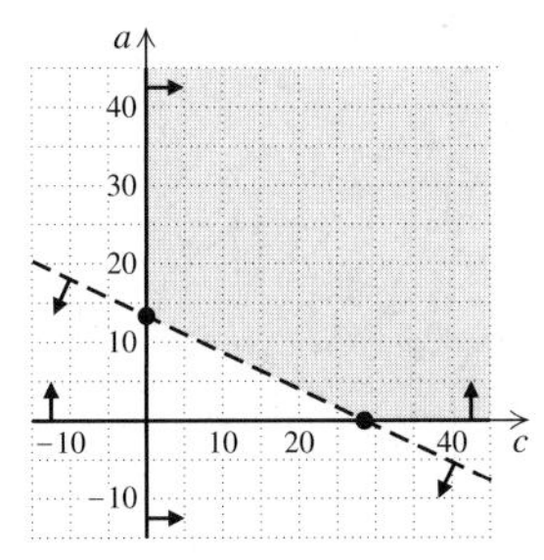

69. (a) $3x + 6y > 2$

(b) $x - 5y \leq 10$

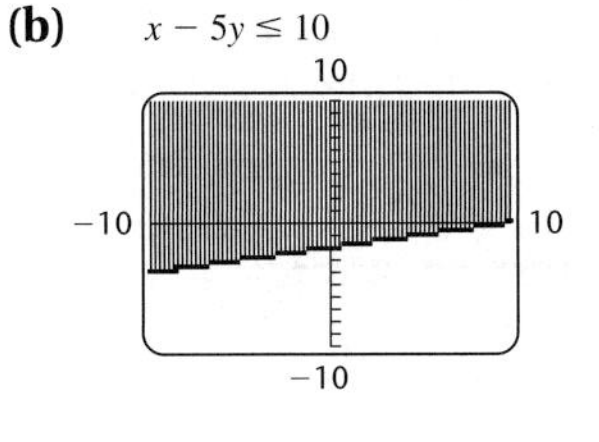

(c) $13x - 25y + 10 \leq 0$ **(d)** $2x + 5y > 0$

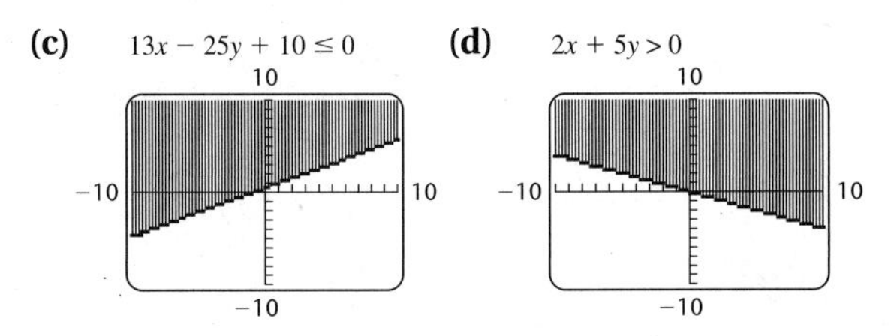

Exercise Set 4.5, pp. 258–260

1. Maximum 84 when $x = 0, y = 6$; minimum 0 when $x = 0, y = 0$ **3.** Maximum 76 when $x = 7, y = 0$; minimum 16 when $x = 0, y = 4$ **5.** Maximum 5 when $x = 3, y = 7$; minimum -15 when $x = 3, y = -3$
7. Chili: 40 orders: burritos: 50 orders
9. Motorcycles: 60; bicycles: 100 **11.** Matching: 5; essay: 15; maximum: 425 **13.** Corporate bonds: $22,000; municipal bonds: $18,000; maximum: $3110
15. Hawaiian Blend: 68; Classic Blend: 39; maximum: $8265 **17.** Knit: 2; worsted: 4 **19.** ▯
21. [1.1], [1.2] -40 **22.** [1.1] 46 **23.** [1.2] $10x + 5$
24. [1.2] $26t + 20$ **25.** [1.2] $3x - 6$ **26.** [1.2] $2t + 2$
27. ▯ **29.** T3's: 30; S5's: 10
31. Chairs: 25; sofas: 9

Review Exercises: Chapter 4, pp. 261–262

1. [4.1] $\{x \mid x \leq -2\}$, or $(-\infty, -2]$;

2. [4.1] $\{a \mid a \leq -21\}$, or $(-\infty, -21]$;

3. [4.1] $\{y \mid y \geq -7\}$, or $[-7, \infty)$;

4. [4.1] $\left\{y \mid y > -\frac{15}{4}\right\}$, or $\left(-\frac{15}{4}, \infty\right)$;

5. [4.1] $\{y \mid y > -30\}$, or $(-30, \infty)$;

6. [4.1] $\left\{x \mid x > -\frac{3}{2}\right\}$, or $\left(-\frac{3}{2}, \infty\right)$;

7. [4.1] $\{x \mid x < -3\}$, or $(-\infty, -3)$;

8. [4.1] $\left\{y \mid y > -\frac{220}{23}\right\}$, or $\left(-\frac{220}{23}, \infty\right)$;

9. [4.1] $\left\{x \mid x \leq -\frac{5}{2}\right\}$, or $\left(-\infty, -\frac{5}{2}\right]$;

10. [4.1] $\{x \mid x \leq 4\}$, or $(-\infty, 4]$ **11.** [4.1] More than 125 hr

12. [4.1] \$1500 **13.** [4.2] $\{1, 5, 9\}$
14. [4.2] $\{1, 2, 3, 5, 6, 9\}$
15. [4.2]; $(-5, 3]$
16. [4.2]; $(-\infty, \infty)$
17. [4.2] $\{x \mid -7 < x \leq 2\}$, or $(-7, 2]$
18. [4.2] $\left\{x \mid -\frac{5}{4} < x < \frac{5}{2}\right\}$, or $\left(-\frac{5}{4}, \frac{5}{2}\right)$
19. [4.2] $\{x \mid x < -3 \text{ or } x > 1\}$, or $(-\infty, -3) \cup (1, \infty)$
20. [4.2] $\{x \mid x < -11 \text{ or } x \geq -6\}$, or $(-\infty, -11) \cup [-6, \infty)$
21. [4.2] $\{x \mid x \leq -6 \text{ or } x \geq 8\}$, or $(-\infty, -6] \cup [8, \infty)$
22. [4.2] $\left\{x \mid x < -\frac{2}{5} \text{ or } x > \frac{8}{5}\right\}$, or $\left(-\infty, -\frac{2}{5}\right) \cup \left(\frac{8}{5}, \infty\right)$
23. [4.2] $(-\infty, 3) \cup (3, \infty)$ **24.** [4.2] $[-3, \infty)$
25. [4.2] $\left(-\infty, \frac{8}{3}\right]$ **26.** [4.3] $\{-4, 4\}$
27. [4.3] $\{t \mid t \leq -3.5 \text{ or } t \geq 3.5\}$, or $(-\infty, -3.5] \cup [3.5, \infty)$
28. [4.3] $\{-5, 9\}$ **29.** [4.3] $\left\{x \mid -\frac{17}{2} < x < \frac{7}{2}\right\}$, or $\left(-\frac{17}{2}, \frac{7}{2}\right)$
30. [4.3] $\left\{x \mid x \leq -\frac{11}{3} \text{ or } x \geq \frac{19}{3}\right\}$, or $\left(-\infty, -\frac{11}{3}\right] \cup \left[\frac{19}{3}, \infty\right)$
31. [4.3] $\left\{-14, \frac{4}{3}\right\}$ **32.** [4.3] $\varnothing$
33. [4.3] $\{x \mid -12 \leq x \leq 4\}$, or $[-12, 4]$
34. [4.3] $\{x \mid x < 0 \text{ or } x > 10\}$, or $(-\infty, 0) \cup (10, \infty)$
35. [4.3] $\varnothing$
36. [4.4]

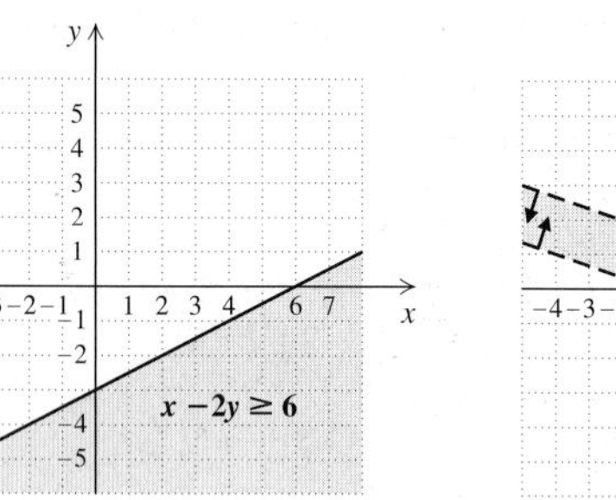

37. [4.4]

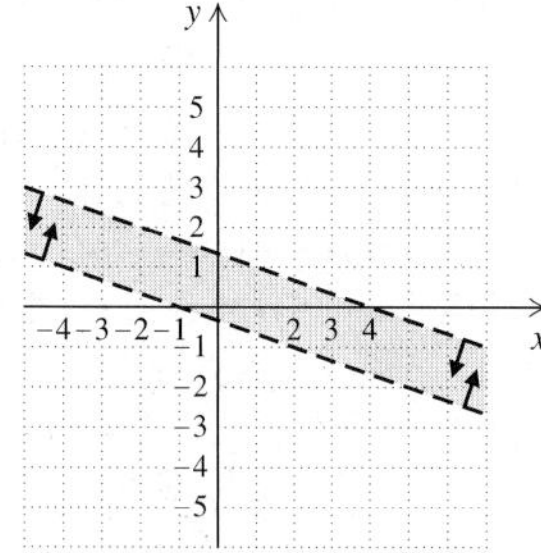

38. [4.4]

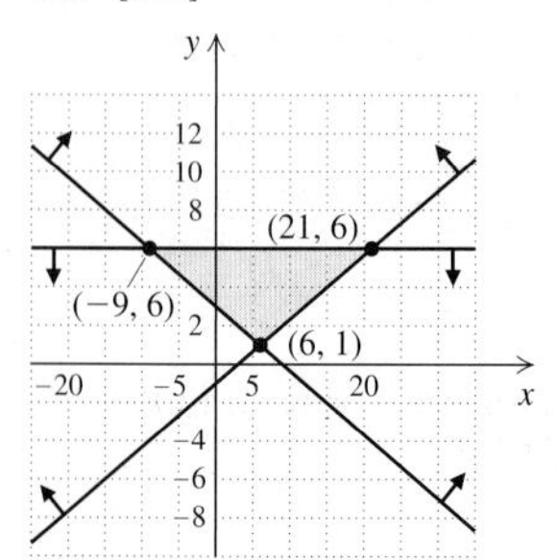

39. [4.5] Maximum 40 when $x = 7$, $y = 15$; minimum 10 when $x = 1$, $y = 3$
40. [4.5] Ohio plant: 120; Oregon plant: 40
41. [4.3] The equation $|x| = p$ has two solutions when p is positive because x can be either p or $-p$. The same equation has no solution when p is negative because no number has a negative absolute value.
42. [4.4] The solution set of a system of inequalities is all ordered pairs that make *all* the individual inequalities true. This consists of ordered pairs that are common to all the individual solution sets, or the intersection of the graphs.
43. [4.3] $\left\{x \mid -\frac{8}{3} \leq x \leq -2\right\}$, or $\left[-\frac{8}{3}, -2\right]$
44. [4.1] False; $-4 < 3$ is true, but $(-4)^2 < 9$ is false.
45. [4.3] $|d - 1.1| \leq 0.03$

Test: Chapter 4, p. 263

1. [4.1] $\{x \mid x < 12\}$, or $(-\infty, 12)$
2. [4.1] $\{y \mid y > -50\}$, or $(-50, \infty)$
3. [4.1] $\{y \mid y \leq -2\}$, or $(-\infty, -2]$
4. [4.1] $\left\{a \mid a \leq \frac{11}{5}\right\}$, or $\left(-\infty, \frac{11}{5}\right]$
5. [4.1] $\left\{x \mid x > \frac{5}{2}\right\}$, or $\left(\frac{5}{2}, \infty\right)$
6. [4.1] $\left\{x \mid x \leq \frac{7}{4}\right\}$, or $\left(-\infty, \frac{7}{4}\right]$
7. [4.1] $\{x \mid x > 1\}$, or $(1, \infty)$ **8.** [4.1] More than $166\frac{2}{3}$ mi
9. [4.1] Less than or equal to 2.5 hr **10.** [4.2] $\{3, 5\}$
11. [4.2] $\{1, 3, 5, 7, 9, 11, 13\}$ **12.** [4.2] $(-\infty, 7]$
13. [4.2] $\{x \mid -1 < x < 6\}$, or $(-1, 6)$
14. [4.2] $\left\{t \mid -\frac{2}{5} < t \leq \frac{9}{5}\right\}$, or $\left(-\frac{2}{5}, \frac{9}{5}\right]$
15. [4.2] $\{x \mid x < 3 \text{ or } x > 6\}$, or $(-\infty, 3) \cup (6, \infty)$
16. [4.2] $\left\{x \mid x < -4 \text{ or } x > -\frac{5}{2}\right\}$, or $(-\infty, -4) \cup \left(-\frac{5}{2}, \infty\right)$
17. [4.2] $\left\{x \mid 4 \leq x < \frac{15}{2}\right\}$, or $\left[4, \frac{15}{2}\right)$
18. [4.3] $\{-9, 9\}$
19. [4.3] $\{a \mid a < -3 \text{ or } a > 3\}$, or $(-\infty, -3) \cup (3, \infty)$

20. [4.3] $\{x \mid -\frac{7}{8} < x < \frac{11}{8}\}$, or $\left(-\frac{7}{8}, \frac{11}{8}\right)$

21. [4.3] $\{t \mid t \leq -\frac{13}{5} \text{ or } t \geq \frac{7}{5}\}$, or $\left(-\infty, -\frac{13}{5}\right] \cup \left[\frac{7}{5}, \infty\right)$

22. [4.3] $\varnothing$

23. [4.2] $\{x \mid x < \frac{1}{2} \text{ or } x > \frac{7}{2}\}$, or $\left(-\infty, \frac{1}{2}\right) \cup \left(\frac{7}{2}, \infty\right)$

24. [4.3] $\{1\}$

25. [4.4]

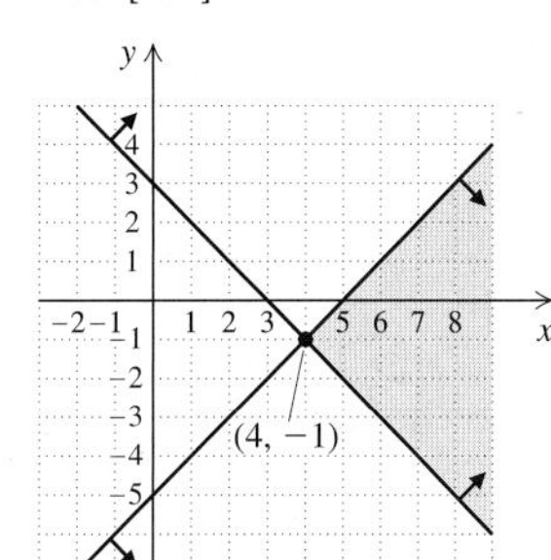

26. [4.4]

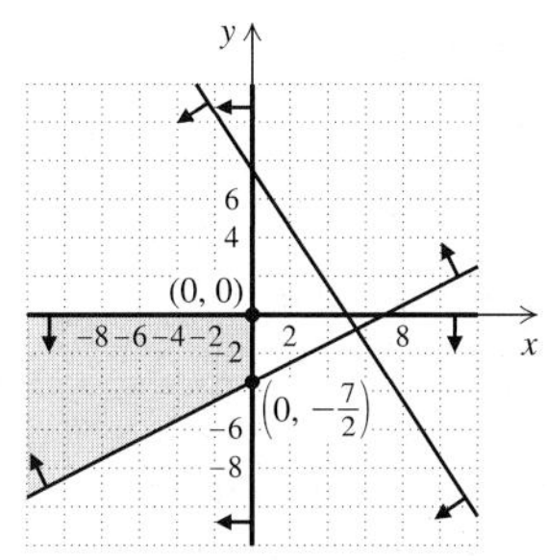

27. [4.5] Maximum 57 when $x = 6$, $y = 9$; minimum 5 when $x = 1$, $y = 0$ **28.** [4.5] Manicures: 35; haircuts: 15; maximum: \$690 **29.** [4.3] $[-1, 0] \cup [4, 6]$ **30.** [4.2] $\left(\frac{1}{5}, \frac{4}{5}\right)$ **31.** [4.3] $|x + 13| \leq 2$

CHAPTER 5

Technology Connection, p. 272

1. The graphs appear to coincide in the given window. Moving the cursor from curve to curve does not change the coordinates.
2. A quick visual inspection can be performed more quickly and easily with this procedure.

Exercise Set 5.1, pp. 273–276

1. 5, 3, 2, 1, 0; 5 **3.** 3, 7, 6, 0; 7 **5.** 5, 6, 2, 1, 0; 6
7. $-4y^3 - 6y^2 + 7y + 19$; $-4y^3$; -4
9. $3x^7 + 5x^2 - x + 12$; $3x^7$; 3
11. $-a^7 + 8a^5 + 5a^3 - 19a^2 + a$; $-a^7$; -1
13. $-9 + 6x - 5x^2 + 3x^4$
15. $3xy^3 + x^2y^2 + 7x^3y - 5x^4$
17. $-7ab + 4ax - 7ax^2 + 4x^6$ **19.** 47; 7
21. -45; $-8\frac{19}{27}$ **23.** -23 **25.** -12 **27.** 9 **29.** 6840
31. 144 ft **33.** \$18,750 **35.** \$8375 **37.** 399
39. 5.2 million **41.** 6.4 million **43.** About 340 mg
45. $M(5) \approx 65$ **47.** 56.5 in^2 **49.** $2a^3 - a + 4$
51. $-6a^2b - 3b^2$ **53.** $10x^2 + 2xy + 15y^2$
55. $12a + 4b - c$ **57.** $-4a^2 - b^2 + 3c^2$
59. $-2x^2 + x - xy - 1$ **61.** $6x^2y - 4xy^2 + 5xy$
63. $9r^2 + 9r - 9$ **65.** $-\frac{5}{24}xy - \frac{27}{20}x^3y^2 + 1.4y^3$
67. $-(5x^3 - 7x^2 + 3x - 9)$, $-5x^3 + 7x^2 - 3x + 9$
69. $-(-12y^5 + 4ay^4 - 7by^2)$, $12y^5 - 4ay^4 + 7by^2$
71. $10x - 9$ **73.** $-4x^2 - 3x + 13$ **75.** $3a - 4b + 3c$
77. $-2x^2 + 6x$ **79.** $-4a^2 + 8ab - 5b^2$
81. $8a^2b + 16ab + 3ab^2$ **83.** $x^4 - x^2 - 1$ **85.** \$9700
87. **89.** [1.3] $7x + 16$ **90.** [1.3] $10a + 59$
91. [1.3] $a^2 + 3a - 4$ **92.** [1.3] $x^2 - x - 2$
93. [1.6] x^9 **94.** [1.6] a^8 **95.**
97. $68x^5 - 81x^4 - 22x^3 + 52x^2 + 2x + 250$
99. $45x^5 - 8x^4 + 208x^3 - 176x^2 + 116x - 25$
101. 494.55 cm^3 **103.** $5x^2 - 8x$ **105.** $8x^{2a} + 7x^a + 7$
107. $x^{5b} + 4x^{4b} + x^{3b} - 6x^{2b} - 9x^b$ **109.**

Technology Connection, p. 283

1. Graphing y_3 with a wide line will produce a screen like the following for each example.

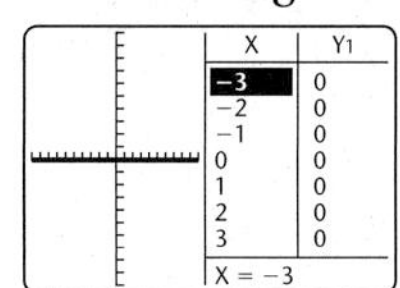

Exercise Set 5.2, pp. 284–286

1. $32a^3$ **3.** $-20x^3y$ **5.** $-10x^5y^6$ **7.** $21x - 7x^2$
9. $20c^3d^2 - 25c^2d^3$ **11.** $6x^2 + 7x - 20$
13. $m^2 - mn - 6n^2$ **15.** $3y^2 - 13xy - 56x^2$
17. $a^4 - 5a^2b^2 + 6b^4$ **19.** $x^3 - 64$ **21.** $x^3 + y^3$
23. $a^4 + 5a^3 - 2a^2 - 9a + 5$
25. $3a^3b^2 - 8a^2b^2 + 3a^3b + 7ab^2 - 2a^2b + 3ab^3 - 6b^3$
27. $x^2 - \frac{3}{4}x + \frac{1}{8}$ **29.** $3x^2 - 1.5xy - 15y^2$
31. $12x^4 - 21x^3 - 17x^2 + 35x - 5$ **33.** $a^2 + 9a + 20$
35. $y^2 - 5y - 24$ **37.** $x^2 + 10x + 25$
39. $x^2 - 4xy + 4y^2$ **41.** $2x^2 + 13x + 18$
43. $100a^2 - 2.4ab + 0.0144b^2$ **45.** $4x^2 - 4xy - 3y^2$
47. $4x^6 - 12x^3y^2 + 9y^4$ **49.** $a^4b^4 + 2a^2b^2 + 1$
51. $16x^2 - 8x + 1$ **53.** $4x^2 - \frac{4}{3}x + \frac{1}{9}$ **55.** $c^2 - 4$
57. $16x^2 - 1$ **59.** $9m^2 - 4n^2$ **61.** $x^6 - y^2z^2$
63. $-m^2n^2 + m^4$, or $m^4 - m^2n^2$ **65.** $x^4 - 1$
67. $a^4 - 2a^2b^2 + b^4$ **69.** $a^2 + 2ab + b^2 - 1$
71. $4x^2 + 12xy + 9y^2 - 16$ **73.** $A = P + 2Pi + Pi^2$
75. **(a)** $t^2 - 2t + 6$; **(b)** $2ah + h^2$; **(c)** $2ah - h^2$
77. **79.** [1.5] $a = \dfrac{d}{b + c}$ **80.** [1.5] $y = \dfrac{w}{x + z}$
81. [1.5] $m = \dfrac{p}{n + 1}$ **82.** [1.5] $s = \dfrac{t}{r + 1}$
83. [3.5] Dimes: 5; nickels: 2; quarters: 6 **84.** [3.3] 8
85. **87.** $x^{4a^2}y^{4ab}$ **89.** $\frac{1}{4}a^{7x}b^{2y+2}$
91. $y^{3n+3}z^{n+3} - 4y^4z^{3n}$

93. $a^2 + 2ac + c^2 - b^2 - 2bd - d^2$
95. $16x^4 + 4x^2y^2 + y^4$ **97.** $x^{4a} - y^{4b}$ **99.** $x^{a^2-b^2}$
101. 0 **103.** (b) and (c) are identities.

Technology Connection, p. 289

1. A table should show that $y_3 = 0$ for any value of x.

Exercise Set 5.3, pp. 291–293

1. $2t(t + 4)$ **3.** $y(y - 5)$ **5.** $y^2(y + 9)$
7. $5x^2(3 - x^2)$ **9.** $4xy(x - 3y)$ **11.** $3(y^2 - y - 3)$
13. $2a(3b - 2d + 6c)$ **15.** $3x^2y^4z^2(3xy^2 - 4x^2z^2 + 5yz)$
17. $-5(x - 7)$ **19.** $-6(y + 12)$ **21.** $-2(x^2 - 2x + 6)$
23. $-3(-y + 8x)$, or $-3(8x - y)$ **25.** $-7(-s + 2t)$, or $-7(2t - s)$ **27.** $-(x^2 - 5x + 9)$
29. $-a(a^3 - 2a^2 + 13)$ **31.** $(b - 5)(a + c)$
33. $(x + 7)(2x - 3)$ **35.** $(x - y)(a^2 - 5)$
37. $(c + d)(a + b)$ **39.** $(b - 1)(b^2 + 2)$
41. $(a - 3)(a^2 - 2)$ **43.** $12x(6x^2 - 3x + 2)$
45. $x^3(x - 1)(x^2 + 1)$ **47.** $(y^2 + 3)(2y^2 + 5)$
49. **(a)** $h(t) = -8t(2t - 9)$; **(b)** $h(1) = 56$ ft
51. $R(n) = n(n - 1)$ **53.** $P(x) = x(x - 3)$
55. $R(x) = 0.4x(700 - x)$ **57.** $N(x) = \frac{1}{6}(x^3 + 3x^2 + 2x)$
59. $H(n) = \frac{1}{2}n(n - 1)$ **61.** **63.** [1.2] -26
64. [1.2] -1 **65.** [1.2] -30 **66.** [1.2] -19
67. [1.4] 56, 58, 60 **68.** [3.5] A: 75; B: 84; C: 63 **69.**
71. $x^5y^4 + x^4y^6 = x^3y(x^2y^3 + xy^5)$
73. $(x^2 - x + 5)(r + s)$
75. $(x^4 + x^2 + 5)(a^4 + a^2 + 5)$ **77.** $x^{1/3}(1 - 7x)$
79. $x^{1/3}(1 - 5x^{1/6} + 3x^{5/12})$ **81.** $3a^n(a + 2 - 5a^2)$
83. $y^{a+b}(7y^a - 5 + 3y^b)$ **85.**

Technology Connection, p. 296

1. They should coincide. **2.** The x-axis
3. Let $y_1 = x^3 - x^2 - 30x$, $y_2 = x(x + 5)(x - 6)$, and $y_3 = y_2 - y_1$. The graphs of y_1 and y_2 should coincide; the graph of y_3 should be the x-axis.
4. Let $y_1 = 2x^2 + x - 15$, $y_2 = (2x + 5)(x - 3)$, and $y_3 = y_2 - y_1$. The graphs of y_1 and y_2 do not coincide; the graph of y_3 is not the x-axis.

Exercise Set 5.4, pp. 301–302

1. $(x + 2)(x + 6)$ **3.** $(t + 3)(t + 5)$ **5.** $(x - 9)(x + 3)$
7. $2(n - 5)(n - 5)$, or $2(n - 5)^2$ **9.** $a(a - 9)(a + 8)$
11. $(x + 9)(x + 5)$ **13.** $(y + 9)(y - 7)$
15. $(t - 9)(t - 5)$ **17.** $(x + 5)(x - 2)$
19. $3(x + 2)(x + 3)$ **21.** $(8 - x)(7 + x)$
23. $y(8 - y)(4 + y)$ **25.** $x^2(x - 5)(x + 16)$
27. Prime **29.** $(p - 8q)(p + 3q)$
31. $(y + 4z)(y + 4z)$, or $(y + 4z)^2$
33. $p^2(p - 79)(p - 1)$ **35.** $x^4(x - 2)(x + 9)$
37. $(3x + 5)(2x - 5)$ **39.** $y(5y + 4)(2y - 3)$
41. $2(4a - 1)(3a - 1)$ **43.** $(5y + 2)(7y + 4)$
45. $2(5t - 3)(t + 1)$ **47.** $4(2x + 1)(x - 4)$
49. $a^4(a + 3)(a - 2)$ **51.** $x^2(7x + 1)(2x - 3)$
53. $4(3a - 4)(a + 1)$ **55.** $(3x + 1)(3x + 4)$
57. $(x + 3)(4x + 3)$ **59.** $-2(2t - 3)(2t + 5)$
61. $xy(6y + 5)(3y - 2)$ **63.** $(24x + 1)(x - 2)$
65. $3x(7x + 3)(3x + 4)$ **67.** $2x^2(6x + 5)(4x - 3)$
69. $(4a - 3b)(3a - 2b)$ **71.** $(2x - 3y)(x + 2y)$
73. $(2x - 7y)(3x - 4y)$ **75.** $(3x - 5y)(3x - 5y)$, or $(3x - 5y)^2$ **77.** $(9xy - 4)(xy + 1)$ **79.**
81. [5.3] $5x(2x^2 - 7x + 1)$ **82.** [5.3] $4t(3t^2 - 10t - 2)$
83. [1.6] $125a^{12}$ **84.** [1.6] $-32x^{10}$ **85.** [2.2] -24
86. [2.2] 880 ft; 960 ft; 1024 ft; 624 ft; 240 ft **87.**
89. $ab^2(2a^3b^4 - 3ab - 20)$ **91.** $\left(x + \frac{4}{5}\right)\left(x - \frac{1}{5}\right)$
93. $(y - 0.1)(y + 0.5)$ **95.** $(x^a + 8)(x^a - 3)$
97. $(bx + a)(dx + c)$ **99.** $a^2(p^a + 2)(p^a - 1)$
101. $[3(x - 7) - 1][2(x - 7) + 5]$, or $(3x - 22)(2x - 9)$
103. 31, -31, 14, -14, 4, -4
105. Since $ax^2 + bx + c = (mx + r)(nx + s)$, from FOIL we know that $a = mn$, $c = rs$, and $b = ms + rn$. If $P = ms$ and $Q = rn$, then $b = P + Q$. Since $ac = mnrs = msrn$, we have $ac = PQ$. **107.**

Exercise Set 5.5, pp. 307–308

1. $(x - 4)^2$ **3.** $(a + 8)^2$ **5.** $2(a + 2)^2$ **7.** $(y - 6)^2$
9. $a(a + 12)^2$ **11.** $2(4x + 3)^2$ **13.** $(5y - 8)^2$
15. $a(a - 5)^2$ **17.** $(0.5x + 0.3)^2$ **19.** $(p - q)^2$
21. $(5a + 3b)^2$ **23.** $4(t - r)^2$ **25.** $(x + 4)(x - 4)$
27. $(p + 7)(p - 7)$ **29.** $(ab + 9)(ab - 9)$
31. $6(x + y)(x - y)$ **33.** $7x(y^2 + z^2)(y + z)(y - z)$
35. $a(2a + 7)(2a - 7)$
37. $3(x^4 + y^4)(x^2 + y^2)(x + y)(x - y)$
39. $a^2(3a + 5b^2)(3a - 5b^2)$ **41.** $\left(\frac{1}{5} + x\right)\left(\frac{1}{5} - x\right)$
43. $(a + b + 3)(a + b - 3)$ **45.** $(x - 3 + y)(x - 3 - y)$
47. $(m - n + 5)(m - n - 5)$
49. $(6 + x + y)(6 - x - y)$
51. $(r - 1 + 2s)(r - 1 - 2s)$
53. $(4 + a + b)(4 - a - b)$
55. $(m - 7)(m + 2)(m - 2)$
57. $(a - 2)(a + b)(a - b)$ **59.** **61.** [1.6] $8a^{12}b^{15}$
62. [1.6] $125x^6y^{12}$ **63.** [5.2] $x^3 + 3x^2y + 3xy^2 + y^3$
64. [5.2] $a^3 + 3a^2 + 3a + 1$ **65.** [3.4] $(2, -1, 3)$
66. [4.3] $\left\{x \mid x \le -\frac{4}{7} \text{ or } x \ge 2\right\}$, or $\left(-\infty, -\frac{4}{7}\right] \cup [2, \infty)$
67. [4.3] $\left\{x \mid -\frac{4}{7} \le x \le 2\right\}$, or $\left[-\frac{4}{7}, 2\right]$
68. [4.1] $\left\{x \mid x < \frac{14}{19}\right\}$, or $\left(-\infty, \frac{14}{19}\right)$ **69.**
71. $-\frac{1}{54}(4r + 3s)^2$ **73.** $(0.3x^4 + 0.8)^2$, or $\frac{1}{100}(3x^4 + 8)^2$
75. $(r + s + 1)(r - s - 9)$ **77.** $(x^{2a} + y^b)(x^{2a} - y^b)$
79. $(5y^a + x^b - 1)(5y^a - x^b + 1)$ **81.** $3(x + 3)^2$

83. $(3x^n - 1)^2$ **85.** $h(2a + h)$
87. (a) $\pi h(R + r)(R - r)$; **(b)** 3,014,400 cm^3 **89.**

Exercise Set 5.6, pp. 312–313

1. $(t + 3)(t^2 - 3t + 9)$ **3.** $(x - 2)(x^2 + 2x + 4)$
5. $(m - 4)(m^2 + 4m + 16)$ **7.** $(2a + 1)(4a^2 - 2a + 1)$
9. $(3 - 2t)(9 + 6t + 4t^2)$ **11.** $(2x + 3)(4x^2 - 6x + 9)$
13. $(y - z)(y^2 + yz + z^2)$ **15.** $\left(x + \frac{1}{3}\right)\left(x^2 - \frac{1}{3}x + \frac{1}{9}\right)$
17. $2(y - 4)(y^2 + 4y + 16)$
19. $8(a + 5)(a^2 - 5a + 25)$ **21.** $r(s + 4)(s^2 - 4s + 16)$
23. $2(y - 3z)(y^2 + 3yz + 9z^2)$
25. $(y + 0.5)(y^2 - 0.5y + 0.25)$
27. $(5c^2 - 2d^2)(25c^4 + 10c^2d^2 + 4d^4)$
29. $3z^2(z - 1)(z^2 + z + 1)$ **31.** $(t^2 + 1)(t^4 - t^2 + 1)$
33. $(p + q)(p^2 - pq + q^2)(p - q)(p^2 + pq + q^2)$
35. $(a^3 + b^4c^5)(a^6 - a^3b^4c^5 + b^8c^{10})$ **37.**
39. [2.2] 224 ft; 288 ft; 320 ft; 288 ft; 128 ft
40. [3.3] 228 ft^2 **41.** [2.2] -2 **42.** [2.3] Slope: $\frac{4}{3}$; y-intercept: $\left(0, -\frac{8}{3}\right)$ **43.** [1.3] $\frac{5}{3}$ **44.** [1.3] $-\frac{7}{2}$ **45.**
47. $(x^{2a} - y^b)(x^{4a} + x^{2a}y^b + y^{2b})$ **49.** $2x(x^2 + 75)$
51. $5\left(xy^2 - \frac{1}{2}\right)\left(x^2y^4 + \frac{1}{2}xy^2 + \frac{1}{4}\right)$
53. $-(3x^{4a} + 3x^{2a} + 1)$ **55.** $(t - 8)(t - 1)(t^2 + t + 1)$
57. $h(2a + h)(a^2 + ah + h^2)(3a^2 + 3ah + h^2)$
59.

Exercise Set 5.7, pp. 316–317

1. $5(m^2 + 2)(m^2 - 2)$ **3.** $(a + 9)(a - 9)$
5. $(4x - 1)(2x + 5)$ **7.** $(a - 5)^2$ **9.** $3(x + 12)(x - 7)$
11. $(3x + 5y)(3x - 5y)$ **13.** $(t^2 + 1)(t^4 - t^2 + 1)$
15. $(x + y + 3)(x - y + 3)$
17. $(7x + 3y)(49x^2 - 21xy + 9y^2)$
19. $(m^3 + 10)(m^3 - 2)$ **21.** $(a + d)(c - b)$
23. $(2c - d)^2$ **25.** $(8 + 3t^2)(3 + t)$
27. $2(x + 3)(x + 2)(x - 2)$
29. $2(3a - 2b)(9a^2 + 6ab + 4b^2)$ **31.** $(6y - 5)(6y + 7)$
33. $(a^4 + b^4)(a^2 + b^2)(a + b)(a - b)$
35. $ab(a + 4b)(a - 4b)$ **37.** $2(a - 3)(a + 3)$
39. $7a(a^3 - 2a^2 + 3a - 1)$ **41.** $(9ab + 2)(3ab + 4)$
43. $p(1 - 4p)(1 + 4p + 16p^2)$
45. $(a - b - 3)(a + b + 3)$ **47.** **49.** [1.3] $\frac{9}{5}$
50. [1.3] $-\frac{13}{7}$
51. [2.3]

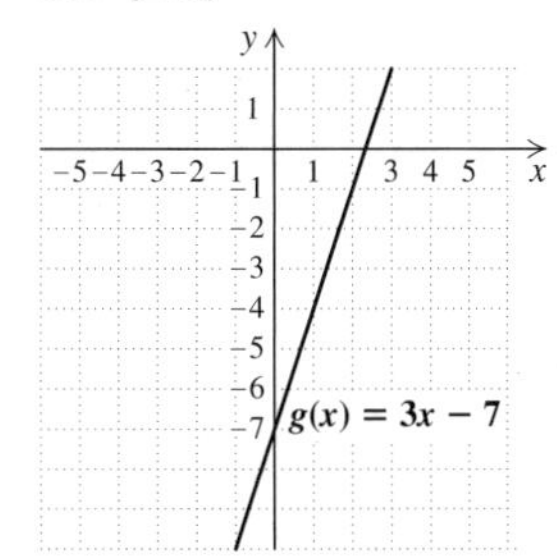

52. [2.3]

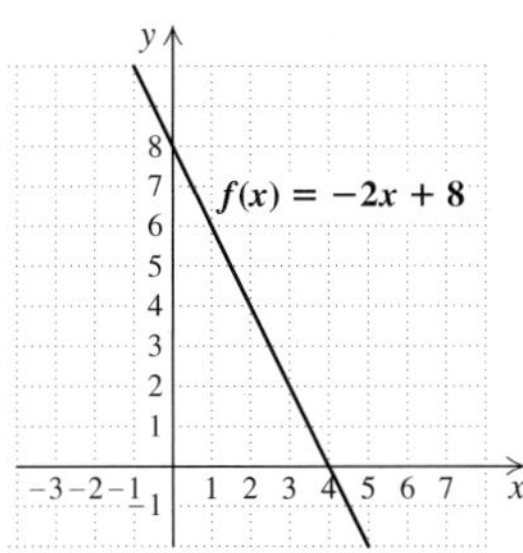

53. [3.3] Correct: 55; incorrect: 20 **54.** [3.3] $\frac{80}{7}$ units
55. **57.** $(12x - 5y^2)(5x - 6y^2)$
59. $(11 + y^2)(2 + y)(2 - y)$
61. $(a + b)(a - b)(a + 7b)(a - 7b)$
63. $(3x^{2s} + 4y^t)(9x^{4s} - 12x^{2s}y^t + 16y^{2t})$
65. $(2x + y - r + 3s)(2x + y + r - 3s)$
67. $6(2t^a + 1)(2t^a - 1)$ **69.** $\left(\frac{x^9}{10} - 1\right)\left(\frac{x^{18}}{100} + \frac{x^9}{10} + 1\right)$
71. $3(x - 3)(x + 2)$ **73.** $3(a + 7)^2$ **75.** 0

Exercise Set 5.8, pp. 325–329

1. $\{-5, 9\}$ **3.** $\{1\}$ **5.** $\{-6\}$ **7.** $\{-5, -4\}$ **9.** $\{0, 8\}$
11. $\{-5, 0, 8\}$ **13.** $\{-4, 4\}$ **15.** $\{-9, 9\}$ **17.** $\left\{\frac{2}{3}, 2\right\}$
19. $\left\{-2, -\frac{3}{4}, 0\right\}$ **21.** $\{-4, 5\}$ **23.** $\left\{-\frac{7}{4}, \frac{4}{3}\right\}$ **25.** $\left\{-\frac{1}{8}, \frac{1}{8}\right\}$
27. $\{-5, -1, 1, 5\}$ **29.** $\{-8, -4\}$ **31.** $\left\{-4, \frac{3}{2}\right\}$
33. $\{-9, -3\}$
35. $\{x \mid x$ is a real number *and* $x \neq -1$ *and* $x \neq 5\}$
37. $\{x \mid x$ is a real number *and* $x \neq -3$ *and* $x \neq 3\}$
39. $\left\{x \mid x \text{ is a real number } \textit{and } x \neq 0 \textit{ and } x \neq \frac{1}{2}\right\}$
41. $\{x \mid x$ is a real number *and* $x \neq 0$ *and* $x \neq 2$ *and* $x \neq 5\}$
43. $-12, 11$ **45.** Length: 12 cm; width: 7 cm **47.** 3 m
49. 3 cm **51.** 10 ft **53.** 16, 18, 20 **55.** Height: 6 ft; base: 4 ft **57.** Height: 16 m; base: 7 m **59.** 41 ft
61. Length: 100 m; width: 75 m **63.** 2 sets **65.** 2 sec
67. 5 sec **69.** **71.** [1.2] $-\frac{5}{8}$ **72.** [1.2] 2
73. [3.3] Faster car: 54 mph; slower car: 39 mph
74. [3.3] Conventional: 36; surround-sound: 42
75. [4.1] $\left\{x \mid x > \frac{10}{3}\right\}$, or $\left(\frac{10}{3}, \infty\right)$ **76.** [3.4] $(2, -2, -4)$
77. **79.** $\left\{-\frac{11}{8}, -\frac{1}{4}, \frac{2}{3}\right\}$ **81.** $\{1\}$
83. Answers may vary. $f(x) = 5x^3 - 20x^2 + 5x + 30$
85. $\{-1, 3\}$; $\{x \mid -2 < x < 4\}$, or $(-2, 4)$
87. 50-yr-olds **89.** About 5.7 sec **91.**
93. $\{-4.00, 1.00\}$ **95.** $\{-3.33, 5.15\}$ **97.**

Review Exercises: Chapter 5, pp. 330–331

1. [5.1] 7, 11, 3, 0; 11 **2.** [5.1] $-5x^3 + 2x^2 + 4x - 7$; $-5x^3$; -5 **3.** [5.1] $-3x^2 + 2x^3 + 3x^6y - 7x^8y^3$
4. [5.1] 0; -6 **5.** [5.1] 4 **6.** [5.1] $-2a^3 + a^2 - 3a - 4$
7. [5.1] $-x^2y - 2xy^2$ **8.** [5.1] $-x^3 + 2x^2 + 5x + 2$
9. [5.1] $-3x^4 + 3x^3 - x + 16$
10. [5.1] $-5xy^2 - 2xy + x^2y$ **11.** [5.1] $9x - 7$
12. [5.1] $-2a + 6b + 7c$ **13.** [5.1] $6x^2 - 7xy + 3y^2$
14. [5.2] $-18x^3y^4$ **15.** [5.2] $x^8 - x^6 + 5x^2 - 3$
16. [5.2] $8a^2b^2 + 2abc - 3c^2$ **17.** [5.2] $4x^2 - 25y^2$
18. [5.2] $4x^2 - 20xy + 25y^2$ **19.** [5.2] $2x^2 + 5x - 3$
20. [5.2] $x^4 + 8x^2y^3 + 16y^6$ **21.** [5.2] $x^3 - 125$
22. [5.2] $x^2 - \frac{1}{2}x + \frac{1}{18}$ **23.** [5.3] $x(6x + 5)$
24. [5.3] $3y^2(3y^2 - 1)$ **25.** [5.3] $3x(5x^3 - 6x^2 + 7x - 3)$
26. [5.4] $(a - 9)(a - 3)$ **27.** [5.4] $(3m + 2)(m + 4)$
28. [5.5] $(5x + 2)^2$ **29.** [5.5] $4(y + 2)(y - 2)$
30. [5.4] $x(x - 2)(x + 7)$ **31.** [5.3] $(a + 2b)(x - y)$

32. [5.3] $(y + 2)(3y^2 - 5)$
33. [5.5] $(a^2 + 9)(a + 3)(a - 3)$
34. [5.3] $4(x^4 + x^2 + 5)$
35. [5.6] $(3x - 2)(9x^2 + 6x + 4)$
36. [5.6] $(0.4b - 0.5c)(0.16b^2 + 0.2bc + 0.25c^2)$
37. [5.3] $y(y^4 + 1)$ **38.** [5.3] $2z^6(z^2 - 8)$
39. [5.6] $2y(3x^2 - 1)(9x^4 + 3x^2 + 1)$
40. [5.5] $4(3x - 5)^2$ **41.** [5.4] $(3t + p)(2t + 5p)$
42. [5.5] $(x + 3)(x - 3)(x + 2)$
43. [5.5] $(a - b + 2t)(a - b - 2t)$ **44.** [5.8] $\{10\}$
45. [5.8] $\left\{\frac{2}{3}, \frac{3}{2}\right\}$ **46.** [5.8] $\left\{0, \frac{7}{4}\right\}$ **47.** [5.8] $\{-4, 4\}$
48. [5.8] $\{-3, 0, 7\}$ **49.** [5.8] $\{-1, 6\}$ **50.** [5.8] -4, 11
51. [5.8] $\left\{x \mid x \text{ is a real number and } x \neq -7 \text{ and } x \neq \frac{2}{3}\right\}$
52. [5.8] 5 units **53.** [5.8] 3, 5, 7; -7, -5, -3
54. [5.8] Length: 8 in.; width: 5 in. **55.** [5.8] 17 ft
56. [5.8] The roots of a polynomial function are the x-coordinates of the points at which the graph of the function crosses the x-axis. **57.** [5.8] The principle of zero products states that if a product is equal to 0, at least one of the factors must be 0. If a product is nonzero, we cannot conclude that any one of the factors is a particular value.
58. [5.6] $2(2x - y)(4x^2 + 2xy + y^2)$
$(2x + y)(4x^2 - 2xy + y^2)$
59. [5.6] $-2(3x^2 + 1)$
60. [5.2], [5.6] $a^3 - b^3 + 3b^2 - 3b + 1$ **61.** [5.2] z^{5n^5}
62. [5.8] $\left\{-1, -\frac{1}{2}\right\}$

Test: Chapter 5, p. 332

1. [5.1] 9 **2.** [5.1] $5x^5y^4 - 2x^4y - 4x^2y + 3xy^3$
3. [5.1] $-4a^3$ **4.** [5.1] 4; 2 **5.** [5.2] $2ah + h^2 - 5h$
6. [5.1] $3xy + 3xy^2$ **7.** [5.1] $-3x^3 + 3x^2 - 6y - 7y^2$
8. [5.1] $7m^3 + 2m^2n + 3mn^2 - 7n^3$ **9.** [5.1] $6a - 8b$
10. [5.1] $2y^2 + 5y + y^3$ **11.** [5.2] $64x^3y^3$
12. [5.2] $12a^2 - 4ab - 5b^2$ **13.** [5.2] $x^3 - 2x^2y + y^3$
14. [5.2] $4x^6 + 20x^3 + 25$ **15.** [5.2] $16y^2 - 72y + 81$
16. [5.2] $x^2 - 4y^2$ **17.** [5.3] $5x^2(3 - x^2)$
18. [5.5] $(y + 5)(y + 2)(y - 2)$
19. [5.4] $(p - 14)(p + 2)$ **20.** [5.4] $(6m + 1)(2m + 3)$
21. [5.5] $(3y + 5)(3y - 5)$
22. [5.6] $3(r - 1)(r^2 + r + 1)$ **23.** [5.5] $(3x - 5)^2$
24. [5.5] $(x^4 + y^4)(x^2 + y^2)(x + y)(x - y)$
25. [5.5] $(y + 4 + 10t)(y + 4 - 10t)$
26. [5.5] $5(2a - b)(2a + b)$ **27.** [5.4] $2(4x - 1)(3x - 5)$
28. [5.6] $2ab(2a^2 + 3b^2)(4a^4 - 6a^2b^2 + 9b^4)$
29. [5.3] $4xy(y^3 + 9x + 2x^2y - 4)$ **30.** [5.8] $\{-3, 6\}$
31. [5.8] $\{-5, 5\}$ **32.** [5.8] $\left\{-7, -\frac{3}{2}, 0\right\}$ **33.** [5.8] $\left\{-\frac{1}{3}, 0\right\}$
34. [5.8] $\{0, 5\}$
35. [5.8] $\{x \mid x \text{ is a real number and } x \neq -1\}$
36. [5.8] Length: 8 cm; width: 5 cm **37.** [5.8] $4\frac{1}{2}$ sec
38. (a) [5.2] $x^5 + x + 1$;
(b) [5.2], [5.7] $(x^2 + x + 1)(x^3 - x^2 + 1)$
39. [5.4] $(3x^n + 4)(2x^n - 5)$

CHAPTER 6

Technology Connection 6.1, p. 338

1. Let $y_1 = (7x^2 + 21x)/(14x)$, $y_2 = (x + 3)/2$, and $y_3 = y_1 - y_2$ (or $y_2 - y_1$). A table or the TRACE feature can be used to show that, except when $x = 0$, y_3 is always 0. As an alternative, let $y_1 = (7x^2 + 21x)/(14x) - (x + 3)/2$ and show that, except when $x = 0$, y_1 is always 0.
2. Let $y_1 = (x + 3)/x$, $y_2 = 3$, and $y_3 = y_1 - y_2$ (or $y_2 - y_1$). Use a table or the TRACE feature to show that y_3 is not always 0. As an alternative, let $y_1 = (x + 3)/x - 3$ and show that y_1 is not always 0.

Exercise Set 6.1, pp. 340–343

1. $\frac{40}{13}$ hr, or $3\frac{1}{13}$ hr **3.** $\frac{2}{3}$; 28; $\frac{163}{10}$ **5.** $-\frac{9}{4}$; does not exist; $-\frac{11}{9}$ **7.** $\frac{4x(x - 3)}{4x(x + 2)}$ **9.** $\frac{(t - 2)(-1)}{(t + 3)(-1)}$ **11.** $\frac{3}{x}$ **13.** $\frac{2}{3t^4}$
15. $a - 5$ **17.** $\frac{3}{5a - 6}$ **19.** $\frac{x - 4}{x + 5}$ **21.** $\frac{5}{x}$ **23.** $-\frac{1}{2}$
25. $\frac{8}{t + 2}$ **27.** $-\frac{1}{1 + 2t}$ **29.** $-\frac{6}{5}$ **31.** $\frac{a - 5}{a + 5}$
33. $\frac{x + 8}{x - 4}$ **35.** $\frac{4 + t}{4 - t}$ **37.** $\frac{7b^2}{6a^4}$ **39.** $\frac{8x^2}{25}$ **41.** $\frac{y + 4}{4}$
43. $\frac{(x + 4)(x - 4)}{x(x + 3)}$ **45.** $-\frac{a + 1}{2 + a}$ **47.** 1
49. $\frac{(x + 5)(2x + 3)}{7x}$ **51.** $c(c - 2)$ **53.** $\frac{a^2 + ab + b^2}{3(a + 2b)}$
55. $\frac{1}{2x + 3y}$ **57.** $6x^4y^7$ **59.** $\frac{5}{x^5}$ **61.** $\frac{(x + 2)(x + 4)}{x^7}$
63. $-\frac{5x + 2}{x - 3}$ **65.** $-\frac{1}{y^3}$ **67.** $\frac{(x + 4)(x + 2)}{3(x - 5)}$
69. $\frac{y(y^2 + 3)}{(y + 3)(y - 2)}$ **71.** $\frac{x^2 + 4x + 16}{(x + 4)^2}$ **73.** $\frac{(2a + b)^2}{2(a + b)}$
75. **77.** [1.2] $-\frac{7}{30}$ **78.** [1.2] $-\frac{13}{40}$ **79.** [1.2] $\frac{5}{14}$
80. [1.2] $\frac{1}{35}$ **81.** [5.1] $4x^3 - 7x^2 + 9x - 5$
82. [5.1] $-2t^4 + 11t^3 - t^2 + 10t - 3$ **83.**
85. (a) $\frac{2x + 2h + 3}{4x + 4h - 1}$; **(b)** $\frac{2x + 3}{8x - 9}$; **(c)** $\frac{x + 5}{4x - 1}$ **87.** $\frac{2s}{r + 2s}$
89. $\frac{6t^2 - 26t + 30}{8t^2 - 15t - 21}$ **91.** $\frac{m - t}{m + t + 1}$
93. $\frac{x^2 + xy + y^2 + x + y}{x - y}$ **95.** $-\frac{2x}{x - 1}$ **97.**
99.

Exercise Set 6.2, pp. 350–352

1. $\frac{4}{a}$ **3.** $-\frac{1}{a^2b}$ **5.** 2 **7.** $\frac{3y + 5}{y - 2}$ **9.** $\frac{1}{x - 4}$

11. $\dfrac{-1}{a+5}$ **13.** $a+b$ **15.** $\dfrac{15}{x}$ **17.** $\dfrac{2}{x+4}$
19. $\dfrac{1}{t^2+4}$ **21.** $\dfrac{1}{m^2+mn+n^2}$ **23.** $\dfrac{2a^2-a+14}{(a-4)(a+3)}$
25. $\dfrac{5x+1}{x+1}$ **27.** $\dfrac{x+y}{x-y}$ **29.** $\dfrac{3x^2+7x+14}{(2x-5)(x-1)(x+2)}$
31. $\dfrac{8x+1}{(x+1)(x-1)}$ **33.** $\dfrac{-x+34}{20(x+2)}$
35. $\dfrac{-a^2+7ab-b^2}{(a-b)(a+b)}$ **37.** $\dfrac{x-5}{(x+5)(x+3)}$
39. $\dfrac{y}{(y-2)(y-3)}$ **41.** $\dfrac{7x+1}{x-y}$
43. $\dfrac{3y^2-3y-29}{(y-3)(y+8)(y-4)}$ **45.** $\dfrac{-2a^2-7a-10}{(a+4)(a-4)(a+6)}$
47. $\dfrac{10a-16}{(a-4)(a-1)(a-2)(a+2)}$ **49.** $\dfrac{2(3t-7)}{t-2}$
51. $\dfrac{-y}{(y+3)(y-1)}$ **53.** $-\dfrac{2y}{2y+1}$
55. $\dfrac{-6x+42}{(x-3)(x-2)(x+1)(x+3)}$ **57.**
59. [1.6] $\dfrac{3y^6z^7}{7x^5}$ **60.** [1.6] $\dfrac{7b^{11}c^7}{9a^2}$ **61.** [1.6] $\dfrac{17s^{12}r^{40}}{5t^{60}}$
62. [2.5] $y=\frac{5}{4}x+\frac{11}{2}$ **63.** [3.5] Dimes: 2 rolls; nickels: 5 rolls; quarters: 5 rolls **64.** [3.3] 30-min tapes: 4; 60-min tapes: 8 **65.** **67.** 420 days **69.** 12 parts
71. $x^4(x^2+1)(x+1)(x-1)(x^2+x+1)(x^2-x+1)$
73. $8a^4, 8a^4b, 8a^4b^2, 8a^4b^3, 8a^4b^4, 8a^4b^5, 8a^4b^6, 8a^4b^7$
75. $\dfrac{x^4+6x^3+2x^2}{(x+2)(x-2)(x+5)}$ **77.** $\dfrac{x^5}{(x^2-4)(x^2+3x-10)}$
79. $\dfrac{9x^2+28x+15}{(x-3)(x+3)^2}$ **81.** $\dfrac{1}{2x(x-5)}$ **83.** $-4t^4$
85.

Technology Connection, p. 356

1. $-1, 0, \frac{2}{3}, 2$

Exercise Set 6.3, pp. 358–360

1. $\dfrac{7a+1}{1-3a}$ **3.** $\dfrac{x^2-1}{x^2+1}$ **5.** $\dfrac{6y+7x}{7y-6x}$ **7.** $\dfrac{x+y}{x}$
9. $\dfrac{x^2(3-y)}{y^2(2x-1)}$ **11.** $\dfrac{1}{a-b}$ **13.** $-\dfrac{1}{x(x+h)}$
15. $\dfrac{(x-4)(x-7)}{(x-5)(x+6)}$ **17.** $\dfrac{4x-7}{7x-9}$ **19.** $\dfrac{a^2-3a-6}{a^2-2a-3}$
21. $\dfrac{x+2}{x+3}$ **23.** $\dfrac{a+1}{2a+5}$ **25.** $\dfrac{-1-3x}{8-2x}$, or $\dfrac{3x+1}{2x-8}$
27. $\dfrac{1}{y+5}$ **29.** $-y$ **31.** $\dfrac{2(5a^2+4a+12)}{5(a^2+6a+18)}$
33. $\dfrac{(2x+1)(x+2)}{2x(x-1)}$ **35.** $\dfrac{(2a-3)(a+5)}{2(a-3)(a+2)}$ **37.** -1
39. $\dfrac{2x^2-11x-27}{2x^2+21x+13}$ **41.** **43.** [1.3] 1 **44.** [1.3] $\frac{11}{4}$
45. [1.5] $y=\dfrac{t-rs}{r}$ **46.** [4.3] $\{-2, 5\}$ **47.** [3.3] First frame: 15 cm; second frame: 29 cm **48.** [1.4] \$34
49. **51.** $\dfrac{5(y+x)}{3(y-x)}$ **53.** $\dfrac{8c}{17}$ **55.** $\dfrac{-3}{x(x+h)}$
57. $\dfrac{2}{(1+x+h)(1+x)}$
59. $\{x \mid x$ is a real number *and* $x \neq \pm 1$ *and* $x \neq \pm 4$ *and* $x \neq \pm 5\}$ **61.** $\dfrac{2+a}{3+a}$
63. $\dfrac{x^4}{81}$; $\{x \mid x$ is a real number *and* $x \neq 3\}$ **65.**
67. \$168.61

Exercise Set 6.4, pp. 366–368

1. $\frac{51}{5}$ **3.** 144 **5.** -2 **7.** 7 **9.** No solution
11. $-\frac{1}{2}$ **13.** No solution **15.** $-\frac{10}{3}$ **17.** -1
19. 2, 3 **21.** -145 **23.** -1 **25.** $-\frac{3}{2}, 5$ **27.** 14
29. $\frac{3}{4}$ **31.** 4 **33.** 3 **35.** No solution **37.** $-\frac{7}{3}$
39. $\frac{1}{7}$ **41.** **43.** [3.5] Multiple-choice: 25; true–false: 30; fill-in: 15 **44.** [3.3] Child's: 118; adult's: 132 **45.** [1.4] 24,640 m^2 **46.** [5.8] 16 and 18
47. [3.1] **(a)** Inconsistent; **(b)** consistent
48. [4.3] $\{x \mid x < -1$ *or* $x > 5\}$, or $(-\infty, -1) \cup (5, \infty)$
49. **51.** $\frac{1}{5}$ **53.** $-\frac{7}{2}$ **55.** 0.0854697 **57.** Yes
59. **61.**

Exercise Set 6.5, pp. 374–377

1. 2 **3.** $-3, -2$ **5.** 6 and 7, -7 and -6 **7.** $3\frac{3}{7}$ hr
9. $9\frac{9}{10}$ hr **11.** $3\frac{3}{4}$ min **13.** Skyler: 12 hr; Jake: 6 hr
15. HQ17: 30 min; EV25: 15 min **17.** Sara: 6 hr; Kate: 3 hr **19.** $1\frac{1}{5}$ hr **21.** 300 min, or 5 hr **23.** 7 mph
25. 5.2 ft/sec **27.** Freight: 66 mph; passenger: 80 mph
29. Express: 45 mph; local: 38 mph **31.** $1\frac{1}{5}$ km/h
33. 2 km/h **35.** 25 mph **37.** 20 mph **39.**
41. [1.6] $5a^4b^6$ **42.** [1.6] $5x^6y^4$ **43.** [1.6] $4s^{10}t^8$
44. [5.1] $-2x^4-7x^2+11x$
45. [5.1] $-8x^3+28x^2-25x+19$
46. [5.1] $11x^4+7x^3-2x^2-4x-10$ **47.**
49. $49\frac{1}{2}$ hr **51.** 11.25 min **53.** 2250 people per hour
55. $14\frac{7}{8}$ mi **57.** 30 mi **59.** $8\frac{2}{11}$ min after 10:30
61. $51\frac{3}{7}$ mph

Exercise Set 6.6, pp. 382–383

1. $\dfrac{17}{3}x^4+3x^3-\dfrac{14}{3}$ **3.** $3a^2+a-\dfrac{3}{7}-\dfrac{2}{a}$
5. $7y-\dfrac{9}{2}-\dfrac{4}{y}$ **7.** $-5x^5+7x^2+1$

9. $1 - ab^2 - a^3b^4$ **11.** $-2pq + 3p - 4q$ **13.** $x + 3$
15. $a - 12 + \dfrac{32}{a+4}$ **17.** $x - 4 + \dfrac{1}{x-5}$ **19.** $y - 5$
21. $y^2 - 2y - 1 + \dfrac{-8}{y-2}$ **23.** $2x^2 - x + 1 + \dfrac{-5}{x+2}$
25. $a^2 + 4a + 15 + \dfrac{70}{a-4}$ **27.** $2y^2 + 2y - 1 + \dfrac{8}{5y-2}$
29. $2x^2 - x - 9 + \dfrac{3x+12}{x^2+2}$ **31.** $4x^2 - 6x + 9, x \neq -\dfrac{3}{2}$
33. $2x - 5, x \neq -\dfrac{2}{3}$ **35.** $x^2 + 1, x \neq -5, x \neq 5$
37. $2x^3 - 3x^2 + 5, x \neq -1, x \neq 1$ **39.**
41. [1.5] $c = \dfrac{ab-k}{d}$ **42.** [1.5] $z = \dfrac{xy-t}{w}$
43. [5.8] 12, 13, 14 **44.** [2.2] $-54a^3$
45. [4.3] $\{x \mid x < -2 \text{ or } x > 5\}$, or $(-\infty, -2) \cup (5, \infty)$
46. [4.3] $\left\{x \mid -\frac{7}{3} < x < 3\right\}$, or $\left(-\frac{7}{3}, 3\right)$ **47.**
49. $a^2 + ab$
51. $a^6 - a^5b + a^4b^2 - a^3b^3 + a^2b^4 - ab^5 + b^6$
53. $-\frac{3}{2}$ **55.** **57.**

Exercise Set 6.7, pp. 387–388

1. $x^2 - 3x + 5 + \dfrac{-12}{x+1}$ **3.** $a + 5 + \dfrac{-4}{a+3}$
5. $x^2 - 5x - 23 + \dfrac{-43}{x-2}$ **7.** $3x^2 - 2x + 2 + \dfrac{-3}{x+3}$
9. $y^2 + 2y + 1 + \dfrac{12}{y-2}$ **11.** $x^4 + 2x^3 + 4x^2 + 8x + 16$
13. $3x^2 + 6x - 3 + \dfrac{2}{x+\frac{1}{3}}$ **15.** 6 **17.** 1 **19.** 54
21. **23.** [1.5] $b = \dfrac{a-9}{c+1}$ **24.** [1.5] $a = \dfrac{bd-8}{c-b}$
25. [5.8] $\{x \mid x \text{ is a real number } and\ x \neq -5\ and\ x \neq 5\}$
26. [5.8] $\left\{x \mid x \text{ is a real number } and\ x \neq -\frac{9}{2}\ and\ x \neq 1\right\}$
27. [2.5] **28.** [2.3]

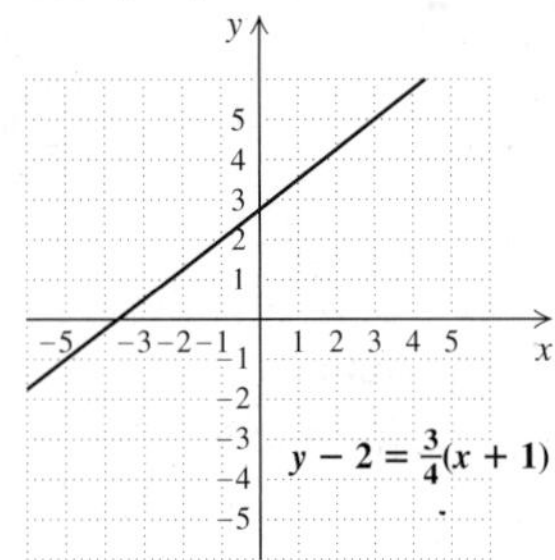

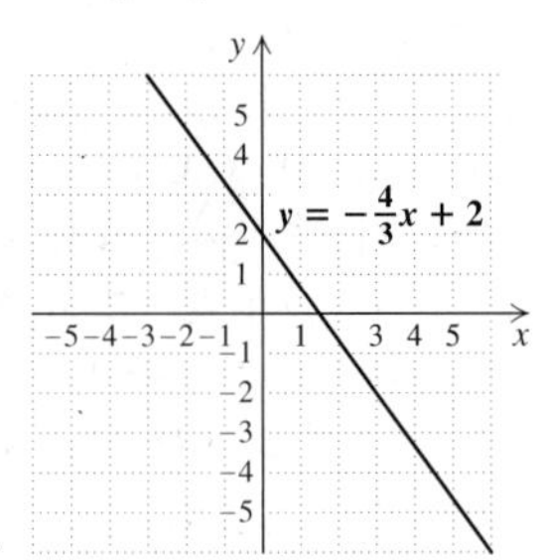

29.
31. **(a)** The degree of R must be less than 1, the degree of $x - r$; **(b)** Let $x = r$. Then

$$\begin{aligned} P(r) &= (r - r) \cdot Q(r) + R \\ &= 0 \cdot Q(r) + R \\ &= R. \end{aligned}$$

33. $0; -\frac{7}{2}, \frac{5}{3}, 4$ **35.** **37.** 0

Exercise Set 6.8, pp. 395–399

1. $d_1 = \dfrac{W_1 d_2}{W_2}$ **3.** $v_1 = \dfrac{2s}{t} - v_2$, or $\dfrac{2s - tv_2}{t}$
5. $r_1 = \dfrac{Rr_2}{r_2 - R}$ **7.** $R = \dfrac{2V}{I} - 2r$, or $\dfrac{2V - 2Ir}{I}$
9. $g = \dfrac{Rs}{s-R}$ **11.** $n = \dfrac{IR}{E - Ir}$ **13.** $q = \dfrac{pf}{p-f}$
15. $t_1 = \dfrac{H}{Sm} + t_2$, or $\dfrac{H + Smt_2}{Sm}$ **17.** $r = \dfrac{Re}{E-e}$
19. $r = 1 - \dfrac{a}{S}$, or $\dfrac{S-a}{S}$ **21.** $a + b = \dfrac{f}{c^2}$
23. $T = -\dfrac{I_f}{I_t} + 1$, or $1 - \dfrac{I_f}{I_t}$, or $\dfrac{I_t - I_f}{I_t}$ **25.** $r_2 = \dfrac{Rr_1}{r_1 - R}$
27. $t_1 = t_2 - \dfrac{v_2 - v_1}{a}$ **29.** $Q = \dfrac{2Tt - 2AT}{A - q}$
31. $k = 7; y = 7x$ **33.** $k = 1.7; y = 1.7x$
35. $k = 6; y = 6x$ **37.** $33\frac{1}{3}$ cm **39.** 241,920,000 cans
41. 32 kg **43.** 3.36 **45.** $k = 60; y = \dfrac{60}{x}$
47. $k = 112; y = \dfrac{112}{x}$ **49.** $k = 9; y = \dfrac{9}{x}$ **51.** 20 min
53. 160 cm^3 **55.** 3.5 hr **57.** $y = \dfrac{2}{3}x^2$ **59.** $y = \dfrac{54}{x^2}$
61. $y = 0.3xz^2$ **63.** $y = \dfrac{4wx^2}{z}$ **65.** 40 W/m^2
67. 220 cm^3 **69.** About 57.42 mph **71.**
73. [2.2] $\mathbb{R}$ **74.** [2.2] $\mathbb{R}$
75. [4.4] **76.** [2.2] $8a^3 - 2a$

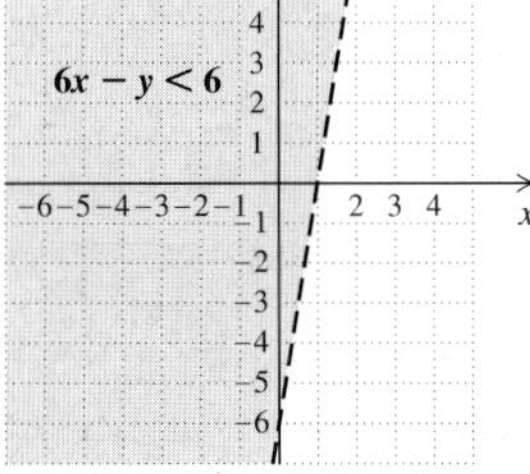

77. [5.6] $(t + 2b)(t^2 - 2bt + 4b^2)$ **78.** [5.8] $-\frac{5}{3}, \frac{7}{2}$
79. **81.** 567 mi **83.** Ratio is $\dfrac{a+12}{a+6}$; percent increase is $\dfrac{6}{a+6} \cdot 100\%$, or $\dfrac{600}{a+6}\%$
85. $t_1 = t_2 + \dfrac{(d_2 - d_1)(t_4 - t_3)}{a(t_4 - t_2)(t_4 - t_3) + d_3 - d_4}$ **87.** Q varies directly as the square of p and inversely as the cube of q.
89. About 1.697 m **91.** $d(s) = \dfrac{28}{s}$; 70 yd

Review Exercises: Chapter 6, pp. 402–403

1. [6.1] **(a)** $-\frac{2}{9}$; **(b)** $-\frac{3}{4}$; **(c)** 0 **2.** [6.2] $48x^3$
3. [6.2] $(x+5)(x-2)(x-4)$ **4.** [6.2] $x+3$
5. [6.2] $\dfrac{1}{x-1}$ **6.** [6.1] $\dfrac{b^2c^6d^2}{a^5}$ **7.** [6.2] $\dfrac{15np+14m}{18m^2n^4p^2}$
8. [6.1] $\dfrac{y-8}{2}$ **9.** [6.1] $\dfrac{(x-2)(x+5)}{x-5}$
10. [6.1] $\dfrac{3a-1}{a-3}$ **11.** [6.1] $\dfrac{(x^2+4x+16)(x-6)}{(x+4)(x+2)}$
12 [6.2] $\dfrac{x-3}{(x+1)(x+3)}$ **13.** [6.2] $\dfrac{x-y}{x+y}$
14. [6.2] $2(x+y)$ **15.** [6.2] $\dfrac{-y}{(y+4)(y-1)}$
16. [6.3] $\dfrac{5}{7}$ **17.** [6.3] $\dfrac{a^2b^2}{2(b^2-ba+a^2)}$
18. [6.3] $\dfrac{(y+11)(y+5)}{(y-5)(y+2)}$ **19.** [6.3] $\dfrac{(14-3x)(x+3)}{2x^2+16x+6}$
20. [6.4] 2 **21.** [6.4] 6 **22.** [6.4] No solution
23. [6.4] 0 **24.** [6.4] $-1, 4$ **25.** [6.5] $5\frac{1}{7}$ hr
26. [6.5] Celeron: 45 sec; Pentium III: 30 sec
27. [6.5] 24 mph **28.** [6.5] Motorcycle: 62 mph; car: 70 mph **29.** [6.6] $4s^2+3s-2rs^2$
30. [6.6] $y^2-5y+25$ **31.** [6.6] $4x+3+\dfrac{-9x-5}{x^2+1}$
32. [6.7] $x^2+6x+20+\dfrac{54}{x-3}$ **33.** [6.7] 341
34. [6.8] $s=\dfrac{Rg}{g-R}$ **35.** [6.8] $m=\dfrac{H}{S(t_1-t_2)}$
36. [6.8] $c=\dfrac{b+3a}{2}$ **37.** [6.8] $t_1=\dfrac{-A}{vT}+t_2$, or $\dfrac{-A+vTt_2}{vT}$ **38.** [6.8] About 202.3 lb **39.** [6.8] 64 L
40. [6.8] $y=\dfrac{\frac{3}{4}}{x}$
41. [6.2], [6.3], [6.4] The least common denominator was used to add and subtract rational expressions, to simplify complex rational expressions, and to solve rational equations. **42.** [6.1], [6.4] A rational *expression* is a quotient of two polynomials. Expressions can be simplified, multiplied, or added, but they cannot be solved for a variable. A rational *equation* is an equation containing rational expressions. In a rational equation, we often can solve for a variable.
43. [6.4] All real numbers except 0 and 13
44. [6.3], [6.4] 45 **45.** [6.5] $9\frac{9}{19}$ sec

Test: Chapter 6, p. 404

1. [6.1] $\dfrac{3}{4(t+1)}$ **2.** [6.1] $\dfrac{x^2-3x+9}{x+4}$
3. [6.2] $(x-3)(x+11)(x-9)$ **4.** [6.2] $\dfrac{25x+x^3}{x+5}$
5. [6.2] $3(a-b)$ **6.** [6.2] $\dfrac{a^3-a^2b+4ab+ab^2-b^3}{(a-b)(a+b)}$
7. [6.2] $\dfrac{-2(2x^2+5x+20)}{(x-4)(x+4)(x^2+4x+16)}$
8. [6.2] $\dfrac{y-4}{(y+3)(y-2)}$ **9.** [6.3] $\dfrac{a(2b+3a)}{5a+b}$
10. [6.3] $\dfrac{(x-9)(x-6)}{(x+6)(x-3)}$ **11.** [6.3] $\dfrac{4x^2-14x+2}{3x^2+7x-11}$
12. [6.4] $-\frac{21}{4}$ **13.** [6.4] 15 **14.** [6.1] 5; 0 **15.** [6.4] $\frac{5}{3}$
16. [6.5] $1\frac{31}{32}$ hr **17.** [6.6] $\dfrac{4b^2c}{a}-\dfrac{5bc^2}{2a}+3bc$
18. [6.6] $y-14+\dfrac{-20}{y-6}$ **19.** [6.6] $6x^2-9+\dfrac{5x+22}{x^2+2}$
20. [6.7] $x^2+9x+40+\dfrac{153}{x-4}$ **21.** [6.7] 449
22. [6.8] $b_1=\dfrac{2A}{h}-b_2$, or $\dfrac{2A-b_2h}{h}$ **23.** [6.5] 5 and 6; -6 and -5 **24.** [6.5] $3\frac{3}{11}$ mph **25.** [6.8] 30 workers
26. [6.8] 637 in^2 **27.** [6.4] $-\frac{19}{3}$
28. [6.4] $\{x \mid x$ is a real number *and* $x \neq 0$ *and* $x \neq 15\}$
29. [6.3], [6.4] x-intercept: $(11, 0)$; y-intercept: $\left(0, -\frac{33}{5}\right)$ **30.** [6.5] Hans: 56 lawns; Franz: 42 lawns

Cumulative Review: Chapters 1–6, pp. 405–406

1. [1.1], [1.2] 10 **2.** [1.7] 5.76×10^9
3. [2.3] Slope: $\frac{7}{4}$; y-intercept: $(0, -3)$
4. [2.5] $y=-\frac{10}{3}x+\frac{11}{3}$ **5.** [3.2] $(-3, 4)$
6. [3.4] $(-2, -3, 1)$ **7.** [3.3] Small: 16; large: 29
8. [3.5] $12, \frac{1}{2}, 7\frac{1}{2}$ **9.** [2.3] \$11.8 million per year
10. [2.5] **(a)** $V(t)=\frac{5}{18}t+15.4$; **(b)** about 19.8 min
11. **(a)** [6.1] $-\frac{1}{2}$; **(b)** [2.2] $\{x \mid x$ is a real number *and* $x \neq 5\}$
12. [5.8] $\frac{1}{4}$ **13.** [5.8] $-\frac{25}{7}, \frac{25}{7}$ **14.** [4.1] $\{x \mid x > -3\}$, or $(-3, \infty)$ **15.** [4.1] $\{x \mid x \geq -1\}$, or $[-1, \infty)$
16. [4.2] $\{x \mid -10 < x < 13\}$, or $(-10, 13)$
17. [4.2] $\left\{x \mid x < -\frac{4}{3}\ or\ x > 6\right\}$, or $\left(-\infty, -\frac{4}{3}\right) \cup (6, \infty)$
18. [4.3] $\{x \mid x < -6.4\ or\ x > 6.4\}$, or $(-\infty, -6.4) \cup (6.4, \infty)$
19. [4.3] $\left\{x \mid -\frac{13}{4} \leq x \leq \frac{15}{4}\right\}$, or $\left[-\frac{13}{4}, \frac{15}{4}\right]$ **20.** [6.4] $-\frac{5}{3}$
21. [6.4] -1 **22.** [6.4] No solution **23.** [6.4] $\frac{1}{3}$
24. [4.3] $1, \frac{7}{3}$ **25.** [4.2] $[7, \infty)$ **26.** [1.5] $n=\dfrac{m-12}{3}$
27. [6.8] $a=\dfrac{Pb}{3-P}$

28. [4.4]

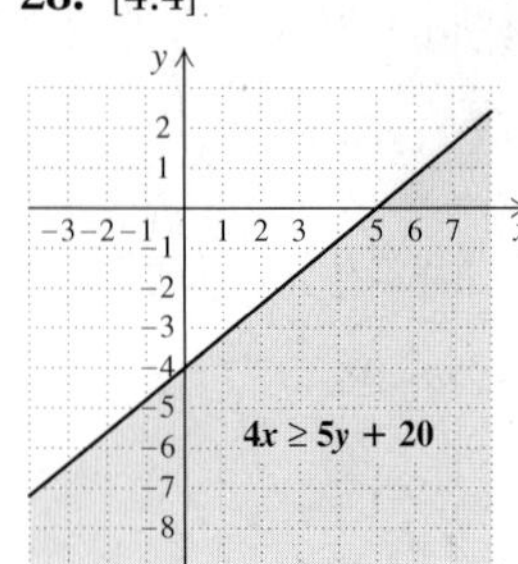

29. [2.3]

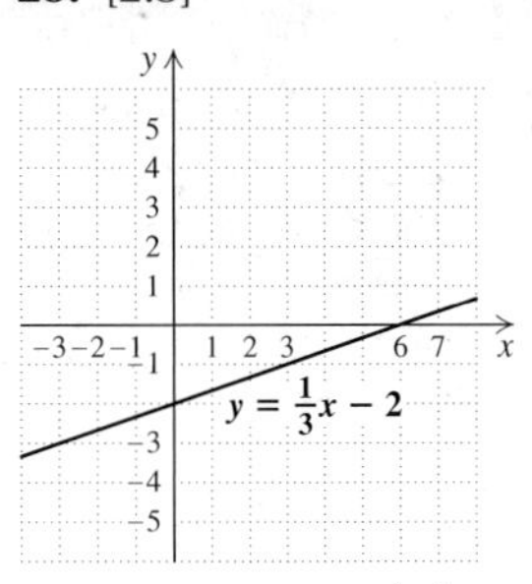

30. [5.1] $-3x^3 + 9x^2 + 3x - 3$ **31.** [5.2] $-15x^4y^4$
32. [5.1] $5a + 5b - 5c$
33. [5.2] $15x^4 - x^3 - 9x^2 + 5x - 2$
34. [5.2] $4x^4 - 4x^2y + y^2$ **35.** [5.2] $4x^4 - y^2$
36. [5.1] $-m^3n^2 - m^2n^2 - 5mn^3$ **37.** [6.1] $\frac{y-6}{2}$
38. [6.1] $x - 1$ **39.** [6.2] $\frac{a^2 + 7ab + b^2}{(a-b)(a+b)}$
40. [6.2] $\frac{-m^2 + 5m - 6}{(m+1)(m-5)}$ **41.** [6.2] $\frac{3y^2 - 2}{3y}$
42. [6.3] $\frac{y-x}{xy(x+y)}$ **43.** [6.6] $9x^2 - 13x + 26 + \frac{-50}{x+2}$
44. [5.3] $2x^2(2x + 9)$ **45.** [5.4] $(x - 6)(x + 14)$
46. [5.5] $(4y - 9)(4y + 9)$
47. [5.6] $8(2x + 1)(4x^2 - 2x + 1)$ **48.** [5.5] $(t - 8)^2$
49. [5.5] $x^2(x - 1)(x + 1)(x^2 + 1)$
50. [5.6] $(0.3b - 0.2c)(0.09b^2 + 0.06bc + 0.04c^2)$
51. [5.4] $(4x - 1)(5x + 3)$ **52.** [5.4] $(3x + 4)(x - 7)$
53. [5.3] $(x^2 - y)(x^3 + y)$
54. [2.6], [5.8] $\{x|x$ is a real number *and* $x \neq 2$ *and* $x \neq 5\}$
55. [6.5] $3\frac{1}{3}$ sec **56.** [5.8] 30 ft **57.** [5.8] All such sets of even integers satisfy this condition. **58.** [6.8] 25 ft
59. [5.2] $x^3 - 12x^2 + 48x - 64$ **60.** [5.8] $-3, 3, -5, 5$
61. [4.2], [4.3] $\{x|-3 \leq x \leq -1$ *or* $7 \leq x \leq 9\}$, or $[-3, -1] \cup [7, 9]$
62. [6.4] All real numbers except 9 and -5
63. [5.8] $-\frac{1}{4}, 0, \frac{1}{4}$

CHAPTER 7

Exercise Set 7.1, pp. 415–416

1. $4, -4$ **3.** $12, -12$ **5.** $9, -9$ **7.** $30, -30$ **9.** $-\frac{7}{6}$
11. 21 **13.** $-\frac{4}{9}$ **15.** 0.3 **17.** -0.07 **19.** $p^2 + 4$; 2
21. $\frac{x}{y+4}$; 3 **23.** $\sqrt{20}$; 0; does not exist; does not exist
25. -3; -1; does not exist; 0 **27.** 1; $\sqrt{2}$; $\sqrt{101}$
29. 1; does not exist; 6 **31.** $6|x|$ **33.** $6|b|$ **35.** $|7 - t|$
37. $|y + 8|$ **39.** $|3x - 5|$ **41.** -5 **43.** -3 **45.** $-\frac{1}{2}$
47. $|y|$ **49.** $7|b|$ **51.** 10 **53.** $|2a + b|$ **55.** $|x^5|$
57. $|a^7|$ **59.** $5t$ **61.** $7c$ **63.** $5 + b$ **65.** $3(x + 2)$, or $3x + 6$ **67.** $5t - 2$ **69.** -4 **71.** $3x$ **73.** 10
75. $4x$ **77.** a^7 **79.** $(x + 3)^5$ **81.** 2; 3; -2; -4
83. 2; does not exist; does not exist; 3 **85.** $\{x|x \geq 5\}$, or $[5, \infty)$ **87.** $\{t|t \geq -3\}$, or $[-3, \infty)$ **89.** $\{x|x \leq 5\}$, or $(-\infty, 5]$ **91.** $\mathbb{R}$ **93.** $\{z|z \geq -\frac{3}{5}\}$, or $[-\frac{3}{5}, \infty)$ **95.** $\mathbb{R}$
97. **99.** [1.6] $a^9b^6c^{15}$ **100.** [1.6] $10a^{10}b^9$
101. [1.6] $\frac{a^6c^{12}}{8b^9}$ **102.** [1.6] $\frac{x^6y^2}{25z^4}$ **103.** [1.6] $2x^4y^5z^2$
104. [1.6] $\frac{5c^3}{a^4b^7}$ **105.**
107. **(a)** 13; **(b)** 15; **(c)** 18; **(d)** 20
109. $\{x|x \geq 0\}$, or $[0, \infty)$;

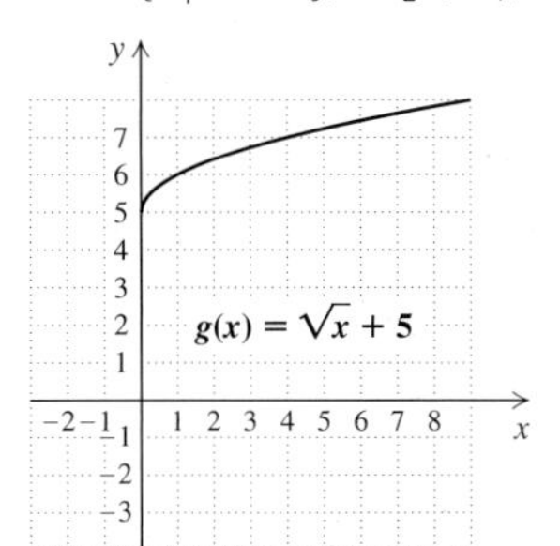

111. $\{x|x \geq 2\}$, or $[2, \infty)$;

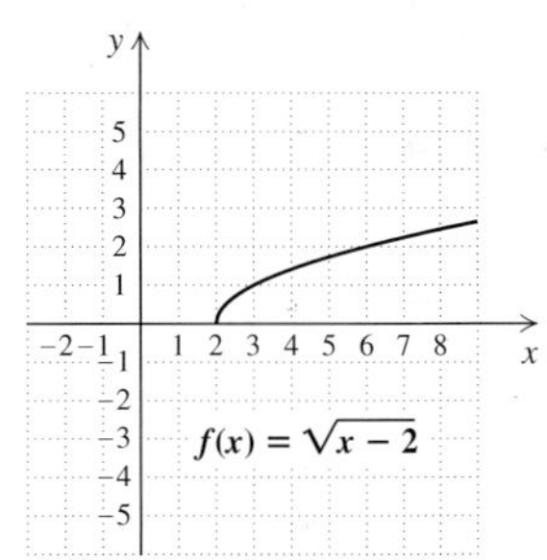

113. $\{x|-4 < x \leq 5\}$, or $(-4, 5]$ **115.**

Technology Connection, p. 418

1. Without parentheses, the expression entered would be $\frac{7^2}{3}$. **2.** For $x = 0$ or $x = 1$, $y_1 = y_2 = y_3$; or $(0, 1)$, $y_1 > y_2 > y_3$; on $(1, \infty)$, $y_1 < y_2 < y_3$.

Technology Connection, p. 420

1. Most graphers do not have keys for radicals of index 3 or higher. On those graphers that offer $\sqrt[x]{\ }$ in a MATH menu, rational exponents still require fewer keystrokes.

Exercise Set 7.2, pp. 420–422

1. $\sqrt[4]{x}$ **3.** 4 **5.** 3 **7.** 3 **9.** $\sqrt[3]{xyz}$ **11.** $\sqrt[5]{a^2b^2}$
13. $\sqrt[3]{a^2}$ **15.** 8 **17.** 343 **19.** 243 **21.** $\sqrt[4]{81^3x^3}$, or $27\sqrt[4]{x^3}$ **23.** $125x^6$ **25.** $20^{1/3}$ **27.** $17^{1/2}$ **29.** $x^{3/2}$
31. $m^{2/5}$ **33.** $(cd)^{1/4}$ **35.** $(xy^2z)^{1/5}$ **37.** $(3mn)^{3/2}$
39. $(8x^2y)^{5/7}$ **41.** $\frac{2x}{z^{2/3}}$ **43.** $\frac{1}{x^{1/3}}$ **45.** $\frac{1}{(2rs)^{3/4}}$ **47.** 4
49. $a^{5/7}$ **51.** $\frac{2a^{3/4}c^{2/3}}{b^{1/2}}$ **53.** $\frac{x^4}{2^{1/3}y^{2/7}}$ **55.** $\left(\frac{8yz}{7x}\right)^{3/5}$
57. $\frac{7x}{z^{1/3}}$ **59.** $\frac{5ac^{1/2}}{3}$ **61.** $5^{7/8}$ **63.** $3^{3/4}$ **65.** $4.1^{1/2}$
67. $10^{6/25}$ **69.** $a^{23/12}$ **71.** 64 **73.** $\frac{m^{1/3}}{n^{1/8}}$ **75.** $\sqrt[3]{a}$
77. x^5 **79.** x^3 **81.** a^5b^5 **83.** $\sqrt[4]{3x}$ **85.** $\sqrt{3a}$

87. $\sqrt[8]{x}$ **89.** a^3b^3 **91.** x^8y^{20} **93.** $\sqrt[12]{xy}$ **95.**
97. [5.2] $11x^4 + 14x^3$ **98.** [5.2] $-3t^6 + 28t^5 - 20t^4$
99. [5.2] $15a^2 - 11ab - 12b^2$
100. [5.2] $49x^2 - 14xy + y^2$ **101.** [1.4] \$93,500
102. [5.8] 0, 1 **103.** **105.** $\sqrt[10]{x^5y^3}$ **107.** $\sqrt[4]{2xy^2}$
109. 524 cycles per second **111.** $2^{7/12} \approx 1.498 \approx 1.5$
113. **(a)** 1.8 m; **(b)** 3.1 m; **(c)** 1.5 m; **(d)** 5.3 m
115. 10 mg

Technology Connection, p. 424

1. The graph of $y = 0$ (the x-axis)

Exercise Set 7.3, pp. 427–429

1. $\sqrt{70}$ **3.** $\sqrt[3]{10}$ **5.** $\sqrt[4]{72}$ **7.** $\sqrt{30ab}$ **9.** $\sqrt[5]{18t^3}$
11. $\sqrt{x^2 - a^2}$ **13.** $\sqrt[3]{0.1x^2}$ **15.** $\sqrt[4]{x^3 - 1}$ **17.** $\sqrt{\dfrac{7x}{6y}}$
19. $\sqrt[7]{\dfrac{5x - 15}{4x + 8}}$ **21.** $5\sqrt{2}$ **23.** $2\sqrt{7}$ **25.** $2\sqrt{2}$
27. $3\sqrt{22}$ **29.** $6a^2\sqrt{b}$ **31.** $2x\sqrt[3]{y^2}$ **33.** $-2x^2\sqrt[3]{2}$
35. $f(x) = 5x\sqrt[3]{x^2}$ **37.** $f(x) = |7(x - 3)|$, or $7|x - 3|$
39. $f(x) = |x - 1|\sqrt{5}$ **41.** $ab^2\sqrt{a}$ **43.** $xy^2z^3\sqrt[3]{x^2z}$
45. $-2ab^2\sqrt[5]{a^2b}$ **47.** $abc\sqrt[5]{ab^3c^4}$ **49.** $3x^2\sqrt[4]{10x}$
51. $5\sqrt{3}$ **53.** $2\sqrt{35}$ **55.** 2 **57.** $18a^3$ **59.** $a\sqrt[3]{10}$
61. $3x^3\sqrt{5x}$ **63.** $s^2t^3\sqrt[3]{t}$ **65.** $(x + 5)^2$
67. $2ab^3\sqrt[4]{3a}$ **69.** $x(y + z)^2\sqrt[5]{x}$ **71.**
73. [6.2] $\dfrac{12x^2 + 5y^2}{64xy}$ **74.** [6.2] $\dfrac{2a + 6b^3}{a^4b^4}$
75. [6.2] $\dfrac{-7x - 13}{2(x - 3)(x + 3)}$ **76.** [6.2] $\dfrac{-3x + 1}{2(x - 5)(x + 5)}$
77. [1.6] $3a^2b^2$ **78.** [1.6] $3ab^5$ **79.**
81. **(a)** 20 mph; **(b)** 37.4 mph; **(c)** 42.4 mph
83. $r^{10}t^3\sqrt{rt}$ **85.** $a^3b^6\sqrt[3]{ab^2}$
87. $f(x) = h(x);\ f(x) \neq g(x)$

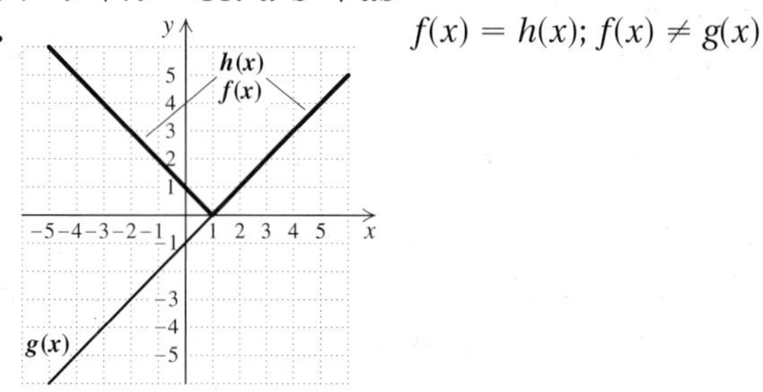

89. $\{x \mid x \leq -1 \text{ or } x \geq 4\}$, or $(-\infty, -1] \cup [4, \infty)$ **91.** 10
93.

Exercise Set 7.4, pp. 434–435

1. $\dfrac{5}{6}$ **3.** $\dfrac{4}{3}$ **5.** $\dfrac{7}{y}$ **7.** $\dfrac{5y\sqrt{y}}{x^2}$ **9.** $\dfrac{3a\sqrt[3]{a}}{2b}$ **11.** $\dfrac{2a}{bc^2}$
13. $\dfrac{ab^2}{c^2}\sqrt[4]{\dfrac{a}{c^2}}$ **15.** $\dfrac{2x}{y^2}\sqrt[5]{\dfrac{x}{y}}$ **17.** $\dfrac{xy}{z^2}\sqrt[6]{\dfrac{y^2}{z^3}}$ **19.** $\sqrt{5}$
21. 3 **23.** $y\sqrt{5y}$ **25.** $2\sqrt[3]{a^2b}$ **27.** $\sqrt{2ab}$
29. $2x^2y^3\sqrt[4]{y^3}$ **31.** $\sqrt[3]{x^2 + xy + y^2}$ **33.** $\dfrac{\sqrt{35}}{7}$
35. $\dfrac{2\sqrt{15}}{5}$ **37.** $\dfrac{2\sqrt[3]{6}}{3}$ **39.** $\dfrac{\sqrt[3]{75ac^2}}{5c}$ **41.** $\dfrac{y\sqrt[3]{180x^2y}}{6x^2}$
43. $\dfrac{\sqrt[3]{2xy^2}}{xy}$ **45.** $\dfrac{\sqrt{14a}}{6}$ **47.** $\dfrac{3\sqrt{5y}}{10xy}$ **49.** $\dfrac{\sqrt{5b}}{6a}$
51. $\dfrac{5}{\sqrt{35x}}$ **53.** $\dfrac{2}{\sqrt{6}}$ **55.** $\dfrac{52}{3\sqrt{91}}$ **57.** $\dfrac{7}{\sqrt[3]{98}}$
59. $\dfrac{7x}{\sqrt{21xy}}$ **61.** $\dfrac{2a^2}{\sqrt[3]{20ab}}$ **63.** $\dfrac{x^2y}{\sqrt{2xy}}$ **65.**
67. [6.1] $\dfrac{3(x - 1)}{(x - 5)(x + 5)}$ **68.** [6.1] $\dfrac{7(x - 2)}{(x + 4)(x - 4)}$
69. [6.1] $\dfrac{a - 1}{a + 7}$ **70.** [6.1] $\dfrac{t + 11}{t + 2}$ **71.** [1.6] $125a^9b^{12}$
72. [1.6] $225x^{10}y^6$ **73.** **75.** **(a)** 1.62 sec; **(b)** 1.99 sec; **(c)** 2.20 sec **77.** $9\sqrt[3]{9n^2}$ **79.** $\dfrac{-3\sqrt{a^2 - 3}}{a^2 - 3}$, or $\dfrac{-3}{\sqrt{a^2 - 3}}$
81. Step 1: $\sqrt[n]{a} = a^{1/n}$, by definition; Step 2: $\left(\dfrac{a}{b}\right)^n = \dfrac{a^n}{b^n}$, raising a quotient to a power; Step 3: $a^{1/n} = \sqrt[n]{a}$, by definition **83.** $(f/g)(x) = 3x$, where x is a real number and $x > 0$ **85.** $(f/g)(x) = \sqrt{x + 3}$, where x is a real number and $x > 3$

Exercise Set 7.5, pp. 440–442

1. $5\sqrt{7}$ **3.** $3\sqrt[3]{5}$ **5.** $13\sqrt[3]{y}$ **7.** $7\sqrt{2}$
9. $13\sqrt[3]{7} + \sqrt{3}$ **11.** $21\sqrt{3}$ **13.** $23\sqrt{5}$ **15.** $9\sqrt[3]{2}$
17. $(1 + 6a)\sqrt{5a}$ **19.** $(x + 2)\sqrt[3]{6x}$ **21.** $3\sqrt{a - 1}$
23. $(x + 3)\sqrt{x - 1}$ **25.** $3\sqrt{7} - 7$ **27.** $4\sqrt{6} - 4\sqrt{10}$
29. $2\sqrt{15} - 6\sqrt{3}$ **31.** -6 **33.** $a + 2a\sqrt{3}$ **35.** 19
37. -19 **39.** $14 + 6\sqrt{5}$ **41.** -6
43. $2\sqrt[3]{9} + 3\sqrt[3]{6} - 2\sqrt[3]{4}$ **45.** $3x + 2\sqrt{3xy} + y$
47. $\dfrac{3 - \sqrt{5}}{2}$ **49.** $\dfrac{12 + 2\sqrt{3} + 6\sqrt{5} + \sqrt{15}}{33}$
51. $\dfrac{a - \sqrt{ab}}{a - b}$ **53.** -1
55. $\dfrac{24 - 3\sqrt{10} - 4\sqrt{14} + \sqrt{35}}{27}$ **57.** $\dfrac{3\sqrt{6} + 4}{2}$
59. $\dfrac{3}{5\sqrt{7} - 10}$ **61.** $\dfrac{2}{14 + 2\sqrt{3} + 3\sqrt{2} + 7\sqrt{6}}$
63. $\dfrac{x - y}{x + 2\sqrt{xy} + y}$ **65.** $a\sqrt[4]{a}$ **67.** $b\sqrt[10]{b^9}$
69. $xy\sqrt[6]{xy^5}$ **71.** $3a^2b\sqrt[4]{ab}$ **73.** $xyz\sqrt[6]{x^5yz^2}$
75. $\sqrt[15]{x^7}$ **77.** $\sqrt[15]{\dfrac{a^7}{b^2}}$ **79.** $\sqrt[10]{xy^3}$ **81.** $\sqrt[12]{(2 + 5x)^5}$
83. $\sqrt[12]{5 + 3x}$ **85.** $x\sqrt[6]{xy^5} - \sqrt[15]{x^{13}y^{14}}$
87. $2m^2 + m\sqrt[4]{n} + 2m\sqrt[3]{n^2} + \sqrt[12]{n^{11}}$ **89.** $\sqrt[4]{2x^2} - x^3$
91. $x^2 - 7$ **93.** $27 - 10\sqrt{2}$ **95.** $8 + 2\sqrt{15}$ **97.**
99. [6.4] 8 **100.** [6.4] $\frac{15}{2}$ **101.** [5.8] Length: 20 units;

width: 5 units **102.** [5.8] $-5, 4$ **103.** [5.8] $\frac{1}{5}, 1$
104. [5.8] $\frac{1}{7}, 1$ **105.** **107.** Radicands; indices
109. Denominators **111.** $f(x) = -6x\sqrt{5+x}$
113. $f(x) = (x + 3x^2)\sqrt[4]{x-1}$
115. $ac^2\left[(3a + 2c)\sqrt{ab} - 2\sqrt[3]{ab}\right]$
117. $9a^2(b+1)\sqrt[6]{243a^5(b+1)^5}$ **119.** $1 - \sqrt{w}$
121. $(\sqrt{x} + \sqrt{5})(\sqrt{x} - \sqrt{5})$
123. $(\sqrt{x} + \sqrt{a})(\sqrt{x} - \sqrt{a})$ **125.** $2x - 2\sqrt{x^2 - 4}$
127. $\frac{b^2 + \sqrt{b}}{1 + b + b^2}$ **129.** $\frac{x - 19}{x + 10\sqrt{x+6} + 31}$

Technology Connection, p. 445

1. The x-coordinates of the points of intersection should approximate the solutions of the examples.

Exercise Set 7.6, pp. 448–449

1. 22 **3.** $\frac{9}{2}$ **5.** 11 **7.** -3 **9.** 29 **11.** 13
13. 0, 64 **15.** No solution **17.** -64 **19.** 81
21. No solution **23.** $\frac{80}{3}$ **25.** 57 **27.** $-\frac{5}{3}$ **29.** 1
31. $\frac{106}{27}$ **33.** 4 **35.** 3, 7 **37.** $\frac{80}{9}$ **39.** -1
41. No solution **43.** 2, 6 **45.** 2 **47.** 4 **49.**
51. [1.4] Height: 7 in.; base: 9 in. **52.** [3.3] 8 60-sec commercials **53.** [6.5] Elaine: 6 hr; Gonzalo: 12 hr
54. [6.5] $\frac{165}{52}$ hr, or about 3.2 hr
55. [4.4] **56.** [2.3]

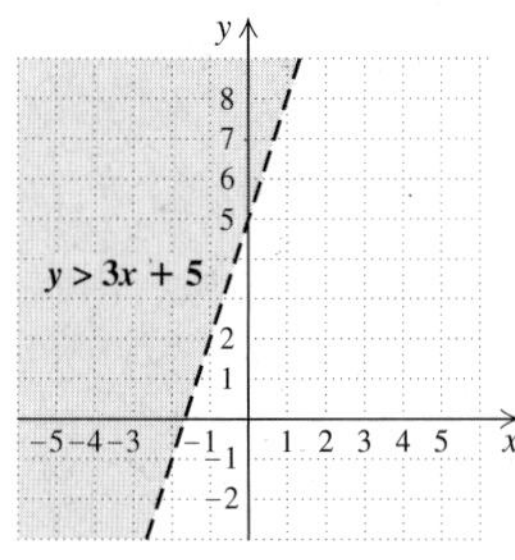

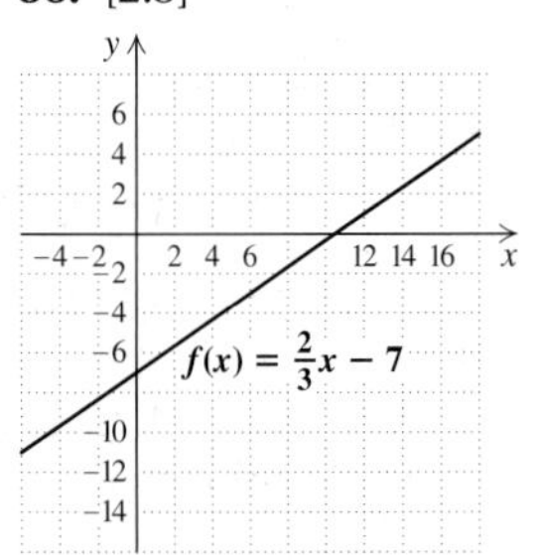

57. **59.** 230°C **61.** $t = \frac{1}{9}\left(\frac{S^2 \cdot 2457}{1087.7^2} - 2617\right)$
63. 4480 rpm **65.** $r = \frac{v^2h}{2gh - v^2}$ **67.** 72.25 ft
69. $\frac{2504}{125}, \frac{2496}{125}$ **71.** 0 **73.** 1, 8 **75.** $\left(\frac{1}{36}, 0\right), (36, 0)$
77. **79.**

Exercise Set 7.7, pp. 455–458

1. $\sqrt{34}$; 5.831 **3.** $9\sqrt{2}$; 12.728 **5.** 5 **7.** $\sqrt{31}$; 5.568
9. $\sqrt{11}$; 3.317 **11.** 1 **13.** $\sqrt{325}$ ft; 18.028 ft
15. $\sqrt{14{,}500}$ ft; 120.416 ft **17.** 20 in. **19.** $\sqrt{13{,}000}$ m; 114.018 m **21.** $a = 5$; $c = 5\sqrt{2} \approx 7.071$ **23.** $a = 7$; $b = 7\sqrt{3} \approx 12.124$ **25.** $a = 5\sqrt{3} \approx 8.660$; $c = 10\sqrt{3} \approx 17.321$ **27.** $a = \frac{13\sqrt{2}}{2} \approx 9.192$; $b = \frac{13\sqrt{2}}{2} \approx 9.192$ **29.** $b = 14\sqrt{3} \approx 24.249$; $c = 28$
31. $3\sqrt{3} \approx 5.196$ **33.** $13\sqrt{2} \approx 18.385$
35. $\frac{19\sqrt{2}}{2} \approx 13.435$ **37.** $\sqrt{10561}$ ft ≈ 102.767 ft
39. $h = 2\sqrt{3}$ ft ≈ 3.464 ft **41.** 8 ft **43.** $(0, -4), (0, 4)$
45. **47.** [1.6] -47 **48.** [1.6] 5
49. [5.3] $x(x-3)(x+3)$ **50.** [5.3] $7a(a-2)(a+2)$
51. [4.3] $\left\{-\frac{2}{3}, 4\right\}$ **52.** [4.3] $\left\{-\frac{4}{3}, 10\right\}$ **53.**
55. $\sqrt{75}$ cm **57.** 4 gal; the total area of the doors and windows is 164 ft^2 or more **59.** 36.15 ft by 36.15 ft by 36.15 ft

Exercise Set 7.8, pp. 464–465

1. $5i$ **3.** $i\sqrt{13}$, or $\sqrt{13}i$ **5.** $3i\sqrt{2}$, or $3\sqrt{2}i$
7. $i\sqrt{3}$, or $\sqrt{3}i$ **9.** $9i$ **11.** $10i\sqrt{3}$, or $10\sqrt{3}i$ **13.** $-7i$
15. $4 - 2\sqrt{15}i$ **17.** $(2 + 2\sqrt{3})i$ **19.** $(6\sqrt{2} - 5)i$
21. $12 + 11i$ **23.** $4 + 5i$ **25.** $-4 - 5i$ **27.** $5 + 5i$
29. -54 **31.** 56 **33.** -35 **35.** $-\sqrt{42}$ **37.** $-5\sqrt{6}$
39. $-6 + 14i$ **41.** $-20 - 24i$ **43.** $-11 + 23i$
45. $40 + 13i$ **47.** $8 + 31i$ **49.** $7 - 61i$
51. $-15 - 30i$ **53.** $39 - 37i$ **55.** $-3 - 4i$
57. $5 + 12i$ **59.** $21 + 20i$ **61.** $\frac{6}{5} + \frac{3}{5}i$ **63.** $\frac{6}{29} + \frac{15}{29}i$
65. $-\frac{7}{9}i$ **67.** $-\frac{3}{4} - \frac{5}{4}i$ **69.** $1 - 2i$ **71.** $-\frac{23}{58} + \frac{43}{58}i$
73. $\frac{6}{25} - \frac{17}{25}i$ **75.** $-i$ **77.** 1 **79.** -1 **81.** i
83. -1 **85.** $-125i$ **87.** 0 **89.** **91.** [2.6] 1
92. [2.6] 1 **93.** [2.6] 50 **94.** [2.6] 0 **95.** [5.8] $-\frac{4}{3}, 7$
96. [4.3] $\left\{x \mid -\frac{29}{3} < x < 5\right\}$, or $\left(-\frac{29}{3}, 5\right)$ **97.**
99. $-9 - 27i$ **101.** $50 - 120i$ **103.** $\frac{250}{41} + \frac{200}{41}i$
105. 8 **107.** $\frac{3}{5} + \frac{9}{5}i$ **109.** 1

Review Exercises: Chapter 7, pp. 468–469

1. [7.1] $\frac{7}{6}$ **2.** [7.1] -0.5 **3.** [7.1] 5
4. [7.1] $\left\{x \mid x \geq \frac{7}{2}\right\}$, or $\left[\frac{7}{2}, \infty\right)$ **5.** [7.1] $7|a|$ **6.** [7.1] $|c + 8|$
7. [7.1] $|x - 3|$ **8.** [7.1] $|2x + 1|$ **9.** [7.1] -2
10. [7.4] $-\frac{4x^2}{3}$ **11.** [7.3] $|x^3y^2|$, or $|x^3|y^2$ **12.** [7.3] $2x^2$
13. [7.2] $(5ab)^{4/3}$ **14.** [7.2] $8a^4\sqrt{a}$ **15.** [7.2] x^3y^5
16. [7.2] $\sqrt[3]{x^2y}$ **17.** [7.2] $\frac{1}{x^{2/5}}$ **18.** [7.2] $7^{1/6}$
19. [7.3] $f(x) = 5|x - 3|$ **20.** [7.3] $\sqrt{15xy}$
21. [7.3] $3a\sqrt[3]{a^2b^2}$ **22.** [7.3] $-6x^5y^4\sqrt[3]{2x^2}$
23. [7.4] $y\sqrt[3]{6}$ **24.** [7.4] $\frac{5\sqrt{x}}{2}$ **25.** [7.4] $\frac{2a^2\sqrt[4]{3a^3}}{c^2}$
26. [7.5] $7\sqrt[3]{x}$ **27.** [7.5] $3\sqrt{3}$ **28.** [7.5] $(2x + y^2)\sqrt[3]{x}$
29. [7.5] $15\sqrt{2}$ **30.** [7.5] $-43 - 2\sqrt{10}$ **31.** [7.5] $\sqrt[4]{x^3}$
32. [7.5] $\sqrt[12]{x^5}$ **33.** [7.5] $a^2 - 2a\sqrt{2} + 2$
34. [7.5] $-2\sqrt{6} + 6$ **35.** [7.5] $\frac{6}{3 + \sqrt{6}}$ **36.** [7.6] 21
37. [7.6] -126

38. [7.6] 4 **39.** [7.6] 14 **40.** [7.7] $5\sqrt{2}$ cm; 7.071 cm
41. [7.7] $\sqrt{24}$ ft; 4.899 ft **42.** [7.7] $a = 10$; $b = 10\sqrt{3} \approx 17.321$ **43.** [7.8] $-2i\sqrt{2}$, or $-2\sqrt{2}i$
44. [7.8] $-2 - 9i$ **45.** [7.8] $1 + i$ **46.** [7.8] 29
47. [7.8] i **48.** [7.8] $9 - 12i$ **49.** [7.8] $\frac{13}{25} - \frac{34}{25}i$
50. [7.1] An absolute-value sign must be used to simplify $\sqrt[n]{x^n}$ when n is even, since x may be negative. If x is negative while n is even, the radical expression cannot be simplified to x, since $\sqrt[n]{x^n}$ represents the principal, or positive, root. When n is odd, there is only one root, and it will be positive or negative depending on the sign of x. Thus there is no absolute-value sign when n is odd.
51. [7.8] Every real number is a complex number, but there are complex numbers that are not real. A complex number $a + bi$ is not real if $b \neq 0$. **52.** [7.6] 3
53. [7.8] $-\frac{2}{5} + \frac{9}{10}i$

Test: Chapter 7, p. 469

1. [7.3] $5\sqrt{3}$ **2.** [7.4] $-\frac{2}{x^2}$ **3.** [7.1] $10|a|$
4. [7.1] $|x - 4|$ **5.** [7.3] $x^2y\sqrt[5]{x^2y^3}$ **6.** [7.4] $\left|\frac{5x}{6y^2}\right|$, or $\frac{5|x|}{6y^2}$
7. [7.3] $\sqrt[3]{10xy^2}$ **8.** [7.4] $\sqrt[5]{x^2y^2}$ **9.** [7.5] $xy\sqrt[4]{x}$
10. [7.5] $\sqrt[20]{a^3}$ **11.** [7.5] $5\sqrt{2}$ **12.** [7.5] $(x^2 + 3y)\sqrt{y}$
13. [7.5] $14 - 19\sqrt{x} - 3x$ **14.** [7.2] $\sqrt[6]{(2a^3b)^5}$
15. [7.2] $(7xy)^{1/2}$ **16.** [7.1] $\{x | x \le 2\}$, or $(-\infty, 2]$
17. [7.5] $27 + 10\sqrt{2}$ **18.** [7.5] $\sqrt{6} - \sqrt{3}$ **19.** [7.6] 7
20. [7.6] No solution **21.** [7.7] Leg: 7 cm; hypotenuse: $7\sqrt{2}$ cm $\approx$ 9.899 cm **22.** [7.7] $\sqrt{10{,}600}$ ft $\approx$ 102.956 ft
23. [7.8] $5i\sqrt{2}$, or $5\sqrt{2}i$ **24.** [7.8] $10 + 2i$
25. [7.8] -24 **26.** [7.8] $15 - 8i$ **27.** [7.8] $-\frac{13}{53} - \frac{19}{53}i$
28. [7.8] i **29.** [7.6] 3 **30.** [7.8] $-\frac{17}{4}i$

CHAPTER 8

Technology Connection, p. 481

1. The right-hand x-intercept should be an approximation of $4 + \sqrt{23}$.
2. x-intercepts should be approximations of $-3 + \sqrt{5}$ and $-3 - \sqrt{5}$ for Example 8; approximations of $(-5 + \sqrt{37})/2$ and $(-5 - \sqrt{37})/2$ for Example 10(b)
3. A grapher can give only rational-number approximations of the two irrational solutions. An *exact* solution cannot be found with a grapher.
4. The graph of $y = 4x^2 + 9$ has no x-intercepts.

Exercise Set 8.1, pp. 481–483

1. $\pm\sqrt{3}$ **3.** $\pm\frac{2}{5}i$ **5.** $\pm\sqrt{\frac{2}{3}}$, or $\pm\frac{\sqrt{6}}{3}$ **7.** $-7, 3$
9. $-5 \pm 2\sqrt{2}$ **11.** $1 \pm 7i$ **13.** $\frac{-3 \pm \sqrt{14}}{2}$
15. $-7, 13$ **17.** $1, 9$ **19.** $-4 \pm \sqrt{13}$ **21.** $-14, 0$
23. $x^2 + 8x + 16, (x + 4)^2$ **25.** $x^2 - 6x + 9, (x - 3)^2$
27. $x^2 - 24x + 144, (x - 12)^2$ **29.** $t^2 + 9t + \frac{81}{4}, \left(t + \frac{9}{2}\right)^2$
31. $x^2 - 3x + \frac{9}{4}, \left(x - \frac{3}{2}\right)^2$ **33.** $x^2 + \frac{2}{3}x + \frac{1}{9}, \left(x + \frac{1}{3}\right)^2$
35. $t^2 - \frac{5}{3}t + \frac{25}{36}, \left(t - \frac{5}{6}\right)^2$ **37.** $x^2 + \frac{9}{5}x + \frac{81}{100}, \left(x + \frac{9}{10}\right)^2$
39. $-7, 1$ **41.** $5 \pm \sqrt{47}$ **43.** $-7, -1$ **45.** $3, 7$
47. $\frac{-5 \pm \sqrt{13}}{2}$ **49.** $3 \pm i$ **51.** $-2 \pm 3i$ **53.** $-\frac{1}{2}, 3$
55. $-\frac{3}{2}, -\frac{1}{2}$ **57.** $-\frac{3}{2}, \frac{5}{3}$ **59.** $\frac{-2 \pm \sqrt{2}}{2}$ **61.** $\frac{5 \pm \sqrt{61}}{6}$
63. 10% **65.** 18.75% **67.** 4% **69.** About 10.7 sec
71. About 6.3 sec **73.** **75.** [1.2] 28 **76.** [1.2] -92
77. [7.3] $3\sqrt[3]{10}$ **78.** [7.3] $4\sqrt{5}$ **79.** [7.1] 5
80. [7.1] 7 **81.** **83.** ± 18 **85.** $-\frac{7}{2}, -\sqrt{5}, 0, \sqrt{5}, 8$
87. Barge: 8 km/h; fishing boat: 15 km/h **89.**
91. ,

Exercise Set 8.2, pp. 488–489

1. $\frac{-7 \pm \sqrt{61}}{2}$ **3.** $3 \pm \sqrt{7}$ **5.** $-\frac{1}{2} \pm \frac{\sqrt{7}}{2}i$ **7.** $2 \pm 3i$
9. $3 \pm \sqrt{5}$ **11.** $\frac{-4 \pm \sqrt{19}}{3}$ **13.** $\frac{-1 \pm \sqrt{17}}{2}$
15. $\frac{-9 \pm \sqrt{129}}{24}$ **17.** $\frac{2}{5}$ **19.** $\frac{-11 \pm \sqrt{41}}{8}$ **21.** $5, 10$
23. $\frac{13 \pm \sqrt{509}}{10}$ **25.** $2 \pm \sqrt{5}i$ **27.** $2, -1 \pm \sqrt{3}i$
29. $\frac{5 \pm \sqrt{37}}{6}$ **31.** $5 \pm \sqrt{53}$ **33.** $\frac{7 \pm \sqrt{85}}{2}$ **35.** $\frac{3}{2}, 6$
37. $-5.31662479, 1.31662479$ **39.** $0.7639320225, 5.236067978$ **41.** $-1.265564437, 2.765564437$ **43.**
45. [3.3] Kenyan: 30 lb; Peruvian: 20 lb
46. [3.3] Cream-filled: 46; glazed: 44 **47.** [7.3] $9a^2b^3\sqrt{2a}$
48. [7.3] $4a^2b^3\sqrt{6}$ **49.** [6.3] $\frac{3(x + 1)}{3x + 1}$
50. [6.3] $\frac{4b}{3ab^2 - 4a^2}$ **51.** **53.** $(-2, 0), (1, 0)$
55. $4 - 2\sqrt{2}, 4 + 2\sqrt{2}$ **57.** $-1.1792101, 0.3392101$
59. $\frac{-5\sqrt{2} \pm \sqrt{34}}{4}$ **61.** $\frac{1}{2}$ **63.**

Exercise Set 8.3, pp. 494–497

1. First part: 12 km/h; second part: 8 km/h **3.** 35 mph
5. Cessna: 150 mph, Beechcraft: 200 mph; or Cessna: 200 mph, Beechcraft: 250 mph
7. To Hillsboro: 10 mph; return trip: 4 mph
9. About 11 mph **11.** 12 hr **13.** About 3.24 mph
15. $r = \frac{1}{2}\sqrt{\frac{A}{\pi}}$ **17.** $r = \frac{-\pi h + \sqrt{\pi^2h^2 + 2\pi A}}{2\pi}$

19. $s = \sqrt{\dfrac{kQ_1Q_2}{N}}$ **21.** $g = \dfrac{4\pi^2 l}{T^2}$

23. $c = \sqrt{d^2 - a^2 - b^2}$ **25.** $t = \dfrac{-v_0 + \sqrt{v_0^2 + 2gs}}{g}$

27. $n = \dfrac{1 + \sqrt{1 + 8N}}{2}$ **29.** $h = \dfrac{V^2}{12.25}$

31. $t = \dfrac{-b \pm \sqrt{b^2 - 4ac}}{2a}$ **33. (a)** 10.1 sec; **(b)** 7.49 sec; **(c)** 272.5 m **35.** 2.9 sec **37.** 0.87 sec **39.** 2.5 m/sec

41. 7% **43.** **45.** [1.2] -104 **46.** [7.8] $2i\sqrt{11}$

47. [6.1] $\dfrac{x + y}{2}$ **48.** [6.1] $\dfrac{a^2 - b^2}{b}$ **49.** [7.4] $\dfrac{1 + \sqrt{5}}{2}$

50. [7.4] $\dfrac{1 - \sqrt{7}}{5}$ **51.**

53. $t = \dfrac{-10.2 \pm 6\sqrt{-A^2 + 13A - 39.36}}{A - 6.5}$

55. $l = \dfrac{w + w\sqrt{5}}{2}$ **57.** $\pm\sqrt{2}$

59. $n = \pm\sqrt{\dfrac{r^2 \pm \sqrt{r^4 + 4m^4r^2p - 4mp}}{2m}}$

61. $A(S) = \dfrac{\pi S}{6}$

Exercise Set 8.4, pp. 501–502

1. Two irrational **3.** Two imaginary **5.** Two irrational **7.** One rational **9.** Two imaginary **11.** Two rational **13.** Two rational **15.** Two rational **17.** Two irrational **19.** Two imaginary **21.** Two irrational

23. $x^2 + 4x - 21 = 0$ **25.** $x^2 - 6x + 9 = 0$

27. $x^2 + 7x + 10 = 0$ **29.** $3x^2 - 14x + 8 = 0$

31. $6x^2 - 5x + 1 = 0$ **33.** $x^2 - 0.8x - 0.84 = 0$

35. $x^2 - 7 = 0$ **37.** $x^2 - 18 = 0$ **39.** $x^2 + 9 = 0$

41. $x^2 - 10x + 29 = 0$ **43.** $x^2 - 4x - 6 = 0$

45. $x^3 - 4x^2 - 7x + 10 = 0$ **47.** $x^3 - 2x^2 - 3x = 0$

49. **51.** [1.6] $81a^8$ **52.** [1.6] $16x^6$

53. [5.8] $(-1, 0), (8, 0)$ **54.** [5.8] $(2, 0), (4, 0)$

55. [3.3] 6 30-sec commercials

56. [2.3]

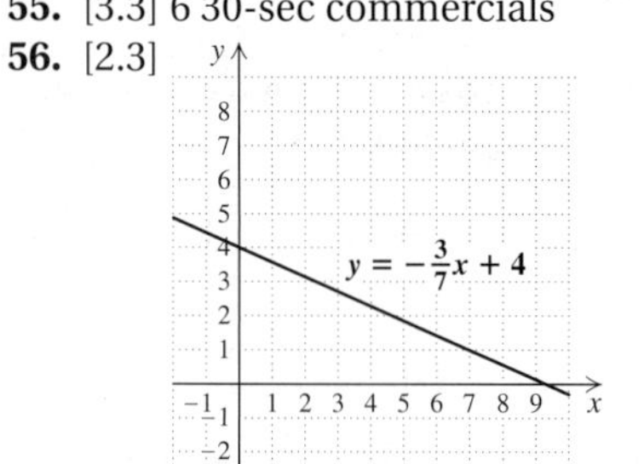

57.

59. $a = 1, b = 2, c = -3$ **61. (a)** $-\frac{3}{5}$; **(b)** $-\frac{1}{3}$

63. (a) $9 + 9i$; **(b)** $3 + 3i$ **65.** The solutions of $ax^2 + bx + c = 0$ are $x = \dfrac{-b \pm \sqrt{b^2 - 4ac}}{2a}$. When there is one solution, $b^2 - 4ac$ must be 0, so $x = \dfrac{-b \pm 0}{2a} = \dfrac{-b}{2a}$. **67.** $a = 8, b = 20, c = -12$

69. $x^4 - 8x^3 + 21x^2 - 2x - 52 = 0$ **71.**

Exercise Set 8.5, pp. 507–508

1. $\pm 1, \pm 3$ **3.** $\pm\sqrt{3}, \pm 3$ **5.** $\pm\dfrac{\sqrt{5}}{3}, \pm 1$ **7.** $9 + 4\sqrt{5}$

9. $\pm 2\sqrt{2}, \pm 3$ **11.** No solution **13.** $-\frac{1}{2}, \frac{1}{3}$ **15.** $-\frac{4}{5}, 1$

17. $-27, 8$ **19.** 729 **21.** 1 **23.** No solution **25.** $\frac{12}{5}$

27. $\left(\frac{4}{25}, 0\right)$ **29.** $\left(\dfrac{3 + \sqrt{33}}{2}, 0\right), \left(\dfrac{3 - \sqrt{33}}{2}, 0\right), (4, 0), (-1, 0)$ **31.** $(-243, 0), (32, 0)$ **33.** No x-intercepts

35.

37. [2.3]

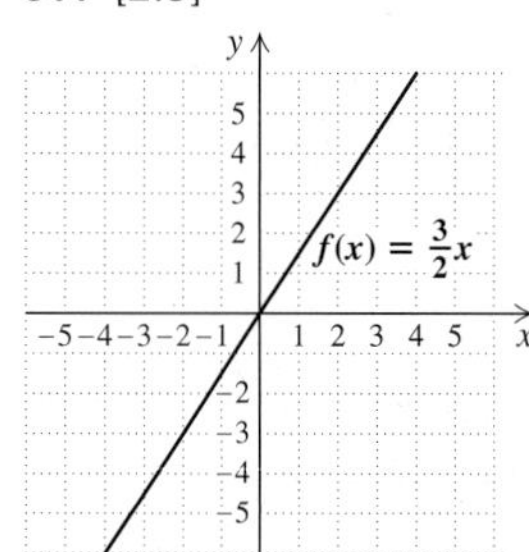

38. [2.3]

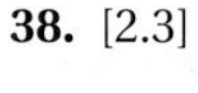

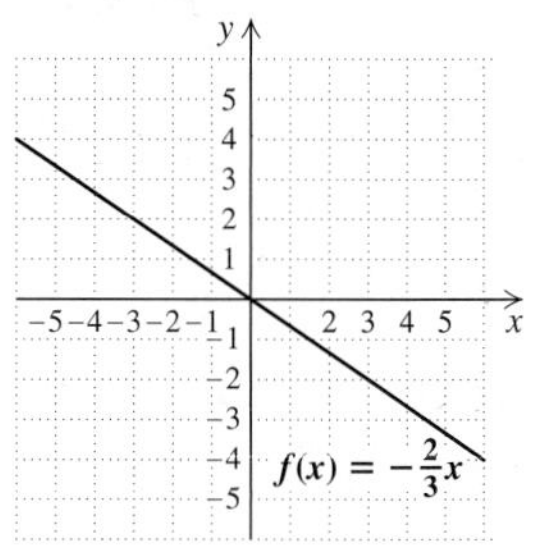

39. [2.1]

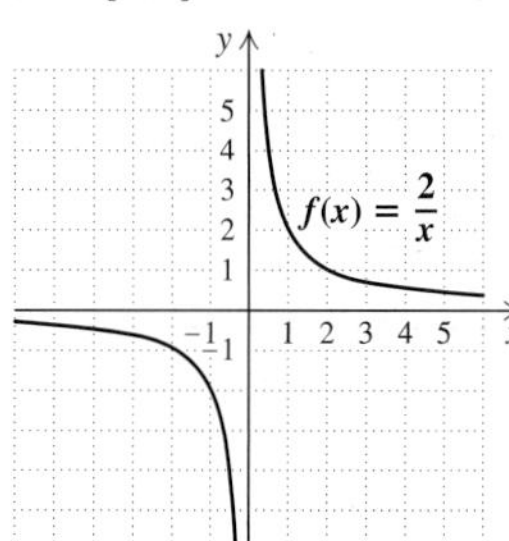

40. [2.1]

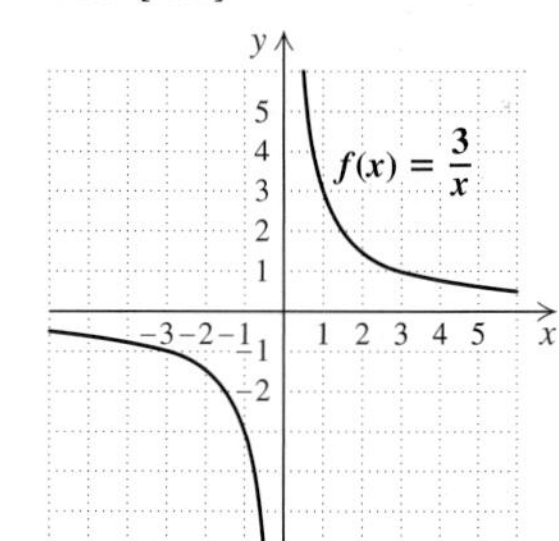

41. [3.3] A: 4 L; B: 8 L **42.** [5.2] $a^2 + a$ **43.**

45. $\pm\sqrt{\dfrac{7 \pm \sqrt{29}}{10}}$ **47.** $-2, -1, 5, 6$ **49.** $\frac{100}{99}$

51. $-5, -3, -2, 0, 2, 3, 5$ **53.** 1, 3 **55.**

57.

Exercise Set 8.6, pp. 515–516

1.

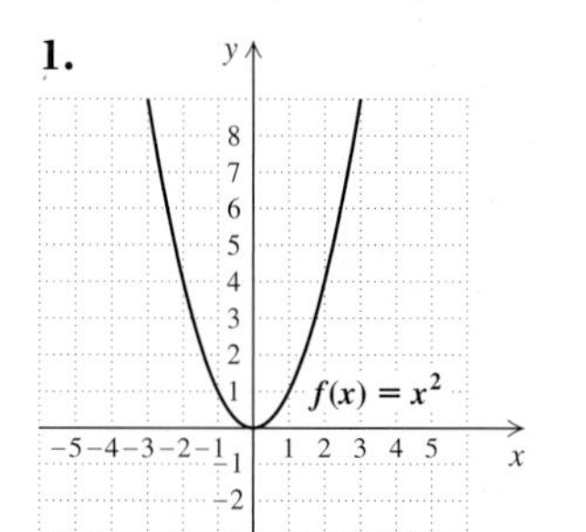

3.

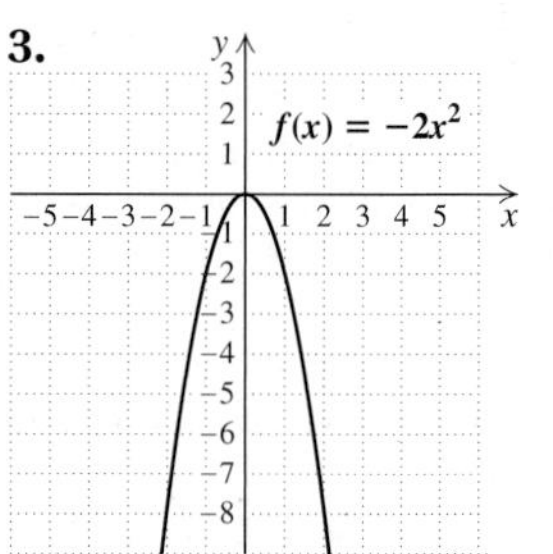

5.

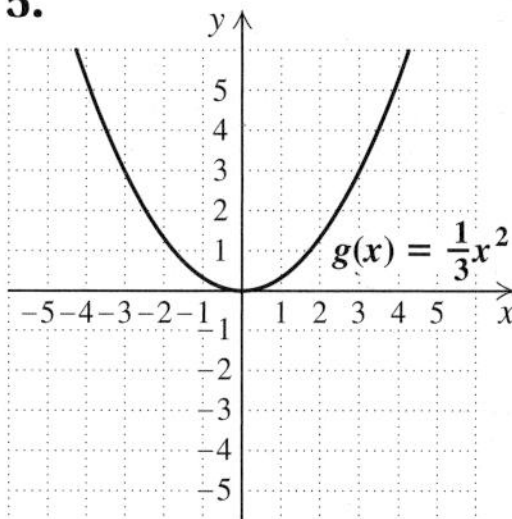

7.

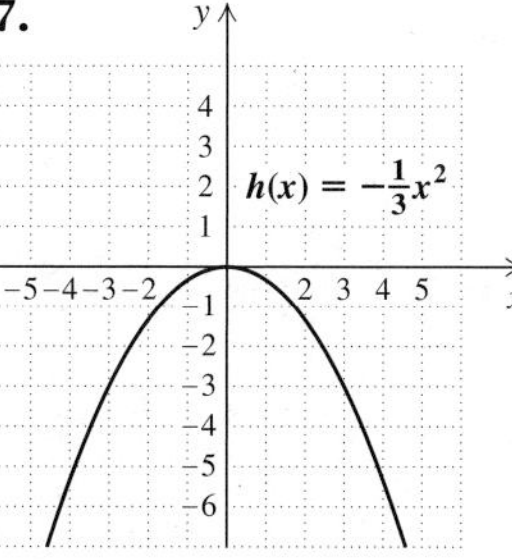

9.

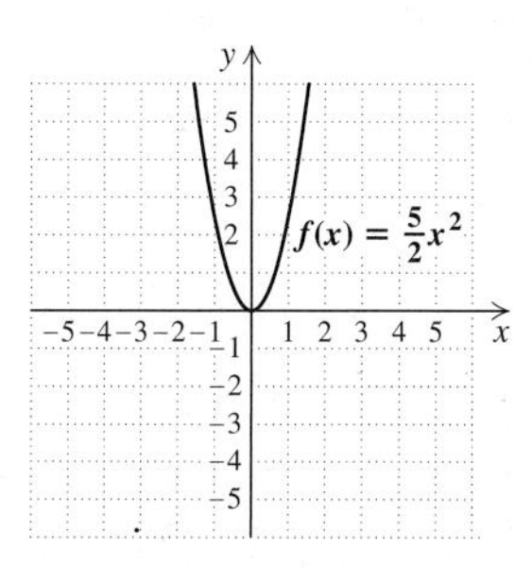

11. Vertex: $(-4, 0)$;
axis of symmetry: $x = -4$

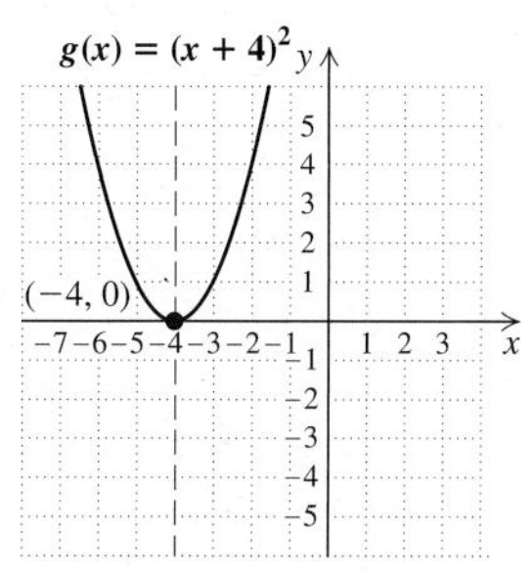

13. Vertex: $(1, 0)$;
axis of symmetry: $x = 1$

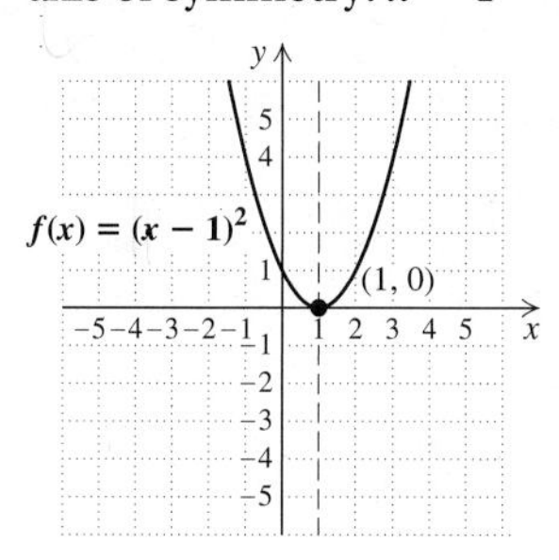

15. Vertex: $(3, 0)$;
axis of symmetry: $x = 3$

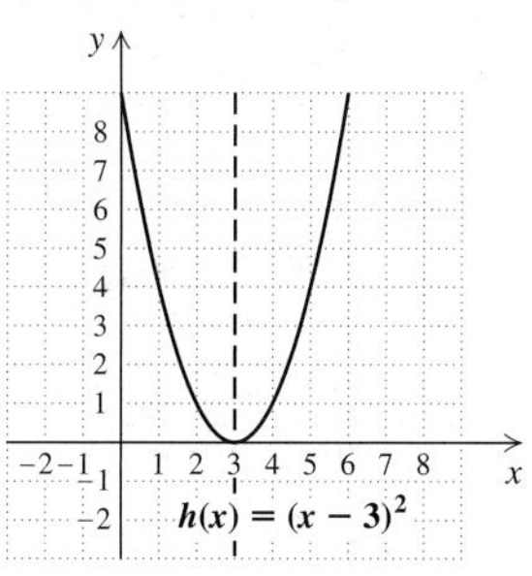

17. Vertex: $(-4, 0)$;
axis of symmetry: $x = -4$

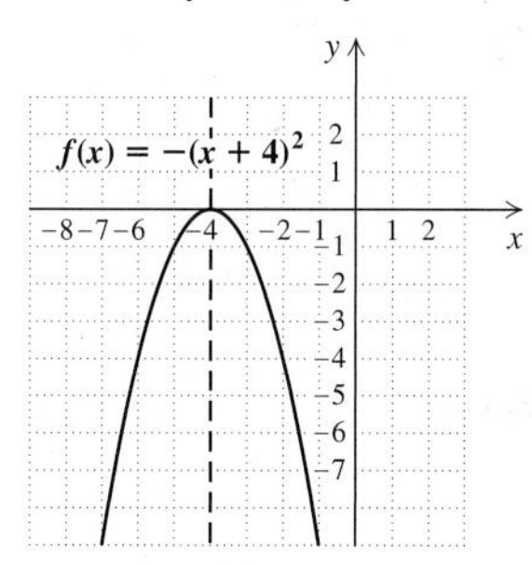

19. Vertex: $(1, 0)$;
axis of symmetry: $x = 1$

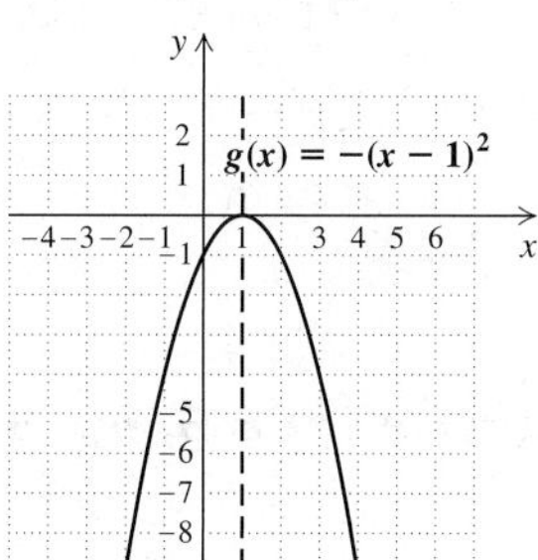

21. Vertex: $(-1, 0)$;
axis of symmetry: $x = -1$

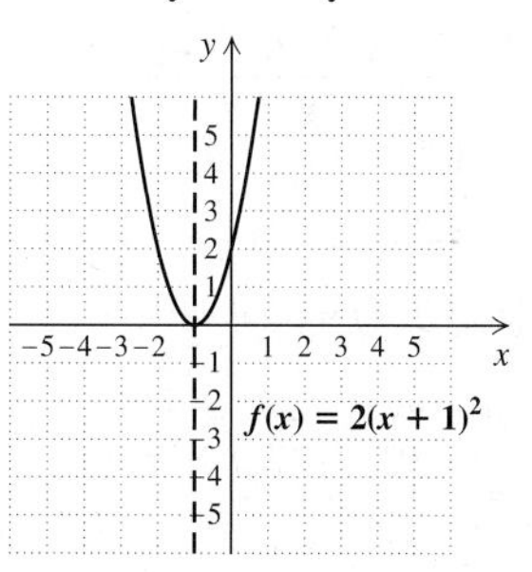

23. Vertex: $(4, 0)$;
axis of symmetry: $x = 4$

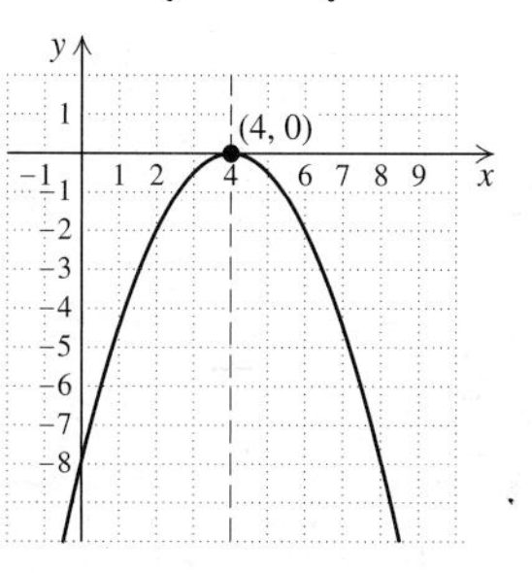

$h(x) = -\frac{1}{2}(x - 4)^2$

25. Vertex: $(1, 0)$;
axis of symmetry: $x = 1$

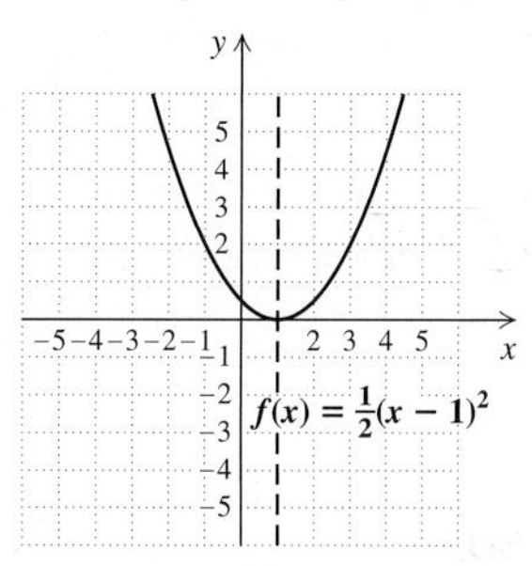

27. Vertex: $(-5, 0)$;
axis of symmetry: $x = -5$

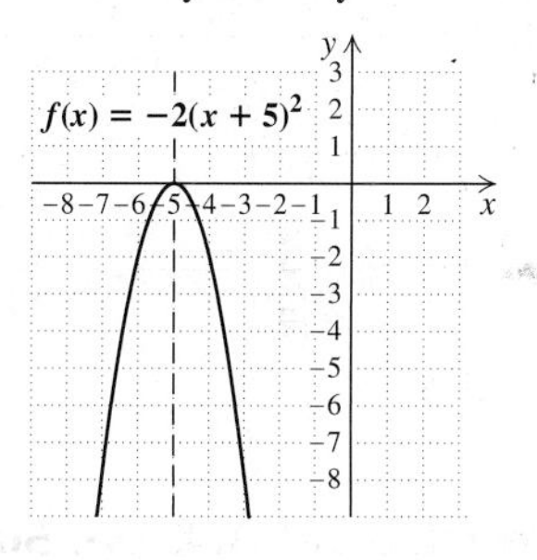

29. Vertex: $\left(\frac{1}{2}, 0\right)$;
axis of symmetry: $x = \frac{1}{2}$

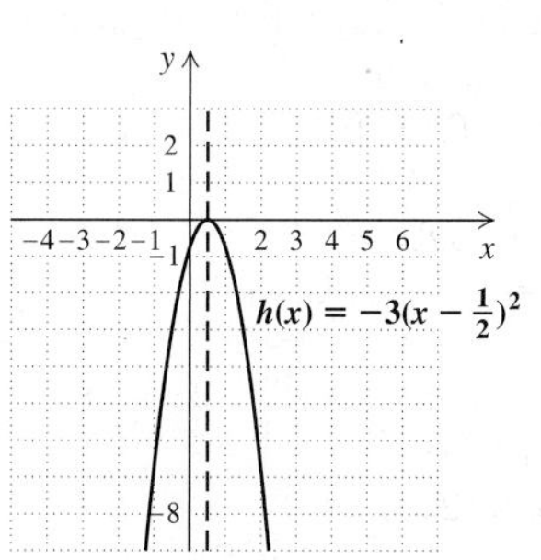

31. Vertex: $(5, 2)$;
axis of symmetry: $x = 5$;
minimum: 2

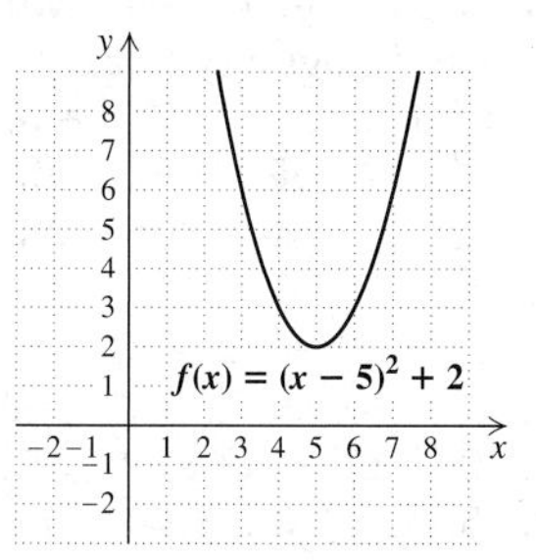

33. Vertex: $(-1, -3)$;
axis of symmetry: $x = -1$;
minimum: -3

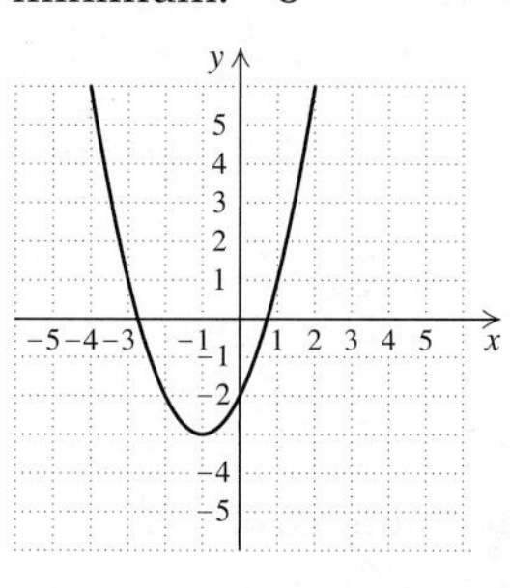

$f(x) = (x + 1)^2 - 3$

35. Vertex: $(-4, 1)$;
axis of symmetry: $x = -4$;
minimum: 1

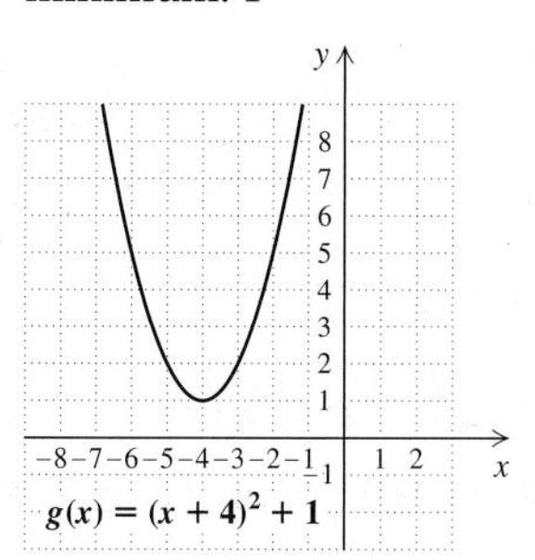

37. Vertex: $(1, -3)$; axis of symmetry: $x = 1$; maximum: -3

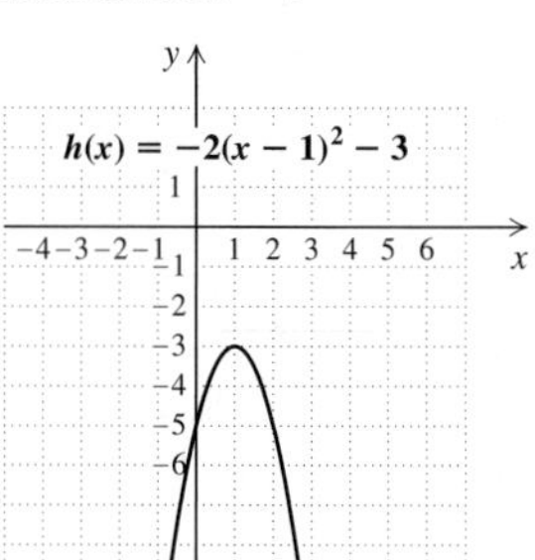

39. Vertex: $(-4, 1)$; axis of symmetry: $x = -4$; minimum: 1

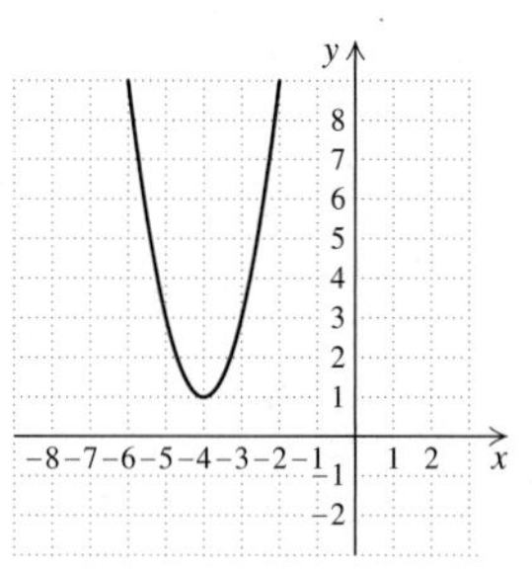

$f(x) = 2(x + 4)^2 + 1$

41. Vertex: $(1, 4)$; axis of symmetry: $x = 1$; maximum: 4

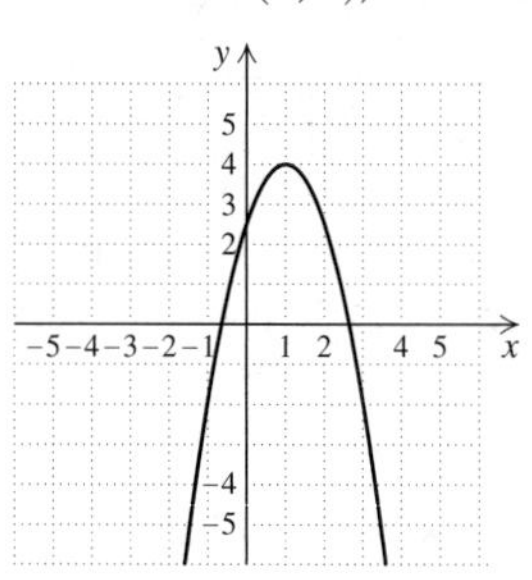

$g(x) = -\frac{3}{2}(x - 1)^2 + 4$

43. Vertex: $(9, 7)$; axis of symmetry: $x = 9$; minimum: 7 **45.** Vertex: $(-6, 11)$; axis of symmetry: $x = -6$; maximum: 11 **47.** Vertex: $\left(-\frac{1}{4}, -13\right)$; axis of symmetry: $x = -\frac{1}{4}$; minimum: -13 **49.** Vertex: $(-4.58, 65\pi)$; axis of symmetry: $x = -4.58$; minimum: 65π **51.**

53. [2.4]

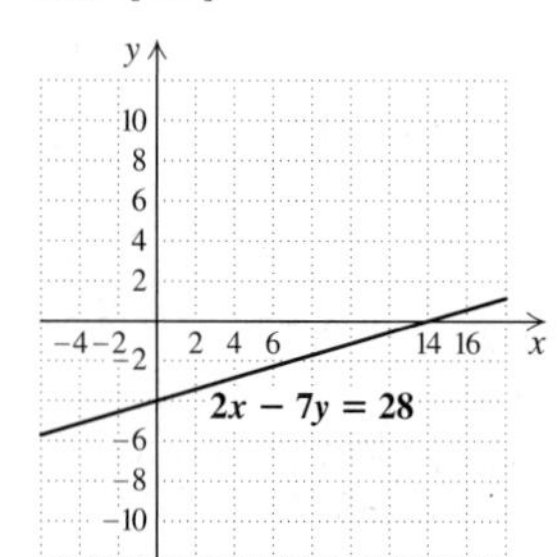

54. [2.4]

$6x - 3y = 36$

55. [3.2] $(-5, -1)$ **56.** [3.2] $(-1, 2)$
57. [8.1] $x^2 + 5x + \frac{25}{4}$ **58.** [8.1] $x^2 - 9x + \frac{81}{4}$ **59.**
61. $f(x) = \frac{3}{5}(x - 4)^2 + 1$ **63.** $f(x) = \frac{3}{5}(x - 3)^2 - 1$
65. $f(x) = \frac{3}{5}(x + 2)^2 - 5$ **67.** $g(x) = -2(x - 5)^2$
69. $f(x) = 2(x + 4)^2$ **71.** $g(x) = -2(x - 3)^2 + 8$
73. $F(x) = 3(x - 5)^2 + 1$

75.

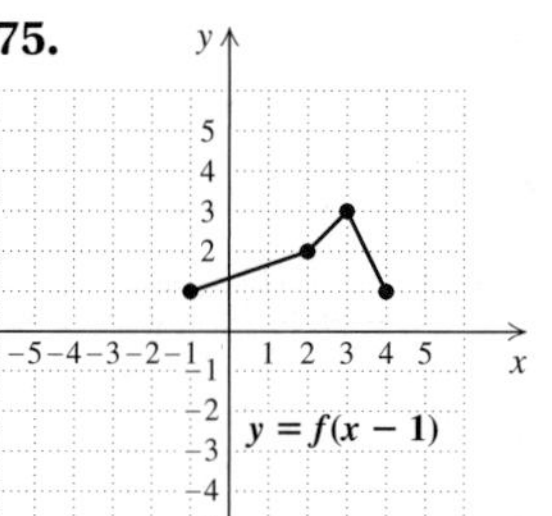

77.

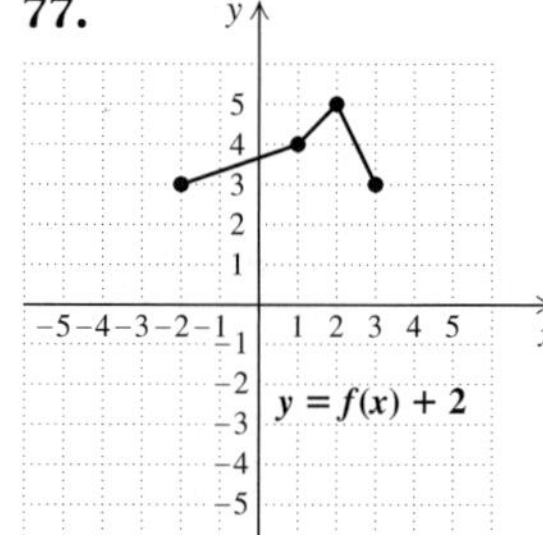

79.

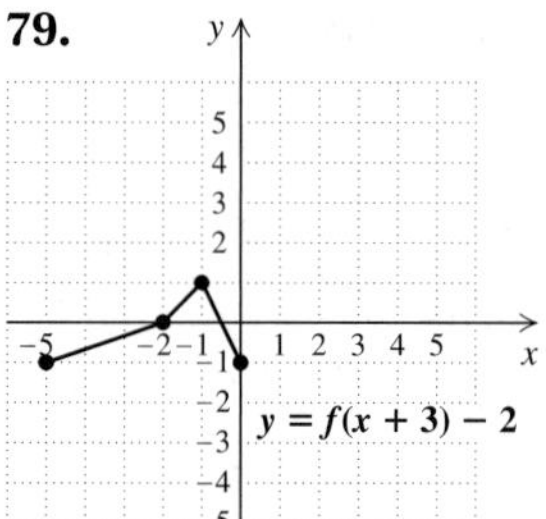

81. **83.** ,

Exercise Set 8.7, pp. 522–523

1. (a) Vertex: $(-2, 1)$; axis of symmetry: $x = -2$;
(b)

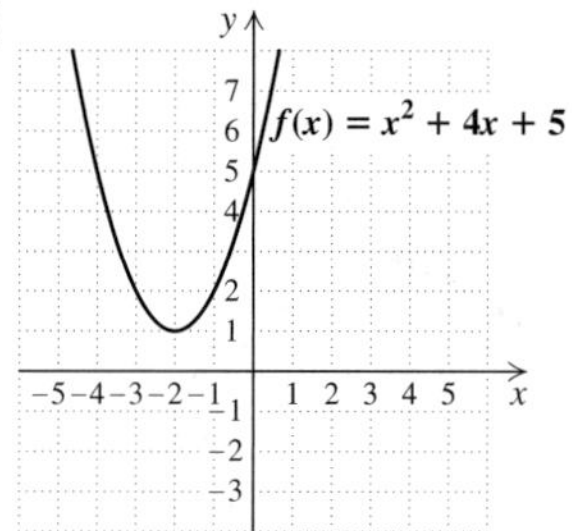

3. (a) Vertex: $(3, 4)$; axis of symmetry: $x = 3$;
(b)

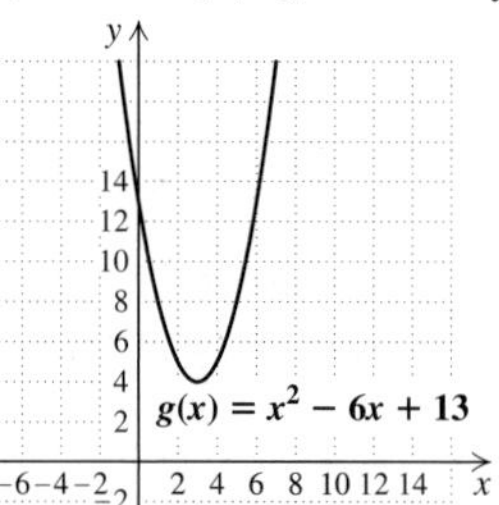

5. **(a)** Vertex: $(-4, 4)$; axis of symmetry: $x = -4$;
(b)

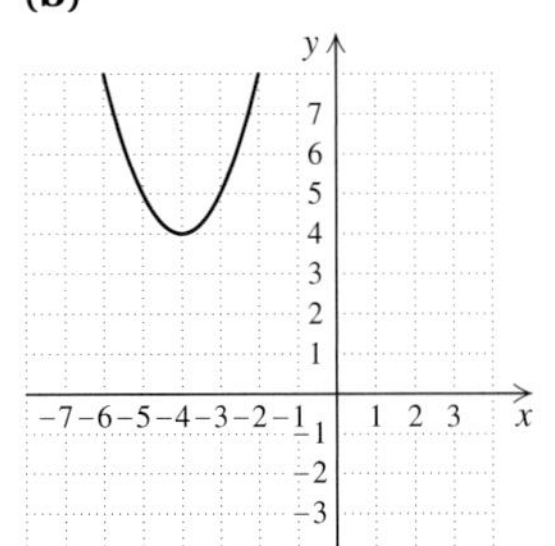

$f(x) = x^2 + 8x + 20$

7. **(a)** Vertex: $(4, -7)$; axis of symmetry: $x = 4$;
(b)

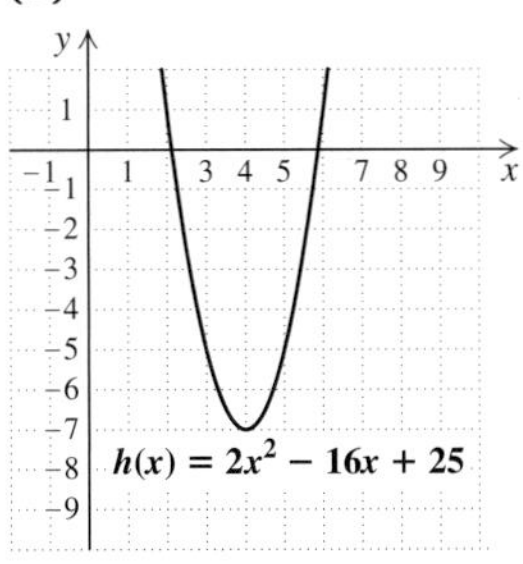

9. **(a)** Vertex: $(1, 6)$; axis of symmetry: $x = 1$;
(b)

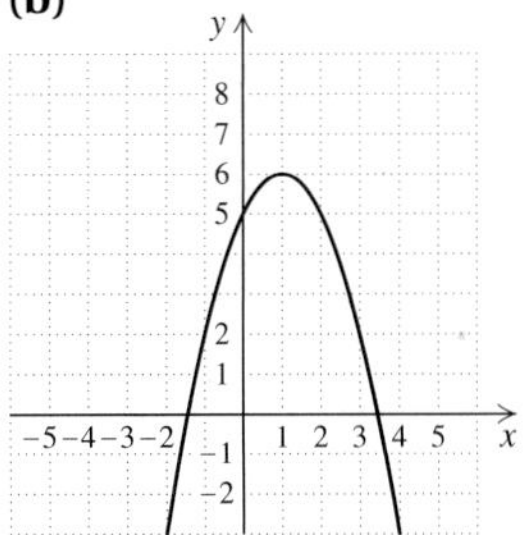

$f(x) = -x^2 + 2x + 5$

11. **(a)** Vertex: $\left(-\frac{3}{2}, -\frac{49}{4}\right)$; axis of symmetry: $x = -\frac{3}{2}$;
(b)

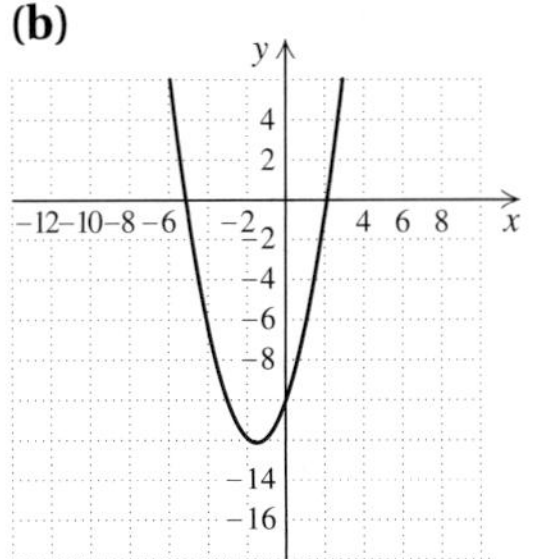

$g(x) = x^2 + 3x - 10$

13. **(a)** Vertex: $(4, 2)$; axis of symmetry: $x = 4$;
(b)

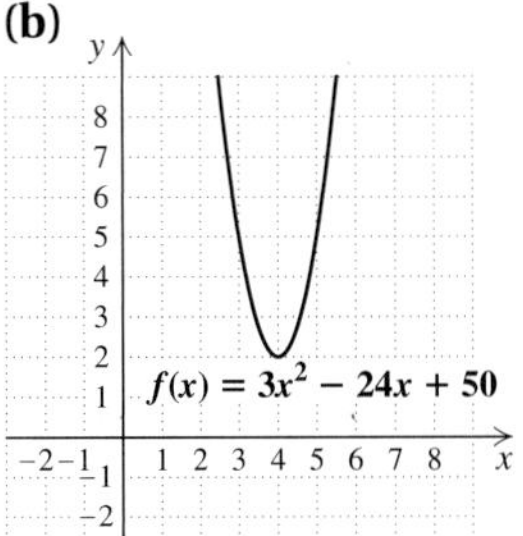

15. **(a)** Vertex: $\left(-\frac{7}{2}, -\frac{49}{4}\right)$; axis of symmetry: $x = -\frac{7}{2}$;
(b)

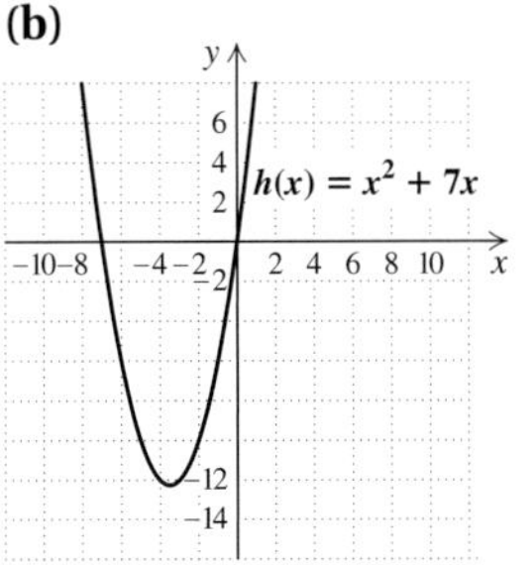

17. **(a)** Vertex: $(-1, -4)$; axis of symmetry: $x = -1$;
(b)

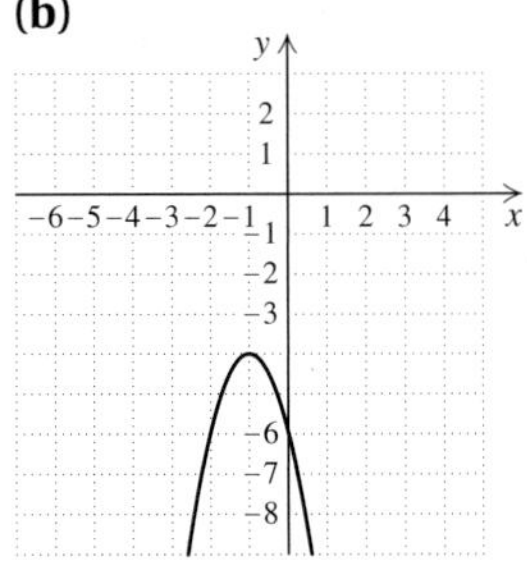

$f(x) = -2x^2 - 4x - 6$

19. **(a)** Vertex: $(2, -5)$; axis of symmetry: $x = 2$;
(b)

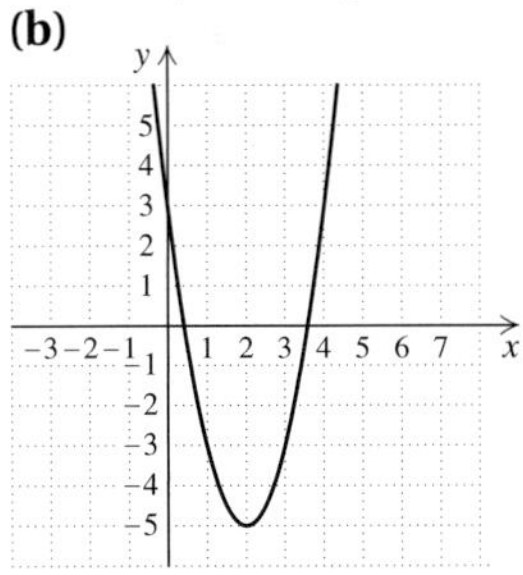

$g(x) = 2x^2 - 8x + 3$

21. **(a)** Vertex: $\left(\frac{5}{6}, \frac{1}{12}\right)$; axis of symmetry: $x = \frac{5}{6}$;
(b)

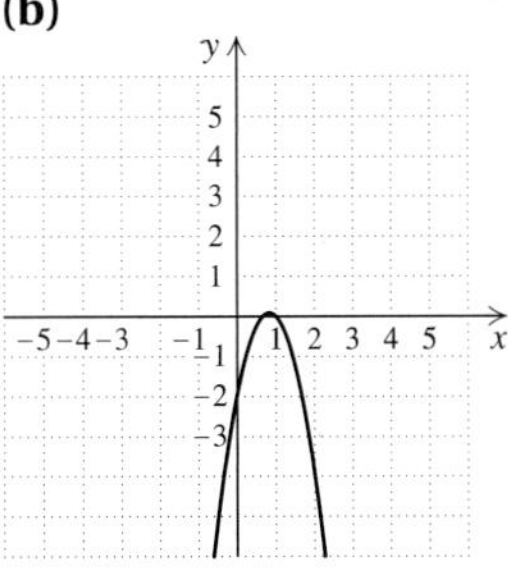

$f(x) = -3x^2 + 5x - 2$

23. **(a)** Vertex: $\left(-4, -\frac{5}{3}\right)$; axis of symmetry: $x = -4$;
(b)

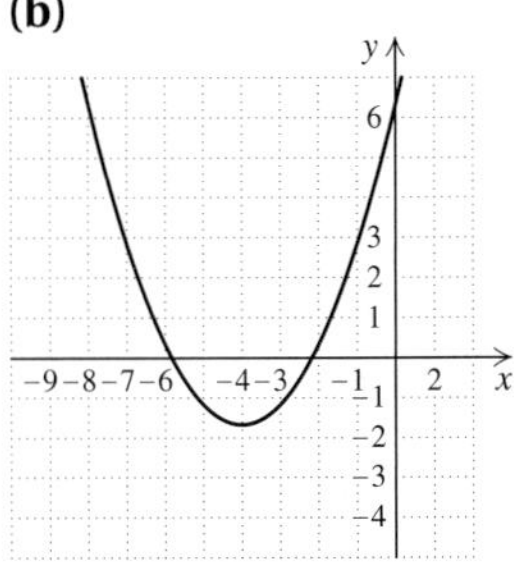

$h(x) = \frac{1}{2}x^2 + 4x + \frac{19}{3}$

25. $(3 - \sqrt{6}, 0), (3 + \sqrt{6}, 0); (0, 3)$ **27.** $(-1, 0), (3, 0); (0, 3)$ **29.** $(0, 0), (9, 0); (0, 0)$ **31.** $(2, 0); (0, -4)$ **33.** No x-intercept; $(0, 3)$ **35.** **37.** [3.2] $(2, -2)$ **38.** [3.2] $(7, 1)$ **39.** [3.4] $(3, 2, 1)$ **40.** [3.4] $(1, -3, 2)$ **41.** [7.6] 5 **42.** [7.6] 4 **43.** **45.** **(a)** Minimum: -6.953660714; **(b)** $(-1.056433682, 0)$, $(2.413576539, 0)$; $(0, -5.89)$ **47.** **(a)** $-2.4, 3.4$; **(b)** $-1.3, 2.3$ **49.** $f(x) = m\left(x - \dfrac{n}{2m}\right)^2 + \dfrac{4mp - n^2}{4m}$

51. $f(x) = \frac{5}{16}x^2 - \frac{15}{8}x - \frac{35}{16}$, or $f(x) = \frac{5}{16}(x - 3)^2 - 5$

53.

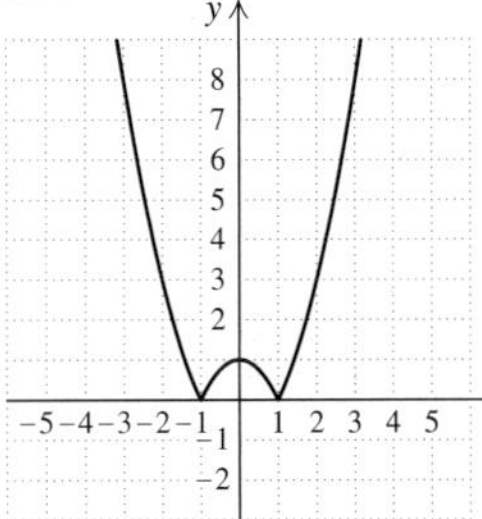

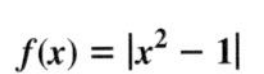

$f(x) = |x^2 - 1|$

55.

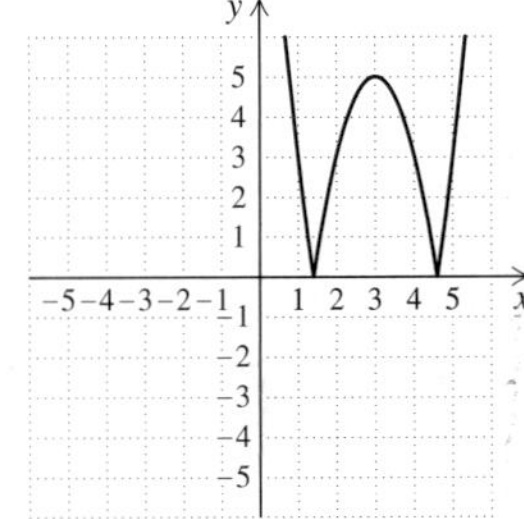

$f(x) = |2(x - 3)^2 - 5|$

Exercise Set 8.8, pp. 528–531

1. \$4; 3 mos. after January **3.** 11 days after the concert was announced; about 62 **5.** 180 ft by 180 ft **7.** 200 ft^2; 10 ft by 20 ft. (The barn serves as a 20-ft side.) **9.** 3.5 in. **11.** 81; 9 and 9 **13.** -16; 4 and -4 **15.** 25; -5 and -5 **17.** Not quadratic **19.** Quadratic **21.** Not quadratic **23.** Not quadratic **25.** $f(x) = 2x^2 + 3x - 1$ **27.** $f(x) = -\frac{1}{4}x^2 + 3x - 5$ **29.** **(a)** $A(s) = \frac{3}{16}s^2 - \frac{135}{4}s + 1750$; **(b)** about 531 **31.** $h(d) = -0.0068d^2 + 0.8571d$ **33.** **35.** [6.2] $\dfrac{x - 9}{(x + 9)(x + 7)}$ **36.** [6.1] $\dfrac{(x - 3)(x + 1)}{(x - 7)(x + 3)}$ **37.** [6.1] $\dfrac{(t + 2)(t + 8)}{(t - 3)(t + 1)}$ **38.** [6.2] $\dfrac{2t(t + 2)}{(t - 7)(t - 3)(t + 7)}$

39. [4.1] $\{x|x < 8\}$, or $(-\infty, 8)$
40. [4.1] $\{x|x \geq 10\}$, or $[10, \infty)$ **41.**
43. The radius of the circular portion of the window and the height of the rectangular portion should each be $\frac{24}{\pi + 4}$ ft. **45.** 30 **47.** 78.4 ft **49.**

Technology Connection, p. 536

1. $\{x|-0.78 \leq x \leq 1.59\}$, or $[-0.78, 1.59]$
2. $\{x|x \leq -0.21 \text{ or } x \geq 2.47\}$, or $(-\infty, -0.21] \cup [2.47, \infty)$
3. $\{x|x < -1.26 \text{ or } x > 2.33\}$, or $(-\infty, -1.26) \cup (2.33, \infty)$
4. $\{x|x > -1.37\}$, or $(-1.37, \infty)$

Exercise Set 8.9, pp. 539–540

1. $(-4, 3)$, or $\{x|-4 < x < 3\}$ **3.** $(-\infty, -7] \cup [2, \infty)$, or $\{x|x \leq -7 \text{ or } x \geq 2\}$ **5.** $(-1, 2)$, or $\{x|-1 < x < 2\}$
7. $[-5, 5]$, or $\{x|-5 \leq x \leq 5\}$ **9.** $\varnothing$
11. $(-2, 6)$, or $\{x|-2 < x < 6\}$
13. $(-\infty, -2) \cup (0, 2)$, or $\{x|x < -2 \text{ or } 0 < x < 2\}$
15. $(-3, -1) \cup (2, \infty)$, or $\{x|-3 < x < -1 \text{ or } x > 2\}$
17. $(-\infty, -3) \cup (-2, 1)$, or $\{x|x < -3 \text{ or } -2 < x < 1\}$
19. $(-\infty, -3)$, or $\{x|x < -3\}$ **21.** $(-\infty, -1] \cup (5, \infty)$, or $\{x|x \leq -1 \text{ or } x > 5\}$ **23.** $\left[-\frac{2}{3}, 2\right)$, or $\left\{x\middle|-\frac{2}{3} \leq x < 2\right\}$
25. $(-\infty, -6)$, or $\{x|x < -6\}$ **27.** $(-\infty, -1] \cup [2, 5)$, or $\{x|x \leq -1 \text{ or } 2 \leq x < 5\}$ **29.** $(-\infty, -3) \cup [0, \infty)$, or $\{x|x < -3 \text{ or } x \geq 0\}$ **31.** $(0, \infty)$, or $\{x|x > 0\}$
33. $(-\infty, -4) \cup [1, 3)$, or $\{x|x < -4 \text{ or } 1 \leq x < 3\}$
35. $\left(0, \frac{1}{4}\right)$, or $\left\{x\middle|0 < x < \frac{1}{4}\right\}$ **37.** **39.** [1.6] $-8a^9b^6c^{12}$
40. [1.6] $25a^8b^{14}$ **41.** [1.6] $\frac{1}{32}$ **42.** [1.6] $\frac{1}{81}$
43. [5.2] $3a^2 + 6a + 3$ **44.** [2.2] $5a + 7$ **45.**
47. $\left(-1 - \sqrt{6}, -1 + \sqrt{6}\right)$, or $\left\{x\middle|-1 - \sqrt{6} < x < -1 + \sqrt{6}\right\}$ **49.** $\{0\}$
51. **(a)** $(10, 200)$, or $\{x|10 < x < 200\}$; **(b)** $[0, 10) \cup (200, \infty)$, or $\{x|0 \leq x < 10 \text{ or } x > 200\}$
53. $\{n|n \text{ is an integer and } 12 \leq n \leq 25\}$ **55.** $f(x) = 0$ for $x = -2, 1, 3$; $f(x) < 0$ for $(-\infty, -2) \cup (1, 3)$, or $\{x|x < -2 \text{ or } 1 < x < 3\}$; $f(x) > 0$ for $(-2, 1) \cup (3, \infty)$, or $\{x|-2 < x < 1 \text{ or } x > 3\}$ **57.** $f(x)$ has no zeros; $f(x) < 0$ for $(-\infty, 0)$, or $\{x|x < 0\}$; $f(x) > 0$ for $(0, \infty)$, or $\{x|x > 0\}$
59. $f(x) = 0$ for $x = -1, 0$; $f(x) < 0$ for $(-\infty, -3) \cup (-1, 0)$, or $\{x|x < -3 \text{ or } -1 < x < 0\}$; $f(x) > 0$ for $(-3, -1) \cup (0, 2) \cup (2, \infty)$, or $\{x|-3 < x < -1 \text{ or } 0 < x < 2 \text{ or } x > 2\}$ **61.**

Review Exercises: Chapter 8, pp. 543–544

1. [8.1] $\pm\sqrt{\frac{7}{2}}$, or $\pm\frac{\sqrt{14}}{2}$ **2.** [8.1] $0, -\frac{5}{14}$ **3.** [8.1] 3, 9
4. [8.2] $\frac{5 \pm i\sqrt{11}}{2}$ **5.** [8.2] 3, 5 **6.** [8.2] $\frac{-9 \pm \sqrt{85}}{2}$
7. [8.2] $-0.3722813233, 5.3722813233$ **8.** [8.2] $-\frac{1}{4}, 1$
9. [8.1] $x^2 - 12x + 36$; $(x - 6)^2$ **10.** [8.1] $x^2 + \frac{3}{5}x + \frac{9}{100}$; $\left(x + \frac{3}{10}\right)^2$ **11.** [8.1] $3 \pm 2\sqrt{2}$ **12.** [8.1] 10%
13. [8.1] 6.7 sec **14.** [8.3] About 153 mph
15. [8.3] 6 hr **16.** [8.4] Two irrational
17. [8.4] Two imaginary **18.** [8.4] $x^2 - 5 = 0$
19. [8.4] $x^2 + 8x + 16 = 0$ **20.** [8.5] $(-3, 0), (-2, 0), (2, 0), (3, 0)$ **21.** [8.5] $-5, 3$ **22.** [8.5] $\pm\sqrt{2}, \pm\sqrt{7}$
23. [8.6]

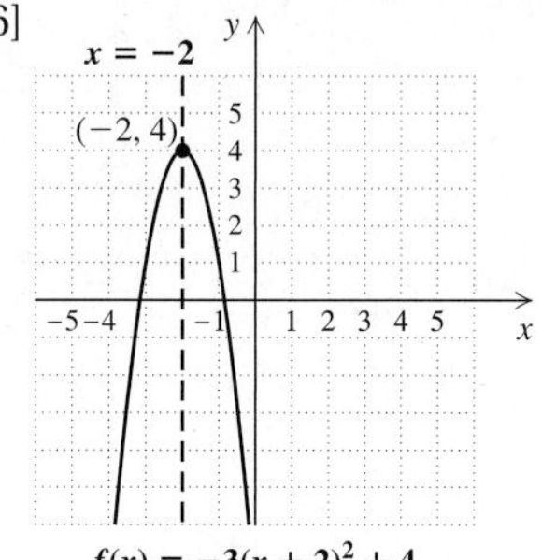

24. [8.7] **(a)** Vertex: $(3, 5)$; axis of symmetry: $x = 3$;
(b)

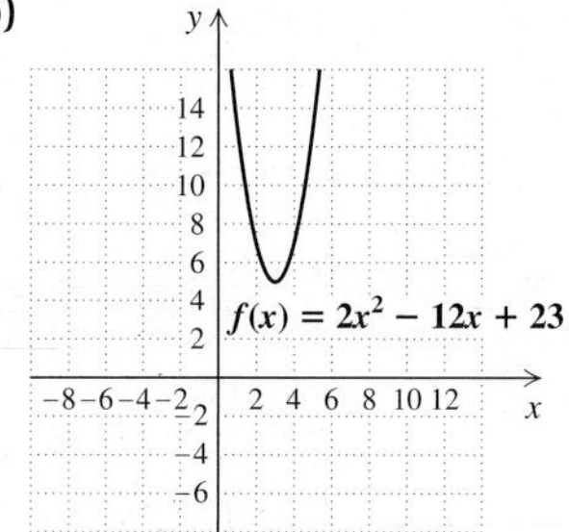

25. [8.7] $(2, 0), (7, 0)$; $(0, 14)$ **26.** [8.3] $p = \frac{9\pi^2}{N^2}$
27. [8.3] $T = \frac{1 \pm \sqrt{1 + 24A}}{6}$ **28.** [8.8] Quadratic
29. [8.8] Not quadratic **30.** [8.8] 56.25 ft^2; 7.5 ft by 7.5 ft
31. [8.8] **(a)** $f(x) = \frac{37}{50}x^2 - \frac{31}{2}x + 189$; **(b)** 175 million
32. [8.9] $(-1, 0) \cup (3, \infty)$, or $\{x|-1 < x < 0 \text{ or } x > 3\}$
33. [8.9] $(-3, 5]$, or $\{x|-3 < x \leq 5\}$
34. [8.7], [8.8] The x-coordinate of the maximum or minimum point lies halfway between the x-coordinates of the x-intercepts. **35.** [8.2], [8.4] Yes; if the discriminant is a perfect square, then the solutions are rational numbers, p/q and r/s. (Note that if the discriminant is 0, then $p/q = r/s$.) Then the equation can be written in factored form, $(qx - p)(sx - r) = 0$.
36. [8.5] Four; let $u = x^2$. Then $au^2 + bu + c = 0$ has at most two solutions, $u = m$ and $u = n$. Now substitute x^2 for u and obtain $x^2 = m$ or $x^2 = n$. These equations yield the solutions $x = \pm\sqrt{m}$ and $x = \pm\sqrt{n}$. When $m \neq n$, the maximum number of solutions, four, occurs.

37. [8.1], [8.2], [8.7] Completing the square was used to solve quadratic equations and to graph quadratic functions by rewriting the function in the form $f(x) = a(x - h)^2 + k$. **38.** [8.7] $f(x) = \frac{7}{15}x^2 - \frac{14}{15}x - 7$
39. [8.4] $h = 60, k = 60$ **40.** [8.5] 18, 324

Test: Chapter 8, pp. 544–545

1. [8.1] $\pm\frac{4\sqrt{3}}{3}$ **2.** [8.2] 2, 9 **3.** [8.2] $\frac{-1 \pm i\sqrt{3}}{2}$
4. [8.2] $1 \pm \sqrt{6}$ **5.** [8.5] $-2, \frac{2}{3}$ **6.** [8.2] -4.192582404, 1.192582404 **7.** [8.2] $-\frac{3}{4}, \frac{7}{3}$ **8.** [8.1] $x^2 + 14x + 49$; $(x + 7)^2$ **9.** [8.1] $x^2 - \frac{2}{7}x + \frac{1}{49}$; $\left(x - \frac{1}{7}\right)^2$
10. [8.1] $-5 \pm \sqrt{10}$ **11.** [8.3] 16 km/h **12.** [8.3] 2 hr
13. [8.4] Two imaginary **14.** [8.4] $3x^2 + 5x - 2 = 0$
15. [8.5] $(-3, 0), (-1, 0), \left(-2 - \sqrt{5}, 0\right), \left(-2 + \sqrt{5}, 0\right)$
16. [8.6]

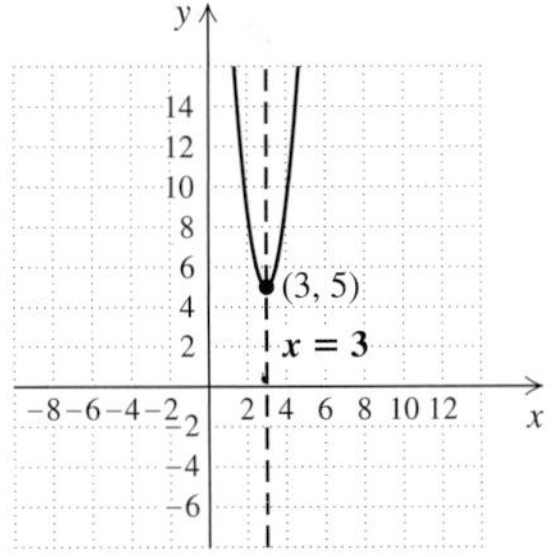

$f(x) = 4(x - 3)^2 + 5$
Minimum: 5

17. [8.7] **(a)** $(-1, -8), x = -1$;
(b)

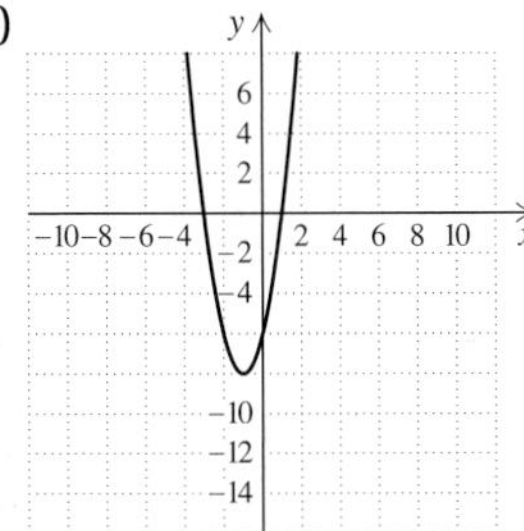
$f(x) = 2x^2 + 4x - 6$

18. [8.7] $(-2, 0), (3, 0)$; $(0, -6)$ **19.** [8.3] $r = \sqrt{\frac{3V}{\pi} - R^2}$
20. [8.8] Not quadratic **21.** [8.8] Minimum \$129/cap when 325 caps are built **22.** [8.8] $f(x) = \frac{1}{5}x^2 - \frac{3}{5}x$
23. [8.9] $[-6, 1]$, or $\{x \mid -6 \le x \le 1\}$
24. [8.9] $(-1, 0) \cup (1, \infty)$, or $\{x \mid -1 < x < 0 \text{ or } x > 1\}$
25. [8.4] $\frac{1}{2}$ **26.** [8.4] $x^4 - 14x^3 + 67x^2 - 114x + 26 = 0$; answers may vary.
27. [8.4] $x^6 - 10x^5 + 20x^4 + 50x^3 - 119x^2 - 60x + 150 = 0$; answers may vary

CHAPTER 9

Technology Connection, p. 550

1. A table shows that $y_2 = y_3$. $y_1 = 7x + 3$; $y_2 = y_1^2$; $y_3 = (7x + 3)^2$

X	Y2	Y3
5	1444	1444
6	2025	2025
7	2704	2704
8	3481	3481
9	4356	4356
10	5329	5329
11	6400	6400

X = 5

A similar table shows that for $y_2 = x^2$ and $y_4 = y_2(y_1)$, we have $y_3 = y_4$.
A graph can also be used: $y_2 = y_1^2$; $y_3 = (7x + 3)^2$

Technology Connection, p. 557

1. Graph each pair of functions in a square window along with the line $y = x$ and determine whether the first two functions are reflections of each other across $y = x$. For further verification, examine a table of values for each pair of functions. **2.** Yes; most graphers do not require that the inverse relation be a function.

Exercise Set 9.1, pp. 558–560

1. $(f \circ g)(1) = 12$; $(g \circ f)(1) = 9$; $(f \circ g)(x) = 4x^2 + 4x + 4$; $(g \circ f)(x) = 2x^2 + 7$
3. $(f \circ g)(1) = 20$; $(g \circ f)(1) = 22$; $(f \circ g)(x) = 15x^2 + 5$; $(g \circ f)(x) = 45x^2 - 30x + 7$
5. $(f \circ g)(1) = 8$; $(g \circ f)(1) = \frac{1}{64}$; $(f \circ g)(x) = \frac{1}{x^2} + 7$; $(g \circ f)(x) = \frac{1}{(x + 7)^2}$ **7.** $f(x) = x^2$; $g(x) = 7 + 5x$
9. $f(x) = \sqrt{x}$; $g(x) = 2x + 7$ **11.** $f(x) = \frac{2}{x}$; $g(x) = x - 3$ **13.** $f(x) = \frac{1}{\sqrt{x}}$; $g(x) = 7x + 2$
15. $f(x) = \frac{1}{x} + x$; $g(x) = \sqrt{3x}$ **17.** Yes **19.** No
21. Yes **23.** No **25.** **(a)** Yes; **(b)** $f^{-1}(x) = x + 4$
27. **(a)** Yes; **(b)** $f^{-1}(x) = x - 3$ **29.** **(a)** Yes; **(b)** $g^{-1}(x) = x - 5$ **31.** **(a)** Yes; **(b)** $f^{-1}(x) = \frac{x}{4}$

33. **(a)** Yes; **(b)** $g^{-1}(x) = \dfrac{x+1}{4}$ **35.** **(a)** No

37. **(a)** Yes; **(b)** $f^{-1}(x) = \dfrac{1}{x}$ **39.** **(a)** Yes;
(b) $f^{-1}(x) = \dfrac{3x-1}{2}$ **41.** **(a)** Yes; **(b)** $f^{-1}(x) = \sqrt[3]{x+5}$

43. **(a)** Yes; **(b)** $g^{-1}(x) = \sqrt[3]{x} + 2$ **45.** **(a)** Yes;
(b) $f^{-1}(x) = x^2, x \geq 0$ **47.** **(a)** Yes; **(b)** $f^{-1}(x) = \sqrt{\dfrac{x-1}{2}}$

49.

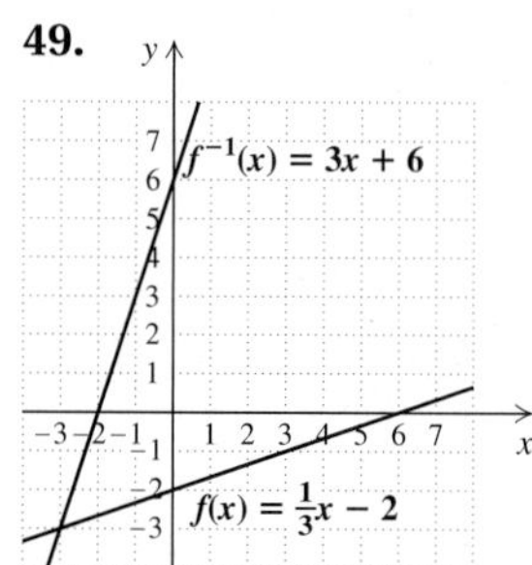

51.

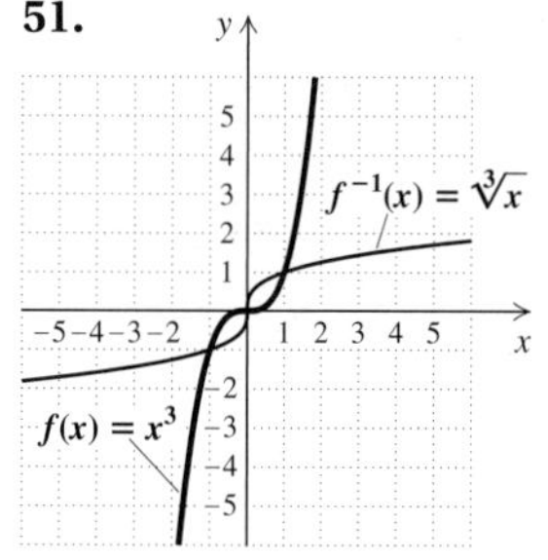

53.

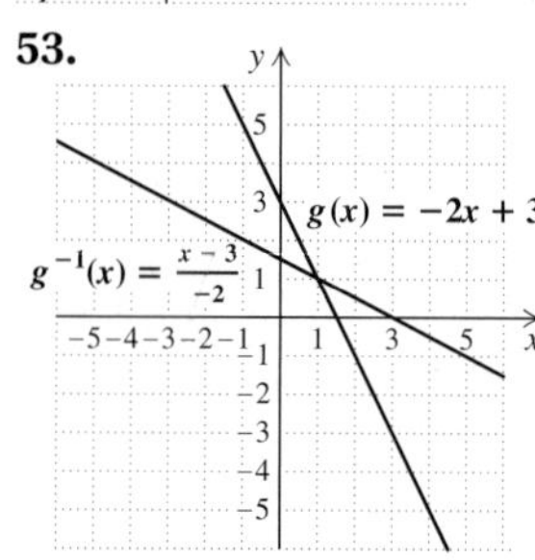

55.

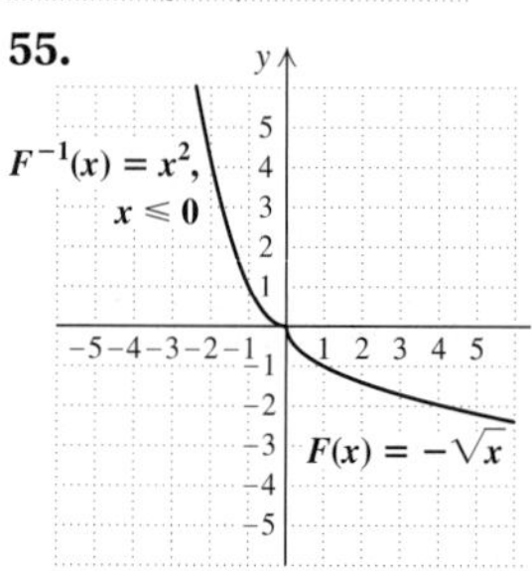

57.

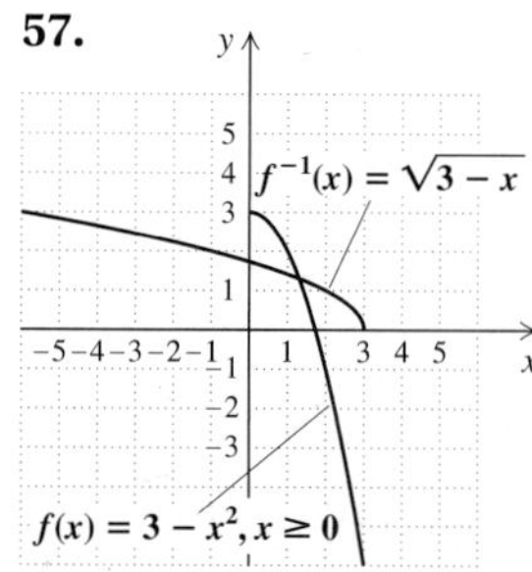

59. **(1)** $(f^{-1} \circ f)(x) = f^{-1}(f(x)) = f^{-1}\left(\frac{4}{5}x\right) = \frac{5}{4}\left(\frac{4}{5}x\right) = x$;
(2) $(f \circ f^{-1})(x) = f(f^{-1}(x)) = f\left(\frac{5}{4}x\right) = \frac{4}{5}\left(\frac{5}{4}x\right) = x$

61. **(1)** $(f^{-1} \circ f)(x) = f^{-1}(f(x)) = f^{-1}\left(\dfrac{1-x}{x}\right)$

$$= \frac{1}{\left(\dfrac{1-x}{x}\right) + 1} = \frac{1}{\dfrac{1-x+x}{x}} = x;$$

(2) $(f \circ f^{-1})(x) = f(f^{-1}(x)) = f\left(\dfrac{1}{x+1}\right)$

$$= \frac{1 - \left(\dfrac{1}{x+1}\right)}{\left(\dfrac{1}{x+1}\right)} = \frac{\dfrac{x+1-1}{x+1}}{\dfrac{1}{x+1}} = x$$

63. **(a)** 40, 42, 46, 50; **(b)** $f^{-1}(x) = x - 32$; **(c)** 8, 10, 14, 18
65. **67.** [1.6] $a^{13}b^{13}$ **68.** [1.6] $x^{10}y^{12}$ **69.** [7.2] 81
70. [7.2] 125 **71.** [1.5] $y = \frac{3}{2}(x+7)$
72. [1.5] $y = \dfrac{10-x}{3}$ **73.**

75.

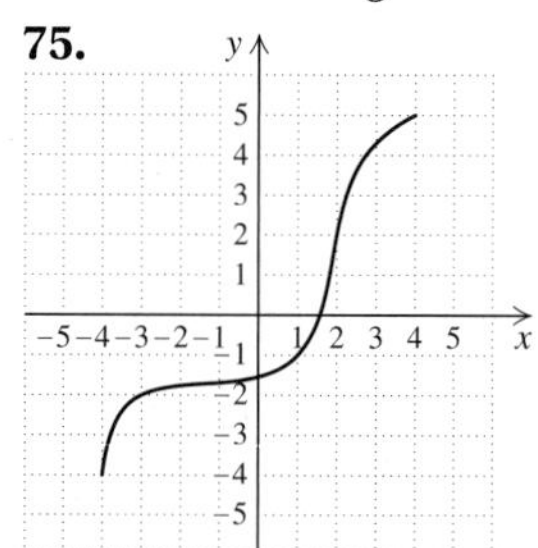

77. $g(x) = \dfrac{x}{2} + 20$ **79.**

81. Suppose that $h(x) = (f \circ g)(x)$. First, note that for $I(x) = x$, $(f \circ I)(x) = f(I(x)) = f(x)$ for any function f.

(i) $((g^{-1} \circ f^{-1}) \circ h)(x) = ((g^{-1} \circ f^{-1}) \circ (f \circ g))(x)$
$= ((g^{-1} \circ (f^{-1} \circ f)) \circ g)(x)$
$= ((g^{-1} \circ I) \circ g)(x)$
$= (g^{-1} \circ g)(x) = x$

(ii) $(h \circ (g^{-1} \circ f^{-1}))(x) = ((f \circ g) \circ (g^{-1} \circ f^{-1}))(x)$
$= ((f \circ (g \circ g^{-1})) \circ f^{-1})(x)$
$= ((f \circ I) \circ f^{-1})(x)$
$= (f \circ f^{-1})(x) = x.$

Therefore, $(g^{-1} \circ f^{-1})(x) = h^{-1}(x)$. **83.** Yes **85.** No
87. **(1)** C; **(2)** A; **(3)** B; **(4)** D **89.**

Technology Connection, p. 563

1.

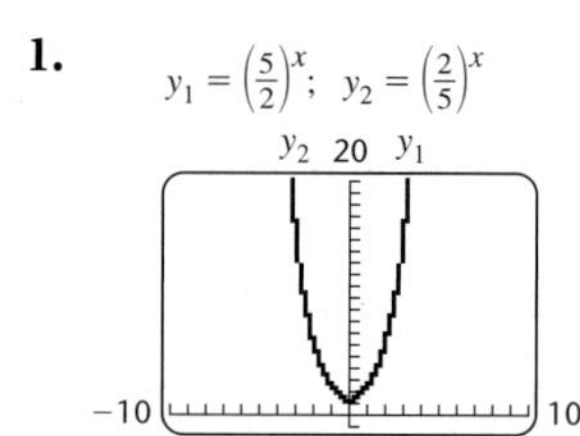

2.

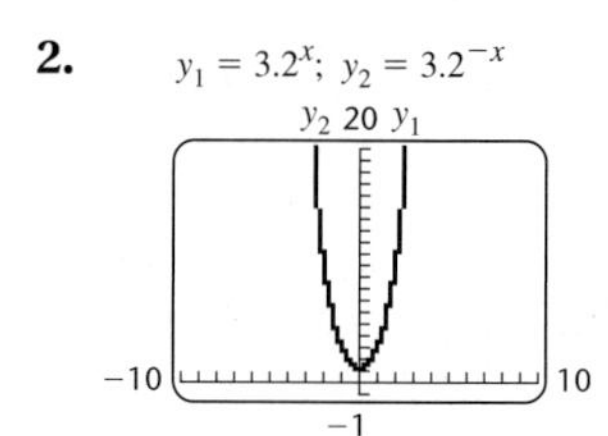

3.

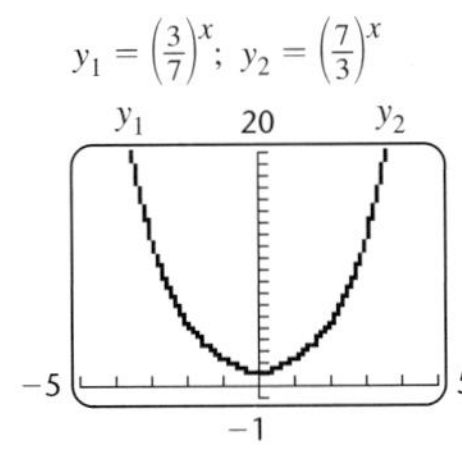

4.

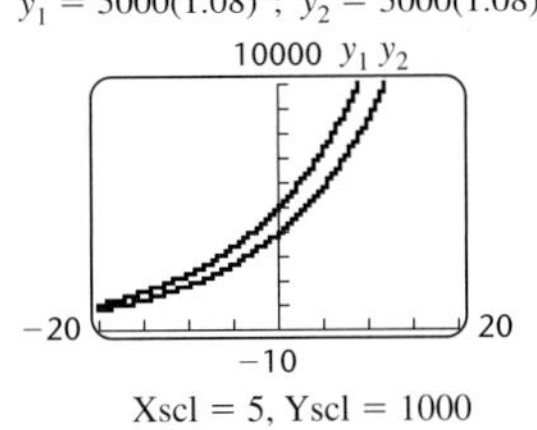

17.

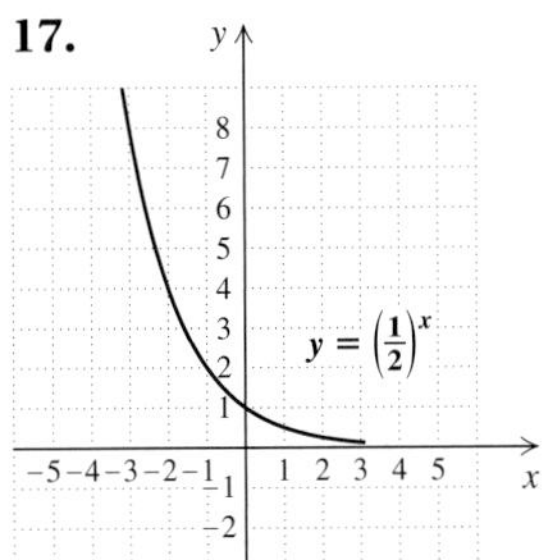

19.

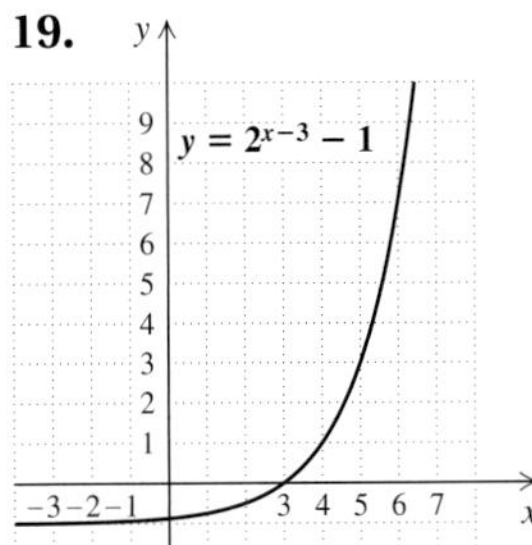

Exercise Set 9.2, pp. 567–568

1.

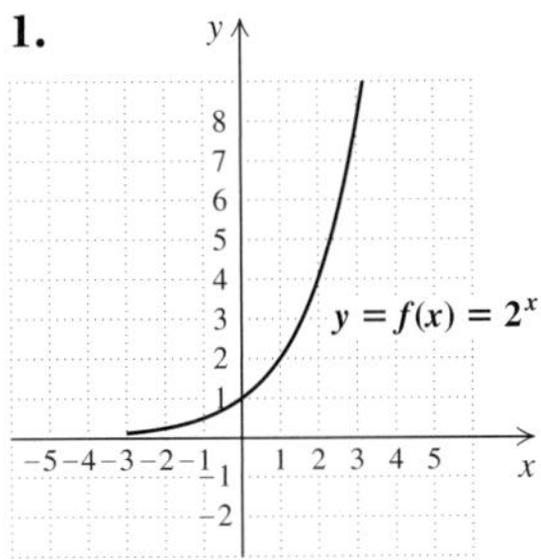

3.

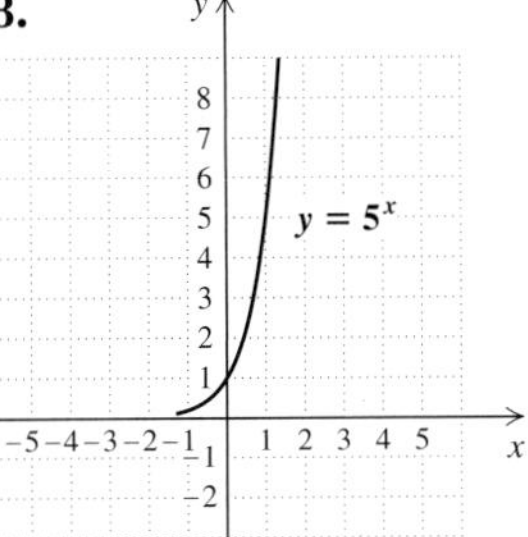

5.

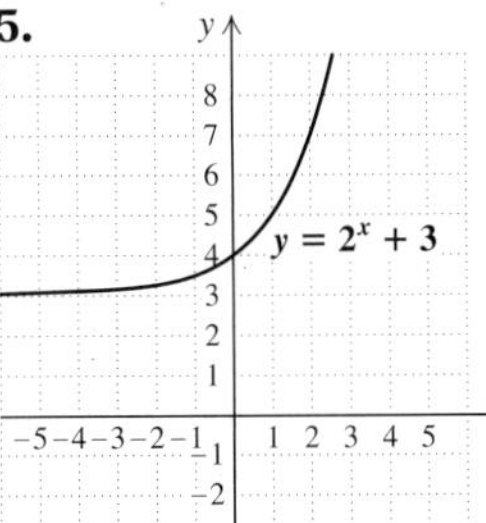

7.

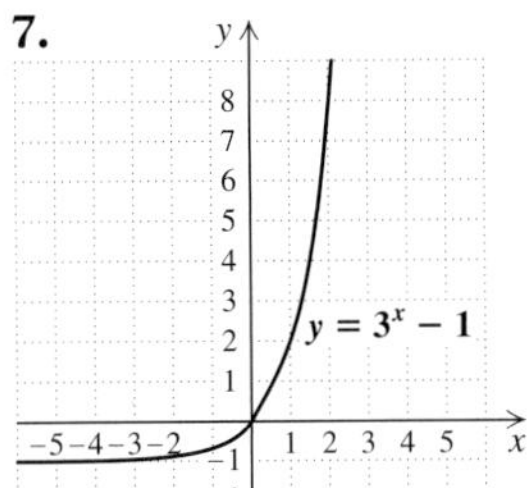

9.

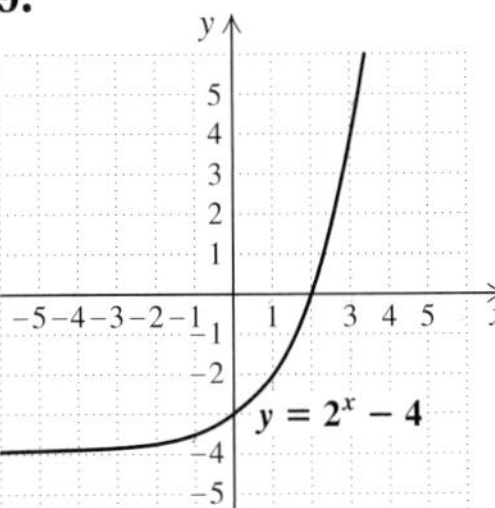

11.

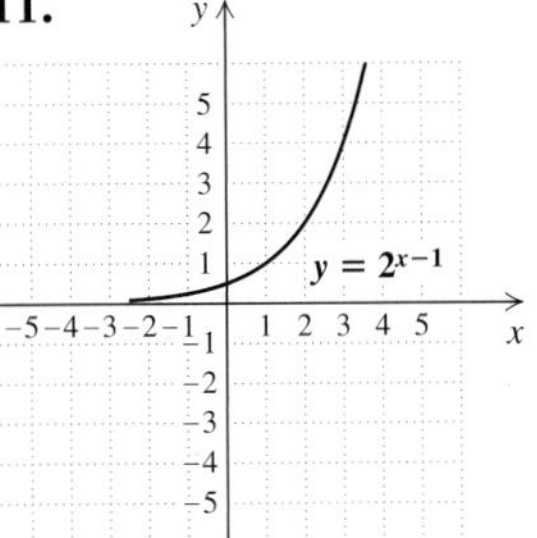

13.

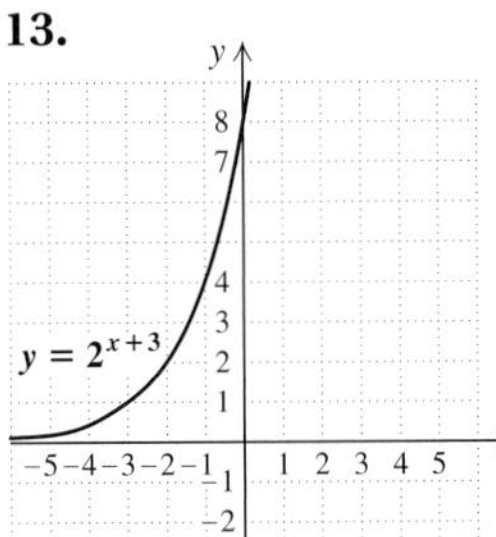

15.

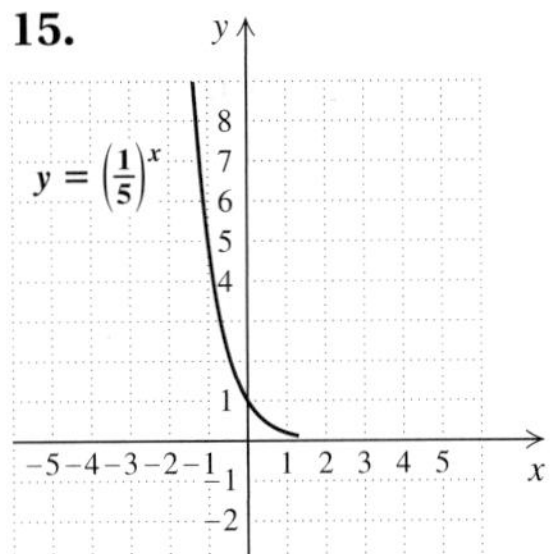

21.

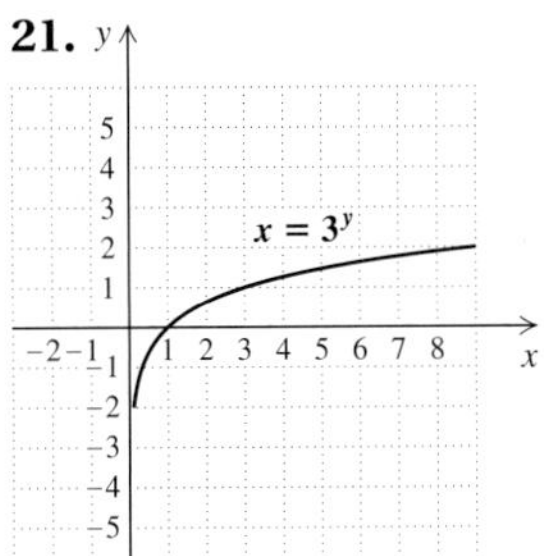

23.

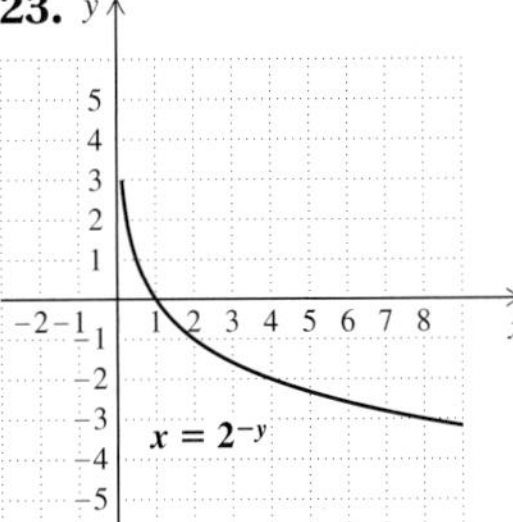

25.

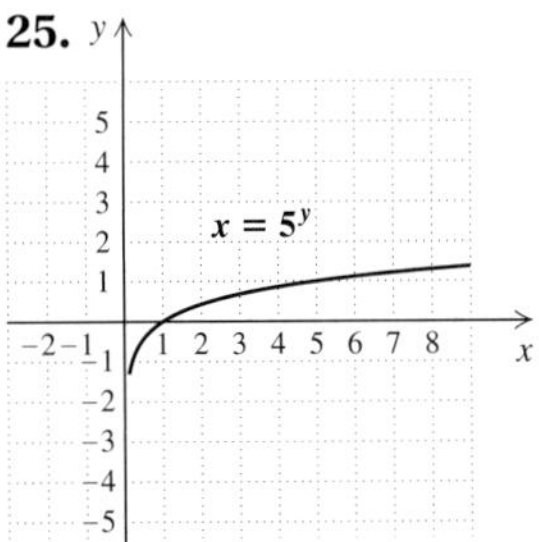

27.

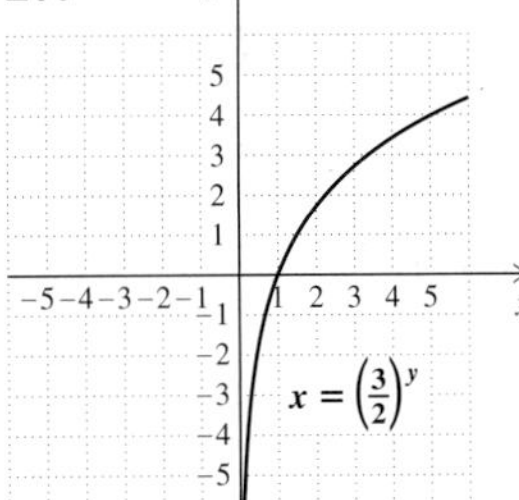

29.

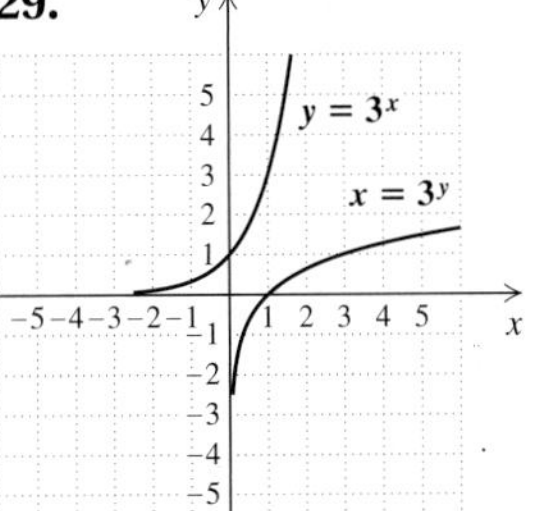

31.

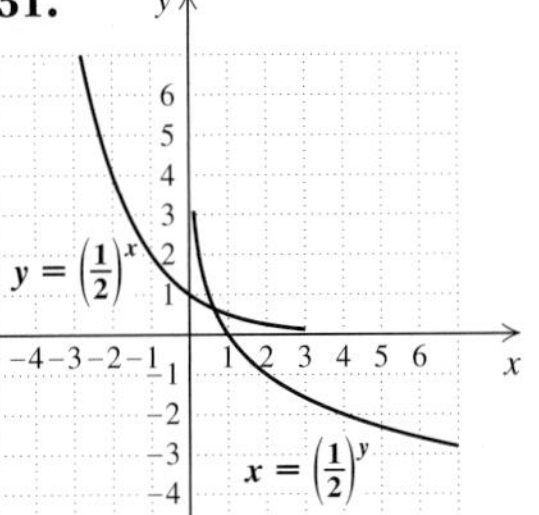

33. (a) About 6.4 billion; about 6.8 billion; about 7.3 billion;

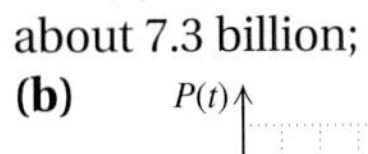

(b)

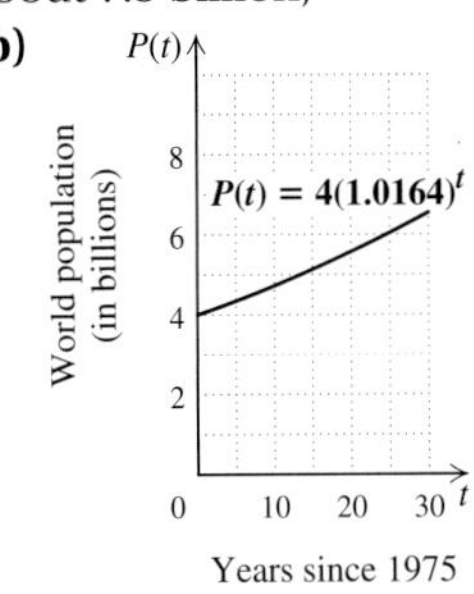

35. **(a)** About 44,079 whales; about 12,953 whales; **(b)**

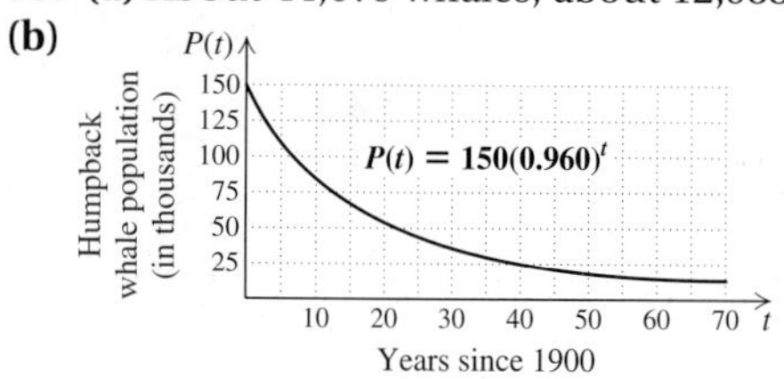

37. **(a)** 250,000; 166,667; 49,383; 4335; **(b)**

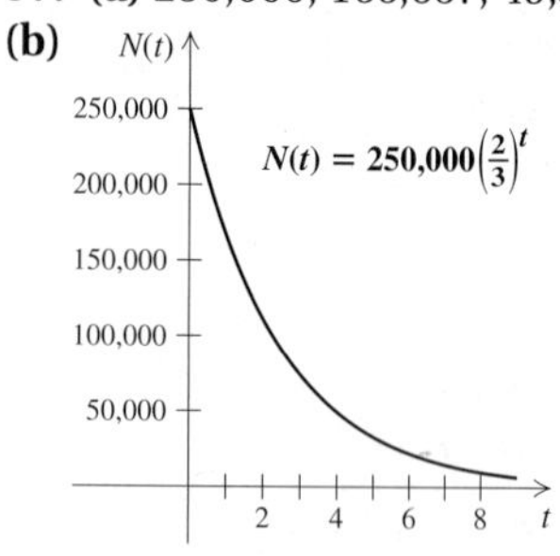

39. **(a)** 0.3 million, or 300,000; 12.1 million; 490.6 million; 3119.5 million; **(b)**

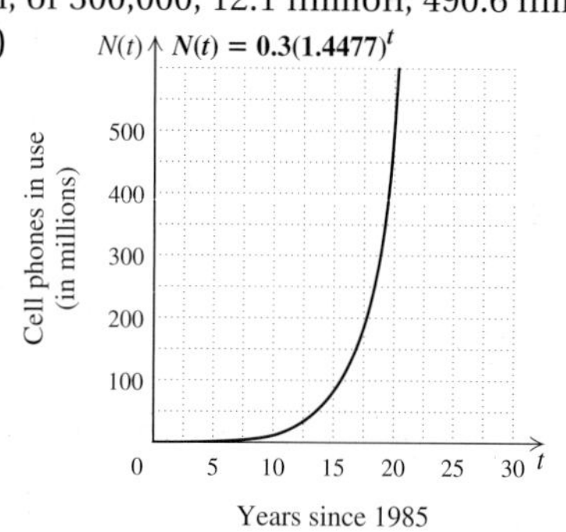

41.

43. [1.6] $\frac{1}{25}$ **44.** [1.6] $\frac{1}{32}$ **45.** [7.2] 100 **46.** [7.2] $\frac{1}{125}$ **47.** [1.6] $5a^6b^3$ **48.** [1.6] $6x^4y$ **49.**

51. $\pi^{2.4}$

53.

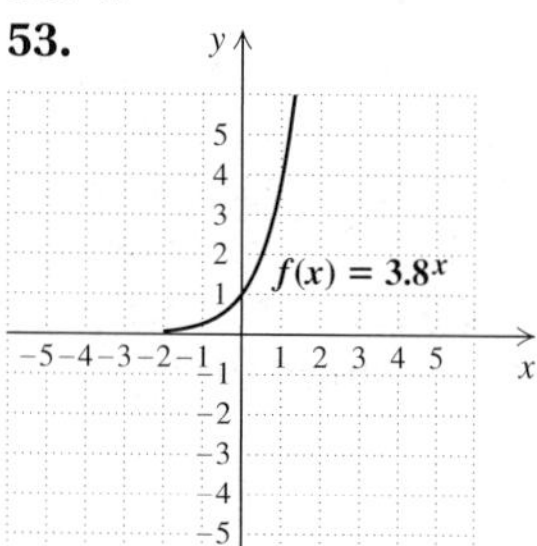

55.

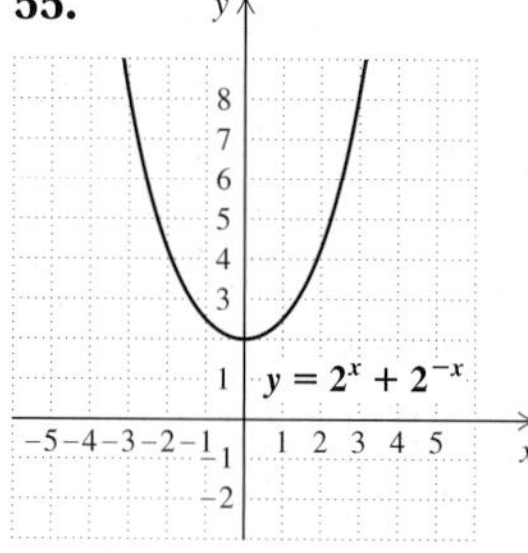

57.

$y = |2^x - 2|$

59.

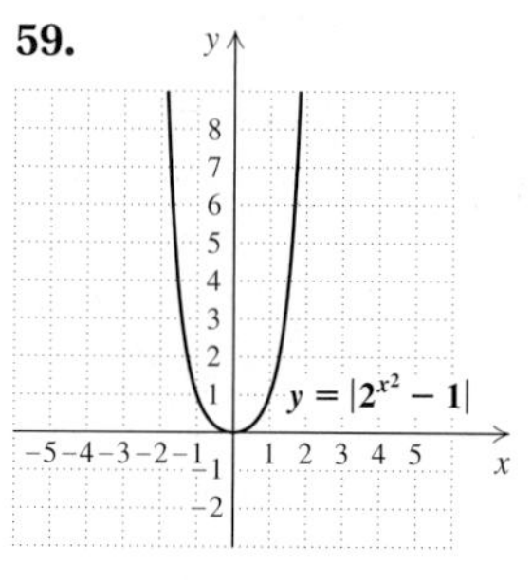

61.

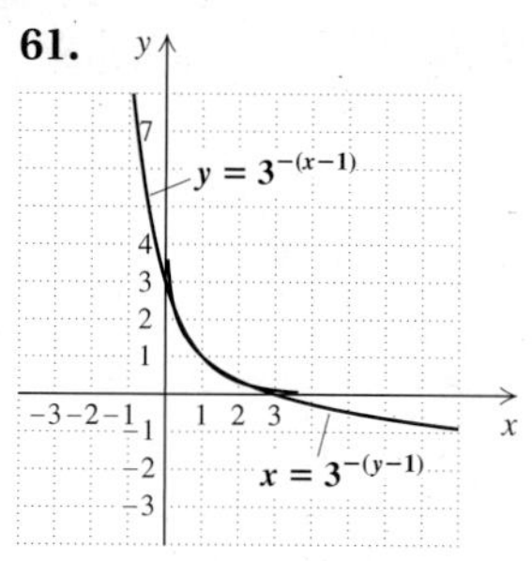

63. $A(t) = 169.3393318(2.535133248)^t$, where $A(t)$ is total sales, in millions of dollars, t years after 1997; \$288,911,061,500 **65.** 19 wpm, 66 wpm, 110 wpm

Exercise Set 9.3, pp. 575–576

1. 2 **3.** 3 **5.** 4 **7.** -2 **9.** -1 **11.** 4 **13.** 1 **15.** 0 **17.** 7 **19.** -1 **21.** $\frac{1}{2}$ **23.** $\frac{3}{2}$ **25.** $\frac{2}{3}$ **27.** 7

29.

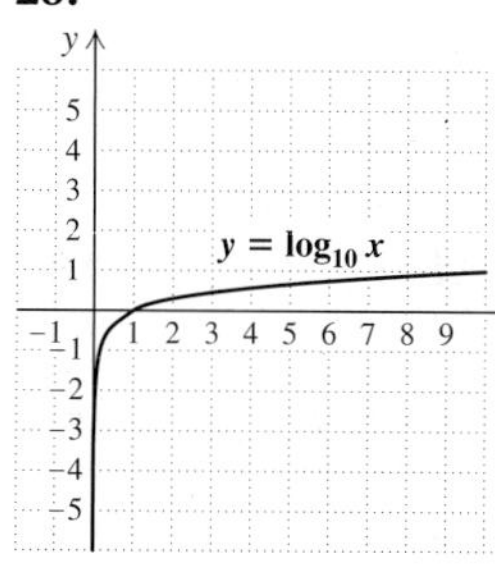

31.

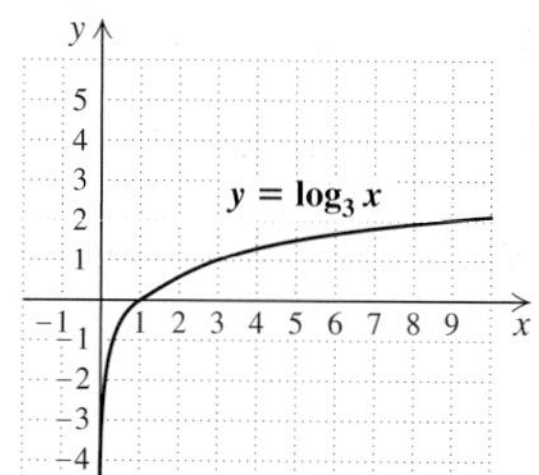

33.

$f(x) = \log_6 x$

35.

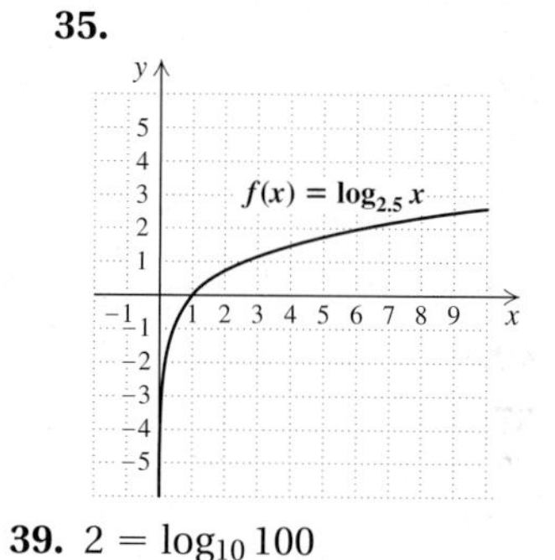

37.

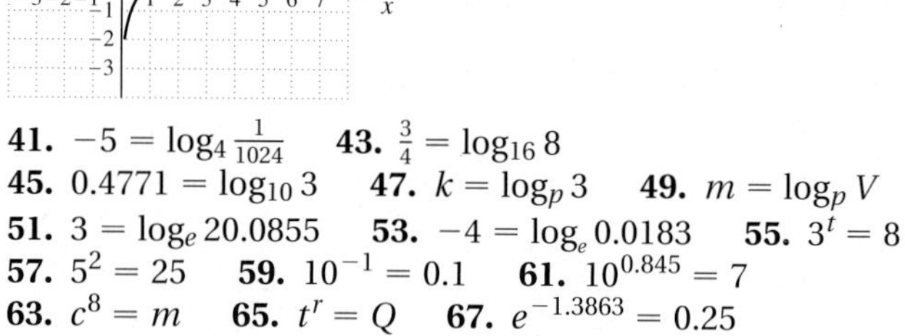

39. $2 = \log_{10} 100$

41. $-5 = \log_4 \frac{1}{1024}$ **43.** $\frac{3}{4} = \log_{16} 8$ **45.** $0.4771 = \log_{10} 3$ **47.** $k = \log_p 3$ **49.** $m = \log_p V$ **51.** $3 = \log_e 20.0855$ **53.** $-4 = \log_e 0.0183$ **55.** $3^t = 8$ **57.** $5^2 = 25$ **59.** $10^{-1} = 0.1$ **61.** $10^{0.845} = 7$ **63.** $c^8 = m$ **65.** $t^r = Q$ **67.** $e^{-1.3863} = 0.25$

69. $r^{-x} = T$ **71.** 9 **73.** 4 **75.** 2 **77.** 2 **79.** 7
81. 9 **83.** $\frac{1}{9}$ **85.** 4 **87.** **89.** [1.6] x^8
90. [1.6] a^{12} **91.** [1.6] a^7b^8 **92.** [1.6] x^5y^{12}
93. [6.3] $\dfrac{x(3y-2)}{2y+x}$ **94.** [6.3] $\dfrac{x+2}{x+1}$ **95.**

97.

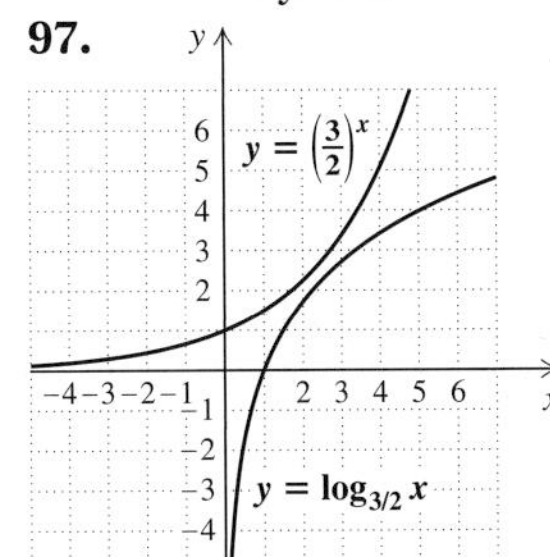

99.

$y = \log_3 |x + 1|$

101. 25 **103.** $-\frac{7}{16}$ **105.** 3 **107.** 1 **109.** 1
111.

Exercise Set 9.4, pp. 582–583

1. $\log_3 81 + \log_3 27$ **3.** $\log_4 64 + \log_4 16$
5. $\log_c r + \log_c s + \log_c t$ **7.** $\log_a(5 \cdot 14)$, or $\log_a 70$
9. $\log_c(t \cdot y)$ **11.** $8 \log_a r$ **13.** $6 \log_c y$ **15.** $-3 \log_b C$
17. $\log_2 53 - \log_2 17$ **19.** $\log_b m - \log_b n$
21. $\log_a \dfrac{15}{3}$, or $\log_a 5$ **23.** $\log_b \frac{36}{4}$, or $\log_b 9$ **25.** $\log_a \frac{7}{18}$
27. $5 \log_a x + 7 \log_a y + 6 \log_a z$
29. $\log_b x + 2 \log_b y - 3 \log_b z$
31. $4 \log_a x - 3 \log_a y - \log_a z$
33. $\log_b x + 2 \log_b y - \log_b w - 3 \log_b z$
35. $\frac{1}{2}(7 \log_a x - 5 \log_a y - 8 \log_a z)$
37. $\frac{1}{3}(6 \log_a x + 3 \log_a y - 2 - 7 \log_a z)$ **39.** $\log_a x^7z^3$
41. $\log_a x$ **43.** $\log_a \dfrac{y^5}{x^{3/2}}$ **45.** $\log_a(x-2)$ **47.** 1.953
49. -0.369 **51.** -1.161 **53.** $\frac{3}{2}$ **55.** Cannot be found
57. 3.114 **59.** 9 **61.** m **63.**
65. [2.2]

$f(x) = \sqrt{x} - 3$

66. [2.2]

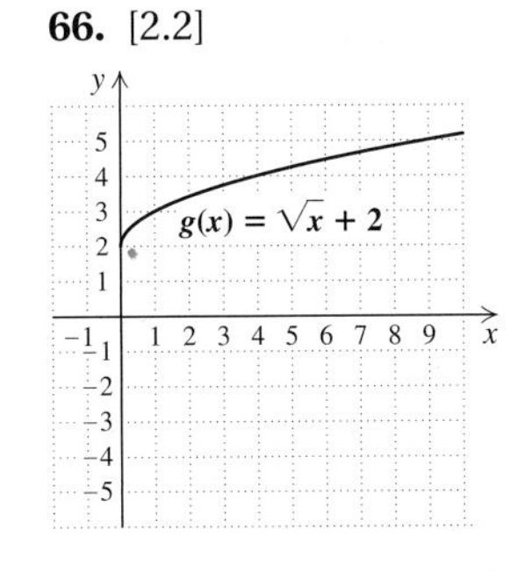

67. [2.2]

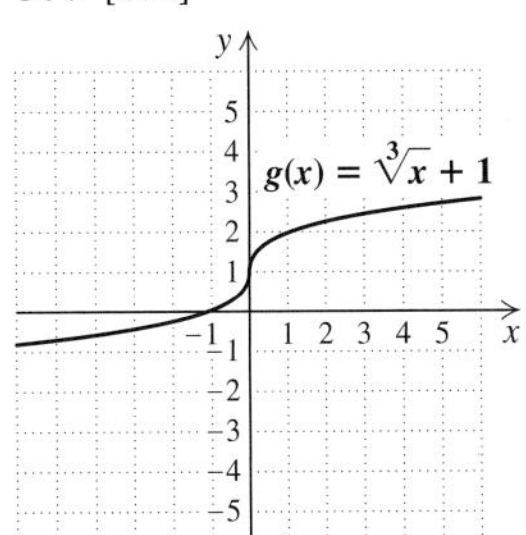

68. [2.2]

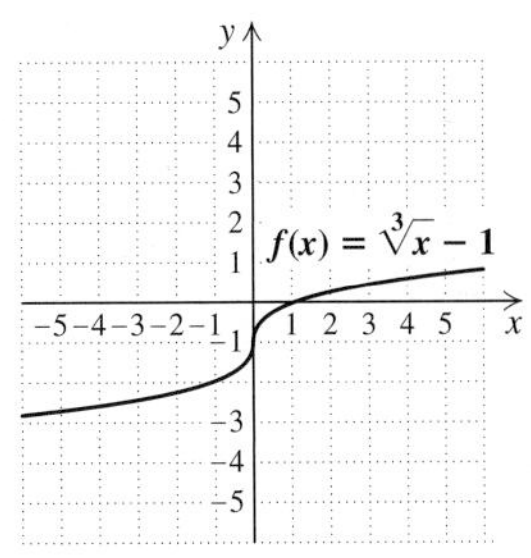

69. [1.6] $a^{17}b^{17}$ **70.** [1.6] $x^{11}y^6z^8$ **71.**
73. $\log_a(x^6 - x^4y^2 + x^2y^4 - y^6)$
75. $\frac{1}{2}\log_a(1-s) + \frac{1}{2}\log_a(1+s)$ **77.** $\frac{10}{3}$ **79.** -2
81. True

Technology Connection, p. 588

1.

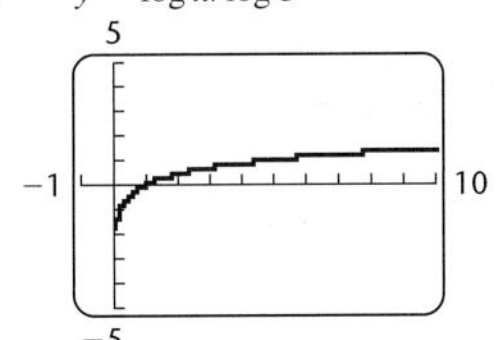

2. y = log x/log 7

3. y = log (x+2)/log 5

4. y = log x/log 7 + 2

Exercise Set 9.5, pp. 589–590

1. 0.7782 **3.** 1.8621 **5.** 3 **7.** -0.2782 **9.** 1.7986
11. 199.5262 **13.** 1.4894 **15.** 0.0011 **17.** 1.6094
19. 4.0431 **21.** -5.0832 **23.** 96.7583 **25.** 15.0293
27. 0.0305 **29.** 109.9472 **31.** 2.5237 **33.** 6.6439
35. 2.1452 **37.** -2.3219 **39.** -2.3219 **41.** 3.5471
43.

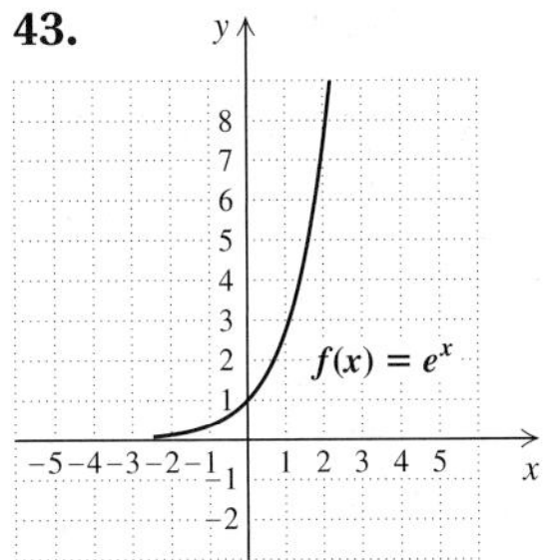

Domain: $\mathbb{R}$; range: $(0, \infty)$

45.

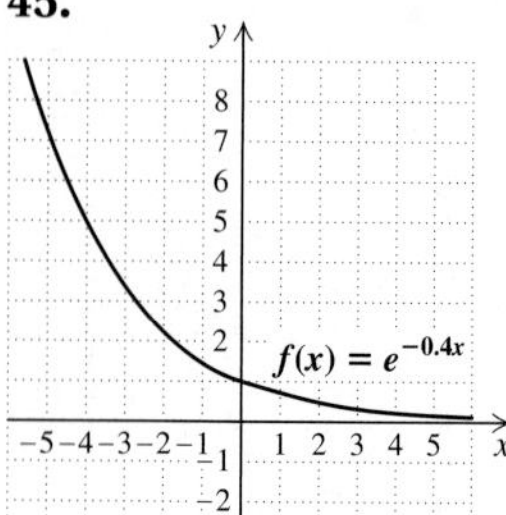

Domain: $\mathbb{R}$; range: $(0, \infty)$

47.

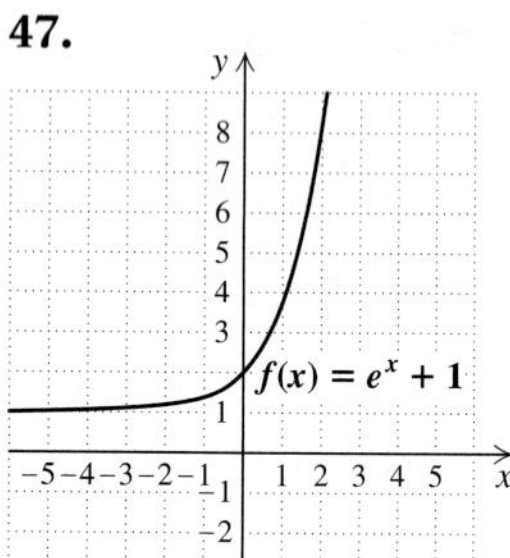

Domain: $\mathbb{R}$; range: $(1, \infty)$

49.

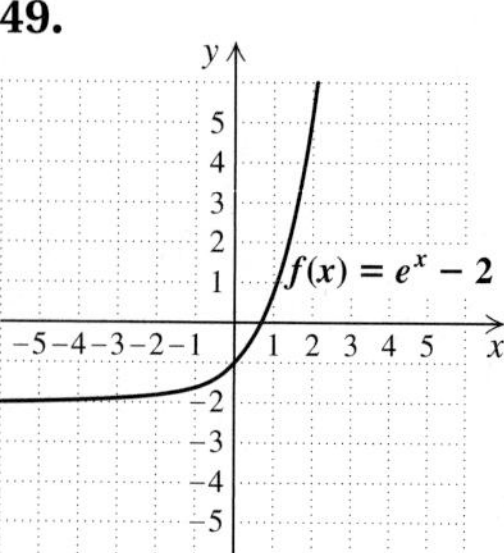

Domain: $\mathbb{R}$; range: $(-2, \infty)$

51.

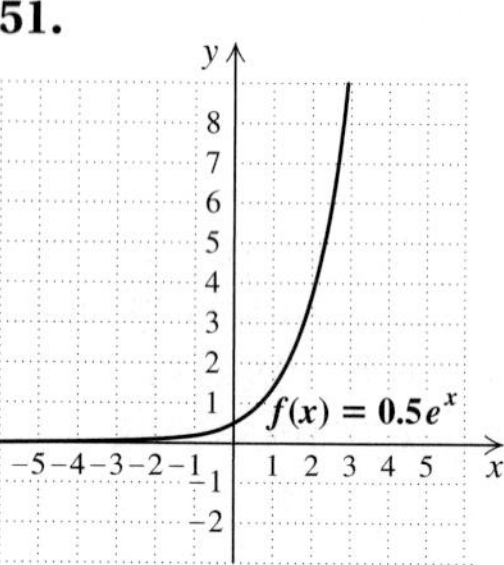
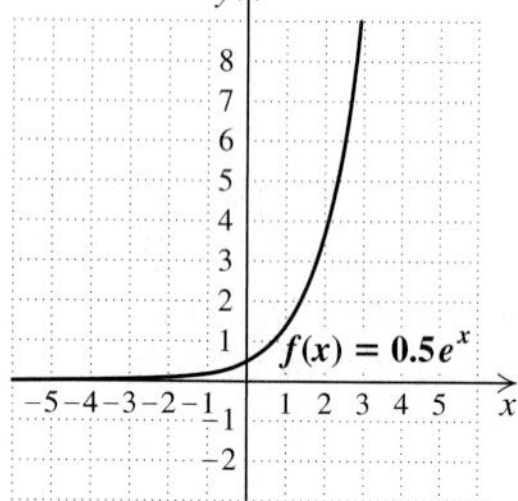

Domain: $\mathbb{R}$; range: $(0, \infty)$

53.

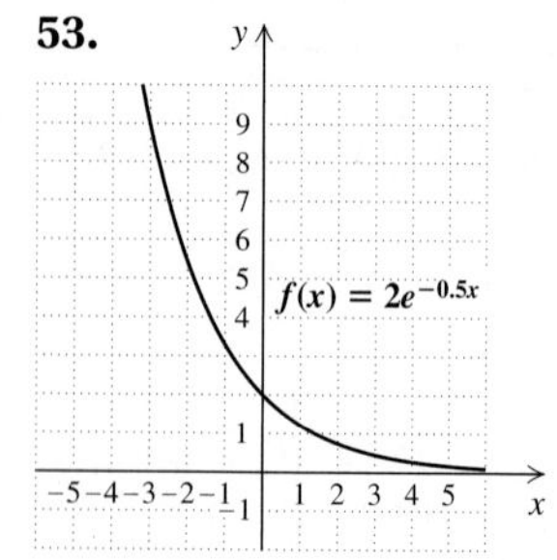

Domain: $\mathbb{R}$; range: $(0, \infty)$

55.

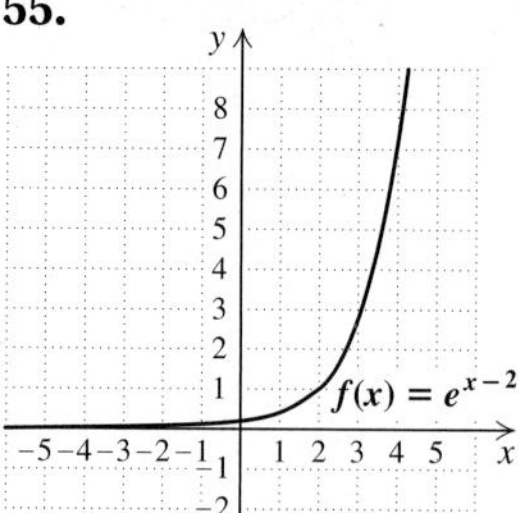

Domain: $\mathbb{R}$; range: $(0, \infty)$

57.

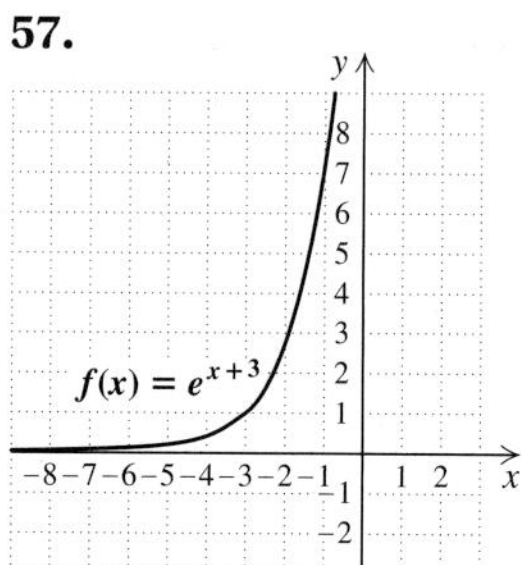

Domain: $\mathbb{R}$; range: $(0, \infty)$

59.

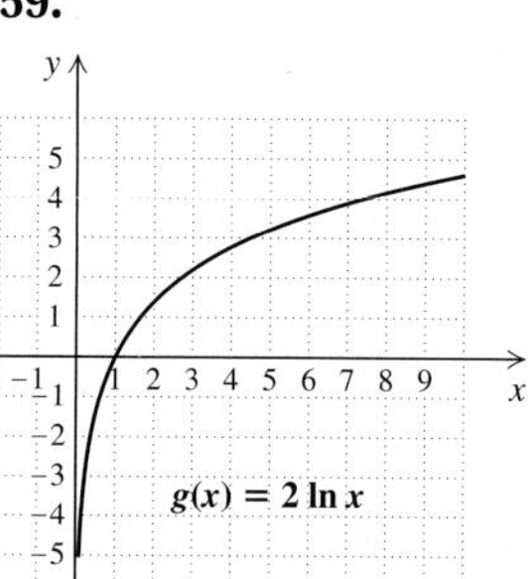
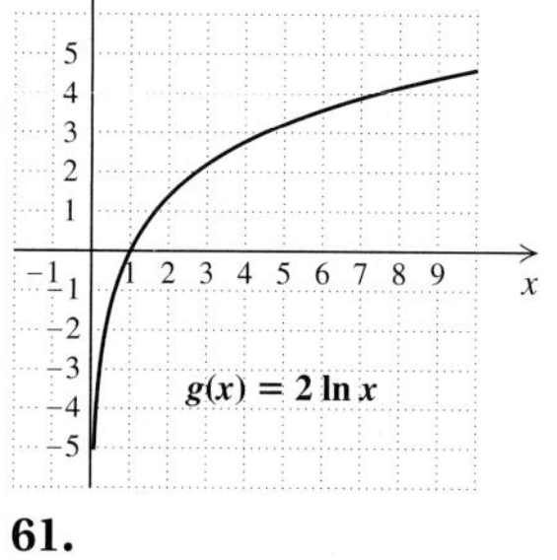

Domain: $(0, \infty)$; range: $\mathbb{R}$

61.

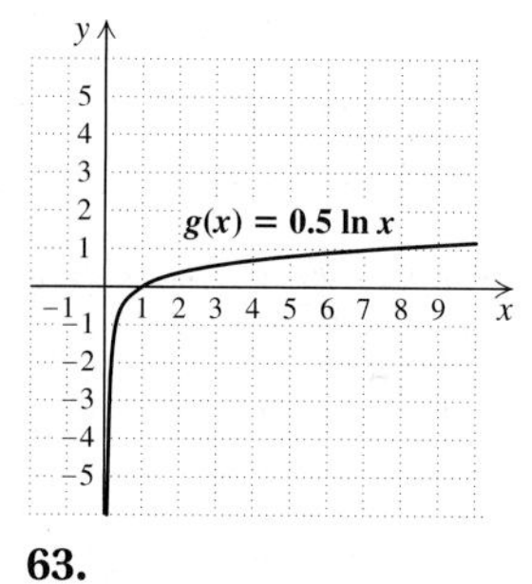

Domain: $(0, \infty)$; range: $\mathbb{R}$

63.

$g(x) = \ln x + 3$

Domain: $(0, \infty)$; range: $\mathbb{R}$

65. Domain: $(0, \infty)$; range: $\mathbb{R}$

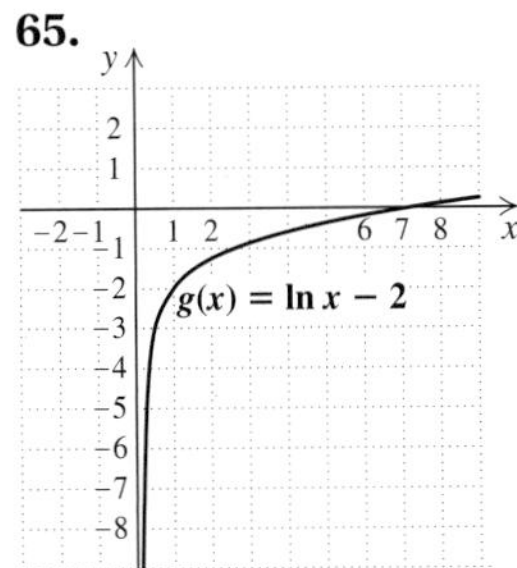

67. Domain: $(-1, \infty)$; range: $\mathbb{R}$

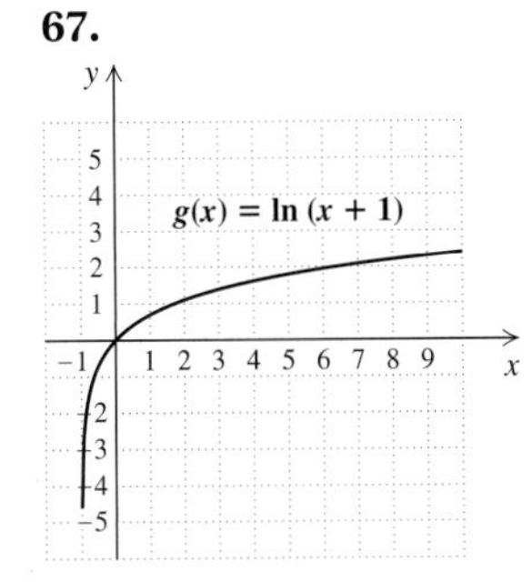

69. Domain: $(3, \infty)$; range: $\mathbb{R}$

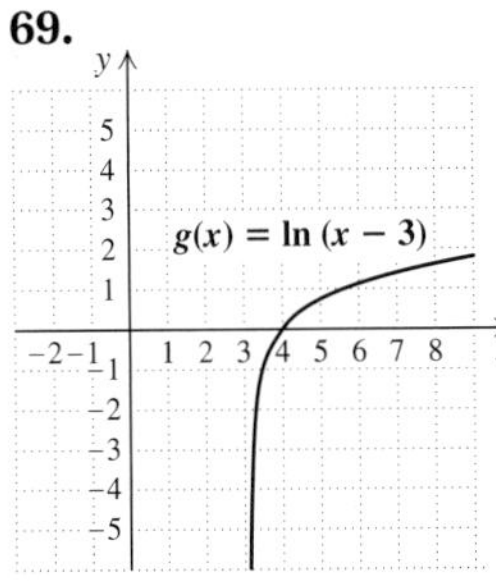

71. **73.** [5.8] $-\frac{5}{2}, \frac{5}{2}$ **74.** [5.8] $0, \frac{7}{5}$ **75.** [1.3] $\frac{15}{17}$
76. [1.3] $\frac{9}{13}$ **77.** [8.5] 16, 256 **78.** [8.5] $\frac{1}{4}, 9$ **79.**
81. 2.452 **83.** 1.442 **85.** $\log M = \dfrac{\ln M}{\ln 10}$
87. 1086.5129 **89.** 4.9855
91. **(a)** Domain: $\{x \mid x > 0\}$, or $(0, \infty)$; range: $\{y \mid y < 0.5135\}$, or $(-\infty, 0.5135)$; **(b)** $[-1, 5, -10, 5]$;
(c) $y = 3.4 \ln x - 0.25e^x$

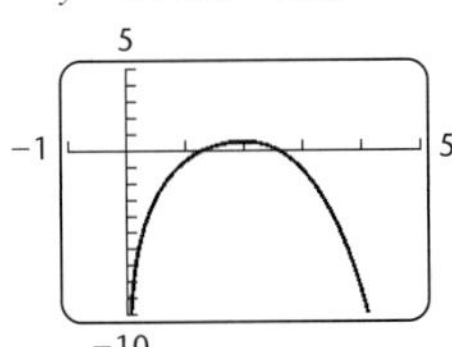

93. **(a)** Domain: $\{x \mid x > 0\}$, or $(0, \infty)$; range: $\{y \mid y > -0.2453\}$, or $(-0.2453, \infty)$; **(b)** $[-1, 5, -1, 10]$;
(c) $y = 2x^3 \ln x$ **95.**

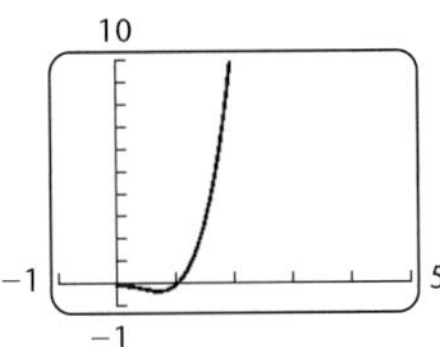

Technology Connection, p. 594

1. 0.38 **2.** −1.96 **3.** 0.90 **4.** −1.53 **5.** 0.13, 8.47
6. −0.75, 0.75

Exercise Set 9.6, pp. 595–596

1. 4 **3.** 3 **5.** 2 **7.** $\frac{4}{3}$ **9.** 0 **11.** 2 **13.** −4, 1
15. $\dfrac{\log 15}{\log 2} \approx 3.907$ **17.** $\dfrac{\log 13}{\log 4} - 1 \approx 0.850$
19. $\ln 100 \approx 4.605$ **21.** $\dfrac{\ln 0.08}{-0.07} \approx 36.082$
23. $\dfrac{\log 3}{\log 3 - \log 2} \approx 2.710$ **25.** $\dfrac{\log 4}{\log 5 - \log 4} \approx 6.213$
27. $\dfrac{\log 65}{\log 7.2} \approx 2.115$ **29.** 125 **31.** 2 **33.** 1000
35. $\frac{1}{10{,}000}$ **37.** $e \approx 2.718$ **39.** $e^{-3} \approx 0.050$ **41.** −4
43. 10 **45.** No solution **47.** $\frac{83}{15}$ **49.** 1 **51.** 5
53. $\frac{17}{2}$ **55.** 4 **57.** **59.** [6.8] $y = 9x$
60. [6.8] $y = \dfrac{21.35}{x}$ **61.** [8.3] $L = \dfrac{8T^2}{\pi^2}$
62. [8.3] $c = \sqrt{\dfrac{E}{m}}$ **63.** [6.5] $1\frac{1}{5}$ hr **64.** [6.5] $9\frac{3}{8}$ min
65. **67.** No solution **69.** −4 **71.** 2 **73.** $\pm\sqrt{34}$
75. $10^{100{,}000}$ **77.** 1, 100 **79.** $\frac{1}{100{,}000}$, 100,000 **81.** $-\frac{1}{3}$
83. 38 **85.** 1

Exercise Set 9.7, pp. 604–608

1. **(a)** 2003; **(b)** 2.2 yr **3.** **(a)** 4.2 yr; **(b)** 9.0 yr
5. **(a)** 20,114; **(b)** about 58 **7.** 4.9
9. 10^{-7} moles per liter **11.** 65 dB
13. $10^{-1.5}$ or about 3.2×10^{-2} W/m^2
15. **(a)** $P(t) = P_0e^{0.06t}$; **(b)** \$5309.18, \$5637.48; **(c)** 11.6 yr
17. **(a)** $P(t) = 283.75e^{0.013t}$, where t is the number of years after 2000 and $P(t)$ is in millions; **(b)** 302.81 million; **(c)** 2010 **19.** **(a)** 86.4 min; **(b)** 300.5 min; **(c)** 20 min
21. **(a)** 2000; **(b)** 2452; **(c)**

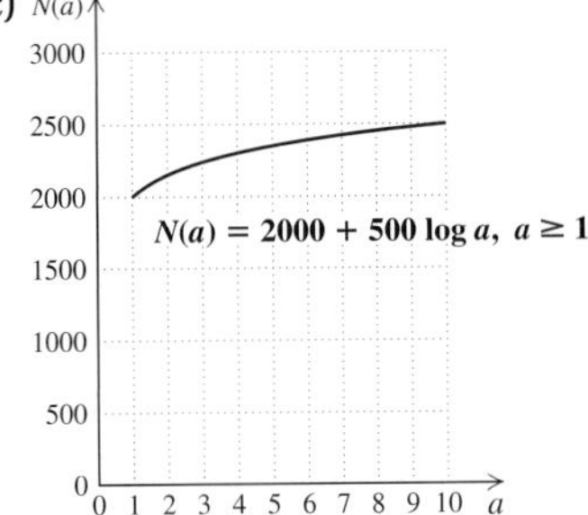

(d) \$1,000,000 thousand, or \$1,000,000,000
23. **(a)** $N(t) = 17e^{0.534t}$, where t is the number of years since 1995; **(b)** about 419 **25.** 2002
27. **(a)** $k \approx 0.094$; $W(t) = 17.5e^{-0.094t}$, where t is the number of years since 1996 and $W(t)$ is in millions of tons; **(b)** 6.8 million tons; **(c)** 2173

29. About 2103 yr **31.** About 7.2 days
33. 69.3% per year
35. **(a)** $k \approx 0.135$; $V(t) = 640{,}500e^{0.135t}$, where t is the number of years since 1996; **(b)** about \$2.47 million; **(c)** 5.1 yr; **(d)** 2004 **37.**
39. [8.7]

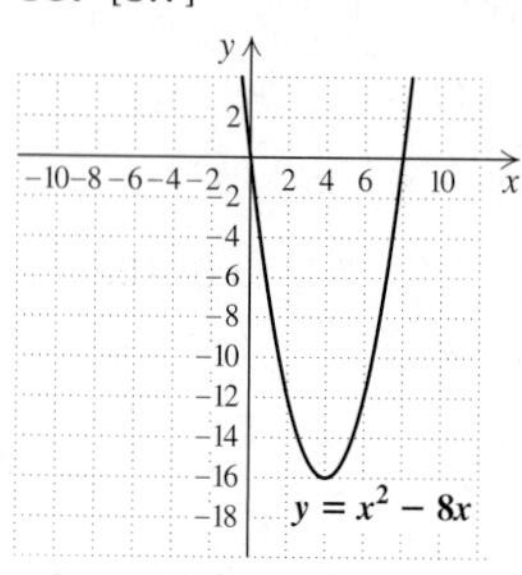

40. [8.7]

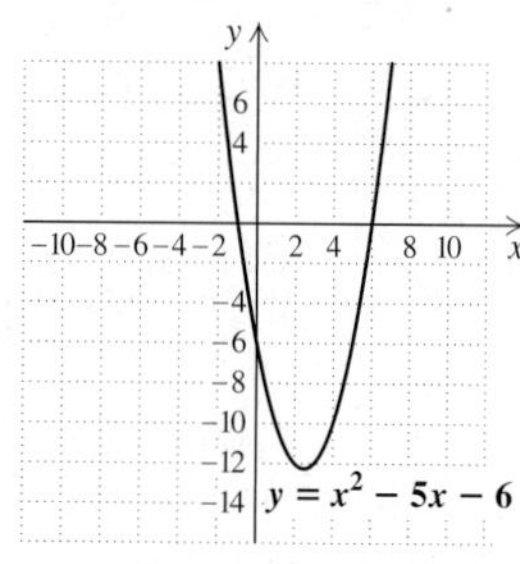

41. [8.7]

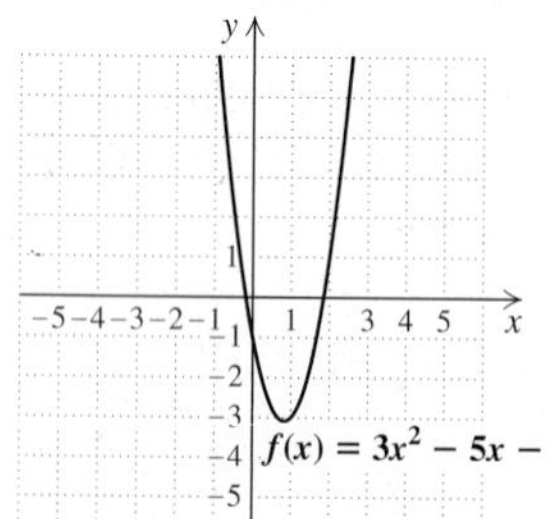

42. [8.7]

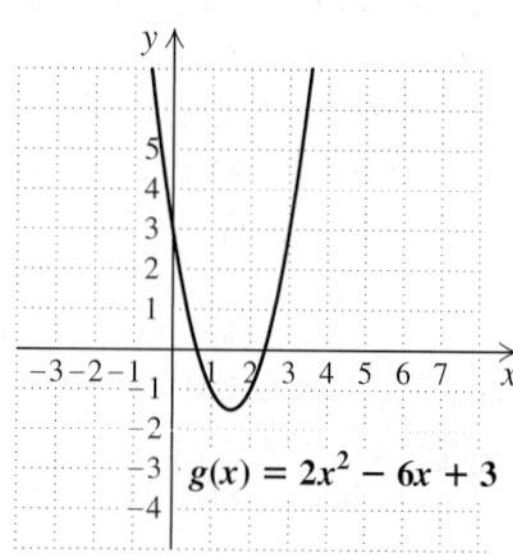

43. [8.1] $4 \pm \sqrt{23}$ **44.** [8.1] $-5 \pm \sqrt{31}$ **45.**
47. \$13.4 million **49.** $P(t) = 39e^{0.302t}$, where t is the number of years after 2001 and $P(t)$ is a percent.
51. About 80,922 yr, or with rounding, about 80,792 yr

Review Exercises: Chapter 9, pp. 612–613

1. [9.1] $(f \circ g)(x) = 4x^2 - 12x + 10$; $(g \circ f)(x) = 2x^2 - 1$
2. [9.1] $f(x) = \sqrt{x}$; $g(x) = 3 - x$ **3.** [9.1] No
4. [9.1] $f^{-1}(x) = x + 3$ **5.** [9.1] $g^{-1}(x) = \dfrac{2x - 1}{3}$
6. [9.1] $f^{-1}(x) = \dfrac{\sqrt[3]{x}}{3}$
7. [9.2]

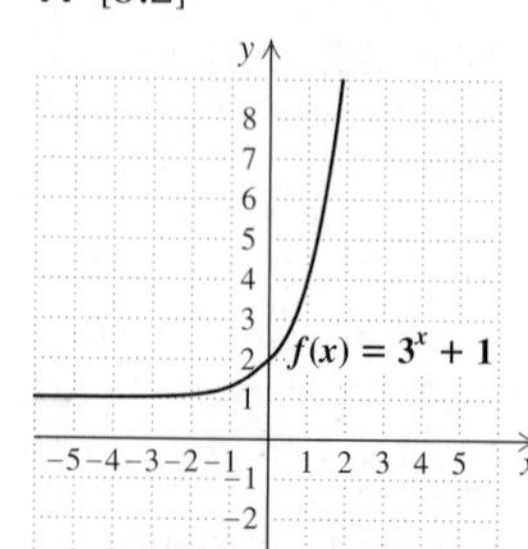

8. [9.2]

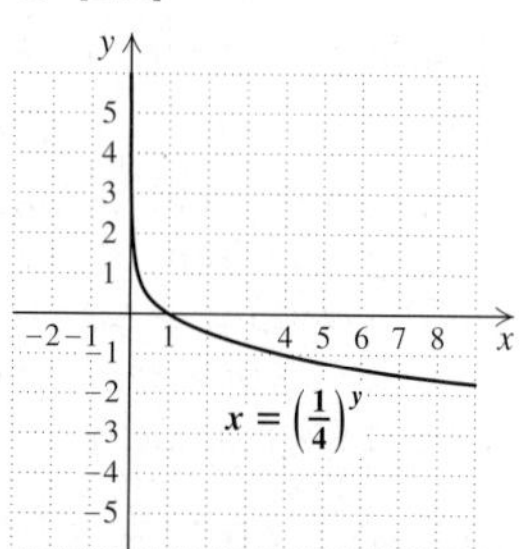

9. [9.3]

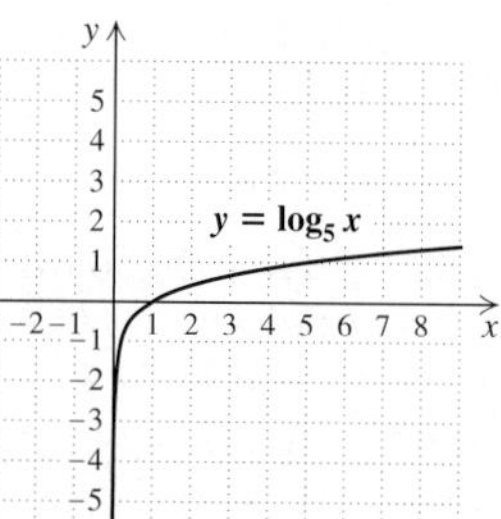

10. [9.3] 2

11. [9.3] -1 **12.** [9.3] $\frac{1}{2}$ **13.** [9.3] 12
14. [9.3] $\log_{10} \frac{1}{100} = -2$ **15.** [9.3] $\log_{25} 5 = \frac{1}{2}$
16. [9.3] $16 = 4^x$ **17.** [9.3] $1 = 8^0$
18. [9.4] $4 \log_a x + 2 \log_a y + 3 \log_a z$
19. [9.4] $3 \log_a x - (\log_a y + 2 \log_a z)$, or $3 \log_a x - \log_a y - 2 \log_a z$
20. [9.4] $\frac{1}{4}(2 \log z - 3 \log x - \log y)$
21. [9.4] $\log_a(8 \cdot 15)$, or $\log_a 120$ **22.** [9.4] $\log_a \frac{72}{12}$, or $\log_a 6$ **23.** [9.4] $\log \dfrac{a^{1/2}}{bc^2}$ **24.** [9.4] $\log_a \sqrt[3]{\dfrac{x}{y^2}}$
25. [9.4] 1 **26.** [9.4] 0 **27.** [9.4] 17 **28.** [9.4] 6.93
29. [9.4] -3.2698 **30.** [9.4] 8.7601 **31.** [9.4] 3.2698
32. [9.4] 2.54995 **33.** [9.4] -3.6602 **34.** [9.5] 1.9138
35. [9.5] 61.5177 **36.** [9.5] -2.9957 **37.** [9.5] 0.3753
38. [9.5] 0.4307 **39.** [9.5] 1.7097
40. [9.5] Domain: $\mathbb{R}$; range: $(-1, \infty)$

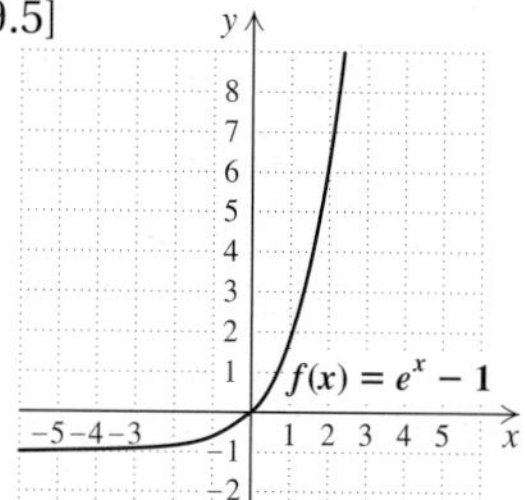

41. [9.5] Domain: $(0, \infty)$; range: $\mathbb{R}$

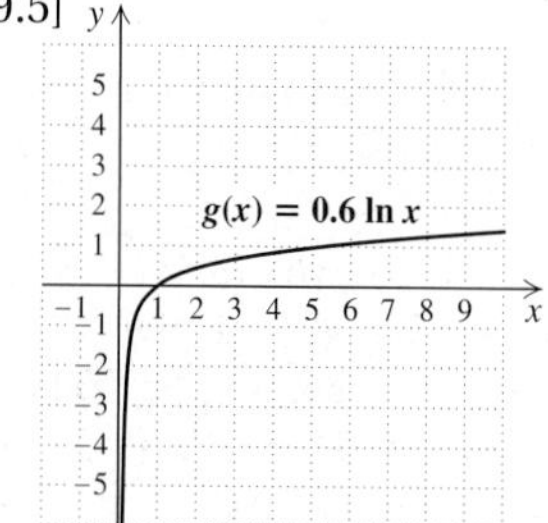

42. [9.6] 5 **43.** [9.6] -2 **44.** [9.6] $\frac{1}{9}$ **45.** [9.6] 2
46. [9.6] $\frac{1}{10{,}000}$ **47.** [9.6] $e^{-2} \approx 0.1353$ **48.** [9.6] $\frac{7}{2}$
49. [9.6] $-5, 1$ **50.** [9.6] $\dfrac{\log 8.3}{\log 4} \approx 1.5266$
51. [9.6] $\dfrac{\ln 0.03}{-0.1} \approx 35.0656$ **52.** [9.6] $e^{-3} \approx 0.0498$
53. [9.6] 4 **54.** [9.6] 8 **55.** [9.6] 20 **56.** [9.6] $\sqrt{43}$

57. [9.7] **(a)** 62; **(b)** 46.8; **(c)** 35 months
58. [9.7] **(a)** 6.6 yr; **(b)** 3.1 yr **59.** [9.7] **(a)** 0.426; $C(t) = 667e^{0.426t}$; **(b)** about 47,230; **(c)** 2001
60. [9.7] 23.105% per yr **61.** [9.7] 8.25 yr
62. [9.7] 3463 yr **63.** [9.7] 6.6 **64.** [9.7] 90 dB
65. [9.3] Negative numbers do not have logarithms because logarithm bases are positive, and there is no power to which a positive number can be raised to yield a negative number. **66.** [9.6] Taking the logarithm on each side of an equation produces an equivalent equation because the logarithm function is one-to-one. If two quantities are equal, their logarithms must be equal, and if the logarithms of two quantities are equal, the quantities must be the same. **67.** [9.6] e^{e^3} **68.** [9.6] $-3, -1$
69. [9.6] $\left(\frac{8}{3}, -\frac{2}{3}\right)$

Test: Chapter 9, pp. 613–614

1. [9.1] $(f \circ g)(x) = 2 + 6x + 4x^2$; $(g \circ f)(x) = 2x^2 + 2x + 1$ **2.** [9.1] $f(x) = \dfrac{1}{x}$; $g(x) = 2x^2 + 1$ **3.** [9.1] No **4.** [9.1] $f^{-1}(x) = \dfrac{x+3}{4}$
5. [9.1] $g^{-1}(x) = \sqrt[3]{x} - 1$
6. [9.2]

7. [9.2]

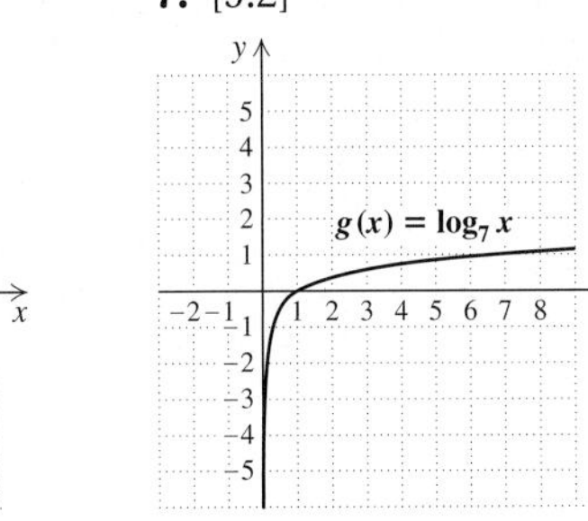

8. [9.3] 3 **9.** [9.3] $\frac{1}{2}$ **10.** [9.3] 18
11. [9.3] $\log_4 \frac{1}{64} = -3$ **12.** [9.3] $\log_{256} 16 = \frac{1}{2}$
13. [9.3] $49 = 7^m$ **14.** [9.3] $81 = 3^4$
15. [9.4] $3 \log a + \frac{1}{2} \log b - 2 \log c$
16. [9.4] $\log_a\left(z^2 \sqrt[3]{x}\right)$ **17.** [9.4] 1 **18.** [9.4] 23
19. [9.4] 0 **20.** [9.4] -0.544 **21.** [9.4] 1.079
22. [9.4] 1.204 **23.** [9.5] 1.0899 **24.** [9.5] 0.1585
25. [9.5] -3.3524 **26.** [9.5] 121.5104 **27.** [9.5] 2.0959
28. [9.5] Domain: $\mathbb{R}$; range: $(3, \infty)$

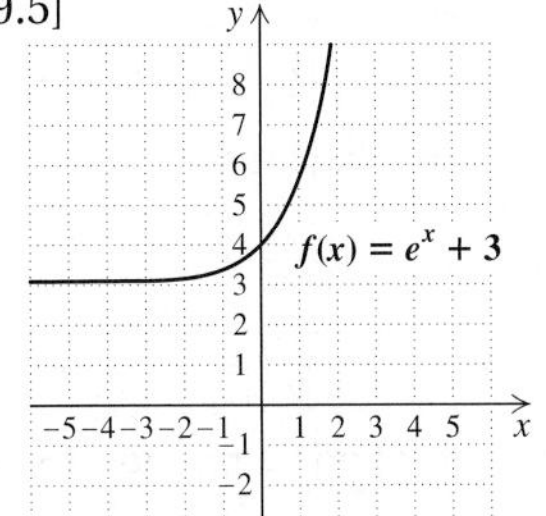

29. [9.5] Domain: $(4, \infty)$; range: $\mathbb{R}$

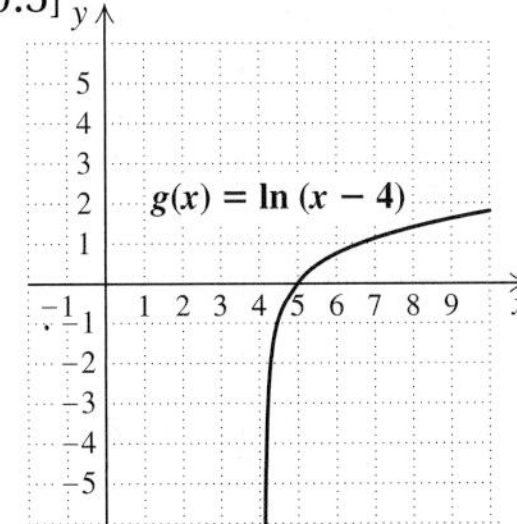

30. [9.6] -5 **31.** [9.6] 5 **32.** [9.6] 2 **33.** [9.6] 10,000
34. [9.6] $\frac{1}{3}$ **35.** [9.6] $\dfrac{\log 1.2}{\log 7} \approx 0.0937$
36. [9.6] $e^{1/4} \approx 1.2840$ **37.** [9.6] 4
38. [9.7] **(a)** 2.46 ft/sec; **(b)** 984,262
39. [9.7] **(a)** $P(t) = 30e^{0.015t}$, where $P(t)$ is in millions and t is the number of years after 2000; **(b)** 31.4 million; 34.9 million; **(c)** 2034; **(d)** 46.2 yr
40. [9.7] **(a)** $k \approx 0.027$; $C(t) = 60e^{0.027t}$; **(b)** 681.5; **(c)** 2024
41. [9.7] 4.6% **42.** [9.7] 4684 yr **43.** [9.7] $10^{-4.5}$ W/m^2
44. [9.7] 7.0 **45.** [9.6] $-309, 316$ **46.** [9.4] 2

Cumulative Review: Chapters 1–9: pp. 615–617

1. [1.1], [1.6] 2 **2.** [1.2] 6 **3.** [1.6] $\dfrac{y^{12}}{16x^8}$
4. [1.6] $\dfrac{20x^6z^2}{y}$ **5.** [1.6] $\dfrac{-y^4}{3z^5}$ **6.** [1.3] $-4x - 1$
7. [1.1] 25 **8.** [1.3] $\frac{11}{2}$ **9.** [3.2] $(3, -1)$
10. [3.4] $(1, -2, 0)$ **11.** [5.8] $-2, 5$ **12.** [6.4] $\frac{9}{2}$
13. [6.4], [8.2] $\frac{5}{8}$ **14.** [7.6] $\frac{3}{4}$ **15.** [7.6] $\frac{1}{2}$ **16.** [8.1] $\pm 5i$
17. [8.5] 9, 25 **18.** [8.5] $\pm 2, \pm 3$ **19.** [9.3] 8
20. [9.3] 7 **21.** [9.6] $\frac{3}{2}$ **22.** [9.6] $\dfrac{\log 7}{5 \log 3} \approx 0.3542$
23. [9.6] $\frac{80}{9}$ **24.** [8.9] $(-\infty, -5) \cup (1, \infty)$, or $\{x \mid x < -5 \text{ or } x > 1\}$ **25.** [8.2] $-3 \pm 2\sqrt{5}$
26. [4.3] $\{x \mid x \le -3 \text{ or } x \ge 6\}$, or $(-\infty, -3] \cup [6, \infty)$
27. [6.8] $a = \dfrac{Db}{b - D}$ **28.** [6.8] $q = \dfrac{pf}{p - f}$
29. [1.5] $B = \dfrac{3M - 2A}{2}$, or $B = \frac{3}{2}M - A$ **30.** [3.7] 2
31. [3.7] 3
32. [5.8] $\left\{x \mid x \text{ is a real number and } x \neq -\frac{1}{3} \text{ and } x \neq 2\right\}$
33. **(a)** [2.3] 1 million per year; **(b)** [2.5] $E(t) = 2.4 + t$; **(c)** [2.5] 7.4 million **34.** [1.4] Length: 36 m; width: 20 m
35. [1.4] *A*: 15°; *B*: 45°; *C*: 120° **36.** [6.5] $5\frac{5}{11}$ min
37. [3.3] Thick and Tasty: 6 oz; Light and Lean: 9 oz
38. [6.5] $2\frac{7}{9}$ km/h **39.** [8.8] -49; -7 and 7
40. [9.7] 78 **41.** [9.7] 67.5 **42.** [9.7] $P(t) = 19e^{0.015t}$
43. [9.7] 20.5 million; 22.7 million **44.** [9.7] 46.2 yr
45. [8.6] 18 **46.** [5.1] $7p^2q^3 + pq + p - 9$
47. [5.1] $8x^2 - 11x - 1$ **48.** [5.2] $9x^4 - 12x^2y + 4y^2$

49. [5.2] $10a^2 - 9ab - 9b^2$ **50.** [6.1] $\dfrac{(x+4)(x-3)}{2(x-1)}$

51. [6.3] $\dfrac{1}{x-4}$ **52.** [6.1] $\dfrac{a+2}{6}$

53. [6.2] $\dfrac{7x+4}{(x+6)(x-6)}$ **54.** [5.3] $x(y - 2z + w)$

55. [5.6] $(1-5x)(1+5x+25x^2)$

56. [5.4] $2(3x-2y)(x+2y)$ **57.** [5.3] $(x^3+7)(x-4)$

58. [5.5] $2(m+3n)^2$

59. [5.5] $(x-2y)(x+2y)(x^2+4y^2)$ **60.** [2.2] -12

61. [6.6] $x^3 - 2x^2 - 4x - 12 + \dfrac{-42}{x-3}$

62. [1.7] 1.8×10^{-1} **63.** [7.4] $2y^2\sqrt[3]{y}$

64. [7.4] $14xy^2\sqrt{x}$ **65.** [7.2] $81a^8b\sqrt[3]{b}$

66. [7.5] $\dfrac{6+\sqrt{y}-y}{4-y}$ **67.** [7.4] $\sqrt[10]{(x+5)^3}$

68. [7.8] $12 + 4\sqrt{3}i$ **69.** [7.8] $8 + i$

70. [9.1] $f^{-1}(x) = \dfrac{x-7}{-2}$, or $f^{-1}(x) = \dfrac{7-x}{2}$

71. [2.5] $f(x) = -5x - 3$ **72.** [2.5] $y = \frac{1}{2}x + 7$

73. [2.4]

$5x = 15 + 3y$

74. [8.7]

$y = 2x^2 - 4x - 1$

75. [9.3]

$y = \log_3 x$

76. [9.1]

$y = 3^x$

77. [4.4]

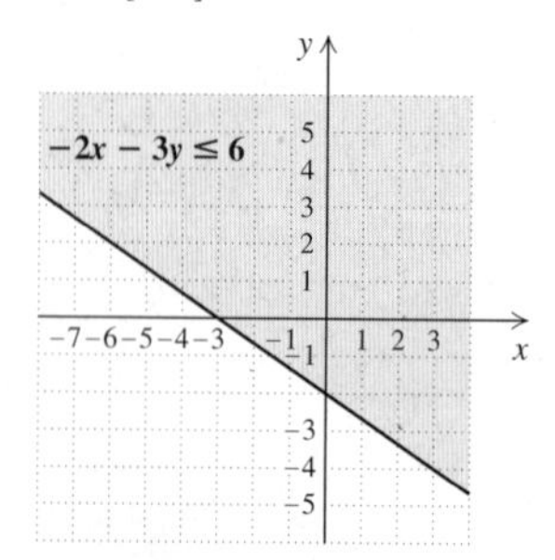

78. [8.7]

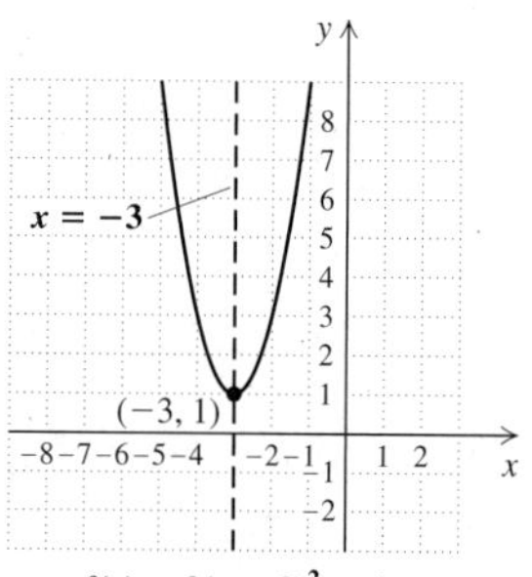

$f(x) = 2(x+3)^2 + 1$
Minimum: 1

79. [9.5] Domain: $\mathbb{R}$; range: $(0, \infty)$

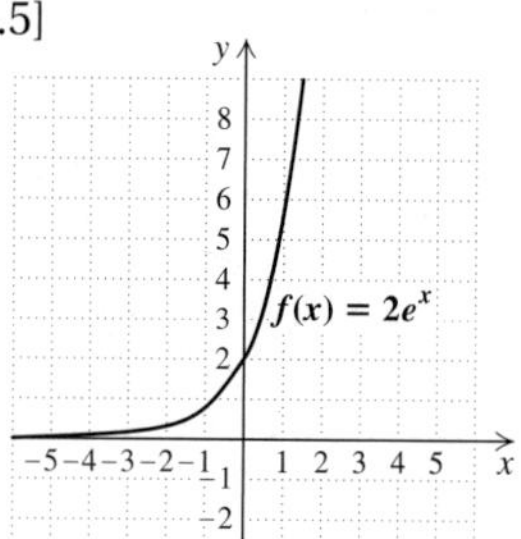

80. [9.4] $2\log a + 3\log c - \log b$ **81.** [9.4] $\log\left(\dfrac{x^3}{y^{1/2}z^2}\right)$

82. [9.3] $a^x = 5$ **83.** [9.3] $\log_x t = 3$ **84.** [9.5] -1.2545

85. [9.5] 776.2471 **86.** [9.5] 2.5479 **87.** [9.5] 0.2466

88. [6.4] All real numbers except 1 and -2

89. [9.6] $\frac{1}{3}, \frac{10{,}000}{3}$ **90.** [8.3] 35 mph

CHAPTER 10

Technology Connection, p. 628

1.

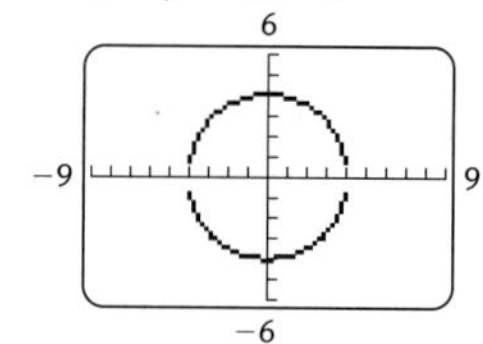

2.

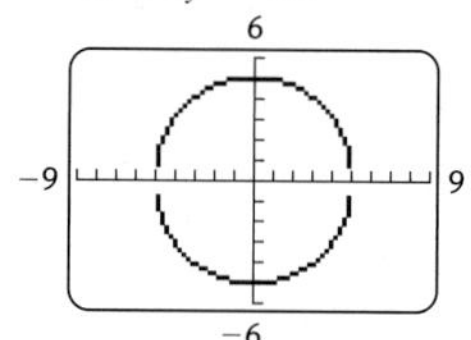

3.

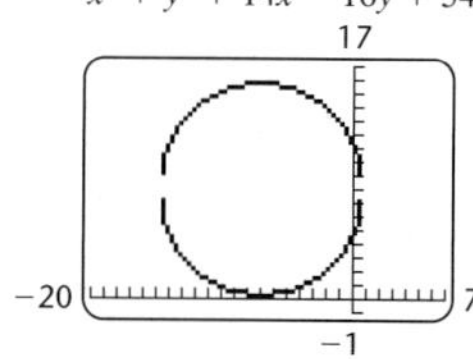

4.

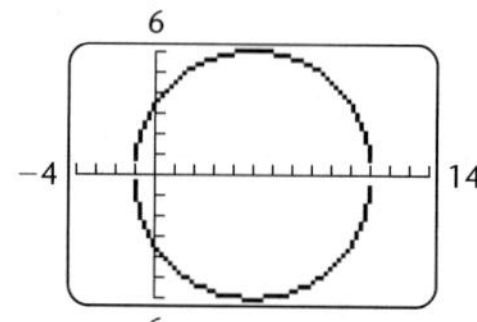

Exercise Set 10.1, pp. 628–631

1.

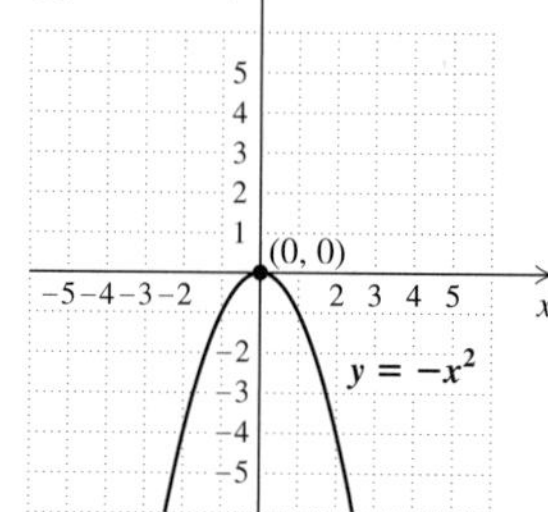

3.

$y = -x^2 + 4x - 5$

(2, −1)

5.

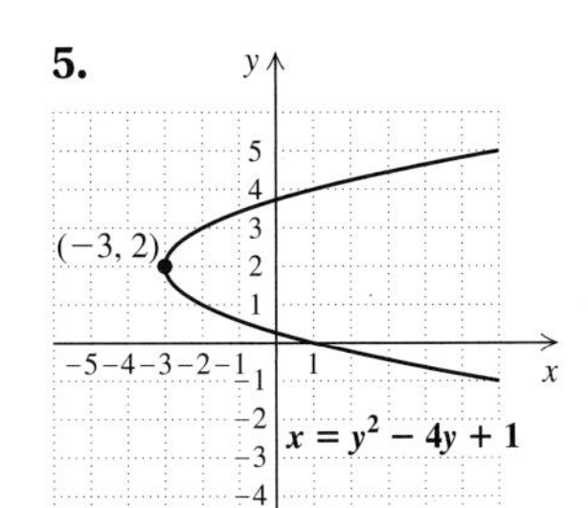

7.

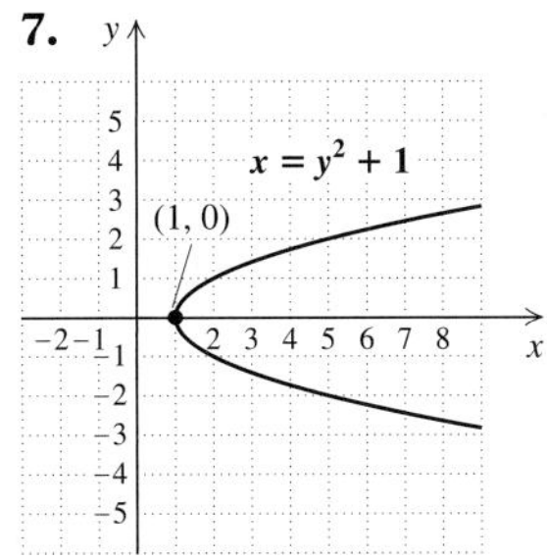

9.

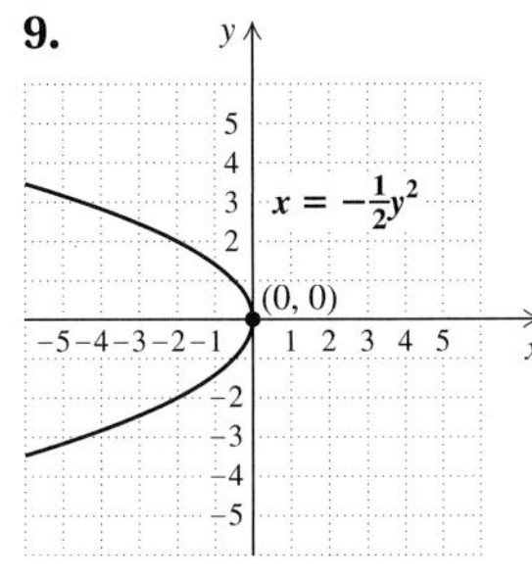

11.

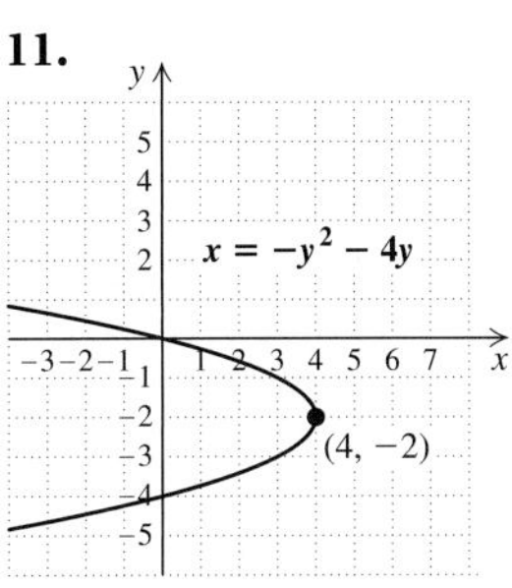

13.

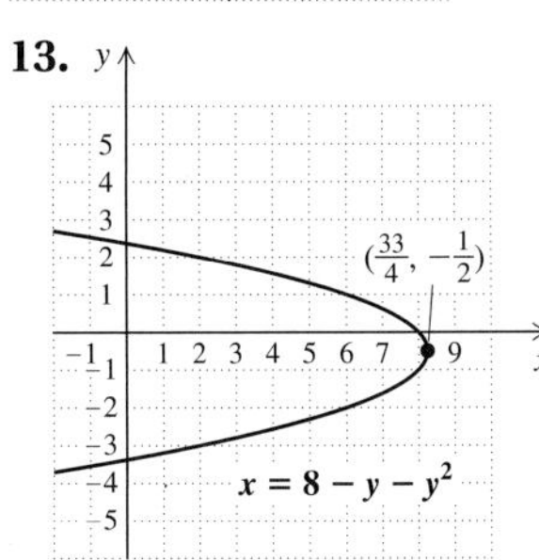

15.

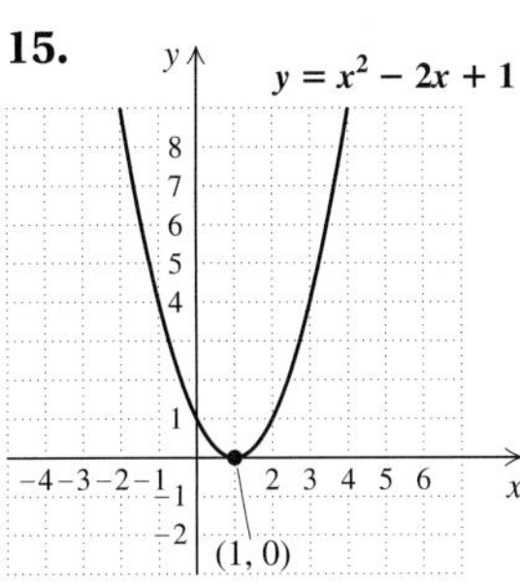

17.

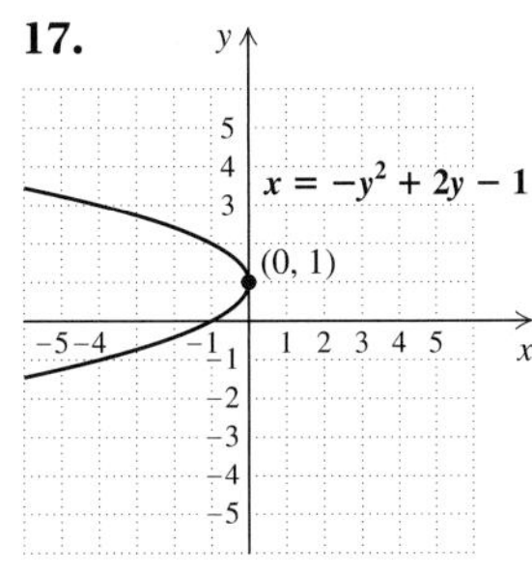

19.

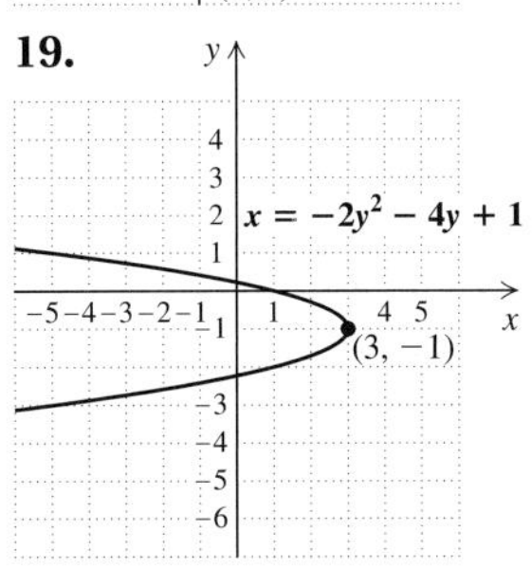

21. 5 **23.** $\sqrt{18} \approx 4.243$ **25.** $\sqrt{200} \approx 14.142$
27. 17.8 **29.** $\dfrac{\sqrt{41}}{7} \approx 0.915$ **31.** $\sqrt{8} \approx 2.828$
33. $\sqrt{17 + 2\sqrt{14} + 2\sqrt{15}} \approx 5.677$ **35.** $\sqrt{s^2 + t^2}$
37. $(1, 4)$ **39.** $\left(\frac{7}{2}, \frac{7}{2}\right)$ **41.** $(-1, -3)$ **43.** $(-0.25, -0.3)$
45. $\left(-\frac{1}{12}, \frac{1}{24}\right)$ **47.** $\left(\dfrac{\sqrt{2} + \sqrt{3}}{2}, \dfrac{3}{2}\right)$ **49.** $x^2 + y^2 = 36$
51. $(x - 7)^2 + (y - 3)^2 = 5$
53. $(x + 4)^2 + (y - 3)^2 = 48$

55. $(x + 7)^2 + (y + 2)^2 = 50$ **57.** $x^2 + y^2 = 25$
59. $(x + 4)^2 + (y - 1)^2 = 20$
61. $(0, 0)$; 7 **63.** $(-1, -3)$; 2

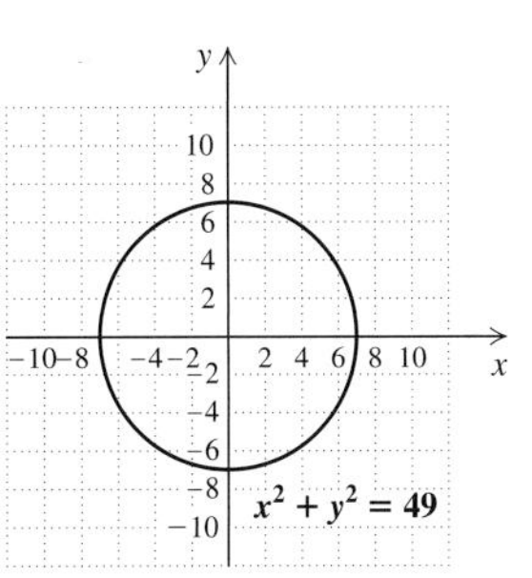

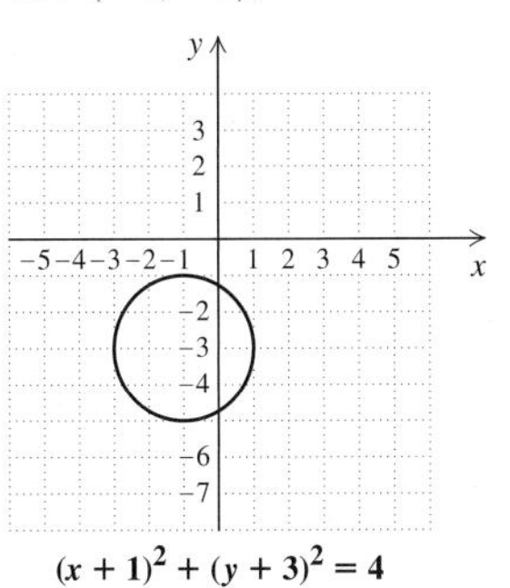

$(x + 1)^2 + (y + 3)^2 = 4$

65. $(4, -3)$; $\sqrt{10}$ **67.** $(0, 0)$; $\sqrt{7}$

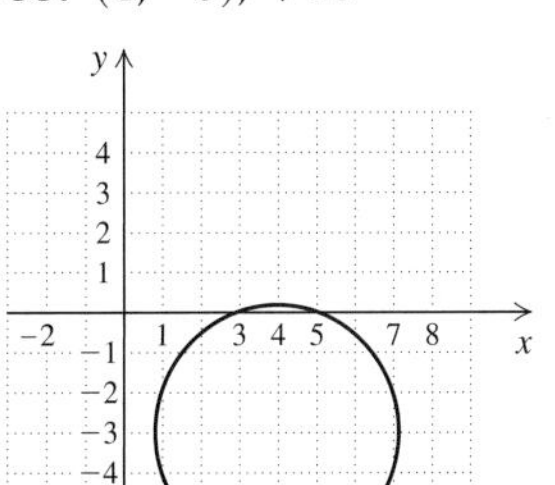

$(x - 4)^2 + (y + 3)^2 = 10$

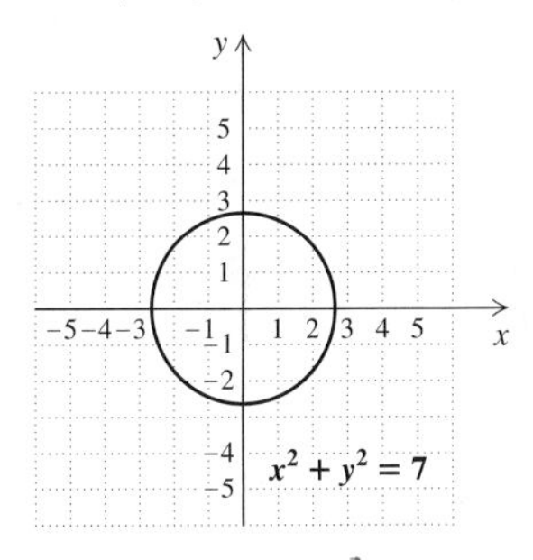

69. $(5, 0)$; $\frac{1}{2}$ **71.** $(-4, 3)$; $\sqrt{40}$, or $2\sqrt{10}$

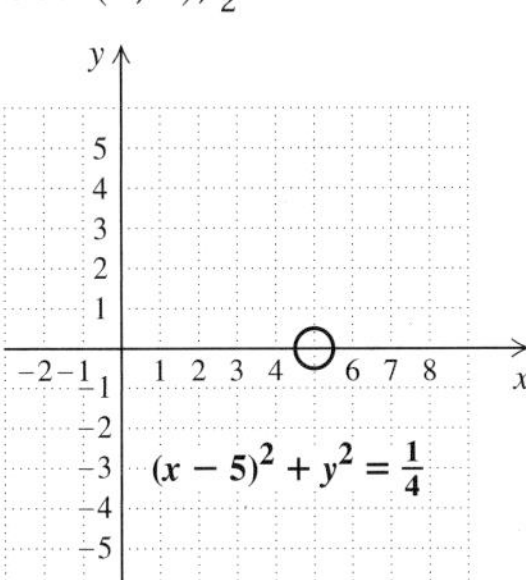

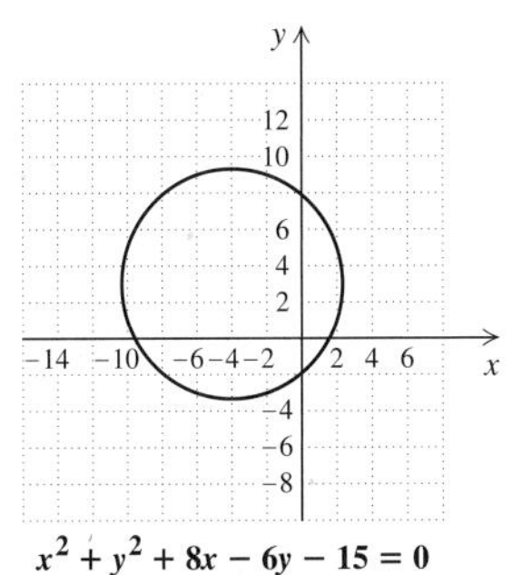

$x^2 + y^2 + 8x - 6y - 15 = 0$

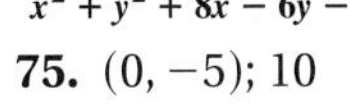

73. $(4, -1)$; 2 **75.** $(0, -5)$; 10

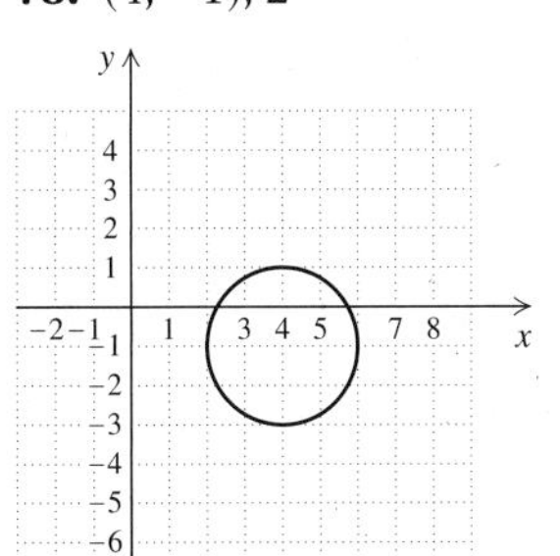

$x^2 + y^2 - 8x + 2y + 13 = 0$

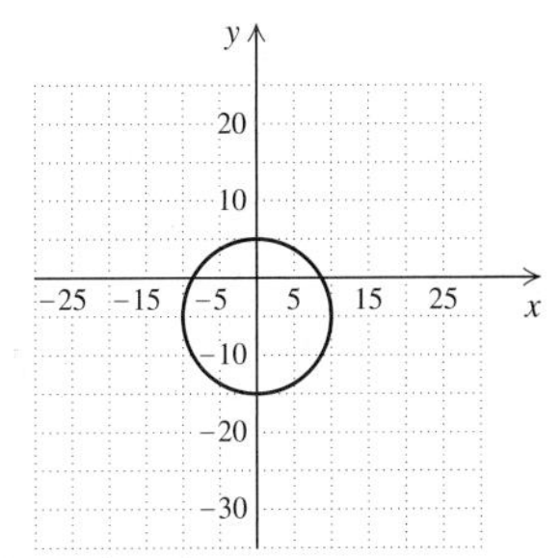

$x^2 + y^2 + 10y - 75 = 0$

77. $\left(-\frac{7}{2}, \frac{3}{2}\right)$; $\sqrt{\frac{98}{4}}$, or $\frac{7\sqrt{2}}{2}$

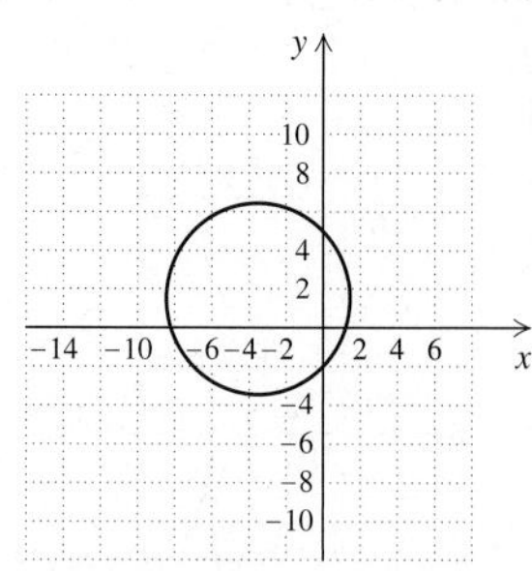

79. $(0, 0)$; $\frac{1}{6}$

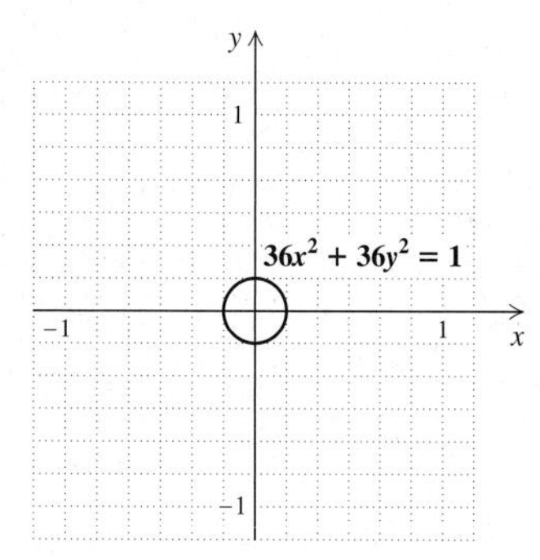

81. ▯ **83.** [6.4] $-\frac{2}{3}$ **84.** [6.4] $\frac{25}{6}$ **85.** [5.8] 4 in.
86. [1.4] 2640 mi **87.** [3.2] $\left(\frac{35}{17}, \frac{5}{34}\right)$ **88.** [3.2] $\left(0, -\frac{9}{5}\right)$
89. ▯ **91.** $(x - 3)^2 + (y + 5)^2 = 9$
93. $(x - 3)^2 + y^2 = 25$ **95.** $(0, 4)$ **97.** **(a)** $(0, -8467.8)$; **(b)** 8487.3 mm **99.** **(a)** $(0, -1522.8)$; **(b)** 1524.9 cm
101. 29 cm **103.** **(a)** $-2.4, 3.4$; **(b)** $-1.3, 2.3$
105. Let $P_1 = (x_1, y_1)$, $P_2 = (x_2, y_2)$, and $M = \left(\frac{x_1 + x_2}{2}, \frac{y_1 + y_2}{2}\right)$. Let $d(AB)$ denote the distance from point A to point B.

(i)
$$d(P_1M) = \sqrt{\left(\frac{x_1 + x_2}{2} - x_1\right)^2 + \left(\frac{y_1 + y_2}{2} - y_1\right)^2} = \frac{1}{2}\sqrt{(x_2 - x_1)^2 + (y_2 - y_1)^2};$$

$$d(P_2M) = \sqrt{\left(\frac{x_1 + x_2}{2} - x_2\right)^2 + \left(\frac{y_1 + y_2}{2} - y_2\right)^2} = \frac{1}{2}\sqrt{(x_1 - x_2)^2 + (y_1 - y_2)^2} = \frac{1}{2}\sqrt{(x_2 - x_1)^2 + (y_2 - y_1)^2} = d(P_1M).$$

(ii)
$$d(P_1M) + d(P_2M) = \frac{1}{2}\sqrt{(x_2 - x_1)^2 + (y_2 - y_1)^2} + \frac{1}{2}\sqrt{(x_2 - x_1)^2 + (y_2 - y_1)^2} = \sqrt{(x_2 - x_1)^2 + (y_2 - y_1)^2} = d(P_1P_2).$$

107. ▯, ▯

Exercise Set 10.2, pp. 636–637

1.

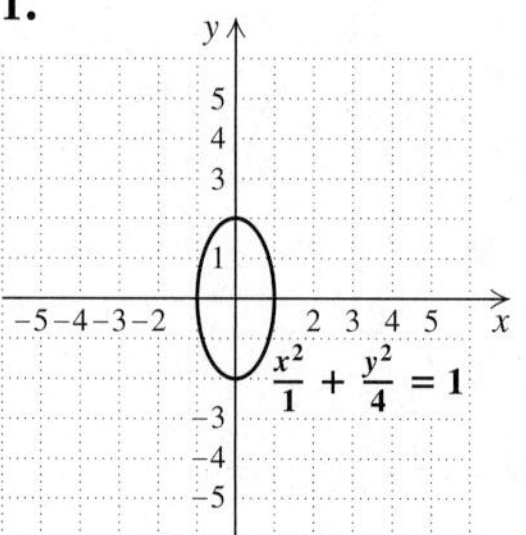

3.

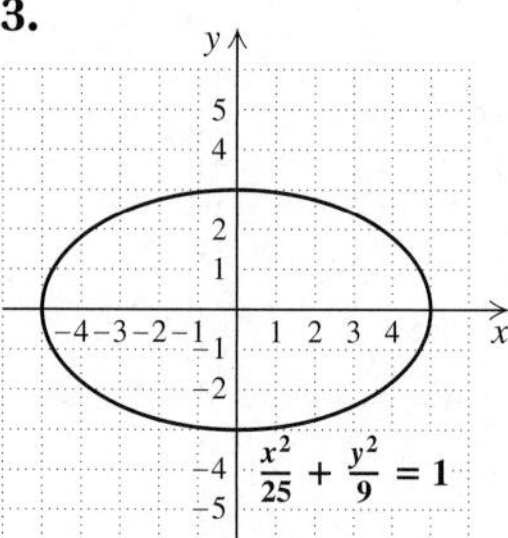

5.

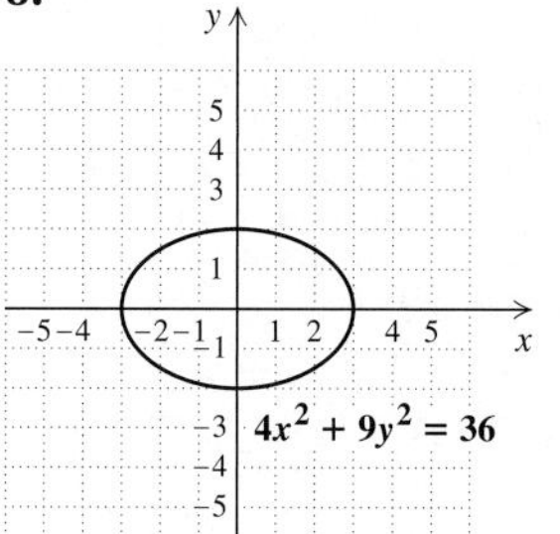

7.

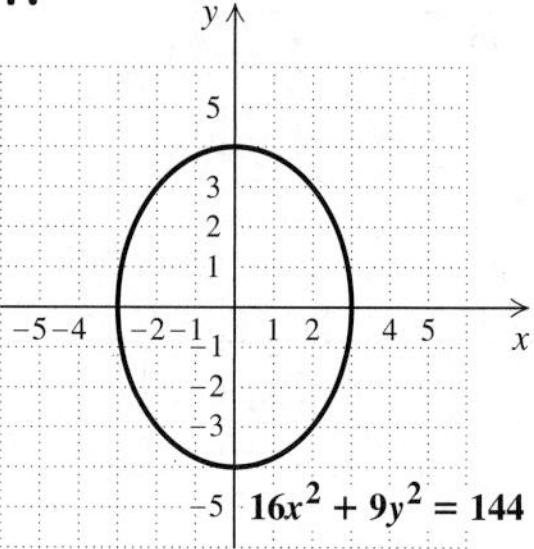

9.

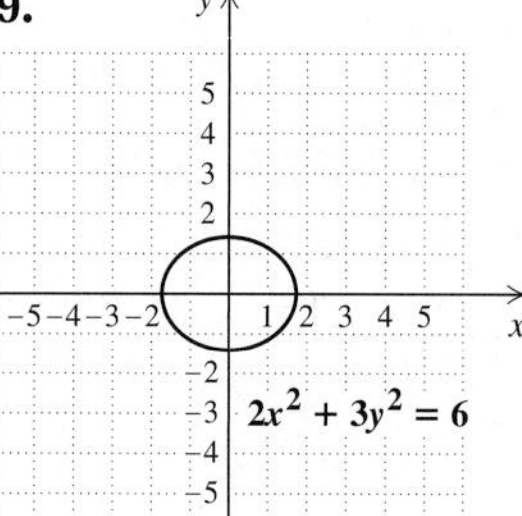

11.

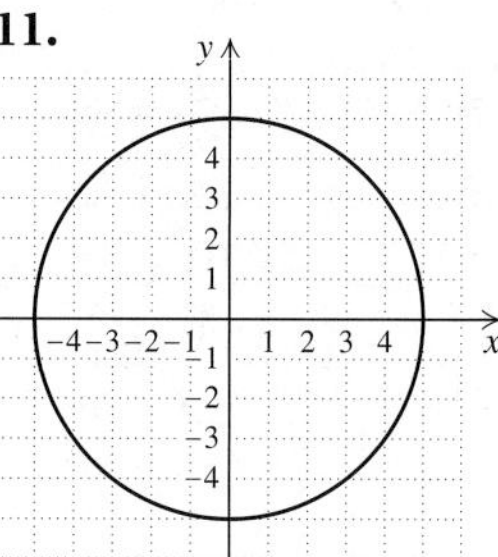

$5x^2 + 5y^2 = 125$

13.

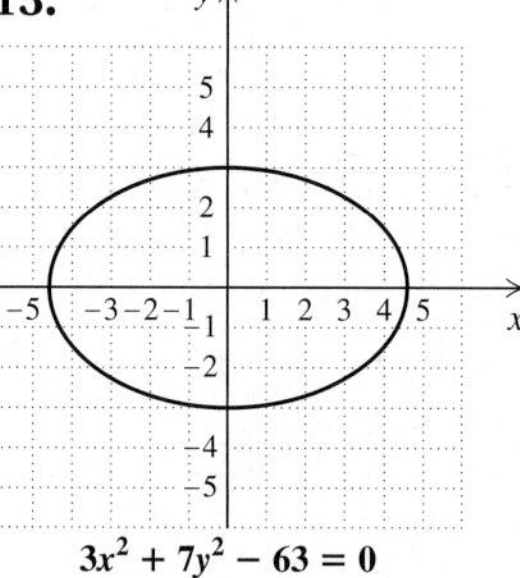

$3x^2 + 7y^2 - 63 = 0$

15.

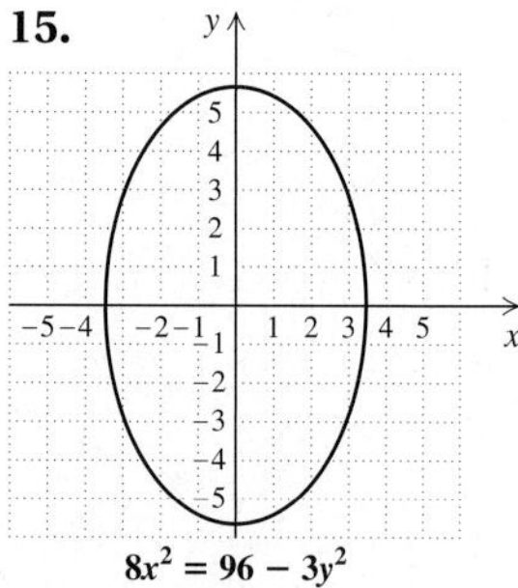

$8x^2 = 96 - 3y^2$

17.

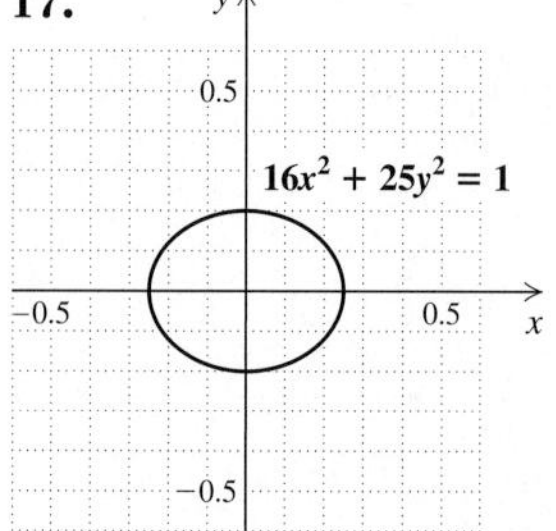

19.

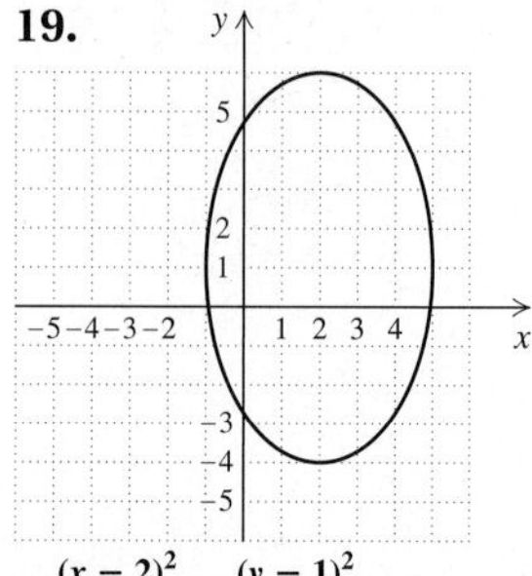

$\frac{(x - 2)^2}{9} + \frac{(y - 1)^2}{25} = 1$

21.

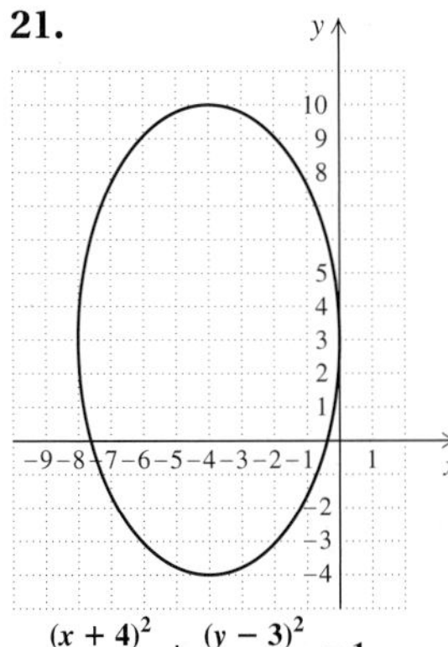

$\frac{(x+4)^2}{16} + \frac{(y-3)^2}{49} = 1$

23.

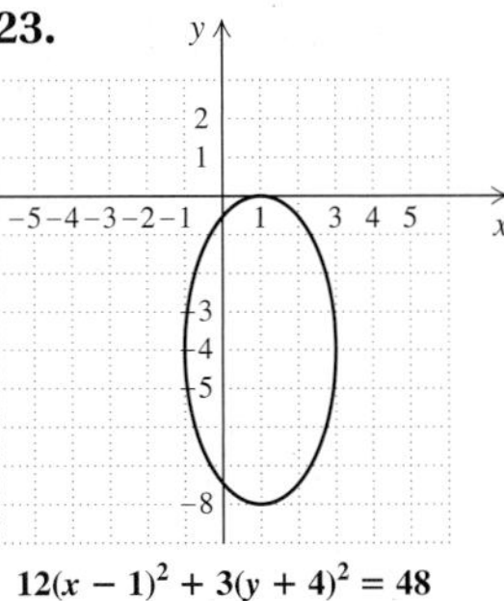

$12(x-1)^2 + 3(y+4)^2 = 48$

25.

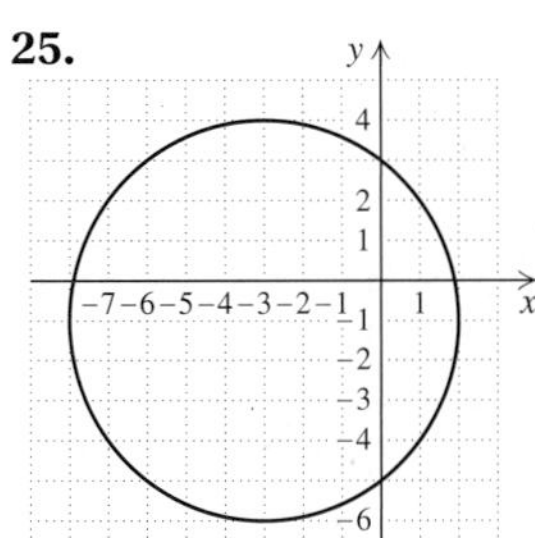

$4(x+3)^2 + 4(y+1)^2 - 10 = 90$

27. 📓 **29.** [6.4] $\frac{16}{9}$

30. [6.4] $-\frac{19}{8}$ **31.** [8.2] $5 \pm \sqrt{3}$ **32.** [8.2] $3 \pm \sqrt{7}$

33. [7.6] $\frac{3}{2}$ **34.** [7.6] No solution **35.** 📓

37. $\frac{x^2}{81} + \frac{y^2}{121} = 1$ **39.** $\frac{(x-2)^2}{16} + \frac{(y+1)^2}{9} = 1$

41. 2.134×10^8 mi **43.**

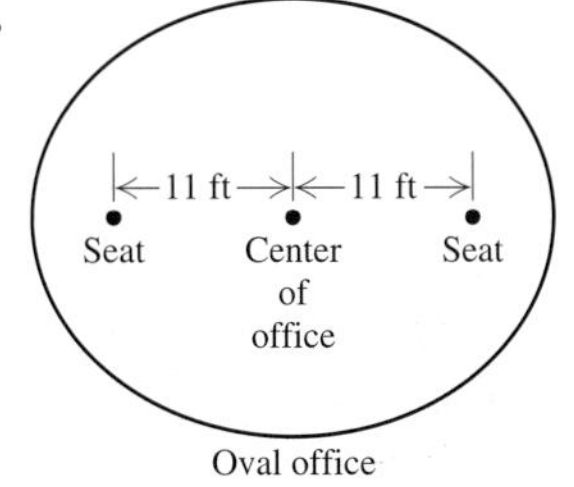

Oval office

45. Length: 700 yd; width: 200 yd

47.

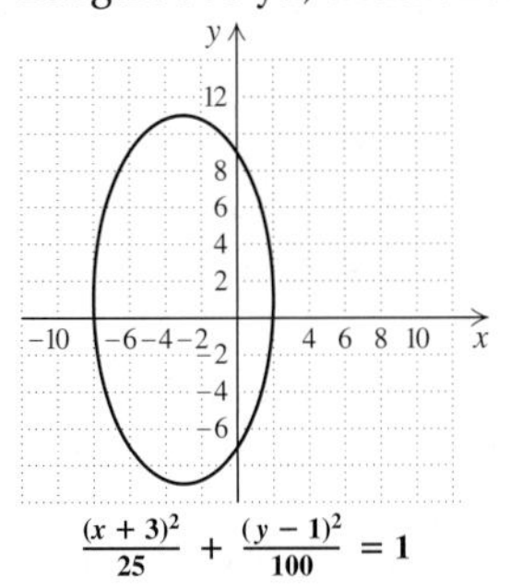

$\frac{(x+3)^2}{25} + \frac{(y-1)^2}{100} = 1$

Technology Connection, p. 643

1. $y_1 = \frac{\sqrt{15x^2 - 240}}{2}$; $y_2 = -\frac{\sqrt{15x^2 - 240}}{2}$

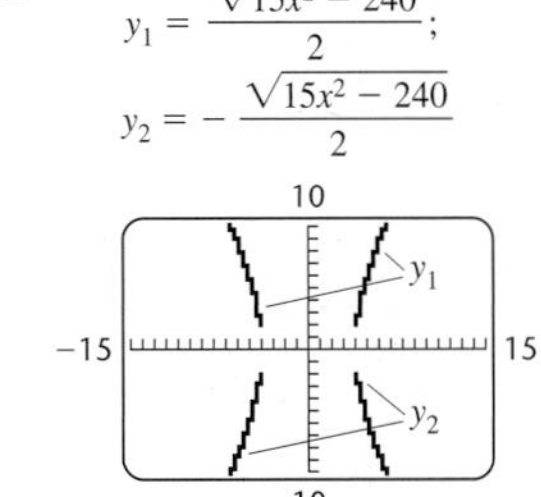

2. $y_1 = \sqrt{\frac{16x^2 - 64}{3}}$; $y_2 = -\sqrt{\frac{16x^2 - 64}{3}}$

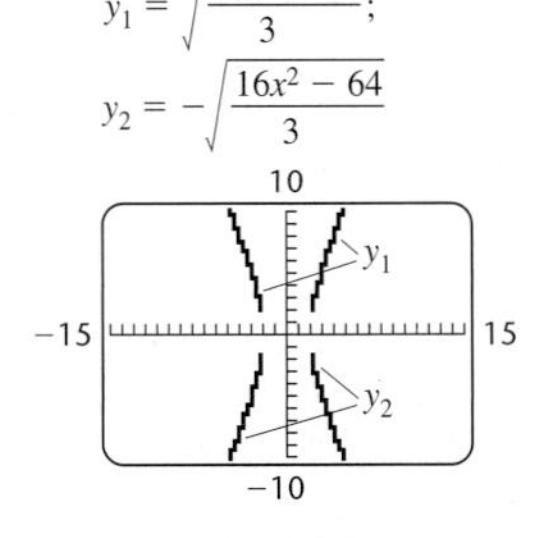

3. $y_1 = \frac{\sqrt{5x^2 + 320}}{4}$; $y_2 = -\frac{\sqrt{5x^2 + 320}}{4}$

4. $y_1 = \sqrt{\frac{9x^2 + 441}{45}}$; $y_2 = -\sqrt{\frac{9x^2 + 441}{45}}$

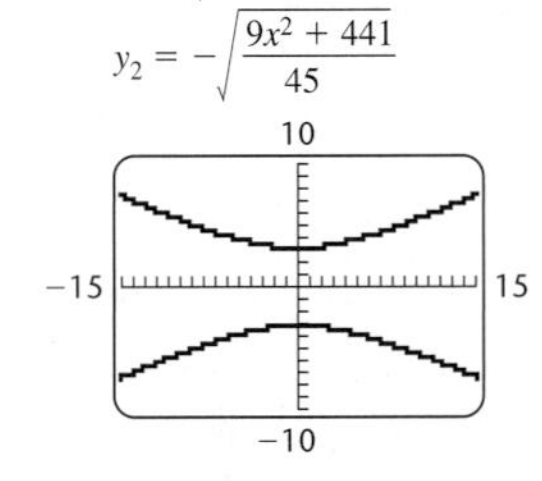

Exercise Set 10.3, pp. 647–648

1.

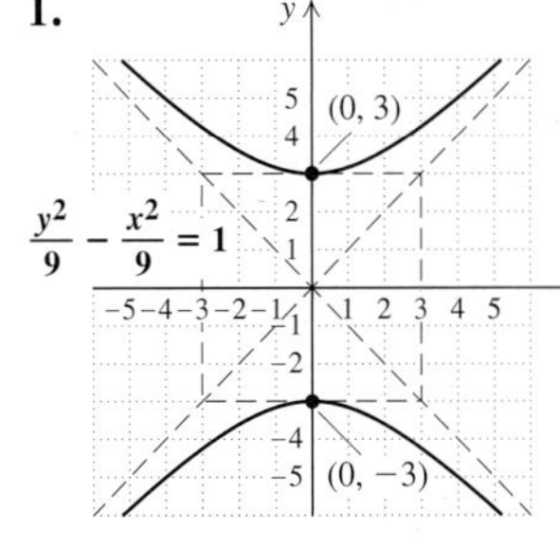

$\frac{y^2}{9} - \frac{x^2}{9} = 1$

3.

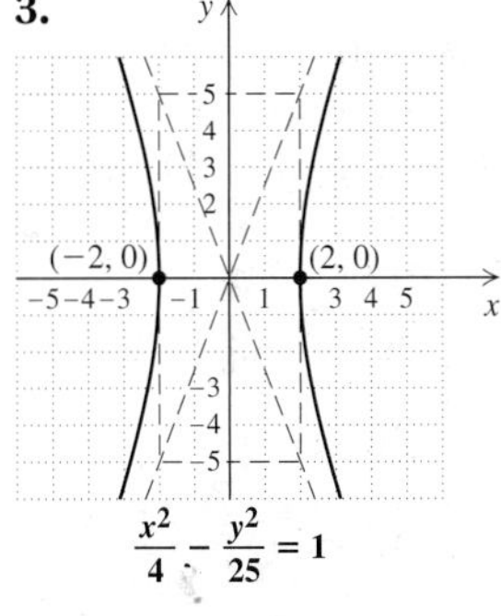

$\frac{x^2}{4} - \frac{y^2}{25} = 1$

5.

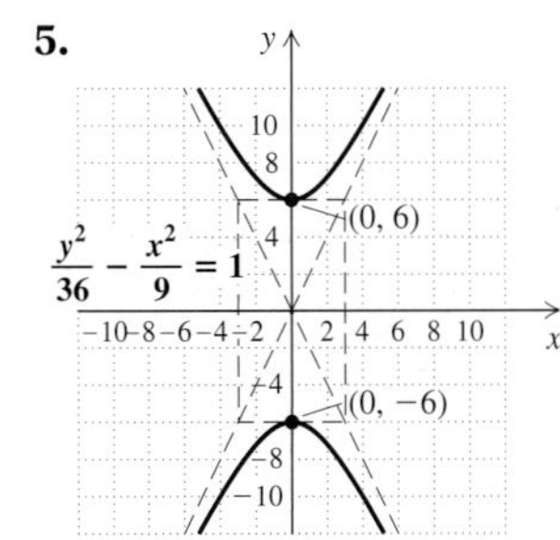

$\frac{y^2}{36} - \frac{x^2}{9} = 1$

7.

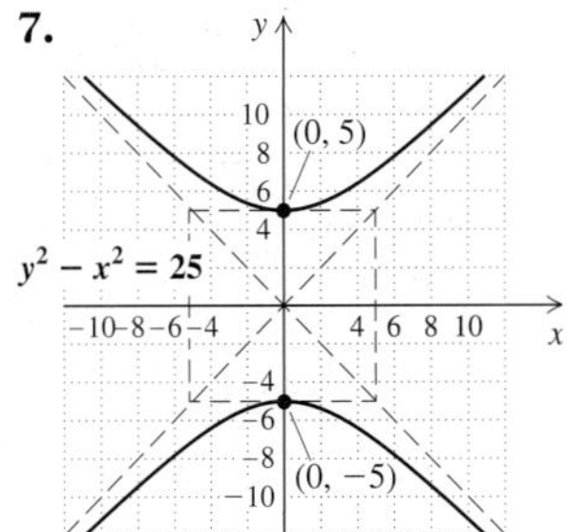

$y^2 - x^2 = 25$

9.

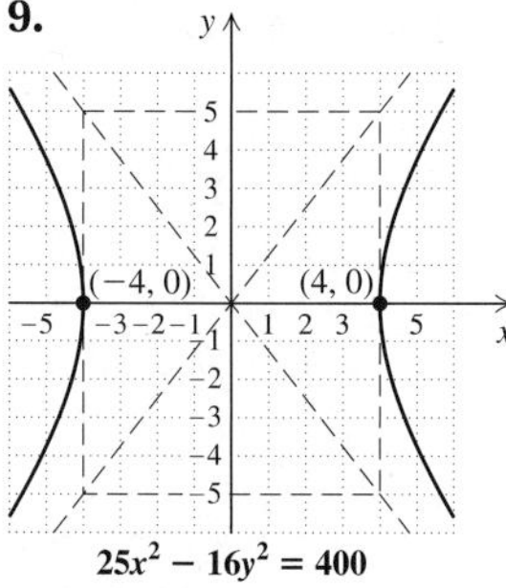

$25x^2 - 16y^2 = 400$

11.

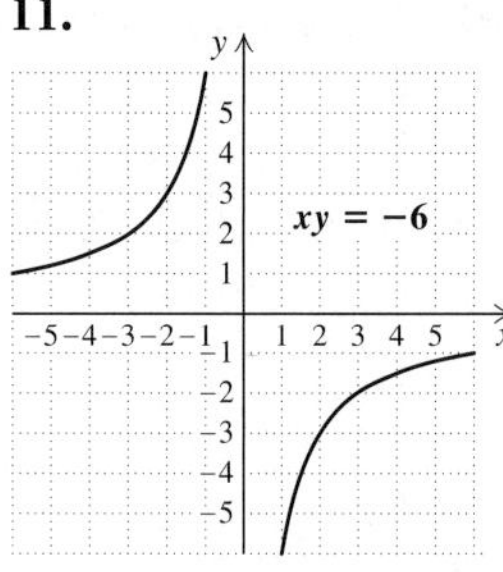

13.

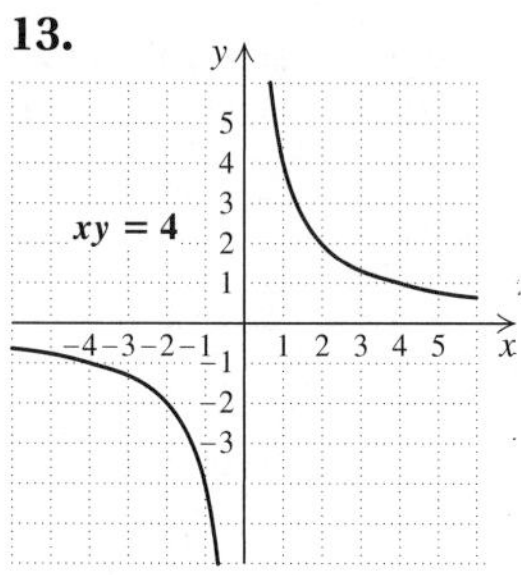

15.

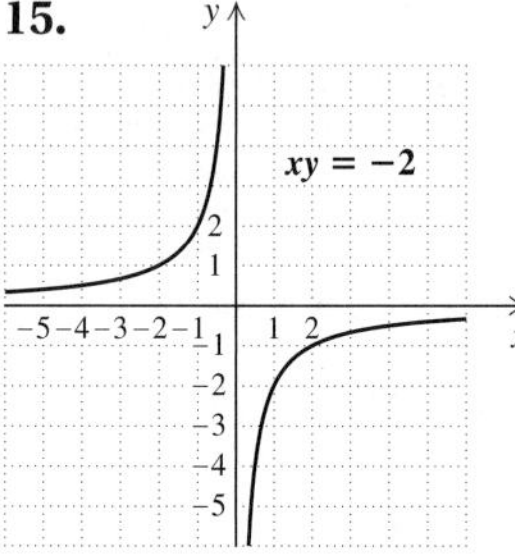

17.

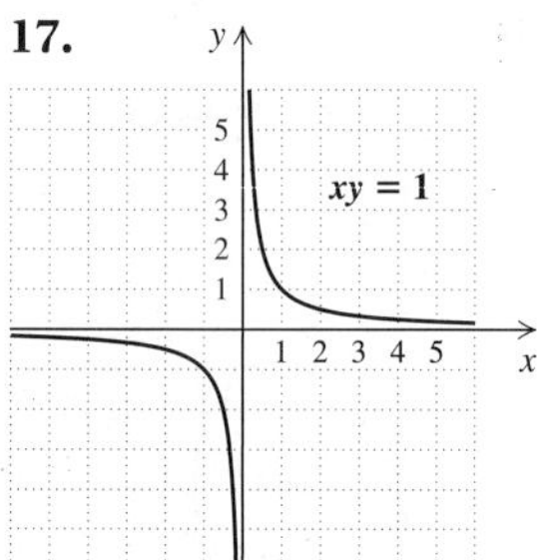

19. Circle **21.** Ellipse

23. Hyperbola **25.** Circle **27.** Ellipse
29. Hyperbola **31.** Parabola **33.** Hyperbola
35. Circle **37.** Ellipse **39.** [icon] **41.** [3.2] $\left(-\frac{22}{3}, \frac{37}{9}\right)$
42. [3.2] $(9, -4)$ **43.** [5.8] $-3, 3$ **44.** [5.8] $-1, 1$
45. [1.4] \$35 **46.** [1.4] 69 **47.** [icon] **49.** $\frac{y^2}{36} - \frac{x^2}{4} = 1$

51. C: $(5, 2)$; V: $(-1, 2)$, $(11, 2)$; asymptotes: $y - 2 = \frac{5}{6}(x - 5)$, $y - 2 = -\frac{5}{6}(x - 5)$

$\frac{(x-5)^2}{36} - \frac{(y-2)^2}{25} = 1$

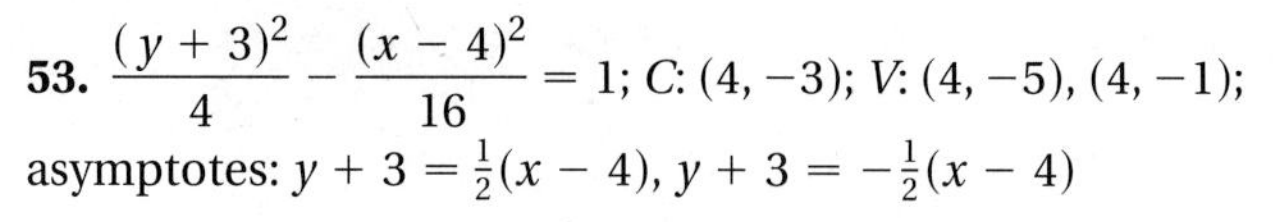

53. $\frac{(y+3)^2}{4} - \frac{(x-4)^2}{16} = 1$; C: $(4, -3)$; V: $(4, -5)$, $(4, -1)$; asymptotes: $y + 3 = \frac{1}{2}(x - 4)$, $y + 3 = -\frac{1}{2}(x - 4)$

$8(y + 3)^2 - 2(x - 4)^2 = 32$

55. $\frac{(x+3)^2}{1} - \frac{(y-2)^2}{4} = 1$; C: $(-3, 2)$; V: $(-4, 2)$, $(-2, 2)$; asymptotes: $y - 2 = 2(x + 3)$, $y - 2 = -2(x + 3)$

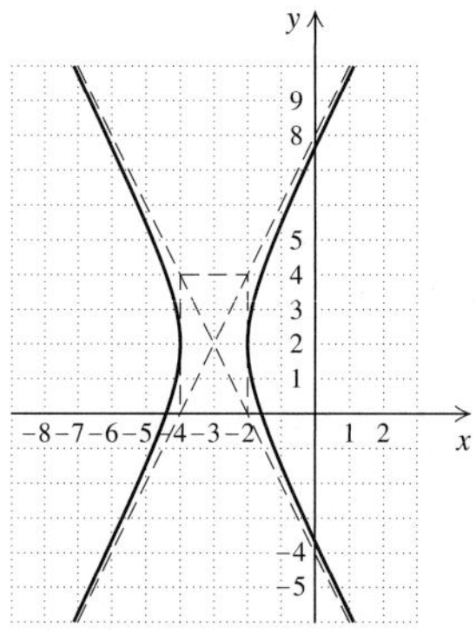

$4x^2 - y^2 + 24x + 4y + 28 = 0$

57. [icon]

Technology Connection, p. 651

1. $(-1.50, -1.17)$; $(3.50, 0.50)$
2. $(-2.77, 2.52)$; $(-2.77, -2.52)$

Technology Connection, p. 653

1. $y_1 = \sqrt{(20 - x^2)/4}$; $y_2 = -\sqrt{(20 - x^2)/4}$; $y_3 = 4/x$

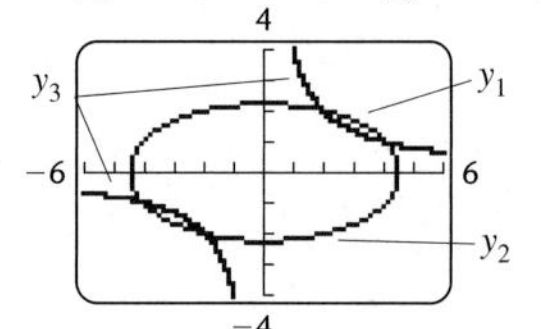

Exercise Set 10.4, pp. 655–657

1. $(-4,-3),(3,4)$ **3.** $(2,0),(0,3)$ **5.** $(2,4),(1,1)$
7. $\left(\frac{11}{4},-\frac{9}{8}\right),(1,-2)$ **9.** $\left(\frac{13}{3},\frac{5}{3}\right),(-5,-3)$
11. $\left(\dfrac{3+\sqrt{7}}{2},\dfrac{-1+\sqrt{7}}{2}\right),\left(\dfrac{3-\sqrt{7}}{2},\dfrac{-1-\sqrt{7}}{2}\right)$
13. $\left(-3,\frac{5}{2}\right),(3,1)$ **15.** $(1,-7),(-7,1)$
17. $(3,0),\left(-\frac{9}{5},\frac{8}{5}\right)$ **19.** $(1,4),(4,1)$ **21.** $(0,0),(1,1)$,
$\left(-\dfrac{1}{2}+\dfrac{\sqrt{3}}{2}i,-\dfrac{1}{2}-\dfrac{\sqrt{3}}{2}i\right),\left(-\dfrac{1}{2}-\dfrac{\sqrt{3}}{2}i,-\dfrac{1}{2}+\dfrac{\sqrt{3}}{2}i\right)$
23. $(-3,0),(3,0)$ **25.** $(-4,-3),(-3,-4),(3,4),(4,3)$
27. $\left(\dfrac{4i\sqrt{35}}{7},\dfrac{6\sqrt{21}}{7}\right),\left(-\dfrac{4i\sqrt{35}}{7},\dfrac{6\sqrt{21}}{7}\right)$,
$\left(\dfrac{4i\sqrt{35}}{7},-\dfrac{6\sqrt{21}}{7}\right),\left(\dfrac{4i\sqrt{35}}{7},-\dfrac{6\sqrt{21}}{7}\right)$
29. $(-\sqrt{2},-\sqrt{14}),(-\sqrt{2},\sqrt{14}),(\sqrt{2},-\sqrt{14}),(\sqrt{2},\sqrt{14})$
31. $(-2,-1),(-1,-2),(1,2),(2,1)$
33. $(-3,-2),(-2,-3),(2,3),(3,2)$ **35.** $(2,5),(-2,-5)$
37. $(3,2),(-3,-2)$ **39.** $(-3,4),(3,4),(0,-5)$
41. Length: 8 cm; width: 6 cm
43. Length: 5 in.; width: 4 in.
45. Length: 12 ft; width: 5 ft **47.** 13 and 12
49. \$3750, 6% **51.** Length: $\sqrt{3}$ m; width: 1 m
53. **55.** [1.2] -16 **56.** [1.2] -32 **57.** [1.2] 1
58. [1.2] $-\frac{1}{4}$ **59.** [1.2] 44 **60.** [1.2] 28 **61.**
63. 61.52 cm and 38.48 cm **65.** $4x^2+3y^2=43$
67. $\left(\frac{1}{3},\frac{1}{2}\right),\left(\frac{1}{2},\frac{1}{3}\right)$ **69.** Length: 24.8 cm; height: 18.6 cm
71. 30

Review Exercises: Chapter 10, p. 659

1. [10.1] 4 **2.** [10.1] 5 **3.** [10.1] $\sqrt{90.1}\approx 9.492$
4. [10.1] $\sqrt{9+4a^2}$ **5.** [10.1] $(4,6)$ **6.** [10.1] $\left(-3,\frac{5}{2}\right)$
7. [10.1] $\left(\dfrac{3}{4},\dfrac{\sqrt{3}-\sqrt{2}}{2}\right)$ **8.** [10.1] $\left(\frac{1}{2},2a\right)$
9. [10.1] $(-2,3),\sqrt{2}$ **10.** [10.1] $(5,0),7$
11. [10.1] $(3,1),3$ **12.** [10.1] $(-4,3),\sqrt{35}$
13. [10.1] $(x+4)^2+(y-3)^2=48$
14. [10.1] $(x-7)^2+(y+2)^2=20$
15. [10.1], [10.3] Circle

$4x^2+4y^2=100$

16. [10.2], [10.3] Ellipse

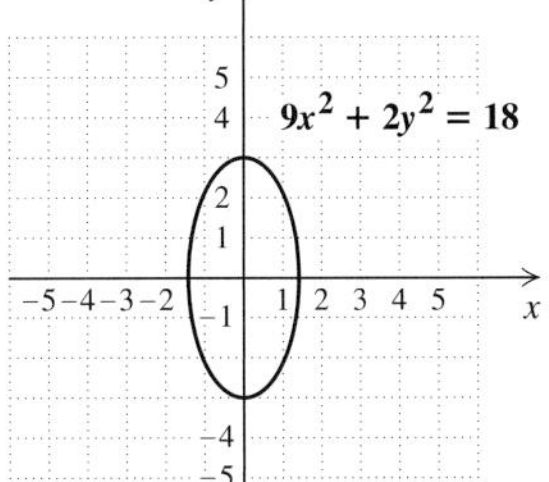
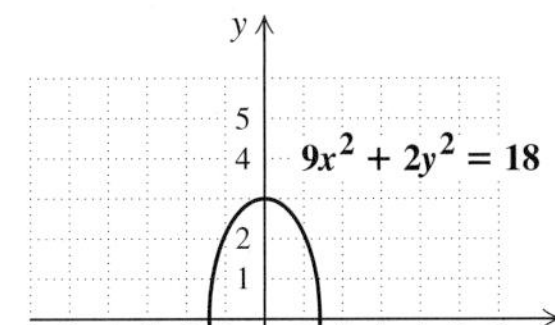

17. [10.1], [10.3] Parabola

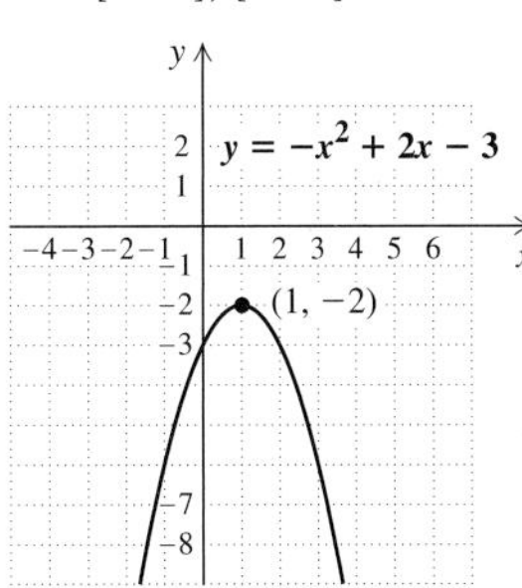

18. [10.3] Hyperbola

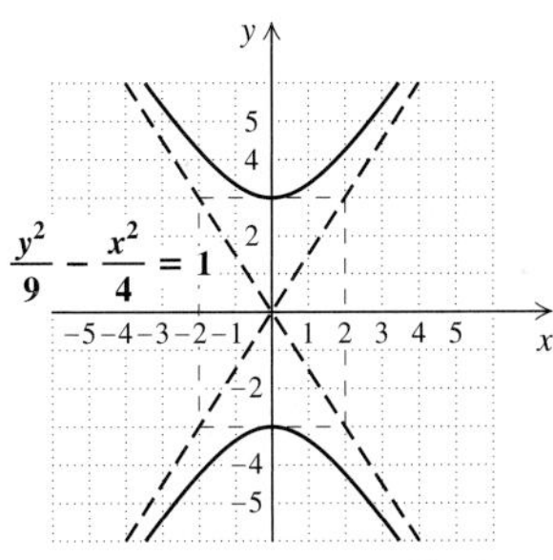

19. [10.3] Hyperbola

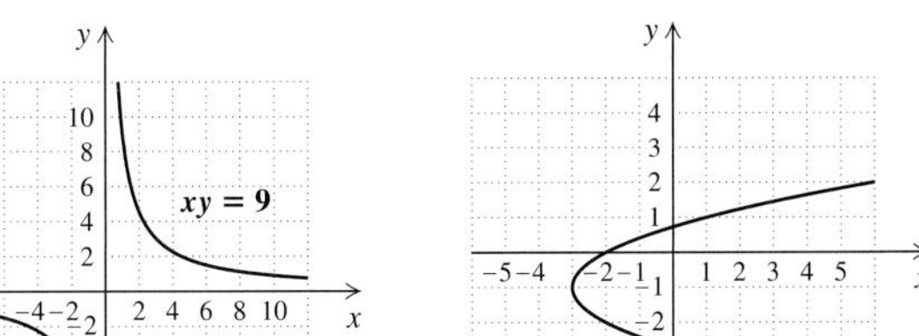

20. [10.1], [10.3] Parabola

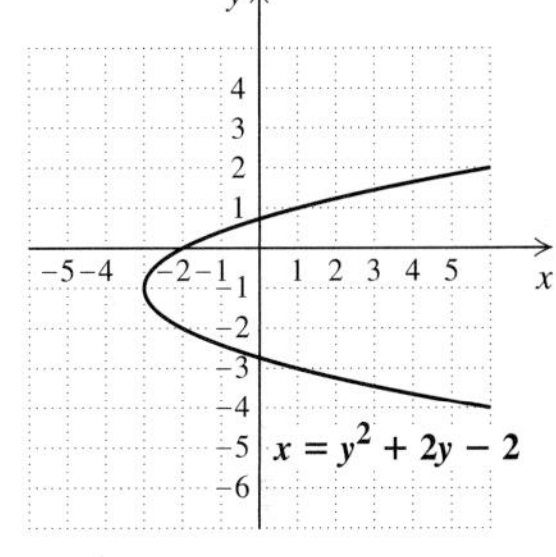

21. [10.2], [10.3] Ellipse

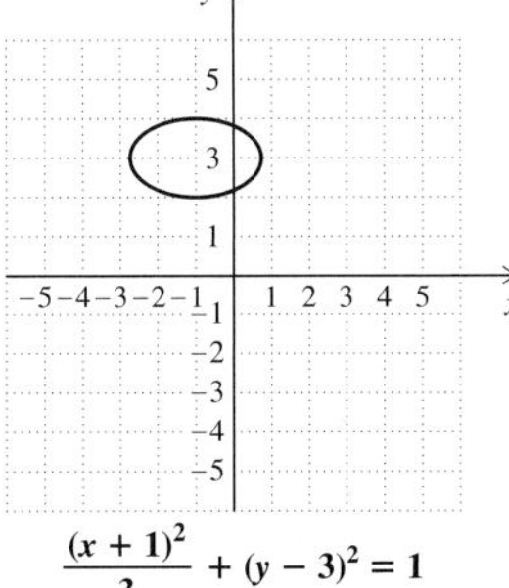

22. [10.1], [10.3] Circle

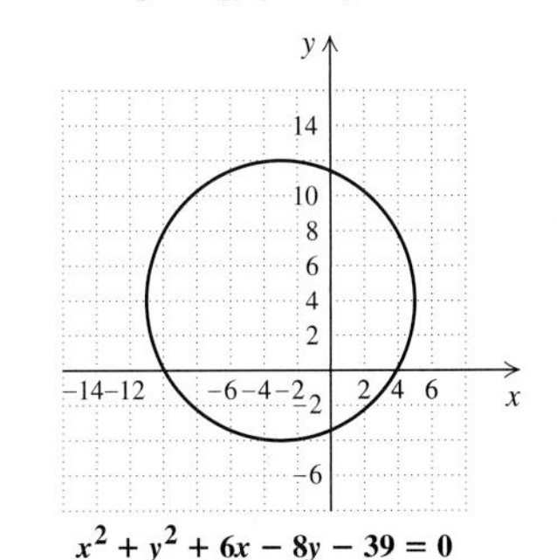

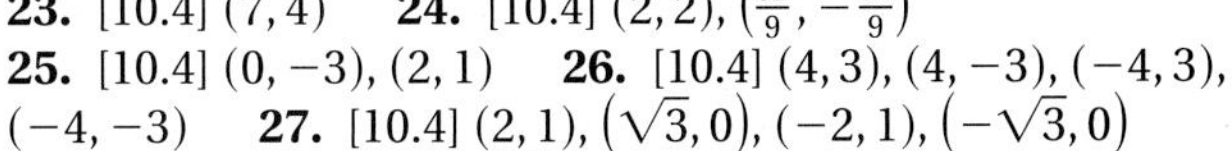

23. [10.4] $(7,4)$ **24.** [10.4] $(2,2),\left(\frac{32}{9},-\frac{10}{9}\right)$
25. [10.4] $(0,-3),(2,1)$ **26.** [10.4] $(4,3),(4,-3),(-4,3),(-4,-3)$ **27.** [10.4] $(2,1),(\sqrt{3},0),(-2,1),(-\sqrt{3},0)$
28. [10.4] $(3,-3),\left(-\frac{3}{5},\frac{21}{5}\right)$ **29.** [10.4] $(6,8),(6,-8),(-6,8),(-6,-8)$ **30.** [10.4] $(2,2),(-2,-2),(2\sqrt{2},\sqrt{2}),(-2\sqrt{2},-\sqrt{2})$ **31.** [10.4] Length: 12 m; width: 7 m
32. [10.4] Length: 12 in.; width: 9 in.
33. [10.4] 32 cm, 20 cm **34.** [10.4] 3 ft, 11 ft
35. [10.1], [10.3] The graph of a parabola has one branch whereas the graph of a hyperbola has two branches. A hyperbola has asymptotes, but a parabola does not. **36.** [10.1], [10.2], [10.3] Function notation is not used in this chapter because many of the relations are not functions. Function notation could be used for vertical parabolas and for hyperbolas that have the axes as asymptotes.
37. [10.4] $(-5,-4\sqrt{2}),(-5,4\sqrt{2}),(3,-2\sqrt{2}),(3,2\sqrt{2})$

38. [10.1] $(0, 6), (0, -6)$
39. [10.1], [10.4] $(x - 2)^2 + (y + 1)^2 = 25$
40. [10.2] $\frac{x^2}{49} + \frac{y^2}{9} = 1$ **41.** [10.1] $\left(\frac{9}{4}, 0\right)$

Test: Chapter 10, p. 660

1. [10.1] $9\sqrt{2} \approx 12.728$ **2.** [10.1] $2\sqrt{9 + a^2}$
3. [10.1] $\left(-\frac{1}{2}, \frac{7}{2}\right)$ **4.** [10.1] $(0, 0)$ **5.** [10.1] $(-2, 3)$, 8
6. [10.1] $(-2, 3)$, 3
7. [10.1], [10.3] Parabola

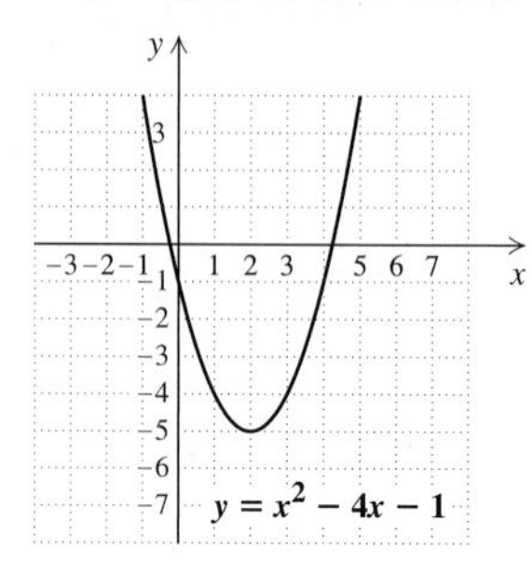

8. [10.1], [10.3] Circle

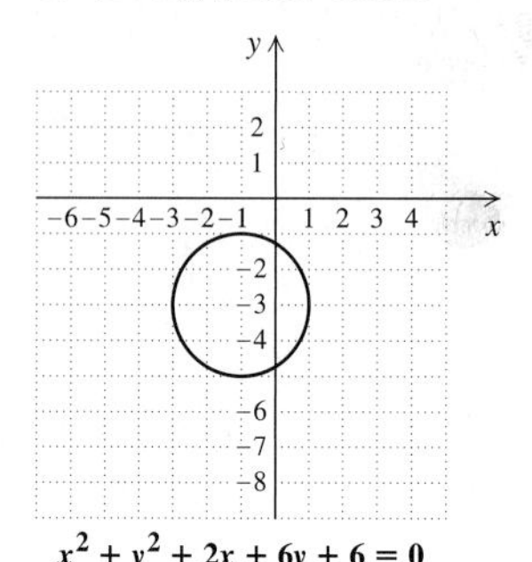

9. [10.3] Hyperbola

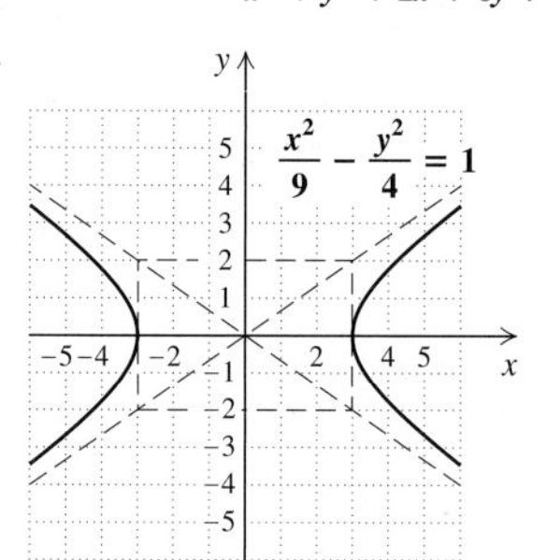

10. [10.2], [10.3] Ellipse

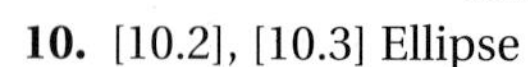

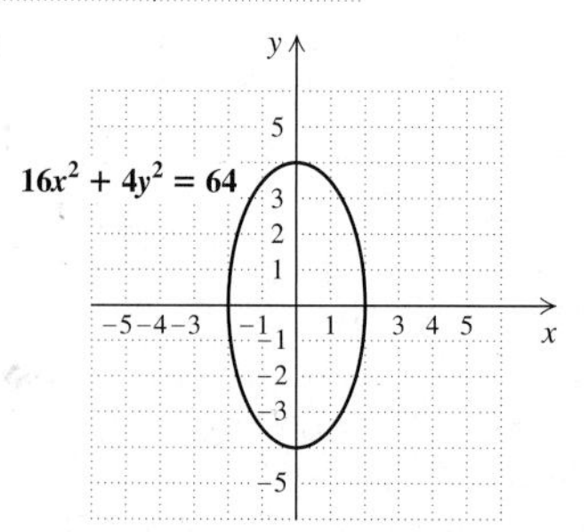

11. [10.3] Hyperbola

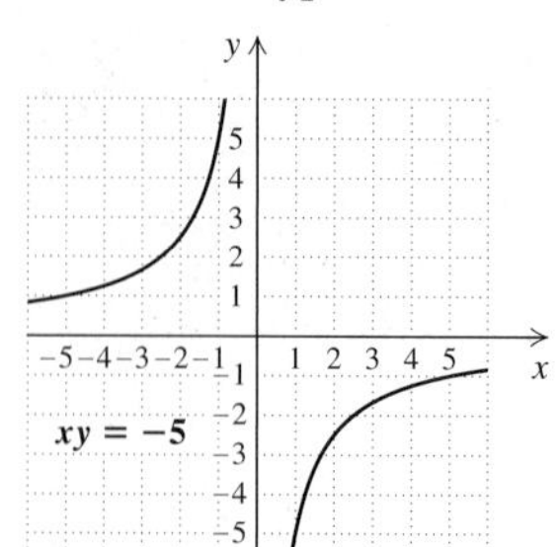

12. [10.1], [10.3] Parabola

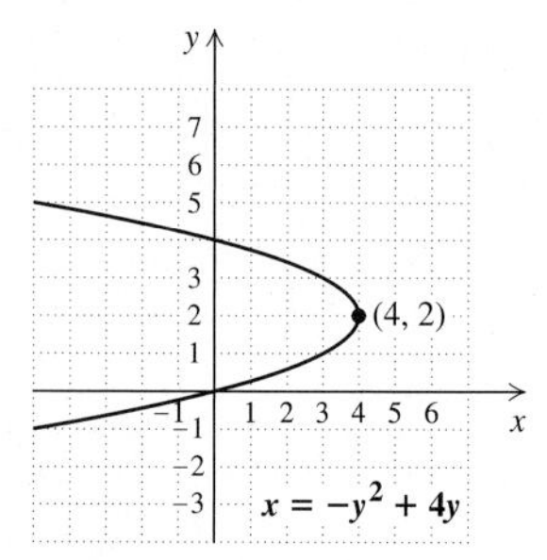

13. [10.4] $(0, 3), (4, 0)$ **14.** [10.4] $(4, 0), (-4, 0)$
15. [10.4] $(3, 2), (-3, -2)$ **16.** [10.4] $\left(\sqrt{6}, 2\right), \left(\sqrt{6}, -2\right), \left(-\sqrt{6}, 2\right), \left(-\sqrt{6}, -2\right)$ **17.** [10.4] 2 by 11
18. [10.4] $\sqrt{5}$ m, $\sqrt{3}$ m **19.** [10.4] Length: 16 ft; width: 12 ft **20.** [10.4] \$1200, 6%
21. [10.2] $\frac{(x - 6)^2}{25} + \frac{(y - 3)^2}{9} = 1$ **22.** [10.1] $\left(0, -\frac{31}{4}\right)$
23. [10.4] 9

CHAPTER 11

Exercise Set 11.1, pp. 666–668

1. 3, 8, 13, 18; 48; 73 **3.** $\frac{1}{2}, \frac{2}{3}, \frac{3}{4}, \frac{4}{5}; \frac{10}{11}; \frac{15}{16}$
5. $-1, 0, 3, 8; 80; 195$ **7.** $2, 2\frac{1}{2}, 3\frac{1}{3}, 4\frac{1}{4}; 10\frac{1}{10}; 15\frac{1}{15}$
9. $-1, 4, -9, 16; 100; -225$ **11.** $-2, -1, 4, -7; -25; 40$
13. 9 **15.** 364 **17.** -23.5 **19.** -363 **21.** $\frac{441}{400}$
23. $2n - 1$ **25.** $(-1)^{n+1}$ **27.** $(-1)^n \cdot n$
29. $(-1)^n \cdot 2 \cdot (3)^{n-1}$ **31.** $\frac{n}{n + 1}$ **33.** 5^n
35. $(-1)^n \cdot n^2$ **37.** 4 **39.** 30
41. $\frac{1}{2} + \frac{1}{4} + \frac{1}{6} + \frac{1}{8} + \frac{1}{10} = \frac{137}{120}$
43. $3^0 + 3^1 + 3^2 + 3^3 + 3^4 = 121$
45. $\frac{1}{2} + \frac{2}{3} + \frac{3}{4} + \frac{4}{5} + \frac{5}{6} + \frac{6}{7} + \frac{7}{8} + \frac{8}{9} = \frac{15{,}551}{2520}$
47. $(-1)^2 2^1 + (-1)^3 2^2 + (-1)^4 2^3 + (-1)^5 2^4 + (-1)^6 2^5 + (-1)^7 2^6 + (-1)^8 2^7 + (-1)^9 2^8 = -170$
49. $(0^2 - 2 \cdot 0 + 3) + (1^2 - 2 \cdot 1 + 3) + (2^2 - 2 \cdot 2 + 3) + (3^2 - 2 \cdot 3 + 3) + (4^2 - 2 \cdot 4 + 3) + (5^2 - 2 \cdot 5 + 3) = 43$
51. $\frac{(-1)^3}{3 \cdot 4} + \frac{(-1)^4}{4 \cdot 5} + \frac{(-1)^5}{5 \cdot 6} = -\frac{1}{15}$ **53.** $\sum_{k=1}^{5} \frac{k + 1}{k + 2}$
55. $\sum_{k=1}^{6} k^2$ **57.** $\sum_{k=2}^{n} (-1)^k k^2$ **59.** $\sum_{k=1}^{\infty} 5k$
61. $\sum_{k=1}^{\infty} \frac{1}{k(k + 1)}$ **63.** **65.** [1.1] 77 **66.** [1.1] 23
67. [5.2] $x^3 + 3x^2y + 3xy^2 + y^3$
68. [5.2] $a^3 - 3a^2b + 3ab^2 - b^3$
69. [5.2] $8a^3 - 12a^2b + 6ab^2 - b^3$
70. [5.2] $8x^3 + 12x^2y + 6xy^2 + y^3$ **71.**
73. 1, 3, 13, 63, 313, 1563 **75.** 1, 2, 4, 8, 16, 32, 64, 128, 256, 512, 1024, 2048, 4096, 8192, 16,384, 32,768, 65,536
77. $S_{100} = 0; S_{101} = -1$ **79.** $i, -1, -i, 1, i; i$
81. 11th term

Exercise Set 11.2, pp. 675–677

1. $a_1 = 2, d = 4$ **3.** $a_1 = 6, d = -4$ **5.** $a_1 = \frac{3}{2}, d = \frac{3}{4}$
7. $a_1 = \$5.12, d = \$0.12$ **9.** 47 **11.** -41
13. $-\$1628.16$ **15.** 27th **17.** 102nd **19.** 82
21. 5 **23.** 28 **25.** $a_1 = 8; d = -3; 8, 5, 2, -1, -4$

27. $a_1 = 1; d = 1$ **29.** 780 **31.** 31,375 **33.** 2550 **35.** 918 **37.** 1030 **39.** 42; 420 **41.** 1260 **43.** \$31,000 **45.** 722 **47.** **49.** [6.2] $\frac{13}{30x}$ **50.** [6.2] $\frac{23}{36t}$ **51.** [9.3] $a^k = P$ **52.** [9.3] $e^a = t$ **53.** [10.1] $x^2 + y^2 = 81$ **54.** [10.1] $(x + 2)^2 + (y - 5)^2 = 18$ **55.** **57.** $S_n = n^2$ **59.** \$8760, \$7961.77, \$7163.54, \$6365.31, \$5567.08; \$4768.85, \$3970.62, \$3172.39, \$2374.16, \$1575.93 **61.** Let d = the common difference. Since p, m, and q form an arithmetic sequence, $m = p + d$ and $q = p + 2d$. Then $\frac{p + q}{2} = \frac{p + (p + 2d)}{2} = p + d = m$. **63.** 156,375

Exercise Set 11.3, pp. 684–686

1. 2 **3.** -1 **5.** $-\frac{1}{2}$ **7.** $\frac{1}{5}$ **9.** $\frac{3}{m}$ **11.** 192 **13.** 80 **15.** 648 **17.** \$2331.64 **19.** $a_n = 3^{n-1}$ **21.** $a_n = (-1)^{n-1}$ **23.** $a_n = \frac{1}{x^n}$ **25.** 762 **27.** $\frac{547}{18}$ **29.** $\frac{1 - x^8}{1 - x}$, or $(1 + x)(1 + x^2)(1 + x^4)$ **31.** \$5134.51 **33.** $\frac{64}{3}$ **35.** $\frac{49}{4}$ **37.** No **39.** No **41.** $\frac{43}{99}$ **43.** \$25,000 **45.** $\frac{7}{9}$ **47.** $\frac{830}{99}$ **49.** $\frac{5}{33}$ **51.** $\frac{5}{1024}$ ft **53.** 155,797 **55.** \$43,318.94 **57.** 25 min **59.** 3100.35 ft **61.** 20.48 in. **63.** **65.** [5.2] $x^3 + 3x^2y + 3xy^2 + y^3$ **66.** [5.2] $a^3 - 3a^2b + 3ab^2 - b^3$ **67.** [3.2] $\left(-\frac{63}{29}, -\frac{114}{29}\right)$ **68.** [3.4] $(-1, 2, 3)$ **69.** **71.** $\frac{x^2[1 - (-x)^n]}{1 + x}$ **73.** 512 cm^2 **75.** ,

Exercise Set 11.4, pp. 694–696

1. 40,320 **3.** 3,628,800 **5.** 210 **7.** 720 **9.** 28 **11.** 210 **13.** 190 **15.** 595 **17.** $m^5 + 5m^4n + 10m^3n^2 + 10m^2n^3 + 5mn^4 + n^5$ **19.** $x^6 - 6x^5y + 15x^4y^2 - 20x^3y^3 + 15x^2y^4 - 6xy^5 + y^6$ **21.** $x^{10} - 15x^8y + 90x^6y^2 - 270x^4y^3 + 405x^2y^4 - 243y^5$ **23.** $729c^6 - 1458c^5d + 1215c^4d^2 - 540c^3d^3 + 135c^2d^4 - 18cd^5 + d^6$ **25.** $x^3 - 3x^2y + 3xy^2 - y^3$ **27.** $x^9 + \frac{18x^8}{y} + \frac{144x^7}{y^2} + \frac{672x^6}{y^3} + \frac{2016x^5}{y^4} + \frac{4032x^4}{y^5} + \frac{5376x^3}{y^6} + \frac{4608x^2}{y^7} + \frac{2304x}{y^8} + \frac{512}{y^9}$ **29.** $a^{10} - 5a^8b^3 + 10a^6b^6 - 10a^4b^9 + 5a^2b^{12} - b^{15}$ **31.** $9 - 12\sqrt{3}t + 18t^2 - 4\sqrt{3}t^3 + t^4$ **33.** $x^{-8} + 4x^{-4} + 6 + 4x^4 + x^8$ **35.** $15a^4b^2$ **37.** $-64{,}481{,}508a^3$ **39.** $1120x^{12}y^2$ **41.** $-1{,}959{,}552u^5v^{10}$ **43.** y^8 **45.** **47.** [9.6] 4 **48.** [9.6] $\frac{5}{2}$ **49.** [9.6] 5.6348 **50.** [9.6] ± 5 **51.** **53.** Consider a set of 5 elements, $\{A, B, C, D, E\}$. List all the subsets of size 3:

$$\{A, B, C\}, \{A, B, D\}, \{A, B, E\}, \{A, C, D\}, \{A, C, E\}, \{A, D, E\}, \{B, C, D\}, \{B, C, E\}, \{B, D, E\}, \{C, D, E\}$$

There are exactly 10 subsets of size 3 and $\binom{5}{3} = 10$, so there are exactly $\binom{5}{3}$ ways of forming a subset of size 3 from a set of 5 elements. **55.** $\binom{8}{5}(0.15)^3(0.85)^5 \approx 0.084$ **57.** $\binom{8}{6}(0.15)^2(0.85)^6 + \binom{8}{7}(0.15)(0.85)^7 + \binom{8}{8}(0.85)^8 \approx 0.89$ **59.** $\frac{55}{144}$ **61.** $\frac{\sqrt[3]{q}}{2p}$ **63.** 8

Review Exercises: Chapter 11, p. 697

1. [11.1] 1, 5, 9, 13; 29; 45 **2.** [11.1] $0, \frac{1}{5}, \frac{1}{5}, \frac{3}{17}; \frac{7}{65}; \frac{11}{145}$ **3.** [11.1] $a_n = -2n$ **4.** [11.1] $a_n = (-1)^n(2n - 1)$ **5.** [11.1] $-2 + 4 + (-8) + 16 + (-32) = -22$ **6.** [11.1] $-3 + (-5) + (-7) + (-9) + (-11) + (-13) = -48$ **7.** [11.1] $\sum_{k=1}^{5} 4k$ **8.** [11.1] $\sum_{k=1}^{5} \frac{1}{(-2)^k}$ **9.** [11.2] 85 **10.** [11.2] $\frac{8}{3}$ **11.** [11.2] $d = 1.25, a_1 = 11.25$ **12.** [11.2] -544 **13.** [11.2] 8580 **14.** [11.3] $1024\sqrt{2}$ **15.** [11.3] $\frac{2}{3}$ **16.** [11.3] $a_n = 2(-1)^n$ **17.** [11.3] $a_n = 3\left(\frac{x}{4}\right)^{n-1}$ **18.** [11.3] 4095 **19.** [11.3] $-4095x$ **20.** [11.3] 12 **21.** [11.3] $\frac{49}{11}$ **22.** [11.3] No **23.** [11.3] No **24.** [11.3] \$40,000 **25.** [11.3] $\frac{5}{9}$ **26.** [11.3] $\frac{46}{33}$ **27.** [11.2] \$17.80 **28.** [11.2] 903 **29.** [11.3] \$22,521.92 **30.** [11.3] 6 m **31.** [11.4] 5040 **32.** [11.4] 56 **33.** [11.4] $190a^{18}b^2$ **34.** [11.4] $x^4 - 8x^3y + 24x^2y^2 - 32xy^3 + 16y^4$ **35.** [11.3] For a geometric sequence with $|r| < 1$, as n gets larger, the absolute value of the terms gets smaller, since $|r^n|$ gets smaller. **36.** [11.4] The first form of the binomial theorem draws the coefficients from Pascal's triangle; the second form uses factorial notation. The second form avoids the need to compute all preceding rows of Pascal's triangle, and is generally easier to use when only one term of an expression is needed. When several terms of an expansion are needed and n is not large (say, $n \leq 8$), it is often easier to use Pascal's triangle. **37.** [11.3] $\frac{1 - (-x)^n}{x + 1}$ **38.** [11.4] $x^{-15} + 5x^{-9} + 10x^{-3} + 10x^3 + 5x^9 + x^{15}$

Test: Chapter 11, p. 698

1. [11.1] 1, 7, 13, 19, 25; 91 **2.** [11.1] $a_n = 4(\frac{1}{3})^n$
3. [11.1] $1 + (-1) + (-5) + (-13) + (-29) = -47$
4. [11.1] $\sum_{k=1}^{5} (-1)^{k+1}k^3$ **5.** [11.2] -46 **6.** [11.2] $\frac{3}{8}$
7. [11.2] $a_1 = 31.2; d = -3.8$ **8.** [11.2] 2508
9. [11.3] $\frac{9}{128}$ **10.** [11.3] $\frac{2}{3}$ **11.** [11.3] $(-1)^{n+1}3^n$
12. [11.3] $511 + 511x$ **13.** [11.3] 1 **14.** [11.3] No
15. [11.3] $\frac{\$25{,}000}{23} \approx \$1086.96$ **16.** [11.3] $\frac{85}{99}$
17. [11.2] 63 **18.** [11.2] \$17,100 **19.** [11.3] \$8981.05
20. [11.3] 36 m **21.** [11.4] 78
22. [11.4] $x^{10} - 15x^8y + 90x^6y^2 - 270x^4y^3 + 405x^2y^4 - 243y^5$ **23.** [11.4] $220a^9x^3$
24. [11.2] $n(n + 1)$
25. [11.3] $\dfrac{1 - \left(\dfrac{1}{x}\right)^n}{1 - \dfrac{1}{x}}$, or $\dfrac{x^n - 1}{x^{n-1}(x - 1)}$

Cumulative Review: Chapters 1–11, pp. 699–701

1. [1.6] $-45x^6y^{-4}$, or $\dfrac{-45x^6}{y^4}$ **2.** [1.2] 6.3
3. [1.3] $-3y + 17$ **4.** [1.1] 280 **5.** [1.1], [1.2] $\frac{7}{6}$
6. [5.1] $3a^2 - 8ab - 15b^2$ **7.** [5.1] $13x^3 - 7x^2 - 6x + 6$
8. [5.2] $6a^2 + 7a - 5$ **9.** [5.2] $9a^4 - 30a^2y + 25y^2$
10. [6.2] $\dfrac{4}{x + 2}$ **11.** [6.1] $\dfrac{x - 4}{3(x + 2)}$
12. [6.1] $\dfrac{(x + y)(x^2 + xy + y^2)}{x^2 + y^2}$ **13.** [6.3] $x - a$
14. [5.5] $(2x - 3)^2$ **15.** [5.6] $(3a - 2)(9a^2 + 6a + 4)$
16. [5.3] $(a^2 - b)(a + 3)$ **17.** [5.7] $3(y^2 + 3)(5y^2 - 4)$
18. [2.2] 20 **19.** [6.6] $7x^3 + 9x^2 + 19x + 38 + \dfrac{72}{x - 2}$
20. [1.3] $\frac{2}{3}$ **21.** [6.4] $-\frac{6}{5}, 4$ **22.** [4.2] $\mathbb{R}$, or $(-\infty, \infty)$
23. [3.2] $(1, -1)$ **24.** [3.4] $(2, -1, 1)$ **25.** [7.6] 2
26. [8.5] $\pm 2, \pm 5$ **27.** [10.4] $(\sqrt{5}, \sqrt{3}), (\sqrt{5}, -\sqrt{3}), (-\sqrt{5}, \sqrt{3}), (-\sqrt{5}, -\sqrt{3})$ **28.** [9.6] $\dfrac{\ln 8}{\ln 5} \approx 1.2920$
29. [9.6] 1005 **30.** [9.6] $\frac{1}{16}$ **31.** [9.6] $-\frac{1}{2}$
32. [4.3] $\{x \mid -2 \le x \le 3\}$, or $[-2, 3]$ **33.** [8.1] $\pm i\sqrt{2}$
34. [8.2] $-2 \pm \sqrt{7}$ **35.** [8.10] $\{y \mid y < -5 \text{ or } y > 2\}$, or $(-\infty, -5) \cup (2, \infty)$ **36.** [5.8] $-6, 8$
37. [7.6] No solution **38.** [10.4] 5 ft by 12 ft
39. [4.1] More than 4 **40.** [1.4] 65, 66, 67 **41.** [3.3] $11\frac{3}{7}$
42. [3.3] \$2.68 herb: 10 oz; \$4.60 herb: 14 oz
43. [6.5] 350 mph **44.** [6.5] $8\frac{2}{5}$ hr or 8 hr, 24 min
45. [8.6] 20 **46.** [8.9] 1250 ft^2

47. [2.4]

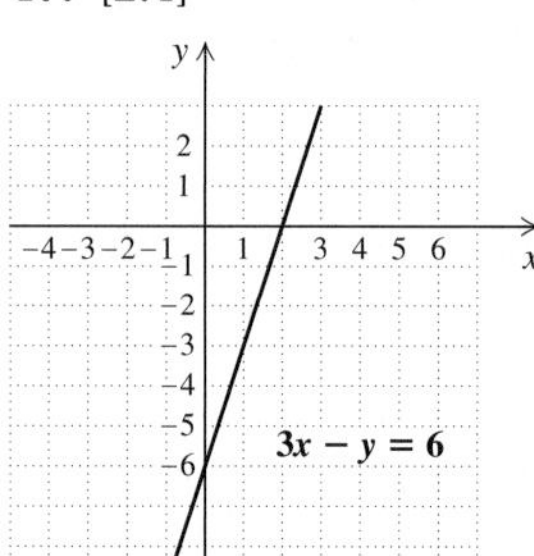

48. [10.2]

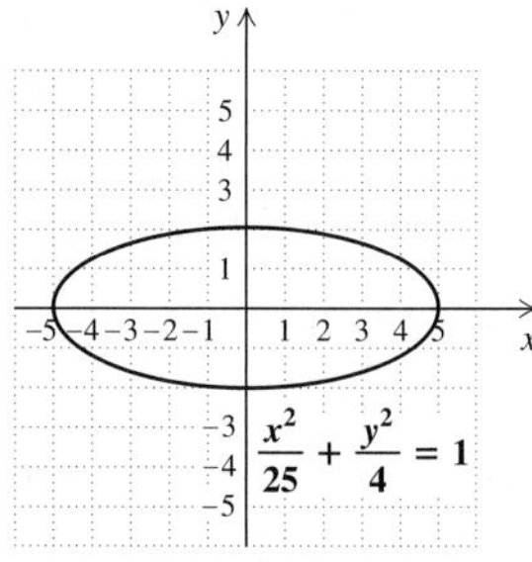

49. [9.3]

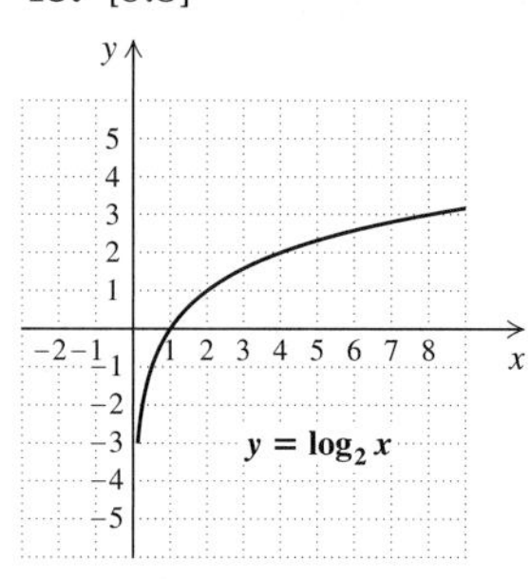

50. [4.4]

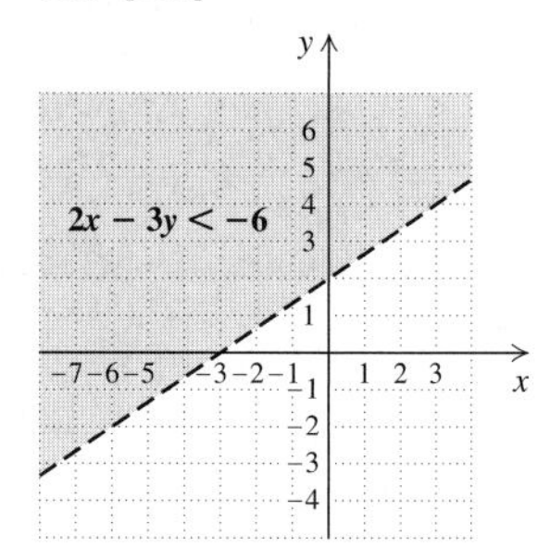

51. [8.7]

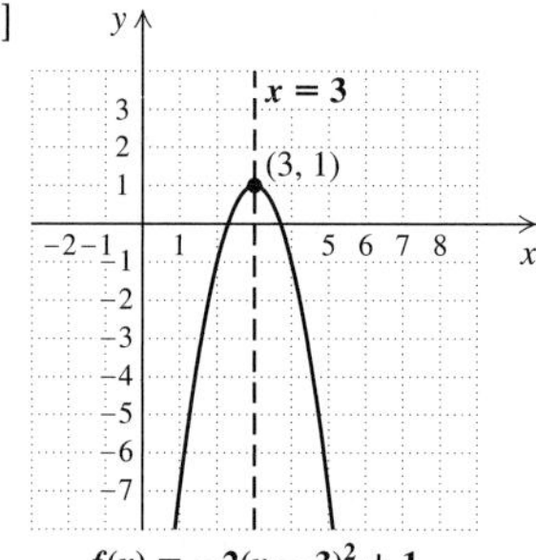

52. [1.5] $r = \dfrac{V - P}{-Pt}$, or $\dfrac{P - V}{Pt}$ **53.** [6.8] $R = \dfrac{Ir}{1 - I}$
54. [2.5] $y = 3x - 3$ **55.** [4.2] $\{x \mid x \le \frac{5}{3}\}$, or $(-\infty, \frac{5}{3}]$
56. [5.8] $\{x \mid x$ is a real number *and* $x \ne 1\}$, or $(-\infty, 1) \cup (1, \infty)$ **57.** [1.7] 6.8×10^{-12}
58. [7.3] $8x^2\sqrt{y}$ **59.** [7.2] $125x^2y^{3/4}$
60. [7.5] $\dfrac{\sqrt[3]{5xy}}{y}$ **61.** [7.5] $\dfrac{1 - 2\sqrt{x} + x}{1 - x}$
62. [7.8] $26 - 13i$ **63.** [8.4] $x^2 - 50 = 0$
64. [10.1] $(2, -3); 6$ **65.** [9.4] $\log_a = \dfrac{\sqrt[3]{x^2} \cdot z^5}{\sqrt{y}}$
66. [9.3] $a^5 = c$ **67.** [9.5] 3.7541 **68.** [9.5] 0.0003
69. [9.5] 8.6442 **70.** [9.5] 0.0277
71. [9.6] **(a)** $k \approx 0.261$; $C(t) = 0.12e^{0.261t}$; **(b)** about 48.6 million **72.** [10.1] 5 **73.** [11.2] -121
74. [11.2] 875 **75.** [11.3] $16(\frac{1}{4})^{n-1}$
76. [11.4] $13{,}440a^4b^6$ **77.** [11.3] $74.88671875x$

78. [11.3] \$168.95 **79.** [6.4] All real numbers except 0 and −12 **80.** [9.6] 81 **81.** [8.6] y gets divided by 8 **82.** [7.8] $-\frac{7}{13} + \frac{2\sqrt{30}}{13}i$ **83.** [3.5] 84 yr

APPENDIX

Exercise Set I.1, p. 706

1. 3 **2.** 9 **3.** 33.8 **4.** 5.446808511 **5.** 137.921536 **6.** 45.6976 **7.** 1.777777778 **8.** 8

Exercise Set I.2, p. 708

1. $C = 65.0008$ cm; $A = 336.2232$ cm^2
2. **(a)** \$84.72; **(b)** \$50.83; **(c)** \$37.07; **(d)** \$73.29
3. 659; −21; 4,945,843; 274,828,929
4. −1.06329696; −1.06329696; $2.250728506 \times 10^{12}$; $2.250728506 \times 10^{12}$

Exercise Set 2.1, p. 710

1. (a)

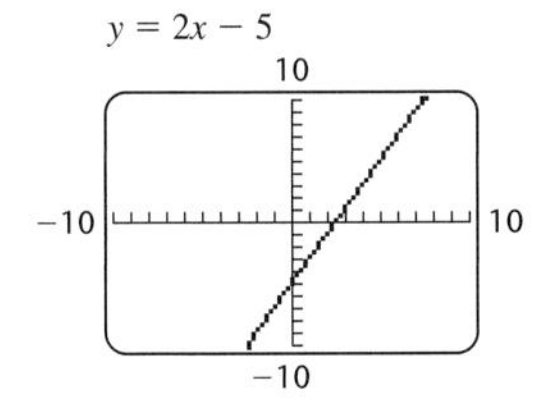

(b)

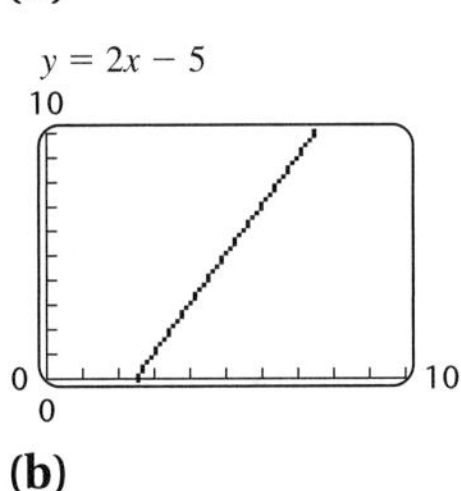

2. (a)

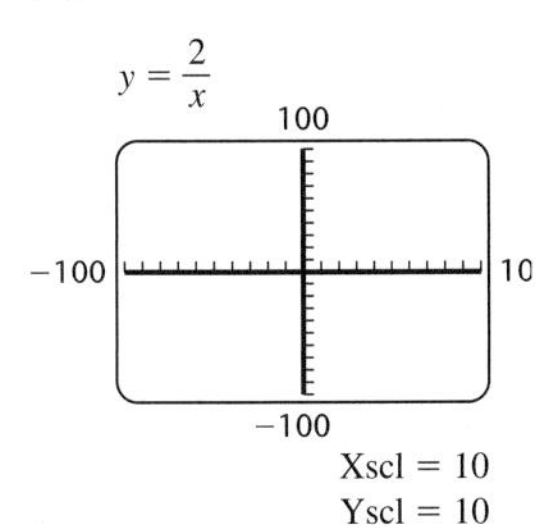

(b)

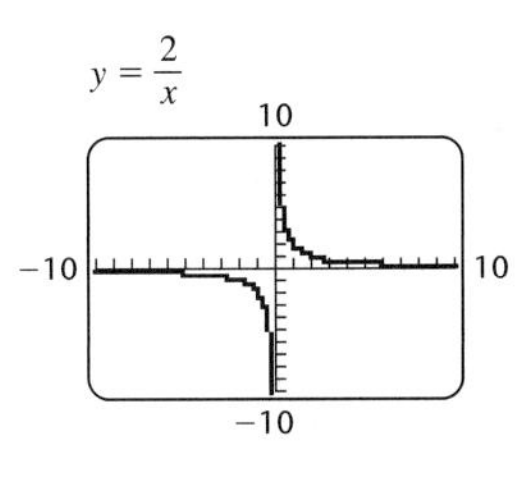

3.

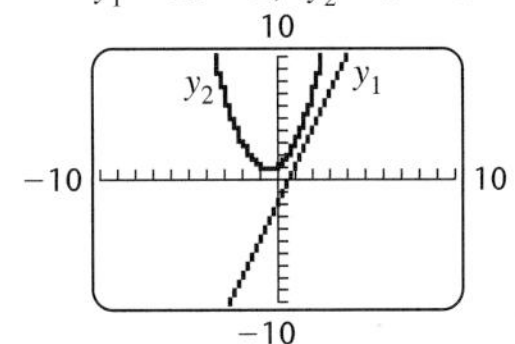

(a) The grapher will graph the entire equation $y_1 = 3x - 2$ first, and then $y_2 = x^2 + x + 1$. **(b)** The grapher will appear to draw both graphs at the same time. **4.** If we use ZDecimal and TRACE, y is 4.52 when $x = -1.2$.

Exercise Set 2.5, p. 713

1. Yes **2.** No **3.** No **4.** **(a)** $f(x) = 0.231x + 67.31$; **(b)** 76.55 **5.** **(a)** $f(x) = 75.93023256x + 8323.372093$; **(b)** about \$9082.67 **6.** **(a)** $f(x) = 3.43x + 87.28$; **(b)** about 258.8 million tons

Exercise Set 4.4, p. 715

1.

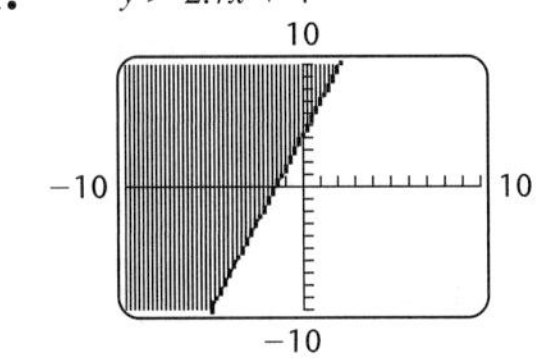

2.

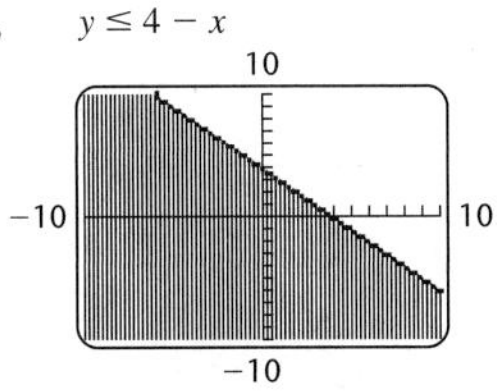

3.

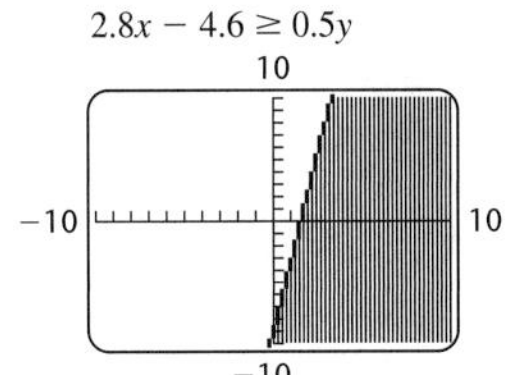

4.

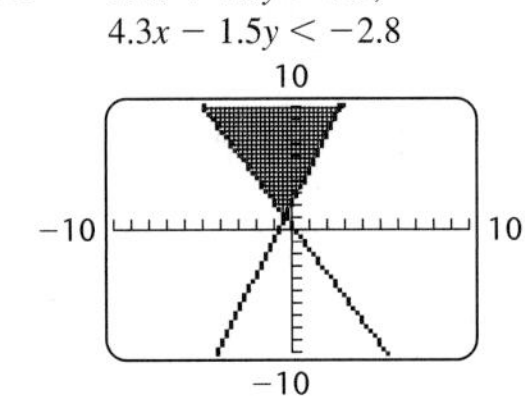

Exercise Set 8.8, p. 716

1. Maximum of −5.22 at $x = 0.88$
2. Minimum of 2.55 at $x = -0.40$
3. Minimum of −440.82 at $x = -383.33$

Index

A

Abscissa, 71
Absolute value, 11, 14, 65, 233
  and distance, 11, 234
  equations with, 233–236
  on a grapher, 85
  inequalities with, 237–239
  and radical expressions, 411, 413, 466
Absolute-value principle, 234, 238, 261
Acceleration, average, 396
Addition
  associative law, 18, 65
  commutative law, 18, 65
  of complex numbers, 460
  of exponents, 47, 54, 66, 419, 466
  of functions, 127, 135
  of logarithms, 577, 611
  of polynomials, 270
  of radical expressions, 436, 467
  of rational expressions, 343, 345, 401
  of real numbers, 13, 65
Addition principle
  for equations, 23, 65
  for inequalities, 214, 260
Additive inverse, 13
  of polynomials, 271
Air resistance, 116
Algebra of functions, 127, 135
Algebraic expression, 2
  evaluating, 4
  least common multiple, 345, 401
  translating to, 3
And, 226, 260
Angles
  complementary, 149
  supplementary, 148
Angstrom (Å), 62
Applications, *see* Applied problems; Formulas. *See also the Index of Applications.*
Applied problems, 1, 10, 33–38, 43–46, 59–63, 67–69, 85, 89–91, 98–100, 102–105, 111, 115–117, 119, 123–125, 128, 131–133, 136–139, 141–144, 148–150, 158–171, 179–186, 190, 191, 196, 198, 201–203, 205–209, 218–222, 232, 241, 250, 251, 256–263, 268, 273–276, 285, 288, 291, 292, 302, 308, 312, 316, 322–328, 331, 332, 340, 341, 351, 359, 360, 367–377, 383, 388–390, 392, 393, 395–399, 403–406, 416, 421, 422, 428, 435, 441, 448, 449, 451, 452, 455–458, 468, 469, 479, 480, 482, 483, 489–497, 501, 507, 524–531, 539, 540, 543–545, 558, 559, 566–568, 595–608, 612–617, 619, 629–631, 637, 648, 654–657, 659–661, 668, 673, 674, 676, 677, 682–686, 695, 697–701. *See also the Index of Applications for specific applications.*
Approximating solutions, quadratic equations, 487
Area formulas, 65. *See also the Index of Applications.*
Arithmetic progressions, *see* Arithmetic sequences.
Arithmetic sequences, 668, 669, 696
  common difference, 669
  *n*th term, 669, 696
  sum of *n* terms, 671, 696
Arithmetic series, 671
Ascending order, 268
Associative laws, 18, 65
Asymptote of an exponential function, 562
Asymptotes of a hyperbola, 639, 640, 642, 648
Average acceleration, 396
Average speed, 377
Axes, 70
Axis of a hyperbola, 639
Axis of symmetry, 509, 510, 511, 512, 520, 541, 542, 621

B

Base, 5
  changing, 586, 611
Base-10 logarithms, 583
Binomial, 267
Binomial coefficient, 692, 696
Binomial expansion
  using factorial notation, 693, 696
  using Pascal's triangle, 690
  $(r + 1)$st term, 694, 696
Binomial theorem, 690, 693, 696
Binomials, multiplication of, 278, 280–283
Birth weight, projected, 45
Branches of a hyperbola, 639
Break-even analysis, 197. *See also the Index of Applications.*
Break-even point, 199

C

Calculator. *See also* Graphers.
  logarithms on, 584, 586
  square roots on, 409
Canceling, 338
Carbon dating, 603, 611
Carry out, problem solving, 30, 32, 65
Cartesian coordinate system, 70
Center
  circle, 625
  ellipse, 632, 634
  hyperbola, 639, 648
Change-of-base formula, 586, 611
Changing the sign, 14, 271, 344
Check the answer in problem solving, 30, 32, 65
Check the solution, 23, 318, 363, 444, 503, 591, 593
  by evaluating, 349
Circle, 625
  area, 39, 65
  center, 625
  circumference, 39
  equation, 626, 644, 658
  graph of, 626
  radius, 625
Circumference, 39
Clearing decimals, 155
Clearing fractions, 155, 362
Closed interval, 213
Coefficients, 267
  binomial, 692, 696
Collaborative Corner, 39, 64, 92, 134, 159, 180, 223, 233, 252, 277, 286, 361, 377, 400, 423, 450, 517, 532, 569, 609, 638, 687
Collecting like terms, 24, 270
Columns of a matrix, 187
Combined variation, 395
Combining like terms, 24, 270
  radical, 436, 467
Common difference, 669
Common factor, 287
Common logarithms, 583
  on a calculator, 584
Common multiple, least, 345, 401
Common ratio, 678
Commutative laws, 18, 65
Complementary angles, 149
Completing the square, 475–479, 540
  in graphing, 517
Complex numbers, 458
  addition, 460
  conjugate, 462
  division, 462
  multiplication, 461
  powers of i, 463
  subtraction, 460
Complex rational expression, 352
  simplifying, 354, 356, 401
Composite function, 548, 549, 610
  and inverses, 557, 610
Compound inequalities, 224
Compound interest, 479, 566, 599, 602, 610, 611. *See also the Index of Applications.*
Conditional equation, 26
Conic sections, 620. *See also* Circle; Ellipse; Hyperbola; Parabola.
  classifying from equations, 646
Conjugate
  of a complex number, 462
  of a radical expression, 437
Conjunction, 224
CONNECTED mode on a grapher, 133
Connecting the concepts, 25, 122, 156, 172, 249, 266, 287, 361, 443, 514, 560, 643, 688
Consistent system of equations, 147, 155, 176, 177, 178, 204
Constant, 2
  of proportionality, 391, 393
Constant function, 106
Constraints, 253
Contradiction, 26
Contrast on a grapher, 704
Coordinate system, 70
Coordinates, 71
Corner principle, 254, 261
Correspondence, 79, 83
Cost, 197
Counting numbers, 6
Cramer's rule, 192, 193, 195, 204
Cube, volume, 5, 65
Cube roots, 412, 466
Cube-root function, 412
Cubes, factoring sum and difference of, 309, 311, 329
Cubic polynomial, 267
Cursor, 704
Cylinder, volume, 43, 65

D

Data point, 525
Decay model, exponential, 603, 611
Decay rate, 603
Decibel, 596
Decimal notation, 8. *See also* Decimals.
  converting to scientific notation, 57
Decimals. *See also* Decimal notation.
  clearing, 155
  repeating, 8
  terminating, 8
Decomposition of functions, 550
Degree
  of a polynomial, 267
  of a polynomial equation, 317
  of a term, 266
Demand, 200
Denominators, rationalizing, 431, 437
Density, 43
Dependent equations, 147, 155, 178, 204
Dependent variable, 84
Depreciation, straight-line, 677
Descartes, René, 70
Descending order, 268
Determinant, 191, 192, 194, 195, 204
Diagonals of a polygon, 292
Difference, common, 669
Difference
  of cubes, factoring, 309, 311, 329
  of functions, 127, 135
  of logarithms, 578, 611
  of rational expressions, 343, 345, 401
  of two squares, factoring, 305, 311, 329
Dimensions of a matrix, 191
Direct variation, 391, 402
Discriminant, 498, 541
  and x-intercepts, 499
Disjoint sets, 226
Disjunction, 227

Distance
on a number line, 236
between points
in the plane, 624, 658
on a vertical or horizontal line, 623
Distance formula, 624, 658
Distance of a free-fall, 480, 542
Distance traveled, 166. *See also the Index of Applications.*
Distributive law, 19, 65
Division
of complex numbers, 462
with exponential notation, 47, 54, 66, 419, 466
of functions, 127, 135
by a monomial, 378
of polynomials, 378–382
with radical expressions, 429, 466
of rational expressions, 339, 401
of real numbers, 15, 65
and reciprocals, 16
remainder, 380
with scientific notation, 59
synthetic, 383–386
by zero, 17
Domain
of a function, 78, 79, 84, 85, 229, 320
and graphs, 128, 335
of a relation, 83
of a sum, difference, product, or quotient of functions, 129
Doppler effect, 390
DOT mode on a grapher, 133
Doubling time, 599, 600
Dummy variable, 84

E

e, 585, 611
Editing entries on grapher, 705
Element
of a matrix, 187
of a set, 7
Eliminating a variable, 153
Elimination method, 153
using matrices, 186
Ellipse, 632
applications, 635, 636
center, 632, 634
equation of, 632, 634, 646, 658
foci, 632
graph of, 632, 633
intercepts, 632, 633
vertices, 632
Empty set, 27
Entering expressions on a grapher, 704
Entries of a matrix, 187
Equality, principle of
exponential, 574, 590, 611
logarithmic, 591, 611
Equation, 2. *See also* Formulas.
with absolute value, 233–236
addition principle, 23, 65
of a circle, 626, 644, 658
conditional, 26
contradiction, 26
distance, 166
of an ellipse, 632, 634, 645, 658
equivalent, 22, 25
exponential, 590
and functions, 84
on a grapher, 707
graphs, 72
linear, 72–74, 93, 94
nonlinear, 74, 75
of a hyperbola, 639, 642, 645, 646, 648, 658
identity, 26
linear, 26, 114, 122, 135, 172
logarithmic, 573
multiplication principle, 23, 65
nonlinear, 74
of a parabola, 621, 644, 658
point–slope, 117, 118, 122, 135
polynomial, 317
quadratic, 317
in quadratic form, 502–506
radical, 443
rate, 166
rational, 362
reducible to linear, 158
reducible to quadratic, 502–506
related, 242
slope–intercept, 96, 122, 135
solution set, 27
solutions of, 3, 72
solving, *see* Solving equations
systems of, 142, 649. *See also* Systems of equations.
time, 166
writing from solutions, 499
Equilibrium point, 200
Equivalent equations, 22, 25
Equivalent expressions, 18, 25
Equivalent inequalities, 214
Escape velocity, 389
Evaluating
checking by, 349
determinants, 192, 194, 204
expressions, 4
polynomial functions, 268
Even root, 413, 466
Exponential decay mode, 603, 611
Exponential equality, principle of, 574, 590, 611
Exponential equations, 590. *See also* Exponential functions.
converting to logarithmic equations, 572
solving, 590
Exponential functions, 562, 571, 610. *See also* Exponential equations.
graphs of, 562, 587, 610
asymptotes, 562
inverse of, 570
Exponential growth model, 600, 611
Exponential notation, 5. *See also* Exponents.
on a calculator, 51
and division, 47, 54, 66, 419, 466
and multiplication, 47, 54, 66, 419, 466
and raising a power to a power, 51, 54, 66
and raising a product to a power, 52, 54, 66
and raising a quotient to a power, 53, 54, 66
Exponents, 5. *See also* Exponential notation.
addition of, 47, 54, 66, 419, 466
irrational, 561
laws of, 54, 66, 419, 466
multiplying, 51, 54, 66, 419, 466
negative, 49, 50, 54, 66, 418
of one, 5
rational, 416, 466
rules for, 54, 66, 419, 466
subtracting, 47, 54, 66, 419, 466
variables as, 562
of zero, 48, 54, 66
Expression
algebraic, 2
equivalent, 18, 25
evaluating, 4
radical, 409
rational, 334
Extrapolation, 85

F

f-stop, 390
Factor, 4, 19
Factorial notation, 691, 696

Factoring, 19
common factors, 287
completely, 288
difference of cubes, 309, 311, 329
difference of squares, 305, 311, 329
by grouping, 289, 306
perfect-square trinomials, 304
polynomials, 287–290, 293–301, 303–307, 309–311, 313–316, 329, 330
the radicand, 424
solving equations by, 318, 472
strategy, 313, 330
sum of cubes, 309, 329
trinomials, 293–301, 304, 329, 330
Factors, 4, 19
and negative exponents, 50
Familiarization in problem solving, 30, 32, 65
Feasible region, 254
Finite sequence, 662
Finite series, 664
First coordinate, 71
Fitting a function to data, 525
Fixed costs, 197
Foci of an ellipse, 632
Focus, 632
FOIL method
of factoring, 297, 329
of multiplication, 280
Formula(s), 39. *See also the Index of Applications.*
area, 65
average acceleration, 396
change-of-base, 586, 611
circumference, 39, 65
compound interest, 479, 566, 610
density, 43
distance between points, 624, 658
distance traveled, 65, 166
Doppler effect, 390
escape velocity, 389
f-stop, 388
free-fall distance, 480, 542
for inverses of functions, 553, 610
as mathematical models, 42
midpoint, 625, 658
motion, 31, 371, 402
perimeter of a square, 65
quadratic, 484, 541
simple interest, 40, 42, 65
solving, 40–42, 65, 388–391, 491–494, 541
taxable interest rate, 395
volume, 65
Fraction, sign of, 15, 65
Fractional equations, *see* Rational equations
Fractional exponents, *see* Rational exponents
Fractional expressions, *see* Rational expressions
Fractional notation, 8
Fractions, clearing, 155
Free-fall distance, 480, 542
Functions, 79
addition of, 127, 135
algebra of, 127, 135
composite, 548, 549, 610
constant, 106
cube-root, 412
decomposition, 550
dependent variable, 84
division of, 127, 135
domain, 78, 79, 84, 85
dummy variable, 84
and equations, 84
evaluating, 268, 269, 283, 707
exponential, 562, 571, 610
graphs of, 80–83
greatest integer, 91
horizontal-line test, 552
independent variable, 84
input, 83
inverse of, 553, 557, 610
linear, 93
logarithmic, 570, 571
maximum/minimum value, 716
multiplication of, 127, 135
notation, 83
objective, 253
one-to-one, 552
output, 83
polynomial, 268
quadratic, 485
range, 78, 79
rational, 334
related, 242
roots of, 321
sequence, 662
square-root, 409
subtraction of, 127, 135
value, 84
vertical-line test, 82, 135
zeros of, 321, 534

G

General term of a sequence, 663
Geometric progressions, *see* Geometric sequences
Geometric sequences, 677, 678, 696. *See also* Geometric series.
common ratio, 678
nth term, 678, 696
sum of n terms, 680, 696
Geometric series, 679
infinite, 680
limit, 681, 696
Golden rectangle, 497
Graphers, 14, 704
and absolute value, 85
alpha keys, 705
and binomial coefficient, 692
and circles, 628
and composition of functions, 550
connected mode, 133
contrast, 704
cursor, 704
and determinants, 195
DOT mode, 133
editing entries, 705
and ellipses, 634
entering data, 526, 712
entering expressions, 704
and equations, 707
evaluating functions, 269, 707
and exponential equations, 594
and exponential functions, 563, 566
and exponential notation, 51
and factorial notation, 692
GraphStyle, 245
G-T mode, 283, 510
home screen, 705
horizontal mode, 510
and hyperbolas, 643
and inequalities, 217, 239, 245, 714
intersect, 113, 147
and inverses, 557
linear regression, 125, 711
and logarithmic equations, 594
and logarithmic functions, 588
logic option, 232
maximum/minimum function values, 716
menus, 705
and negative numbers, 705
and perpendicular lines, 120
pixels, 704
programs, 707
quadratic regression, 526, 531
recalling values, 476, 707
regression, 125, 526, 531, 711
ROOT option, 505
row-equivalent operations, 189
scale of an axis, 563, 708

and scientific notation, 59
SEQUENCE mode, 663, 668
SEQUENTIAL mode, 710
SHADE option, 245
SIMULTANEOUS mode, 105, 710
and solving equations, 113
systems of, 147, 651
split screen, 283, 510
squaring axes, 557, 628, 634, 643
standard viewing window, 709
storing values, 476, 707
subtraction, 705
tables, 75
test key, 217, 232
trace, 71, 269, 709
value, 709
vars, 230
window, 71, 708
Y-VARS, 230
zdecimal, 709
zero, 321
ZOOM menu, 709
zsquare, 120

Graphs
of circles, 626, 644
and completing the square, 517
of cube-root function, 412
and domains, 128, 337
of ellipses, 632, 633, 645
of equations, 72. *See also* Graphs of linear equations; Graphs of nonlinear equations.
systems of, 144
of exponential functions, 562, 587, 610
of functions, 80–83
and their inverses, 554
vertical-line test, 82, 135
of horizontal lines, 106, 107
of hyperbolas, 640, 645, 646
of inequalities, 212
in one variable, 212
systems of, 246
in two variables, 242–246, 261
using intercepts, 108
of inverses of functions, 554
of linear equations, 72–75, 93, 94
using intercepts, 108
using slope, 96, 118
of logarithmic functions, 570, 587, 610
of nonlinear equations, 74
of parabolas, 508–514, 517–521, 541, 542, 622, 644
of polynomial functions, 321
of quadratic functions, 508–514, 517–521, 541, 542
intercepts, 521
and rational equations, 365
using slope, 96, 118
and solving equations, 110, 112, 113
systems of, 144
of square-root function, 410
translations, 93
of vertical lines, 107, 108
x-intercept, 108, 109, 521
y-intercept, 94, 521

GraphStyle, 245
Greater than (>), 12
Greater than or equal to (≥), 12
Greatest common factor, 287
Greatest integer function, 91
Grouping, factoring by, 289, 306
trinomials, 300, 330
Growth model, exponential, 600, 611
Growth rate, 119
exponential, 600

H

Half-life, 603
Half-open interval, 213
Hang time, 492
Harmonic mean, 400
Home screen, 705
Horizontal line, 106, 135
Horizontal-line test, 552
Hyperbola, 639
applications, 642
asymptotes, 639, 640, 642, 648
axis, 639
branches, 639
center, 639, 648
equation of, 639, 645, 646, 648, 658
nonstandard form, 642
graphs of, 640
vertices, 639, 648
Hypotenuse, 323

I

i, 459
powers of, 463
Identity, 26
Imaginary numbers, 459
Inconsistent system of equations, 147, 155, 176, 177, 178, 204
Independent equations, 147, 204
Independent variable, 84
Indeterminate answer, 17
Index of a radical expression, 413
Index of summation, 665
Inequalities, 12, 212
with absolute value, 237–239
addition principle for, 214, 260
with "and," 226
compound, 224
conjunction, 224
disjunction, 227
equivalent, 214
graphs of, 212
one variable, 212
systems of, 246
two variables, 242–246, 261, 714
linear, 242
multiplication principle for, 216, 260
with "or," 228
polynomial, 532, 536, 542
quadratic, 532
rational, 537, 538, 542
solution, 212, 242
solution set, 212
solving, 212, 214–218, 225–229, 536, 538, 542
systems of, 246
Infinite sequence, 662
Infinite series, 664, 680
Infinity (∞), 214
Input, 83
Integers, 6
Intercepts
of an ellipse, 632, 633
graphing using, 108
of a parabola, 521
x-, 108, 109
y-, 94, 108, 109
Interest. *See also the Index of Applications.*
compound, 479, 566, 599, 602, 610, 611
simple, 40, 42
Interest rate, taxable, 395
Interpolation, 85
INTERSECT feature on a grapher, 113, 147
Intersection of sets, 224, 260
Interval notation, 213
Inverse relation, 551. *See also* Inverses of functions.
Inverse variation, 392
Inversely proportional, 393
Inverses
additive, 13. See also Opposites.
of functions, 553, 557, 610

Inverses (*continued*)
multiplicative, 16. *See also* Reciprocals.
Invert and multiply, 339
Irrational numbers, 8
as exponents, 561
Isosceles right triangle, 452, 455, 467

J

Joint variation, 394, 402

L

Largest common factor, 287
Law(s)
associative, 18, 65
commutative, 18, 65
distributive, 19, 65
of exponents, 54, 66, 419, 466
of opposites, 13, 65
of reciprocals, 16, 65
Leading coefficient, 267
Leading term, 267
Least common denominator (LCD), 346
Least common multiple (LCM), 345, 401
Legs of a right triangle, 323
Less than (<), 12
Less than or equal to (≤), 12
Light year, 62
Like radicals, 436
Like terms, 24, 270
Limit of an infinite geometric series, 681, 696
Linear equation in one variable, 26
Linear equation in three variables, 172
Linear equations in two variables, 74. *See also* Linear functions.
graphs of, 72–74, 93, 94
point–slope, 117, 118, 122, 135
slope–intercept, 96, 122, 135
standard form, 114, 122, 135
systems of, 142
Linear functions, 93
Linear inequality, 242
systems of, 246
Linear polynomial, 267
Linear programming, 253
Linear regression, 125, 711
Lines
equations of, *see* Linear equations
horizontal, 106, 135
parallel, 120, 135
perpendicular, 121, 135
slope of, 95, 135
vertical, 107, 135
Lithotripter, 635
ln, *see* Natural logarithms
Loan payment, 569
log, *see* Common Logarithms
Logarithm, *see* Logarithms
Logarithm functions, *see* Logarithmic functions
Logarithmic equality, principle of, 591, 611
Logarithmic equations, 573
converting to exponential equations, 572
solving, 573, 587, 592, 610
Logarithmic functions, 570, 571. *See also* Logarithms.
graphs, 570
Logarithms, 570, 571, 610. *See also* Logarithmic functions.
base a of a, 575, 611
of the base to a power, 581, 611
on a calculator, 584, 586
change-of-base formula, 586, 611
common, 583
difference of, 578, 611
Naperian, 585
natural, 585
of one, 575, 611
of powers, 578, 611
of products, 577, 611
properties of, 611
of quotients, 578, 611
sum of, 577, 611
LOGIC option on a grapher, 232
Loudness of sound, 596, 611

M

Mathematical model, 42
Matrices, 187. *See also* Matrix.
Matrix, 187
columns, 187
determinant of, 191, 192, 194, 195, 204
dimensions, 191
elements (or entries), 187
row-equivalent, 190
rows, 187
and solving systems of equations, 187
square, 191
Maximum on a grapher, 716
Maximum value of a quadratic function, 512, 523, 541, 542
Mean, harmonic, 398
Menus on a grapher, 705
Midpoint formula, 625, 658
Mil, 63
Minimum on a grapher, 716
Minimum value of a quadratic function, 512, 523, 541, 542
Mixture problems, 160
Model, mathematical, 42
Monomials, 266, 267
as divisors, 378
multiplying, 277
Motion formula, 31, 371, 402
Motion problems, 166, 371, 490
Multiple, least common, 345, 401
Multiplication
associative law, 18, 65
of binomials, 278, 280–283
commutative law, 18, 65
of complex numbers, 461
with exponential notation, 47, 54, 66, 419, 466
of exponents, 51, 54, 66, 419, 466
of functions, 127, 135
by one, 335, 353
of polynomials, 277–283
of radical expressions, 423, 437, 466
of rational expressions, 335, 401
of real numbers, 15, 65
with scientific notation, 58
Multiplication principle
for equations, 23, 65
for inequalities, 216, 260
Multiplicative inverse, 16. *See also* Reciprocals.

N

nth term
arithmetic sequence, 669, 696
geometric sequence, 678, 696
Naperian logarithms, 585
Natural logarithms, 585
Natural numbers, 6
Negative exponents, 49, 54, 66
and factors, 50, 54
Nested evaluation, 388
Nonlinear equations, 74
systems of, 649
Notation
decimal, 8
exponential, 5
factorial, 691
fractional, 8
function, 83
interval, 213

radical, 409
scientific, 56, 66
set, 7, 213
sigma, 665
summation, 665
Number line, 7
Numbers
complex, 459
counting, 6
imaginary, 459
integers, 6
irrational, 8
natural, 6
rational, 7
real, 8
whole, 6
Numerators, rationalizing, 433, 437

O

Objective function, 253
Odd root, 413, 466
One
as an exponent, 5
logarithm of, 575, 611
multiplying by, 335, 353
One-to-one function, 552
Open interval, 213
Operations
order of, 5
row-equivalent, 189, 190, 204
Opposites, 13, 25, 271
and absolute value, 14
law of, 13, 65
and subtraction, 14, 271
Or, 228, 260
Order of operations, 5
Ordered pair, 71
Ordered triple, 172
Ordinate, 71
Origin, 71
Output, 83

P

Pairs, ordered, 71
Parabola, 620
applications, 620
axis of symmetry, 509, 510, 511, 512, 520, 541, 542
equations of, 621, 644, 658
graphs of, 508–514, 517–521, 541, 542, 622
intercepts, 521
vertex, 509, 520, 542
Parallel lines, 120, 135
Parallelogram, area, 39, 65
Parsec, 62, 67
Partial sum, 664
Pascal's triangle, 689
and binomial expansion, 691
Pendulum, period, 435, 492
Perfect cube, 424
Perfect nth power, 424
Perfect square, 424
Perfect-square trinomial, 303
factoring, 304, 329
Perimeter, square, 34
Period of a pendulum, 435, 492
Perpendicular lines, 121, 135
pH, 598, 611
Pi (π), 39
Plane, 70
Plotting points, 71
Point–slope equation, 117, 118, 122, 135
Points, plotting, 71
Polynomial, 266. *See also* Polynomial function.
addition, 270
additive inverse, 271
ascending order, 268
binomial, 267
coefficients, 267
combining (or collecting) like terms, 270
cubic, 267
degree of, 267
descending order, 268
division of, 378–382
synthetic, 383–386
factoring
common factors, 287
completely, 288
difference of cubes, 309, 311, 329
difference of squares, 305, 311, 329
by grouping, 289, 300, 306, 330
perfect-square trinomials, 304, 329
strategy, 313, 330
sum of cubes, 309, 311, 329
trinomials, 293–301, 304, 329, 330
leading coefficient, 267
leading term, 267
linear, 267
monomial, 266, 267
multiplication, 277–283
in one variable, 267
opposite, 271
perfect-square trinomial, 304, 329
prime, 288
quadratic, 267
in several variables, 267
subtraction, 271
terms, 266
trinomial, 267
Polynomial equation, 317
degree, 317
standard form, 318
Polynomial function, 268. *See also* Polynomial.
evaluating, 268
and graphs, 321
roots, 321
zeros, 321
Polynomial inequality, 532, 536, 542
Population growth, *see the Index of Applications*
Power raised to a power, 51, 54, 66, 419, 466
Power rule
for exponents, 51, 54, 66, 419, 466
for logarithms, 578, 611
Powers, 5
of i, 463
logarithm of, 581, 611
principle of, 443, 467
Prime polynomial, 288
Principal square root, 409
Principle
absolute-value, 234, 238, 261
addition
for equations, 23, 65
for inequalities, 214, 260
corner, 254, 261
of exponential equality, 574, 590, 611
of logarithmic equality, 591, 611
multiplication
for equations, 23, 65
for inequalities, 216, 260
of powers, 443, 467
of square roots, 451, 467, 473, 474, 540
of zero products, 318, 330
Problem solving, five-step strategy, 30, 32, 65
Product. *See also* Multiplication.
of functions, 127, 135
logarithm of, 577, 611
raising to a power, 52, 54, 66
of rational expressions, 335, 401
of solutions, quadratic equation, 502
of sums and differences, 282
of two polynomials, 277–283

Product rule
  for exponents, 47, 54, 66, 419, 466
  for logarithms, 577, 611
  for radicals, 423, 466
Profit, 197
Progression, 662. *See also* Sequence.
  arithmetic, 669. *See also* Arithmetic sequences.
  geometric, 678. *See also* Geometric sequences.
Projection of a function, 81
Properties of logarithms, 611
Proportional, 391, 393
Pure imaginary numbers, 460
Pythagorean theorem, 323, 451, 467

Q

Quadrants, 71
Quadratic equations, 317
  approximating solutions, 487
  discriminant, 498, 451
  product of solutions, 502
  reducible to quadratic, 502–506
  solutions, nature of, 498, 541
  solving
    by completing the square, 475–479, 540
    by factoring, 318, 472
    by principle of square roots, 451, 467, 473, 540
    by quadratic formula, 484, 541
  standard form, 318
  sum of solutions, 502
  writing from solutions, 499
Quadratic form, equation in, 502–506
Quadratic formula, 484, 541
Quadratic functions, 485
  fitting to data, 525
  graphs of, 508–514, 517–521, 541, 542
  maximum value of, 512, 523, 541, 542
  minimum value of, 512, 523, 541, 542
Quadratic inequalities, 532
Quadratic polynomial, 267
Quadratic regression on a grapher, 526, 531
Quotient
  of functions, 127, 135
  logarithm of, 578, 611
  raising to a power, 53, 54, 66
  of rational expressions, 339, 401
  root of, 429, 466
Quotient rule
  for exponents, 47, 54, 66
  for logarithms, 578, 611
  for radicals, 429, 466

R

Radical equations, 443
Radical expressions, 409
  and absolute value, 411
  addition of, 436, 467
  conjugates, 437
  dividing, 429, 466
  index, 413
  like, 436
  multiplying, 423, 437, 466
  product rule, 423, 466
  quotient rule, 429, 466
  radicand, 409
  rationalizing denominators or numerators, 431, 433, 437
  simplifying, 420, 424, 439
  subtraction, 436
Radical sign, 409
Radicals, *see* Radical expressions
Radicand, 409
  factoring,424
Radii, 625
Radius, 625
Raising a power to a power, 51, 54, 66, 419, 466
Raising a product to a power, 52, 54, 66, 419, 466
Raising a quotient to a power, 53, 54, 66
Range
  of a function, 78, 79
  of a relation, 83
Rate
  of change, 98
  equation, 166
  exponential decay, 603
  exponential growth, 600
Ratio, common, 678
Rational equations, 362
  and graphs, 365
Rational exponents, 416, 466
  and simplifying radical expressions, 439
Rational expression, 334
  addition, 343, 345, 401
  complex, 352
  dividing, 339, 401
  least common denominator (LCD), 346
  multiplying, 335, 401
  multiplying by one, 335, 353
  reciprocal of, 339
  simplifying, 336
  subtraction, 343, 345, 401
Rational function, 334
Rational inequalities, 537, 538, 542
Rational numbers, 7
  as exponents, 416, 466
Rationalizing
  denominators, 431, 437
  numerators, 433, 438
Real numbers, 8
  addition, 13, 65
  division, 15, 65
  multiplication, 15, 65
  subtraction, 14
Recalling values on a grapher, 476, 707
Reciprocals, 16
  and division, 16
  and exponential notation, 49
  law of, 16, 65
  of rational expressions, 339
Rectangle, area, 40, 65
Rectangle, golden, 497
Recursively defined sequence, 667
Reducible to linear equations, 158
Reducible to quadratic equations, 502–506
Reflection, 555, 563
Regression on a grapher, 125, 526, 531, 711
Related equation, 242
Related function, 242
Relation, 83
  inverse, 551
Remainder, 380
Remainder theorem, 387
Removing a factor equal to 1, 337, 486
Repeating decimals, 8
Revenue, 197
Right circular cylinder, volume, 65
Right triangle, 323
  isosceles, 452, 455, 467
  30°–60°–90°, 453, 455, 467
Rise, 95, 96
Root of a function, 321
ROOT option on a grapher, 505
Roots
  cube, 412, 466
  even, 413, 466
  multiplying, 423, 437, 466
  odd, 413, 466
  and quotients, 429, 466
  square, 408, 466

Roster notation, 7
Rounding, 58
Row-equivalent matrices, 190
Row-equivalent operations, 189, 190, 204
Rows of a matrix, 187
Run, 95, 96

S
Salvage value, 677
Scale of axis on a grapher, 563, 708
Scientific notation, 56, 66
  on a calculator, 59
  converting to decimal notation, 57
  dividing with, 59
  multiplying with, 58
  in problem solving, 59
Second coordinate, 71
Sequence, 666
  arithmetic, 669, 696
  finite, 662
  general term, 663
  geometric, 678, 696
  infinite, 662
  partial sum, 664
  recursively defined, 667
  terms, 663
Sequence mode, 663
Sequential mode, 710
Series
  arithmetic, 671
  finite, 664
  geometric, 679
  infinite, 664
  sigma (summation) notation, 665
Set-builder notation, 7, 213
Sets
  disjoint, 226
  element of, 7
  empty, 27
  intersection, 224, 260
  notation, 7, 213
  of numbers, 9
  of solutions, 27, 212
  subset, 9
  union, 227, 260
SHADE option on a grapher, 245
Sigma notation, 665
Sign of a fraction, 15, 65
Significant digits, 58
Signs, changing, 14, 271, 344
Similar terms, 24, 270
Simple interest, 40, 42
Simplifying
  complex rational expression, 354, 356, 401
  radical expressions, 420, 424, 439
  rational expressions, 336
Simultaneous mode on a grapher, 105, 710
Slope, 95, 135
  graphing using, 96, 118
  of a horizontal line, 106, 135
  and parallel lines, 120, 135
  and perpendicular lines, 121, 135
  as rate of change, 98
  undefined, 107, 135
  of a vertical line, 107, 135
  zero, 106, 135
Slope–intercept equation, 96, 122, 135
Solution sets
  of equations, 27
  of inequalities, 212
Solutions
  approximating, 487
  of equations, 3, 72
  of inequalities, 212, 242
  nature of, quadratic equations, 498, 541
  of systems of equations, 144, 172
  writing equations from, 499
Solving equations
  with absolute value, 233–236
  using addition principle, 23
  exponential, 590
  by factoring, 318, 472
  graphically, 110, 112, 113
  logarithmic, 573, 592
  using multiplication principle, 23
  quadratic, 485, 541
    by completing the square, 475–479, 540
    by factoring, 318, 472
    by principle of square roots, 451, 467, 473, 474, 540
    by quadratic formula, 484, 541
  in quadratic form, 502–506
  radical, 444, 446, 467
  rational, 362, 391
  reducible to quadratic, 502–506
  systems, *see* Systems of equations
Solving formulas, 40–42, 65, 388–391
Solving inequalities, 212
  with absolute value, 237–239
  in one variable, 214–218, 225–229
  polynomial, 536, 542
  rational, 538, 542
Solving problems, five-step strategy, 30, 32, 65
Solving systems of equations, *see* Systems of equations
Sound, loudness, 596, 611
Speed, average, 377
Split screen on a grapher, 283
Square
  area, 5, 65
  perimeter, 34
Square matrix, 191
Square roots, 408, 466
  and absolute value, 411
  on a calculator, 409
  of negative numbers, 408, 458
  principal, 409
  principle of, 451, 467, 473, 474, 540
Square-root function, 409
Squares
  of binomials, 281
  differences of, 305
  perfect, 424
  sum of, 311
Squaring axes on a grapher, 557, 628, 634, 643
Standard form
  of circle equations, 626
  of ellipse equations, 632
  of hyperbola equations, 639, 648, 658
  of linear equations, 114, 122, 135, 172
  of polynomial equations, 318
Standard viewing window, 709
State the answer, 30, 32, 65
Storing values on a grapher, 476, 707
Straight-line depreciation, 677
Strategy
  factoring, 313, 330
  problem-solving, 30, 32, 65
Study tip, 6, 17, 35, 49, 70, 80, 93, 118, 143, 152, 166, 174, 212, 234, 243, 257, 293, 313, 340, 348, 362, 408, 429
Subscripts, 41
Subset, 9
Substituting, 4
Substitution method, 151
Subtraction
  additive inverses and, 14, 271
  of complex numbers, 460
  of exponents, 47, 54, 66, 419, 466
  of functions, 127, 135
  of logarithms, 578, 611
  opposites and, 14, 271
  of polynomials, 271
  of radical expressions, 436
  of rational expression, 343, 345, 401

Subtraction (*continued*)
of real numbers, 14
Sum
of cubes, factoring, 309, 311, 329
of functions, 127, 135
of logarithms, 577, 611
of polynomials, 271
of rational expressions, 343, 345, 401
of sequences
arithmetic, n terms, 671, 696
geometric, n terms, 680, 696
partial, 664
of solutions, quadratic equation, 502
of squares, 311
Sum and difference, product of, 282
Summation notation, 665
Supplementary angles, 148
Supply and demand, 200. *See also the Index of Applications.*
Symmetry, axis of, 509, 510, 511, 512, 541, 542, 621
Synthetic division, 383–386
Systems of equations in three variables, 172
consistent, 176, 177, 178
dependent equations, 178
inconsistent, 176, 177, 178
solution of, 172
solving, 173, 174, 203
Systems of equations in two variables, 142
consistent, 147, 155, 204
dependent equations, 147, 155, 204
graphs of, 144
inconsistent, 147, 155, 204
independent equations, 147, 204
nonlinear, 649
solution of, 144
solving
Cramer's rule, 192, 193, 195, 204
elimination method, 153
graphically, 144
using matrices, 187
substitution method, 151
System of inequalities, 246

T
Tables on a grapher, 75
Taxable interest rate, 397
Technology Connection, 14, 51, 59, 71, 75, 85, 94, 96, 113, 120, 131, 147, 189, 195, 217, 230, 235, 239, 245, 269, 272, 281, 283, 289, 296, 321, 338, 344, 356, 365, 386, 411, 414, 418, 420, 424, 425, 445, 476, 481, 486, 487, 499, 505, 509, 510, 512, 525, 526, 536, 550, 557, 563, 566, 571, 588, 594, 628, 634, 643, 651, 653, 663, 692
Term, *see* Terms
Terminating decimals, 8
Terms, 24, 266
coefficients, 267
combining (or collecting) like, 24
degree of, 266
leading, 267
like, 24, 270
of a polynomial, 266
of a sequence, 663
similar, 24
Test key on a grapher, 217
Test points, 243, 534, 537
Theorem
binomial, 690, 693, 696
Pythagorean, 323, 451, 467
remainder, 387
30°–60°–90° right triangle, 453, 455, 467
Time equation, 166
Total cost, 197
Total profit, 197
Total revenue, 197
Total-value problems, 160
Trace on a grapher, 71, 269, 709
Trade-in value, 677
Translating
to algebraic expressions, 3
in problem solving, 30, 32, 65, 218
Translations of graphs, 93
Trapezoid, area, 41, 65
Trial-and-error factoring, 293–297
Triangle
area, 4, 65
right, 323
Trinomials, 267
factoring, 293–301, 329, 330
perfect-square, 304, 329
Triple, ordered, 172

U
Undefined answer, 17
Undefined slope, 107, 135
Union, 227, 260

V
Value of a function, 84
Value on a grapher, 709
Variable, 2
dependent, 84
dummy, 84
eliminating, 153
in the exponent, 562
independent, 84
substituting for, 4
Variable costs, 197
Variation
combined, 395
direct, 391, 402
inverse, 392, 402
joint, 394, 402
Variation constant, 391, 393
VARS on a grapher, 230
Vertex of a parabola, 509, 520, 542. *See also* Vertices.
Vertical line, 107, 135
Vertical-line test for functions, 82, 135
Vertices. *See also* Vertex.
of an ellipse, 632
of a hyperbola, 639, 648
Volume formulas, 65

W
Waiting time, 45
Whispering gallery, 636
Whole numbers, 6
Window on a grapher, 71, 708
standard, 709
Work problems, 368, 401

X
x-intercept, 108, 109, 521
x, y-coordinate system, 70

Y
y-intercept, 94, 108, 109, 521
Y-VARS, 230

Z
ZDECIMAL, 709
Zero
division by, 16
as an exponent, 48, 54, 66
of a function, 321, 534
on a grapher, 321
slope, 106, 135
Zero products, principle of, 318, 330
ZSQUARE option on a grapher, 120

Index of Applications

In addition to the applications highlighted below, there are other applied problems and examples of problem solving in the text. An extensive list of their locations can be found under the heading "Applied problems" in the index at the back of the book.

Astronomy

Astronomy, 59, 62, 63, 68, 353, 360, 389, 396, 637
Escape velocity, 398, 449

Biology

Bacteria in mud, 62
Cell biology, 669
Gender, 171
Growth of bacteria, 567, 606
Newborn calves, 524
Veterinary medicine, 268

Business

Adjusted wages, 38
Advertising, 184, 209, 606
Airplane production, 259, 260
Biscuit production, 259
Book sales, 171
Camcorder production, 327
Catering, 103, 169, 250
Cellular phone use, 99
Coffee blends, 169, 171, 259, 489
Computer hits, 102
Computer manufacturing, 202
Computer printers, 375
Computer production, 203, 262
Consumer demand, 123
Cost projections, 111
Cover charges, 531
Crop yield, 531
Cycle production, 258
Daily wages, 682
Data processing, 403
Depreciation of a computer, 125
Dinner prices, 367
Dog food production, 203
Donuts, 489
Earning plans, 219
Earnings, 359
Furniture production, 260
Granola blends, 169, 191
Grape growing, 259
Hotel management, 375
Hourly wages, 673, 697
Inventory, 196
Lens production, 184
Livestock feed, 169
Loudspeaker production, 203
Lunchtime profits, 258
Mail order, 374
Making change, 170
Manufacturing, 222, 292
Manufacturing caps, 202
Manufacturing lamps, 202
Manufacturing radios, 198
Maximizing profit, 258, 259, 260, 263, 528
Minimizing cost, 259, 260, 262, 528, 545
Minimum wage, 137
Mixed nuts, 169
Monthly earnings, 686
Newspaper delivery, 375
Office supplies, 62
On-line travel plans, 98
Operating expenses, 125
Overseas travel growth, 102
Paid admissions, 250
Photo developing, 340
Pricing, 37
Printing, 375
Printing and engraving, 62
Publishing, 222
Purchasing, 33, 143, 160, 497
Radio airplay, 149
Railing sales, 657
Real estate, 142, 160, 170
Restaurant management, 185
Retail sales, 90, 148
Sales, 169

Sales of cotton goods, 103
Sales of DVD players, 568
Sales of food, 196
Sales of pharmaceuticals, 149
Salvage value, 100, 104, 567, 613
Seller's supply, 123
Soft-drink production, 136
Spaces in a parking lot, 416
Sport coat production, 202
Steel manufacturing, 449
Straight-line depreciation, 677
Supply and demand, 608
Tea blends, 161
Telemarketing, 184
Television sales, 328
Teller work, 170
Textile production, 259
Ticket revenue, 186
Ticket sales, 528
Total cost, 274, 292
Total profit, 275, 292, 539
Total revenue, 274, 292
Trade-in value, 104
Trail mix, 191
Value of a copier, 668
Wages, 221, 261, 285
Weekly pay, 103
Work rate, 396, 397
Yo-yo production, 203

Chemistry

Automotive maintenance, 169, 171
Biochemistry, 171, 497
Carbon dating, 603, 607, 613, 614
Chemistry, 607
Dating fossils, 422
Food science, 170
Ink remover, 169
Mixing fertilizers, 164
Nontoxic floor wax, 149
Nontoxic scouring powder, 150
pH of liquids, 598, 605, 613
Temperature conversions, 89, 125, 222
Temperature of liquids, 232
Wood stains, 171

Construction/Engineering

Air resistance, 116
Architecture, 170, 182, 528, 654
Auditorium design, 676, 698
Automotive repair, 449
Box construction, 328
Bridge design, 531
Bridge expansion, 457
Cabinet making, 327
Carpentry, 323
Contracting, 458
Doorway construction, 630
Ferris wheel design, 631
Home construction, 136, 607
Home remodeling, 158
Home restoration, 374
Insulation, 196
Landscaping, 103
Lumber production, 149
Milling, 258
Molding plastics, 528
Norman window, 531
Painting, 375, 448, 458
Paving, 375
Plumbing, 607
Road pavement messages, 422
Roofing, 458
Stained glass window design, 528
Welding rates, 184
Work rate, 102

Consumer Applications

Automobile pricing, 184
Cellular phone charges, 115, 125
Checking-account rates, 221
Converting dress sizes, 232
Copying costs, 116
Cost of a FedEx delivery, 115
Cost of a movie ticket, 103
Cost of renting a truck, 103
Cost of a road call, 115
Cost of a taxi ride, 103, 158
Customer service complaints, 613
Dress sizes, 558, 559
e-filers, 615
Gas mileage, 258
Home prices, 38
Insurance benefits, 222
Insurance claims, 221
Medical insurance, 125
Milk consumption, 133
Minimizing tolls, 231
Moving costs, 221
Phone rates, 221
Pizza prices, 530
Service call, 263
Telephone charges, 115
Truck rentals, 221
Tuition cost, 136
Van rental, 139, 263
Volume and cost, 399
Wedding costs, 222

Economics

Accumulated savings, 676
Amount owed, 685
Banking, 44
Break-even analysis, 198, 201 203
Compounding interest, 285, 496, 543
Consumer Price Index, 614
Economics, 104
Finance, 63
Financial planning, 360
Interest, 396, 482
Interest compounded annually, 482, 496, 566, 599, 701
Interest compounded continuously, 602, 606, 613, 614
Interest rate, 251
Investing, 44, 45, 222, 258, 262, 685
Investment growth, 479
Investments, 169, 185, 191, 656, 660
Loan amount, 206
Loan repayment, 682, 697
Portfolio value, 613
Real estate taxes, 421
Simple interest, 40, 42
Spending, 676
Stock prices, 528
Straight-line depreciation, 677
Student loan repayment, 605
Student loans, 163, 169
Supply and demand, 201–203
Taxable interest, 395

Environment

Coal demand, 607
Composting, 528
Firefighting, 637
Home maintenance, 34
Household waste, 250
Hydrology, 526
Lead pollution, 397
Logging, 406
Meteorology, 116
National park land, 124
Natural gas demand, 103
Oil demand, 606
Recycling, 123, 375, 567, 605
Solid-waste generation, 232
Ultraviolet index, 393, 397
Use of aluminum cans, 396
Waste generation, 403
Water from melting snow, 392
Wind chill, 116, 428
Yard waste, 607
Zebra mussel populations, 568, 600

Geometry

Agriculture, 367
Angles in a triangle, 36, 37, 67, 184, 205, 615
Antenna wires, 326
Area of a circle, 659
Area of a parallelogram, 44
Area of a rectangle, 40, 656, 657, 660
Area of a square, 326, 331, 660
Area of a trapezoid, 41
Area of a triangle, 38, 448
Box design, 657
Camping tent, 457
Complementary angles, 149
Computer parts, 656
Computer screens, 657
Design, 150
Diagonal of a cube, 497
Diagonals of a polygon, 292, 540
Dimensions of a rug, 656
Distance over water, 456
Equilateral triangle, 89
Fenced-in land, 524
Fencing, 45
Framing, 326, 359
Garden design, 327, 331, 528, 615, 656, 659
Gardening, 45
Golden rectangle, 497
Guy wire, 452, 455
HDTV dimensions, 655, 657
Ladder location, 327
Landscaping, 251, 326
Luggage size, 251
Maximizing area, 544
Minimizing area, 531
Parking lot design, 327
Patio design, 528
Perimeter, 38, 292, 316, 629, 656, 659, 699
Sailing, 326
Supplementary angles, 148
Surface area, 274, 275, 291, 406, 497
Television sets, 456
Tent design, 326
Tile design, 656
Vegetable garden, 456
Volume of carpet, 308
Volume of a display, 276
Warning dye, 403

Health/Life Sciences

AIDS cases in the U. S. population, 90, 568
Audiology, 605
Blood alcohol level, 90
Cholesterol level, 182
Dentistry, 637
Energy expenditure, 91
Hair growth, 103
Health care, 398
Heart attacks and cholesterol, 89
Heart transplants, 606
Life expectancy of females in the United States, 103, 123, 125
Life expectancy of males in the United States, 123
Medicine, 274, 605
Mumps cases, 607
Nutrition, 185
Obstetrics, 185
Pregnancy, 91
Prescription drugs, 351
Projected birth weight, 45
Public health, 606
Tattoo removal, 119
Women giving birth, 131

Miscellaneous

Ages, 150, 186
Audiotapes, 351
Coin value, 190, 191
Counting spheres in a pile, 292
Cutting firewood, 375
Electing officers, 273
Filling a bog, 376
Filling a pool, 374
Filling a tank, 374, 495
Filling a tub, 376
Fundraising, 149
High-fives, 292
Home appliances, 352
Keyboarding speed, 568
Mowing, 368, 404
Number of handshakes, 540, 668
Portrait of Dr. Gachet, 608
President's office, 637
Pumping rate, 397
Repainting a car, 370
Sharing raffle tickets, 186
Sighting to the horizon, 449
Stacking paper, 686
Technology in U. S. schools, 150
Telephone pole storage, 674, 676
Test questions, 367
Value of coins, 285, 353
Waiting time, 45
Waxing a car, 375
Wood cutting, 375

Physics

Acoustics, 390
Atmospheric drag, 398
Atmospheric pressure, 608
Average acceleration, 396, 399
Average speed, 377, 396
Current and resistance, 397
Density, 43, 46
Doppler effect, 390
Downward speed, 496
Drag force, 398
Electricity, 396
Falling distance, 273, 493, 495
Fireworks, 327, 332
Free-falling objects, 480, 482, 543
Height of a baseball, 291, 312
Height of a rocket, 291, 302
Height of thrown object, 299, 539
Hooke's law, 396
Intensity of light, 398
Intensity of a signal, 398
Mass of a particle, 422
Motion of a spring, 241
Music, 351, 421, 422
Ohm's law, 396
Optics, 388
Period of a pendulum, 435, 492
Power of a motor, 631
Pressure at sea depth, 124, 232
Prize tee shirts, 322, 327
Rebound distance, 685, 686, 697, 698
Relative aperture, 397
Resistance, 396
Safety flares, 327
Sound level, 596, 613, 614
Special relativity, 497
Speed of a skidding car, 428
Stopping distance of a car, 398
Telecommunications, 60
Tension of a musical string, 399
Trajectory of a launched object, 531
Volume of gas, 398
Volume and pressure, 397
Wavelength and frequency, 397
Weight of a coin, 44
Weight of salt, 45
Weight on Mars, 397

Social Sciences

Archaeology, 89, 607, 630
Aspects of love, 90

Electing officers, 273
History, 185
PAC contributions, 123
Spread of a rumor, 605
Walking speed, 614

Sports/Hobbies/Entertainment

Archery, 530
Band formations, 676
Baseball, 451, 456, 695
Basketball floor widths, 251
Basketball scoring, 149, 185, 648
Bicycling, 371, 494
Boating, 375, 376, 377
Bungee jumping, 496, 683
Canoeing, 376, 494
Court dimensions, 149
Display of a sports card, 324
Exercise, 171
Gardening, 676
Golf distance finder, 399
Hang time, 492, 496
Hockey rankings, 149
Hockey wins and losses, 251
Kayaking, 375
League schedules, 496
Music, 605
NASCAR attendance, 274
Number of games in a league, 292
Paddleboats, 495
Record in the 400-m run, 123
Records in the 1500-m run, 123
Records in the men's 200-dash, 218
Records in the women's 100-m dash, 232
Rowing, 495
Running rate, 102
Running speed, 99
Skateboarding, 605
Skiing, 630
Skiing rate, 102
Skydiving, 328
Snowboarding, 630
Softball, 455
Speaker placement, 456
Sports salaries, 608
Swimming, 36
Value of a sports card, 607
Video rentals, 149
Walking, 378

Statistics/Demographics

Cellular phones, 568, 604
Compact discs, 604
Computer virus spread, 601
Crying rate, 185
Exam scores, 316
Forgetting, 606, 612
Fuel economy, 171
PAC contriibutions, 123
Population growth, 90, 567, 606, 614, 616, 685
Semester average, 396
Shrinking population, 685
Test scores, 37, 38, 67, 68, 158, 208, 256, 258, 316, 616
Voting attitudes, 89
Weekly allowance, 397
Widows and divorcees, 695
Work experience, 150
Working mothers, 85

Transportation

Air travel, 132, 494, 543, 629, 700
Aircraft departures, 208
Airline routes, 291
Airplane seating, 149
Aviation, 36, 376, 377
Boating, 36, 170, 482, 544
Bus travel, 376
Canoeing, 170
Car rentals, 158
Car speed, 494
Car travel, 170
Car trips, 494
Cost of a speeding ticket, 105
Cruising altitude, 37
Daily accidents, 274
Design of a van, 656
Driving, 328
Elevators, 251
Escalators, 376
Fishing boat, 616
Highway fatalities, 328
License fees, 92
Marine travel, 167
Median age of cars, 136
Moped speed, 376
Motorcycle travel, 490
Moving sidewalks, 36, 375
Navigation, 328, 494
Parking fees, 115
Point of no return, 170
Rate of descent, 102
Shipping, 376
Train speed, 376, 617
Train travel, 166, 170, 205, 376
Travel by car, 377
Tugboat, 373

VIDEOTAPE AND CD INDEX

Text/Video/CD Section	Exercise Numbers
Section 1.1	35, 41, 47
Section 1.2	21, 31, 49, 69, 85, 125, 135
Section 1.3	7, 17, 61, 71
Section 1.4	5
Section 1.5	23
Section 1.6	9, 49, 77
Section 1.7	5, 55
Section 2.1	1, 43
Section 2.2	7, 9, 41(b), 41(d), 41(f)
Section 2.3	31, 41
Section 2.4	5, 15, 21, 37, 51, 63, 67, 69
Section 2.5	15, 23, 53, 71
Section 2.6	17, 51, 55
Section 3.1	3, 9
Section 3.2	13
Section 3.3	37
Section 3.4	1, 29, 31
Section 3.5	5
Section 3.6	9
Section 3.7	3, 9, 11, 19
Section 3.8	9
Section 4.1	63
Section 4.2	11, 23
Section 4.3	none
Section 4.4	3, 13, 45
Section 4.5	12
Section 5.1	3, 37, 69
Section 5.2	23, 45, 59
Section 5.3	15, 27
Section 5.4	55
Section 5.5	9, 29, 41
Section 5.6	17, 31
Section 5.7	15, 45
Section 5.8	1, 37
Section 6.1	1, 45
Section 6.2	37
Section 6.3	7, 21, 33

Text/Video/CD Section	Exercise Numbers
Section 6.4	12, 33
Section 6.5	10, 21, 24
Section 6.6	13, 36
Section 6.7	7
Section 6.8	1, 9, 31, 43, 61, 63
Section 7.1	1, 33, 37, 69, 75, 89
Section 7.2	11, 23, 27, 39, 49, 55, 65, 85
Section 7.3	1, 9, 21, 39, 43, 51, 65
Section 7.4	25, 31, 61
Section 7.5	25, 37, 71, 81
Section 7.6	34
Section 7.7	15, 17, 27
Section 7.8	17, 73, 78
Section 8.1	63
Section 8.2	9
Section 8.3	25
Section 8.4	7, 43
Section 8.5	none
Section 8.6	39
Section 8.7	7
Section 8.8	9, 29
Section 8.9	17, 33
Section 9.1	17, 19
Section 9.2	18, 21, 35
Section 9.3	3, 7, 15, 27, 39, 43, 55, 59, 73, 77
Section 9.4	7
Section 9.5	11, 27
Section 9.6	23, 39
Section 9.7	7
Section 10.1	53
Section 10.2	13, 21, 41
Section 10.3	31, 41
Section 11.1	7, 23, 39
Section 11.2	17, 41
Section 11.3	5, 37
Section 11.4	39

FOR EXTRA PRACTICE

The following sections feature extra practice problems available when you register for MyMathLab.com. For your convenience, these exercises are also printed in the *Instructor's Resource Guide*. Please contact your instructor for assistance.

Section 1.2	Operations and Properties of Real Numbers
Section 1.4	Introduction to Problem Solving
Section 1.6	Properties of Exponents
Section 2.1	Graphs
Section 3.2	Solving by Substitution or Elimination
Section 3.3	Solving Applications: Systems of Two Equations
Section 3.4	Systems of Equations in Three Variables
Section 3.5	Solving Applications: Systems of Three Equations
Section 4.1	Inequalities and Applications
Section 4.3	Absolute-Value Equations and Inequalities
Section 4.4	Inequalities in Two Variables
Section 5.3	Common Factors and Factoring by Grouping
Section 5.4	Factoring Trinomials
Section 5.5	Factoring Perfect-Square Trinomials and Differences of Squares
Section 5.6	Factoring Sums or Differences of Cubes
Section 5.7	Factoring: A General Strategy
Section 5.8	Applications of Polynomial Equations
Section 6.1	Rational Expressions and Functions: Multiplying and Dividing
Section 6.2	Rational Expressions and Functions: Adding and Subtracting
Section 6.3	Complex Rational Expressions
Section 6.4	Rational Equations
Section 6.5	Solving Applications Using Rational Equations
Section 6.6	Division of Polynomials
Section 6.7	Synthetic Division
Section 7.1	Radical Expressions and Functions
Section 7.2	Rational Numbers as Exponents
Section 7.3	Multiplying Radical Expressions
Section 7.4	Dividing Radical Expressions
Section 7.5	Expressions Containing Several Radical Terms
Section 7.6	Solving Radical Equations
Section 7.7	Geometric Applications
Section 8.1	Quadratic Equations
Section 8.2	The Quadratic Formula
Section 8.3	Applications Involving Quadratic Equations
Section 8.5	Equations Reducible to Quadratic
Section 8.6	Quadratic Functions and Their Graphs
Section 8.7	More About Graphing Quadratic Functions
Section 8.9	Polynomial and Rational Inequalities
Section 9.6	Solving Exponential and Logarithmic Equations
Section 10.4	Nonlinear Systems of Equations
Section 11.4	The Binomial Theorem

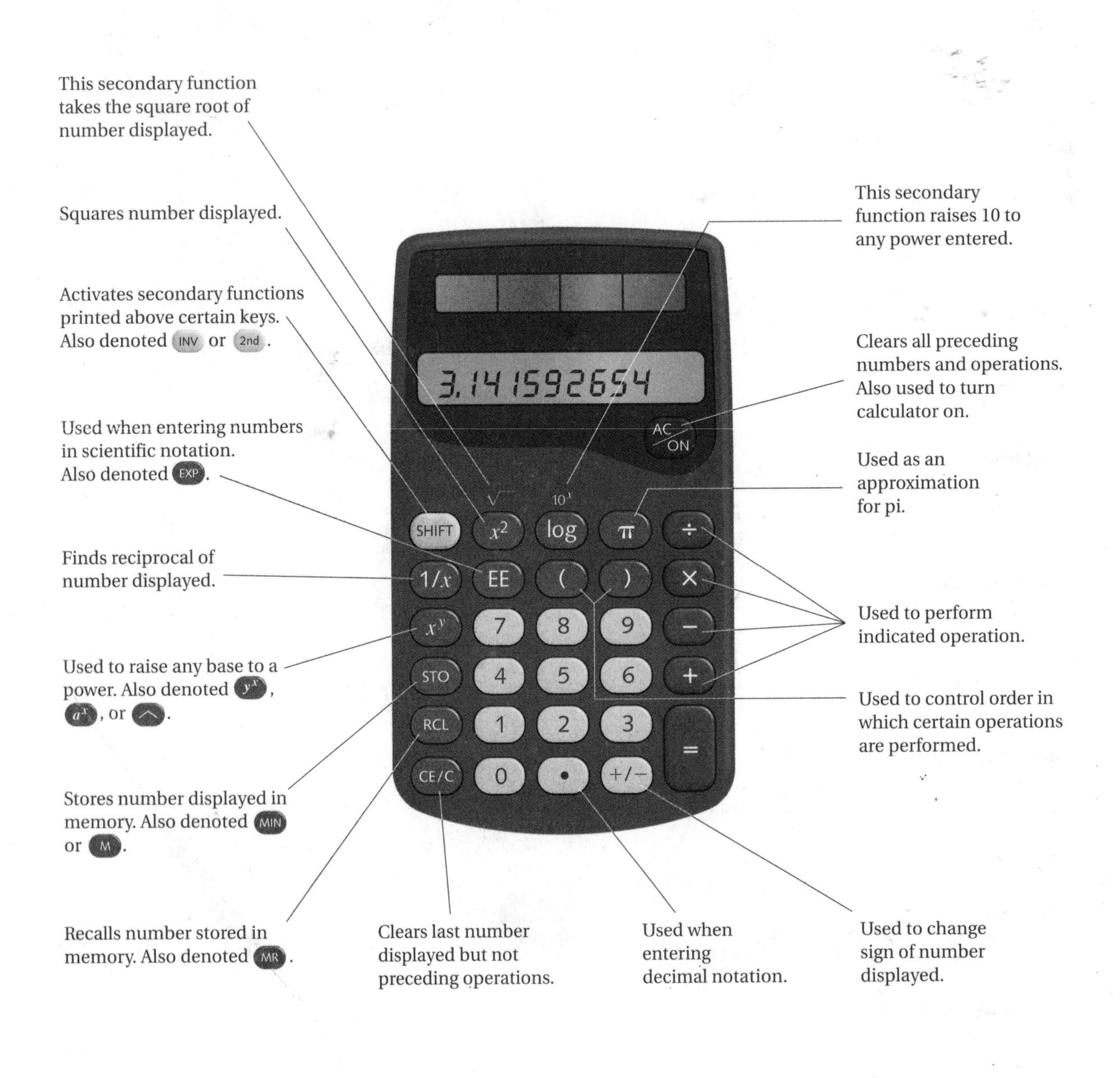

This secondary function takes the square root of number displayed.
Squares number displayed.
Activates secondary functions printed above certain keys. Also denoted INV or 2nd.
Used when entering numbers in scientific notation. Also denoted EXP.
Finds reciprocal of number displayed.
Used to raise any base to a power. Also denoted y^x, a^x, or ^.
Stores number displayed in memory. Also denoted MIN or M.
Recalls number stored in memory. Also denoted MR.
Clears last number displayed but not preceding operations.
Used when entering decimal notation.
Used to change sign of number displayed.
This secondary function raises 10 to any power entered.
Clears all preceding numbers and operations. Also used to turn calculator on.
Used as an approximation for pi.
Used to perform indicated operation.
Used to control order in which certain operations are performed.
3.141592654
AC ON
√
10^x
SHIFT
x^2
log
π
÷
1/x
EE
(
)
×
x^y
7
8
9
−
STO
4
5
6
+
RCL
1
2
3
=
CE/C
0
•
+/−